D1393612

# ILLUSTRATED ENCYCLOPEDIA OF THE UNIVERSE

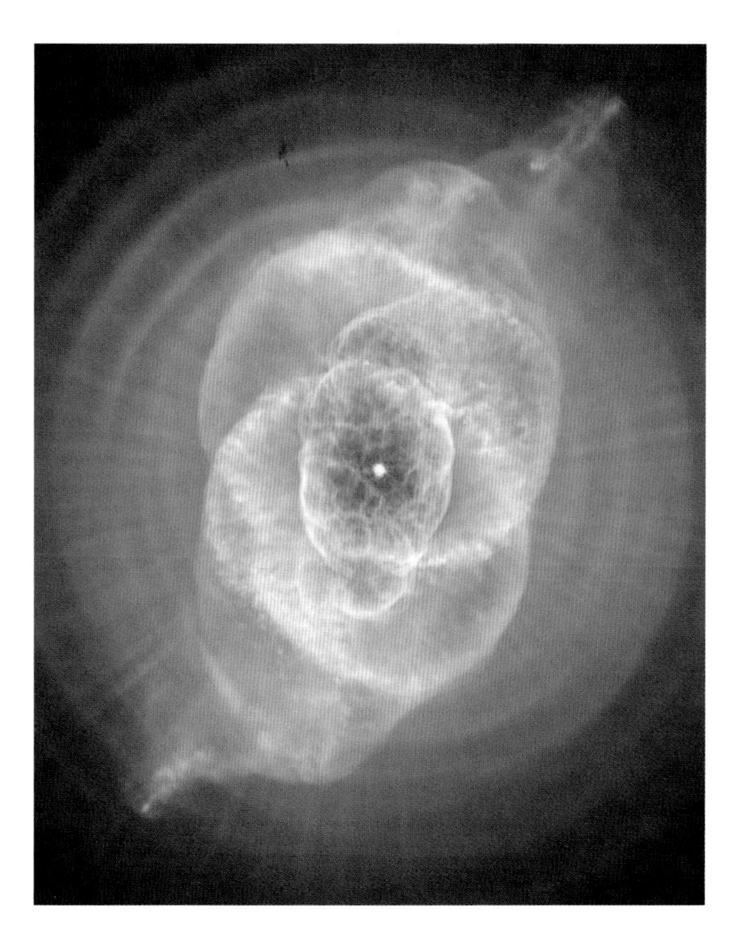

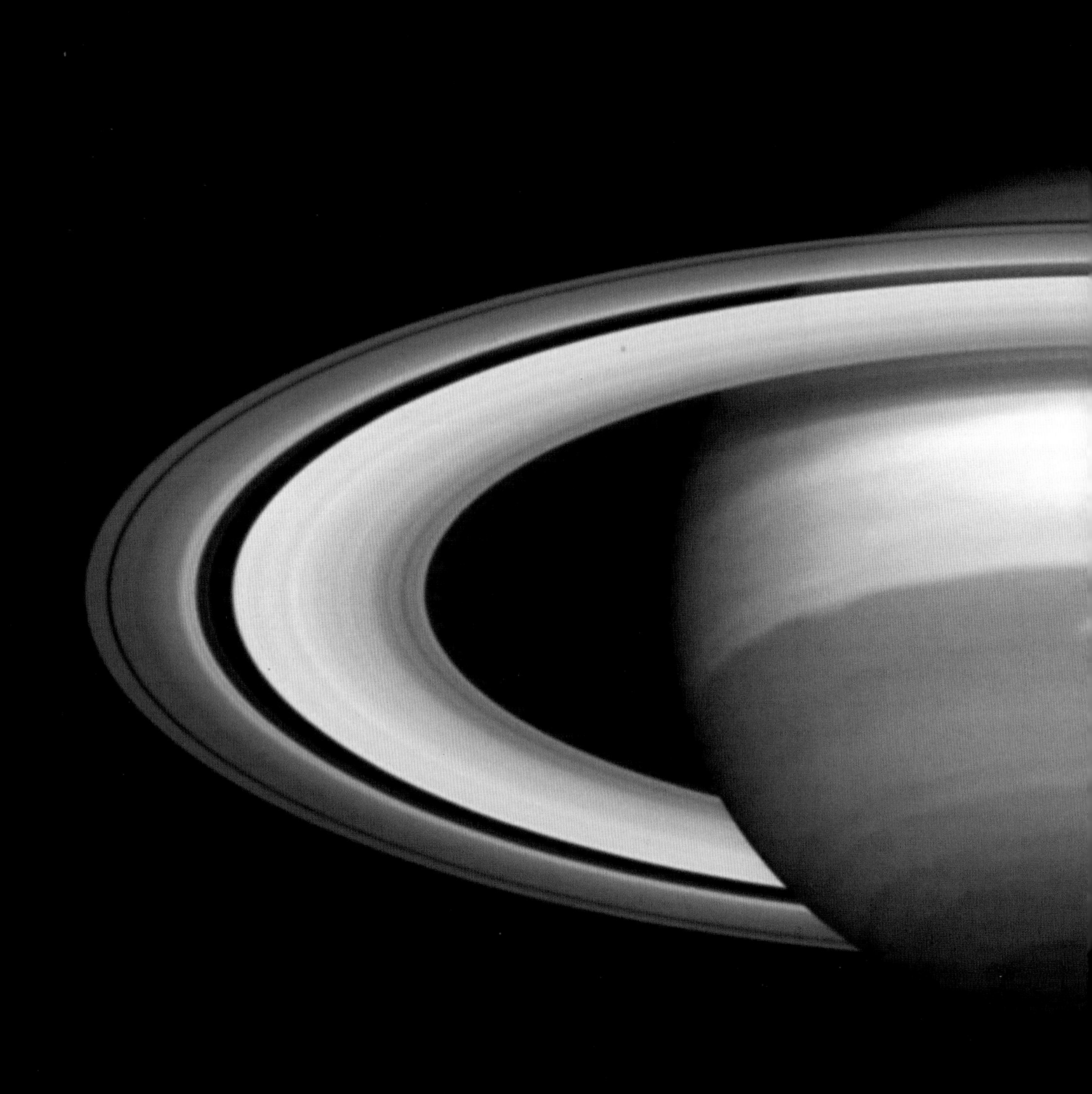

ILLUSTRATED ENCYCLOPEDIA OF THE

# UNIVERSE

GENERAL EDITOR **MARTIN REES**

LONDON, NEW YORK, MELBOURNE,
MUNICH, AND DELHI

**SENIOR EDITOR** Peter Frances

**SENIOR ART EDITORS** Mabel Chan, Peter Laws

**PROJECT EDITORS** Georgina Garner, Rob Houston, Gill Pitts, David Summers

**PROJECT ART EDITORS** Dave Ball, Sunita Gahir, Alison Gardner, Mark Lloyd

**DESIGNERS** Kenny Grant, Jerry Udall

**EDITORS** Joanna Chisholm, Ben Hoare, Jane Simmonds, Giles Sparrow, Nikky Twyman, Miezan van Zyl

**DESIGN ASSISTANT** Marilou Prokopiou

**DTP DESIGNERS** John Goldsmid, Martin Nilsson, Simon Longstaff

**INDEXER** Hilary Bird

**PICTURE RESEARCHER** Louise Thomas

**ILLUSTRATORS** Antbits, Combustion Design and Advertising, Fanatic Design, JP Map Graphics, Moonrunner Design, Pikaia Imaging, Planetary Visions, Precision Illustration

**PRODUCTION CONTROLLER** Heather Hughes

**MANAGING EDITOR** Liz Wheeler

**MANAGING ART EDITOR** Philip Ormerod

**PUBLISHING DIRECTOR** Jonathan Metcalf

**ART DIRECTOR** Bryn Walls

First published as *Universe* in Great Britain in 2005
This edition published in 2011
by Dorling Kindersley Limited
80 Strand, London WC2R 0RL

A Penguin Company



2 4 6 8 10 9 7 5 3 1

002 – UD141 – Jan/11



A CIP catalogue record for this book is available from the British Library

ISBN 978 1 4053 6331 0

Colour reproduction by GRB Editrice, s.r.l., Italy
Printed and bound in China by Leo Paper Products

See our complete catalogue at
www.dk.com

# CONTENTS

ABOUT THIS BOOK 6

A SHORT TOUR OF THE UNIVERSE 8
BY MARTIN REES

## INTRODUCTION

**WHAT IS THE UNIVERSE?** 20
- THE SCALE OF THE UNIVERSE 22
- CELESTIAL OBJECTS 24
- MATTER 28
- RADIATION 32
- GRAVITY, MOTION, AND ORBITS 36
- SPACE AND TIME 38
- EXPANDING SPACE 42

**THE BEGINNING AND END OF THE UNIVERSE** 44
- THE BIG BANG 46
- OUT OF THE DARKNESS 50
- LIFE IN THE UNIVERSE 52
- THE FATE OF THE UNIVERSE 54

**THE VIEW FROM EARTH** 56
- THE CELESTIAL SPHERE 58
- CELESTIAL CYCLES 60
- PLANETARY MOTION 64
- STAR MOTION AND PATTERNS 66
- LIGHTS IN THE SKY 70
- NAKED-EYE ASTRONOMY 72
- BINOCULAR ASTRONOMY 74
- TELESCOPE ASTRONOMY 76

**EXPLORING SPACE** 80
ANCIENT ASTRONOMY 82
EARLY SCIENTIFIC ASTRONOMY 84
THE COPERNICAN REVOLUTION 86
THE INFINITE UNIVERSE 90
SHAPING SPACE 92
SPACE-AGE ASTRONOMY 94
EARLY SPACECRAFT 98
JOURNEYS TO THE MOON 102
IN EARTH'S ORBIT 106
BEYOND EARTH 108

## GUIDE TO THE UNIVERSE

**THE SOLAR SYSTEM** 114
THE HISTORY OF THE SOLAR SYSTEM 116
THE FAMILY OF THE SUN 118
THE SUN 120
MERCURY 124
VENUS 128
EARTH 138
THE MOON 148
MARS 160
JUPITER 176
SATURN 186
URANUS 196
NEPTUNE 200
DWARF PLANETS 204
THE KUIPER BELT AND THE OORT CLOUD 206
ASTEROIDS 208
COMETS 214
METEORS AND METEORITES 220

**THE MILKY WAY** 224
THE MILKY WAY 226
STARS 230
THE LIFE CYCLES OF STARS 232
STAR FORMATION 236
MAIN-SEQUENCE STARS 246
OLD STARS 250
STELLAR END POINTS 262
MULTIPLE STARS 270
VARIABLE STARS 278
STAR CLUSTERS 284
EXTRA-SOLAR PLANETS 290

**BEYOND THE MILKY WAY** 292
TYPES OF GALAXY 294
GALAXY EVOLUTION 298
ACTIVE GALAXIES 310
GALAXY CLUSTERS 316
GALAXY SUPERCLUSTERS 324

## THE NIGHT SKY

**THE CONSTELLATIONS** 328
THE HISTORY OF CONSTELLATIONS 330
MAPPING THE SKY 332
*Guide to the Constellations* 338

**MONTHLY SKY GUIDE** 410
USING THE SKY GUIDES 412
JANUARY 414
FEBRUARY 420
MARCH 426
APRIL 432
MAY 438
JUNE 444
JULY 450
AUGUST 456
SEPTEMBER 462
OCTOBER 468
NOVEMBER 474
DECEMBER 480

**GLOSSARY** 486

**INDEX** 494

**ACKNOWLEDGMENTS** 510

# ABOUT THIS BOOK

*Universe* is divided into three main sections. The **INTRODUCTION** is an overview of the basic concepts of astronomy. **GUIDE TO THE UNIVERSE** looks, in turn, at the Solar System, the Milky Way (our home galaxy), and the regions of space that lie beyond. Finally, **THE NIGHT SKY** is a guide to the sky for the amateur skywatcher.

## INTRODUCTION

This section is about the Universe and astronomy as a whole. It is subdivided into four parts. WHAT IS THE UNIVERSE? looks at different kinds of objects in the Universe and the forces governing how they behave and interact. THE BEGINNING AND END OF THE UNIVERSE covers the origin and history of the Universe, while THE VIEW FROM EARTH explains what we see when we look at the sky. EXPLORING SPACE is a history of humankind's study of the Universe and our journeys beyond the Earth.

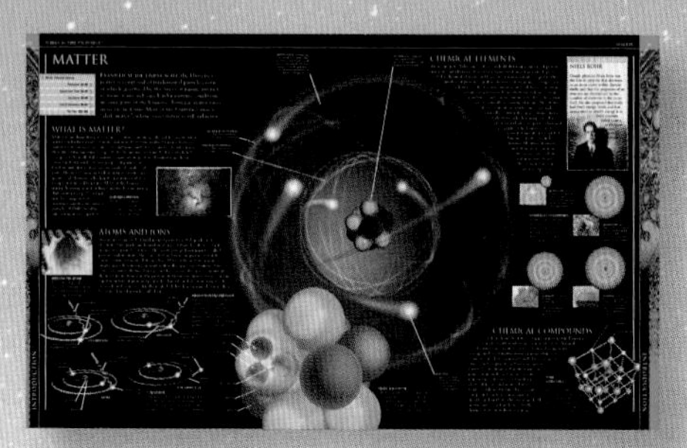

**◁ WHAT IS THE UNIVERSE?**
This section begins by looking at some basic questions about the size and shape of the Universe. It goes on to explain concepts such as matter and radiation, the motion of objects in space, and the relationship between time and space.

**THE BEGINNING AND END OF THE UNIVERSE ▷**
The Universe is thought to have originated in an event known as the Big Bang. This section describes the Big Bang in detail and looks at how the Universe came to be the way it is now, as well as how it might end.

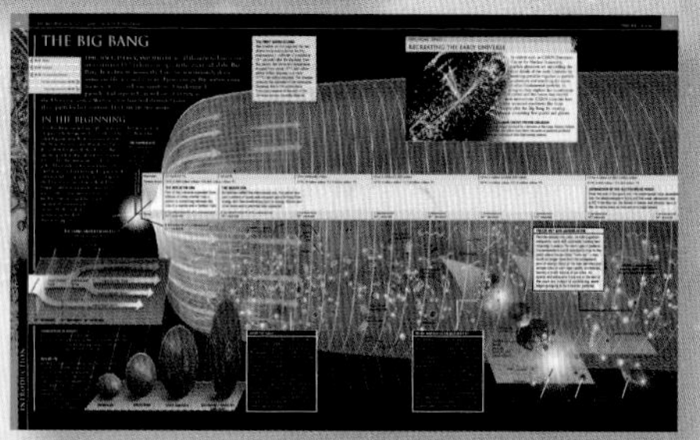

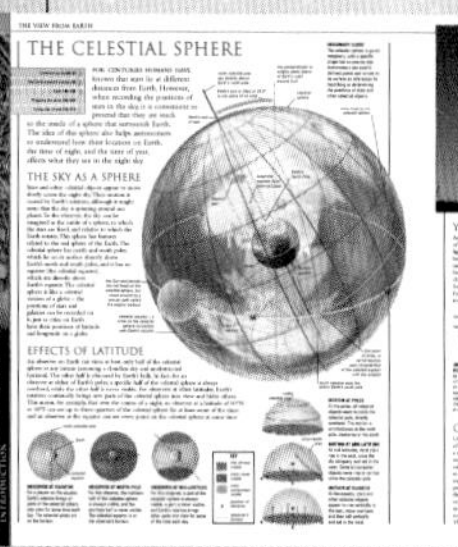

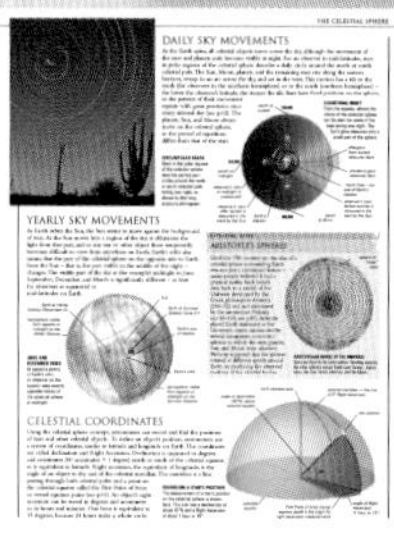

**◁ THE VIEW FROM EARTH**
This section presents a simple model for making sense of the changing appearance of the sky. It also contains practical advice on looking at the sky with the naked eye, telescopes, and binoculars.

**EXPLORING SPACE ▷**
This section is about our exploration of the Cosmos, from the discoveries of ancient civilizations to the findings of modern astronomers. It also looks at the history of both manned and unmanned spaceflight.

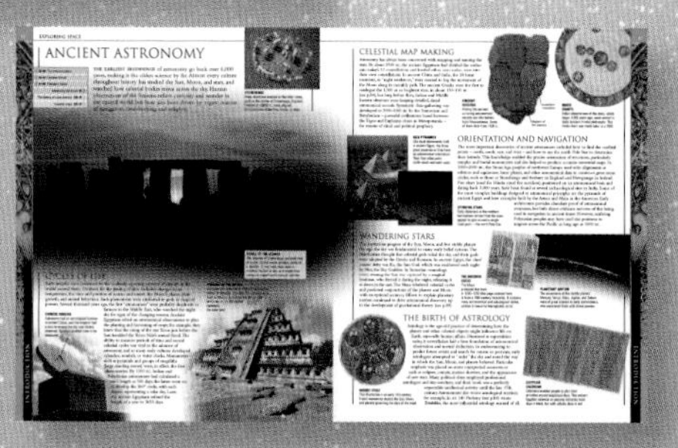

## GUIDE TO THE UNIVERSE

This part of the book focuses on specific regions of space, starting from the Sun and then moving outwards to progressively more distant reaches of the Universe. It is divided into three sections, covering the Solar System, the Milky Way, and features beyond the Milky Way. In each section, introductory pages describe features in a general way and explain the processes behind their formation. These pages are often followed by detailed profiles of actual features (such as individual stars), usually arranged in order of their distance from Earth.

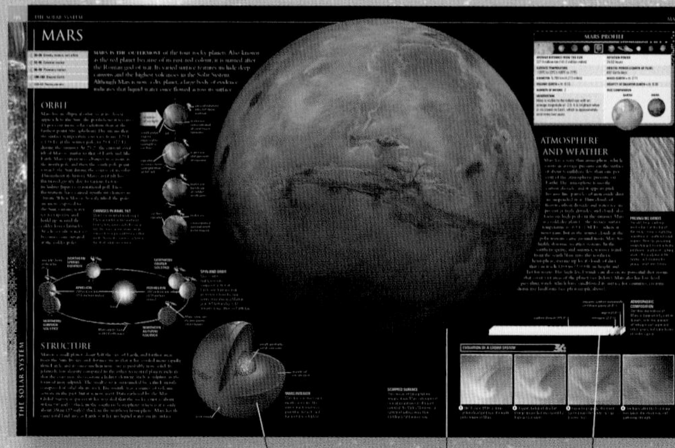


**△ THE SOLAR SYSTEM**
This section is about the Sun and the many bodies in orbit around it. It covers the eight planets one by one and then looks at asteroids, comets, and meteors, as well as the remote regions on the margins of the Solar System. For most planets, profiles of individual surface features or moons are also included.

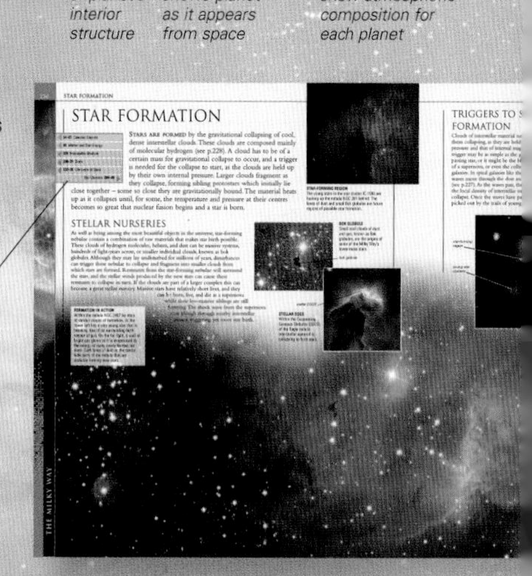


**THE MILKY WAY ▷**
The subject of this section is the Milky Way and the stars, nebulae, and planets that it contains. Pages such as those shown here describe how particular types of features are formed.

## THE NIGHT SKY

This section is an atlas of the night sky. It is divided into two parts. The first (THE CONSTELLATIONS) is a guide to the 88 regions into which astronomers divide the sky. It contains illustrated profiles of all the constellations, arranged according to their position in the sky with the most northerly ones first and the southernmost last. The second part (the MONTHLY SKY GUIDE) is a month-by-month guide, containing a summary of the highlights for each month, detailed star charts, and charts showing the positions of the planets.

**THE CONSTELLATIONS ▷**
Each constellation profile is illustrated with a chart, two locator maps, and one or more photographs. A more detailed guide to the section can be found on pp.332–33.

## THEMED PANELS

Three types of colour-coded panels are used to present a more detailed focus on selected subjects. These panels appear both on explanatory pages and in feature profiles.

EXPLORING SPACE

SEARCHING FOR A PLANET

If the distant fixed stars are photographed on two nights, a few days apart, and there is a planet in the field of view, this planet will move with respect to the background. If a blink comparator is then used to view the two photographic plates in quick succession, the bodies that have moved appear to blink. On 23 January

MYTHS AND STORIES

ASTROLOGY AND THE ECLIPTIC

Astrology is the study of the positions and movements of the Sun, Moon, and planets in the sky in the belief that these influence human affairs. At one time, when astronomy was applied mainly to devising calendars, astronomy and astrology were intertwined, but their aims and methods have now diverged. Astrologers pay little attention to constellations, but measure the positions of the Sun and planets in sections of the ecliptic that they call "Aries" and "Taurus", for example. However, these sections no longer match the constellations of Aries, Taurus, and so on.

STARGAZER

JOHANNES KEPLER

German astronomer Johannes Kepler (1571–1630) discovered the laws of planetary motion (see p.87). His first law states that planets orbit the Sun in elliptical paths. The next states that the closer a planet comes to the Sun, the faster it moves, while his third law describes the link between a planet's average distance and its orbital period. Newton used Kepler's

**◁ EXPLORING SPACE**
This type of feature is used to describe the study of space, either from the Earth's surface or from spacecraft. Individual panels describe particular discoveries or investigations.

**MYTHS AND STORIES ▷**
As well as being studied scientifically, objects in the night sky have featured in myths, superstitions, and folklore, which form the subject of this type of panel.

**◁ BIOGRAPHY**
Profiles of notable astronomers and pioneers of spaceflight, as well as a brief summary of their achievements, appear in this type of panel.

## CONTRIBUTORS

**Martin Rees** General editor

**Robert Dinwiddie**
What is the Universe?
The Beginning and End of the Universe
The View From Earth
The Solar System

**Philip Eales** The Milky Way

**David Hughes**
Exploring Space
The Solar System

**Iain Nicolson** Glossary

**Ian Ridpath** The Night Sky

**Giles Sparrow**
Exploring Space
Beyond the Milky Way

**Pam Spence** The Milky Way

**Carole Stott** The Solar System

**Kevin Tildsley** The Milky Way

*name or astronomical catalogue number of feature (features without a popular name are identified by number)*

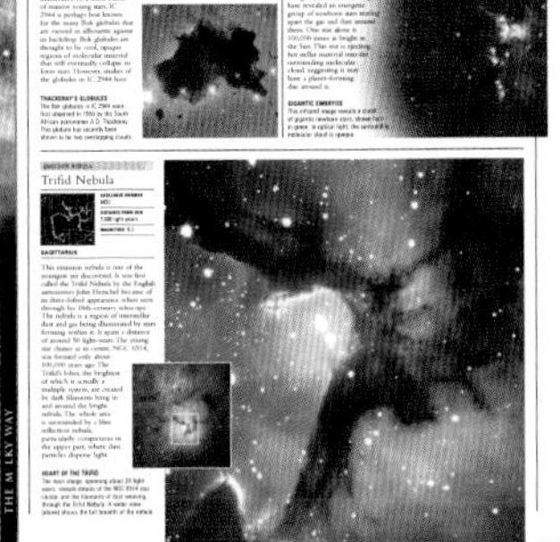

EMISSION NEBULA

Carina Nebula

| | |
|---|---|
| CATALOGUE NUMBER | NGC 3372 |
| DISTANCE FROM SUN | 8,000 light-years |
| MAGNITUDE | 1 |

CARINA

*locator map shows constellation in which feature can be found and its position within the constellation*

*table of summary information (varies between sections)*

*selected features are described in double-page feature profiles*

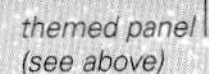

*themed panel (see above)*

**△ FEATURE PROFILES**
Throughout the Guide to the Universe, introductory pages are often followed by profiles of a selection of specific objects. For example, the introduction to star-formation (left) is followed by profiles of actual star-forming regions in the Milky Way (above).

**BEYOND THE MILKY WAY ▷**
This section looks at features found beyond our own galaxy, including other galaxies and galaxy clusters and superclusters, the largest known structures in the Universe.

*locator shows constellation in context*

*artwork of figure depicted by constellation*

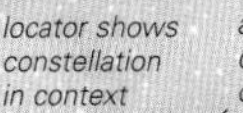

**△ MONTHLY SKY GUIDE**
This section includes two charts for each month of the year, for observers in northern and southern latitudes. The section is described in more detail on pp.412–13.

*chart on this page shows view looking north, with view to south on facing page*

*lines on chart show reference points for observers at different latitudes*

## THE GREEK ALPHABET

Astronomers use a convention for naming some stars in which Greek letters are assigned to stars according to their brightness. These letters appear on some of the charts in this book.

| | | | |
|---|---|---|---|
| α | alpha | ν | nu |
| β | beta | ξ | xi |
| γ | gamma | ο | omicron |
| δ | delta | π | pi |
| ε | epsilon | ρ | rho |
| ζ | zeta | σ | sigma |
| η | eta | τ | tau |
| θ | theta | υ | upsilon |
| ι | iota | φ | phi |
| κ | kappa | χ | chi |
| λ | lambda | ψ | psi |
| μ | mu | ϖ | omega |

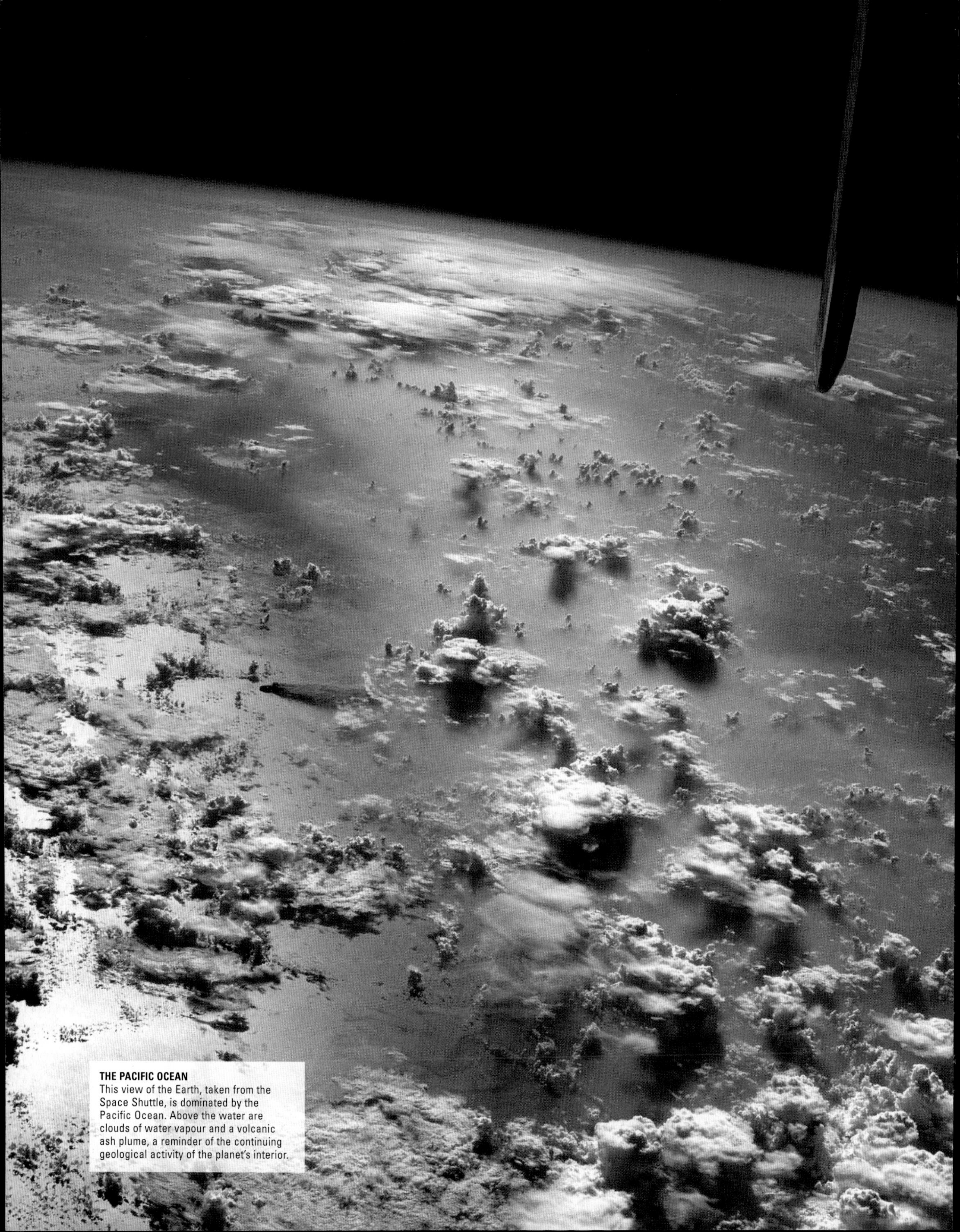

**THE PACIFIC OCEAN**
This view of the Earth, taken from the Space Shuttle, is dominated by the Pacific Ocean. Above the water are clouds of water vapour and a volcanic ash plume, a reminder of the continuing geological activity of the planet's interior.

# A SHORT TOUR OF THE UNIVERSE

**KENNEDY SPACE CENTER**
Many of humankind's first ventures into space were set in motion on the launchpads of Kennedy Space Center. This remains the busiest launch and landing site of America's space programme, and it is also the main base for the Space Shuttle.

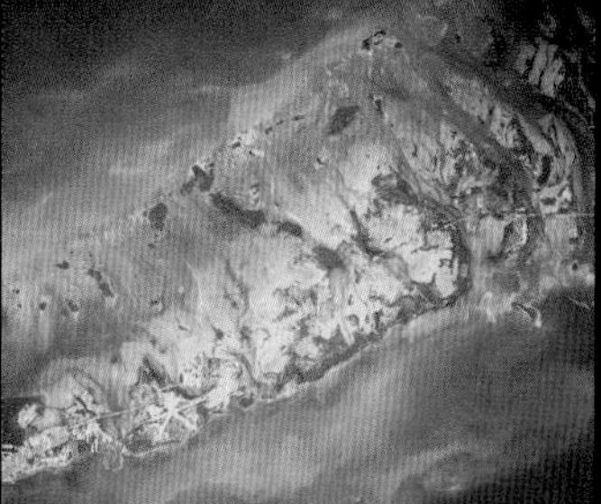

**THE FLORIDA COAST**
The islands and reefs of the Florida Keys are seen here from Earth orbit. The reefs are partly made of living organisms, in the form of corals. To date, life has not been found anywhere other than on Earth, but the search for alien life will be perhaps the most fascinating quest of the 21st century.

THE NIGHT SKY has always evoked mystery and wonder. Since antiquity, astronomers have tried to understand the patterns of the "fixed stars", and the motions of the Moon and planets. The motive was partly a practical one but there has always been a more "poetic" motivation, too – a quest to understand our place in nature. Modern science reveals a cosmos far vaster and more varied than our ancestors could have envisioned.

No continents on Earth remain to be discovered. The exploratory challenge has now broadened to the cosmos. Humans have walked on the Moon; unmanned spacecraft have beamed back views of all the planets; and some people now living may one day walk on Mars.

The stars, fixed in the "vault of heaven", were a mystery to the ancients. They are still unattainably remote, but we know that many of them are shining even more brightly than the Sun. Within the last decade, we have learned something remarkable that was long suspected: many stars are, like our Sun, encircled by orbiting planets. The number of known planetary systems already runs into hundreds – there could altogether be a billion in our galaxy. Could some of these planets resemble the Earth, and harbour life? Even intelligent life?

All the stars visible to the unaided eye are part of our home galaxy – a structure so vast that light takes a hundred thousand years to cross it. But this galaxy, the Milky Way, is just one of billions visible through large telescopes. These galaxies are hurtling apart from each other, as though they had all originated in a "big bang" 13 or 14 billion years ago. But we don't know what banged, nor why it banged.

The beauty of the night sky is a common experience of people from all cultures – indeed it is something that we share with all generations since prehistoric times. Our modern perception of the "cosmic environment" is even grander. Astronomers are now setting Earth in a cosmic context. They seek to understand how the cosmos developed its intricate complexity – how the first galaxies, stars, and planets formed and how on at least one planet atoms assembled into creatures able to ponder their origins. This book sets humanity's concept of the cosmos in its historic context, and presents the latest discoveries and theories. It is a beautiful "field guide" to our cosmic habitat: it should enlighten and delight anyone who has looked up at the stars with wonder, and wished to understand them better. MARTIN REES

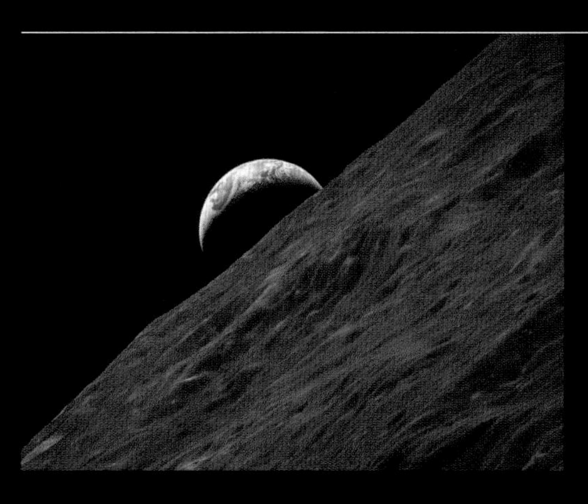

**THE MOON**
**1.3 light-seconds from Earth**
The Earth is seen here rising above the horizon of its satellite, the Moon. Our home planet's delicate biosphere contrasts with the sterile moonscape on which the Apollo astronauts left their footprints.

# The Sun

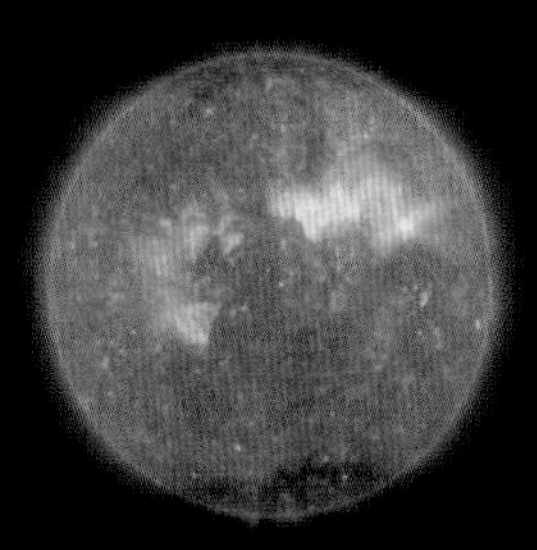

**OUR LOCAL STAR**
The Sun dominates the Solar System. Our chief source of heat and light, it also holds the Earth and the rest of the planets in their orbits. This ultraviolet image reveals the dynamic activity in the ultra-hot corona above the Sun's visible surface.

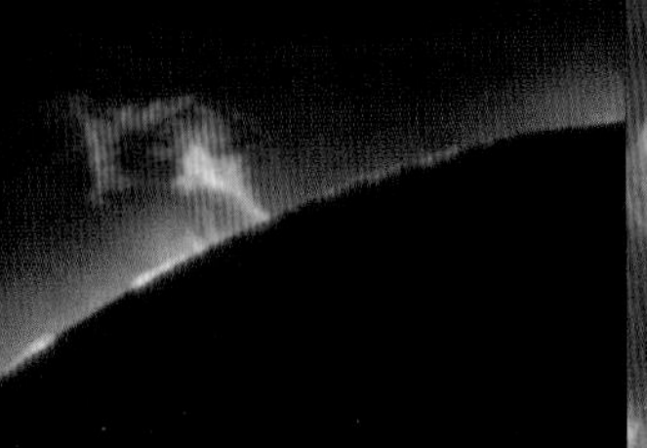

**A SOLAR FLARE**
The Sun usually appears to the unaided eye as a bright but featureless disc. However, during a total solar eclipse, when light from the disc is blocked out by the Moon, violent flares in the outer layers of the atmosphere can be seen more clearly.

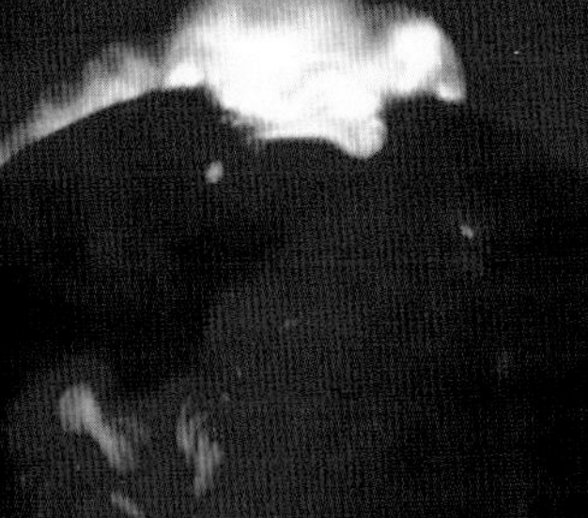

**ULTRA-HOT CORONA**
The gas in the Sun's corona is heated to several million degrees, causing it to emit X-rays, which show up in this image taken by the Japanese YOHKOH satellite. The dark areas are regions of low-density gas that emit a stream of particles, known as the solar wind, into space.

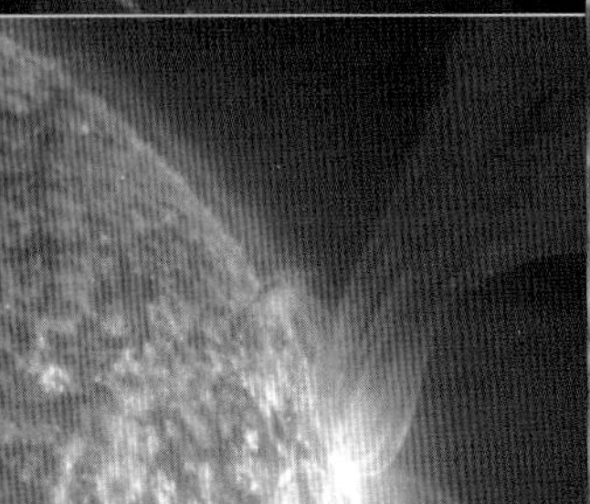

**PROMINENCES**
In the corona, electrified gas called plasma forms into huge clouds known as prominences, flowing through the Sun's magnetic field. As the prominence in this image erupts, it hurls plasma out of the Sun's atmosphere and into space.

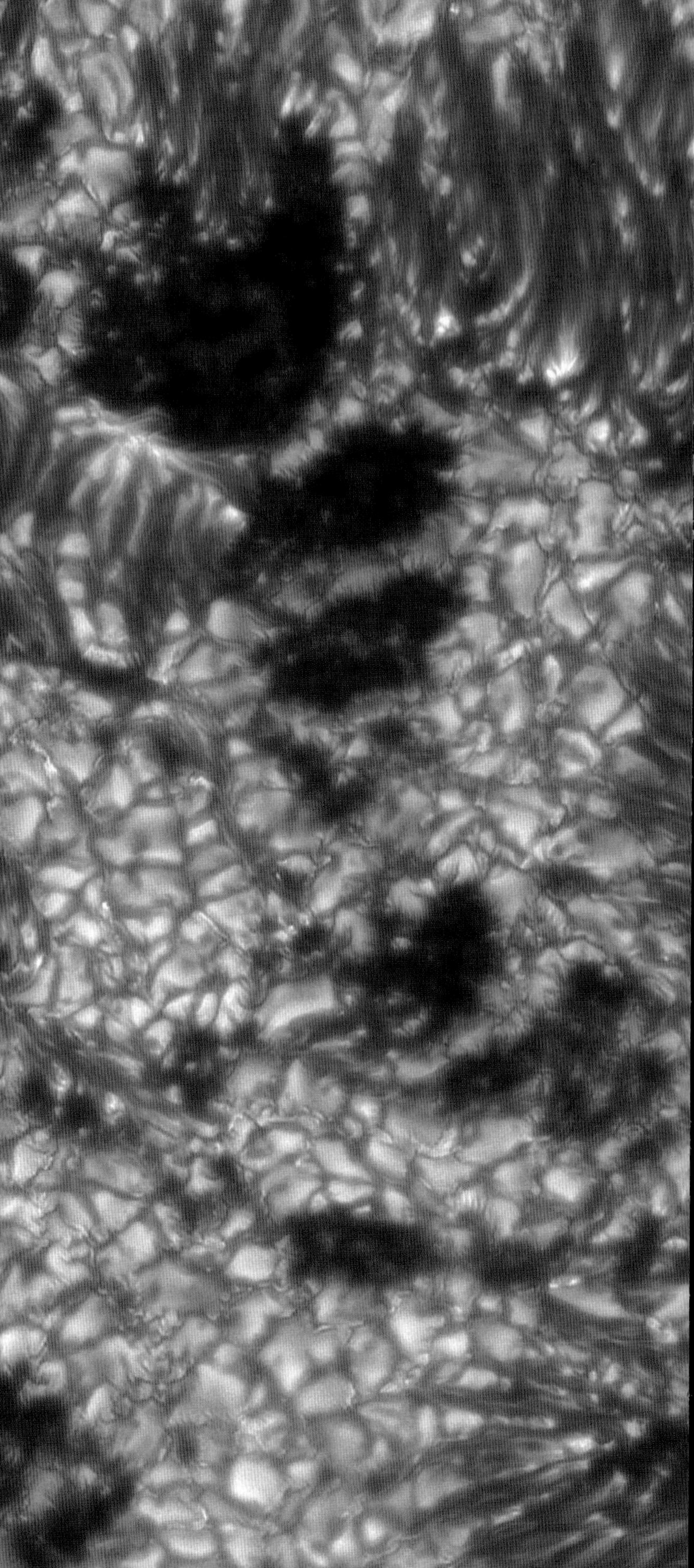

**SUNSPOTS ON THE SOLAR SURFACE**
**8 light-minutes from Earth**

These regions, cooler and darker than the rest of the Sun's surface, are sustained by strong magnetic fields. Some sunspots are large enough to engulf the Earth. Sunspot numbers vary in cycles that take about 11 years to complete, and peaks in the cycle coincide with disturbances, such as aurorae, in our own atmosphere.

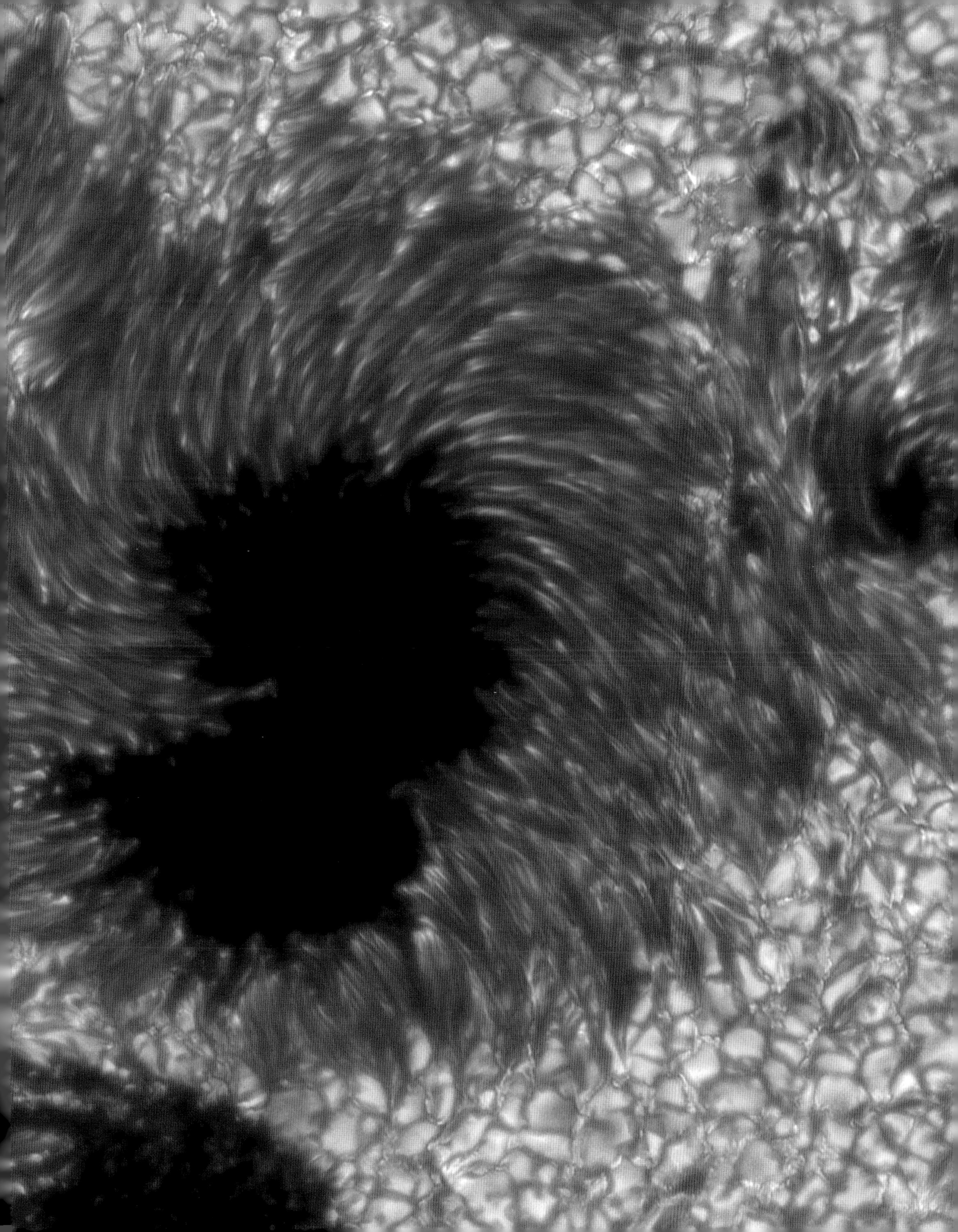

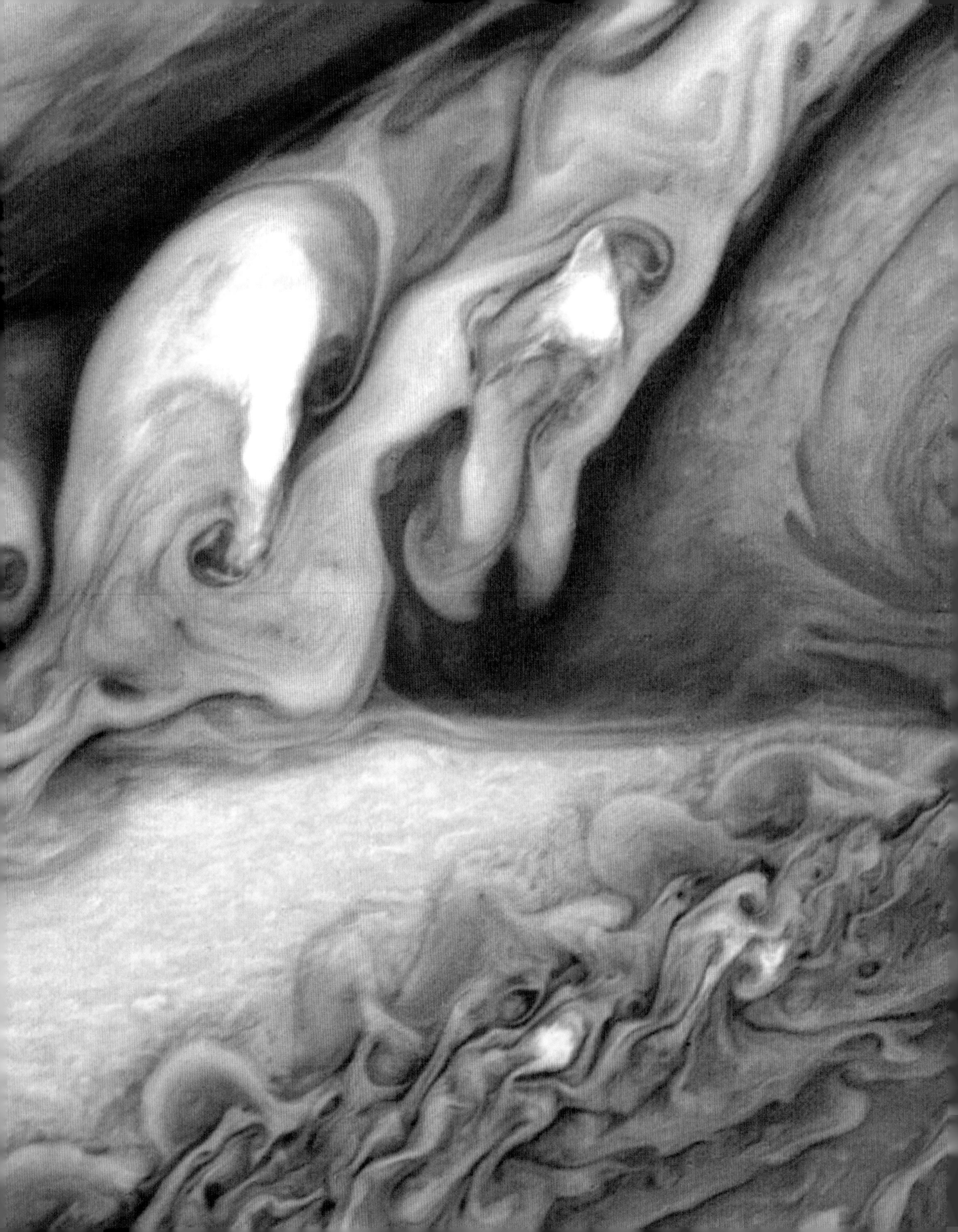

**CANYONS ON MARS**
**4 light-minutes from Earth**

Mars is one of the Solar System's four inner rocky planets. This image (with exaggerated vertical scale) shows part of the Valles Marineris, a vast canyon system.

**SATURN AND ITS RINGS**
**71 light-minutes from Earth**

There are rings of dust and ice particles in near-circular orbits around all four of the giant gas planets, but those around Saturn are especially beautiful. This close-up was taken by the Cassini spacecraft.

**IO**
**34 light-minutes from Earth**

Jupiter has 63 known moons – and there are almost certainly others yet to be discovered. Io, Jupiter's innermost moon, is seen here moving in front of the turbulent face of the planet.

**433 EROS**
**3.8 light-minutes from Earth**

A vast number of asteroids are in independent orbit around the Sun. Eros is marked by the impact of much smaller bodies. This image was taken by the NEAR–Shoemaker craft from only 100km (60 miles) above the surface.

# The planets

**JUPITER'S GREAT RED SPOT**
**34 light-minutes from Earth**

The gas giant Jupiter is more massive than all the other planets in the Solar System combined. Its mysterious swirling vortex, the Great Red Spot, has been known since the 17th century, but our knowledge of Jupiter improved greatly when the planet was visited by unmanned spacecraft in the 1970s. This image of the Great Red Spot was taken in 1979 by Voyager 1, using filters that exaggerate its colours.

# Stars and galaxies

**HE CENTRE OF OUR GALAXY**
**5,000 light-years from Earth**

he centre of our galaxy, the Milky Way, is thought to arbour a black hole as eavy as 3 million Suns. This mage reveals flare-ups in X-ay activity close to the event orizon, the point of no return or any objects or light that pproach the black hole.

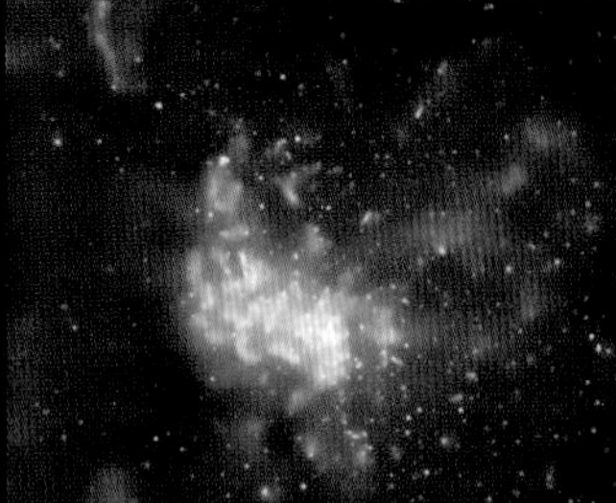

**ENTAURUS A**
**5 million light-years rom Earth**

Not all galaxies exist in solation; occasionally they nteract. Centaurus A is ar more "active" than our own galaxy. It has an even bigger black hole than the Milky Way's, and its gravity may have captured and "cannibalized" a smaller neighbour.

**HE WHIRLPOOL GALAXY**
**1 million light-years rom Earth**

he Whirlpool is involved in nother case of galaxy nteraction. A spinning, disc-ike galaxy, viewed face-on, ts spiral structure may have been induced by the ravitational pull of a smaller atellite galaxy (at the top of his picture).

**THE ORION NEBULA**
**1,500 light-years from Earth**

The Orion Nebula is a vast cloud of glowing dusty gas within the Milky Way, inside which new stars are forming. The nebula contains bright blue stars much younger than the Sun, and some protostars whose nuclear power sources have not yet ignited.

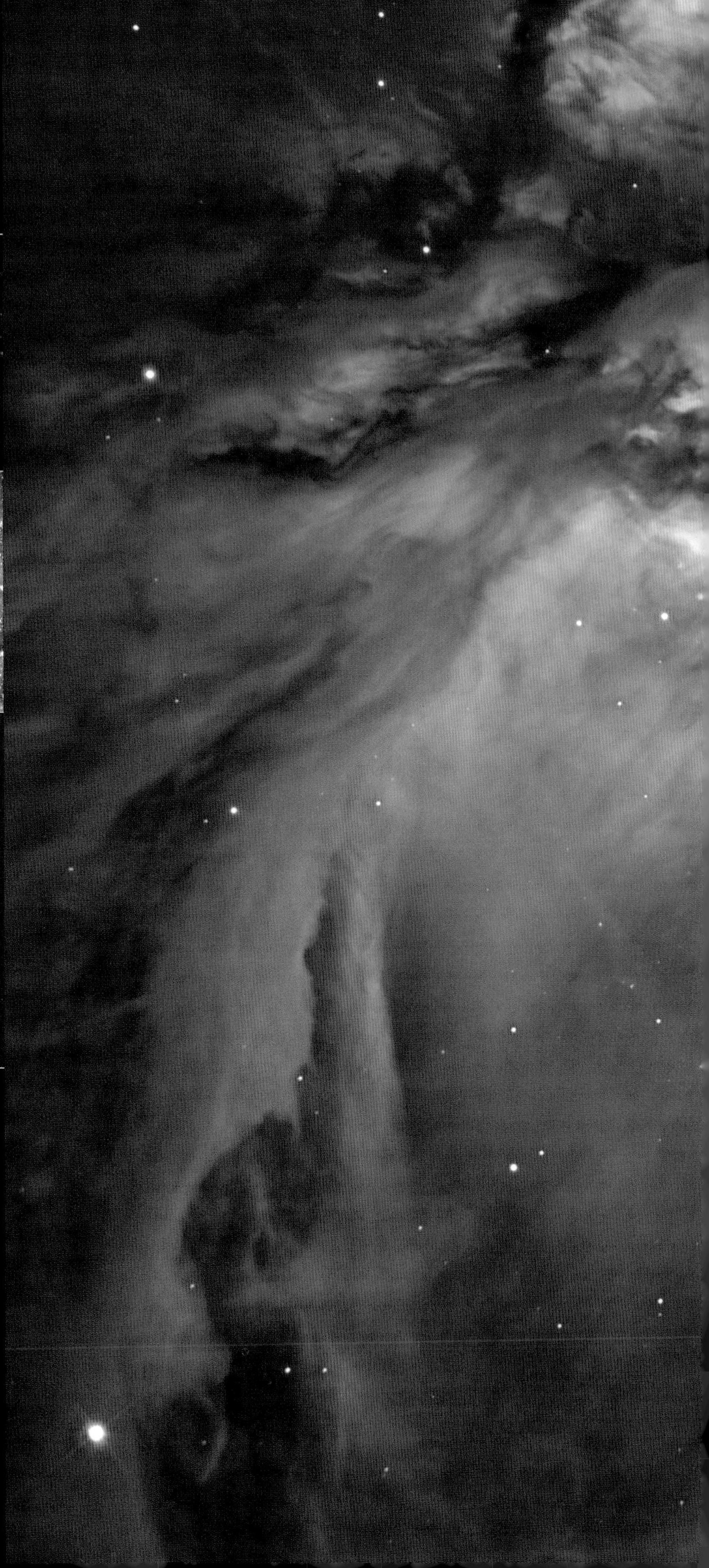

# The limits of time and space

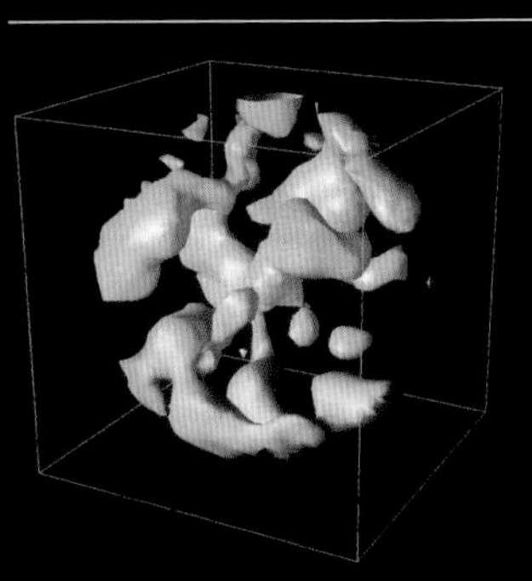

**GALAXY SUPERCLUSTERS**
This image, generated by plotting the positions of 15,000 galaxies, depicts the main "topographic" features of our cosmic environment out to 700 million light-years from Earth. The yellow blobs are superclusters of galaxies, which are interspersed with black voids.

**LARGE-SCALE STRUCTURES**
This view of the sky, in infrared light, shows how galaxies outside the Milky Way are distributed in clusters and filamentary structures. The galaxies are colour-coded according to brightness, with bright ones in blue and faint ones in red.

**THE HUBBLE ULTRA-DEEP FIELD**
This image was created by aiming the Hubble Space Telescope at an area of sky 100 times smaller than the full Moon and taking a long exposure. It reveals hundreds of ultra-distant galaxies, some of them nearly as old as the Universe itself.

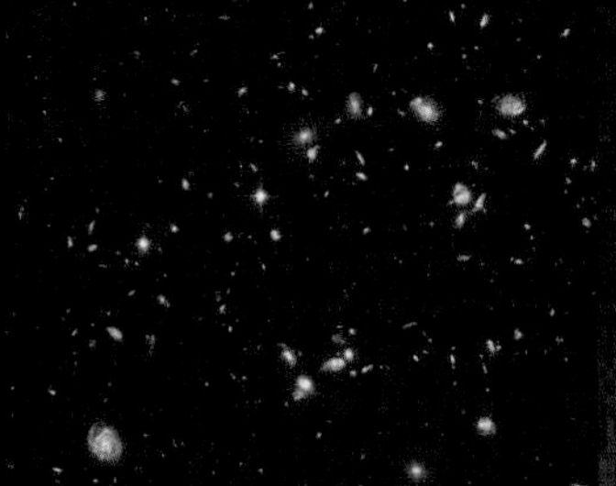

**DISTANT CLUSTER OF GALAXIES**
**8.8 billion light-years from Earth**
Galaxies commonly occur with other galaxies in groups called clusters. The Milky Way, for example, is part of a cluster called the Local Group. The cluster shown here is one of the most distant known to astronomers. It is so far away that the light now reaching us from it set out 8.8 billion years ago, well before the Earth was formed. Superimposed on the picture is an X-ray image (shown in purple), which reveals hot, diffuse gas that pervades the cluster.

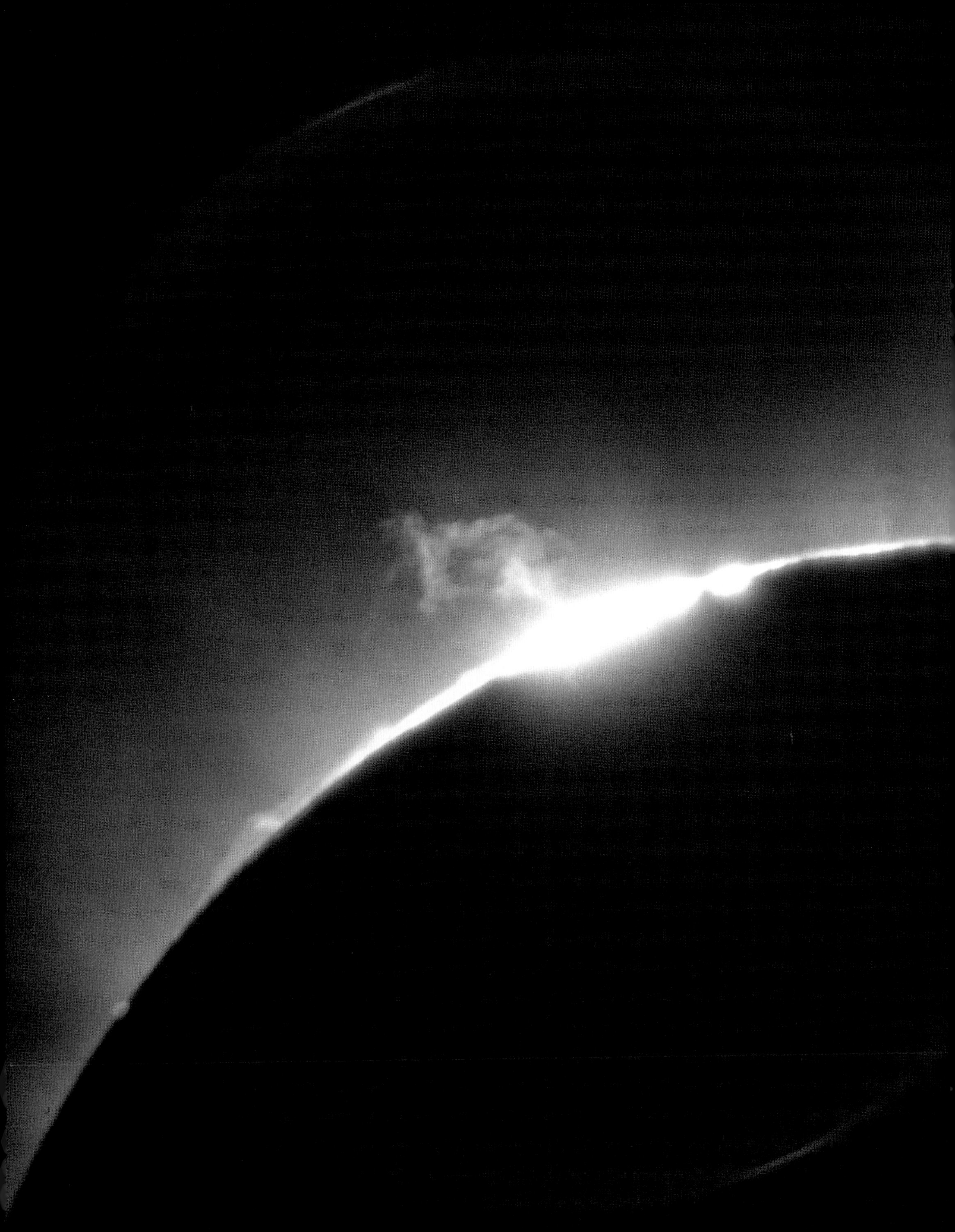

# INTRODUCTION

*"There are grounds for cautious optimism that we may now be near the end of the search for the ultimate laws of nature."*

Stephen Hawking

THE UNIVERSE IS ALL OF EXISTENCE – all of space and time and all the matter and energy within it. The Universe is unknowably vast, and ever since it formed, it has been expanding, carrying distant regions apart at speeds up to, and in some cases possibly exceeding, the speed of light. The Universe encompasses everything from the smallest atom to the largest galaxy cluster, and yet it seems that all are governed by the same basic laws. All visible matter (which is only a small percentage of the total matter) is built from the same subatomic blocks, and the same fundamental forces govern all interactions between these elements. Knowledge of these cosmic operating principles – from general relativity to quantum physics – informs cosmology, the study of the Universe as an entity. Cosmologists hope to answer questions such as "How big is the Universe?", "How old is it?", and "How does it work, on the grandest scale?".

**BOW SHOCK AROUND A STAR**
This mysterious image from the Orion Nebula shows how matter and radiation interact on a stellar scale. A star surrounded by gas and dust has met a fierce wind of particles blowing from a bright young star (out of frame). Around the star, a crescent-shaped gaseous bow shock has built up, like water flowing past the prow of a boat.

# WHAT IS THE UNIVERSE?

# THE SCALE OF THE UNIVERSE


Celestial objects 24–27
Expanding space 42–43
The fate of the Universe 54–55
The family of the Sun 118–19
The Milky Way 226–29
Beyond the Milky Way 292–325


**EVERYTHING IN THE UNIVERSE** is part of something larger. The scale of the Earth and its moon may be relatively easy for the human mind to grasp, but the nearest star is unimaginably remote, and the farthest galaxies are billions of times more distant yet. Cosmologists, who study of the size and structure of the Universe, use mathematical models to build a picture of the Universe's vast scale.

## THE SIZE OF THE UNIVERSE

Cosmologists may never determine exactly how big the Universe is. It could be infinite. Alternatively, it might have a finite volume, but even a finite Universe would have no centre or boundaries and would curve in on itself. So paradoxically, an object travelling off in one direction would eventually reappear from the opposite direction. What is certain is that the Universe is expanding and has been doing so since its origins in the Big Bang, 13.7 billion years ago (see p.46). By studying the patterns of radiation left from the Big Bang, cosmologists can estimate the minimum size of the Universe, should it turn out to be finite. Some parts must be separated by at least tens of billions of light-years. As a light-year is the distance that light travels in a year, (9,460 billion km, or 5,878 billion miles), the Universe is bewilderingly big.

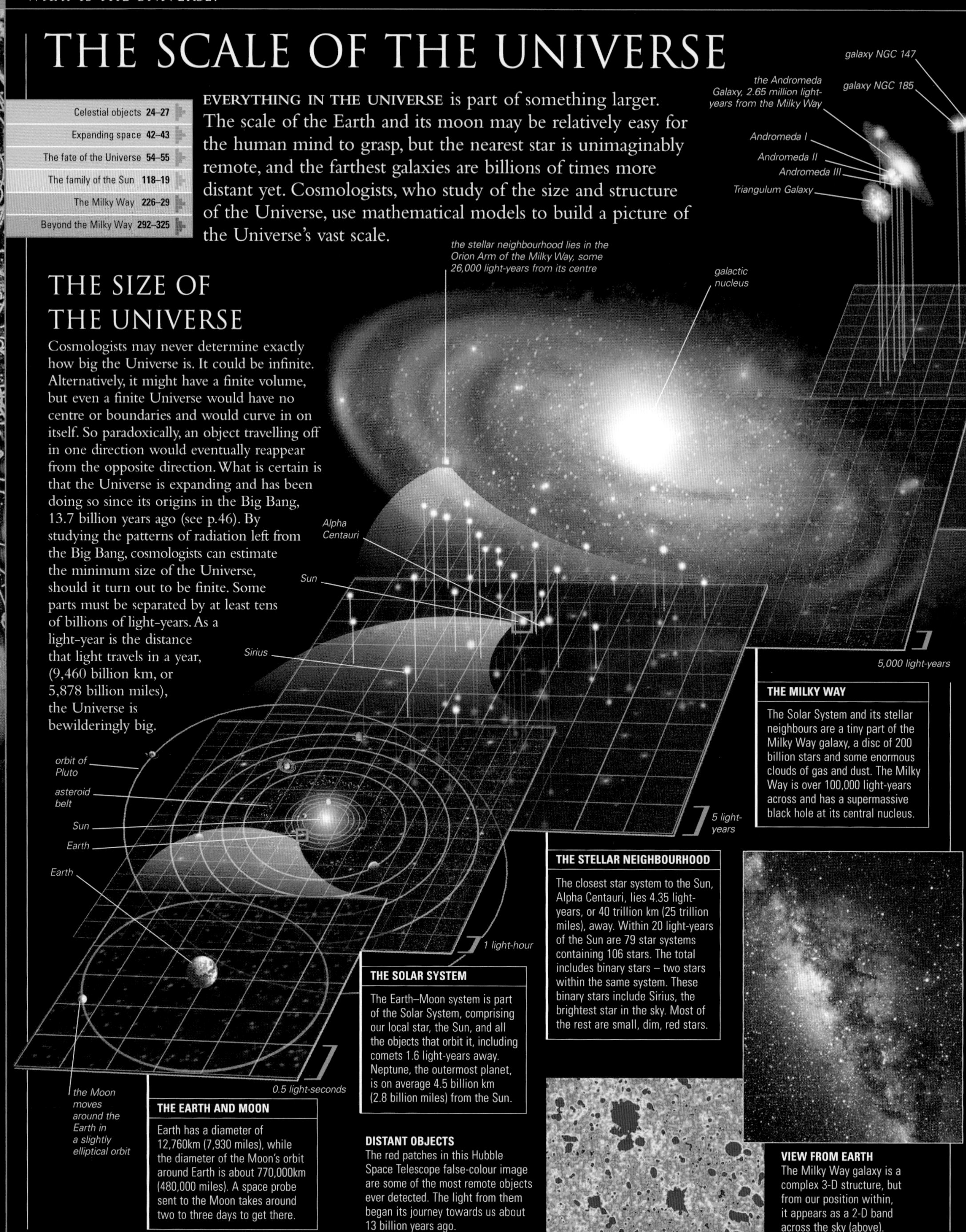


**THE MILKY WAY**

The Solar System and its stellar neighbours are a tiny part of the Milky Way galaxy, a disc of 200 billion stars and some enormous clouds of gas and dust. The Milky Way is over 100,000 light-years across and has a supermassive black hole at its central nucleus.

**THE STELLAR NEIGHBOURHOOD**

The closest star system to the Sun, Alpha Centauri, lies 4.35 light-years, or 40 trillion km (25 trillion miles), away. Within 20 light-years of the Sun are 79 star systems containing 106 stars. The total includes binary stars – two stars within the same system. These binary stars include Sirius, the brightest star in the sky. Most of the rest are small, dim, red stars.

**THE SOLAR SYSTEM**

The Earth–Moon system is part of the Solar System, comprising our local star, the Sun, and all the objects that orbit it, including comets 1.6 light-years away. Neptune, the outermost planet, is on average 4.5 billion km (2.8 billion miles) from the Sun.

**THE EARTH AND MOON**

Earth has a diameter of 12,760km (7,930 miles), while the diameter of the Moon's orbit around Earth is about 770,000km (480,000 miles). A space probe sent to the Moon takes around two to three days to get there.

**DISTANT OBJECTS**
The red patches in this Hubble Space Telescope false-colour image are some of the most remote objects ever detected. The light from them began its journey towards us about 13 billion years ago.

**VIEW FROM EARTH**
The Milky Way galaxy is a complex 3-D structure, but from our position within, it appears as a 2-D band across the sky (above).

**THE LOCAL GROUP OF GALAXIES**

The Milky Way is one of a cluster of galaxies, called the Local Group, that occupies a region 10 million light-years across. The Local Group contains 46 galaxies, only one of which – the Andromeda Galaxy – is bigger than the Milky Way. Most others are small (dwarf) galaxies.

**THE LOCAL SUPERCLUSTER**

The Local Group of galaxies, together with some nearby galaxy clusters, such as the giant Virgo Cluster, is contained within a vast structure called the Virgo Supercluster. It is 100 million light-years across and (if dwarf galaxies are included) contains tens of thousands of galaxies.

**DISTANT GALAXY CLUSTER**

The vast galaxy cluster Abell 2218 (left) is visible from Earth even though it is more than 2 billion light-years away.

**LARGE-SCALE STRUCTURE**

Galaxy superclusters clump into knots or extend as filaments that can be billions of light-years long, with large voids separating them. However, at the largest scale, the density of galaxies, and thus all visible matter, in the Universe is uniform.

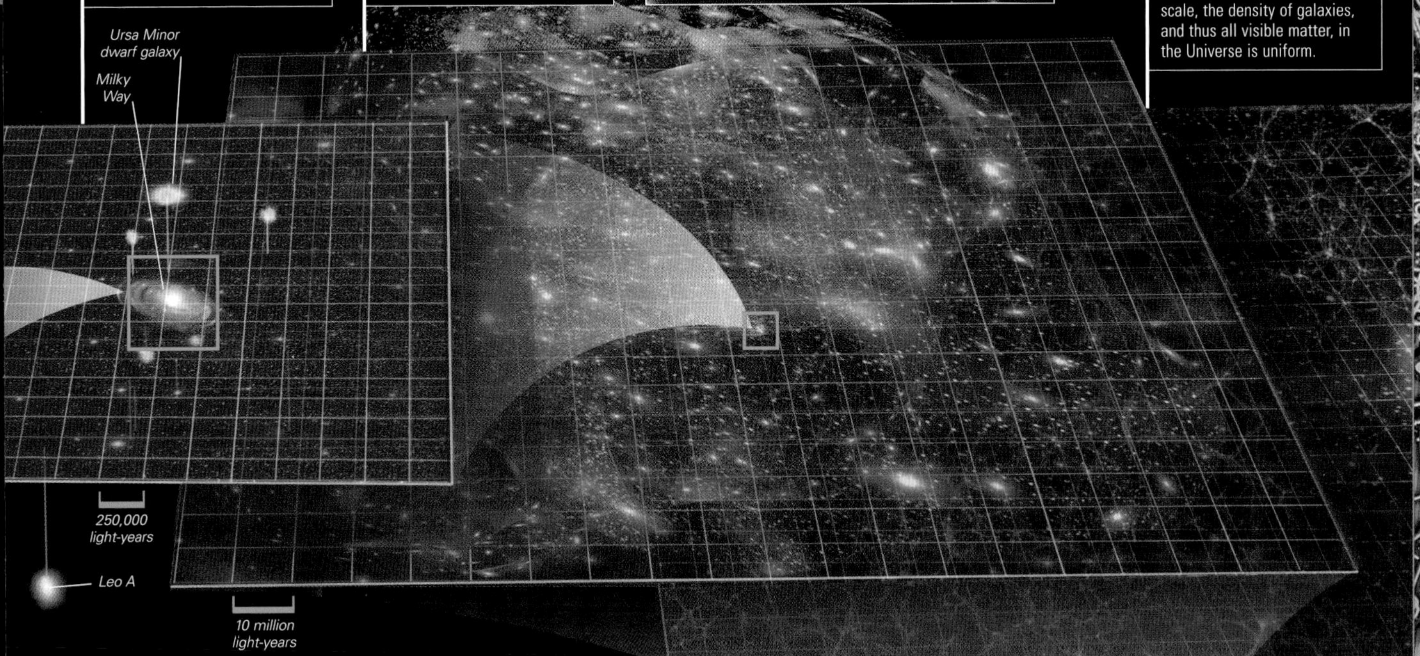

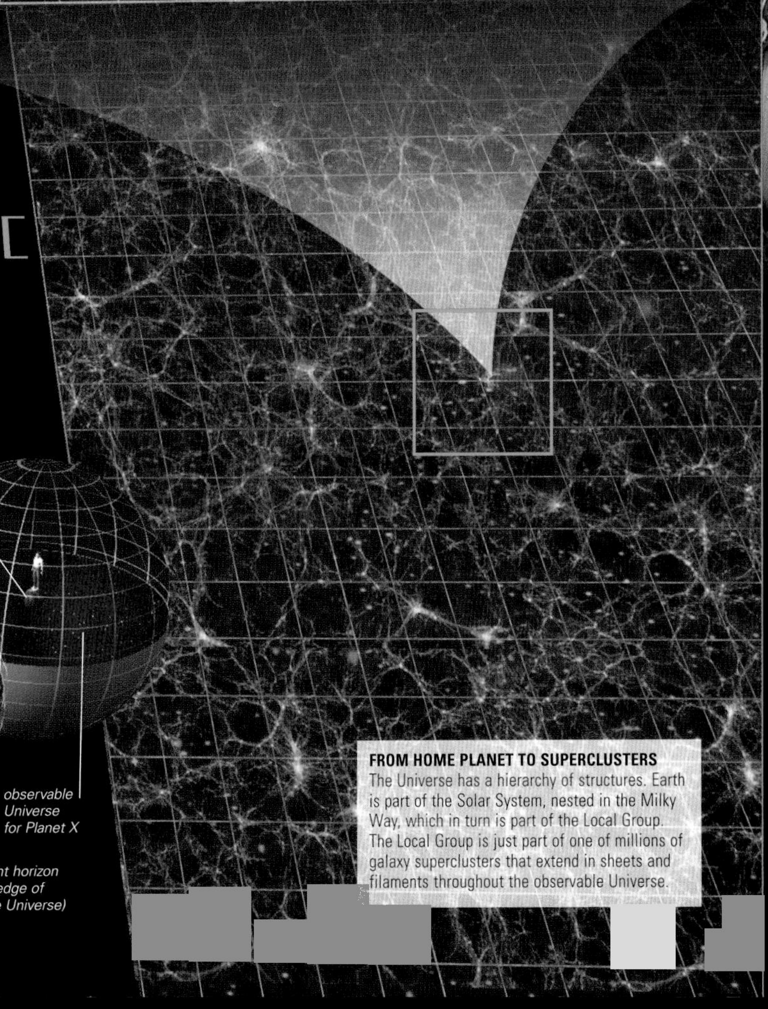

# THE OBSERVABLE UNIVERSE

Although the Universe has no edges and may be infinite, the part of it that scientists have knowledge of is bounded and finite. Called the observable Universe, it is the spherical region around Earth from which light has had time to reach us since the Universe began. The boundary that separates this region from the rest of the Universe is called the cosmic light horizon. Light reaching Earth from an object very close to this horizon must have been travelling for most of the age of the Universe, that is, approximately 13.7 billion years. This light must have travelled a distance of around 13.7 billion light-years to reach Earth. Such a distance can be defined as a "lookback" or "light-travel-time" distance between Earth and the distant object. However, the true distance is much greater, because since the light arriving at Earth left the object, the object has been carried further away by the Universe's expansion (see p.43).

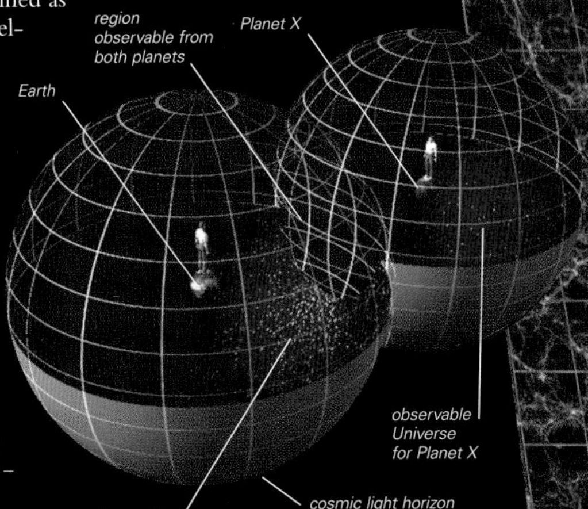

**OVERLAPPING OBSERVABLE UNIVERSES**

Earth and Planet X – an imaginary planet with intelligent life, located tens of billions of light-years away – would have different observable Universes. These may overlap, as shown here, or they may not.

**FROM HOME PLANET TO SUPERCLUSTERS**

The Universe has a hierarchy of structures. Earth is part of the Solar System, nested in the Milky Way, which in turn is part of the Local Group. The Local Group is just part of one of millions of galaxy superclusters that extend in sheets and filaments throughout the observable Universe.

# CELESTIAL OBJECTS

The family of the Sun 118–21
Stars 230–31
The life cycles of stars 232–35
Extra-solar Planets 290–91
Types of galaxy 294–95
Galaxy clusters 316–17
Galaxy superclusters 324–25

THE UNIVERSE CONSISTS of energy, space, and matter. Some of the matter drifts through space as single atoms or simple gas molecules. Other matter clumps into islands of material, from dust motes to giant suns, or implodes to form black holes. Gravity binds all of these objects into the great clouds and discs of material known as galaxies. Galaxies in turn fall into clusters and finally form the biggest celestial objects of all – superclusters.

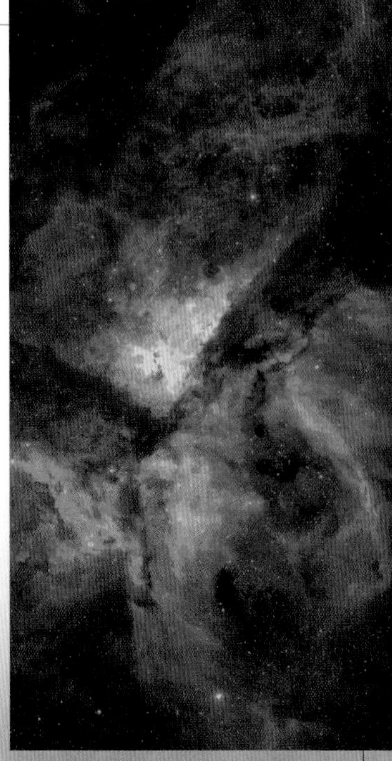

**STAR-FORMING NEBULA**
The Carina Nebula, a giant cloud of gas, is a prominent feature of the sky in the southern hemisphere and is visible to the naked eye. Different colours in this image represent temperature variations in the gas.

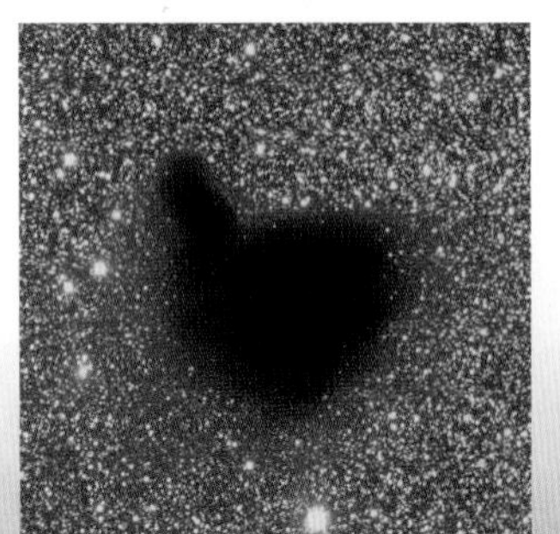

**DARK NEBULA**
A globule of dust and dense gas, Barnard 68 is an example of a dark nebula. The thick dust obscures the rich star field behind it.

## GAS, DUST, AND PARTICLES

Much of the ordinary matter of the Universe exists as a thin and tenuous gas within and around galaxies and as an even thinner gas between galaxies. The gas is made mainly of hydrogen and helium atoms, but some clouds inside galaxies contain atoms of heavier chemical elements and simple molecules. Mixed in with the galactic gas clouds is dust – tiny solid particles of carbon or substances such as silicates (compounds of silicon and oxygen). Within galaxies, the gas and dust make up what is called the interstellar medium. Visible clumps of this medium, many of them the sites of star formation, are called nebulae. Some, called emission nebulae, produce a brilliant glow as their constituent atoms absorb radiant energy from stars and re-radiate it as light. In contrast, dark nebulae are visible only as smudges that block out starlight. Particles of matter also exist in space in the form of cosmic rays – highly energetic subatomic particles travelling at high speed through the cosmos.

**GLOWING GAS**
This ocean of glowing gas is an active region of star formation in the Omega Nebula, an emission nebula. Clouds of gas and dust may give birth to stars and planets, but they are also cast off by dying stars, eventually to be recycled into the next stellar generation.

## STARS AND BROWN DWARFS

The Universe's light comes mainly from stars – hot balls of gas that generate energy through nuclear fusion in their cores. Stars form from the condensation of clumps of gas and dust in nebulae, and sometimes occur in pairs or clusters. Depending on their initial mass, stars vary in colour, surface temperature, brightness, and life span. The most massive stars, known as giants and supergiants, are the hottest and brightest, but last for only a few million years. Low-mass stars (the most numerous) are small, dim, red, and may live for billions of years; they are called red dwarfs. Smaller yet are brown dwarfs. These are failed stars, not massive or hot enough to sustain the type of fusion that occurs in stars, and emit only a dim glow. Brown dwarfs may account for a lot of the ordinary matter in the Universe.

**BROWN DWARF**
The dot at the right of centre in this picture is a brown dwarf called Gliese 229b. The bigger, brighter object is the red-dwarf star Gliese 229, around which it orbits.

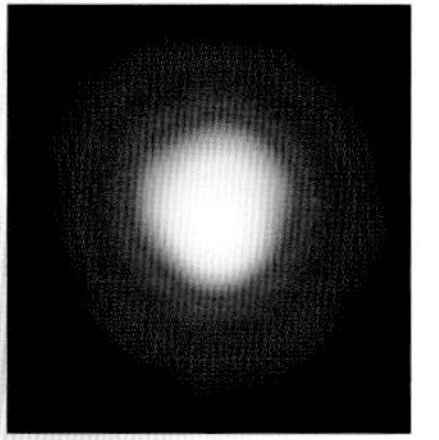

**SUPERGIANT**
The supergiant star Betelgeuse appears here as a disc because it is so big, even though it is 425 light-years away.

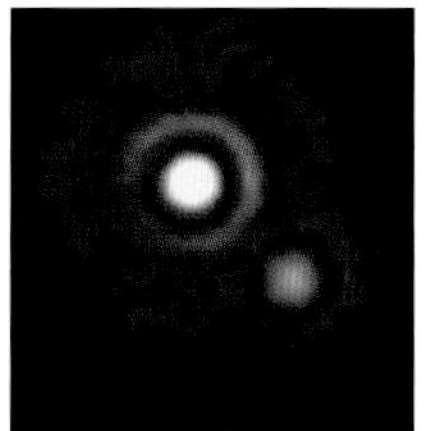

**DOUBLE STAR**
Izar is a binary, or double star, consisting of a bright yellow-orange primary star and a dimmer, bluish companion.

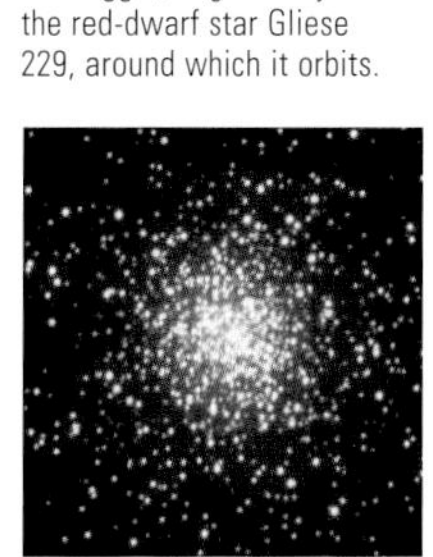

**GLOBULAR CLUSTER**
Star clusters such as M3, above, are ancient objects that orbit galaxies. M3 has about half a million stars.

## STAR REMNANTS

Stars do not last forever. Even the smallest, longest-lived red dwarfs eventually fade away. Stars of medium mass, such as the Sun, expand into large, low-density stars called red giants before they blow off most of their outer layers. They then collapse to form white dwarf stars that gradually cool and fade. The expanding shells of blown-off matter surrounding such stars are called planetary nebulae (although they have nothing to do with planets). More massive stars have yet more spectacular ends, disintegrating in explosions called supernovae. The expanding shell of ejected matter may be seen for thousands of years and is called a supernova remnant. Not all the star's material is blown off, however. Part of the core collapses to a compact, extremely dense object called a neutron star. The most massive stars of all collapse to black holes (see p.26).

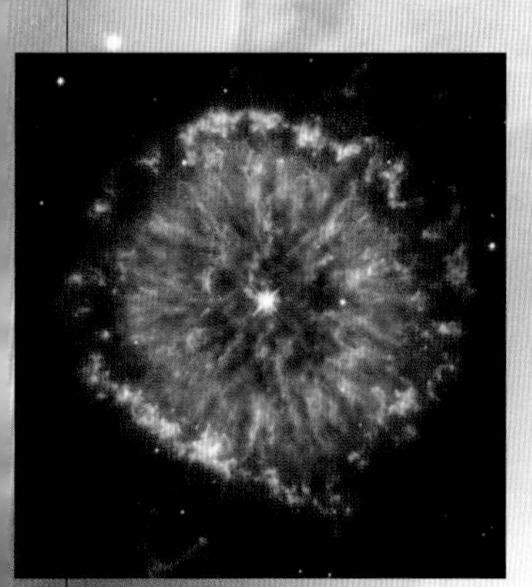

**PLANETARY NEBULA**
This glowing cloud of gas, called NGC 6751, was ejected several thousand years ago from the hot, white dwarf star visible at its centre.

**SUPERNOVA REMNANT**
The Veil Nebula is the shock wave from a star that exploded 5,000–15,000 years ago. It is 2,600 light-years away, and its material may one day form new stars.

## PLANETS AND SMALLER BODIES

The Solar System (our own star, the Sun, and everything that orbits it) is thought to have formed from dust and gas that condensed into a spinning disc called a protoplanetary disc. The central material became the Sun, while the outer matter formed planets and other small, cold objects. A planet is a sphere orbiting a star and, unlike brown dwarfs, producing no nuclear fusion. As planets and protoplanetary discs are found orbiting stars elsewhere in our galaxy, it is probable that the Solar System is typical, and that planets are common in the Universe. In the Solar System, the planets are either gas giants, such as Jupiter, or smaller, rocky bodies, such as Earth and Mars. Yet smaller objects fall into several categories. Moons are objects that orbit planets or asteroids. Asteroids are rocky bodies from about 50m (160ft) to 1,000km (600 miles) across. Comets are chunks of ice and rock, a few kilometres in diameter, that orbit in the far reaches of the Solar System. Ice dwarfs are similar but are up to several hundred kilometres across. Meteoroids are the remains of shattered asteroids or dust from comets.

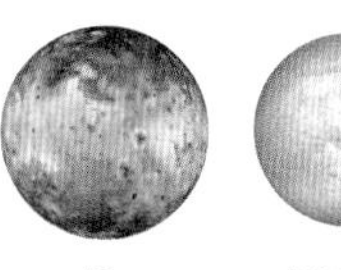

IO

EUROPA

GANYMEDE

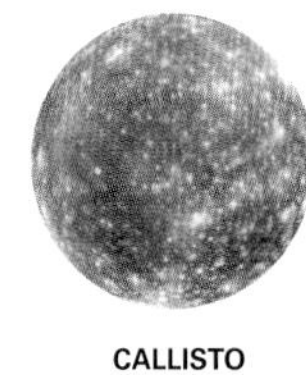

CALLISTO

**GALILEAN MOONS**
Other than Earth's own Moon, these four large moons orbiting the planet Jupiter were the first ever discovered, by Galileo Galilei in 1610 (see pp.88–89).

**PLANET EARTH**
Our home planet seems unusual in having surface water and supporting life. We do not know how rare this is in the Universe.

*coma*

*gas tail*

*dust tail*

**COMET IKEYA–ZHANG**
A few comets travel in orbits that bring them close to the Sun. Frozen chemicals in the comet then vaporize to produce a glowing coma (head) and long tails of dust and gas. This bright comet was visible in 2002.

# GALAXIES

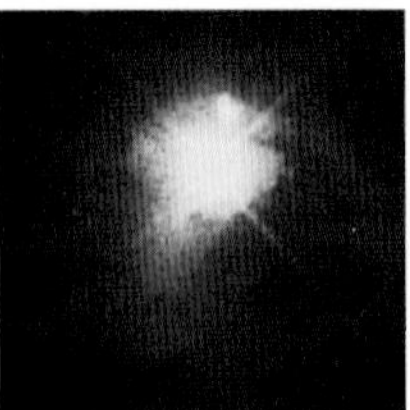

**QUASAR**
Some, if not most, galaxies are thought to have been quasars earlier in their life. Quasars are extremely luminous galaxies powered by matter falling into a massive, central black hole.

The Solar System occupies just a tiny part of an enormous, disc-shaped structure of stars, gas, and dust called the Milky Way galaxy. Until around a hundred years ago, our galaxy was thought to comprise the whole Universe; few people imagined that anything might exist outside of the Milky Way. Today, we know that just the observable part of our Universe contains more than 100 billion separate galaxies. They vary in size from dwarf galaxies, a few hundred light-years across and containing a few million stars, to giants spanning several hundred thousand light-years and containing several trillion stars. As well as stars, galaxies contain clouds of gas, dust, and dark matter (see opposite), all held together by gravity. They come in five shapes: spiral, barred spiral, elliptical (spherical to rugby-ball-shaped), lenticular (lens-shaped), and irregular. Astronomers identify galaxies by their number in one of several databases of celestial objects. For example, NGC 1530 indicates galaxy 1530 in a database called the New General Catalogue (NGC).

**SPIRAL GALAXY**
This image taken by the Spitzer Space Telescope shows a nearby spiral galaxy called M81. The sensor captured infrared radiation, rather than visible light, and the image highlights dust in the galaxy's nucleus and spiral arms.

**BARRED SPIRAL**
In a barred spiral galaxy, such as NGC 1530, above, the spiral arms radiate from the ends of the central bar-like structure, rather than from the nucleus.

# BLACK HOLES

A black hole is a region of space containing, at its centre, some matter squeezed into a point of infinite density, called a singularity. Within a spherical region around the singularity, the gravitational pull is so great that nothing, not even light, can escape. Black holes can therefore be detected only from the behaviour of material around them; those discovered so far typically have a disc of gas and dust spinning around the hole, throwing off hot, high-speed jets of material or emitting radiation (such as X-rays) as matter falls into the hole. There are two main types of black hole – supermassive and stellar. Supermassive black holes, which can have a mass equivalent to billions of suns, exist in the centres of most galaxies, including our own. Their exact origin is not yet understood, but they may be a by-product of the process of galaxy formation. Stellar black holes form from the collapsed remains of exploded supergiant stars (see p.263), and may be very common in all galaxies.

**STELLAR BLACK HOLE**
The black hole SS 433 is situated in the centre of this false-colour X-ray image. It is detectable because it is sucking in matter from a nearby star and blasting out material and X-ray radiation, visible here as two bright yellow lobes.

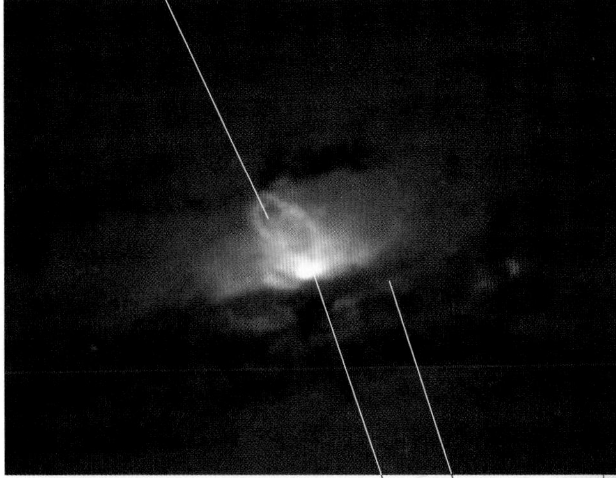


**GALACTIC BLACK HOLE**
A huge bubble of hot gas rises from a disc of dust spinning around what is thought to be a supermassive black hole in the centre of a nearby galaxy, NGC 4438.

**HICKSON COMPACT GROUP**
This cluster includes a face-on spiral galaxy in the centre of the image, two closer oblique spirals, and an elliptical galaxy at lower right.

# GALAXY CLUSTERS

Galaxies are bound by gravity to form clusters of about 20 to several thousand. Clusters vary from 3 to 30 million light-years across. Some have a concentrated central core and a well-defined spherical structure; others are irregular in shape and structure. The cluster of galaxies that contains our own galaxy is called the Local Group. The neighbouring Virgo Cluster is a large, irregular cluster of several hundred galaxies, lying 50 million light-years away. Chains of a dozen or so galaxy clusters are linked loosely by gravity and make up superclusters, which can be up to 200 million light-years in extent. Superclusters in turn are arranged in broad sheets and filaments, separated by voids of about 100 million light-years across. The sheets and voids form a network that permeates the entire observable Universe.

**RICH CLUSTER**
One of the most massive galaxy clusters known, Abell 1689 is thought to contain hundreds of galaxies (coloured gold here).

# DARK MATTER AND DARK ENERGY

There is far more matter in the Universe than that contained in stars and other visible objects. The invisible mass is called "dark matter". Its composition is unknown. Some might take the form of MACHOs (massive compact halo objects) – dark, planet-like bodies – or WIMPs (weakly interacting massive particles) – exotic subatomic entities that scarcely interact with ordinary matter. Evidence for dark matter includes the motion of galaxies in clusters. They move faster than can be explained by the gravity of visible matter – there must be further mass present. Even if all the dark matter deduced from observations is included, the density of the Universe is not sufficient to satisfy theories of its evolution. To find a solution, cosmologists have proposed the existence of "dark energy", a force that counteracts gravity and causes the Universe to expand faster (see p.54). The exact nature of dark energy is still speculative.

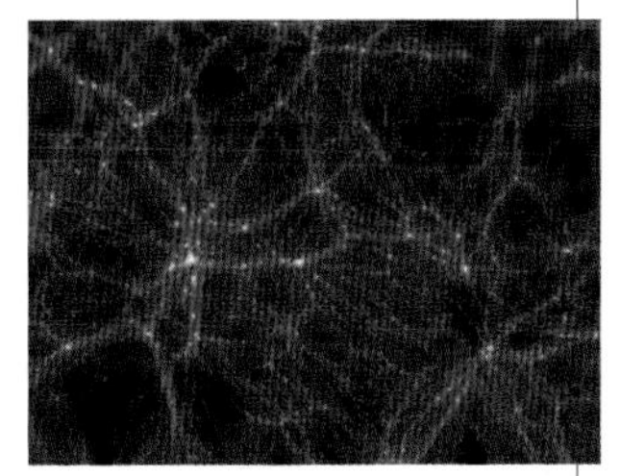

**DARK-MATTER DISTRIBUTION.**
This image from a computer simulation shows the way in which dark matter (red clumps and filaments) must be distributed within the galaxy superclusters in our local Universe.

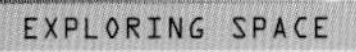

EXPLORING SPACE

## THE SEARCH FOR DARK MATTER

To find dark matter, scientists are investigating some of the several forms it could take. Underground detectors search for evasive particles, such as WIMPs and neutrinos. Neutrinos are so tiny, they were once thought to be massless, but they do have a minute mass. There are so many neutrinos in the cosmos that their combined mass could account for 1–2 per cent of the Universe's dark matter. WIMPs, if detected, could account for far more.

**NEUTRINO DETECTOR**
Neutrinos are extremely difficult to detect. This instrument is full of oil during operation. The numerous photomultiplier tubes detect flashes of light as neutrinos collide with the oil.

**DISTORTED GALAXY**
Nicknamed "the Tadpole", this galaxy lies 420 million light-years away. Like any galaxy, it is a vast, spinning wheel of matter bound together by gravity. In clusters, gravity can also rip galaxies apart. The streamer of stars emerging from this galaxy is thought to have been torn out by the gravity of a smaller galaxy passing close by.

# MATTER

24–27 Celestial objects
Radiation 32–35
Space and time 38–41
The Big Bang 46–49
Out of the darkness 50–51
The Sun 120–23

**EXAMINED AT THE TINIEST SCALE**, the Universe's matter is composed of fundamental particles, some of which, governed by various forces, group together to form atoms and ions. In addition to these well-understood types of matter, other forms exist. Most of the Universe's mass consists of this "dark matter", whose exact nature is still unknown.

**STRUCTURE OF A CARBON ATOM**
At the centre of an atom is the nucleus, which contains protons and neutrons. Electrons move around within two regions, called shells, surrounding the nucleus. The shells appear fuzzy because electrons do not move in defined paths.

**EMPTY SPACE**
*Most of an atom is empty – the protons, neutrons, and electrons are all shown here much larger than their real size relative to the whole atom*

**OUTER ELECTRON SHELL**
*Region in which four electrons orbit*

**INNER ELECTRON SHELL**
*Region within which two electrons orbit*

## WHAT IS MATTER?

Matter is anything that possesses mass – that is, anything affected by gravity. Most matter on Earth is made of atoms and ions. Elsewhere in the Universe, however, matter exists under a vast range of conditions and takes a variety of forms, from thin interstellar medium (see p.228) to the matter in infinitely dense black holes (see p.263). Not all of this matter is made of atoms, but all matter is made of some kind of particle. Certain types of particle are fundamental – that is, they are not made of smaller sub-units. The most common particles within ordinary matter are quarks and electrons, which make up atoms and ions and form all visible matter. Most of the Universe's matter, however, is not ordinary matter, but dark matter (see p.27), perhaps composed partly of neutrinos, theoretical WIMPs (weakly interacting massive particles), or both.

**LUMINOUS MATTER**
These illuminated gas clouds in interstellar space are made of ordinary matter, composed of atoms and ions.

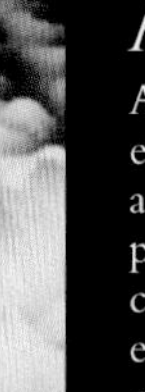

**IMAGING THE ATOM**
This image of gold atoms on a grid of green carbon atoms was made by a scanning-tunnelling microscope.

## ATOMS AND IONS

Atoms are composed of fundamental particles called quarks and electrons. The quarks are bound in groups of three by gluons, which are massless particles of force. The quark groups form particles called protons and neutrons. These are clustered in a compact region at the centre of the atom called the nucleus. Most of the rest of an atom is empty space, but moving around within this space are electrons. These carry a negative electrical charge and have a very low mass – nearly all the mass in an atom is in the protons and neutrons. Atoms always contain equal numbers of protons (positively charged) and electrons (negatively charged) and so are electrically neutral. If they lose or gain electrons, they become charged particles called ions.

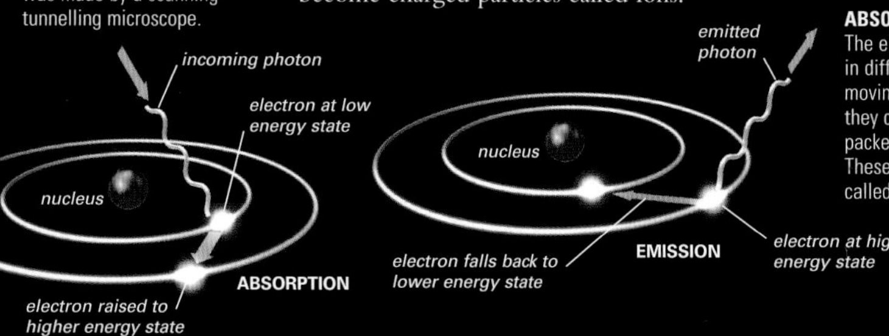


**ABSORPTION AND EMISSION**
The electrons in atoms can exist in different energy states. By moving between energy states they can either absorb or emit packets, or quanta, of energy. These energy packets are called photons.

incoming high-energy photon
inner-shell electron
nucleus
electron in outer shell
neutron
proton
ATOM (NEUTRAL, NO CHARGE)
ejected electron (charge -1)
nucleus
empty shell
ION (CHARGE +1)

**IONIZATION**
One way an atom may become a positive ion is by the electron absorbing energy from a high-energy photon and, as a result, being ejected, along with its charge, from the atom.

red quark
gluon
green quark
blue quark
neutron
proton

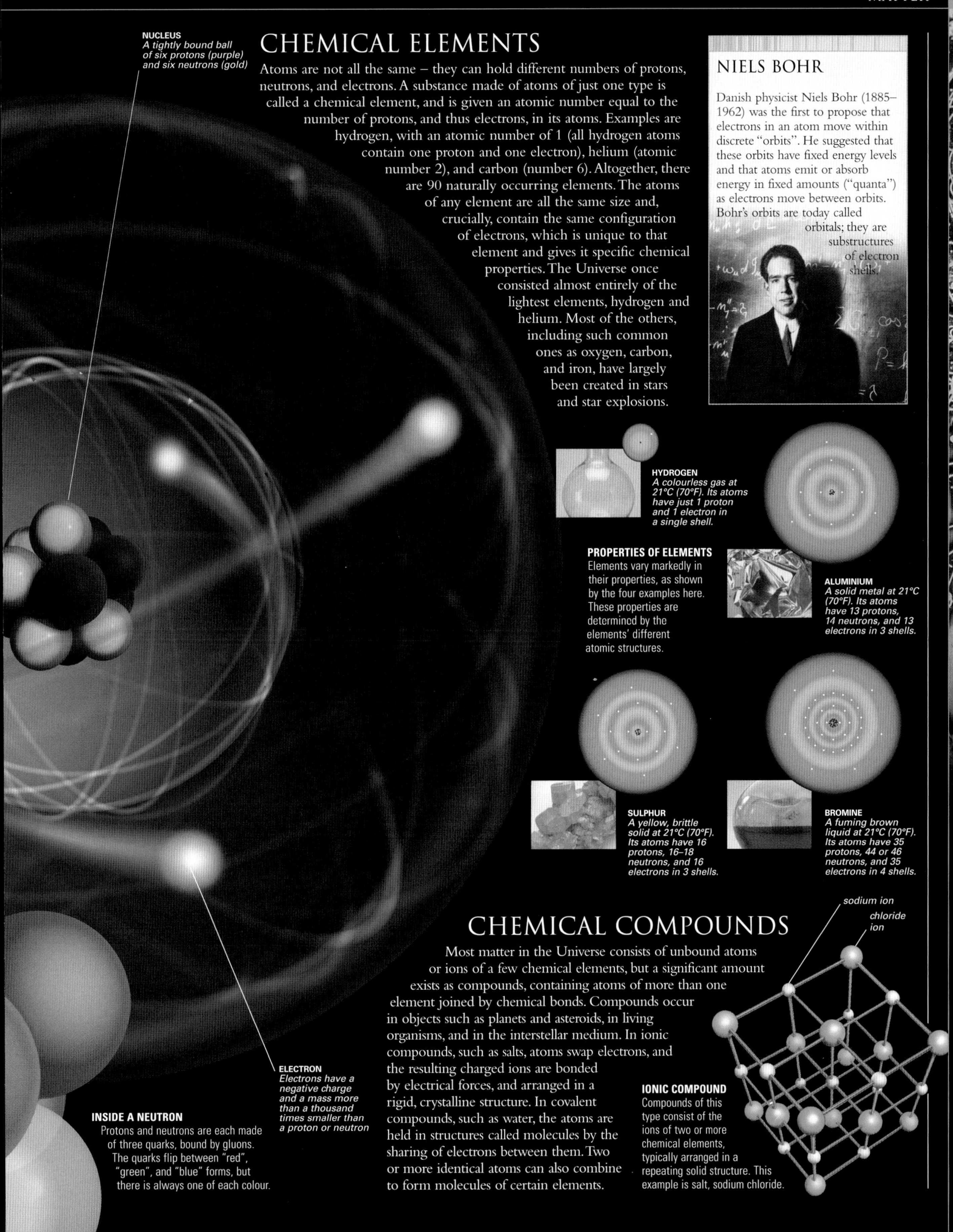

**NUCLEUS**
*A tightly bound ball of six protons (purple) and six neutrons (gold)*

# CHEMICAL ELEMENTS

Atoms are not all the same – they can hold different numbers of protons, neutrons, and electrons. A substance made of atoms of just one type is called a chemical element, and is given an atomic number equal to the number of protons, and thus electrons, in its atoms. Examples are hydrogen, with an atomic number of 1 (all hydrogen atoms contain one proton and one electron), helium (atomic number 2), and carbon (number 6). Altogether, there are 90 naturally occurring elements. The atoms of any element are all the same size and, crucially, contain the same configuration of electrons, which is unique to that element and gives it specific chemical properties. The Universe once consisted almost entirely of the lightest elements, hydrogen and helium. Most of the others, including such common ones as oxygen, carbon, and iron, have largely been created in stars and star explosions.

## NIELS BOHR

Danish physicist Niels Bohr (1885–1962) was the first to propose that electrons in an atom move within discrete "orbits". He suggested that these orbits have fixed energy levels and that atoms emit or absorb energy in fixed amounts ("quanta") as electrons move between orbits. Bohr's orbits are today called orbitals; they are substructures of electron shells.

**HYDROGEN**
*A colourless gas at 21°C (70°F). Its atoms have just 1 proton and 1 electron in a single shell.*

**PROPERTIES OF ELEMENTS**
Elements vary markedly in their properties, as shown by the four examples here. These properties are determined by the elements' different atomic structures.

**ALUMINIUM**
*A solid metal at 21°C (70°F). Its atoms have 13 protons, 14 neutrons, and 13 electrons in 3 shells.*

**SULPHUR**
*A yellow, brittle solid at 21°C (70°F). Its atoms have 16 protons, 16–18 neutrons, and 16 electrons in 3 shells.*

**BROMINE**
*A fuming brown liquid at 21°C (70°F). Its atoms have 35 protons, 44 or 46 neutrons, and 35 electrons in 4 shells.*

# CHEMICAL COMPOUNDS

Most matter in the Universe consists of unbound atoms or ions of a few chemical elements, but a significant amount exists as compounds, containing atoms of more than one element joined by chemical bonds. Compounds occur in objects such as planets and asteroids, in living organisms, and in the interstellar medium. In ionic compounds, such as salts, atoms swap electrons, and the resulting charged ions are bonded by electrical forces, and arranged in a rigid, crystalline structure. In covalent compounds, such as water, the atoms are held in structures called molecules by the sharing of electrons between them. Two or more identical atoms can also combine to form molecules of certain elements.

*sodium ion*
*chloride ion*

**IONIC COMPOUND**
Compounds of this type consist of the ions of two or more chemical elements, typically arranged in a repeating solid structure. This example is salt, sodium chloride.

**ELECTRON**
*Electrons have a negative charge and a mass more than a thousand times smaller than a proton or neutron*

**INSIDE A NEUTRON**
Protons and neutrons are each made of three quarks, bound by gluons. The quarks flip between "red", "green", and "blue" forms, but there is always one of each colour.

# STATES OF MATTER

Ordinary matter exists in four states, called solid, liquid, gas, and plasma. These differ in the energy of the matter's particles (molecules, atoms, or ions) and in the particles' freedom to move relative to one another. Substances can transfer between states, by losing or gaining heat energy, for instance. The constituents of a solid are locked by strong bonds and can hardly move, whereas in a liquid they are bound only by weak bonds and can move freely. In a gas, the particles are bound very weakly and move with greater freedom, occasionally colliding. A gas becomes a plasma when it is so hot that collisions start to knock electrons out of its atoms. A plasma therefore consists of ions and electrons moving extremely energetically. Because stars are made of plasma, it is the most common state of ordinary matter in the Universe; the gaseous state is the second most common.

**SOLID, LIQUID, AND GAS**
On Earth, water can sometimes be found as a liquid, in solid form (ice or snow), and as a gas (water vapour), all in close proximity.

# FORCES INSIDE MATTER

The bonds that link the constituents of solids, liquids, gases, and plasma are based on the electromagnetic (EM) force. This force attracts particles of unlike electrical charge and repels like charges. It is one of three forces that control the small-scale structure of matter. The others are the strong nuclear force, composed of fundamental and residual parts; and the weak nuclear force or interaction. Together with a fourth force, gravity, these are the fundamental forces of nature. The EM, weak, and strong forces are mediated by force-carrier particles, which belong to a group of particles called the bosons. The EM force, as well as binding atoms in solids and liquids, also holds electrons within atoms. The strong force holds together protons, neutrons, and atomic nuclei. The weak force brings about radioactive decay and other nuclear interactions.

**PLASMA**
Plasma exists naturally in stars but can also be artificially created. In a plasma ball, electricity is induced to flow from a charged metal ball through a gas to the surface of a glass sphere, creating plasma streamers.

neutron
red down quark
fundamental strong nuclear force
gluon, the force particle
green up quark
blue down quark

**FUNDAMENTAL STRONG NUCLEAR FORCE**
Also known as the colour force, this force binds quarks within protons and neutrons. It controls the quarks' "colour" property, and as it operates, the quarks constantly change "colour" by exchanging virtual gluons (the force-carrier particles).

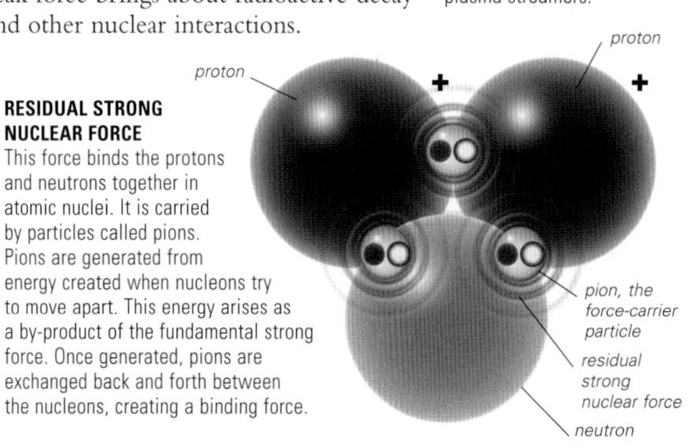


**RESIDUAL STRONG NUCLEAR FORCE**
This force binds the protons and neutrons together in atomic nuclei. It is carried by particles called pions. Pions are generated from energy created when nucleons try to move apart. This energy arises as a by-product of the fundamental strong force. Once generated, pions are exchanged back and forth between the nucleons, creating a binding force.

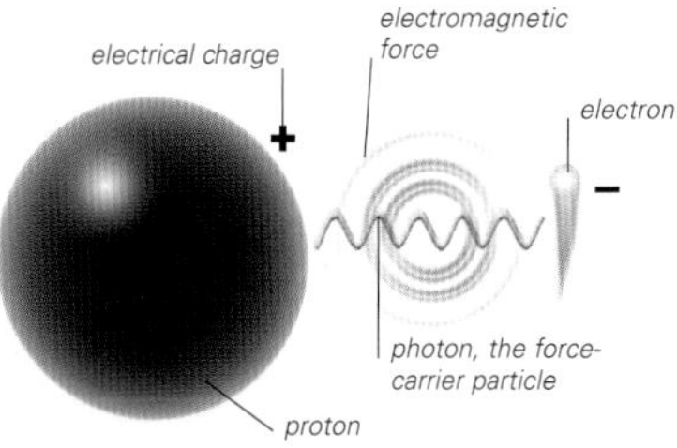


**ELECTROMAGNETIC FORCE**
Within an atom, the electromagnetic (EM) force holds the electrons within the shells surrounding the nucleus. It attracts the negatively electrically charged electrons towards the positively charged nucleus and keeps electrons apart. The force carrier for the EM force is the photon.

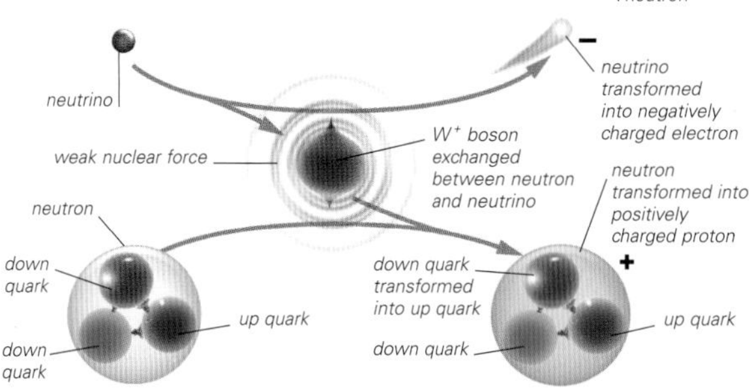


**WEAK INTERACTION, OR WEAK NUCLEAR FORCE**
This force governs radioactive decay, among other interactions. Its force-carrier particles are $W^+$, $W^-$, and $Z^0$ bosons. Here, a $W^+$ boson controls the changing of a neutrino into an electron and the transformation of a down quark into an up quark, converting a neutron into a proton.

## STEVEN WEINBERG

The American physicist Steven Weinberg (b.1933) is best known for his theory that two of the fundamental forces – the weak interaction and the electromagnetic force – are unified, or work in an identical way, at extremely high energy levels, such as those existing just after the Big Bang (see p.46). Weinberg's so-called electroweak theory was confirmed by particle accelerator experiments in 1973. He and his colleagues received the 1979 Nobel prize for physics.

# PARTICLE PHYSICS

For some decades, physicists have directed research towards a better understanding of matter and the four fundamental forces. Part of the purpose has been to clarify what happened in the early Universe, shortly after the Big Bang. Research is centred on smashing particles together in devices called particle accelerators. These experiments have identified hundreds of different particles (most of them highly unstable), which differ in their masses, electric charges, other properties such as "spin", and in the fundamental forces they are subject to. Known particles, and their interactions, are currently explained by a theory called the standard model of particle physics. This explains the properties of most of the particles (see table, right). One exception is the graviton, a hypothetical particle thought to carry the force of gravity. The graviton does not fit into the scheme, because the best theory of gravity (general relativity, see pp.40–41) is incompatible with aspects of the standard model. New theories such as string theory (see panel, below) attempt to unite gravity with particle physics.

**SPRAY OF PARTICLES**
This image from a detector within a particle accelerator shows a spray of light particles shooting to the right, following collision of two higher-mass particles on the left.

## CLASSIFICATION OF PARTICLES

Physicists distinguish composite particles, which have an internal structure, from fundamental particles, which do not. They also divide particles into fermions and bosons. Fermions (leptons, quarks, and baryons) are the building blocks of matter. Bosons (gauge bosons and mesons) are primarily force-carrier particles.

### FUNDAMENTAL PARTICLES

Leptons and quarks form matter, while gauge bosons carry forces. Quarks feel the strong nuclear force, but leptons do not.

**LEPTONS**

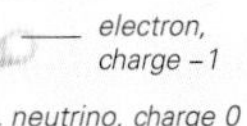


Six different leptons exist, but the 2 above are the only stable ones and are those that occur in ordinary matter.

**QUARKS**

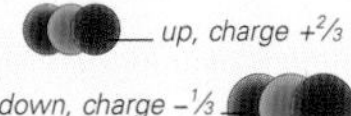


There are 6 "flavours" of quark, but only 2 occur in ordinary matter: "up" and "down". Each can exist in any of 3 "colours".

**GAUGE BOSONS**

These are force-carrier particles. Some shown are hypothetical.

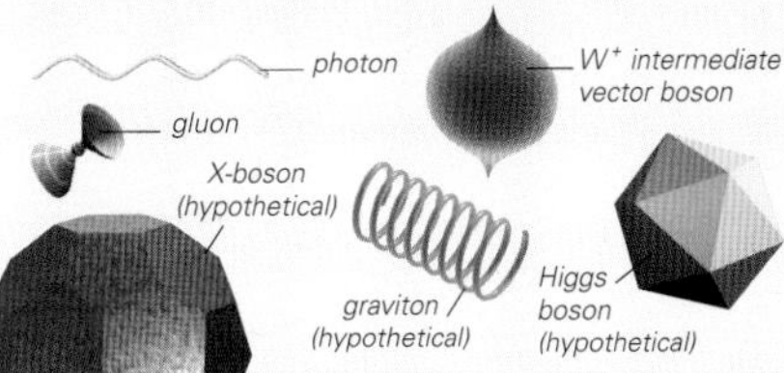


### COMPOSITE PARTICLES

Also known as hadrons, these are composed of quarks, antiquarks, or both, bound by gluons.

**BARYONS**
Relatively large-mass particles containing 3 quarks.

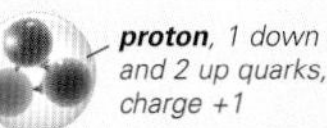


**MESONS**
Particles containing a quark and an antiquark.

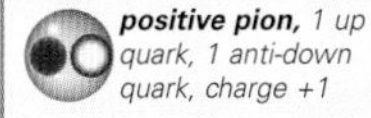


Hundreds of other baryons and mesons exist.

### ANTIPARTICLES

Most particles have an antimatter equivalent that has the same mass, but whose charge and other properties are opposite.

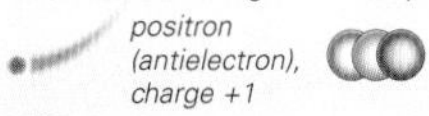
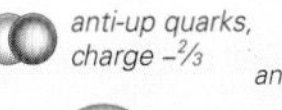


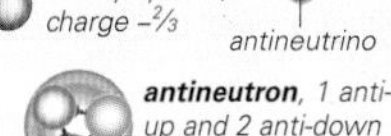


### EXOTIC PARTICLES

Further particles have been hypothesized that do not have a place in this particle classification. They include magnetic monopoles and WIMPs (weakly interacting massive particles).

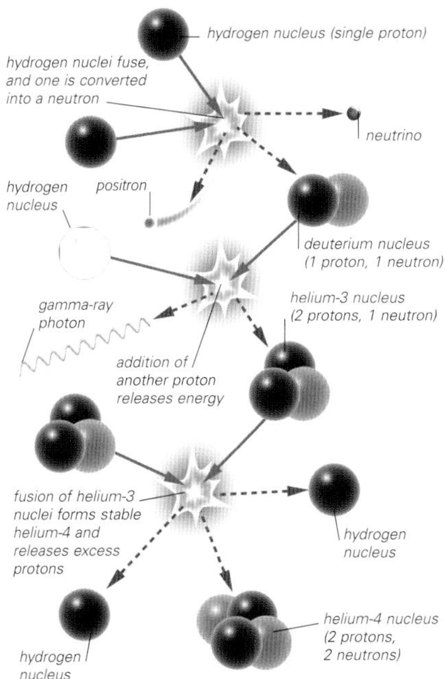


**FUSION REACTION IN THE SUN**
In stars the size of the Sun or smaller, the dominant energy-producing fusion process is called the proton–proton chain. This chain of high-energy collisions fuses hydrogen nuclei (free protons), via several intermediate stages, into helium-4 nuclei. Energy is released in the form of gamma-ray photons and in the movement energy of the helium nuclei. Positrons and neutrinos are also produced.

# NUCLEAR FISSION AND FUSION

Twentieth-century physicists learned that atomic nuclei are not immutable but can break up or join together. In nature, unstable atomic nuclei can spontaneously disassemble, giving off particles and energy, measured as radioactivity. Similarly, in the artificial process of nuclear fission, large nuclei are intentionally split into smaller parts, with huge energy release. On a cosmic scale, a more important phenomenon is nuclear fusion. In this process, atomic nuclei join, forming a larger nucleus and releasing energy. Fusion powers stars and has created the atoms of all chemical elements heavier than beryllium. The most common fusion reaction in stars joins hydrogen nuclei (protons) into helium nuclei. In this and other fusion reactions, the products of the reaction have a slightly lower mass than the combined mass of the reactants. The lost mass converts into huge amounts of energy, in accordance with Einstein's famous equation $E=mc^2$ that links energy (E), mass (m), and the speed of light (c) (see p.39).

**THE HEAT OF FUSION**
All solar energy comes from nuclear fusion in the Sun's core. The energy gradually migrates to the Sun's surface and into space through heat transfer by convection, conduction, and radiation.

EXPLORING SPACE

## STRING THEORY

For decades, physicists have sought a "theory of everything" (see Quantum gravity, p.39) that will unify the four fundamental forces of nature and provide an underlying scheme for how particles are constructed. A leading contender is string theory, which proposes that each fundamental particle consists of a tiny vibrating filament called a string. The vibrational modes, or frequencies, of these strings lend particles their varied properties. Although it sounds bizarre, many leading physicists are enthusiastic adherents of string theory.

LOW-FREQUENCY STRING

HIGH-FREQUENCY STRING

**VIBRATING STRINGS**
A string is closed, like a loop, or open, like a hair. The two closed strings shown here are vibrating at different resonant frequencies, just as the strings on a guitar have rates at which they prefer to vibrate.

# RADIATION

28–31 Matter
Light and gravity 40
The birth of astrophysics 90
Telescope technology 93
Stars 230–31


RADIATION IS ENERGY IN THE FORM of waves or particles that are emitted from a source and can travel through space and some types of matter. Electromagnetic (EM) radiation includes light, X-rays, and infrared radiation. Particulate radiation consists of fast-moving charged particles such as cosmic rays and particles emitted in radioactive decay. EM radiation is vastly more significant in astronomy.

## ELECTROMAGNETIC RADIATION

Energy in the form of EM radiation is one of the two main components of the Universe, the other being matter (see p.28). This type of radiation is produced by the motion of electrically charged particles, such as electrons. A moving charge gives rise to a magnetic field. If the motion is constant, then the magnetic field varies and in turn produces an electric field. By interacting with each other, the two fields sustain one another and move through space, transferring energy. As well as visible light, EM radiation includes radio waves, microwaves, infrared (heat) radiation, ultraviolet radiation, X-rays, and gamma rays. All these phenomena travel through space at the same speed – called the velocity of light. This speed is very nearly 300,000km (186,000 miles) per second or 1 billion kph (670 million mph).

electric field strength
amplitude
magnetic field strength
wavelength

**HOW WAVES TRAVEL**
An EM wave consists of oscillating electrical and magnetic fields arranged perpendicular to each other, and carrying energy forward.

## WAVE-LIKE BEHAVIOUR

In most situations, EM radiation acts as a wave – a disturbance moving energy from one place to another. It has properties such as wavelength (the distance between two successive peaks of the wave) and frequency (the number of waves passing a given point each second). This wave-like nature is shown by experiments such as the double-slit test (see below), in which light waves diffract (spread out) after passing through a slit and also interfere with each other as their peaks and troughs overlap. The forms of EM radiation differ only in wavelength, but this affects other properties, such as penetrating power and ability to ionize atoms (see p.28).

**INTERFERING WAVES**
The slit experiment is analogous to disturbing two points on the surface of a liquid. The ripples interfere to corrugate the liquid's surface.

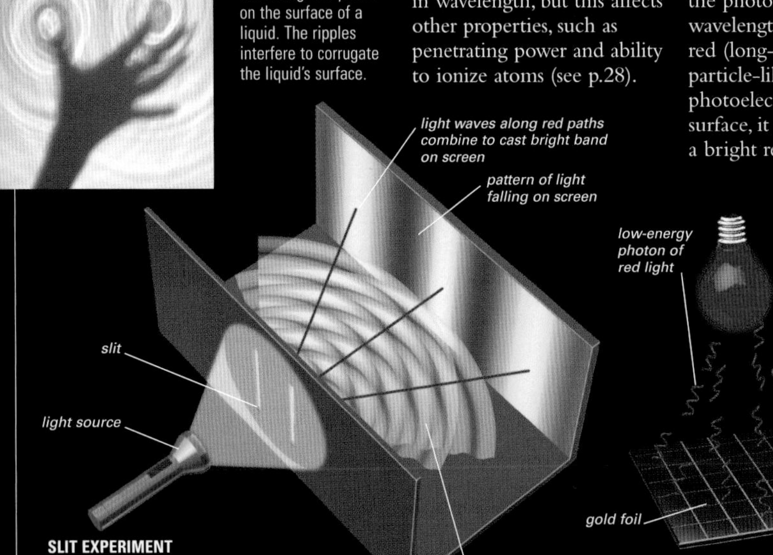


**SLIT EXPERIMENT**
If light is shone at a card containing two slits, diffraction spreads the light waves out like ripples emanating in arcs from each slit. The two wave trains then interfere to produce a light and dark pattern on the screen.

## PARTICLE-LIKE BEHAVIOUR

EM radiation behaves mainly like a wave, but it can also be considered to consist of tiny packages or "quanta" of energy called photons. These have no mass but carry a fixed amount of energy. The energy in a photon depends on its wavelength – the shorter the wavelength, the more energetic the photon. For example, photons of blue (short-wavelength) light are more energetic than photons of red (long-wavelength) light. A classic demonstration of light's particle-like properties is provided by a phenomenon called the photoelectric effect (see below). If a blue light is shone at a metal surface, it causes electrons to be ejected from the metal, whereas even a bright red light has no such effect.

low-energy photon of red light
gold foil
high-energy photon of blue light
low-energy electron
ultra-high-energy ultraviolet photon
electron ejected at higher energy

**RED LIGHT**
When red light is shone on a metal surface, no electrons are ejected, even if the light is intensely bright.

**BLUE LIGHT**
Blue light shining on the same surface causes electrons to be ejected, because the blue photons are more energetic.

**ULTRAVIOLET LIGHT**
Shining ultraviolet light on the metal surface causes electrons to be ejected at very high energy.

# ANALYSING LIGHT

The radiation output of celestial objects is a mixture of wavelengths. When passed through a prism, the light is split into its component wavelengths, giving a record called a spectrum. A star's spectrum usually contains dark lines called absorption lines, caused by photons being absorbed at certain wavelengths by atoms in the star's atmosphere. They can be used to establish what chemical elements are present. The spectrum of a nebula can also reveal its composition. When heated by radiation from a nearby star, its atoms emit their own light. The resulting spectrum, called an emission spectrum, consists of a series of bright lines characteristic of different elements.

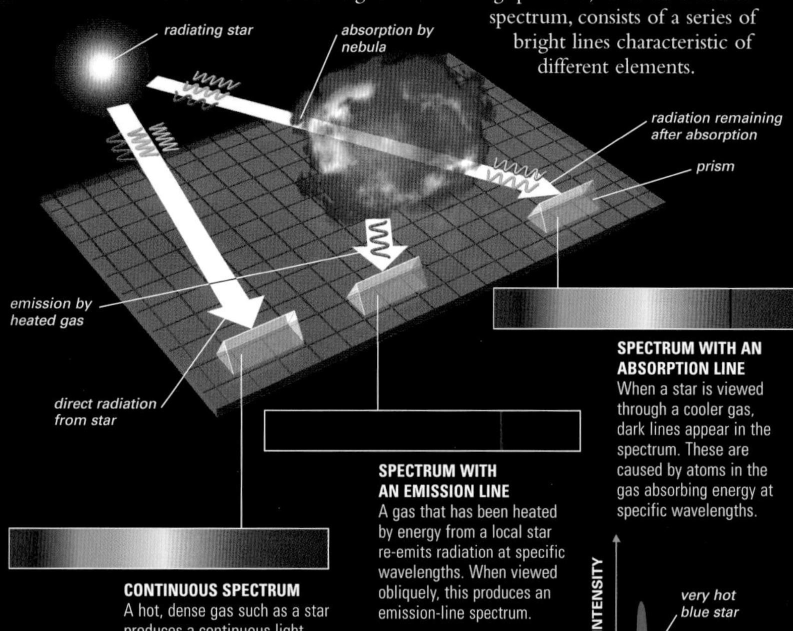


**SPECTRUM WITH AN ABSORPTION LINE**
When a star is viewed through a cooler gas, dark lines appear in the spectrum. These are caused by atoms in the gas absorbing energy at specific wavelengths.

**SPECTRUM WITH AN EMISSION LINE**
A gas that has been heated by energy from a local star re-emits radiation at specific wavelengths. When viewed obliquely, this produces an emission-line spectrum.

**EMISSION NEBULA**
This nebula glows as its gas is heated by nearby stars. The emitted light consists of photons of a few specific wavelengths. These photons were emitted by the gas's atoms as their electrons settled to lower energy states.

**CONTINUOUS SPECTRUM**
A hot, dense gas such as a star produces a continuous light spectrum from its surface, with all different light wavelengths (colours) represented.

**RADIATION FROM HOT OBJECTS**
Not only is the total radiation greater from hotter objects, but the wavelength of peak intensity is also shorter (towards the blue end of the light spectrum). Astronomers can calculate the temperature of stars by measuring the peak of the star's spectrum.

INTENSITY
very hot blue star
hot yellow star, such as the Sun
cooler red star
Earth
WAVELENGTH

# RED SHIFT AND BLUE SHIFT

The spectrum of radiation received by an observer shifts if the source of the radiation is moving relative to the observer – and celestial objects are always moving. Astronomers can detect the shifts by measuring the position of spectral lines, which occur at characteristic places. Observers watching an object moving away see its spectral lines shifted towards longer wavelengths (a red shift). For an approaching object, the lines are shifted to the shorter wavelengths (a blue shift). The greater the relative velocity between source and observer, the greater the shift. Distant galaxies show large red shifts, indicating they are receding at enormous speeds; these are called cosmological red shifts.

## LOUIS DE BROGLIE

The French physicist Louis de Broglie (1892–1987) received the Nobel prize in 1929. He found that particles of matter, such as electrons, have wave-like properties. The dual nature of matter and light (each has both particle-like and wave-like properties) is called wave-particle duality.

galaxy receding from observer 1 and approaching observer 2

wavefront of emitted radiation

wavefronts spread out

wavefronts bunched up

OBSERVER 1

OBSERVER 2

RED-SHIFTED SPECTRUM LINE

BLUE-SHIFTED SPECTRUM LINE

**SHIFTING WAVELENGTHS**
Shifts occur because of a phenomenon called the Doppler effect. The wavefronts of light from a receding object are stretched out, increasing their wavelength, while those of an approaching object are squashed up.

# ACROSS THE SPECTRUM

Celestial objects can emit radiation right across the EM spectrum, from radio waves, through visible light to gamma rays. Some complex objects, such as galaxies and supernova remnants, shine at nearly all these wavelengths. Cool objects tend to radiate photons with less energy and are therefore only visible at longer wavelengths. Towards the gamma-ray end of the spectrum, photons are increasingly powerful. High-energy X-rays and gamma rays originate only from extremely hot sources, such as the gas of galaxy clusters (see p.317) or violent events, such as the swallowing of matter by black holes (see p.263). To detect all this radiation and form images, astronomers need a range of instruments – each type of radiation has different properties and must be collected and focused in a particular way. Radiation at many wavelengths does not penetrate to Earth's surface, and is detectable only by orbiting observatories above the atmosphere.

parabolic dish reflects all incoming radio waves to the subreflector

receiver

subreflector focuses the radio waves onto receiver

primary reflector dish

Sun shield

**RADIO WAVES** TELESCOPE ARRAY
Radio waves can be many metres long. To create sharp images from such long waves, astronomers collect and focus them using telescopes with huge dish antennae. They might use a single dish or an entire array. The Very Large Array (right), in New Mexico, is the world's largest array. It consists of 27 dishes, each 25m (82ft) across, moving on a Y-shaped railway network. Their data combine to form a single, fine-detailed image, the dishes effectively forming one giant, 27km (16-mile) antenna.

**MICROWAVES** SPACE PROBE
Most microwaves are absorbed by Earth's atmosphere, so microwave observatories must be placed in space. Launched in 2001, the Wilkinson Microwave Anisotropy Probe (WMAP, above) is a NASA spacecraft with a goal to map the cosmic microwave background radiation (see p.50) across the whole sky. This is the oldest electromagnetic radiation in the Universe, released soon after the Big Bang. The probe was placed in a stable orbit around the Sun 1.5 million km (900,000 miles) from Earth.

**INFRARED** MOUNTAINTOP TELESCOPE
Little infrared radiation from space reaches sea level on Earth, but some penetrates down to the height of mountaintops. Some infrared telescopes, such as NASA's Spitzer Space Telescope (see p.245), have been launched into space, but most infrared astronomy is conducted from mountain observatories. This one, the United Kingdom Infrared Telescope (UKIRT) is at 4,194m (13,760ft) in Hawaii. Like optical telescopes, it uses mirrors to collect and focus the radiation. With a 3.8m (12.5ft) mirror, UKIRT achieves great brightness and resolution. It can pick up dim galaxies, brown dwarfs, nebulae, and interstellar molecules glowing in the infrared, and it can peer into star-forming nebulae to image the young stars shining within.

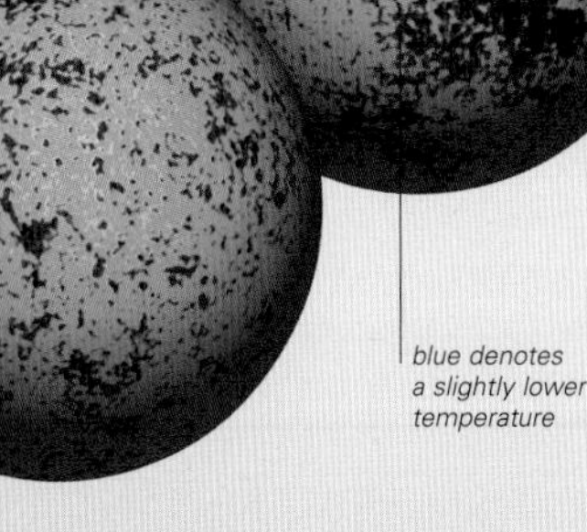

*red denotes a fractionally higher temperature*

*blue denotes a slightly lower temperature*

**RADIO WAVES** GALAXY
In this map of the Andromeda Galaxy produced by a radio telescope, red and yellow indicate the highest-intensity radio-wave emissions. To produce such an image, a radio dish must scan an area of sky. As it points at each location in the sky in turn, the telescope gradually builds a picture by recording the radio intensity at each location. The resolution is low because radio waves are so long. Radio emissions are produced by hydrogen clouds in galaxies, or by synchrotron radiation from active galaxies (see p.310) and black holes (see p.263).

**MICROWAVES** UNIVERSE
The lack of microwave sources in the nearby Universe is fortunate, because it reduces difficulties in observing the cosmic background radiation, which reaches Earth at microwave wavelengths. The pattern of microwaves from the whole sky, as measured by WMAP, is here projected onto two hemispheres.

**INFRARED** GALACTIC CENTRE
This infrared image penetrates to the central region of the Milky Way galaxy, which in visible light is hidden behind thick clouds of dust. The core of the galaxy is at upper left. The reddening of the stars in that area and along the galactic plane is caused by dust scattering.

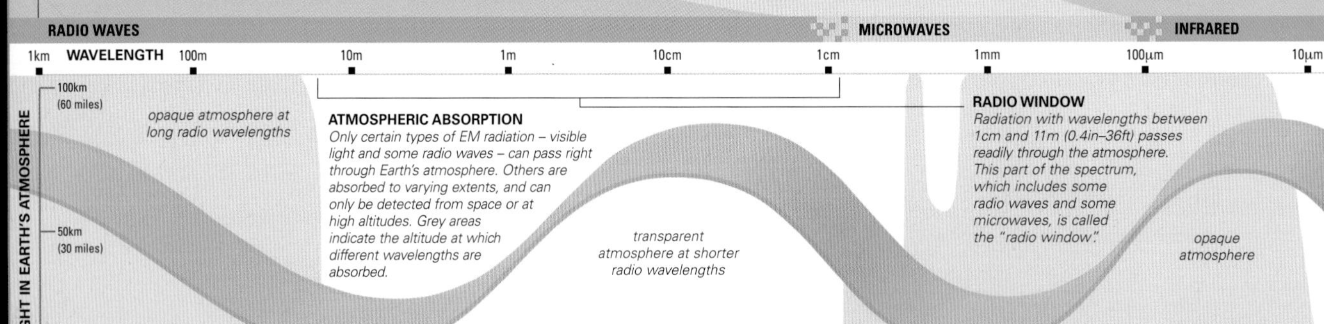
RADIO WAVES
MICROWAVES
INFRARED
1km
WAVELENGTH
100m
10m
1m
10cm
1cm
1mm
100μm
10μm
HEIGHT IN EARTH'S ATMOSPHERE
100km (60 miles)
50km (30 miles)
0
opaque atmosphere at long radio wavelengths
ATMOSPHERIC ABSORPTION
Only certain types of EM radiation – visible light and some radio waves – can pass right through Earth's atmosphere. Others are absorbed to varying extents, and can only be detected from space or at high altitudes. Grey areas indicate the altitude at which different wavelengths are absorbed.
transparent atmosphere at shorter radio wavelengths
RADIO WINDOW
Radiation with wavelengths between 1cm and 11m (0.4in–36ft) passes readily through the atmosphere. This part of the spectrum, which includes some radio waves and some microwaves, is called the "radio window."
opaque atmosphere

EXPLORING SPACE

## IMAGES FROM INVISIBLE RADIATION

Astronomers have developed telescopes that can gather information from EM radiation other than visible light, but they still face a problem of how to visualize the invisible. The most popular technique uses computers to create "false-colour" images – pictures that show the object in particular wavelengths of radiation, sometimes colour-coded, but often just using varying intensities of a single colour. These images of Kepler's Supernova (see p.269) show radiation in visible light, infrared, and two different wavelengths of X-rays, revealing the temperature of different regions and the overall structure.

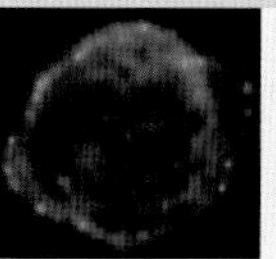

1 High-energy (short-wavelength) X-ray image from Chandra Observatory.

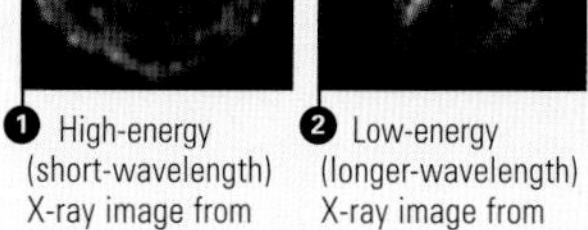

2 Low-energy (longer-wavelength) X-ray image from Chandra Observatory.

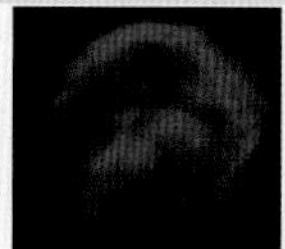

3 Image in infrared radiation, taken by the Spitzer Space Telescope.

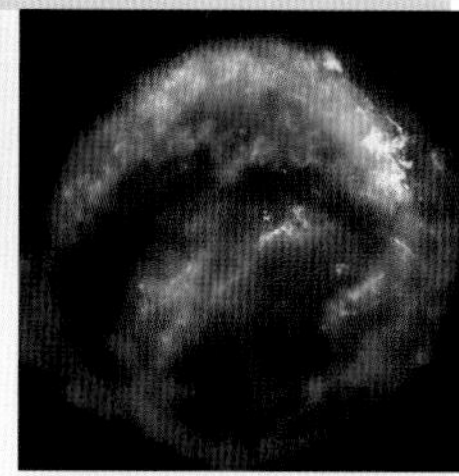

**COMBINED IMAGE**
The false-coloured images are combined with a Hubble image of the remnant in visible light.

**VISIBLE LIGHT** OPTICAL TELESCOPE
Optical telescopes with the largest mirrors achieve the brightest, sharpest images and the greatest power (see p.76). They range from those of amateur astronomers, such as this example with a 21.5cm (8.5in) mirror, to those of large observatories, with mirrors up to 10m (33ft) wide. Planned telescopes include the 30m (98ft) California Extremely Large Telescope (CELT) and the 100m (330ft) Overwhelmingly Large Telescope (OWL).

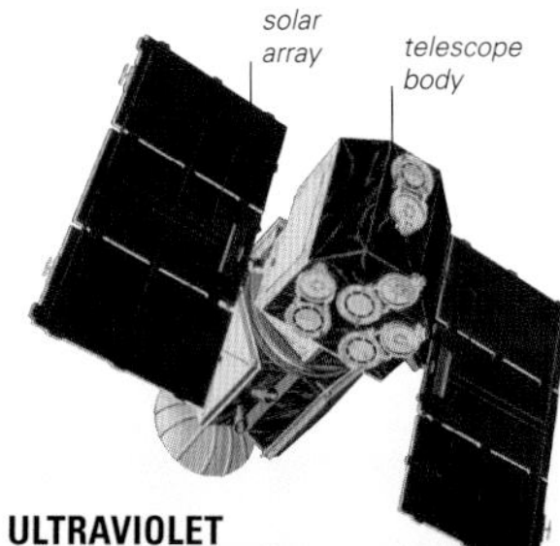


**ULTRAVIOLET**
ORBITING TELESCOPE
NASA's Extreme Ultraviolet Explorer satellite surveyed sources of extreme (very-short wavelength) ultraviolet radiation during the 1990s. Ultraviolet light originates from hot sources such as white dwarfs, neutron stars, and Seyfert galaxies (see p.310).

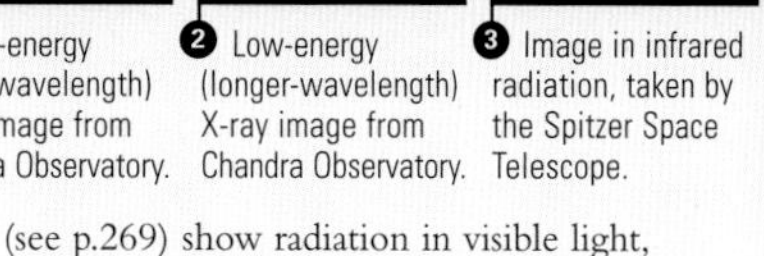


**X-RAYS**
ORBITING OBSERVATORY
X-rays are highly energetic and so powerful that they pass through conventional mirrors. To focus X-rays, telescopes, such as the Chandra X-ray Observatory (above), use a nest of curved "grazing incidence" mirrors of polished metal. X-rays glance off these mirrors, like ricocheting bullets, towards the focal point.

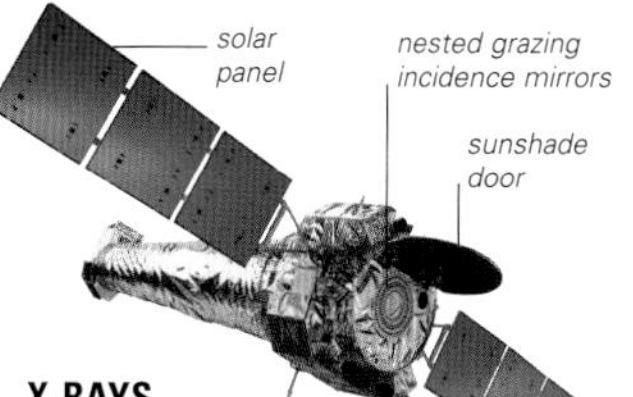


**GAMMA-RAY**
ORBITING OBSERVATORY
Gamma rays are the most energetic EM waves, released by the most violent cosmic events. The Compton Gamma Ray Observatory orbited Earth in the 1990s to study gamma rays from phenomena such as supernovae, pulsars, and gamma-ray bursts.

**VISIBLE LIGHT** GALAXY
The spiral galaxy M90, which lies 30 million light-years away, is shown here as it appears to human eyes through a large telescope. This galaxy is similar in size to the Milky Way. The image was taken at Kitt Peak National Observatory in Arizona, USA.

**ULTRAVIOLET** GALAXY
This image of spiral galaxy M74 is a composite of visible light and ultraviolet images. The high-energy ultraviolet emission is in blue and white and picks out extremely hot, young stars in the spiral arms and in the galactic nucleus.

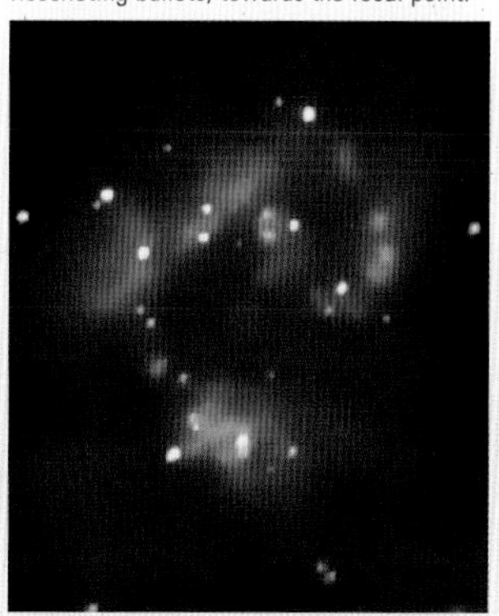

**X-RAYS** GALAXY
The orange-pink regions in this Chandra Observatory image of two colliding galaxies (called the Antennae, see p.307) are X-ray-emitting "superbubbles" of hot gas. The point X-ray sources (bright spots) are black holes and neutron stars.

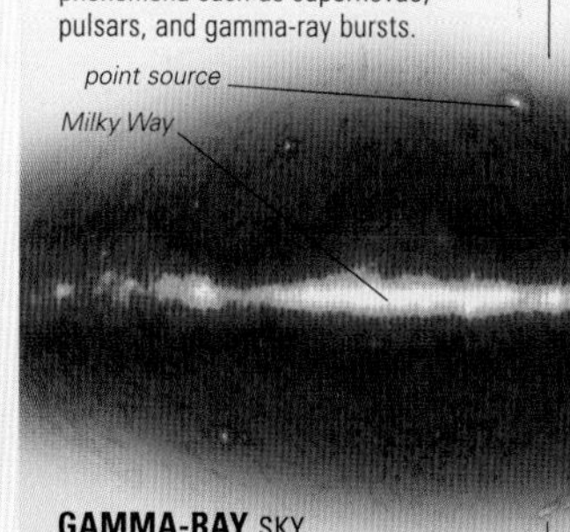


**GAMMA-RAY** SKY
Gamma rays are too powerful to focus, so sharp images are impossible. This view of the sky shows the Milky Way as a bright band. Point sources may be neutron stars or hypernovae (see p.51). The image comes from the Energetic Gamma Ray Experiment Telescope (EGRET) on the Compton Observatory.

VISIBLE | ULTRAVIOLET | X-RAYS | GAMMA RAYS

1µm | 100nm | 10nm | 1nm | 0.1nm | 0.01nm | 0.001nm | 0.0001nm | 0.00001nm

transparent atmosphere

**OPTICAL WINDOW**
*Wavelengths of radiation between 300 and 1100nm (nanometres) pass easily through the atmosphere (the visible light spectrum extends from 400 to 700 nanometres).*

opaque atmosphere

# GRAVITY, MOTION, AND ORBITS


Space and time 38–39
Planetary motion 64–65
The Copernican revolution 86–87
In Earth's orbit 106–107
Beyond Earth 108–109
The family of the Sun 118–19


GRAVITY IS THE ATTRACTIVE FORCE that exists between every object in the Universe, the force that both holds stars and galaxies together and causes a pin to drop. Gravity is weaker than nature's other fundamental forces, but because it acts over great distances, and between all bodies possessing mass, it has played a major part in shaping the Universe. Gravity is also crucial in determining orbits and creating phenomena such as planetary rings and black holes.

**DISCS AND RINGS**
The disc- and ring-like structures common in celestial objects are maintained by gravity. Examples include Saturn's rings (pictured), spiral galaxies, and the discs around black holes. Every particle in Saturn's rings is held in orbit through gravitational interactions with billions of other particles and with Saturn itself.

## NEWTONIAN GRAVITY

The scientific study of gravity began with Galileo Galilei's demonstration, in about 1590, that objects of different weight fall to the ground at exactly the same, accelerating rate. In 1665 or 1666, Isaac Newton realized that whatever force causes objects to fall might extend into space and be responsible for holding the Moon in its orbit. By analysing the motions of several heavenly objects, Newton formulated his law of universal gravitation. It stated that every body in the Universe exerts an attractive force (gravity) on every other body and described how this force varies with the masses of the bodies and their separation. To this day, Newton's law remains applicable for understanding and predicting the movements of most astronomical objects.

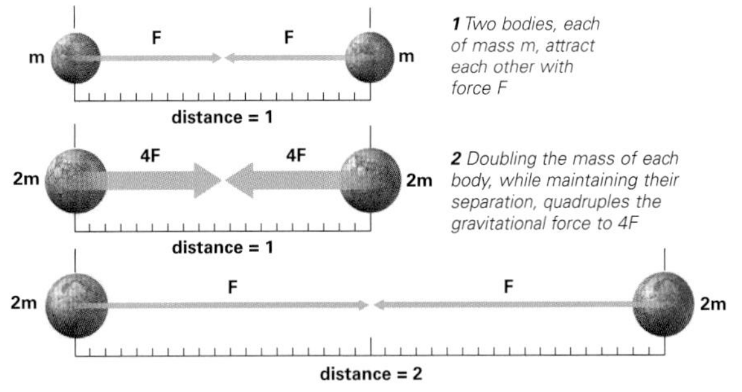


*1 Two bodies, each of mass m, attract each other with force F*

*2 Doubling the mass of each body, while maintaining their separation, quadruples the gravitational force to 4F*

*3 Doubling the separation between the bodies reduces the force by a factor of 4, back to F*

**MASS AND DISTANCE**
Any two bodies are attracted by a force of gravity proportional to the mass of one multiplied by the mass of the other. The force is also inversely proportional to the square of their separation.

### ISAAC NEWTON

The English mathematician and physicist Isaac Newton (1642–1727) was one of the greatest ever scientific intellects. As well as his discoveries in the fields of gravity and motion, he co-discovered the mathematical technique of calculus. In 1705, Newton became the first scientist to be knighted for his work.

## NEWTON'S LAWS OF MOTION

From his studies of gravity and the motions of heavenly bodies, and again extending concepts first developed by Galileo, Newton formulated his three laws of motion. Before Galileo and Newton, people thought that an object in motion could continue moving only if a force acted on it. In his first law of motion, Newton contradicted this idea: he stated that an object remains in uniform motion or at rest unless a net force acts on it (a net force is the sum of all forces acting on an object). Newton's second law states that a net force acting on an object causes it to accelerate (change its velocity) at a rate that is directly proportional to the magnitude of the force. It also states that the smaller the mass of an object, the higher the acceleration it experiences for a given force. Newton's third law states that for every action there is an equal and opposite reaction – for example, Earth's gravitational pull on the Moon is matched by the pull of the Moon on Earth.

**FIRST LAW OF MOTION**
The first law states that an object remains in a state of rest or moves at a constant speed in a straight line unless acted on by a net force.

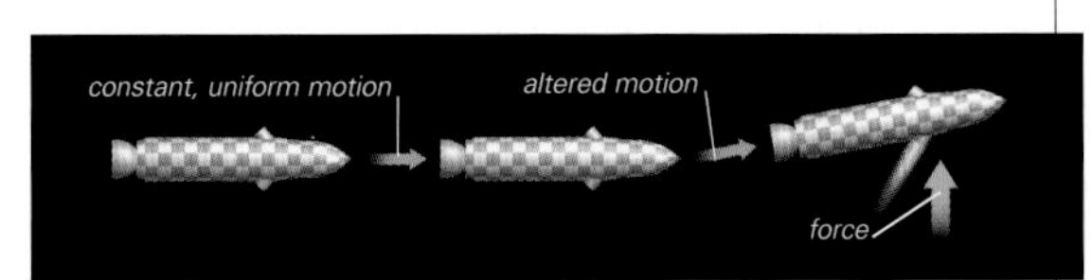


**SECOND LAW OF MOTION**
When an object of low mass and one of greater mass are subjected for a force of the same magnitude, the low-mass object accelerates at a higher rate.

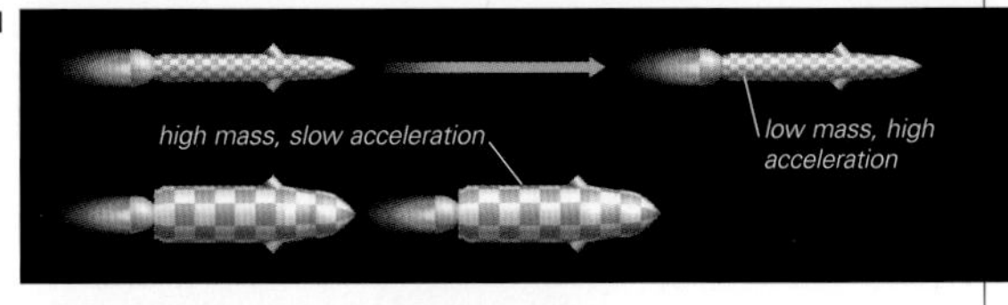


**THIRD LAW OF MOTION**
To every action there is an equal and opposite reaction. The forward thrust of a rocket is a reaction to the backward blast of combusted fuel.



## WEIGHT AND FREE FALL

The size of the gravitational force acting on an object is called its weight. An object's rest mass (measured in kilograms or pounds) is constant, while its weight (measured in newtons) varies according to the local strength of gravity. A mass of 1kg (2.2lb) weighs 9.8 newtons on Earth, but only 1.65 newtons on the Moon. Weight can be measured, and the feeling of weight experienced, only when the gravity producing it is resisted by a second, opposing, force. A person standing on Earth feels weight not so much from the pull of gravity as from the opposing push of the ground on his or her feet. In contrast, a person orbiting Earth is actually falling towards Earth under gravity. Such a person is in "free fall" and experiences apparent weightlessness. This is due not to lack of gravity but to the absence of a force opposing gravity.

**WEIGHTLESSNESS**
Astronauts in training must frequently experience apparent weightlessness. Here, a plane is plunging sharply from high altitude, putting the trainee astronauts inside into free fall.

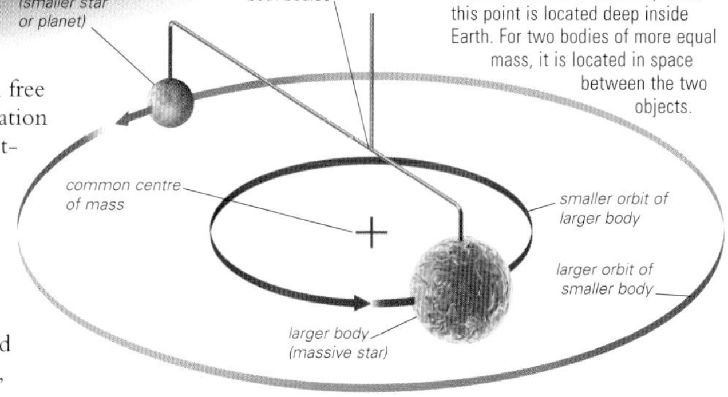

COMMON CENTRE OF MASS

In an orbital system of two bodies, the smaller body does not simply orbit the larger one. Instead, both revolve around the joint centre of mass. In the Earth–Moon system, this point is located deep inside Earth. For two bodies of more equal mass, it is located in space between the two objects.

smaller body (smaller star or planet)
pivot: centre of rotation of both bodies
common centre of mass
smaller orbit of larger body
larger orbit of smaller body
larger body (massive star)

## SHAPES OF ORBITS

When one object is in orbit around another object of higher mass, it is in free fall towards the larger body. It experiences a constant gravitational acceleration towards the larger object that deflects what would otherwise be its straight-line motion into a curved trajectory. The direction of its motion, and the direction of acceleration both constantly change, producing its curved path. All closed orbits in nature have the shape of an ellipse (a stretched circle). The degree to which an ellipse varies from a perfect circle is called its eccentricity. Many Solar System orbits (such as the Moon's around Earth) are not very eccentric – that is, they are almost circular. Others, such as Pluto's orbit around the Sun, are much more eccentric and highly elongated. Some celestial bodies follow open, non-returning orbits, along curves with shapes called parabolas and hyperbolas.

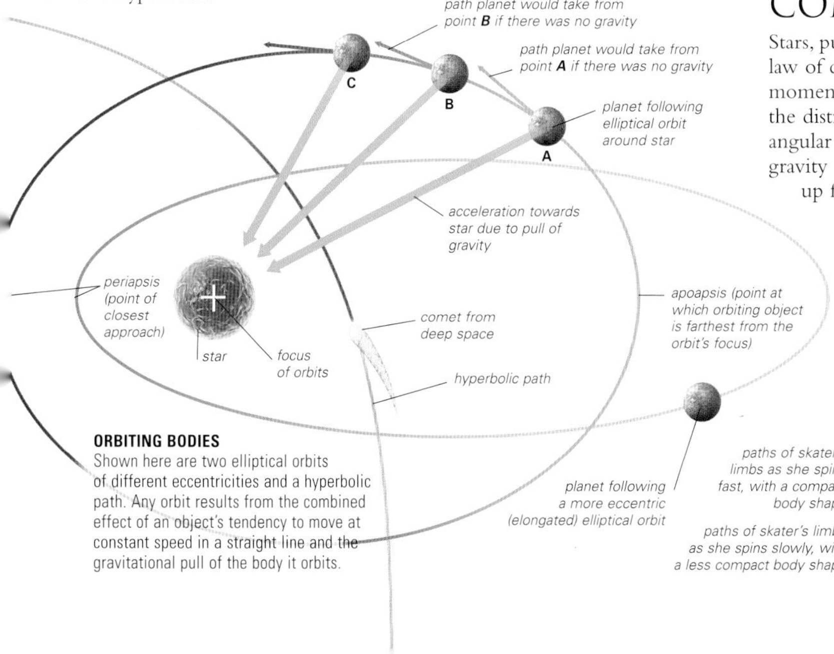

path planet would take from point **B** if there was no gravity
path planet would take from point **A** if there was no gravity
C
B
A
planet following elliptical orbit around star
acceleration towards star due to pull of gravity
periapsis (point of closest approach)
star
focus of orbits
comet from deep space
hyperbolic path
apoapsis (point at which orbiting object is farthest from the orbit's focus)
planet following a more eccentric (elongated) elliptical orbit

ORBITING BODIES

Shown here are two elliptical orbits of different eccentricities and a hyperbolic path. Any orbit results from the combined effect of an object's tendency to move at constant speed in a straight line and the gravitational pull of the body it orbits.

## COMPACT, ROTATING BODIES

Stars, pulsars, galaxies, and planets all rotate, governed by the law of conservation of angular momentum. An object's angular momentum is related to its rotational energy, which depends on the distribution of mass in the object and on how fast it spins. The angular momentum of any rotating object stays constant, so if gravity causes the object to contract, its spin rate increases to make up for the redistribution of mass. Compact, fast-rotating objects therefore tend to form from diffuse, slowly-rotating ones.

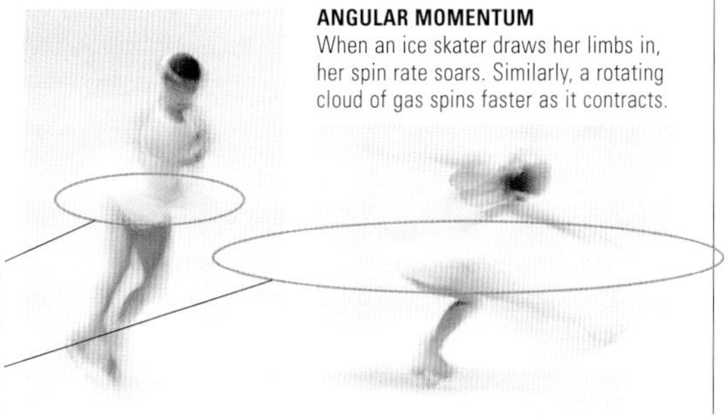

ANGULAR MOMENTUM

When an ice skater draws her limbs in, her spin rate soars. Similarly, a rotating cloud of gas spins faster as it contracts.

paths of skater's limbs as she spins fast, with a compact body shape
paths of skater's limbs as she spins slowly, with a less compact body shape

# SPACE AND TIME


32–33 Radiation
36–37 Gravity, motion, and orbits
Expanding space 42–43
Shaping space 92–93
Space-age astronomy 94–97
The family of the Sun 118–21


**MOST PEOPLE SHARE SOME COMMON-SENSE NOTIONS** about the world. One is that time passes at the same rate for everyone. Another is that the length of a rigid object does not change. In fact, such ideas, which once formed a bedrock for the laws of physics, are an illusion: they apply only to the restricted range of situations with which people are most familiar. In fact, time and space are not absolute, but stretch and warp depending on relative viewpoint. What is more, the presence of matter distorts both space and time to produce the force of gravity.

## PROBLEMS IN NEWTON'S UNIVERSE

Problems with the Newtonian view of space and time (see p.36) first surfaced towards the end of the 19th century. Up to that time, scientists assumed that the positions and motions of objects in space should all be measurable relative to some non-moving, absolute "frame of reference", which they thought was filled with an invisible medium called "the ether". However, in 1887, an experiment to measure Earth's motion through this ether, by detecting variations in the velocity of light sent through it in different directions, produced some unexpected results. First, it failed to confirm the existence of the ether. Second, it indicated that light always travels at the same speed relative to an observer, whatever that observer's own movements. This finding suggested that light does not follow the same rules of relative motion that govern everyday objects such as cars and bullets. If a person were to chase a bullet at half of the bullet's speed, the rate at which the bullet moved away from him or her would halve. However, if a person were to chase a light wave at half the speed of light, the wave would continue moving away from him or her at exactly the same velocity.

**CONSTANT SPEED OF LIGHT**
Light leaves both the ceiling lights and the headlights of the moving cars at the same speed relative to its source. Paradoxically, light from both sources reaches an observer standing in the tunnel, again, at exactly the same speed.

### ALBERT EINSTEIN

The work of German-born mathematician and physicist Albert Einstein made him the most famous scientist of the 20th century. Although he won the Nobel prize for his work on the particle-like properties of light (see p.32), he is more famous for his theories of special relativity (1905) and general relativity (1915). These theories introduced a revolutionary new way of thinking about space, time, mass, energy, and gravity.

## SPECIAL RELATIVITY

In 1905, Albert Einstein rejected the idea that there is any absolute or "preferred" frame of reference in the Universe. In other words, everything is relative. He also rejected the idea that time is absolute, suggesting that it need not pass at the same rate everywhere. To replace the old ideas, he formed the special theory of relativity, called "special" because it is restricted to frames of reference in constant, unchanging motion (because they are not being accelerated by a force, see p.36). He based the entire theory on two principles. The first principle, called the principle of relativity, states that the same laws of physics apply equally in all constantly moving frames of reference. The second principle states that the speed of light is constant and independent of the motion of the observer or source of light. Einstein recognized that this second principle conflicts with accepted notions about how velocities add together; further, that combining it with the first principle seems to lead to perplexing, non-intuitive results. He perceived, however, that human intuition about time and space could be wrong.

direction of Observer 2's motion

path of Observer 1's ball, as seen by Observer 2

Observer 1

path of Observer 1's ball, as seen by Observer 1

direction of Observer 1's motion

**VIEWPOINT ONE**
From Observer 1's point of view, the green ball within his or her own frame of reference travels up and down. If Observer 1 looks across at the red ball in a frame of reference in relative motion, the ball seems to follow an arc.

**MOVING FRAMES OF REFERENCE**
Here we see two travellers – effectively two moving reference frames – passing each other. Each tosses a ball up. By the principle of relativity, the laws of physics apply in each reference frame, so each traveller observes the behaviour of the two balls as predicted by those laws. Although the two travellers see different motions for each ball, neither traveller's point of view is superior to the other's – both are equally valid, and there is no preferred frame of reference.

# EFFECTS OF SPECIAL RELATIVITY

The results that flow from the principles of special relativity are remarkable. Using thought experiments, Einstein showed that for the speed of light to be the same in all reference frames, measures of space and time in one frame must be transformed to those in another. These transformations show that when an object moves at high speed relative to an observer, the observer sees less of its length – an effect called Lorentz contraction. Also, time for such an object appears to run more slowly – an effect called time dilation. So measurements of time and space vary between moving reference frames. Einstein also showed that an object gains mass when its energy increases and loses it when its energy decreases. This led him to realise that mass and energy have an equivalence, which he expressed in the famous equation E (energy) = m (mass) x $c^2$ (the speed of light squared).

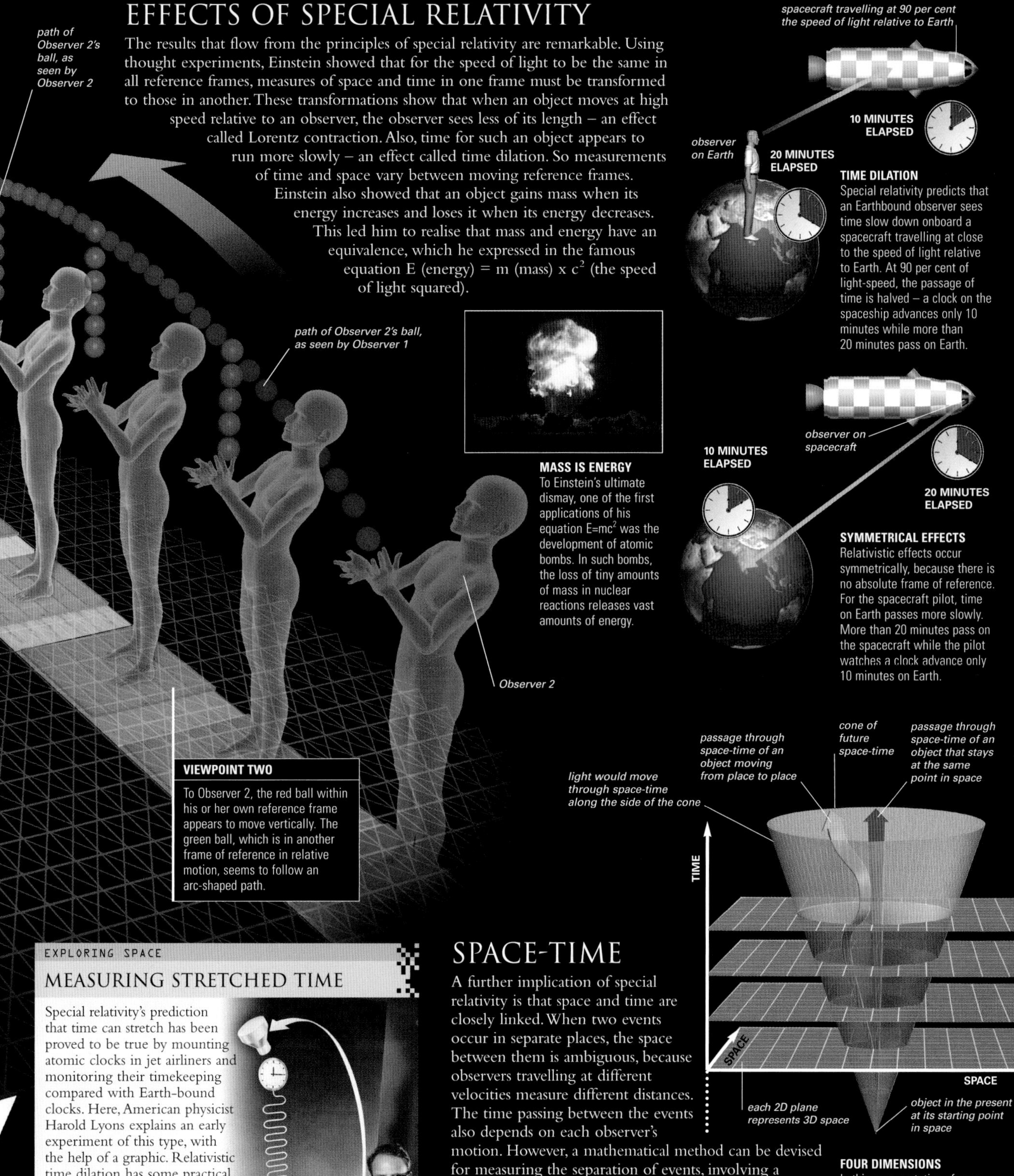


**MASS IS ENERGY**
To Einstein's ultimate dismay, one of the first applications of his equation $E=mc^2$ was the development of atomic bombs. In such bombs, the loss of tiny amounts of mass in nuclear reactions releases vast amounts of energy.

**TIME DILATION**
Special relativity predicts that an Earthbound observer sees time slow down onboard a spacecraft travelling at close to the speed of light relative to Earth. At 90 per cent of light-speed, the passage of time is halved – a clock on the spaceship advances only 10 minutes while more than 20 minutes pass on Earth.

**SYMMETRICAL EFFECTS**
Relativistic effects occur symmetrically, because there is no absolute frame of reference. For the spacecraft pilot, time on Earth passes more slowly. More than 20 minutes pass on the spacecraft while the pilot watches a clock advance only 10 minutes on Earth.

**VIEWPOINT TWO**
To Observer 2, the red ball within his or her own reference frame appears to move vertically. The green ball, which is in another frame of reference in relative motion, seems to follow an arc-shaped path.

EXPLORING SPACE

## MEASURING STRETCHED TIME

Special relativity's prediction that time can stretch has been proved to be true by mounting atomic clocks in jet airliners and monitoring their timekeeping compared with Earth-bound clocks. Here, American physicist Harold Lyons explains an early experiment of this type, with the help of a graphic. Relativistic time dilation has some practical consequences. The atomic clocks in global positioning system (GPS) satellites run about 7.2 microseconds a day slower than Earth-bound atomic clocks, so their data must be adjusted to maintain accuracy.

# SPACE-TIME

A further implication of special relativity is that space and time are closely linked. When two events occur in separate places, the space between them is ambiguous, because observers travelling at different velocities measure different distances. The time passing between the events also depends on each observer's motion. However, a mathematical method can be devised for measuring the separation of events, involving a combination of space and time, that gives values that all observers can agree on. This led to the idea that events in the Universe should no longer be described in three spatial dimensions, but rather in a four-dimensional world, incorporating time, called space-time. In this system, the separation between any two events is described by a value called a space-time interval.

**FOUR DIMENSIONS**
In this representation of space-time, time moves upwards into the future, while the three spatial dimensions are reduced to two-dimensional planes. The cone represents the effective limits of space-time for any object – its boundary is defined by the speed of light.

# ACCELERATING MOTION

Having completed his study of relativity in the special case of reference frames in uniform motion (inertial reference frames), Einstein turned his attention to changing, or accelerated motion. In particular, he examined the link between gravity and acceleration. This led him to formulate a proposition called the principle of equivalence, which describes gravity and acceleration as different perspectives of the same thing. Specifically, Einstein stated that it was impossible for any experiment to tell the difference between being at rest in a uniform gravitational field and being accelerated. He illustrated this idea using thought experiments involving scientists sealed into boxes under various conditions of acceleration and gravity. Starting from the principle of equivalence, by 1915 Einstein had gone on to develop his most complex and major masterpiece, the general theory of relativity, which provided a new description of gravity.

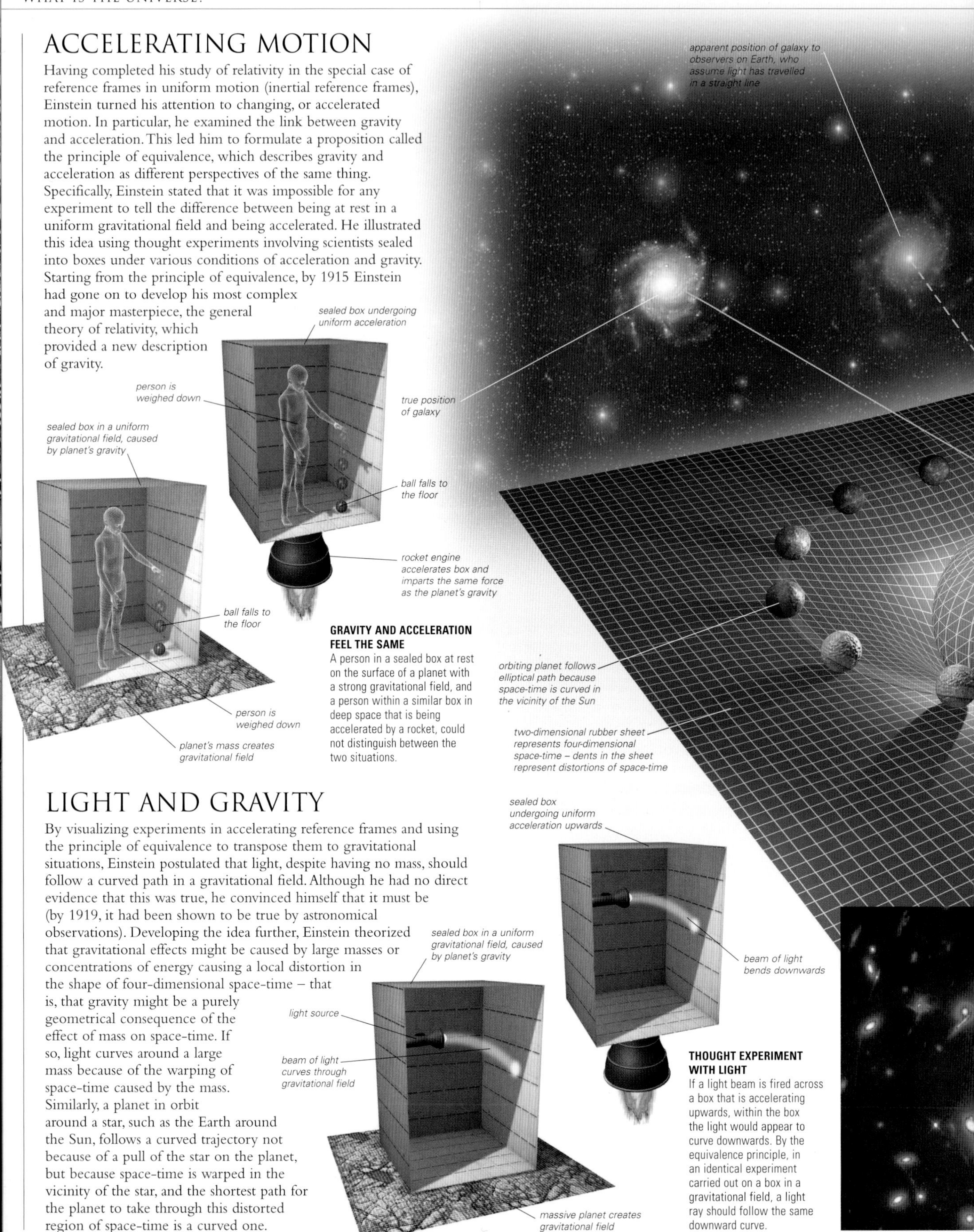


**GRAVITY AND ACCELERATION FEEL THE SAME**
A person in a sealed box at rest on the surface of a planet with a strong gravitational field, and a person within a similar box in deep space that is being accelerated by a rocket, could not distinguish between the two situations.

# LIGHT AND GRAVITY

By visualizing experiments in accelerating reference frames and using the principle of equivalence to transpose them to gravitational situations, Einstein postulated that light, despite having no mass, should follow a curved path in a gravitational field. Although he had no direct evidence that this was true, he convinced himself that it must be (by 1919, it had been shown to be true by astronomical observations). Developing the idea further, Einstein theorized that gravitational effects might be caused by large masses or concentrations of energy causing a local distortion in the shape of four-dimensional space-time – that is, that gravity might be a purely geometrical consequence of the effect of mass on space-time. If so, light curves around a large mass because of the warping of space-time caused by the mass. Similarly, a planet in orbit around a star, such as the Earth around the Sun, follows a curved trajectory not because of a pull of the star on the planet, but because space-time is warped in the vicinity of the star, and the shortest path for the planet to take through this distorted region of space-time is a curved one.

**THOUGHT EXPERIMENT WITH LIGHT**
If a light beam is fired across a box that is accelerating upwards, within the box the light would appear to curve downwards. By the equivalence principle, in an identical experiment carried out on a box in a gravitational field, a light ray should follow the same downward curve.

# GENERAL RELATIVITY AT WORK

Einstein encapsulated his theory of how mass distorts space-time in his set of "field equations". Physicists have used these equations to find that it is in the strongest gravitational fields – where massive, dense objects distort space-time most strongly – that reality departs farthest from that predicted by Newton (see p.36). For instance, Mercury is so close to the Sun that it always moves in a strong gravitational field (or in strongly curved space-time). Its orbit is distorted in a way that Newton could not account for, but which general relativity explains perfectly (see p.124). General relativity also provides a framework for models of the Universe's structure, development, and eventual fate. It predicts that the Universe must either be expanding or contracting. Before the introduction of general relativity, space and time were thought of only as an arena in which events took place. After general relativity, physicists realized that space and time are dynamic entities that can be affected by mass, forces, and energy.

**DENTED SPACE-TIME**
Space-time can be visualized as a rubber sheet in which massive objects make dents. In this view, planets orbit the Sun because they roll around the dent it produces. Similarly, light passing by a massive object has its straight-line path deflected by following the local curvature of space-time. Remember, however, that it is 4-dimensional space-time, not just space, that is warped.

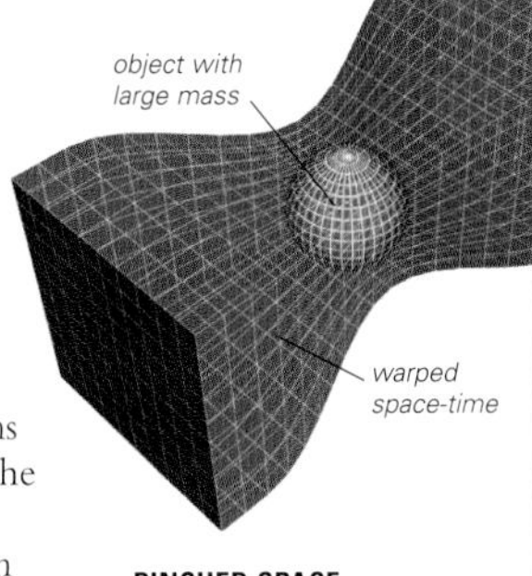

**PINCHED SPACE**
Instead of a two-dimensional sheet, four-dimensional space-time can also be visualized as a three-dimensional volume that is narrowed or "pinched in" around large masses.

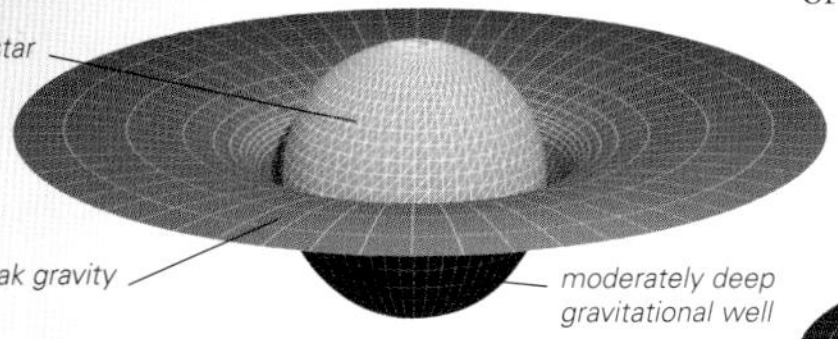

**WHITE DWARF STAR**
A white dwarf is a very dense, planet-sized star that can be thought of as producing a smaller but deeper dent in space-time than does a star like the Sun.

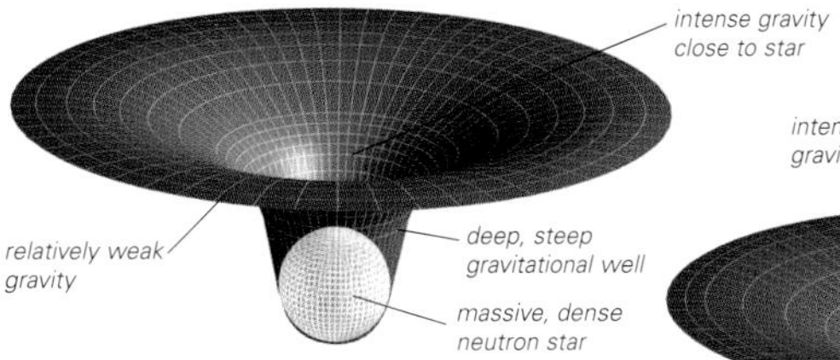

**NEUTRON STAR**
A neutron star is an exceedingly dense stellar remnant that makes a very deep dent in space-time. A neutron star significantly deflects light passing by, but cannot capture it.

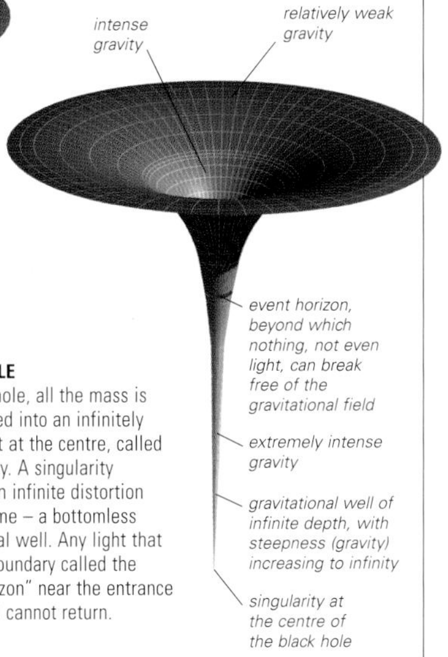

**BLACK HOLE**
In a black hole, all the mass is concentrated into an infinitely dense point at the centre, called a singularity. A singularity produces an infinite distortion in space-time – a bottomless gravitational well. Any light that passes a boundary called the "event horizon" near the entrance to this well cannot return.

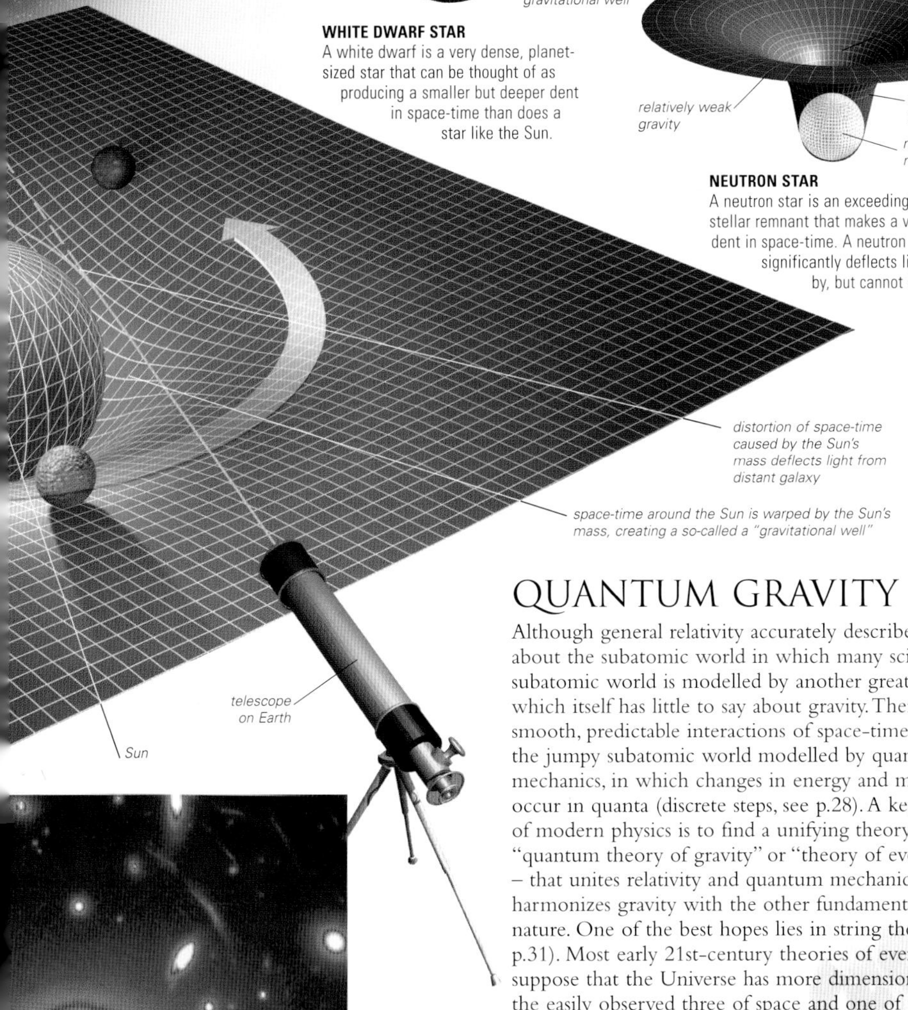

# QUANTUM GRAVITY

Although general relativity accurately describes the Universe on a large scale, it has little to say about the subatomic world in which many scientists believe gravity must originate. This subatomic world is modelled by another great theory of physics, called quantum mechanics, which itself has little to say about gravity. There is, it seems, little in common between the smooth, predictable interactions of space-time and matter predicted by general relativity and the jumpy subatomic world modelled by quantum mechanics, in which changes in energy and matter occur in quanta (discrete steps, see p.28). A key goal of modern physics is to find a unifying theory – a "quantum theory of gravity" or "theory of everything" – that unites relativity and quantum mechanics, and harmonizes gravity with the other fundamental forces of nature. One of the best hopes lies in string theory (see p.31). Most early 21st-century theories of everything suppose that the Universe has more dimensions than the easily observed three of space and one of time.

**MULTIDIMENSIONAL SPACE-TIME**
These figures, called Calabi–Yau spaces, are purported to hold six or more dimensions "curled up" within them. By incorporating one of these tiny objects at each point in space-time, string theorists envisage ten or more dimensions.

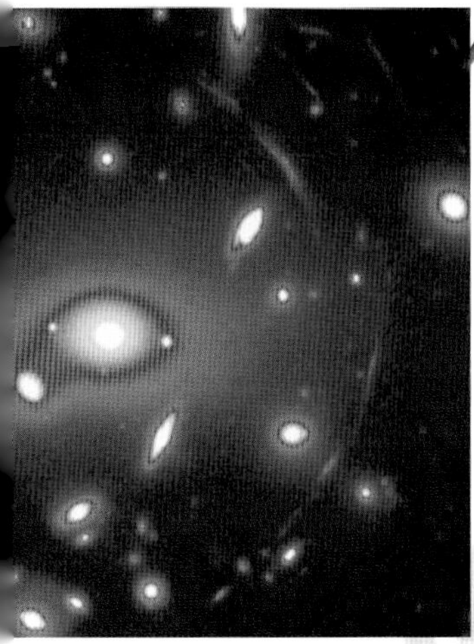

**GRAVITY BENDING LIGHT**
The effect of gravity on the path of light is not obvious unless an observer looks deep into space at the Universe's most massive objects – clusters of galaxies. This image shows galaxies as white blobs. Their combined gravity bends light so much that the images of more distant galaxies appear as blue streaks, stretched and squashed by the galaxy cluster's gravity.

# EXPANDING SPACE

22–23 The scale of the Universe
32–35 Radiation
The Big Bang 46–49
The fate of the Universe 54–55
Galaxy clusters 316–17

A CRUCIAL PROPERTY of the Universe is that it is expanding. It must be growing, because distant galaxies are quickly receding from Earth and more distant ones are receding even faster. Assuming that the Universe has always been expanding, it must once have been smaller and denser – a fact that strongly supports the Big Bang theory of its origin.

## MEASURING EXPANSION

The rate of the Universe's expansion can be calculated by comparing the distances to remote galaxies and the speeds at which they are receding. The galaxies' velocities are measured by examining the red shifts in their light spectra (see p.33). Their distances are calculated by detecting a class of stars called Cepheid variables in the galaxies and measuring the stars' cycles of magnitude variation (see pp.278, 303). The result is a number known as the Hubble constant – an expression of the Universe's expansion rate. The value of the constant has been debated by cosmologists, but is currently thought to be about 80,000kph (50,000mph) per million light-years. This means, for example, that two galaxies situated 1 billion light-years apart are receding from each other at 80 million kph (50 million mph). On a familiar timescale, this is actually a very gradual expansion – an increase in the galaxies' distance of 1 per cent takes tens of millions of years.

recessional velocity (measured by red shift)
distance from Earth (measured by variable stars)

**HUBBLE CONSTANT**
The recession velocity of remote galaxies rises with distance, and this relationship forms a straight line on a graph. Estimates of the line's slope yield values of the Hubble constant.

cones represent two possible histories of the expanding Universe
15 billion years ago, size of Universe is zero – a possible Big Bang occurs
Universe expanding at a constant rate in the past
12 billion years ago, size of Universe is zero
Universe expanding at a faster rate then slowing down
present day

**AGE OF THE UNIVERSE**
Cosmologists can estimate the age of the Universe by extrapolating its expansion rate backwards to the point at which the size of the observable Universe was zero. Depending on how the expansion rate has changed, estimates for the Universe's age range from 12 to 15 billion years. The current best estimate is 13.7 billion years.

## THE NATURE OF EXPANSION

Several notable features have been established about the Universe's expansion. First, although all distant galaxies are moving away, neither Earth nor any other point in space is at the centre of the Universe. Rather, everything is receding from everything else, and there is no centre. Second, at a local scale, gravity dominates over cosmological expansion and holds matter together. The scale at which this happens is surprisingly large – even entire clusters of galaxies resist expansion and hold together. Third, it is incorrect to think of galaxies and galaxy clusters moving away from each other "through" space. A more accurate picture is that of space itself expanding and carrying objects with it. Finally, the expansion rate almost certainly varies. Cosmologists are greatly interested in establishing how the expansion rate may change in future. The future rate of expansion will decide the eventual fate of the Universe (see pp.54–55).

**LOCAL GRAVITY**
The galaxies above are not moving apart. They will continue to collide despite cosmological expansion. Galaxy clusters are also held together by gravity.

the Universe 6 billion years ago was much smaller
galaxies close together
free gas and dust not yet absorbed into galaxies
6 BILLION YEARS AGO
3 BILLION YEARS AGO
some galaxies evolve into spiral shapes
PRESENT DAY
galaxies becoming less crowded
galaxy cluster, bound by gravity, does not expand
3 BILLION YEARS IN THE FUTURE

# TIME AND EXPANDING SPACE

**PEERING INTO DEEP SPACE**
This Hubble "deep-field" photograph shows a jumble of galaxies viewed at different distances. Each appears as it existed billions of years ago.

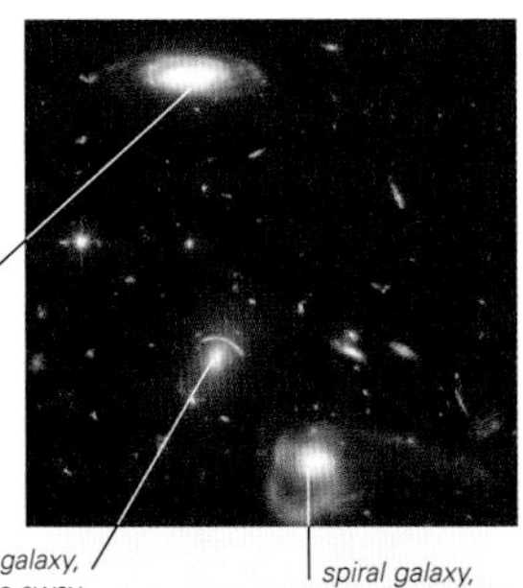


The continued expansion of space, combined with the constant speed of light, turns the Universe into a giant time machine. The light from a remote galaxy has taken billions of years to reach Earth, so astronomers see the galaxy as it was billions of years ago. In effect, the deeper astronomers look into space, the farther they peer into the Universe's history. In the remotest regions, they see only incompletely formed galaxies as they looked soon after the Big Bang. The most dim and distant of these galaxies is receding from Earth at speeds approaching the speed of light. Should astronomers observe such objects for millions of years, they would see them evolving more slowly than if they were closer and not being carried away so fast. At greater distances yet, beyond the observable Universe (see p.23), there may exist other objects that have moved away so fast that light from them has never reached Earth.

## EDWIN HUBBLE

The American astronomer Edwin Hubble (1889–1953) is famous for being the first to prove that the Universe is expanding. He showed the direct relationship between the recession speeds of remote galaxies and their distances from Earth, now known as Hubble's Law. Hubble is also noted for his earlier proof that galaxies are external to the Milky Way and for his system of galaxy classification. The Hubble Space Telescope and the Hubble constant are both named after him.

# LOOKBACK DISTANCE

The expansion of space complicates the expression of distances to very remote objects, particularly those that we now observe as they existed more than 5 billion years ago. When astronomers describe the distance to such faraway objects, by convention they use the "lookback" or "light-travel-time" distance. This is the distance that light from the object has travelled through space to reach us today, and it tells us how long ago the light left the object. But because space has expanded in the interim, the distance of the galaxy when the light began its journey towards Earth is less than the lookback distance. Conversely, the true distance to the remote object (called the "comoving" distance) is greater than the lookback distance. These distinctions need to be remembered when, for example, a galaxy is stated as being 10 billion light-years away.

**DIVERGING WORLDS**
An object described as being 11 billion light-years away (lookback distance) has a greater true distance (comoving distance), due to the effects of the Universe's expansion.

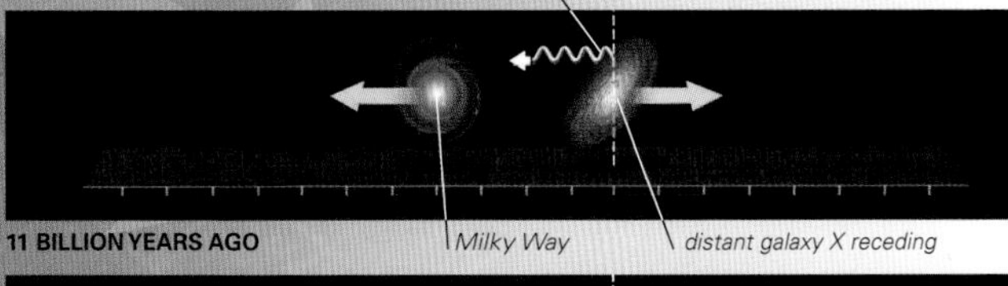


*1. Eleven billion years ago, a photon of light departs distant galaxy X travelling towards the Milky Way. The two galaxies are separated by 4 billion light-years of space.*

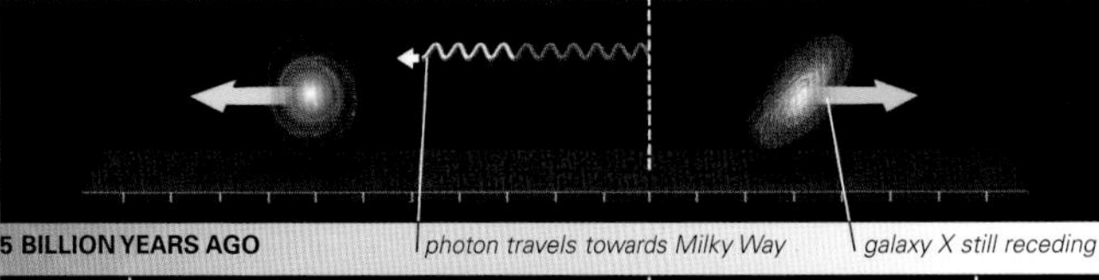


*2. Six billion years later, the photon has not yet reached its destination, because space has expanded, carrying the galaxies much farther apart.*

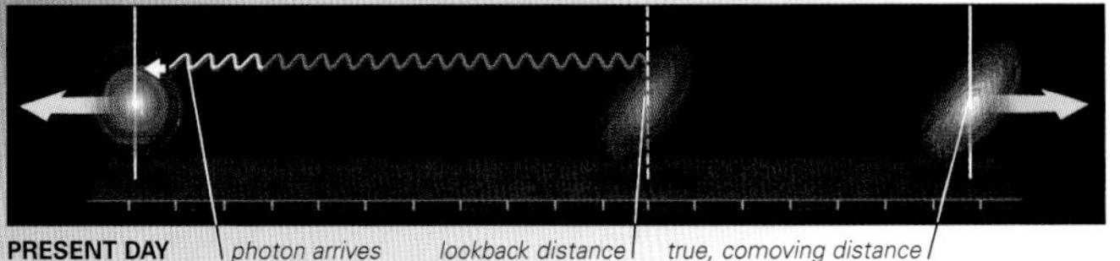


*3. The photon reaches the Milky Way, where an observer sees X as it was 11 billion years ago, 11 billion light-years away (lookback distance). Meanwhile, X's true (comoving) distance has increased to 18 billion light-years.*

**ACCELERATING EXPANSION**
This is a conceptual interpretation of how a region of space may have changed over a 9-billion-year period. As space has expanded, so the galaxies within it have been carried apart, evolving as they go. This interpretation shows expansion speeding up – a scenario gaining support from cosmologists.

*"Some say the world will end in fire,*
*Some say in ice."*
Robert Frost

THE STORY OF THE UNIVERSE can be traced back to its very first instants, according to the Big Bang theory of its origins. In the Big Bang model, the Universe was once infinitely small, dense, and hot. The Big Bang began a process of expansion and cooling that continues today. It was not an explosion of matter into space, but an expansion of space itself, and in the beginning, it brought time and space into existence. The Big Bang model does not explain all features of the Universe, however, and it continues to be refined. Nonetheless, scientists use it as a framework for mapping the continuing evolution of the Universe, through events such as the decoupling of matter and radiation (when the first atoms were formed and the Universe became transparent) and the condensation of the first galaxies and the first stars. Study of the Big Bang and the balance between the Universe's gravity and a force called dark energy can even help predict how the Universe will end.

**CRADLE OF STAR BIRTH**
This pillar of gas and dust is the Cone Nebula, one of the most active cradles of star formation in the Milky Way. The clouds of material giving birth to these stars were once parts of stars themselves. The recycling of material in the life cycles of stars has been key to the Universe's enrichment and evolution.

# THE BEGINNING AND END OF THE UNIVERSE

# THE BIG BANG

28–31 Matter
32–35 Radiation
42–43 Expanding space
The fate of the Universe 54–55
Space-age astronomy 94–97

**TIME, SPACE, ENERGY, AND MATTER** are all thought to have come into existence 13.7 billion years ago, in the event called the Big Bang. In its first moments, the Universe was infinitely dense, unimaginably hot, and contained pure energy. But within a tiny fraction of a second, vast numbers of fundamental particles had appeared, created out of energy as the Universe cooled. Within a few hundred thousand years, these particles had combined to form the first atoms.

## IN THE BEGINNING

The Big Bang was not an explosion in space, but an expansion of space, which happened everywhere. Physicists do not know what happened in the first instant after the Big Bang, known as the Planck era, but at the end of this period, they believe that gravity split from the other forces of nature, followed by the strong nuclear force (see p.30). Many believe this event triggered "inflation" – a short but rapid expansion. If inflation did occur, it helps to explain why the Universe seems so smooth and flat. During inflation, a fantastic amount of mass-energy came into existence, in tandem with an equal but negative amount of gravitational energy. By the end of inflation, matter had begun to appear.

**THE FIRST MICROSECOND**
The timeline on this page and the next shows some events during the first microsecond (1 millionth of a second or $10^{-6}$ seconds) after the Big Bang. Over this period, the Universe's temperature dropped from about $10^{34}$°C (ten billion trillion trillion degrees) to a mere $10^{13}$°C (ten trillion degrees). The timeline refers to the diameter of the observable Universe: this is the approximate historical diameter of the part of the Universe we can currently observe.

**THE PLANCK ERA**
No current theory of physics can describe what happened in the Universe during this time.

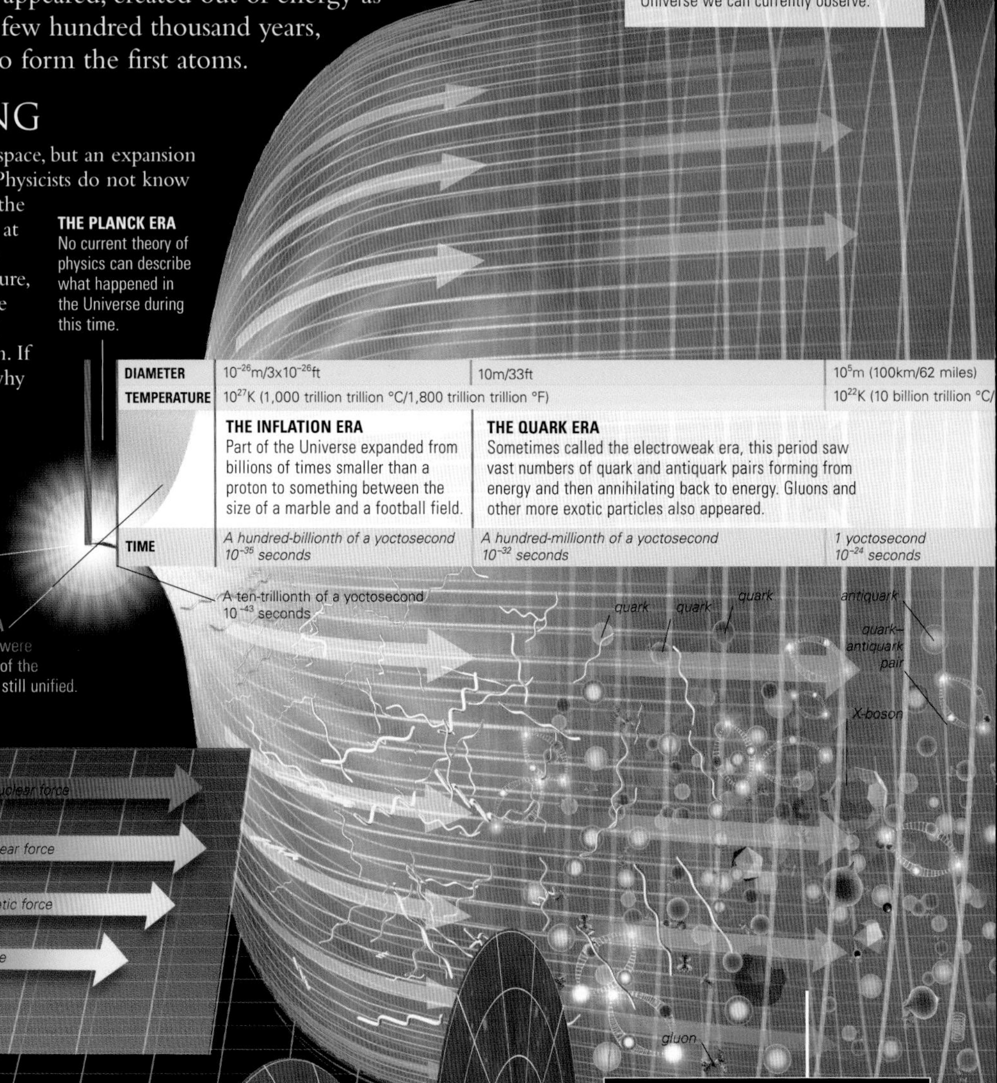

singularity at the start of time

A ten-trillionth of a yoctosecond $10^{-43}$ seconds

**THE GRAND UNIFIED THEORY ERA**
During this era, matter and energy were completely interchangeable. Three of the fundamental forces of nature were still unified.

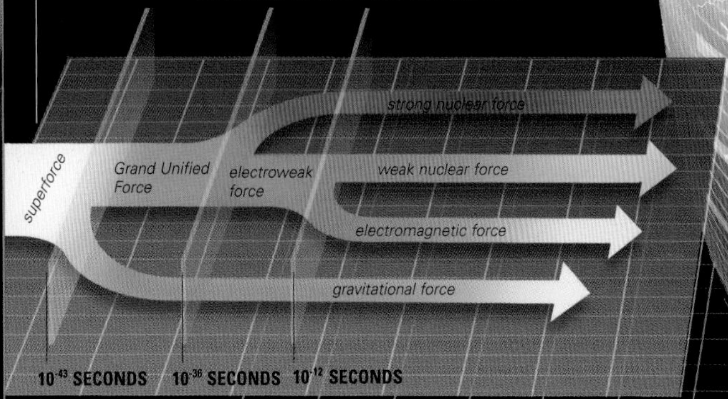

| | | | |
|---|---|---|---|
| DIAMETER | $10^{-26}$m/3x$10^{-26}$ft | 10m/33ft | $10^5$m (100km/62 miles) |
| TEMPERATURE | $10^{27}$K (1,000 trillion trillion °C/1,800 trillion trillion °F) | | $10^{22}$K (10 billion trillion °C/ |
| | **THE INFLATION ERA** Part of the Universe expanded from billions of times smaller than a proton to something between the size of a marble and a football field. | **THE QUARK ERA** Sometimes called the electroweak era, this period saw vast numbers of quark and antiquark pairs forming from energy and then annihilating back to energy. Gluons and other more exotic particles also appeared. | |
| TIME | *A hundred-billionth of a yoctosecond $10^{-35}$ seconds* | *A hundred-millionth of a yoctosecond $10^{-32}$ seconds* | *1 yoctosecond $10^{-24}$ seconds* |

quark
quark
quark
antiquark
quark–antiquark pair
X-boson
gluon

superforce
Grand Unified Force
electroweak force
strong nuclear force
weak nuclear force
electromagnetic force
gravitational force
$10^{-43}$ SECONDS
$10^{-36}$ SECONDS
$10^{-12}$ SECONDS

**SEPARATION OF FORCES**
Physicists believe that at the exceedingly high temperatures present just after the Big Bang, the four fundamental forces were unified. Then, as the Universe cooled, the forces separated, or "froze out", at the time intervals shown here.

**INFLATION**
In a Big Bang without inflation, what are now widely spaced regions of the Universe could never have become so similar in density and temperature. Inflation theory proposes that our observable Universe is derived from a tiny homogeneous patch of the original Universe. The effect of inflation is like expanding a wrinkled sphere – after the expansion, its surface appears smooth and flat.

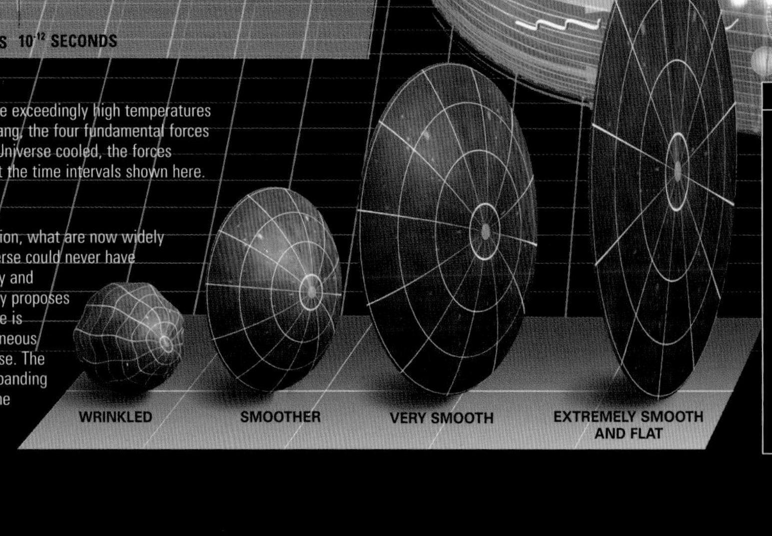

WRINKLED
SMOOTHER
VERY SMOOTH
EXTREMELY SMOOTH AND FLAT

**PARTICLE SOUP**
About $10^{-32}$ seconds after the Big Bang, the Universe is thought to have been a "soup" of fundamental particles and antiparticles. These were continually formed from energy as particle–antiparticle pairs, which then met and annihilated back to energy. Among these particles were some that still exist today as constituents of matter or as force carrier particles. These include quarks and their antiparticles (antiquarks), and bosons such as gluons (see pp.30–31). Other particles may have been present that no longer exist or are hard to detect – perhaps some gravitons (hypothetical gravity-carrying particles) and Higgs bosons, also hypothetical, which impart mass to other particles.

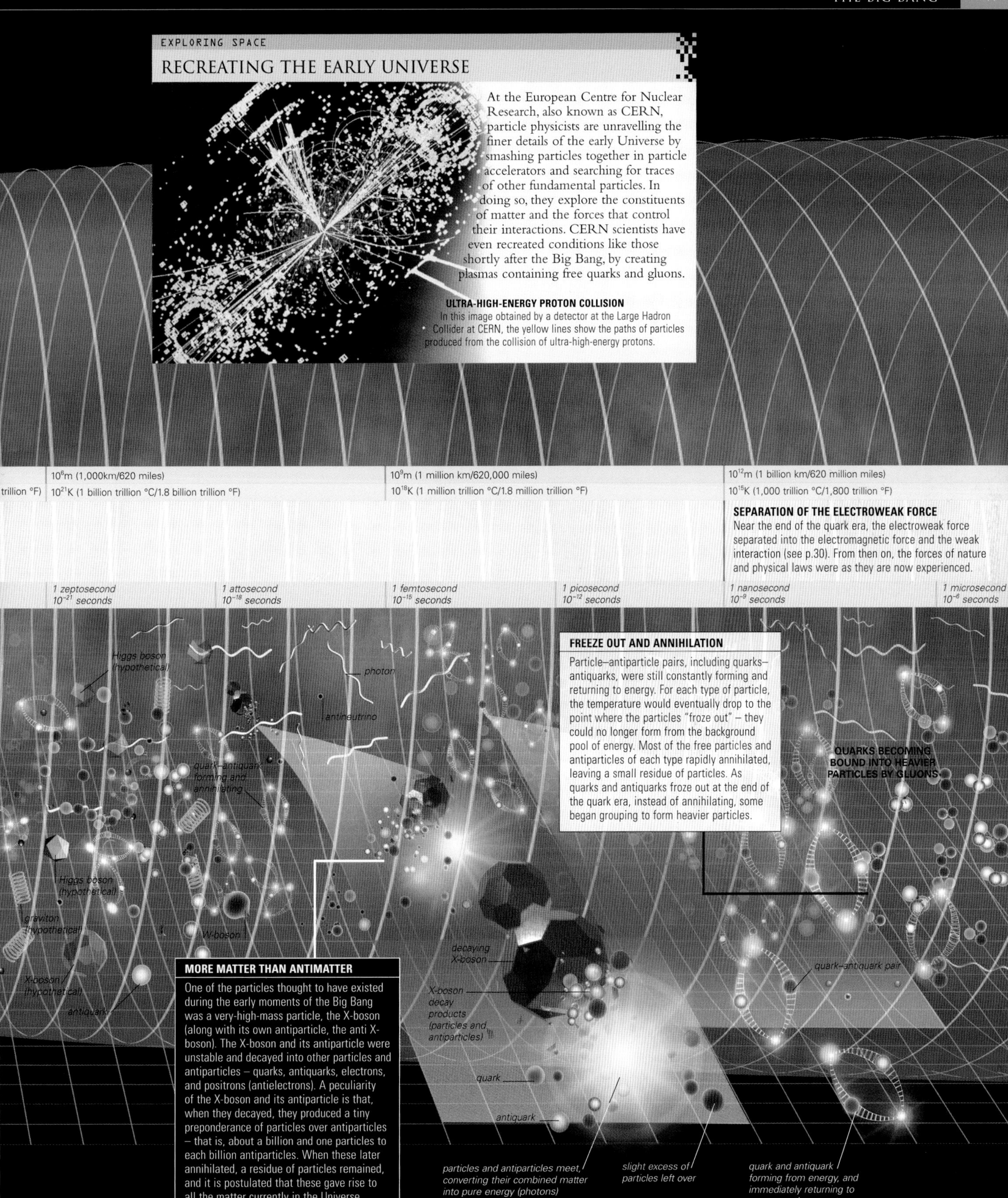


EXPLORING SPACE

## RECREATING THE EARLY UNIVERSE

At the European Centre for Nuclear Research, also known as CERN, particle physicists are unravelling the finer details of the early Universe by smashing particles together in particle accelerators and searching for traces of other fundamental particles. In doing so, they explore the constituents of matter and the forces that control their interactions. CERN scientists have even recreated conditions like those shortly after the Big Bang, by creating plasmas containing free quarks and gluons.

**ULTRA-HIGH-ENERGY PROTON COLLISION**
In this image obtained by a detector at the Large Hadron Collider at CERN, the yellow lines show the paths of particles produced from the collision of ultra-high-energy protons.

**SEPARATION OF THE ELECTROWEAK FORCE**
Near the end of the quark era, the electroweak force separated into the electromagnetic force and the weak interaction (see p.30). From then on, the forces of nature and physical laws were as they are now experienced.

**FREEZE OUT AND ANNIHILATION**
Particle–antiparticle pairs, including quarks–antiquarks, were still constantly forming and returning to energy. For each type of particle, the temperature would eventually drop to the point where the particles "froze out" – they could no longer form from the background pool of energy. Most of the free particles and antiparticles of each type rapidly annihilated, leaving a small residue of particles. As quarks and antiquarks froze out at the end of the quark era, instead of annihilating, some began grouping to form heavier particles.

**MORE MATTER THAN ANTIMATTER**
One of the particles thought to have existed during the early moments of the Big Bang was a very-high-mass particle, the X-boson (along with its own antiparticle, the anti X-boson). The X-boson and its antiparticle were unstable and decayed into other particles and antiparticles – quarks, antiquarks, electrons, and positrons (antielectrons). A peculiarity of the X-boson and its antiparticle is that, when they decayed, they produced a tiny preponderance of particles over antiparticles – that is, about a billion and one particles to each billion antiparticles. When these later annihilated, a residue of particles remained, and it is postulated that these gave rise to all the matter currently in the Universe.

# THE EMERGENCE OF MATTER

About 1 microsecond ($10^{-6}$ or one millionth of a second) after the Big Bang, the young Universe contained, in addition to vast quantities of radiant energy, or photons, a seething "soup" of quarks, antiquarks, and gluons. Also present were the class of fundamental particles called leptons (mainly electrons, neutrinos, and their antiparticles) forming from energy and then annihilating back to energy. The stage was set for the next processes of matter formation that led to our current Universe. First, quarks and gluons met to make heavier particles – particularly protons and a smaller number of neutrons. Next, the neutrons combined with some of the protons to form atomic nuclei, mainly those of helium. The remaining protons, destined to form the nuclei of hydrogen atoms, stayed uncombined. Finally, after half a million years, the Universe cooled sufficiently for electrons to combine with the free protons and helium nuclei – so forming the first atoms.

### THE NEXT HALF-MILLION YEARS

The timeline on these two pages shows events from 1 microsecond to 500,000 years after the Big Bang. The temperature dropped from $10^{13}$°K (10 trillion °C/18 trillion °F) to 2,500°C (4,500°F). Today's observable Universe expanded from 100 billion km (about 50 light-hours) to many millions of light-years wide.

## GEORGE GAMOW

Influenced by the original "Big Bang" concept of Georges Lemaître (see p.96), Ukrainian-American physicist George Gamow (1904–1968) played a major role in developing the "hot Big Bang" theory. This, supplemented by inflation, is the mainstream theory today. With his students Alpher and Herman, Gamow studied details of the theory, estimating the present cosmic temperature as 5K above absolute zero.

| | | | |
|---|---|---|---|
| **DIAMETER** | 100 billion km/60 billion miles | 1,000 billion km/600 billion miles | 10 light-years (1 light-year = 9.46 trillion km/5.88 trillic |
| **TEMPERATURE** | $10^{13}$K (10 trillion °C/18 trillion °F) | $10^{12}$K (1,000 billion °C/1,800 billion °F) | $10^{10}$K (10 billion °C/18 billion °F) |
| | **HADRON ERA** Around the beginning of this era, quarks and antiquarks began combining to form particles called hadrons. These included baryons (protons and neutrons), antibaryons, and mesons. | **LEPTON ERA** During this era, leptons (electrons, neutrinos, and their antiparticles) were very numerous. By its end, the electrons annihilated with positrons (antielectrons). | **NUCLEOSYNTHESIS ERA** Neutrons gradually converted into protons as the Universe cooled, but when there was about one neutron for every seven protons, most remaining neutrons combined with protons to make helium nuclei, each with two protons and two neutrons. |
| **TIME** | *1 microsecond $10^{-6}$ seconds – 1 millionth of a second* | *1 millisecond $10^{-3}$ seconds – 1 thousandth of a second* | *1 second* |

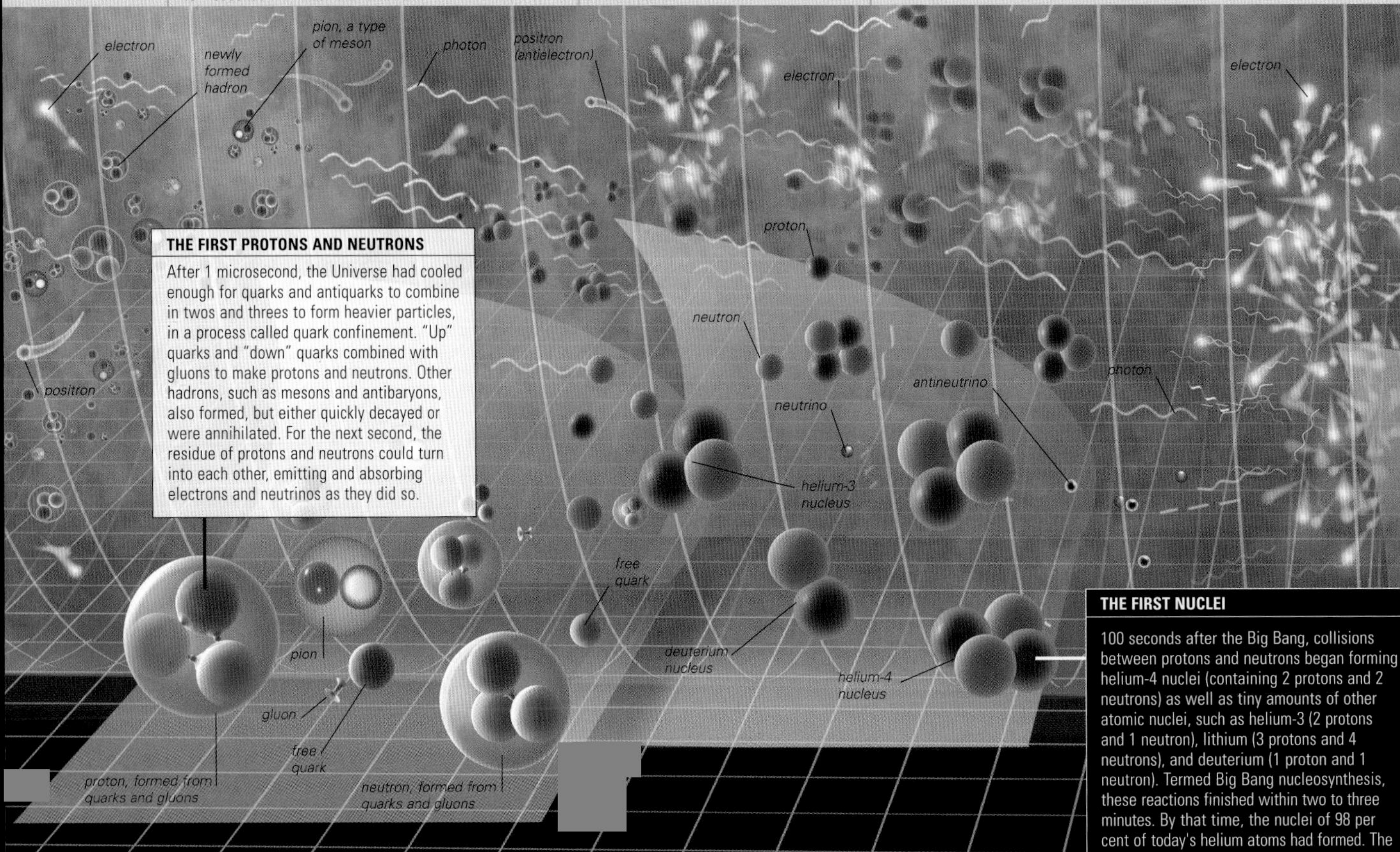


### THE FIRST PROTONS AND NEUTRONS

After 1 microsecond, the Universe had cooled enough for quarks and antiquarks to combine in twos and threes to form heavier particles, in a process called quark confinement. "Up" quarks and "down" quarks combined with gluons to make protons and neutrons. Other hadrons, such as mesons and antibaryons, also formed, but either quickly decayed or were annihilated. For the next second, the residue of protons and neutrons could turn into each other, emitting and absorbing electrons and neutrinos as they did so.

### THE FIRST NUCLEI

100 seconds after the Big Bang, collisions between protons and neutrons began forming helium-4 nuclei (containing 2 protons and 2 neutrons) as well as tiny amounts of other atomic nuclei, such as helium-3 (2 protons and 1 neutron), lithium (3 protons and 4 neutrons), and deuterium (1 proton and 1 neutron). Termed Big Bang nucleosynthesis, these reactions finished within two to three minutes. By that time, the nuclei of 98 per cent of today's helium atoms had formed. The reactions also mopped up all the free neutrons.

## EVIDENCE FOR THE BIG BANG

The strongest evidence for the Big Bang is the radiation it left, called the cosmic microwave background radiation (CMBR). George Gamow (see panel, opposite) predicted the radiation's existence in 1948. Its detection in the 1960s was confirmation, for most cosmologists, of the Big Bang theory. Other observations help support the theory.

**BACKGROUND RADIATION** The spectrum of the CMBR, discovered by Arno Penzias and Robert Wilson (below), indicates a uniformly hot early Universe.

**EXPANSION** If the Universe is expanding and cooling, it must once have been much smaller and hotter.

**BALANCE OF ELEMENTS** Big Bang theory exactly predicts the proportion of light elements (hydrogen, helium, and lithium) seen in the Universe today.

**GENERAL RELATIVITY** Einstein's theory predicts that the Universe must either be expanding or contracting – it cannot stay the same size.

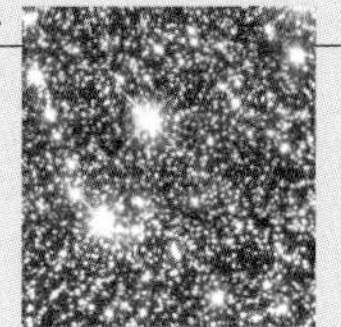

**DARK NIGHT SKY** If the Universe were both infinitely large and old, Earth would receive light from every part of the night sky and it would look bright – much brighter even than the densest star field (above). The fact that it is not is called Olbers' paradox. The Big Bang resolves the paradox by proposing that the Universe has not always existed.

10,000 light-years

$10^8$K (100 million °C/180 million °F)

**OPAQUE ERA**
During this relatively lengthy era, the ocean of matter particles (comprising mainly electrons, protons, and helium nuclei) were in a continual state of interaction with photons (radiant energy), making the Universe "foggy".

*200 seconds*

**BALANCE OF ELEMENTS**
At the end of the Opaque Era, many more free protons existed than helium nuclei, or other atomic nuclei. The scene was set for the first atoms to form. When they did, about nine hydrogen atoms were made for each helium atom. A few lithium and deuterium (heavy hydrogen) atoms also formed.

100 million light-years

3,000K (2,700°C/4,900°F)

**MATTER ERA**
At the start of our present era, photons were free to travel through the Universe. Most electrons were bound to atoms until the first stars formed, reheating matter.

*300,000 years*

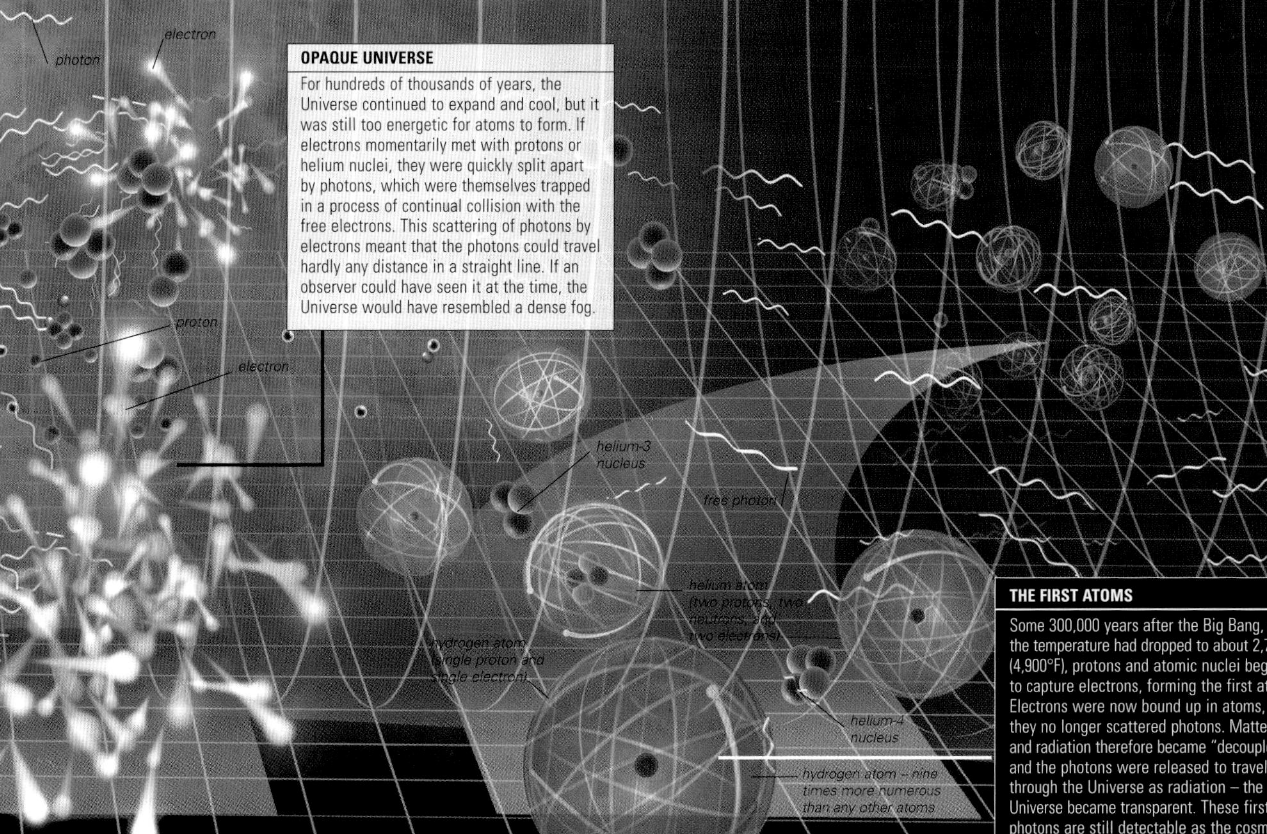


**OPAQUE UNIVERSE**
For hundreds of thousands of years, the Universe continued to expand and cool, but it was still too energetic for atoms to form. If electrons momentarily met with protons or helium nuclei, they were quickly split apart by photons, which were themselves trapped in a process of continual collision with the free electrons. This scattering of photons by electrons meant that the photons could travel hardly any distance in a straight line. If an observer could have seen it at the time, the Universe would have resembled a dense fog.

**THE FIRST ATOMS**
Some 300,000 years after the Big Bang, when the temperature had dropped to about 2,700°C (4,900°F), protons and atomic nuclei began to capture electrons, forming the first atoms. Electrons were now bound up in atoms, so they no longer scattered photons. Matter and radiation therefore became "decoupled", and the photons were released to travel through the Universe as radiation – the Universe became transparent. These first free photons are still detectable as the cosmic microwave background radiation (CMBR).

# OUT OF THE DARKNESS

28–31 Matter
32–35 Radiation
Stars 230–31
Stellar end points 262–63
Galaxy evolution 298–99
Galaxy superclusters 324–25

THE PERIOD FROM THE BIRTH of atoms, 300,000 years after the Big Bang, to the ignition of the first stars, hundreds of millions of years later, is known as the "dark ages" of the Universe. What happened in this era, and the subsequent "cosmic renaissance" as starlight filled the Universe, is an intricate puzzle. Astronomers are solving it by analysing the relic radiation of the Big Bang and using the world's most powerful telescopes to peer to the edges of the Universe.

## THE AFTERMATH OF THE BIG BANG

At an age of 350,000 years, the Universe was full of photons of radiation streaming in all directions, and of atoms of hydrogen and helium, neutrinos, and other dark matter. Although it was still hot, at 2,500°C (4,900°F), and full of radiation, astronomers see no light if they try to peer back to that moment. The reason is that as the Universe has expanded, it has stretched the wavelengths of radiation by a factor of a thousand. The photons reach Earth not as visible light, but as low-energy photons of cosmic microwave background radiation (CMBR). Their wavelengths, once characteristic of the fireball of the Universe, is now that of a cold object with a temperature of -270°C (-454°F) – only 3°C (5°F) above absolute zero.

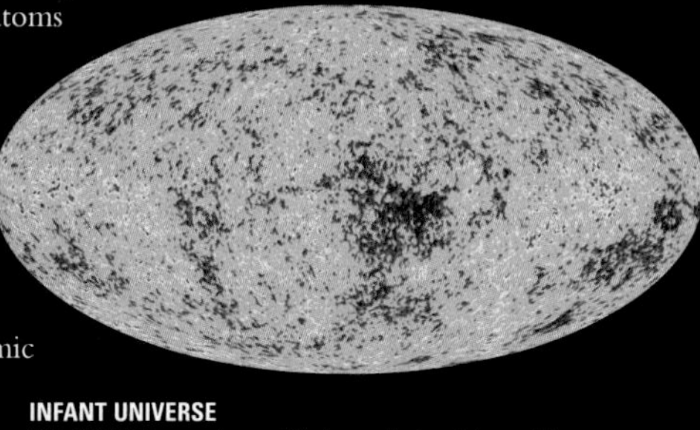

**INFANT UNIVERSE**
This WMAP image (see p.34), is an all-sky picture of the minute fluctuations in the temperature of the CMBR, which relate to early irregularities in matter density. In effect, it is an image of the infant Universe.

## THE DARK AGES

Earth will never receive visible light from the period before the first stars ignited, a few hundred million years after the Big Bang, but cosmologists can reconstruct what happened during that time using other data, such as those of the CMBR. The CMBR reveals tiny fluctuations in the density of matter at the time the first atoms formed. Cosmologists think that gravity, working on these ripples, caused the matter to begin forming into clumps and strands. These irregularities in the initial cloud of matter probably laid the framework of present-day large-scale objects, such as galaxy superclusters (see pp.324–25). The development of such structures over billions of years has been simulated with computers. These simulations rely on assumptions about the density and properties of matter, including dark matter, in the infant Universe, as well as the influence of dark energy (a force opposing gravity, see p.54). Some simulations closely resemble the distribution of matter seen in the Universe today.

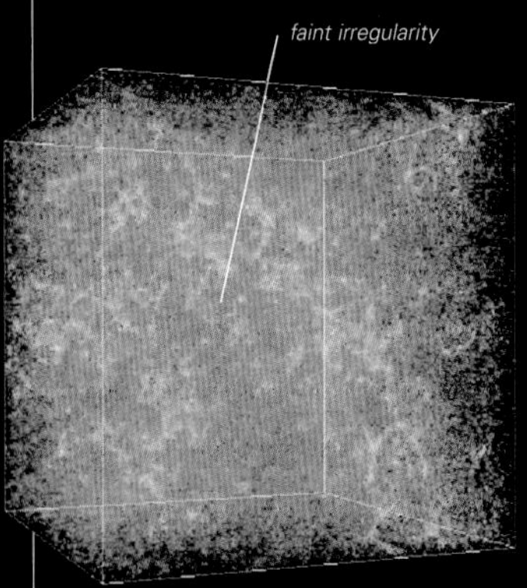


**UNIVERSE AT 500,000 YEARS OLD**
This computer simulation of the development of structure in the Universe starts with matter almost uniformly dispersed in a cube that is 140 million light-years high, wide, and deep.

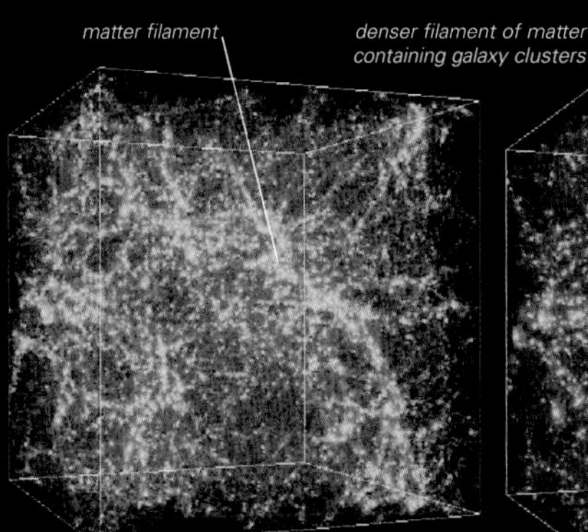


**1.3 BILLION YEARS OLD**
A billion years later, considerable clumping and filament formation has occurred. To compensate for the cosmic expansion since the previous stage, the cube has been scaled to size.

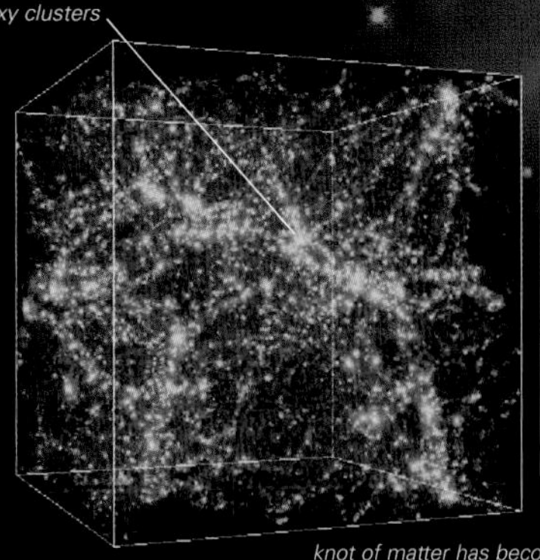


**5 BILLION YEARS OLD**
A further 4 billion years later, and (again after rescaling) the matter has condensed into some intricate filamentous structures interspersed with sizeable bubbles or voids of empty space.

**13.7 BILLION YEARS OLD**
The matter distribution in the simulation now resembles the sort of galaxy-supercluster structure seen in the local Universe (within a few billion light-years).

# EARLY GALAXIES

Astronomers are still trying to pinpoint when the very first stars ignited and in what types of early galactic structures this may have occurred. Recent infrared studies, with instruments such as the Spitzer Space Telescope and Very Large Telescope, have revealed what seem to be very faint galaxies, with extremely high red shifts, existing as little as 500 million years after the Big Bang. Their existence indicates that well-developed precursor knots and clumps of condensing matter may have existed as little as 100 to 300 million years after the Big Bang. It is within these structures that the first stars probably formed.

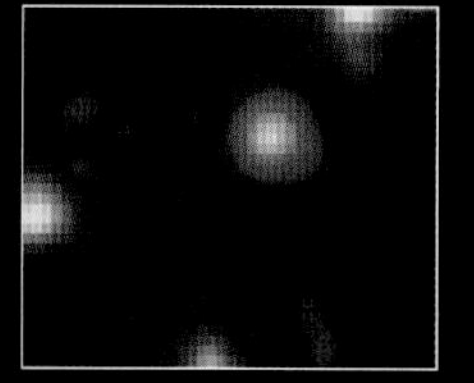

**EARLY GALAXY IN INFRARED**
The purple glow in this image is an active galactic nucleus. It is seen as it was only 700 million years after the Big Bang.

# THE FIRST STARS

The first stars, which may have formed only 200 million years after the Big Bang, were made almost entirely of hydrogen and helium, as virtually no other elements were present. Physicists think that star-forming nebulae that lacked heavy elements condensed into larger clumps than those of today. Stars forming from these clumps would have been very large and hot, with perhaps 100 to 1,000 times the mass of the Sun. Many would have lasted only a few million years before dying as supernovae. Ultraviolet light from these stars may have triggered a key moment in the Universe's evolution – the re-ionization of its hydrogen, turning it from a neutral gas back into the ionized (electrically charged) form seen today. Alternatively, radiation from quasars (see p.310) may have re-ionized the Universe.

**DEATH OF MEGASTARS**
The first, massive stars may have exploded as "hypernovae" – events associated today with black-hole formation and violent bursts of gamma rays. These artist's impressions depict one model of hypernova development.

200-solar-mass "megastar"

gamma-ray jet

core collapses into star's own black hole

star sheds outer shells of matter

**IONIZING POWER OF STARS**
These young, high-mass stars in the Orion Nebula ionize the gas around them, causing it to glow. Ionized hydrogen between galaxy clusters today may have been created by the far fiercer radiation of the first generation of stars and hypernovae.

# COSMIC CHEMICAL ENRICHMENT

During the course of their lives and deaths, the first massive stars created and dispersed new chemical elements into space and into other collapsing protogalactic clumps. A zoo of new elements, such as carbon, oxygen, silicon, and iron, was formed from nuclear fusion in the hot cores of these stars. Elements heavier than iron, such as barium and lead, were formed during their violent deaths. Second- and third-generation stars, smaller than the primordial megastars, later formed from the enriching interstellar medium. These stars created more of the heavier elements and returned them to the interstellar medium via stellar winds and supernovae explosions. Galactic mergers and the stripping of gas from galaxies (see p.317) led to further intergalactic mixing and dispersion. These processes of recycling and enrichment of the cosmos continue today. In the Milky Way galaxy, the new heavier elements have been essential to the formation of objects from rocky planets to living organisms.

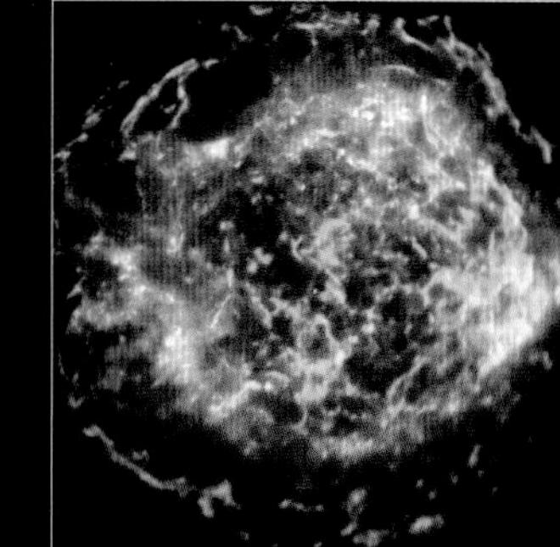

**STARDUST**
Supernova remnant Cassiopeia A is a sphere of enriched material expanding into space. Elements heavier than iron have mostly been made and dispersed by supernovae.

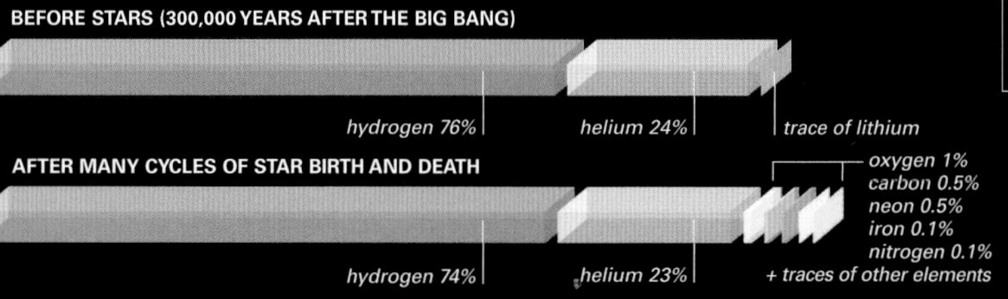


**COMPOSITION OF THE UNIVERSE**
The early Universe consisted of hydrogen and helium, with a trace of lithium. Today it still consists mainly of hydrogen and helium, but stellar processes have boosted the contribution from other chemical elements to more than 2 per cent.

# LIFE IN THE UNIVERSE

29 Chemical compounds
Telescope technology 93
The hunt goes on 97
Beyond Earth 108–11
Life on Earth 141

THE ONLY KNOWN LIFE IN THE COSMOS is that on Earth. Life on Earth is so ubiquitous, however, and the Universe so enormous, that many scientists think there is a very good chance that life also exists elsewhere. Much depends on whether the development of life on Earth was a colossal fluke – the product of an extremely improbable series of events – or, as many believe, not so unexpected given what is suspected about primordial conditions on the planet.

## LIVING ORGANISMS

What exactly constitutes a living organism? Human ideas on this are heavily reliant on the study of life on Earth, as scientists have no experience of the potential breadth of life beyond. Nonetheless, biologists are agreed on a few basic features that distinguish life from non-life anywhere in the cosmos – as a bare minimum, a living entity must be able to replicate itself and, over time, to evolve. Beyond that, the definition of life is not universally agreed. As an illustration, there is uncertainty about whether viruses are living. Although they self-replicate, viruses lack some characteristics that most biologists consider essential to life; notably they do not exist as cells or possess their own biochemical machinery. It is also uncertain that other characteristics common to life on Earth, such as carbon chemistry or the use of liquid water, must inevitably be a feature of extraterrestrial life. Disagreements over such matters add complexity to discussions of the likelihood of life beyond Earth.

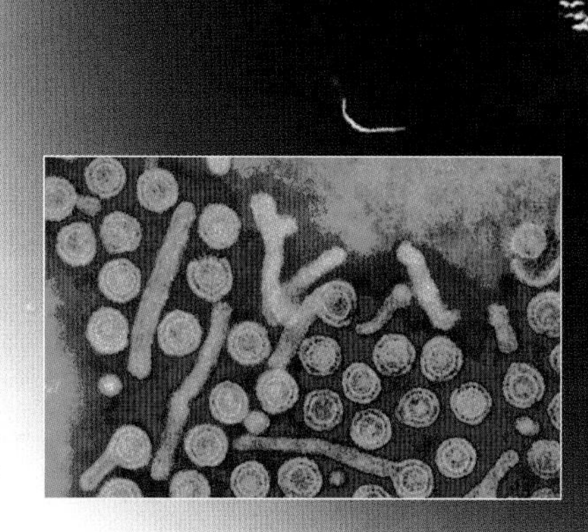

**VIRUS PARTICLES**
Viruses, such as this hepatitis virus, are on the border between living and non-living matter. They self-replicate but can do so only by hijacking the metabolic machinery of animal, plant, or bacterial cells.

**SUBZERO LIFE FORM**
This so-far-unclassified life form was found living deep in the Antarctic ice sheet. Life can exist in a wider range of conditions than once thought.

## ORIGINS OF LIFE

Most scientists agree that the beginnings of life on Earth were linked to the accumulation of simple organic (carbon-containing) molecules in a "primordial soup" in Earth's oceans not long after their formation. The molecules originated from reactions of chemicals in Earth's atmosphere, stimulated by energy, perhaps from lightning. Within the soup, over millions of years the organic compounds reacted to form larger and more complex molecules, until a molecule appeared with the capacity to replicate itself. By its nature, this molecule – a rudimentary gene – became more common. Through mutations and the mechanism of natural selection, variants of this gene developed more sophisticated survival adaptations, eventually evolving into a bacteria-like cell – the precursor of all other life on Earth.

Many evolutionary biologists would say that the decisive event was the appearance of the self-replicator, after which living organisms would inevitably follow.

**STROMATOLITES**
Some of the earliest remains of life are fossil stromatolites – mineral mounds built billions of years ago in shallow seas by cyanobacteria (blue-green algae). Stromatolites still grow on the Australian coast (left).

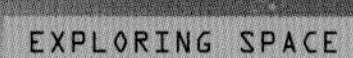

EXPLORING SPACE

### RECREATING PRIMORDIAL EARTH

In 1953, American chemist Stanley Miller (b.1930) recreated what he thought was Earth's primordial atmosphere in a flask. He sent sparks, simulating lightning, into the gas mixture, which lacked oxygen. The result was many different amino acids – some of the basic building blocks of life.

**STANLEY MILLER**
Here, Stanley Miller recreates the experiment he first conducted as a graduate student. It showed that amino acids could have formed in Earth's oxygen-free early atmosphere.

# HOW RARE IS LIFE?

Until about 30 years ago, the ranges of conditions thought essential to life, such as those of temperature and humidity, were thought to be narrow. Since then, scientists have found extremophiles (organisms that thrive in extreme conditions) living in adverse environments on Earth. Organisms may live deep in ice sheets or in boiling-hot water around vents in the ocean floor. Some exist in communities divorced from sunlight and live on energy from chemical sources. Bacteria are even found living 3km (2 miles) deep in the Earth's crust, living on hydrogen, which they convert to water. Extremophiles have encouraged the idea that life can exist in a wide range of conditions. Some scientists are still hopeful that extraterrestrial life will be found in the Solar System, although exploration of the most likely location, Mars, has proved negative so far. Beyond the Solar System, many scientists think that life must be widespread. At these remote distances, scientists are most interested in whether intelligent, contactable life exists. In the 1960s, American radio astronomer Frank Drake (b.1930) developed an equation for predicting the number of civilizations in the galaxy capable of interstellar communication. Because few of the factors in the equation can be estimated accurately, applying it (see panel, right) can have any outcome from less than one to millions, depending on the estimated values. Nevertheless, it is not unreasonable to suggest that at least a few such civilizations may exist in the Milky Way.

**LIFE ON EUROPA?**
Jupiter's moon Europa is covered with ice. There may be a liquid ocean underneath, possibly containing water, with the possibility of life. In 2012, NASA plans to send a probe to scan Europa and two other Jovian moons.

## ALIEN CIVILIZATIONS?

Applying the Drake Equation involves estimating factors, such as the the fraction of stars that develop planets, then multiplying all the factors. The example below uses only moderately optimistic estimates (some are just guesses).

**RATE OF STAR BIRTH** A fair estimate would be 50 new stars per year in the Milky Way.

*50% of new stars develop planets*

**STARS WITH PLANETS** Perhaps 50 per cent of these stars develop planetary systems.

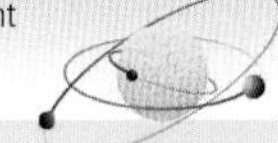

*0.4 planets will be habitable*

**HABITABLE PLANETS** On average maybe only 0.4 planets per system are habitable.

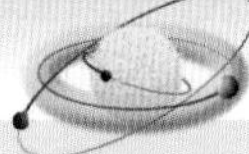

*90% of habitable planets develop life*

**PLANETS WITH LIFE** Life may well develop on 90% of habitable planets.

*90% of life-bearing planets bear only simple life* *10%*

**INTELLIGENT LIFE** Possibly about 10% of new instances of life develop intelligence.

*90% of intelligent life never talks to the stars* *10%*

**COMMUNICATING LIFE** Possibly only 10% of such life develops interstellar communications.

*some civilizations die before contact*

**LIFE SPAN OF CIVILIZATION** These civilizations might, on average, last 10,000 years.

*900 civilizations alive today*

**CONCLUSION**
Using the estimates above, one might expect there to be about 50 x 0.5 x 0.4 x 0.9 x 0.1 x 0.1 x 10,000 = 900 alien civilizations in our galaxy that, in theory, we should be able to communicate with. However, some of the estimates may be wildly wrong.

**RECOGNIZING LIFE**
If humans ever encounter extraterrestrial life, it is by no means certain that we would immediately recognize it. Not everyone would see life, rather than just discoloration, in this algal bloom growing in the North Atlantic.

# LOOKING FOR LIFE

Attempts to identify extraterrestrial life forms follow a number of approaches. Within the Solar System, scientists analyse images of planets and moons for signs of life and send probes to feasible locations, such as Mars and Saturn's moon Titan (see p.110). Outside the Solar System, the main focus of the search is SETI (the search for extraterrestrial intelligence) – a set of programs that involve scanning the sky for radio signals that look like they were sent by aliens. A search has also begun for Earth-like planets around nearby stars (see p.291). Finally, CETI (communication with extraterrestrial intelligence) involves broadcasting the presence of humans by sending signals towards target stars. In 1974, a CETI message in binary code was sent towards the M13 star cluster, 21,000 light-years away. In 1999, the more elaborate "Encounter 2001" message was sent from a Ukrainian radio telescope towards some nearby Sun-like stars. Even if aliens pick up this message, we can expect no reply for at least a century.

**ARECIBO DISH**
The Arecibo Telescope in Puerto Rico is the world's largest single-dish radio telescope. It has been used extensively for SETI and in one CETI attempt.

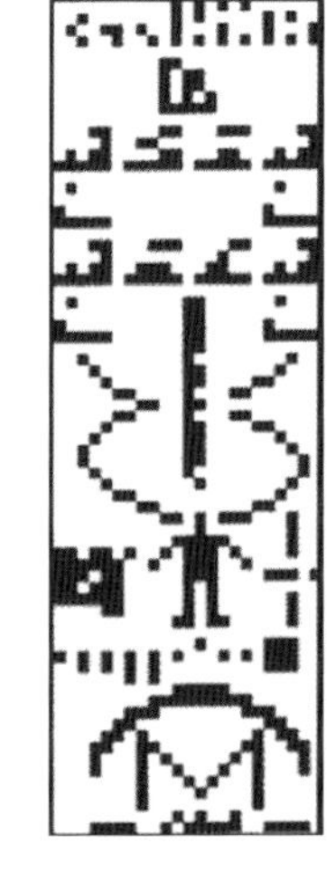

**MESSAGE TO ALIENS**
The Arecibo Telescope message contains symbols of a human body, DNA, the Solar System, and the Arecibo dish itself.

# THE FATE OF THE UNIVERSE

22–23 The scale of the Universe
24–27 Celestial objects
28–31 Matter
38–41 Space and time
46–49 The Big Bang

ALTHOUGH IT IS POSSIBLE THAT THE UNIVERSE will last forever, the types of structures that exist in it today, such as planets, stars, and galaxies, almost certainly will not. At some distant point in the future, our galaxy and others will either be ripped apart, suffer a long, protracted, cold death, or be crushed out of existence in a reverse of the Big Bang. Which of these fates befalls the Universe depends to a considerable extent on the nature of dark energy – a mysterious, gravity-opposing force recently found to be playing a major part in the Universe's large-scale behaviour.

## BIG CRUNCH AND BIG CHILL

Until recently, cosmologists assumed that the Universe's expansion rate (see pp.42–43) must be slowing, due to the "braking" effects of gravity. They also believed that a single factor – the Universe's mass-energy density – would decide which of two basic fates awaited it. Cosmologists measure the density of both mass and energy together since Einstein demonstrated that mass and energy are equivalent and interchangeable (see p.39). They calculated that if this density was above a critical value, gravity would eventually cause the Universe to stop expanding and collapse in a fiery, all-annihilating implosion (a "Big Crunch"). If, however, the Universe's density was below or exactly on the critical value, the Universe would expand forever, albeit with its expansion rate gradually slowed by gravity. In this case, the Universe would end in a lengthy, cold death (a "Big Chill"). Research aimed at resolving this issue found that the Universe has properties suggesting that it is extremely close to being "flat" (opposite), with a density of exactly the critical value. Even though some of the mass-energy in the Universe needed to render it flat seemed hard to locate, its density must be near the critical value, and so its most likely fate was eternal expansion. However, in the late 1990s, models of the fate of the Universe were thrown into confusion by new findings indicating that the Universe's expansion is not slowing down at all.

**FOUR POTENTIAL FATES**
Depending on the average density of the Universe and the future behaviour of dark energy, the Universe has a number of possible different fates. Four alternatives, of differing likelihood, are depicted here.

**BIG CHILL**
If the Universe has a mass-energy density close to or just less than the critical value, and should the effects of dark energy tail off, the Universe might continue to expand at a rate that slowly decreases but never comes to a complete halt. Over unimaginably long periods of time, it suffers a lingering cold death or "Big Chill".

**MODIFIED BIG CHILL**
If the effects of dark energy continue as they do at present, the Universe will expand at an increasing rate whatever its density. Structures that are not bound by gravity will fly apart, ultimately at speeds faster than the speed of light (space itself can expand at such speed, although matter and radiation cannot). This scenario will also end in a lingering cold death or Big Chill.

**BIG RIP**
If the strength of dark energy increased, it could overcome all the fundamental forces and totally disintegrate the Universe in a "Big Rip". This could happen 20–30 billion years from now. First galaxies would be torn apart, then solar systems. A few months later, stars and planets would explode, followed shortly by atoms. Time would then stop.

## DARK ENERGY

The new findings (see above) came from studies of supernovae in remote galaxies. The apparent brightness of these exploding stars can be used to calculate their distance, and by comparing their distances with the red shifts of their home galaxies, scientists can calculate how fast the Universe was expanding at different times in its history. The calculations showed that the expansion of the Universe is accelerating and that some repulsive force is opposing gravity, causing matter to fly apart. This force has been called dark energy, and its exact nature is uncertain, though it appears similar to a gravity-opposing force, the "cosmological constant", proposed by Albert Einstein as part of his theory of general relativity (see p.40–41). The existence of dark energy also accounts for the missing mass-energy in the Universe required to make it flat (above), and modifies the number of possible fates for the Universe.

SUPERNOVA DISCOVERY

**SUPERNOVAE CLUES**
Type Ia supernovae, like that depicted here, all have the same intrinsic brightness. Hence their apparent brightness reveals their distance.

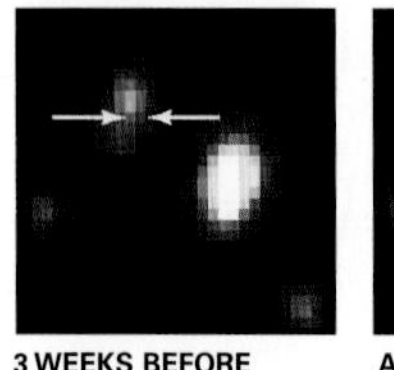

3 WEEKS BEFORE

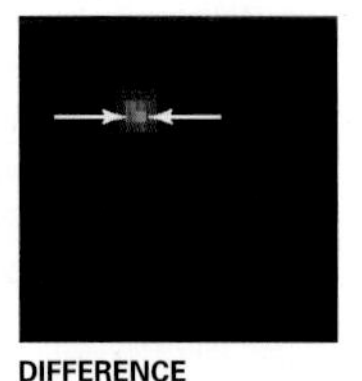

AFTER SUPERNOVA

DIFFERENCE

**DARK ENERGY DOMINANCE**
Dark energy provides 70 per cent of the mass-energy density of the Universe. Atom-based matter (in stars and the interstellar medium) and neutrinos contribute just 5 per cent.

DARK ENERGY
neutrinos 0.3%, stars 0.5%, heavy elements 0.03%
dark energy about 70%
dark matter about 25%
free hydrogen and helium 4%

# THE GEOMETRY OF SPACE

Cosmologists base their ideas on the fate of the Universe partly on mathematical models. These indicate that, depending on its mass-energy density, the Universe has three possible geometries, each with a different space-time curvature that can be represented by a 2-D shape. Before the discovery of dark energy, there was a correspondence between these geometries and the fate of the Universe. A positively curved or "closed" universe was envisaged to end in a Big Crunch and a negatively curved or "open" universe in a Big Chill. A "flat" universe would also end in a Big Chill but one in which the Universe's expansion eventually slows to a virtual standstill. With the discovery of dark energy, the correspondence no longer holds. If dark energy remains constant in intensity, any type of Universe may expand forever. If dark energy is capable of reversing, any type of universe could end in a Big Crunch. Currently, the most favoured view is that the Universe is flat and will undergo an accelerating expansion. A cataclysmic "Big Rip" scenario, in which increasing dark energy tears the Universe apart, is less likely.

**FLAT UNIVERSE**
If the density of the Universe is exactly on a critical value, it is "flat". In a flat universe, parallel lines never meet. The 2-D analogy is a plane. The Universe is thought to be flat or nearly flat.

**CLOSED UNIVERSE**
If the Universe is denser than a critical value, it is positively curved or "closed" and is finite in mass and extent. In such a universe, parallel lines converge. The 2-D analogy is a spherical surface.

**OPEN UNIVERSE**
If the Universe is less dense than a critical value, it is negatively curved or "open" and infinite. The 2D analogy of such a universe is a saddle-shaped surface on which parallel lines diverge.



**BIG CRUNCH**
In this version of doomsday, all matter and energy collapse to an infinitely hot, dense singularity, in a reverse of the Big Bang. This scenario currently looks the least likely unless the effect of dark energy reverses in future. Even if it did happen, the earliest it could do so would be tens of billions of years from now.

# A COLD DEATH

If the Universe peters out in a Big Chill, its death will take a long time. Over the next $10^{12}$ (1 trillion) years, galaxies will exhaust their gas in forming new stars. About $10^{25}$ (10 trillion trillion) years in the future, most of the Universe's matter will be locked up in star corpses such as black holes and burnt-out white dwarfs circling and falling into the supermassive black holes at the centres of galaxies. At $10^{32}$ (1 followed by 32 zeros) years from now, protons will start decaying to radiation (photons), electrons, positrons, and neutrinos. All matter not in black holes will fall apart. After another $10^{67}$ years, black holes will start evaporating by emitting particles and radiation, and in about $10^{100}$ years, even supermassive black holes will evaporate. The utterly cold, dark Universe will then be nothing but a diffuse sea of photons and fundamental particles.

**FINAL SURVIVORS**
At a very distant stage of the Big Chill, all the Universe's matter, even that in black holes, will have decayed or evaporated to radiation. Apart from some very long-wavelength photons, the only constituents of the Universe will be neutrinos, electrons, and positrons.



**FATE OF GALAXIES**
A trillion years from now, the Universe will contain just old, fading, galaxies. All their gas and dust will be used up and most of the stars will be dying.

*"Mortal as I am, I know that I am born for a day. But when I follow at my pleasure the serried multitude of the stars in their circular course, my feet no longer touch the Earth."*

Ptolemy

OBJECTS IN THE UNIVERSE – galaxies, stars, planets, nebulae – are scattered across three dimensions of space and one of time. Viewed from widely separated locations in the Universe, their relative positions look completely different. To find objects in space, study their movements, and make celestial maps, astronomers need an agreed reference frame, and for most purposes the frame used is Earth itself. The prime element of this Earth-based view is the celestial sphere – an imaginary shell around Earth to which astronomers pretend the stars are attached. Apparent movements of celestial objects on this sphere can be related to the actual movements of Earth, the planets (as they orbit the Sun), the Moon (as it orbits Earth), and the stars as they move within the Milky Way. Understanding the celestial sphere, and conventions for naming and finding objects on it, are essential first steps in astronomy.

**MOVEMENT ON THE SKY**
This photograph, obtained over a four-hour period from the Las Campanas Observatory in Chile, looks towards the south celestial pole. The circular, clockwise star trails across the sky are a feature of the Earth-based view of the cosmos, as they result solely from the Earth's rotation.

# THE VIEW FROM EARTH

# THE CELESTIAL SPHERE


Celestial cycles 60–63
The Earth-centred cosmos 85
Earth's orbit 138
Mapping the sky 332–37
Using the sky guides 412–13


FOR CENTURIES, humans have known that stars lie at different distances from Earth. However, when recording the positions of stars in the sky, it is convenient to pretend that they are all stuck to the inside of a sphere that surrounds Earth. The idea of this sphere also helps astronomers to understand how their location on Earth, the time of night, and the time of year affect what they see in the night sky.

## THE SKY AS A SPHERE

To an observer on Earth, the stars appear to move slowly across the night sky. Their motion is caused by Earth's rotation, although it might seem that the sky is spinning around our planet. To the observer, the sky can be imagined as the inside of a sphere, known as the celestial sphere, to which the stars are fixed, and relative to which the Earth rotates. This sphere has features related to the real sphere of the Earth. It has north and south poles, which lie on its surface directly above Earth's North and South Poles, and it has an equator (the celestial equator), which sits directly above Earth's equator. The celestial sphere is like a celestial version of a globe – the positions of stars and galaxies can be recorded on it, just as cities on Earth have their positions of latitude and longitude on a globe.



**IMAGINARY GLOBE**
The celestial sphere is purely imaginary, with a specific shape but no precise size. Astronomers use exactly defined points and curves on its surface as references for describing or determining the positions of stars and other celestial objects.

## EFFECTS OF LATITUDE

An observer on Earth can view, at best, only half of the celestial sphere at any instant (assuming a cloudless sky and unobstructed horizon). The other half is obscured by Earth's bulk. In fact, for an observer at either of Earth's poles, a specific half of the celestial sphere is always overhead, while the other half is never visible. For observers at other latitudes, Earth's rotation continually brings new parts of the celestial sphere into view and hides others. This means, for example, that over the course of a night, an observer at a latitude of 60°N or 60°S can see up to three-quarters of the celestial sphere for at least some of the time; and an observer at the equator can see every point on the celestial sphere at some time.



**OBSERVER AT EQUATOR**
For a person on the equator, Earth's rotation brings all parts of the celestial sphere into view for some time each day. The celestial poles are on the horizon.

**OBSERVER AT NORTH POLE**
For this observer, the northern half of the celestial sphere is always visible, and the southern half is never visible. The celestial equator is on the observer's horizon.

**OBSERVER AT MID-LATITUDE**
For this observer, a part of the celestial sphere is always visible, a part is never visible, and Earth's rotation brings other parts into view for some of the time each day.





**MOTION AT NORTH POLE**
At the poles, all celestial objects seem to circle the celestial pole, directly overhead. The motion is anticlockwise at the North Pole, clockwise at the south.

**MOTION AT MID-LATITUDE**
At mid-latitudes, most stars rise in the east, cross the sky obliquely, and set in the west. Some (circumpolar) objects never rise or set but circle the celestial pole.

**MOTION AT EQUATOR**
At the equator, stars and other celestial objects appear to rise vertically in the east, move overhead, and then fall vertically and set in the west.

**CIRCUMPOLAR STARS**
Stars in the polar regions of the celestial sphere describe perfect part-circles around the north or south celestial pole during one night, as shown by this long-exposure photograph.

# DAILY SKY MOVEMENTS

As the Earth spins, all celestial objects move across the sky, although the movements of the stars and planets become visible only at night. For an observer in mid-latitudes, stars in polar regions of the celestial sphere describe a daily circle around the north or south celestial pole. The Sun, Moon, planets, and the remaining stars rise along the eastern horizon, sweep in an arc across the sky, and set in the west. This motion has a tilt to the south (for observers in the northern hemisphere) or to the north (southern hemisphere) – the lower the observer's latitude, the steeper the tilt. Stars have fixed positions on the sphere, so the pattern of their movement repeats with great precision once every sidereal day (see p.62). The planets, Sun, and Moon always move on the celestial sphere, so the period of repetition differs from that of the stars.

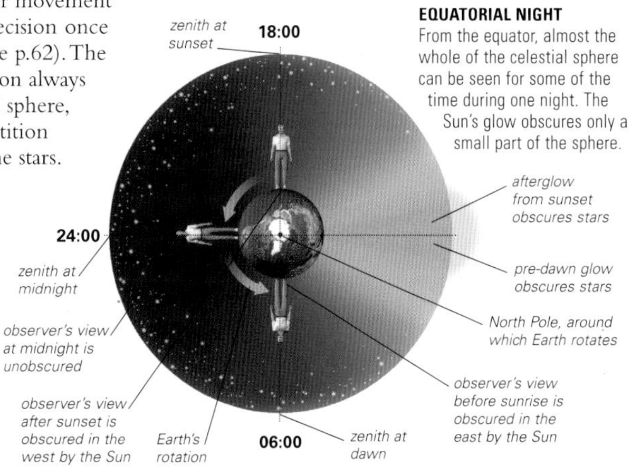


**EQUATORIAL NIGHT**
From the equator, almost the whole of the celestial sphere can be seen for some of the time during one night. The Sun's glow obscures only a small part of the sphere.

# YEARLY SKY MOVEMENTS

As Earth orbits the Sun, the Sun seems to move against the background of stars. As the Sun moves into a region of the sky, its glare washes out the light from that part, and so any star or other object there temporarily becomes difficult to view from anywhere on Earth. Earth's orbit also means that the part of the celestial sphere on the opposite side to Earth from the Sun – that is, the part visible in the middle of the night – changes. The visible part of the sky at, for example, midnight in June, September, December, and March is significantly different – at least for observers at equatorial or mid-latitudes on Earth.

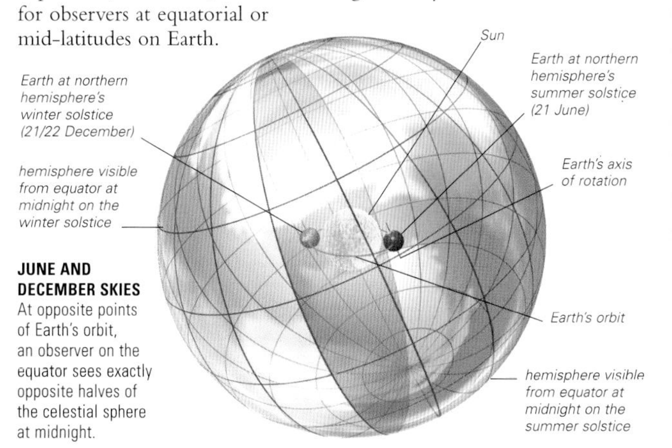


**JUNE AND DECEMBER SKIES**
At opposite points of Earth's orbit, an observer on the equator sees exactly opposite halves of the celestial sphere at midnight.

EXPLORING SPACE

## ARISTOTLE'S SPHERES

Until the 17th century AD, the idea of a celestial sphere surrounding Earth was not just a convenient fiction – many people believed it had a physical reality. Such beliefs date back to a model of the Universe developed by the Greek philosopher Aristotle (384–322 BC) and elaborated by the astronomer Ptolemy (AD 85–165, see p.85). Aristotle placed Earth stationary at the Universe's centre, surrounded by several transparent, concentric spheres to which the stars, planets, Sun, and Moon were attached. Ptolemy supposed that the spheres rotated at different speeds around Earth, so producing the observed motions of the celestial bodies.

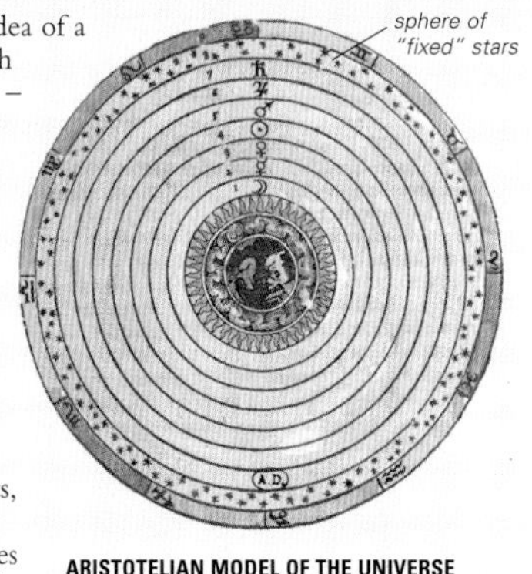


**ARISTOTELIAN MODEL OF THE UNIVERSE**
Stars are fixed to the outer sphere. Working inwards, the other spheres around Earth carry Saturn, Jupiter, Mars, the Sun, Venus, Mercury, and the Moon.

# CELESTIAL COORDINATES

Using the celestial sphere concept, astronomers can record and find the positions of stars and other celestial objects. To define an object's position, astronomers use a system of coordinates, similar to latitude and longitude on Earth. The coordinates are called declination and right ascension. Declination is measured in degrees and arcminutes (60 arcminutes = 1 degree/1°) north or south of the celestial equator, so is equivalent to latitude. Right ascension, the equivalent of longitude, is the angle of an object to the east of the celestial meridian. The meridian is a line passing through both celestial poles and a point on the celestial equator called the first point of Aries or vernal equinox point (see p.61). An object's right ascension can be stated in degrees and arcminutes or in hours and minutes. One hour is equivalent to 15°, because 24 hours make a whole circle.

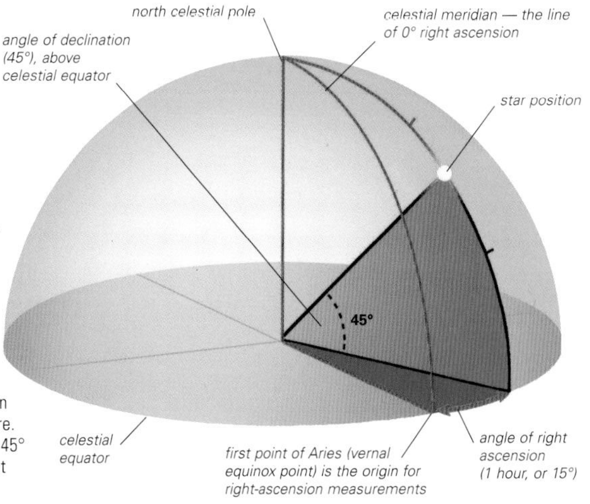


**RECORDING A STAR'S POSITION**
The measurement of a star's position on the celestial sphere is shown here. This star has a declination of about 45° (sometimes written +45°) and a right ascension of about 1 hour, or 15°.

# CELESTIAL CYCLES


58–59 The celestial sphere
Ancient astronomy 82–83
The Sun 120–23
Earth 138–41
The Moon 148–53
Mapping the sky 332–37


TO AN OBSERVER ON EARTH, celestial events occur within the context of cycles determined by the motions of Earth, Sun, and Moon. These cycles provide us with some of our basic units for measuring time, such as days and years. They include the apparent daily motions of all celestial objects across the sky, the annual apparent movement of the Sun against the celestial sphere, the seasonal cycle, and the monthly cycle of lunar phases. Other related cycles produce the dramatic but predictable events known as lunar and solar eclipses.

**THE SUN'S ANALEMMA**
To produce this image, the Sun was photographed, above a sundial, at the same time of day on 37 occasions throughout one year. The vertical change in its position is due to Earth's tilt. The horizontal drift is due to Earth changing its speed on its elliptical orbit around the Sun. The resulting figure-of-eight pattern is called an analemma.

MYTHS AND STORIES

### ASTROLOGY AND THE ECLIPTIC

Astrology is the study of the positions and movements of the Sun, Moon, and planets in the sky in the belief that these influence human affairs. At one time, when astronomy was applied mainly to devising calendars, astronomy and astrology were intertwined, but their aims and methods have now diverged. Astrologers pay little attention to constellations, but measure the positions of the Sun and planets in sections of the ecliptic that they call "Aries" and "Taurus", for example. However, these sections no longer match the constellations of Aries, Taurus, and so on.

**STARGAZER**
This 17th-century illustration, taken from a treatise written in India on the zodiac, depicts a stargazer using an early form of mounted telescope.

## THE SUN'S CELESTIAL PATH

As the Earth travels round the Sun, to an observer on Earth the Sun seems to trace a path across the celestial sphere known as the ecliptic. Because of the Sun's glare, this movement is not obvious, but the Sun moves a small distance each day against the background of stars. The band of sky extending for 9 degrees (see p.59) on either side of the Sun's path is called the zodiac and incorporates parts or all of 24 constellations (see p.68). Of these, the Sun passes through 13 constellations, of which 12 form the "signs of the zodiac", well-known to followers of astrology (see panel, left). The Sun spends a variable number of days in each of these 13 constellations. However, the Sun currently passes through each constellation on dates very different from traditional astrological dates. For example, someone born between 21 March and 19 April, is said to have the sign Aries, although the Sun currently passes through Aries between 19 April and 23 May. This disparity is partly caused by a phenomenon called precession.

## PRECESSION

The Earth's axis of rotation is tilted to the ecliptic plane by 23.5°. The tilt is crucial in causing seasons (see opposite). At present, the axis points at a position on the northern celestial sphere (the north celestial pole) close to the star Polaris, but this will not always be so. Like a spinning top, Earth is executing a slow "wobble", which alters the direction of its axis over a 25,800-year cycle. The wobble, called precession, is caused by the gravity of the Sun and Moon. It also causes the south celestial pole, the celestial equator, and two other reference points on the celestial sphere, called the equinox points, to change their locations gradually. The coordinates of stars and other "fixed" objects, such as galaxies (see p.59), therefore change, so astronomers must quote them according to a standard "epoch" of around 50 years. The current standard was exactly correct on 1 January 2000.

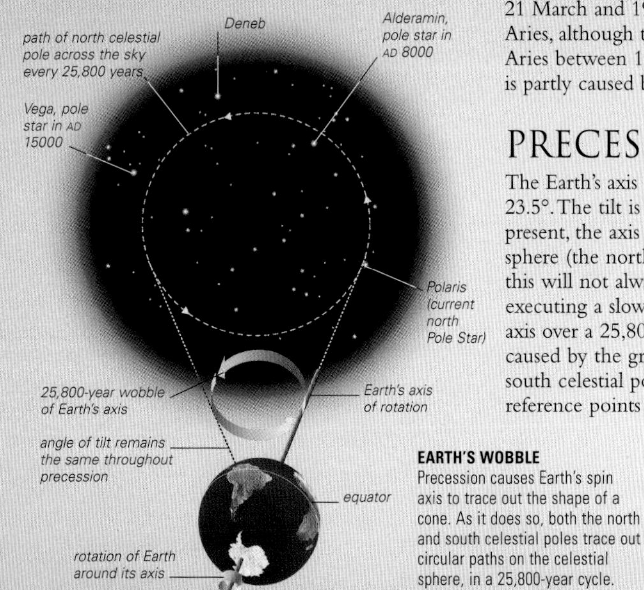


**EARTH'S WOBBLE**
Precession causes Earth's spin axis to trace out the shape of a cone. As it does so, both the north and south celestial poles trace out circular paths on the celestial sphere, in a 25,800-year cycle.

**ISLAMIC ZODIAC**
This Islamic depiction of part of the celestial sphere includes several constellations that are also well-known zodiacal "star signs", such as Scorpius and Leo. The illustration decorates a 19th-century manuscript from India that brought together Islamic, Hindu, and European knowledge of astronomy.

**MIDNIGHT SUN**
This multiple-exposure photograph (below) shows the path of the Sun around midnight near the summer solstice in Iceland. Since the photograph was taken in polar latitudes, Earth's angle of tilt ensures the Sun does not set.

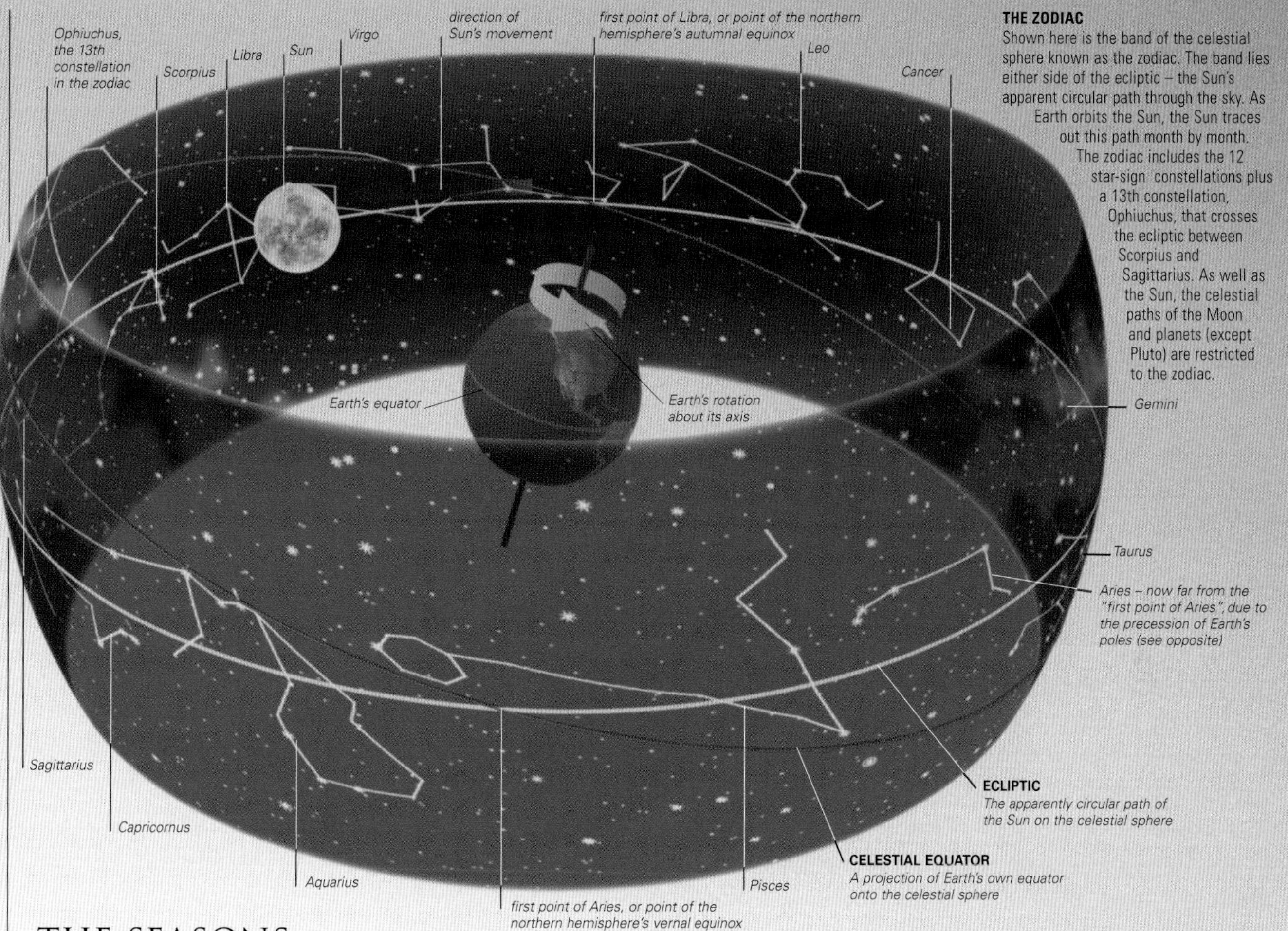


**THE ZODIAC**
Shown here is the band of the celestial sphere known as the zodiac. The band lies either side of the ecliptic – the Sun's apparent circular path through the sky. As Earth orbits the Sun, the Sun traces out this path month by month. The zodiac includes the 12 star-sign constellations plus a 13th constellation, Ophiuchus, that crosses the ecliptic between Scorpius and Sagittarius. As well as the Sun, the celestial paths of the Moon and planets (except Pluto) are restricted to the zodiac.

# THE SEASONS

Earth's orbit around the Sun takes 365.25 days and provides a key unit of time, the year. Earth's seasons result from the tilt of its axis relative to its orbit. Due to Earth's tilt, one or other of its hemispheres is normally pointed towards the Sun. The hemisphere that tilts towards the Sun receives more sunlight and is therefore warmer. Each year, the northern hemisphere reaches its maximum tilt towards the Sun around 21 June – summer solstice in the northern hemisphere and winter solstice in the southern hemisphere. For some time around this date, the north polar region is sunlit all day, while the south polar region is in darkness. Conversely, around 21 December, the situation is reversed. Between the solstices are the equinoxes, when Earth's axis is broadside to the Sun and the periods of daylight and darkness are equal for all points on Earth. Earth's tilt also defines the tropics. The Sun is overhead at midday on the Tropic of Cancer (23.5°N) around 21 June, above the Tropic of Capricorn (23.5°S) near to 21 December, and directly above the equator at midday during the equinoxes.

**SOLSTICES AND EQUINOXES**
At the solstices, in June and December, one hemisphere has its longest day, the other its shortest. At the equinoxes, in March and September, the length of day and night are equal for everywhere on Earth.

Earth on 20 or 21 March, the northern hemisphere's vernal or spring equinox
Earth on 21 or 22 December, the northern hemisphere's winter solstice
midday sun overhead at Tropic of Cancer
midday sun overhead at Tropic of Capricorn
Sun
Earth's orbit
Earth on 21 or 22 June, the northern hemisphere's summer solstice
Earth on 22 or 23 September, the northern hemisphere's autumnal equinox

23.5° angle of tilt
axis of spin
Tropic of Cancer, 23.5°N
solar radiation
direction of Earth's spin
Tropic of Capricorn, 23.5°S

**SUNLIGHT INTENSITY**
The intensity of solar radiation is greatest within the tropics. Towards the poles, the Sun's rays impinge at an oblique angle. They must pass through a greater thickness of atmosphere, and they are spread over a wider area of ground.

# MEASURING DAYS

**SOLAR TIME**
Solar time is the way of gauging time from the Sun's apparent motion across the sky, as measured by a sundial. One solar day is subdivided into 24 hours.

Every day, Earth rotates once, and most locations on its surface pass from sunlight to shadow and back, producing the day–night cycle. However, there are two possible definitions for what constitutes a day, and only one of these, the solar day, lasts for exactly 24 hours. A solar day is defined by the apparent movement of the Sun across the sky produced by Earth's rotation. It is the length of time the Sun takes to return to its highest point in the sky from the same point the previous day. The other type of day, the sidereal day, is defined by Earth's rotation relative to the stars. It is the length of time a star takes to return to its highest point in the sky on successive days. A sidereal day is 4 minutes shorter than a solar day.

1 APRIL, 20:00

8 APRIL, 20:00

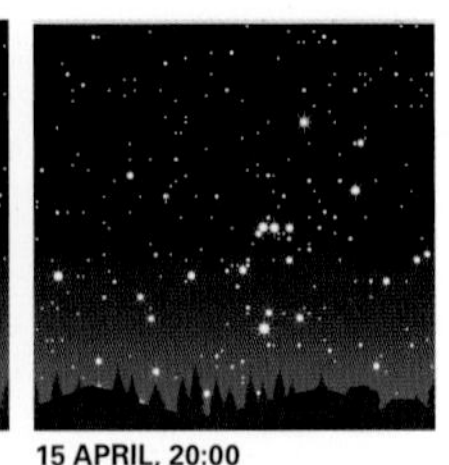
15 APRIL, 20:00

**SIDEREAL TIME**
The distinctive constellation Orion (see pp.374–75), here pictured as if from 50°N, appears lower in the sky at the same solar time each day, as the daily 4-minute difference between solar and sidereal time mounts up.

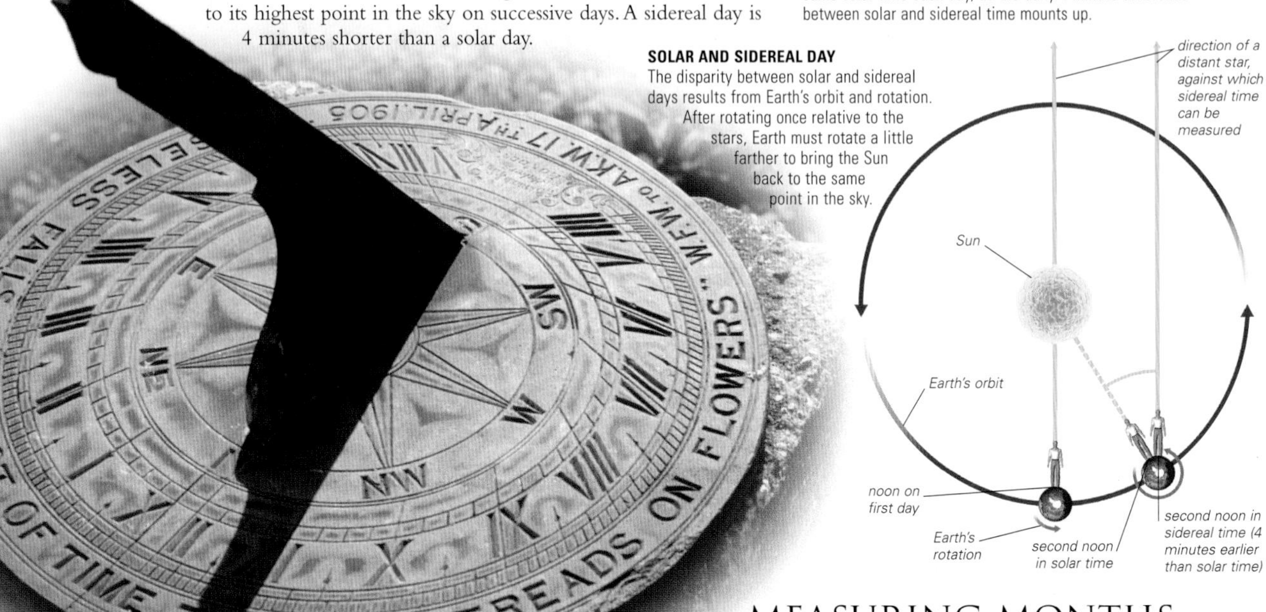


**SOLAR AND SIDEREAL DAY**
The disparity between solar and sidereal days results from Earth's orbit and rotation. After rotating once relative to the stars, Earth must rotate a little farther to bring the Sun back to the same point in the sky.

# MEASURING MONTHS

The concept of a month is based on the Moon's orbit around Earth. During each of the Moon's orbits, the angle between Earth, the Moon, and the Sun continuously changes, giving rise to the Moon's phases. The phases cycle through new Moon (when the Moon is between Earth and the Sun), crescent, quarter, and gibbous, to full Moon (when the Earth lies between the Moon and the Sun). A complete cycle of the Moon's phases takes 29.5 solar days and defines a lunar month. However, Earth's progress around the Sun complicates the expression of a month, just as it confuses the measurement of a day. The Moon in fact takes only 27.3 days to orbit Earth with reference to the background stars. Astronomers call this period a sidereal month. The disparity results because Earth's progress around the Sun alters the angles between the Earth, Sun, and Moon. After one full orbit of Earth (a sidereal month), the Moon must orbit a bit farther to return to its original alignment with Earth and the Sun (a lunar month).

sunlight
1. waxing crescent
2. first quarter
3. waxing gibbous
4. full moon
5. waning gibbous
6. last quarter
7. waning crescent
8. new moon

**CHANGING ANGLES**
During each lunar orbit, the angle between Earth, the Moon, and the Sun changes. The part of the Moon's sunlit face seen by an observer on Earth changes in a cyclical fashion.

1. WAXING CRESCENT

2. FIRST QUARTER

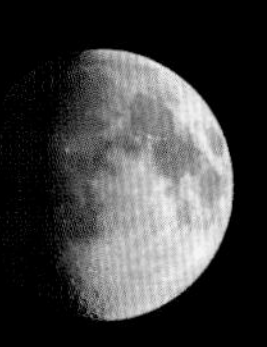
3. WAXING GIBBOUS

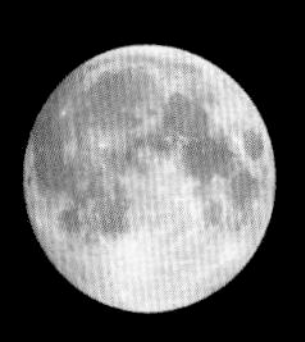
4. FULL MOON

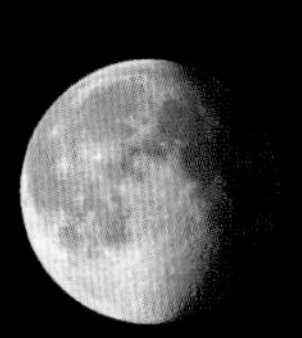
5. WANING GIBBOUS

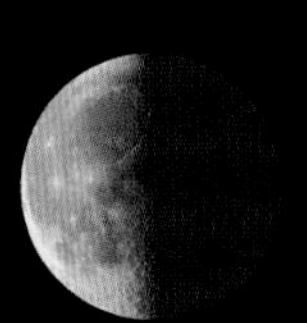
6. LAST QUARTER

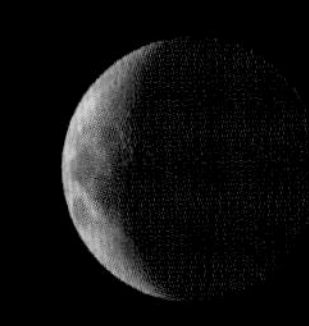
7. WANING CRESCENT

8. NEW MOON

MYTHS AND STORIES

## EVIL PORTENTS

Astronomers have predicted eclipses reliably since about 700 BC, but that has not stopped doomsayers and astrologers from reading evil omens into these routine celestial events. They have often prophesied disasters associated with eclipses, and although they meet with no more than occasional success, some people listen. The Incas below, for instance, are pictured as awestruck by an eclipse, in a European atlas of 1827. Eclipses may not be useful for predicting the future, but accounts of past eclipses are of great value to today's historians, who can calculate the dates of events with great precision if the historical accounts include records of eclipses.

# LUNAR ECLIPSES

As the Moon orbits the Earth, it occasionally moves into Earth's shadow – an occurrence called a lunar eclipse – or blocks sunlight from reaching a part of Earth's surface – a solar eclipse. Eclipses do not happen every month, because the plane of the Moon's orbit around Earth does not coincide with the plane of Earth's orbit around the Sun. Nevertheless, an eclipse of some kind occurs several times each year. Lunar eclipses occur two or three times a year, always during full Moon. Astronomers classify lunar eclipses into three different types. In a penumbral eclipse, the Moon passes through Earth's penumbra (part-shadow), leading to only a slight dimming of the Moon. In a partial eclipse, a portion of the Moon passes through Earth's umbra (full shadow), while in a total eclipse the whole Moon passes through the umbra.

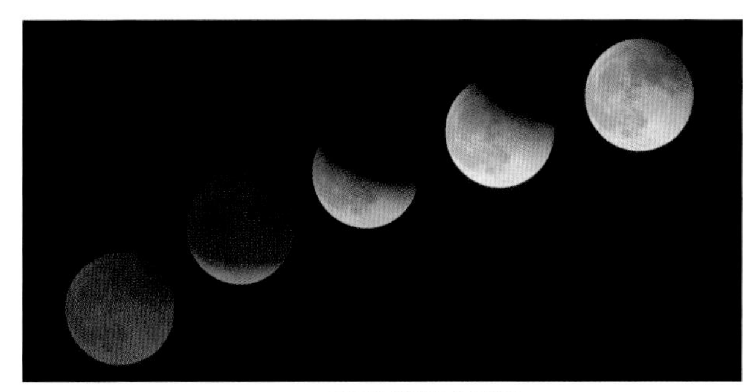

**TOTAL LUNAR ECLIPSE**
This composite photograph shows stages of a total lunar eclipse. The moon appears red at the eclipse's peak (bottom left), because a little red light is bent towards it by refraction in Earth's atmosphere.

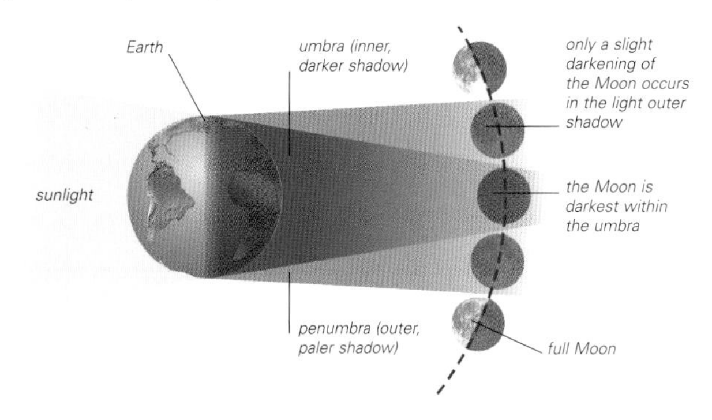


**MECHANICS OF A LUNAR ECLIPSE**
Earth's shadow consists of the penumbra, within which some sunlight is blocked out, and the umbra, or full shadow. In a total eclipse, the Moon passes through the penumbra, umbra, and then the penumbra again.

# SOLAR ECLIPSES

An eclipse of the Sun occurs when the Moon blocks sunlight from reaching part of the Earth. During a total eclipse, viewers within a strip of Earth's surface, called the path of totality, witness the Sun totally obscured for a few moments by the Moon. Outside this area is a larger region where viewers see the Sun only partly obscured. More common are partial eclipses, which cause no path of totality. A third type of solar eclipse is the annular eclipse, occurring when the Moon is farther from Earth than average, so that its disc is too small to cover the Sun's disc totally. At the peak of an annular eclipse, the Moon looks like a dark disc inside a narrow ring of sunlight. Solar eclipses happen two or three times a year, but total eclipses occur only about once every 18 months. During the period of totality, the Sun's corona (its hot outer atmosphere) becomes visible.

**TOTALITY PATHS**
The part of Earth's surface over which the Moon's full shadow will sweep during a total solar eclipse, called the path of totality, can be predicted precisely. Below are the paths for eclipses up to 2015.

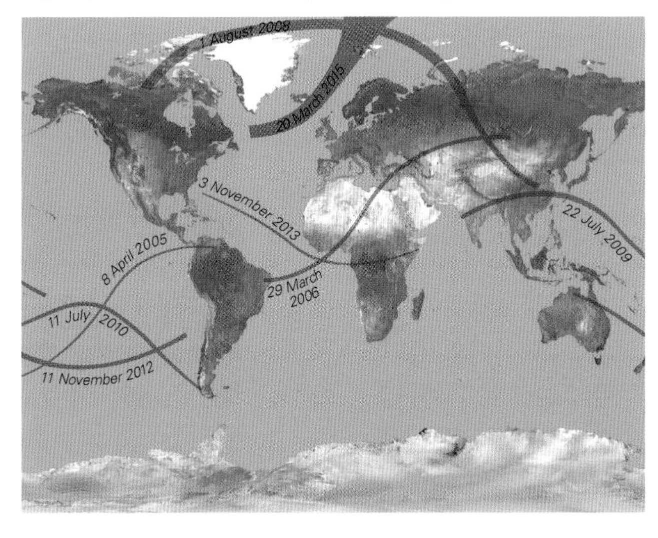


**ECLIPSE SEQUENCE**
This multiple exposure photograph depicts more than 20 stages of a total solar eclipse, seen in Mexico in 1991. At the centre can be seen the corona around the fully eclipsed Sun.

**BAILY'S BEADS**
At the beginning and end of a total solar eclipse, the Moon's rough, cratered surface breaks a thin slice of Sun into patches of light called "Baily's Beads"

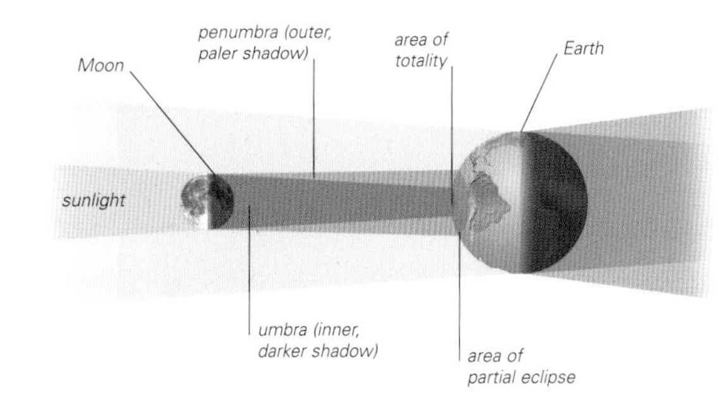


**MOON SHADOW**
The shadow cast by the Moon during a total solar eclipse consists of the central umbra (associated with the area of totality) and the penumbra (area of partial eclipse).

# PLANETARY MOTION


60–63 Celestial cycles
Naked-eye astronomy 72–73
Binocular astronomy 74–75
Laws of planetary motion 87
Using the sky guides 412–13


THE PLANETS IN THE SOLAR SYSTEM are much closer to Earth than are the stars, so as they orbit the Sun they appear to wander across the starry background. This sky motion is influenced by Earth's own solar orbit, which changes the point of view of Earth-bound observers. The planets closest to Earth move round on the celestial sphere more rapidly than the more distant planets; this is partly due to perspective and partly because the closer a planet is to the Sun, the faster is its orbital speed.

**ALWAYS NEAR THE SUN**
The Moon and Venus appear close together here in the dawn sky. Venus is only ever visible in the eastern sky for up to a few hours before dawn, or in the western sky after dusk – it is never seen in the middle of the night. This is because it orbits closer to the Sun than Earth and so never strays far from the Sun in the sky.

## INFERIOR AND SUPERIOR PLANETS

In terms of their motions in the sky as seen from Earth, the planets are divided into two groups. The inferior planets, Mercury and Venus, are those that orbit closer to the Sun than does Earth. They never move far from the Sun on the celestial sphere – the greatest angle by which the planets stray from the Sun (called their maximum elongation) is 28° for Mercury and 45° for Venus. Because they are close to Earth and orbiting quickly, both planets move rapidly against the background stars. They also display phases, like the Moon's (see p.62), because there is some variation in the angle between Earth, the planet, and the Sun. All the other planets, from Mars outwards, are called superior planets. These are not "tied" to the Sun on the celestial sphere, and so can be seen in the middle of the night. Apart from Mars, the superior planets are too far from Earth to display clear phases, and they move slowly on the celestial sphere – the farther they are from the Sun, the slower their movement.

### JOHANNES KEPLER

The German astronomer Johannes Kepler (1571–1630) discovered the laws of planetary motion (see p.87). His first law states that planets orbit the Sun in elliptical paths. The next states that the closer a planet comes to the Sun, the faster it moves, while his third law describes the link between a planet's distance from the Sun and its orbital period. Newton used Kepler's laws to formulate his theory of gravity.

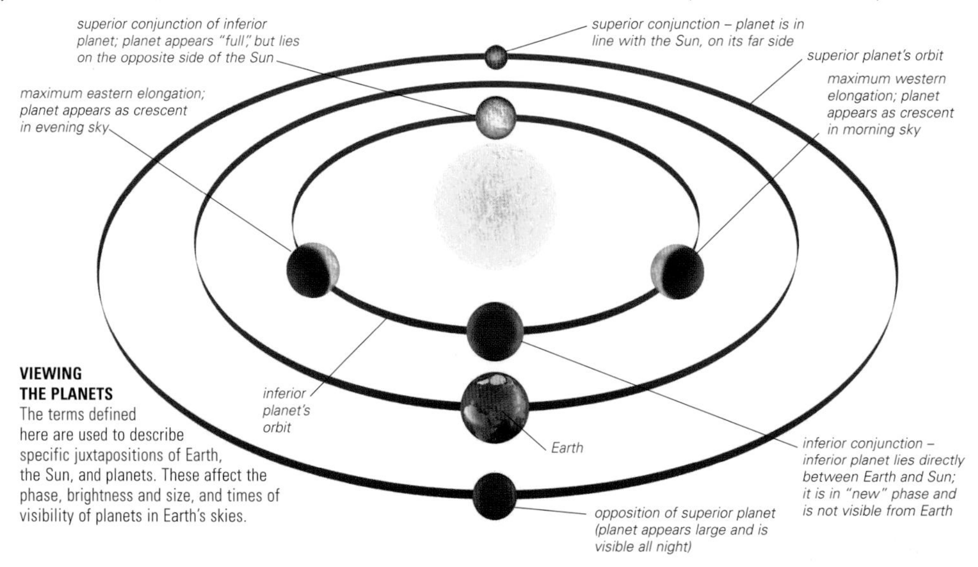


**VIEWING THE PLANETS**
The terms defined here are used to describe specific juxtapositions of Earth, the Sun, and planets. These affect the phase, brightness and size, and times of visibility of planets in Earth's skies.

path of Mars across sky
ecliptic plane
Mars's orbit inclined relative to ecliptic plane
Mars
Sun
Earth's orbit
Earth

**ZIGZAG ON THE SKY**
In retrograde motion, a planet may perform a loop or a zigzag on the sky, depending on the angle of its orbit relative to Earth's.

## RETROGRADE MOTION

The planets generally move through the sky from west to east against the background of stars, night by night. However, periodically, a planet moves from east to west for a short time – a phenomenon called retrograde motion. Retrograde motion is an effect of changing perspective. Superior planets such as Mars show retrograde motion when Earth "overtakes" the other planet at opposition (when Earth moves between the superior planet and the Sun). The inferior planets Mercury and Venus show retrograde motion either side of inferior conjunction. They overtake Earth as they pass between Earth and the Sun.

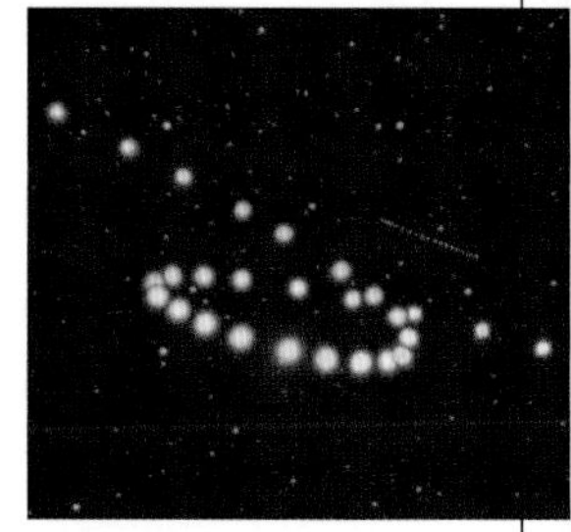

**MARS LOOPING THE LOOP**
This composite of photographs taken over several months shows a retrograde loop in Mars's motion against the background stars. The additional short dotted line is produced by Uranus.

INTRODUCTION

# ALIGNMENTS IN THE SKY

Because all the planets orbit the Sun roughly in the same plane (see pp.118–19), they never stray from the band in the sky called the zodiac (see p.61). It is not uncommon for several of the planets to be in the same part of the sky at the same time, often arranged roughly in a line. Such events, called planetary conjunctions, are of no deep significance, but can be a spectacular sight. Another type of alignment, called a transit, occurs when an inferior planet comes directly between Earth and the Sun, passing across the Sun's disc. A pair of Venus transits, eight years apart, occur about once a century or so, while Mercury transits happen about 12 times a century. In earlier times, these transits allowed astronomers to obtain more accurate data on distances in the Solar System. A final type of alignment is an occultation – one celestial body passing in front of, and hiding, another. Occultations of one planet by another, such as Venus occulting Jupiter, occur only a few times a century; in contrast, occultations of one or other of the bright planets by the Moon occur 10 or 11 times a year.

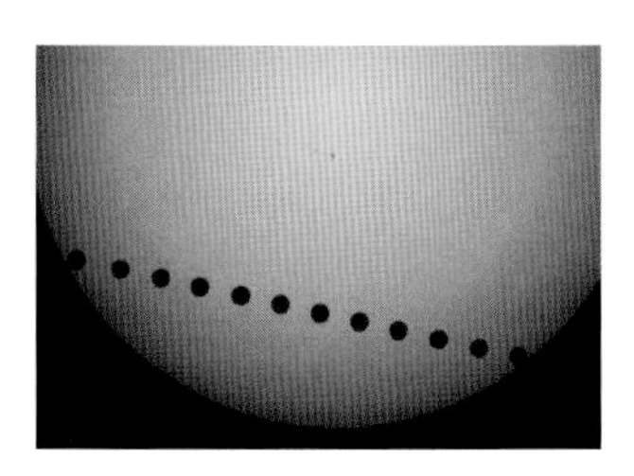

**VENUS'S PATH ACROSS THE SUN'S DISC**
This composite photograph of Venus's 2004 transit spans just over five hours. During this time, astronomers gathered data on the Sun's changing light to use as a model to look for Earth-sized planets orbiting other stars.

**TRANSIT OF VENUS**
This photograph of the 2004 Venus transit shows our nearest planetary neighbour as a dark circle close to the edge of the Sun's disc. This was the first Venus transit since 1882. Another will occur in 2012, but after that no more are expected until 2117.

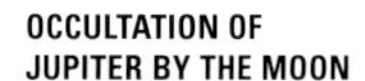

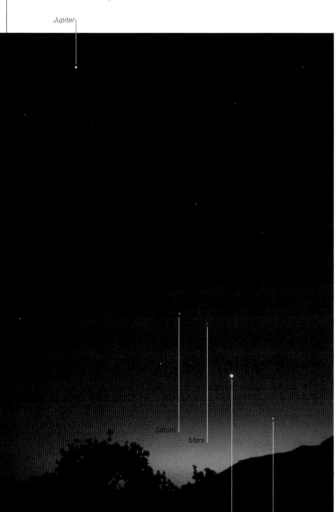


**PLANETARY CONJUNCTION, APRIL 2002**
The conjunction shown here, involving all five naked-eye planets, was visible after sunset for several evenings in April 2002. Although the planets appear close, they are separated by tens or hundreds of millions of kilometres.

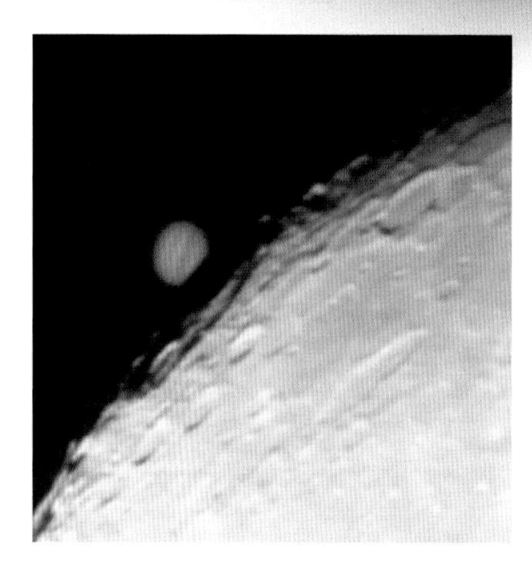

**OCCULTATION OF JUPITER BY THE MOON**
This occultation occurred on 26 January 2002 and was visible above a latitude of 55°N. Here, the planet sinks out of sight beyond the dark far wall of the lunar crater Bailly. Occultations by the Moon tend to run in series, when for a period the planet and Moon wander into alignment as seen from Earth. An occultation then occurs each sidereal month, until eventually the planet and Moon drift out of alignment again.

MYTHS AND STORIES

## CONSTANTINE'S SKY SIGN

On 27 October, AD 312, the Christian Roman Emperor Constantine fought a famous battle at the Milvian Bridge, outside Rome, emerging victorious against his rival, Maxentius. It is said that, on the eve of the battle, Constantine saw a symbol in the sky, consisting of the Greek letters chi ($\chi$) and rho ($\rho$) – the first two letters of the word "Christ". Hearing a voice saying "In this sign you shall win", he ordered his soldiers to put the symbol on their shields and attributed his victory to taking this step. Historians now know of a conjunction of bright planets on the date of the battle, and some suggest that these planets, the Moon, and some nearby stars formed the chi-rho symbol.

**CONJUNCTION OF AD 312**
The four brightest planets (yellow) and several bright stars (blue-white) were close in the southwestern sky after sunset on the eve of the battle. The Moon (far left) may have added to the impression of a chi-rho pattern on the sky.

**THE BATTLE OF MILVIAN BRIDGE**
In this French woodcut depicting the battle, Constantine and his followers look at the sign in the sky as their adversaries perish.

# STAR MOTION AND PATTERNS


58–59 The celestial sphere
Stars 230–31
The history of constellations 330–31
Mapping the sky 332–37


STARS MAY SEEM TO BE FIXED to the celestial sphere, but in fact their positions are changing, albeit very slowly. There are two parts to this motion: a tiny, yearly wobble of a star's position in the sky, called parallax shift; and a continuous directional motion, called proper motion. To record the motion of stars, and properties such as their colour and brightness, each star needs a name. Naming systems and catalogues have their roots in the constellations, which were invented to describe the patterns formed by stars in the sky.

**STAR COLOURS**
Although at first glance they all look white, stars differ in their colours, that is in the mixture of light wavelengths they emit. This is a long-exposure photograph of the bright stars of Orion, taken while changing the camera's focus. Each star looks white when sharply focused, but when its light is spread out, its true colour is revealed.

EXPLORING SPACE

### HIPPARCOS

Hipparcos is a European Space Agency satellite that between 1989 and 1993 performed surveys of the stars. Its name is short for High Precision Parallax Collecting Satellite and was chosen to honour the Greek astronomer Hipparchus. Its mission has resulted in two catalogues. The Hipparcos catalogue records the position, parallax, proper motions, brightness, and colour of over 118,000 stars, to a high level of precision. The Tycho catalogue records over 1 million stars with measurements of lower accuracy.

**HIPPARCOS SATELLITE**
The satellite spun slowly in space, scanning strips of the sky as it rotated. It measured the motion of each star about 100 to 150 times.

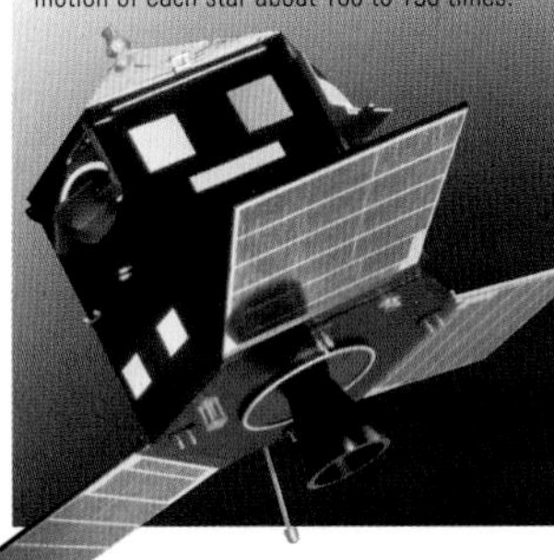

## PARALLAX SHIFT

Parallax shift is an apparent change in the position of a relatively close object against a more distant background as the observer's location changes. When an observer takes two photographs of a nearby star from opposite sides of Earth's orbit around the Sun, the star's position against the background of stars moves slightly. When the observer measures the size of this shift, knowing the diameter of Earth's orbit, she or he can calculate the star's distance using trigonometry. Until recently, this technique was limited to stars within a few hundred light-years of Earth, because the shifts of distant stars were too small to measure accurately. However, by using accurate instruments carried in satellites, much greater precision is possible: those carried in the Hipparcos satellite (see panel, left) have allowed calculation of star distances up to a few thousand light-years from Earth. For more distant stars, the shift is vanishingly small, and so other methods must be used for estimating their distances.

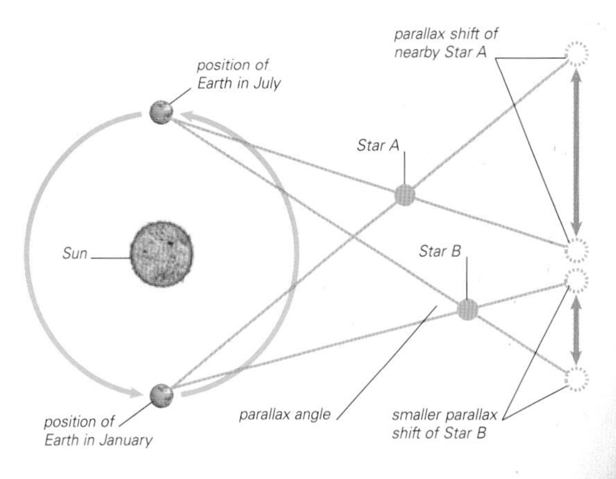


**MEASURING DISTANCE USING PARALLAX**
When Star A is observed from opposite sides of Earth's orbit, its apparent shift in position is greater than that of more distant Star B. From the shift, an observer can calculate the parallax angle between the star and the two positions of Earth. The star's distance can be determined from this angle.

## PROPER MOTION OF STARS

All stars in our galaxy are moving at different velocities relative to the Solar System, to the galactic centre, and to each other. This motion gives rise to an apparent angular movement across the celestial sphere called a star's proper motion – measured in degrees per year. Most stars are so distant that their proper motions are negligible. About 200 have proper motions of more than 1 arcsecond a year – or 1 degree of angular movement in 3,600 years. Barnard's star (see p.365) has the fastest proper motion, moving at 10.3 arcseconds per year. It takes 180 years to travel the diameter of the full Moon in the sky. If astronomers know both the proper motion of a star and its distance, they can calculate its transverse velocity relative to Earth – that is, its velocity at right angles to the line of sight from Earth. The other component of a star's velocity relative to Earth is called its radial velocity (its velocity towards or away from Earth), measured by shifts in the star's spectrum (see p.33).

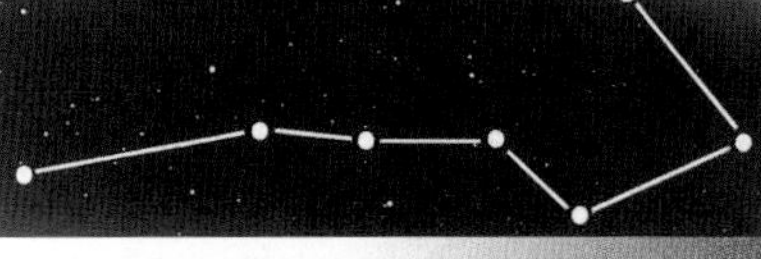

THE PLOUGH IN 100,000 BC

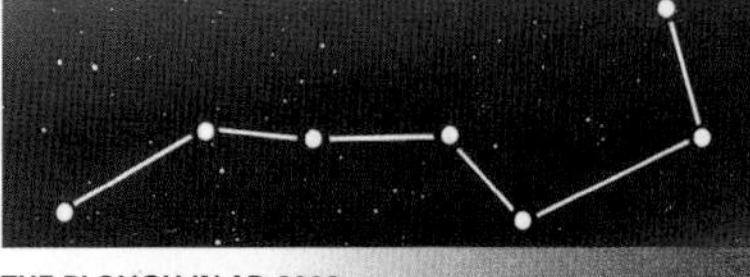

THE PLOUGH IN AD 2000

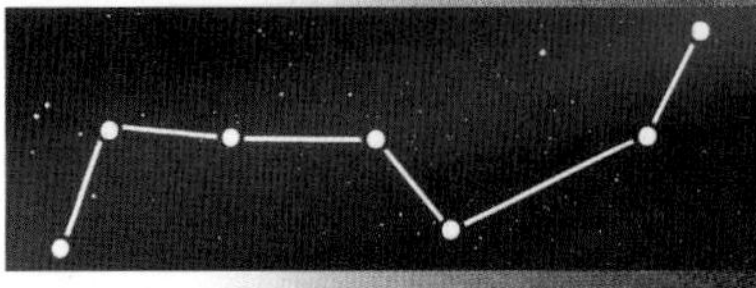

THE PLOUGH IN AD 100,000

**CHANGING SHAPE**
The shape of the star pattern known as the Plough gradually changes due to the proper motions of its stars. Five stars are moving in unison as a group, but the two stars on the ends are moving independently.

# THE BRIGHTNESS OF STARS

A star's brightness in the sky depends on its distance from Earth and on its intrinsic brightness, which is related to its luminosity (the amount of energy it radiates into space per second, see p.231). To compare how stars would look if they were all at the same distance, astronomers use a measure of intrinsic brightness called the absolute magnitude scale. This scale uses high positive numbers to denote dim stars and negative numbers for the brightest ones. A star's brightness as seen from Earth, on the other hand, is described by its apparent magnitude. Again, the smaller the number of a star's apparent magnitude, the brighter the star. Stars with an apparent magnitude of +6 are only just detectable to the naked eye, whereas the apparent magnitude of most of the 50 brightest stars is between +2 and 0. The four brightest (including the brightest star of all, Sirius) have negative apparent magnitudes.

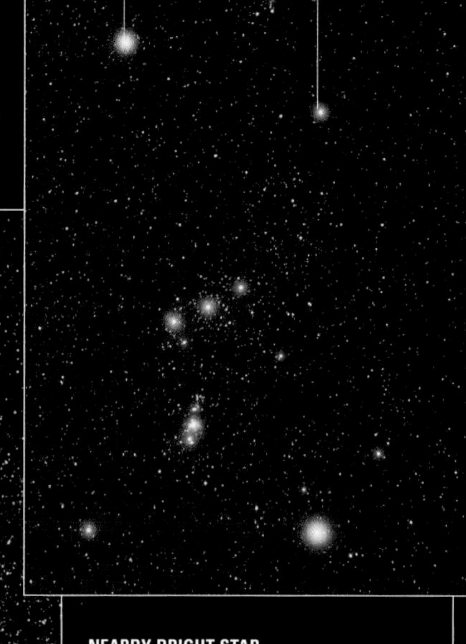
Betelgeuse
Bellatrix

**INTRINSICALLY BRIGHT STAR**
The stars Betelgeuse and Bellatrix mark the shoulders of Orion. Betelgeuse is noticeably brighter (apparent magnitude 0.45) than Bellatrix (1.64), despite being twice as distant. It is a red, high-luminosity supergiant, whereas Bellatrix is a much less luminous giant.

*Alpha Centauri*

*Hadar (Beta Centauri)*

**NEARBY BRIGHT STAR**
In the constellation Centaurus, the triple star system Alpha ($\alpha$) Centauri is a little brighter (apparent magnitude -0.01) than the binary star Hadar, or Beta ($\beta$) Centauri (0.61). The reason for Alpha Centauri's brightness is its proximity – it is our closest stellar neighbour. The blue giant stars that make up Hadar are much more luminous than the stars in Alpha Centauri, but they are about 120 times farther away.

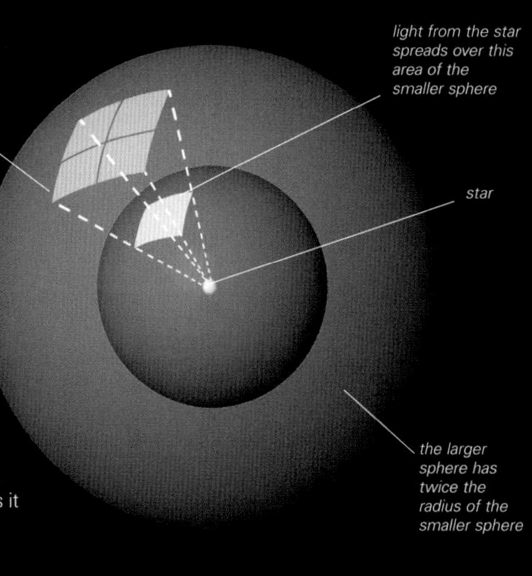
when the light reaches the larger sphere, it is spread over four times the area (the square of the distance, or 2x2)
light from the star spreads over this area of the smaller sphere
star
the larger sphere has twice the radius of the smaller sphere

**THE INVERSE SQUARE RULE**
The apparent brightness of a star drops in proportion to the square of its distance from the observer – a rule called the inverse square law. This happens because light energy is spread out over a progressively larger area as it travels away from the star.

# CONSTELLATIONS

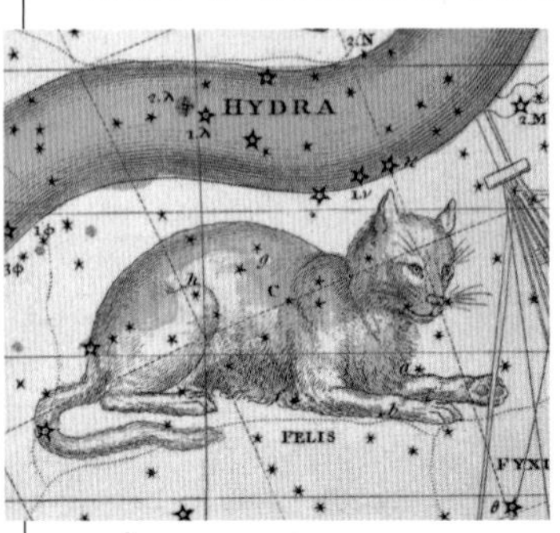


**LOST CONSTELLATIONS**
Some constellations have proved short-lived. In the 19th century, Felis, the cat, was incorporated into what is now part of the constellation of Hydra. It appeared on several star charts but was not officially adopted.

Since ancient times, people have seen imaginary shapes among groups of stars in the night sky. Using lines, they have joined the stars in these groups to form figures called constellations and named these constellations after the shapes they represent. Each constellation has a Latin name, which in most cases is either that of an animal, for example, Leo (the lion), an object, such as Crater (the cup), or a mythological character, such as Hercules. Some constellations, such as Orion (the Hunter), are easy to recognize; others, such as Pisces (the Fishes) are less distinct. Since 1930, an internationally agreed system has divided the celestial sphere into 88 irregular areas, each containing one of these figures. In fact, from an astronomical point of view, the word "constellation" is now applied to the area of the sky containing the figure rather than to the figure itself. All stars inside the boundaries of a constellation area belong to that constellation, even if they are not connected to the stars that produce the constellation figure. Within some constellations are some smaller, distinctive groups of stars known as asterisms; these include Orion's belt (a line of three bright stars in Orion) and the Plough or Big Dipper (a group of seven stars in the constellation Ursa Major).

A few asterisms cut across constellation boundaries. For example, most of the "Square of Pegasus" asterism is in Pegasus, but one of its corners is in Andromeda.

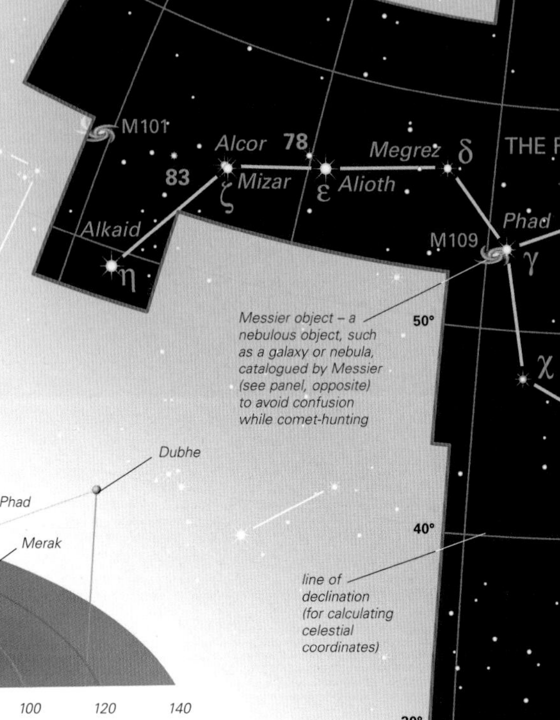


**STAR CHART**
This star chart of Ursa Major (the Great Bear) shows the constellation figure (the pattern of lines joining bright stars) and labels many of the stars, as well as objects such as galaxies, lying within the constellation's boundaries.

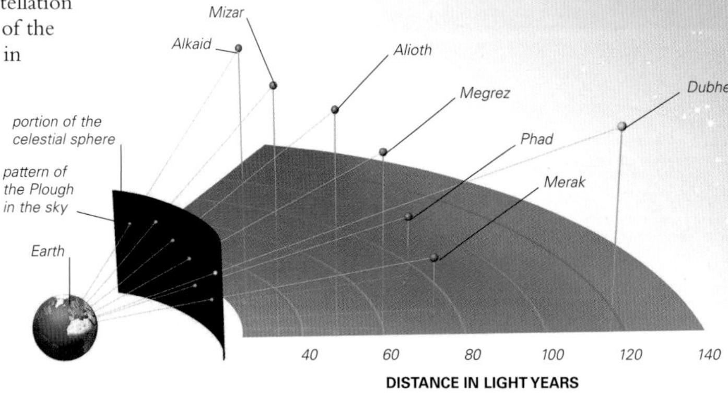


**LINE-OF-SIGHT EFFECT**
A star pattern such as the Plough in Ursa Major is a two-dimensional view of what may be a widely-scattered sample of stars. The stars might seem to lie in the same plane, but they are at different distances from Earth. If we could view the stars from elsewhere in space, they would form a totally different pattern.

EXPLORING SPACE

## BAYER'S SYSTEM

Johann Bayer ascribed Greek letters to the stars in a constellation, roughly in order of decreasing brightness. Regulus, the brightest star in the constellation of Leo, was given the name Alpha (α) Leonis, the second brightest (Denebola) was called Beta (β) Leonis, and so on. In some cases, Bayer used other ordering systems. The Plough in Ursa Major is lettered by following the stars from west to east.

**BAYER'S MAP OF URSA MAJOR**
The seven stars of the Plough can be seen in the upper left area of this chart from Bayer's *Uranometria*.

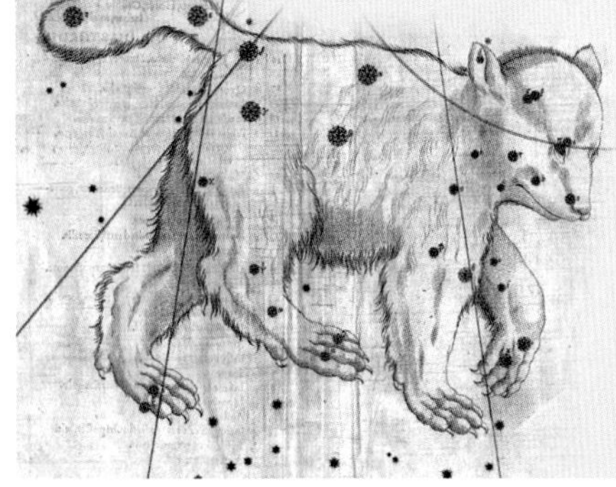

# NAMING THE STARS

Most of the brighter stars in the sky have ancient names of Babylonian, Greek, or Arabic origin. The name Sirius, for example, comes from a Greek word meaning "scorching". The first systematic naming of stars was introduced by Johann Bayer in 1603 (see panel, left, and p.331). Bayer distinguished up to 24 stars in each constellation by labelling them with Greek letters, after which he resorted to using Roman lower case letters, a to z. In 1712, English astronomer John Flamsteed (1646–1719) introduced another system, in which stars are numbered in order of their right ascension (see p.59) from west to east across their constellation. Stars are usually named by linking their Bayer letter or Flamsteed number with the genitive form (possessive case) of the constellation name – so 56 Cygni denotes the star that is 56th closest to the western edge of the constellation Cygnus. Since the 18th century, numerous further catalogues have identified and numbered many more faint stars, and specialized systems have been devised for cataloguing variable, binary, and multiple stars.

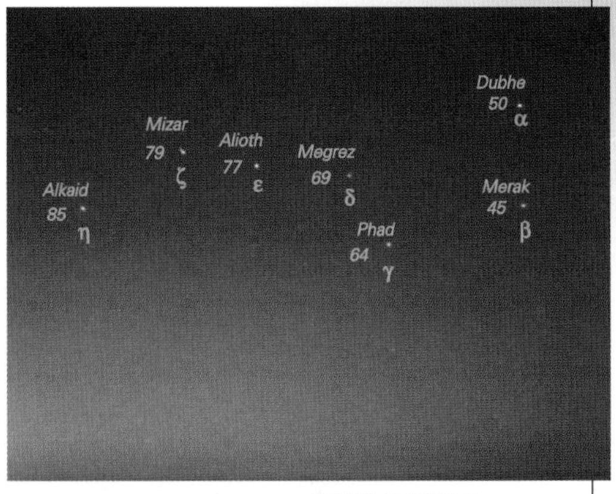


**SYSTEMS OF BAYER AND FLAMSTEED**
This photo of the Plough in Ursa Major shows the ancient name of each star, its Bayer designation, and its Flamsteed number. For example the star Alkaid can also be called Eta (η) Ursae Majoris (Bayer) or 85 Ursae Majoris (Flamsteed).

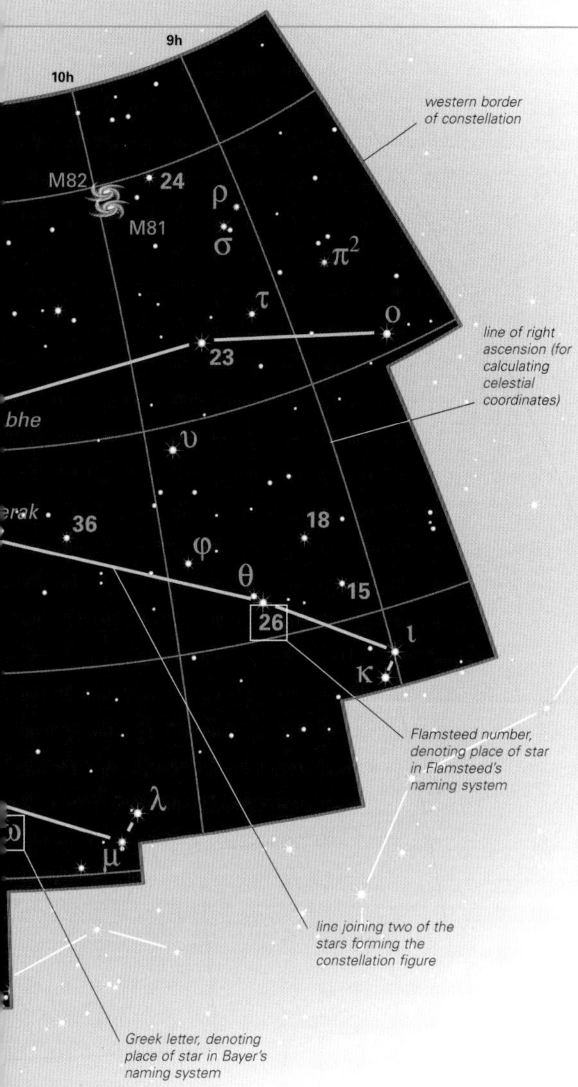


# CATALOGUES OF NEBULOUS OBJECTS

Besides individual stars, various other types of object, such as star clusters, nebulae, and galaxies, have practically fixed positions on the celestial sphere. Most of these objects appear as no more than hazy blurs in the sky, even through a telescope. The first person to catalogue such objects was a French astronomer, Charles Messier (see panel, below), in the 18th century. He compiled a list of 110 hazy objects, though none of these are from the southern polar skies – that is because Messier carried out his observations from Paris, and anything in declination below 40°S was below his horizon. In 1888, a much larger catalogue called the *New General Catalogue of Nebulae and Star Clusters* (NGC) was published, and this was later expanded by what is called the *Index Catalogue* (IC). To this day, the NGC and IC are important catalogues of nebulae, star clusters, and galaxies. Their current versions cover the entire sky and provide data on more than 13,000 objects, all identified by NGC or IC numbers. In addition, several hundred specialist astronomical catalogues are in use, covering different types of objects, parts of the sky, and regions of the electromagnetic spectrum. Many catalogues are now maintained as computer databases accessible over the Internet.

NGC 2841, A SPIRAL GALAXY

**NEW GENERAL CATALOGUE**

More than 150 New General Catalogue (NGC) objects lie within the constellation Ursa Major. Two are shown here, both spiral galaxies in a region around the Great Bear's forelegs, not far from Theta (θ) Ursae Majoris. NGC 2841 has delicate, tightly wound arms, within which astronomers have recorded many supernovae explosions. NGC 3079 has an active central region, from which rises a lumpy bubble of hot gas, 3,500 light-years wide, driven by star formation.

NGC 3079, A SPIRAL GALAXY VIEWED EDGE-ON

## CHARLES MESSIER

The French comet-hunter Charles Messier (1730–1817) compiled a catalogue of 110 nebulous-looking objects in the sky that could be mistaken for comets. Not all of them were discovered by himself – many were spotted by another Frenchman, Pierre Méchain, and yet others had been found years earlier by astronomers such as Edmond Halley. Messier's first true discovery was M3, a globular star cluster in Canes Venatici. Ironically, Messier is more famous for his catalogue of non-comets than he is for the real comets he discovered.

M81, A SPIRAL GALAXY (SEE P.304)

M82, AN IRREGULAR GALAXY (SEE P.304)

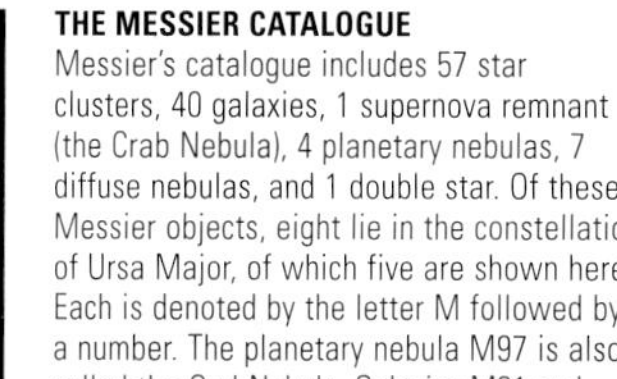

**THE MESSIER CATALOGUE**

Messier's catalogue includes 57 star clusters, 40 galaxies, 1 supernova remnant (the Crab Nebula), 4 planetary nebulas, 7 diffuse nebulas, and 1 double star. Of these Messier objects, eight lie in the constellation of Ursa Major, of which five are shown here. Each is denoted by the letter M followed by a number. The planetary nebula M97 is also called the Owl Nebula. Galaxies M81 and M82 are neighbours in the sky and can be viewed simultaneously with a good pair of binoculars. M109 lies close to the star Phad – Gamma (γ) Ursae Majoris – in the Plough.

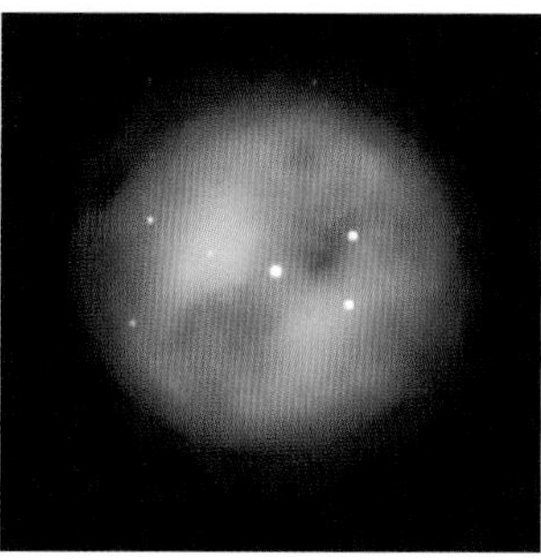

M97, A PLANETARY NEBULA

M108, A SPIRAL GALAXY

M109, A BARRED SPIRAL GALAXY

# LIGHTS IN THE SKY


32–35 Radiation
60–63 Celestial cycles
Naked-eye astronomy 72–73
Earth's atmosphere and weather 139
Comets 214–15
Meteors and meteorites 220–21


**AS WELL AS STARS, GALAXIES, NEBULAE,** and Solar System objects, other phenomena can cause lights to appear in the night sky. In the main, these originate in light or particles of matter reaching Earth in various indirect ways from the Sun, but some are generated by Earth-bound processes. Amateur stargazers need to be aware of these sources of nocturnal light to avoid confusion with astronomical phenomena.

## AURORAE

The aurora borealis (northern lights) and aurora australis (southern lights) appear when charged particles from the Sun, carried to Earth in the solar wind (see pp.122–23), become trapped by Earth's magnetic field. They are then accelerated into regions above the north and south magnetic poles, where they excite particles of gas in the upper atmosphere, 100–400km (60–250 miles) above Earth's surface. The appearance and location of aurorae change in response to the solar wind. They are most often visible at high latitudes, towards Earth's magnetic poles, but may be seen at lower latitudes during disturbances in the solar wind, such as after mass ejections from the Sun (see pp.122–23).

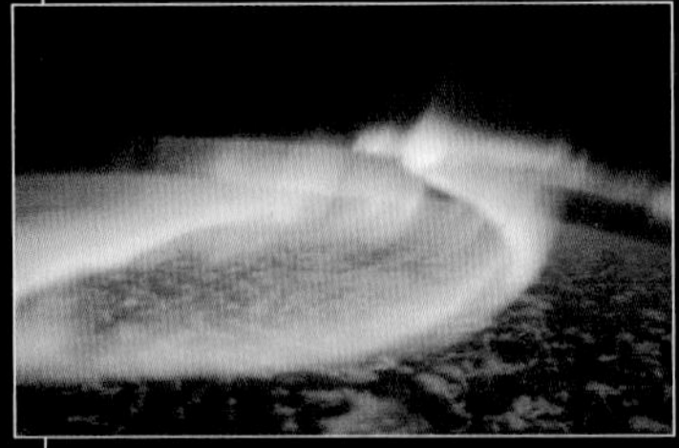

**AURORA FROM THE SPACE SHUTTLE**
This photograph of the aurora australis was taken from the Space Shuttle Discovery during a 1991 mission. A study of the aurora's features was one of the mission tasks.

**AURORA BOREALIS**
A colourful display of the northern lights is visible here over silhouetted trees near Fairbanks, Alaska, USA. The colours stem from light emission by different atmospheric gases.

## ICE HALOES

Atmospheric haloes are caused by ice crystals high in Earth's atmosphere refracting light. Light either from the Sun or the Moon (that is, reflected sunlight) can cause haloes. The most common halo is a circle of light with a radius of 22° around the Moon or Sun. Also present may be splashes of light, called moon dogs or sun dogs (parhelia), arcs, and circles of light that seem to pass through the Sun or Moon. All these phenomena result from the identical angles between the faces of atmospheric ice crystals. Even if the crystals are not all aligned, they tend to deflect light in some directions more strongly than in others.

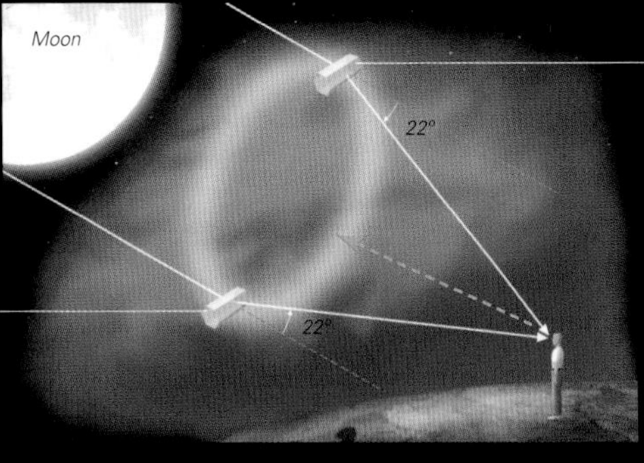


**OBSERVING A 22° HALO**
This halo is formed when ice crystals in the atmosphere refract light from the Moon to the observer on Earth by an angle of 22°. A light ray is refracted through this angle as it passes through two faces of an ice crystal.

**HALO AND MOON DOGS**
This photograph taken in Arctic Canada shows several refraction phenomena. The patches of light on either side of the Moon, called moon dogs, are caused by horizontal ice crystals in the atmosphere refracting light. The band of light running through the moon dogs is called a parhelic circle. Also visible is a circular 22° halo.

SEEING THE ZODIACAL LIGHT
The zodiacal light is most distinct just before dawn in autumn, far from any light pollution. It is near the horizon and forms a rough triangle.

## ZODIACAL LIGHT

A faint glow is sometimes visible in the eastern sky before dawn or occasionally in the west after sunset. Called zodiacal light, it is caused by sunlight scattered off interplanetary dust particles in the plane of the Solar System – the ecliptic plane (see p. 64). The mixture of wavelengths in the light is the same as that in the Sun's spectrum. A related phenomenon is called the gegenschein (German for "counterglow"). It is sometimes perceivable on a dark night, far from any light pollution, as a spot on the celestial sphere directly opposite the Sun's position in the sky. The dust particles in space responsible for both zodiacal light and gegenschein are thought to be from asteroid collisions and comets and have diameters of about 1mm (0.04in).

THE GEGENSCHEIN
This faint, circular glow, 10° across, is most often spotted at midnight, in an area above the southern horizon (for northern-hemisphere viewers).

## NOCTILUCENT CLOUDS

Clouds at extremely high altitude (around 80km/50 miles, high) in Earth's atmosphere can shine at night by reflecting sunlight long after the Sun has set. These "noctilucent" (night-shining) clouds are seen after sunset or before dawn. It is thought that they consist of small, ice-coated particles that reflect sunlight. Noctilucent clouds are most often seen between latitudes between 50° and 65° north and south, from May to August in northern latitudes and November to February in southern latitudes. They may also form at other latitudes and times of year.

SHINING CLOUDS
Noctilucent clouds are silvery-blue and usually appear as interwoven streaks. They are only ever seen against a partly lit sky background, the clouds occupying a sunlit portion of Earth's atmosphere.

## MOVING LIGHTS AND FLASHES

Many phenomena can cause moving lights and flashes across the sky. Rapid streaks of light are likely to be meteors or shooting stars – that is, dust particles entering and burning up in the atmosphere. A bigger, but very rare variant is a fireball – simply a larger meteor burning up. Slower-moving, steady, or flashing lights are more likely to be aircraft, satellites, or orbiting spacecraft. Large light flashes are usually electrical discharges connected with lightning storms. In recent years, meteorologists have named two new types of lightning – "red sprites" and "blue jets". Both are electrical discharges between the tops of thunderclouds and the ionosphere above.

BLUE JETS
These cone-shaped discharges are 50–60km (31–37 miles) high, 10km (6 miles) wide at the top, and result from lightning in the atmosphere ionizing nitrogen atoms, causing them to glow blue as they re-emit light. In the past, blue jets may have been reported as UFOs.

PATH OF THE ISS
As the International Space Station (ISS) orbits Earth, it is visible from the ground because it reflects sunlight. This photograph of the Space Station was taken using a 60-second camera exposure, which indicates how quickly the spacecraft moves across the night sky.

MYTHS AND STORIES

### UFO SIGHTINGS

Every year there are reports of unidentified flying objects (UFOs). Most of these can be accounted for by natural phenomena such as meteors, aurorae, ball lightning, by unusual clouds, or by man-made objects such as satellites and aircraft. After excluding such causes, there are still unexplained cases. It would be unscientific to dismiss the possibility that these UFOs are signs of extraterrestrial visitors without further investigation – just as it would be to accept it before excluding less exotic explanations.

FLYING SAUCER?
This object, suggestive of a flying saucer, is actually a lenticular cloud. Clouds like this are usually formed by vertical air movements around the sides or summits of mountains.

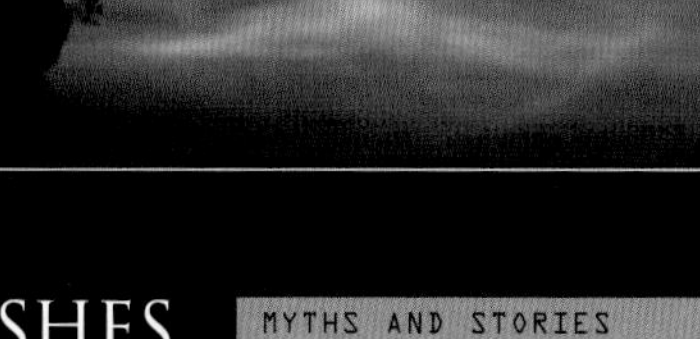

# NAKED-EYE ASTRONOMY

58–59 The celestial sphere
60–63 Celestial cycles
64–65 Planetary motion
70–71 Lights in the sky
Mapping the sky 332–37
Monthly sky guide 410–85

OPTICAL INSTRUMENTS ARE NOT NECESSARY to gain a foothold in astronomy – our ancestors did without them for thousands of years. Today's naked-eye observer, equipped with a little foreknowledge and some basic equipment, can still appreciate the constellations, observe the brightest deep-sky objects, and trace the paths of the Moon and planets in the night sky.

## PREPARING TO STARGAZE

To get the most from stargazing, some preparation is needed. The human eye takes some 20 minutes to adjust to darkness and, as the pupil opens, more detail and fainter objects become visible. Look at a planisphere or monthly sky chart (see pp.410–85) to see what is currently in the sky. A good location is one shielded from street lights, and ideally away from their indirect glow. Try to avoid all artificial light – if necessary, use a torch with a red filter. Keep a notebook or a prepared report form to record observations, especially if looking for particular phenomena, such as meteors. To see faint stars and deep-sky objects, avoid nights when a bright Moon washes out the sky. Even on a dark, cloudless night, air turbulence can affect the observing quality or "seeing" – the best nights are often those which do not suddenly get colder at sunset.

**SEEING AND TWINKLE**
Variable "seeing" is caused by warm air currents rising from the ground at nightfall. These telescope images of Jupiter show the range of seeing from poor to fine, but seeing also limits the visibility of stars with the naked eye and determines the amount of "twinkle".

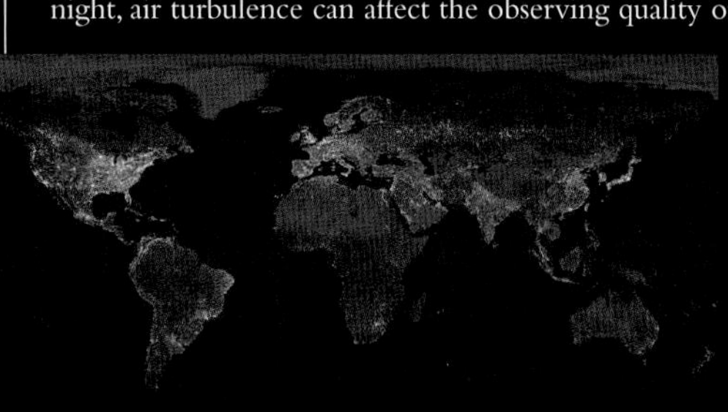

**LIGHT POLLUTION**
This composite satellite image shows the extent of artificial lighting on Earth. In industrialized regions, it is almost impossible to find truly dark skies.

**GOOD STREET LIGHTING**
In some countries, non-essential street lights are switched off late at night. Elsewhere, shades are installed, which project all the light downwards, preventing it from leaking into the sky. Such measures can increase the light on the street, save energy, and preserve the night sky for stargazers.

**PLANISPHERE**
A planisphere is a useful tool for any amateur astronomer. The user rotates the discs so that the time and date markers on the edge match up correctly, and the window reveals a map of the sky at that moment. A single planisphere is useful only for a limited range of latitudes, so be sure to get one with the correct settings.

**THE MOON AND VENUS**
Solar System objects such as the Moon and Venus can be spectacular sights even with the unaided eye. This beautiful twilight pairing was photographed in January 2004.

# MEASUREMENTS ON THE SKY

Distances between objects in the sky are often expressed as degrees of angle. All the way around the horizon measures 360°, while the angle from horizon to zenith (the point directly overhead) is 90°. The Sun and Moon both have an angular diameter of 0.5°, while an outstretched hand can be used to estimate other distances. When studying star charts, bear in mind that one hour of right ascension (RA) along the celestial equator is equivalent to 15° of declination (see p.59), but right ascension circles get tighter towards the celestial poles, so at 60°N an hour's difference in RA is equivalent to only 7.5° of declination.

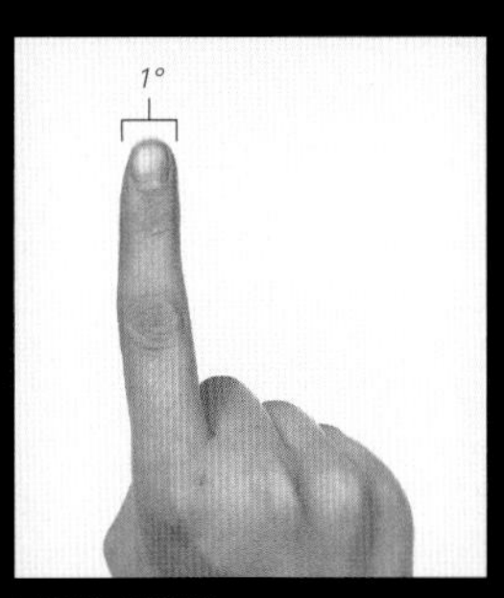


**FINGER WIDTH**
Held out at arm's length, a typical adult index finger blocks out roughly one degree of the sky – enough to cover the Moon twice over.

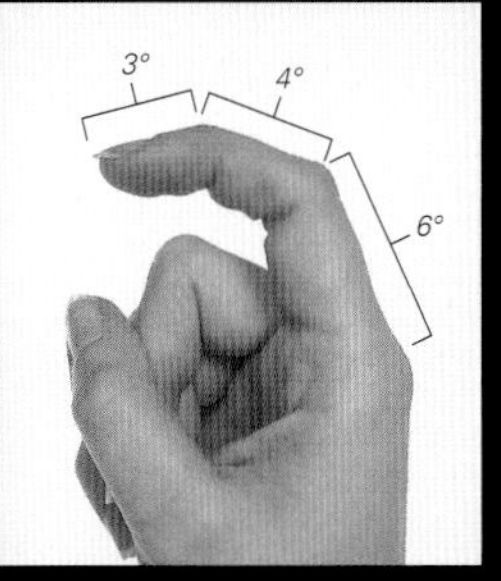


**FINGER JOINTS**
The finger joints provide measures for distances of a few degrees. A side-on fingertip is about 3° wide, the second joint 4°, and the third 6°.

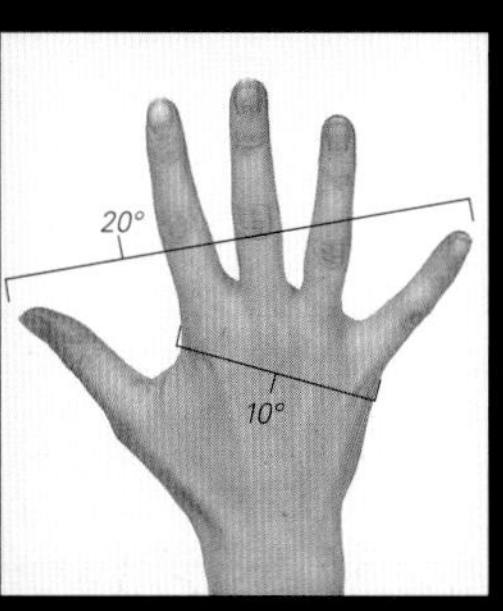


**HAND SPANS**
The hand (not including the thumb), is about 10° across at arm's length, while a stretched hand-span covers 20° of sky.

# STAR-HOPPING

The best way to learn the layout of the night sky is to first find a few bright stars and constellations, then work outwards into more obscure areas. Two key regions are the Plough (the brightest seven stars in the constellation Ursa Major, close to the north celestial pole) and the area around the brilliant constellation Orion, including the Winter Triangle (see p.420) on the celestial equator. By following lines between certain stars in these constellations, one can find other stars and begin to learn the sky's overall layout. The Plough is a useful pointer, since two of its stars align with Polaris, the star that marks the north celestial pole. Because the sky seems to revolve around the celestial poles, Polaris is the one fixed point in the northern sky (there is no bright south Pole Star). Other useful keystones are the Summer Triangle (see p.450), comprising the northern stars Vega, Deneb, and Altair, and the Southern Cross (see p.421) and False Cross (see p.427) in the far south.

**STAR HOPS FROM THE PLOUGH**
A line through Dubhe and Merak along one side of the Plough points straight to Polaris in one direction, and (allowing for the curvature of the sky), towards the bright star Regulus in Leo in the other direction. Following the curve of the Plough's handle, meanwhile, leads to the bright red star Arcturus in Boötes and eventually to Spica in Virgo.

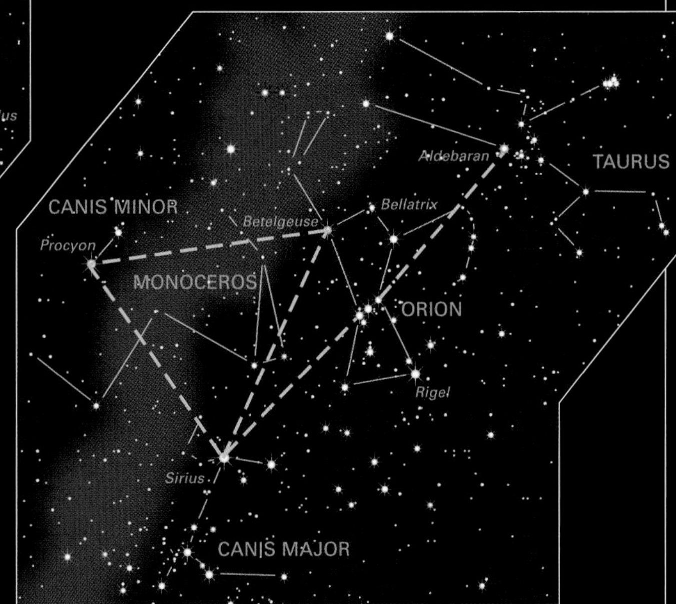


**ORION'S BELT AND THE WINTER TRIANGLE**
The distinctive line of three bright stars forming Orion's belt points in one direction towards the red giant Aldebaran in Taurus, and in the other towards Sirius, the brightest star in the sky, in Canis Major. Sirius, Betelgeuse (on Orion's shoulder), and Procyon (in Canis Minor) make up the equilateral Winter Triangle.

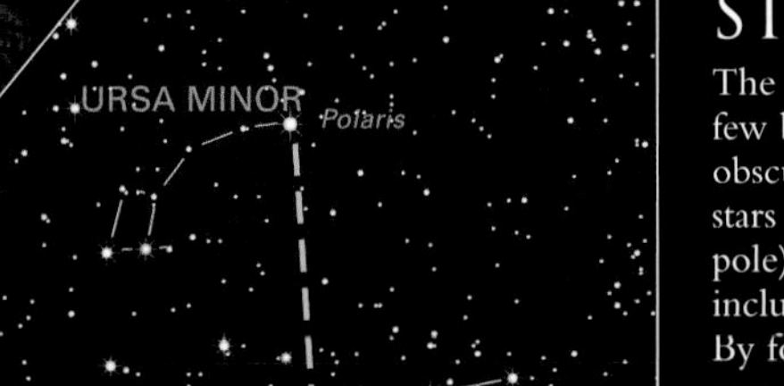

# BINOCULAR ASTRONOMY

60–63 Celestial cycles
72–73 Naked-eye astronomy
Telescope astronomy 76–79
Mapping the sky 332–37
Monthly sky guide 410–85

FOR MOST NEWCOMERS to astronomy, the most useful piece of equipment is a pair of binoculars. As well as being easy and comfortable to use, binoculars (unlike telescopes) allow stargazers to see images the right way up. A range of fascinating astronomical objects can be observed through them.

## BINOCULAR CHARACTERISTICS

Binoculars are like a combination of two low-powered telescopes. The two main designs, called porro-prism and roof-prism, differ in their optics, but either can be useful for astronomy. More important when choosing binoculars are the two main numbers describing their optical qualities; for example, 7x50 or 12x70. The first figure is the magnification. For a newcomer, a magnification of 7x or 10x is usually adequate – with a higher magnification, it can be difficult to locate objects in the sky. The second figure is the aperture, or diameter of the objective lenses, measured in millimetres. This number expresses the binoculars' light-gathering power, which is important in seeing faint objects. For night-sky viewing, an aperture of at least 50mm (2in) is preferable.

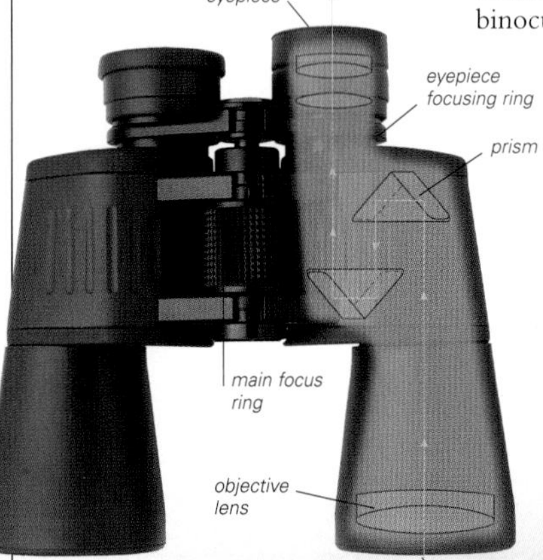


**STANDARD BINOCULARS**
These typically have 50mm (2in) objective lenses and a magnification of 7x or 10x. This pair has a porro-prism design.

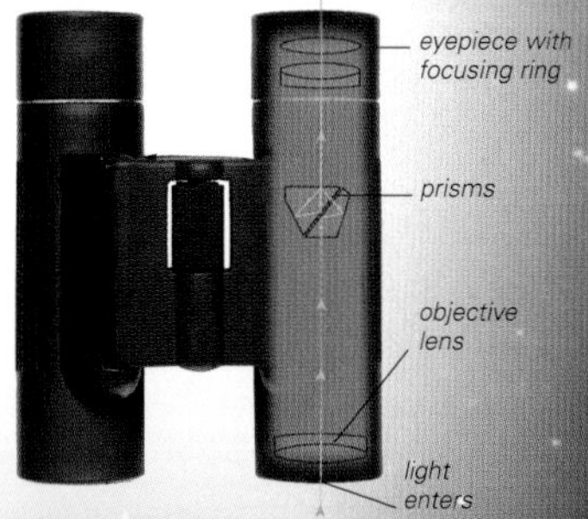


**COMPACT BINOCULARS**
These are lightweight but their objective lenses are rather small for astronomy. This pair has a roof-prism design.

EXPLORING SPACE

### BINOCULAR FINDS

Astronomers make some important discoveries using binoculars. The Arizonan astronomer Peter Collins uses binoculars to search for the stellar outbursts known as novae (see p.278). To make the method effective, he memorizes thousands of star positions. Comets are also frequently first seen by binocular enthusiasts. Japanese astronomer Hyakutake Yuji spotted Comet Hyakutake (see p.217) in 1996 using a pair of giant (25x100mm) binoculars.

**PETER COLLINS**

**IDYLLIC SKYGAZING**
The modest magnifying power of binoculars is more than enough to reveal many of the sky's most interesting objects. Wilderness camping is a good way to get away from light pollution.

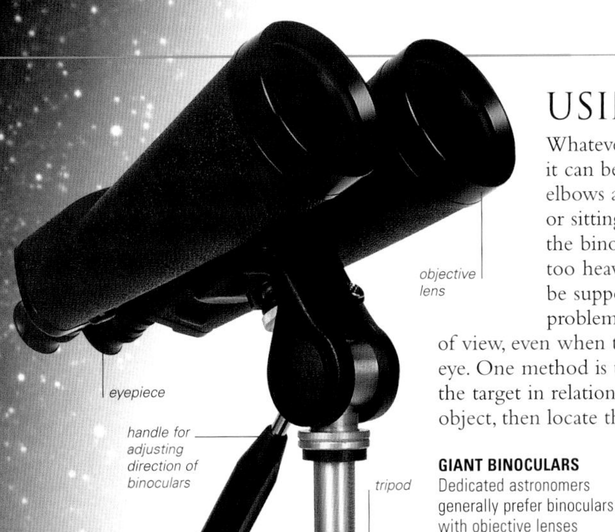


**GIANT BINOCULARS**
Dedicated astronomers generally prefer binoculars with objective lenses of 70mm (2.8in) and magnifications of 15–20x.

# USING BINOCULARS

Whatever size of binoculars astronomers choose, it can be difficult to keep them steady. Placing elbows against something solid, such as a wall, or sitting down in a deckchair, can help to stop the binoculars wobbling. Giant binoculars are too heavy to hold steady in the hands, so should be supported on a tripod. Another common problem is finding the target object in the field of view, even when the object is visible to the naked eye. One method is to establish the position of the target in relation to an easier-to-locate object, then locate the easier object and finally navigate to the target object. Alternatively, work upwards from a recognizable feature on the horizon.

**KEEPING YOUR BINOCULARS STEADY**
Sitting and placing the elbows on the knees can support the weight of binoculars and keep them steady.

**HOW TO FOCUS A PAIR OF BINOCULARS**
A pair of binoculars is not immediately in perfect focus for every user, as users' eyesight differs. To fix this, follow the instructions below.

**1 IDENTIFY FOCUSING RING**
Find which eyepiece can be rotated to focus independently (usually the right). Look through with your eye closed on that side.

**2 FOCUS LEFT EYEPIECE**
Rotate the binoculars' main, central focusing ring, which moves both eyepieces, until the left-eyepiece image comes into sharp focus.

**3 CLOSE LEFT EYE, OPEN RIGHT EYE**
Now open only the other eye (in this example, the right), and use the eyepiece focusing ring to bring the image into focus.

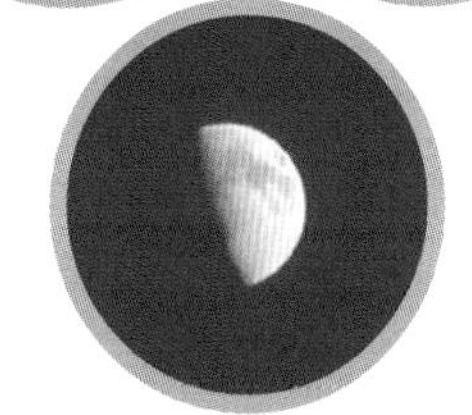

**4 FOCUS AND THEN USE BOTH EYES**
Both eyepieces should now be in focus, so now you can open both eyes and start observing.

# BINOCULAR FIELD OF VIEW

The size of the circular area of sky seen through binoculars is called the field of view and is usually expressed as an angle. The field of view is closely related to magnification – the higher the magnification, the lower the field of view. A typical field of view of a pair of medium-power binoculars (10x) is 6–8°. This offers a good compromise between adequate magnification and a field of view wide enough to see most of a large object such as the Andromeda Galaxy (see pp.302–303). For viewing larger areas yet, lower-power binoculars (5–7x), with a field of view of at least 9°, are more suitable. Conversely, for looking at more compact objects, such as Jupiter and its moons, binoculars with higher magnification, and a field of view of 3° or even less, are better to use.

**M31 VIEWED THROUGH BINOCULARS**
This is how the Andromeda Galaxy (M31, above) appears through medium- to low-magnification binoculars, with a field of view of about 8°.

**M31 VIEWED THROUGH A TELESCOPE**
Here the central part of the Andromeda Galaxy is shown as you might see it through very-high-magnification binoculars, or a small telescope, with a field of view of about 1.5°.

# BINOCULAR OBJECTS

A striking first object for a novice binocular user is the Orion Nebula (see p.239). Other choices might be the Andromeda Galaxy (above), and the fabulous star clouds and nebulae in the Sagittarius and Scorpius regions of the Milky Way, including the Lagoon nebula (see p.241). For viewers south of 50°N, an excellent binocular object is the Omega Centauri star cluster (see p.288). To find these, all that is needed is some star charts (see pp.410–85) or astronomy software (see p.78). Also try observing the Moon, Jupiter and its moons, and the phases of Venus.

**THE MILKY WAY**
Shown here is a dense region of the Milky Way in Sagittarius, as seen through low-power binoculars with a field of view of 12°.

**ORION NEBULA**
This appears as a blue-green smudge in Orion, shown here as it appears in medium-power binoculars with a field of view of 8°.

**THE PLEIADES**
This spectacular star cluster in Taurus is seen here as it appears through high-power binoculars with a field of view of about 3°.

# TELESCOPE ASTRONOMY


34–35 Across the spectrum
59 Celestial coordinates
Galileo's discoveries 88
Newton's reflecting telescope 89
Instrumental progress 91
Mapping the sky 332–33
Monthly sky guide 410–85


TELESCOPES ARE THE ULTIMATE astronomical instruments – artificial eyes that capture and focus much larger amounts of light than the human eye can take in, bringing faint objects within the limits of human vision. They come in a variety of forms – the simplest have evolved little from those in use four centuries ago, while the most advanced amateur instruments offer high-performance optics and computerized controls that were the preserve of professionals just a few years ago.

## TELESCOPE DESIGNS

A telescope's function is to collect light from distant objects, bring it to a focus, and then magnify it. There are two basic ways of doing this, using either a lens or a concave mirror. A lens refracts the light passing through it, bending it inwards to a focal point somewhere behind it. A curved mirror reflects light rays back onto converging paths that come to a focus somewhere in front of it. Both mirrors and lenses can focus light rays that enter the telescope only on a near-parallel path, but fortunately all astronomical objects are distant enough for this to be the case. Once the captured rays have passed the focus, they begin to diverge again, at which point they are captured by an eyepiece, which returns the rays to parallel directions, magnifying them in the process. Because light rays entering the eyepiece have crossed over as they pass through the focus, the image in the eyepiece is usually inverted. The different operating principles of lens-based telescopes (refractors) and mirror-based ones (reflectors) dictate contrasting designs.

refracted light
piggyback finder scope
objective lens
altazimuth mount
focused light
90° eyepiece – a sliding tube allows it to move in and out to focus.

**REFRACTING TELESCOPE**
Refractors are typically long tubes with one or more objective lenses at one end (see panel, below) and an eyepiece at the other.

eyepiece
secondary mirror
lightweight tube
primary (objective) mirror
Dobsonian altazimuth mount

**NEWTONIAN REFLECTOR TELESCOPE**
In this popular design, converging light rays from the primary mirror are reflected out of the side of the telescope by a flat secondary mirror. Here, they come to a focus and are magnified by the eyepiece.

**OBSERVING ON THE MOVE**
An equatorial-mounted refractor with a sturdy tripod is an ideal telescope for transporting in the back of a car and setting up at a dark site in the countryside.

light enters
correcting lens
light enters
convex secondary mirror
finder
concave primary mirror
eyepiece
equatorial "wedge" mount
light enters

**SCHMIDT–CASSEGRAIN TELESCOPE**
In this compact reflector design, a convex secondary mirror directs light to the eyepiece through a hole in the primary mirror. By bouncing the light back on itself, the length of the telescope tube is reduced.

## MOUNTINGS

The way a telescope is mounted can greatly affect its performance. The most common types of mounting are the altazimuth – which allows the instrument to pivot in altitude (up and down) and azimuth (parallel to the horizon) – and the equatorial, which aligns the telescope's movement with lines of right ascension and declination in the sky (see p.59). Altazimuth mountings are simple to set up, but because objects in the sky are constantly changing their altitude and azimuth, tracking objects requires continued adjustment of both. Equatorial mounts are heavier and take longer to set up but, once aligned to a celestial pole, allow a user to follow objects round the sky with movement in a single axis.

**ALTAZIMUTH MOUNTING**
A good altazimuth mount has control rods that allow gradual adjustment of the telescope's direction without the need to touch the telescope tube and shake the image in the eyepiece. A very simple type of altazimuth mount, called the Dobsonian, is often used on very large reflectors.

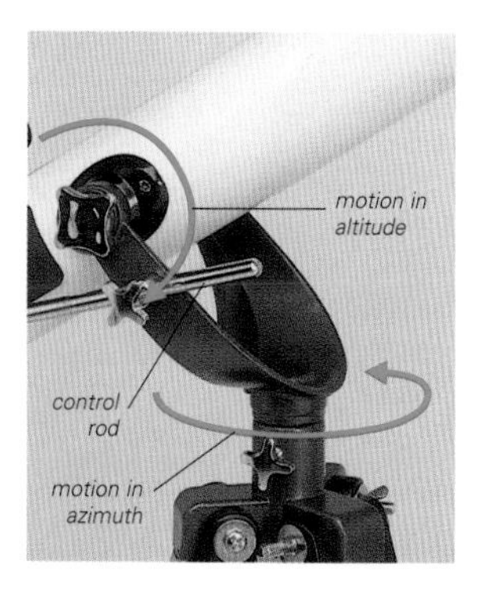


**EQUATORIAL MOUNTING**
An equatorial mount can slew the telescope to odd positions, so it is often balanced with a heavy counterweight. Once the mount axis is aligned to the celestial pole, the telescope's movements are limited to parallels of declination and right ascension. A clock drive can turn the telescope to keep pace with the sky.

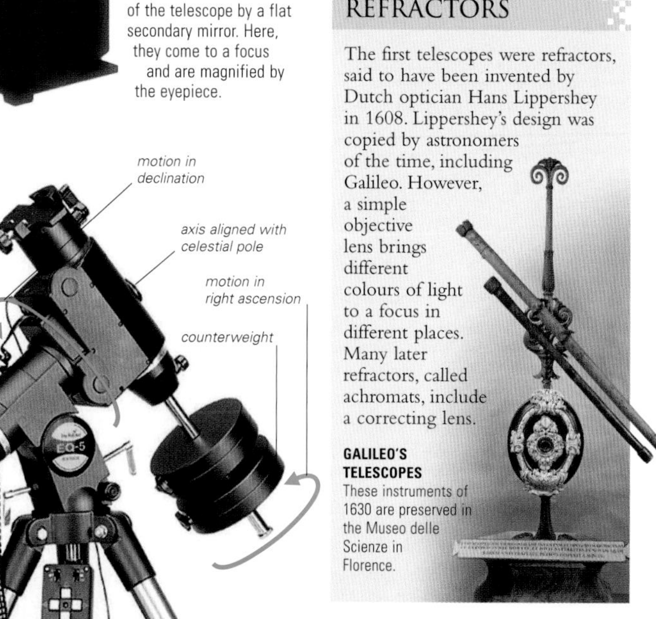


EXPLORING SPACE

### THE FIRST REFRACTORS

The first telescopes were refractors, said to have been invented by Dutch optician Hans Lippershey in 1608. Lippershey's design was copied by astronomers of the time, including Galileo. However, a simple objective lens brings different colours of light to a focus in different places. Many later refractors, called achromats, include a correcting lens.

**GALILEO'S TELESCOPES**
These instruments of 1630 are preserved in the Museo delle Scienze in Florence.

INTRODUCTION

## APERTURE AND MAGNIFICATION

Two major factors affect the image in a telescope eyepiece – aperture and magnification. Aperture is the diameter of the telescope's primary mirror or objective lens. The aperture of a telescope is linked to its "light grasp" – the amount of light it can collect. Doubling the aperture quadruples the light grasp, enabling the telescope to make fainter objects visible. Magnification is limited by a telescope's aperture, but controlled by the strength of the telescope's eyepiece. Eyepiece power is identified by its focal length – the distance at which it focuses parallel rays of light. Eyepieces with a shorter focus give greater magnification. Objective lenses and primary mirrors also have a focal length, and dividing this focal length by that of the eyepiece gives the overall magnification of that combination. An eyepiece's design also affects the amount of visible sky – the field of view.

**LARGER APERTURE**
This image of the open cluster M35 was taken through a telescope with a 100mm (4in) objective lens. The light grasp is 200 times that of a human eye and four times that of the 50mm (2in) telescope below. Stars are visible to roughly magnitude 12.5.

**7MM NAGLER EYEPIECE**

**7MM PLÖSSL EYEPIECE**

**EYEPIECE COMPARISON**
These two images show the open star cluster NGC 884 through the same telescope with different 7mm eyepieces. The Nagler eyepiece is designed for an extremely wide field of view. The Plössl has a more typical, narrower field.

**SMALL APERTURE**
This image of M35 was taken at the same magnification through a telescope with an objective lens 50mm (2in) in diameter. The light grasp is roughly 50 times that of a human eye, and stars are visible down to roughly magnitude 11.0.

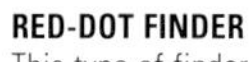

## FINDING DEVICES

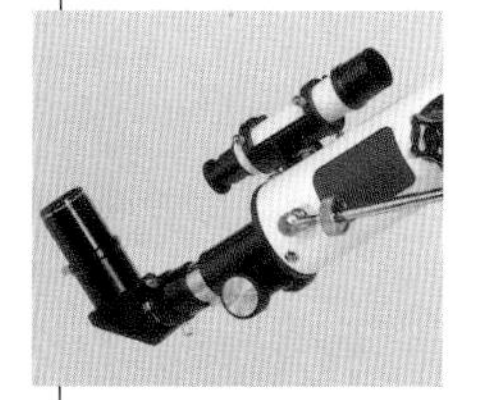

**OPTICAL FINDER**
An object aligned with the centre of an optical finder's cross hairs should also be visible in the main telescope, at least through a low-powered eyepiece.

Another vital element of any telescope is a finder scope. Even at the lowest magnifications, the view through a telescope eyepiece is too narrow for navigation, and simply pointing the telescope in the direction of a celestial object is unlikely to locate it. Finder scopes attach to the side of a telescope and are aligned in precisely the same direction. The most popular finders are simply low-powered telescopes with a targeting device (called a "reticle"), such as cross hairs. Other types project an accurate target reticle onto an unmagnified view of the sky, indicating exactly where the telescope is pointing.

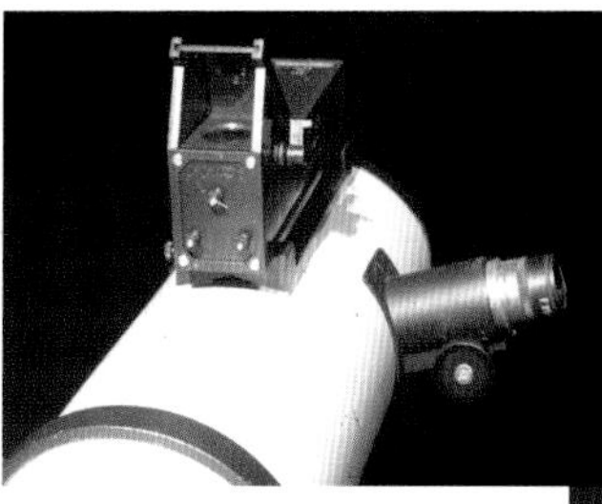

**REFLEX FINDER**
This device uses mirrors to project illuminated circles onto the view of the sky. The circle diameters correspond to different fields of view in the sky.

**RED-DOT FINDER**
This type of finder uses an LED and mirrors to project a red dot onto the sky, showing precisely where the instrument is pointing.

# THE COMPUTERIZED TELESCOPE

Computers simplify the hard work of tracking down and observing faint deep-sky objects or obscure comets and asteroids. A basic knowledge of how to use a telescope is still essential, but once an instrument is properly set up computerized databases and mechanical drives can make locating and tracking simple. Most serious amateur telescopes today come with hand-held controllers and can interface with desktop PCs or hand-held PDA computers. Built-in databases of object locations and orbits can be updated over the Internet. Many are even equipped with GPS satellite navigation equipment. With GPS, a telescope can identify its location on Earth's surface and use its drives to compensate for the movement of the sky without the user having to align its equatorial mount precisely.

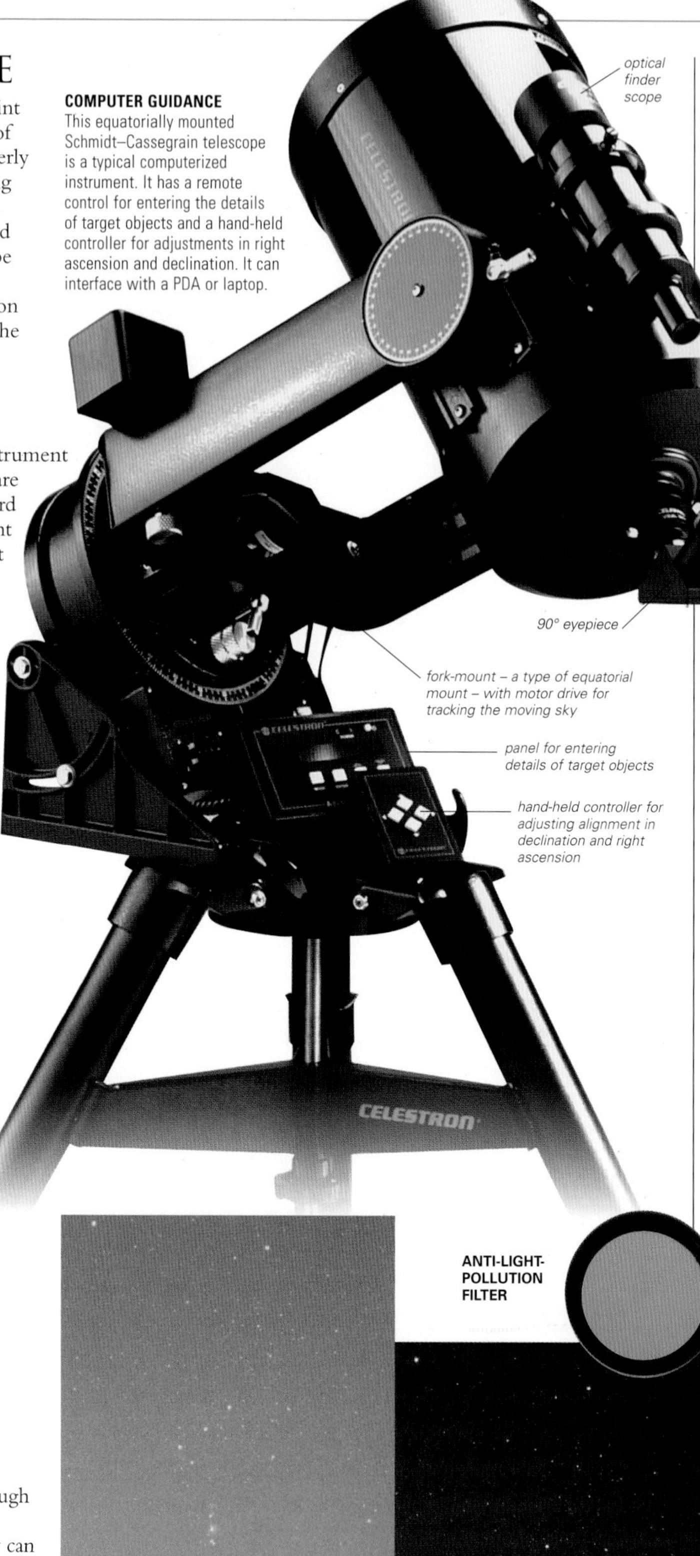


**COMPUTER GUIDANCE**
This equatorially mounted Schmidt–Cassegrain telescope is a typical computerized instrument. It has a remote control for entering the details of target objects and a hand-held controller for adjustments in right ascension and declination. It can interface with a PDA or laptop.

# ASTRONOMICAL SOFTWARE

Computerized telescopes are supplied with software to control the instrument from a computer, but other packages are also available. Planning software allows a user to schedule their observations and offers facilities to record notes and store images. Desktop planetarium software recreates the night sky on a computer screen, at any location or time from the distant past to the far future. Some programs, such as the free Celestia, can even simulate the view from elsewhere in the Solar System or the Galaxy.

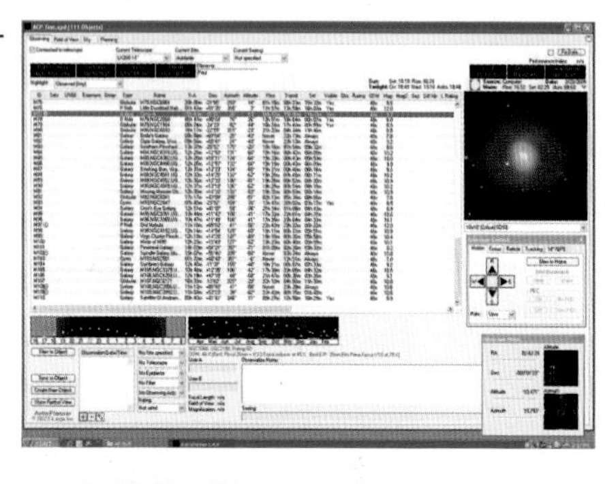

**TELESCOPE CONTROL**
Astroplanner is an observation planning and logging program that allows a user to specify a list of objects they want to observe, then calculates the best schedule for seeing them. It can also simulate the view through the eyepiece, and even steer the telescope from one object to another.

**NIGHT-SKY SIMULATORS**
Desktop Universe is one of several planetarium programs on the market. Others include Starry Night, TheSky, and RedShift. This program's unique feature is that its map is a real picture of the sky – a mosaic of 20,000 CCD images featuring more than 55 million stars and 1 million galaxies – all linked to a comprehensive database of information about each object.

# USING FILTERS

Filters can make a huge difference to the view through a telescope. They are glass discs that screw onto the eyepiece and filter out certain colours of light. They can cut out light pollution, enhance contrast on the surface of planets or other objects, or emphasize particular wavelengths emitted by certain phenomena. However, so-called "sun filters" that attach to an eyepiece are dangerous – they crack easily and should never be relied upon for looking at the Sun.

**COLOURED FILTERS**
These filters darken and block their "complimentary" colour, enhancing contrast in planetary features – blue will darken red features, yellow blocks purple, and red blocks blue.

**ANTI-LIGHT-POLLUTION FILTER**

**FILTERING OUT LIGHT POLLUTION**
Many street lamps emit yellow light with a very narrow range of wavelengths, making the sky glow orange (above). A light-pollution filter can cut it out while leaving most of the light from distant stars unaffected (right).

**TRIPOD PHOTOGRAPHY**
The simplest astrophotographs are just long-exposure pictures taken with a tripod-mounted camera. With sensitive film, one can photograph bright constellations in 30 seconds. Longer exposures show star trails caused by Earth's spin.

# ASTROPHOTOGRAPHY

While many amateur astronomers enjoy sketching the objects they see through their telescope, photography is the ultimate way of recording the sky. Single-lens reflex (SLR) film cameras are the most versatile camera for astrophotography – the main requirement is that the camera has a function to keep the shutter open and the film exposed for as long as required. A cable release to trigger the shutter without touching the camera will also help to avoid shake and blurring of the image. The camera can be used independently, piggybacked onto a telescope for guidance, or attached to the telescope's "prime focus" in place of an eyepiece. The best results require specialized, fast film, which reacts rapidly to even low light levels. However, there is always a trade-off to be made, since fast films have larger light-sensitive grains and will result in a grainier picture. Some astrophotographers go to extreme lengths to get the best results, chilling their film and the entire camera to make it more responsive.

*piggyback-mounted 35mm SLR camera with telephoto lens*

*Schmidt–Cassegrain telescope*

*cable shutter release*

*equatorial mount with motor drive*

**PIGGYBACK MOUNTING**
Fixing a camera to an equatorial telescope allows easy tracking of objects and long-exposure photography without star trails. A telephoto lens is ideal for capturing wide-field images of whole constellations. Once the telescope is aligned, its drive keeps the camera pointed at the target. A user without a motor drive can track the target manually while looking through the eyepiece, although this requires some skill.

**PRIME-FOCUS CAMERA MOUNTING**
With the right adaptor, an SLR camera body can be attached directly to the telescope's eyepiece tube. The telescope acts as a giant telephoto lens, focusing light onto the camera's film plane. It allows much greater magnification (and therefore a narrower field of view). A telescope with some form of automatic guidance is a must, since manual tracking through the camera eyepiece is almost impossible.

# DIGITAL ASTROPHOTOGRAPHY

Most digital cameras are limited to exposures too short to capture stars. However, specialist digital equipment, such as charge-coupled device (CCD) detectors, opens up possibilities beyond the reach of film. CCDs can be more sensitive than film, and because they "count" the number of photons hitting each part of the chip, a user can combine the data from several exposures. Image-processing software can then enhance the images and extract data. Even simple "webcams" can image the night sky – with the right software. However, even the highest-resolution CCD cannot yet capture the same resolution as film.

**CCD CAMERA AND CHIP**
A typical CCD camera is a simple cylinder that attaches to the eyepiece tube, with cables linking it to a computer.

*CCD assembly*

MEADE LX90 EMC

**DETAIL OF A CCD**

**POST-PROCESSING**
Registax is a free image-processing program that "stacks" frames from webcams. Detailed images of planets can be produced by selecting data from each frame.

*raw image*

*data selection*

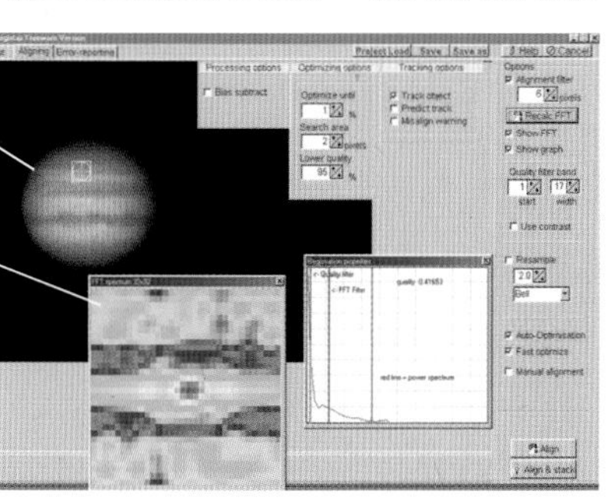

**HIGH-QUALITY AMATEUR PHOTOGRAPH**
This image of NGC 1977, the Ghost Nebula in Orion, was made through an amateur 300mm (12.5in) telescope and combines four 90-minute exposures. A digital stacking program was then used to add the images together.

*"Mankind will not remain on Earth forever, but in its quest for light and space will at first timidly penetrate beyond the confines of the atmosphere, and later will conquer for itself all the space near the Sun."*

Konstantin E. Tsiolkovsky

THE HISTORY OF ASTRONOMY links prehistoric stone circles with 21st-century space probes. It is a story of successive revolutions – in the understanding of our place in the cosmos, and in the ways we learn more about it. Geometry enabled the ancient Greeks to make the first measurements of the Sun, Earth, and Moon, while accurate instruments overthrew their Earth-centred ideas in the 17th century. The telescope opened up the sky, presenting many new mysteries for explanation by Isaac Newton's theory of gravitation. Spectroscopy offered a new method for studying the properties of stars and ultimately revealed the existence of other galaxies and the expansion of the Universe. In the last few decades, the development of spaceflight has allowed orbiting telescopes to probe deep space, while the journeys of a few people and machines beyond the Earth's immediate orbit have put our planet truly into perspective for the first time.

**WORLD IN MOTION**
A Space Shuttle hurtles around the Earth – which itself orbits the Sun at more than 100,000kph (62,000mph) – yet the illusion of stillness is complete. It is easy to see why the ancients thought that the Universe revolved around them.

# EXPLORING SPACE

# ANCIENT ASTRONOMY

58–59 The celestial sphere
60–63 Celestial cycles
64–65 Planetary motion
Measuring distances 84
The history of constellations 330–31
Mapping the sky 332–37

THE HISTORY OF ASTRONOMY stretches back over 6,000 years, making it the oldest science by far. Almost every culture throughout history has studied the Sun, Moon, and stars and watched how celestial bodies move across the sky. Human observations of the heavens reflect curiosity and wonder in the natural world, but have also been driven by urgent reasons of navigation, timekeeping and religion.

**STONEHENGE**
Many structures erected in Neolithic times, such as the circles at Stonehenge, England, finished in *c.*2600 BC, were aligned with positions of the Sun, Moon, or stars.

**RIDDLE OF THE STONES**
The purpose of Stonehenge, pictured here at sunset on the winter solstice, remains a mystery. It may have been used to worship the Sun or sky, or it might have served as a giant astronomical calendar.

## CYCLES OF LIFE

Early peoples were fascinated by the endlessly repeating patterns of change in the world around them. Evidence for the passing of time included changes in air temperature, the time and position of sunrise and sunset, the Moon's phases, plant growth, and animal behaviour. Such phenomena were attributed to gods or magical powers. Several thousand years ago, the first "astronomers" were probably shepherds or farmers in the Middle East, who watched the night sky for signs of the changing seasons. Ancient Egyptians relied on astronomical observations to plan the planting and harvesting of crops; for example, they knew that the rising of the star Sirius just before the Sun heralded the River Nile's annual flood. The ability to measure periods of time and record celestial cycles was vital to the advance of astronomy, and so many early cultures developed calendars, sundials, or water clocks. Monuments such as pyramids and groups of megaliths (large standing stones) were, in effect, the first observatories. By 1000 BC, Indian and Babylonian astronomers had calculated a year's length as 360 days; the latter went on to develop the 360° circle, with each degree representing a solar day. Later, the ancient Egyptians refined the length of a year to 365.25 days.

**CHINESE SUNDIAL**
Astronomy had an astrological function in ancient China, and the emperor had a duty to ensure the sky was closely observed. Sundials enabled time to be measured.

**PYRAMID OF THE NICHES**
Astronomical cycles have often influenced the design of temples, including this pyramid at El Tajin in Mexico, built from the 9th to 13th centuries AD. Its 365 niches represent the solar year.

## CELESTIAL MAP MAKING

Astronomy has always been concerned with mapping and naming the stars. By about 3500 BC, the ancient Egyptians had divided the zodiac into today's 12 constellations and herded other, non-zodiac stars into their own constellations. In ancient China and India, the 28 lunar mansions, or "night residences," were created to log the movement of the Moon along its monthly path. The ancient Greeks were the first to catalogue the 1,000 or so brightest stars, in about 150–100 BC (see p.84), but long before then, Indian and Middle Eastern observers were keeping detailed, dated astronomical records. Systematic data-gathering was developed in 3000–1000 BC by the Sumerians and Babylonians – powerful civilizations based between the Tigris and Euphrates rivers in Mesopotamia – for reasons of ritual and political prophecy.

**ANCIENT RECORDS**
Among the earliest surviving astronomical records are clay tablets from Mesopotamia. Some of them date from 1100 BC.

**HINDU CHART**
Indian observations of the stars, which began 4,000 years ago, were central to Vedic (ancient Hindu) philosophy. This chart was made later, in *c.* AD 1500.

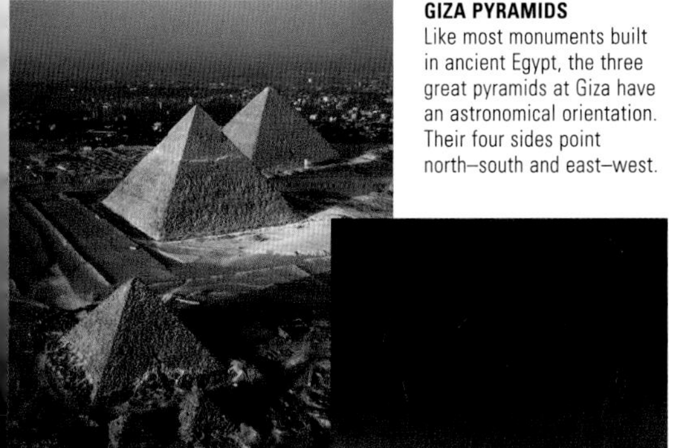

**GIZA PYRAMIDS**
Like most monuments built in ancient Egypt, the three great pyramids at Giza have an astronomical orientation. Their four sides point north–south and east–west.

## ORIENTATION AND NAVIGATION

The most important discoveries of ancient astronomers included how to find the cardinal points – north, south, east, and west – and how to use the north Pole Star to determine their latitude. This knowledge enabled the precise orientation of structures, particularly temples and burial monuments and also helped to produce accurate terrestrial maps. In 3000–2000 BC, the Stone Age peoples of northwest Europe used solar alignments at solstices and equinoxes, lunar phases, and other astronomical data to construct great stone circles, such as those at Stonehenge and Avebury in England and Newgrange in Ireland. Fire altars (used for Hindu ritual fire sacrifices), positioned on an astronomical basis and dating back 3,000 years, have been found at several archaeological sites in India. Some of the most complex buildings designed to astronomical principles are the pyramids of ancient Egypt and later examples built by the Aztecs and Maya in the Americas. Early architecture provides abundant proof of astronomical awareness, but little direct evidence survives of this being used in navigation in ancient times. However, seafaring Polynesian peoples may have used star positions to migrate across the Pacific as long ago as 1000 BC.

**SPINNING STARS**
Early observers in the northern hemisphere noticed that the stars appear to spin around a single fixed point – the north Pole Star.

## WANDERING STARS

The mysterious progress of the Sun, Moon, and five visible planets through the sky was fundamental to many early belief systems. The Babylonians thought that celestial gods ruled the sky, and their gods were adopted by the Greeks and Romans. In ancient Egypt, the chief cosmic deity was Ra, the Sun God, which was swallowed each night by Nut, the Sky Goddess. In Sumerian cosmology, every evening the Sun was captured by a magical boatman, who ferried it during the night, releasing it at dawn in the east. The Maya tabulated celestial cycles and predicted conjunctions of the planets and Moon with exceptional accuracy. Efforts to explain planetary motion continued to drive astronomical discovery up to the development of gravitational theory (see p.89).

**THE DRESDEN CODEX**
The Maya produced this book in 1200–1250 (the page pictured here is from a 19th-century facsimile). It contains many astronomical and astrological tables, written in beautiful hieroglyphic script.

**PLANETARY MOTION**
The movements of the visible planets Mercury, Venus, Mars, Jupiter, and Saturn were of great interest to early astronomers, who associated them with divine powers.

## THE BIRTH OF ASTROLOGY

Astrology is the age-old practice of determining how the planets and other celestial objects might influence life on Earth, especially human affairs. Dismissed as superstition today, it nevertheless laid a firm foundation of astronomical observation and mental deduction. In endeavouring to predict future events and search for omens or portents, early astrologers attempted to "order" the sky and noted the way in which the Sun, Moon, and planets behaved. Particular emphasis was placed on more unexpected occurrences such as eclipses, comets, meteor showers, and the appearance of new stars. Many political elites employed professional astrologers and sky-watchers, and their work was a perfectly respectable intellectual activity until the late 17th century. Astronomers also wrote astrological treatises; for example, in AD 140 Ptolemy (see p.85) wrote *Tetrabiblos*, the most influential astrology manual of all.

**WEEKLY CYCLE**
This illustration in an early 14th-century French manuscript depicts the Sun, Moon, and planets governing the days of the week.

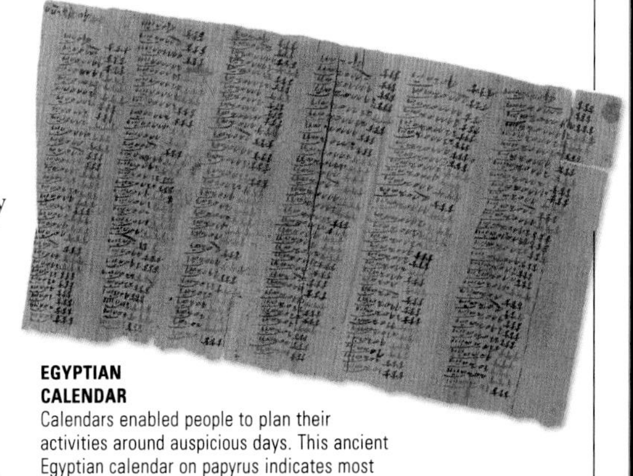

**EGYPTIAN CALENDAR**
Calendars enabled people to plan their activities around auspicious days. This ancient Egyptian calendar on papyrus indicates most days in black, but with unlucky days in red.

# EARLY SCIENTIFIC ASTRONOMY


22–23 The scale of the Universe
58–59 The celestial sphere
60–63 Celestial cycles
64–65 Planetary motion
The Copernican revolution 86–87
The history of constellations 330–31


THE RISE OF ANCIENT GREECE was a watershed in the development of astronomy as a rational science. Greek academics formulated complex physical laws and modelled the Universe. After the foundation of Islam, Arabic scholars, driven by religious requirements, developed ever more accurate methods for observing the heavens.

**ASTRONOMERS ON MOUNT ATHOS**
Classical Greek scholars, depicted at work in this 15th-century German manuscript, established astronomy as a disciplined science. Their experimental approach to studying the sky is immediately apparent.

### ARISTOTLE

One of the most influential of all Western philosophers, Aristotle (384–322 BC) was born on the Chalcidic peninsula in northern Greece and studied at the famous Athens Academy under Plato. He was an adviser to Alexander the Great, and he founded his own school, the Lyceum, in 335 BC. Aristotle believed the Universe was governed by physical laws, which he tried to explain through deduction and logic.

## SEEDS OF SCIENTIFIC THOUGHT

The ancient Greek world saw unprecedented political and intellectual freedom and produced astonishing scientific progress. For 700 years, from about 500 BC to AD 200, Greek philosophers sought answers to some of the fundamental astronomical questions. They still believed in a divine master plan and had some ideas that seem strange to us today – Heraclitus (540–500 BC), for example, suggested that the stars were lit each night, and the Sun each morning, like oil lamps. But Plato (*c.*427–347 BC) argued that geometry was the basis of all truth, thus providing the necessary impetus for deducing exactly how the cosmos worked. Aristotle (see panel, left) fixed the Earth at the centre of the Universe and suggested that the planets were eternal bodies moving on perfectly circular orbits. Eudoxus (*c.*408–355 BC) charted the northern constellations, and Hipparchus (*c.*190–120 BC) sorted the stars into six orders of brightness for the first time.

**HIPPARCHUS**
The star magnitude system in use today is based on the one originally devised by Hipparchus in the 2nd century BC.

## MEASURING DISTANCES

Advances in geometry and trigonometry enabled the Greeks to measure astronomical distances with reasonable precision. In about 500 BC, Pythagoras (*c.*580–500 BC) proposed that the Sun, Earth, Moon, and planets were all spherical, a notion that Aristotle confirmed, in the case of the Earth, by observing the shape of its shadow during a lunar eclipse. In about 250 BC, Eratosthenes (*c.*276–194 BC) noted that sunlight fell straight down a well at Syene, Egypt, at noon on midsummer's day. By comparing this data with that from Alexandria, trigonometry gave the Earth's circumference to an accuracy of within 5 per cent. Hipparchus (see above) and Aristarchus (320–230 BC) measured the Earth–Moon distance by timing lunar eclipses. Unfortunately the Earth–Sun distance was not measured accurately.

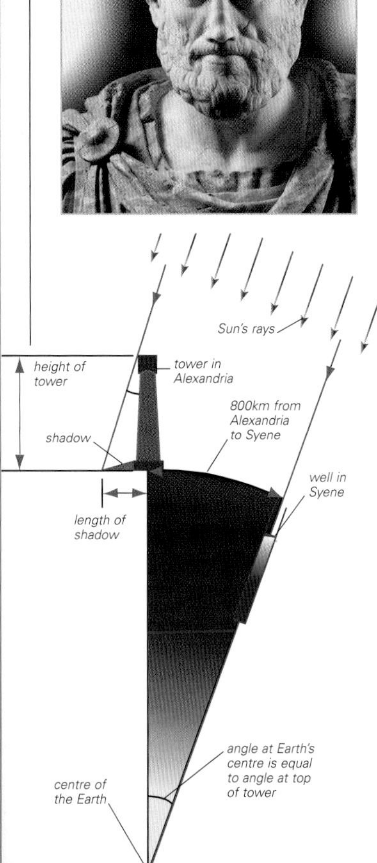


**ESTIMATING THE EARTH'S CIRCUMFERENCE**
Eratosthenes used the fact that the Sun's rays are nearly parallel to calculate the Earth's size. When the Sun was directly overhead at Syene on the Tropic of Cancer (see p.61), it cast a shadow at an angle of 7° in Alexandria. The 800km (500 miles) between these places was therefore 0.0194 (or 7/360) of the Earth's circumference.

# THE EARTH-CENTRED COSMOS

It was only natural that people should assume the cosmos was geocentric, or Earth-centred. After all, we do not perceive the Earth's motion as it spins through space, and the stars in the firmament appear to be fixed. A few early thinkers, such as Aristarchus in 280 BC, disputed this system and placed the Sun at the centre, but their idea did not gain credence. Instead, Ptolemy (see panel, below) refined Aristotle's cosmological world view to create a linear sequence of uniform orbits. His order was: Moon, Mercury, Venus, Sun, Mars, Jupiter, Saturn, and, finally, the stars. This was based on perceived speed; for example, the Moon went round the sky every month, the Sun every year, and Saturn every 29.5 years. In order to reconcile the mistaken concept of circular (as opposed to elliptical) orbits with actual planetary observations, Ptolemy was forced to argue that each planet also revolved in a small circle, or epicycle, while it was orbiting the Earth.

**PTOLEMAIC SYSTEM**
This plate from *The Celestial Atlas*, published in Amsterdam in 1660–61, shows Ptolemy's system of orbits, in which the Moon, Sun, planets, and stars circle the Earth. Diagrams of the Universe resembled this one until the Renaissance.

## PTOLEMY

Ptolemy (*c.* AD 100–170) lived and worked in the great metropolis of Alexandria, Egypt, which was then part of the Greek empire. He was one of the last – and the greatest – of the ancient Greek astronomers. His Earth-centred model of the Universe, outlined in the treatise *Almagest*, dominated astronomical theory for 1,400 years. Ptolemy also made a catalogue of 1,022 stars in 48 constellations, based on earlier work by Hipparchus.

# ARABIC ASTRONOMY

After the decline of the ancient Greek city-states, the most important advances in astronomy were achieved by Arabic scholars. The period of Arabic dominance lasted for about 800 years, from the foundation of Islam in AD 622 until the 15th century. Astronomers working in the Middle East, Central Asia, North Africa, and Moorish Spain translated Greek and Sanskrit (ancient Indian) texts into Arabic and assimilated their astronomical knowledge. Islamic rules for daily and monthly worship, and the need to find the direction of Mecca in order to pray and orientate mosques, meant that there was now an urgent requirement to determine time and position extremely accurately. Sophisticated spherical trigonometry, trigonometric functions, and algebra were developed, and the astrolabe (a Greek invention, used as a clock) was vastly improved. The approach of great Islamic astronomers, such as al-Battânî (*c.* AD 850–929) and Ulugh Beg (1394–1449), was founded on patient observation. To this end, numerous observatories were built, of which the grandest were at Baghdad in Iraq, Samarkand in Uzbekistan, and Maraghah in Iran. They housed large astronomical instruments, including wall-mounted quadrants, which were used to measure the altitude of astronomical objects as they crossed a north–south plane (meridian).

**ULUGH BEG'S OBSERVATORY**
Located near Samarkand in Central Asia, this observatory was built in about 1420. It contained a huge meridian arc or quadrant, partly hewn into the rock, which was used for taking celestial sightings.

**ASTROLABE**
An astrolabe uses stellar, solar, and lunar positions to tell the time and find latitude. The brass model pictured here was made in 17th-century Persia.

# THE COPERNICAN REVOLUTION


24–27 Celestial objects
36–37 Gravity, motion, and orbits
38 Problems in Newton's Universe
66–67 Star motion and patterns
The stellar cosmos 90
The family of the Sun 118–19


**FOR CENTURIES, ASTRONOMERS** based their theories on the assumption that Earth lay at the centre of the Universe. In the 16th century, this privileged position was undermined by the suggestion that the Earth was just one of several planets circling the Sun. This revolution was accompanied by huge technological advances, particularly the invention of telescopes, ushering in a new era of research and great discoveries.

**MEASURING TIME**
Medieval astronomy was usually inaccurate by later standards. For example, it was hampered by the crudeness of the period's mechanical clocks, such as this cathedral clock in Strasbourg, France. They could keep time only to the nearest 15 minutes or so.

## MEDIEVAL ASTRONOMY

After the fall of Rome in AD 476, astronomers in Europe, the Middle East, and Asia left behind their relative cultural isolation and exchanged ideas more freely. This was due partly to the growth of trade during the medieval period and also to the spread of Islam (see p.85). Observational astronomy flourished, with emphasis on planetary conjunctions, solar and lunar eclipses, and the appearances of comets or new stars. Many medieval astronomers were university teachers who earned a living through teaching. Despite limited resources, they made notable advances, including a new, precise catalogue of naked-eye star positions devised by Ulugh Beg (1394–1449) of Mongolia – the first since the days of Hipparchus, 1,600 years previously (see p.84).

## A SUN-CENTRED UNIVERSE

It is possible to date the birth of modern astronomy to 1543, when Nicolaus Copernicus (see panel, right) published a ground-breaking treatise about a Sun-centred, or heliocentric, Universe. He was dissatisfied with the inaccuracy of the geocentric models of planetary orbits, in which the Earth was at the centre, which had dominated astronomy since Ptolemy's work in the second century AD (see p.85). His simpler theory of a central Sun and an orbiting Earth explained many previously puzzling observations, because there were now two types of planet: those inside the Earth's orbit; and those beyond. The sequence Mercury, Venus, Earth, Mars, Jupiter, and Saturn was now in order of increasing orbital periods and planet–Sun distances. The "lantern of the Universe", as Copernicus called it, took its rightful place at the centre of the cosmos. Although this meant the Earth must move at high velocity, the theory had already been proposed by Aristarchus (see p.85) and others.

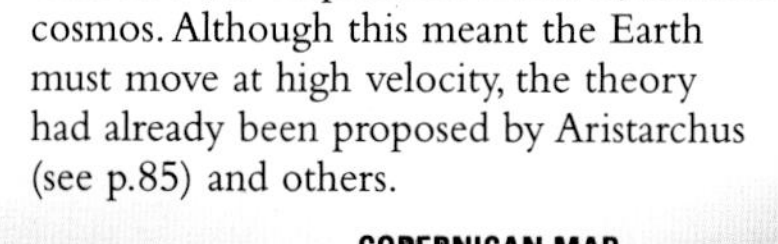

### NICOLAUS COPERNICUS

Born in Torun, Poland, Copernicus (1473–1543) studied theology, law, and medicine at university. In 1503, he became the canon of Frauenberg Cathedral. This post provided financial security and left him plenty of time to indulge his passion for astronomy. He described his idea of a Sun-centred universe in his book *On the Revolution of the Heavenly Spheres*, published in the year of his death.

**COPERNICAN MAP**
This map by Andreas Cellarius demonstrates the Copernican theory of the Earth and other planets circling the Sun, with the zodiac stars beyond.

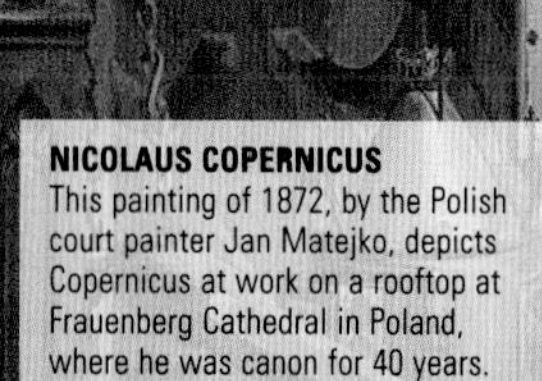

**NICOLAUS COPERNICUS**
This painting of 1872, by the Polish court painter Jan Matejko, depicts Copernicus at work on a rooftop at Frauenberg Cathedral in Poland, where he was canon for 40 years.

# THE NOTION OF SPACE

The Greeks and subsequent astronomers placed the stars just beyond Saturn, but this caused a problem if Copernicus's theory was correct. If, as he argued, the Earth orbited the Sun, why did the stars not have a reciprocal movement? Copernicus had no choice but to banish stars to a vast distance so that their motion would not be apparent to human observers. He also suggested for the first time that bright stars were closer than faint ones. However, others wondered why God would have created such an enormous, seemingly useless space between the orbit of Saturn and the stars. Tycho Brahe (1546–1601), a Danish nobleman, thought that the future of astronomy depended on the estimation of true distances and the accurate recording of planetary positions at different times. To this end, he greatly improved standards of observation by making precisely calibrated measuring instruments. In 1572, his patient observations of a supernova (exploding star) in the constellation Cassiopeia convinced him that stars were not all a fixed distance away, but were changeable objects that existed in "space". It required courage, or the backing of powerful patrons, to put forward such ideas in the age of the Inquisition – the feared tribunal of the Roman Catholic Church that tried and punished those who challenged the orthodox view of the Universe.

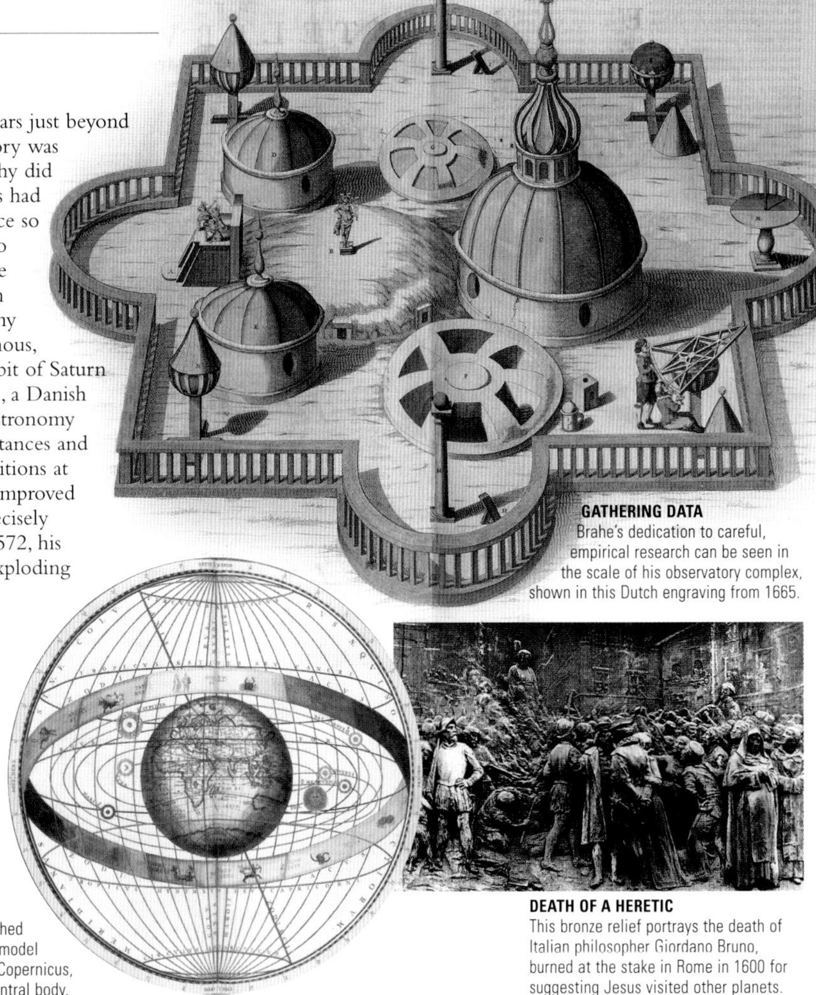

**GATHERING DATA**
Brahe's dedication to careful, empirical research can be seen in the scale of his observatory complex, shown in this Dutch engraving from 1665.

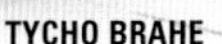

**TYCHO BRAHE**
A towering figure of Renaissance astronomy, Tycho Brahe lived and worked at Uraniborg observatory on the island of Hven, Denmark.

**BRAHE'S SYSTEM**
This detail of a print published in 1660–61 shows Brahe's model of planetary orbits. Unlike Copernicus, he retained Earth as the central body.

**DEATH OF A HERETIC**
This bronze relief portrays the death of Italian philosopher Giordano Bruno, burned at the stake in Rome in 1600 for suggesting Jesus visited other planets.

# LAWS OF PLANETARY MOTION

The Copernican Sun-centred model of the universe still described planetary orbits in terms of epicycles – each planet followed a small circle as it revolved around the central body. This view had prevailed since Ptolemy (see p.85), but the actual form of the orbits was a mystery. Two things were needed to solve the problem. The first was the accurate planetary data obtained by Tycho Brahe (above). The second was faith in the veracity of that data, coupled with tenacity and mathematical genius. The latter was provided by German astronomer Johannes Kepler (1571–1630), Tycho's successor after 1601. Basing his work on observations of Mars's celestial path, Kepler eventually formulated three laws of planetary motion. First, in 1609, he revealed that orbits are elliptical, not circular or epicyclic. The Sun is at one focus of the ellipse (an ellipse is a loop drawn around two foci; a circle is an ellipse with a single focus). Second, he showed that the line between the Sun and a planet sweeps out equal areas of space in equal times, so a planet's orbital speed is slower in the outer reaches of its orbit. Third, in 1619, Kepler also proved the link between the size and period of an orbit. Kepler held mystical views – he believed the cosmos was permeated by a musical chord, with each planet producing a tone in proportion to its speed. Like everyone else before Newton's discovery of gravity (see p.89), he had no idea what pushed the planets along their orbits.

**JOHANNES KEPLER**

**HEAVENS ABOVE**
This amusing woodcut, designed by Camille Flammarion in Paris in 1888, mocks the ignorance of medieval and Renaissance astronomers. A scholar is shown poking his head through the shell-like celestial sphere in a desperate attempt to understand what makes the heavens move.

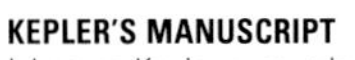

**KEPLER'S MANUSCRIPT**
Johannes Kepler expounded his theories of planetary motion in the treatise *Epitome of Copernican Astronomy*, first published in 1619.

# GALILEO'S DISCOVERIES

Concrete proof that the Copernican model of the Universe (see p.86) was correct was provided by Galileo Galilei (see panel, below). In 1608, a Flemish spectacle-maker invented the telescope, and the news spread rapidly. Galileo quickly constructed several of his own in 1609, thereby becoming the founder of telescopic astronomy. His devices had a magnification of up to x30. Within months, he had discovered that the Moon is mountainous and Venus has phases, saw sunspots for the first time, and noted four of Jupiter's moons. He even suggested that the stars were distant suns in their own right. Above all, he realized that the phases of Venus can be explained only if the planets orbit the Sun, rather than the Earth. Galileo's other achievements included showing that the acceleration of a falling object is independent of its composition, and that a pendulum's swing period does not depend on its amplitude.

**GALILEO THE TEACHER**
A brilliant communicator, Galileo published his findings in Italian as well as Latin, to make them more accessible. However, his belligerent debating style made many enemies.

*wooden tube is covered in paper*

*copper binding*

**TELESCOPE**
Galileo's simple telescopes (this is a replica) suffered from poor glass and a very narrow field of view, but brought new worlds within reach.

## GALILEO GALILEI

Galileo (1564–1642) was born into a professional family in Pisa, Italy. He held the mathematics professorship at Padua University from 1592 to 1610. Practical in approach, he used telescopic observations to prove his controversial theories, but was condemned by the Roman Catholic Church. At his trial in Rome in 1633 (left), the Inquisition placed him under house arrest for life.

**MOON SKETCHES**
Telescopes enabled Galileo to make detailed drawings of the surface of the Moon, revealing its uneven topography.

**SIDEREUS NUNCIUS**
These pages are taken from Galileo's book, *The Starry Messenger*, published in Venice in 1610. The short work outlined his recent discoveries, sending shock waves throughout the scientific community.

# GREAT OBSERVATORIES

The emergence of a new generation of observatories during the late 17th and 18th centuries was a watershed in the history of astronomy. Observatories had formerly been the preserve of a select band of private individuals, including Tycho Brahe in Denmark (see p.87) and Johannes Hevelius in Danzig, Prussia (now Gdansk, Poland), or of university professors, such as Galileo (see above) and Giovanni Cassini of Bologna, Italy. Now these were augmented by well-funded royal institutions. Louis XIV founded the Académie des Sciences in Paris in 1666, and the French Royal Observatory, completed in 1672. In England, Charles II's foundation of the Royal Society in 1660 led to the building of the Royal Greenwich Observatory, where John Flamsteed (1646–1719), the first Astronomer Royal, began work in 1676. Soon astronomers were tackling the three major challenges of the time: working out the size of the Solar System; measuring the distance to the stars; and finding the latitude and longitude of places on land and sea. This pursuit was partly driven by the need for better navigation in an age of expanding European empires and by the desire for national prestige. From the early 1700s, observatories multiplied rapidly, opening in Berlin, Prussia (1711), Jaipur, India (1726), Uppsala, Sweden (1730), Vilnius, Prussia (1753), Washington, USA (1838), and Pulkovo, Russia (1839). Much effort was devoted to mapping the sky and tracing movements of celestial objects.

**MAPPING STARS**
Observatories such as that at Greenwich used pendulum clocks and more accurate transit instruments (inset) to measure the exact times and angles of stars as they crossed a north–south line, or meridian.

**PARIS ROYAL OBSERVATORY**
This building opened in 1672, heralding a competitive age in which science and astronomy were part of national prestige.

# RULES OF ATTRACTION

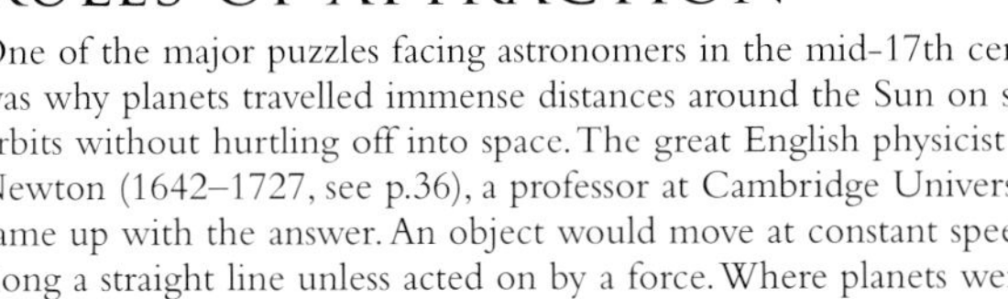

One of the major puzzles facing astronomers in the mid-17th century was why planets travelled immense distances around the Sun on stable orbits without hurtling off into space. The great English physicist Isaac Newton (1642–1727, see p.36), a professor at Cambridge University, came up with the answer. An object would move at constant speed along a straight line unless acted on by a force. Where planets were concerned, this force was solar gravity. He soon realized that gravity was universal. It controlled the path of a falling object (such as the apple that Newton famously saw fall from a tree in 1666), the Moon circling the Earth, and a comet travelling towards the Sun from the distant recesses of space. Gravity explained all three of Kepler's laws (see p.87). It also explained the height of the tides. After Newton's death, the return of Halley's Comet in 1758 (see p.90) proved that gravity did indeed apply to the edge of the Solar System, and the use of gravitational theory made it possible to calculate the mass of the Earth and the Sun.

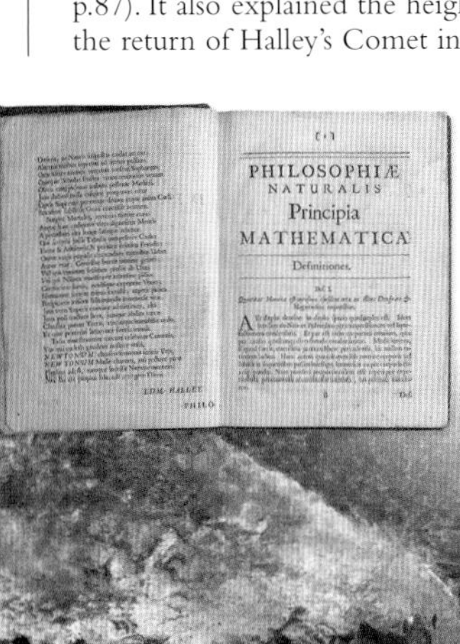


**NEWTON'S MASTER WORK**
In 1687, Newton published the first clear mathematical definition of gravity, in his treatise *Philosophiae Naturalis: Principia Mathematica.* It revolutionized astronomy and physics.

# NEWTON'S REFLECTING TELESCOPE

In 1666, Newton found that a prism breaks up white light into a rainbow of colours. Unfortunately, telescope lenses do the same, and the blue light at one end of the spectrum is focused at a different spot to the red light at the other. This is called chromatic aberration. It creates haloes of coloured light around the object being viewed, causing serious loss of image quality. One solution is to use curved mirrors to focus all the light to the same point, regardless of its colour. In 1663, the Scotsman James Gregory (1638–75) designed a reflecting telescope with a large concave primary mirror and a smaller concave secondary one that bounced light back through a hole in the primary mirror to a magnifying lens at the rear. Newton modified the design, using a flat secondary mirror to reflect the captured light to a side-mounted eyepiece. Unveiled in 1672, Newton's telescope won huge acclaim.

**CRUCIAL EXPERIMENT**
These notes made by Newton record his experiment on the behaviour of light when it is refracted by prisms.

**NEWTON AND LIGHT**
This coloured engraving depicts the 25-year-old Newton in his university rooms, using a glass prism to investigate the nature of light. He was the first scientist to understand colour.

eyepiece

sliding focus

**NEWTON'S TELESCOPE**
Compared to earlier telescopes, Newton's version was very compact, measuring just 30cm (12in). It also eliminated the problem of chromatic aberration. The model shown here is a replica.

**DIVINE POWERS**
Newton is using compasses to study a diagram in this print made in 1795 by the English artist William Blake (1757–1827). Blake's setting of a god-like Newton in a fantastical landscape criticizes Newton for confining the wonder and beauty of the created Universe within the restricting laws of science.

# THE INFINITE UNIVERSE

22–23 The scale of the Universe
28–29 Matter
32–35 Radiation
Matter and star energy 92
Comets 214–19
Stars 230–31

**BY THE MID-19TH CENTURY,** astronomy had evolved from an essentially mathematical science into a discipline incorporating the new knowledge and techniques of physicists and chemists. Rapid technological progress, particularly the invention of photography and ever more powerful telescopes, enabled astronomers to study celestial bodies in far greater detail. They began to classify different objects and study how they behaved.

## THE AGE OF ENLIGHTENMENT

The 18th and 19th centuries saw great progress in science due to improved scientific methods. In astronomy it was a period of consolidation, based on achievements in the measurement and classification of celestial bodies. The orbits of comets were of great interest after the work of Englishman Edmond Halley (1656–1742), who showed that "his" comet returned to the Sun every 76 years. Orbits of planets were determined more precisely than before, and their surface features, including Jupiter's Great Red Spot and the polar caps of Mars, were observed. In 1728, the Earth's velocity was calculated. William Herschel (see panel, right) discovered the seventh planet – Uranus – in 1781, and the Italian Giuseppe Piazzi (1746–1826) identified the first asteroid – Ceres – in 1801. Urbain Le Verrier (see p.118), Pierre-Simon Laplace (1749–1827), both of France, and Englishman John Adams (1819–92) applied Newtonian gravitational theory to predict positions for an unseen planet beyond Uranus. Neptune was eventually discovered in 1846.

**HALLEY'S COMET**
In 1696, Halley said that the three historic comets of 1531, 1607, and 1682 were all the same object. He correctly predicted its return in 1758.

**EDMOND HALLEY**
Halley realized that some comets were permanent members of the Solar System, regularly returning to the Sun.

### WILLIAM HERSCHEL

Born in Hanover, Germany, in 1738, Herschel emigrated to England in 1757, earning his living as a music teacher and conductor. His interest turned to astronomy, and by the late 1770s he was making the best metallic mirrors and telescopes in the world. In 1781, he fortuitously discovered the planet Uranus. A year later he was appointed court astronomer. Royal grants enabled the construction of a large telescope, which Herschel and his sister Caroline (1750–1848) used to survey the sky. He died in 1822.

## THE STELLAR COSMOS

Improvements in telescopes throughout the 18th century enabled astronomy to become the study of a dynamic, infinite Universe. Previously, stars were just fixed points of light, their distance unknown, but in the 1710s Halley had discovered them to be moving, triggering concerted efforts to understand their behaviour. Many stars were found to be twins orbiting a common centre of mass under the influence of Newtonian gravity, and in the 1780s variable stars such as Algol and Delta (δ) Cephei were investigated. In 1781, the Frenchman Charles Messier (1730–1817) published a catalogue of 103 "fuzzy bodies", or nebulae (clouds of dust and gas). Herschel extended this work, spending many hours counting stars of different brightnesses in an ambitious attempt to survey the entire sky. In 1783, he deduced that the Sun is travelling towards the star Lambda (λ) Herculis and, mistakenly, concluded that it lies at the centre of a galactic stellar system. During the first two decades of the 19th century, Laplace developed mathematical analysis to produce the first model of the Solar System's origins.

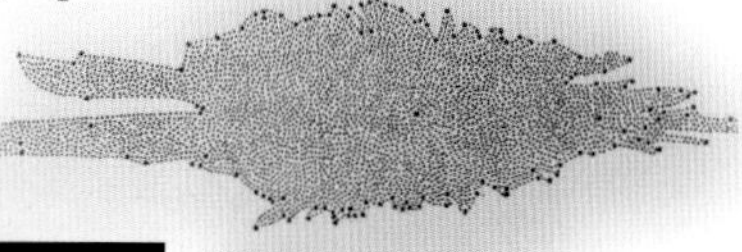

**MAPPING THE MILKY WAY**
This sectional map of the Milky Way was produced by Herschel in 1785. He wrongly assumed that its stars are uniformly distributed and that his telescope could see to the very edge of the galaxy.

**EXPANDING HORIZONS**
In a kind of 19th-century "space race", astronomers competed to find information about stars, nebulae, and galaxies. This image of the Omega Nebula (M17) is based on a drawing made in the 1860s.

## THE BIRTH OF ASTROPHYSICS

During the 19th century, astronomers continued to apply developments in mathematics, physics, and chemistry to understand the make-up and behaviour of planets, comets, and stars. A new field – astrophysics – emerged, which transformed objects in space into scientific entities. Astronomy had been concerned mainly with their movement, whereas astrophysics examined their basic parameters such as radius, mass, temperature, and composition. In 1815, the German optician Joseph Fraunhofer (1787–1826) studied the spectrum created by the Sun's light, and noticed dark lines crossing it. Later, these were found to be produced by chemicals in the Sun, which absorb light. As each chemical absorbs light differently, the patterns of lines in spectra reveal which are present. In the 1860s, spectroscopy (the study of spectra) led British astronomer William Huggins (1824–1910) to discover that stars contain the same elements as the Earth.

**WILLIAM HUGGINS**
Huggins was among the first astronomers to use spectroscopy.

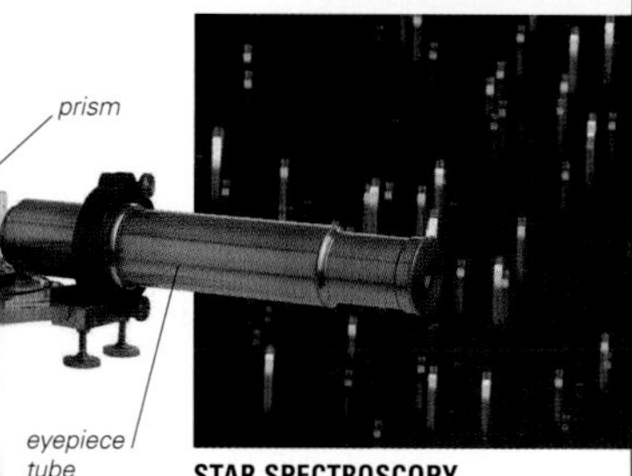


**SPECTROSCOPE**
A spectroscope (this is a 19th-century model) uses a prism to split white light into its constituent wavelengths, producing a spectrum.

**STAR SPECTROSCOPY**
Photographs of the spectra of stars, such as these in the Hyades cluster, enable the stars' chemical composition and temperature to be established.

# INSTRUMENTAL PROGRESS

Telescopes grew larger and larger between the early 18th century – when astronomers such as Edmond Halley (see left) struggled to use unwieldy, inaccurate instruments with small-diameter objective lenses – and the mid-19th century. They doubled in lens diameter about every 40 years, which meant that the distance to the farthest viewable object also doubled and the number of objects that could be seen increased by a factor of eight. Telescope mounts became more sophisticated, and lens quality improved. Refracting telescopes were revolutionized in 1758, when English optician John Dolland (1706–61) introduced the doublet lens, which focused blue and red light at the same point. Huge reflecting telescopes were built, by Herschel among others, and in 1824, the first equatorially mounted telescope was completed, in which one axis of the instrument was aligned with the north Pole Star; a clockwork mechanism moved the machine around this axis to track stars. In the 1840s, photography started to replace pencil and paper for recording data. Photographic plates could be exposed for hours, enabling a record to be made of much fainter objects than those visible to the unaided human eye.

**MICROMETER**
This device measures distances between objects, such as stars, in a telescope's field of view. It uses a cross hair that can be minutely adjusted with a finely turned screw.

**REFLECTING TELESCOPE**
As the century progressed, the size of the objective mirrors increased steadily. This 2.1m (7ft) wooden model magnified 200 times.

**GIANT TELESCOPE**
Instead of concentrating on focal length alone, Herschel perfected ways of making larger mirrors. In 1789, he installed a telescope at Slough, England, with a 1.25m (4ft) mirror. Its massive light-gathering power enabled him to peer deeper into space.

**THE HERSCHELS AT WORK**
This 19th-century illustration shows William Herschel observing the sky with his sister Caroline, who collaborated on many projects and discovered eight comets of her own.

# SHAPING SPACE

22–23 The scale of the Universe
28–31 Matter
42–43 Expanding space
Stars 230–31
Main-sequence stars 246–47
Beyond the Milky Way 292–325

AT THE BEGINNING of the 20th century, there was thought to be just one galaxy, with the Sun at the centre. By the 1930s, astronomers realized that there were billions of galaxies, millions of light-years apart, and that the Universe was expanding. They also started to understand the sources of stellar energy.

ALBERT EINSTEIN

## MATTER AND STAR ENERGY

By 1900, due to the discovery of radioactive decay four years earlier, it was realized that the Earth was over 1 billion years old. This great age was consistent with earlier estimates made by geologists and with Charles Darwin's estimate of the time needed for natural selection to do its work. It remained a mystery how the Sun could have kept shining for so long. One idea was that the Sun was fuelled by meteorites falling into it; another was that the Sun was slowly shrinking. Neither of these theories could account for such a long-lived Sun. However, in 1905, the German-American Albert Einstein (1879–1955) proposed that $E = mc^2$, and so energy, $E$, could be produced by destroying mass, $m$ (see pp.38–39). In the 1920s, the British astrophysicist Arthur Eddington (1882–1944) suggested that the energy of the Sun – and other stars – comes from nuclear fusion. German-American physicist Hans Bethe (1906–2005) then worked out the basic fusion process (see p.31). It is a nuclear reaction that does indeed liberate energy from matter. Astrophysicists now have a detailed view of how stars are fuelled and estimate that Sun-like stars can shine for 10 billion years.

ARTHUR EDDINGTON

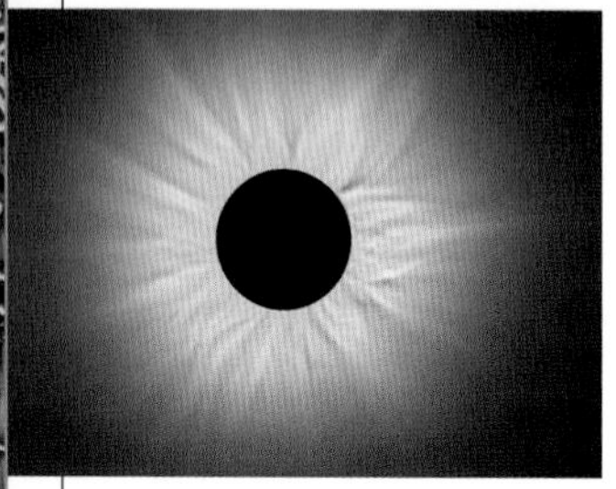

**PROVING RELATIVITY**
Eddington measured starlight deflected by the Sun during a solar eclipse in 1919, and proved Einstein's theory regarding the degree to which light could be bent by astronomical forces.

## THE GREAT DEBATE

In 1920, there was a public debate between American astronomers Harlow Shapley and Heber Curtis (see panel, below) over the form of our galaxy in particular and the extent of the Universe in general. Shapley believed that there was only one "big galaxy", with the Earth two-thirds of the way towards the edge. He said that its diameter was about 300,000 light-years – 10 times larger than Curtis's estimate. This value was obtained by using Cepheid variable stars (which periodically increase and decrease in brightness) as distance indicators (see p.278). Curtis, on the other hand, was convinced that many nebulous objects in the sky were not in our galaxy, but were actually galaxies just like ours. He proposed that these "island universes" were spread evenly across the sky, but that some were obscured by dust in our galactic disc, explaining their distribution above and below the plane of the Milky Way.

### HARLOW SHAPLEY AND HEBER CURTIS

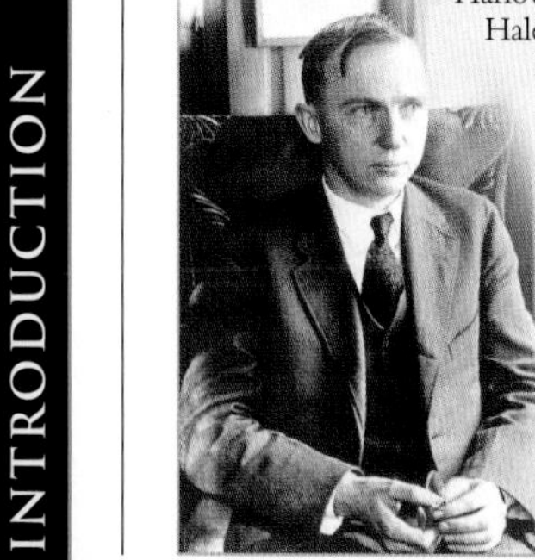

SHAPLEY

CURTIS

Harlow Shapley (1885–1972) directed the Hale Observatory at Harvard University, USA, from 1921 to 1952. Shapley changed our concept of the galaxy, showing that the Sun orbits the central nucleus, 30,000 light-years away. Heber Curtis (1872–1942) used the Lick refracting telescope on Mount Hamilton, California, USA, to measure stellar velocities, then turned to galaxy photography. He argued that there are vast numbers of galactic "islands" and that galactic discs are dusty, explaining why external galaxies appear to avoid the plane of the Milky Way.

# THE EXPANDING UNIVERSE

Huge new American telescopes revolutionized 20th-century astronomy. The Hooker Telescope (below) was so large that it could detect the Cepheid stars in the Andromeda Nebula (M31). Their faintness indicated that its distance to Earth was about 10 times the diameter of the Milky Way galaxy. Heber Curtis (opposite) had been right: Andromeda was not a "nebula", but a galaxy. Other spiral nebulae were clearly galaxies, too. Using the Hooker Telescope, the American Edwin Hubble (1889–1953) showed that galaxies are "building blocks" for a universe far vaster than previously imagined. In about 1927, he investigated galaxies' spectral features and found that, not only was light from most galaxies shifted towards the red end of the spectrum, indicating that they were moving away from us, but this recessional velocity was larger for fainter (more distant) galaxies. The Universe was actually expanding, so must have been much smaller in the past. The gradient of the velocity–distance graph indicated the age of the Universe. Astronomers also realized that something had started the expansion, sowing the seeds of the Big Bang theory (see p.96).

**EDWIN HUBBLE**
Hubble, pictured in front of the famous telescope at Mount Wilson, radically improved understanding of the Universe's structure, proving beyond doubt that it is expanding.

# TELESCOPE TECHNOLOGY

Observational astronomy reached a watershed in the last decades of the 19th century. In America, new hi-tech observatories and university departments were established, often funded by millionaire businessmen. Astronomers needed to examine very faint, distant objects, but the limited strength and relatively low transparency of large discs of glass meant that telescope objective lenses could not be made larger than about 1m (3.3ft) in diameter – the size of the Yerkes refracting telescope, which opened in Williams Bay, Wisconsin, in 1897. So refractors were left behind: a new age of huge reflector telescopes dawned. The 2.5m (8ft) Hooker Telescope, built on Mount Wilson, California, became operational in 1918. In 1948, the 5.2m (17ft) Hale Telescope entered service on Mount Palomar, California. In 1993, the first Keck Telescope, with a 10m (33ft) mirror made of a mosaic of smaller segments, came into service. Coupled with this progressive increase in telescope size was a massive increase in the sensitivity of detectors. Photographic plates became more sensitive and were used extensively for mapping the sky and producing permanent records of stellar spectra and positions. Another crucial development was radio astronomy. Radio telescopes, the first of which was made in 1937, work by picking up long-wave radio emissions from deep space.

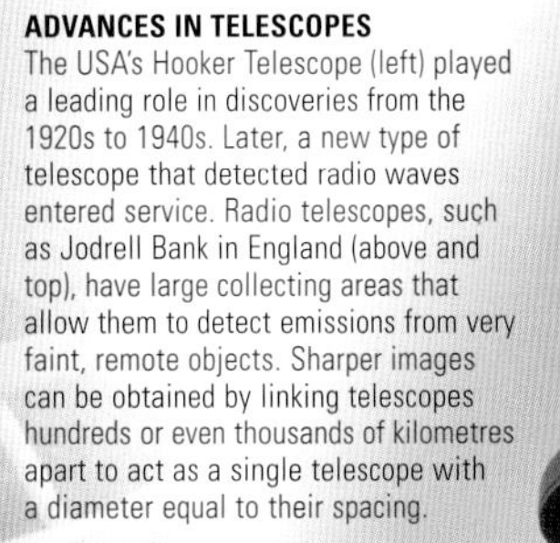

**ADVANCES IN TELESCOPES**
The USA's Hooker Telescope (left) played a leading role in discoveries from the 1920s to 1940s. Later, a new type of telescope that detected radio waves entered service. Radio telescopes, such as Jodrell Bank in England (above and top), have large collecting areas that allow them to detect emissions from very faint, remote objects. Sharper images can be obtained by linking telescopes hundreds or even thousands of kilometres apart to act as a single telescope with a diameter equal to their spacing.

**HALE TELESCOPE**
Like all large 20th-century telescopes, the Hale (visible through its open dome) was built at altitude so that atmospheric absorption and variability has less effect on image quality.

# SPACE-AGE ASTRONOMY

44–55 The beginning and end of the Universe
Beyond Earth 108–111
The interstellar medium 228
Stellar end points 262–63
Extra-solar planets 290–91
Beyond the Milky Way 292–325

WITH THE START of the Space Age (the era of spaceflight) in the 1950s, travel beyond Earth and contact with planets, comets, and asteroids became a reality (see pp.98–111). The dawning of radio, infrared, ultraviolet, X-ray, and gamma-ray astronomy offered humankind new perspectives on the Universe, revealing the existence of new and exotic objects that previously were not even imagined.

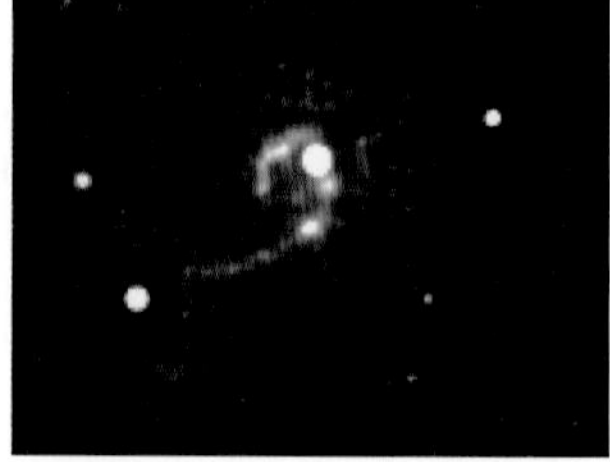

**QUASAR**
The discovery and recognition of quasars, such as HE1013-2136 above, caused a revolution in our understanding of the distant, early Universe.

## EXTREME STARS

Relationships between the mass, size, and luminosity of stars first became apparent from the Hertzsprung–Russell diagram (see p.230) around 1911, and led to the recognition of giants and dwarfs. Since then, many types of extreme stars have been found. In 1915, W. S. Adams identified Sirius B (see p.264) as the first known white dwarf – a star of the Sun's mass squeezed into the Earth's volume. In 1931, Indian astrophysicist S. Chandrasekhar, using new models for the behaviour of subatomic particles, discovered an upper limit to a white dwarf's mass – 1.4 solar masses. Above this, an expiring stellar core will collapse into a superdense neutron star just a few kilometres across, blowing apart the rest of its star in a supernova explosion (see p.262). Spinning neutron stars are seen from Earth as pulsating radio sources: astronomers Jocelyn Bell-Burnell and Antony Hewish in the UK discovered the first radio pulsar in 1967. A visual neutron star was found two years later. At the other extreme, recent years have seen the discovery of low-mass stars called brown dwarfs, which are too cool to generate energy by nuclear processes.

**BROWN DWARF**
This image from the Gemini North telescope shows a faint brown-dwarf companion (lower left) orbiting the Sun-like star 15 Sagittae. The dwarf is thought to have at least 48 times the mass of Jupiter.

**GEMINI NORTH**
Giant telescopes and computer technology have extended our view of the Universe. The 8m (26ft) Gemini North telescope in Hawaii is linked to a twin instrument thousands of kilometres away (see image, below).

### STEPHEN HAWKING

Despite severe physical disability due to motor neurone disease, the English theoretical physicist Stephen Hawking (b.1942) has made a great contribution to cosmology. He postulated that mini black holes formed just after the Big Bang and, by combining quantum theory and general relativity, he has shown that black holes could create and emit subatomic particles – so-called Hawking radiation.

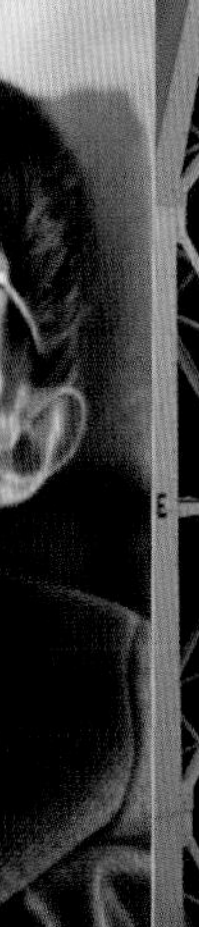
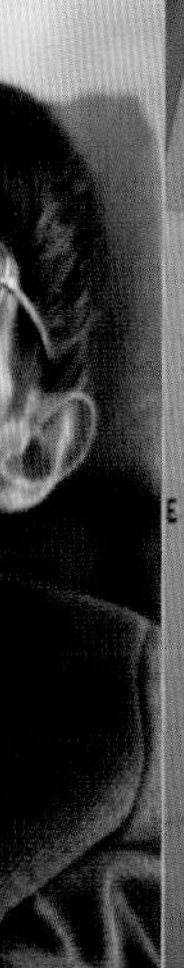
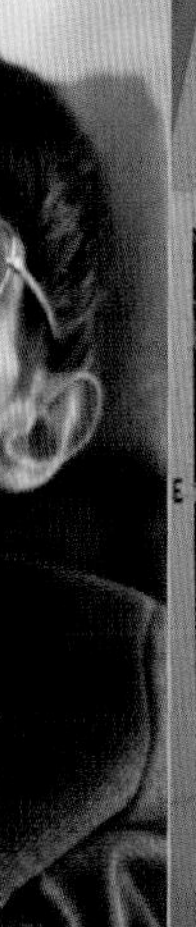

## BLACK HOLES

The possibility of black holes (see p.26) was suggested as early as 1783 by English astronomer John Michell, who considered the idea of an object so massive that its own light could not escape its gravity. The idea reappeared in 1916 as a result of Albert Einstein's theory of general relativity, but black holes remained a theoretical curiosity until the 1960s, when the launch of the first X-ray astronomy satellites led to the discovery of X-ray binary stars such as Cygnus X-1 (see p.268). X-ray binaries require a massive, compact source of energy, which could be provided only by a black hole. The discovery of stellar-mass black holes also opened the way for the acceptance of quasars – compact, highly red-shifted objects – as distant, violent galaxies powered by supermassive black holes at their cores (see p.310).

# INSIDE STARS

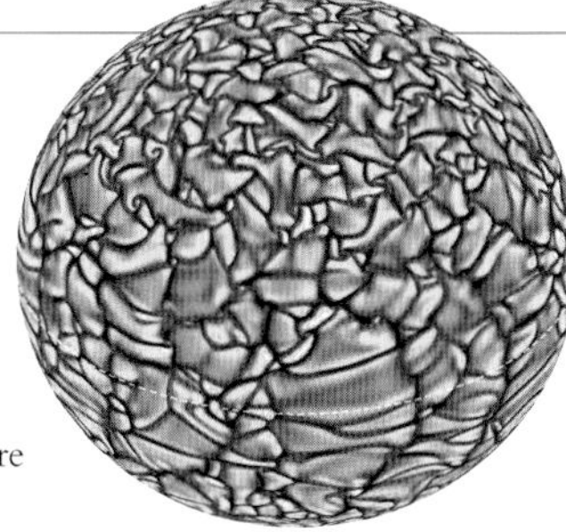
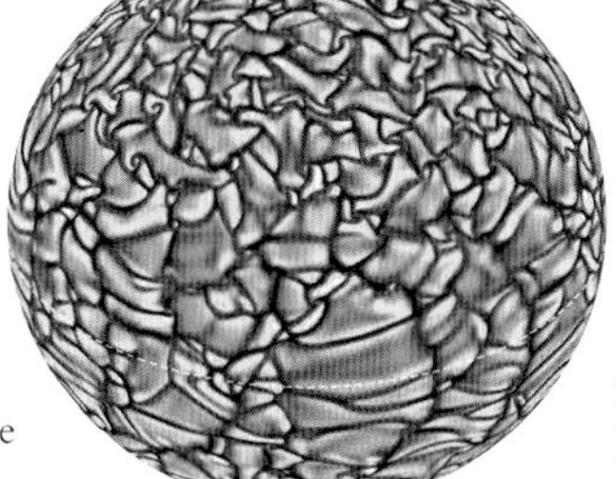

A MAP OF THE SUN
Using helioseismology, astronomers have mapped the movement of the Sun's surface, revealing huge convection cells such as these.

With optical telescopes, astronomers can see to a depth of about 500km (300 miles) below the solar surface. Unfortunately, since the Sun has a radius of 700,000km (430,000 miles), a huge volume cannot be observed directly. During the 1920s, astrophysicists calculated that the solar centre has a temperature of about 15 million °C (27 million °F) and a density around 150 times that of water. In 1939, German-born Hans Bethe showed how nuclear processes acting under these extreme conditions would slowly convert hydrogen into helium, releasing huge amounts of energy by destroying mass. English and German-born astronomers Fred Hoyle and Martin Schwarzschild extended this model in the 1950s, showing how helium was transformed into carbon and oxygen in giant stars. Soon astrophysicists had found mechanisms for the manufacture of even heavier elements, such as cobalt and iron, in the most massive stars. In the 1960s, the first neutrino detectors were used to detect subatomic particles released by nuclear fusion in the Sun, and since the 1970s, the technique of helioseismology has allowed astronomers to monitor the way that sound waves resonate through the solar interior, helping to establish its detailed structure.

SOLAR FURNACE
Studies of solar sound waves have revealed that they may drive spicules – tendrils of hot gas that rise out of the Sun's surface, then fade and fall back within a few minutes.

# THE INTERSTELLAR MEDIUM

The discovery of large quantities of gas and dust between the stars (see p.228) has been a triumph of radio astronomy. In 1944, Dutch astronomer Hendrick van de Hulst predicted that interstellar hydrogen would emit radio waves with a distinctive 21cm wavelength, and this was confirmed in 1951 by American physicists Harold Ewen and Edward Purcell. Soon radio telescopes were being used to map the distribution of this neutral atomic hydrogen in the arms of our galaxy and others. From 1974, surveys of the 2.6mm wavelength, which acts as a tracer for molecular hydrogen, led to the discovery of huge molecular clouds, the birthplaces of stars.

IMAGE PROCESSING
Computers can reveal hidden detail, such as structure in faint gas clouds.

DUST AND GAS
Radio and infrared astronomy have shown that interstellar material (shown here in the Orion Nebula) is common throughout the Galaxy, while atomic physics has explained its characteristic effects on light.

GEMINI SOUTH
The huge Gemini South optical/infrared telescope in the Chilean Andes has a perfect view of southern skies. Computer-controlled actuators constantly adjust the shape of its mirror to account for distortions caused by its great weight.

## ASTRONOMY FROM SPACE

The Space Age started on 4 October 1957, when the USSR launched their Sputnik 1 satellite (see p.99), and astronomers were quick to take advantage of the ability to observe from beyond Earth's atmosphere. Rocket-borne detectors had already picked up intriguing signals at unusual wavelengths during their brief trips out of the atmosphere, and the first orbiting observatory, Ariel 1, was launched by Britain in 1961, equipped with an ultraviolet telescope. Other satellites, such as the US Explorer and Uhuru series, swiftly mapped the major sources of ultraviolet, infrared, and X-ray radiation. Meanwhile, space probes spread out across the Solar System, returning information about the interplanetary environment and charting the planets with a variety of cameras, radar mappers, and other instruments.

**TITAN FROM ABOVE**
This view of the surface of Saturn's cloud-wrapped moon, Titan, was returned by the Huygens lander during its descent in January 2005.

**MAPPING WITH RADAR**
The Magellan probe orbited Venus for three years in the early 1990s and surveyed its surface with radar, charting its elevation, surface roughness, and reflectivity, as seen in this exaggerated image.

## EXTRA-SOLAR PLANETS

Great improvements in spectroscopy have enabled astronomers to measure slight variations in stellar velocities produced by the gravitational influence of orbiting planets. In 1995, Swiss astronomers Michel Mayor and Didier Queloz of the Geneva Observatory discovered a planet with a mass just under half that of Jupiter, orbiting the star 51 Pegasi, 48 light-years away. Hundreds of extra-solar planets are now known, and a whole new field of astronomy has opened. Astronomers have been surprised by their findings – the majority of the new planetary systems are very different from the Solar System. Some have giant planets orbiting very close to a parent star, while others have planets in orbits that are highly elliptical. Systems like our own seem to be rare.

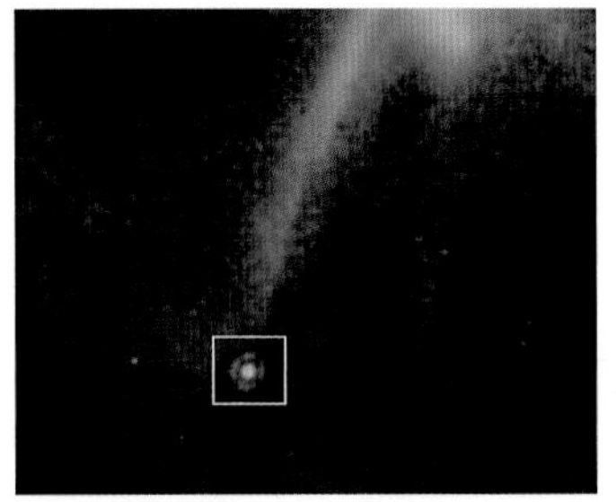

**POSSIBLE PLANET**
This Hubble Space Telescope photograph of a faint object orbiting a young binary star system, 450 light-years away in Taurus, was hailed in 1998 as the first image of an extrasolar planet. Alternatively, it could be a small brown dwarf.

## THE ORIGIN OF THE UNIVERSE

The discovery during the 1920s that the Universe is expanding triggered a new wave of cosmological thinking. In 1931, Belgian priest and astronomer Georges Lemaître suggested that all the material in the Universe had started as a single, highly condensed sphere – the origin of the Big Bang theory (see p.46). In 1948, the Austrian Thomas Gold, and the Britons Hermann Bondi and Fred Hoyle, proposed a rival theory of continuous creation, in which material was being created all the time to fill in the gaps left by the expansion. Fortunately, the theories could be tested by observation, and the evidence has come down on the side of the Big Bang. In 1980, Alan Guth at Stanford University, USA, extended the Big Bang cosmology by introducing inflation, helping to resolve several of the major problems with the theory. However, some important questions are still to be answered.

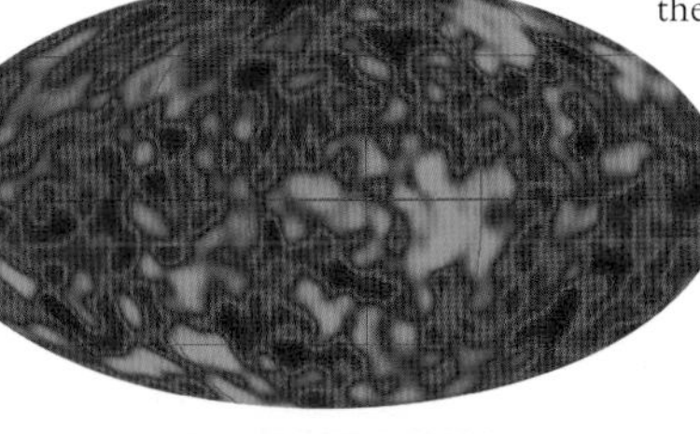

**MICROWAVE BACKGROUND RADIATION**
The discovery in 1964 of radiation from the Big Bang, cooled by time and red-shifting to just above absolute zero, was the clinching evidence for the Big Bang.

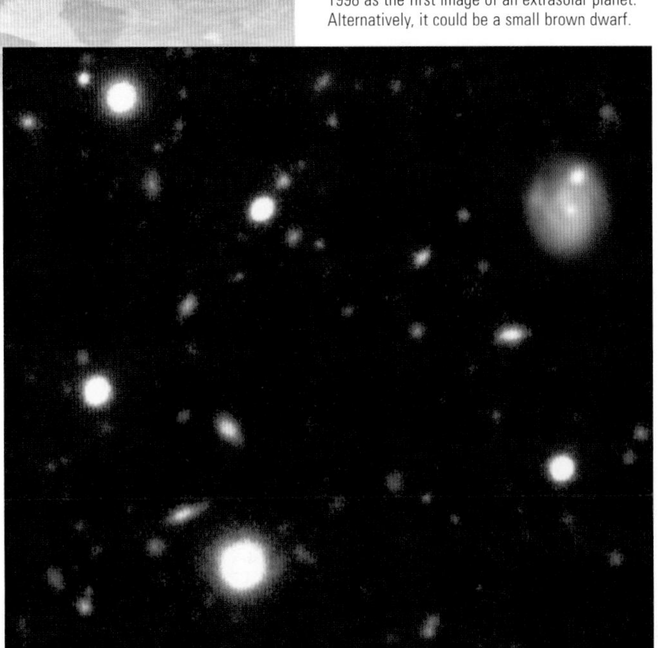

**HUBBLE DEEP FIELD**
Images of the faintest, most distant, and earliest galaxies, such as this, coupled with radio surveys of distant galaxies, have shown that galaxies were more densely packed in earlier times – powerful evidence in favour of the Big Bang.

**RADIO RECEIVER**
The central platform of the Arecibo radio telescope hangs some 140m (460ft) above the dish. It weighs around 900 tons and houses the telescope's receiver equipment.

**ARECIBO**
The Arecibo radio telescope in Puerto Rico is the world's largest, some 305m (1,000ft) across. An icon of modern astronomy, it has also been used to beam signals to the stars (see p.53).

# THE HUNT GOES ON

The more astronomers learn about the Universe, the more there is to learn. In the 1970s, it was recognized that the Universe contains much more material than is, as yet, visible: dark matter affects the rotation of galaxies; missing mass is predicted by the Big Bang theory but not yet observed; and exotic "dark energy" appears to control the expansion rate of the Universe (see p.54). In contrast to these grand cosmological mysteries stands the more immediate, yet just as significant, search for life in the Universe (see pp.52–53), encompassed in the new science of astrobiology. Over the last few centuries, the size of the largest telescope mirrors has doubled approximately every 40 years, putting more and more of the Universe within their grasp. Soon today's 10m (33ft) mirrors may be replaced by even larger giants, and observations made from above the Earth's atmosphere should become ever more productive. The 2.5m (8.2ft) Hubble Space Telescope will be dwarfed by the 6.5m (21.3ft) mirror of the James Webb Space Telescope, due for launch in 2011. This new generation of telescopes may finally put the very first generation of stars and galaxies clearly within our view. And as astronomers look ever farther with ever-more sensitive instruments, they will inevitably be confronted with new mysteries and exotic bodies that cannot yet be imagined.

**LIFE FROM MARS?**
In 1997, NASA scientists suggested that a meteorite originating from Mars held chemical traces of ancient life, and perhaps even microscopic fossils, within it. Their findings are still controversial.

**MARS ROVER**
Robot vehicles on the surface of Mars have proved that it once had substantial oceans, bolstering the theory that life might have evolved there, too.

# EARLY SPACECRAFT

36–37 Gravity, motion, and orbits
Journeys to the Moon 102–105
Putting space to work 106
Atmosphere and weather 140

THE LATE 20TH CENTURY saw a revolution in our understanding of the Universe, triggered by the development of spaceflight. For the first time, instead of just looking into space, humans and their machines travelled in it. The early days of space exploration were filled with setbacks and hazards, but once these obstacles were overcome, progress was rapid.

## ROCKET DREAMERS

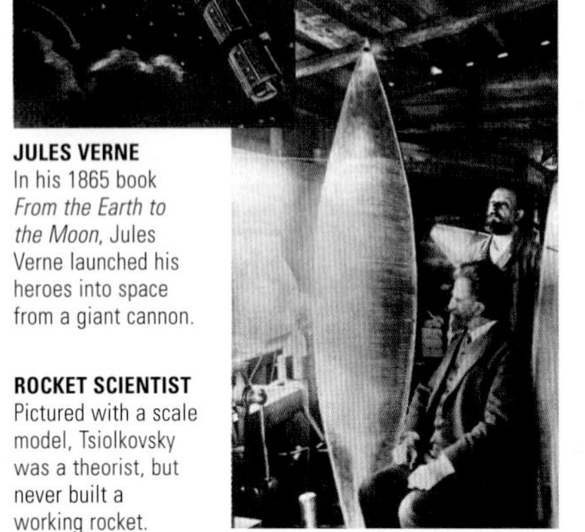

**JULES VERNE**
In his 1865 book *From the Earth to the Moon*, Jules Verne launched his heroes into space from a giant cannon.

**ROCKET SCIENTIST**
Pictured with a scale model, Tsiolkovsky was a theorist, but never built a working rocket.

The idea of space travel is as old as storytelling itself, but with little idea of the laws of physics or the nature of space, fantasists often relied on absurd and comical means to carry their fictional space travellers beyond Earth. However, Newton's laws of motion and universal gravitation, coupled with the fact that space is a vacuum, meant that only one form of propulsion was truly capable of carrying people through space – the rocket. First developed as fireworks and weapons in medieval China, rockets contain the fuel needed for propulsion, and are pushed forward by the exhaust gases from their engines. Although French writer Cyrano de Bergerac (1619-55) had suggested using rockets to reach the Moon in the 17th century, the first person to seriously consider the realities of space travel was Russian schoolteacher Konstantin Tsiolkovsky (1857-1935), who worked out many of the principles of liquid-fuelled, multi-stage rockets, and published them in the 1890s.

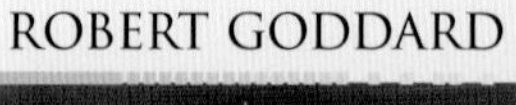

### ROBERT GODDARD

Physics professor Robert Goddard (1882–1945) braved ridicule to prove the rocket's potential for space travel. He began to develop signal rockets for the US Navy before the First World War, but did his most important work in the 1920s and 1930s, developing the first liquid-fuel rockets, along with stabilizing and navigation systems, and even rocket-borne cameras.

## LIFT-OFF

The principles of rocketry were well established by the early 20th century, but there were major problems in turning spaceflight into a practical reality. The greatest was fuel efficiency. American engineer Robert Goddard (see panel, above) had the idea of using efficient liquid fuels, and in 1926 he launched a 3m- (10ft-) tall rocket fuelled by liquid oxygen and gasoline. Goddard and Tsiolkovsky's ideas were adopted by enthusiasts worldwide, including the Society for Space Travel (VfR) in Germany. During the 1930s, many of the VfR's scientists were recruited by the Nazi government into a military programme that continued through the Second World War. The culmination of their efforts, the V2 rocket-propelled missile, came too late to save Germany from defeat, but proved that rocket weapons were the way of the future.

**GERMAN ROCKETRY**
German rocket theorist Hermann Oberth (left), stands alongside members of the VfR and a prototype rocket.

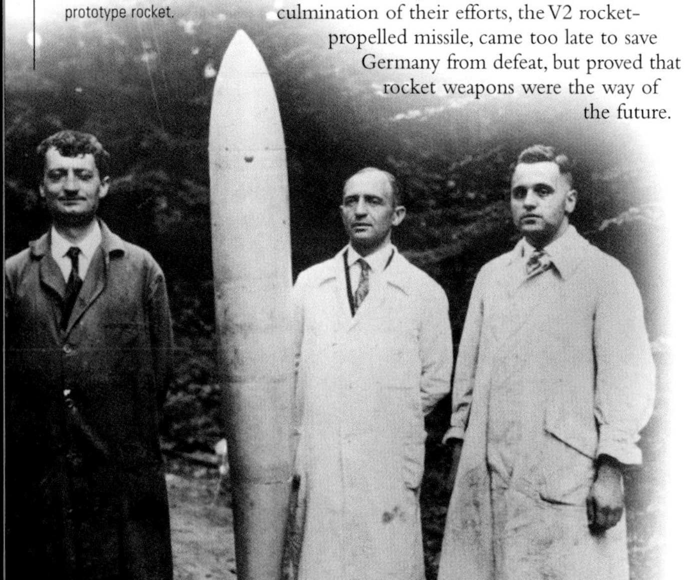

## THE SPACE RACE

In the aftermath of the Second World War, both major powers wanted rocket technology for themselves. Most of the scientists fled to the West, but the Soviet Union captured the V2 factories. The reason for such interest became clear as relations rapidly deteriorated and the Cold War began. Both sides believed that rocket-propelled inter-continental ballistic missiles (ICBMs) would be the answer to delivering the ultimate weapon – a nuclear warhead. It was in this context that the US and the Soviet Union began their space programmes. Both sides saw that missile technology could be used to reach Earth's orbit and they recognized that such launches would demonstrate the power of their rockets and provide great propaganda benefits.

**WHITE SANDS**
In the late 1940s, US scientists launched captured V2s from White Sands Missile Range in New Mexico.

**COLD WARRIORS**
Soviet rockets were often displayed in military parades – a clear reminder of their primary use as ICBMs.

**ADAPTABLE WEAPON**
The V2 rocket was all the more terrifying because of its mobility in the field. V2s could be transported by road, set up, and launched within hours, making launch sites difficult to strike.

## INTO ORBIT

In the early 1950s, both the US and the Soviet Union announced plans to launch satellites during the International Geophysical Year of 1957–58. The Soviet programme was able to proceed in total secrecy, using its massive R-7 ballistic missiles as launch vehicles, while the American scientists suffered under public scrutiny. German scientist Wernher von Braun's plans to launch a satellite using the US Redstone ICBM were shelved, in favour of the US Navy's research rocket programme Vanguard. As the Vanguard programme approached a November launch target, the Soviet promises were all but forgotten, until, on 4 October 1957, they announced the successful launch of Sputnik 1 into orbit. Tracking stations soon picked up its radio signals and confirmed that the Soviet Union had seized the lead in the Space Race. US humiliation was compounded on 6 December, when a Vanguard launch ended in a fire on the launch pad. Von Braun's project was immediately resurrected, and the first US satellite, Explorer 1, successfully entered orbit less than two months later, on 31 January 1958.

**SOVIET PROPAGANDA**
The Soviet Union was quick to capitalize on its lead in the Space Race. Propaganda posters hailed Soviet superiority over the West, while Americans were left to fear the implications of their rivals' mastery of space.

**SPUTNIK 1**
The first Soviet satellite was little more than a metal ball with antennae, a radio transmitter, and batteries, but its simple radio signal shook the world.

**DOG IN SPACE**
A month after Sputnik 1 was launched, on 3 November 1957, the Soviet Union launched Sputnik 2, a more sophisticated satellite carrying a passenger – a dog called Laika. With no means of returning the satellite to Earth, she was to die in orbit.

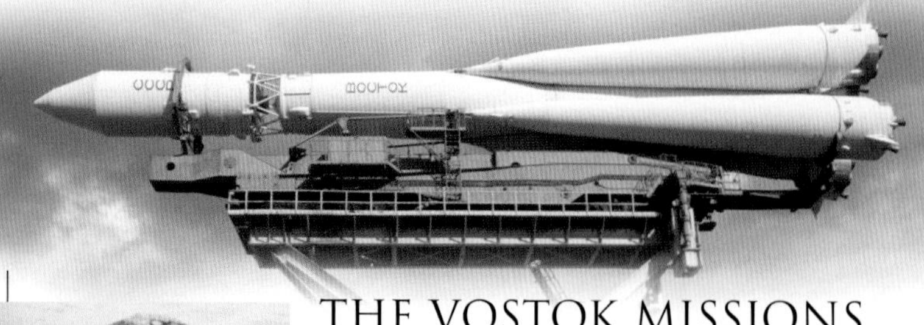

# THE VOSTOK MISSIONS

By the end of the 1950s, the next great target of the Space Race was clear – which of the superpowers would be the first to put a person into orbit? The Soviets had an obvious power advantage, as Sputnik 2, their second satellite, already weighed half a tonne, and they had managed to send probes beyond Earth's immediate neighbourhood. On 12 April 1961, they announced to a shocked world that Colonel Yuri Gagarin had become the first man in space, aboard Vostok 1. Gagarin returned to Earth a hero of the Soviet Union, and the US space programme, now managed by NASA, was again left trailing behind. Later Vostok missions broke new ground by putting the first woman into space, increasing orbit times, and even putting more than one spacecraft in orbit at the same time.

**VOSTOK 1**
Although it was not revealed at the time, Gagarin ejected from Vostok 1's re-entry capsule (shown here) and parachuted to the ground.

## YURI GAGARIN

Born near Smolensk, Russia, Yuri Gagarin (1934–68) joined the Soviet Air Force and became a fighter pilot after graduation from college. Gagarin's first words in space were "I see Earth – it's so beautiful!" Hailed as a hero on his return from space, Gagarin fell into depression and drink problems. He died in a jet crash in March 1968.

**VOSTOK 2**
The second Vostok mission, launched on 6 August 1961, saw Gherman Titov spend an entire day in orbit.

## SERGEI KOROLEV

Sergei Pavlovich Korolev (1906–66) was a Ukrainian-born engineer and mastermind of the early Soviet space programme. While training as an aircraft designer and working on jet-engine research, he built and launched the first Soviet liquid-fuelled rocket in 1933. After working on the Russian equivalent of the V2 during the Second World War, Korolev was put in charge of the new space programme, designing the first Russian ICBM (used to launch Sputnik), the Vostok rocket, and the Vostok, Voskhod, and Soyuz spacecraft.

**ABOVE THE EARTH**
US astronaut Ed White became the first American to walk in space, in June 1965, during the Gemini 4 mission (see p.102). Though tethered to the spacecraft, White also carried a manoeuvring "gun" capable of firing brief jets of nitrogen to push himself around.

# PROJECT MERCURY

NASA's answer to Vostok was the Mercury programme, which launched six astronauts into space between 1961 and 1963. Unlike the Soviet programme, American efforts were carried out in the full glare of publicity. The Mercury space capsules were tiny and lightweight, partly because they had to be carried by the small Redstone rockets during early flights. NASA was developing a new launch vehicle, the Atlas, to launch Mercury, but the capsule had to begin testing before the larger rocket was ready. Stung by the surprise launch of Vostok 1, NASA raced to retaliate and successfully put their first man in space on 5 May. Launched on a Redstone rocket, Alan Shepard's (see p.104) Freedom 7 mission did not have sufficient speed to go into orbit, but it did reach a maximum altitude of 185km (115 miles) during its 15-minute flight. After a second sub-orbital flight, the Atlas rocket was finally ready in late 1961. After a number of tests, John Glenn became the first American in orbit on 20 February 1962.

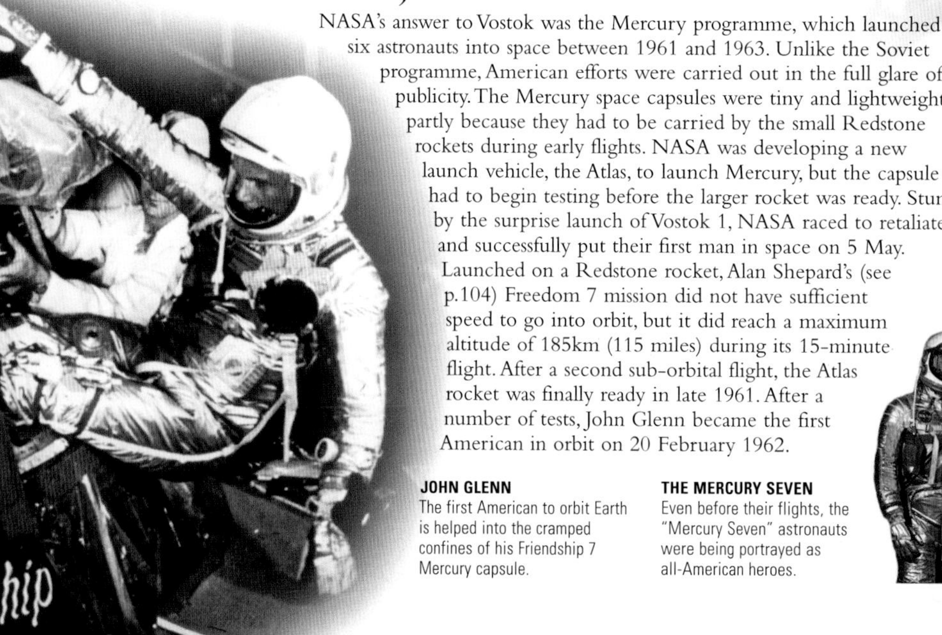

**JOHN GLENN**
The first American to orbit Earth is helped into the cramped confines of his Friendship 7 Mercury capsule.

**THE MERCURY SEVEN**
Even before their flights, the "Mercury Seven" astronauts were being portrayed as all-American heroes.

## NEXT STEPS

Eager to maintain its lead in the Space Race, the Soviet Union now took a major risk. The Americans had already announced their planned two-man Gemini missions (see p.102), and in an effort to pre-empt them, Soviet spaceflight director Korolev (see panel, opposite) planned to launch a three-man mission. This was a big challenge, as work had already begun on the Soyuz spacecraft intended to go to the Moon. In the end, the Soviet engineers came up with an ingenious compromise. Voskhod was effectively a modified Vostok capsule with just enough room to carry a crew of three. The flight of Voskhod 1 on 12 October 1964 was a success, beating the first crewed Gemini missions by five months. Voskhod 2, launched days before the first Gemini test flight, was even more successful. During the flight, Alexei Leonov became the first person to walk in space, and the Soviets snatched yet another propaganda coup.

**BEYOND VOSTOK**
Cosmonauts Pavel Belyaeyev and Alexei Leonov, attired in their spacesuits, inside the tight confines of the Voskhod 2 capsule.

**WALKING IN SPACE**
This historic sequence of images captures Alexei Leonov's space walk of 18 March 1965. Leonov spent about 20 minutes floating in space, tethered to the Voskhod spacecraft.

## ROBOT EXPLORERS

While public attention was largely focused on the manned space programmes, a second Space Race was continuing in parallel – one that would have important consequences for our understanding of the Solar System. Each superpower tried to outdo the other with new "firsts" in the exploration of other worlds. Robotic space exploration had a patchy early history, with numerous failures, either on launch pads, en-route to targets, or during attempted landings. Notable successes included the Soviet Lunik (or Luna) 1, which in January 1959 became the first object to escape Earth's gravity and enter orbit around the Sun; Lunik (Luna) 2, which hit the Moon in September of that year; and Lunik (Luna) 3, which returned the first photographs of the Moon's far side. NASA's Pioneer 5 became the first probe deliberately sent into interplanetary space, entering an orbit between Earth and Venus in 1960, while the US Mariner probes 2 and 4 successfully beat Soviet efforts to reach the other planets, flying past Venus and Mars in 1962 and 1965 respectively.

**MOON LANDER**
The US Ranger missions crashed into the Moon, sending back pictures up to the moment of impact.

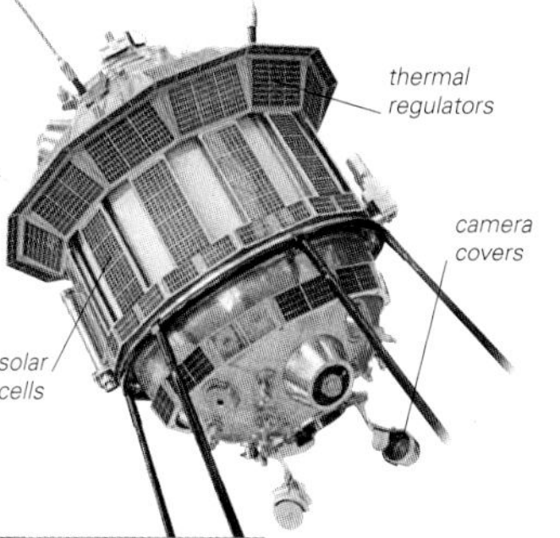


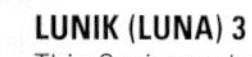

**LUNIK (LUNA) 3**
This Soviet probe used a television camera to return the first pictures of the Moon's far side to Earth.

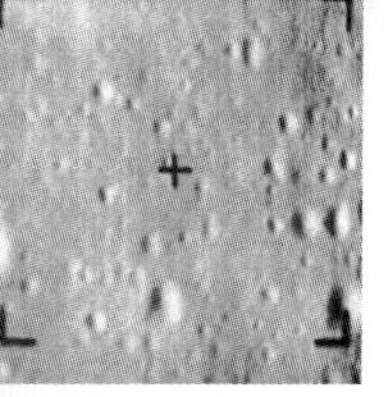

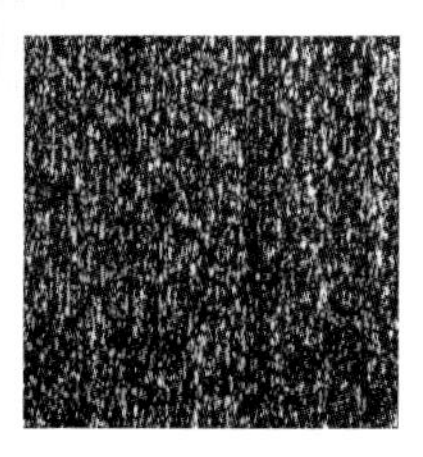

**MOON'S SURFACE**
These unique images of the Moon's surface were captured in the moments just before the US Ranger 8 probe crash-landed.

# JOURNEYS TO THE MOON

36–37 Gravity, motion, and orbits
98 The space race
101 Next steps
In Earth's orbit 106–07
The Moon 148–59

THE APOLLO MISSIONS to the Moon have often been described as humankind's greatest technical achievement. Vast quantities of manpower and resources were invested into a programme motivated as much by propaganda as science. The missions revealed much about the Moon, and their technological spin-offs helped shape the modern world.

## A MAN ON THE MOON

On 25 May 1961, US President John F. Kennedy made an announcement that shocked the world. At a time when the United States had still to put an astronaut in orbit around the Earth, he pledged that they would land people on the Moon by the end of the decade. The scale of the task was daunting. Astronauts had ventured little more than 300km (190 miles) from the Earth's surface, and now Kennedy gave NASA the task of sending them some 400,000km (250,000 miles), landing them on the surface of an unknown world, and returning them safely. However, if it could be done, it would send a powerful message that the United States was now the superior space power. NASA immediately began to investigate possible ways to land on the Moon. The mission design they settled on would involve using the largest rocket ever built, designed by German rocket scientist Wernher von Braun (see panel, below), to send three linked spacecraft to the Moon – only one of which would make it back to Earth. The name Apollo, taken from the Greek god of the Sun, was suggested by NASA spaceflight director Dr Abe Silverstein.

**"BEFORE THIS DECADE IS OUT"** President Kennedy lays down his historic challenge before the Houses of Congress.

**APOLLO ROCKET** The Apollo programme called for the construction of the most powerful rocket ever built – the three-stage Saturn V.

### WERNHER VON BRAUN

Fascinated by spaceflight from childhood, Wernher von Braun (1912–1977) joined the German VfR rocket society in his teens. After working on V2 rockets, von Braun fled west at the end of World War II, and found work in the US rocket programme. Von Braun's rockets were used for both the Apollo programme and the early American satellite launches.

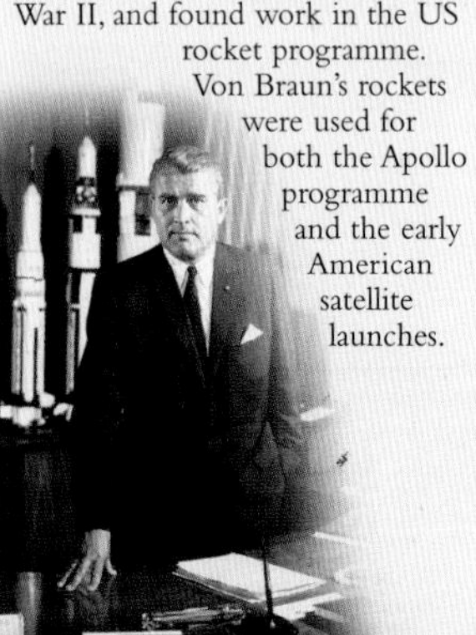

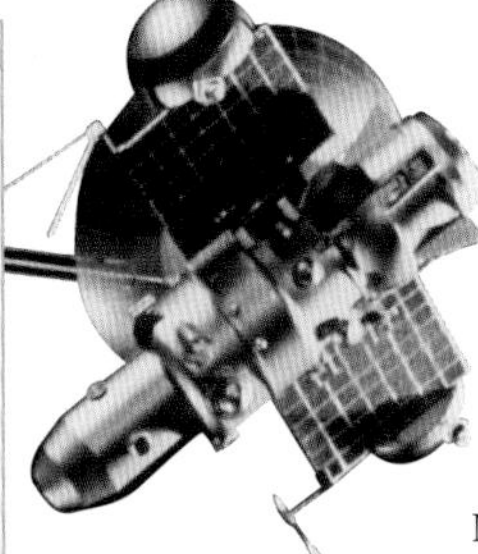

**SOVIET SETBACKS** Although officially denied, the Soviets had a secret Moon programme that was eventually intended to include manned missions. However, problems with their launch vehicles and spacecraft (based on this Zond 3 vehicle) soon saw them trailing behind.

## PROJECT GEMINI

Each Apollo mission would involve a number of rendezvous, docking, and undocking manoeuvres in space – operations that NASA and its astronauts had never tried before. It would also require a minimum of seven days just to get to the Moon and back. To gain experience in longer-duration spaceflight and to try out the delicate operations needed by Apollo, NASA announced that the Mercury programme would be cut short, and immediately replaced with a series of two-man missions called Project Gemini. In total, there were 10 crewed Gemini missions between 1964 and 1966. A number of these involved rendezvous between spacecraft in orbit, space walks, and even dockings with unmanned Agena target vehicles. The spacecraft, composed of three modules, was also a major step forward. While the two astronauts spent the mission inside a re-entry module just 50 per cent bigger than a Mercury capsule, their air and water supplies, electrical equipment and experiments were mostly kept in a separate equipment module. A third retrograde module contained rocket engines for manoeuvring the spacecraft in orbit and slowing it down prior to re-entry.

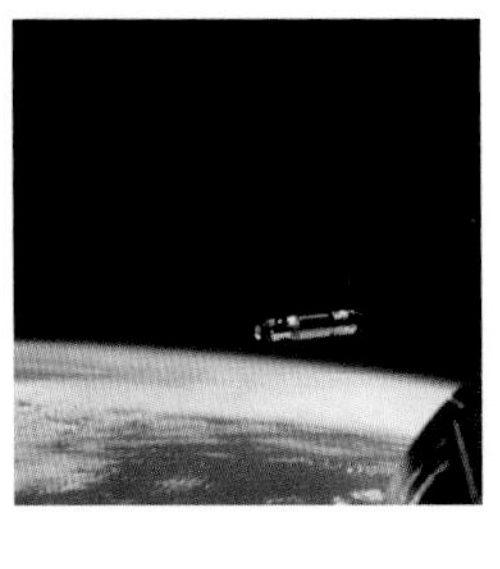

**DOCKING REHEARSALS** This sequence of pictures shows the rendezvous between a Gemini spacecraft and an unmanned Agena docking target.

**LUNAR ORBITER**
This spacecraft made the first comprehensive surveys of the Moon, ending the reliance on lunar maps compiled by Earth-based telescopes.

## PAVING THE WAY

A key challenge for the Apollo planners was a lack of knowledge about the Moon itself. At the time, the history of the Moon, the nature of its craters, and the properties of its surface were little understood – there were serious worries that the lunar soil might be so powdery that spacecraft would sink into it. To answer such concerns, NASA planned a series of robot spacecraft that would survey the Moon in detail and attempt both crash- and soft landings. The first wave were the Ranger spacecraft, four of which crashed into the Moon in 1964 and 1965. Sending back photographs up to the moment of impact, the probes confirmed that the Moon was heavily cratered, at even the smallest scales. This implied that the craters were caused by impacts. In 1966 a second phase began, with the Lunar Orbiter and Surveyor spacecraft. The Orbiters photographed the Moon from as close as 40km (25 miles), looking for interesting sites for the manned landings, while the Surveyors made a series of soft landings, establishing that the lunar surface was stable.

### JIM LOVELL

After serving as a naval test pilot in the 1950s, Jim Lovell (b.1928) joined NASA's astronaut programme and flew as commander on the Gemini 7 and 12 missions, as well as Apollo 8. Selected as a reserve commander for the Apollo 11 lunar landing, Lovell is most famous as commander of the Apollo 13 mission, which survived a mid-flight accident to make it back to Earth intact (see p.105).

## APOLLO IN ORBIT

By late 1966, the Apollo programme was moving ahead rapidly. The enormous Saturn V rockets were under construction, and the spacecraft were ready for testing. But tragedy struck in January 1967, when the crew of Apollo 1 were killed by a fire in their capsule during a launch rehearsal. In the aftermath, Apollo missions 2 and 3 were cancelled, while missions 4, 5, and 6 were reassigned to automated test launches. It was not until October 1968, with the launch of Apollo 7, that NASA astronauts returned to space. This mission into Earth's orbit was swiftly followed by Apollo 8. Launched for the first time by a Saturn V rocket, this spacecraft's mission was altered to include an orbit around the Moon over Christmas, after NASA heard rumours that a Soviet launch to send a manned vehicle around the Moon was imminent. The rival launch never happened, and after Apollo 9 and Apollo 10, which tested all the mission equipment in lunar orbit, NASA was finally ready to attempt a landing on the Moon.

**EARLY TRAGEDY**
The Apollo 1 mission was tragically aborted after a fire in the oxygen-filled capsule killed the three crewmen.

**TO THE MOON**
The first Saturn V rocket blasts free of its launch tower, marking the beginning of the Apollo 8 mission.

**TOUCHING DISTANCE**
The "Charlie Brown" command module speeds over the lunar surface during the Apollo 10 mission. This picture was taken from the lunar module "Snoopy" during a test separation.

# THE FIRST MOON LANDING

**TRIP TO THE MOON**
Neil Armstrong, Michael Collins, and Buzz Aldrin board a van for the 13km (8-mile) trip to the launch pad at the start of their epic voyage to the Moon.

Apollo 11 blasted off from Cape Kennedy (now Cape Canaveral) in Florida on 16 July 1969 and entered orbit around the Moon three days later. Neil Armstrong and Edwin "Buzz" Aldrin then climbed aboard the lunar module "Eagle" for the descent to the surface, leaving Michael Collins aboard the command and services modules (CSM) "Columbia" in lunar orbit. Eagle touched down safely on the lava plain known as the Mare Tranquillitatis (Sea of Tranquillity), and six hours after landing, Neil Armstrong left the module and stepped down the ladder, touching the surface at 2.56am on 20 July. Armstrong and Aldrin stayed on the surface for some 21 hours, completing one moonwalk during which they planted a flag and a commemorative plaque, set up a number of experiments, collected rock samples, and had a telephone conversation with US president Richard Nixon.

**READY FOR LAUNCH**
Huge crowds gathered in Florida to witness the launch of Apollo 11, while millions more watched on television around the world.

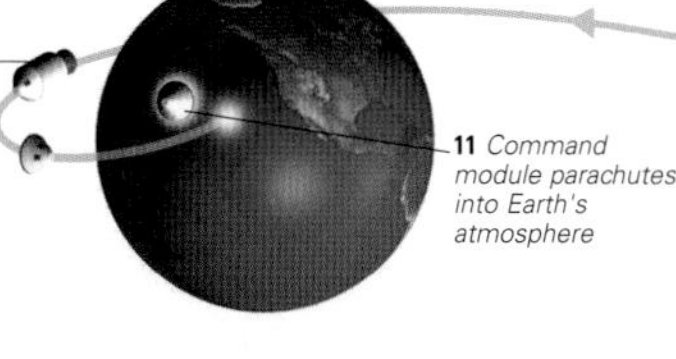

1 Saturn V launches from Florida
2 Lower rocket stages discarded
3 Engine burn puts spacecraft on course for Moon
4 CSM and lunar modules separate and re-join in orbit
5 Lunar module, last rocket stage, and CSM separate
6 Lunar module lands while CSM orbits above
7 Lunar module launches from surface
8 Rendezvous with CSM in orbit
9 Mid-course correction of CSM
10 Command and services modules separate
11 Command module parachutes into Earth's atmosphere

**MISSION PROFILE**
Each Apollo mission involved a number of complex orbital manoeuvres and dockings between the individual spacecraft. If anything had gone wrong, the astronauts would have been doomed, stranded on the lunar surface or trapped in orbit around the Moon.

## MOON MISSIONS

Over the three years following Apollo 11, NASA targeted further landings at regions of particular interest, ensuring that, by the end of the Apollo programme, astronauts had brought back samples from most types of lunar terrain.

**APOLLO 11** Landed on the Moon on 21 July 1969. This picture shows Neil Armstrong's first footprint on the Moon – a mark that could remain unchanged for longer than the human race survives.

**APOLLO 12** Landed on the Moon on 19 November 1969, in the Oceanus Procellarum (Ocean of Storms) region. Its objective was to deploy a wide range of scientific instruments to perform a detailed scientific lunar exploration.

**APOLLO 13** Launched on 11 April 1970, but mission aborted after a mid-flight accident. The astronauts returned to Earth after a 143-hour trip round the Moon.

**APOLLO 14** Landed on 5 February 1971 at Fra Mauro crater, the intended Apollo 13 landing site. It was the only mission to carry a "lunar cart" for sample collection. Apollo 14 commander Alan Shepard famously hit two golf balls on the Moon.

**LUNAR ROVER**

**APOLLO 15** Landed on 30 July 1971 near Hadley Rille, a volcanic feature. This was the first mission to carry a lunar roving vehicle to extend the range of exploration.

**APOLLO 16** Landed on 21 April 1972 near the Descartes Crater in the previously unvisited lunar highlands region, helping to resolve many of the outstanding questions about their origin.

**APOLLO 17** Landed on 11 December 1972 near the Taurus–Littrow valley. In this picture, Eugene Cernan, the last man on the Moon, salutes the US flag.

SPECIAL EDITION
LIFE
TO THE MOON AND BACK

**ONE SMALL STEP**
This magazine cover features a famous photograph of Buzz Aldrin on the Moon (Neil Armstrong is reflected in his spacesuit's visor).

**WALKING ON THE MOON**
Buzz Aldrin is pictured tending to the passive seismic experiment (a moonquake detector) deployed near the Apollo 11 landing site.

## APOLLO 11 CREW

Neil Armstrong, Buzz Aldrin, and Michael Collins (each born in 1930) were all experienced test or combat pilots before joining NASA's astronaut programme. Each had his first spaceflight during the Gemini missions – Collins and Aldrin as pilots, Armstrong as a commander. Lauded as heroes on their return from the Moon, each went on to a successful career beyond NASA.

**LUNAR ROVER**
The lunar roving vehicle travelled at up to 18kph (11mph). Apollo 17 commander Eugene Cernan described it as "the finest machine I ever had the pleasure to drive".

## LUNAR MODULE

The rickety-looking lunar module was the most vital element of each Apollo mission. Since it would never have to fly in an atmosphere, the designers were free to choose a strictly functional shape. Although the module was 9.5m (31ft) wide and 7m (23ft) tall, the cabin was still so cramped that the astronauts had to stand up during flight. A large rocket engine on the underside of the upper section was used to slow the capsule as it approached the lunar surface, and also provided the thrust to blast the module away from the surface and into lunar orbit.

**EN ROUTE TO THE MOON**
An Apollo lunar module is pictured above the Earth during the complex redocking operation.

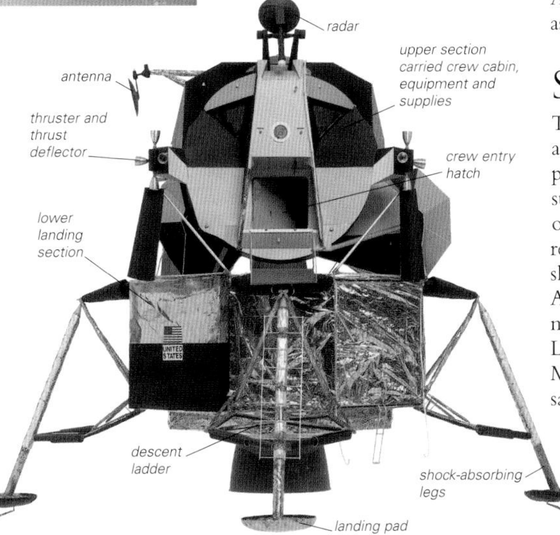


## LATER MISSIONS

NASA originally planned ten Apollo lunar landings, but only six were completed. Although the Apollo 12 was a success, Apollo 13 became famous for all the wrong reasons, when electrical failure and loss of oxygen, for a time, put the crew in grave danger. The last three Apollos carried a lunar roving vehicle, considerably extending the area that astronauts could explore. Fading public interest and NASA budget cuts led to the cancellation of the last three planned missions. The last gasps of Apollo were the Skylab space station (see p.106), which used a Saturn V rocket, and the Apollo–Soyuz mission, a rendezvous between American and Soviet astronauts in Earth orbit.

**MISSION BADGE**
Apollo 13's mission badge depicts a chariot pulling the Sun across the sky.

**APOLLO–SOYUZ**
Apollo astronaut Donald "Deke" Slayton (left) and cosmonaut Alexei Leonov enjoy their historic orbital rendezvous.

## SCIENCE LESSONS

The Apollo missions taught astronomers a great deal about the chemistry and history of the Moon (see pp.148–49), and the rocks collected are still being studied around the world. The record of cratering on the Moon, and the radiometric dating of samples, revealed the period of intense bombardment that shaped the Solar System during its first billion years. Although the Soviet Union never attempted a manned lunar landing, they did put a series of Lunokhod rovers on the Moon and returned small samples of dust to Earth.

**PRECIOUS ROCKS**
The Apollo missions brought back some 380kg (840lb) of rock, and each piece is still a very valuable scientific sample.

**EXPERIMENTS**
Some of the scientific equipment left on the Moon is still functional today.

# IN EARTH'S ORBIT

36–37 Gravity, motion, and orbits
96 Looking back in time
99 Into orbit
103 Apollo in orbit
Atmosphere and weather 140

ALTHOUGH MUCH OF THE FOCUS of space exploration has been on going to distant planets and moons, the great majority of missions have gone no farther than Earth's orbit, and it is here that the advent of spaceflight has had most impact. Nearby space is now crowded with Earth-orbiting satellites with both scientific and commercial purposes.

**SOVIET STATION**
An artwork of the Salyut 6 space station shows a Soyuz module docked at the near end and a supply module at the far end.

## EARLY SPACE STATIONS

Beaten by the US in the race to the Moon, the Soviet Union changed the direction of its space programme in the late 1960s, focusing on establishing semi-permanent outposts in Earth's orbit. The early Soviet space stations of the 1970s, called Salyuts, were cylinders some 13m (43ft) long and a maximum of 4m (13ft) wide, where a crew of three lived for several weeks in cramped conditions. Although the space race was winding down, NASA still felt the need to compete and, in 1973, launched the Skylab station. Skylab was visited by three separate crews through the course of a year, and NASA briefly held the record for spaceflight duration. However, as NASA turned its attention to the Space Shuttle (see opposite), space stations were left to the Soviets. The more advanced Salyut 6 and Salyut 7 were much larger than their predecessors and could be extended with modules launched from Earth. By the mid-1980s, cosmonauts were remaining in orbit for months at a time, carrying out valuable scientific experiments during their stays.

**SKYLAB**
NASA's first space station, Skylab, floats above Earth. The longest mission to this "orbital workshop" lasted 84 days.

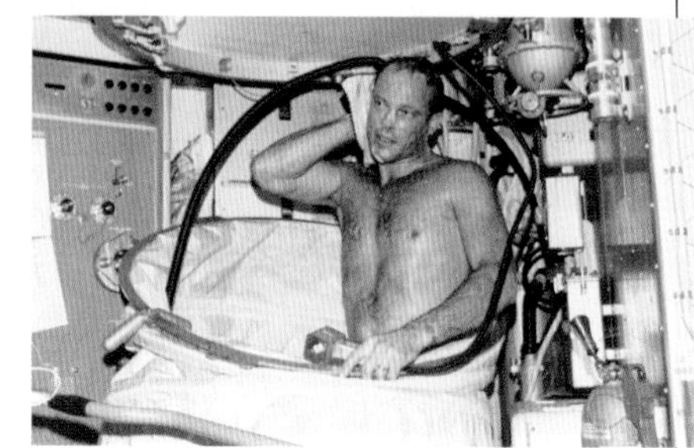

**INSIDE SKYLAB**
Despite the cramped conditions, Skylab's designers did find room for an ingenious collapsible shower, here used by astronaut Jack Lousma.

## PUTTING SPACE TO WORK

Alongside exploration for political and scientific reasons, the past few decades have seen the rise of the practical exploitation of space. Private companies and smaller governments can now afford to launch satellites, and the world has been transformed by the results. The communications satellite arose from the work of Arthur C. Clarke (see panel, below) and others, and orbiting radio transmitters are today also used in the Global Positioning System, which allows users to find their position anywhere on Earth, to within a few metres. The potential for observing Earth from orbit became apparent when the first astronauts reported clearly seeing land features from space – much to the surprise of their ground controllers. Today, a variety of Earth-observing satellites rings the Earth, ranging from weather satellites monitoring entire hemispheres, to spy satellites capable of seeing detail less than 1m (3ft) across. Even more sophisticated are remote-sensing satellites. Viewing Earth at a variety of wavelengths, they can gather huge amounts of information about the ground below, such as the direction of ocean currents, the location of mineral deposits, and even the health of crops.

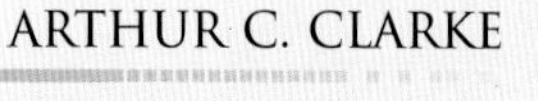

### ARTHUR C. CLARKE

Once described as "the prophet of the Space Age", British science-fiction author Arthur C. Clarke (b.1917) has invented or predicted several concepts vital to the modern use of space. Most influential of all was his 1945 suggestion that geostationary satellites, orbiting Earth once a day, could be used for communications.

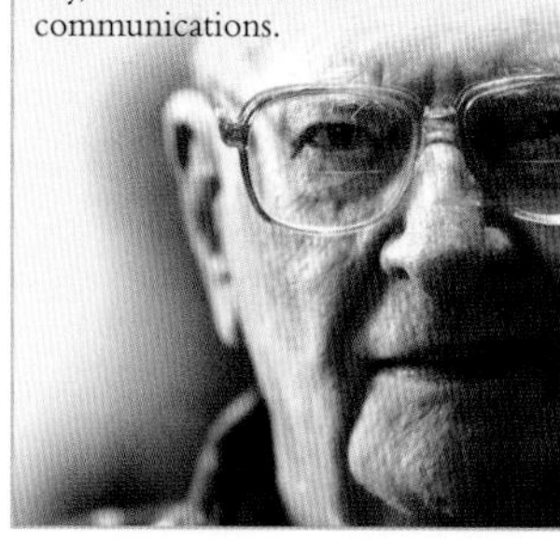

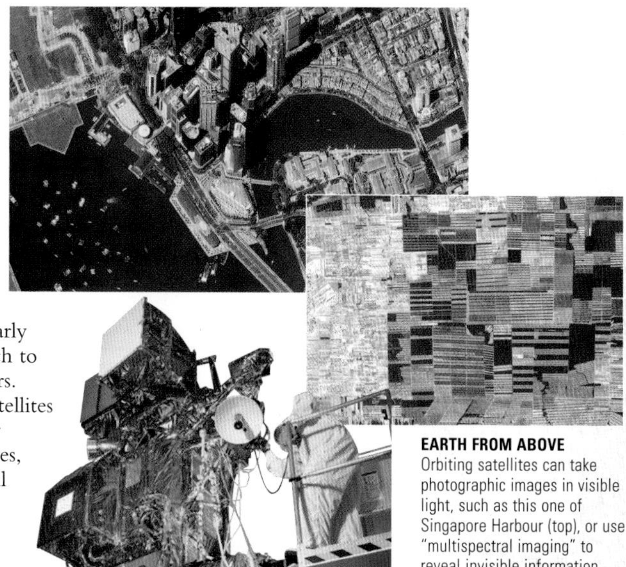

**EARTH FROM ABOVE**
Orbiting satellites can take photographic images in visible light, such as this one of Singapore Harbour (top), or use "multispectral imaging" to reveal invisible information, such as crop types (above).

**LANDSAT SATELLITE**
A technician works on a Landsat Earth remote-sensing satellite before launch. The Landsats have revealed a great deal about the nature of our planet.

**PERFECT SKIES**
Probably the most famous of all orbiting observatories, the Hubble Space Telescope, launched in 1990, hangs above the atmosphere and its distorting influence.

## ORBITING OBSERVATORIES

Many of the greatest discoveries and most spectacular images of the distant Universe have come from satellites. Earth's atmosphere creates a problem for astronomers, as it filters out most types of electromagnetic radiation. However, running a telescope in space presents unique problems. Not only must it be remotely controlled and the pictures returned to Earth, but it must also operate in a hostile environment. The extreme temperature fluctuations between sunlight and darkness can easily distort delicate telescope optics, as the entire instrument expands and contracts, so clever design and insulation are necessary. Even then, the operating life of a telescope in orbit is limited. Despite solar-power cells, most satellites need a fuel supply to reorient themselves in space and coolant to protect their sensitive electronics.

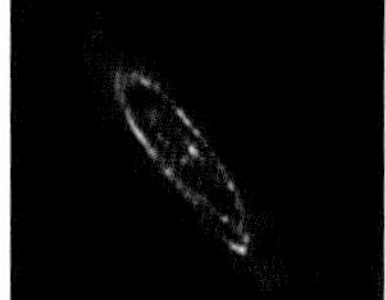

**ABOVE THE NOISE**
The Infrared Space Observatory (above) captures hidden details such as these dust rings in the Andromeda Galaxy.

## MANNED SPACEFLIGHT

Since the Apollo Moon missions, manned spaceflight has continued to develop. Despite two high-profile tragedies, in 1986 and 2003, NASA's Space Shuttle, the first reusable spacecraft, turned space travel into an almost routine affair. Since the first flight, in 1981, the Shuttles have completed more than 100 missions, carrying experiments, launching satellites, and observing the Earth and space. Meanwhile, the Soviet Union developed the principle of a modular, expandable space station with Mir (1986–2001). The collapse of the Soviet Union made international collaboration possible, while spiralling costs made it a necessity. NASA's plans for a permanent space station developed into a huge international project, and, when the International Space Station is completed, it seems that the Space Shuttle will finally be retired after more than 25 years of service. Currently, the US has ambitious plans for humans to return to the Moon and then move on to Mars.

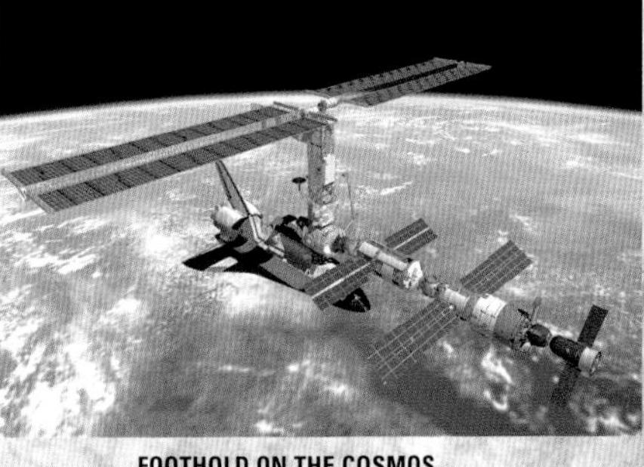

**FOOTHOLD ON THE COSMOS**
When completed, the International Space Station (ISS) will be the length of a football pitch and weigh 455 tonnes (500 tons). The ISS will serve as the basis for human operations in Earth's orbit for at least the first quarter of the 21st century.

### DENNIS TITO

Forty years after Yuri Gagarin became the first human in space, American businessman Dennis Tito (b.1941) set a new landmark in 2001 in the exploration of space when he became the world's first space tourist. As part of a programme set up by the Russian space agency, he paid $20m for a six-day stay on the International Space Station.

**THE SPACE SHUTTLE**
The orbiting space plane, rocket-launched but able to glide back to Earth intact at the end of each mission, is the most important element of NASA's ground-breaking Space Shuttle system.

# BEYOND EARTH


Mariner 10 126
Missions to Venus 130
Missions to Mars 162
Saturn 186–95
Asteroids 208–13
Comets 214–19


ALTHOUGH HUMANS HAVE NOT YET VENTURED beyond the Moon, our agents – automated space probes – have gone much farther afield. Robotic explorers have visited all the planets except Pluto, and have also surveyed dozens of satellites and a handful of minor bodies. During their travels, they have transformed our view of the Solar System, revealing other worlds almost as complex as our own.

**FIRST TO VENUS**
The Mariner 2 spacecraft weighed just 202kg (445lb), with a hexagonal body 1.5m (5ft) wide. It was powered by a span of solar panels 5m (16ft) across.

**PIONEER PROBES**
NASA's Pioneer probes, including Pioneer 6 (top) and the later Pioneer Venus (above), paved the way for automated space exploration.

## THE FIRST STEPS

The first probe to leave the Earth's influence and go into its own orbit around the Sun was Russia's Lunik 1 (also known as Luna 1), which, in 1959, missed its target, the Moon, and became the first interplanetary probe by accident. It was soon followed by more deliberate efforts. NASA's Pioneers 5 to 9 were launched successfully into orbits between Earth and Venus between 1960 and 1968, and several of these solar-powered probes are still sending back scientific data. Just as they had raced to put satellites and humans into space, the Cold War superpowers competed to be the first to reach the planets. However, failures in Soviet probes, and a prudent NASA policy of building its probes in identical pairs, allowed the US to take the lead in this race. Mariner 2 became the first probe to fly past Venus, in December 1962, measuring the planet's extreme surface temperature and confirming its unusually slow rotation. Mariner 4 flew past Mars in July 1965, measuring its atmosphere and photographing its cratered southern highlands.

**ON TO MARS**
Mariner 4 was more compact than Mariner 2, but 60kg (130lb) heavier. As the Sun's energy is weaker out at the Martian orbit, it required an array of four solar panels.

**PRECIOUS PICTURES**
These two images, out of the 21 returned from Mars by Mariner 4, show topographic features of the planet's surface.

## THE INNER SOLAR SYSTEM

Being closer to the Sun, both Venus and Mercury are travelling faster than the Earth, so any probe must pick up speed in order to enter their orbits. Nevertheless, an armada of spacecraft overcame this technical challenge and visited Venus in the years after the first fly-by. Several Soviet Venera probes attempted to land on the hostile surface, only to be destroyed during descent. In 1967, Venera 4, equipped with tank-like shielding, successfully sent back signals from Venus's surface. It was eight years later that Venera 9 returned the first pictures of the Venusian landscape. Both NASA and the Soviets also launched orbiters to survey the planet from space, and in 1978 the Pioneer Venus mission mapped the planet with radar and released atmospheric probes. In 1989, Magellan, a NASA orbiter equipped with sophisticated radar, studied the planet in unprecedented detail over four years. Mercury's orbital speed presents a larger problem: to date, only Mariner 10 has imaged its surface.

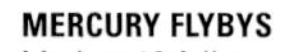

**MERCURY FLYBYS**
Mariner 10 followed a flight path that allowed it to make three close approaches to Mercury in successive orbits. Even then it could only photograph half the surface.

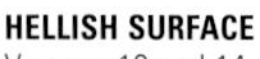

**SHIELDED LANDER**
The Venera 13 lander had a disc-shaped atmospheric brake and landing cushions.

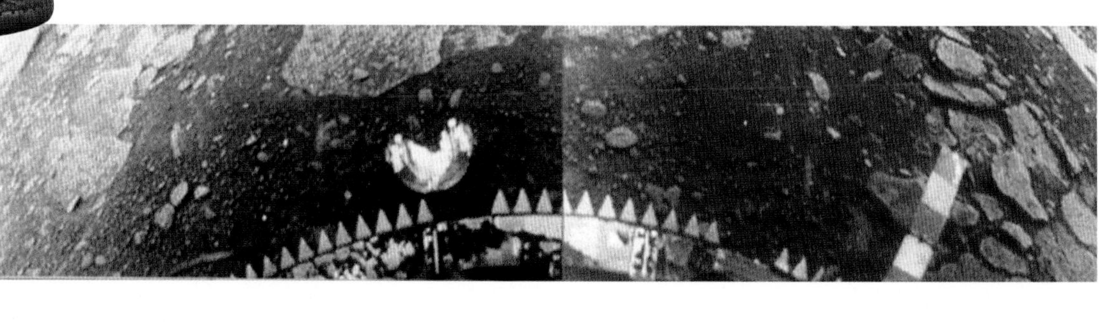

**HELLISH SURFACE**
Veneras 13 and 14 returned the first colour images from Venus's surface, in March 1981. These were the only landers on which all of the cameras worked properly.

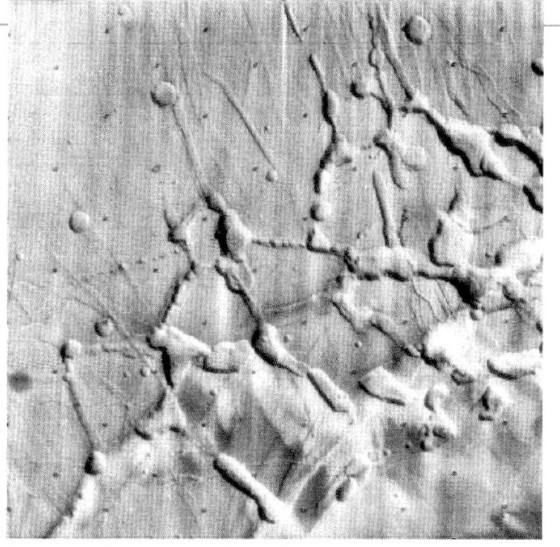

**COLLAPSE FEATURES**
These features, photographed by Mariner 9 near the Martian south pole, are thought to have been caused by landslips as underground ice thawed or evaporated.

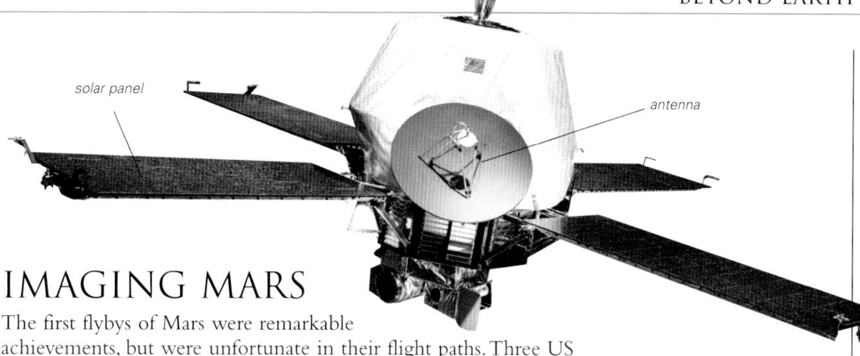


**IN MARTIAN ORBIT**
Mariner 9 weighed just under a tonne on launch, almost half of which was fuel for the rocket motor that slowed it down to put it into orbit around Mars.

## IMAGING MARS

The first flybys of Mars were remarkable achievements, but were unfortunate in their flight paths. Three US probes returned pictures of 10 per cent of the Martian surface and yet missed all the evidence of volcanism and water that makes the planet so fascinating. NASA's luck changed in 1971, when Mariner 9 became the first spacecraft to orbit Mars. It carried out a photographic survey that revealed the Valles Marineris system (see p.166–67), the towering volcanoes of the Tharsis region, and the first signs of liquid-eroded canyons. With interest in the Red Planet rejuvenated, NASA's Viking mission launched twin spacecraft to Mars, each consisting of an orbiter and a lander. The Viking orbiters created the first detailed photographic atlas of Mars, while the landers sent back pictures and weather reports from the surface, and also carried out a number of experiments, including a controversial test for microbial life.

### CARL SAGAN

US astronomer Carl Sagan (1934–96) worked on many of NASA's space probes and pioneered serious thought about life in other worlds. At NASA, he championed the idea that the Pioneer and Voyager probes should carry a record of humanity with them – a plaque for the Pioneers and a gold laser disk for the Voyagers.

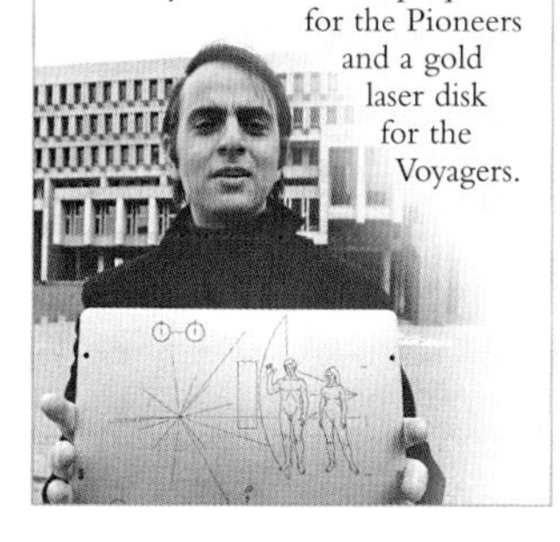

## THE GRAND TOUR

The 1970s presented a rare opportunity, as Jupiter, Saturn, Uranus, and Neptune came into an alignment that would allow a spacecraft to travel from one to another using a "gravitational slingshot" effect. A so-called "Grand Tour" mission would take little more than a decade. In 1977, NASA launched Voyagers 1 and 2, initially targeted at just Jupiter and Saturn. Voyager 1 passed Jupiter in March 1979 and Saturn in November 1980, making the first close approach to its moon Titan (see p.194). Voyager 2 followed a few months later, and, once Voyager 1 had accomplished its mission, NASA technicians put their backup plan into action, swinging the second probe around Saturn onto a course for Uranus, in 1986, and Neptune, in 1989. This extremely successful mission offered us our first and (so far) only glimpses of the cold outer giants, their moons and rings. Both spacecraft are still travelling beyond Pluto, towards the outer Solar System.

**PLUCKY PROBES**
The Voyager flyby probes were the most successful ever launched, returning more than 80,000 images of the outer planets.

**TO THE OUTER PLANETS**
The Voyagers were preceded by Pioneer 10, which reached Jupiter in 1973, and Pioneer 11, which reached Saturn in 1979. Their images, like this one of Saturn, were the best of the time but were soon bettered by Voyager.

**IO'S VOLCANOES**
The active volcanoes on Io, one of Jupiter's moons, were among Voyager 1's greatest discoveries. Voyager 2 was able to follow up with a much closer look.

# EXPLORERS ON MARS

Despite the success of the Viking landers, probes did not return to Mars until the late 1990s, following a series of ill-fated Soviet and US missions. In 1997, NASA's Mars Global Surveyor swung into orbit around the planet, equipped with the latest cameras, while a lander, Mars Pathfinder, deployed a wheeled rover called Sojourner, which sampled rocks and soil. Many new spacecraft followed this advance guard, with each new discovery making the possibility of life on Mars seem more plausible. A number of artificial satellites now orbit the planet, using remote-sensing techniques to probe the Martian subsoil, while a pair of robust Mars Exploration Rovers, Spirit and Opportunity, have discovered the clinching proof that Mars had widespread and long-lived oceans in its past. More missions are being planned to further the search for existing water and life on Mars.

**MARS PATHFINDER**
The Sojourner rover was intended to operate for only seven days but actually lasted for 83 days, travelling 100m (330ft).

**LAYERED ROCK**
An image from NASA's Spirit rover shows a layered rock dubbed Tetl. Scientists are hoping to discover whether the rock's layering is volcanic or sedimentary in origin.

**GUSEV PANORAMA**
This mosaic image shows the landscape in the Gusev crater, landing site of the Spirit rover.

# THROUGH THE MOONS OF JUPITER

The Voyager flybys of Jupiter's huge moons revealed them to be fascinating worlds worth closer inspection. The Galileo probe, destined to orbit Jupiter, was launched in 1989, but did not arrive until 1995. However, the results proved worth the wait. The orbiter made the first close flyby of an asteroid while en route, then deployed a probe into Jupiter's atmosphere, before beginning a mission that would surpass all expectations. Galileo studied the volcanoes of Io and the weather systems of Jupiter itself, while bolstering the evidence for an ocean beneath Europa's icy crust and on the outer Jovian moons Ganymede and Callisto.

**VOYAGE BEGINS**
The Space Shuttle *Atlantis* prepares to release Galileo (still attached to a booster-rocket stage) in October 1989.

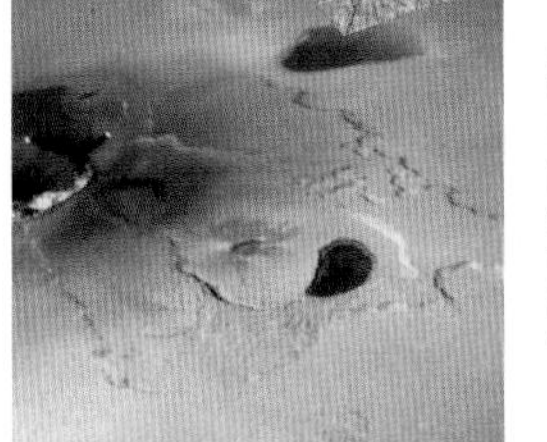

**OCEANS AND VOLCANOES**
Galileo's high-resolution images revolutionized accepted ideas about the icy crust of Europa (above) and the volcanic calderas of Io (left).

# CASSINI AND BEYOND

Galileo was followed by an even more ambitious mission to Saturn. Cassini, an enormous probe weighing 5.6 tonnes, was launched in 1997. After a complex journey during which it flew twice past Venus, once past Earth, and also swung close to Jupiter, it arrived in orbit around Saturn in 2004. Onboard, it carried a European-built lander called Huygens, which parachuted into the atmosphere of the moon Titan in January 2005, returning pictures during its descent and revealing a world in which liquid methane appears to play the same role that water does on Earth. Cassini will monitor Saturn and also fly close to many of Saturn's fascinating and varied satellites. It is likely to pave the way for even more ambitious future orbiter missions, such as Prometheus (formerly the Jupiter Icy Moons Orbiter), a nuclear-powered spacecraft intended to study Jupiter's moons in unprecedented detail. Prometheus could be the first in a new generation of nuclear space probes that could open up the Solar System to faster and even deeper exploration.

**EXPLORING TITAN**
Cassini was fitted with cameras to pierce the haze of Titan's atmosphere (right), while the Huygens lander provided a close-up view (above).

**GIANT PROBE**
The bus-sized Cassini probe awaits launch in its "clean room" in 1997.

# COMETS AND ASTEROIDS

While most probes have been targeted at planets, scientists have not neglected the minor members of the Solar System. Comets and asteroids, largely unchanged since they formed 4.5 billion years ago, can hold fascinating clues about the creation of the Solar System and even the origin of life itself. In 1985 and 1986, as Halley's Comet moved past the Sun for the first time in 76 years, a fleet of probes set out to meet it. The Near Earth Asteroid Rendezvous (NEAR) carried out a flyby of the Main-belt asteroid Mathilde in 1997, before surveying the large asteroid Eros for an entire year. More ambitious missions have followed. NASA's Stardust mission collected material from the tail of Comet Wild-2 in 2004 for eventual return to Earth; the European Rosetta mission will put a lander on a comet for the first time; and the New Horizons probe will fly past Pluto before exploring the Kuiper Belt, the home of many comets (see p.206).

**NEW HORIZONS**
Seen here under construction, this probe will use a boost from Jupiter's gravity to reach Pluto in just nine years.

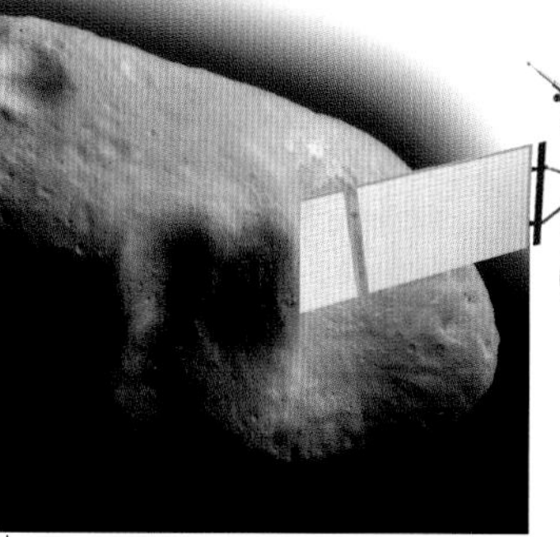

**ROSETTA**
This European Space Agency probe will orbit Comet Churyumov-Gerasimenko in 2014.

**EROS**
The 33km- (20-mile-) long near-earth asteroid Eros was studied intensively by the NEAR probe in 2000 and 2001.

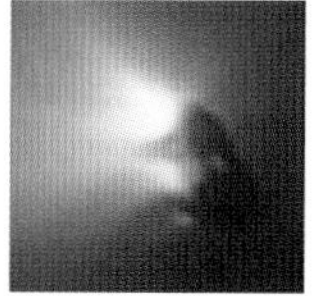

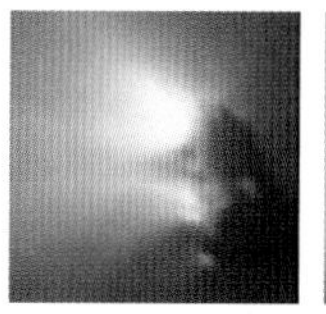

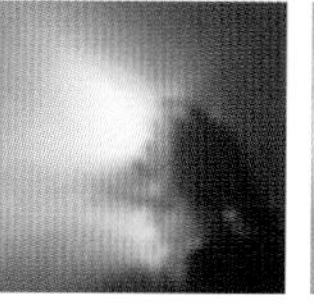

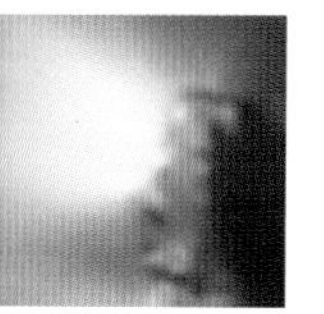

**COMET FLYBYS**
A sequence of images captures the European Giotto probe's close encounter with the nucleus of Halley's Comet in 1986.

# MAN'S FUTURE IN SPACE

It is inevitable that crewed spacecraft will one day venture into the Solar System. While the US and China have plans for a return to the Moon, the US is also considering the possibility of a mission to Mars. Although these plans may be abandoned, they will be replaced by others, and in the meantime space scientists are gathering the knowledge they will need to make long-duration space travel a reality. Russia gained unique experience of prolonged zero gravity with the Mir space station missions, and the crew of the International Space Station are a valuable source of information on new space medicines. Elsewhere, ground-based experiments, such as Biosphere 2, have provided useful information on how astronauts might produce their own food, water, and oxygen on other planets and how an isolated crew might cope in the confines of a spacecraft for many months.

**SMART-1**
ESA's SMART-1 lunar probe was launched to test new propulsion technologies.

## VALERI POLIAKOV

Russian cosmonaut Valeri Poliakov (b.1942) has specialized in the challenges of long-duration spaceflight. After qualifying as a doctor, he carried out research into space medicine alongside his cosmonaut training in the 1970s. His first flight aboard Mir, in 1988, lasted 240 days, while his second, in 1994, lasted an epic 437 days – a spaceflight duration record sure to stand for many years to come.

**BIOSPHERE 2**
Two "crews" lived in the self-contained Biosphere 2 in the early 1990s for up to two years. Based in the Arizona desert, its research continues and the facility is now open to the public.

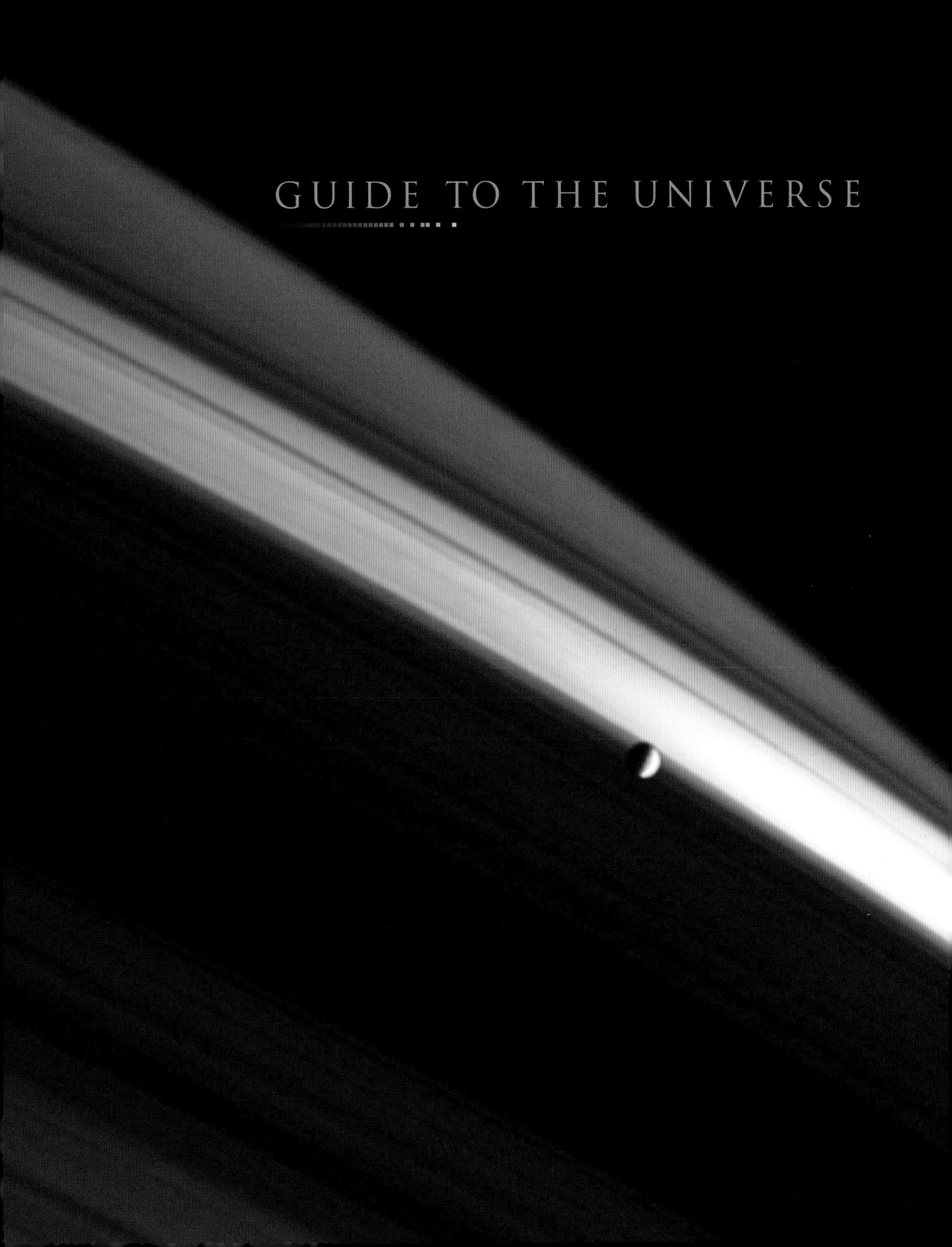
GUIDE TO THE UNIVERSE

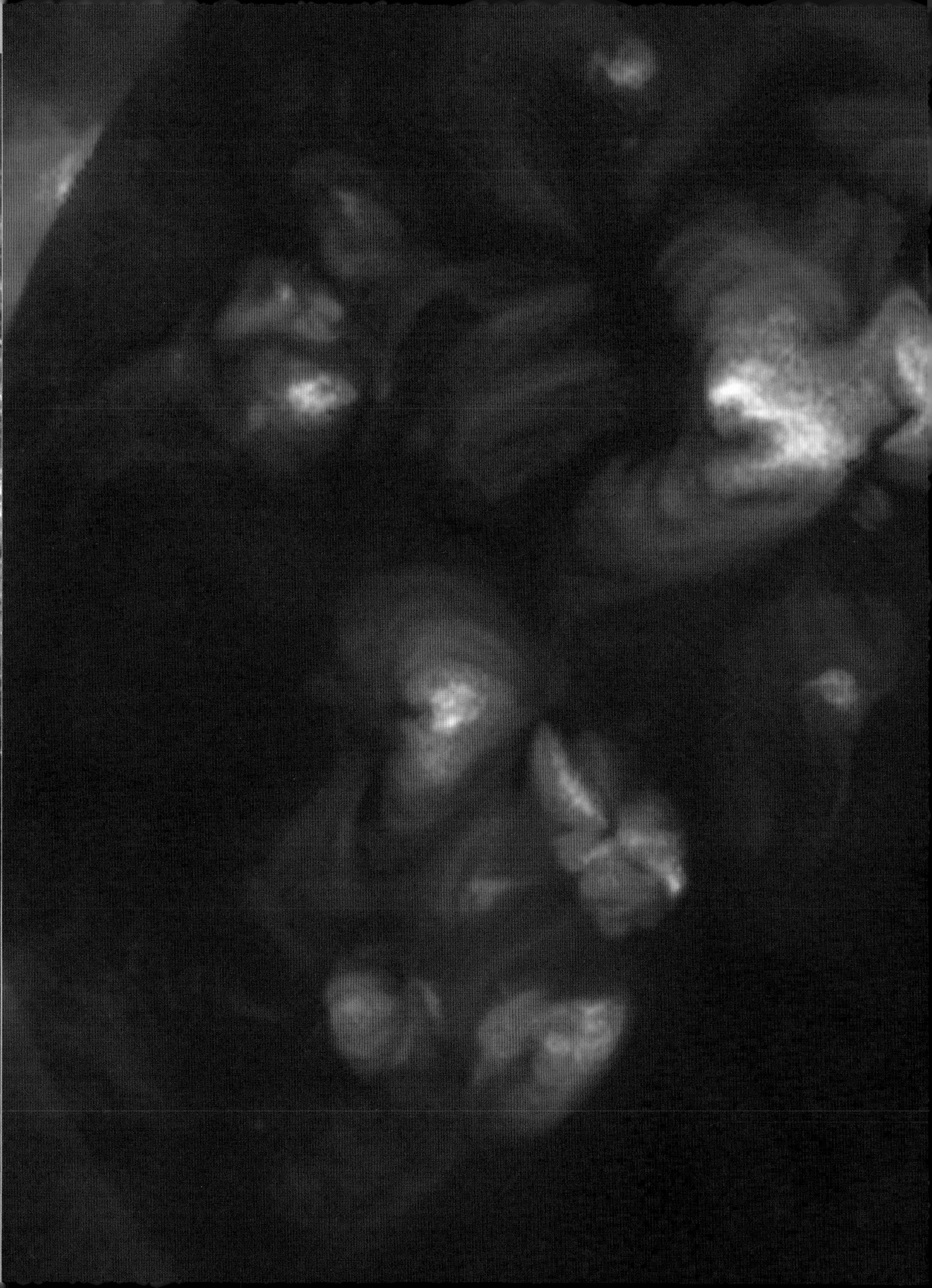

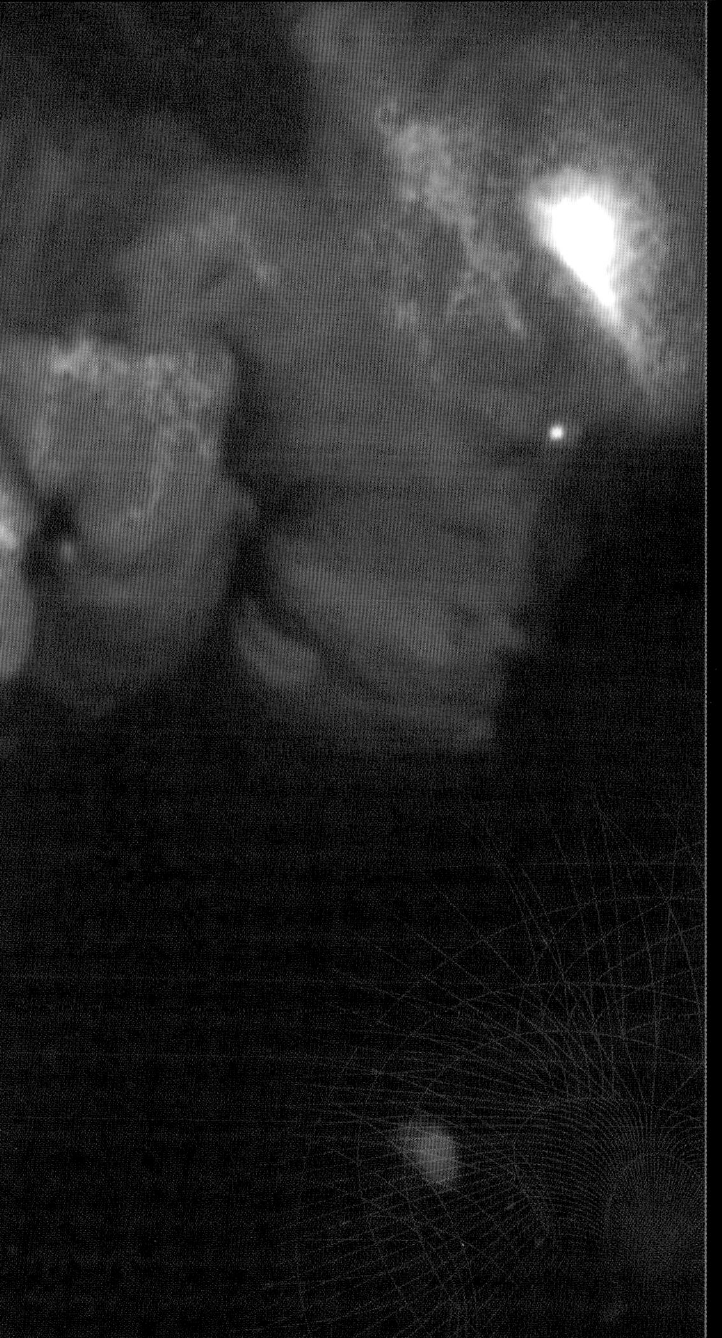

*"How vast those Orbs must be, and how inconsiderable this Earth, the Theatre upon which all our mighty Designs, all our Navigations, and all our Wars are transacted, is when compared to them."*

Christiaan Huygens

THE SOLAR SYSTEM IS the region of space that falls within the gravitational influence of the Sun, an ordinary yellow star that has shone steadily for almost 5 billion years. After the Sun itself, the most significant objects in the Solar System are the planets – a group of assorted rocky, gaseous, and icy worlds that follow independent, roughly circular orbits around their central star. Most of the planets are orbited in turn by moons, while a huge number of smaller lumps of rock and ice also follow their own courses around the Sun – though largely confined in a few relatively crowded zones. Myriad tiny particles flow around all these larger bodies – ranging from fragments of atoms blown out by the Sun to motes of dust and ice left in the wake of comets. Our local corner of the Universe has been studied intensely from the time of the first stargazers to the modern era of space probes, yet it is still a source of wonder and surprise.

**SOLAR FLARE**
On the broiling surface of the Sun, a cataclysmic release of magnetic energy triggers a solar flare – a violent outburst of radiation and high-energy particles that will reach Earth within hours.

# THE SOLAR SYSTEM

# THE HISTORY OF THE SOLAR SYSTEM

22–23 The scale of the Universe
24–27 Celestial objects
32–35 Radiation
36–37 Gravity, motion, and orbits
64–65 Planetary motion
86–89 The Copernican revolution

THE SOLAR SYSTEM IS THOUGHT to have begun forming about 4.6 billion years ago from a gigantic cloud of gas and dust, called the solar nebula. This cloud contained several times the mass of the present-day Sun. Over millions of years, it collapsed to a flat, spinning disc, which had a dense, hot central region. The central part of the disc eventually became the Sun, while the planets and everything else in the Solar System formed from a portion of the remaining material.

## THE FORMATION OF THE SOLAR SYSTEM

No one knows for certain what caused the great cloud of gas and dust, the solar nebula from which the Solar System formed, to start collapsing. What is certain is that gravity somehow overcame the forces associated with gas pressure that would otherwise have kept it expanded. As it collapsed, the cloud flattened into a pancake-shaped disc with a bulge at its centre. Just as an ice skater spins faster as she pulls in her arms, the disc began to rotate faster and faster as it contracted. The central region also became hotter and denser. In the parts of the disc closest to this hot central region, only rocky particles and metals could remain in solid form. Other materials were vaporized. In due course, these rocky and metallic particles gradually came together to form planetesimals (small bodies of rock, up to several kilometres in diameter) and eventually the inner rocky planets – Mercury, Venus, Earth, and Mars. In the cooler outer regions of the disc, a similar process occurred, but the solid particles that came together to form planetesimals contained large amounts of various ices, such as water, ammonia, and methane ices, as well as rock. These materials were destined eventually to form the cores of the gas-giant planets – Jupiter, Saturn, Uranus, and Neptune.

**1 SOLAR NEBULA FORMS**
The solar nebula started as a huge cloud of cold gas and dust, many times larger than our present Solar System. Its initial temperature would have been about -230°C (-382°F). From the start, the solar nebula was probably spinning very slowly.

**SIX STEPS TO FORM A SOLAR SYSTEM**
Shown here is an outline of the nebular hypothesis – the most widely accepted theory for how the Solar System formed. It provides a plausible explanation for many of the basic facts about the Solar System. For example, it explains why the orbits of most of the planets lie roughly in the same plane and why the planets all orbit in the same direction.

### PIERRE-SIMON DE LAPLACE

Pierre Laplace (1749–1827) was a French mathematician who developed the nebular hypothesis – the idea, originally proposed by the German philosopher Immanuel Kant, that the Solar System originated from the contraction of a huge gaseous nebula. Today, this hypothesis provides the most widely accepted theory for how the Solar System formed. Another of Laplace's contributions to science was to analyse the complex forces of gravitational attraction between the planets. He investigated how these might affect the stability of the Solar System and concluded that the system is inherently stable.

**6 REMAINING DEBRIS**
Radiation from the Sun blew away most of the remaining gas and other unaccreted material in the planetary Solar System. Some of the leftover planetesimals in the outer part of the disc formed the vast and remote Oort Cloud of comets.

*inner Solar System*

*frozen cometary nuclei*

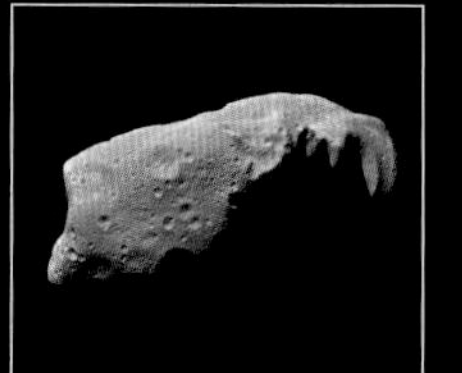

**IDA**
The ring of planetesimals between Mars and Jupiter failed to form a planet, possibly because of the gravitational influence of Jupiter. Instead they formed a belt of asteroids, which include this asteroid, Ida.

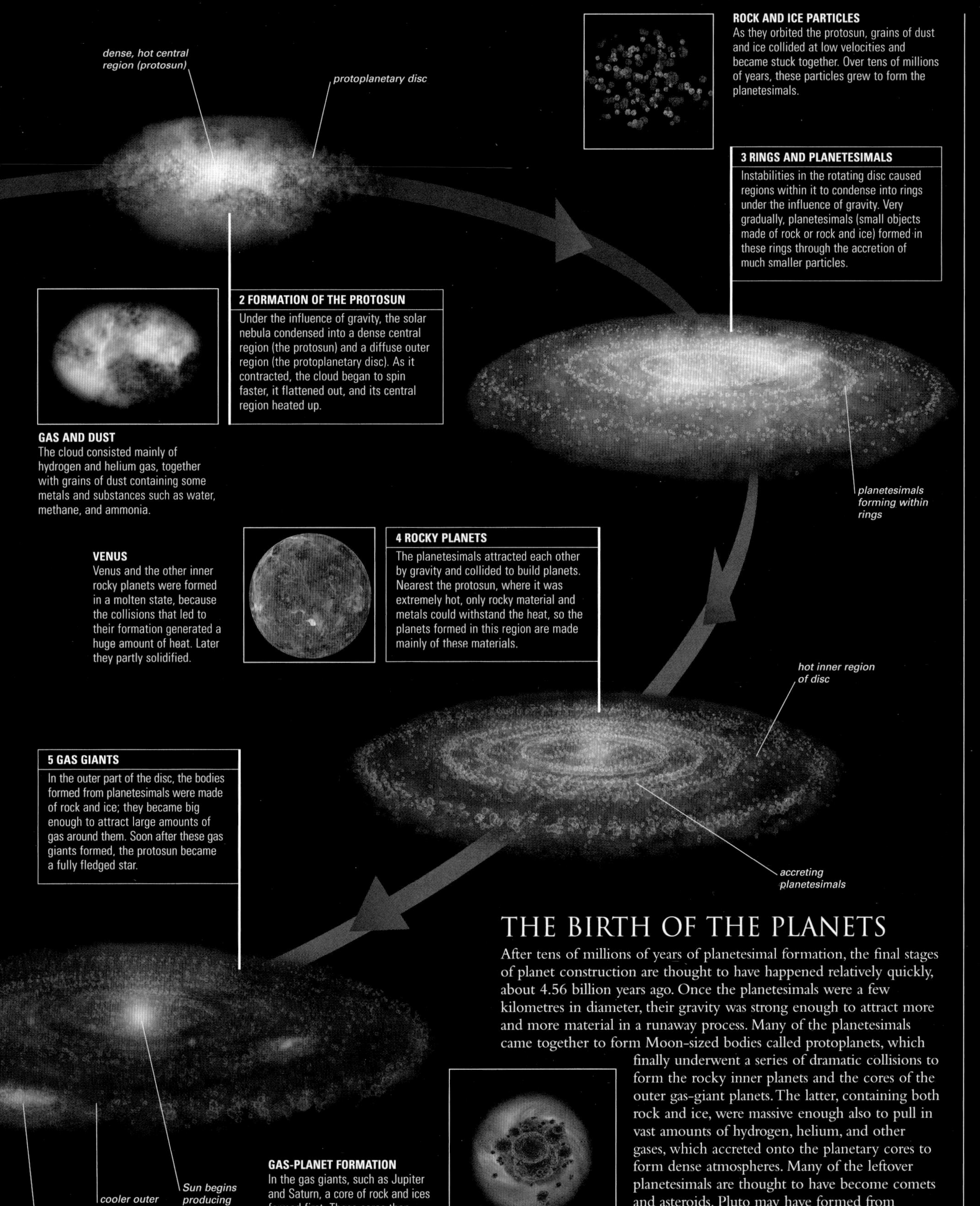


**ROCK AND ICE PARTICLES**
As they orbited the protosun, grains of dust and ice collided at low velocities and became stuck together. Over tens of millions of years, these particles grew to form the planetesimals.

**3 RINGS AND PLANETESIMALS**
Instabilities in the rotating disc caused regions within it to condense into rings under the influence of gravity. Very gradually, planetesimals (small objects made of rock or rock and ice) formed in these rings through the accretion of much smaller particles.

**2 FORMATION OF THE PROTOSUN**
Under the influence of gravity, the solar nebula condensed into a dense central region (the protosun) and a diffuse outer region (the protoplanetary disc). As it contracted, the cloud began to spin faster, it flattened out, and its central region heated up.

**GAS AND DUST**
The cloud consisted mainly of hydrogen and helium gas, together with grains of dust containing some metals and substances such as water, methane, and ammonia.

**VENUS**
Venus and the other inner rocky planets were formed in a molten state, because the collisions that led to their formation generated a huge amount of heat. Later they partly solidified.

**4 ROCKY PLANETS**
The planetesimals attracted each other by gravity and collided to build planets. Nearest the protosun, where it was extremely hot, only rocky material and metals could withstand the heat, so the planets formed in this region are made mainly of these materials.

**5 GAS GIANTS**
In the outer part of the disc, the bodies formed from planetesimals were made of rock and ice; they became big enough to attract large amounts of gas around them. Soon after these gas giants formed, the protosun became a fully fledged star.

# THE BIRTH OF THE PLANETS

After tens of millions of years of planetesimal formation, the final stages of planet construction are thought to have happened relatively quickly, about 4.56 billion years ago. Once the planetesimals were a few kilometres in diameter, their gravity was strong enough to attract more and more material in a runaway process. Many of the planetesimals came together to form Moon-sized bodies called protoplanets, which finally underwent a series of dramatic collisions to form the rocky inner planets and the cores of the outer gas-giant planets. The latter, containing both rock and ice, were massive enough also to pull in vast amounts of hydrogen, helium, and other gases, which accreted onto the planetary cores to form dense atmospheres. Many of the leftover planetesimals are thought to have become comets and asteroids. Pluto may have formed from material not used in the gas giants or may have been captured by the Solar System at a later time.

**GAS-PLANET FORMATION**
In the gas giants, such as Jupiter and Saturn, a core of rock and ices formed first. These cores then attracted, and became enveloped by, large amounts of gas.

# THE FAMILY OF THE SUN

22–23 The scale of the Universe
24–27 Celestial objects
32–35 Radiation
36–37 Gravity, motion, and orbits
64–65 Planetary motion
86–89 The Copernican revolution
The Milky Way 226–29

THE SOLAR SYSTEM CONSISTS OF the Sun, eight recognized planets, over 140 moons, and countless small bodies such as comets and asteroids. Its inner region contains the Sun and the rocky planets – Mercury, Venus, Earth, and Mars. Beyond this lie a ring of asteroids, called the Main Belt, and the gas giant planets Jupiter, Saturn, Uranus, and Neptune. Next is a huge region populated by Pluto and other ice dwarfs and finally a vast cloud of comets. In total, the Solar System is about 15,000 billion km (9,300 billion miles) across; the planets occupy a zone extending just 6 billion km (3.75 billion miles) from the Sun.

## URBAIN LE VERRIER

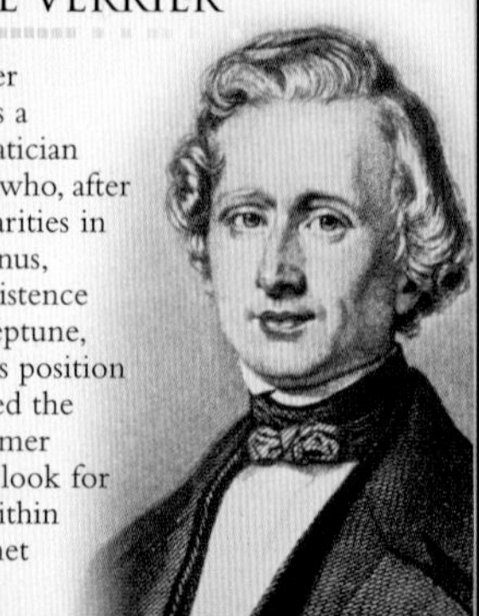

Urbain Le Verrier (1811–1877) was a French mathematician and astronomer who, after studying irregularities in the orbit of Uranus, predicted the existence of the planet Neptune, and calculated its position in 1846. He asked the German astronomer Johann Galle to look for Neptune, and within an hour the planet had been found.

## ORBITS IN THE SOLAR SYSTEM

Most orbits of objects in the Solar System have the shapes of ellipses (stretched circles). However, for most of the planets, these ellipses are close to being circular. Only Mercury has an orbit that differs very markedly from being circular. All the planets and nearly all asteroids orbit the Sun in the same direction, which is also the direction in which the Sun spins on its own axis. The orbital period (the time it takes a planet to orbit the Sun) increases with distance from the Sun, from 88 Earth days for Mercury to nearly 165 Earth years for Neptune, following a mathematical relationship first discovered by the German astronomer Johannes Kepler in the early 17th century (see p.64). As well as having longer orbits to complete, the planets farther from the Sun move much more slowly.

**THE SUN**
*The Sun's diameter at the equator is 1.4 million km (864,900 miles), and its equatorial rotation period is about 25 Earth days*

**EARTH**
*Orbits the Sun in 365.26 Earth days at an average distance of 149.6 million km (92.9 million miles)*

**JUPITER**
*Orbits the Sun in 11.86 Earth years at an average distance of 778.4 million km (483.4 million miles)*

**URANUS**
*Orbits the Sun in 84.01 Earth years at an average distance of 2.9 billion km (1.8 billion miles)*

**MERCURY**
*Orbits the Sun in 88 Earth days at an average distance of 57.9 million km (36.0 million miles)*

**PLANET ORBITS**
All the orbits of the planets, and the Asteroid Belt, lie roughly in a flat plane known as the ecliptic plane. Only Mercury orbits at a significant angle to this plane (7.0°). The planets and their orbits are not shown to scale here.

# THE GAS GIANTS

The four large planets immediately beyond the Asteroid Belt are called the gas giants. These planets have many properties in common. Each has a core composed of rock and ice. This is surrounded by a liquid or semi-solid mantle containing hydrogen and helium, or, in the case of Uranus and Neptune, a combination of methane, ammonia, and water ices. Each has a deep, often stormy atmosphere composed mainly of hydrogen and helium. All four have a significant magnetic field, but Jupiter's is exceptional, being 20,000 times stronger than that of Earth. Each of the gas giants is orbited by a large number of moons, several dozen in the case of Jupiter. Finally, all four gas giants have ring systems made of grains of rock or ice. These rings may have been present since the planets formed, or they may be the fragmented remains of moons that were broken up by the gas giants' powerful gravitational fields.

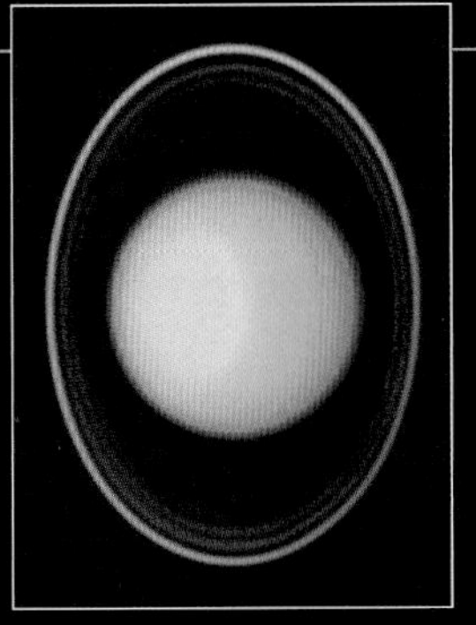

**URANUS AND RINGS**
Uranus has 11 major rings and a blue coloration caused by the presence of methane in its atmosphere (this is a Hubble infrared image). Its spin axis is tilted right over on the side.

# THE ROCKY PLANETS

The four inner planets of the Solar System are also called the rocky planets. They are much smaller than the gas giants, have few or no moons, and no rings. All four were born in a molten state due to the heat of the collisions that led to their formation. While molten, the materials from which they are made became separated into a metallic core and a rocky mantle and crust. Throughout their later history, all these planets suffered heavy bombardment by meteorites that left craters on their surfaces, although on Earth these craters have largely become hidden by various geological processes. In some other respects, the rocky planets are quite diverse. For example, Venus has a dense atmosphere consisting mainly of carbon dioxide, while Mars has a thin atmosphere composed of the same gas. In contrast, Mercury has virtually no atmosphere and Earth's is rich in nitrogen and oxygen.

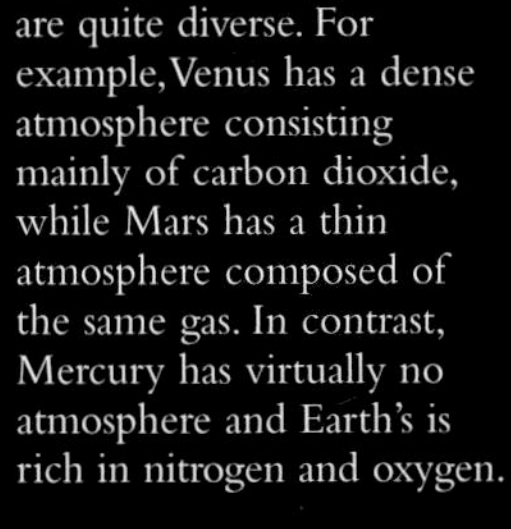

**GOSSES BLUFF CRATER**
This impact crater in a central desert region of Australia resulted from an asteroid 1km (0.6 miles) wide that smashed into the Earth's surface 142 million years ago.

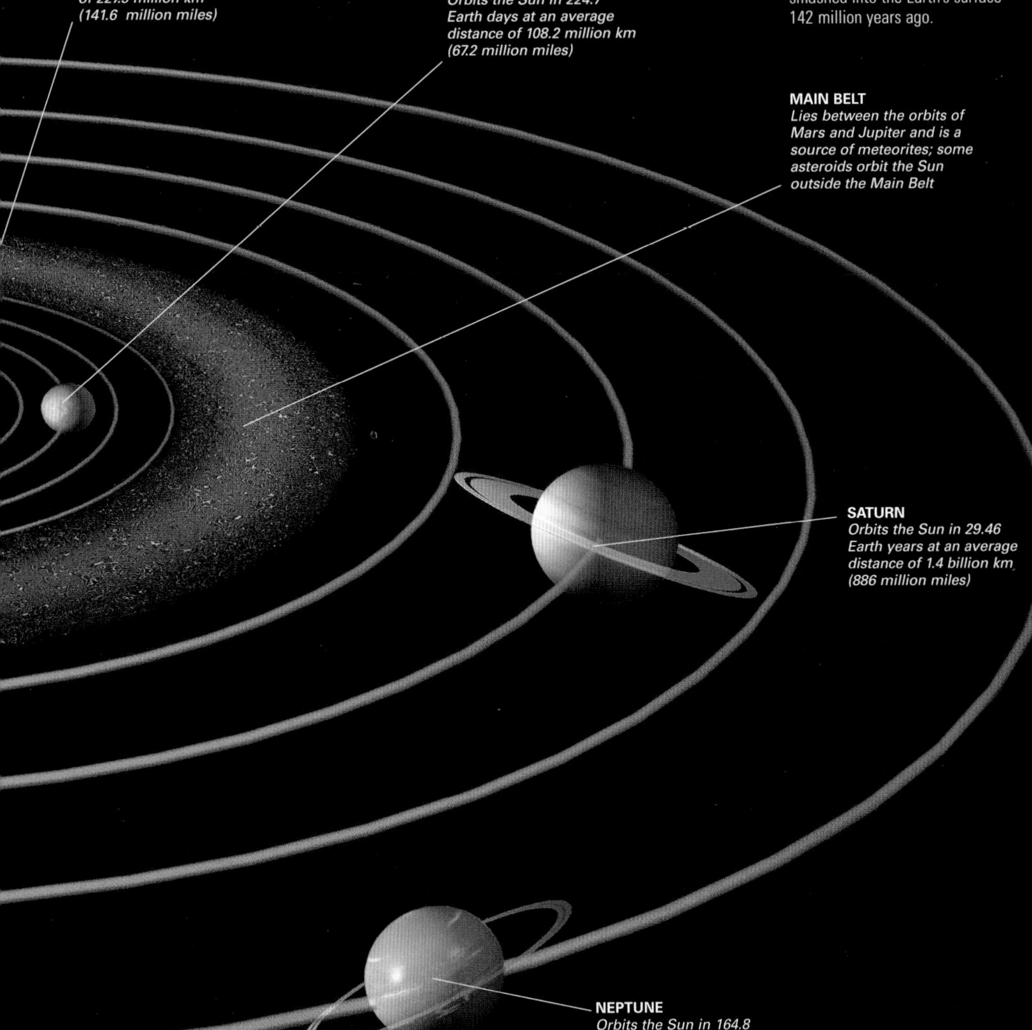
MARS
Orbits the Sun in 687 Earth days at an average distance of 227.9 million km (141.6 million miles)
VENUS
Orbits the Sun in 224.7 Earth days at an average distance of 108.2 million km (67.2 million miles)
MAIN BELT
Lies between the orbits of Mars and Jupiter and is a source of meteorites; some asteroids orbit the Sun outside the Main Belt
SATURN
Orbits the Sun in 29.46 Earth years at an average distance of 1.4 billion km (886 million miles)
NEPTUNE
Orbits the Sun in 164.8 Earth years at an average distance of 4.5 billion km (2.8 billion miles)

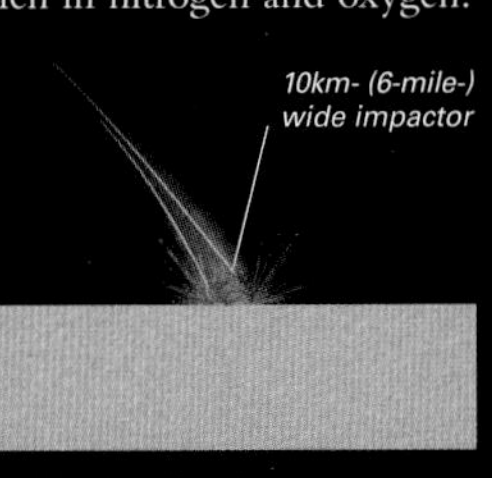
10km- (6-mile-) wide impactor

IMPACTOR STRIKES

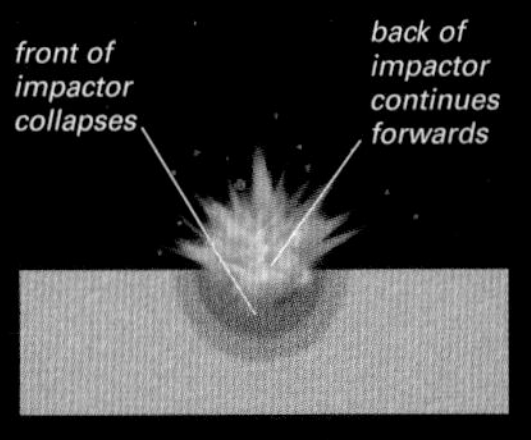
front of impactor collapses
back of impactor continues forwards

EXPLOSION ON IMPACT

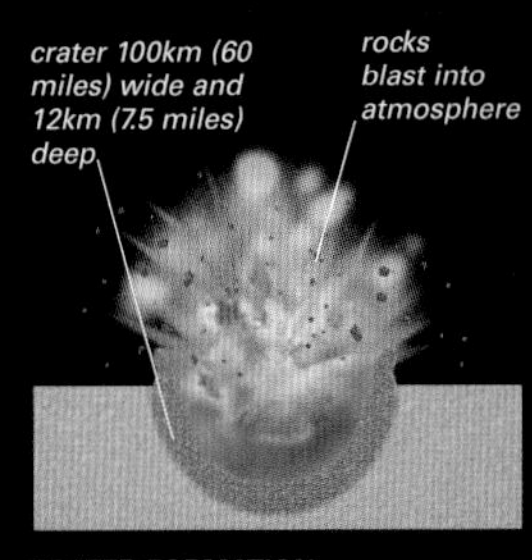
crater 100km (60 miles) wide and 12km (7.5 miles) deep
rocks blast into atmosphere

CRATER FORMATION

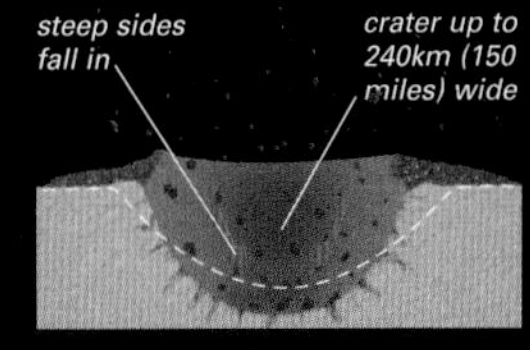
steep sides fall in
crater up to 240km (150 miles) wide

CRATER COLLAPSE

**DEEP IMPACT**
This sequence shows what typically happens when a 10km- (6-mile-) wide projectile hits a rocky planet or moon. The crater formed is much larger than the impactor. The latter usually vaporizes on impact, though some melted or shattered remnants may be left at the site.

# THE SUN

31 Nuclear fission and fusion
32–35 Radiation
63 Solar eclipses
Stars 230–31
The life cycles of stars 232–35
Star formation 236–37
Main-sequence stars 246–47

THE SUN IS A 4.6-BILLION-YEAR-OLD main-sequence star. It is a huge sphere of exceedingly hot plasma (ionized gas) containing 750 times the mass of all the Solar System's planets put together. In its core, nuclear reactions produce helium from hydrogen and generate colossal amounts of energy. This energy is gradually carried outwards until it eventually escapes from the Sun's surface.

## INTERNAL STRUCTURE

The Sun has three internal layers, although there are no sharp boundaries between them. At the centre is the core, where temperatures and pressures are extremely high. In the core, nuclear fusion turns the nuclei of hydrogen atoms (protons) into helium nuclei at the rate of about 600 million tons per second. Released as by-products of the process are energy, in the form of photons of electromagnetic (EM) radiation, and neutrinos (particles with no charge and almost no mass). The EM radiation travels out from the core through a slightly cooler region, the radiative zone. It takes about 1 million years to find its way out of this zone, as the photons are continually absorbed and re-emitted by ions in the plasma. Further out, the energy wells up in a convective zone – where huge flows of rising hot plasma occur next to areas of falling cooler plasma – and is transferred to a surface layer called the photosphere. There it escapes as heat, light, and other forms of radiation.

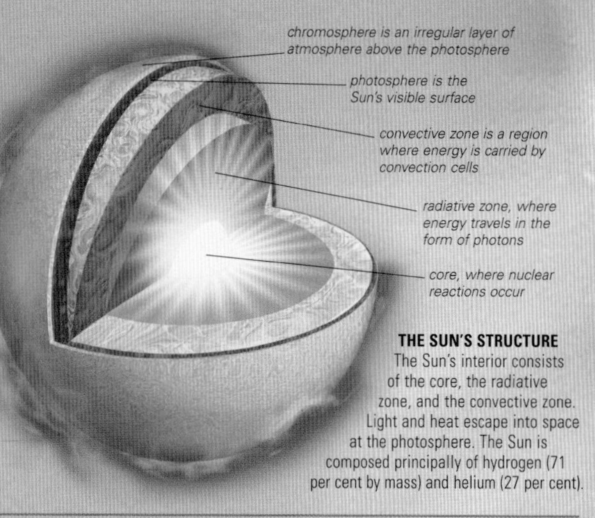


**THE SUN'S STRUCTURE**
The Sun's interior consists of the core, the radiative zone, and the convective zone. Light and heat escape into space at the photosphere. The Sun is composed principally of hydrogen (71 per cent by mass) and helium (27 per cent).

### SUN PROFILE

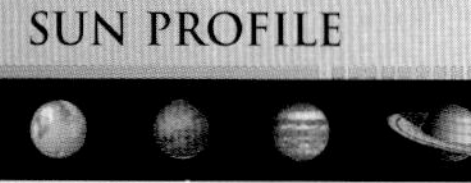

| | |
|---|---|
| **AVERAGE DISTANCE FROM EARTH** 149.6 million km (93.0 million miles) | **ROTATION PERIOD (POLAR)** 34 Earth days |
| **SURFACE TEMPERATURE** 5,500°C (9,932°F) | **ROTATION PERIOD (EQUATORIAL)** 25 Earth days |
| **CORE TEMPERATURE** 15 million °C (27 million °F) | **MASS (EARTH = 1)** 333,000 |
| **DIAMETER AT EQUATOR** 1.4 million km (864,900 miles) | **SIZE COMPARISON** |
| **OBSERVATION** The Sun has an apparent magnitude of -26.7 and should never be observed directly with the naked eye or any optical instrument. It can be observed safely only through special solar filters. | |

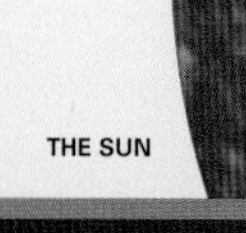


**VIOLENT SUN**
This composite image taken by the SOHO observatory shows both the Sun's surface and its corona. When the corona image was taken, billions of tons of matter were being blasted through it into space.

## STUDYING THE SUN FROM SPACE

Since 1960, a series of space probes and satellites have been launched by NASA and other organizations with the aim of collecting data about the Sun. Some of the most important missions are listed below.

**1960–68 PIONEERS 5 TO 9 (USA)** These were a series of probes that successfully orbited the Sun and studied the solar wind, solar flares, and the interplanetary magnetic field.

HELIOS 1

**1974, 1976 HELIOS 1 AND 2 (USA AND GERMANY)** The two Helios probes were put into orbits that involved high-velocity passes close to the Sun's surface. They measured the solar wind and the Sun's magnetic field.

**1980 SOLAR MAXIMUM MISSION (USA)** This studied the Sun at its most active, collecting X-rays, gamma rays, and ultraviolet radiation produced by flares, sunspots, and prominences.

SOLAR SATELLITE

**1990 ULYSSES (USA AND EUROPE)** The first space probe to be sent into an orbit over the Sun's poles, Ulysses has studied the solar wind and the Sun's magnetic field over its polar regions.

**1991 YOHKOH (JAPAN, USA, AND UK)** Yohkoh was an Earth-orbiting satellite that for 10 years observed high-energy radiation (X-rays and gamma rays) produced by solar flares, as well as pre-flare conditions.

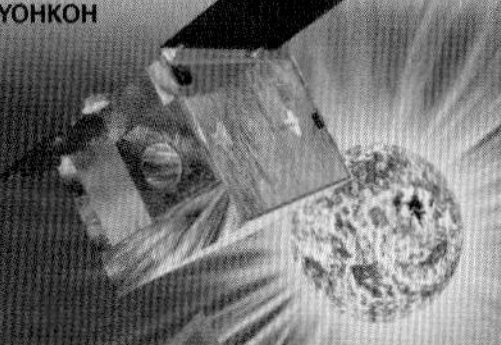

YOHKOH

SOHO

**1995 SOHO (USA AND EUROPE)** This solar observatory follows a special "halo" orbit about 1.5 million km (930,000 miles) from Earth in the direction of the Sun. SOHO (solar and heliospheric observatory) studies the Sun's interior, as well as events at its surface.

**1998 TRACE (USA)** Trace is a satellite in Earth's orbit that studies the corona and a thinner layer in the Sun's atmosphere called the transition region. The objective of TRACE (transition region and coronal explorer) is to better understand the connection between the Sun's magnetic field and coronal heating.

TRACE

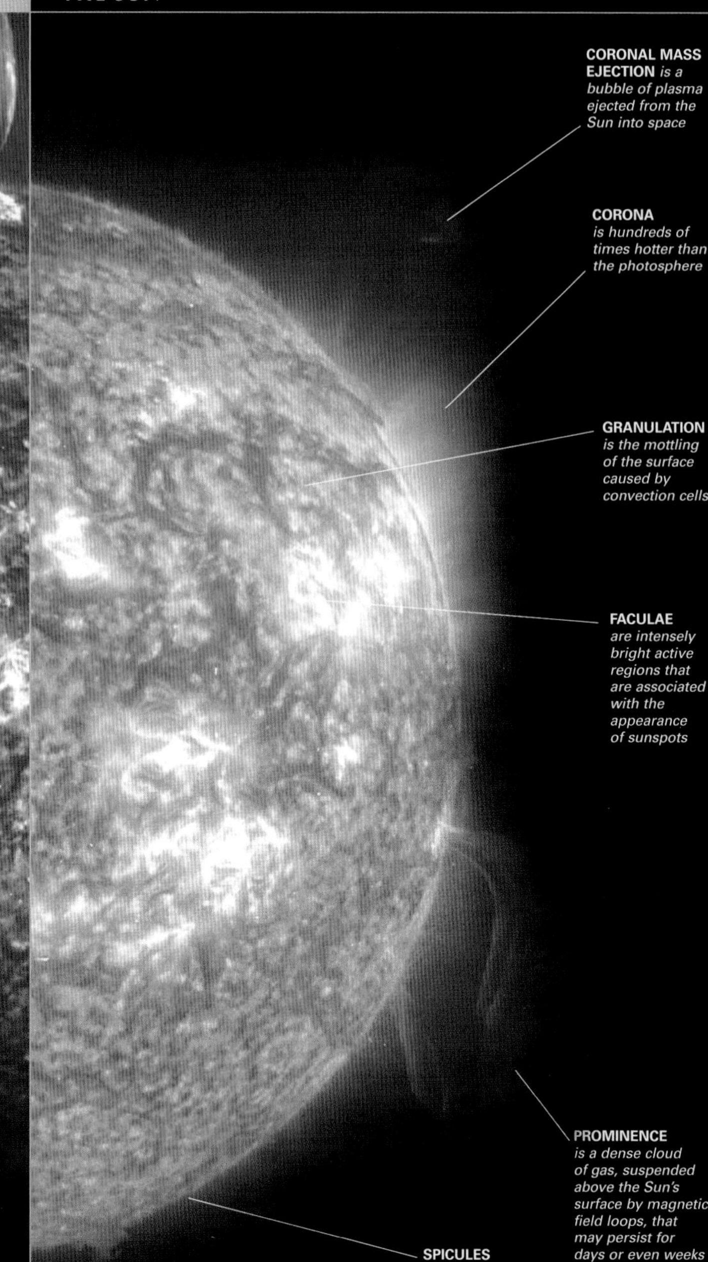
CORONAL MASS EJECTION is a bubble of plasma ejected from the Sun into space
CORONA is hundreds of times hotter than the photosphere
GRANULATION is the mottling of the surface caused by convection cells
FACULAE are intensely bright active regions that are associated with the appearance of sunspots
PROMINENCE is a dense cloud of gas, suspended above the Sun's surface by magnetic field loops, that may persist for days or even weeks
SPICULES are short-lived jets of gas that are 10,000km (6,000 miles) long

**SOLAR ACTIVITY**
This ultraviolet image of the Sun was obtained by an instrument onboard the SOHO solar observatory. It shows the Sun's chromosphere (the layer just above the photosphere) and various protuberances, including a huge solar prominence, as well as a number of active regions on the solar surface. The image also shows a coronal mass ejection with a bright central area of ultraviolet emission.

# SURFACE

The visible surface of the Sun is called the photosphere. It is a layer of plasma (ionized gas) about 100km (60 miles) thick and appears granulated or bubbly. The bumps, which are about 1,000km (600 miles) wide, are the upper surfaces of convection cells that bring hot plasma up from the Sun's interior. Other significant features of the photosphere are sunspots, which are cooler regions that appear dark against their brighter, hotter surroundings. Sunspots and related phenomena, such as solar flares (tremendous explosions on the Sun's surface), and plasma loops, are thought to have a common underlying cause – they are associated with strong magnetic fields or disturbances in these fields. The magnetic fields result from the fact that the Sun is a rotating body that consists largely of electrically charged particles (ions in its plasma). Different parts of the Sun's convective zone rotate at different rates (faster at the equator than the poles), causing the magnetic field lines to become twisted and entangled over time. Sunspots are caused by concentrations of magnetic field lines inhibiting the flow of heat from the interior where they intersect the photosphere. Other types of disturbance are caused by twisted field lines popping out of the Sun's surface, releasing tremendous energy, or by plasma erupting as loops along magnetic field lines. The amount of sunspot and related activity varies from a minimum to a maximum over an 11-year cycle.

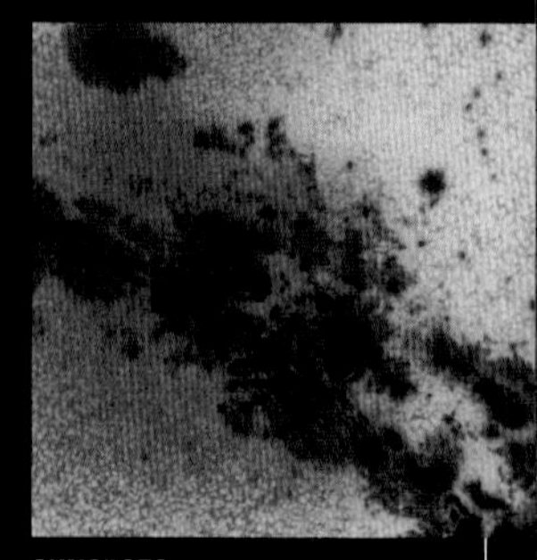

**SUNSPOTS**
Each sunspot has a dark central region, the umbra, and a lighter periphery, the penumbra. Away from the sunspots, the Sun's surface looks granulated. Each granule is the top of a convection cell in the Sun's interior.

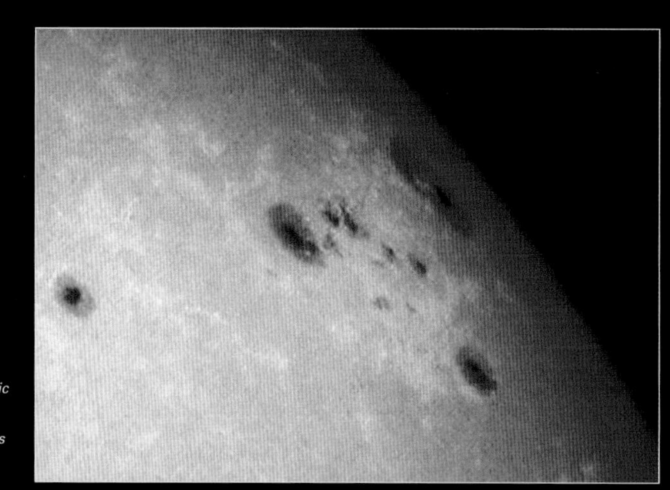

**PHOTOSPHERE**
The base of the photosphere has a temperature of 5,700°C (10,300°F) but its upper layers are cooler and emit less light. Here, the edge of the Sun's disc looks darker because light from it has emanated from these cooler regions.

**FIRST OBSERVATION OF A SOLAR QUAKE**

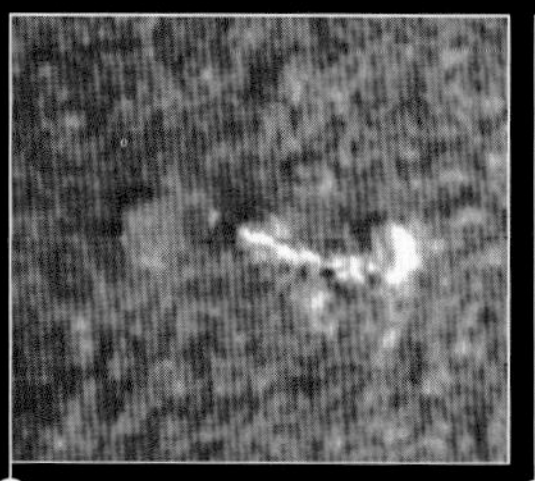

1 In July 1996, by analysing data obtained by an instrument on the SOHO observatory, scientists recorded a solar quake for the first time.

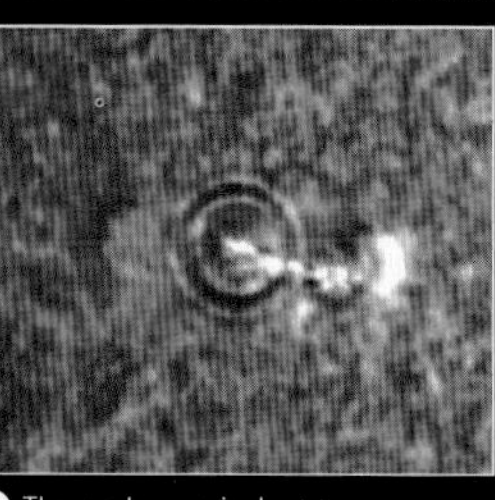

2 The quake, equivalent to an earthquake of magnitude 11, was caused by a solar flare, visible as the white blob with a "tail" to its left.

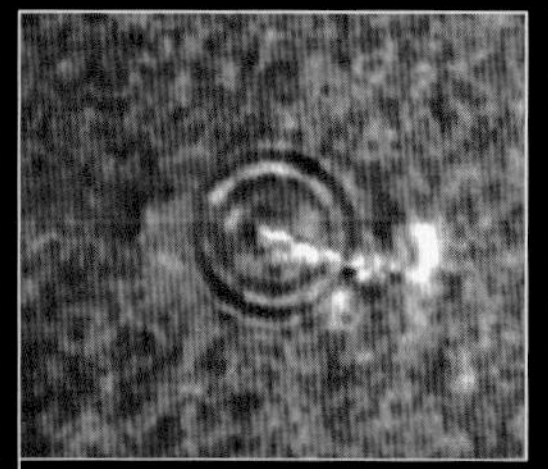

3 The seismic waves looked like ripples on a pond but were 3km (2 miles) high and reached a maximum speed of 400,000kph (248,600mph).

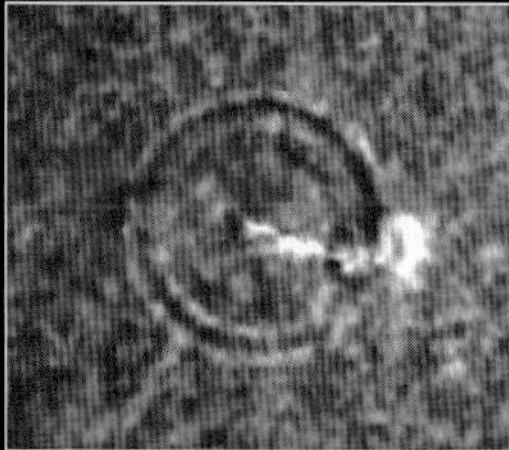

4 Over the course of an hour, the waves travelled a distance equal to 10 Earth diameters before fading into the fiery background.

## JOSEPH VON FRAUNHOFER

A German physicist and optical instrument maker, Joseph von Fraunhofer (1787–1826) is best known for his investigation of dark lines in the Sun's spectrum. Now known as Fraunhofer lines, they correspond to wavelengths of light absorbed by chemical elements in the outer parts of the Sun's atmosphere. Fraunhofer's observations were later used to help determine the composition of the Sun and other stars.

# ATMOSPHERE

As well as forming its visible surface, the photosphere is the lowest layer of the Sun's atmosphere. Above it are three more atmospheric layers. The orangey-red chromosphere lies above the photosphere and is about 2,000km (1,200 miles) deep. From the bottom to the top, its temperature rises from 4,500°C (8,100°F) to about 20,000°C (36,000°F). The chromosphere contains many flamelike columns of plasma called spicules, each rising up to 10,000km (6,000 miles) high along local magnetic field lines and lasting for a few minutes. Between the chromosphere and the corona is a thin, irregular layer called the transition region, within which the temperature rises from 20,000°C (36,000°F) to about 1 million °C (1.8 million °F). Scientists are studying this region in an attempt to understand the cause of the temperature increase. The outermost layer of the solar atmosphere, the corona, consists of thin plasma. At a great distance from the Sun, this blends with the solar wind, a stream of charged particles (mainly protons and electrons) flowing away from the Sun across the Solar System. The corona is extremely hot, 2 million °C (3.6 million °F), for reasons that are not entirely clear, although magnetic phenomena are believed to be a major cause of the heating. Coronal mass ejections (CMEs) are huge bubbles of plasma, containing billions of tons of material, that are occasionally ejected from the Sun's surface through the corona into space. CMEs can disturb the solar wind, which results in changes to aurorae in Earth's atmosphere (see p.70).

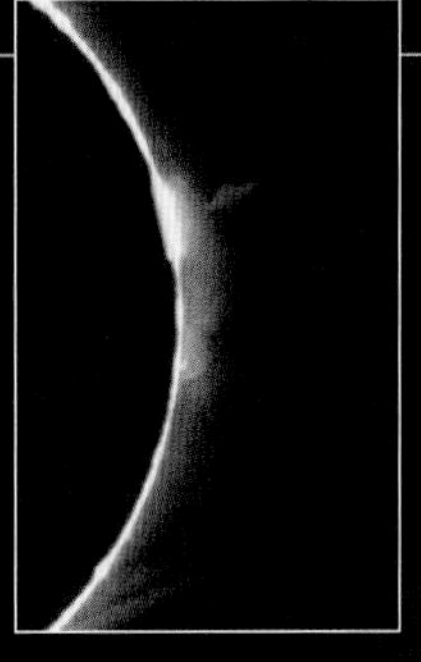

**CHROMOSPHERE**
The Sun's chromosphere is visible here as an irregular, thin red arc adjacent to the much brighter photosphere. Also apparent is a flame-like protuberance from the chromosphere into the corona.

**CORONA**
The outermost layer of the Sun, the corona extends outwards into space for millions of kilometres from the chromosphere. It is most easily observed during a total eclipse of the Sun, as here.

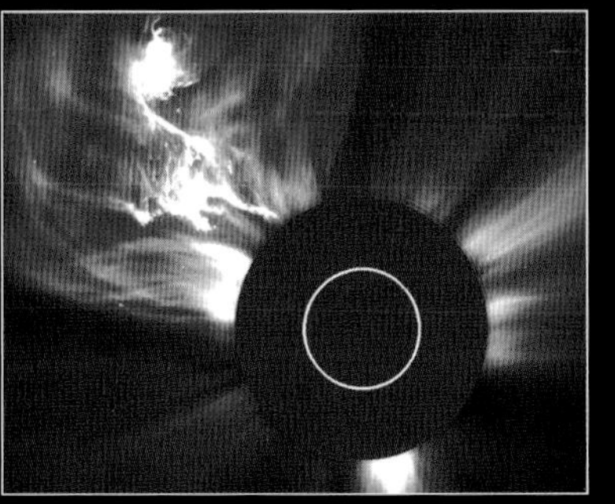

**CORONAL MASS EJECTION**
This image of a coronal mass ejection (top left) was taken by the SOHO solar observatory, using a coronagraph – an instrument that blocks direct sunlight by means of an occulter (the central smooth red area in the image). The white circle represents the occulted disc of the Sun.

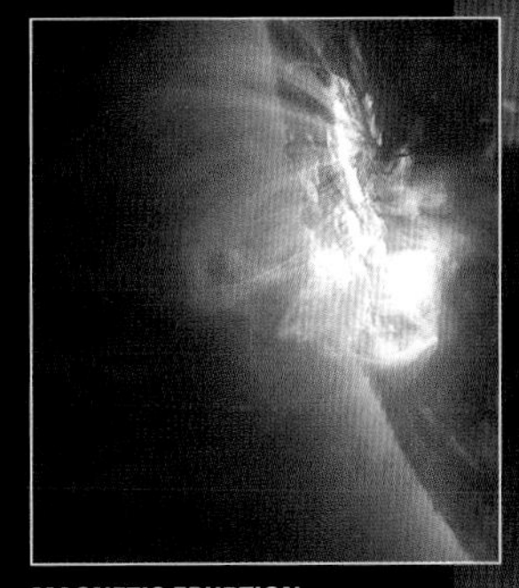

**MAGNETIC ERUPTION**
Hot plasma explodes into the atmosphere, following magnetic field lines. In this TRACE image, colours represent temperature, with blue being the coolest and red, hottest.

**NORTHERN LIGHTS**
When charged particles from the solar wind reach Earth, they can cause aurorae. This photograph of the aurora borealis was taken in Manitoba, Canada.

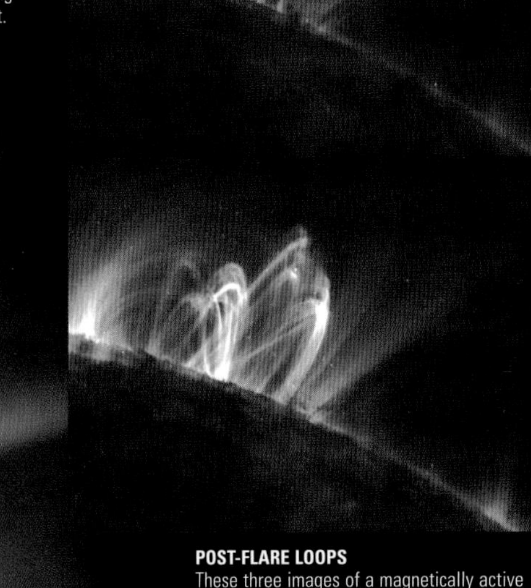

**POST-FLARE LOOPS**
These three images of a magnetically active solar region, taken by the TRACE satellite, span a period of 2.5 hours. The loops in the Sun's corona probably followed a solar flare and consist of plasma heated to exceedingly high temperatures along magnetic field lines.

# MERCURY

36–37 Gravity, motion, and orbits
64–65 Planetary motion
108–11 Beyond Earth
118–19 The family of the Sun

MERCURY IS THE SECOND SMALLEST planet in the Solar System, the closest planet to the Sun, and the richest in iron. The surface environment is extremely harsh. There is hardly any shielding atmosphere, and the temperature rises to a blistering 430°C (800°F) during the day then plummets to an air-freezing -180°C (-290°F) at night. No other planet experiences such a wide range of temperatures. Its surface has been churned up by meteoritic bombardment and is dark and dusty.

## ORBIT

With the exception of Pluto, Mercury has the most eccentric of all the planetary orbits. At perihelion it is only 46 million km (28.6 million miles) from the Sun, but at aphelion it is 69.8 million km (43.3 million miles) away. The plane of Mercury's equator coincides with the plane of its orbit (in other words, its axis of rotation is almost vertical). This means that the planet has no seasons, and that some craters close to the poles never receive any sunlight and are permanently cold. The orbit is inclined at 7° to the plane of the Earth's orbit. Because Mercury orbits inside the Earth's orbit, it displays phases, just like the Moon (see p.62).

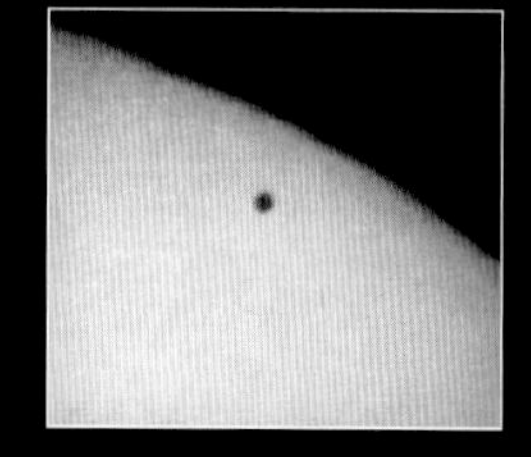

**TRANSIT OF MERCURY**
Mercury passes directly between the Earth and the Sun about 13 times a century, appearing as a small black dot silhouetted against the disc of the Sun.

**SPIN AND ORBIT**
Mercury rotates three times in two orbits (in other words, there are three Mercurian "days" in two Mercurian "years"). This unusual spin–orbit coupling means that for an observer standing on Mercury there would be an interval of 176 Earth days between one sunrise and the next.

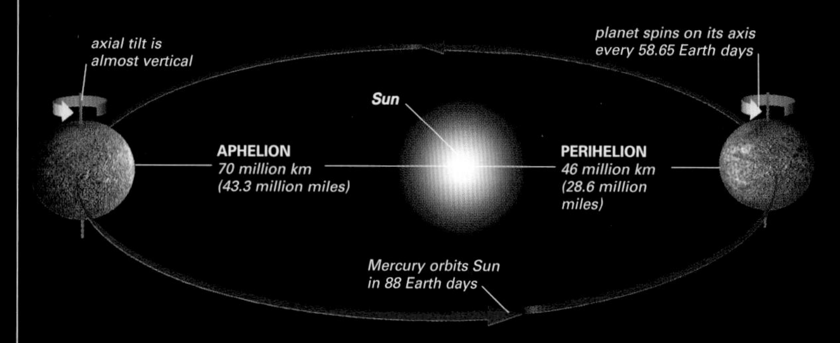


EXPLORING SPACE

## EINSTEIN AND MERCURY

Mercury's perihelion position moves slightly more than Isaac Newton's theories of motion predict. In the 19th century, it was proposed that a planet (called Vulcan) inside Mercury's orbit produced this effect. In his general theory of relativity of 1915, the German physicist Albert Einstein suggested that space near the Sun was curved and correctly predicted the exact amount by which the perihelion would move.

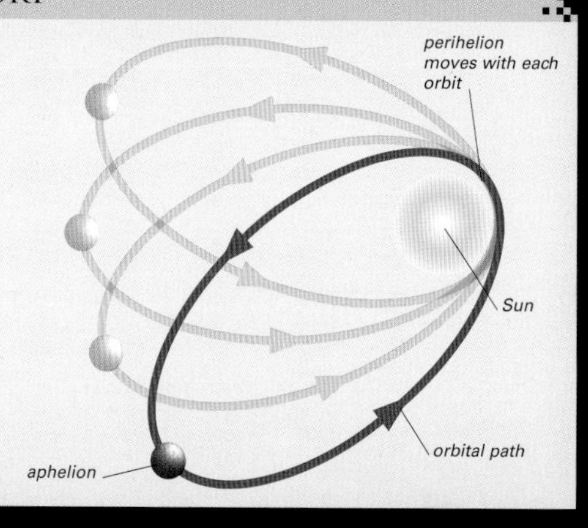


**MERCURY'S WOBBLY ORBIT**
Mercury's perihelion advances by about 1.55° every century, which is 0.012° more than is expected given the gravitational influence of nearby planets.

**POCKMARKED PLANET**
Mercury's heavily cratered surface resembles the highland areas of the Moon. The planet also has large expanses of younger, smooth, lightly cratered plains, rather like the lunar maria.

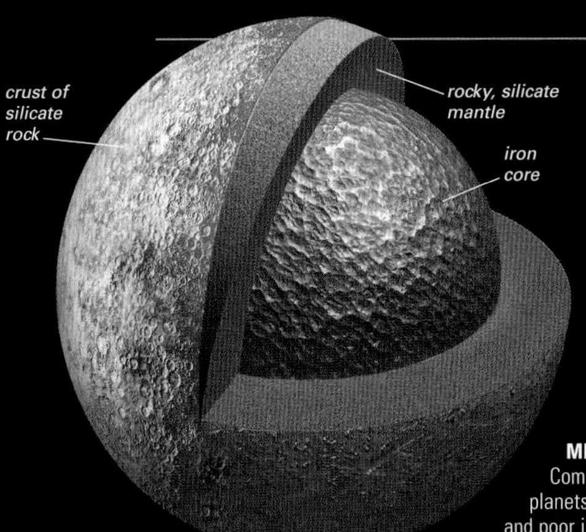


## STRUCTURE

The very high density of Mercury indicates that it is rich in iron. This iron sank to the centre some 4 billion years ago, producing a huge core, 3,600km (2,235 miles) in diameter. There is a possibility that a thin layer of the outer core is still molten. The solid rocky mantle is about 550km (340 miles) thick and makes up most of the outer 25 per cent of the planet. This outer mantle has slowly cooled, and during the last billion years volcanic eruptions and lava flows have ceased, making the planet tectonically inactive. The mantle and the thin crust mainly consist of the silicate mineral anorthosite, just like the old lunar highlands. There are no iron oxides. Unlike on other planets, it seems that all the iron has gone into the core, which produces a magnetic field with a strength that is about one per cent of Earth's magnetic field.

**MERCURY INTERIOR**
Compared to the other rocky planets, Mercury is very rich in metals and poor in heat-producing radioactive elements. Its huge iron core is probably solid.

## ATMOSPHERE

Mercury has a very thin temporary atmosphere because the planet's mass is too small for an atmosphere to persist. Mercury is very close to the Sun, so daytime temperatures are extremely high, reaching 430°C (806°F). The escape velocity is less than half that of Earth's, so hot, light elements in the atmosphere, such as helium, quickly fly off into space. All the atmospheric gases therefore need constant replenishment. Mercury's atmosphere was analysed by an ultraviolet spectrometer onboard the Mariner 10 spacecraft in 1974. Oxygen, helium, and hydrogen were detected in this way, and subsequently atmospheric sodium, potassium, and calcium have been detected by Earth-based telescopes. The hydrogen and helium are captured from the solar wind of gas that is constantly escaping from the Sun. The other elements originate from the planet's surface and are intermittently kicked up into the tenuous atmosphere by the impact of ions from Mercury's magnetosphere and micrometeorite particles from the Solar System dust cloud. The atmospheric gases are much denser on the cold night-side of the planet than on the hot day-side, as the molecules have less energy to escape.

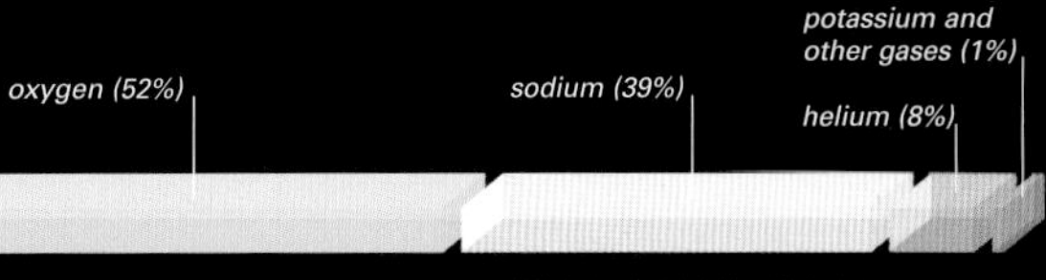


**ATMOSPHERIC COMPOSITION**
Oxygen is the most abundant gas, followed by sodium and helium. However, loss and regeneration of the gases is continuous, and the atmospheric composition can vary drastically over time.

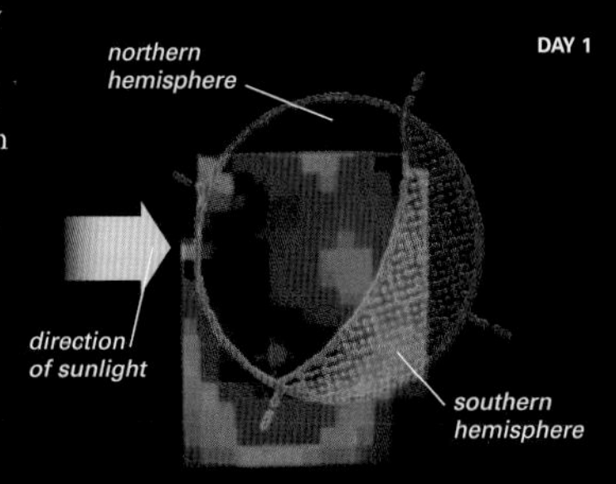


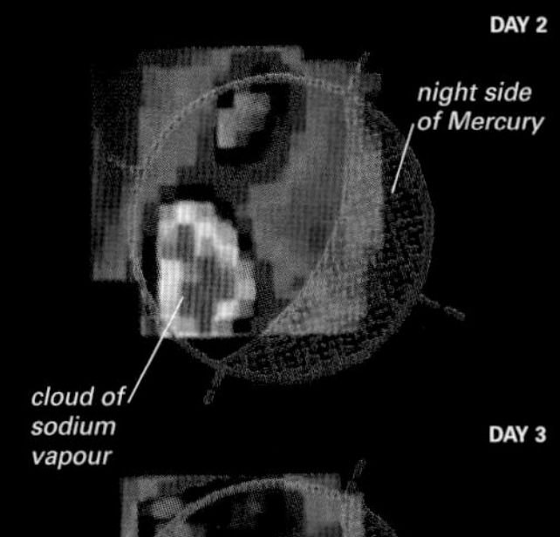


**TEMPORARY ATMOSPHERE**
Thin clouds of sodium suddenly appear over some regions of the planet and then just as quickly disappear. The clouds might be produced by meteorite impacts – the freshly cratered surface releases sodium vapour when it is next heated by sunlight. Another possibility is that ionized particles, which would cause aurorae in Earth's atmosphere, actually hit Mercury's surface and release sodium from the regolith.

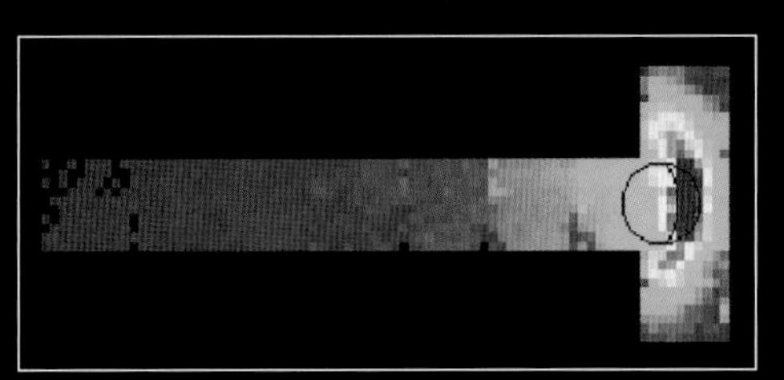

**MERCURY'S SODIUM TAIL**
Pressure exerted by sunlight pushes atoms away from Mercury, and this effect is strongest at perihelion. In 2001, the McMath-Pierce solar telescope in Arizona, USA, imaged a sunward sodium half-corona together with the escape of this sodium gas "downwind", forming a long, comet-like tail.

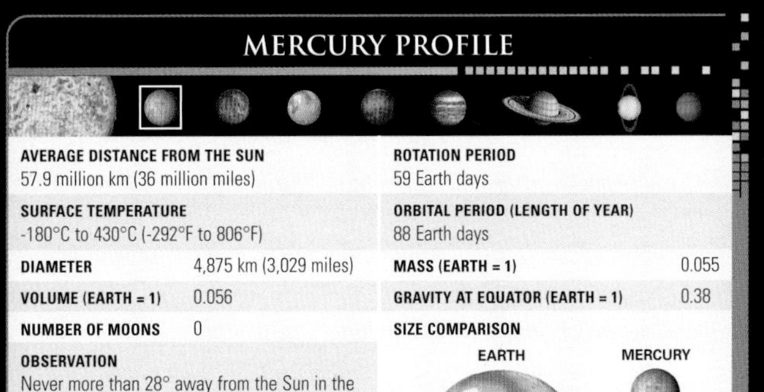

### MERCURY PROFILE

| | | | |
|---|---|---|---|
| **AVERAGE DISTANCE FROM THE SUN** 57.9 million km (36 million miles) | | **ROTATION PERIOD** 59 Earth days | |
| **SURFACE TEMPERATURE** -180°C to 430°C (-292°F to 806°F) | | **ORBITAL PERIOD (LENGTH OF YEAR)** 88 Earth days | |
| **DIAMETER** | 4,875 km (3,029 miles) | **MASS (EARTH = 1)** | 0.055 |
| **VOLUME (EARTH = 1)** | 0.056 | **GRAVITY AT EQUATOR (EARTH = 1)** | 0.38 |
| **NUMBER OF MOONS** | 0 | **SIZE COMPARISON** | EARTH / MERCURY |
| **OBSERVATION** Never more than 28° away from the Sun in the sky, Mercury is always seen at dawn or dusk. It is the most difficult of the nearby planets to spot and is visible only for a few days each month. | | | |

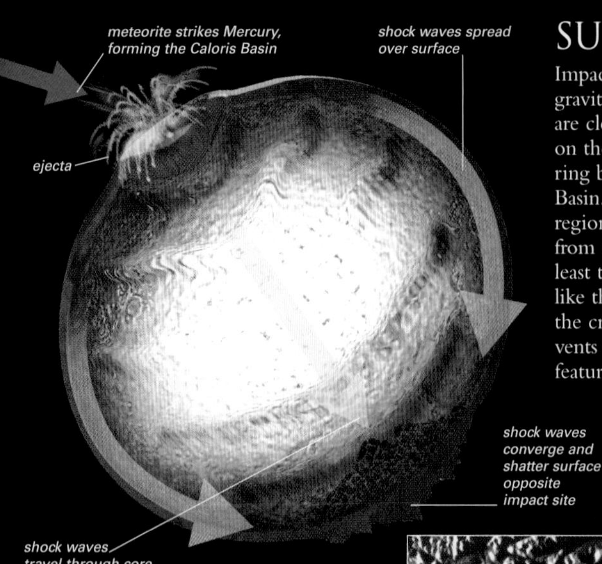


**IMPACT SHOCK WAVES**
A few minutes after the formation of the Caloris Basin, the shock waves generated by the impact came to a focus on the opposite side of the planet. This caused a massive upheaval over an area of 250,000 square kilometres (96,500 square miles), raising ridges up to 1.8km (1.1 miles) high and 5–10km (3–6 miles) across. Crater rims were broken into small hills and depressions.

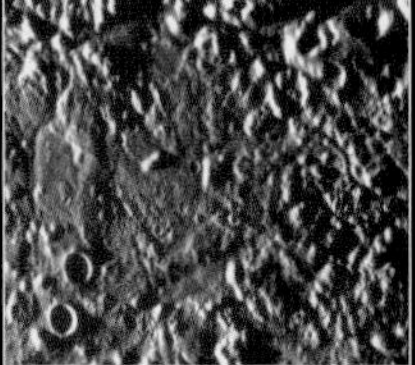

**CHAOTIC TERRAIN OPPOSITE THE CALORIS BASIN**

## SURFACE FEATURES

Impact craters cover Mercury's visible surface. As the surface gravity is about twice that of the Moon, the ejecta blankets are closer to the parent craters and thicker than those found on the Moon. Large meteorite impacts have produced multi-ring basins. A particularly impressive example is the Caloris Basin. On the opposite side of the planet to the basin is a region of strange terrain produced by earthquakes resulting from the impact (see left). The craters are interspersed by at least two generations of flat plains of solidified basaltic lava, like the lunar maria. Fluid lava oozed gently out of vents in the crust and pooled in depressions. Eventually, most of the vents were covered by the lava. Unfortunately, many of the features typical of volcanic activity cannot be identified due to the low resolution of the images taken by the Mariner 10 spacecraft. Mercury's surface also has several ridges, which are up to 1–3km (0.6–1.9 miles) high and 500km (310 miles) long.

**SURFACE COMPOSITION**
In this false-colour mosaic, yellow represents areas of the silicate crust that have been exposed by cratering, while the blue regions are younger volcanic rocks.



**RIDGE FORMATION**
Mercury uniquely has steep, cliff-like north–south ridges stretching all over the surface. There are two probable causes. Tidal forces slowed the planet's rotation, changing its shape from ovoidal to spherical. Also, as Mercury cooled it shrank, decreasing its diameter by 0.1 per cent. The surface was compressed, and parts of the crust were pushed over adjacent areas.

## GEOGRAPHY

Mariner 10's flight path meant that only one hemisphere was seen, but astronomers expect the other side of Mercury to be similar. The 20° longitude meridian passes exactly through the centre of a small crater that has been named Hun Kal, which means "20" in the Mayan language. Other craters have been named after famous artists, authors, and musicians, such as Michelangelo, Dickens, and Beethoven. Most of the plains are named after the word for the planet Mercury in various languages.

EXPLORING SPACE

### MARINER 10

This is the only spacecraft to visit Mercury. On its way, it flew close to Venus, and this deflected the orbit so that, in late March 1974, the spacecraft passed by a half-lit, shadowy planet. At its closest, it was just 707km (439 miles) away. About 2,000 TV pictures were sent back to Earth. Mariner 10 flew by Mercury again on 21 September and 16 March 1975. Unfortunately, the same regions were seen again.

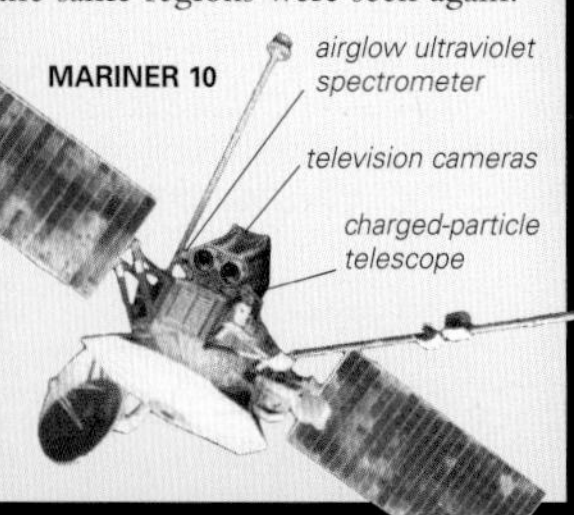


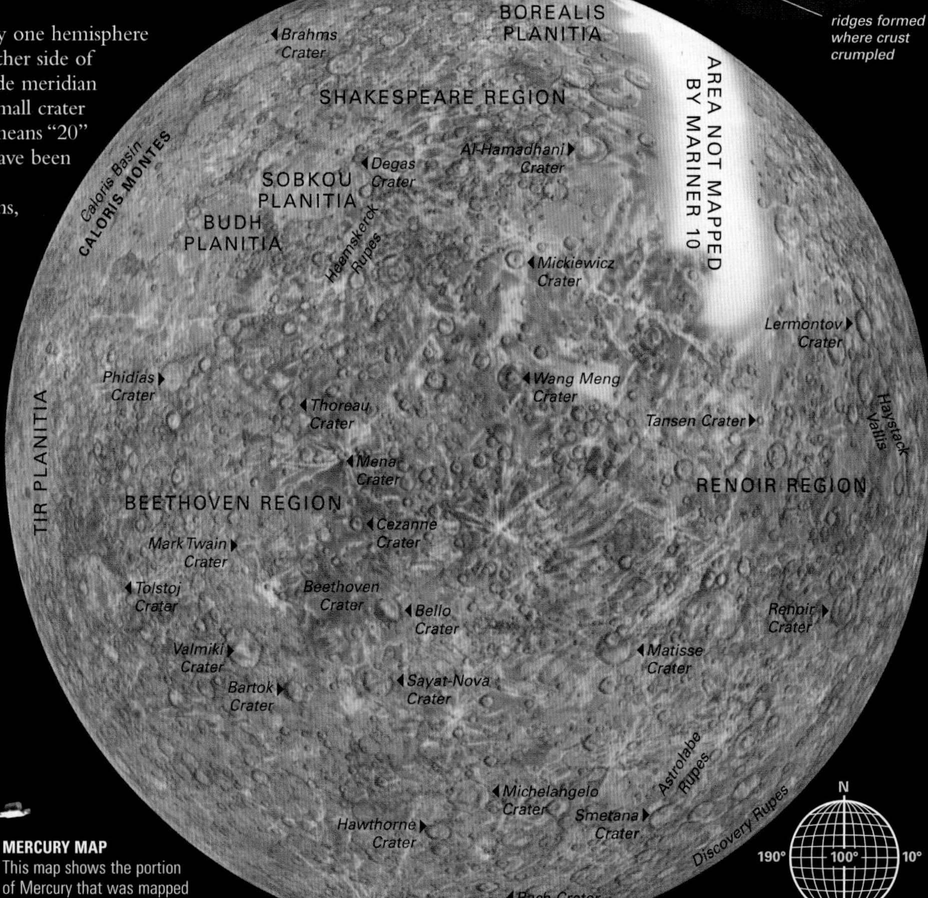


**MERCURY MAP**
This map shows the portion of Mercury that was mapped by Mariner 10.

# FEATURES OF MERCURY

Mercury is covered with impact craters ranging in size from small, bowl-shaped craters to a basin that is a quarter of the diameter of the planet. Its flat plains (called planitiae) were formed when lava flooded low-lying regions. In the last billion years, the impact rate has greatly decreased, volcanism has ceased, and the surface has changed little.

**SOUTH POLE**
The temperature is permanently freezing in Mercury's south polar region because it receives very little sunlight.

SHAKESPEARE REGION

## Caloris Basin

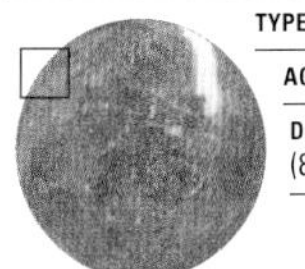

**TYPE** Impact crater
**AGE** 4 billion years
**DIAMETER** 1,350km (840 miles)

The formation of this huge multi-ring basin, which is larger than the US state of Texas, was a major event in the early history of the planet. It is similar to the Orientale Basin on the Moon. The asteroid responsible for creating the basin was probably about 100km (60 miles) across. Ejecta were thrown more than 1,000km (620 miles) beyond the outer rim, producing many radial ridges. The tremendous impact that produced Caloris led to seismic waves being focused on the opposite side of the planet, causing an earthquake. The waves were then reflected back to the basin and fractured the surrounding rocks. Caloris was probably produced towards the end of the period of massive bombardment. Subsequently, the basin floor filled with lava, which cooled and fractured in a polygonal fashion. The basin is now about 2km (1.2 miles deep). The centre of the Caloris Basin lies to the left of the mosaic shown right; the Mariner mission did not photograph it. The name "Caloris" is derived from the Latin for heat, and as the Sun is overhead at perihelion, the basin is one of the two hottest places on Mercury.

**BASIN FLOOR**
The lava-filled basin floor is wrinkled with ridges and furrows and pitted with impact craters of varying sizes.

**RECENT IMPACTS**
The results of smaller and more recent impacts are seen in this mosaic of images of the basin taken by Mariner 10.

RENOIR REGION

## Discovery Rupes

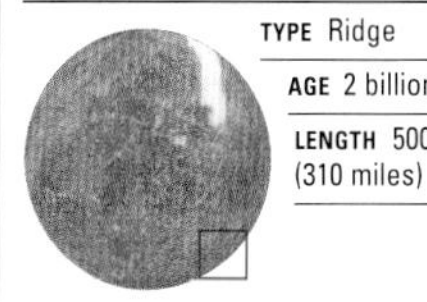

**TYPE** Ridge
**AGE** 2 billion years
**LENGTH** 500km (310 miles)

This cliff-like ridge (*rupes* in Latin), which in places is 2km (1.2 miles) high, is younger than both the craters and the volcanic plains that it crosses. Discovery runs in a northeast–southwest direction and, at 500km (310 miles) in length, it is the longest cliff on Mercury. It was formed when part of the rocky crust cracked and was lifted up as the planet cooled and shrank. Discovery is one of 16 cliff systems that have been discovered on Mercury to date.

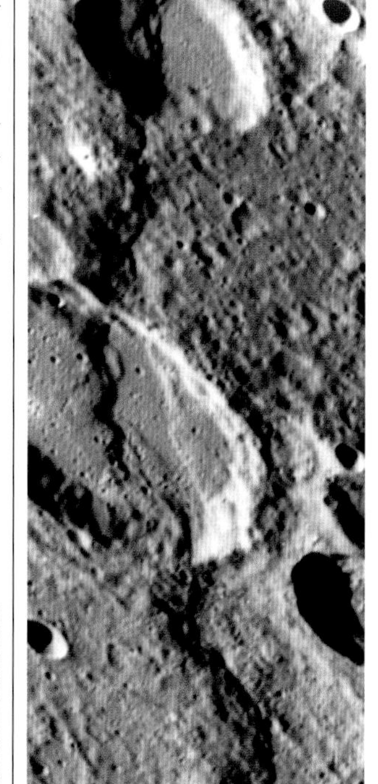

**DISCOVERY RUPES CUTTING THROUGH IMPACT CRATERS**

SHAKESPEARE REGION

## Brahms Crater

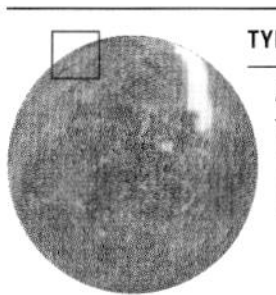

**TYPE** Impact crater
**AGE** 3.5 billion years
**DIAMETER** 97km (60 miles)

A large, mature, complex crater to the north of the Caloris Basin, Brahms has a prominent central mountainous peak about 20km (12 miles) across. The walls have slipped inwards, forming a series of elaborate, concentric, stair-like terraces and a highly irregular rim. This structure is typical of a crater this size – craters with diameters less than 10km (6 miles) are bowl-shaped, and craters with diameters greater than 130km (80 miles) develop central rings (see Bach, right). Radial hills of ejecta surround Brahms.

**CENTRAL PEAK**
This 3km- (2-mile-) high mountain was produced when the subsurface rock rebounded after being struck by an asteroid.

SHAKESPEARE REGION

## Degas Crater

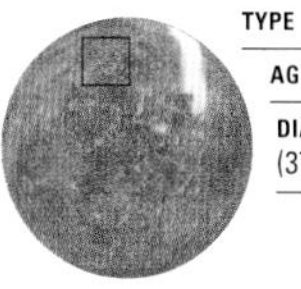

**TYPE** Impact crater
**AGE** 500 million years
**DIAMETER** 60km (37 miles)

This bright ray crater is relatively young, and it overlies the slightly larger Brontë crater to its north. The rays extend out radially for several hundred kilometres, crossing all other features. These highly reflective wispy streaks were caused by the fine pulverized rock ejected from Degas churning up the soil surface on impact. It will take about a billion years for the solar wind to erase them.

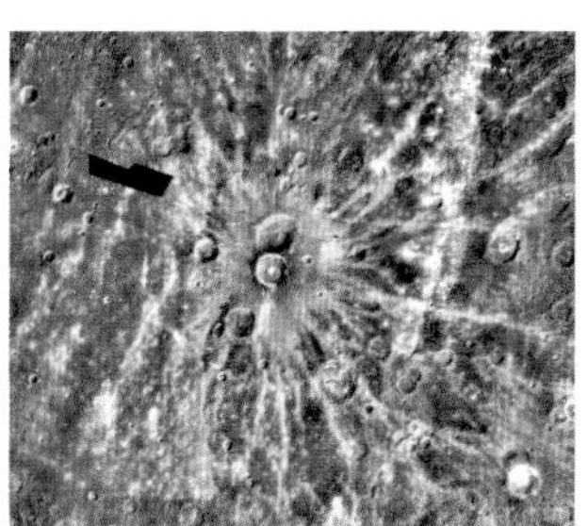

**RAY CRATER**

BEETHOVEN REGION

## Bach Crater

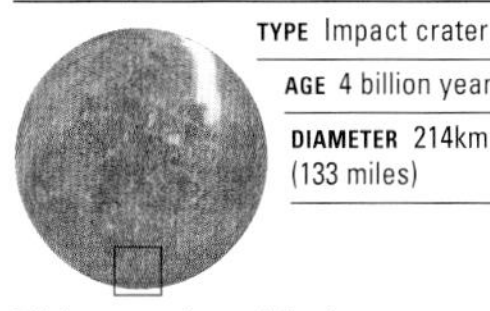

**TYPE** Impact crater
**AGE** 4 billion years
**DIAMETER** 214km (133 miles)

This two-ringed basin represents an intermediate class of craters, between the slightly smaller ones with central mountainous peaks and larger ones with multiple rings. The prominent inner ring is half the width of the outer, and the overall circularity is impressive. Bach was formed towards the end of the period of heavy bombardment. Lava later flooded the crater, producing the smooth floor.

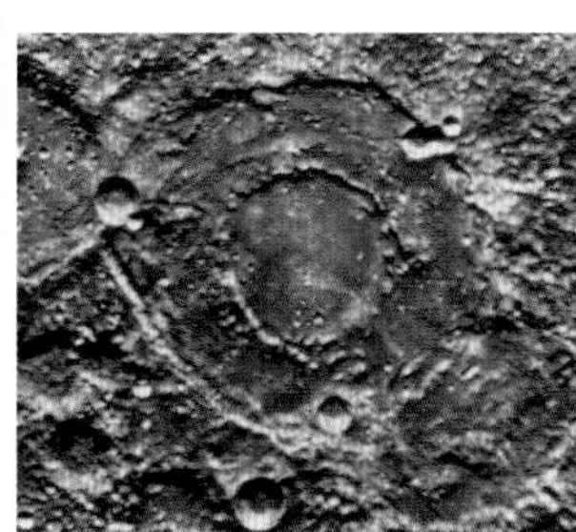

**TWO-RINGED BASIN**

# VENUS

36–37 Gravity, motion, and orbits
64–65 Planetary motion
108–09 Beyond Earth
118–19 The family of the Sun

VENUS IS THE SECOND PLANET FROM THE SUN and Earth's inner neighbour. The two planets are virtually identical in size and composition but these are very different worlds. An unbroken blanket of dense clouds permanently envelops Venus. Underneath lies a gloomy, lifeless, dry world with a scorching surface, hotter than that of any other planet. Radar has penetrated the clouds and revealed a landscape dominated by volcanism.

## ORBIT

Venus's orbital path is the least elliptical of all the planets. It is almost a perfect circle so there is little difference between the planet's aphelion and perihelion distances. Venus takes 224.7 Earth days to complete one orbit. As it orbits the Sun, Venus spins extremely slowly on its axis – slower than any other planet. It takes 243 Earth days for just one spin, which means that a Venusian day is longer than a Venusian year. However, the time between one sunrise and the next on Venus is 117 Earth days. This is because the planet is travelling along its orbit as it spins, and so any one spot on the surface faces the Sun every 117 Earth days. Venus's slow spin is also in the opposite direction from most other planets. Venus does not have seasons as it moves through its orbit. This is because of its almost circular path and the planet's small axial tilt. Venus's orbit lies inside that of the Earth, and about every 19 months Venus moves ahead of Earth on its inside track and passes between our planet and the Sun. At this close encounter, Venus is within 100 times the distance of the Moon.

**SPIN AND ORBIT**
Venus is tipped over by 177.4°. This means its spin axis is tilted by just 2.6° from the vertical. As a result, neither of the planet's hemispheres nor poles points notably towards the Sun during the course of an orbit.

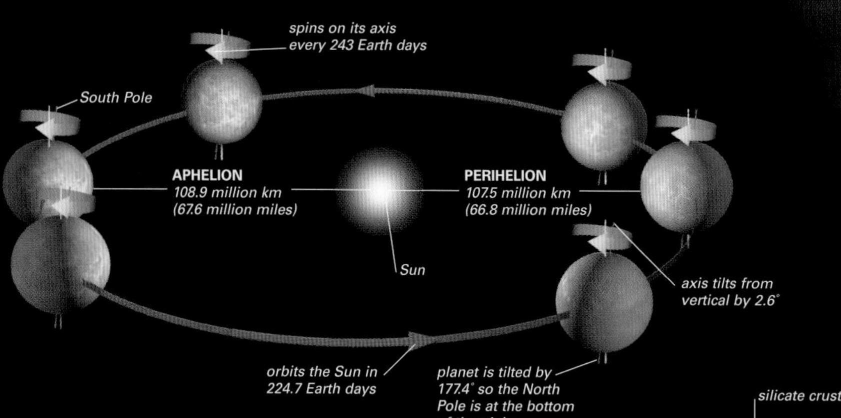


## STRUCTURE

Venus is one of the four terrestrial planets and the most similar of the group to Earth. It is a dense, rocky world just smaller than Earth and with less mass. Venus's Earth-like size and density leads scientists to believe that its internal structure, its core dimensions, and the thickness of its mantle are also similar to Earth's. So, Venus's metal core is thought to have a solid inner part and a molten outer part, like Earth's core. In contrast to Earth, Venus has no detectable magnetic field. The planet spins extremely slowly compared to Earth, far too slowly to produce the circulation of the molten core that is needed to generate a magnetic field. Venus's internal heat – generated early in the planet's history and from radioactive decay in the mantle – is lost through the crust by conduction and volcanism. Heat melts the subsurface mantle material, and magma is released onto the surface.

silicate crust
rocky mantle
molten iron and nickel outer core
solid iron and nickel inner core

**VENUS INTERIOR**
Venus was formed from the same material as Earth about 4.5 billion years ago and has differentiated into distinct layers in a similar way to Earth. A substantial part of the core has solidified; the exact amount still molten is unknown.

**TERRIBLE BEAUTY**
Venus's thick, reflective clouds enable the planet to shine brightly so that from a distance it looks beguiling and beautiful, which is why it was named after the Roman goddess of love and beauty. Close up, it is a different story; no human could survive on this planet.

## VENUS PROFILE

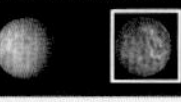

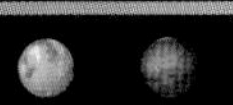

**AVERAGE DISTANCE FROM THE SUN**
108.2 million km (67.2 million miles)

**ROTATION PERIOD**
243 Earth days

**SURFACE TEMPERATURE**
464°C (867°F)

**ORBITAL PERIOD (LENGTH OF YEAR)**
224.7 Earth days

**DIAMETER** 12,104 km (7,521 miles)

**MASS (EARTH = 1)** 0.82

**VOLUME (EARTH = 1)** 0.86

**GRAVITY AT EQUATOR (EARTH = 1)** 0.9

**NUMBER OF MOONS** 0

**OBSERVATION**
Venus is the brightest planet in Earth's sky and is surpassed in brightness only by the Sun and the Moon. Its maximum magnitude is -4.7. It is seen in the early morning or early evening sky.

**SIZE COMPARISON**

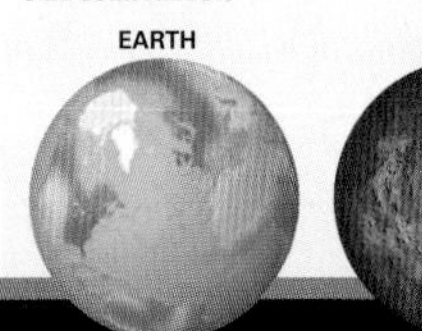


carbon dioxide 96.5%

nitrogen and trace gases 3.5%

## ATMOSPHERE

**COMPOSITION OF ATMOSPHERE**
Along with carbon dioxide and nitrogen, Venus's atmosphere contains traces of other gases, such as water vapour, sulphur dioxide, and argon.

Venus's carbon-dioxide-rich atmosphere stretches up from the ground for about 80km (50 miles). A deck of clouds with three distinct layers lies within the atmosphere. The lowest layer is the densest and contains large droplets of sulphuric acid. The middle layer contains fewer droplets, and the top layer has small droplets. Close to the planet's surface, the atmosphere moves very slowly and turns with the planet's spin. Higher up, in the cloudy part of the atmosphere, fierce winds blow westwards. The clouds speed round Venus once every four Earth days. The clouds reflect the majority of sunlight reaching Venus back into space, and so this is an overcast, orange-coloured world. Venus's equator receives more solar heat than its polar regions. Yet, the surface temperature at the equator and the poles varies by only a few degrees from 464°C (867°F), as do the day and night temperatures. The initial difference generates cloud-top winds that transfer the heat to the polar regions in one large circulation cell. As a result Venus has no weather.

about 80 per cent of sunlight reflects away

cloud deck stretches from about 45km (28 miles) to about 70km (43 miles) above the ground

carbon dioxide in atmosphere holds in heat

reflected light means cloud surface is bright and easy to see

thick layers of sulphuric acid clouds stop most sunlight reaching the surface

infrared radiation is absorbed by carbon dioxide and cannot escape into space

20 per cent of sunlight reaches rocky surface

**GREENHOUSE EFFECT**
Venus's thick cloud layers trap heat and help produce the planet's high surface temperature in the same way that glass traps heat in a greenhouse. Only 20 per cent of sunlight reaches the surface. Once there, it warms up the rock. Heat in the form of infrared radiation is then released, but it cannot escape and adds to the warming process.

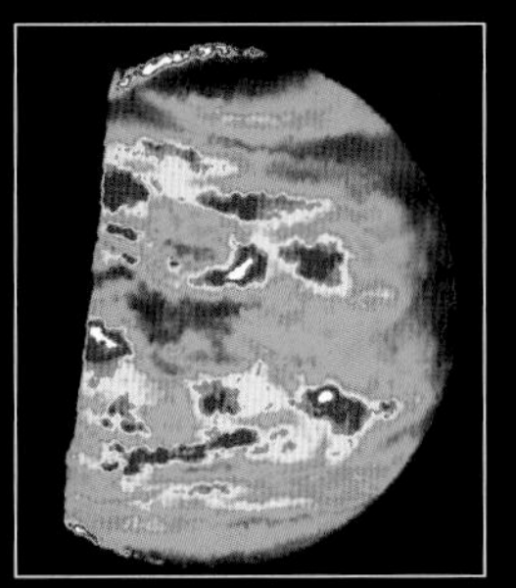

**MIDDLE CLOUD LAYER**
In this infrared image of Venus, heat from the lower atmosphere shines through the sulphuric acid clouds. The colours indicate the relative transparency of the middle cloud layer: white and red are thin clouds; black and blue are thick.

## MISSIONS TO VENUS

Over 20 probes have investigated Venus. The first was the USA's Mariner 2, which made the first successful flyby of a planet in December 1962. Since then, probes have orbited Venus, plunged into its atmosphere, and landed on its scorching hot surface.

**1967 VENERA 4 (USSR)** Sixteen different Venera probes travelled to Venus between 1961 and 1983. Venera 4 parachuted through Venus's atmosphere in October 1967 and returned the information that it is primarily composed of carbon dioxide.

VENERA 4

**1970 VENERA 7 (USSR)** This was the first probe to make a controlled landing on the surface. An instrument capsule landed on the night side and measured the temperature.

**1975 VENERA 9 AND 10 (USSR)** The first image of the surface came from Venera 9. Its lander touched down on 22 October 1975 and returned an image of rocks and soil. Venera 10 did the same three days later.

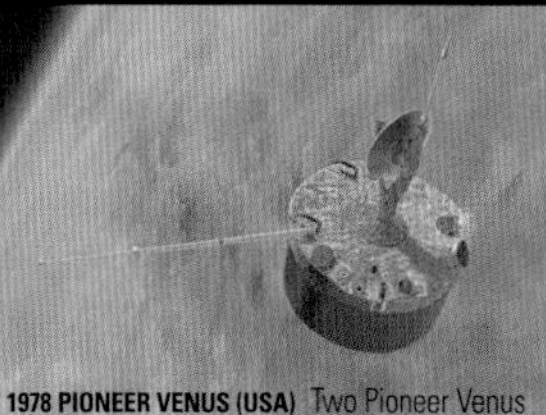

**1978 PIONEER VENUS (USA)** Two Pioneer Venus probes, each with several components, arrived in 1978. An orbiter collected data that was used to make the first global map of Venus, and probes studied the atmosphere.

**1981 VENERA 13 (USSR)** Venera 13 survived on the surface for 2 hours 7 minutes on 1 March 1982. It took the first colour images and analysed a soil sample. Flat slabs of rock and soil can be seen beyond the edge of the probe in the image below.

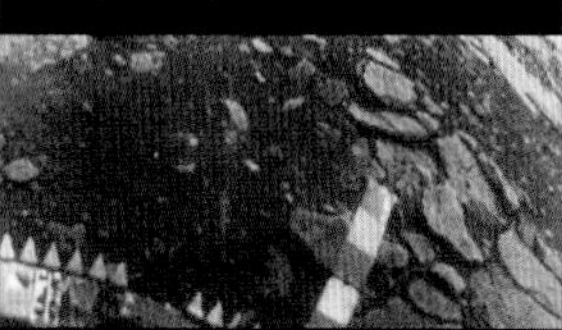

**1990 MAGELLAN (USA)** Between September 1990 and October 1994, Magellan made four 243-day mapping cycles of Venus. It collected gravity data on the fourth cycle.

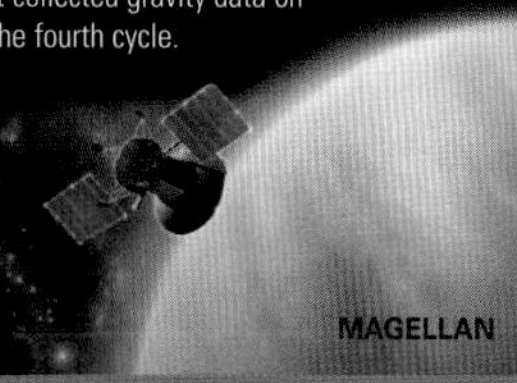

MAGELLAN

# TECTONIC FEATURES

Venus could be expected to have global features similar to those on Earth, but it differs in one key respect: it does not have moving plates. This means that its surface tends to move up or down rather than sideways. Yet, Venus displays many familiar, Earth-like features formed by a range of tectonic processes, as well as some unfamiliar ones, such as arachnoids (see below). Venus has hundreds of volcanoes, from large, shallow-sloped shield volcanoes such as Maat Mons, to small nameless domes. About 85 per cent of the planetary surface is low-lying volcanic plains consisting of vast areas of flood lava. There has been volcanic activity as recently as about 500 million years ago, and it is possible that some of the volcanoes could be active. Other features are a result of the crust pulling apart or compressing. There are troughs, rifts, and chasms, as well as mountain belts such as Maxwell Montes, ridges, and rugged highland regions. Venus's highest mountains and biggest volcanoes are comparable in size to the largest on Earth, but overall this planet has less variation in height.

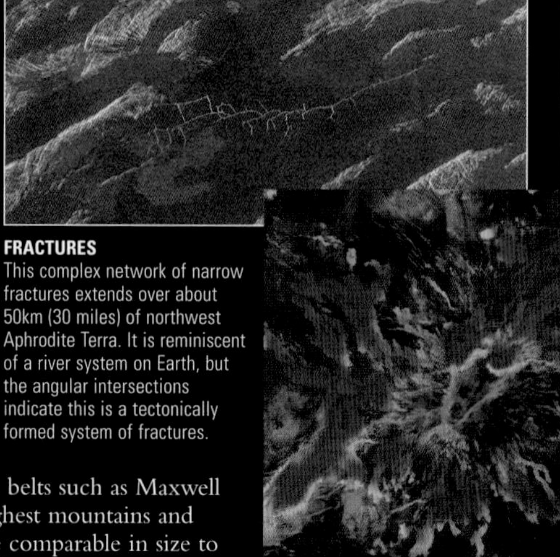

**FRACTURES**
This complex network of narrow fractures extends over about 50km (30 miles) of northwest Aphrodite Terra. It is reminiscent of a river system on Earth, but the angular intersections indicate this is a tectonically formed system of fractures.

**LAVA FLOWS**
Solidified lava flows spread out for hundreds of kilometres in all directions from one of Venus's many volcanoes. The colours represent levels of heat radiation.

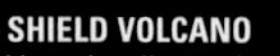

**SHIELD VOLCANO**
Venus's tallest volcano, Maat Mons, rises to almost 5km (3 miles) above the surrounding terrain and is 8km (5 miles) above the planet's mean surface level.

**ARACHNOID**
This spider-like volcanic feature has a central circular depression (or dome) surrounded by a raised rim with radiating ridges and valleys.

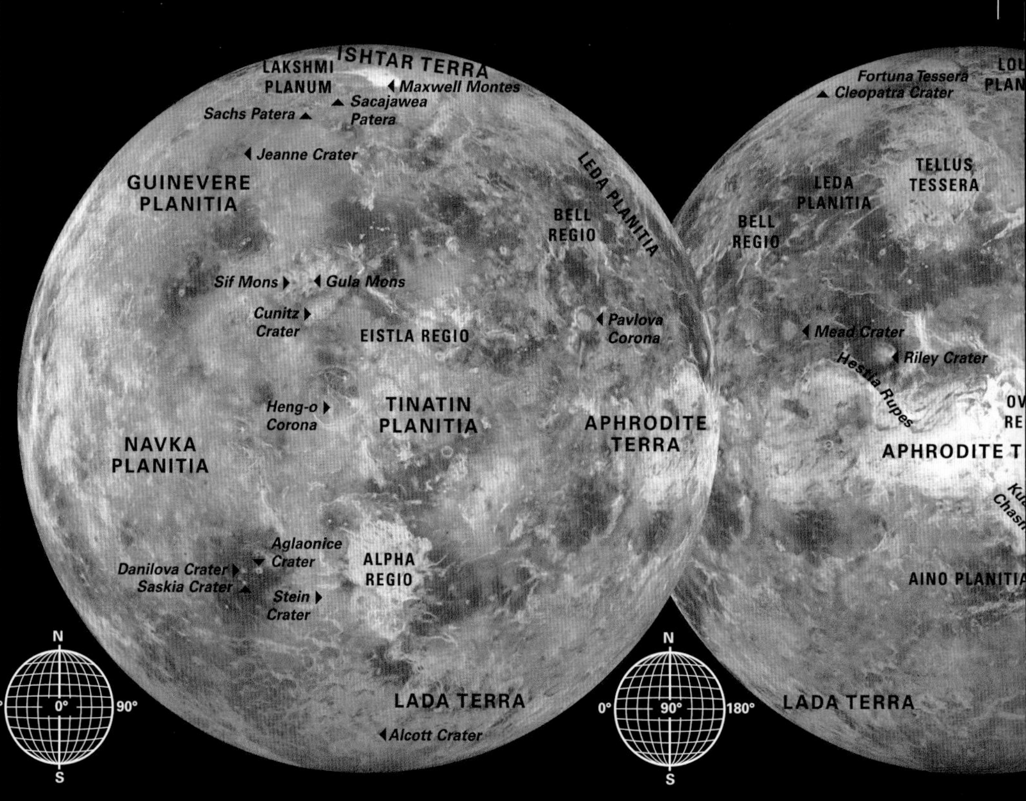


**VENUS MAPS**
These four views combine to show the complete surface of Venus. They have been labelled to show surface features, such as mountains, craters, highland regions, upland areas, and lowland plains.

## IMPACT CRATERS

Although many hundreds of impact craters have been identified on Venus, this total is low compared to that for the Moon and Mercury. There were more craters in the past, but they were wiped out by resurfacing due to volcanic activity about 500 million years ago. Venus's craters have some characteristics not seen elsewhere in the Solar System, because its dense atmosphere and high temperature affect the incoming impactor and crater ejecta. Ejecta can, for example, be blown by winds and form fluid-like flows. And some potential impactors are too small to reach the surface intact. They break up in the atmosphere, and either a resultant shock wave pulverizes the surface or a blanket of fine material formed by the break-up produces a dark halo mark before a crater forms. Wind has also modified the surface, creating wind streaks and what may be sand dunes.

**WIND STREAK**
A 35km- (22-mile-) long tail of material has been created on the northeast side of this small volcano by prevailing winds.

**UNUSUAL CRATER**
This small crater, about 6km (4 miles) across, has terraced walls and lobes of ejecta radiating out from the rim to give it an unusual starfish-like appearance.

**DARK HALO**
A dark halo surrounds a bright feature that appears to be a cluster of small impacts, ejecta, and debris formed by an impactor that broke up in the atmosphere.

**WIND EROSION**
Impact debris thrown 500km (300 miles) to the northeast of Mead Crater has been modified by surface winds. Wind streaks are visible, but it is not known if these are bright streaks on dark terrain, or vice versa.

## GEOGRAPHY

Present maps of Venus are based on data collected by the Magellan probe (see panel, opposite), with additional information from previous missions. The colouring of the maps below and Magellan images is based on the colours recorded by Venera 13 and 14. The orange colour is due to the atmosphere filtering out the blue light. The following terminology is used for the surface features: lowland plains are termed planitia; high plains, planum; extensive landmasses, terra; mountain ranges, montes; and mountains or volcanoes, mons. A chasma is a deep, elongated, steep-sided depression. The features are all named after women, both historical and mythological, with the exception of Maxwell Montes (see p.132).

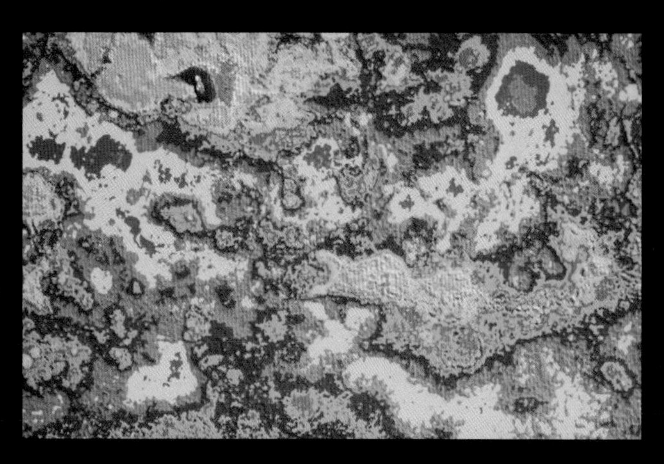

**TOPOGRAPHIC MAP**
This relief map, based on Pioneer Venus data, covers the surface area from approximately 78°N to 63°S. High land is coloured yellow, with the highest of all in red. The green-coloured massifs of Ishtar Terra (top) and Aphrodite Terra (right) stand out from the surrounding lowland shown in blue.

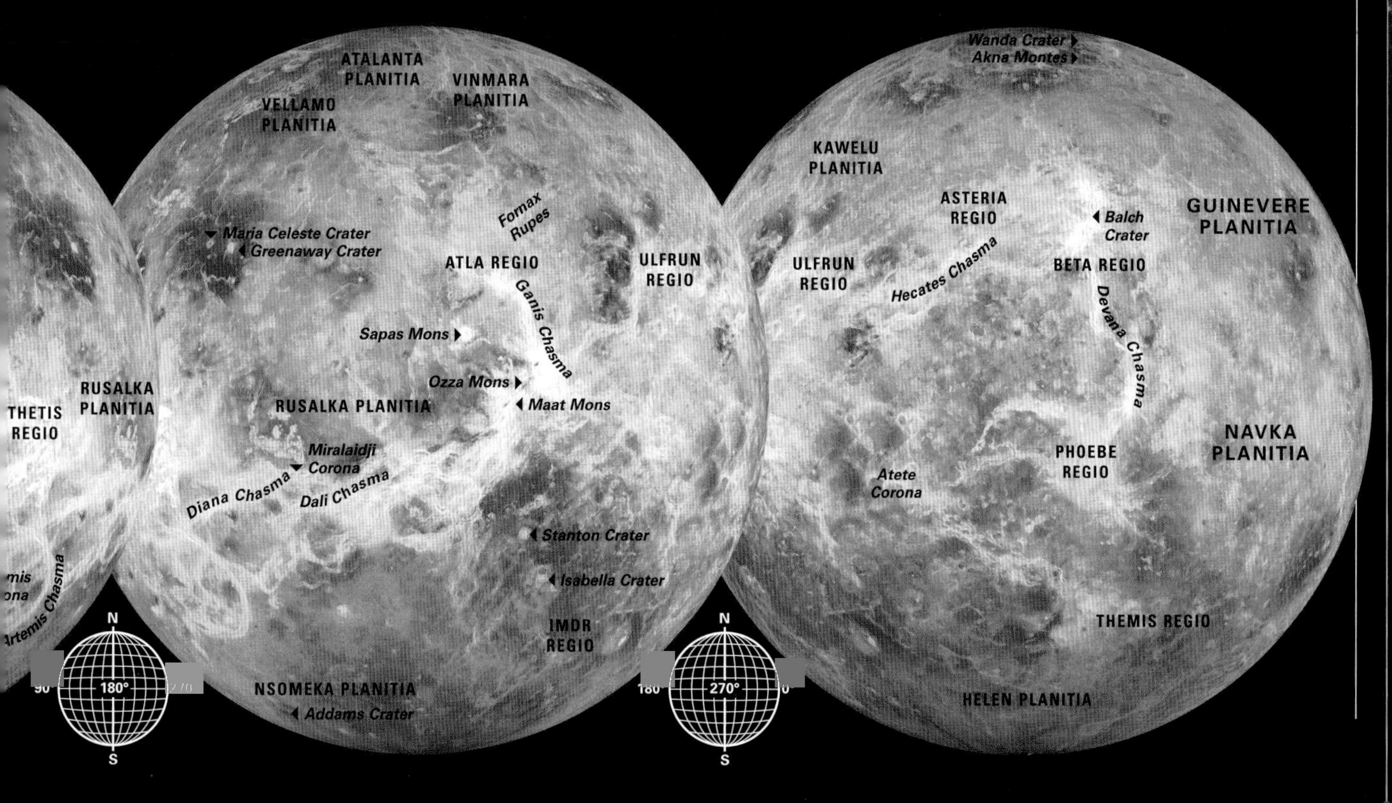

# TECTONIC FEATURES

Thanks to space-probe exploration, astronomers have a full and detailed picture of Venus's varied landscape. The planet has three main highland regions, termed terra. They are Aphrodite, which dominates the equatorial zone, and Lada and Ishtar. Over 20 smaller, upland areas, termed regio, are found around the planet. Extensive lowland plains, termed planitia, complete the global picture. Volcanic activity is evident across most of the surface but the volcanoes are not randomly distributed. There are more in the uplands, particularly in Atla and Beta Regio, than in the highlands or plains.

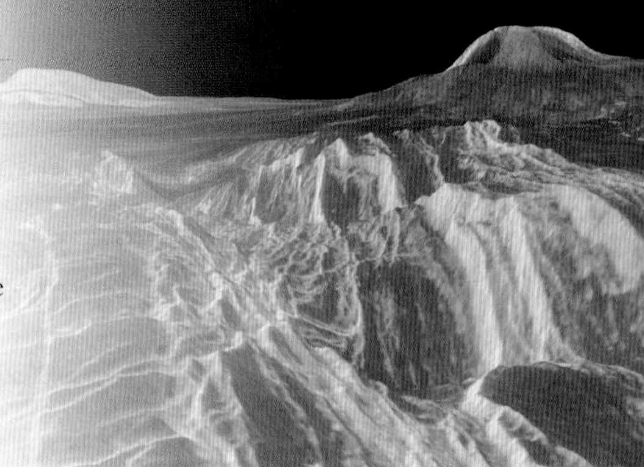

**VOLCANIC TERRAIN**
This view across western Eistla Regio is typical of the Venusian surface. The volcanoes on the skyline are Sif Mons (left) and Gula Mons.

ISHTAR TERRA

## Ishtar Terra

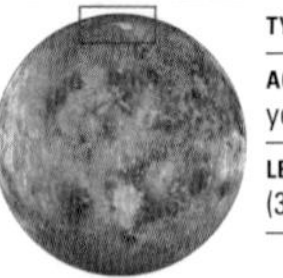

**TYPE** Highland terrain
**AGE** Under 500 million years
**LENGTH** 5,610km (3,485 miles)

Ishtar Terra is a large plateau about the size of Australia, which stands 3.3km (2 miles) above the surrounding lowlands. It is the nearest thing on Venus to the continents on Earth. Its western region is the elevated plateau Lakshmi Planum, which is bounded at the northwest by the Akna Montes and the Freyja Montes, and to the south by the Danu Montes. The steep-sided Maxwell Montes range makes up the eastern part of Ishtar Terra, along with a deformed area, Fortuna Tessera, to the mountain range's north and east. The plateau was possibly formed as areas of planetary crust were driven together. It is likely that beneath Ishtar there is cooled, thickened crust that is kept up by a rising region of mantle.

**LAVA CHANNEL**
Running for well over 2,000km (1,200 miles), this lava channel is unusually long.

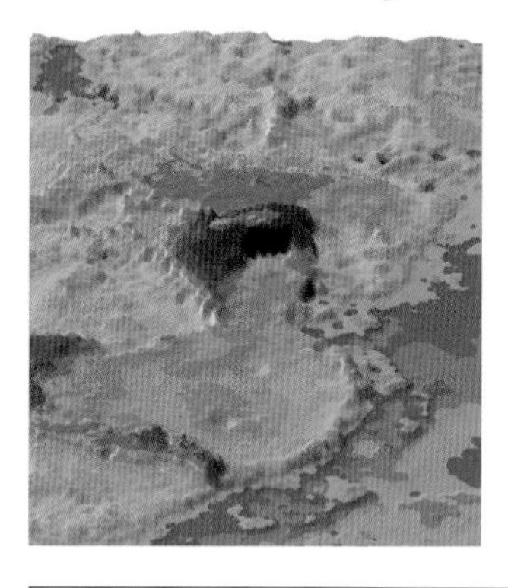

**MASSIVE PLATEAU**
Looking eastwards across Ishtar Terra, this false-coloured view, created from Pioneer-Venus 1 data, highlights the varying height of the terrain. Blue represents the lowest elevation, and red is the highest.

ISHTAR TERRA

## Akna Montes

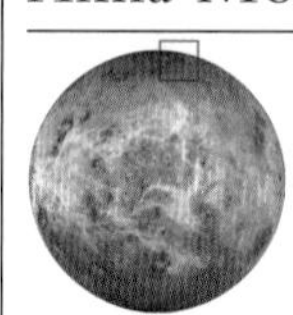

**TYPE** Mountain range
**AGE** Under 500 million years
**LENGTH** 830km (515 miles)

Forming the western border of Lakshmi Planum, Akna Montes is a ridge belt that appears to be the result of folding due to northwest–southeast compression. The mountain building is thought to have occurred after the plains formed, as the plains in this region seem to be deformed.

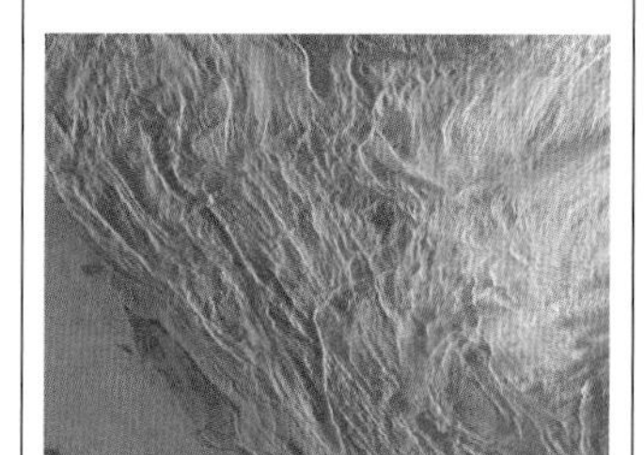

**FOLDING DUE TO COMPRESSION**

ISHTAR TERRA

## Fortuna Tessera

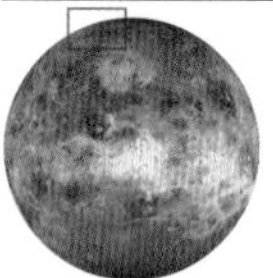

**TYPE** Ridged terrain
**AGE** Under 500 million years
**LENGTH** 2,801km (1,739 miles)

Fortuna Tessera is an area of north–south trending ridges about 1,000km (600 miles) wide. The distinctive pattern made by the region's intersecting ridges and grooves led to this type of terrain originally being called parquet terrain, after its resemblance to woodblock flooring, although it is now termed tessera. The image shown here is a view looking westwards across about 250km (155 miles) of Fortuna Tessera towards the slopes of Maxwell Montes (coloured in blue).

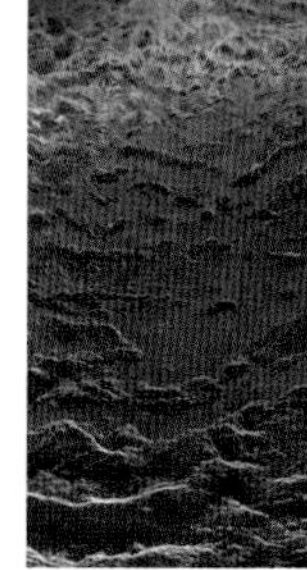

**RIDGES**

ISHTAR TERRA

## Lakshmi Planum

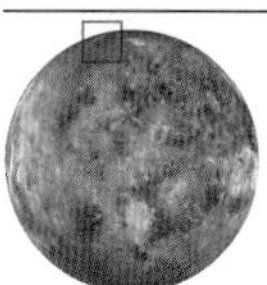

**TYPE** Volcanic plain
**AGE** Under 500 million years
**LENGTH** 2,345km (1,456 miles)

The western part of Ishtar Terra consists of Lakshmi Planum. This is a smooth plateau, 4km (2.5 miles) high, formed by extensive volcanic eruptions. The plateau is encircled by curving mountain belts – the Danu, Akna, Freyja, and Maxwell Montes – and steep escarpments such as Vesta Rupes to its southwest. This massive plain covers an area that is about twice the size of Earth's Tibetan Plateau (see pp.144–45). Two large volcanic features, the Colette Patera and Sacajawea Patera (see opposite), which punctuate the otherwise relatively smooth plain, were identified in Venera 15 and 16 data. Their floors lie over 2.5km (1.5 miles) below the plateau level. There are just a few planums on Venus, all named after goddesses. Lakshmi is the Indian goddess of love and war.

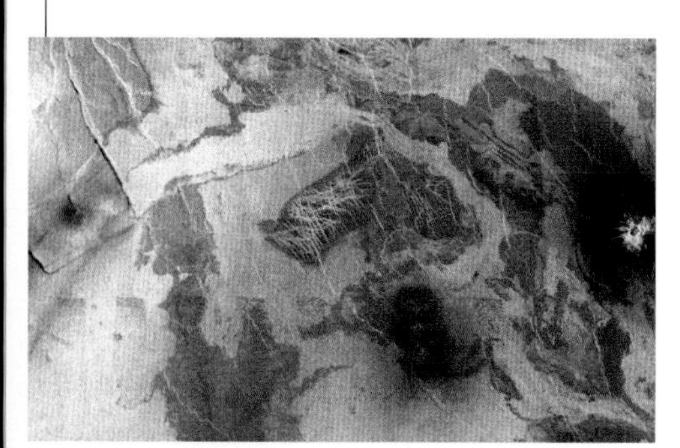

**LAVA FLOWS**
The eastern Lakshmi region is covered by lava flows. The dark flows are smooth, and the light ones are rough in texture. A bright impact crater can be seen on the right.

ISHTAR TERRA

## Maxwell Montes

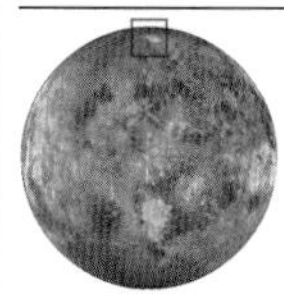

**TYPE** Mountain range
**AGE** Under 500 million years
**LENGTH** 797km (495 miles)

The Maxwell Montes mountain range forms the eastern boundary of Lakshmi Planum. It is the highest point on Venus, rising over 10km (6 miles) above the surrounding lowlands. In its higher regions, the ridges, which are 10–20km (6–12 miles) apart, have a sawtooth pattern. The mountains fall away to Fortuna Tessera to the east. The western side is a complex of grooves and ridges and is particularly steep – Magellan data revealed that the southwestern flank has a slope of 35°. The mountain range was formed by compression and crustal foreshortening, which produced folding and thrust faulting. Venusian mountain ranges are usually named after goddesses, but Maxwell Montes is named after the British physicist James Clerk Maxwell, a pioneer of electromagnetic radiation.

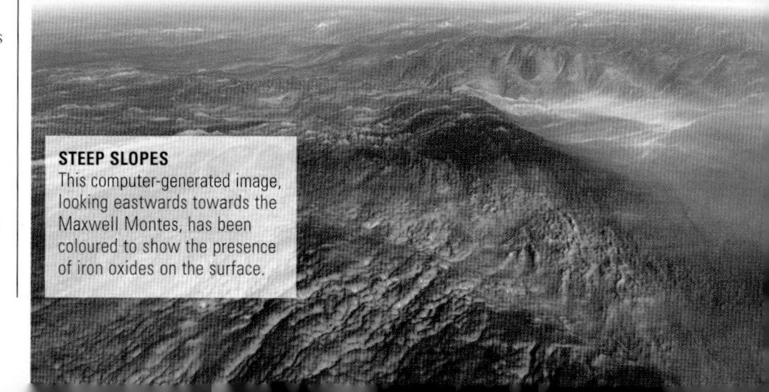

**STEEP SLOPES**
This computer-generated image, looking eastwards towards the Maxwell Montes, has been coloured to show the presence of iron oxides on the surface.

ISHTAR TERRA

## Sacajawea Patera

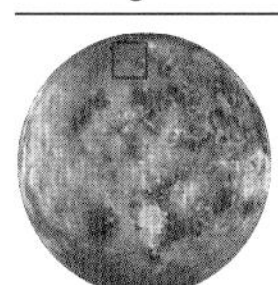

**TYPE** Caldera

**AGE** Under 500 million years

**DIAMETER** 233km (145 miles)

Sacajawea Patera is an elliptically shaped volcanic caldera on Lakshmi Planum. It is thought to have formed when a large underground chamber was drained of magma and then collapsed. The resulting caldera then sagged. The depression is about 1.2km (0.75 miles) deep and is enclosed by a zone of concentric troughs and scarps that extend up to 100km (60 miles) in length and are 0.5–4km (0.3–2.5 miles) apart. They are thought to have formed as the caldera sagged.

Sacajawea was a Shoshoni Indian woman, born in 1786, who worked as an interpreter.

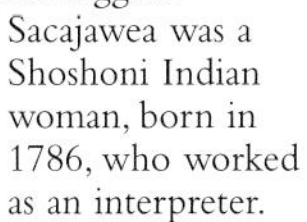

**SAG-CALDERA**
Bright linear scarps extend out from Sacajawea Patera's eastern edge.

GUINEVERE PLANITIA

## Sachs Patera

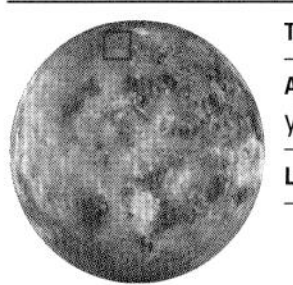

**TYPE** Caldera

**AGE** Under 500 million years

**LENGTH** 65km (40 miles)

Sachs Patera is about 130m (420ft) deep and is surrounded by scarps spaced 2–5km (1–3 miles) apart. A second, separately produced arc-shaped set of scarps lies to the north (top in the image below) of the main caldera. Solidified lava flows extend 10–25km (6–16 miles) to the north and northwest of those scarps.

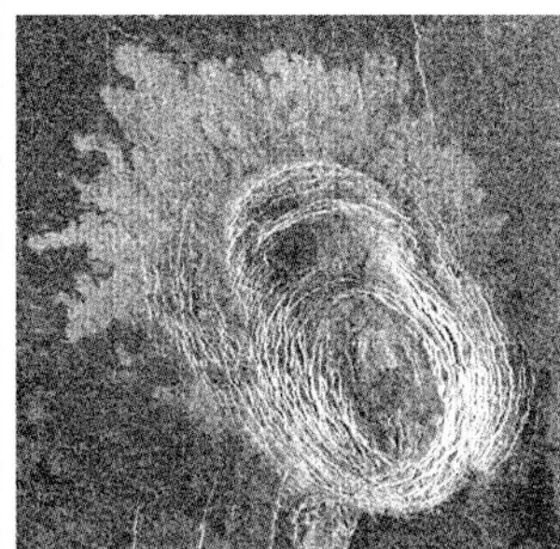

**SCARPS AROUND SACHS PATERA**

BETA REGIO

## Beta Regio

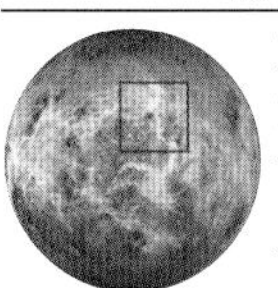

**TYPE** Volcanic highland

**AGE** Under 500 million years

**LENGTH** 2,869km (1,781 miles)

Beta Regio is a large highland region dominated by Rhea Mons and Theia Mons. Rhea, which lies 800km (500 miles) to the north of Thea, was originally thought to be a volcano but Magellan data revealed it to be an uplifted massif cut through by a rift valley, the Devana Chasma (right). Theia Mons is a volcano superimposed onto the rift.

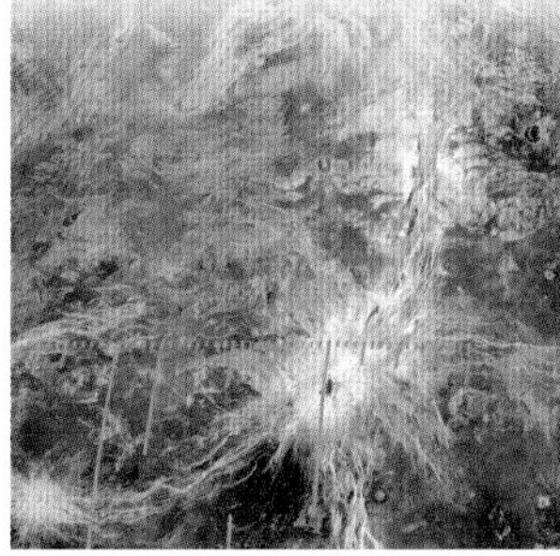

**RHEA AND THEIA MONS**

BETA REGIO

## Devana Chasma

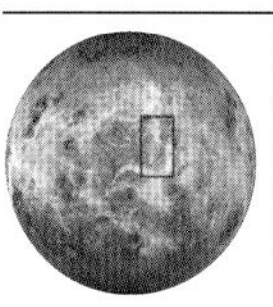

**TYPE** Fault

**AGE** Under 500 million years

**LENGTH** 4,600km (2,860 miles)

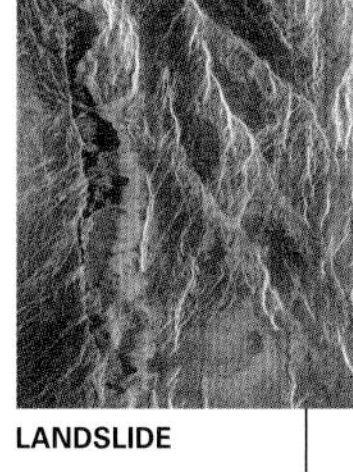

**LANDSLIDE**

Devana Chasma is a large fault that cuts through Beta Regio (left). This major rift valley extends in a north–south direction and was produced as the planet's crust pulled apart and the surface sank to form a valley floor with steep sides. It is similar to the Great Rift Valley on Earth (see p.142). Devana Chasma slices through Rhea Mons and Theia Mons. The fault is over 2km (1.2 miles) deep and about 80km (50 miles) wide near Rhea Mons. Elsewhere it is broader, as much as 240km (150 miles) wide. To the south of Theia Mons, it continues to the highland region Phoebe Regio and reaches depths of 6km (3.7 miles). Faults and grabens cut through and fan out from parts of the rift valley.

**WESTERN EISTLA REGIO**
Lava flows extending for hundreds of kilometres fill the foreground of this image. In the distance, Gula Mons (left) and Sif Mons (right) rise above the plain, about 730km (450 miles) apart.

EISTLA REGIO

## Eistla Regio

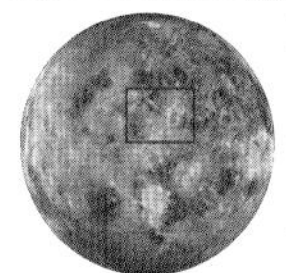

**TYPE** Volcanic highland

**AGE** Under 500 million years

**LENGTH** 8,015km (4,977 miles)

Eistla Regio is one of Venus's smaller upland areas, which are located in the lower basin land separating the major highland areas. Eistla Regio lies in the equatorial region to the west of the major highland, Aphrodite Terra. It is a series of broad crustal rises, each of which is several thousand kilometres across. The landscape was seen for the first time in the 1980s, when data collected by the Pioneer Venus Orbiter was used to produce the first accurate topographic map of Venus. Prominent features, such as the volcanoes Sif Mons and Gula Mons (right) and their lava flows, were clearly visible in the west of the region. Eistla Regio was also the first of the equatorial highlands imaged in the 1990s by Magellan, which revealed more detail of the broad volcanic rises and rift zones. An unusual type of volcanic dome, unique to Venus, is found within Eistla Regio. The domes are circular, flat-topped mounds of lava and so are often referred to as pancake domes. It is believed that when the lava erupted through the surface it was highly viscous and so didn't flow freely. Cracks and pits in the domes are caused by cooling and withdrawal of lava.

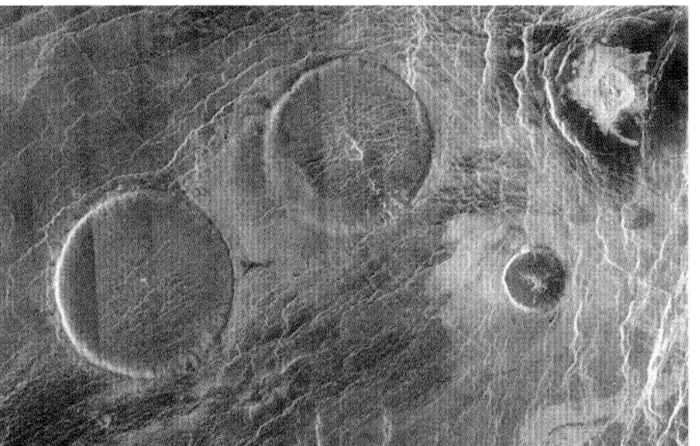

**PANCAKE DOMES**
The two large, flat pancake domes are each about 60km (37 miles) across and less than 1km (0.6 miles) in height.

EISTLA REGIO

## Gula Mons

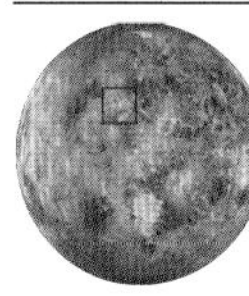

**TYPE** Shield volcano

**AGE** Under 500 million years

**HEIGHT** 3km (2 miles)

Gula Mons is the larger of the two volcanoes that dominate the highland rise of western Eistla Regio (left). At its widest, it measures about 400km (250 miles) across. This shield volcano is encircled by hundreds of kilometres of lava flows. It does not have a caldera at its summit but a fracture line that is 150km (93 miles) long. The volcano is also at the centre of an array of crustal fractures. A particularly prominent one, Guor Linear, is a rift system that extends for at least 1,000km (600 miles) from the southeast flank.

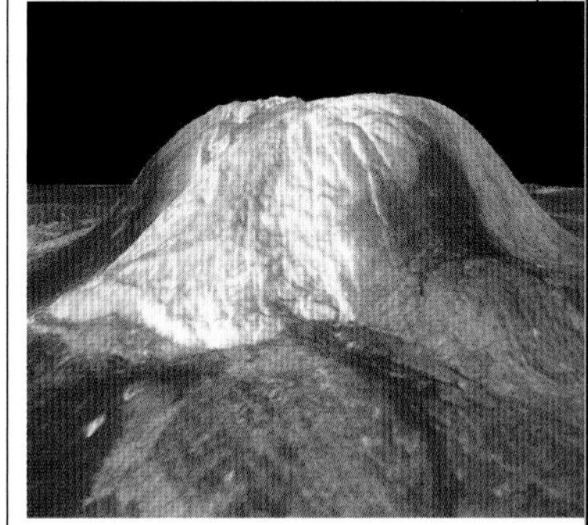

**SOUTHWEST SLOPES OF SUMMIT**

ATLA REGIO

## Sapas Mons

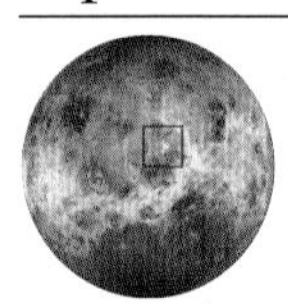

**TYPE** Shield volcano

**AGE** Under 500 million years

**HEIGHT** 1.5km (1 mile)

Rising 1.5km (1 mile) above the surrounding terrain and with a diameter of about 217km (135 miles), Sapas Mons is one of Venus's shield volcanoes. These are shaped like a shield or inverted plate, with a broad base and gently sloping sides, and are like those found on Earth's Hawaiian Islands. Sapas Mons is located in the Atla Regio, a broad volcanic rise just north of Venus's equator with an average elevation of 3km (2 miles). The region is believed to have formed as a result of large volumes of molten rock welling up from the planet's interior. It is home to some particularly large shield volcanoes, which are linked by complex systems of fractures. These include Ozza Mons, which is 6km (4 miles) high, and the largest Venusian volcano, Maat Mons, which is 8km (5 miles) high. Sapas Mons is covered in lava flows and grew in size as the layers of lava accumulated. The flows near the summit appear bright in Magellan radar images, suggesting that these are rougher than the dark flows farther down the volcano's flanks. The flows commonly overlap, and many originate from the flanks rather than the summit. The summit has two mesas with flat to slightly convex tops. Nearby are groups of pits up to 1km (0.6 miles) wide that are believed to have formed when underground chambers of magma drained away and the surface collapsed. The shield volcanoes are in the main named after goddesses: Sapas was a Phoenician goddess; Ozza, a Persian one; and Maat, an Egyptian.

**CRATER ON EASTERN FLANK**
Bright lava flows from Sapas Mons have stopped short of an impact crater on the volcano's eastern side. The flows, which are tens of kilometres long, cover some of the ejecta and so are younger than the crater.

**DOUBLE SUMMIT**
In this Magellan image of Sapas Mons taken from directly overhead, the two flat-topped mesas, which give the volcano the appearance of a double summit, appear dark against the bright lava flows. The area shown covers about 650km (400 miles) from top to bottom.

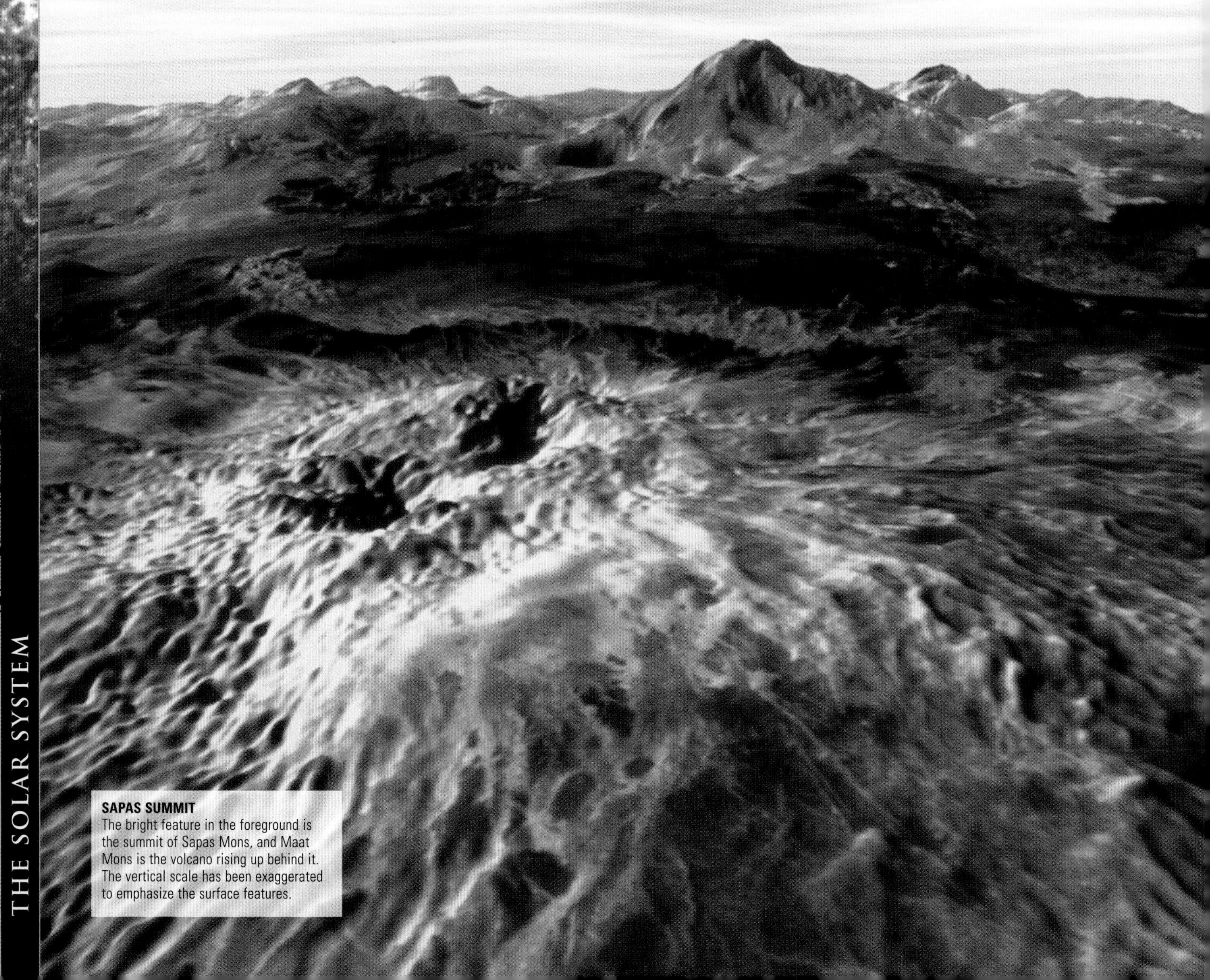

**SAPAS SUMMIT**
The bright feature in the foreground is the summit of Sapas Mons, and Maat Mons is the volcano rising up behind it. The vertical scale has been exaggerated to emphasize the surface features.

**LINEAR RIDGES**
Ridges 30–60km (20–40 miles) long lie along a northern slope of Ovda Regio. Dark lava, or possibly windblown dirt, fills the spaces between the ridges.

APHRODITE TERRA

# Ovda Regio

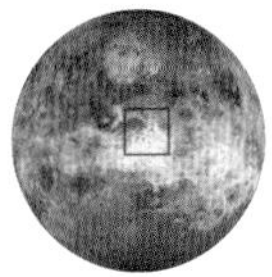

**TYPE** Highland terrain

**AGE** Under 500 million years

**DIAMETER** 5,280km (3,279 miles)

Ovda Regio is a highland area in Venus's equatorial region. It forms the western part of Aphrodite Terra, Venus's most extensive highland system, which rises 3km (2 miles) above the mean surface level. Ovda Regio is one of a handful of highland regions on Venus that displays a type of complex ridge terrain known as tessera, a form of terrain, that was first identified in images taken by Veneras 15 and 16. Tesserae are raised plateau-shaped regions with chaotic and complex patterns of crisscrossing lines. In places, the planet's crust has been fractured into kilometre-sized blocks. Elsewhere there are folds, faults, and belts of ridges and grooves hundreds of kilometres long. These are best seen along Ovda Regio's boundaries, where curving ridges and troughs have developed. There is also evidence that volcanic activity has played its part in the shaping of this landscape. Magma, which may have welled up from the planet's interior, has flowed across part of the region, and ridges formed by compression have filled with lava. Ovda Regio is named after a Marijian (Russian) forest goddess.

**HIGHLANDS AND LOWLANDS**
Tessera ridges run between the Ovda Regio highland (right) and lowland lava flows (left). Some of the highland depressions have been partially filled by smooth material.

APHRODITE TERRA

# Miralaidji Corona

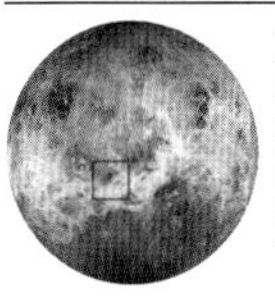

**TYPE** Corona

**AGE** Under 500 million years

**DIAMETER** 300km (186 miles)

This large volcanic feature was formed by a plume of magma rising under the Venusian surface. The magma partially melted the crustal rock, which was raised up above the surrounding land to produce the corona, a blister-like formation with radial faulting. The coronae on Venus range in size from about 50 to 2,600km (30 to 1,600 miles) across and are circular to elongate in shape. They are named after fertility goddesses. Miralaidji is an Aborigine fertility goddess.

RADIAL FAULTING

APHRODITE TERRA

# Dali Chasma

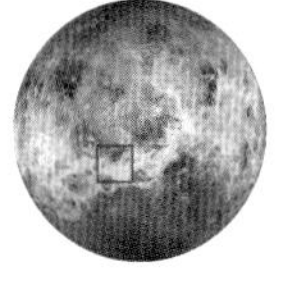

**TYPE** Fault

**AGE** Under 500 million years

**LENGTH** 2,077km (1,291 miles)

The Dali Chasma lies in western Aphrodite Terra. It is a system of canyons and deep troughs coupled with high mountains that makes a broad, curving cut through more than 2,000km (1,200 miles) of the planet's surface. Along with the Diana Chasma system, it connects the Ovda and Thetis highland regions with the large volcanoes at Atla Regio. The mountain ranges associated with the canyons rise for 3–4km (2–2.5 miles) above the surrounding terrain. The canyons are 2–3km (1.2–2.5 miles) deep.

**TROUGHS**
In this view along the Dali Chasma, part of the raised rim of the 1,000km- (600-mile-) wide Latuna Corona is visible on the left.

APHRODITE TERRA

# Artemis Corona

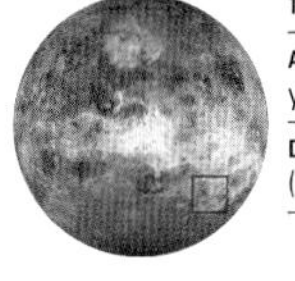

**TYPE** Corona

**AGE** Under 500 million years

**DIAMETER** 2,600km (1,614 miles)

Artemis is more than twice as big as the next largest corona on Venus, Heng-o. A near-circular trough, Artemis Chasma, which has a raised rim, marks its boundary. Within it are complex systems of fractures, volcanic flows, and small volcanoes. Artemis, like other coronae, could have been formed by hot material welling up under the surface. But its large size and the surrounding trough mean that other forces, such as the pulling apart of the crust and surface, were involved.

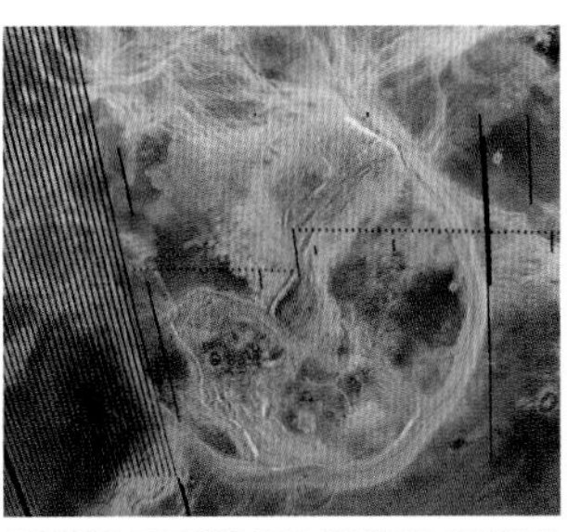

ARTEMIS CORONA AND ARTEMIS CHASMA

LADA TERRA

# Lada Terra

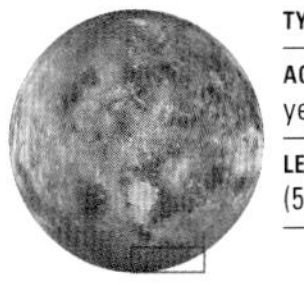

**TYPE** Highland terrain

**AGE** Under 500 million years

**LENGTH** 8,615km (5,350 miles)

Lada is the second largest of three highland regions on Venus. It is in the south-polar region of the planet, largely south of latitude 50°S, and comparatively little is known about it. Lada Terra includes some typical tessera terrain of crisscrossing troughs and ridges. Volcanic activity has also affected the area. Lada includes three large coronae (blister-like features), called Quetzalpetlatl, Eithinoha, and Otygen. Lava has flowed over and cut through the northern part of the region. All three terras on Venus are named after goddesses of love: Aphrodite is named after the Greek goddess; Ishtar (see p.132), the Babylonian goddess; and Lada, the Slavic goddess.

**LAVA CHANNEL**
Part of a 1,200km- (745-mile-) long channel carved through Lada Terra by high-temperature, very fluid lava runs from west to east across the centre of this image.

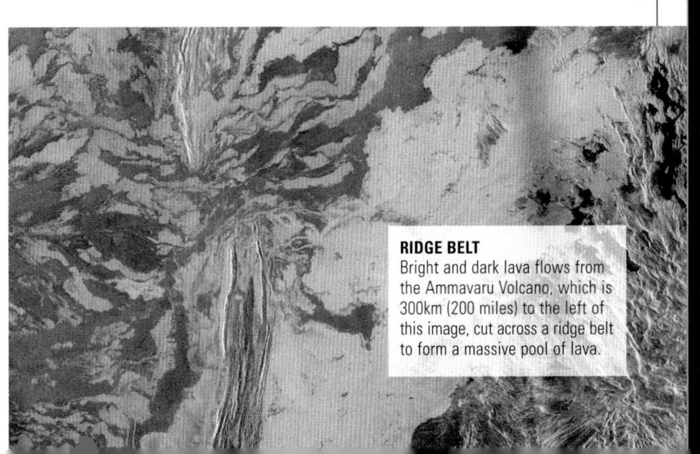

**RIDGE BELT**
Bright and dark lava flows from the Ammavaru Volcano, which is 300km (200 miles) to the left of this image, cut across a ridge belt to form a massive pool of lava.

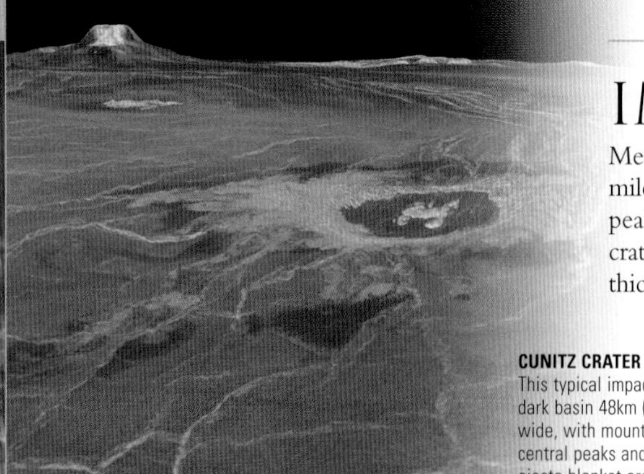

# IMPACT CRATERS

Meteorite impact craters on Venus range in size from 7km (4 miles) to 270km (168 miles) across. The largest are multiringed, those of intermediate size have central peaks, and the smaller ones have smooth floors. The smallest of all – simple, bowl-like craters that are common on the Moon and Mars – are scarce on Venus, because the thick atmosphere filters out the small asteroids that would create them. Venusian craters are young and in many cases in pristine condition. The last volcanic resurfacing of Venus could have occurred as recently as 500 million years ago, so its craters must have mostly formed since then, and there has been little geological activity or weathering to affect them. Individual craters on Venus are named after women of note or are given female first names.

**CUNITZ CRATER**
This typical impact crater has a dark basin 48km (30 miles) wide, with mountainous central peaks and a bright ejecta blanket around it.

ISHTAR TERRA

## Wanda Crater

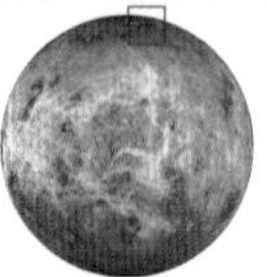

**TYPE** Central-peak crater
**AGE** Under 500 million years
**DIAMETER** 21.6km (13.4 miles)

Wanda Crater is in the northern part of the Akna Montes mountain range. It was mapped first in 1984, by the Venera 15 and 16 missions, and Magellan studied it again a few years later. The crater has a large, rugged peak in the centre of its smooth, lava-flooded floor. About one-third of all Venusian craters have such peaks. Material from the mountain ridges seems to have collapsed into the crater's western edge.

**CENTRAL PEAK**

ISHTAR TERRA

## Cleopatra Crater

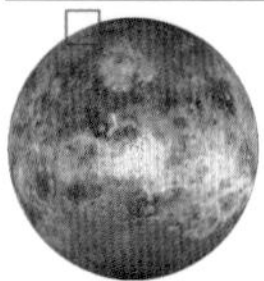

**TYPE** Double-ring crater
**AGE** Under 500 million years
**DIAMETER** 105km (65 miles)

Cleopatra Crater is named after the legendary Egyptian queen. It is located on Maxwell Montes, Venus's highest mountain range, and stands out as a smooth, eye-like feature against the rough mountainous terrain. Cleopatra was imaged by the Venera 15 and 16 spacecraft and the Arecibo radio telescope in the mid-1980s. It was one of several circular features that resembled both an impact crater and a volcanic feature. The data of the time revealed a feature of apparently great depth, without the rim deposits typical of impact craters. As a result, Cleopatra was classified as a volcanic caldera. However, high-resolution images from Magellan revealed an inner basin and rough ejecta deposits, providing conclusive proof that Cleopatra is an impact crater.

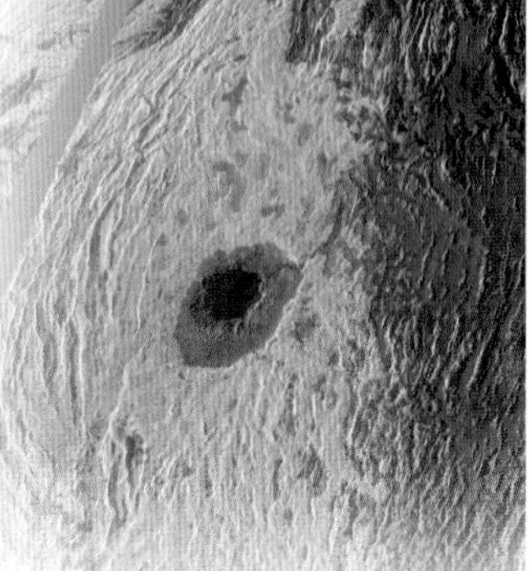

**MYSTERY CRATER**
The dark inner basin, the rim, and the surrounding ejecta revealed in this Magellan image from 1990 convinced astronomers that Cleopatra is an impact crater.

GUINEVERE PLANITIA

## Jeanne Crater

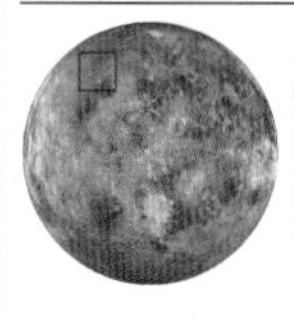

**TYPE** Central-peak crater
**AGE** Under 500 million years
**DIAMETER** 19.4km (12 miles)

An asteroid travelling from the southwest smashed into the Guinevere Planitia obliquely and created Jeanne Crater. Ejecta pushed out of the impact basin produced a distinctive triangular shape. Lobes formed to the northwest of the crater as molten material produced by the impact flowed downhill.

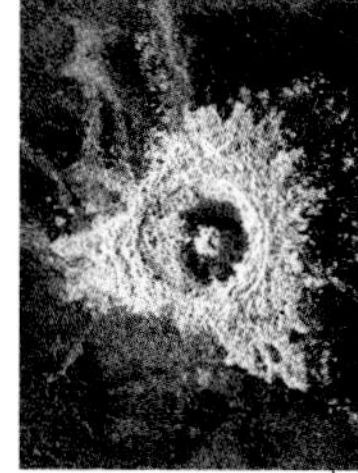

**TRIANGULAR EJECTA**

BETA REGIO

## Balch Crater

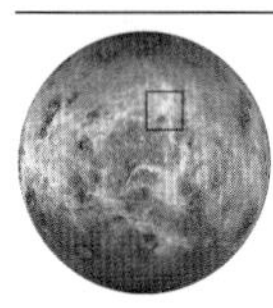

**TYPE** Central-peak crater
**AGE** Under 500 million years
**DIAMETER** 40km (25 miles)

Most impact craters on Venus have remained unchanged since they were formed and have sharply defined rims. However, a relatively small number have been modified by volcanic eruptions and other kinds of tectonic activity. Balch Crater is one of these. Its circular form was split in two as the land was wrenched apart during the formation of a deep rift valley. The rift, which is up to 20km (12.4 miles) wide, created a division that runs from north to south through the crater's centre. The western half of the crater remains intact, but most of the eastern half was destroyed. A central peak and an ejecta blanket are visible in the western half. The crater was initially called Somerville, but is now named after American economist and Nobel laureate Emily Balch.

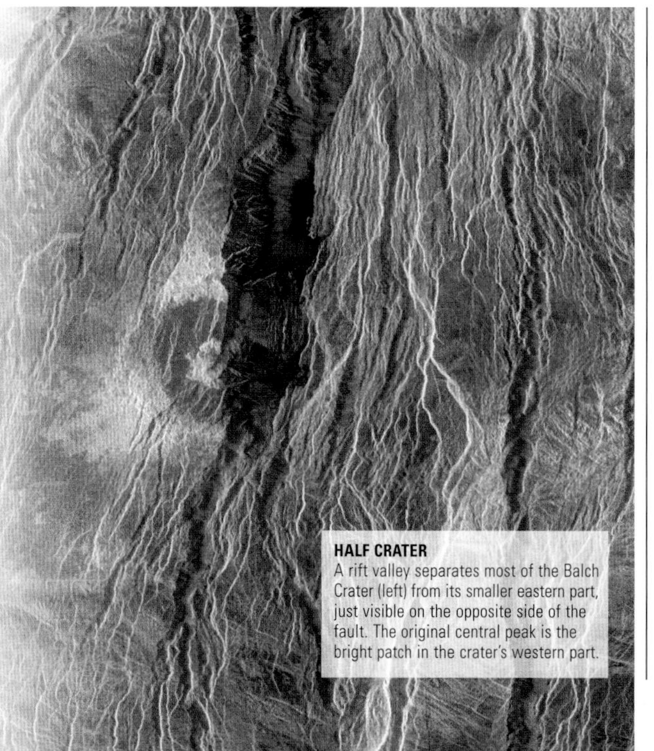

**HALF CRATER**
A rift valley separates most of the Balch Crater (left) from its smaller eastern part, just visible on the opposite side of the fault. The original central peak is the bright patch in the crater's western part.

APHRODITE TERRA

## Riley Crater

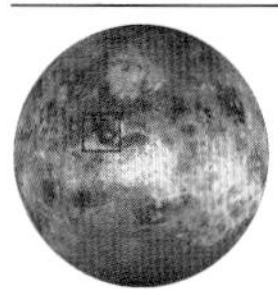

**TYPE** Central-peak crater
**AGE** Under 500 million years
**DIAMETER** 25km (16 miles)

Riley Crater, named after 19th-century botanist Margaretta Riley, is one of the few Venusian craters to have been precisely measured. Comparison of images from different angles shows that the 25km- (16-mile-) wide crater's floor lies 580m (1,880ft) below the surrounding plain, the rim is 620m (2,009ft) above it, and the peak is 536m (1,737ft) high.

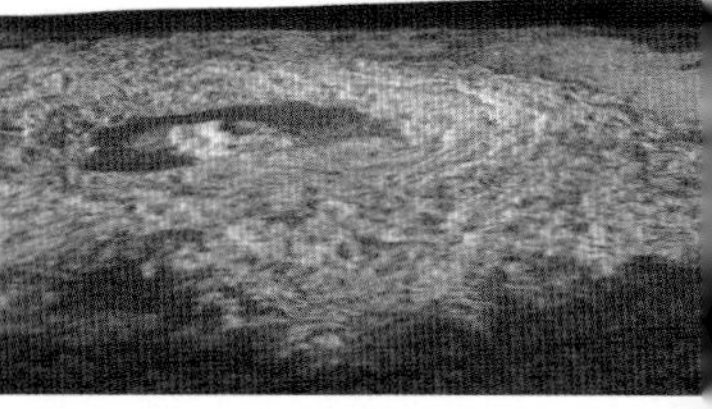

**OBLIQUE VIEW OF RILEY CRATER**

APHRODITE TERRA

## Mead Crater

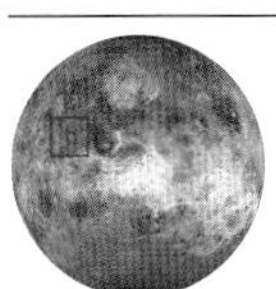

**TYPE** Multi-ringed crater

**AGE** Under 500 million years

**DIAMETER** 270km (168 miles)

Mead is the largest impact crater on Venus – although compared to craters on the Moon and Mercury, it is not very large. Mead is a multiringed crater whose inner ring is the rim of the crater basin. This encloses a smooth, flat floor, which hides a possible central peak. The crater floor was flooded at the time of impact as a result of impact melt or by volcanic lava being released from below the surface. This explains why a crater of Mead's size is so shallow; there is a drop of only about 1km (0.6 miles) between the crater rim and the crater centre.

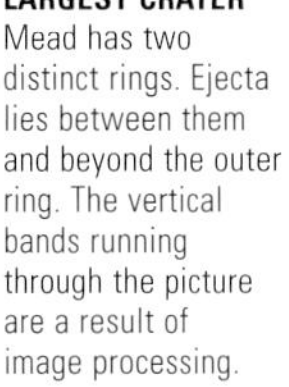

**LARGEST CRATER** Mead has two distinct rings. Ejecta lies between them and beyond the outer ring. The vertical bands running through the picture are a result of image processing.

LAVINIA PLANITIA

## Saskia Crater

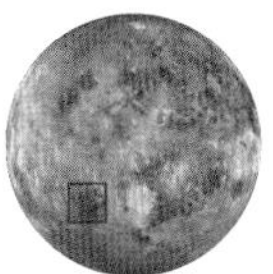

**TYPE** Central-peak crater

**AGE** Under 500 million years

**DIAMETER** 37.1km (23 miles)

Saskia is a middle-sized crater, and its ejecta pattern is typical for its size. The ejecta blanket extends all the way around the crater's basin, suggesting that the impacting body smashed into the surface at a high angle. The crater has central peaks, formed as the planet's surface recoiled after being pushed down by the energy released during the impact. The original crater rim has collapsed and formed terraced walls. The incoming object must have been about 2.5km (1.6 miles) across to produce a crater of this size. Images of Saskia and other craters, such as the similarly sized Danilova (48.8km/30.3 miles wide) and Aglaonice (63.7km/39.6 miles wide), which lie within a few hundred kilometres of Saskia, have been produced from radar data collected by Magellan. Raw radar images (such as the one above) do not show features as they would appear to the naked eye. Instead, brightness varies according to the smoothness of the surface – rough areas appear light, while smooth ones look dark.

**THREE CRATERS** Saskia lies at the lower left of this 500km- (300-mile-) wide segment of Lavinia Planitia. Above it are the Danilova and Aglaonice craters.

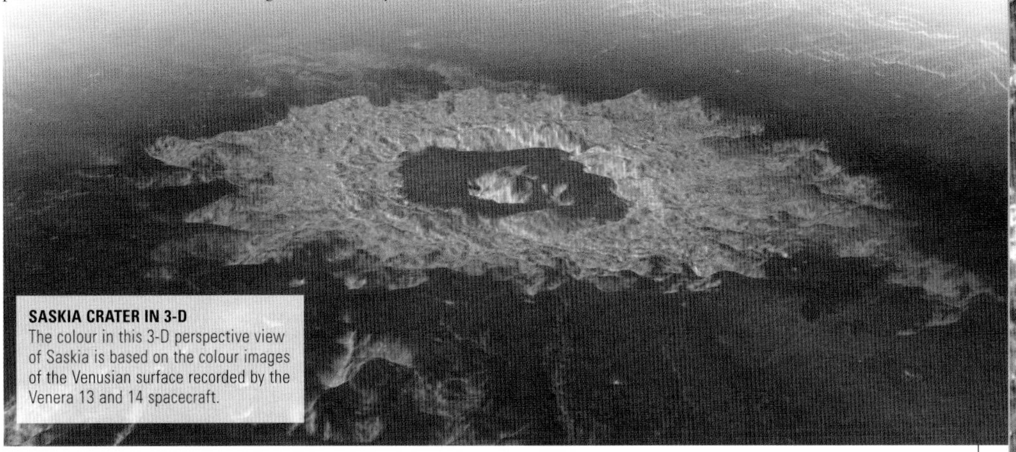

**SASKIA CRATER IN 3-D** The colour in this 3-D perspective view of Saskia is based on the colour images of the Venusian surface recorded by the Venera 13 and 14 spacecraft.

LAVINIA PLANITIA

## Stein Crater Field

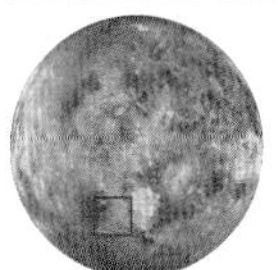

**TYPE** Crater field

**AGE** Under 500 million years

**DIAMETER** 14km (8.7 miles), 11km (6.8 miles), and 9km (5.6 miles)

Small asteroids heading for Venus's surface can be broken up by the planet's dense atmosphere. The resulting fragments continue heading towards the surface, striking it simultaneously within a relatively small area to form a crater field. The Stein field consists of three small craters. The two smallest ones overlap. Material ejected by all three craters extends mainly to the northeast, suggesting that the fragments struck from the southwest. Material melted by the impacts has formed flow deposits, also lying to the northeast.

**STEIN TRIPLETS**

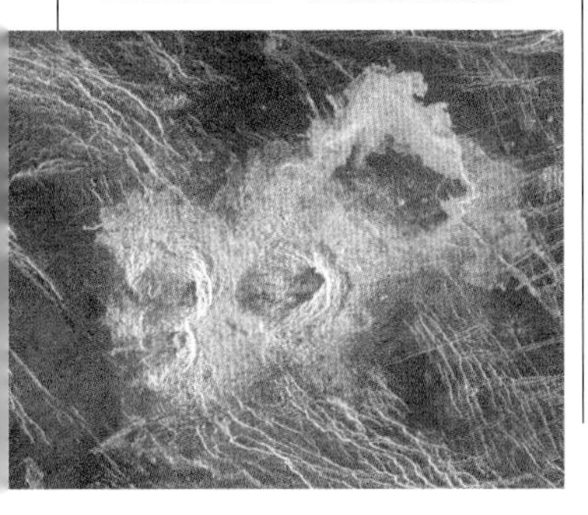

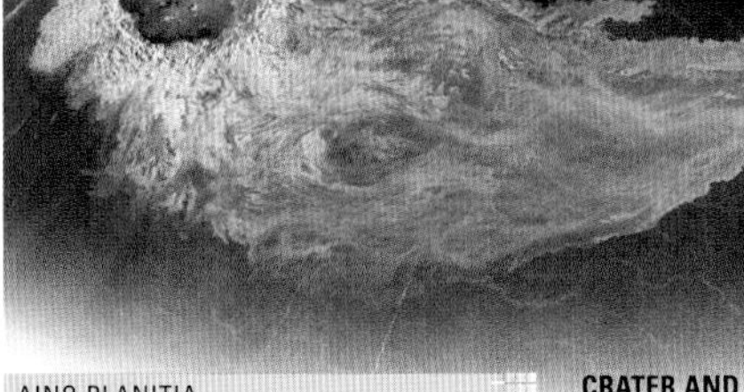

**CRATER AND OUTFLOW** A 600-km- (372-mile) long, radar-bright flow of once-molten debris stretches to the east of Addams Crater.

AINO PLANITIA

## Addams Crater

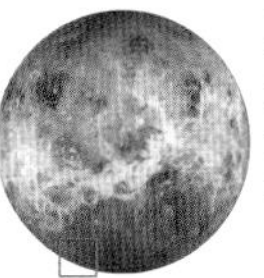

**TYPE** Central-peak crater

**AGE** Under 500 million years

**DIAMETER** 87km (54 miles)

The large, circular Addams Crater measures almost 90km (55 miles) across but it is its long tail that makes this crater unusual. An asteroid has hit the ground from the northwest and created a crater basin with an ejecta blanket stretching out beyond about three-quarters of the crater rim. Additionally, impact-melt ejecta and lava extend out from about a third of the rim, creating a mermaid-style tail to the east. The molten material flowed downhill for about 600km (372 miles) from the impact site. The Magellan spacecraft found this area to be radar bright – that is, it bounced back a large portion of the radio waves that Magellan transmitted to it, which suggests it has a rugged surface. Venus's high surface temperature of about 464°C (867°F) allows ejecta to remain molten for a longer time than if it were on Earth. However, once the material cools below about 1,000°C (1,800°F) it becomes so viscous it stops flowing. The crater is named after the American social reformer Jane Addams.

LADA TERRA

## Alcott Crater

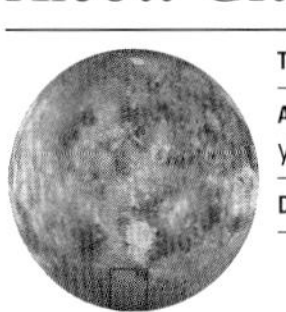

**TYPE** Degraded crater

**AGE** Under 500 million years

**DIAMETER** 66km (41 miles)

Alcott is one of the few craters on Venus that has been modified by volcanic activity not associated with the crater's production. Many craters have floors flooded with lava that came up through the surface as the crater basin was formed. In Alcott's case, lava erupted elsewhere and then flowed over the crater. About half of the crater's rim is still visible, along with ejecta from the original impact lying to the south and east. A channel where lava once flowed touches the southwest edge of the crater.

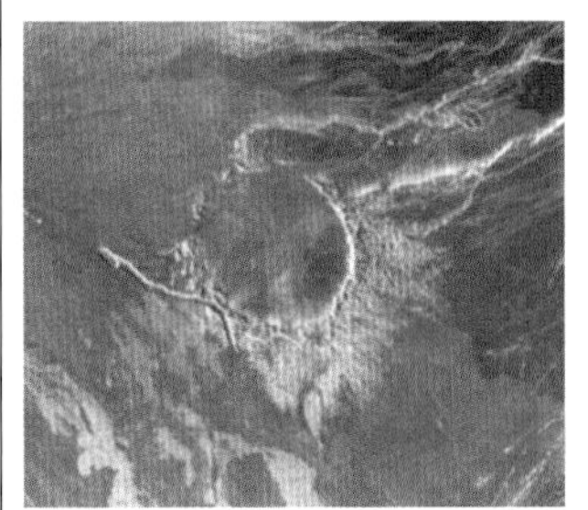

**MODIFIED BY LAVA**

# EARTH

36–37 Gravity, motion, and orbits
52–53 Life in the Universe
60–63 Celestial cycles
64–65 Planetary motion
70–71 Lights in the sky
119 Rocky planets

**EARTH IS THE THIRD-CLOSEST PLANET** to the Sun. The largest of the four rocky planets, it formed approximately 4.56 billion years ago. Earth's internal structure is similar to that of its planetary neighbours, but it is unique in the Solar System in that it has abundant liquid water at its surface, an oxygen-rich atmosphere, and it is known to support life. Earth's surface is in a state of constant dynamic change as a result of processes occurring within its interior and in its oceans and atmosphere.

## ORBIT

Earth orbits the Sun at an average speed of 108,000kph (67,000mph), in an anticlockwise direction when viewed from above its North Pole. Like the other planets, Earth orbits the Sun along an elliptical path. As a result, about 7 per cent more solar radiation currently reaches Earth's surface in January than in July. The plane of Earth's orbit around the Sun is called the ecliptic plane. Earth's spin axis is not perpendicular to this plane but is tilted at an angle of 23.5°. The eccentricity of Earth's elliptical orbit around the Sun (the degree to which it varies from a perfect circle) changes over a cycle of about 100,000 years, and its axial tilt varies over a 42,000-year cycle. Combined with a third cycle – a wobble in the direction in which the spin axis points in space, called precession (see p.60) – these variations are believed to play a part in causing long-term cycles in Earth's climate, such as ice ages.

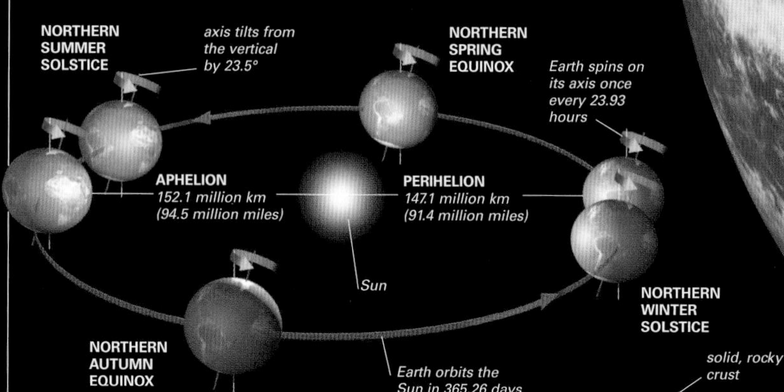


**SPIN AND ORBIT**
Earth is about 3 per cent nearer the Sun at perihelion (in January) than at aphelion (in July). Its axial tilt combined with its elliptical orbit gives rise to seasons (see p.61).

## STRUCTURE

Earth's rotation causes its equatorial regions to bulge out slightly, by about 21km (13 miles) compared to the poles. Internally, Earth has three main layers. The central core has a diameter of about 7,000km (4,350 miles) and is made mainly of iron with a small amount of nickel. It has a central solid part, which has a temperature of about 4,700°C (8,500°F), and an outer liquid part. Surrounding the core is the mantle, which contains rocks rich in magnesium and iron and is about 2,800km (1,700 miles) deep. Earth's crust consists of many different types of rocks and minerals, predominantly silicates, and is differentiated into continental crust and a thinner oceanic crust.

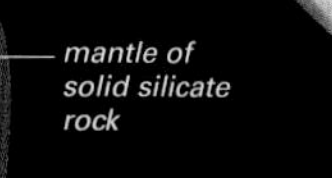


**EARTH INTERIOR**
At Earth's centre is a hot dense core. Surrounding the core are the mantle and the thin, rocky outer crust, which supports Earth's biosphere, with its oceans, atmosphere, plants, and animals.

THE SOLAR SYSTEM

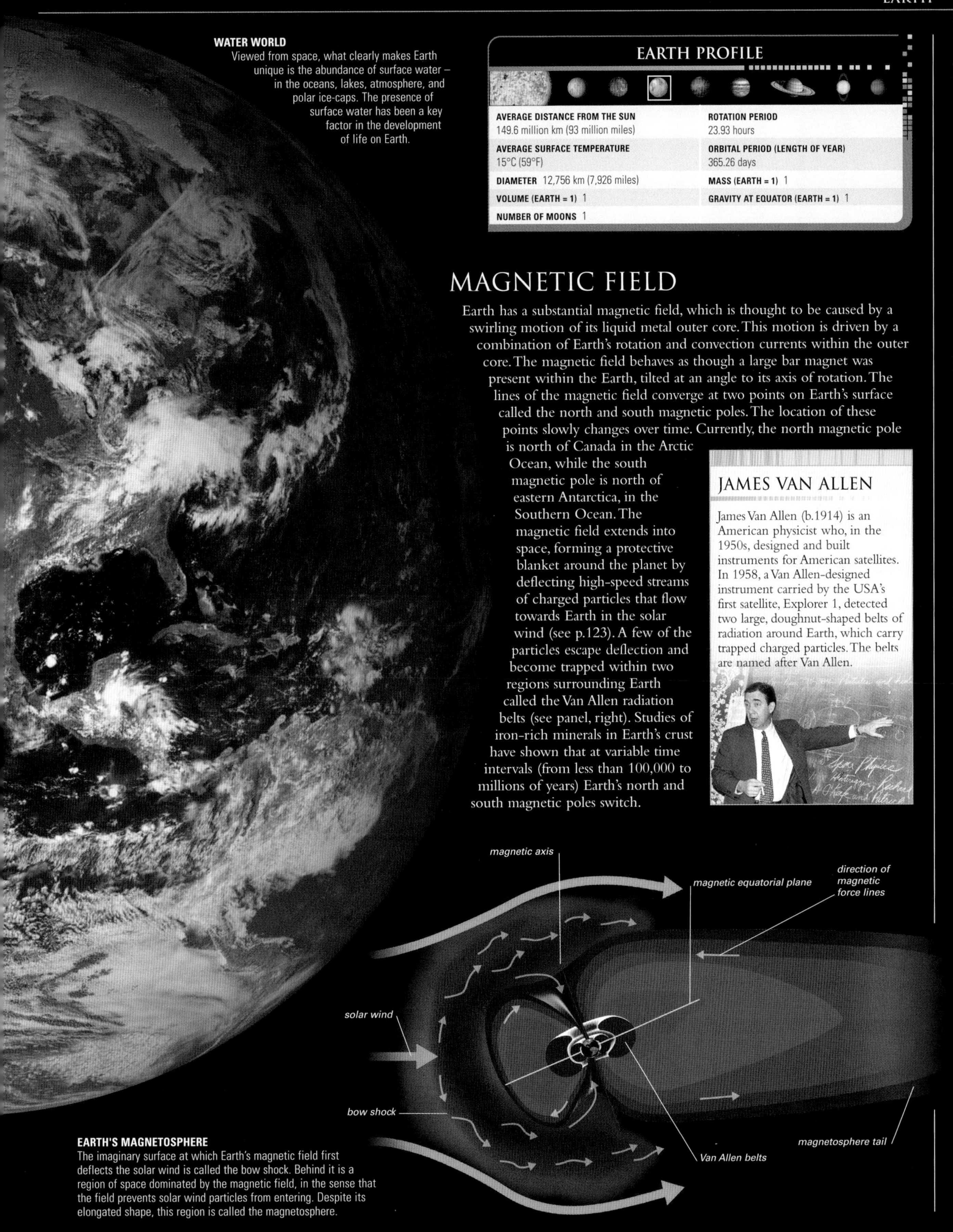


**WATER WORLD**
Viewed from space, what clearly makes Earth unique is the abundance of surface water – in the oceans, lakes, atmosphere, and polar ice-caps. The presence of surface water has been a key factor in the development of life on Earth.

## EARTH PROFILE

| | |
|---|---|
| **AVERAGE DISTANCE FROM THE SUN** 149.6 million km (93 million miles) | **ROTATION PERIOD** 23.93 hours |
| **AVERAGE SURFACE TEMPERATURE** 15°C (59°F) | **ORBITAL PERIOD (LENGTH OF YEAR)** 365.26 days |
| **DIAMETER** 12,756 km (7,926 miles) | **MASS (EARTH = 1)** 1 |
| **VOLUME (EARTH = 1)** 1 | **GRAVITY AT EQUATOR (EARTH = 1)** 1 |
| **NUMBER OF MOONS** 1 | |

# MAGNETIC FIELD

Earth has a substantial magnetic field, which is thought to be caused by a swirling motion of its liquid metal outer core. This motion is driven by a combination of Earth's rotation and convection currents within the outer core. The magnetic field behaves as though a large bar magnet was present within the Earth, tilted at an angle to its axis of rotation. The lines of the magnetic field converge at two points on Earth's surface called the north and south magnetic poles. The location of these points slowly changes over time. Currently, the north magnetic pole is north of Canada in the Arctic Ocean, while the south magnetic pole is north of eastern Antarctica, in the Southern Ocean. The magnetic field extends into space, forming a protective blanket around the planet by deflecting high-speed streams of charged particles that flow towards Earth in the solar wind (see p.123). A few of the particles escape deflection and become trapped within two regions surrounding Earth called the Van Allen radiation belts (see panel, right). Studies of iron-rich minerals in Earth's crust have shown that at variable time intervals (from less than 100,000 to millions of years) Earth's north and south magnetic poles switch.

## JAMES VAN ALLEN

James Van Allen (b.1914) is an American physicist who, in the 1950s, designed and built instruments for American satellites. In 1958, a Van Allen-designed instrument carried by the USA's first satellite, Explorer 1, detected two large, doughnut-shaped belts of radiation around Earth, which carry trapped charged particles. The belts are named after Van Allen.

**EARTH'S MAGNETOSPHERE**
The imaginary surface at which Earth's magnetic field first deflects the solar wind is called the bow shock. Behind it is a region of space dominated by the magnetic field, in the sense that the field prevents solar wind particles from entering. Despite its elongated shape, this region is called the magnetosphere.

# ATMOSPHERE AND WEATHER

Earth is surrounded by the atmosphere, a layer of gases many hundreds of kilometres thick. This atmosphere is thought to have arisen partly from gases spewed out by ancient volcanoes, although its oxygen content – so vital to most forms of life – was created mainly by plants. Through the effects of gravity, the atmosphere is densest at Earth's surface and rapidly thins with altitude. With increasing altitude, there are also changes in temperature and a progressive drop in atmospheric pressure. For example, at a height of 30km (19 miles), the pressure is just 1 per cent that at sea level. Within the lowest layer of the atmosphere, the troposphere, continual changes occur in temperature, air flow (wind), humidity, and precipitation, known as weather. The basic cause of weather is the fact that Earth absorbs more of the Sun's heat at the equator than the poles. This produces variations in atmospheric pressure, which create wind systems. The winds drive ocean currents and cause masses of air with different temperatures and moisture content to circulate over the planet's surface. Earth's rotation plays a part in causing this atmospheric circulation because of the Coriolis effect (below).

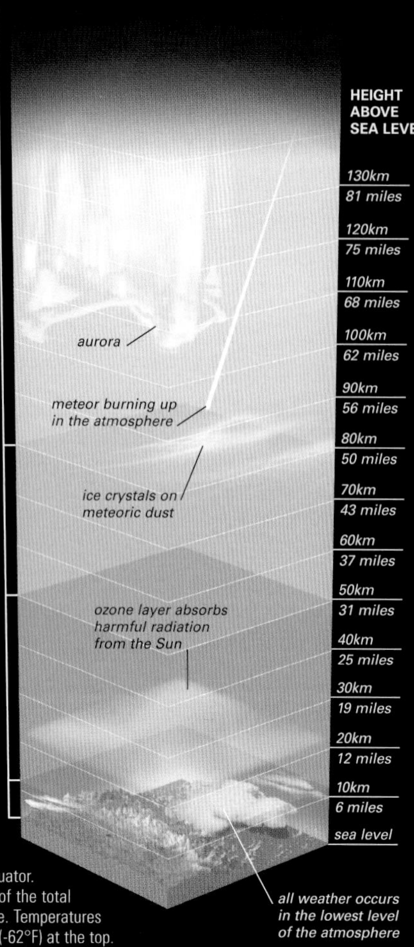

**THERMOSPHERE**
The thermosphere extends to over 600km (375 miles) above the Earth's surface. Temperature rises rapidly in the lower thermosphere due to absorption of solar energy and then increases gradually with altitude, reaching as high as 1,700°C (3,100°F).

**MESOSPHERE**
This layer extends up to about 80km (50 miles). Temperatures fall through the mesosphere to as low as -93°C (-135°F).

**STRATOSPHERE**
The stratosphere is a calm layer stretching up to about 50km (30 miles) above sea level. The temperature rises to -3°C (27°F) at the top of this layer.

**TROPOSPHERE**
This layer extends to 8km (5 miles) above the poles and 16km (10 miles) above the equator. It contains 75 per cent of the total mass of the atmosphere. Temperatures fall to as low as -52°C (-62°F) at the top.

HEIGHT ABOVE SEA LEVEL
130km / 81 miles
120km / 75 miles
110km / 68 miles
100km / 62 miles
90km / 56 miles
80km / 50 miles
70km / 43 miles
60km / 37 miles
50km / 31 miles
40km / 25 miles
30km / 19 miles
20km / 12 miles
10km / 6 miles
sea level

aurora
meteor burning up in the atmosphere
ice crystals on meteoric dust
ozone layer absorbs harmful radiation from the Sun
all weather occurs in the lowest level of the atmosphere

**ATMOSPHERIC LAYERS**
The four main layers in Earth's atmosphere are distinguished by different temperature characteristics. No boundary exists at the top of the atmosphere. Its upper regions progressively thin out and merge with space.

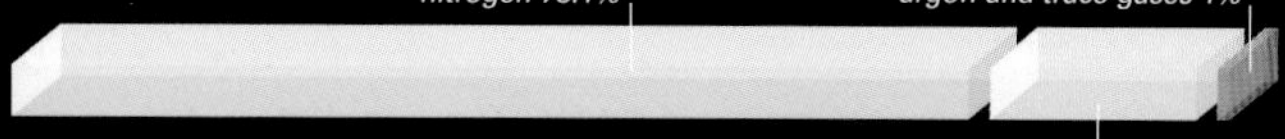

nitrogen 78.1%
argon and trace gases 1%
oxygen 20.9%

**COMPOSITION OF ATMOSPHERE**
Nitrogen and oxygen make up 99 per cent of dry air by volume. About 0.9 per cent is argon, and the rest consists of tiny amounts of other gases. The atmosphere also contains variable amounts (up to 4 per cent) of water vapour.

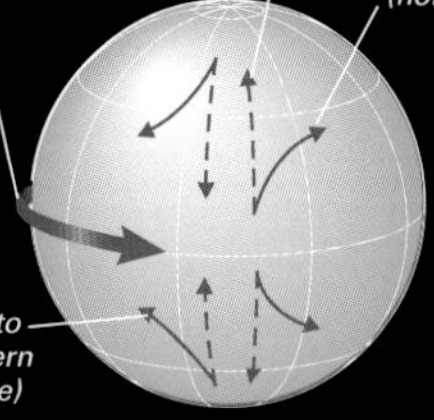

initial direction of moving air
deflection to right (northern hemisphere)
direction of spin
deflection to left (southern hemisphere)

**THE CORIOLIS EFFECT**
The Coriolis effect causes deflections of air moving across Earth's surface. It is a consequence of the fact that objects at different latitudes move at different speeds around Earth's spin axis.

# PLATE TECTONICS

Earth's crust and the top part of its mantle are joined in a structure called the lithosphere. This is broken up into several solid structures called plates, which "float" on underlying semi-molten regions of the mantle and move relative to each other. Most plates carry both oceanic crust and some thicker continental crust, although a few carry only oceanic crust. The scientific theory concerning the motions of these plates is called plate tectonics, and the phenomena associated with the movements are called tectonic features. Most tectonic features, which include ocean ridges, deep sea trenches, high mountain ranges, and volcanoes, result from processes occurring at plate boundaries. Their nature depends on the type of crust on either side of the boundary and whether the plates are moving towards or away from each other. Tectonic features occurring away from plate boundaries include volcanic island chains, such as the Hawaiian islands. These are caused by magma (molten rock) upwelling from "hotspots" in the mantle, causing a series of volcanoes to form on the overlying plate.

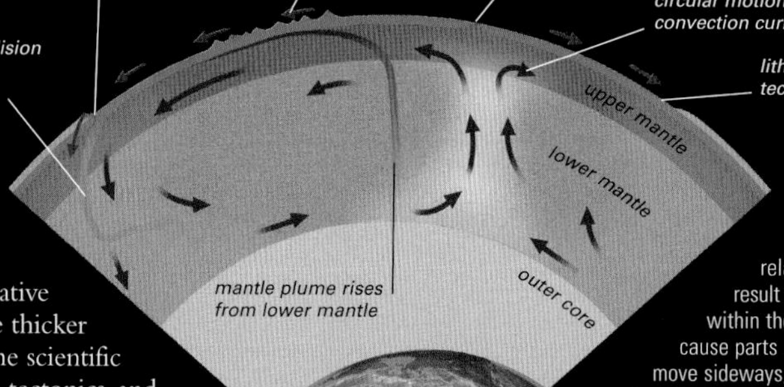

destructive boundary, where tectonic plates converge
plate dragged along by convection current
constructive boundary, where plates diverge and new crust is created
circular motion of convection current
lithospheric tectonic plate
plate in collision descends into mantle
upper mantle
lower mantle
outer core
mantle plume rises from lower mantle

**MOVING PLATES**
Earth's plates move relative to each other as a result of convection currents within the mantle. The currents cause parts of the mantle to rise, move sideways, and then sink again, dragging the plates along as they do so.

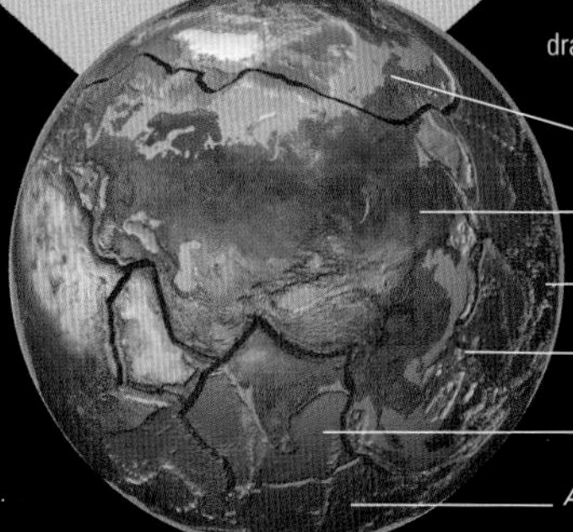

North American Plate
Eurasian Plate
Pacific Plate
Plate boundary
Indian Plate
Australian Plate

**TECTONIC PLATES**
Earth's surface is broken into seven large plates, such as the Eurasian plate, and many smaller ones, such as the Indian plate. Each continent is embedded in one or more plates.

## SURFACE FEATURES

From space, the flatter areas of Earth's land surface (apart from the areas dominated by ice) appear either dark green or various shades of yellow-brown. The green areas are forests and grasslands, which comprise a major component of Earth's biosphere (the planet's life-sustaining regions). The yellow-brown areas are mainly deserts, which have been created over long periods by various weathering and erosional processes. Like the other rocky planets, Earth has suffered many thousands of meteorite impacts over its history (see p.119). But, because Earth's surface is so dynamic, the evidence for most of these impacts has disappeared, removed by erosion or covered up by depositional processes.

**SANDY DESERT**
Deserts cover about 20 per cent of Earth's land surface, but only a small proportion are occupied by sand dunes, like these in the Sahara Desert.

## WATER

Water is a dominant feature of Earth's surface. Overall, about 97 per cent of the water is in oceans (which cover 75 per cent of the surface), 2 per cent is in ice-sheets and glaciers, less than 1 per cent is in ground water (underground and in rocks), and the rest is in rivers, lakes, and the atmosphere. The presence of liquid water has been key to the development of life on Earth, and the heat capacity of the oceans has been important in keeping the planet's temperature relatively stable. Liquid water is also responsible for most of the erosion and weathering of Earth's continents, a process unique in the Solar System, although it is believed to have occurred on Mars in the past.

**RAINFOREST**
Forests cover 30 per cent of Earth's land surface and range from the cold, dark boreal forest of the far north to the dense rainforests of the humid tropics.

### KINGDOMS OF LIFE

**Biologists use various systems for classifying living organisms, but the most widely used is the five-kingdom system. This classifies organisms mainly on the basis of their cell structure and method of obtaining nutrients and energy. However, not all scientists accept this system as satisfactory, and some have proposed switching to an eight-kingdom system or one with 30 kingdoms grouped into three superkingdoms.**

**ANIMALS**

Animals are multicellular organisms that contain muscles or other contractile structures allowing some method of movement. They acquire nutrients, and so gain energy, by ingesting food. Many animals, including mammals, are vertebrates (they possess a backbone), but a far larger number are invertebrates (without a backbone).

**VERTEBRATE**

**PLANTS**

Plants are multicellular organisms that obtain energy from sunlight through the process of photosynthesis. Their cells contain special pigments for absorbing light energy and are enclosed by cell walls made of cellulose.

**FLOWERING PLANT**

**FUNGI**

Fungi acquire nutrients by absorption from other living organisms or dead and decaying organic material. They have no means of locomotion. They range from yeasts (microscopic unicellular organisms) to multicellular forms with large fruiting bodies, such as mushrooms.

**TOADSTOOL**

**PROTISTS**

Protists are microscopic, mainly single-celled organisms whose cells contain nuclei. Some gain energy from sunlight, others ingest food like animals.

**PARAMECIUM**

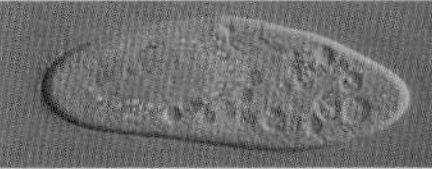

**MONERANS**

Monerans are the simplest, smallest, most primitive, and most abundant organisms on Earth. The two main groups are bacteria and blue-green algae (cyanobacteria). Monerans are single-celled but their cells contain no distinct nucleus. Most reproduce by splitting in two.

**MYCOBACTERIUM**

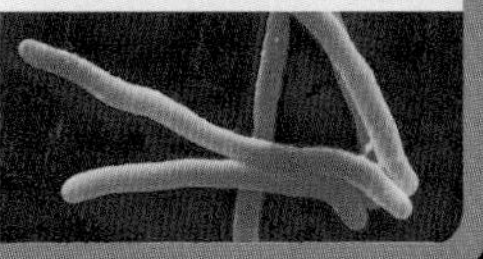

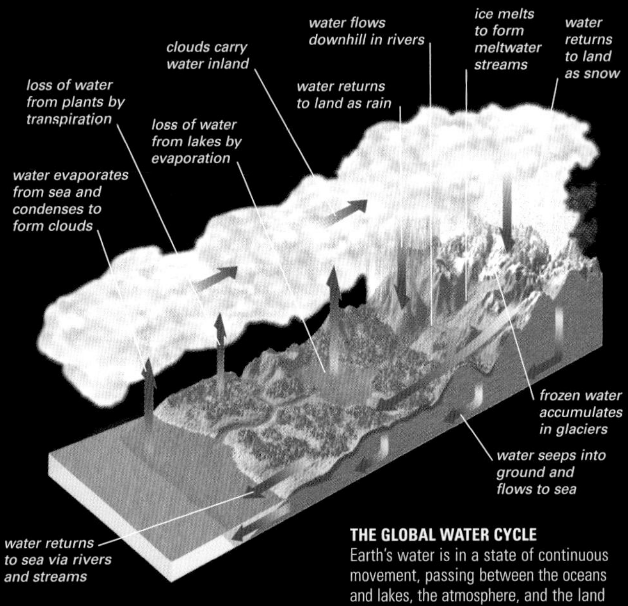

water flows downhill in rivers
ice melts to form meltwater streams
water returns to land as snow
clouds carry water inland
loss of water from plants by transpiration
water returns to land as rain
loss of water from lakes by evaporation
water evaporates from sea and condenses to form clouds
frozen water accumulates in glaciers
water seeps into ground and flows to sea
water returns to sea via rivers and streams

**THE GLOBAL WATER CYCLE**
Earth's water is in a state of continuous movement, passing between the oceans and lakes, the atmosphere, and the land in a cycle of connected processes.

## LIFE ON EARTH

Evidence in ancient rocks points to the presence of simple, bacteria-like organisms on Earth some 3.8 billion years ago. However, the prevailing scientific view is that life started on Earth long before that, as a result of complex chemical reactions in the oceans or atmosphere. These reactions eventually led to the appearance of a self-replicating and self-repairing molecule, a precursor of DNA (deoxyribonucleic acid). Once life, in this rudimentary form, had started, processes such as mutation and natural selection inevitably led, over the vast expanses of geological time, to a collection of life-forms of increasing diversity and complexity. Life spread from the seas to the land and to every corner of the planet. Currently, Earth is teeming with life in astonishing abundance and diversity.

# TECTONIC FEATURES

Most of Earth's tectonic features are associated with plate boundaries. At constructive (or divergent) boundaries, plates move apart and new crust is added. Examples are mid-ocean ridges and the Great Rift Valley. At destructive (or convergent) boundaries, two plates push against each other, producing a range of features, depending on the nature of the crust on each plate. Many plate boundaries are associated with an increased frequency of volcanism, earthquakes, or both.

**THE SAN ANDREAS FAULT**
This fault in California, USA, known for producing earthquakes, marks a transform boundary where two plates push past each other.

AFRICA *east* and ASIA *southwest*

## Great Rift Valley

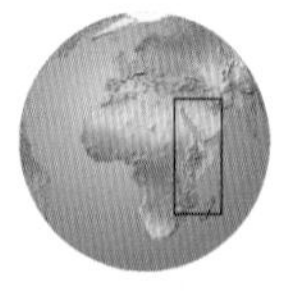

**LOCATION** Extending from Mozambique northwards through East Africa, the Red Sea, and into Lebanon

**TYPE** Series of rift faults

**LENGTH** 8,500km (5,300 miles)

The Great Rift Valley provides an example of the geological process of rifting – the stretching and tearing apart of a section of continental crust by a plume of hot magma pushing up underneath it. Rifting is associated with the development of a constructive plate boundary, which is formed as ascending magma creates new crust and pushes the plates on either side of the rift apart. The main section of the Great Rift Valley runs (in two branches) through east Africa. Over tens of millions of years, rifting in this region has caused extensive faulting, the collapse of large chunks of crust, and associated features such as volcanism and a series of lakes in the subsided sections. As rifting continues, it is anticipated that a large area of eastern Africa will eventually split off as a separate island. A northern arm of the rift valley runs up the Red Sea and eventually reaches Lebanon, in the north. This coincides with a divergent boundary that is pushing Arabia away from Africa.

**OL DOINYO LENGAI**
This active volcano in northern Tanzania sits in the middle of the east African part of the Great Rift Valley.

**THE NORTHERN RED SEA**
The Gulf of Aqaba (centre right), a branch of the Red Sea, forms part of the northern arm of the Great Rift Valley. The Gulf of Suez (centre) is a side-branch of the rift.

**BLACK SMOKERS**
Hydrothermal vents are underwater geysers located near mid-ocean ridges. The hot water spewed out by some vents, called "black smokers", is discoloured by the dark mineral iron sulphide.

ATLANTIC OCEAN

## Mid-Atlantic Ridge

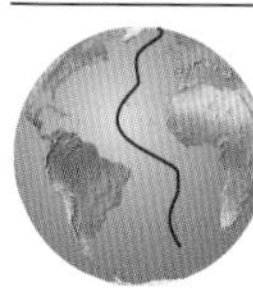

**LOCATION** Extending from the Arctic Ocean to the Southern Ocean

**TYPE** Slow spreading mid-ocean ridge

**LENGTH** 16,000km (10,000 miles)

The Mid-Atlantic Ridge is the longest mountain chain on Earth and one of its most active volcanic regions, albeit mainly underwater. The ridge sits on top of the Mid-Atlantic Rise, a bulge that runs the length of the Atlantic Ocean floor. Both rise and ridge coincide with plate boundaries that divide the North and South American plates, on the west, from the Eurasian and African plates, on the east. These are constructive plate boundaries, where new ocean crust is formed by magma upwelling from Earth's mantle. As this crust forms, the plates on either side are pushed away from the ridge at a rate of 1–10cm (0.4–4in) a year, widening the Atlantic basin. The discovery in the 1960s of this spreading of the Atlantic sea floor – evidenced by the fact that crustal material near the ridge is younger than that farther away – led to general acceptance of the theory of continental drift. The ridge is a site of extensive earthquake activity and volcanism, along with many seamounts (isolated underwater mountains). Where the volcanoes break the ocean surface, they have formed islands such as Iceland and the Azores.

**SURTSEY**
Between 1963 and 1967, a massive and dramatic submarine eruption, from a section of the Mid-Atlantic Ridge to the south of Iceland, produced the new island of Surtsey.

**FIERY ARENAL**
Arenal is one of the most active volcanoes in Costa Rica – a region where the small Cocos Plate is subducted under the neighbouring Caribbean Plate.

PACIFIC OCEAN

## Pacific Ring of Fire

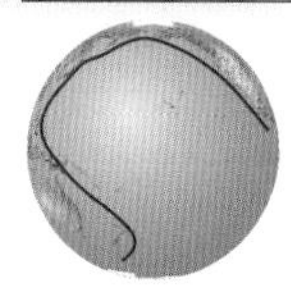

**LOCATION** Pacific Ocean rim, from Chile to New Zealand

**TYPE** Series of destructive boundaries

**LENGTH** 32,000km (20,000 miles)

The Ring of Fire is a huge arc of volcanic and seismic (earthquake) activity around the rim of the Pacific Ocean. It stretches from the western coasts of South America and North America, across the Aleutian Islands of Alaska, and down the eastern edge of Asia, to the northeast of Papua New Guinea, and finally to New Zealand. More than half of the world's active volcanoes above sea level are part of the ring. The Ring of Fire results from the Pacific Plate and other smaller plates in the Pacific colliding with neighbouring plates along a series of destructive plate boundaries. The main driving force for this activity is the creation of new crust by a large mid-ocean ridge in the eastern Pacific (the East Pacific Rise). Here, new material is continually added to the Pacific and Nazca plates, and to the small Cocos Plate, forcing them towards the edges of the Pacific.

Across much of its northern and western edges, the oceanic crust of the Pacific Plate is subducted (forced underneath) by the oceanic crust of other plates, forming deep-sea trenches. This predisposes these regions to earthquakes, and the subducted crust also melts at depth to create hot magma, which reaches the surface through volcanoes. The result has been the formation of many highly volcanic island arcs in these regions – examples being the Aleutian Islands, the Kurile Islands, the islands of Japan, and the Mariana Islands.

On the eastern side of the Pacific, the situation is somewhat different. Here, parts of the Pacific, Nazca, and Cocos plates are being subducted below continental crust. Deep-sea trenches have also formed here, but instead of island arcs, the plate collisions have led to the formation of large mountain ranges, interspersed with volcanoes, along much of the western coast of the Americas. These include the Cascade Range in Washington State, USA, home of the active volcano Mount St. Helens, and the Andes in South America, Earth's longest and most active land mountain range.

**MOUNT RUAPEHU**
At the southwest corner of the Ring of Fire is New Zealand. Here, steam rises from the country's tallest volcano, Ruapehu, between eruptions that occurred in 1995 and 1996.

**THE ANDES**
On the western edge of South America, subduction of the Nazca Plate under the South American Plate has created the Andes, another highly active region.

**OKMOK VOLCANO**
The volcanic Aleutian Islands were created as the Pacific Plate was driven under the oceanic crust of the North American Plate. This volcano is on the island of Umnak.

**MOUNT FUJI**
In the northwest Pacific, the subduction of the Pacific Plate under the Eurasian Plate is responsible for creating the islands of Japan, the site of volcanoes such as Mount Fuji, which last erupted in 1707.

ASIA *south*

## Himalayas

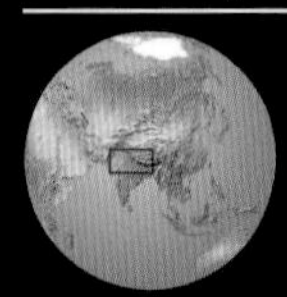

**LOCATION** Running southeast from northern Pakistan and India across Nepal to Bhutan

**TYPE** Continent–continent collision

**LENGTH** 3,800km (2,400 miles)

**TIBETAN RANGE**
The Kailas Range is a central region of the Himalayas, close to the border between Tibet and India. Here the mountains are viewed from the Tibetan Plateau, which is itself about 5 km (3 miles) above sea level.

The Himalayas are the highest mountain range on Earth, as well as one of the youngest. If the neighbouring Karakoram Range is included, the Himalayas contain Earth's 14 highest mountain peaks, each with an altitude of over 8km (5 miles), including its highest mountain, Mount Everest. These peaks are still being uplifted at the rate of some 50cm (20in) a century by the continent–continent collision that originally formed them. However, the mountains are weathered and eroded at almost the same rate, with the debris carried away by great rivers, such as the Ganges and Indus to the south.

The collision that brought about both the Himalayas and the Tibetan Plateau to its north occurred between 50 and 30 million years ago when tectonic plate movements caused India – at that time an island continent – to crash into Southeast Asia. For millions of years before the collision, the floor of the ocean between India and Asia (called Tethys) was consumed by subduction under the Eurasian Plate. But once the ocean closed, first the continental margins between India and Asia, and finally the continents themselves, collided. The crust from both was thickened, deformed, and metamorphosed, and parts of both continents and the floor of the Tethys Ocean were pushed up to form the Himalayas. Today, because the Himalayas are still rising, earthquakes and accompanying landslides remain a common occurrence.

The mountains form a number of distinct ranges. Travelling northwards, from the high plains of the Ganges, the first of these are the Siwalik Hills, a line of gravel deposits carried down from the high mountains. Here, there are subtropical forests of bamboo and other vegetation. Farther north are the Lesser Himalayas, which rise to heights of about 5,000m (3,000ft) and are traversed by numerous deep gorges formed by swift-flowing streams. Farthest north are the Great Himalayas, between 6,000 and 8,800m (20,000 and 29,000ft) tall and containing the highest peaks. This region is heavily glaciated and contains lakes filled with glacial meltwater.

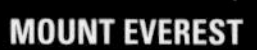

**EASTERN HIMALAYAS**
In this satellite view of an eastern region of the Himalayas, which extends into China, the snow-covered high-altitude regions are clearly delineated.

**MOUNT EVEREST**
At 8,850m (29,035ft), Everest is the highest peak on Earth. Satellite studies show that it is still being uplifted by a few millimetres a year.

**GLACIAL LAKES**
Many of the higher areas of the Himalayas are covered in glaciers and dotted with lakes dammed by glacial moraines. In the left foreground is the Tsho Rolpa Glacier Lake in northeast Nepal, which, at 4,600m (15,092ft), is one of the highest lakes on Earth.

**ROOF OF THE WORLD**
In this photograph, taken from a NASA Space Shuttle, the snow-covered Himalayas, on the left, are bordered by Earth's largest upland region, the vast and lake-spattered Tibetan Plateau.

# FEATURES FORMED BY WATER

Some of the most obvious and striking features of Earth's surface are large bodies and flows of liquid water, such as oceans, seas, lakes, and rivers. In addition to these are landforms caused by the erosional or depositional power of liquid water, which include gorges, river valleys, and coastal features ranging from beaches to eroded headlands. Ice, too, has had a major impact on Earth's appearance. Ice-formed features include existing bodies of ice, such as glaciers and ice-sheets, and landforms, such as U-shaped valleys, sculpted by the movement of past glaciers.

**GRAND CANYON**
Carved over millions of years by the Colorado River, the Grand Canyon is Earth's largest gorge.

NORTH AMERICA *northeast*

## Great Lakes

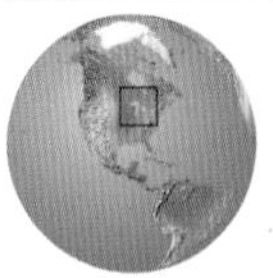

**LOCATION** Straddling the border of the USA and Canada

**TYPE** System of freshwater lakes

**AREA** 244,767 square km (94,480 square miles)

The Great Lakes of North America are a system of five connected lakes that together form the largest body of fresh water on Earth. The lakes – named, from west to east, Superior, Michigan, Huron, Erie, and Ontario – contain 20 per cent of Earth's surface fresh water and drain a basin of approximately 751,100 square km (289,900 square miles). They are connected to each other by short rivers, a strait, and canals, and drain into the Atlantic Ocean via the St. Lawrence River. The Great Lakes began to form at the end of the last ice age when glacier-carved basins were filled with meltwater left by a retreating ice-sheet. Originally, several of today's lakes were united in one huge lake, but following post-glacial uplift in the region, they took on their present form about 10,000 years ago. The lake surfaces vary in height above sea level, from 183m (600ft) at Lake Superior to 75m (246ft) at Lake Ontario. Sprinkled across the lakes are thousands of islands, including Isle Royale on Superior, which is itself big enough to hold several lakes.

**NIAGARA FALLS**
The greatest drop in water level within the Great Lakes system is at Niagara Falls, between lakes Erie and Ontario. Here, the water plunges a spectacular 51m (167ft).

**LAKES HURON AND SUPERIOR**
In this photograph taken from a NASA Space Shuttle, the largest lake, Superior, is on the right, and appears partly iced over. Lake Huron is on the left.

SOUTH AMERICA *north*

## Amazon River

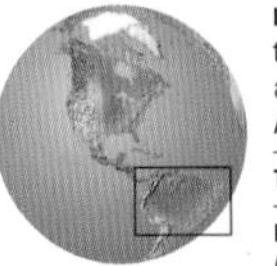

**LOCATION** Flows from the Peruvian Andes, across Brazil to the Atlantic Ocean

**TYPE** River

**LENGTH** 6,430km (3,995 miles)

The Amazon is the greatest river on Earth, whether measured by the area of the planet's land surface that it drains or by the volume of water that it discharges every year. Overall, the Amazon accounts for nearly 20 per cent of all river water discharged into Earth's oceans. The source of the Amazon has been established as a headwater of the River Apurímac, a tributary of the Ucayali, high in the Andes of southern Peru. The Ucayali flows north from this area, turns east, and joins another major tributary, the River Marañón, where it becomes the Amazon proper. The river then meanders for thousands of kilometres across the Amazon Basin, a vast flat area that contains Earth's largest rainforest, merging with numerous tributaries along the way. Just east of Manaus, at its confluence with the River Negro, the Amazon is already 16km (10 miles) wide, while still 1,600km (1,000 miles) from the sea. At its mouth, the Amazon discharges into the Atlantic Ocean at the incredible rate of about 770 billion litres (170 billion gallons of water) every hour.

**BRAIDING**
Over its course, the Amazon frequently braids into channels, creating many temporary islands.

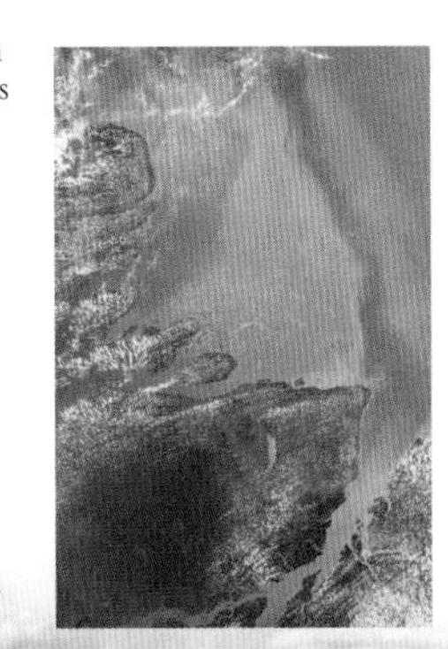

**MEETING THE ATLANTIC**
The mouth of the Amazon occupies the whole top part of this image, which covers an area of tens of thousands of square kilometres. Rio Pará, the estuary of a separate major river, the Tocantins, can be seen at bottom.

**MEANDERING TRIBUTARY**
The Tigre is a tributary of the Amazon in Peru. Here, it meanders through the Peruvian rainforest, over 3,000km (1,860 miles) from the Amazon's mouth.

THE VOLGA DELTA
The huge triangular delta of the River Volga is visible in the bottom of this image, with the Caspian Sea stretching out beyond it to the south.

ASIA *west*

## Caspian Sea

**LOCATION** On the borders of Azerbaijan, Iran, Kazakhstan, Russia, and Turkmenistan

**TYPE** Saline inland sea

**AREA** 371,000 square km (143,000 square miles)

The Caspian Sea is the largest inland body of water on Earth. It contains salty rather than fresh water, so can be appropriately described either as a salt lake or as an inland sea. The Caspian was once joined, via another inland sea, the Black Sea, to the Mediterranean. However, several million years ago it was cut off from those other seas when water levels fell during an ice age. The sea has no outflow other than by evaporation, but it receives considerable inputs of water from the River Volga (supplying three-quarters of its inflow) and from the Ural, Terek, and several other rivers. Its surface level has changed throughout history in line with discharges from the Volga, which in turn have depended on rainfall levels in the Volga's vast catchment basin in Russia. Today, the Caspian Sea contains about 78,200 cubic km (18,800 cubic miles) of water – about one-third of Earth's inland surface water. Its salinity (saltiness) varies from 1 per cent in the north, where the Volga flows in, to about 20 per cent in Kara-Bogaz-Bol Bay, a partially cut-off area on its eastern shore.

**OIL EXTRACTION**
Some of Earth's largest oil reserves underlie the Caspian Sea. The greatest concentration of proven reserves and extraction facilities are in its northeastern section.

ANTARCTICA

## Antarctic Ice-sheet

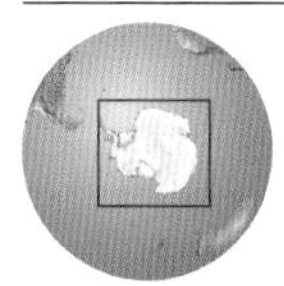

**LOCATION** Covering most of Antarctica

**TYPE** Continental ice-sheet

**AREA** 13.7 million sq. km (5.3 million sq. miles)

Earth's largest glacier, the Antarctic Ice-sheet, is an immense mass of ice that covers almost the whole of the continent of Antarctica and holds over 70 per cent of Earth's fresh water. The ice-sheet has two distinct parts, separated by a range of mountains called the Transantarctic Range. The West Antarctic Ice-sheet has a maximum ice thickness of 3.5km (2.2 miles), and its base lies mainly below sea level. The larger East Antarctic Ice-sheet is over 4.5km (2.8 miles) thick in places with a base above sea level. Both parts of the ice-sheet are domed, being slightly higher at their centres and sloping gently down towards their edges. A few areas around the edges of the ice-sheets, such as some regions within the Transantarctic Range, are known to be rich sources of meteorites (see pp.220–21). Meteorites continually fall onto the ice-sheet and become buried in it. But in a few places, where there is an upward flow of ice and some evaporation, they concentrate again at the surface. For some years, there have been concerns that the West Antarctic Ice-sheet is shrinking due to global warming. Scientists agree that the West Antarctic Ice-sheet has been showing a general pattern of retreat for over 10,000 years, but think there is only a small risk of it collapsing within the next few centuries.

**SATELLITE VIEW**
This radar image shows the whole of Antarctica, with the larger, eastern section of its ice-sheet on the left. The grey area around its coast is partly ice-shelf and partly sea-ice.

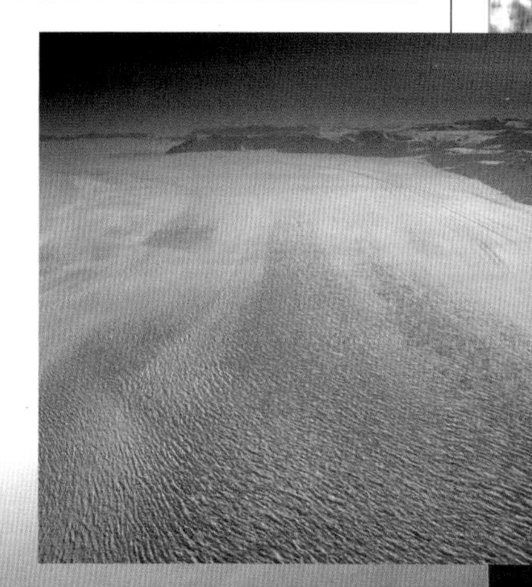

**THE BEARDMORE GLACIER**
This huge glacier drains the East Antarctic Ice-sheet into the Ross Ice-shelf. At 415km (260 miles) in length, it is one of the longest glaciers on Earth.

**THE LARSEN ICE-SHELF**
Around the coast of Antarctica, glaciers and ice-streams merge to form platforms of floating ice called ice-shelves. These are home to large colonies of penguins.

# THE MOON

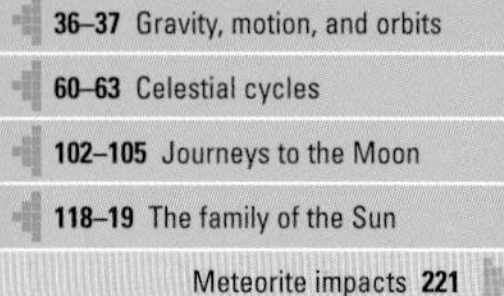


**EVEN THOUGH IT HAS ONLY** 1.2 per cent of the mass of Earth, the Moon is still the fifth-largest planetary satellite in the Solar System. When full, it is the brightest object in our sky after the Sun, and its gravity exerts a strong influence over our planet. However, the Moon is too small to retain a substantial atmosphere, and geological activity has long since ceased, so it is a lifeless, dusty, and dead world. Twelve men have walked on its surface and over 380kg (838lb) of lunar rock have been collected, but scientists are still not sure exactly how the Moon formed.

## ORBIT

The Moon has an elliptical orbit around the Earth, so the distance between the two bodies varies. At its closest to Earth (perigee), the Moon is 10 per cent nearer than when at its farthest point (apogee). The Moon takes 27.32 Earth days to spin on its axis, which is the same time it takes to orbit the Earth. This is known as synchronous rotation (see right) and keeps one side of the Moon permanently facing Earth – although eccentricities in the Moon's orbit called librations allow a few regions of the far side to come into view. Because the Earth is moving around the Sun, the Moon takes 29.53 Earth days to return to the same position relative to the Sun in Earth's sky, completing its cycle of phases (see p.62). This is also the length of a lunar day (the time between successive sunrises on the Moon).

**SYNCHRONOUS ROTATION**
For each orbit of Earth, the Moon spins once on its axis. As a result, it always keeps the same face towards Earth.

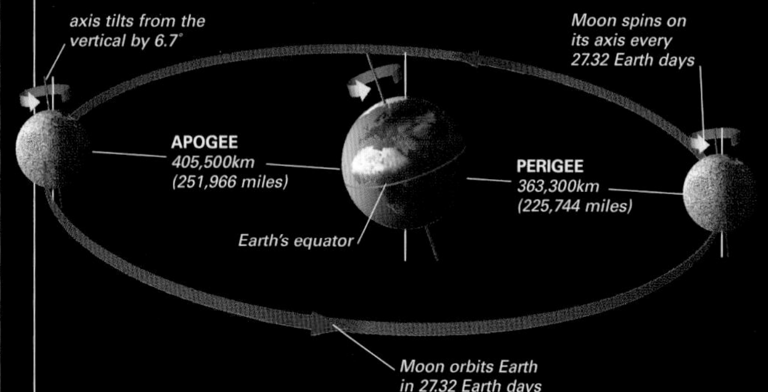


**SPIN AND ORBIT**
The Moon's orbital path is tilted at an angle to Earth's equator, causing its path across the sky to vary in an 18-year cycle. Tidal forces mean that the Moon is slowing down Earth's rotation, while the Moon moves away from the Earth at a rate of about 3cm (1in) each year.

## STRUCTURE

The lunar crust is made of calcium-rich, granite-like rock. It is about 48km (30 miles) thick on the near side and 74km (46 miles) thick on the far side. Because of the Moon's history of meteorite bombardment, the crust is severely cracked. The cracking extends to a depth of 25km (15 miles); below that, the crust is completely solid. The Moon's rocky mantle is rich in silicate minerals but poor in metals such as iron. The upper mantle is solid, rigid, and stable. Radioactive decay of minor components of the lunar rock means that the temperature increases with depth. The lower mantle lies about 1,000km (600 miles) below the crust, and here the rock gradually becomes partially molten. The average density of the Moon indicates that it might have a small iron core. The Apollo missions measured the velocities of shock waves travelling through the Moon, but the results proved inconclusive. Further seismic evidence is needed to confirm the existence of a metallic core.

**MOON INTERIOR**
The density of the Moon is much less than that of the whole Earth, but is similar to that of Earth's mantle. It is possible that the Moon is entirely made of solid rock and has no metallic core at all.

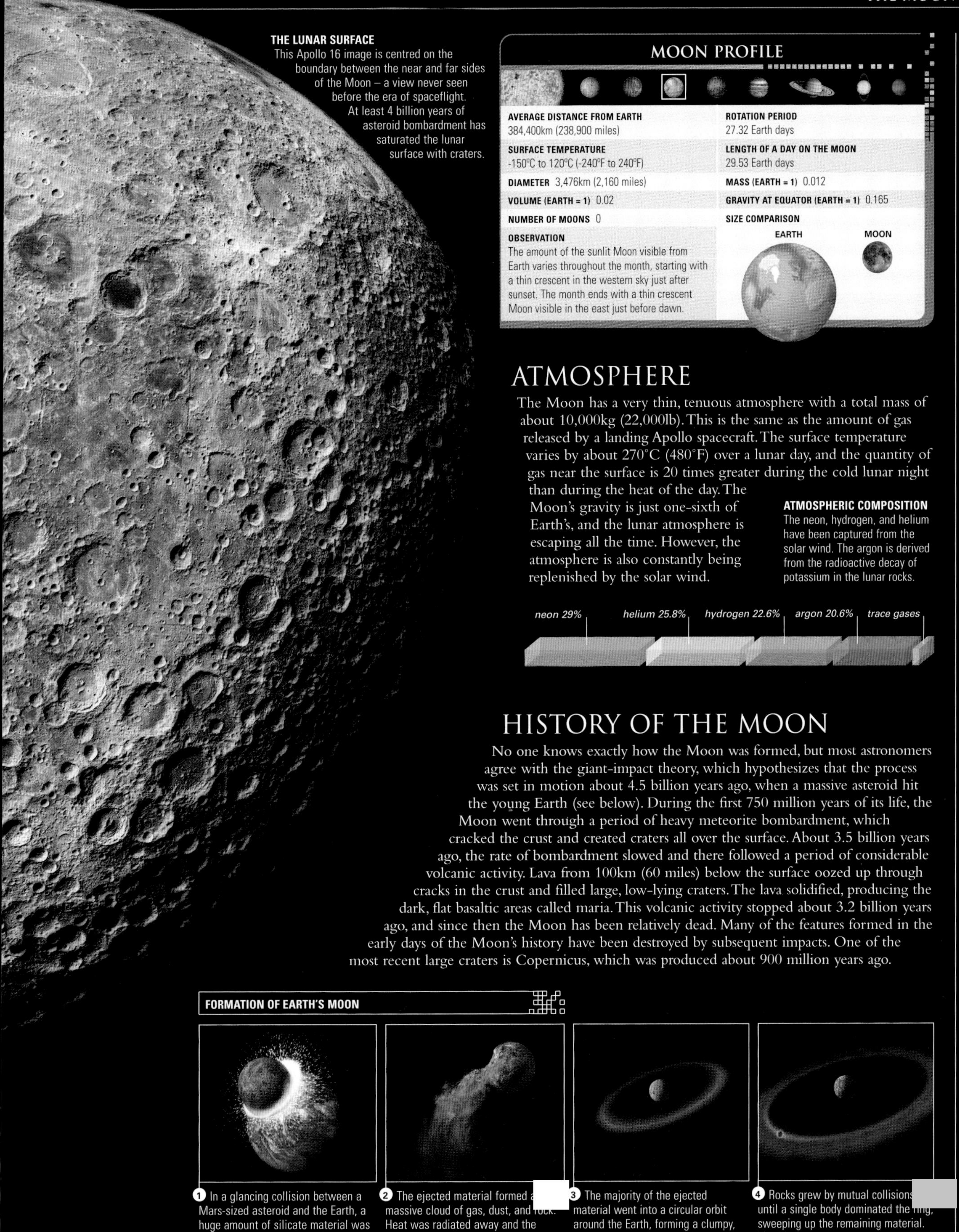

**THE LUNAR SURFACE**
This Apollo 16 image is centred on the boundary between the near and far sides of the Moon – a view never seen before the era of spaceflight. At least 4 billion years of asteroid bombardment has saturated the lunar surface with craters.

## MOON PROFILE

| | |
|---|---|
| **AVERAGE DISTANCE FROM EARTH** 384,400km (238,900 miles) | **ROTATION PERIOD** 27.32 Earth days |
| **SURFACE TEMPERATURE** -150°C to 120°C (-240°F to 240°F) | **LENGTH OF A DAY ON THE MOON** 29.53 Earth days |
| **DIAMETER** 3,476km (2,160 miles) | **MASS (EARTH = 1)** 0.012 |
| **VOLUME (EARTH = 1)** 0.02 | **GRAVITY AT EQUATOR (EARTH = 1)** 0.165 |
| **NUMBER OF MOONS** 0 | **SIZE COMPARISON** EARTH MOON |
| **OBSERVATION** The amount of the sunlit Moon visible from Earth varies throughout the month, starting with a thin crescent in the western sky just after sunset. The month ends with a thin crescent Moon visible in the east just before dawn. | |

# ATMOSPHERE

The Moon has a very thin, tenuous atmosphere with a total mass of about 10,000kg (22,000lb). This is the same as the amount of gas released by a landing Apollo spacecraft. The surface temperature varies by about 270°C (480°F) over a lunar day, and the quantity of gas near the surface is 20 times greater during the cold lunar night than during the heat of the day. The Moon's gravity is just one-sixth of Earth's, and the lunar atmosphere is escaping all the time. However, the atmosphere is also constantly being replenished by the solar wind.

**ATMOSPHERIC COMPOSITION**
The neon, hydrogen, and helium have been captured from the solar wind. The argon is derived from the radioactive decay of potassium in the lunar rocks.

neon 29% | helium 25.8% | hydrogen 22.6% | argon 20.6% | trace gases

# HISTORY OF THE MOON

No one knows exactly how the Moon was formed, but most astronomers agree with the giant-impact theory, which hypothesizes that the process was set in motion about 4.5 billion years ago, when a massive asteroid hit the young Earth (see below). During the first 750 million years of its life, the Moon went through a period of heavy meteorite bombardment, which cracked the crust and created craters all over the surface. About 3.5 billion years ago, the rate of bombardment slowed and there followed a period of considerable volcanic activity. Lava from 100km (60 miles) below the surface oozed up through cracks in the crust and filled large, low-lying craters. The lava solidified, producing the dark, flat basaltic areas called maria. This volcanic activity stopped about 3.2 billion years ago, and since then the Moon has been relatively dead. Many of the features formed in the early days of the Moon's history have been destroyed by subsequent impacts. One of the most recent large craters is Copernicus, which was produced about 900 million years ago.

**FORMATION OF EARTH'S MOON**

1 In a glancing collision between a Mars-sized asteroid and the Earth, a huge amount of silicate material was jetted away from the Earth's mantle.

2 The ejected material formed a massive cloud of gas, dust, and rock. Heat was radiated away and the cloud quickly began to cool.

3 The majority of the ejected material went into a circular orbit around the Earth, forming a clumpy, dense, doughnut-shaped ring.

4 Rocks grew by mutual collisions until a single body dominated the ring, sweeping up the remaining material. The Moon was born.

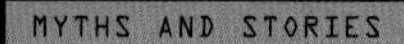

MYTHS AND STORIES

## WEREWOLVES

Many myths and old folk tales attribute strange powers to the Moon. Some say that a full Moon can turn people mad (the origin of the word "lunacy"), and many cultures, from Eurasia to the Americas, share a belief that when the Moon is full some humans can be transformed into vicious werewolves. The superstition is widespread and ancient – even the Babylonian King Nebuchadnezzar (*c.*630–*c.*562BC) imagined that he had become a werewolf.

# LUNAR INFLUENCES

Although the Moon is much smaller than Earth, its gravity still exerts an influence. The Moon's gravitational attraction is felt most strongly on the side of the Earth facing the Moon, and this pulls water in the oceans towards it. Inertia (the tendency of objects with mass to resist forces acting upon them) attempts to keep the water in place, but because the gravitational force is greater a bulge of water is pulled towards the Moon. On the opposite side of the Earth, the water's inertia is stronger than the Moon's gravity, so a second bulge of water is created. As the Earth rotates, the bulges sweep over the planet's surface creating daily changes in sea level called tides. The time of the high tide changes according to the Moon's position in the sky. The height of the tides changes during the lunar cycle, but the actual height also depends on local geography. In shallow coastal bays, the tidal range can be huge.

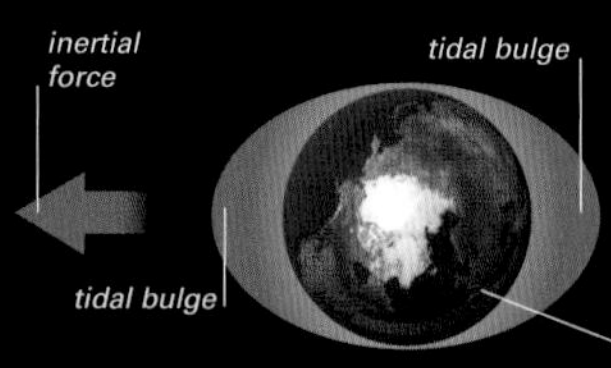

**TIDAL BULGES**
Gravitational interaction between the Earth and Moon creates two bulges in Earth's oceans (exaggerated here). As the Earth spins on its axis, the bulges of water sweep over the surface, creating tides.

**TIDAL RANGE**
The magenta in this satellite image of Morecambe Bay on the northwest coast of England reveals the inlets and mudflats that are left exposed at low tide.

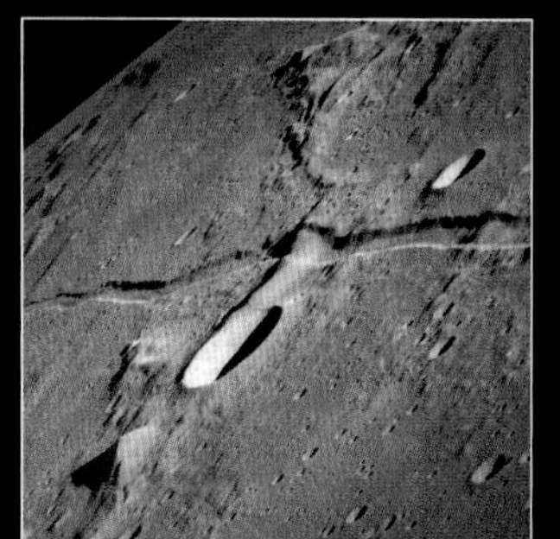

**LAVA TUBE**
Over 5km (3 miles) wide and hundreds of kilometres long, this rille is a collapsed tube-like structure through which lava once flowed. Moonquakes caused by nearby impacts may have caused the roof to fall in.

# SURFACE FEATURES

The surface of the Moon has been pulverized by meteorites and is covered by a rough, porous blanket of rubble several metres thick. This debris ranges in size from particles of dust to huge lumps of rock tens of metres across. The soil (or regolith) consists of fine-grained, fragmented bedrock, the size of the grains getting progressively larger with depth. As there is no wind or rain, the surface material does not move far, and its composition can change considerably from place to place. The thickness also varies – in young mare regions it is about 5m (16ft) thick, but this increases to 10m (32ft) in the old highlands. Micrometeorite impacts continuously erode exposed rocks, and they are also damaged by cosmic rays and solar-flare particles. The topmost layer of soil is saturated with hydrogen ions absorbed from the solar wind.

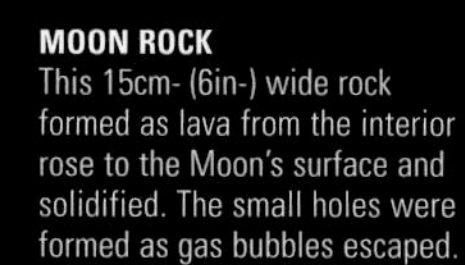

**MOON ROCK**
This 15cm- (6in-) wide rock formed as lava from the interior rose to the Moon's surface and solidified. The small holes were formed as gas bubbles escaped.

**TRACKS IN THE SOIL**
Lunar Rover tyre tracks lead away from the Apollo 15 module "Falcon", nestling near Hadley Rille in 1971. Over a million or more years, they will eventually be erased by meteorite bombardment.

## CRATERS

The vast majority of lunar craters are produced by impacts. Asteroids usually strike the Moon at velocities of about 72,000kph (45,000mph). The resulting crater is about 15 times larger than the impacting body. Unless the asteroid nearly skims the surface on entry, the resultant crater is circular. Three types are formed. Those smaller than 10km (6 miles) across are bowl-shaped, having a depth of around 20 per cent of the diameter. Craters between 10km and 150km (6–90 miles) in diameter have outer walls that have slumped into the initial crater pit. There is often a central mountainous peak produced by the recoil of the underlying stressed rocks. The crater depth is a few kilometres, and much excavated material falls back into the crater just after the impact. Craters wider than 150km (90 miles) contain concentric rings of mountains, created as rebounding material rippled out from the centre before solidifying. Such craters were so deep that hot magma flooded to the surface and filled the bottom of the crater with lava.

**SUNRISE OVER COPERNICUS CRATER**

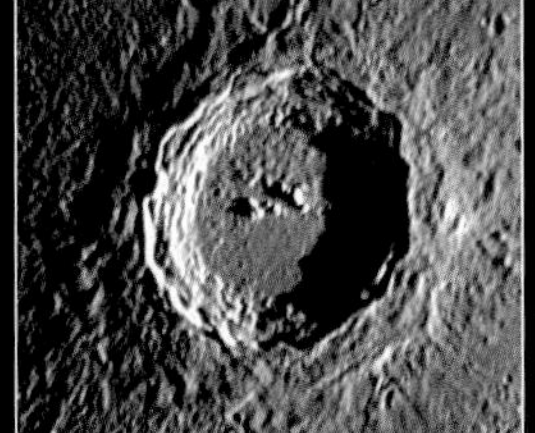

1 Just after dawn in the crater, the low eastern Sun casts long shadows, which emphasize the variation in height between the floor and rim.

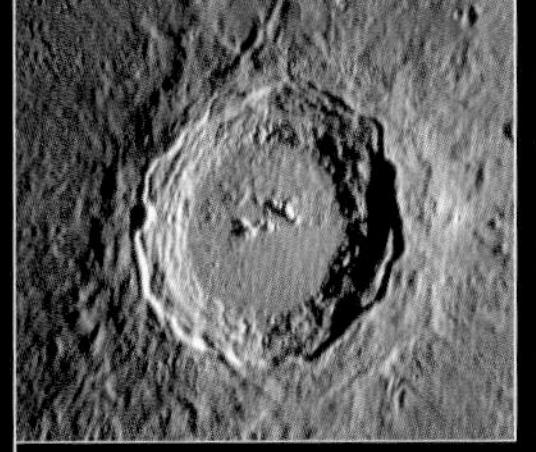

2 Halfway through the morning, small shadows enhance the ejecta blanket outside the crater. The temperature inside the crater is rising.

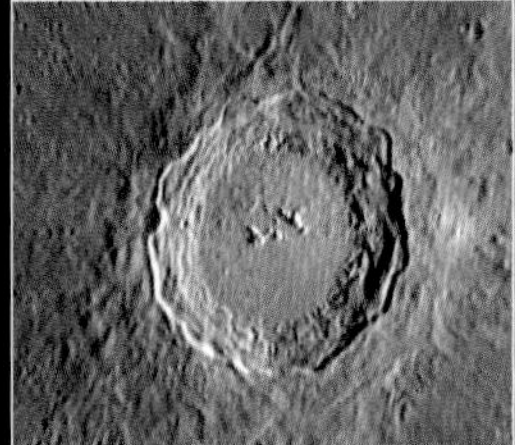

3 At noon, the Sun is overhead, and the scene appears much flatter and washed out. The temperature is now more than 100°C (212°F).

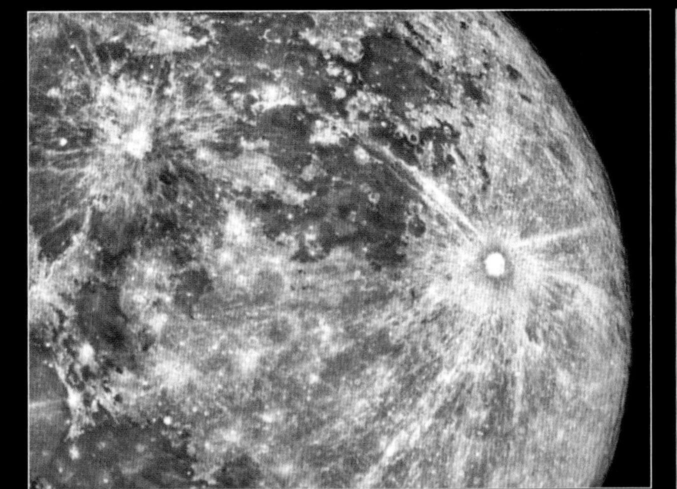

**RAY CRATERS**
Material ejected from a crater during an impact is often confined to narrow jets. Where this material hits the surface, it ploughs up the lunar soil, and this disturbed region then reflects more sunlight than its surroundings. From Earth, these appear as rays. The rays around Tycho Crater (far right) extend for thousands of kilometres.

### EUGENE SHOEMAKER

Gene Shoemaker (1928–1997) was an American astrogeologist who studied terrestrial and lunar meteorite impact craters and dreamt of going to the Moon. Addison's disease prevented that. Instead, he taught the Apollo astronauts to be field geologists. In 1969, he joined a team at Palomar, USA, searching for near-Earth asteroids. After Shoemaker died, some of his ashes were carried to the Moon aboard the Lunar Prospector space probe in 1999.

## MAPPING THE MOON

Some ancient Greeks thought the Moon was like the Earth and that its dark areas were water. This belief continued to the 17th century, when the dark patches were given aquatic names such as mare (sea) and oceanus (ocean) on the first proper maps. Palus Putredinis (the Marsh of Decay) and Sinus Iridum (the Bay of Rainbows) are evocative examples. Italian astronomer Galileo Galilei (see p.88) was the first to realize that the height of surface features could be added to maps by noting how the shadow lengths changed during the lunar day. The first photographic atlas appeared in 1897, but the real leap forward came with the advent of spaceflight. In 1959, the Soviet Union sent the Luna 3 space probe behind the Moon to photograph the far side. NASA's five Lunar Orbiter spacecraft imaged 99 per cent of lunar surface in 1966–67, paying special attention to potential Apollo landing sites. More recently, Clementine and Lunar Prospector surveyed our satellite's mineral composition in the 1990s.

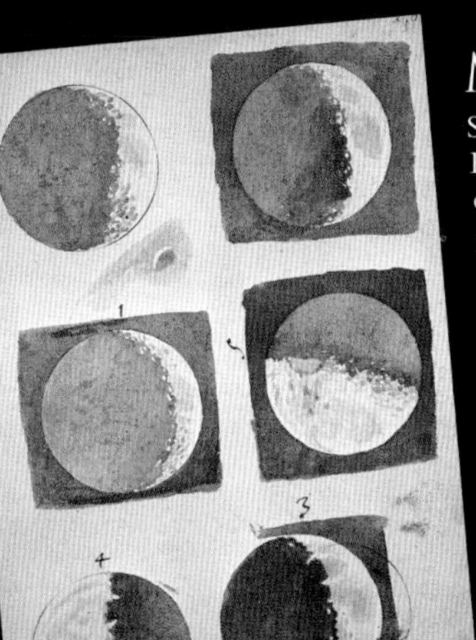

**GALILEO SKETCHES**
Galileo's first telescopic observations of the Moon were made on 30 November 1609. The pictures, published in *Sidereus Nuncius* in 1610, emphasized the roughness of the surface.

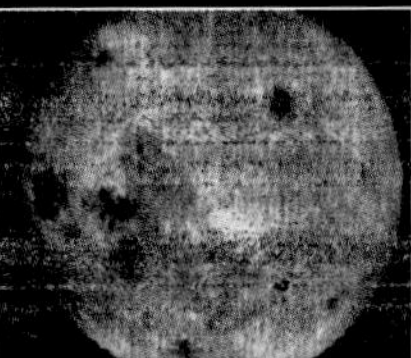

**LUNA 3**
On 7 October 1959, the Soviet Union's Luna 3 space probe imaged the far side of the Moon. It had never been seen before.

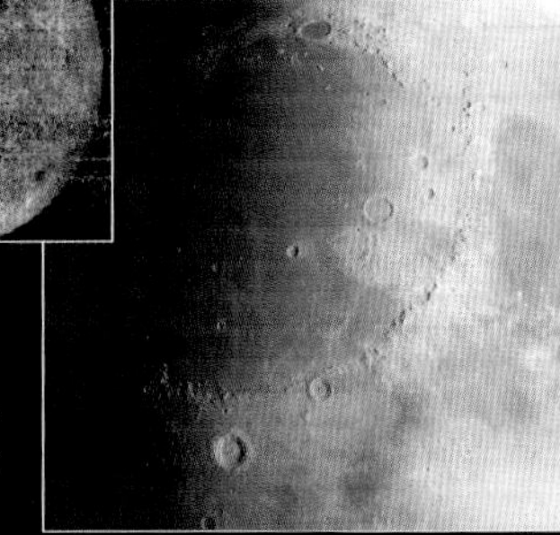

**LUNAR ORBITER IV**
This superb wide-angle image of the half-lit Mare Imbrium was one of 546 images taken by NASA's orbiter on 11–26 May 1967 from a height of about 4,000km (2,485 miles).

**SMART-1**
During its approach phase, ESA's SMART-1 spacecraft took this image of an illuminated region of the far side, near the lunar north pole, on 12 November 2004, from a distance of about 60,000km (37,250 miles).

# THE NEAR AND FAR SIDES OF THE MOON

The Moon's spin and orbital periods became locked together very early in its existence, when it was much closer to Earth than it is now, and the surface was still molten due to the heating produced by massive early impacts. As a result, the Earth's influence has led to noticeable differences in the appearance of the two sides. The far side is on average about 5km (3 miles) higher, with respect to the Moon's centre of mass, than the near side, and its low-density crust is 26km (16 miles) thicker. Since the near side is lower, volcanic magma has more easily found its way to the surface here, pouring from volcanic fissures into the low-lying regions of the largest craters and solidifying to form the lunar seas. By contrast, the far side – forever facing outwards from Earth – seems to have suffered slightly heavier bombardment and cratering than the near side.

**TOPOGRAPHIC VIEW**
In the mid-1990s, the US Clementine mission used a laser ranging instrument to measure accurate heights on the lunar surface. The far side is on the left, the near on the right. Blue is low, green medium, and red high. The large blue area is the 2,400km- (1,490-mile-) wide South Pole–Aitken Basin, whose existence was confirmed by Clementine.

**NEAR SIDE**
Many features on the Moon's near side have classically inspired names. Landing sites of the six crewed spacecraft and most of the probes that reached the Moon between 1959 and 1999 (see panel, opposite) are marked on this map.

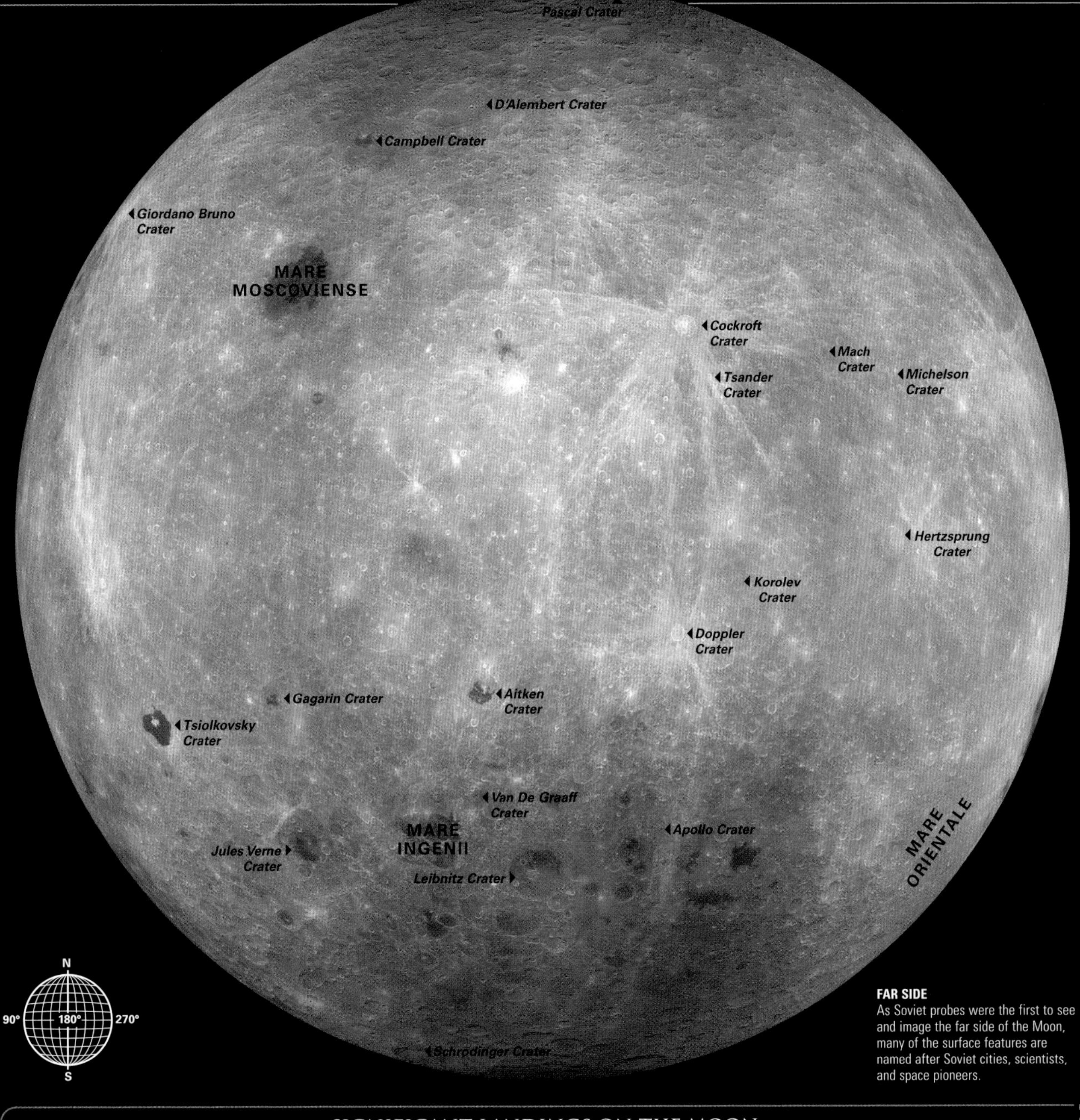


**FAR SIDE**
As Soviet probes were the first to see and image the far side of the Moon, many of the surface features are named after Soviet cities, scientists, and space pioneers.

## SIGNIFICANT LANDINGS ON THE MOON

Between them, automated space probes and human explorers have studied a wide range of terrains on the near side of the Moon. At first, just crashing a probe into the Moon at all was a significant achievement, but by the time of the Apollo missions, landings were being targeted at particular areas to answer specific questions about the Moon's geology and history.

| MISSION | DATE OF ARRIVAL | TYPE | ACHIEVEMENT |
|---|---|---|---|
| Luna 2 (USSR) | 13 September 1959 | Impact | Makes first crash-landing on the Moon |
| Ranger 7 (USA) | 31 July 1964 | Impact | Takes first close-up photos of surface |
| Ranger 8 (USA) | 20 February 1965 | Impact | Takes 7,137 good-quality photos |
| Ranger 9 (USA) | 24 March 1965 | Impact | Finds volcanic vents |
| Luna 9 (USSR) | 3 February 1966 | Lander | Makes first soft landing |
| Surveyor 1 (USA) | 2 June 1966 | Lander | Acquires data on radar reflectivity of lunar surface |
| Luna 13 (USSR) | 24 December 1966 | Lander | Successfully uses mechanical soil probe |
| Surveyor 3 (USA) | 20 April 1967 | Lander | Images future Apollo 12 landing site |
| Surveyor 5 (USA) | 11 September 1967 | Lander | Analyses magnetic properties of soil |
| Surveyor 6 (USA) | 10 November 1967 | Lander | Takes nearly 30,000 photos |
| Surveyor 7 (USA) | 10 January 1968 | Lander | Analyses surface near Tycho Crater |
| Apollo 11 (USA) | 20 July 1969 | Manned | Lands first astronauts on the Moon |
| Apollo 12 (USA) | 19 November 1969 | Manned | Makes first pinpoint landing |
| Luna 16 (USSR) | 20 September 1970 | Lander | Makes first automated sample return |
| Luna 17 (USSR) | 17 November 1970 | Rover | Carries first robotic lunar rover |
| Apollo 14 (USA) | 5 February 1971 | Manned | Carries "lunar cart" for sample collection |
| Apollo 15 (USA) | 30 July 1971 | Manned | Carries first manned lunar rovers |
| Luna 20 (USSR) | 21 February 1972 | Lander | Makes automated sample return |
| Apollo 16 (USA) | 21 April 1972 | Manned | Explores central highlands |
| Apollo 17 (USA) | 11 December 1972 | Manned | Makes longest stay on Moon (75 hours) |
| Luna 21 (USSR) | 15 January 1973 | Rover | Explores Posidonius Crater |
| Luna 24 (USSR) | 14 August 1976 | Lander | Returns sample from Mare Crisium |
| Hiten (Japan) | 10 April 1993 | Impact | Crashes into Furnerius region |
| Lunar Prospector (USA) | 31 July 1999 | Impact | Orbiter makes controlled crash near the south pole to look for evidence of water |

# FEATURES OF THE MOON

From afar, the Moon is clearly divided into two types of terrain. There are large, dark plains called maria (the Latin for "seas") and also brighter, undulating, heavily cratered highland regions. The whole surface was initially covered with craters, most of which were produced during a time of massive bombardment. The rate at which asteroids have been striking the Moon has decreased over the last 4 billion years. Around 4 billion years ago, the Moon was also volcanically active. Lava rose to the surface through cracks and fissures, filling the lower parts of the large craters to produce the dark plains. The plains reflect only about 4 per cent of the sunlight that hits them, whereas the mountains reflect about 11 per cent.

**MOSAIC OF THE NORTH POLE**
The lunar North Pole is partially hidden from view from Earth and is best imaged by orbiting spacecraft. Galileo took a series of photographs of the region on 7 December 1992 on its way to Jupiter.

NEAR SIDE *northern hemisphere*

## Aristarchus Crater

**TYPE** Impact crater

**AGE** About 300 million years

**DIAMETER** 37km (23 miles)

This young crater has a series of nested terraces, which were produced by concentric slices of rock in the wall slipping downwards. This both widened the crater and made it considerably shallower, as the initially deep central region was filled with material from the rim. Aristarchus was mapped by the Apollo Infrared Scanning Radiometer. During the night, the temperature in the crater is about 30°C (54°F) higher than that of the surrounding terrain. Young craters contain many large boulders. These take a long time to heat up during the day and also a long time to cool down at night. As time passes the boulders are broken up by small impacting asteroids, so this thermal difference eventually disappears.

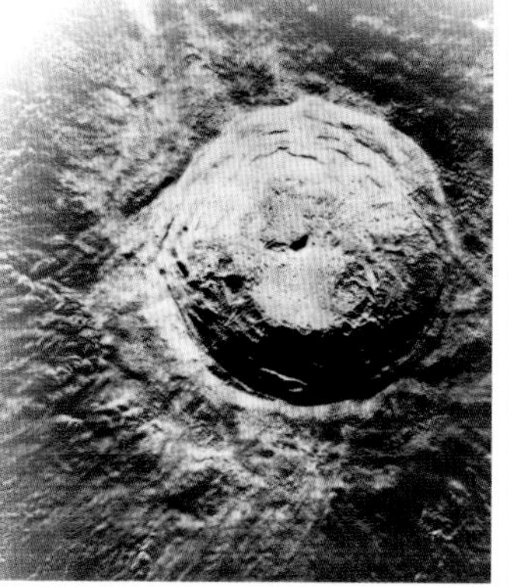

**LUNAR ORBITER 5 IMAGE**
This view of Aristarchus, taken from directly above, underlines the crater's circularity and reveals the extensive surrounding blanket of hummocky ejecta.

NEAR SIDE *northern hemisphere*

## Mare Crisium

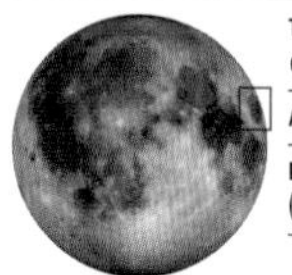

**TYPE** Lava-filled impact crater (sea)

**AGE** 3.9 billion years

**DIAMETER** 563km (350 miles)

Mare Crisium has an extremely smooth floor, which varies in height by less than 90m (290ft). The lava that flooded Crisium had extremely low viscosity and became like a still pond before it solidified. The Soviet Luna 24 probe was the last mission to bring back rock samples from the Moon. In 1976, it returned to Earth with a core of rock weighing 170g (6oz), which was collected from Crisium's floor.

**OVAL CRATER**
The Mare Crisium, which can be seen with the naked eye from Earth, is nearly circular in shape. Over 95 per cent of lunar craters are completely circular.

NEAR SIDE *northern hemisphere*

## Montes Apenninus

**TYPE** Mountain range

**AGE** 3.9 billion years

**LENGTH** 401km (249 miles)

The Lunar Apennine mountains form a ring around the southwestern edge of the Mare Imbrium impact basin. They consist of crustal blocks rising more than 3km (1.9 miles) above the flat lava plain, pushed up by the shockwave from the Imbrium impact. The mountain chain stretches for some 600km (375 miles), though its southern end is partially buried beneath lava flows.

**LUNAR MOUNTAINS**
The Apennines lie in the lower right of this Apollo 15 image. The dark area to their left is Palus Putredinis.

NEAR SIDE *northern hemisphere*

## Mare Tranquillitatis

**TYPE** Sea

**AGE** 3.6 billion years

**DIAMETER** 873km (542 miles)

The surfaces of lunar maria are much darker than highland rock and are also considerably younger. This means that they are relatively smooth and contain only a few impact craters. Their low reflectivity is due to the chemistry of the very fluid lava that flooded them. The Mare Tranquillitatis (Latin for "Sea of Tranquillity") lies just north of the lunar equator and joins onto the southeast part of the Mare Serenitatis (Sea of Serenity). Together, the two seas form one of the Moon's most prominent features. The basin in which the "sea" formed is very ancient, predating the formation of the Imbrium Basin 3.9 billion years ago. It overlaps with other basins at several points, but only flooded with lava about 3.6 billion years ago. The Sea of Tranquillity was famously the landing place of US astronauts Neil Armstrong and Buzz Aldrin on their 1969 Apollo 11 mission.

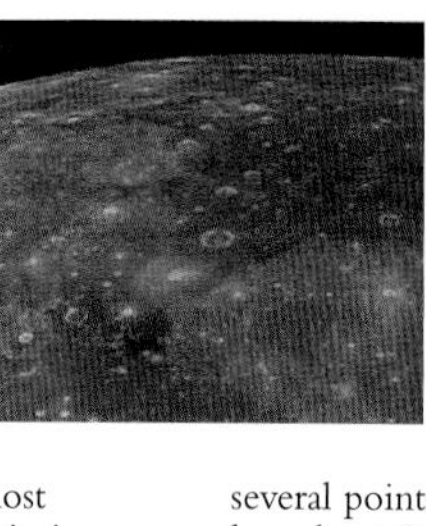

**RICH IN TITANIUM**
This Galileo image has been colour-coded according to the titanium content of the rock. The blue Tranquillitatis region is rich in titanium, whereas the orange Serenitatis region at the lower right is titanium-poor.

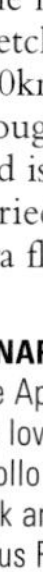

**BEFORE TOUCHDOWN**
The flat, desolate plain of the Sea of Tranquillity stretches away to the north in this view from the Apollo 11 lunar module taken just before landing.

NEAR SIDE *northern hemisphere*

## Copernicus Crater

**TYPE** Impact crater

**AGE** 900 million years

**DIAMETER** 91km (57 miles)

This young ray crater has massive terraced walls. The crater floor is below the general level of the surrounding plain, and lies 3.7km (2.3 miles) below the top of the surrounding walls. Copernicus is an intermediate-sized crater with high central peaks. These mountains were formed when the rock directly below the crater rebounded after being compressed by the explosion caused by the impacting asteroid. The vicinity of Copernicus is peppered with secondary craters formed by boulders thrown out during the impact. Fine, light grey rock particles ejected during the crater's formation were collected by the Apollo 12 astronauts near their landing site. Such particles were responsible for forming the rays that surround the crater. The high reflectivity of the rays is due to the ejecta churning up the lunar regolith (rough material reflects more light than smooth material).

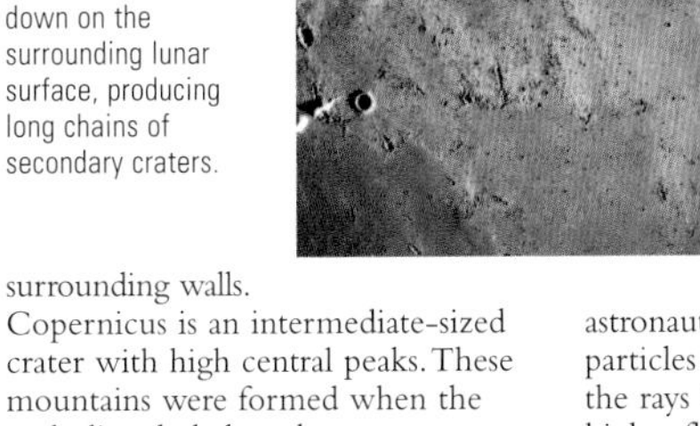

**CRATER CHAINS**
The material excavated by an impact showers down on the surrounding lunar surface, producing long chains of secondary craters.

**LUNAR ORBITER 2 IMAGE**
Copernicus Crater's terraced walls and central peaks were revealed by NASA's second Lunar Orbiter in 1966.

NEAR SIDE *southern hemisphere*

## Alphonsus Crater

**TYPE** Impact crater

**AGE** 4.0 billion years

**DIAMETER** 117km (80 miles)

NASA's Ranger 9 spacecraft was deliberately crash-landed into the Alphonsus Crater on 24 March 1965, taking television pictures as it approached. The crater formed in an impact, but the dark patches and fractures Ranger 9 found on its floor are thought to be a result of volcanic activity – probably explosive eruptions. Because of these features, Alphonsus was considered a possible landing site for later Apollo missions.

THREE MINUTES BEFORE IMPACT

NEAR SIDE *southern hemisphere*

## Rupes Altai

**TYPE** Cliff

**AGE** 4.2 billion years

**LENGTH** 507km (315 miles)

Altai is by far the longest cliff on the Moon. It is about 1.8km (1.1 miles) high. The energy that is released during an impact does more than just excavate a crater and lift material out to form walls and an ejecta blanket. Violent seismic shock waves radiate away from the impact point. An obstacle such as a mountain can halt these waves and the lunar crust may then buckle, forming a long cliff. Altai was created by the Nectaris impact.

**ALTAI ESCARPMENT**
Altai is a cliff that runs north–south. So, when the Moon is at first quarter, Altai is struck by the early morning sunlight, and it appears as a dark line of shadow.

NEAR SIDE *southern hemisphere*

## Humboldt Crater

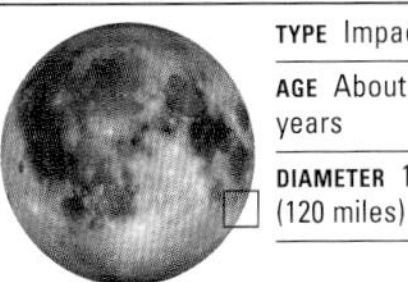

**TYPE** Impact crater

**AGE** About 3.8 billion years

**DIAMETER** 189km (120 miles)

This crater is remarkable because its lava-filled floor is crisscrossed with a series of radial and concentric fractures (or rilles). On closer inspection, some look like collapsed tubes through which lava once flowed, others like rift valleys. Lunar volcanic activity lasted for over 500 million years. Lava would seep up into a crater and then cool, shrink, crack, and sink. It would then be covered by more lava. The final basaltic infill would have many layers.

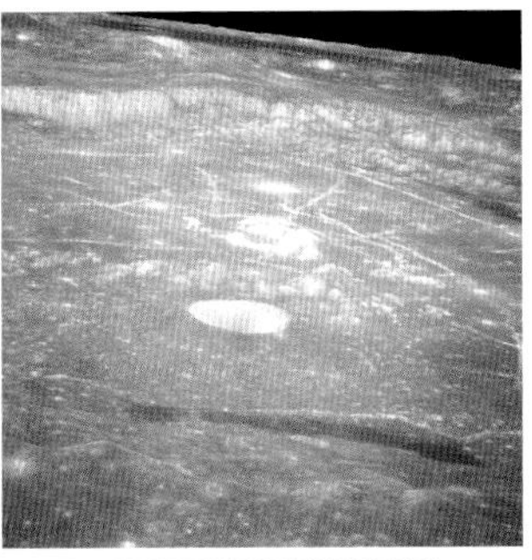

SOUTHEAST VIEW INTO HUMBOLDT

NEAR SIDE *southern hemisphere*

## Tycho Crater

**TYPE** Impact crater

**AGE** 100 million years

**DIAMETER** 85km (52 miles)

Lying in the southern highlands, Tycho is one of the most perfect walled craters on the Moon, with a central mountain peak towering 3km (1.8 miles) above a rough infilled inner region. Surveyor 7 landed on the north rim of Tycho's ejecta blanket in January 1968. About 21,000 photographs were taken, and the soil was chemically analysed. The highland soil was found to be mainly made of calcium-aluminium silicates, in contrast to the maria material, which is iron-magnesium silicate.

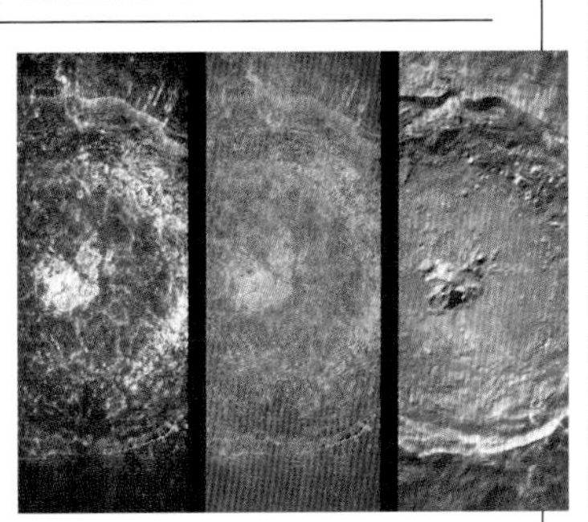

**THREE FILTERED IMAGES OF TYCHO**
The Ultraviolet/Visual camera onboard the Clementine spacecraft was fitted with a series of filters. Differing colour combinations revealed the variability in the physical and chemical structure of the crater rock.

**YOUNGEST LARGE LUNAR CRATER?**
Although Tycho is one of the youngest lunar craters (Giordano Bruno may be younger), it still formed in the age of the dinosaurs.

NEAR SIDE *northern hemisphere*

## Taurus-Littrow Valley

**TYPE** Valley

**AGE** About 3.85 billion years

**LENGTH** 30km (18.6 miles)

**HIGHLAND MASSIFS**
The flat-based Taurus-Littrow Valley can be seen in the centre of this image, nestling between the rugged, block-like mountains known prosaically as the North, South, and East massifs.

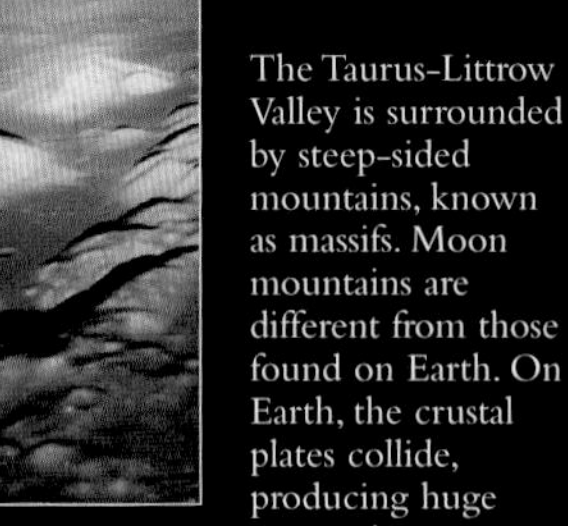

In December 1972, the last manned mission to the Moon landed in the dark-floored Taurus-Littrow Valley at the edge of the basalt-filled Mare Serenitatis. The range of geological features was impressive, and the Apollo 17 astronauts found three distinct types of rock in the region. One piece of crushed magnesium olivine was 4.6 billion years old and had crystallized directly from the melted shell of the just-formed Moon. The nearby Serenitatis crater was produced about 3.9 billion years ago, and much of the basaltic rock dates from that time, when the crater was flooded with lava. The third type of rock was found on the top of nearby hills. This was barium-rich granite and had been ejected from one of the surrounding large craters. Most of the material near the landing site was extremely dark and consisted of cinders and ash ejected billions of years ago from nearby volcanic vents and fissures.

The Taurus-Littrow Valley is surrounded by steep-sided mountains, known as massifs. Moon mountains are different from those found on Earth. On Earth, the crustal plates collide, producing huge mountain ranges like the Alps and Himalayas. These new mountains are subsequently eroded by rain and ice. The Moon's crust is not broken into plates. Nothing moves. All the Moon mountains are produced by impacts, and the mountains around the Taurus-Littrow Valley are the remains of old crater walls. Part of the valley floor just to the north of South

### HARRISON SCHMITT

Harrison "Jack" Schmitt (b.1935) was born in New Mexico, USA. He studied geology at Caltech and Harvard University. While working for the US Geological Survey, he joined a team instructing astronauts in the art of field geology. In June 1965, Schmitt was selected as a scientist-astronaut by NASA and was later chosen to be the lunar module pilot for Apollo 17. In December 1972, he became the first and only geologist to walk on the Moon. One of the highlights of the Apollo 17 mission was his discovery of orange glass within the lunar rock.

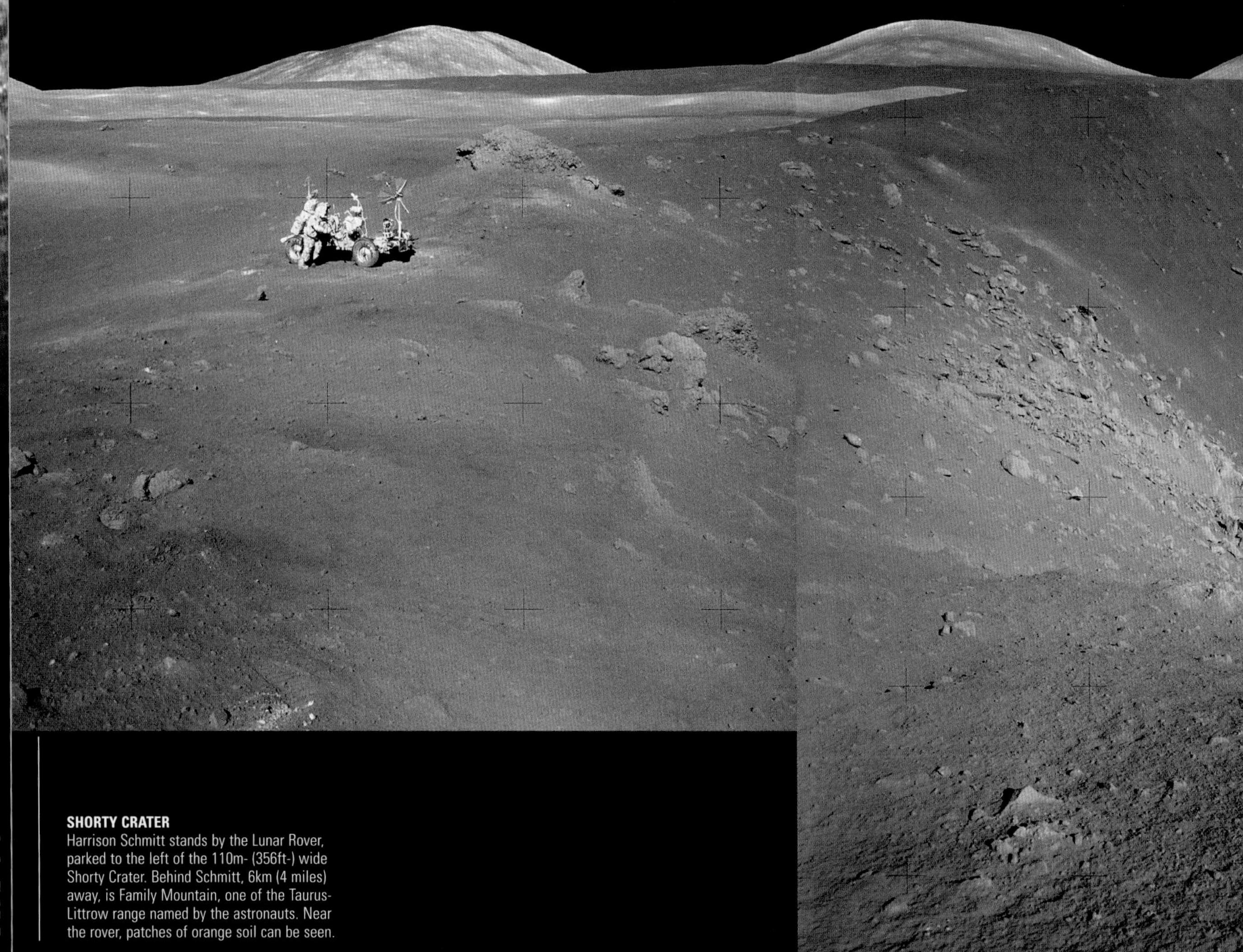

**SHORTY CRATER**
Harrison Schmitt stands by the Lunar Rover, parked to the left of the 110m- (356ft-) wide Shorty Crater. Behind Schmitt, 6km (4 miles) away, is Family Mountain, one of the Taurus-Littrow range named by the astronauts. Near the rover, patches of orange soil can be seen.

Massif was covered with a light mantle of regolith a few metres thick. This had been produced by a rock avalanche, possibly triggered when the area was bombarded by boulders ejected when the nearby Tycho Crater was formed. As the Moon is being continuously bombarded by asteroids, the number of craters per unit area increases with time. There are relatively few craters on the Taurus-Littrow valley floor, which was taken to indicate that the surface is even younger than the Apollo 12 landing site. One crater in the valley, Shorty, was once thought to be a volcanic vent, but more detailed analysis of its raised rim and central mound indicated that, like millions of other lunar craters, it was produced by an impacting asteroid.

**SPLIT ROCK**
This house-sized boulder was ejected from an impact crater in the Mare Serenitatis and then rolled down into the valley. Scoop marks can be clearly seen where some samples have been taken from its surface.

EXPLORING SPACE

## MOON GLASS

The lunar regolith contains large amounts of volcanic glass. This occurs as glazings on rock fragments and also as tiny teardrop- and dumbbell-shaped droplets. Colours range from green and wine-red through to orange and opaque. The orange glass found near Shorty was typical of high-titanium lunar glasses, but it was also rich in zinc.

**ORANGE SOIL IN SHORTY CRATER**
The glassy orange surface soil was excavated by an impact about 20 million years ago. It was actually formed about 3.6 billion years ago.

FAR SIDE *northern hemisphere*

## Pascal Crater

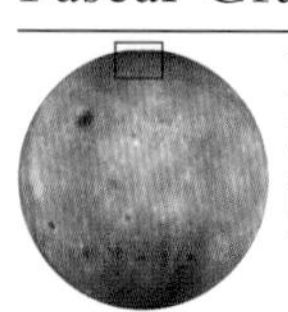

**TYPE** Impact crater

**AGE** About 4.1 billion years

**DIAMETER** 115km (71 miles)

This is one of 300 lunar craters named after mathematicians. It honours the Frenchman Blaise Pascal. The image below was taken in 2004, with the camera looking directly down into the crater. The Sun is low in the sky, below the bottom of the picture. The tiny craters around Pascal are bowl-shaped and young, with circular rims much sharper than the older rim of Pascal. The larger crater's rim was initially eroded by slumping and rock slides, and is now being worn down further by more recent impacts.

PASCAL AND ITS YOUNGER NEIGHBOURS

**DARK FLOOR**
If Tsiolkovsky had been formed earlier in lunar history, the volcanic activity would have been greater and more of the crater floor would have been filled with lava.

FAR SIDE *southern hemisphere*

## Tsiolkovsky Crater

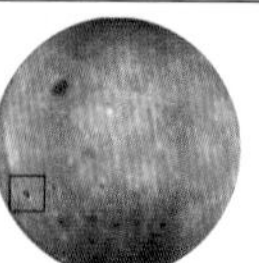

**TYPE** Impact crater

**AGE** About 4.2 billion years

**DIAMETER** 198km (123 miles)

Only half the size of Mare Crisium, this far-side crater is special because only half the interior basin has been filled with lava. The central peak is also unusually offset from the centre of the crater. There have been extensive rock avalanches down the southern rim of the crater. The first images of the lunar far side was obtained in October 1959 by the Soviet spacecraft Luna 3. Resolution was low, but the features that could be seen were nevertheless given names, such as Mare Moscoviense and Sinus Astronautarum. Only a few craters could be made out, including this one. Konstantin Tsiolkovsky was a Russian rocketry pioneer who not only designed a liquid hydrogen/liquid oxygen rocket but also suggested the multi-stage approach to spaceflight. The crater was pencilled in as a possible landing site for one of the post-Apollo 17 missions, which were cancelled.

**ORBITER 3 IMAGE**
The crest of the rim of Tsiolkovsky Crater runs to the upper right of this image. The diagonal banding to its right is probably the result of a large avalanche down the slope of the rim.

FAR SIDE *southern hemisphere*

## Van de Graaff Crater

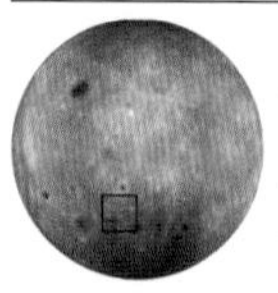

**TYPE** Double impact crater

**AGE** About 3.6 billion years

**LENGTH** 250km (155 miles)

Less than one per cent of the lunar craters are non-circular. Van de Graaff is typical of such irregular craters, which are produced on the rare occasions when the impacting asteroid hits the surface at an angle of less than 4°. Van de Graaff is also special because it is both magnetic and has the highest concentration of natural radiation. Most of the ancient lunar magnetic field decayed away over 3 billion years ago. However, there are still a few magnetic anomalies (magcons), of which Van de Graaff and nearby Aitken are the strongest. Magcons were discovered by small magnetometer sub-satellites released by Apollos 15 and 16.

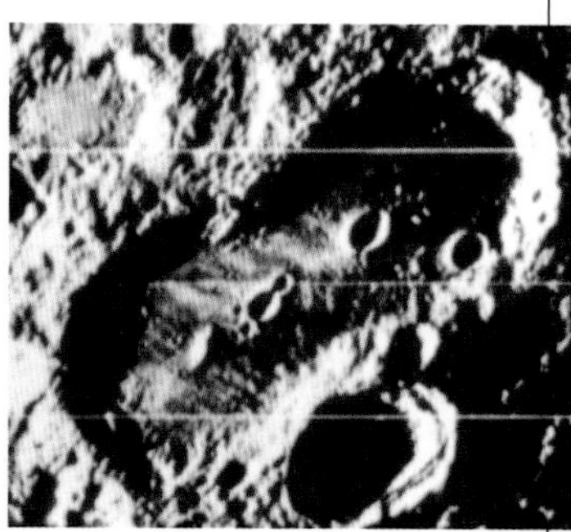

IRREGULARLY SHAPED CRATER

FAR SIDE *southern hemisphere*

## Korolev Crater

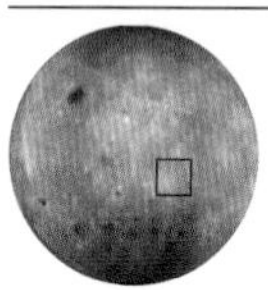

**TYPE** Ringed impact crater

**AGE** About 3.7 billion years

**DIAMETER** 405km (250 miles)

Sergei Korolev (see p.100) led the Soviet space effort in the 1950s and 1960s and was responsible for the early Sputnik and Vostok spacecraft. He has two craters named after him, one on the Moon and the other on Mars. Korolev is one of only 10 craters on the lunar far side that are more than 200km (125 miles) across. It is double-ringed and pocked with smaller craters. The outer ring is 405km (252 miles) in diameter. The inner ring is much less distinct. It is only half the height of the outer ring and its diameter is half that of the outer ring. Together with Hertzsprung and Apollo, Korolev forms a trio of huge ringed formations on the lunar far side. The lunar crust varies in thickness, and it reaches its maximum thickness of 107km (66 miles) in the region around the Korolev Crater.

**INDESTRUCTIBLE?**
This Orbiter 1 image shows that later impacts have done little to obliterate the huge Korolev Crater.

EXPLORING SPACE

### NUCLEAR CRATER

It is very difficult to estimate the relationship between the size of a crater and the size of the asteroid that produced it. Usually the crater is about 20 times bigger. Only in controlled nuclear explosions can an exact relationship between energy release and crater size be established. Sedan Crater in the Nevada Desert, USA (below), is bowl shaped and 368m (1,200ft) across. It was produced by a subsurface nuclear blast equivalent to 100 kilotons of TNT in July 1962. It is very similar to small lunar impact craters such as those within Korolev.

FAR SIDE *southern hemisphere*

## Mare Orientale

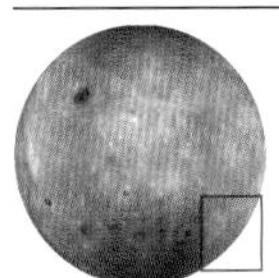

**TYPE** Multi-ring basin

**AGE** 3.8 billion years

**DIAMETER** 900km (560 miles)

This multi-ring basin is half the size of the near side's Mare Imbrium. It lies on the eastern limb of the far side, and from Earth the Montes Rook, the innermost eastern portion of the three distinct rings, can be clearly seen. This giant lunar bull's-eye was formed by a massive asteroid, and two theories have been proposed to explain the rings. The first has the impact excavating a deep transient crater. The cracked inner walls of this crater would have had been unable to support the weight of surrounding crust, so, the rock slumped into the hole, guided by a series of concentric fault systems that account for the rings that remain. Not only was most of the crater filled in, but the break-up of subsurface rock allowed lava from way below the lunar surface to seep up and fill in the central regions. However, the highland crust is about 60km (37 miles) thick, and rock from below that depth should have been excavated, but this deep rock has not been found. Alternatively, the seismic shocks generated by the massive impact could have briefly turned the surrounding rocks into a fluidized powder. Tsunami-type waves moved out through the pulverized rock but quickly became frozen, resulting in three clearly visible mountain rings.

**ROOK AND CORDILLERA MOUNTAINS**
Orientale is surrounded by two huge circular mountain ranges. The outer range is called Montes Cordillera (above right) and the inner one is called Montes Rook (lower left).

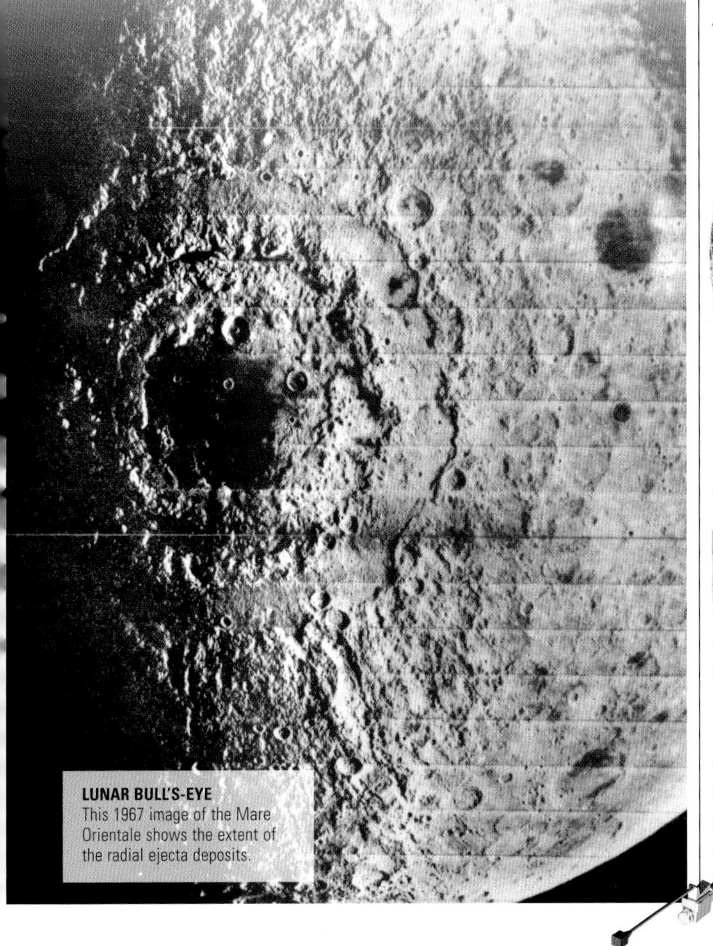

**LUNAR BULL'S-EYE**
This 1967 image of the Mare Orientale shows the extent of the radial ejecta deposits.

FAR SIDE *southern hemisphere*

## South Pole–Aitken Basin

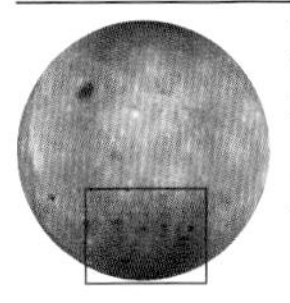

**TYPE** Impact crater

**AGE** 3.9 billion years

**DIAMETER** 2,500km (1,550 miles)

The South Pole–Aitken Basin is an immense impact crater, lying almost entirely on the far side of the Moon. It stretches from just above the South Pole to beyond the Aitken Crater, which is close to the centre of the far side. South Pole–Aitken is a staggering 2,500km (1,550 miles) in diameter, and is over 12km (7.4 miles) deep. It is one of the largest craters in the Solar System, and is comparable in size to the Chryse Basin on Mars. It is about 70 per cent of the diameter of Moon. The asteroid that produced it would have been over 100km (60 miles) across.

Even though the basin was first discovered in 1962, detailed investigation only started when the Galileo spacecraft imaged the Moon in 1992, while on its way to Jupiter. The South Pole–Aitken Basin looked darker than the rest of the far-side highland rocks, indicating that the lower-crustal rocks at the bottom of the deep crater were richer in iron than normal lunar surface material. Iron oxide and titanium oxide abound. Impact geophysicists are convinced that a normal impact could not have produced a crater this large without digging up large amounts of rock from the mantle that lies below the lunar crust. It may be that the crater was produced by a low-velocity collision, with the impactor coming into the surface at a low angle. Huge amounts of material would have been blasted from the lunar surface and would have moved off around the Moon's orbit. In the subsequent 10 million years, this debris would have collided with the Moon, producing many new craters.

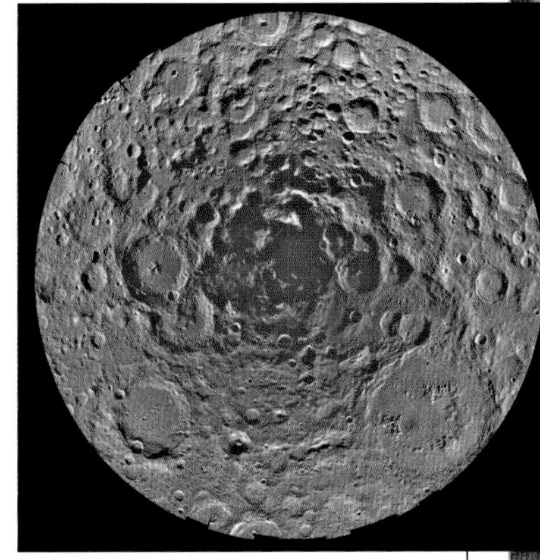

**SOUTH POLE**
The massively cratered, cold lunar South Pole can only be glimpsed tangentially from Earth. NASA's Clementine mission provided the first detailed map of the region in 1994.



**LARGEST KNOWN IMPACT CRATER**
The dashed circle outlines the vast South Pole–Aitken Basin. The laser ranger instrument on the Clementine spacecraft produced an accurate contour map of the region. The purple areas are the deepest, and the white areas the highest.

EXPLORING SPACE

### LOOKING FOR WATER

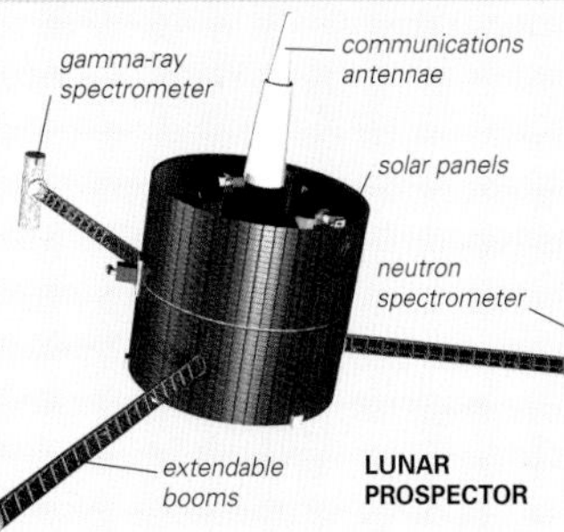


**LUNAR PROSPECTOR**

Because the South Pole–Aitken Basin is depressed well below the lunar surface, it contains many dark regions that are not heated by sunlight. Water seeping up from cracks in the mantle, or released by an impact, will not be able to escape from these "cold traps". In 1998, NASA's Lunar Prospector probe found hydrogen in these traps. One explanation is that it came from the break-up of water ice hidden in the polar shadows.

# MARS

37–37 Gravity, motion, and orbits
60–61 Celestial cycles
64–65 Planetary motion
108–11 Beyond Earth
118–19 The family of the Sun

MARS IS THE OUTERMOST of the four rocky planets. Also known as the red planet because of its rust-red colour, it is named after the Roman god of war. Its varied surface features include deep canyons and the highest volcanoes in the Solar System. Although Mars is now a dry planet, a large body of evidence indicates that liquid water once flowed across its surface.

## ORBIT

Mars has an elliptical orbit, so at its closest approach to the Sun (the perihelion) it receives 45 per cent more solar radiation than at the farthest point (the aphelion). This means that the surface temperature can vary from -125°C (-195°F), at the winter pole, to 25°C (77°F) during the summer. At 25.2°, the current axial tilt of Mars is similar to that of Earth and, like Earth, Mars experiences changes in seasons as the North Pole, and then the South Pole, points towards the Sun during the course of its orbit. Throughout its history, Mars's axial tilt has fluctuated greatly due to various factors, including Jupiter's gravitational pull. These fluctuations have caused significant changes in climate. When Mars is heavily tilted, the poles are more exposed to the Sun, causing water ice to vaporize and build up around the colder lower latitudes. At a lesser tilt, water ice becomes concentrated at the colder poles.

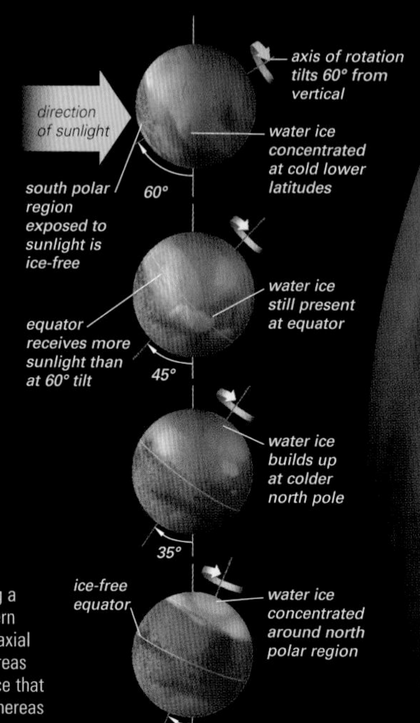


**CHANGES IN AXIAL TILT**
Water-ice distribution during a Martian winter in the northern hemisphere varies with the axial tilt. The translucent white areas shown here represent thin ice that melts during the summer, whereas the thick white ice remains.

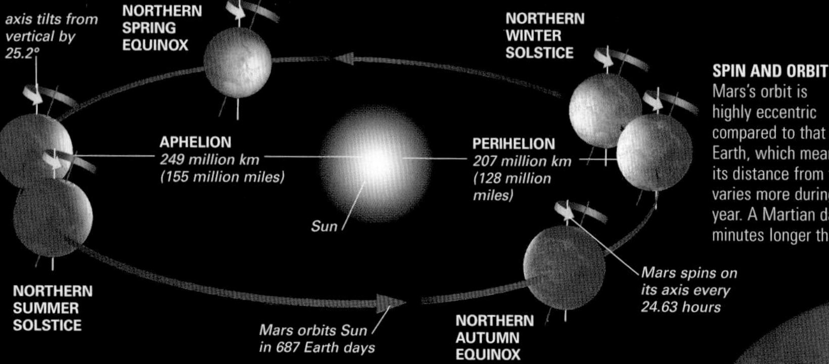


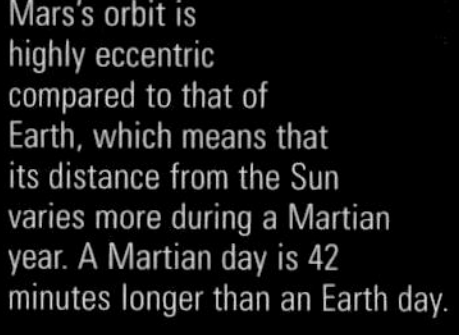


**SPIN AND ORBIT**
Mars's orbit is highly eccentric compared to that of Earth, which means that its distance from the Sun varies more during a Martian year. A Martian day is 42 minutes longer than an Earth day.

## STRUCTURE

Mars is a small planet, about half the size of Earth, and farther away from the Sun. Its size and distance mean that it has cooled more rapidly than Earth, and its once-molten iron core is probably now solid. Its relatively low density compared to the other terrestrial planets indicates that the core may also contain a lighter element, such as sulphur, in the form of iron sulphide. The small core is surrounded by a thick mantle, composed of solid silicate rock. The mantle was a source of volcanic activity in the past, but it is now inert. Data gathered by the Mars Global Surveyor spacecraft has revealed that the rocky crust is about 80km (50 miles) thick in the southern hemisphere, whereas it is only about 35km (22 miles) thick in the northern hemisphere. Mars has the same total land area as Earth, as it has no liquid water on its surface.

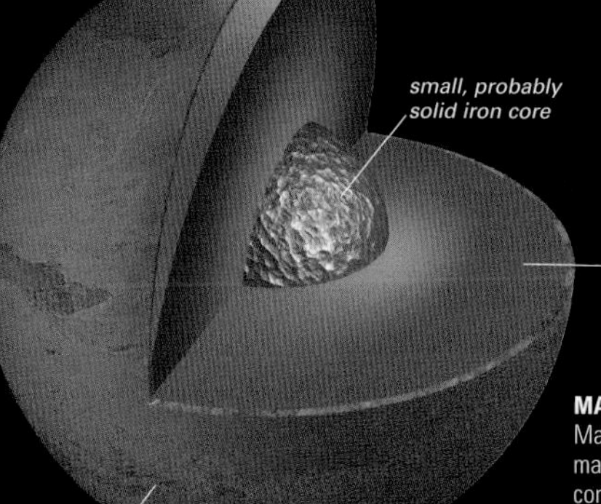


**MARS INTERIOR**
Mars has a distinct crust, mantle, and core. The core is much smaller in proportion to Earth, and has probably solidified.

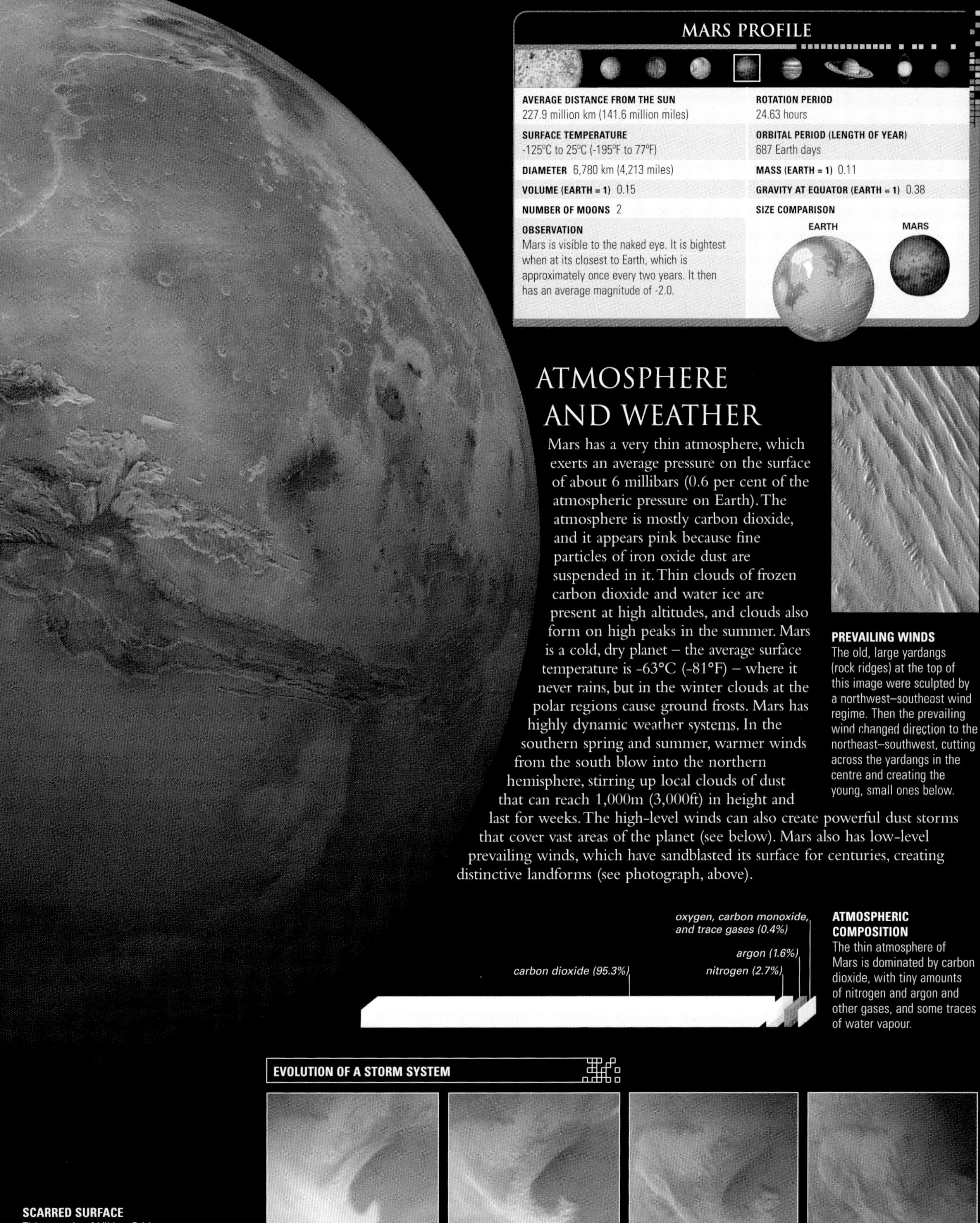

## MARS PROFILE

| AVERAGE DISTANCE FROM THE SUN 227.9 million km (141.6 million miles) | ROTATION PERIOD 24.63 hours |
|---|---|
| SURFACE TEMPERATURE -125°C to 25°C (-195°F to 77°F) | ORBITAL PERIOD (LENGTH OF YEAR) 687 Earth days |
| DIAMETER 6,780 km (4,213 miles) | MASS (EARTH = 1) 0.11 |
| VOLUME (EARTH = 1) 0.15 | GRAVITY AT EQUATOR (EARTH = 1) 0.38 |
| NUMBER OF MOONS 2 | SIZE COMPARISON EARTH MARS |
| OBSERVATION Mars is visible to the naked eye. It is bightest when at its closest to Earth, which is approximately once every two years. It then has an average magnitude of -2.0. | |

# ATMOSPHERE AND WEATHER

Mars has a very thin atmosphere, which exerts an average pressure on the surface of about 6 millibars (0.6 per cent of the atmospheric pressure on Earth). The atmosphere is mostly carbon dioxide, and it appears pink because fine particles of iron oxide dust are suspended in it. Thin clouds of frozen carbon dioxide and water ice are present at high altitudes, and clouds also form on high peaks in the summer. Mars is a cold, dry planet – the average surface temperature is -63°C (-81°F) – where it never rains, but in the winter clouds at the polar regions cause ground frosts. Mars has highly dynamic weather systems. In the southern spring and summer, warmer winds from the south blow into the northern hemisphere, stirring up local clouds of dust that can reach 1,000m (3,000ft) in height and last for weeks. The high-level winds can also create powerful dust storms that cover vast areas of the planet (see below). Mars also has low-level prevailing winds, which have sandblasted its surface for centuries, creating distinctive landforms (see photograph, above).

**PREVAILING WINDS**
The old, large yardangs (rock ridges) at the top of this image were sculpted by a northwest–southeast wind regime. Then the prevailing wind changed direction to the northeast–southwest, cutting across the yardangs in the centre and creating the young, small ones below.

*carbon dioxide (95.3%)*
*nitrogen (2.7%)*
*argon (1.6%)*
*oxygen, carbon monoxide, and trace gases (0.4%)*

**ATMOSPHERIC COMPOSITION**
The thin atmosphere of Mars is dominated by carbon dioxide, with tiny amounts of nitrogen and argon and other gases, and some traces of water vapour.

**EVOLUTION OF A STORM SYSTEM**

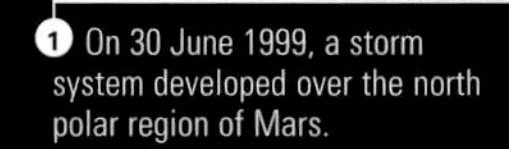

1 On 30 June 1999, a storm system developed over the north polar region of Mars.

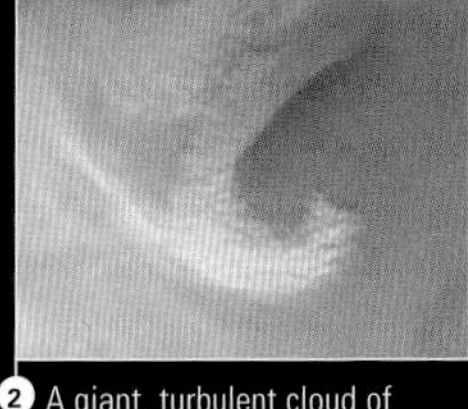

2 A giant, turbulent cloud of orange-brown dust was raised by high surface winds.

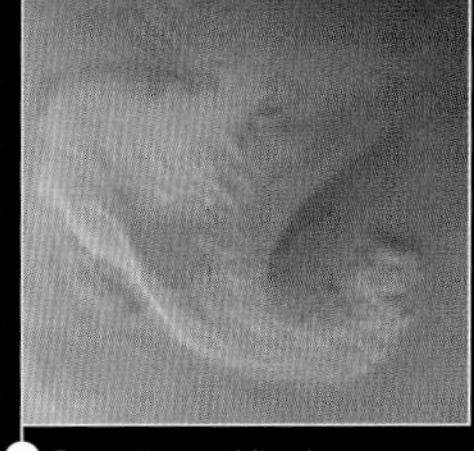

3 Expanding rapidly, the storm swirled over the white ice cap (centre, top).

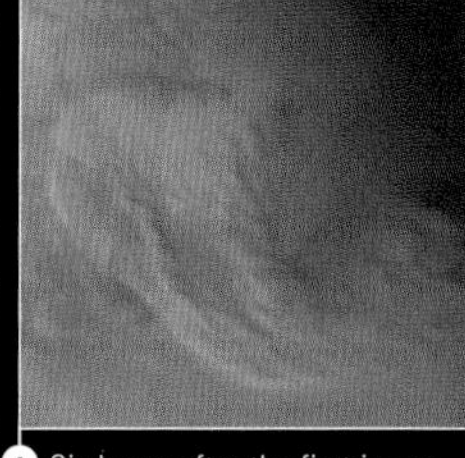

4 Six hours after the first image was taken, the storm was still gathering strength.

**SCARRED SURFACE**
This mosaic of Viking Orbiter images shows Mars's distinct red coloration and reveals the vast extent of the Valles Marineris, a system of valleys more than 4,000km (2,500 miles) long.

## MISSIONS TO MARS

**Numerous spacecraft have been sent to Mars since the first missions were undertaken by the USA and the Soviet Union in the 1960s, with varying success due to technical difficulties. A selection of successful missions is described below.**

**1976 VIKING 1 AND 2 (USA)** These two craft each consisted of an orbiter and a lander. The orbiters sent back images, while the landers descended to two different sites and sent back analyses of the soil and atmosphere, as well as images.

**VIKING ORBITER**

**1997 MARS PATHFINDER (USA)** This mission sent a stationary lander and a free-ranging robot called Sojourner to the surface of Mars. They landed in an ancient floodplain and sent back pictures and analyses of soil samples.

**SOJOURNER**

**1997 MARS GLOBAL SURVEYOR (USA)** Orbiting at an average altitude of 380km (235 miles), the Global Surveyor mapped the entire planet at high resolution. It provided further evidence that water has flowed on Mars in the past.

**GLOBAL SURVEYOR**

**2003 MARS EXPRESS (EUROPE)** This orbiting spacecraft is imaging the entire surface of Mars as well as mapping its mineral composition and studying the Martian atmosphere.

**MARS EXPRESS**

**2004 MARS EXPLORATION ROVERS (USA)** Twin rovers Spirit and Opportunity landed on opposite sides of the planet and studied rocks and soil, looking for evidence of how liquid water affected Mars in the past.

**MARS ROVER**

# SURFACE FEATURES

Mars's surface features have been formed and shaped by meteorite impacts, by the wind (see p.161), and by volcanism and faulting (see Tectonic Features, below). Scientists also believe that water once flowed on and below the surface of Mars (see opposite), carving out features such as valleys and outflow channels. The craters formed during a period of intense meteorite bombardment about 3.9 billion years ago. They are found mainly in the southern hemisphere, which is geologically older than the northern hemisphere, and include the vast Hellas Basin (see p.173), but small craters are found all over Mars. Martian craters are flatter than those on the Moon and show signs of erosion by wind and water; indeed, some have almost been obliterated.

**ROCK GARDEN**
Landers have revealed Mars to be a rocky as well as a sandy, dusty place. Sojourner took these images of part of an area nicknamed the Rock Garden in 1997.

**IMPACT CRATER**
The Herschel impact crater, located in the southern highlands, is about 300km (185 miles) across. This image has been false coloured to show altitude. The lowest areas are the dark blue floors of smaller craters. The Herschel Crater floor is mostly at 1,000m (3,240ft) and the highest parts of the rim (pale pink) are at about 3,000m (9,720ft).

# TECTONIC FEATURES

Billions of years ago, when Mars was a young planet, internal adjustments created the large-scale features seen on its surface today. Internal forces created raised areas on the surface, such as the Tharsis Bulge, and stretched and split the surface to create rift valleys, such as the vast Valles Marineris (see pp.166–67). Landslides, wind, and water have since modified the rift valleys. Volcanic activity dates back billions of years and persisted for much of Mars's history. The planet may still be volcanically active today, although no such activity is expected. Lava eruptions of the past formed today's giant volcanoes, including Olympus Mons (see p.165).

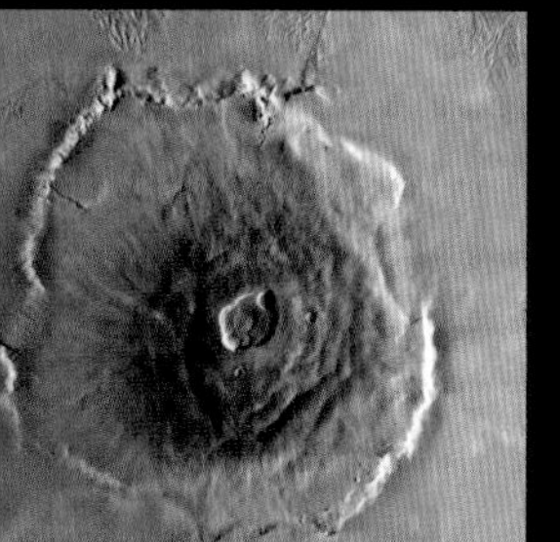

**OLYMPUS MONS**
This mosaic of images of Olympus Mons taken by Viking 1 in 1978 looks deceptively flat – the volcano stands 24km (15 miles) above the surrounding plain.

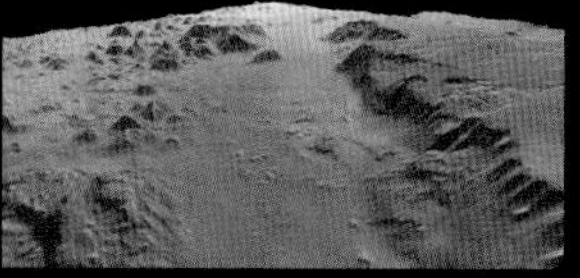

**VALLES MARINERIS**
The Valles Marineris is a complex system of canyons that cuts across Mars at an average depth of 8km (5 miles). If it was on Earth, it would stretch across North America.

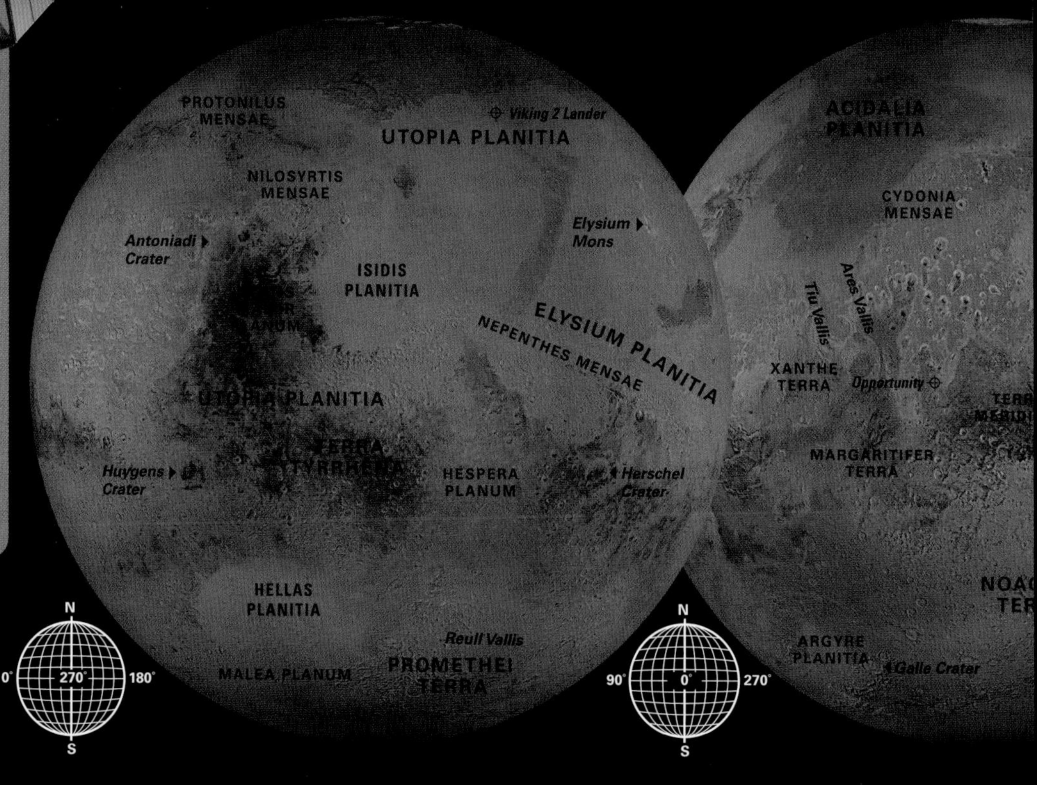


**MARS MAPS**
These four views combine to show the complete surface of Mars. They have been labelled to show large-scale features, as well as the landing sites of some of the spacecraft sent to explore its surface.

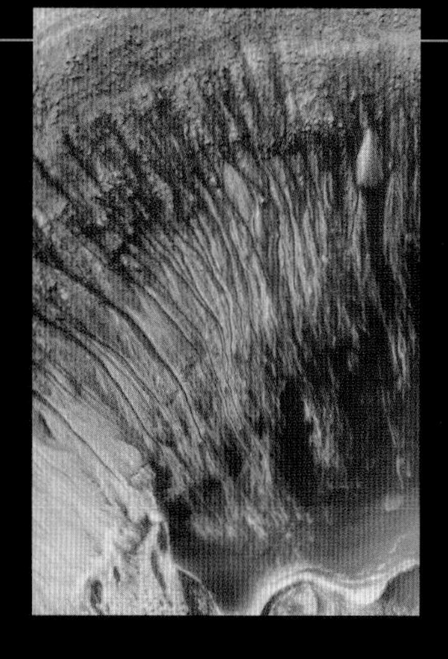

# WATER ON MARS

Scientists have been hoping to establish whether water is present on Mars, as this is essential for the development of life. Liquid water is not present, for this is a cold planet, where water can exist only as ice or vapour. The latter can form low-lying mists and fogs and freezes into a thin layer of white water ice on the rocks and soil when the temperature falls. However, dry river-beds, valleys, and ancient floodplains bear witness to the presence of large amounts of fast-flowing water on the surface 3–4 billion years ago, when Mars was a warmer, wetter world with a thicker atmosphere. Some of that water remains today in the form of ice, which is present both underground and in the polar ice caps. The ice caps wax and wane with the Martian seasons, and are composed of varying amounts of water ice and frozen carbon dioxide.

**THE CASE FOR FLOWING WATER**
High-resolution images from Mars Global Surveyor, such as this one of gullies and finger-like deposits of debris in a crater wall, suggest that liquid water has flowed on Mars in recent geological times.

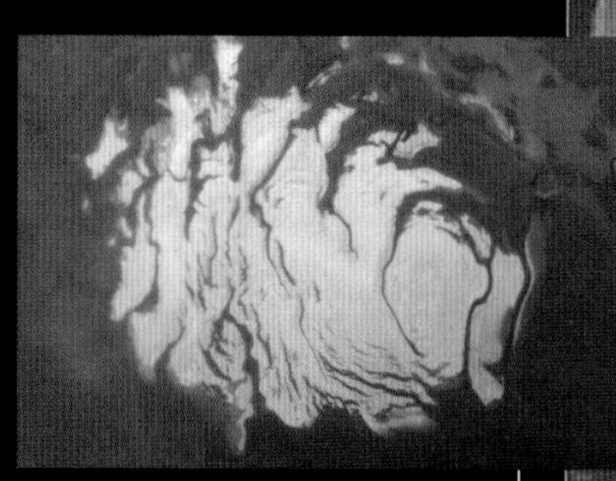

**SOUTH POLAR ICE CAP**
During the summer, the south polar ice cap shrinks, and when photographed in April 2000 it measured just 420km (260 miles) across. The cap is 85 per cent frozen carbon dioxide and 15 per cent water ice in late summer.

# MOONS

Mars has two small, dark moons called Phobos and Deimos, which were discovered by the American astronomer Asaph Hall in August 1877. Deimos, the smaller of the two, is 15km (9.3 miles) long and Phobos is 26.8km (16.6 miles) long. Both are irregular "potato-shaped" rocky bodies and are probably asteroids that were captured in Mars's early history. They both bear the scars of meteorite battering. Deimos orbits Mars at a distance of 23,500km (14,580 miles). Phobos is only 9,380km (5,830 miles) from Mars and getting closer; eventually it will be so close that it will either be torn apart by Mars's gravity field or will collide with the planet.

**MOONS' ORBIT**
Phobos and Deimos both follow near-circular orbits around Mars, and both exhibit synchronous rotation. From Mars, Phobos rises and sets three times every Martian day.

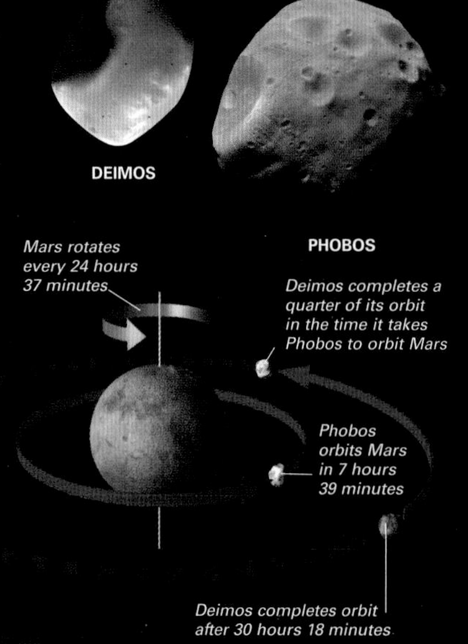


# GEOGRAPHY

The first reliable maps of Mars were made in the late 19th century when astronomers drew what they observed through their telescopes. Today's maps are based on data collected by space probes such as Mars Global Surveyor, which obtained 100,000 photos of Mars and completed a survey of the planet, and Mars Express, which is imaging the entire surface. The following terminology is used for the surface features: lowland plains are termed planitia; high plains, planum; extensive landmasses, terra; and mountains or volcanoes, mons. A chasma is a deep, elongated, steep-sided depression, and a labyrinthus is a system of intersecting valleys or canyons. Individual names are allocated depending on the type of feature. Large valleys (vallis) are named after Mars in various languages and small ones after rivers. Large craters are named after past scientists, writers, and others who have studied Mars; smaller craters are named after villages. Other features are named after the nearest albedo feature on the early maps.

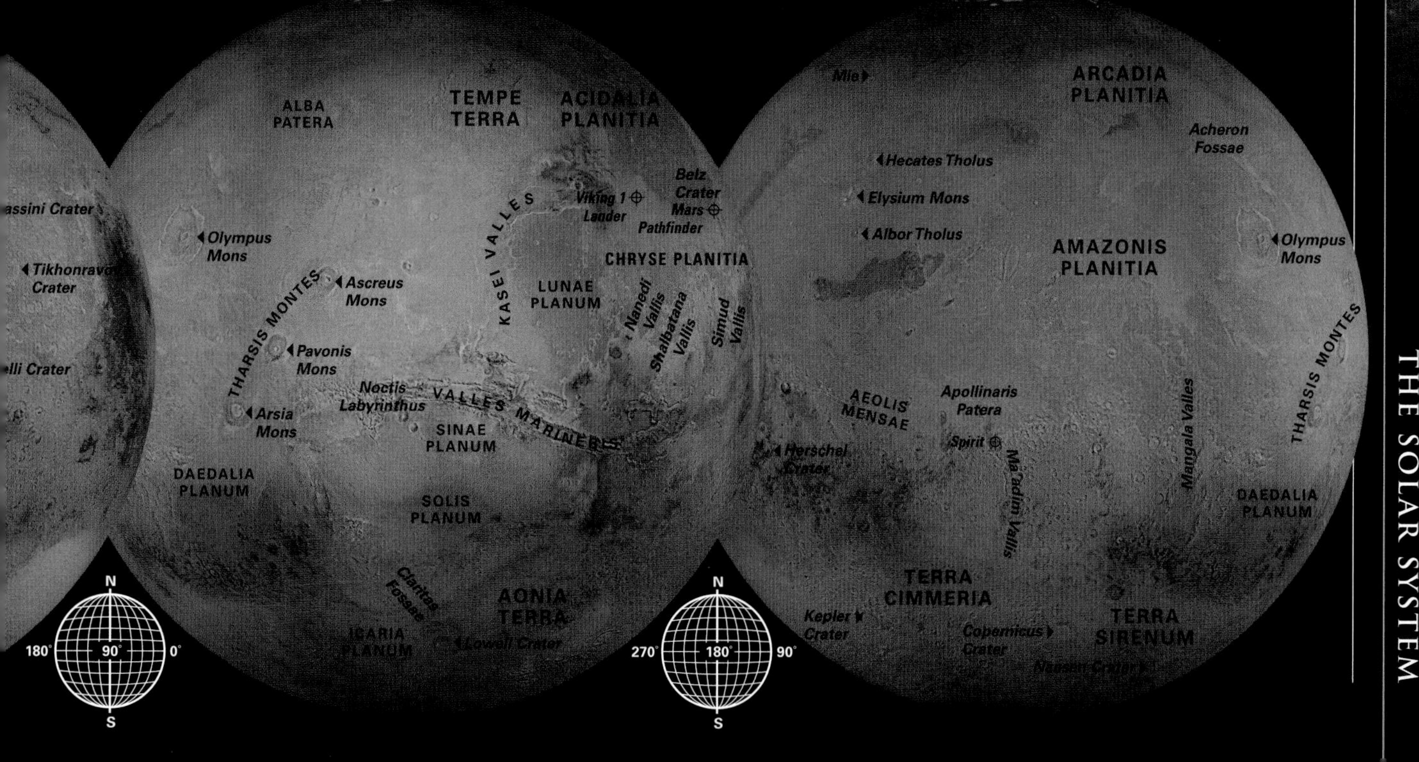

# TECTONIC FEATURES

Mars has two areas of markedly different terrain. Much of the northern hemisphere is characterized by relatively smooth and low-lying volcanic plains. The older southern landscape is typically cratered highland. The boundary between the two is an imaginary circle tilted by about 30° to the equator. The planet's major tectonic features are found within a region that extends roughly 30° each side of the equator. It contains Mars's main volcanic centre, the Tharsis region, and the Valles Marineris, the vast canyon system that slices across the centre of the planet.

**WESTERN FLANK OF OLYMPUS MONS**
Tectonic features on Mars take on familiar forms but are on a much grander scale than those on Earth. This escarpment on the side of Olympus Mons is 7km (4.3 miles) high.

THARSIS MONTES

## Pavonis Mons

**TYPE** Shield volcano
**AGE** 300 million years
**DIAMETER** 375km (235 miles)

A huge bulge in the western hemisphere, commonly known as the Tharsis Bulge, contains volcanoes of various sizes and types, from large shields to smaller domes. Olympus Mons dominates the region. But three other volcanoes, which anywhere else would be considered enormous, are also found here. The three form a line and together make the Tharsis Montes mountain range. Pavonis Mons, situated on the equator, is the middle of the three. It is a shield volcano with a broad base and sloping sides and is similar to those found in Hawaii on Earth. The volcano's summit stands 7km (4.3 miles) above the surrounding plain and has a single caldera within a larger, shallow depression. Hundreds of narrow lava flows are seen to emanate from the rim of the caldera, and others can be traced back to pits situated close by.

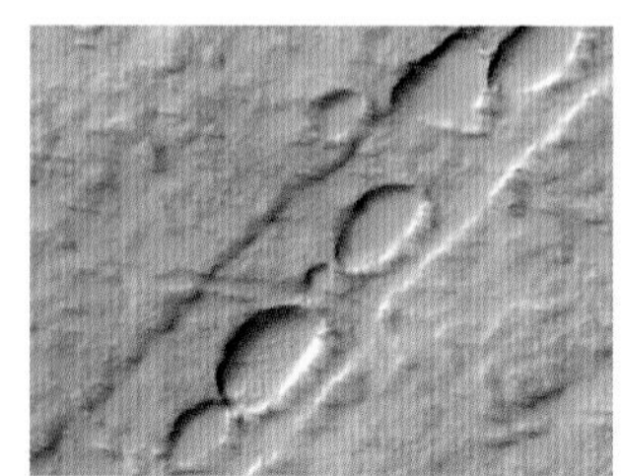

**PIT CHAIN**
A chain of pits lies in a shallow trough on the lower east flank. The pits and trough formed either because the ground was moved apart by tectonic forces or was uplifted by molten rock.

**CHANNELS**
These deep channels on the volcano's southern flank may have started out as subsurface lava tubes whose roofs collapsed as pits developed over them.

**SUMMIT DEPRESSION**
The summit caldera lies within a shallow depression that is almost twice the caldera's size and has faulted sides.

THARSIS MONTES

## Ascraeus Mons

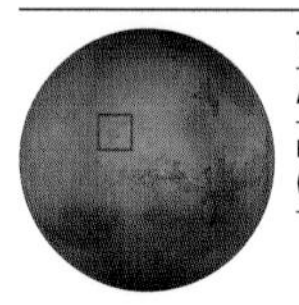

**TYPE** Shield volcano
**AGE** 100 million years
**DIAMETER** 460km (285 miles)

Ascraeus Mons is the northernmost of the three Tharsis Montes volcanoes. The three lie on the crest of the Tharsis Bulge and form a line in a southwest–northeast direction. The line marks the position of a major rift zone, long since buried under lava. The three volcanoes grew by the gradual build-up of thousands of individual and successive lava flows that came to the surface through the rift zone. Ascraeus is the tallest of the three, rising about 18km (11 miles) above the surrounding plain. It has a large number of lines and channels all round the rim of the caldera, showing the paths taken by flowing lava.

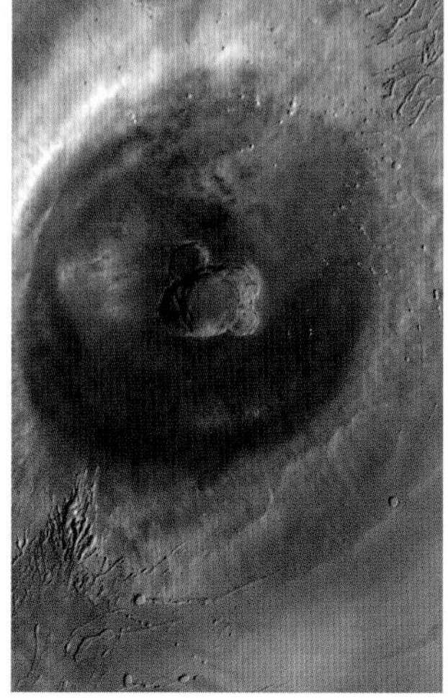

**CALDERA**
The caldera on the summit is made up of eight major depressions and has a nested appearance (centre). Its deepest point is over 3km (1.9 miles) below the rim.

THARSIS MONTES

## Arsia Mons

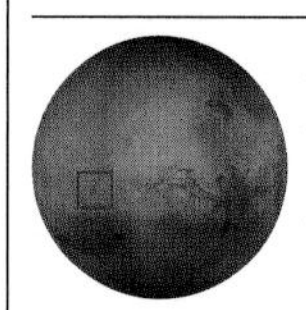

**TYPE** Shield volcano
**AGE** 700 million years
**DIAMETER** 475km (295 miles)

Arsia Mons is second only to the mighty Olympus Mons in terms of volume. It is the southernmost of the three Tharsis Montes volcanoes, and its summit rises more than 9km (5.6 miles) above the surrounding plain. Like the other two, it has a summit caldera bigger than any known on Earth. Arsia Mons measures 120km (75 miles) across and is surrounded by arc-shaped faults. Lava flows fan out down the volcano's shallow slopes. The lava is of basalt-like composition and of low viscosity, and the flows are shorter nearer the summit than on the lower flanks.

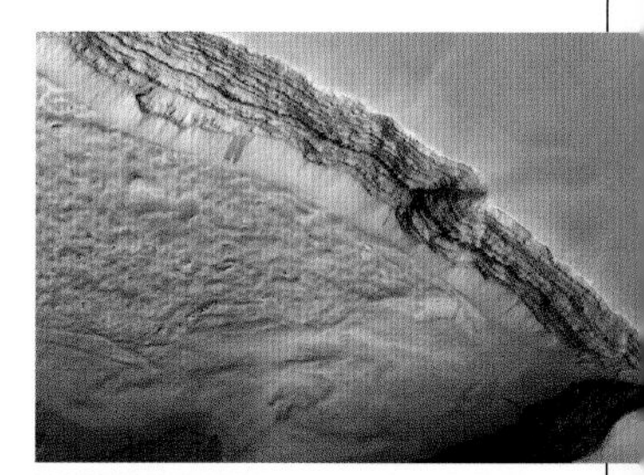

**LAYERED OUTCROP**
This outcrop of layered rock lies in a pit on the volcano's lower west flank. The layers are thought to consist mostly of volcanic rock formed by successive lava flows.

**CLOUDY SUMMIT**
Water-ice clouds hang over the volcano's summit – a common sight every Martian afternoon in the Tharsis region.

THARSIS REGION

## Olympus Mons

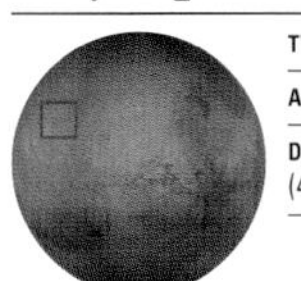

TYPE Shield volcano

AGE 30 million years

DIAMETER 648km (403 miles)

Olympus Mons is unquestionably the largest volcano in the Solar System. Its height, of about 24km (15 miles), makes it the tallest, and its volume is over 50 times that of any shield volcano on Earth. Olympus is one of the giant shield volcanoes of the Tharsis region, which is home to the greatest number of volcanoes on Mars, including the planet's youngest. Volcanoes evolve over long periods of time and can be inactive for hundreds of millions of years. Olympus Mons is considered the youngest of the shield volcanoes.

The summit's complex caldera is surrounded by a surface of wide terraces formed by lava flows, crossed by thinner flows. These are encircled by a huge scarp, up to 6km (3.7 miles) high. Vast plains, termed aureoles, extend from the north and west of the summit, like petals from a flower. These regions of gigantic ridges and blocks, whose origins remain unexplained, extend outwards for up to 1,000km (620 miles).

**LANDSLIDE** The scarp surrounding the summit has experienced several major landslides.

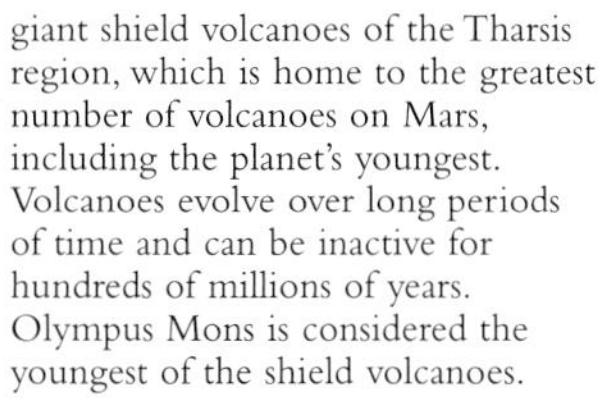

**LAVA FLOWS** These lava flows and a collapsed lava tube (top left) on the southwest flank have been peppered by tiny impact craters.

EXPLORING SPACE

### MARTIAN METEORITES

Solidified basaltic lava covers the Tharsis region. Pieces of lava that had flowed on the Martian surface as recently as 180 million years ago are now on Earth. Impactors hit Mars and ejected them, and after journeys lasting millions of years they fell to Earth as meteorites. They include the Shergotty meteorite (right), which landed in Shergahti, India, on 25 August 1865.

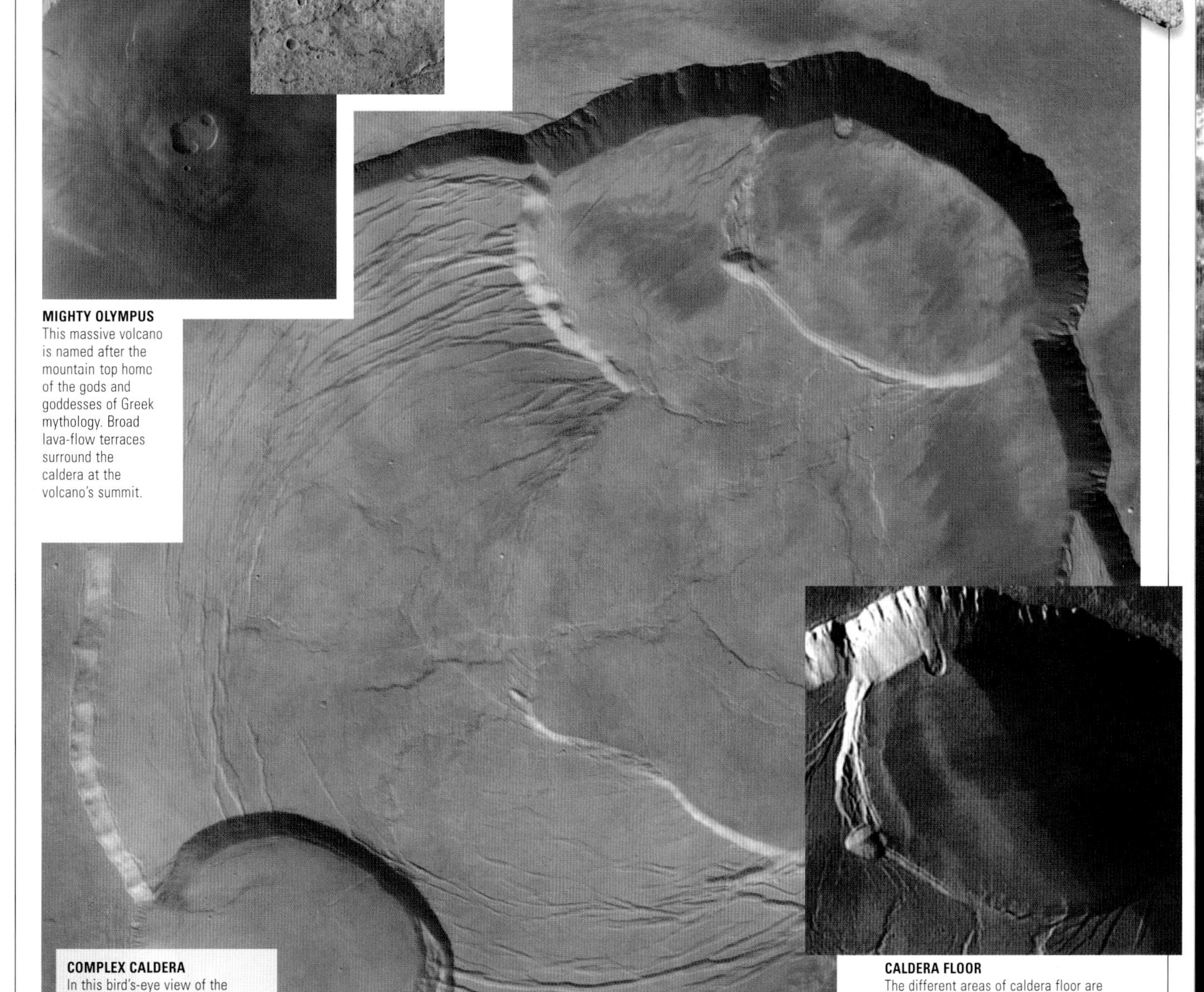

**MIGHTY OLYMPUS** This massive volcano is named after the mountain top home of the gods and goddesses of Greek mythology. Broad lava-flow terraces surround the caldera at the volcano's summit.

**COMPLEX CALDERA** In this bird's-eye view of the 52km- (32-mile-) wide nested caldera on the summit of Olympus Mons, five roughly circular areas of caldera floor can be seen.

**CALDERA FLOOR** The different areas of caldera floor are associated with different periods of volcanic activity. The area above is about 200 million years old. The largest central area, which is marked by ring-shaped faults (see image, left) is more recent, at 140 million years old.

VALLES MARINERIS

## Valles Marineris

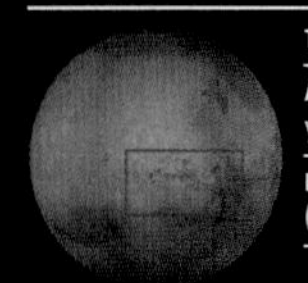

**TYPE** Canyon system

**AGE** About 3.5 billion years

**LENGTH** Over 4,000km (2,500 miles)

Valles Marineris is the largest feature formed by tectonic activity on Mars. It consists of a system of canyons that stretches for over 4,000km (2,500 miles), is up to 700km (430 miles) wide, and averages 8km (5 miles) in depth. The Grand Canyon in Arizona, USA, is dwarfed by comparison; it is only about one-tenth as long and one-fifth as deep. Valles Marineris lies just south of the Martian equator, and the system trends, very roughly, west to east. The trend follows a set of fractures that radiates from the Tharsis Bulge at Marineris's western end.

The origins of the system date back a few billion years to when the canyons were formed by faulting. This contrasts with the Grand Canyon, which is a primarily water-eroded canyon. But water, as well as wind, has played its part in the development of the Marineris system. Buffeting winds, flowing water, and the collapse of unstable walls have all widened and deepened the canyons.

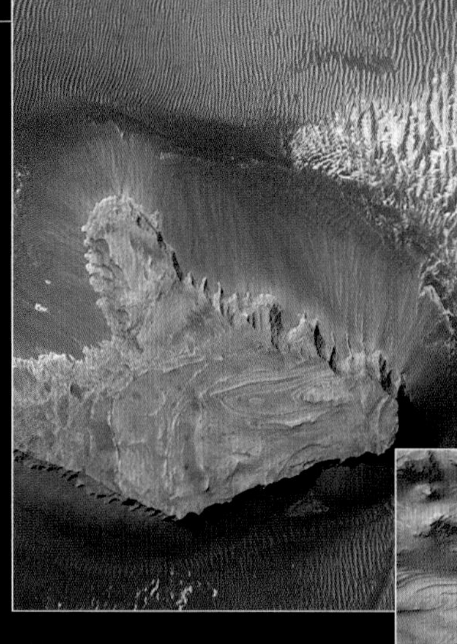

**MESA**
This small mesa (a flat-topped hill) lies in northwestern Candor Chasma in the central Valles Marineris. Light-toned outcrops of layered sedimentary rock are exposed on the top. These may have formed from material deposited in a lake in the chasma. Darker windblown ripples cover the surrounding plains.

The Noctis Labyrinthus region marks the western end of the system. This is a roughly triangular area of intersecting rift valleys that form a maze-like arrangement. The eastern end of Valles Marineris is bounded by chaotic terrain of irregular appearance. Here, smaller canyons and depressions give way to outflow canyons. These carried ancient rivers of water out of Marineris towards the lowland region, Chryse Planitia, to the north. This whole area has seen extensive water erosion; millions of cubic kilometres of material have been removed by water action.

The system's canyons are described as chasma (plural, chasmata) and are given identifying names. The main chasma in the western part of the system is Ius. The central complex is made up of three parallel canyons,

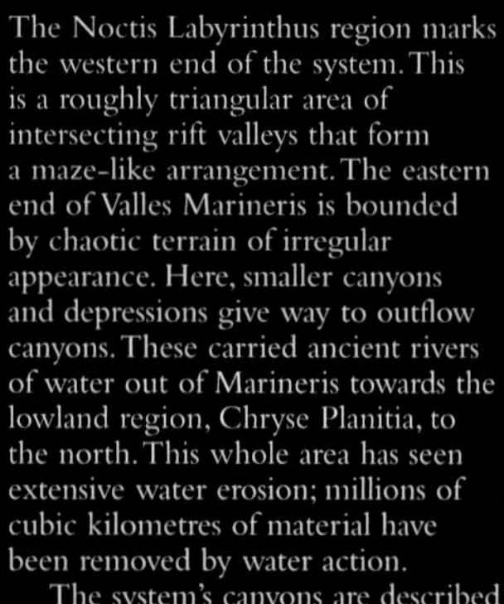

**LAYERED DEPOSITS**
A detail of the floor of western Candor Chasma shows layered sedimentary rock. Up to 100 layers have been counted, each about 10m (33ft) thick. The layers may be made from material deposited in an impact crater before the chasma formed.

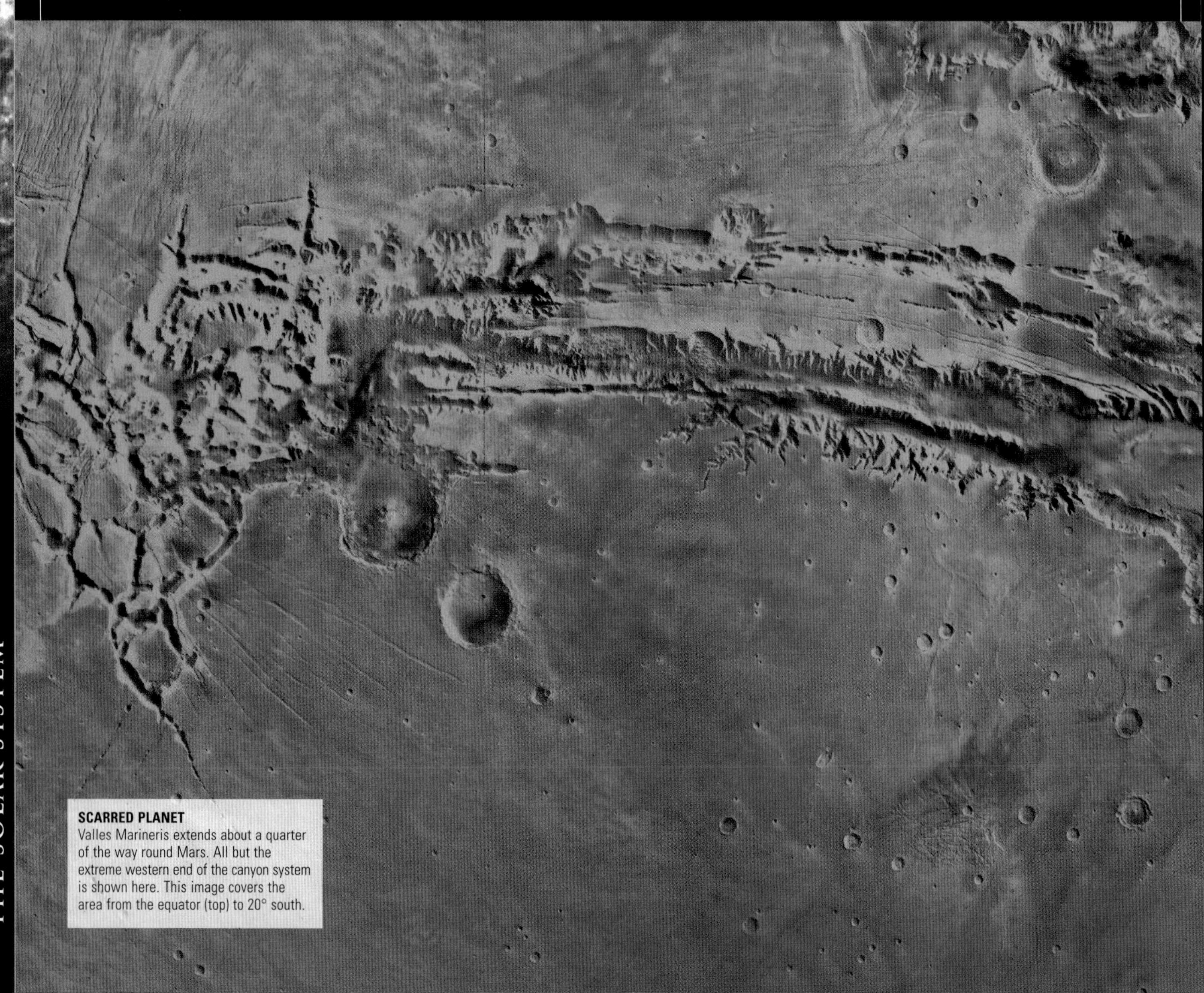

**SCARRED PLANET**
Valles Marineris extends about a quarter of the way round Mars. All but the extreme western end of the canyon system is shown here. This image covers the area from the equator (top) to 20° south.

Ophir and Candor and, to their south, Melas Chasma. The long Coprates Chasma stretches out to the east, where it meets the broader Eos Chasma. The name of the whole system, Valles Marineris, means "Valleys of the Mariner". The Mariner in this case is the Mariner 9 mission that mapped the entire surface of Mars and took the first close-up images of this area. More recent craft, such as Mars Global Surveyor and Mars Express, have provided more detailed coverage. For example, their surveys have revealed layered rock in the canyon walls that could be a profile of the different lava flows that built the plains that the canyons cut through. Rocks on the floor may have formed from windborne dust layers or by deposits in ancient lakes that once filled the canyons.

**LOOKING WEST THROUGH OPHIR CHASMA**
Over billions of years, Ophir Chasma has widened as its walls have collapsed and slumped downwards, covering the floor with debris.

**EASTERN EOS CHASMA**
Water flowed through this broad chasma, out of the Valles Marineris and into a series of valleys and channels.

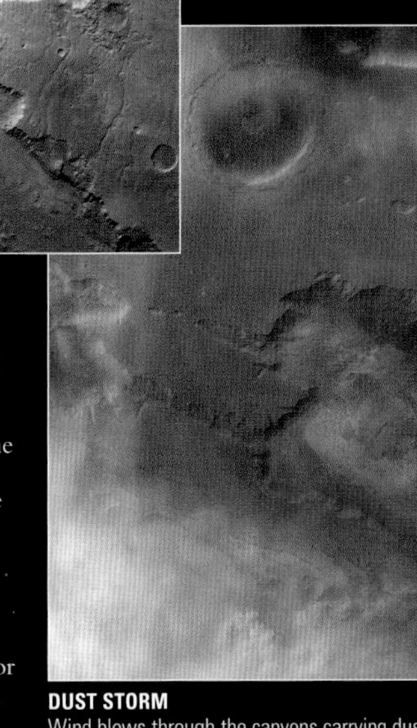

**DUST STORM**
Wind blows through the canyons carrying dust. The pinkish dust cloud at the bottom of this image is moving north across the junction of Ius Chasma and Melas Chasma. The higher, bluish-white clouds are water ice.

EXPLORING SPACE

## MARS EXPRESS'S STEREO CAMERA

The High Resolution Stereo Camera onboard Mars Express began its two-year programme to map the entire Martian surface in January 2004. Its nine charge-coupled device sensors record data one line at a time. Downward, backward, and forward views are used to build up 3-D images. The Super Resolution Channel provides more detailed information.

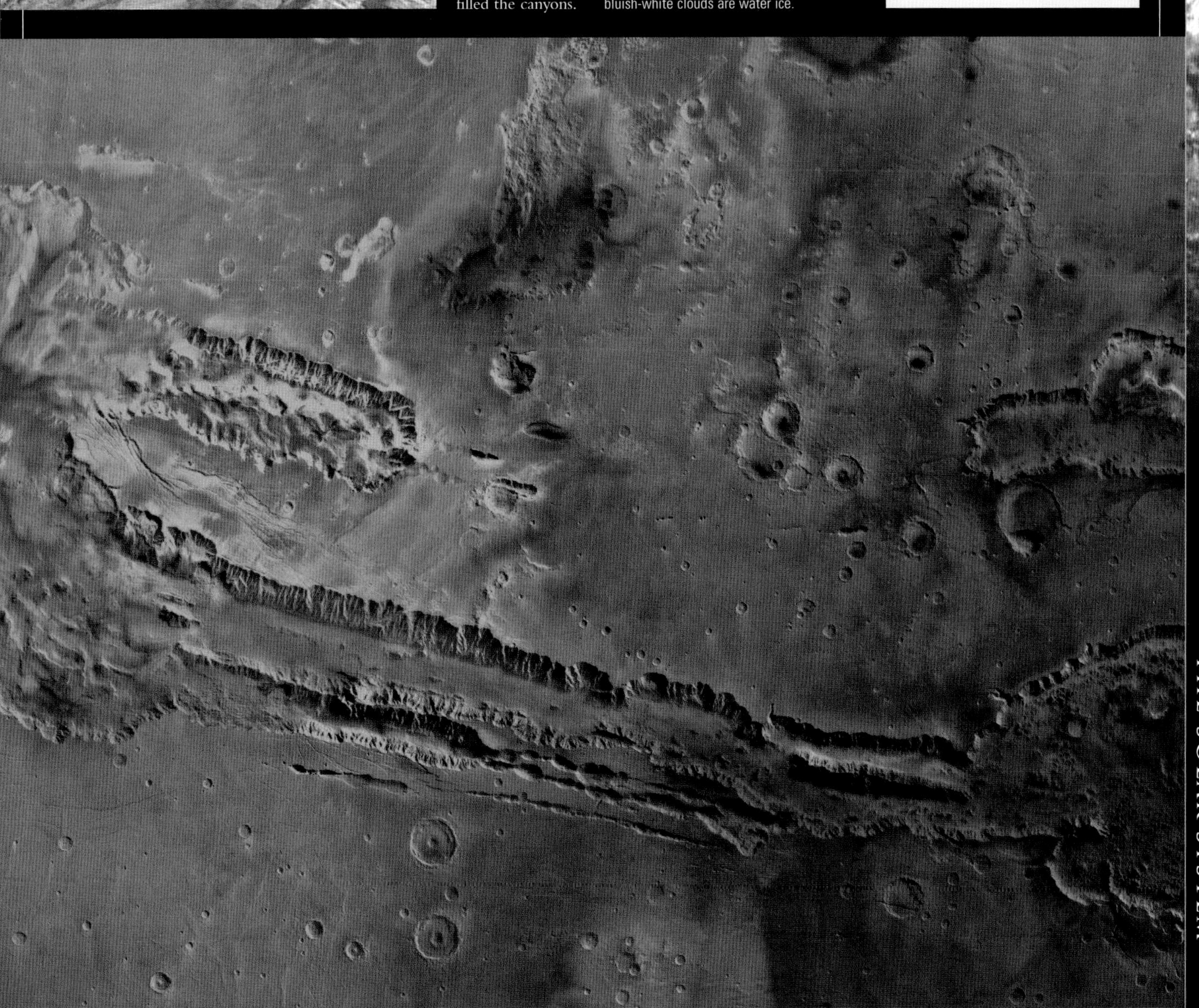

THARSIS REGION

## Acheron Fossae

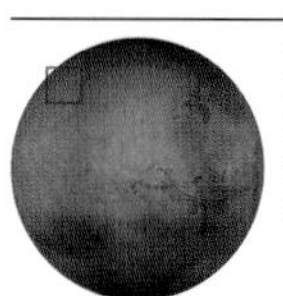

**TYPE** Fault system

**AGE** Over 3.5 billion years

**LENGTH** 1,120km (695 miles)

Acheron Fossae is a relatively high area that has seen intense tectonic activity in the past. It marks the northern edge of the Tharsis Bulge and is located about 1,000km (620 miles) north of Olympus Mons. Acheron is part of a network of fractures that radiates out from the Tharsis Bulge – a huge region of uplift and volcanic activity. It can be compared to the Great Rift Valley on Earth (see p.142), where continental plates have spread apart. The huge curved faults in the Tharsis Bulge were produced in the process of the Tharsis uplift, and crustal cracks formed as hot mantle material pushed upwards onto the surface rock layer. The brittle Martian crust broke along zones of weakness when the tension of the uplift became too strong. The eroded walls, the subdued nature of the high-standing hills, and the wind etching on the flat surfaces all confirm this is ancient terrain. This type of feature gets its name from the Latin *fossa*, meaning "trough". In Greek mythology, Acheron is the river that flows into Hades, the Underworld.

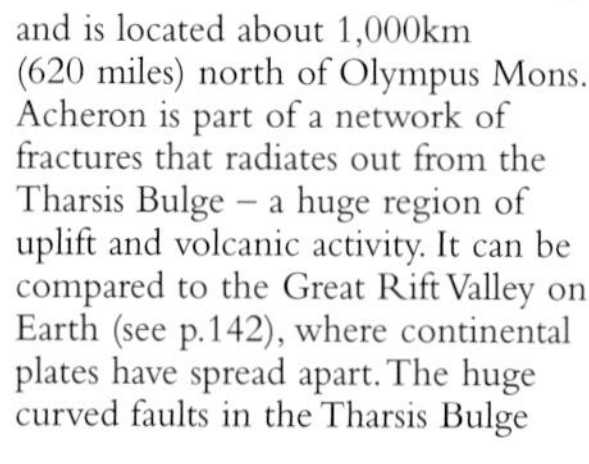

**BROKEN CRUST**
In this perspective view across the highly deformed area of Acheron Fossae, curved faults trending to the northwest dominate the scene.

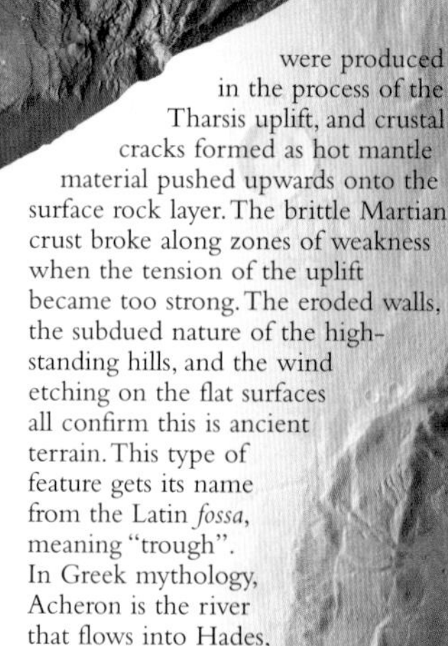

**STRESSED LANDSCAPE**
A fault system cutting across an ancient impact crater is evidence of the stress felt by the Martian crust. The crater floor has since been resurfaced by material from outside the area.

**GRABENS AND HORSTS**
The planetary crust has fallen between parallel faults to form grabens up to 1.7km (1 mile) deep; remnants of the pre-existing heights are termed horsts.

**CLIFF FACE**
This is a close-up of the bright, steep slopes of a scarp or cliff. Dark streaks on the cliff face may be formed as the dust mantle that covers the region gives way and produces a dust avalanche.

TERRA CIMMERIA

## Apollinaris Patera

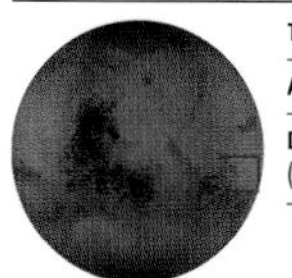

**TYPE** Patera volcano

**AGE** 900 million years

**DIAMETER** 296km (184 miles)

This is an example of a type of volcano that was first identified on Mars. Known as patera volcanoes, they have very gentle slopes (with angles as low as 0.25°). Apollinaris Patera is one of the largest on the planet, situated on the northern edge of Cimmeria Terra, a few degrees south of the equator. It is the only major volcano that is isolated from the two major volcanic regions of Tharsis, to the northeast, and Elysium, to the northwest. Apollinaris is a broad, roughly shield-shaped volcano, reminiscent of an upturned saucer. It is only about 5km (3 miles) high and has a caldera about 80km (50 miles) across. It appears to have been formed by both effusive and explosive activity. Lava flows are clearly visible beyond the summit. A cliff surrounding the caldera area is visible on the northern side, but has disappeared on the opposite side. It is buried under a fan of material whose surface is marked by broad channels. The fan material could have formed from flowing lava or volcanic rock fragments.

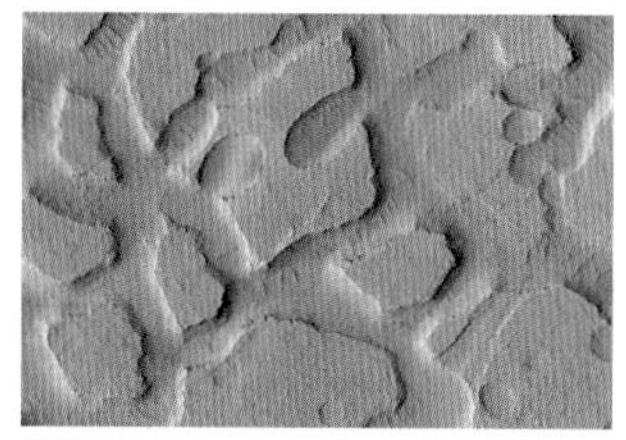

**MESAS AND TROUGHS**
A group of mesas was created by pitting and erosion of the surface in an area north of Apollinaris Patera. Windblown dust has filled the troughs between the mesas.

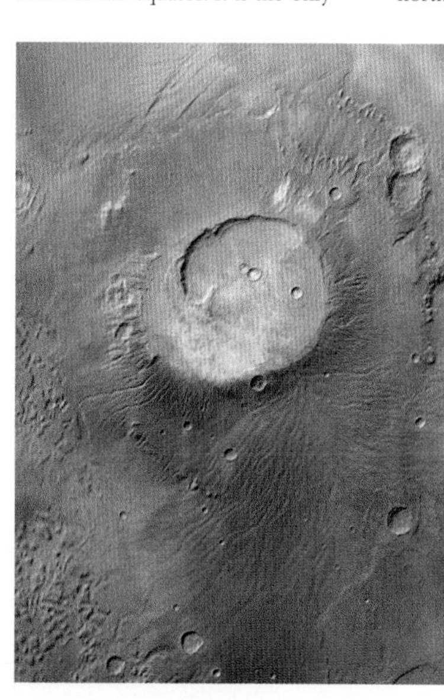

**SPLIT-LEVEL CALDERA**
The caldera has two different floor levels. It is partially hidden here by a patch of blue-white clouds. The summit area is pocked with impact craters.

AONIA TERRA

## Claritas Fossae

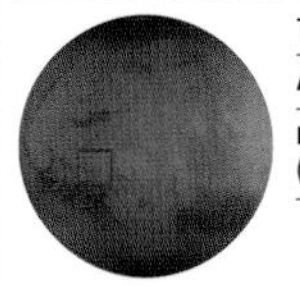

**TYPE** Fault system

**AGE** Over 3.5 billion years

**LENGTH** 2,050km (1,275 miles)

Claritas Fossae is a series of roughly northwest-to-southeast-trending linear fractures, which forms the southern end of the Tharsis Bulge. It is located south of the equator at the western end of the Valles Marineris. The region is about 150km (95 miles) wide at its northern end and 550km (340 miles) wide in the south. Individual fractures range from a few to tens of kilometres across. They formed as a result of enormous stresses associated with the formation of the Tharsis Bulge. As the crust pulled apart, blocks of crust dropped between two faults to form features called grabens. Crustal blocks that remained in place or were thrown up are termed horsts. Claritas Fossae separates two volcanic plains: that of Solis Planum to the east and Daedalia Planum to the west.

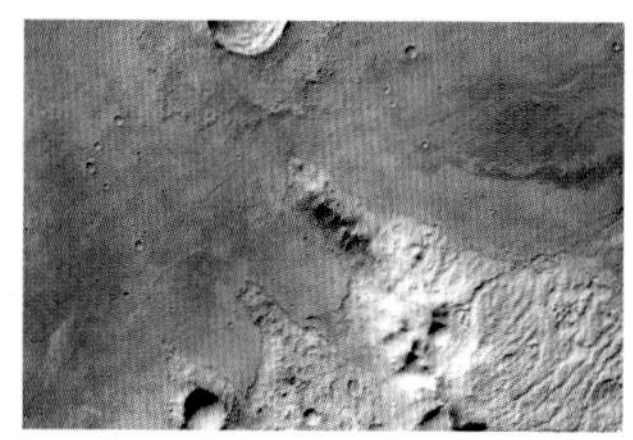

**LAVA BLANKET**
The eastern part of Claritas Fossae (bottom) meets the western part of Solus Planum (top). The lava from Solus has flowed over some of the older fractured terrain of Claritas and surrounds some of the higher ground.

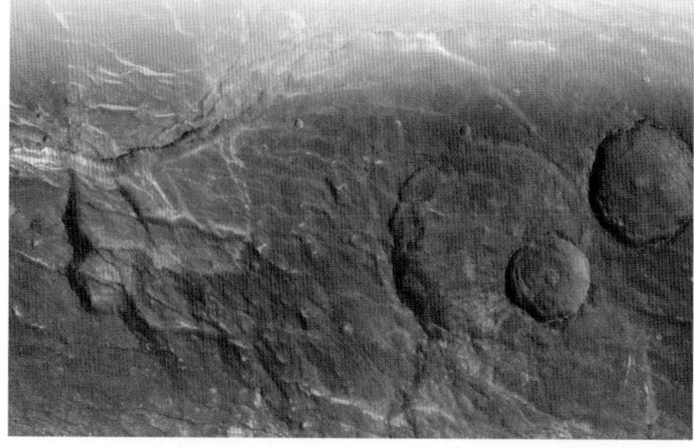

**ANCIENT TERRAIN**
A striking series of fractures runs round the edge of a lower-lying area that has been flooded with lava. The ancient impact craters appear to have had their floors resurfaced.

# FEATURES FORMED BY WATER

Both liquid and solid water have formed and shaped surface features on Mars. Giant channel-like valleys emerge fully formed out of the landscape. Some of these valleys were cut by fast-moving water during catastrophic floods, others were formed by water flowing more gradually through networks of river valleys, and others still were carved by glaciers. Other surface features suggest that Mars once had seas. The rivers and seas have long since vanished, but water ice remains, most markedly in the two ice plateaus that cap the planet.

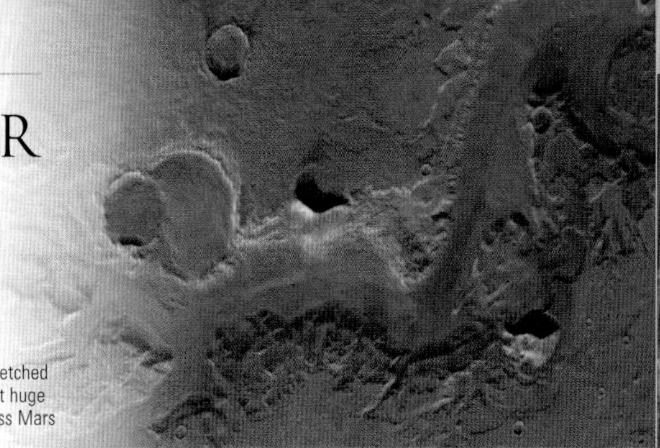

**REULL VALLIS**
Long, wide river channels are etched into the surface, revealing that huge volumes of water flowed across Mars billions of years ago.

PLANUM BOREUM

## North Polar Region

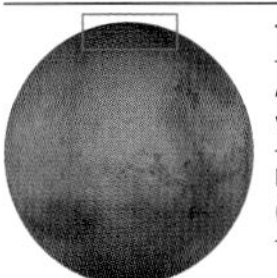

**TYPE** Polar ice-cap

**AGE** Under 2.5 billion years

**DIAMETER** 1,100 km (685 miles)

Two bright, white polar caps stand out against the otherwise dark surface of Mars. The one roughly centred on the North Pole is officially named Planum Boreum – the Northern Plain – although it is generally referred to as the North Polar Cap. Both this and its southern counterpart are easy to detect from Earth, but spacecraft have also flown over the poles allowing monitoring of daily, seasonal, and longer-term change.

The North Polar Cap is an ice-dominated mound that stands several kilometres above the surrounding terrain. It consists of a virtually permanent cap of water ice, which is either covered by or free of a deposit of carbon-dioxide ice, depending on the time of the Martian year. The cap is roughly circular but – as is also the case for the South Polar Cap – its bright ice forms a distinctive swirling, loosely spiral pattern when seen from above (see p.171).

**POLAR POLYGONS**
Polygon-shaped structures, similar to those found in Earth's polar regions, pattern parts of Mars's polar landscape. On Earth, they form as a result of stresses induced by repeated freezing and thawing of water.

The entire region is in darkness for about six months, during the Martian winter. This is when carbon dioxide in the atmosphere condenses into frost and snow, and not only covers the water-ice cap, but also the surrounding region, down to latitudes of about 65° north. When spring turns to summer and the Sun is permanently in the polar sky, its warmth evaporates the carbon dioxide and turns some of the water ice directly into vapour. The polar cap shrinks until just water-ice remains.

The cap is not made exclusively of ice but consists of layers of ice and layers of dusty sediment. Frost grains form around small particles of dust during winter dust storms in much the same way that hailstones form on Earth. These cover the ground until the frost is evaporated in the warmer months, leaving a layer of dust. The metres-deep layers take millions of years to form, building up at the rate of about 1mm (0.04in) per year. A study of these layers will reveal the history of the Martian climate.

**CUTTING THROUGH THE CAP**
A 14.5km- (9-mile-) wide slice of polar cap, at one end of the Chasma Boreale, reveals older, darker layers that contain sand. These are covered by lighter-toned, more uniformly bedded layers, which have no sand and may be a mixture of ice and dust.

**CLIFFS NEAR THE NORTH POLE**
This close-up view of the Martian North Polar Cap shows water ice close to cliffs about 2km (1.2 miles) high. Dark material in the caldera-like structures and dune fields could be volcanic ash.

UTOPIA PLANITIA

## Utopia Planitia

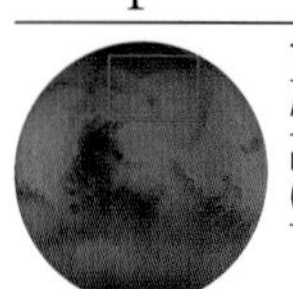

**TYPE** Lowland plain

**AGE** 2–3.5 billion years

**DIAMETER** 3,200km (2,000 miles)

Utopia is one of the enormous lava-covered plains of the northern hemisphere. The giant Elysium volcanoes are at its eastern perimeter. From above, it is possible to see that complex albedo patterns, polygonal fractures, and craters mark the vast rolling plain. Down on the surface, the landscape is uniformly flat and rock-strewn. At least that is the view in northeastern Utopia, where Viking 2 landed on 3 September 1976. Angular boulders of basaltic rock cover the landing site, close to Mie Crater. Small holes in the rocks are a result of bursting bubbles of volcanic gas. A thin layer of frost was also seen by the craft, first in mid-1977, when it covered the surface for about 100 Earth days, and then when it built up again in May 1979, one Martian year (23 Earth months) later.

These giant polygons are not unique to Utopia; they are also seen on other northern plains such as Acidalia and Elysium (below). They are polygon-shaped chunks of flat-lying land separated by huge cracks, or troughs, and are reminiscent of mud cracks seen in dried-up ponds on Earth. Earth's polygonal patterns are book- to table-sized; the Martian ones are on a far grander scale, the size of a town or small city. The land areas are 5–20km (3–12 miles) across, and the cracks between them are hundreds of metres wide. Earth's mud cracks form as ground dries through water evaporation. The Martian cracks could have a similar origin. Certainly, Mars has experienced the large-scale floods of water required. And dust-covered cemented rock found by Viking 2 seemed to be held together by salts left behind as briny water vaporized. However, it has also been suggested that the polygons formed in other ways – for example, in cooling lava.

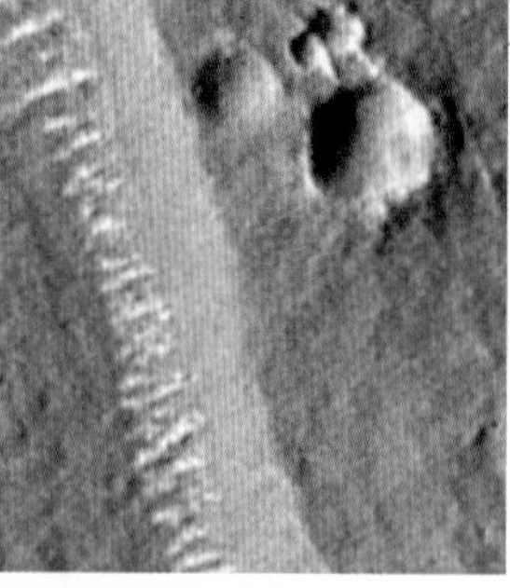

**POLYGON TROUGH FLOOR**
This close-up of one of the huge surface cracks that isolate polygon-shaped areas of land reveals bright, evenly spaced windblown ripples of sediment.

**ICE ON THE ROCKS**
A coating of water ice covers the volcanic rocks and soil at the Viking 2 landing site. The ice layer is very thin, no more than a fraction of a millimetre (one thousandth of an inch) thick.

LUNAR PLANUM

## Kasei Vallis

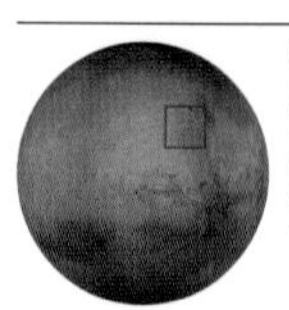

**TYPE** Outflow channel

**AGE** 3–3.5 billion years

**LENGTH** 1,780km (1,105 miles)

Kasei, which takes its name from the Japanese word for Mars, is the largest outflow channel. Not only is it long, but parts of its upper reaches are over 200km (125 miles) across and in places it is over 3km (2 miles) deep. The catastrophic flooding that formed Kasei was greater than any other known flood event on Mars, or Earth. Kasei originates in Lunar Planum, directly north of central Valles Marineris (see p.166), then flows across the ridged plain to Chryse Planitia. Along its route lie streamlined islands, isolated as the water flow split and then rejoined.

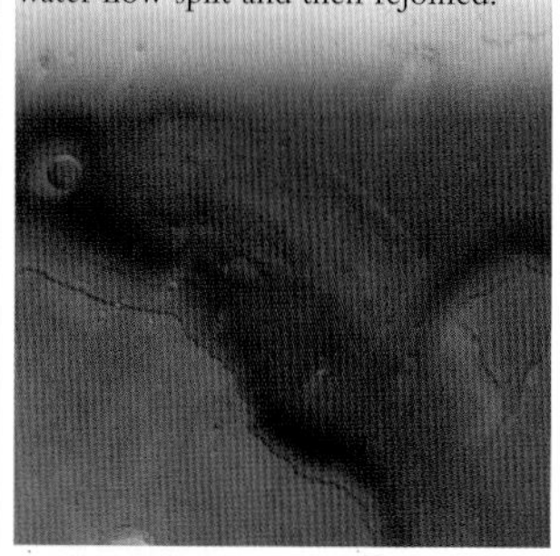

**GLACIAL CHANNEL**
This part of Kasei Vallis was probably carved by glaciers or outflows from lakes beneath glaciers. The dark colour indicates sediments. The view is hampered by atmospheric dust and haze.

ELYSIUM PLANITIA

## Elysium Planitia

**TYPE** Lowland plain

**AGE** Under 2.5 billion years

**DIAMETER** 3,000km (1,860 miles)

The Elysium Planitia is an extensive lava-covered plain just north of the equator. It has been suggested that an area almost directly south of the great volcano Elysium Mons is a dust-covered frozen sea. It is dominated by irregular blocky shapes that look like the rafts of segmented sea-ice seen off the coast of Earth's Antarctica. These "ice plateaus" are surrounded by bare rock. They formed when water flooded through a series of fractures in the Martian crust, creating a sea similar in size to Earth's North Sea. As the water froze, floating pack ice broke into rafts. These were later covered in dust from the nearby volcanoes, and this coating protected them. Unprotected ice between the rafts vaporized into the atmosphere, leaving bare rock around the ice plateaus.

**ICE PLATEAUS AND IMPACT CRATERS**
The darker-toned ice plateaus are a few tens of kilometres across. The relatively small number of impact craters in this area suggests a young surface.

XANTHE TERRA

## Nanedi Vallis

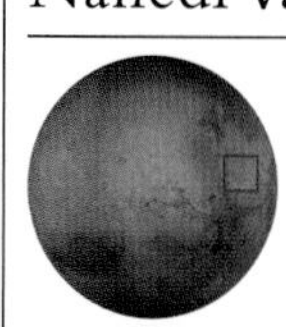

**TYPE** Outflow channel

**AGE** 2–3.5 billion years

**LENGTH** 508km (315 miles)

This major outflow channel lies in a relatively flat area. There is no visible source for the channel in the south, but its snake-like route northwards, across the cratered plains of Xanthe Terra, is clearly seen before the channel comes to a sudden stop. Nanedi Vallis appears to have undergone different stages of flow. Initially, the meandering river almost created some ox-bow lakes. Then, areas of riverbed drained and became the terraces now seen stranded between the main channel and the cratered plain above. A gully down the centre of the channel indicates a final flow of water.

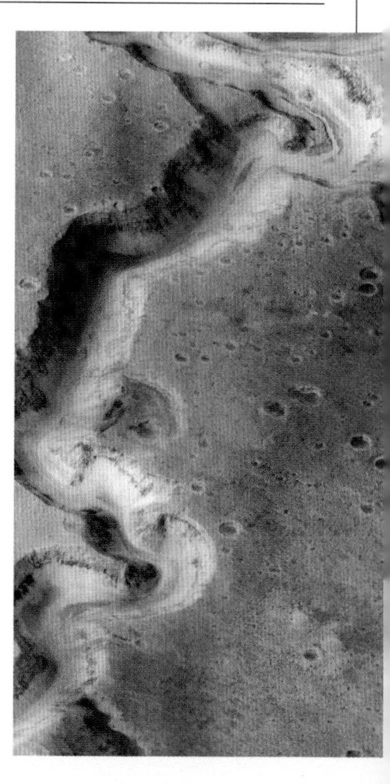

**TERRACING**
Nanedi Vallis formed by water flow over a long period. Terracing is evident in this image, and a portion of the narrow central channel is just visible (top right).

TERRA MERIDIANI

## Meridiani Planum

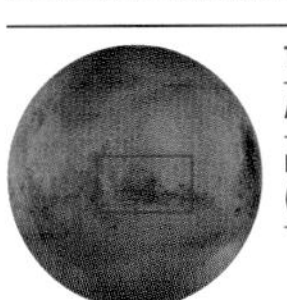

**TYPE** Highland plain
**AGE** Over 3.5 billion years
**DIAMETER** 1,100km (680 miles)

In the westernmost portion of Terra Meridiani and just south of the equator lies the high plain Meridiani Planum. It does not stand out in the global view of Mars but achieved prominence as the landing site and exploration ground for the Opportunity rover. The plain is about 15° due west of Schiaparelli Crater (see p.172). Smaller impact craters pepper the area. They range from Airy, just 41km (26 miles) across, to much smaller bowl-shaped craters, such as the 22m- (72-ft-) wide Eagle Crater where Opportunity landed.

**UNIQUE METEORITE**
This basketball-sized rock has an iron-nickel composition. It is not a Martian rock but a meteorite – the first to be found on a planet other than Earth.

Volcanic basalt is found within the area but the region is of greatest interest because it contains ancient layered sedimentary rock that includes the mineral hematite. Some of this mineral, which on Earth almost always forms in liquid water, is exposed and easily found on the surface. The hematite could have been produced from iron-rich lavas but it is believed that water was involved. This area is dry now but it was once soaking wet and could well have been the site of an ancient lake or sea about 3.7 billion years ago. Eroded layered outcrops beyond the landing site support this theory and point to a deep and long-lasting volume of water as large as Earth's Baltic Sea. At this time, Mars must have been a much warmer and wetter place than it is today.

**HEATSHIELD AND SCORCH MARK**
Subsurface pale dirt was spattered onto the plain when Opportunity landed. The remains of the craft's discarded heat shield are at left and centre.

PROMETHEI TERRA

## Reull Vallis

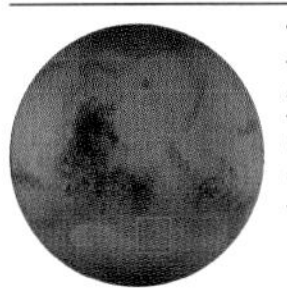

**TYPE** Outflow channel
**AGE** 2–3.5 billion years
**DIAMETER** 945km (587 miles)

Reull Vallis is one of the larger channels of the southern hemisphere. It extends across the northern part of Promethei Terra, to the east of Hellas Basin (see p.173). Reull is thought to have had a complex evolution as it exhibits the characteristics of all three channel types seen on Mars. In the collapsed region at the southern base of the volcano Hadriaca Patera, for example, it is a fully formed outflow channel. But small tributaries also feed into the main channel, as they would in a runoff channel. And the main channel has the features of a fretted channel – a wide, flat floor and steep walls. Reull Vallis takes its name from the Gaelic word for planet.

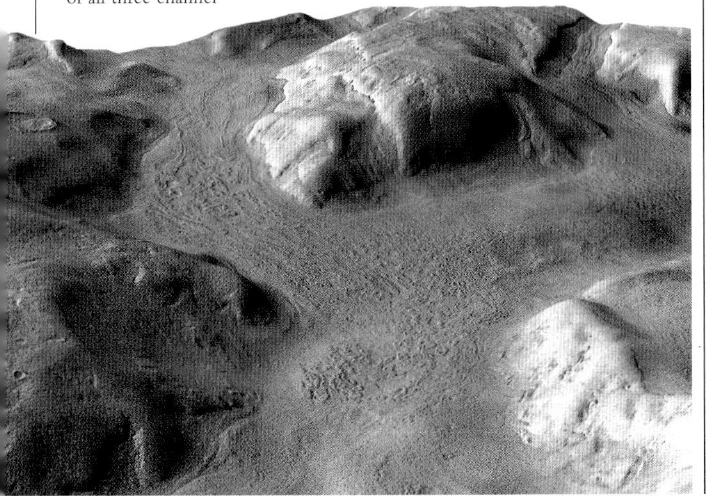

**MERGING CHANNELS**
Reull Vallis (top left) is joined by a tributary, Teviot Vallis (right). The parallel structures in the fretted channel floor were possibly caused by a glacial flow of loose debris mixed with ice.

PLANUM AUSTRALE

## South Polar Region

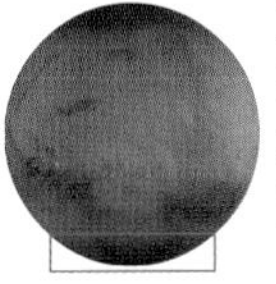

**TYPE** Polar ice cap
**AGE** Under 2.5 billion years
**DIAMETER** 1,450km (900 miles)

The South Polar Cap, known formally as Planum Australe (Southern Plain), is an ice-dominated mound, several kilometres high. It consists of three different parts. First is the bright polar cap that is roughly centred on the South Pole. This is a permanent cap of water ice with a covering of carbon-dioxide ice. Next are the scarps made primarily of water ice, which fall away from the cap to the surrounding plains. Thirdly, hundreds of square kilometres of permafrost encircle the region. The permafrost is water ice mixed into the soil and frozen to the hardness of solid rock. The South Polar Cap shrinks and grows with the seasons like the North Polar Cap (see p.169). Yet surprisingly the southern cap does not get warm enough in the summer to lose its carbon-dioxide ice covering. Dust storms that block out the Sun may keep the cap cooler than expected.

**WATER-ICE SCARPS**
This late summer view, when temperatures are at a high of about -130°C (-200°F), shows the water-ice cap. Scarps edge the ice, revealing that the cap consists of layered deposits.

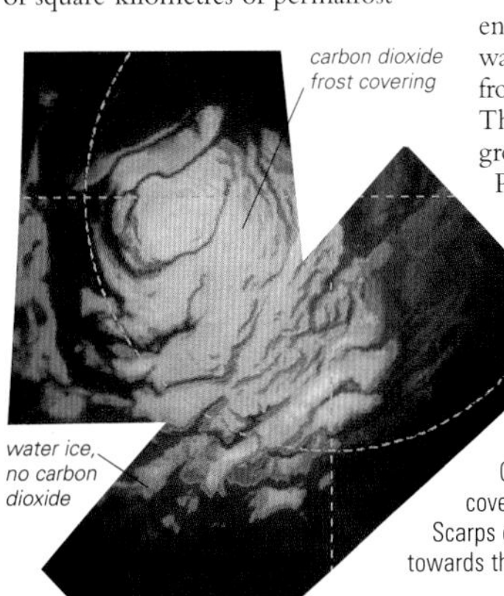


**SOUTH POLAR CAP**
Carbon-dioxide frost (shown as pink) covers over the water-ice cap (green-blue). Scarps of water ice at the edge of the cap slope towards the surrounding plains.

# IMPACT CRATERS

The Martian surface is scarred by tens of thousands of craters, about 1,000 of which have been given names. They range from simple bowl craters, less than 5km (3 miles) across, to basins hundreds of kilometres wide. The oldest craters are found in the southern hemisphere and have been eroded throughout their lifetimes. Their floors have been filled and their rims degraded, and the craters have become characteristically shallow. Smaller, fresher-looking craters have formed on top of them. The ejecta has been distributed by flowing across the surface rather than being flung through the air.

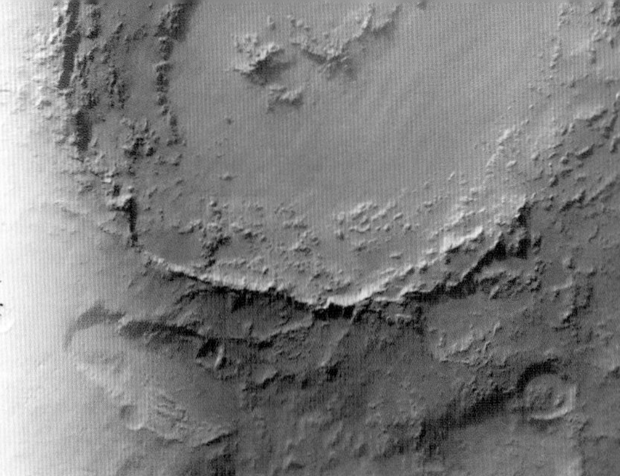

**ANCIENT GEOLOGICAL FEATURE**
Large impact craters such as Hale (right) have had their central peaks and terraced walls continuously eroded for up to 4 billion years.

CHRYSE PLANITIA

## Belz Crater

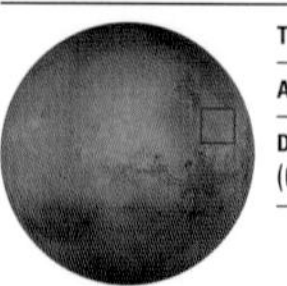

**TYPE** Ejecta-flow crater
**AGE** Under 4 billion years
**DIAMETER** 10.2km (6.3 miles)

Ejecta-flow craters, also known as rampart or splosh craters, are found only on Mars. They are small, with a raised rampart wall and ejecta that has apparently flowed away from the crater. Belz is a perfect example. When such craters form, subsurface water or water ice melted by the heat of the impact is released. This then allows the debris to flow across the surface.

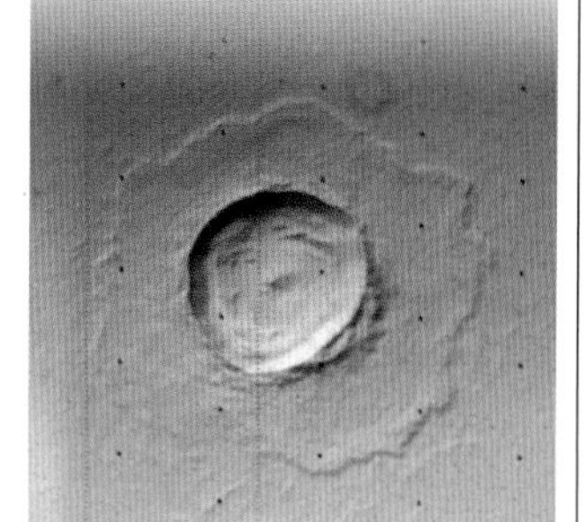

**SPLOSH CRATER**

ARABIA TERRA

## Tikhonravov Crater

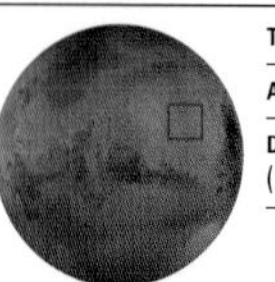

**TYPE** Large crater
**AGE** About 4 billion years
**DIAMETER** 386km (240 miles)

Tikhonravov is almost unrecognizable as a crater. Part of it is buried under deposits, and smaller impact craters have formed on top. Two of these near Tikhonravov's centre have dark patches of sand within them that appear to give the crater eyes.

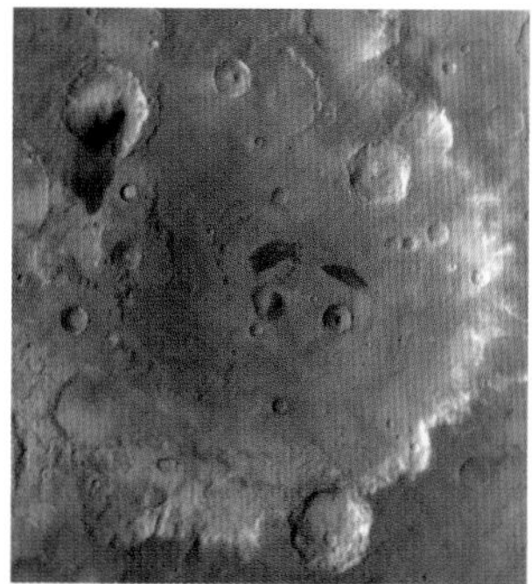

**TIKHONRAVOV'S EYEBROWS**

ARABIA TERRA

## Schiaparelli Crater

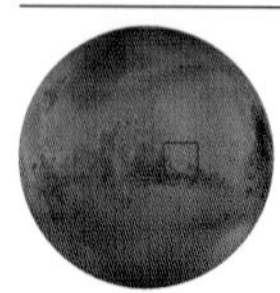

**TYPE** Large crater
**AGE** About 4 billion years
**DIAMETER** 471km (293 miles)

This crater takes its name from the astronomer Giovanni Schiaparelli (see p.220), who spent much of his working life studying Mars. It is a highly circular crater, as are most Martian craters, although a significant number are elliptical – a rarity on the Moon and Mercury. Schiaparelli straddles the equator and is the largest crater in the Arabia Terra. It is an old crater, formed by an impacting body when the planet was young, and shows signs of degradation. The rim has been smoothed down and in parts is completely missing. Any central peak in the crater has been obliterated. Material has been deposited within the crater, and smaller craters have formed across the whole area. Wind continues to shape the landscape by erosion and by moving surface material.

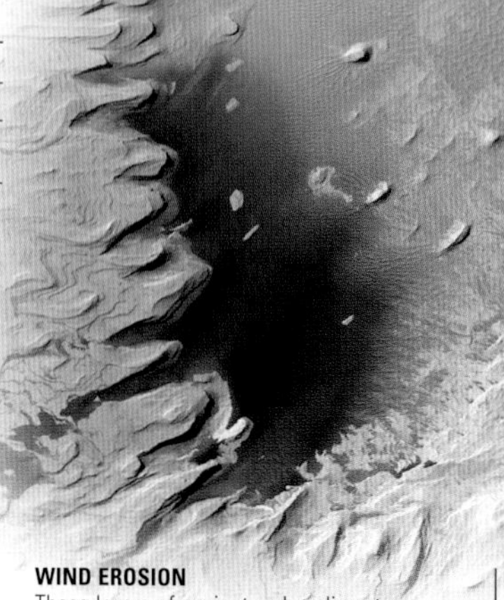

**WIND EROSION**
These layers of ancient rock sediments on the floor of an impact crater lying within the northwestern rim of Schiaparelli have been eroded and exposed by the wind.

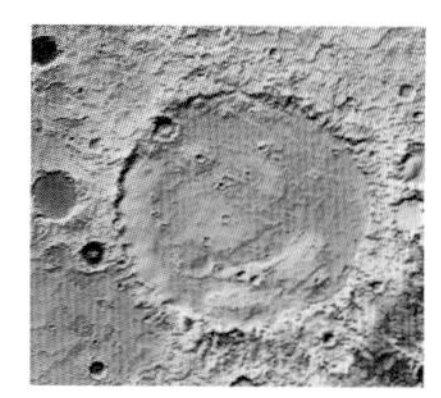

**SHALLOW CRATER**
Here colour is used to indicate altitude. The crater floor is at the same height as much of the surrounding terrain. Higher deposits are in green. The degraded rim (yellow) is only about 1.2km (0.75 miles) above the floor.

TERRA TYRRHENA

## Huygens Crater

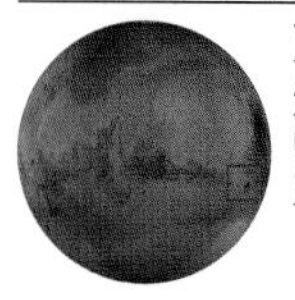

**TYPE** Multi-ringed crater
**AGE** About 4 billion years
**DIAMETER** 470km (292 miles)

Huygens is one of the largest impact craters in the heavily cratered southern highlands of Mars. It was formed during the period of intense bombardment within the first 500 million years of the planet's early history. The age of craters such as Huygens is determined by counting the number of craters that overlay their rims. Huygens has a second ring inside its mountainous rim. This has been filled by material carried into the ring. The rim is heavily eroded, and markings on it suggest that surface water has run off it at some time. The pattern of markings is reminiscent of dendritic drainage systems on Earth, which from above look like the trunk and branches of a tree. Dark material within this crater's drainage channels was either carried by the draining water or by the wind.

**MESAS**
Smooth-topped hills (mesas) on the crater floor are left behind as a former smooth layer of material is eroded to reveal a more rugged surface.

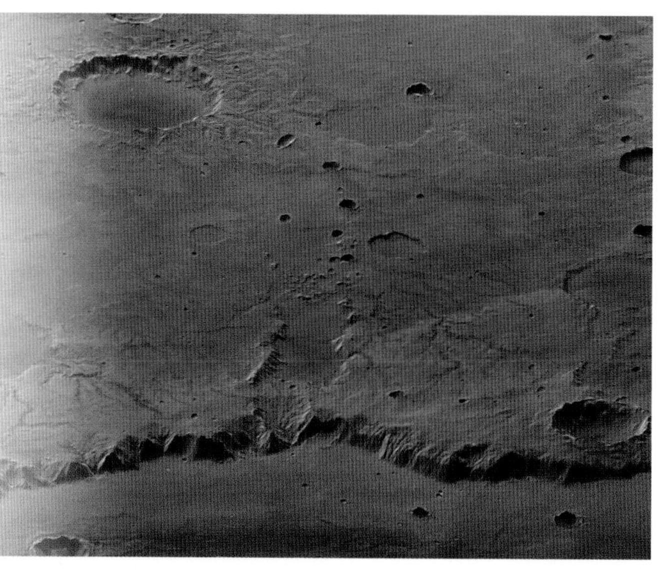

**EASTERN RIM**
In this perspective view across Huygens's eastern rim (foreground) to the surrounding terrain, a branch-like network of drainage channels flows away from the rim, and small, more recently formed craters can be seen.

**ANCIENT BASIN**
The original crater floor has been covered by volcanic and wind-borne deposits, and it also shows signs of change by water and glacial ice. Dust storms continue to shape the surface.

HELLAS PLANITIA

## Hellas Planitia

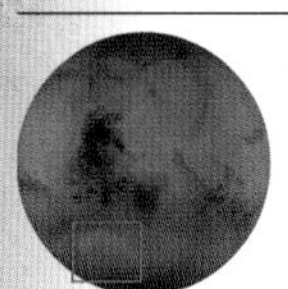

**TYPE** Basin

**AGE** About 4 billion years

**DIAMETER** 2,200km (1,365 miles)

The Hellas Basin is the largest impact crater on Mars and possibly the largest in the Solar System. It is the dominant surface feature in the southern hemisphere. It is not immediately apparent that Hellas is an impact crater. Indeed, its official name, Hellas Planitia, indicates that it is a large, low-lying plain. This designation dates from over a century ago, when the Martian surface was observed only through Earth-based telescopes and the true nature of this vast, shallow feature was not known. Hellas is the Greek word for Greece.

Particularly large craters that have been subsequently altered are termed basins. They are analogous to the maria on Earth's Moon. The term basin is also applied to the second-largest Martian crater, Isidis Planitia, and the third-largest, Argyre Planitia (below). Over the past 3.5–4 billion years, Hellas Basin has had its floor filled by lava and its features changed by wind, water, and fresh crater formation. Despite all this, some of its original features are still visible. Its overall shape and the remains of its rim can still be seen, as can inward-facing, arc-shaped cliffs lying up to several hundred kilometres beyond the rim. These are possibly the remnants of multiple rings.

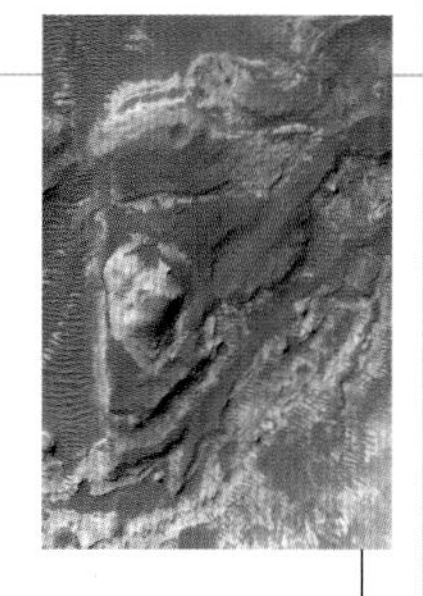

**ROCK OUTCROPS**
Layered sedimentary rocks, which formed long after Hellas, lie in an eroded region northeast of the crater basin. Darker windblown ripples mark the surface.

**ERODED RIM**
This perspective view shows the northern rim – the mountain range formed around the crater as the planet's crust was lifted up at the time of impact. Whole portions of the rim are missing to the northeast and southwest.

SIRENUM TERRA

## Nansen Crater

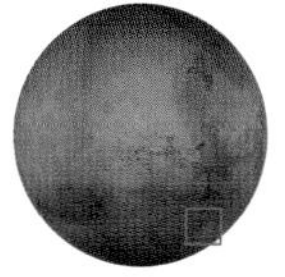

**TYPE** Large crater

**AGE** About 4 billion years

**DIAMETER** 81km (50 miles)

Martian impact craters were first identified in 22 images returned by Mariner 4 in 1966. Nansen Crater was among the first and was named after the Norwegian explorer Fridtjof Nansen. New craters continue to be added to the list as a result of surveys by spacecraft. The Viking orbiter recorded this image of Nansen in 1976. The crater shows signs of erosion; its walls have been nibbled by the wind. Smaller, sharply defined craters have punctuated the surrounding terrain. A more recent crater has formed inside Nansen. Its central dark floor could be volcanic basalt.

**CRATER WITHIN A CRATER**

ARGYRE PLANITIA

## Argyre Planitia

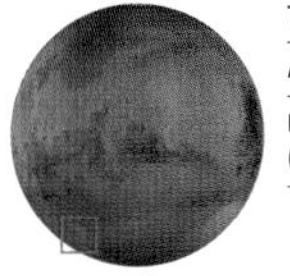

**TYPE** Basin

**AGE** About 4 billion years

**DIAMETER** 800km (500 miles)

Argyre is the third-largest crater on Mars. Its floor has been flooded by volcanic lava, and it has been heavily eroded by wind and water. It is speculated that in the distant past water drained into the basin from the south polar ice cap. Channels entering the basin at its southeastern edge and others leading out from its northern edge reveal the water's route. The path cuts through the mountain ranges that define the basin: the Charitum Montes to the south and the Nereidum Montes to the north.

**FROST IN THE SOUTHERN HILLS**
Frost (mainly of carbon dioxide) covers an area of cratered terrain in the Charitum Montes in early June 2003, at which time the south polar frost cap had been retreating southwards for about a month.

**CRATER DUNE FIELD**
Argyre's floor and rugged highland rim contain smaller craters. Some of these show signs of erosion. This one lying in the northwestern part of Argyre Basin contains a dark dune field.

AONIA TERRA

## Lowell Crater

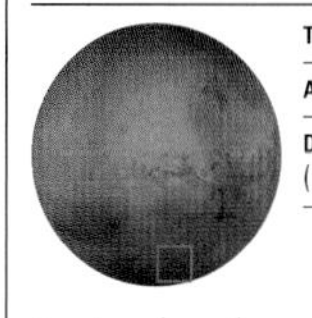

**TYPE** Multi-ringed crater

**AGE** About 4 billion years

**DIAMETER** 203km (126 miles)

Erosion has changed Lowell since its formation early in Mars's history. The edges of both its outer rim and inner ring have been smoothed out, and its fine-grained ejecta soil has been blown about. The crater's appearance continues to undergo long-term changes but it also changes on a short-term basis. Frost covers the crater's face in the winter months as the frost line extends north from the south polar region.

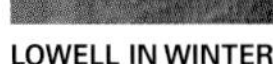

**LOWELL IN WINTER**

MERIDIANI PLANUM

# Endurance Crater

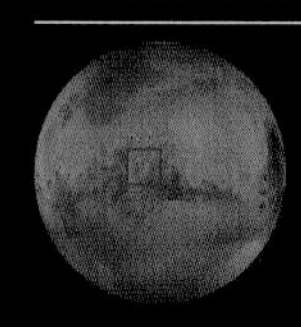

**TYPE** Bowl crater

**AGE** Under 4 billion years

**DIAMETER** 130m (420ft)

This small and inconspicuous crater has been explored and investigated to a greater extent than any other crater on Mars. In early 2004, it did not even have a name but by the end of that year its rim, slopes, and floor had all been imaged and examined by the robotic rover Opportunity. The small craft just happened to land within roving distance of this football-field-sized crater when it made its scheduled landing in the Meridiani Planum in Mars's northern hemisphere.

Endurance, named after the ship that carried Irish-born British explorer Ernest Shackleton to the Antarctic, is an almost circular crater bounded by a rim of rugged cliffs. Its inner walls slope down to the crater floor, 20–30m (66–100ft) below. Layers of bedrock line the crater, some of which are exposed; loose material and sand dunes cover the rest of the floor.

Opportunity spent approximately six months exploring Endurance. The rover started by travelling round the southern third of the crater's rim; here it crossed a region named Karatepe and travelled along the edge of Burns Cliff. It then retraced its route to enter the crater on its southwestern limb.

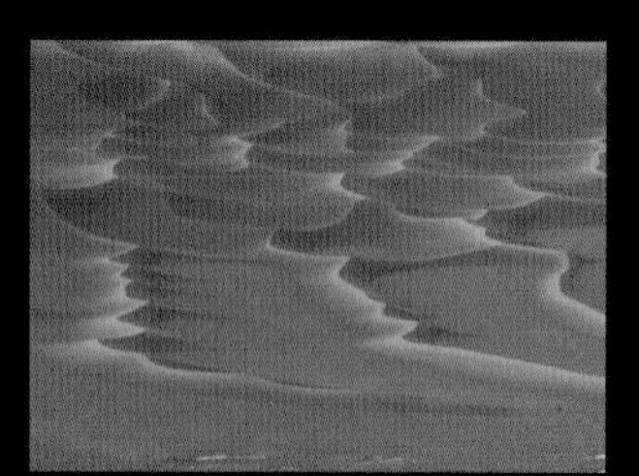

**SAND DUNES**
The centre of the crater floor is covered by small sand dunes. The reddish dust has formed flowing tendrils, which are a few centimetres to a metre or so deep.

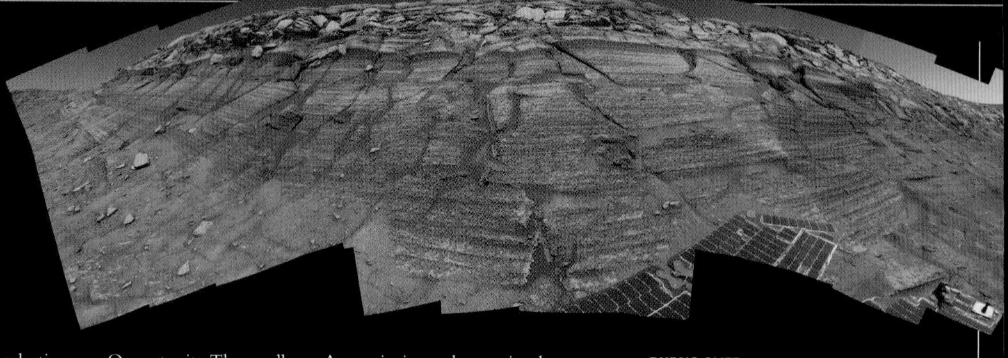

**BURNS CLIFF**
This portion of the crater's southern inner wall is called Burns Cliff. Forty-six Opportunity images taken in November 2004 combine to make this 180° view. The wide-angle camera makes rock walls bulge unrealistically toward the viewer.

Opportunity made its way down the inner slope, examining rocks and soil along its route. It headed towards the crater's centre but got less than halfway before doubling back; any farther and it might have got stuck in the sandy terrain. It then exited the crater to

**DRAMATIC PANORAMA**
This approximately true-colour view across Endurance Crater was taken by Opportunity's panoramic camera as the rover perched on the western rim. A dune field lies in the centre of the crater.

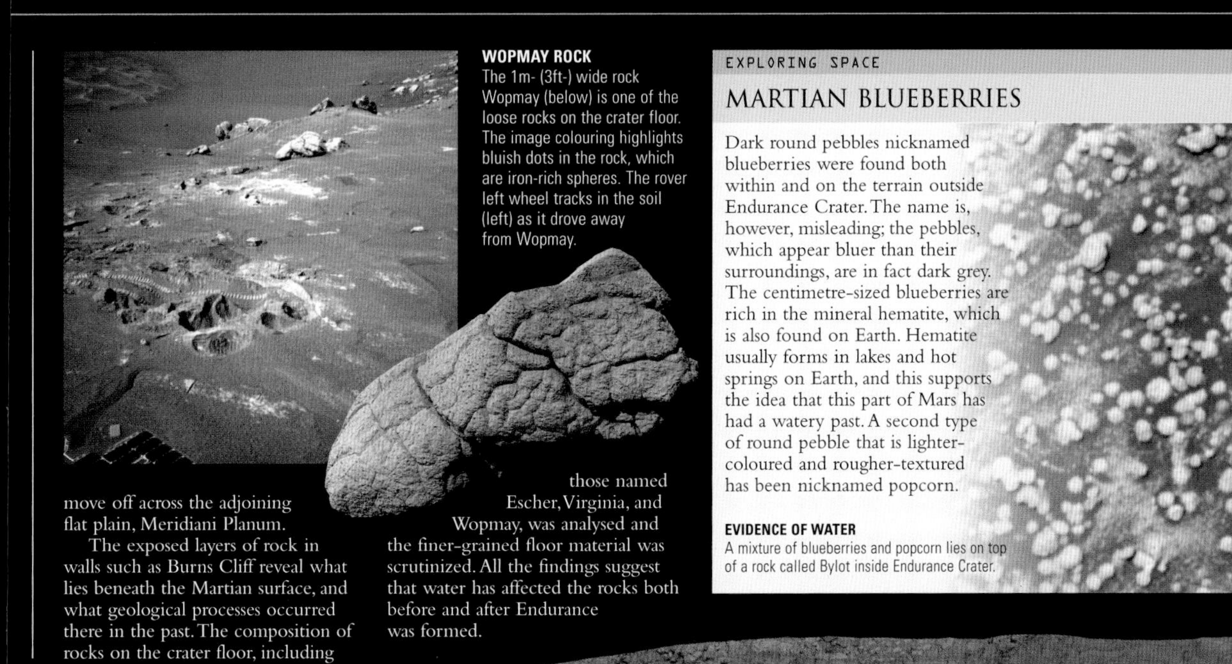

**WOPMAY ROCK**
The 1m- (3ft-) wide rock Wopmay (below) is one of the loose rocks on the crater floor. The image colouring highlights bluish dots in the rock, which are iron-rich spheres. The rover left wheel tracks in the soil (left) as it drove away from Wopmay.

move off across the adjoining flat plain, Meridiani Planum.

The exposed layers of rock in walls such as Burns Cliff reveal what lies beneath the Martian surface, and what geological processes occurred there in the past. The composition of rocks on the crater floor, including those named Escher, Virginia, and Wopmay, was analysed and the finer-grained floor material was scrutinized. All the findings suggest that water has affected the rocks both before and after Endurance was formed.

EXPLORING SPACE

## MARTIAN BLUEBERRIES

Dark round pebbles nicknamed blueberries were found both within and on the terrain outside Endurance Crater. The name is, however, misleading; the pebbles, which appear bluer than their surroundings, are in fact dark grey. The centimetre-sized blueberries are rich in the mineral hematite, which is also found on Earth. Hematite usually forms in lakes and hot springs on Earth, and this supports the idea that this part of Mars has had a watery past. A second type of round pebble that is lighter-coloured and rougher-textured has been nicknamed popcorn.

**EVIDENCE OF WATER**
A mixture of blueberries and popcorn lies on top of a rock called Bylot inside Endurance Crater.

# JUPITER

36–37 Gravity, motion, and orbits
64–65 Planetary motion
108–11 Beyond Earth
119 Gas giants

JUPITER IS THE LARGEST AND MOST MASSIVE of all the planets. It has almost 2.5 times the mass of the other eight planets combined, and over 1,300 Earths could fit inside it. Jupiter bears the name of the most important of all the Roman gods (known as Zeus in Greek mythology). The planet has the largest family of moons in the Solar System, its members named after Jupiter's lovers, descendants, and attendants.

## ORBIT

Jupiter is the fifth planet from the Sun. It lies approximately five times as far away as Earth, but its distance from the Sun is not constant. Its orbit is elliptical and there is a difference of 76.1 million km (47.3 million miles) between its aphelion and perihelion distances. Jupiter's spin axis tilts by 3.1°, and this means that neither of the planet's hemispheres point markedly towards or away from the Sun as it moves round its orbit. Consequently, Jupiter does not have obvious seasons. The planet spins quickly about its axis, more quickly than any other planet. Its rapid spin throws material in its equatorial region outward. The result is a bulging equator and a slightly squashed appearance.

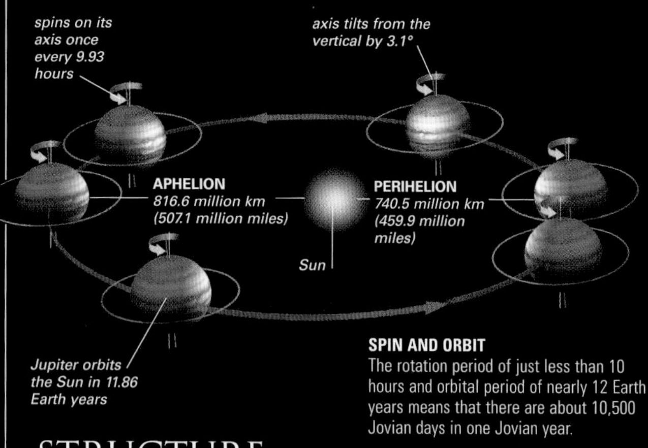


**SPIN AND ORBIT**
The rotation period of just less than 10 hours and orbital period of nearly 12 Earth years means that there are about 10,500 Jovian days in one Jovian year.

## STRUCTURE

Although it is the most massive planet (318 times the mass of the Earth), Jupiter's great size means that its density is low. Its composition is more like the Sun's than any other planet in the Solar System. Jupiter's hydrogen and helium is in a gaseous form in the outer part of the planet, where the temperature is about -110°C (-166°F). Closer to the centre, the pressure, density, and temperature increase. The state of the hydrogen and helium changes accordingly. By about 7,000km (4,350 miles) deep, at about 2,000°C (3,600°F), hydrogen acts more like a liquid than a gas. By 14,000km (8,700 miles), at about 5,000°C (9,000°F), hydrogen has compacted to metallic hydrogen and acts like a molten metal. Deep inside, at a depth of about 60,000km (37,260 miles), is a solid core of rock, metal, and hydrogen compounds. The core is small compared to Jupiter's great size but is about 10 times the mass of Earth.

gaseous hydrogen and helium
outer layer of liquid hydrogen and helium
inner layer of metallic hydrogen
core of rock, metal, and hydrogen compounds

**JUPITER INTERIOR**
At the heart of Jupiter lies a relatively small, dense, and probably solid core. The core is surrounded by layers of metallic, liquid, and gaseous material, which is predominantly hydrogen.

**GAS GIANT**
Jupiter's surface is not solid. Each light or dark band and every big or small swirl or spot is a part of the planet's cloudy atmosphere.

## JUPITER PROFILE

| | |
|---|---|
| **AVERAGE DISTANCE FROM THE SUN** 778.3 million km (483.6 million miles) | **ROTATION PERIOD** 9.93 hours |
| **CLOUD-TOP TEMPERATURE** -110°C (-160°F) | **ORBITAL PERIOD (LENGTH OF YEAR)** 11.86 Earth years |
| **DIAMETER** 142,984km (88,846 miles) | **MASS (EARTH = 1)** 318 |
| **VOLUME (EARTH = 1)** 1,321 | **CLOUD-TOP GRAVITY (EARTH = 1)** 2.53 |
| **NUMBER OF MOONS** 63 | **SIZE COMPARISON** EARTH JUPITER |
| **OBSERVATION** Jupiter is bright and easy to spot. It has a maximum magnitude of -2.9. Even at its faintest it is brighter than Sirius, the brightest star in the sky. Jupiter is best seen at opposition, which occurs once every 13 months. | |

# MAGNETIC FIELD

Jupiter has a magnetic field – it is as if the planet has a large bar magnet deep inside it. The field is generated by electric currents within the thick layer of metallic hydrogen, and the axis joining the magnetic poles is tilted at about 11° to the spin axis. The field is stronger than that of any other planet. It is about 20,000 times stronger than Earth's magnetic field and has great influence on the volume of space surrounding Jupiter. Solar wind particles (see p.123) streaming from the Sun plough into the field. They are slowed down and rerouted to spiral along the field's magnetic lines of force. Some particles enter Jupiter's upper atmosphere around its magnetic poles. They collide with the atmospheric gases, which radiate and produce aurorae. Other charged particles (plasma) are trapped and form a disc-like sheet around Jupiter's magnetic equator. Electric currents flow within this sheet. High-energy particles are trapped and form radiation belts, similar to, but much more intense than, the Van Allen belts round Earth (see p.139). Jupiter's magnetic field is shaped by the solar wind, forming a vast region called the magnetosphere. Its size varies with changes in pressure of the solar wind, but the tail is thought to have a length of about 600 million km (370 million miles).

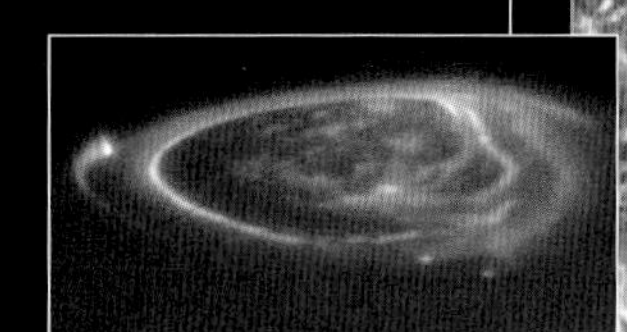

**AURORA**
This striking electric-blue aurora centred on Jupiter's north magnetic pole was imaged by the Hubble Space Telescope in 1998.

solar wind deflected
axis of rotation
direction of magnetic force lines
plasma sheet
axis of magnetic field
northern horn
solar wind
bow shock
southern horn
turbulence
solar wind deflected
tail
magnetic equatorial plane
radiation belt
magnetosheath

**JUPITER'S MAGNETOSPHERE**
Jupiter's magnetosphere – the bubble-like region round Jupiter dominated by the planet's magnetic field – is enormous; it is one thousand times the volume of the Sun, and its tail stretches away from the planet as far as Saturn's orbit. This slice through the magnetosphere reveals its structure.

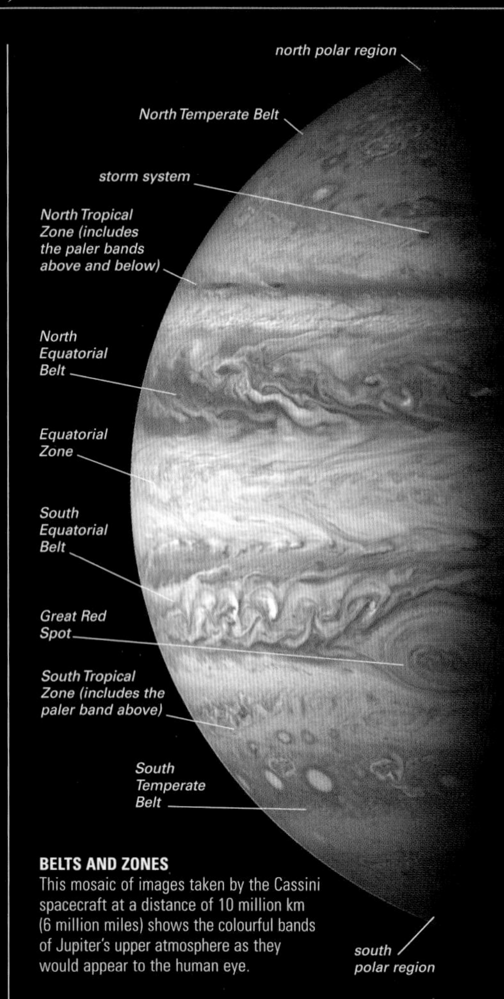


**BELTS AND ZONES**
This mosaic of images taken by the Cassini spacecraft at a distance of 10 million km (6 million miles) shows the colourful bands of Jupiter's upper atmosphere as they would appear to the human eye.

# ATMOSPHERE

Jupiter's atmosphere is dominated by hydrogen, with helium being the next most common gas. The rest is made up of simple hydrogen compounds – such as methane, ammonia, and water – and more complex ones such as ethane, acetylene, and propane. It is these compounds that condense to form the different-coloured clouds of the upper atmosphere and help give Jupiter its distinctive banded appearance. The temperature of the atmosphere increases towards the planet's interior. As gases condense at different temperatures, different types of clouds form at specific altitudes. All the while, the gas in Jupiter's equatorial region is heated by the Sun, and this rises and moves towards the polar regions. Cooler air flows from the polar regions at a lower altitude to take its place, creating in effect a large circulation cell. This hemisphere-wide circulation transfer would be straightforward if Jupiter was stationary. It is not – it rotates, and speedily at that, and a force known as the Coriolis effect (see p.140) deflects the north–south flow into an east–west flow. As a result, the large circulation cell is split into many smaller cells of rising and falling air. These are seen on Jupiter's surface as alternating bands of colour. Jupiter's white bands of cool rising air are called zones. The red-brown bands of warmer falling air are known as belts.

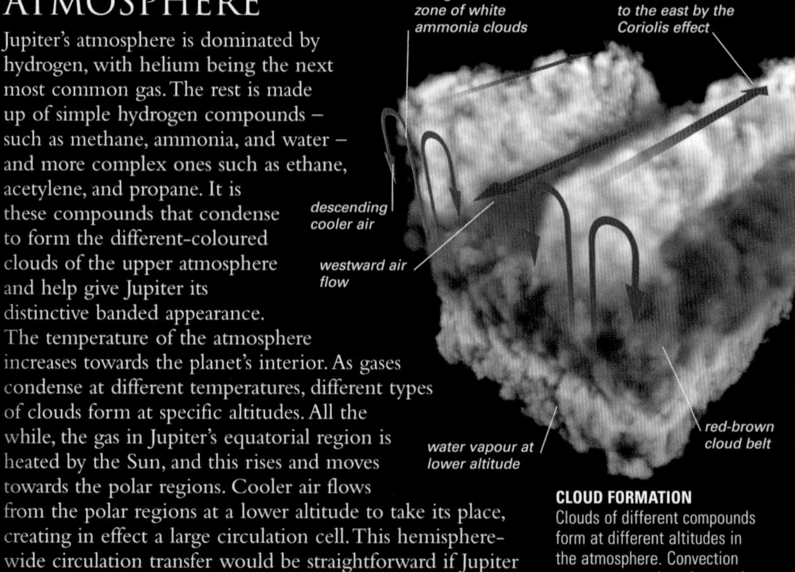


**CLOUD FORMATION**
Clouds of different compounds form at different altitudes in the atmosphere. Convection currents move the mixture of gas upwards. Water is first to reach the altitude where it is cool enough to condense to form clouds. Higher up, where it is cooler, red-brown ammonium hydrosulphide clouds form, and highest of all, where it is coolest, are the white ammonia clouds.

hydrogen (89.8%)

helium with traces of methane and ammonia (10.2%)

**COMPOSITION OF ATMOSPHERE**
Hydrogen dominates, but it is the trace compounds that colour Jupiter's upper atmosphere.

# MOONS

Jupiter has over 60 known moons, over two-thirds of which have been discovered since January 2000. Only 38 of the moons have been given names, and several have yet to have their orbit confirmed. The recent discoveries are typically irregularly shaped rocky bodies a few kilometres across, and are thought to be captured asteroids. By contrast, Jupiter's four largest moons are spherical bodies that were formed at the same time as Jupiter. Collectively known as the Galilean Moons, they were the first moons to be discovered after Earth's Moon (see p.180). As they orbit Jupiter, passing between it and the Sun, their shadows sweep across the planetary surface; from within the shadow, the Sun appears eclipsed. A triple eclipse happens just once or twice a decade.

**TRIPLE ECLIPSE**
Three shadows were cast onto Jupiter's surface on 28 March 2004 as its three largest moons passed between the planet and the Sun. Io is the white circle in the centre, its shadow to its left. Ganymede is the blue circle at upper right, and its shadow lies on Jupiter's left edge. Callisto's shadow is on the upper-right edge, but the moon itself is out of view, to the right of the planet.

**JUPITER'S MOONS**

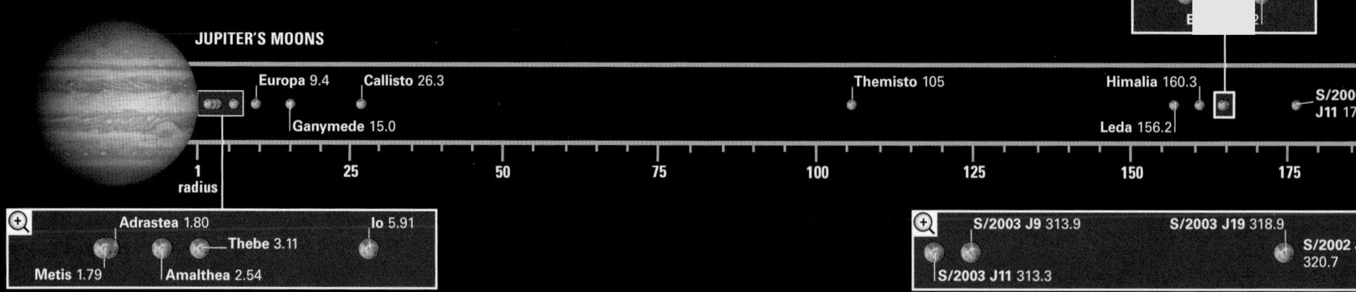


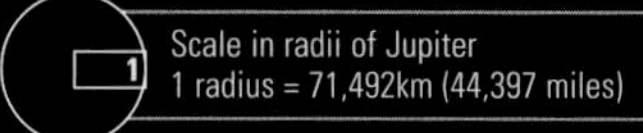

Scale in radii of Jupiter
1 radius = 71,492km (44,397 miles)

Moons are not to scale and increase in size for magnification purposes only

# WEATHER

Jupiter has no notable seasons, and the planet's temperature is virtually uniform. Its polar regions have similar temperatures to its equatorial regions because of internal heating. Jupiter radiates about 1.7 times more heat than it absorbs from the Sun. The excess is infrared heat left from when the planet was formed. Most of Jupiter's weather occurs in the part of its atmosphere that contains its distinct white and red-brown cloud layers and is dominated by clouds, winds, and storms. The rising warm air and descending cool air within the atmosphere produce winds, which are channelled around the planet, both to the east and west, by Jupiter's fast spin. The wind speed changes with latitude; winds within the equatorial region are particularly strong and reach speeds in excess of 400kph (250mph). The solar and infrared heat, the wind, and Jupiter's spin combine to produce regions of turbulent motion, including circular and oval cloud structures, which are giant storms. The smallest of these storms are like the largest hurricanes on Earth. They can be relatively short-lived and last for just days at a time, but others endure for years. Jupiter's most prominent feature, the Great Red Spot, is an enormous high-pressure storm that may have first been sighted from Earth over 340 years ago.

**THE GREAT RED SPOT**
This giant storm, which is bigger than Earth, is constantly changing its size, shape, and colour. It rotates anticlockwise every six to seven days.

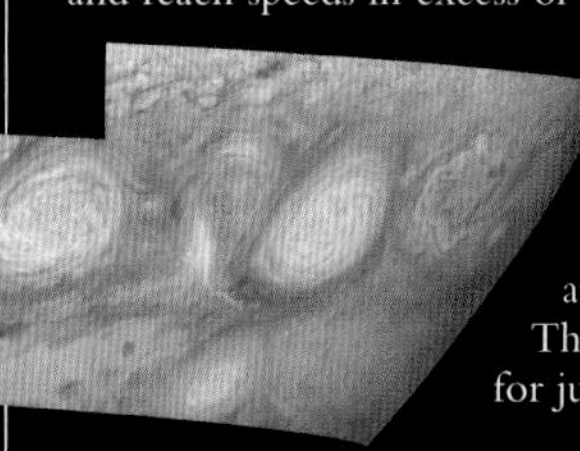

**WHITE OVALS**
Two white oval storms, formed in 1938, appear purple in this false-colour image. A turbulent system lies between them. White ovals remain fixed in latitude, in this case 30°S, but drift in longitude.

EXPLORING SPACE

## DEATH OF A COMET

Comet Shoemaker–Levy 9 was discovered orbiting Jupiter in March 1993. Unusually, it wasn't a single object but a string of 22 cometary chunks. Astronomers calculated that in July 1992 the comet had been torn apart by Jupiter's gravitational pull, and they realized that the fragments were on a collision course with the planet. In July 1994, the fragments hurtled into Jupiter's atmosphere, each impact being followed by an erupting fireball of hot gas.

**HEADING FOR DESTRUCTION**
Shoemaker–Levy 9 fragments orbit Jupiter in May 1994, just weeks before they slammed into the planet's atmosphere. A cloud of gas and dust surrounds each fragment.

# RINGS

Jupiter's ring system was revealed for the first time in an image taken by Voyager 1 in 1979. It is a thin, faint system made of dust-sized particles knocked off Jupiter's four inner moons. The system consists of three parts. The main ring is flat and is about 7,000km (4,350 miles) wide and less than 30km (18 miles) thick. Outside this is the flat gossamer ring, which is 850,000km (528,000 miles) wide and stretches beyond Amalthea to Thebe's orbit. On the inside edge of the main ring is the 20,000km- (12,400-mile-) thick doughnut-shaped halo. Its tiny dust grains reach down to Jupiter's cloud tops.

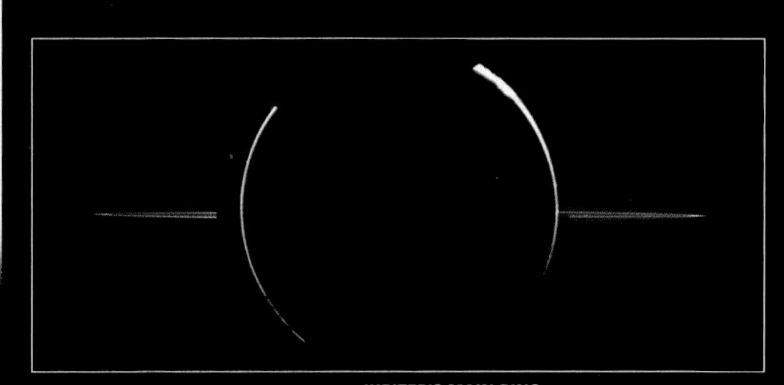

**JUPITER'S MAIN RING**
Jupiter's main ring was imaged by Galileo with the Sun behind the planet. From this position, small particles within the ring and in Jupiter's upper atmosphere stand out. The halo and gossamer ring are revealed only if the main ring is overexposed.

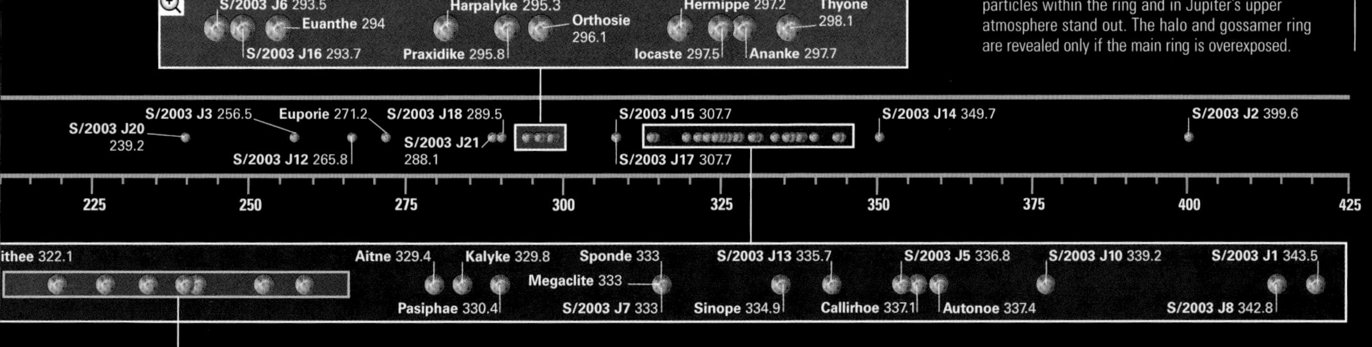

# JUPITER'S MOONS

Jupiter's moons fall into three categories: the four inner moons; the four large Galilean moons; and the rest, the small outer moons. The inner and Galilean moons orbit in the usual direction, that is the same direction as Jupiter's spin – anticlockwise if viewed from above the north pole. Most of the outer moons travel in the opposite direction, suggesting that they originated from an asteroid that fragmented after it was captured by Jupiter's gravitational field.

**SO NEAR AND YET SO FAR**
Io, one of the largest of Jupiter's 63 moons, appears close to its planet, but the two are almost three times the diameter of Jupiter apart.

INNER MOON

## Metis

**DISTANCE FROM JUPITER** 127,960km (79,460 miles)

**ORBITAL PERIOD** 6 hours 58 minutes

**DIAMETER** 40km (25 miles)

Metis, the closest moon to Jupiter, is irregular in shape and lies within the planet's main ring. It was discovered on 4 March 1979 by the Voyager 1 probe. Metis is named after the first wife of Zeus, who was swallowed by him when she became pregnant.

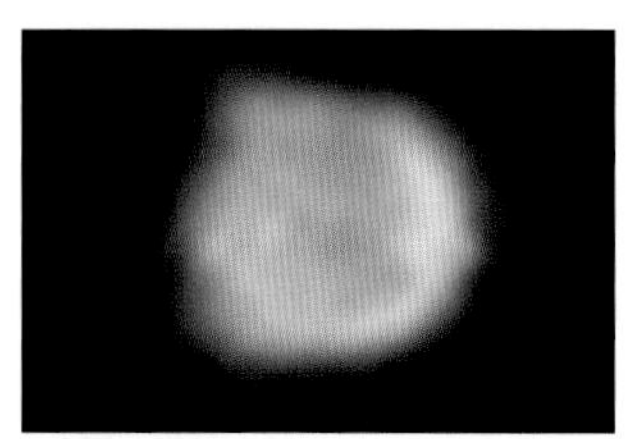

**JUPITER'S CLOSEST MOON**

INNER MOON

## Adrastea

**DISTANCE FROM JUPITER** 128,980km (80,100 miles)

**ORBITAL PERIOD** 7 hours 9 minutes

**LENGTH** 26km (16 miles)

**ADRASTEA**

The small, irregularly shaped Adrastea is the second moon out from Jupiter and lies within its main ring. For each orbit of Jupiter, Adrastea spins once on its axis, so the same side of the moon always faces the planet. This synchronous rotation is also exhibited by Adrastea's three closest neighbours, Metis, Amalthea, and Thebe. Adrastea was discovered by Voyager 2 in July 1979, and is named after a nymph of Crete into whose care, according to Greek mythology, the infant Zeus was entrusted.

INNER MOON

## Thebe

**DISTANCE FROM JUPITER** 221,900km (137,800 miles)

**ORBITAL PERIOD** 16 hours 5 minutes

**LENGTH** 110km (68 miles)

The most distant of the inner moons, Thebe is named after an Egyptian king's daughter who was a grand-daughter of Io. The moon, which was discovered on 5 March 1979 by Voyager 1, lies within the outer part of the Gossamer Ring (see p.179).

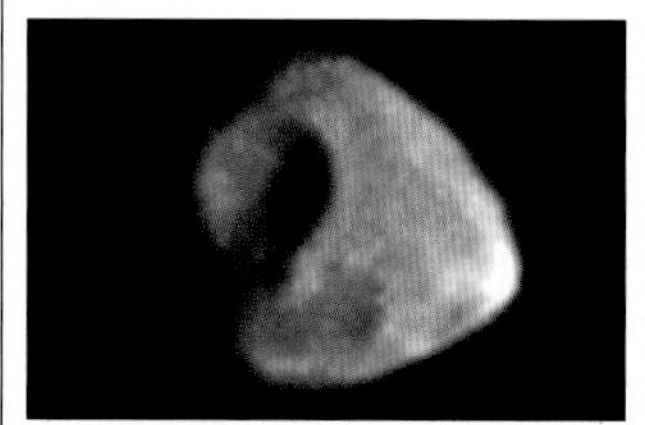

**THEBE SHOWING IMPACT CRATER**

INNER MOON

## Amalthea

**DISTANCE FROM JUPITER** 181,300km (112,590 miles)

**ORBITAL PERIOD** 11 hours 46 minutes

**LENGTH** 262km (163 miles)

The largest of Jupiter's inner moons and the third from the planet, Amalthea is named after the nurse of newborn Zeus. The irregularly shaped moon lies within the Gossamer Ring and is believed to be a source of ring material. Meteoroids from outside the Jovian system collide with Amalthea and the other inner moons, chipping off flecks of dust, which then become part of the ring system. Amalthea's unexpected discovery on 9 September 1892, over 280 years after the four, much larger Galilean moons had been discovered, was headline news.

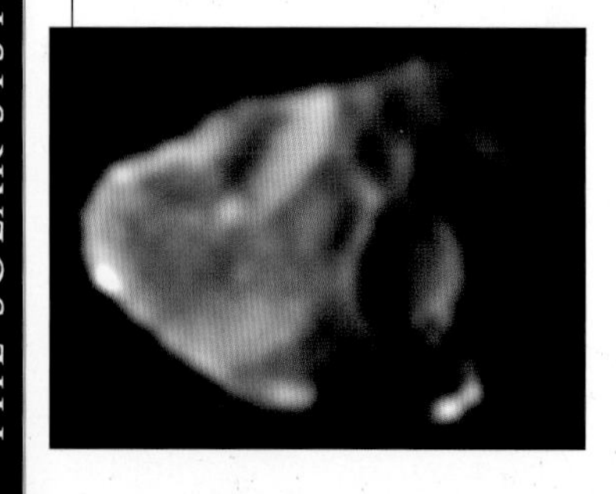

**BARNARD'S TELESCOPE**
Amalthea was the last of Jupiter's moons to be discovered by direct visual observation (as opposed to photography). Its discoverer, the American Edward Barnard, used a 91cm (36in) refractor telescope, which is preserved at the Lick Observatory, California, USA.

**BATTERED SURFACE**
The circular feature in this image is Pan, which, with a diameter of about 90km (56 miles), is the largest impact crater on Amalthea. The bright spot below Pan is associated with another, smaller crater, Gaea (bottom).

OUTER MOON

## Themisto

**DISTANCE FROM JUPITER** 7.5 million km (4.66 million miles)

**ORBITAL PERIOD** 130 Earth days

**DIAMETER** 8km (5 miles)

In November 2000, astronomers at the Mauna Kea Observatory, Hawaii, carried out a systematic search for new moons and identified 11 small moons. Observations recorded on subsequent nights revealed that one of the 11, since named Themisto, was a moon that had been discovered by American astronomer Charles Kowal on 30 September 1975, but then lost.

**THEMISTO REDISCOVERED**
This digital image is one of a series that shows Themisto (highlighted) and its changing position against the background stars, which led to its rediscovery in November 2000.

GALILEAN MOON

## Europa

**DISTANCE FROM JUPITER** 670,900km (416,630 miles)

**ORBITAL PERIOD** 3.55 Earth days

**DIAMETER** 3,122km (1,939 miles)

Europa is an ice-covered ball of rock, which has been studied for about 400 years but whose intriguing nature was only fully revealed once the Galileo space probe started its study in 1996. The probe was named after the Italian scientist Galileo Galilei (see p.88), who observed Europa, along with the three other moons that collectively bear his name, in January 1610, from Padua, Italy. The German astronomer Simon Marius (1573–1624) is believed to

**DAYTIME TEMPERATURE**
Infrared observations reveal heat radiation from Europa's surface at midday. Temperatures at the equator (shown here as yellow) reach about -140°C (-225°F). Farther away from the equator, the surface temperatures are even lower.

have observed the moons first, but it was Galileo who published his findings and brought the moons to the attention of the scientific and wider community.

Jupiter's fourth-largest satellite is a fascinating world. It is a little smaller than Earth's Moon, but much brighter, as its icy surface reflects five times as much light. A liquid sea may lie below Europa's water-ice crust, which is just tens of kilometres thick. This watery layer, which is estimated to be 80–170km (50–105 miles) deep and to contain more liquid than Earth's oceans combined, could be a haven for life. Below lies a rocky mantle surrounding a metallic core.

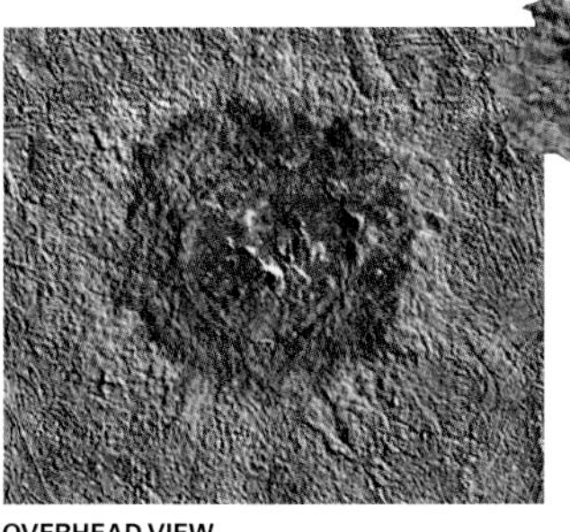

OVERHEAD VIEW

TERRAIN MODEL

**PWYLL CRATER**
This three-dimensional model of the 26km- (16-mile-) wide Pwyll Crater (above) was made by combining images (see example, left) taken from different angles and then applying colour. Unusually, the crater floor (blue) is the same height as the moon's surface, and the central peak (red) is much higher than the crater's rim.

The surface appears to be geologically young and consists of smooth ice plains, disrupted terrain, and regions criss-crossed by dark linear structures that can be thousands of kilometres long.

The mottled appearance of the disrupted terrain comes from crust that has broken up and floated into new positions. Round or oblong, city-sized dark spots freckle the surface. Known as lenticulae, these form as large globules of warm and slushy ice push up from underneath and briefly melt the surface ice. Exactly how the dark lines formed is unclear, but volcanically heated water and ice and other kinds of tectonic activity were involved. Tidal forces fractured the crust, and liquid or icy water erupted through the crack to freeze almost instantly on the surface.

In Greek myth, Europa was the girl who was seduced by Zeus in the form of a white bull and carried off to Crete.

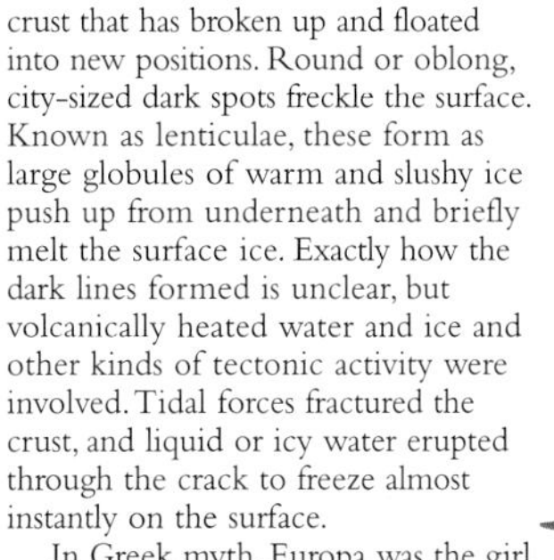

fractures in crust

Pwyll Crater

**IMAGE OF THE FAR SIDE**
This is how the far side of Europa would appear to the human eye. Bright plains in the polar areas (top and bottom) sandwich a darker, disrupted region of the crust.

EXPLORING SPACE

## KEEP EUROPA TIDY

After a six-year journey from Earth, the Galileo space probe spent eight years studying the Jovian system and made 11 close flybys of Europa. The decision was made to destroy the probe because NASA wanted to avoid an impact with Europa and the potential contamination of its subsurface ocean, which could possibly harbour life. With little fuel left, Galileo was put on a collision course with Jupiter. The probe disintegrated in the planet's atmosphere on 21 September 2003.

GALILEO

**ICY SURFACE**
This area of Europa's northern hemisphere shows features typical of the moon's icy surface. Brown grooves and ridges slice across a blue-grey water-ice surface freckled by lenticulae. The colours in this mosaic of images taken by Galileo have been enhanced to reveal detail.

GALILEAN MOON

## Io

**DISTANCE FROM JUPITER** 421,600km (261,800 miles)

**ORBITAL PERIOD** 1.77 Earth days

**DIAMETER** 3,643km (2,262 miles)

Io is a little larger and denser than Earth's Moon, and orbits Jupiter at a distance only slightly greater than the Moon's from Earth. But there the similarities end. Io is a highly coloured world of volcanic pits, calderas and vents, lava flows, and high-reaching plumes. The moon's nature was revealed first by the two Voyager probes and then more fully explored by the Galileo mission. Prior to Voyager 1's arrival in March 1979, scientists expected to find a cold, impact-cratered moon. Instead, it found the most volcanic body in the Solar System.

Io has a thin silicate crust that surrounds a molten silicate layer. Below this lies a comparatively large iron-rich core that extends about halfway to the surface. Io orbits Jupiter quickly, every 42.5 hours or so. As it orbits, it is subjected to the strong gravitational pull of Jupiter on one side and the lesser pull of Europa on the other. Io's surface flexes as a consequence of the varying strength and direction of the pull it experiences. The flexing is accompanied by friction, which produces the heat that keeps part of Io's interior molten. It is this material that erupts through the surface and constantly renews it.

The evidence of such volcanism is seen all over Io. Over 80 major active volcanic sites and more than 300 vents have been identified. Features known as plumes are also found at the surface; these fast-moving and long-lived columns of cold gas and frost grains are more like geysers than volcanic explosions. They are created as superheated sulphur dioxide shoots through fractures in Io's crust. The material in the plumes falls slowly back to the surface as snow and leaves circular or oval frost deposits. Plume material also spreads into space surrounding Io and supplies a doughnut-shaped body of material that has formed along Io's orbital path. Temperatures at the volcanic hotspots can be over 1,230°C (2,240°F), the highest surface temperatures in the Solar System outside the Sun. Elsewhere the surface is cold, reaching just -153°C (-244°F).

Simon Marius (see p.180) suggested the names of the Galilean moons. Io is named after one of Zeus's loves whom he changed into a cow to hide her from his jealous wife. Hera was not fooled and sent a gadfly to torment Io forever. Other surface features are named after people and places from the Io myth or from Dante's *Inferno*, or after fire, sun, volcano and thunder gods, goddesses, and heroes.

**JUPITERSHINE**
Sunlight reflected off Jupiter illuminates Io's western side. The eastern side is in shadow but for a burst of light beyond the limb where the plume of the volcano Prometheus is lit. The yellowish sky is produced by sodium atoms surrounding Io scattering the sunlight.

**VOLCANIC ACTIVITY**
In this colour-enhanced Galileo image, the dark spots on Io's surface are active volcanic centres. The dark eruptive area of Prometheus at centre left is encircled by a pale yellow ring of sulphur-dioxide snow deposited by the volcano's plume.

**TOHIL MONS**
Non-volcanic mountains are also found on Io. Here, the sunlit peak of the 300km- (185-mile-) wide Tohil Mons rises 5.4km (3.4 miles) above Io's surface.

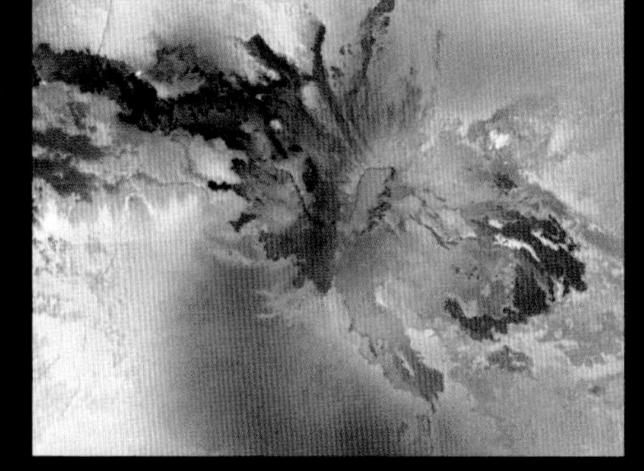

**CULANN PATERA**
Colourful lava flows stream away from the irregularly shaped green-floored volcanic crater of Culann Patera (right of centre). The reasons for the varied colours are uncertain. The diffuse red material is thought to be a compound of sulphur deposited from a plume of gas. The green deposits may be formed when sulphur-rich material coats warm silicate lava.

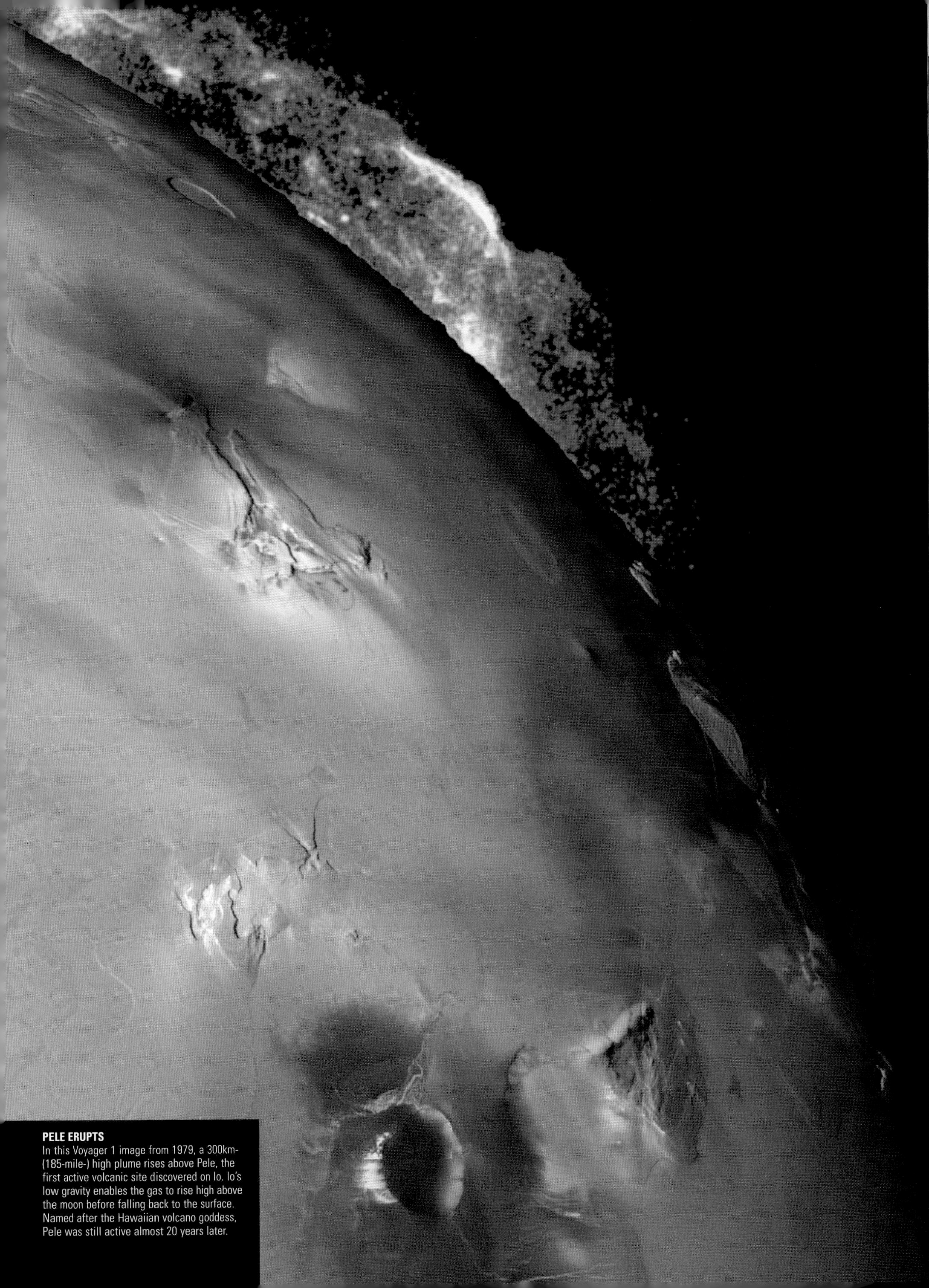

**PELE ERUPTS**
In this Voyager 1 image from 1979, a 300km- (185-mile-) high plume rises above Pele, the first active volcanic site discovered on Io. Io's low gravity enables the gas to rise high above the moon before falling back to the surface. Named after the Hawaiian volcano goddess, Pele was still active almost 20 years later.

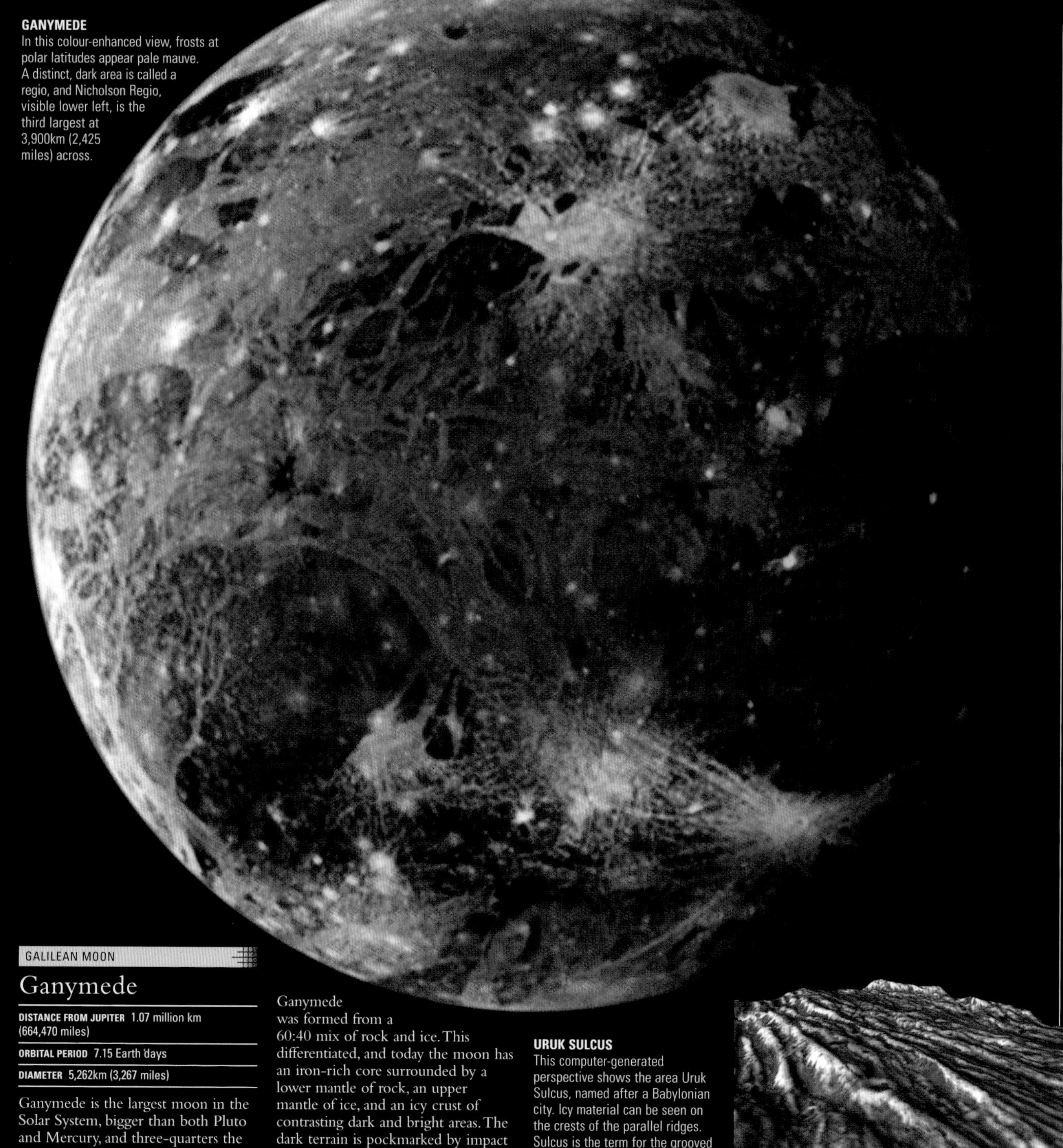

**GANYMEDE**
In this colour-enhanced view, frosts at polar latitudes appear pale mauve. A distinct, dark area is called a regio, and Nicholson Regio, visible lower left, is the third largest at 3,900km (2,425 miles) across.

GALILEAN MOON

## Ganymede

**DISTANCE FROM JUPITER** 1.07 million km (664,470 miles)

**ORBITAL PERIOD** 7.15 Earth days

**DIAMETER** 5,262km (3,267 miles)

Ganymede is the largest moon in the Solar System, bigger than both Pluto and Mercury, and three-quarters the size of Mars. It is named after the beautiful young boy in Greek myth who was taken to Olympus by Zeus and became cupbearer to the Gods.

Ganymede was formed from a 60:40 mix of rock and ice. This differentiated, and today the moon has an iron-rich core surrounded by a lower mantle of rock, an upper mantle of ice, and an icy crust of contrasting dark and bright areas. The dark terrain is pockmarked by impact craters, suggesting that it is an older surface. Circular bright areas termed palimpsests are the smoothed-out and filled-in remains of craters formed on the icy surface in the distant past. The dark terrain is also characterized by long depressions about 7km (4 miles) wide, called furrows. These may have formed as subsurface ice flowed into recently formed craters and material dragged across the surface created the bow-shaped troughs.

The bright terrain, which is rich in water ice with patches of carbon-dioxide ice, is generally smoother and marked by fewer craters. It is crisscrossed by ridges and grooves produced by the tectonic stretching of the moon's surface.

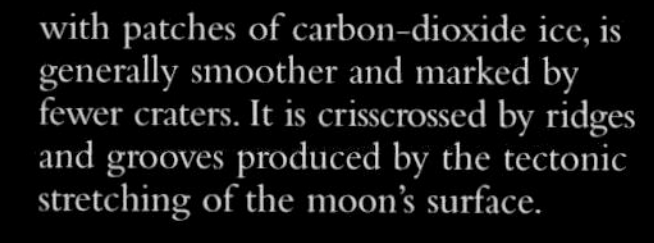

**URUK SULCUS**
This computer-generated perspective shows the area Uruk Sulcus, named after a Babylonian city. Icy material can be seen on the crests of the parallel ridges. Sulcus is the term for the grooved and ridged regions of bright terrain.

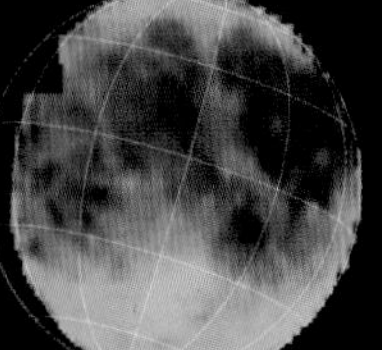

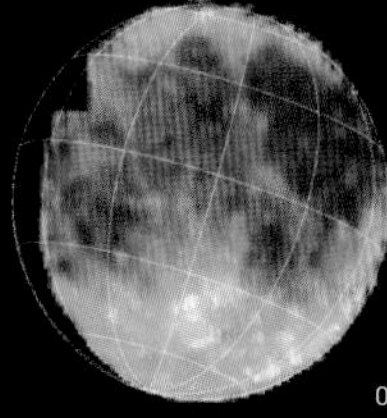

**INFRARED MAPPING**
The infrared image taken by Galileo on the left locates surface water-ice – the brighter the shading, the greater the amount. The colours of the right-hand image indicate the location of minerals (red) and the size of ice grains (shades of blue).

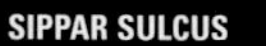

**SIPPAR SULCUS**
This depression within Sippar Sulcus appears to be an old caldera (a volcano's collapsed underground reservoir) containing frozen lava.

GALILEAN MOON

## Callisto

**DISTANCE FROM JUPITER** 1.88 million km (1.17 million miles)

**ORBITAL PERIOD** 16.69 Earth days

**DIAMETER** 4,821km (2,994 miles)

The most distant, second-largest, and darkest of the Galilean moons, Callisto is still brighter than Earth's Moon as its surface contains ice that reflects sunlight. Callisto has undergone little internal change since its formation. Its original mix of rock and ice is only partly differentiated, so that the moon is rockier towards its centre and icier towards its crust. The surface, scarred by craters and multi-ringed structures created by meteorite impacts, bears few signs of geological activity. Callisto does not appear to have been shaped by plate tectonics or cryovolcanism, where ice behaves like volcanic lava; although the ice has eroded the rock in places, causing crater rims to be worn away and sometimes collapse.

**TINDR CRATER**
The partial collapse of the rim of this 76km- (47-mile-) wide crater and its pitted floor are probably the result of erosion by ice.

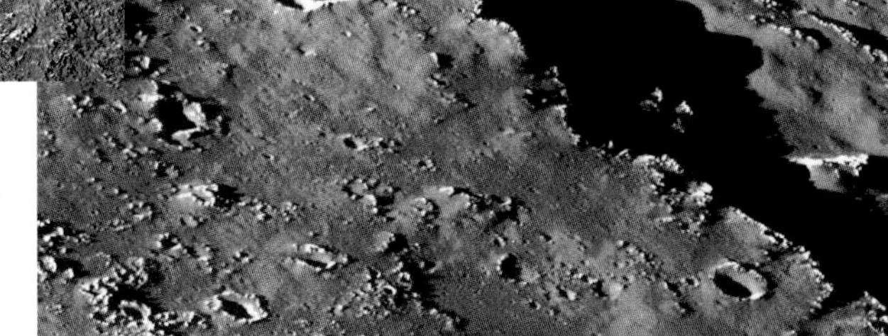

**VALHALLA REGION**
This photograph of part of the Valhalla Basin, lit by sunlight streaming in from the left, shows a 10km- (6-mile-) wide fault scarp, part of Valhalla's ring system. The smallest craters visible are about 155m (510ft) across.

*dark areas lack ice*

*ice on crater rim and floor shines brightly*

**SCARRED SURFACE**
This is the only complete global colour image of Callisto obtained by Galileo. The surface is uniformly cratered, and the bright impact scars are easily visible against its otherwise dark, smooth surface.

The craters are named after heroes and heroines of northern myths, and the large, ringed features, such as the Valhalla Basin (see below), after homes of the gods or heroes. About 2,600km (1,600 miles) across, the Valhalla Basin was probably formed by a large meteorite strike early in Callisto's history, which fractured the cold, brittle crust, allowing ice that was previously below the surface to flood the impact site.

MYTHS AND STORIES

### CALLISTO

Callisto was a beautiful follower of the huntress Artemis, who was seduced by Zeus and bore him a son. According to one myth, Zeus's jealous wife, Hera, turned Callisto into a bear. One day, Callisto came across her son, Arcas, now grown. Fearful for his life, Arcas was only stopped from killing Callisto by Zeus, who raised a whirlwind that carried the pair up into the sky. Callisto became the constellation Ursa Major and Arcas formed Boötes.

**MULTI-RING BASIN**
The multi-ringed Valhalla Basin dominates Callisto's surface. The bright, ice-covered central zone is about 600km (370 miles) across. It is surrounded by rings, which are troughs about 50km (30 miles) apart.

# SATURN

36–37 Gravity, motion, and orbits
64–65 Planetary motion
108–111 Beyond Earth
118–119 The family of the Sun

SATURN IS THE SECOND-LARGEST PLANET and the sixth from the Sun – it is the most distant planet normally visible to the naked eye. A huge ball of gas and liquid, Saturn has a bulging equator and an internal energy source. With a composition dominated by hydrogen, it is the least dense of all the planets. A spectacular system of rings encircles the planet itself, and it also has a large family of moons.

## ORBIT

Saturn takes 29.46 Earth years to complete one orbit of the Sun. It is tilted to its orbital plane by 26.7°, a little more than Earth's axial tilt. This means that as Saturn moves along its orbit, the north and south poles take turns to point towards the Sun. The changing orientation of Saturn to the Sun is seen from Earth by the apparent opening and closing of the planet's ring system. The rings are seen edge-on, for example, at the start of an orbital period. Then, an increasing portion of the rings is seen from above as the North Pole tips towards the Sun. The rings slowly close up and disappear from view as the North Pole starts to tip away, until 14.73 Earth years (half an orbit) later, they appear edge-on again. Now the South Pole tips sunwards and the rings are seen increasingly from below. They close up once again as the South Pole turns away, until they are seen edge-on once more, as the orbit is completed. The strength of the Sun at Saturn is only about 1 per cent of that received on Earth, but it is enough to generate seasonal smog. Saturn is at perihelion at the time the South Pole is facing the Sun.

NORTHERN SPRING EQUINOX
axis tilts from the vertical by 26.7°
spins on its axis every 10.66 hours
NORTHERN SUMMER SOLSTICE
NORTHERN WINTER SOLSTICE
APHELION 1.51 billion km (938 million miles)
PERIHELION 1.35 billion km (838 million miles)
Sun
Saturn orbits the Sun in 29.46 Earth years
NORTHERN AUTUMN EQUINOX

**SPIN AND ORBIT**
Saturn spins on its axis as it orbits. The rapid spin flings material outwards with the result that Saturn is about 10 per cent wider at its equator than its poles. Its bulging equator is bigger than that of any other planet.

## STRUCTURE

Saturn's mass is only 95 times that of Earth's yet 764 Earths could fit inside it. This is because Saturn is composed in the main of the lightest elements, hydrogen and helium, which are in both gaseous and liquid states. Saturn is the least dense of all the planets. If it were possible to put Saturn in an ocean of water, it would float. The planet has no discernible surface: its outer layer is gaseous atmosphere. Inside the planet, pressure and temperature increase with depth and the hydrogen and helium molecules are forced closer and closer together until they become fluid. Deeper still, the atoms are stripped of their electrons and act as a liquid metal. Electric currents within this region generate a magnetic field 71 per cent the strength of Earth's (see p.139).

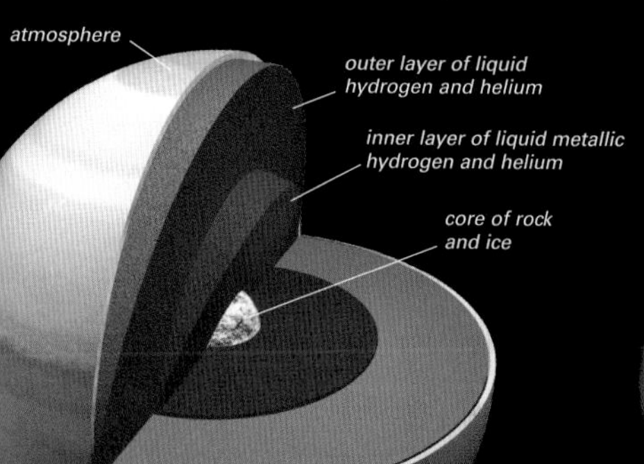


**SATURN'S INTERIOR**
A thin, gaseous atmosphere surrounds a vast shell of liquid hydrogen and helium. The central core is about 10–20 times the mass of Earth.

**RINGLEADER**
Girdled by its bright main rings, Saturn has a hazy, muted appearance in this Voyager 2 image, which shows the planet in its natural colours.

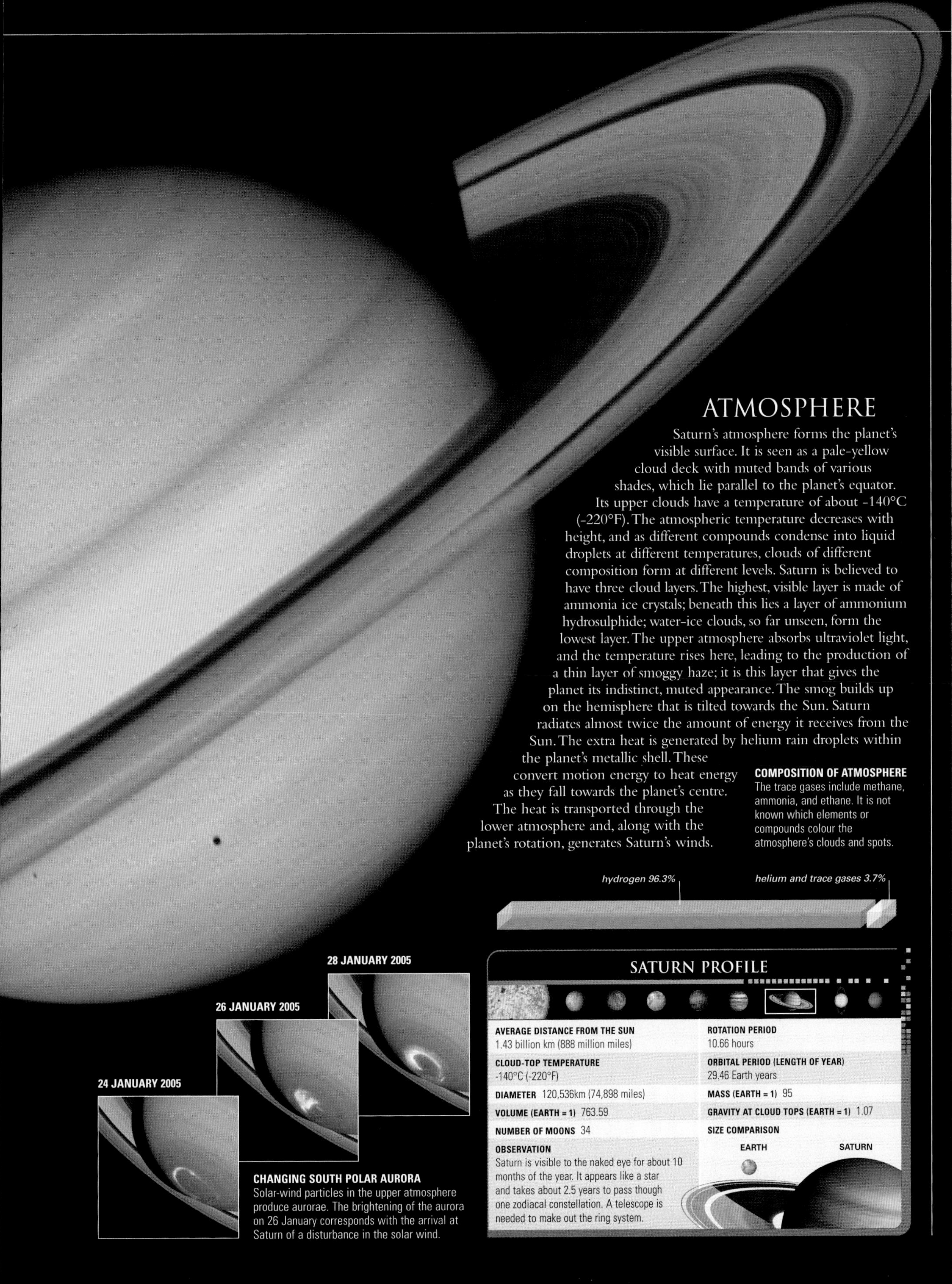

# ATMOSPHERE

Saturn's atmosphere forms the planet's visible surface. It is seen as a pale-yellow cloud deck with muted bands of various shades, which lie parallel to the planet's equator. Its upper clouds have a temperature of about -140°C (-220°F). The atmospheric temperature decreases with height, and as different compounds condense into liquid droplets at different temperatures, clouds of different composition form at different levels. Saturn is believed to have three cloud layers. The highest, visible layer is made of ammonia ice crystals; beneath this lies a layer of ammonium hydrosulphide; water-ice clouds, so far unseen, form the lowest layer. The upper atmosphere absorbs ultraviolet light, and the temperature rises here, leading to the production of a thin layer of smoggy haze; it is this layer that gives the planet its indistinct, muted appearance. The smog builds up on the hemisphere that is tilted towards the Sun. Saturn radiates almost twice the amount of energy it receives from the Sun. The extra heat is generated by helium rain droplets within the planet's metallic shell. These convert motion energy to heat energy as they fall towards the planet's centre. The heat is transported through the lower atmosphere and, along with the planet's rotation, generates Saturn's winds.

**COMPOSITION OF ATMOSPHERE**
The trace gases include methane, ammonia, and ethane. It is not known which elements or compounds colour the atmosphere's clouds and spots.





**CHANGING SOUTH POLAR AURORA**
Solar-wind particles in the upper atmosphere produce aurorae. The brightening of the aurora on 26 January corresponds with the arrival at Saturn of a disturbance in the solar wind.

## SATURN PROFILE

| | |
|---|---|
| **AVERAGE DISTANCE FROM THE SUN** 1.43 billion km (888 million miles) | **ROTATION PERIOD** 10.66 hours |
| **CLOUD-TOP TEMPERATURE** -140°C (-220°F) | **ORBITAL PERIOD (LENGTH OF YEAR)** 29.46 Earth years |
| **DIAMETER** 120,536km (74,898 miles) | **MASS (EARTH = 1)** 95 |
| **VOLUME (EARTH = 1)** 763.59 | **GRAVITY AT CLOUD TOPS (EARTH = 1)** 1.07 |
| **NUMBER OF MOONS** 34 | **SIZE COMPARISON** |

**OBSERVATION**
Saturn is visible to the naked eye for about 10 months of the year. It appears like a star and takes about 2.5 years to pass though one zodiacal constellation. A telescope is needed to make out the ring system.

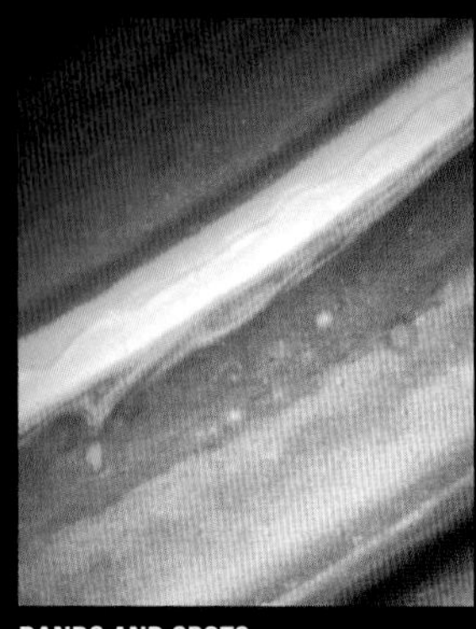

**BANDS AND SPOTS**
Bands of clouds, spots, and ribbon-like features move across Saturn's visible surface. The spots look small but can be thousands of kilometres across.

# WEATHER

Giant upper-atmosphere storms composed of white ammonia ice can be seen from Earth when they rise through the haze. They occur once every 30 years or so, when it is midsummer in the northern hemisphere but, as yet, there is no accepted explanation for the storms. The last of these "Great White Spots" was discovered on 25 September 1990. It spread round the planet, almost encircling the equatorial region over about a month. Smaller, differently coloured oval spots and ribbon-like features have been observed on a more regular basis. In 2004, Cassini revealed a region then dominated by storm activity, nicknamed "storm alley" (see right). Wind speed and direction on the planet are determined by tracking storms and clouds. Saturn's dominant winds blow eastwards, in the same direction as the planet's spin. Near the equator, they reach 1,800kph (1,200mph).

**DRAGON STORM**
The glowing reddish feature (above centre) is a giant, complex thunderstorm, called the Dragon Storm, raging in "storm alley" in the southern hemisphere in 2004. The grey bands are layers of high cloud.

# MOONS

Saturn has 34 confirmed moons. Twenty-five of these have been discovered since 1980, through exploration by the Voyager and Cassini probes and by improved Earth-based observing techniques. Future observations are expected to confirm the presence of more moons. Titan is the largest and was the first to be discovered, in 1655. It is a unique moon, being the only one in the Solar System to have a substantial atmosphere. Saturn's moons are mixes of rock and water ice. Some have ancient, cratered surfaces, and others show signs of resurfacing by tectonics or ice volcanoes. The moons are mostly named after mythological giants. The first to be discovered were named after the Titans, the brothers and sisters of Cronus (Saturn) in Greek mythology. More recent discoveries have Gallic, Inuit, and Norse names.

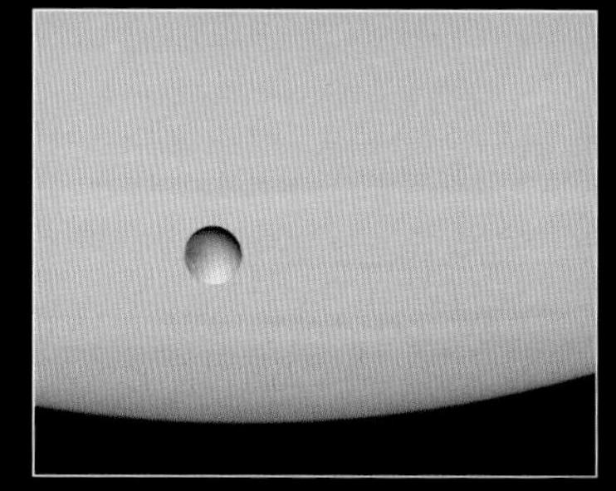

**DIONE**
Cassini produced this image of the moon Dione against the backdrop of Saturn's clouds in December 2005. This natural-colour view reveals the variations in brightness of the moon's icy surface. Dione is Saturn's fourth-largest moon.

**SATURN'S MOONS**

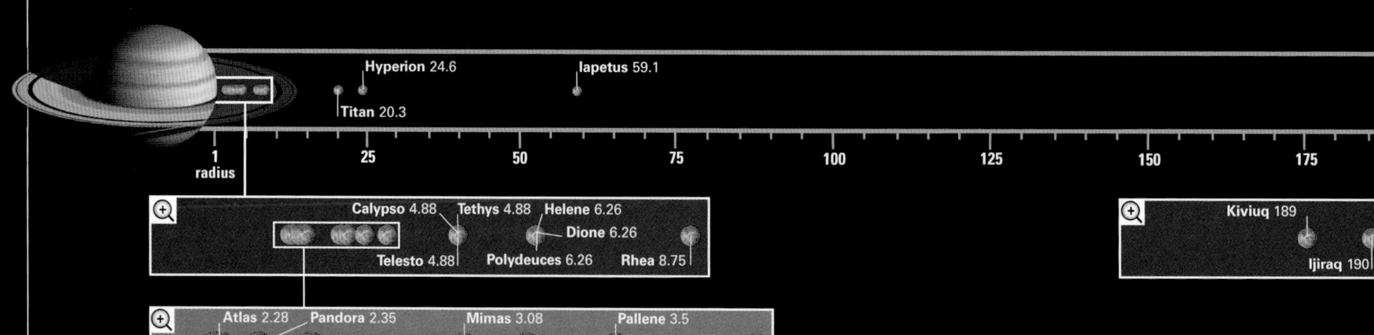

# RINGS

Saturn's rings are the most extensive, massive, and spectacular in the Solar System. They were first observed by Galileo Galilei (see p.88), in 1610, as ear-like lobes either side of Saturn. In 1655, Christiaan Huygens (see p.194) revealed them as a band of material whose appearance changes according to Saturn's position with respect to Earth. The rings are, in fact, collections of separate pieces of dirty water ice that follow individual orbits round Saturn. The pieces range in size from dust grains to boulders several metres across. They are very reflective, so the rings are bright and easy to see. Individual rings are identified by letters, allocated in order of discovery. The readily seen rings are the C, B, and A rings. These are bounded by others, made of tiny particles, that are almost transparent. The thin F ring, the broader G ring, and the diffuse E ring lie outside the main rings. The D ring, inside C, completes the system. The rings change slowly over time, and moons orbiting within the system shepherd particles into rings and maintain gaps such as the Encke gap.

**COMPOSITIONAL DIFFERENCES**
In this ultraviolet image of the outer portion of the C ring (left) and inner B ring (right), red indicates the presence of dirty particles and turquoise indicates purer ice particles.

**A, B, AND C RINGS**
Saturn's sunlit limb is visible through the A, B, and C rings, which range from 5 to 30km (3 to 19 miles) deep. A shadow cast by the rings cuts across the planet's face.

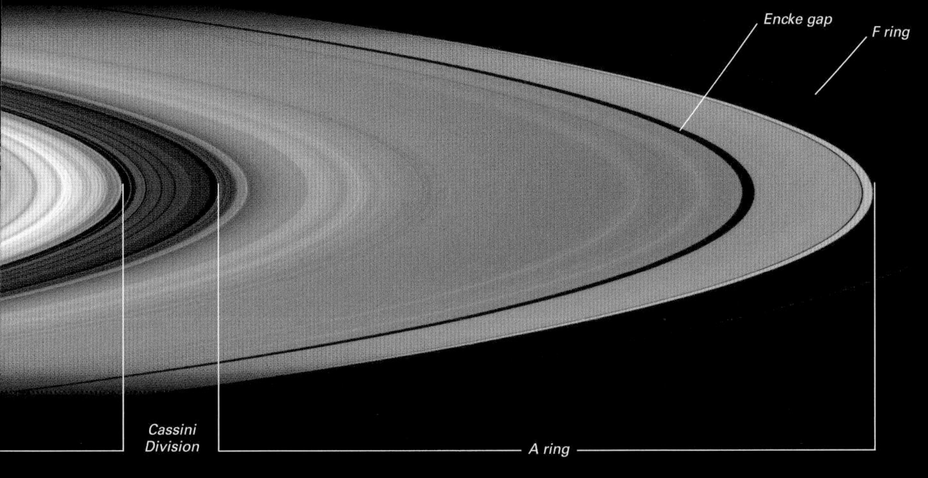


**MAIN RING SYSTEM**
This mosaic of six images shows the main rings in natural colour and reveals the ringlets within the Cassini Division. The distance from the inner edge of C to the F ring is about 65,000km (40,500 miles).

**PROMETHEUS AND THE F RING**
Saturn's innermost moons orbit within the ring system and interact with it. Some act as shepherd moons, confining particles within specific rings. Prometheus (just below the rings in this image) and Pandora work in this way on either side of the F ring.

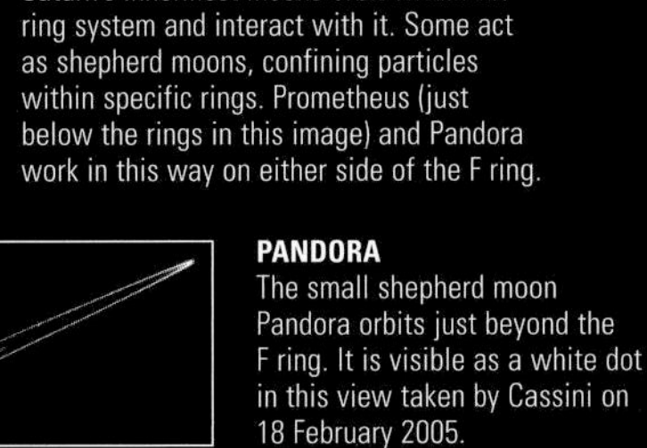

**PANDORA**
The small shepherd moon Pandora orbits just beyond the F ring. It is visible as a white dot in this view taken by Cassini on 18 February 2005.

**THIN RINGS**
Saturn's rings are paper-thin compared to the planet. Cassini imaged them edge-on in February 2005. The bright dot at the left is Dione; the one on the right, within the rings, is Enceladus.

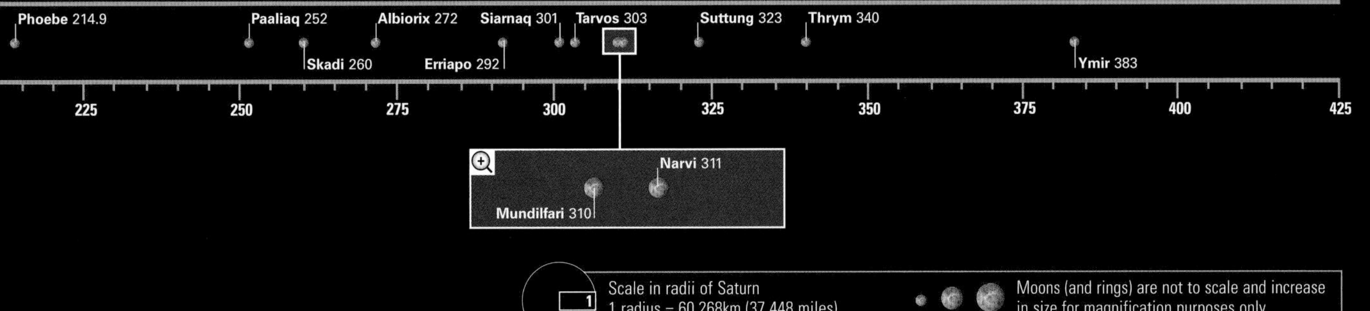

Scale in radii of Saturn
1 radius = 60,268km (37,448 miles)

Moons (and rings) are not to scale and increase in size for magnification purposes only

# SATURN'S MOONS

The moons of Saturn are divided into three groups. The first consists of the major moons, which are large and spherical. The second group, the inner moons, are smaller and irregularly shaped. Members of both these groups orbit within or outside the ring system. The third set of moons lies way beyond the other two – the most distant orbit over 23 million km (14 million miles) from Saturn. These irregularly shaped moons are tiny, just a few kilometres to tens of kilometres across. They have inclined orbits, which suggests that they are captured objects. From Earth, Saturn's moons appear as little more than discs of light, but Voyager and Cassini revealed many of them as worlds in their own right.

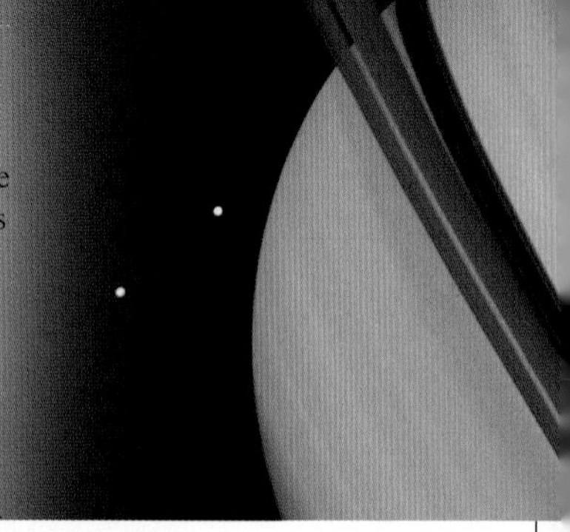

**DWARFED BY SATURN**
Saturn's moons, such as Tethys (top) and Dione (below), are not only small compared to their host planet but, with the exception of Titan, they are all smaller than Earth's Moon.

INNER MOON

## Prometheus

**DISTANCE FROM SATURN** 139,353km (86,539 miles)

**ORBITAL PERIOD** 0.61 Earth days

**LENGTH** 148km (92 miles)

Prometheus is a small, potato-shaped moon orbiting just inside the multi-stranded F ring. It interacts with the ring, and when Cassini imaged ring and moon in 2004 (see image, left), they were connected by a fine thread of material. It is possible that Prometheus is pulling particles out of the F ring.

**SHEPHERD MOON AND RINGLETS**

INNER MOON

## Janus

**DISTANCE FROM SATURN** 151,472km (94,120 miles)

**ORBITAL PERIOD** 0.69 Earth days

**LENGTH** 194km (120 miles)

Heavily cratered and irregularly shaped, Janus orbits Saturn just beyond the F ring and only 50km (30 miles) farther away than its co-orbital moon, Epimetheus. Its existence was first reported in December 1966, and it was named after the Roman god Janus, who could look forward and back at the same time. Yet, it was only confirmed as a moon in February 1980, after Voyager 1 data had been studied.

**BEYOND THE F RING**

INNER MOON

## Epimetheus

**DISTANCE FROM SATURN** 151,422km (94,089 miles)

**ORBITAL PERIOD** 0.69 Earth days

**LENGTH** 138km (86 miles)

Occasionally, moons orbit a planet within about 50km (30 miles) of each other. They are described as co-orbital as they virtually share an orbit. The two moons Epimetheus and Janus (below left), which orbit just beyond the F ring, are such a pair. They swap orbits every four years, taking turns to be slightly closer to Saturn. Epimetheus is a lumpy moon, just 28km (17 miles) longer than it is wide or deep, and it is one of 16 moons that lie within the ring system. Epimetheus is in synchronous rotation – that is, it keeps the same face towards Saturn at all times because its rotation and orbital periods are the same. As it orbits Saturn, it works as a shepherding moon, confining the ring particles within the F ring. Prometheus (left) works in the same way on the inner side of the ring. The existence of Epimetheus was suspected in 1967 but was not confirmed until 26 February 1980. It was one of eight moons discovered in Voyager data that year. The moon is named after a Titan, the family of giants in Greek mythology who once ruled the Earth. Prometheus was one of Epimetheus's five brothers.

**BATTERED SURFACE**
Epimetheus (left) and its co-orbital moon, Janus, are believed to be the remnants of a larger object that was broken apart by an impact.

**CO-ORBITAL MOON**
Epimetheus orbits against the backdrop of Saturn's rings, which are seen nearly edge-on in this view taken by Cassini's narrow-angle camera on 18 February 2005.

INNER MOON

## Pallene

**DISTANCE FROM SATURN** 211,000km (131,000 miles)

**ORBITAL PERIOD** 1.14 Earth days

**LENGTH** 4km (2.5 miles)

Two small moons orbiting between the major moons Mimas and Enceladus were discovered in 2004 in data collected by the Cassini probe. As with all such discoveries, the moons were initially identified by numerical designations (S/2004 S1 and S/2004 S2). The two moons are now known as Methone and Pallene. They were not found by chance but were identified in images taken as part of a search for new moons within this region around Saturn. The contrast of the images was enhanced to increase visibility. Twenty-eight images, including the one at right, acquired over a period of 9.25 hours together make a movie showing Pallene as it progresses along its orbital path around Saturn.

**S/2004 S2**
Pallene is a tiny world, just 4km (2.5 miles) long, which has only been seen as a bright dot. The large, bright object is Saturn, which has been overexposed in an attempt to record new, small, faint moons.

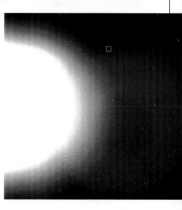

MAJOR MOON

# Mimas

**DISTANCE FROM SATURN** 185,520km (115,208 miles)

**ORBITAL PERIOD** 0.94 Earth days

**DIAMETER** 418km (256 miles)

Mimas is the first of the major moons out from Saturn, and it orbits the planet in the outer part of the ring system. It is in synchronous rotation, so the same side of the moon always faces the planet, in the same way that one side of the Moon always faces Earth. Mimas is a round moon but it is not a perfect sphere – this icy object is about 30km (19 miles) longer than it is wide and deep. Its surface is covered in deep, bowl-shaped impact craters. Many of those greater than about 20km (12 miles) across have central peaks. One crater, Herschel, dwarfs the rest and is the moon's most prominent feature. It is about 130km (80 miles) wide, almost 10km (6 miles) deep, and has a prominent central peak. If the impacting body that formed the crater had been much bigger, it might have smashed the moon apart. The crater is named after the astronomer William Herschel (see p.90), who discovered Mimas on 18 July 1789. It was the sixth of Saturn's moons to be discovered and the first of two discovered by Herschel. Mimas is named after a Titan (see p.188).

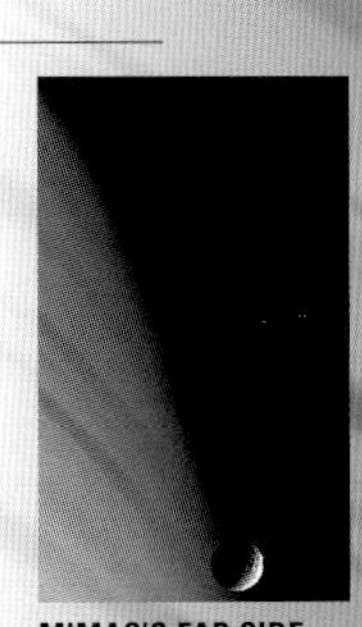

**MIMAS'S FAR SIDE**
Sunlight illuminates part of Mimas as it travels above Saturn. This view shows the side of the moon that always faces away from the planet.

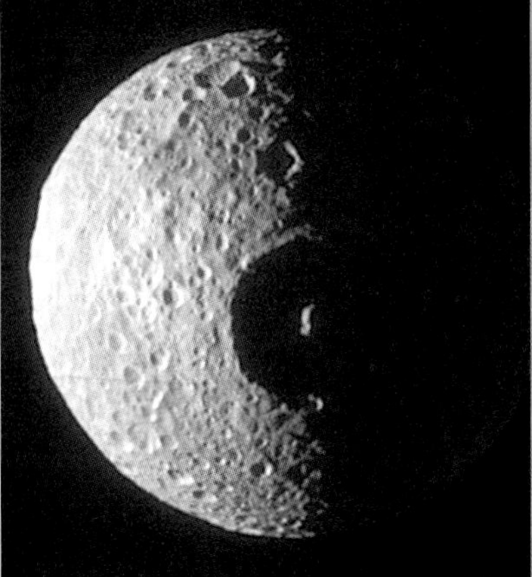

**GIANT CRATER**
The diameter of Herschel Crater is about one third that of Mimas itself. It lies on the moon's leading hemisphere (the side pointing in the direction in which it is moving). Here, the Sun is shining on the central peak from the west.

**TRUE BLUE**
Mimas drifts against the backdrop of Saturn's northern hemisphere in this natural-colour view. Scattering of sunlight in the relatively cloud-free area gives the planet a bluish hue. The dark lines cutting across the atmosphere are shadows cast by Saturn's rings.

MAJOR MOON

## Enceladus

**DISTANCE FROM SATURN** 238,020km (147,898 miles)

**ORBITAL PERIOD** 1.37 Earth days

**DIAMETER** 512km (318 miles)

Enceladus is the 10th moon out from Saturn and orbits within the broad E ring. Its orbit lies within the densest part of the ring, which suggests that Enceladus could be supplying the ring with material. The moon is in synchronous rotation with Saturn. The frosty surface of Enceladus is highly reflective and makes this moon particularly bright, the brightest in the Solar System. The surface terrain suggests that this frigid moon has experienced a long history of tectonic activity and resurfacing. The extent of the geological change is surprising in such a small world – Mimas (see p.191) is about the same size but is inactive. Craters are concentrated in some regions, and elsewhere there are grooves, fractures, and ridges. Images processed to accentuate colour differences have revealed previously unseen detail. The blue colour seen in some fracture walls could be due to the exposure of solid ice, or because the composition or size of particles in the buried ice is different from that on the surface. Enceladus was discovered by William Herschel (see p.90) on 28 August 1789.

**REFLECTIVE SURFACE** The hemisphere of Enceladus that faces away from the Sun can still be seen by Saturnshine, light reflected by Saturn.

**BLUE WALLS** In this image, the walls of a complex fracture system, 85km (53 miles) long, and the narrower fractures that slice through it, appear bluer than the surrounding, heavily cratered terrain.

**SMOOTH PLAINS** This region of smooth plains has a band of chevron-shaped features running across its centre, cut across at the top by a system of crevasses.



**DIVERSE LANDSCAPES** The large number of impact craters indicates that this is one of the oldest terrains on Enceladus. The twisted network of ridges and troughs at top right is younger.

INNER MOON

## Telesto

**DISTANCE FROM SATURN** 294,660km (183,093 miles)

**ORBITAL PERIOD** 1.89 Earth days

**LENGTH** 30km (18.6 miles)

Telesto shares an orbit within the E ring with two other moons: Calypso, which is about the same size as Telesto; and the much larger Tethys (right). Telesto moves along the orbit 60° ahead of Tethys, and Calypso follows 60° behind Tethys. The positions taken on the orbit by the two smaller moons are called the Lagrange points. In these positions, the two small moons can maintain a stable orbit balanced between the gravitational pull of Saturn and that of Tethys. Telesto and Calypso were discovered in 1980; Calypso by Earth-based observation and Telesto in Voyager images. The probe revealed two irregularly shaped moons.

**FIRST-QUARTER MOON** Only half of Telesto's visible hemisphere is lit by the Sun as it orbits below Saturn in this Cassini image taken on 18 January 2005.

**ICY MOON** The pale, icy disc of Tethys was imaged by Cassini on 18 January 2005 as it orbited below Saturn's south polar region, where fierce storms were raging.

MAJOR MOON

## Tethys

**DISTANCE FROM SATURN** 294,660 km (183,093 miles)

**ORBITAL PERIOD** 1.89 Earth days

**DIAMETER** 1,072km (666 miles)

The Italian-French astronomer Giovanni Cassini discovered Tethys on 21 March 1684. Nearly 200 years later, it was discovered that Tethys shares its orbit with two far smaller moons: Telesto (left) and Calypso. Its surface shows that Tethys has undergone tectonic change and resurfacing. Two features stand out. A 400km- (248-mile-) wide impact crater called Odysseus dominates the leading hemisphere. Large but shallow, its original bowl-shape has been flattened by ice flows. The second large feature is the Ithaca Chasma on the side of Tethys facing Saturn. This vast canyon system extends across half of the moon. It may have been formed by tensional fracturing as a result of the impact that produced the Odysseus Crater or when Tethys's interior froze and the moon expanded in size and stretched its surface.

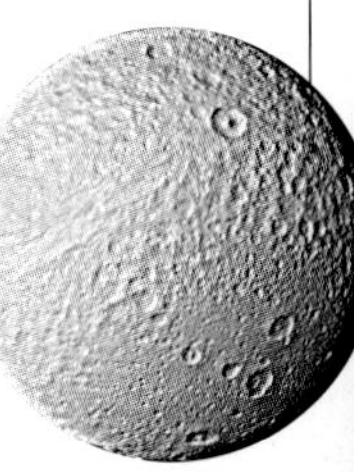

**ITHACA CHASMA** This canyon system, which is up to 4km (2.5 miles) deep, runs from the lower left of the prominent Telemachus Crater (top right).

MAJOR MOON

## Dione

**DISTANCE FROM SATURN** 377,400km (234,505 miles)

**ORBITAL PERIOD** 2.74 Earth days

**DIAMETER** 1,120km (696 miles)

Dione is the most distant moon within Saturn's ring system, but it is not alone in the outer reaches of the E ring. Two other moons, Helene and Polydeuces, follow the same orbit, Helene is ahead of Dione by 60° and Polydeuces follows 60° behind. Helene was discovered in March 1980; Polydeuces was discovered in Cassini data some 24 years later, just after the probe arrived at Saturn. Giovanni Cassini discovered Dione in 1684, on the same day that he discovered Tethys (opposite). Dione has a higher proportion of rock in its rock-ice mix than most of the other moons (only Titan has more), and so it is the second-densest of Saturn's moons. The terrain displays evidence of tectonic activity and resurfacing. There are ridges, faults, valleys, and depressions. There are also craters, which are more densely distributed in some regions than others – Dione's leading face, for example, has more than the trailing face. The largest crater is over 200km (124 miles) across. Dione also has bright streaks on its surface. These wispy features are composed of narrow, bright, icy lines.

**IMPACT CRATERS**
The well-defined central peaks of Dione's largest craters are visible in this Voyager image. Dido Crater lies just left of centre, with Romulus and Remus just above it and Aeneas Crater near the upper limb.

**DIONE'S FAR SIDE**
Impact craters scar the surface of the side of Dione that is permanently turned away from Saturn because it is in synchronous rotation. Areas of wispy terrain are visible on the left of this image.

**CLIFFS OF ICE**
A Cassini close-up of Dione's wispy terrain reveals that it is formed from lines of ice cliffs created by tectonic fractures, rather than deposits of ice and frost as was previously thought.

**ANCIENT SURFACE**
This view of Rhea's trailing side shows a surface pitted by impact craters and, in part, covered by bright, wispy markings. Rhea's surface, along with that of Mimas, is probably the oldest among Saturn's larger moons.

MAJOR MOON

## Rhea

**DISTANCE FROM SATURN** 527,040km (327,487 miles)

**ORBITAL PERIOD** 4.52 Earth days

**DIAMETER** 1,528km (949 miles)

Vast sweeps of ancient cratered terrain cover large parts of Rhea. At first glance, the landscape resembles that seen on Earth's Moon, although Rhea's surface is bright ice. There is some evidence of resurfacing, although not as much as expected for such a large moon. Rhea is Saturn's second-largest moon but other smaller moons, such as its inner neighbours Dione and Tethys, show more resurfacing. It is thought that Rhea froze early in its history and became frigid. Its ice would then have behaved like hard rock. Rhea's craters, for example, are freshly preserved in its icy crust. The craters on other icy moons, such as Jupiter's Callisto (see p.185), have collapsed in the soft, icy crust. Rhea is the 17th moon out from Saturn and the first to lie beyond the ring system. It is named after the Titan Rhea, who was the mother of Zeus in Greek mythology.

**HEAVILY CRATERED**
Rhea's icy surface is heavily cratered, suggesting that it dates back to the period immediately following the formation of the planets. This image shows the region around the moon's North Pole. The largest craters are several kilometres deep.

**FRESH ICE?**
The white areas on the edges of some of these craters are probably fresh ice exposed on steep slopes, or possibly ice and frost deposited by volatile fluids leaking from fractured regions of the moon's surface.

MAJOR MOON

## Titan

**DISTANCE FROM SATURN** 1.22 million km (758,073 miles)

**ORBITAL PERIOD** 15.95 Earth days

**DIAMETER** 5,150km (3,200 miles)

Titan is the second-largest moon in the Solar System after Jupiter's Ganymede (see p.184) and is by far the largest of Saturn's moons. This Mercury-sized body is also one of the most fascinating. A veil of smoggy haze shrouds the moon and permanently obscures the world below. Titan is intriguing, not least because the chemistry of the atmosphere has similarities to that of the young Earth, before life began. The first chance to see the surface and test the atmosphere came in 2005, when Cassini turned its attention to Titan, and Huygens plunged through the atmosphere to the surface (see panel, right).

The nitrogen-rich atmosphere extends for hundreds of kilometres above Titan. Layers of yellow-orange, smog-like haze high within it are the result of chemical reactions triggered by ultraviolet light. Methane clouds form much nearer the surface. These rain methane onto Titan, where it forms rivers and lakes. It then evaporates and forms clouds, and the cycle, which is reminiscent of the water cycle on Earth, continues.

Titan is the densest of Saturn's moons: it is a 50:50 mix of rock and water ice with a surface temperature of -180°C (-292°F). It is a gloomy world because the smog blocks 90 per cent of the incident sunlight. Cassini revealed that its surface is shaped by Earth-like processes – tectonics, erosion, and winds – and perhaps ice volcanism. No liquid methane was detected on the initial flybys but drainage channels and dark elliptical regions, thought to be evaporated lakes, showed where fluids had been. Linear features nicknamed "cat scratches" were also identified.

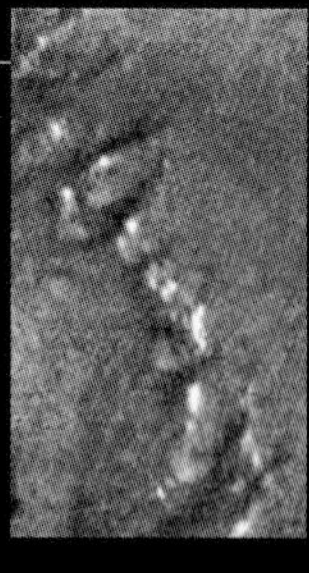

**ISLANDS**
Huygens imaged a dark plain during its descent to the surface. Flow patterns, possibly formed by liquid methane, can be seen around bright "islands".

**TITAN'S ATMOSPHERE**
Infrared and ultraviolet data combined reveals aspects of the atmosphere. Areas where methane absorbs light appear orange and green. The high atmosphere is blue.

**ORANGE AND PURPLE HAZE**
The upper atmosphere consists of separate layers of haze; up to 12 have been detected in this ultraviolet, natural-coloured image of Titan's night-side limb.

EXPLORING SPACE

### THE HUYGENS PROBE

The European probe Huygens travelled to Titan onboard NASA's Cassini spacecraft. Once there, it separated from the larger craft and parachuted into Titan's haze. During its 2.5-hour descent, Huygens tested the atmosphere, measured the speed of the buffeting winds, and took images of the moon's surface. An instrument recorded the first surface touch on 14 January 2005, sending back evidence of a thin, hard crust with softer material beneath.

**HUYGENS AND CASSINI**
The shield-shaped Huygens probe (right) is attached to Cassini's frame in preparation for the launch from Cape Canaveral, Florida, USA, in October 1997.

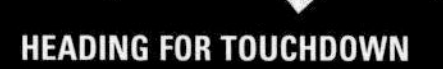

**HEADING FOR TOUCHDOWN**
This is a composite of 30 images taken from altitudes of 13km (8 miles) down to 8km (5 miles) as Huygens descended. River- and channel-like features can be seen at left.

**TITAN REVEALED**
This mosaic of 16 images has been specially processed to remove ground-obscuring haze. The large, bright area at centre right is called Xanadu.

### CHRISTIAAN HUYGENS

As well as being a keen observer of the stars and planets, the Dutch scientist Christiaan Huygens (1629–95) also invented the pendulum clock, constructed first-class telescopes, and proposed the wave theory of light. Huygens discovered Titan, which he referred to as Luna Saturni, on 25 March 1655 while observing Saturn. He also explained the nature of Saturn's rings. The ESA probe to Titan was named in Huygens's honour.

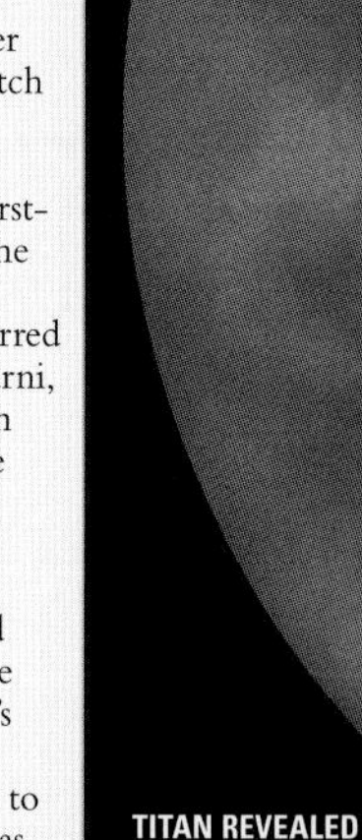

**MOON AND STARS**
During the 82-second exposure required to capture this image, Cassini's cameras kept locked on target as the craft continued to travel. Iapetus is in sharp focus but the stars behind are trails.

MAJOR MOON

# Iapetus

**DISTANCE FROM SATURN** 3.56 million km (2.21 million miles)

**ORBITAL PERIOD** 79.33 Earth days

**DIAMETER** 1,436km (892 miles)

Most of Saturn's inner and major moons orbit in the equatorial plane (also the plane of the rings). Iapetus is an exception, its orbit being inclined by 14.72° to the equatorial plane. Other moons follow orbits with greater inclination but these are the much smaller, outer moons. Iapetus is the 20th moon out from Saturn and the planet's most distant major moon. It is also in synchronous rotation.

Iapetus was discovered by Giovanni Cassini while he was working from Paris on 25 October 1671. He noticed that Iapetus has a naturally dark leading hemisphere and a bright trailing hemisphere. The dark region is called Cassini Regio and is covered in material as dark as coal, in contrast to the icy surface on the bright side. Although the Cassini probe revealed more of the moon's heavily cratered surface, the origin of the dark material remains a mystery. It has been suggested the material erupted from the moon's interior, or that it is ejecta from impacts on a more distant moon, such as Phoebe (right). A unique feature revealed by Cassini has provided another mystery. It is not known whether a 1,300km- (800-mile-) long ridge that coincides almost exactly with the moon's equator is a folded mountain belt or material that erupted through a crack in the surface.

**LANDSLIDE IN CASSINI REGIO**
Land has collapsed down a 15km- (9-mile-) high scarp, which marks the edge of a huge impact crater, into a smaller crater. The long distance travelled by the material along the floor indicates that it could be fine-grained.

**SURFACE COMPOSITION**
False colours represent Iapetus's vastly different surface compositions. Bright blue signifies an area rich in water ice, dark brown indicates a substance rich in organic material, and the yellow region is composed of a mix of ice and organic chemicals.

INNER MOON

# Hyperion

**DISTANCE FROM SATURN** 1.48 million km (919,620 miles)

**ORBITAL PERIOD** 21.28 Earth days

**LENGTH** 370km (230 miles)

Nothing about Hyperion is typical. Firstly, it is an irregularly shaped moon with an average width of about 280km (174 miles). This makes it one of the largest non-spherical bodies in the Solar System. Secondly, it follows an elliptical orbit just beyond Saturn's largest moon, Titan (opposite). And, as it orbits, it rotates in a chaotic manner: its rotation axis wobbles, and the moon appears to tumble as it travels. Its surface is cratered, and there are segments of cliff faces. Its shape and scarred surface suggest Hyperion could be a fragment of a once-larger object that was broken by a major impact. Even Hyperion's discovery was not straightforward. Astronomers in the USA and England found the moon independently and within two days of each other in September 1848.

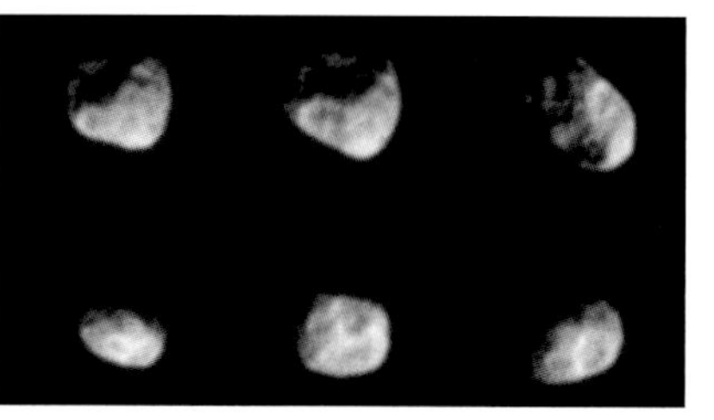

**CHAOTIC TUMBLE**
Cassini viewed Hyperion from various angles as it tumbled along its orbit round Saturn.

OUTER MOON

# Phoebe

**DISTANCE FROM SATURN** 12.95 million km (8.05 million miles)

**ORBITAL PERIOD** 550 Earth days

**LENGTH** 230km (143 miles)

Phoebe was discovered in 1898 and, until 2000, was thought to be Saturn's only outer moon. Fourteen others are now known to exist. Phoebe has a long orbital period and follows a highly inclined orbit, a characteristic of the outer moons. Phoebe's orbit is inclined by 175.3° and so it travels in a retrograde manner (backwards compared to most moons). Half the outer moons orbit this way. Phoebe is by far the largest outer moon; the others are, at most, only 20km (12 miles) across. From Cassini images, it appears to be an ice-rich body coated with a thin layer of dark material.

**CRATER FLOOR**
Debris covers the floor of this impact crater. The streaks inside the crater indicate where loose ejecta has slid down towards the centre.

**BLASTED PHOEBE**
Phoebe's impact craters were revealed by Cassini in June 2004. They are named after the Argonauts in Greek mythology. The largest, Jason, is about 100km (60 miles) across.

# URANUS

36–37 Gravity, motion, and orbits
64–65 Planetary motion
108–11 Beyond Earth
118–19 The family of the Sun

URANUS IS THE THIRD-LARGEST planet and lies twice as far from the Sun as its neighbour Saturn. It is pale blue and featureless, with a sparse ring system and an extensive family of moons. The planet is tipped on its side, and so from Earth the moons and rings appear to encircle it from top to bottom. Uranus was the first planet to be discovered by telescope, but little was known about it until the Voyager 2 spacecraft flew past in January 1986.

**PALE BLUE DISC**
Voyager 2 images have been combined to show the southern hemisphere of Uranus as it would appear to a human onboard the spacecraft.

## ORBIT

Uranus takes 84 Earth years to complete one orbit round the Sun. Its axis of rotation is tipped over by 98°, and the planet moves along the orbital path on its side. Uranus's spin is retrograde, spinning in the opposite direction to most planets. The planet would not have always been like this. Its sideways stance is probably the result of a collision with a planet-sized body when Uranus was young. Each of the poles points to the Sun for 21 years at a time, during the periods centred on the solstices. This means that while one pole experiences a long period of continuous sunlight, the other experiences a similar period of complete darkness. The strength of the sunlight received by the planet is 0.25 per cent of that on Earth. When Voyager encountered Uranus in 1986, its South Pole was pointing almost directly at the Sun. Uranus's equator then became increasingly edge-on to the Sun. After 2007, it will progressively turn away, until the North Pole faces the Sun in 2030.

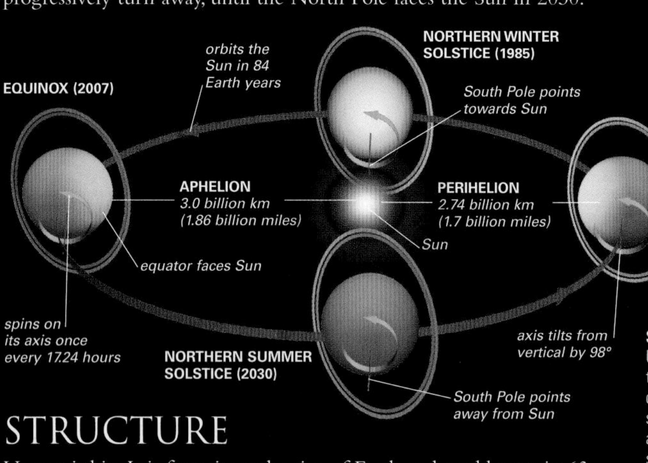


**SPIN AND ORBIT**
Uranus's long orbit and its extreme tilt combine to produce long seasonal differences. Each pole experiences summer when pointing towards the Sun and winter when it is pointing away. At such times, the pole is in the middle of Uranus's disc when viewed from Earth. At the equinoxes, the equator and rings are edge-on to the Sun.

## STRUCTURE

Uranus is big. It is four times the size of Earth and could contain 63 Earths inside it; yet it has only 14.5 times the mass of Earth. So, the material it is made of must be less dense than that of Earth. Uranus is too massive for its main ingredient to be hydrogen, which is the main constituent of the bigger planets, Saturn and Jupiter. It is made mainly of water, methane, and ammonia ices, which are surrounded by a gaseous layer. Electric currents within its icy layer are believed to generate the planet's magnetic field, which is offset by 58.6° from Uranus's spin axis.

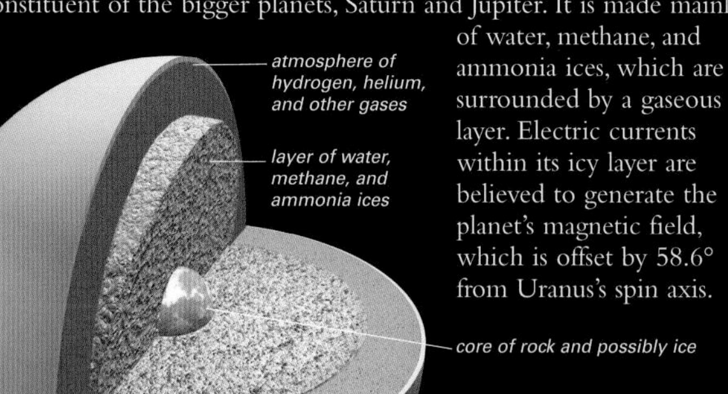


**URANUS INTERIOR**
Uranus does not have a solid surface. The visible surface is its atmosphere. Below this lies a layer of water and ices, which surrounds a small core of rock and possibly ice.

### URANUS PROFILE

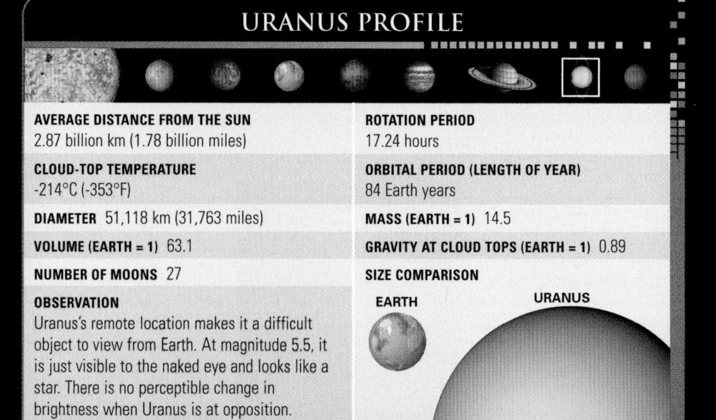

| | |
|---|---|
| **AVERAGE DISTANCE FROM THE SUN** 2.87 billion km (1.78 billion miles) | **ROTATION PERIOD** 17.24 hours |
| **CLOUD-TOP TEMPERATURE** -214°C (-353°F) | **ORBITAL PERIOD (LENGTH OF YEAR)** 84 Earth years |
| **DIAMETER** 51,118 km (31,763 miles) | **MASS (EARTH = 1)** 14.5 |
| **VOLUME (EARTH = 1)** 63.1 | **GRAVITY AT CLOUD TOPS (EARTH = 1)** 0.89 |
| **NUMBER OF MOONS** 27 | **SIZE COMPARISON** EARTH URANUS |
| **OBSERVATION** Uranus's remote location makes it a difficult object to view from Earth. At magnitude 5.5, it is just visible to the naked eye and looks like a star. There is no perceptible change in brightness when Uranus is at opposition. | |

## ATMOSPHERE AND WEATHER

Uranus's blue colour is a result of the absorption of the incoming sunlight's red wavelengths by methane-ice clouds within the planet's cold atmosphere. The cloud-top temperature of -214°C (-353°F) recorded by Voyager 2 appears to be fairly uniform. The action of ultraviolet sunlight on the methane produces haze particles, and these hide the lower atmosphere, making Uranus appear calm. The planet is, however, actively changing. The Voyager 2 data revealed the movement of ammonia and water clouds around Uranus, carried by wind and the planet's rotation. It also revealed that Uranus radiates about the same amount of energy as it receives from the Sun and has no significant internal heat to drive a complex weather system. More recently, observations made using ground-based telescopes have also made it possible for astronomers to track changes in Uranus's atmosphere.

**CLOUDS**
This Keck II telescope infrared image has been processed to show vertical structure. The highest clouds appear white; mid-level ones, bright blue; and the lowest clouds, darker blue. As a by-product, the rings are coloured red.

**COMPOSITION OF ATMOSPHERE**
The atmosphere is made mainly of hydrogen, which extends beyond the visible cloud tops and forms a corona around Uranus.

hydrogen 82.5% | helium 15.2% | methane 2.3%

## RINGS AND MOONS

Uranus has 11 rings that together extend out from 12,400 to 25,600km (7,700–15,900 miles) from the planet. The rings are so widely separated and so narrow that the system has more gap than ring. All but the inner and outer rings are between 1km and 13km (0.6 and 8 miles) wide, and all are less than 15km (9 miles) in height. They are made of charcoal-dark pieces of carbon-rich material measuring from a few centimetres to possibly a few metres across, plus dust particles. The first five rings were discovered in 1977 (see panel, right). The rings do not lie quite in the equatorial plane, nor are they circular or of uniform width. This is probably due to the gravitational influence of small, nearby moons. One of these, Cordelia, lies within the ring system. Uranus has 27 moons. The five major moons were discovered using Earth-based telescopes. Smaller ones have been found since the mid-1980s, through analysis of Voyager 2 data or by using today's improved observing techniques. More discoveries are expected.

**FALSE-COLOUR VIEW OF THE RINGS**
Nine of Uranus's rings are visible in this Voyager 2 image. The faint, pastel lines are due to image enhancement. The brightest, colourless ring (far right) is the outermost ring, epsilon. To its left are five rings in shades of blue-green, then three in off-white.

EXPLORING SPACE

### RINGS DISCOVERED

In March 1977, astronomers onboard the Kuiper Airborne Observatory, an adapted high-flying aircraft, were preparing to observe a rare occultation of a star by Uranus, in order to measure the planet's diameter. Before the star was covered by the planet's disc, it blinked on and off five times. A second set of blinks was recorded after the star appeared from behind the planet. Rings around Uranus had blocked out the star's light.

**KUIPER AIRBORNE OBSERVATORY**
Astronomers and technicians operate an infrared telescope, which looks out to space through an open door in the side of the aircraft.

**URANUS'S MOONS**

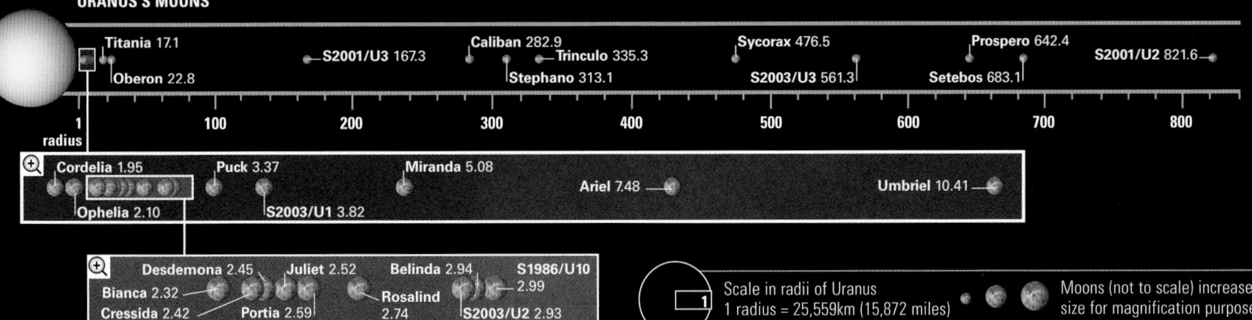

Moons (not to scale) increase in size for magnification purposes

# URANUS'S MOONS

Uranus's moons can be divided into three groups. Moving out from Uranus, they are: the small inner satellites; the five major moons, which orbit in a regular manner; and the small outer moons, many of which follow retrograde orbits. Much of what is known about the moons, and the only close-up views, came from the Voyager 2 flyby in 1985-86. This revealed the major moons to be dark, dense rocky bodies with icy surfaces, featuring impact craters, fractures, and volcanic water-ice flows. The moons are named after characters in the plays of the English dramatist William Shakespeare or in the verse of the English poet Alexander Pope.

**THE VIEW FROM EARTH**
Some of the 27 moons that orbit Uranus can be seen in this infrared image, which was taken by the Hubble Space Telescope in 1998.

INNER MOON

## Cordelia

**DISTANCE FROM URANUS** 49,770km (30,910 miles)

**ORBITAL PERIOD** 0.34 Earth days

**DIAMETER** 40km (25 miles)

Cordelia is the innermost and one of the smallest of Uranus's moons. A team of Voyager 2 astronomers discovered it on 20 January 1986. Cordelia was one of 10 moons that were discovered in the weeks between 30 December 1985 and 23 January 1986 as the Voyager 2 spacecraft flew by Uranus and transmitted images back to Earth. Astronomers had expected to find some more moons in orbit around Uranus. In particular, it was expected that pairs of shepherd moons – moons that are positioned either side of a ring and keep the ring's constituent particles in place – would be found. Surprisingly, just one pair, that of Cordelia and Ophelia, was discovered. Cordelia takes its name from the daughter of Lear in Shakespeare's *King Lear*.

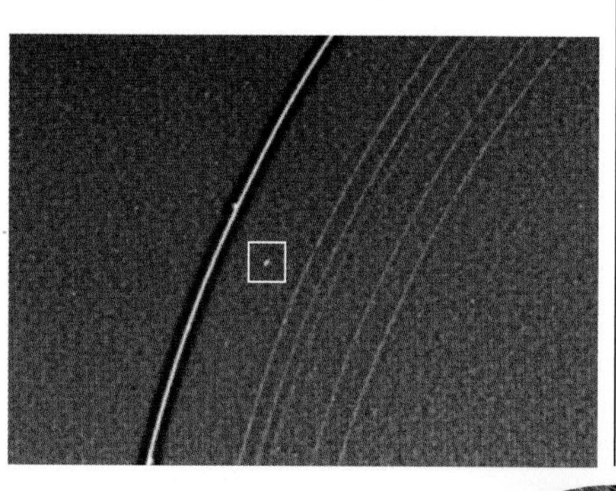

**SHEPHERD MOON**
Cordelia is the innermost of two shepherd moons lying either side of Uranus's bright outer ring.

INNER MOON

## Ophelia

**DISTANCE FROM URANUS** 53,790km (33,400 miles)

**ORBITAL PERIOD** 0.38 Earth days

**DIAMETER** 42km (26 miles)

Ophelia is one of a pair of moons that orbit either side of Uranus's outer ring, the epsilon ring. It was discovered at the same time as its partner, Cordelia, on 20 January 1986. The two are small, not much bigger than the particles that make up the thin, narrow ring. The moon is named after the heroine in Shakespeare's *Hamlet*.

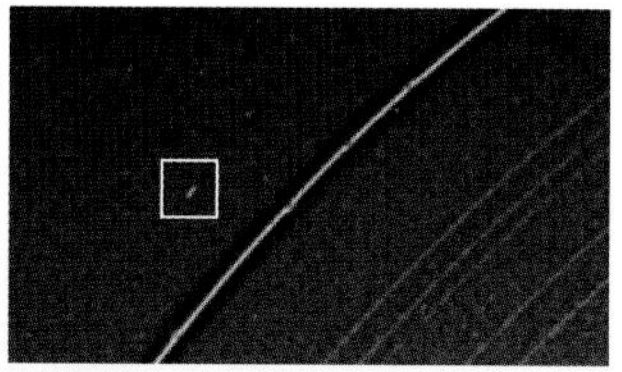

**OPHELIA LIES OUTSIDE THE EPSILON RING**

INNER MOON

## Puck

**DISTANCE FROM URANUS** 86,010km (53,410 miles)

**ORBITAL PERIOD** 0.76 Earth days

**DIAMETER** 162km (101 miles)

Puck was discovered on 30 December 1985 and was the first of the 10 small moons to be found in the Voyager 2 data. It is the second-farthest inner moon from Uranus and was discovered as the probe approached the planet. There was time to calculate that an image could be recorded on 24 January, the day of closest approach. The image (above) revealed an almost circular moon with craters. The largest crater (upper right) is named Lob, after a British Puck-like sprite.

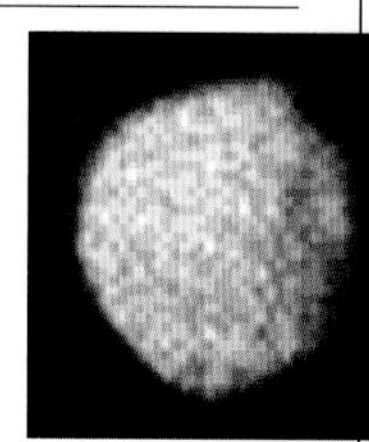

**CRATERED MOON**

MAJOR MOON

## Miranda

**DISTANCE FROM URANUS** 129,390km (80,350 miles)

**ORBITAL PERIOD** 1.41 Earth days

**DIAMETER** 480km (300 miles)

Miranda is the smallest and innermost of Uranus's five major moons, and was discovered by the Dutch-born American astronomer Gerard Kuiper on 16 February 1948. When all five major moons were seen in close-up for the first time, on 24 January 1986, it was Miranda that gave astronomers the biggest surprise. As Voyager 2 passed within 32,000km (19,870 miles) of its surface, the probe revealed a bizarre-looking world, where various surface features butt up against one another in a seemingly unnatural way. One explanation for this strange appearance is that Miranda experienced a catastrophic collision in its past. The moon shattered into pieces and then reassembled in the disjointed way seen today. An alternative theory posits that the moon's evolution was halted before it could be completed. Soon after its formation, dense, rocky material began to sink and lighter material, such as water ice, rose towards the surface. This process then stopped because the necessary internal heat had disappeared. The surface clearly has different types of terrain from different time periods.

**GEOLOGICAL MIX**
On the left lies an ancient terrain of rolling hills and degraded craters; to the right is a younger terrain of valleys and ridges.

**FULL DISC**
The complex terrain of the bright, chevron-shaped Inverness Corona stands out in this south-polar view of Miranda.

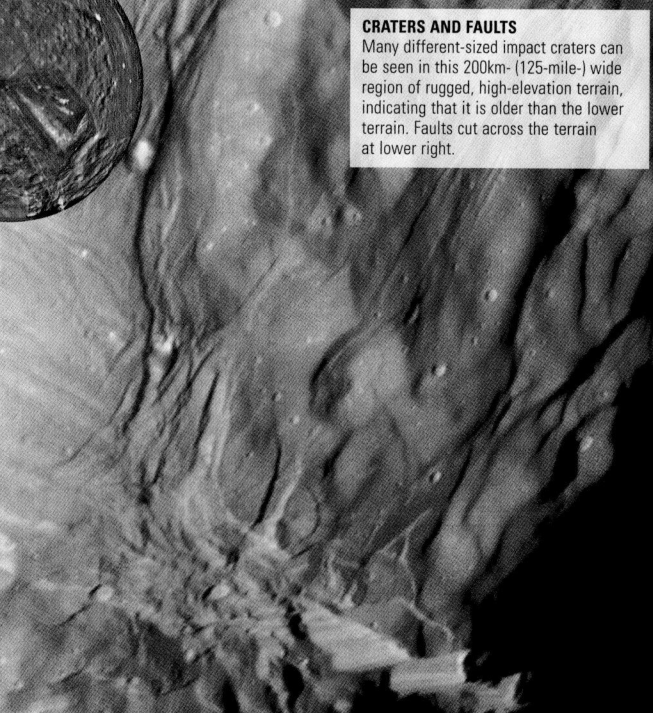

**CRATERS AND FAULTS**
Many different-sized impact craters can be seen in this 200km- (125-mile-) wide region of rugged, high-elevation terrain, indicating that it is older than the lower terrain. Faults cut across the terrain at lower right.

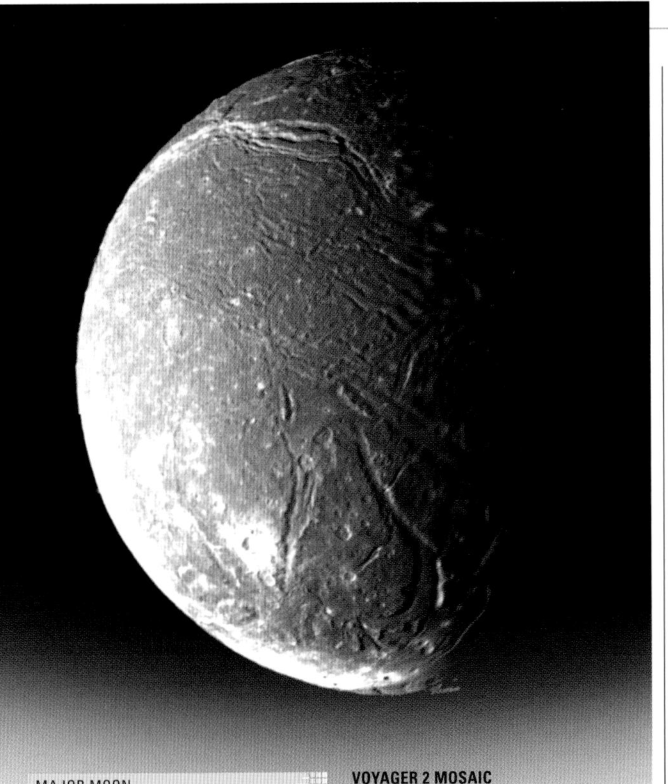

MAJOR MOON

## Ariel

**DISTANCE FROM URANUS** 191,020km (118,620 miles)

**ORBITAL PERIOD** 2.52 Earth days

**DIAMETER** 1,162km (722 miles)

**VOYAGER 2 MOSAIC**
Four Voyager 2 images were combined to produce this view of Ariel. Kachina Chasmata slices across the top, and the Domovoy Crater is on the left, below the centre. Below and to its right is the 50km- (30-mile-) wide Melusine Crater, which is surrounded by bright ejecta.

Ariel and Umbriel (below) were both discovered on 24 October 1851 by the English brewer and astronomer William Lassell (see p.203). Ariel is named after a spirit in Shakespeare's play *The Tempest*. Of the four largest moons, this is the brightest, with the youngest surface. It has impact craters, but these are relatively small – many of them are just 5–10km (3–6 miles) wide. Domovoy, at 71km (44 miles) across, is one of the largest. The sites of any older, larger craters that Ariel once had have been resurfaced. Long faults that formed when Ariel's crust expanded cut across the moon to a depth of 10km (6 miles). One fault, Kachina Chasmata, is 622km (386 miles) long. The floors of such valleys are covered in icy deposits that seeped to the surface from below.

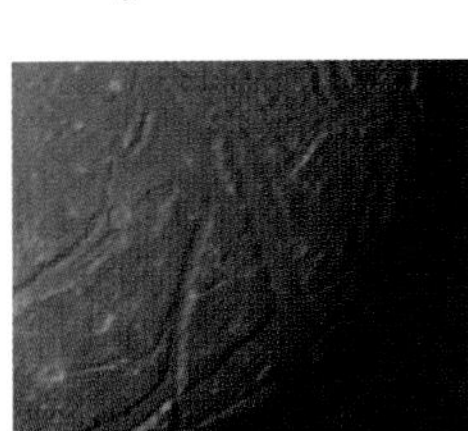

**COMPLEX TERRAIN**
The long, broad valley faults in Ariel's southern hemisphere are filled with deposits and are more sparsely cratered than the surrounding terrain.

MAJOR MOON

## Umbriel

**DISTANCE FROM URANUS** 226,300km (140,530 miles)

**ORBITAL PERIOD** 4.14 Earth days

**DIAMETER** 1,169km (726 miles)

Umbriel is the darkest of Uranus's major moons, reflecting only 16 per cent of the light striking its surface. It is just slightly larger than Ariel, a fact confirmed by the Voyager 2 data. Previous observations had led astronomers to believe that Umbriel was much smaller. This was because of the difficulty in observing such a small, distant moon that reflects little light. Voyager 2 revealed a world covered in craters, many of which are tens of kilometres across. Unlike Ariel, Umbriel appears to have no bright, young ray craters, which means that its surface is older. There is no indication that it has been changed by internal activity. Umbriel's one bright feature, Wunda, is classified as a crater although its nature is unknown.

**SOUTHERN HEMISPHERE**
Umbriel is almost uniformly covered by impact craters. Its one bright feature, the 131km- (81-mile-) wide Wunda at the top of this image, is unfortunately virtually hidden from view.

MAJOR MOON

## Titania

**DISTANCE FROM URANUS** 435,910km (270,700 miles)

**ORBITAL PERIOD** 8.7 Earth days

**DIAMETER** 1,578km (979 miles)

At a little less than half the size of the Moon, Titania is Uranus's largest moon. This rocky world has a grey, icy surface that is covered by impact craters. Icy material ejected when the craters formed reflects the light and stands out on Titania's surface. Large cracks are also visible and are an indication of an active interior. Some of these cut across the craters and appear to be the moon's most recent geological features. They were probably caused by the expansion of water freezing under the crust. There are also smooth regions with few craters that may have been formed by the extrusion of ice and rock. Titania was discovered by the German-born astronomer William Herschel (see p.90) on 11 January 1787, using his homemade 6m (20ft) telescope in his garden in Slough, England (see p.91).

**FULL DISC**
At top right is Titania's largest crater, Gertrude, which is 326km (202 miles) across. Below it, the Messina Chasmata cuts across the moon.

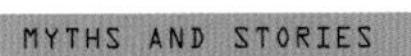

MYTHS AND STORIES

### QUEEN OF THE FAIRIES

Titania and Oberon are the king and queen of the fairies in William Shakespeare's play *A Midsummer Night's Dream*. After a disagreement, Oberon squeezes flower juice into Titania's eyes as she sleeps so that on awakening she will fall in love with the next person she sees. Titania wakes and falls in love with Bottom, the weaver (seen here in a film still from 1999), who has been given an ass's head by the impish sprite Puck.

MAJOR MOON

## Oberon

**DISTANCE FROM URANUS** 583,520km (362,370 miles)

**ORBITAL PERIOD** 13.46 Earth days

**DIAMETER** 1,523km (946 miles)

Oberon was the first Uranian moon to be discovered, as William Herschel observed it before spotting Titania. It has an icy surface pockmarked by ancient impact craters. There are several large craters surrounded by bright ejecta rays. Hamlet, which is just below centre in the Voyager 2 image below, has a diameter of 296km (184 miles). Its floor is partially covered by dark material, and it has a bright central peak. A 6km- (4-mile-) high mountain protrudes from the lower left limb of the moon.

**ICY SURFACE**

OUTER MOON

## Caliban

**DISTANCE FROM URANUS** 7.2 million km (4.5 million miles)

**ORBITAL PERIOD** 579.5 Earth days

**DIAMETER** 96km (60 miles)

Caliban and another small moon, Sycorax, were discovered in September 1997. Both moons follow retrograde and highly inclined orbits. Sycorax is the more distant of the two, at 12.2 million km (7.6 million miles) from Uranus. They were the first of Uranus's irregular moons to be discovered and are believed to be icy asteroids that were captured soon after the planet's formation.

**CALIBAN DISCOVERED**
Caliban lies within the square outline in this image, which was taken using the Hale telescope at Mount Palomar, California, USA. The glow on the right is from Uranus, and the bright dots are background stars.

# NEPTUNE

36–37 Gravity, motion, and orbits
64–65 Planetary motion
109 The grand tour
118–19 The family of the Sun

NEPTUNE IS THE SMALLEST and the coldest of the four gas giants, as well as the most distant from the Sun. It was discovered in 1846, and just one spacecraft, Voyager 2, has been to investigate this remote world. When the probe flew by in 1989, it provided the first close-up view of Neptune and revealed that it is the windiest planet in the Solar System. Voyager 2 also found a set of rings encircling Neptune, as well as six new moons.

## ORBIT

Neptune takes 164.8 Earth years to orbit the Sun, which means that it has completed only one circuit since its discovery in 1846. The planet is tilted to its orbital plane by 28.3°, and as it progresses on its orbit, the north and south poles point sunwards in turn. Neptune is about 30 times farther from the Sun than Earth, and at this distance the Sun is 900 times dimmer. Yet this remote, cold world is still affected by the Sun's heat and light and apparently undergoes seasonal change. Ground-based and Hubble Space Telescope observations show that the southern hemisphere has grown brighter since 1980, and this, as well as an observed increase in the amount, width, and brightness of banded cloud features, has been taken as an indication of seasonal change. However, a longer period of observations is needed to be sure that this seasonal model is correct. The change is slow and the seasons are long. The southern hemisphere is currently in the middle of summer. Once this is over, it is expected to move through autumn, to a colder winter. Then, after 40 years of spring and a gradual increase in temperature and brightness, it will experience summer once more.

**SPIN AND ORBIT**
Neptune's orbit is elliptical but less so than most planets. Only Venus has a more circular orbit. This means there is no marked difference between Neptune's aphelion and perihelion distances.

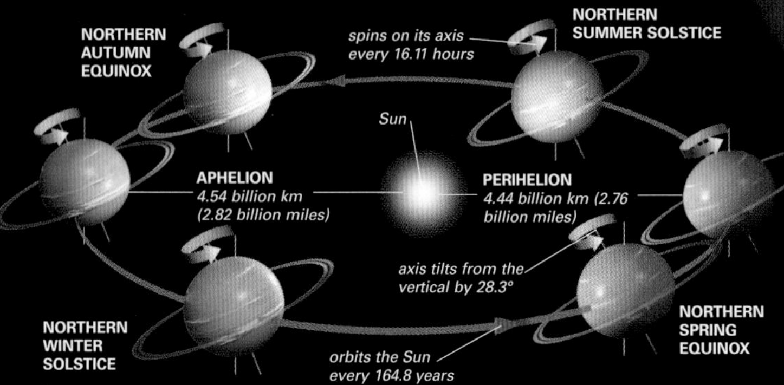


## STRUCTURE

Neptune is very similar in size and structure to Uranus, and neither planet has a discernible solid surface. Like its inner neighbour, Neptune is too massive in relation to its size to be composed mainly of hydrogen. Only about 15 per cent of the planet's mass is hydrogen. Its main ingredient is a mix of water, ammonia, and methane ices that makes up the planet's biggest layer. Neptune's magnetic field, which is tilted by 46.8° to the spin axis, originates in this layer. Above it, lies the atmosphere. This is a shallow, hydrogen-rich layer that also contains helium and methane gas. Below the layer of water and ices, there is a small core of rock and possibly ice. The boundaries between the layers are not clearly defined. The planet rotates quickly on its axis, taking 16.11 hours for one spin, and as a result Neptune has an equatorial bulge. Its polar diameter is 848km (527 miles) less than its equatorial diameter.

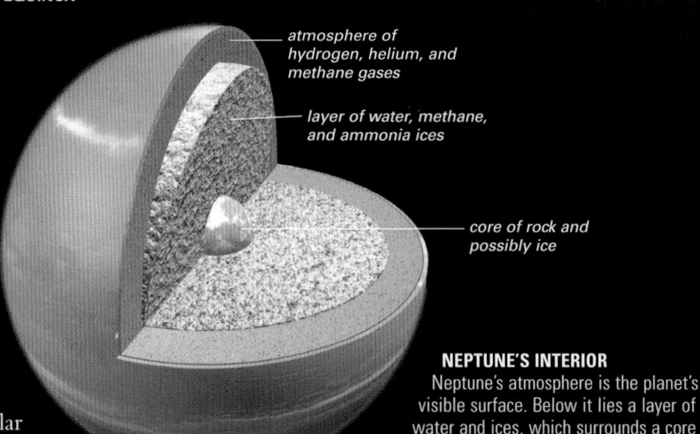


**NEPTUNE'S INTERIOR**
Neptune's atmosphere is the planet's visible surface. Below it lies a layer of water and ices, which surrounds a core of rock and possibly ice.

**THE BLUE PLANET**
This image of Neptune, which was taken by Voyager 2 on 19 August 1989, reveals the planet's dynamic atmosphere. The Great Dark Spot, which is almost as big as Earth, lies in the centre of the planet's disc. A little dark spot and, just above it, the fast-moving cloud feature named the Scooter, are visible on the west limb. A band of cloud stretches across the northern polar region.

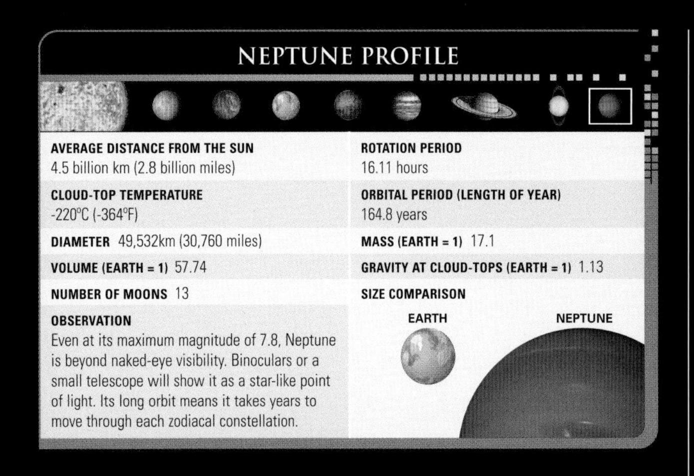

## NEPTUNE PROFILE

| | |
|---|---|
| **AVERAGE DISTANCE FROM THE SUN** 4.5 billion km (2.8 billion miles) | **ROTATION PERIOD** 16.11 hours |
| **CLOUD-TOP TEMPERATURE** -220°C (-364°F) | **ORBITAL PERIOD (LENGTH OF YEAR)** 164.8 years |
| **DIAMETER** 49,532km (30,760 miles) | **MASS (EARTH = 1)** 17.1 |
| **VOLUME (EARTH = 1)** 57.74 | **GRAVITY AT CLOUD-TOPS (EARTH = 1)** 1.13 |
| **NUMBER OF MOONS** 13 | **SIZE COMPARISON** |
| **OBSERVATION** Even at its maximum magnitude of 7.8, Neptune is beyond naked-eye visibility. Binoculars or a small telescope will show it as a star-like point of light. Its long orbit means it takes years to move through each zodiacal constellation. | EARTH NEPTUNE |

# ATMOSPHERE AND WEATHER

Neptune is a perplexing place. For a planet so far from the Sun, it has a surprisingly dynamic atmosphere that exhibits colossal storms and super-fast winds. The heat Neptune receives from the Sun is not enough to drive its weather. The atmosphere may be warmed from below by Neptune's internal heat source, and this is the trigger for larger-scale atmospheric changes. The white bands that encircle the planet are cloud cover, produced when the heated atmosphere rises and then condenses, forming clouds. The winds are most ferocious in the equatorial regions, where they blow westward and reach a staggering 2,160kph (1,340mph). Gigantic, dark, storm-like features accompanied by bright, high-altitude clouds appear and then disappear. One, the Great Dark Spot, was seen by Voyager 2 in 1989. When the Hubble Space Telescope looked for the storm in 1996, it had disappeared.

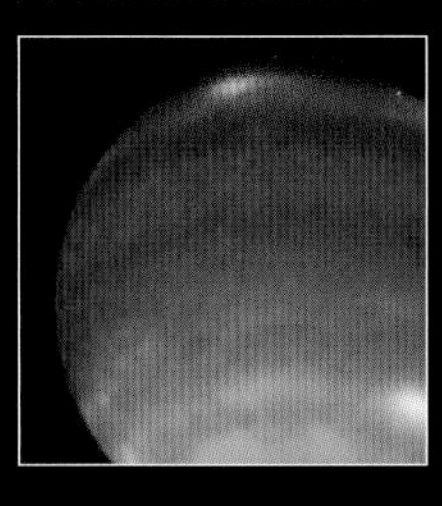

**CLOUDS OVER NEPTUNE**
Neptune's atmosphere lies in bands, which are parallel to the equator. The bright patches are high-altitude clouds, floating above the blue methane layer.

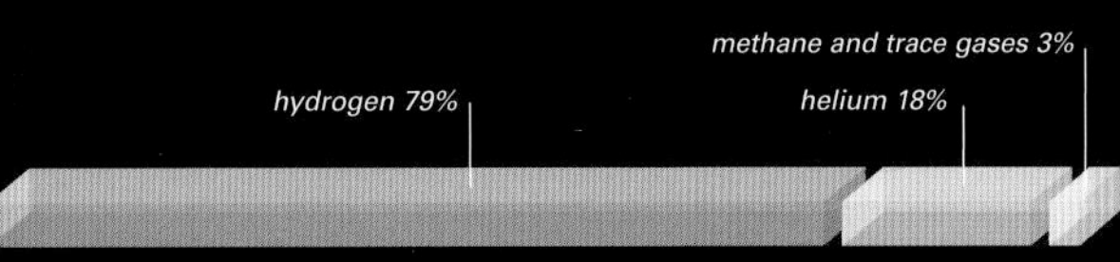


**COMPOSITION OF ATMOSPHERE**
Neptune's atmosphere is made mostly of hydrogen. But it is the methane that gives the planet its deep blue colour, absorbing red light and reflecting blue.

# RINGS AND MOONS

The first indication that Neptune has a ring system came in the 1980s, when stars were seen to blink on and off near the planet's disc. Intriguingly, Neptune seemed to have ring arcs. The mystery was solved when Voyager 2 discovered that Neptune has a ring system with an outer ring so thinly populated that it does not dim starlight but contains three dense regions that do. Neptune has five sparse yet complete rings; moving in from the outer Adams ring, they are Arago, Lassell, Le Verrier, and Galle. A sixth, unnamed partial ring lies within Adams. The rings are made of tiny pieces, of unknown composition, which together would make a body just a few kilometres across. The material is believed to have come from nearby moons. Four of Neptune's 13 moons are within the ring system. It is one of the moons, Galatea, that prevents the arc material from spreading uniformly round the Adams ring. Only one of the 13, Triton, is of notable size. It and Nereid were discovered before the days of space probes. Five small moons have been discovered since 2002, and more will probably be found.

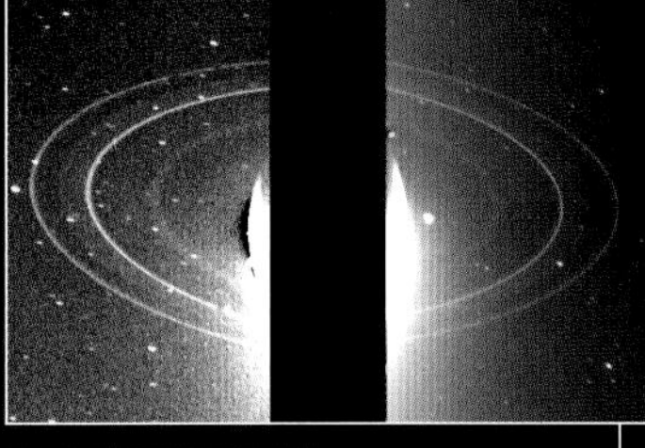

**THE RINGS OF NEPTUNE**
Two Voyager 2 images placed together reveal Neptune's ring system. The two bright rings are Adams and Le Verrier. The faint Galle ring is innermost, and the diffuse band, Lassell, is visible between the two bright ones.

**NEPTUNE'S MOONS**

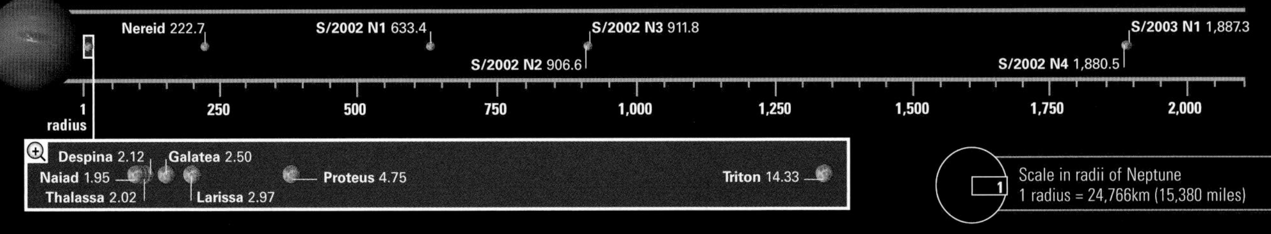

# NEPTUNE'S MOONS

Neptune has only one major moon – Triton. All its other satellites are small and can be described as inner or outer moons depending on whether they are closer to or farther from Neptune than Triton. The six inner moons were discovered by analysis of Voyager 2 data in 1989. The moons are named after characters associated with the Roman god of the sea, Neptune, or his Greek counterpart, Poseidon.

**NEPTUNE AND TRITON**
This image of the crescent moon of Triton below the crescent of Neptune was captured by Voyager 2 as it flew away from the planet.

INNER MOON

## Larissa

**DISTANCE FROM NEPTUNE** 73,458km (45,617 miles)

**ORBITAL PERIOD** 0.55 Earth days

**LENGTH** 216km (134 miles)

Larissa is the fifth moon from Neptune, lying outside the ring system. The moon was first spotted from Earth in 1981, but astronomers eventually decided that it was a ring arc circling Neptune. In late July 1989, a Voyager 2 team of astronomers confirmed that it is, in fact, an irregularly shaped, cratered moon. It was named after a lover of Poseidon.

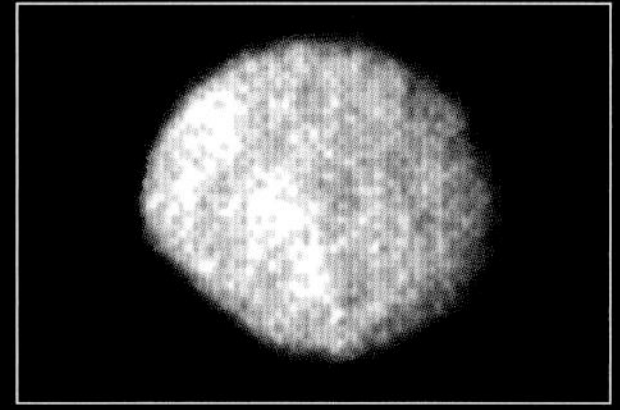

**IRREGULARLY SHAPED MOON**

INNER MOON

## Proteus

**DISTANCE FROM NEPTUNE** 117,647km (73,059 miles)

**ORBITAL PERIOD** 1.12 Earth days

**LENGTH** 440km (273 miles)

The most distant of the inner moons from Neptune, Proteus is also the largest of the six – their size increases with distance. It has an almost equatorial orbit, speeding round Neptune in less than 27 hours. Its visible surface has extensive cratering, but just one major feature stands out – a large, almost circular depression measuring 255km (158 miles) across, with a rugged floor. Proteus was the first of the six inner moons to be discovered by Voyager 2 scientists. It was detected in mid-June 1989, within two months of the probe's closest approach to Neptune, enabling the observing sequence to be changed. The images subsequently recorded by Voyager 2 revealed a grey, irregular but roughly spheroid moon that reflects 6 per cent of the sunlight hitting it. The moon was later named Proteus after a Greek sea god.

**TWO VIEWS**
The first image of Proteus (far right) shows the moon half-lit. The second was taken closer in (the black dots are a processing artefact).

OUTER MOON

## Nereid

**DISTANCE FROM NEPTUNE** 5.5 million km (3.4 million miles)

**ORBITAL PERIOD** 360.1 Earth days

**DIAMETER** 340km (211 miles)

Nereid was discovered on 1 May 1949 by the Dutch-born astronomer Gerard Kuiper, while working at the McDonald Observatory, Texas, USA. Little is still known about this moon. Voyager 2 flew by at a distance of 4.7 million km (2.9 million miles) in 1989 and could take only a low-resolution image. Nereid's outstanding characteristic is its highly eccentric and inclined orbit, which takes the moon out as far as about 9.5 million km (5.9 million miles) from Neptune and to within just 817,200km (507,500 miles) at its closest approach.

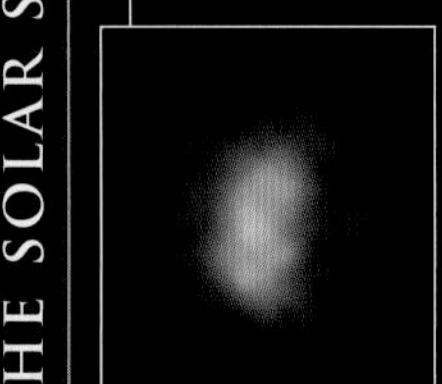

**BEST VIEW**
Voyager 2 revealed Nereid to be a dark moon, reflecting only 14 per cent of the sunlight it receives.

OUTER MOON

## S/2002 N1

**DISTANCE FROM NEPTUNE** 15.7 million km (9.7 million miles)

**ORBITAL PERIOD** 1,874.8 Earth days

**DIAMETER** 48km (30 miles)

S/2002 N1 was discovered by an international team of astronomers who were carrying out a systematic search for new Neptunian moons. Their task wasn't easy because moons as small and as distant as S/2002 N1 are extremely difficult to detect. S/2002 N1 follows a highly inclined and elliptical orbit. The origin of the irregular outer moons, which now number five, is unknown. More may be found, as these moons could be the result of an ancient collision between a former moon and a passing body such as a Kuiper Belt object.

EXPLORING SPACE

### LOOKING FOR NEW MOONS

A team of astronomers announced the discovery of three new moons, including S/2002 N1, on 13 January 2003. They had taken multiple images of the sky around Neptune from two sites in Hawaii and Chile. The images were combined to boost the signal of faint objects. The new moons showed up as points of light against the background of stars, which appeared as streaks of light.

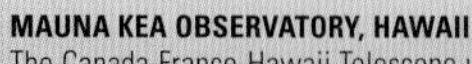

**MAUNA KEA OBSERVATORY, HAWAII**
The Canada-France-Hawaii Telescope used in the search is at Mauna Kea. The other site was the Cerro Tololo Inter-American Observatory in Chile.

MAJOR MOON

## Triton

**DISTANCE FROM NEPTUNE** 354,760km (220,306 miles)

**ORBITAL PERIOD** 5.88 Earth days

**DIAMETER** 2,707km (1,681 miles)

Triton was the first of Neptune's moons to be discovered, just 17 days after the discovery of the planet was announced. William Lassell (see panel, right) used the coordinates published in *The Times* to locate Neptune in early October 1846. On 10 October, he found its biggest moon, using the 61cm (24in) reflecting telescope at his observatory in Liverpool, England. The moon was named Triton after the sea-god son of Poseidon. The Voyager 2 flyby nearly 143 years later revealed most of what is now known about this icy world.

**SMOOTH PLAIN**
The 300km- (185-mile-) wide Ruach Planitia is in the cantaloupe terrain. It may be an old impact crater that has been filled in.

**ICY SURFACE**
The cantaloupe terrain is at the top of this image. The pink colour of the south-polar cap may come from compounds formed when methane ice reacts with sunlight.

Triton is by far the largest of Neptune's moons and is bigger than Pluto. It follows a circular orbit and exhibits synchronous rotation, so the same side always faces Neptune. Peculiarly for such a large moon, Triton is in retrograde motion, travelling in the opposite direction to Neptune's spin. This could be a clue to its origin. Triton may have formed elsewhere in the Solar System and been captured by Neptune. The moon's mix of two parts rock to one part ice is differentiated into a rocky core, a possibly liquid mantle, and an ice crust. Its geologically young, icy surface has few craters and displays a range of features. An area of linear grooves, ridges, and circular depressions is nicknamed the cantaloupe after its resemblance to a melon's skin. Dark patches mark the south-polar region. These form when solar heat turns subsurface nitrogen ice into gas. This erupts through surface vents in geyser-like plumes, which carry dark, possibly carbonaceous dust into the atmosphere before depositing it on Triton's surface.

**POLAR PROJECTION**
Triton's south polar region is seen head on in this image. A band of bluish material extends out from the central polar cap into the equatorial region. It is probably fresh nitrogen frost or snow redistributed by the wind.



**SOUTHERN HEMISPHERE**
Three Voyager 2 images were combined to produce this almost full-disc image of Triton. From a distance, the southern-hemisphere terrain appears mottled.

## WILLIAM LASSELL

English businessman William Lassell (1799–1880) used the profits from his brewery to fund his passion for astronomy. He designed and built large reflecting telescopes that were the finest of the day. He made his observations first from his home in Liverpool, England, then from the island of Malta. In addition to Triton, Lassell discovered the Uranian moons Ariel and Umbriel, and Saturn's moon Hyperion.

# DWARF PLANETS

36–37 Gravity, motion, and orbits
64–65 Planetary motion
108–111 Beyond Earth
118–19 The family of the Sun
Kuiper Belt and Oort Cloud 206–207

A DWARF PLANET is a body that orbits the Sun and has enough mass and gravity to be roughly spherical but has not cleared the region around its orbit and is not a satellite. Dwarf planets have only recently been recognized. When it was first discovered, the best known, Pluto, was treated as a planet, but when larger objects were found in the Kuiper Belt, astronomers agreed to form the new class.

## PLUTO

Pluto is one of the lonely outriders of the Solar System – its average distance from the Sun is 5.9 billion km (3.7 billion miles). It is a cold world made of rock and ice where the surface temperature is about -230°C (-382°F). Pluto is comparatively unknown; the first spacecraft to visit it will not arrive there until 2015. But for decades, astronomers have noted differences from its planetary neighbours. Pluto is small, just 2,304km (1,432 miles) across with a mass of about one-400th that of Earth. Its orbit is far from circular and is inclined by 17.1° to the plane of the planets (see pp.118–19). For 20 years of its 248.6 year orbit it is closer to the Sun than Neptune. Its spin axis is tilted 122° from the vertical, which means Pluto spins in the opposite direction to Earth. Pluto has a nitrogen rich atmosphere, which, with little gravity to retain it, escapes rapidly and is replenished by the evaporation of surface ices. Its largest moon Charon is about 1,180km (730 miles) across. Pluto and Charon keep the same face towards each other at all times. Both spin on their axes every 6.38 Earth days, and Charon orbits Pluto once in the same period. Two small moons, Nix and Hydra, were discovered in May 2005.

**PLUTO'S PATCHY SURFACE**
Hubble images have shown the reflectivity of Pluto's surface to vary. The lighter areas are probably patches of nitrogen frost, methane frost, and water ice.

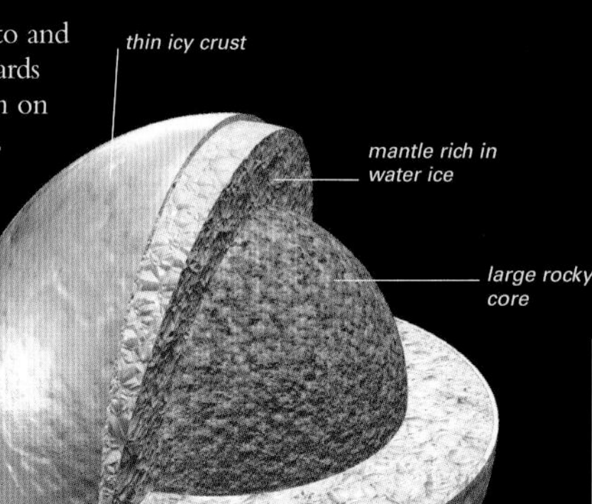


**PLUTO INTERIOR**
Pluto's density indicates that it is probably 70 per cent rock and 30 per cent water ice. Radioactivity in the rock has warmed the interior, ensuring Pluto is spherical and differentiated; rock has sunk to the centre and water ice has formed a surface layer. Also present are ices of nitrogen, carbon monoxide, and methane.

**GROUND-BASED IMAGE**
This is the best view of the Pluto–Charon system that can be obtained from the ground. The image was taken by the Canada–France–Hawaii telescope. Charon was discovered in 1978, when astronomers noticed that Pluto's image became elongated periodically. It was soon realized that this was because Pluto has a moon. Charon orbits Pluto at a distance of 19,600km (12,170 miles).

**COMPOSITION OF ATMOSPHERE**
Nitrogen dominates Pluto's atmosphere. It is estimated that for every million molecules of nitrogen there are about 300 of carbon monoxide and 15 of methane.

**HUBBLE'S VIEW OF PLUTO**
This is one of the best images ever taken of Pluto, (the larger object at the top), and its moon, Charon. To the right the two moons Nix (top) and Hydra (bottom) are also clearly visible.

EXPLORING SPACE

## SEARCHING FOR A PLANET

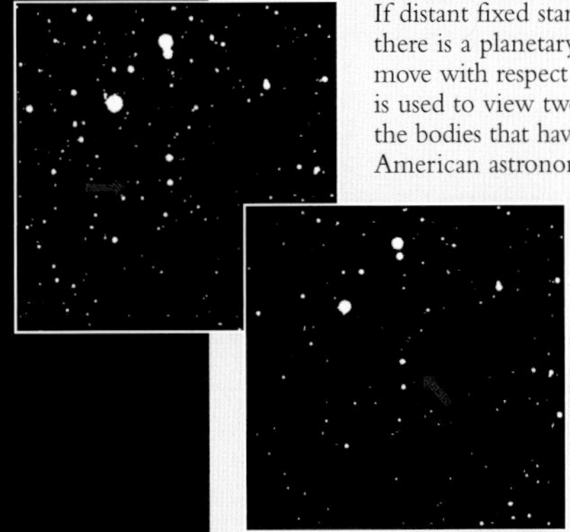

If distant fixed stars are photographed a few days apart, and there is a planetary body in the field of view, that body will move with respect to the starry background. A blink comparator is used to view two photographic plates in quick succession and the bodies that have moved appear to blink. On 23 January 1930, American astronomer Clyde Tombaugh, working at the Lowell Observatory, in Arizona, took an hour-long exposure of the Delta Geminorum region. On 29 January, he imaged the region again and one "star" in his plates (left) had moved – he had discovered Pluto. Astronomers at Palomar Observatory have found tens of Kuiper Belt objects using this technique. Their images are recorded electronically, 1.5 hours apart, and checked by computers. When three images recorded in October 2003 were reanalysed in 2005 they revealed the object we now call Eris.

# OTHER DWARF PLANETS

Eris, named after the Greek goddess of discord and strife, is the largest known dwarf planet. Although larger than Pluto, its size remains uncertain and estimates range from 2,400km (1,500 miles) to around 3,000km (1,850 miles) across. When discovered it was the most distant object ever seen in orbit round the Sun – almost 10 billion miles away. It takes 560 years to complete one circuit of its elongated and highly inclined orbit. Eris is believed to be a mixture of rock and ice, with an icy surface. Ceres, just 960km (596 miles) across, orbits the Sun every 4.6 years within the Main Belt of asteroids. It was the first asteroid found, discovered in 1801, and remains the largest known. It is mainly made of rock, but contains a significant quantity of water ice. Its size and mass mean it is spherical, and like the terrestrial planets, it has a differentiated interior; denser silicate material is in its core and lighter minerals and ice near the surface. Pluto, Eris, and Ceres are the only known dwarf planets, but likely candidates are known and more are expected to be announced in the future.

EXPLORING SPACE

## WHAT IS A PLANET?

In August 2006, the world's largest professional body for astronomers, the International Astronomical Union (IAU), defined a planet to be a celestial body that orbits the Sun, has sufficient mass and gravity to be nearly round, and has cleared the neighbourhood around its orbit. The IAU also decided on a new class of object: dwarf planets. A dwarf planet is similar to a planet, but has not cleared its neighbourhood, and is not a satellite.

**ERIS AND DYSNOMIA**
On 10 September 2005, astronomers using the 10-metre Keck telescope in Hawaii discovered that Eris has a moon (seen to the right of Eris). It has since been named Dysnomia, after the daughter of the goddess Eris.

**VIEWS OF CERES**
When viewed from Earth, Ceres, highlighted in the Hyades star cluster in the constellation of Taurus, looks like a star. When viewed by Hubble (top) its almost-round planetary shape is revealed. This enhanced image also indicates surface regions that could be asteroid impact features.

# THE KUIPER BELT AND THE OORT CLOUD

24–24 Celestial objects
36–37 Gravity, motion, and orbits
Comets 214–15

A FLATTENED BELT OF ICY, comet-like bodies encircles the planetary region of the Solar System. Known as the Kuiper Belt, it begins just beyond the orbit of Neptune, and most of its members take more than 250 years to orbit the Sun. The whole Solar System itself is believed to be surrounded by a huge symmetrical cloud of comets called the Oort Cloud. The comets have randomly inclined, elongated orbits that stretch nearly halfway to the closest stars.

## THE KUIPER BELT

Stretching between 6 and 12 billion km (3.7 and 7.4 billion miles) from the Sun, the Kuiper Belt is a flattened disc of comet-like bodies. Although its existence was suggested by Irish astronomer Kenneth Edgeworth in 1943 and by Gerard Kuiper (see panel, below) in 1951, it was only confirmed during the 1990s with the discovery of an object called 1992 QB1. By 2002, over 600 objects had been found. About half of these are trapped in a 2:3 resonance with Neptune (just like Pluto, see p.204). Two other groups have perihelia ranging between 6.1 and 6.9 billion km (3.8 and 4.3 billion miles); one has circular orbits, the other eccentric orbits. The largest object found so far is the dwarf planet Eris (see p.205), but most range from about 1,000km (600 miles) across to less than 100km (60 miles). There must, however, be vast numbers of them, and they are the main source of short-period comets. Chiron is a member of a group of objects called Centaurs, believed to have been captured from the Kuiper Belt. It is now on a 50.7-year orbit that takes it from just inside the orbit of Saturn to just inside the orbit of Neptune.

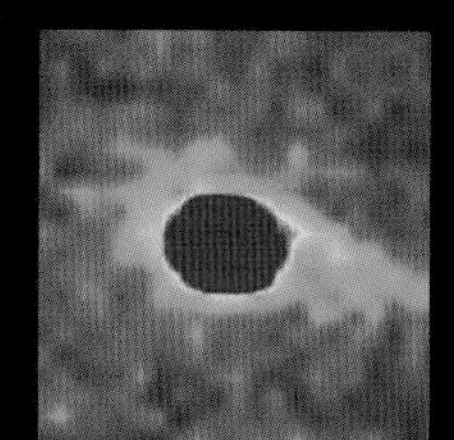

**CHIRON**
Chiron may be an asteroid or it may be a comet. It is 180km (112 miles) across and is surrounded by a coma of dust and gas.

**FLATTENED BELT**
The Kuiper Belt is shaped like a doughnut and stretches far beyond the orbit of Neptune. Pluto orbits largely within the Belt.

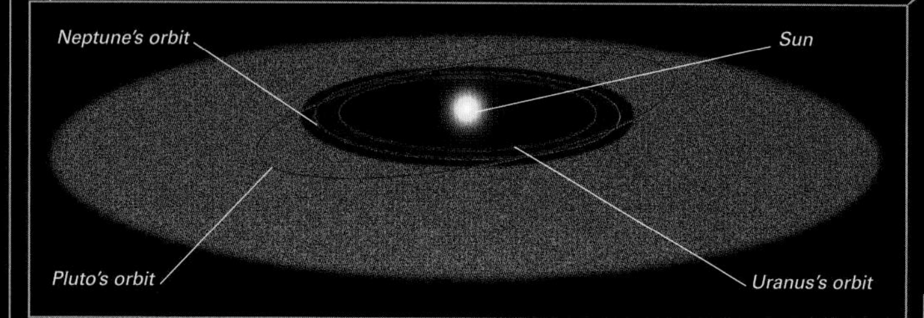


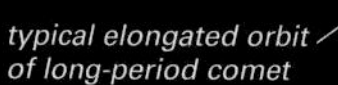


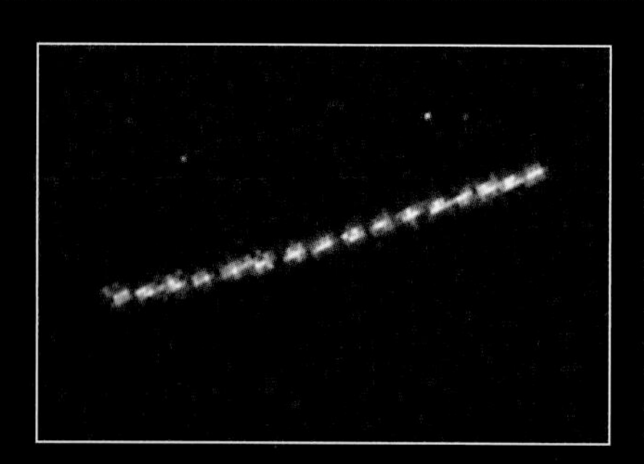

**QUAOAR**
This image has been exposed 16 times and shows Quaoar, discovered in 2002, moving across the sky on its 288-year orbit. With a diameter of about 1,250km (775 miles) – half the size of Pluto – Quaoar is the second-largest known member of the Kuiper Belt.

### GERARD KUIPER

Gerard Kuiper (1905–1973) emigrated from the Netherlands to the USA in 1933. He is acknowledged as the father of modern planetary science, discovering not only the moons Miranda and Nereid but also the atmospheres of Mars and Titan. In 1951, he suggested that short-period comets were sourced from a thick, elliptical disc of rock-and-ice bodies lying just beyond the orbit of Neptune.

# THE OORT CLOUD

The Oort Cloud is a spherical reservoir of long-period comets, which replace those that are destroyed in the inner Solar System. After the formation of the outer planets, many comets were left over and were then scattered, mainly by Neptune, onto orbits that took them towards the stars. Some flew off into the galactic disc, others into the inner Solar System. The remainder formed a huge spherical cloud, about 1.6 light-years in radius, first described by the Estonian-born astrophysicist Ernst Opik and the Dutch astronomer Jan Oort over 50 years ago. This cloud is stirred up by passing stars; the nearest star at the moment being about 4.2 light-years away. Some comets escape the cloud while others are pushed into the inner Solar System, where they start to decay. Here they are seen as long-period comets, their orbits being so elongated that they appear parabolic. The comets that remain in the inner part of the Oort Cloud tend to have orbits closer to the plane of the Solar System. It is estimated that the cloud contains over 1 trillion comets. This vast number is needed in order to explain the rate at which these very long-period comets come close to the Sun and can be detected by Earth-based astronomers. There were more initially, and this number is decreasing exponentially. The total present-day mass of the Oort Cloud is a few times greater than the mass of the Earth.

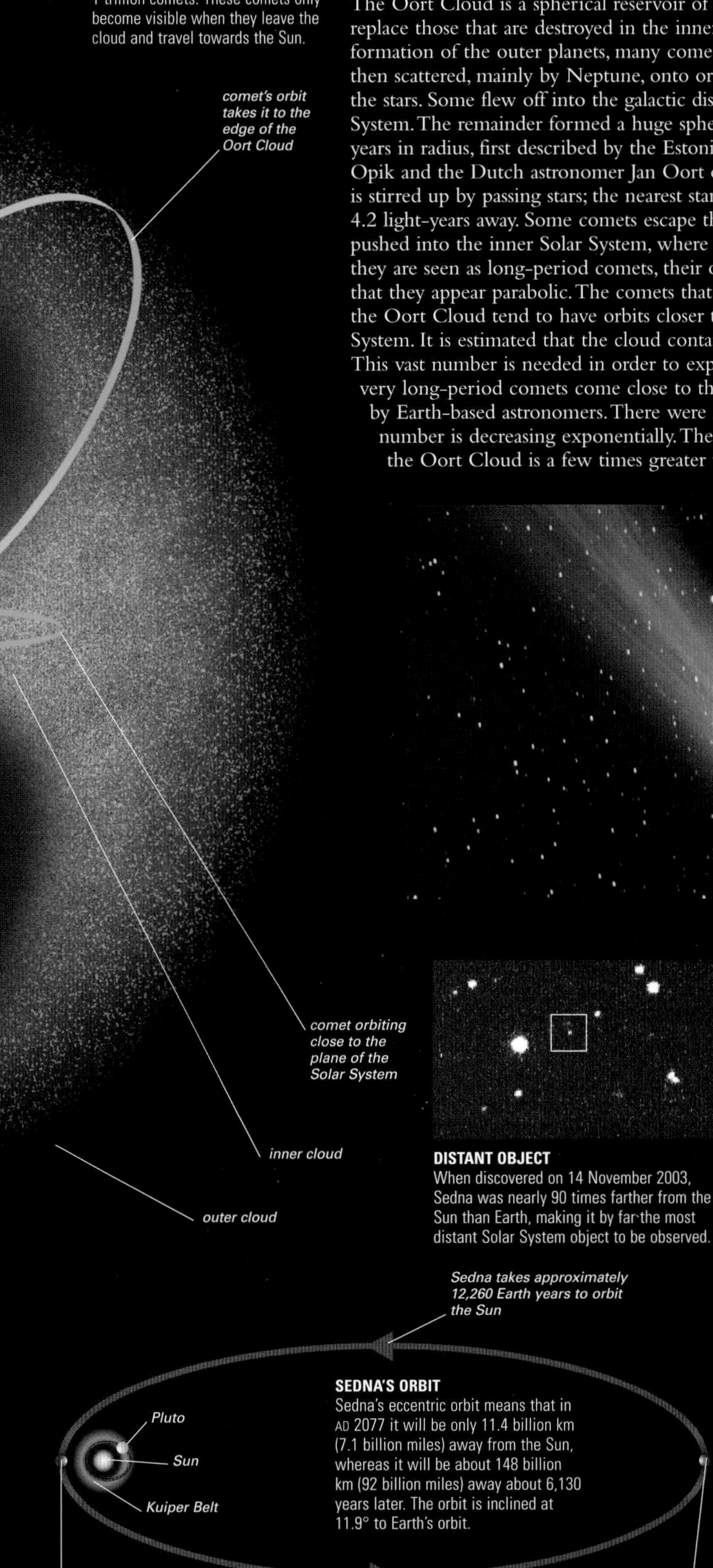
comet's orbit takes it to the edge of the Oort Cloud
comet orbiting close to the plane of the Solar System
inner cloud
outer cloud

**OORT CLOUD**
The Oort Cloud consists of over 1 trillion comets. These comets only become visible when they leave the cloud and travel towards the Sun.

## JAN HENDRIK OORT

Jan Oort (1900–1992) was born in Franeker, the Netherlands. He became one of Europe's top astronomers and was the director of the Leiden Observatory from 1945 to 1970. Oort is mostly remembered for his suggestion that the Solar System is surrounded by the vast symmetrical cloud of comets, which was named after him. He is also famous for pioneering the Dutch radio astronomy telescopes, using radio waves to map the distribution of hydrogen in the galactic plane, and for confirming that the centre of our galaxy is some 30,000 light-years away in the direction of Sagittarius.

**LONG-PERIOD COMET**
Long-period comets, such as Hyakutake, enter the inner Solar System from the Oort Cloud. Passing stars are responsible for pushing them inwards. They have approached the Sun only a few times in their lives.

**DISTANT OBJECT**
When discovered on 14 November 2003, Sedna was nearly 90 times farther from the Sun than Earth, making it by far the most distant Solar System object to be observed.

# SEDNA

Sedna was about 13.5 billion km (8.4 billion miles) away from the Sun when it was discovered in 2003 by three American astronomers, Mike Brown, Chad Trujillo, and David Rabinowitz, who were searching for Kuiper Belt objects. Sedna has an orbit similar to an Oort Cloud comet, but 10 times smaller. Sedna is 1,500 km (930 miles) across – too big to have been formed where it now orbits, out beyond the Kuiper Belt. It must have been formed closer to the Sun and then thrown out by a passing planet. Sedna is the second-reddest body in the Solar System, after Mars. With a surface temperature of -240°C (-400°F), it is extremely cold. The Hubble Space Telescope indicated that Sedna had no moon. Initial observations showed that Sedna's brightness varied every 20 days. If this was due to Sedna's rotation, it would be the third-slowest spinner in the Solar System. A retreating moon would explain this, but no moon has been found thus far. The mystery has yet to be solved.

Sedna takes approximately 12,260 Earth years to orbit the Sun
Pluto
Sun
Kuiper Belt
Sedna's closest approach to the Sun will be in AD 2077
Sedna will reach aphelion in about AD 8207

**SEDNA'S ORBIT**
Sedna's eccentric orbit means that in AD 2077 it will be only 11.4 billion km (7.1 billion miles) away from the Sun, whereas it will be about 148 billion km (92 billion miles) away about 6,130 years later. The orbit is inclined at 11.9° to Earth's orbit.

**RED BODY**
Named after the Inuit goddess of the ocean, Sedna's surface is almost as red as that of Mars, as shown in this artist's impression.

# ASTEROIDS

32–35 Radiation
36–37 Gravity, motion, and orbits
116–17 The history of the Solar System
Meteorites 222–23

ASTEROIDS ARE REMNANTS OF a failed attempt to form a rocky planet that would have been about four times as massive as Earth. They are dry, dusty objects and far too small to have atmospheres. Over 200,000 have been discovered, although over a billion are predicted to exist. The astronomers who discover asteroids have the right to name them.

## ORBITS

Most asteroids are found in a concentration known as the Main Belt, which lies between Mars and Jupiter, about 2.8 times farther from the Sun than Earth. Typically, they take between four and five years to orbit the Sun. The orbits are slightly elliptical and of low inclination. Even though the asteroids are all orbiting in the same direction, collisions at velocities of a few kilometres per second often take place. So, as time passes, asteroids tend to break up. Some asteroids have been captured into rather strange orbits. The Trojans have the same orbital period as Jupiter and tend to be either 60° in front or 60° behind that planet. Then there are the Amor and Apollo asteroid groups (named after individual asteroids), with paths that cross the orbits of Mars and Earth respectively. Aten asteroids have such small orbits that they spend most of their time inside Earth's orbit. These three groups are classed as near-Earth asteroids. They can be dangerous, having the potential to hit Earth and cause a great deal of damage. Fortunately, this happens very rarely.

**ASTEROID PATH**
To picture stars, the Hubble Space Telescope scans the sky, keeping the stars stationary in the image frame. Asteroids, being much closer than the stars and in orbit around the Sun, form streaky trails (the blue line) during the exposure time.

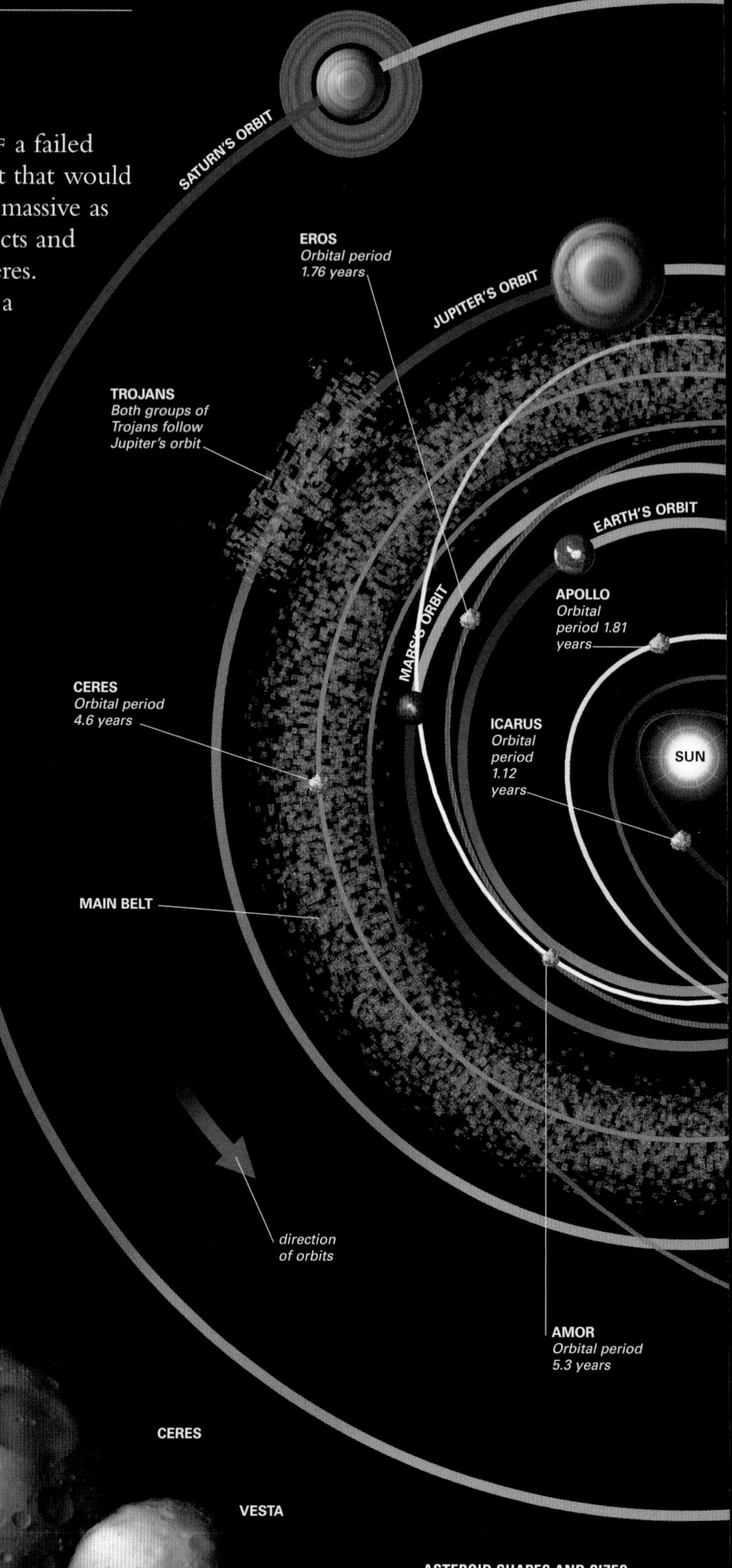

SATURN'S ORBIT
EROS *Orbital period 1.76 years*
JUPITER'S ORBIT
TROJANS *Both groups of Trojans follow Jupiter's orbit*
EARTH'S ORBIT
MARS'S ORBIT
APOLLO *Orbital period 1.81 years*
CERES *Orbital period 4.6 years*
ICARUS *Orbital period 1.12 years*
SUN
MAIN BELT
*direction of orbits*
AMOR *Orbital period 5.3 years*

## STRUCTURES

At the dawn of the Solar System, there existed quite a few asteroids nearly as large as Mars. The radioactive decay of elements within the asteroidal rock melted these large bodies, and, during their fluid stage, gravity pulled them into a spherical shape before they cooled. Many of these have since been broken up or reshaped by collisions with other asteroids. Smaller asteroids, which cooled more efficiently than larger ones, did not reach melting point and retained a uniform rocky-metallic composition and their original irregular shape. There are three main compositional classes of asteroid. The vast majority are either carbonaceous (C-type) or silicaceous (S-type). The next most populated class is metallic (M-type). These classes correspond to carbonaceous chondrite (stony) meteorites, stony-iron meteorites, and iron meteorites.

CERES
VESTA
IDA

**ASTEROID SHAPES AND SIZES**
The largest asteroids, such as Ceres and Vesta, are nearly spherical, whereas smaller asteroids, such as Ida, are irregularly shaped. All asteroids have craters on their surfaces, but some areas have been sandblasted and smoothed by a multitude of minor collisions.

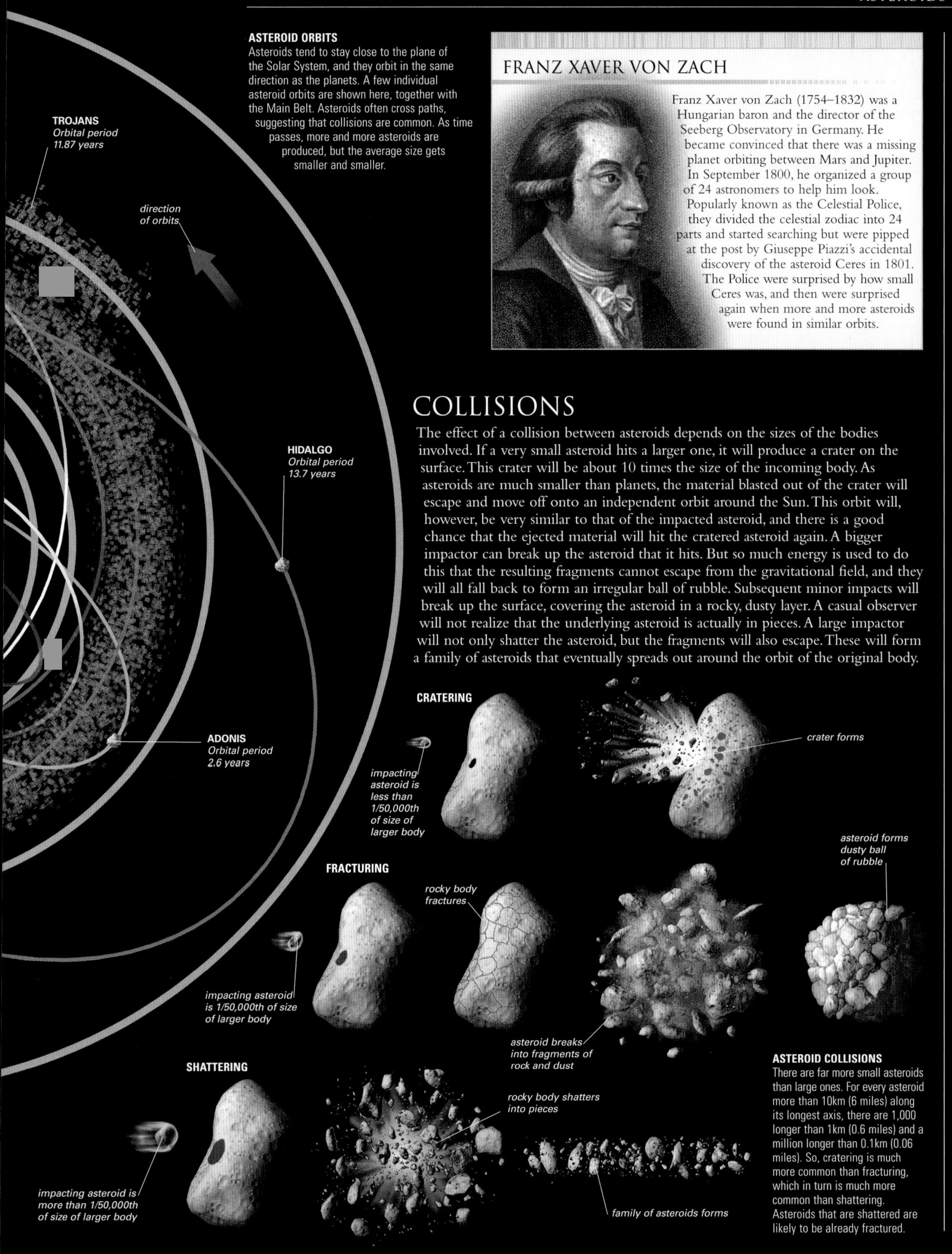
ASTEROID ORBITS
Asteroids tend to stay close to the plane of the Solar System, and they orbit in the same direction as the planets. A few individual asteroid orbits are shown here, together with the Main Belt. Asteroids often cross paths, suggesting that collisions are common. As time passes, more and more asteroids are produced, but the average size gets smaller and smaller.
TROJANS
Orbital period
11.87 years
direction
of orbits
HIDALGO
Orbital period
13.7 years
ADONIS
Orbital period
2.6 years
CRATERING
impacting asteroid is less than 1/50,000th of size of larger body
crater forms
FRACTURING
rocky body fractures
impacting asteroid is 1/50,000th of size of larger body
asteroid breaks into fragments of rock and dust
asteroid forms dusty ball of rubble
SHATTERING
impacting asteroid is more than 1/50,000th of size of larger body
rocky body shatters into pieces
family of asteroids forms

## FRANZ XAVER VON ZACH

Franz Xaver von Zach (1754–1832) was a Hungarian baron and the director of the Seeberg Observatory in Germany. He became convinced that there was a missing planet orbiting between Mars and Jupiter. In September 1800, he organized a group of 24 astronomers to help him look. Popularly known as the Celestial Police, they divided the celestial zodiac into 24 parts and started searching but were pipped at the post by Giuseppe Piazzi's accidental discovery of the asteroid Ceres in 1801. The Police were surprised by how small Ceres was, and then were surprised again when more and more asteroids were found in similar orbits.

## COLLISIONS

The effect of a collision between asteroids depends on the sizes of the bodies involved. If a very small asteroid hits a larger one, it will produce a crater on the surface. This crater will be about 10 times the size of the incoming body. As asteroids are much smaller than planets, the material blasted out of the crater will escape and move off onto an independent orbit around the Sun. This orbit will, however, be very similar to that of the impacted asteroid, and there is a good chance that the ejected material will hit the cratered asteroid again. A bigger impactor can break up the asteroid that it hits. But so much energy is used to do this that the resulting fragments cannot escape from the gravitational field, and they will all fall back to form an irregular ball of rubble. Subsequent minor impacts will break up the surface, covering the asteroid in a rocky, dusty layer. A casual observer will not realize that the underlying asteroid is actually in pieces. A large impactor will not only shatter the asteroid, but the fragments will also escape. These will form a family of asteroids that eventually spreads out around the orbit of the original body.

**ASTEROID COLLISIONS**
There are far more small asteroids than large ones. For every asteroid more than 10km (6 miles) along its longest axis, there are 1,000 longer than 1km (0.6 miles) and a million longer than 0.1km (0.06 miles). So, cratering is much more common than fracturing, which in turn is much more common than shattering. Asteroids that are shattered are likely to be already fractured.

# ASTEROIDS

Mainly moving between the orbits of Mars and Jupiter, asteroids are the remnants of a planet-formation process that failed. Today's asteroid belt contains only about 100 asteroids that are larger than 200km (125 miles) across. But there are 100,000 asteroids greater than about 20km (12.5 miles) across and a staggering 1 billion that are over 2km (1.25 miles) along their longest axis. Ceres, the first asteroid to be discovered, in 1801, is now called a dwarf planet (see p.204–205). Ceres contains about 25 per cent of the mass of all the asteroids combined.

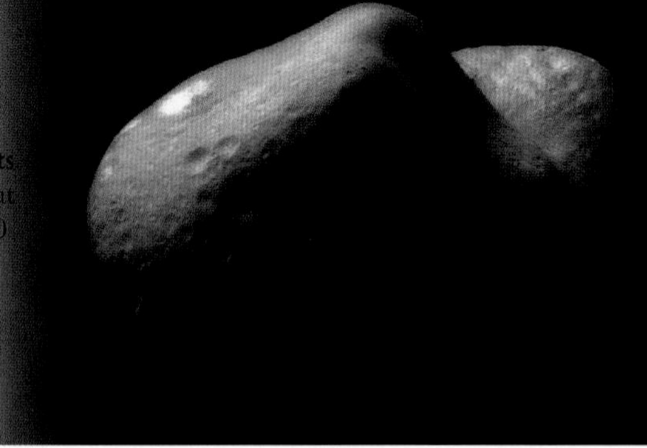

**EROS**
Only asteroids bigger than about 350km (215 miles) in diameter are spherical. Eros is an irregularly shaped fragment of a much larger body.

MAIN-BELT ASTEROID

## 951 Gaspra

**AVERAGE DISTANCE TO SUN** 331 million km (206 million miles)

**ORBITAL PERIOD** 3.29 years

**ROTATION PERIOD** 7.04 hours

**LENGTH** 18 km (11.2 miles)

**DATE OF DISCOVERY** 30 July 1916

Until 1991, asteroids could be glimpsed only from afar. In October of that year, a much closer view was obtained when the Galileo spacecraft flew within 1,600km (1,000 miles) of Gaspra, taking 57 colour images. Gaspra is a silicate-rich asteroid. The surface is very grey, with some of the recently exposed crater edges being bluish and some of the older, low-lying areas appearing slightly red.

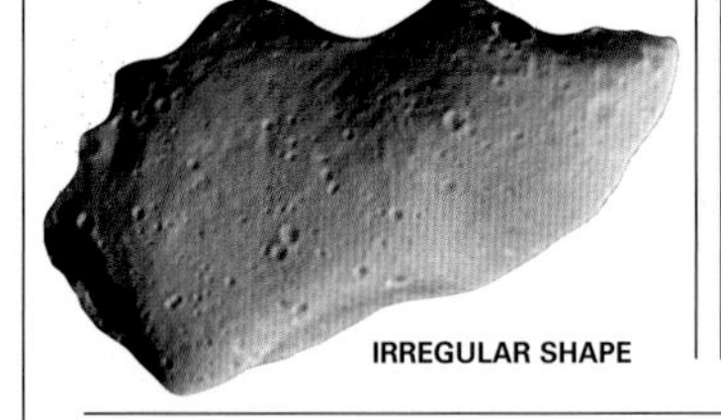

**IRREGULAR SHAPE**

MAIN-BELT ASTEROID

## 5535 Annefrank

**AVERAGE DISTANCE TO SUN** 331 million km (206 million miles)

**ORBITAL PERIOD** 3.29 years

**ROTATION PERIOD** Not known

**LENGTH** 6km (3.7 miles)

**DATE OF DISCOVERY** 23 March 1942

Annefrank orbits in the inner regions of the Main Belt of asteroids and is a member of the Augusta family. On 2 November 2002, Annefrank was imaged by NASA's Stardust spacecraft as it passed within 3,300km (2,050 miles) on its way to Comet Wild 2. Interestingly, Annefrank turned out to be twice as large as had been predicted from Earth-based observations. The brightness that is detected from Earth is proportional to the reflectivity multiplied by the surface area, but astronomers had used too high a value for the reflectivity. The asteroid was named after the famous diarist Anne Frank, who died during the Holocaust.

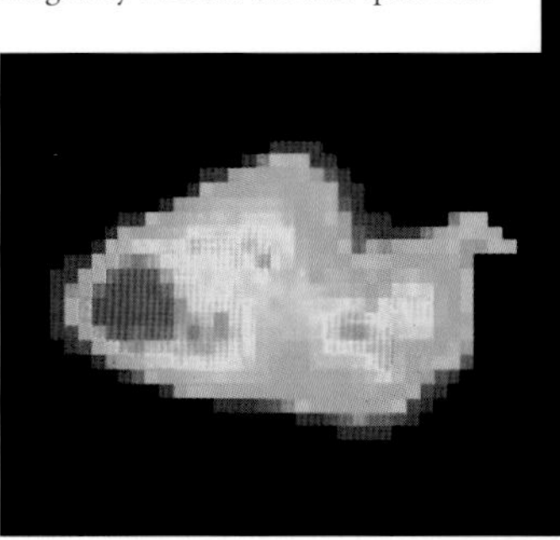

**SURFACE BRIGHTNESS**
False colours (left) are used to highlight differences in brightness over the surface of the asteroid (above). The variations are mainly due to dusty soil layers reflecting different amounts of sunlight in different directions.

NEAR-EARTH ASTEROID

## 4179 Toutatis

**AVERAGE DISTANCE TO SUN** 376 million km (234 million miles)

**ORBITAL PERIOD** 3.98 years

**ROTATION PERIOD** 5.4 and 7.3 days

**LENGTH** 4.26km (2.65 miles)

**DATE OF DISCOVERY** 4 January 1989

Toutatis was named after a Celtic god (who, incidentally, appears in the Asterix comic books). A typical near-Earth asteroid, it sweeps past the planet nearly every four years. In September 2004, it came as close as just four times the distance of the Earth to the Moon. Toutatis is an S-class asteroid, similar to a stony-iron meteorite in composition. It tumbles in space rather like a rugby ball after a botched pass, spinning around two axes, with periods of 5.4 and 7.3 days.

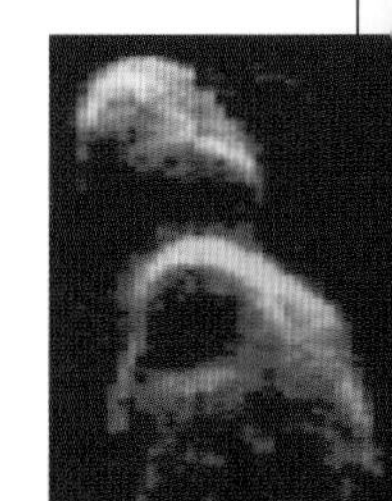

**RADAR IMAGE**

MAIN-BELT ASTEROID

## 4 Vesta

**AVERAGE DISTANCE TO SUN** 353 million km (219 million miles)

**ORBITAL PERIOD** 3.63 years

**ROTATION PERIOD** 5.34 hours

**DIAMETER** 560km (348 miles)

**DATE OF DISCOVERY** 27 March 1807

Vesta is a big asteroid with a surface that reflects a large percentage of the incoming light. This makes it the brightest asteroid in the night sky and the only one visible to the unaided eye. It is also the only asteroid known to have distinctive light and dark areas on its surface. Most asteroids of Vesta's size are expected to be nearly spherical, but Vesta's shape has been distorted by a massive impact, producing a huge crater 460km (285 miles) across and 13km (8 miles) deep. The crater is almost as wide as Vesta itself, and has a mountainous central peak, like some of the craters on the Moon. Some of the fragments of Vesta's crust produced by the cratering process are still trailing Vesta on similar orbits. Others have hit Earth and been recognized as strange basaltic rock meteorites. Six per cent of Earth's recent meteorite falls are Vesta-like in their mineralogical make-up.

The composition is similar to the lava that spews out of Hawaiian volcanoes. In its early life, Vesta melted and then resolidified, with denser material sinking to the core. It now has a layered structure – like the planets – with a low-density crust lying above layers of pyroxene and olivine, and iron. It is thought to be the only remaining differentiated asteroid in the Main Belt. With a density slightly less than that of Mars, Vesta is one of the densest asteroids known.

**METEORITE**
This meteorite, which landed in Western Australia in October 1960, originated from Vesta.

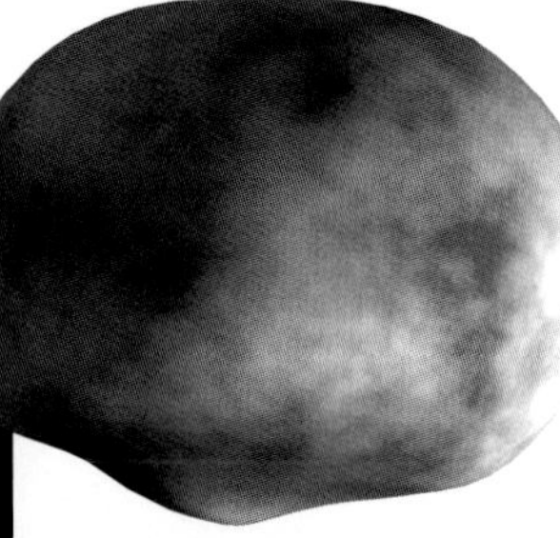

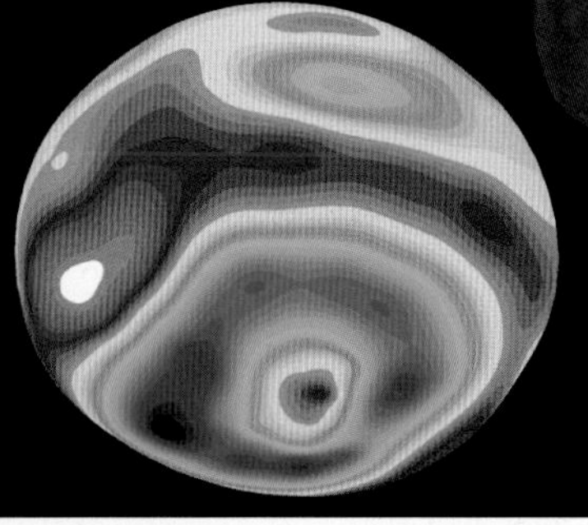

**TWO VIEWS OF VESTA**
Using the 78 images taken by Hubble in 1996, powerful computers created a 3-D model of Vesta (above) and a coloured composite of the surface (left). The latter shows a huge crater with a central peak near Vesta's South Pole.

MAIN-BELT ASTEROID

## 253 Mathilde

**AVERAGE DISTANCE TO SUN** 396 million km (246 million miles)

**ORBITAL PERIOD** 4.31 years

**ROTATION PERIOD** About 418 hours

**LENGTH** 66km (41 miles)

**DATE OF DISCOVERY** 12 November 1885

Mathilde is the biggest asteroid visited by a space probe so far, but, because it spins very slowly, only about half the surface was imaged. It is a primitive carbonaceous asteroid with a density much lower than that of most rocks, suggesting that it is full of holes. Mathilde is probably a compacted pile of rubble.

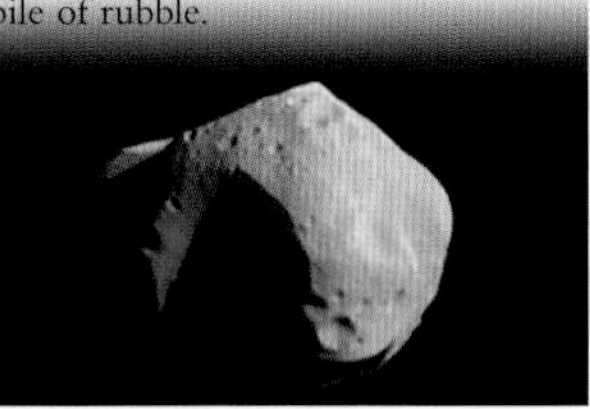

**WEDGE-SHAPED CRATER**

MAIN-BELT ASTEROID

## 243 Ida

**AVERAGE DISTANCE TO SUN** 428 million km (266 million miles)

**ORBITAL PERIOD** 4.84 years

**ROTATION PERIOD** 4.63 hours

**LENGTH** 60km (37 miles)

**DATE OF DISCOVERY** 29 September 1884

Ida was one of 119 asteroids discovered by the Austrian astronomer Johann Palisa, who, together with Max Wolf of Heidelberg, Germany, was a pioneer in the use of photography to produce star maps and hunt for minor planets (another name for asteroids). Ida is a member of the Koronis family. Asteroidal families were discovered by the Japanese astronomer Hirayama Kiyotsugu in 1918. He found that there were groups of asteroids with very similar orbital parameters. The individual members were strung out on one orbit and formed a stream of minor bodies in the inner Solar System (see p.208). Koronis is the most prominent member of Ida's family.

Ida is famous because the Galileo spacecraft imaged it in detail as it flew within 11,000km (6,800 miles) during August 1993, on its way to Jupiter. As Ida spins every 4 hours 36 minutes, Galileo was able to image most of the surface during the flyby. Ida was originally thought to be an S-type asteroid like Gaspra (see opposite), but observations revealed that its density is too low, and it is more likely to be a C-type asteroid. It has about five times more craters per unit area than Gaspra, indicating that its surface is considerably older. The most exciting outcome of the Galileo flyby was the discovery that Ida has its own moon, Dactyl. This binary system is thought to have been formed during the asteroid collision and break-up that created the Koronis family.

**DACTYL**
At just 1.6km (1 mile) long, Dactyl is tiny. Its orbit around Ida is nearly circular, with a radius of about 90km (56 miles) and an orbital period of about 27 hours.

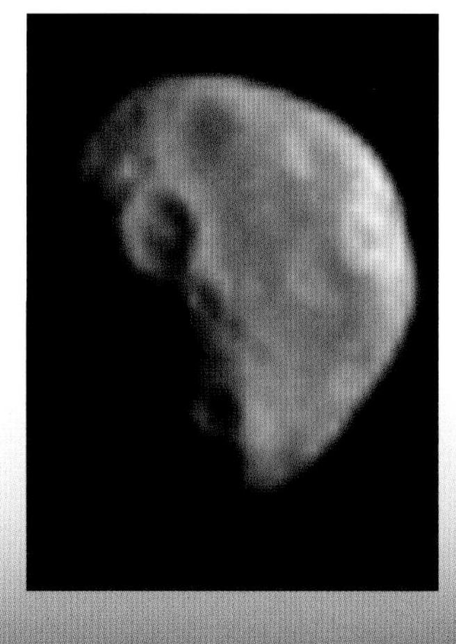

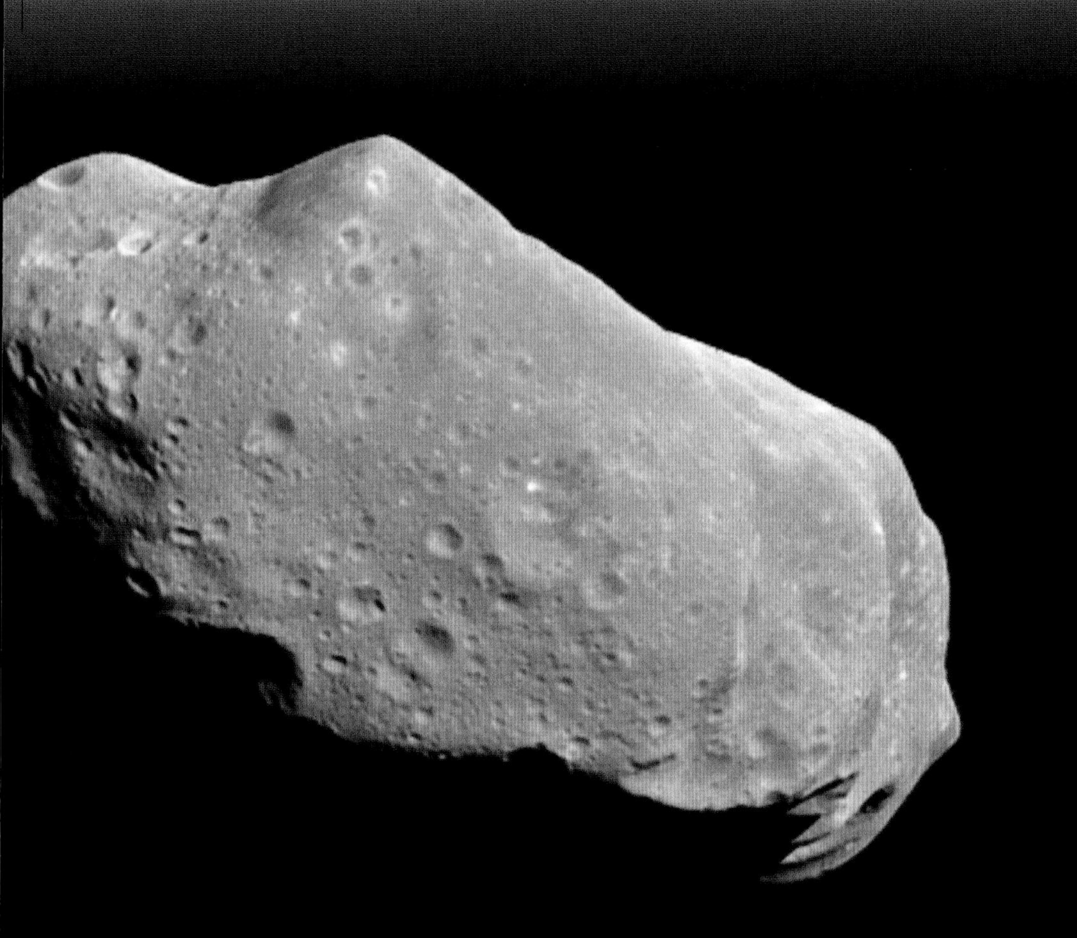

**IDA AND ITS MOON**
Dactyl was the first asteroid satellite to be discovered. Ann Harch, a Galileo mission member, noticed it when examining images that had been stored on the spacecraft when it passed Ida six months earlier.

NEAR-EARTH ASTEROID

## 433 Eros

**AVERAGE DISTANCE TO SUN** 218 million km (136 million miles)

**ORBITAL PERIOD** 1.76 years

**ROTATION PERIOD** 5.27 hours

**LENGTH** 31km (19.25 miles)

**DATE OF DISCOVERY** 13 August 1898

Lying in near-Earth orbit, outside the Main Belt, Eros is usually closer to the Sun than to Mars (see p.208). Its orbit also brings it close to Earth – at the last close approach, in 1975, Eros came within 22 million km (14 million miles) of the planet. The orbit is unstable, and Eros has a one-in-ten chance of hitting either Earth or Mars in the next million years. In 1960, Eros was detected by radar, and infrared measurements taken in the 1970s indicated that the surface was not just bare rock but was covered by a thermally insulating blanket of dust and rock fragments. Eros was the first asteroid to be orbited by a spacecraft and the first to be landed on. It was chosen for close study because it is big and nearby.

On 14 February 2000, the Near Earth Asteroid Rendezvous (NEAR) spacecraft (renamed NEAR Shoemaker in March 2000) went into orbit around Eros. It landed 363 days later. About 160,000 images were recorded. They revealed an irregularly shaped body, which had heavily cratered 2-billion-year-old areas lying next to relatively smooth regions. Even though the gravitational field is very small, several thousand boulders larger than 15m (50ft) across have fallen back to the surface after being ejected by impacts, and some surface dust has rolled down the slopes to form sand dunes a few metres deep. Laser measurements of the NEAR–Eros distance as the spacecraft orbited have not only produced an accurate map of the asteroid's shape but also indicated that the interior is nearly uniform, with a density about the same as that of the Earth's crust. Eros is not a pile of rubble like Mathilde; it is a single, solid lump of rock. The spacecraft's gamma-ray spectrometer worked for two weeks after touchdown. Eros was found to be silicate-rich and highly reflective.

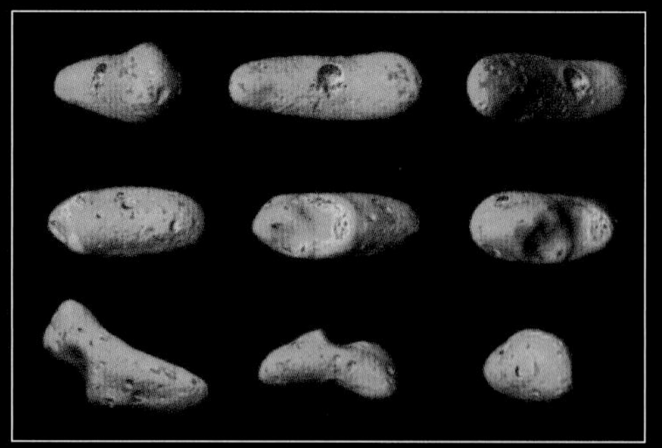

**COMPUTER MODELS**
The gravity on Eros is about 1/2,000th of that on Earth, but varies by nearly a factor of two from place to place. The colours in this image represent the rate at which a rock would roll downhill. It would roll fastest in the red areas, and it wouldn't move at all in the blue areas.

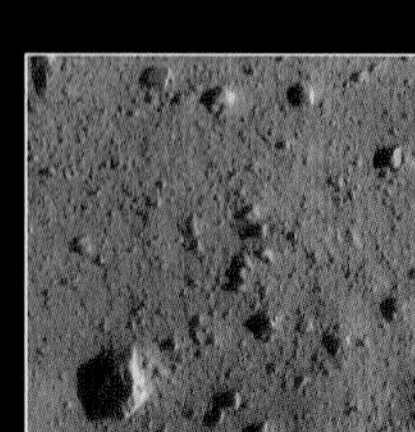

**CLOSING IN**
NEAR Shoemaker took this image from a height of 1,150m (3,770ft) shortly before it touched down on Eros.

**ROCK AND REGOLITH**
Some of the rocks and regolith on Eros's surface have a red colouring. The longer their exposure to minor impacts and the solar wind, the redder they appear.

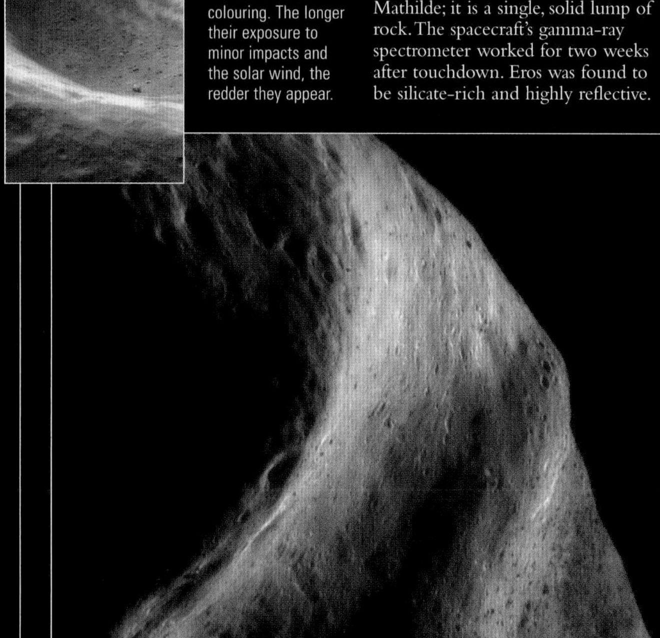

**THE SADDLE**
Four images have been combined to produce this view of the "saddle" region at the south of the asteroid. This 10km- (6-mile-) wide scoop, which has been named Himeros, is relatively boulder-free, unlike the region at the lower right of the frame.

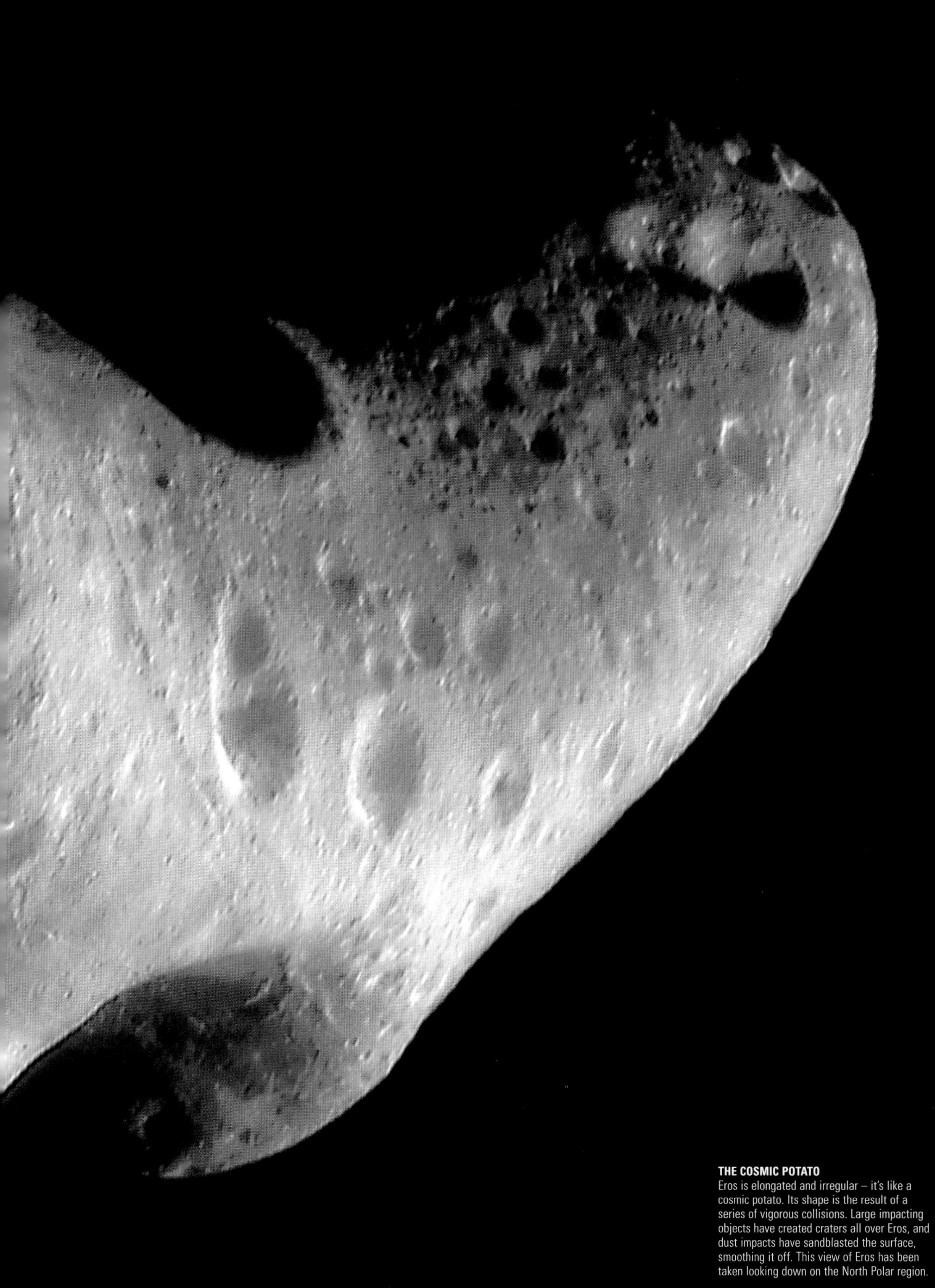

**THE COSMIC POTATO**
Eros is elongated and irregular – it's like a cosmic potato. Its shape is the result of a series of vigorous collisions. Large impacting objects have created craters all over Eros, and dust impacts have sandblasted the surface, smoothing it off. This view of Eros has been taken looking down on the North Polar region.

# COMETS

36–37 Gravity, motion, and orbits
206–207 Kuiper Belt and Oort Cloud
Meteors and Meteorites 220–21

COMETS PRODUCE A STRIKING celestial spectacle when they enter the inner Solar System. Their small nuclei become surrounded by a bright cloud, or coma, of dust and gas about 100,000km (60,000 miles) across. Large comets that get close to the Sun also produce long, glowing tails that can extend many tens of millions of kilometres into space and are bright enough to be seen in Earth's sky.

**SPECTACULAR SIGHT**
Most comets are only discovered when they are bright enough to glow in Earth's sky. Comet Hale–Bopp was discovered in late July 1997 and could be seen for several weeks afterwards. It will return to Earth's sky again around AD 5400.

## ORBITS

Cometary orbits divide into two classes. Short-period comets orbit the Sun in the same direction as the planets. Most have orbital periods of about seven years, and get no further from the Sun than Jupiter. Short-period comets were captured into the inner Solar System by the gravitational influence of Jupiter. If they remain on these small orbits, they will decay quickly. Some, however, will be ejected by Jupiter onto much larger orbits,and then possibly recaptured. Intermediate- and long-period comets have orbital periods greater than 20 years (see p.216). Their orbital planes are inclined at random to the plane of the Solar System. Many of these comets travel huge distances into the interstellar regions. Most of the recorded comets get close to the Sun, where they develop comae and tails and can be easily discovered. There are vast numbers of comets on more distant orbits that are too faint to be found.

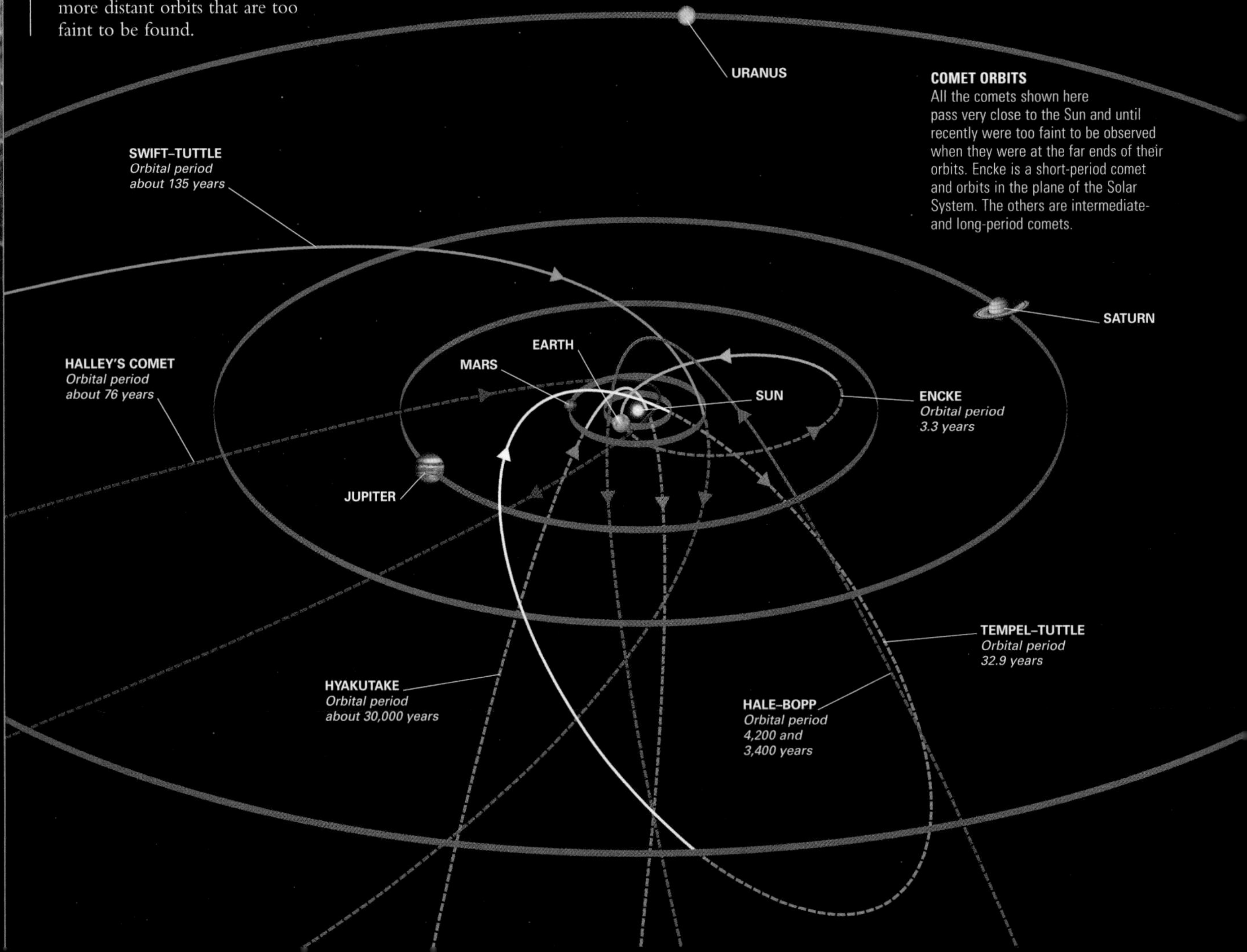

URANUS

**COMET ORBITS**
All the comets shown here pass very close to the Sun and until recently were too faint to be observed when they were at the far ends of their orbits. Encke is a short-period comet and orbits in the plane of the Solar System. The others are intermediate- and long-period comets.

**SWIFT–TUTTLE**
*Orbital period about 135 years*

SATURN

EARTH

MARS

**HALLEY'S COMET**
*Orbital period about 76 years*

SUN

**ENCKE**
*Orbital period 3.3 years*

JUPITER

**TEMPEL–TUTTLE**
*Orbital period 32.9 years*

**HYAKUTAKE**
*Orbital period about 30,000 years*

**HALE–BOPP**
*Orbital period 4,200 and 3,400 years*

## STRUCTURES

The fount of all cometary activity is a low-density, fragile, irregularly shaped, small nucleus that resembles a "dirty snowball". The dirt is silicate rock in the form of small dust particles. The snow is mainly composed of water, but about 1 in 20 molecules are more exotic, being carbon dioxide, carbon monoxide, methane, ammonia, or more complex organic compounds. The nucleus is covered by a thin, dusty layer, which is composed of cometary material that has lost snow from between its cracks and crevices. The snow is converted directly from the solid into the gaseous state by the high level of solar radiation the comet receives when it is close to the Sun.

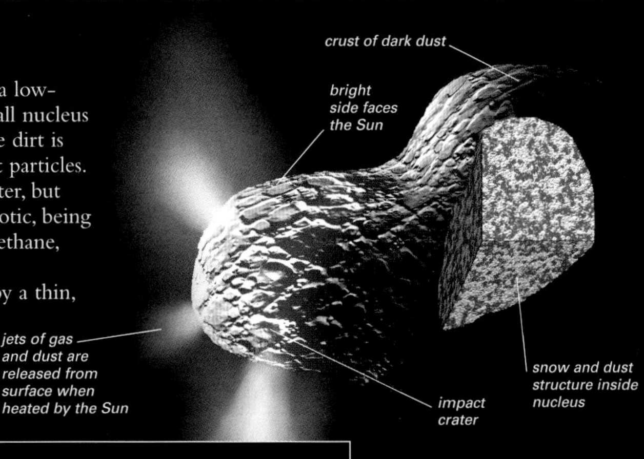


**CROSS-SECTION**
The nucleus has a uniform structure, consisting of many smaller "dirty snowballs". The surface dust layer is only a few centimetres thick and appears dark because it reflects little light. The strength of the whole structure is negligible. Not only do tidal forces pull comets apart, but many simply fragment at random.

**NUCLEUS**
The central part of Comet Borrelly's elongated nucleus has a smooth terrain, but the more "mottled" regions consist of steep-sided hills that are separated by pits and troughs.

### FRED WHIPPLE

Fred Whipple (1906–2004) was an astronomy professor at Harvard, USA, and the director of the Smithsonian Astrophysical Observatory from 1955 to 1973. In 1951, he introduced the "dirty snowball" model of the cometary nucleus, in which the snowball spins. As the Sun heated one side, its heat was slowly transmitted down to the underlying snows, which eventually turn straight to gas. This resulted in a jet force along the cometary orbit which either accelerated or decelerated the nucleus depending on the direction of its spin.

## LIFE CYCLES

A comet spends the vast majority of its life in a dormant, deep-freeze state. Activity is triggered by an increase in temperature. When the comet gets closer to the Sun than the outer part of the Main Belt (see p.208), frozen carbon dioxide and carbon monoxide in the nucleus start to sublime (that is, they pass directly from the solid to the gaseous state). Once the comet is inside the orbit of Mars, it is hot enough for water to join in the activity. The nucleus quickly surrounds itself with an expanding spherical cloud of gas and dust, called a coma. The coma is at its maximum size when the comet is closest to the Sun. A comet that passes through the inner Solar System will lose the equivalent of a 2m- (6ft-) thick layer from its surface. The comet moving away from the Sun is thus smaller than it was on its approach. Mass is lost every time a comet passes perihelion. Borrelly, for example, orbits the Sun about every seven years. If it stays on the same orbit, its 3.2km- (2-mile-) wide nucleus will be reduced to nothing in about 6,000 years. Comets are transient members of the inner Solar System. They are soon dissipated by solar radiation. Large cometary dust particles form a meteoroid stream around the orbit. Gas molecules and small particles of dust are just blown away from the Sun and join the galactic disc.

**CRATER CHAIN**
This 200km- (120-mile-) long chain of impact craters, named Enki Catena, is on Ganymede, the largest of Jupiter's moons. It is likely that Ganymede was struck by 12 or so fragments of a comet that had just been pulled apart by tidal forces as it passed too close to Jupiter.

dust tail is curved
gas tail is straight and narrow
tails are longest close to the Sun
tails shrink as the comet moves away from the Sun
perihelion
Sun
a comet's tail always points away from the Sun
tails grow as the comet travels towards the Sun
naked nucleus
aphelion

**COMETARY TAILS**
As a comet nears the Sun, it develops two tails. The curved tail is formed of dust that is pushed away by solar radiation. The straight tail consists of ionized gas that has been blown away from the coma by the solar wind.

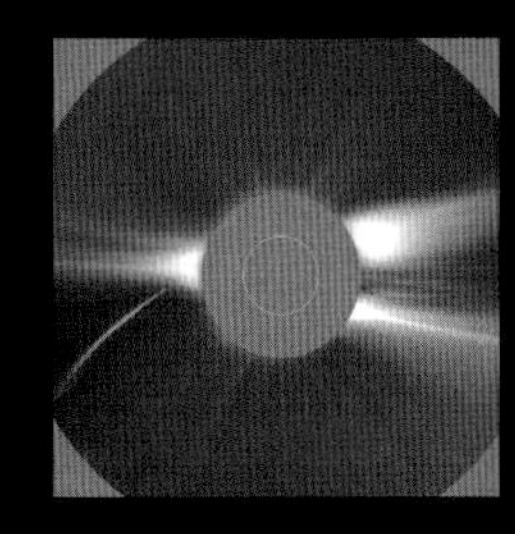

**COMET SOHO-6**
Large numbers of sungrazing comets have been discovered by the SOHO satellite. Here Soho-6 is seen as an orange streak, at left, approaching the masked Sun.

**HALLEY'S TAIL**
These 14 images of Halley's Comet were taken between 26 April and 11 June 1910, around the time it passed perihelion. An impressive tail was produced and dissipated in just seven weeks of its 76-year orbit.

# COMETS

There are billions of comets at the edge of the Solar System, but very few have been observed, for they are bright enough to be seen in Earth's sky only when they travel into the inner Solar System and approach the Sun. Nearly 900 comets have been recorded and their orbits calculated thus far. About 200 of the catalogued comets are periodic, having orbital periods of less than 20 years (short period) or between 20 and 200 years (intermediate period). Most, but not all, comets are named after their discoverers.

**COMET HALE–BOPP**
Caught in the evening sky above Germany in 1997, Hale–Bopp, one of the brightest comets of the 20th century, clearly has two tails.

INTERMEDIATE-PERIOD COMET

## Ikeya–Seki

**CLOSEST APPROACH TO THE SUN** 470,000km (290,000 miles)

**ORBITAL PERIOD** 184 years

**FIRST RECORDED** 8 September 1965

This comet is named after the two amateur Japanese comet hunters, Ikeya Kaoru and Seki Tsutomu, who discovered it independently (and within five minutes of each other) in 1965. On 21 October 1965, as it passed perihelion, the comet was so bright that it was visible in the noon sky only 2 degrees from the Sun. Tidal forces then caused the nucleus to split into three parts. Ikeya–Seki faded quickly as it moved away from the Sun, but the tail grew until it extended over 60 degrees across the sky. At this stage it was 195 million km (121 million miles) from the Sun.

**SUNGRAZER**
Ikeya–Seki is a sungrazer and passed within just 470,000km (290,000 miles) of the Sun's surface in 1965. It is one of over 200 comets in the Kreutz sungrazer family.

LONG-PERIOD COMET

## Great Comet of 1680

**CLOSEST APPROACH TO THE SUN** 940,000km (580,000 miles)

**ORBITAL PERIOD** 9,400 years

**FIRST RECORDED** 14 November 1680

This comet has two great claims to fame. It was the first comet to be discovered by telescope and the first to have a known orbit. Some 70 years after the telescope was invented, the German astronomer Gottfried Kirch found the comet by accident when observing the Moon in 1680. The orbit was calculated by the English mathematician Isaac Newton using his new theory of universal gravity, and the results were published in his masterpiece *Principia* in 1687. The comet is a sungrazer and was seen twice: first, as a morning phenomenon, when it was approaching the Sun; and subsequently in the evenings, when it was receding. Newton was the first to realize that these apparitions were of the same comet. The English physicist Robert Hooke noticed a stream of light issuing from the nucleus. This was the first description of jets of material emanating from active areas.

**GREAT COMETS**
Great comets, such as this 1680 comet, are extremely bright and can be very startling when they appear.

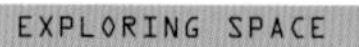

EXPLORING SPACE

### COMET ORBIT

Isaac Newton made observations of the Great Comet of 1680. At the time, a conventional view held that comets travelled in straight lines, passing through the Solar System only once. Based on his observations, Newton realized that he had seen a comet travelling around the Sun on a parabolic curve. In 1687, in the *Principia*, he used his study of comets and other phenomena to confirm his law of universal gravitation. He also showed how to calculate a comet's orbit from three accurate observations of its position. Using Newton's laws, Edmond Halley successfully predicted the return of the comet named after him.

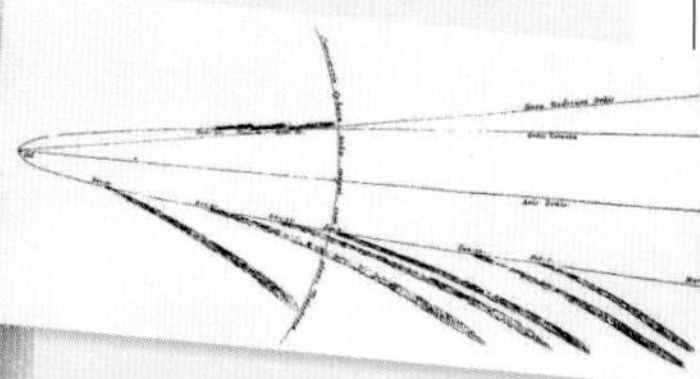

NEWTON'S ORBIT SKETCH

INTERMEDIATE-PERIOD COMET

## Swift–Tuttle

**CLOSEST APPROACH TO THE SUN** 143 million km (88 million miles)

**ORBITAL PERIOD** About 135 years

**FIRST RECORDED** 16 July 1862

After Swift–Tuttle's discovery in 1862, calculations of its orbit established the relationship between comets and meteoroid streams. Every August, the Earth passes through a stream of dust particles that produces the Perseid meteor shower, named after the constellation from which the shooting stars appear to be emanating. In 1866 Giovanni Schiaparelli (see p.220), the director of the Milan Observatory in Italy, calculated the mean orbit of the Perseid meteoroids. He immediately realized that this orbit was very similar to that of Comet Swift–Tuttle, which intersects Earth's path. He concluded that meteoroid streams were produced by the decay of comets, the meteoroids being no more than cometary dust particles, a fraction of a gram in mass, hitting the Earth's upper atmosphere at velocities of about 216,000kph (134,000mph). About the same number of Perseid meteors are seen each year, so the dust must be evenly spread around the cometary orbit. This uniformity takes a long time to come about. Swift–Tuttle must have passed the Sun on the same orbit a few hundred times to produce this effect. Comets are decaying but they have to pass through the inner Solar System a thousand times or so before they are whittled down to nothing.

**PERSEIDS**
It takes about two weeks for the Earth to pass through this meteoroid stream. The peak rate is on 12 August at about 50 visible meteors per hour.

**PERIODIC COMET**
Swift–Tuttle was discovered independently by American astronomers Lewis Swift and Horace Tuttle in 1862. This optical image was taken in 1992, when the comet approached the Sun once again.

LONG-PERIOD COMET

## West

**CLOSEST APPROACH TO THE SUN** 29 million km (18 million miles)

**ORBITAL PERIOD** About 500,000 years

**FIRST RECORDED** 5 November 1975

This comet was one of the first to have a spectrum of hydroxyl (OH) detected. Comet West was discovered by Richard West, an astronomer at the European Southern Observatory, when he examined a batch of photographic plates taken by the 100cm (39.4in) Schmidt telescope at La Silla, Chile. It was on the inner edge of the asteroid belt, on its way towards the Sun. At the time the comet was visible only from the southern hemisphere. During February 1976, the comet not only moved into the northern sky, but also brightened impressively. By the end of February, it was easily visible to the naked eye. It was closest to the Sun on 25 February. Just before it reached perihelion, the nucleus of the comet broke into two. A week or so later it split further and the comet eventually broke into four pieces. These could be seen gradually moving away from each other throughout March, and they all developed a separate tail.

Rocket-borne spectrometers were used to investigate Comet West. These looked at ultraviolet radiation, a region of the spectrum containing hydroxyl bands. These are important because cometary snow contains water molecules that divide into hydrogen (H) and hydroxyl (OH) ions when they are released from the nucleus. By studying the comet with spectrometers, it was possible to measure how much water was lost as it approached perihelion.

**TAIL BANDS**
The striations that can be seen in Comet West's tail are known as synchronic bands. Each band is produced by a puff of dust emitted from the spinning nucleus.

**EARTHGRAZER**
When it passed within Earth's orbit, Hyakutake became one of the brightest comets of the 20th century.

LONG-PERIOD COMET

## Hyakutake

**CLOSEST APPROACH TO THE SUN** 34.4 million km (21.4 million miles)

**ORBITAL PERIOD** About 30,000 years

**FIRST RECORDED** 30 January 1996

This comet became a Great Comet not (like Hale–Bopp) because the nucleus was big, but because on 24 March 1996 it got to within a mere 15 million km (9 million miles) of Earth. It was discovered by the Japanese amateur astronomer Hyakutake Yuji using only a pair of high-powered binoculars. The comet became so bright that large radio-telescope spectrometers could detect minor constituents in the coma, such as a compound of water and deuterium (HDO) and methanol ($CH_3OH$). Hyakutake was the first comet to be observed to emit X-rays. Subsequently, it was found that other comets are also sources of X-rays, the rays being produced when electrons in the coma are captured by ions in the solar wind. On 1 May 1996, the Ulysses spacecraft detected Hyakutake's gas tail when 570 million km (355 million miles) from the nucleus. This is the longest comet tail to be detected thus far. Sections of Hyakutake's gas tail have disconnected due to interactions between magnetic fields in the solar wind and the tail.

**TELESCOPIC VIEW**
In March and April 1996, superb short-exposure photographs of Hyakutake could be obtained using only large telephoto lenses or small telescopes.

SHORT-PERIOD COMET

## Encke

**CLOSEST APPROACH TO THE SUN** 51 million km (32 million miles)

**ORBITAL PERIOD** 3.3 years

**FIRST RECORDED** 17 January 1786

Comet Encke was "discovered" in 1786 (by the French astronomer Pierre Méchain), in 1795 (by the German-born astronomer Caroline Herschel), and in 1805 and 1818-19 (by the French astronomer Jean Louis Pons). These comets were found to be the same only after orbital calculations in 1819 by the German astronomer Johann Encke, who then predicted its return in 1822. Comet Encke is unusual in that, like Halley's Comet, it is not named after its discoverer. It has the shortest period of any known comet and has been seen returning to the Sun on over 59 occasions. The orbit is also shrinking in size, for Encke comes back to perihelion about 2.5 hours sooner than it should. Some astronomers have suggested that this is due to the comet ploughing through a resistive medium in the Solar System. But other comets have returned later than predicted, and the time error has varied from one orbit to the next. Astronomers have realized that the changing orbits were caused by the "jet effect" of gas escaping from the comet's nucleus. The comet receives a push from the expanding gases and, depending on its direction of spin in relation to its orbit, it is either accelerated or decelerated.

**COMA**
Not all comets have tails. Some, such as Encke, just have a dense spherical envelope of gas and dust around the nucleus called the coma. The density of the gas decreases as it flows away from the nucleus. Cometary comae have no boundaries; they just fade away.

INTERMEDIATE-PERIOD COMET

## Halley's Comet

**CLOSEST APPROACH TO THE SUN** 88 million km (55 million miles)

**ORBITAL PERIOD** About 76 years

**FIRST RECORDED** 240 BC

In 1696, Edmond Halley, England's second Astronomer Royal, reported to the Royal Society in London that comets that had been recorded in 1531, 1607, and 1682 had very similar orbits. He concluded that this was the same comet returning to the inner Solar System about every 76 years, moving under the influence of the newly discovered solar gravitational force. What is more, Halley predicted that the comet would return in 1758. Halley's Comet was the first periodic comet to be discovered. This indicated that at least some comets were permanent members of the Solar System.

Orbital analysis has revealed that Halley's Comet has been recorded 30 times, the first known sighting being in Chinese historical diaries of 240 BC. The last appearance, in 1986, was 30 years after the start of the space age, and five spacecraft visited the comet. The most productive was ESA's Giotto mission. This flew to within 600km (370 miles) of the nucleus and took the first ever pictures. Giotto proved that cometary nuclei are large, potato-shaped dirty snowballs and that the majority of the snow is water ice. Halley was about 150 million km (93 million miles) from the Sun when Giotto encountered it. Only about 10 per cent of the surface was actively emitting gas and dust at the time. On average, a comet loses a surface layer about 2m (6.5ft) deep every time it passes through the inner Solar System. At this rate, Halley's Comet will survive for about another 200,000 years.

**NUCLEUS**
Giotto revealed that Halley's nucleus is 15.3 km (9.5 miles) long. The brightest parts of this image are jets of dust streaming towards the Sun.

**HALLEY AGAINST A STAR FIELD**
This photograph was taken from Australia on 11 March 1986, three days before the comet was visited by the Giotto spacecraft.

MYTHS AND STORIES

### CELESTIAL OMEN

Some superstitious people regard comets as portents of death and disaster. Before Edmond Halley's work, all comets were unexpected. They were often compared to flaming swords. England's King Harold II was worried by the appearance of Halley's Comet in 1066. But what was a bad omen for him was a good sign for the Norman Duke William, who conquered Harold at Hastings.

**BAYEUX TAPESTRY**
This crewel embroidery beautifully depicts the coma and tail of Halley's Comet (top left), as seen in 1066. It looks like a primitive rocket spewing out flames.

LONG-PERIOD COMET

## Hale–Bopp

**CLOSEST APPROACH TO THE SUN** 137 million km (85 million miles)

**ORBITAL PERIOD** 4,200 and 3,400 years

**FIRST RECORDED** 23 July 1995

Comet Hale–Bopp was discovered independently and accidentally by the American amateur astronomers Alan Hale and Thomas Bopp, who were looking at Messier objects in the clear skies of the western USA (it was close to M70). Later, after the orbit had been calculated, Hale–Bopp was found to be at a distance of over 1 billion km (620 million miles). This is between the orbits of Jupiter and Saturn and is an almost unprecedented distance for the discovery of a non-periodic comet. The orbit showed that it had been to the inner Solar System before, some 4,200 years ago, but because it passed close to Jupiter a few months after discovery it will return again in about 3,400 years. Hale–Bopp passed perihelion on 1 April 1997. It was one of the brightest comets of the century, not because, like Hyakutake, it came very close to Earth, but simply because it had a huge nucleus, about 35km (22 miles) across.

**TWIN TAILS**
The two tails of Comet Hale–Bopp shine brightly over the Little Ajo mountains in Arizona, USA, shortly after sunset in 1997.

SHORT-PERIOD COMET

## Giacobini–Zinner

**CLOSEST APPROACH TO THE SUN** 155 million km (96 million miles)

**ORBITAL PERIOD** 6.61 years

**FIRST RECORDED** November 1900

This comet was the first to be investigated in situ. The International Comet Explorer spacecraft flew through the tail about 7,800km (4,840 miles) from the nucleus on 11 September 1985. The measurements concentrated on the way in which the plasma in the solar wind interacted magnetically with the expanding atmosphere of the comet. In 1946, Earth crossed the comet's path just 15 days after it had passed. About 2,300 meteors per hour were recorded.

**GIACOBINI–ZINNER IN 1905**

SHORT-PERIOD COMET

## Churyumov–Gerasimenko

**CLOSEST APPROACH TO THE SUN** 194 million km (121 million miles)

**ORBITAL PERIOD** 6.59 years

**FIRST RECORDED** 20 September 1969

In 1969, the Russian astronomer Klim Churyumov was inspecting a photographic plate taken by Svetlana Gerasimenko to see if he could find an image of Comet Comas Solá and made an exciting new discovery instead. Churyumov–Gerasimenko is a typical short-period comet, keeping between the orbits of Mars and Jupiter as it travels round the Sun. It has recently become famous because it is now the target of ESA's Rosetta mission. This orbiting spacecraft was intended to go to Comet Wirtanen, but the launch was delayed due to problems with the Ariane 5 rocket. It finally launched on 2 March 2004. Rosetta will go into orbit around Churyumov–Gerasimenko in November 2014, when it is 790 million km (490 million miles) away from the Sun. The comet's nucleus will be cold and inactive, enabling a small lander, called Philae, to perform a Solar System first by touching down on the surface. Rosetta and Philae will then stay with the comet as it travels into the inner Solar System and will monitor the way in which the activity "switches on".

**SHRINKING TAIL**
This image was taken 82 days after Comet Churyumov-Gerasimenko passed perihelion in 2002. A small tail can still be seen.

**NUCLEUS**
Churyumov–Gerasimenko's nucleus is shaped like a rugby ball and 5km (3 mile) long. It is much smaller than the bright, white central region of the cometary coma.

SHORT-PERIOD COMET

## Borrelly

**CLOSEST APPROACH TO THE SUN** 203 million km (126 million miles)

**ORBITAL PERIOD** 6.86 years

**FIRST RECORDED** 28 December 1904

The flyby of NASA's Deep Space 1 mission on 22 September 2001 revealed that this periodic comet has a nucleus shaped like a bowling pin, about 8km (5 miles) long. Reflecting on average only 3 per cent of the sunlight that hits it, Borrelly has the darkest known surface in the inner Solar System. Any ice in the nucleus is hidden below the hot and dry, mottled, sooty black surface.

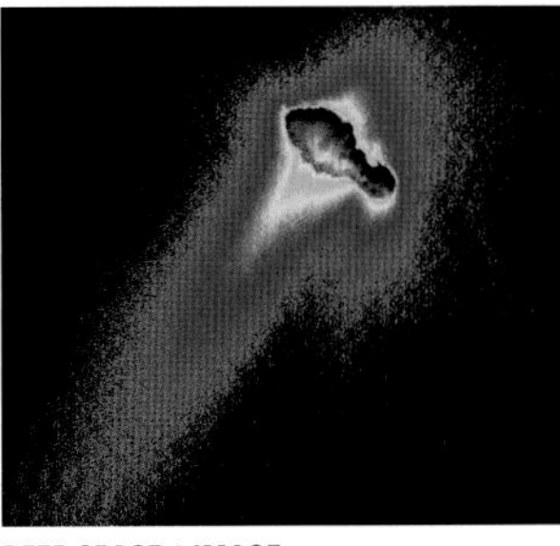

**DEEP SPACE 1 IMAGE**
The production of the jets of gas and dust emanating from Borrelly's nucleus is eroding the surface. There is a possibility that the nucleus will split in two in the future.

SHORT-PERIOD COMET

## Wild 2

**CLOSEST APPROACH TO THE SUN** 236 million km (147 million miles)

**ORBITAL PERIOD** 6.39 years

**FIRST RECORDED** 6 January 1978

Wild 2 is a relatively fresh comet that was brought into an orbit in the inner Solar System as recently as September 1974, when it had a close encounter with Jupiter. It is too faint to be seen with the naked eye as its nucleus is only 5.5km (3.4 miles) long. Wild 2's present path round the Sun takes it very close to the orbits of both Mars and Jupiter. It may oscillate between its present orbit and an orbit with a period of about 30 years that brings it only as close as Jupiter. Wild 2 was chosen as target for NASA's Stardust mission (see panel, below) because the spacecraft could fly by at the relatively low speed of 21,900kph (13,600mph), capturing comet dust on the way.

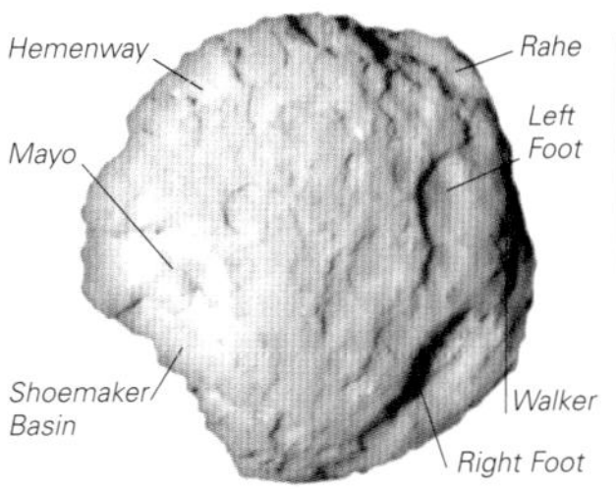


**CLOSE-UP OF NUCLEUS**
The surface of the nucleus is covered by steep-walled depressions hundreds of metres deep. They are mostly named after famous cometary scientists.

EXPLORING SPACE

### THE STARDUST MISSION

The Stardust spacecraft flew by Wild 2 on 2 January 2004. It captured both interstellar dust and dust blown away from the comet's nucleus. Aerogel placed on an extended tennis-racket-shaped collector was used to capture the particles without heating them up or changing their physical characteristics. The craft returned to Earth in 2006 and the collector, stowed in a canister, parachuted to safety in the desert in Utah, USA.

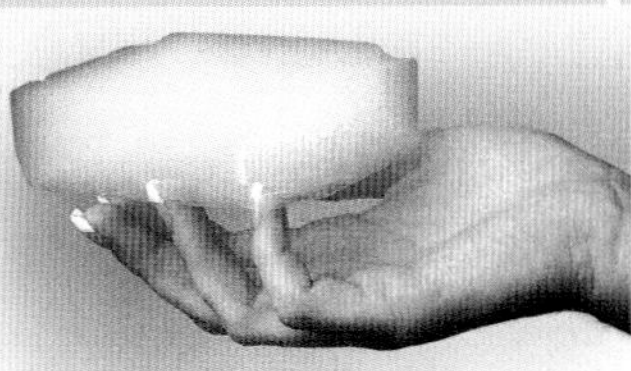

**AEROGEL**
Although it has a ghostly appearance, aerogel is solid. It is a silicon-based sponge-like foam, 1,000 times less dense than glass.

SHORT-PERIOD COMET

## Shoemaker–Levy 9

**ORBITAL DISTANCE FROM JUPITER** 90,000 km (56,000 miles)

**ORBITAL PERIOD AROUND JUPITER** 2.03 years

**FIRST RECORDED** 25 March 1993

Unlike normal comets, this one was discovered in orbit around Jupiter by the American astronomers Gene and Carolyn Shoemaker and David Levy. Even more remarkably, it was in 22 pieces, having been ripped apart on 7 July 1992, when it passed too close to Jupiter. These fragments subsequently crashed into the atmosphere in Jupiter's southern hemisphere in July 1994 (see p.179). Observatories all over the world and the Hubble Space Telescope witnessed the sequence of events. The nucleus was originally just over 1km (0.6 miles) across and was most likely captured by Jupiter in the 1920s.

**SHATTERED NUCLEUS**
The bright streak at the centre of this image (which covers 1 million km/620,000 miles) is the string of nuclei and associated comae.

### CAROLYN SHOEMAKER

After taking up astronomy at the age of 51 after her three children had grown up, Carolyn Shoemaker (b.1929) has now discovered over 800 asteroids and 32 comets. She uses the 46cm (18in) Schmidt wide-angle telescope at the Palomar Observatory in California, USA. Her patience and attention to detail is vital when it comes to inspecting photographic plates that are taken about an hour apart and then studied stereoscopically. Typically, 100 hours of searching are required for each comet discovery. Carolyn was married to Gene Shoemaker (see p.151).

# METEORS AND METEORITES

36–37 Gravity, motion, and orbits
208–11 Asteroids
214–16 Comets
Monthly sky guide 410–85

**POPULARLY KNOWN AS SHOOTING STARS,** meteors are linear trails of light-radiating material produced in Earth's upper atmosphere by the impact of often small, dusty fragments of comets or asteroids called meteoroids. About 1 million visible meteors are produced each day. If the meteor is not completely destroyed by the atmosphere, it will hit the ground and is then called a meteorite. If the meteorite is very large, a crater will be formed by the impact.

## METEOROIDS

Most of the dusty meteoroids responsible for visual meteors come from the decaying surfaces of cometary nuclei. When a comet is close to the Sun, its surface becomes hot, and snow just below the surface is converted into gas. This gas escapes and breaks up the surface of the friable, dusty nucleus and blows small dust particles away from the comet. These dusty meteoroids have velocities that are slightly different to that of their parent comet. This causes them to have slightly different orbits, and as time passes they form a stream of particles all around the original orbit of the comet. This stream is fed by new meteoroids every time the parent comet swings past the Sun. The inner Solar System is full of these streams. Dense streams are produced by large comets that get close to the Sun. Streams with relatively few meteoroids are formed by smaller and more distant comets. As the Earth orbits the Sun, it continually passes in and out of these streams, colliding with some of the meteoroids that they contain. Names are given to some meteor showers that occur at fixed times of year, such as the Leonids (right).

**FIREBALL**
The brightest meteors of all are known as fireballs. They have a magnitude of at least -5, shining more brightly than planets such as Venus and Jupiter.

**LEONID METEOR SHOWER**
Leonid meteors are seen around 17 November every year and are so-called because they appear to pour out of the constellation of Leo. Every 33 years, the shower strengthens into a veritable storm. The woodcut on the right was carved by the Swiss artist Karl Jauslin in 1888; it represents the maximum activity of the 1833 Leonids.

### GIOVANNI SCHIAPARELLI

Giovanni Schiaparelli (1835–1910) was an Italian astronomer who worked at the Brera Observatory in Milan and has two claims to fame. In 1866, he calculated the orbits of the Leonid and Perseid meteoroids and realized that they were similar to the orbits of comets Tempel–Tuttle and Swift–Tuttle respectively. He concluded that cometary decay produced meteoroid streams. In the late 1870s, he went on to map Mars's surface.

## METEORITES

Small extraterrestrial bodies that hit the Earth's atmosphere are completely destroyed during the production of the associated meteor. If, however, the impacting body has a mass of between about 30kg (66lb) and 10,000 tons, only the surface layers are lost during atmospheric entry, and the atmosphere slows down the incoming body until it eventually reaches a "free-fall" velocity of just over 150kph (90mph). The central remnant then hits the ground. The fraction of the incoming body that survives depends on its initial velocity and composition. Meteorites are referred to as "falls" if they are seen to enter and are then picked up just afterwards. Those that are discovered some time later are called "finds". Meteorites are classified as one of three compositional types.

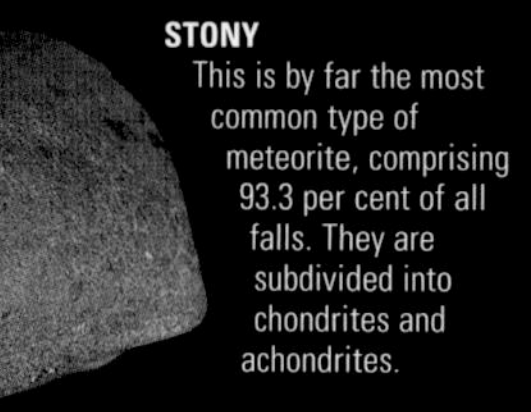

**STONY**
This is by far the most common type of meteorite, comprising 93.3 per cent of all falls. They are subdivided into chondrites and achondrites.

**STONY-IRON**
The rarest meteorites – just 1.3 per cent of meteorite falls – are a mixture of stone and iron-nickel alloy, similar to the composition of the rocky planets.

**IRON**
Iron meteorites make up 5.4 per cent of all falls. They are composed mainly of iron-nickel alloy (consisting of 5–10 per cent nickel by weight) and small amounts of other minerals.

**METEOR TRAIL**
Meteors are randomly occurring narrow streaks of light that shoot across a few tens of degrees of the sky in just a few seconds. A typical meteor will be about 150km (90 miles) away from the observer.

# METEORITE IMPACTS

Earth's atmosphere shields the surface from the vast majority of incoming extraterrestrial bodies. The typical impact velocity at the top of the atmosphere is about 72,000kph (45,000mph), and the leading surface of the meteoroid quickly heats up and starts boiling as a result of hitting air molecules at this speed. Usually the body is so small that it boils away completely. Parts of medium-sized bodies survive to fall as meteorites. A very large body, having a mass greater than about 100,000 tons, is, however, hardly affected by the atmosphere. It punches through the gas like a bullet through tissue paper, energetically slamming into the Earth's surface and gouging out a circular crater that is typically 20 times larger than its own size (see p.119). The enormous energy generated ensures that most of the impactor is vaporized in the process, and seismic shocks and blast waves are produced. The resulting huge earthquake will topple any trees for many kilometres around. The surrounding atmosphere reaches furnace temperatures causing widespread fires. A tsunami will be produced if the impact is in the ocean. An impact crater greater than 20km (12 miles) in diameter is produced on Earth about once every 500,000 years.

**IMPACT CRATER**
About 50,000 years ago, an iron meteorite hit this desert region in Arizona, USA. The resulting crater, called Meteor Crater, is 1.2km (0.75 miles) wide and 170m (550ft) deep. Ejecta produced by the impact can be seen as hummocky deposits lying beyond the crater rim.

MOLDAVITE (GREEN GLASS)

DISC-SHAPED TEKTITE

**IMPACTITES**
These centimetre-sized, glassy bodies are formed when the Earth's rock melts or shatters due to the heat and pressure of an impact.

EXPLORING SPACE

## FINDING METEORITES

The best way to find meteorites is to search exposed glacial ice-fields or sandy deserts free of other large rock. Ideal sites are the eroded blue-ice regions of Antarctica and the Nullarbor Plain in Australia. Since 1976, US, European, and Japanese expeditions have searched for meteorites in Antarctica, and thousands of individual specimens have been recovered. Many are from the same fall, due to the incoming body fragmenting as it passed through Earth's atmosphere.

**NOMAD ROVER IN ANTARCTICA**
In January 2000, the US robot Nomad achieved a first by finding and identifying five meteorites lying on the ice in eastern Antarctica, using just sensors and artificial intelligence.

# METEORITES

Meteorites are mainly pieces of asteroids that have fallen to Earth from space, but a few very rare meteorites have come from the surface of Mars and the Moon. Some meteorites are made up of the primitive material that originally formed rocky planets. These give researchers a glimpse of the conditions at the dawn of the Solar System. Others are fragments of bodies that have differentiated into metallic cores and rocky surfaces, providing an indirect opportunity of studying the deep interior of a rocky planet. Meteorites are named after the place where they landed.

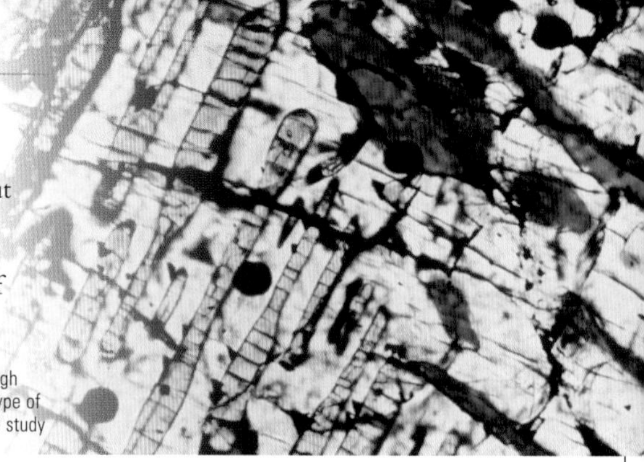

**METEORITE CROSS-SECTION**
By shining polarised light through thin sections of chondrites (a type of stony meteorite), scientists can study their crystalline structure.

NORTH AMERICA *north*

## Tagish Lake

**LOCATION** British Columbia, Canada
**TYPE** Stony
**MASS** About 1kg (2.2lb)
**DATE OF DISCOVERY** 2000

Over 500 fragments of this meteorite rained down onto the frozen surface of Tagish Lake on 18 January 2000. The meteorite was dark red and rich in carbon. Analysis showed that it was extremely primitive, containing many unaltered stellar dust grains that had been part of the cloud of material that formed the Sun and the planets.

**FRAGMENT ENCASED IN ICE**

NORTH AMERICA *southwest*

## Canyon Diablo

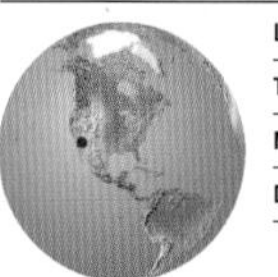

**LOCATION** Arizona, USA
**TYPE** Iron
**MASS** 30 tons
**DATE OF DISCOVERY** 1891

Many pieces of this meteorite, ranging from minute fragments to chunks weighing about 500kg (1,100lb), have been found near Meteor Crater in Arizona, USA. Much more is thought to be buried under one of the crater rims. If a Canyon Diablo meteorite is sawn in half and then one of the faces is polished and etched with acid, a characteristic surface pattern appears.

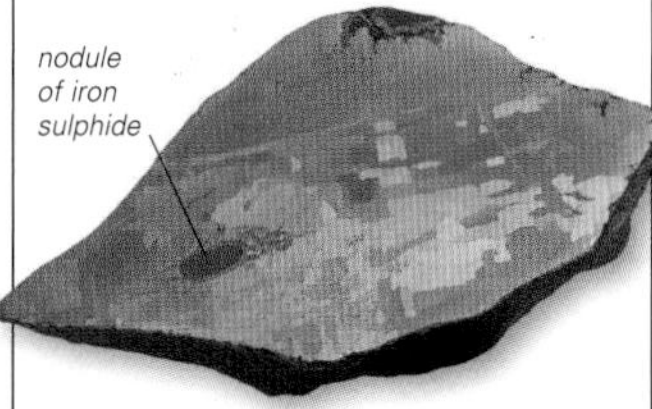


**ACID-ETCHED, POLISHED CROSS-SECTION**

NORTH AMERICA *south*

## Allende

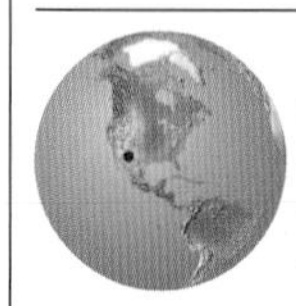

**LOCATION** Chihuahua, Mexico
**TYPE** Stony
**MASS** 2 tons
**DATE OF DISCOVERY** 1969

On 8 February 1969, a fireball was seen streaking across the sky above Mexico. It exploded, and a shower of stones fell over an area of about 150 square km (60 square miles). Two tons of material were speedily collected and distributed among the scientific community. Allende was found to be a very rare type of primitive meteorite. Previously, only gram-sized amounts of this meteorite type were known. As such large samples of Allende were available, destructive analysis was possible. The white calcium- and aluminium-rich crystals were separated out from the surrounding rock. They were found to contain the decay products of radioactive aluminium-26, indicating that these crystals were formed in the outer shells of stars that exploded as supernovae and were then subsequently incorporated into planetary material.

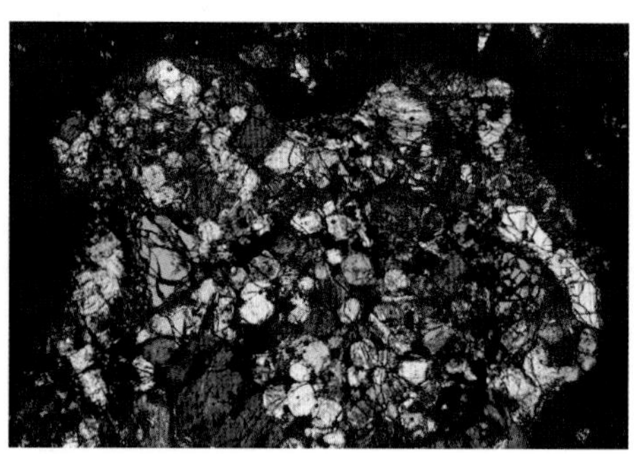

**CHONDRULE**
This thin, magnified section of an Allende meteorite shows one of many spherical, pea-sized chondrules that are locked in the stony matrix. Chondrules are droplets of silicate rock that have cooled extremely rapidly from a molten state.

EUROPE *west*

## Glatton

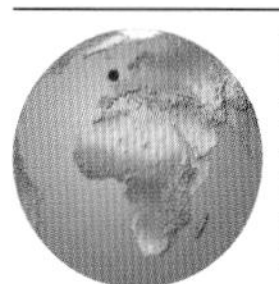

**LOCATION** Cambridgeshire, UK
**TYPE** Stony
**MASS** 767g (27oz)
**DATE OF DISCOVERY** 1991

On 5 May 1991, while planting out a bed of onions just before Sunday lunch, the retired English civil servant Arthur Pettifor heard a loud whining noise. Noticing one of the conifers in his hedge waving about, he got up and looked in the bottom of the hedge. He spotted a small stone that was luke-warm to the touch. If Pettifor had not been gardening, the stony meteorite would never have been found.

**LUCKY FIND**

EUROPE *west*

## Ensisheim

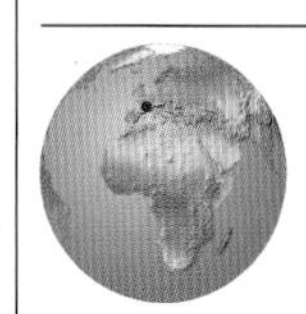

**LOCATION** Alsace, France
**TYPE** Stony
**MASS** 127kg (280lb)
**DATE OF DISCOVERY** 1492

This large stone is the oldest meteorite fall that can be positively dated. It was carefully preserved by being hung from the roof of the parish church of Ensisheim, Alsace. This veneration was due to the fall being regarded by the Holy Roman Emperor Maximilian as a favourable omen for the success of his war with France and his efforts to repel Turkish invasions. Initially, Ensisheim was thought to be a "thunderstone", a rock ejected from a nearby volcano and subsequently struck by lightning. In the early 19th century, it was chemically analysed and found to contain 2.3 per cent nickel. This is very rare in rocks on Earth, and theories of an extraterrestrial origin started to proliferate.

**MEDIEVAL WOODCUT**
The woodcut at the top of this medieval manuscript shows the meteorite falling near Ensisheim after producing a brilliant fireball in the sky on 16 November 1492.

**METEORITE FRAGMENT**
This highly valuable 8kg (17.6lb) sample of the Ensisheim meteorite is kept at the Museum of Paris, France.

AFRICA *north*

## Nakhla

**LOCATION** Alexandria, Egypt
**TYPE** Stony
**MASS** 40kg (88lb)
**DATE OF DISCOVERY** 1911

On 28 June 1911, about 40 stones landed near Alexandria, the largest weighing 1.8kg (4lb). Nakhla is a volcanic, lava-like rock that formed 1,200 million years ago. It is one of over 16 meteorites that have been blasted from the surface of Mars and, after many millions of years in space, fallen to Earth.

*black, glassy fusion crust formed during fall*

**MARTIAN METEORITE**

AFRICA *southwest*

## Hoba West

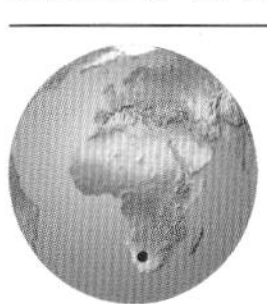

**LOCATION** Grootfontein, Namibia

**TYPE** Iron

**MASS** 66 tons

**DATE OF DISCOVERY** 1920

The largest meteorite to have been found on Earth, Hoba West measures 2.7 x 2.7 x 0.9m (8.9 x 8.9 x 3ft). It consists of 84 per cent iron and 16 per cent nickel. Hoba West has never been moved from where it landed. In the past, enterprising individuals tried to recover this valuable lump of "scrap" metal. To protect it from damage and sample-taking, the Namibian Government has declared it to be a national monument. Hoba West represents the maximum mass that the Earth's atmosphere can slow down to a free-fall velocity. If its parent meteoroid had been much bigger, or the trajectory of the fall steeper, the impact with the ground would have been much faster. This would have led to the destruction of most of the meteorite and the production of a crater in the Earth's surface. Large lumps of surface iron, such as Hoba West, are hard to overlook.

**RUSTING AWAY**
The Hoba West meteorite weighed about 66 tons when it was discovered but it has started to rust away and today weighs less than 60 tons.

**LARGEST KNOWN METEORITE**
A team of scientists from Kings College, London, UK, pose on top of Hoba West in the 1920s. Standing second from the left is Dr L.J. Spencer, who became Keeper of Minerals at the British Museum, London, in the 1930s.

AFRICA *south*

## Cold Bokkeveld

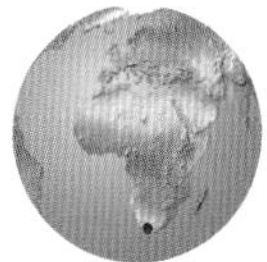

**LOCATION** Western Cape, South Africa

**TYPE** Stony

**MASS** About 4kg (8.8lb)

**DATE OF DISCOVERY** 1838

This meteorite is a perfect example of a stony chondrite, a class of primitive meteorite that makes up almost 90 per cent of those found so far. They consist of silicate, metallic, and sulphide minerals and are thought to represent the material from which the Earth was formed. They contain tiny, spherical chondrules cemented into a rocky matrix. These rocky droplets solidified extremely quickly from a starting temperature of at least 1,400°C (2,600°F). Chondrules contain a mixture of imperfect crystals and glass. Cold Bokkeveld is carbonaceous, which means that it contains compounds of carbon, hydrogen, oxygen, and nitrogen. These are the main constituents of living cells. Carbonaceous chondrites thus contain the building blocks of life.

**WATER FROM STONE**
This tiny chondrule is surrounded by a water-rich matrix (shown as black). Cold Bokkeveld contains about 10 per cent water by mass, which would be released if it was heated.

AUSTRALIA *west*

## Mundrabilla

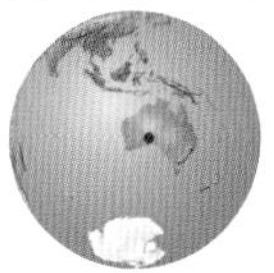

**LOCATION** Nullarbor Plain, Western Australia

**TYPE** Iron

**MASS** About 18 tons

**DATE OF DISCOVERY** 1911

Mundrabilla is on the Trans Australian railway line in a featureless desert. Three small irons were found there in 1911 and 1918. Renewed interest in 1966 led to the discovery of two meteorites weighing 5 and 11 tons. Mundrabilla took many millions of years to solidify, and it offers a rare chance to investigate the formation of alloys at low gravity. A 45kg (100lb) core of one of the meteorites (below) is undergoing computer X-ray analysis by NASA.

**UNDER INVESTIGATION**

ANTARCTICA

## ALH 81005

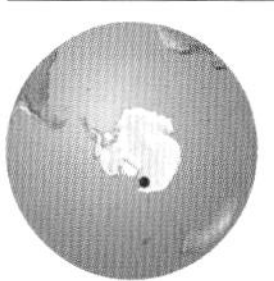

**LOCATION** Allan Hills, Antarctica

**TYPE** Stony

**MASS** 31.4g (1.1oz)

**DATE OF DISCOVERY** 1982

ALH 81005 is a lunar meteorite. About 36 have been discovered, a mere 0.08 per cent of the present total. The cosmic-ray damage they have suffered indicates that they have been blasted from the surface of the Moon by a meteorite impact in the last 20 million years. The main mineral is anorthite (calcium aluminium silicate), which is very rare in asteroids. The composition of these stony meteorites is very similar to that of the lunar-highland rocks brought back to Earth by the Apollo astronauts.

**MOON ROCK**
This golfball-sized rock was found by the US Antarctic Search for Meteorites programme in 1982. It was the first meteorite to be recognized as being of lunar origin.

# THE MILKY WAY

*"A broad and ample road, whose dust is gold,*
*And pavement stars, as stars to thee appear*
*Seen in the galaxy, that milky way*
*Which nightly as a circling zone thou seest*
*Powder'd with stars."*

John Milton

THE SOLAR SYSTEM is part of a vast collection of stars, gas, and dust called the Milky Way galaxy. Galaxies can take various forms, but the Milky Way is a spiral. The Sun and its system of planets lie halfway from the centre, on the edge of one of the spiral arms. For thousands of years, humans have pondered the significance of the pale white band that stretches through the sky. This Milky Way is the light from millions of stars that lie in the disc of the galaxy. Within the Milky Way lie stars at every stage of creation, from the immense clouds of interstellar material, which contain the building material of stars, to the exotic stellar black holes, neutron stars, and white dwarfs, which are the end points of a star's life. Most of the Milky Way's visible mass consists of stellar material, but about 90 per cent of its total mass is made up of invisible "dark matter", which remains a mystery yet to be explained.

**GLOWING PATHWAY**
From Earth, the Milky Way presents a glowing pathway of stars and gas streaking across the night sky. The billions of stars that make up the Milky Way are arranged in a great spiral disc, and from our position, halfway from its centre, we view the disc end-on.

# THE MILKY WAY

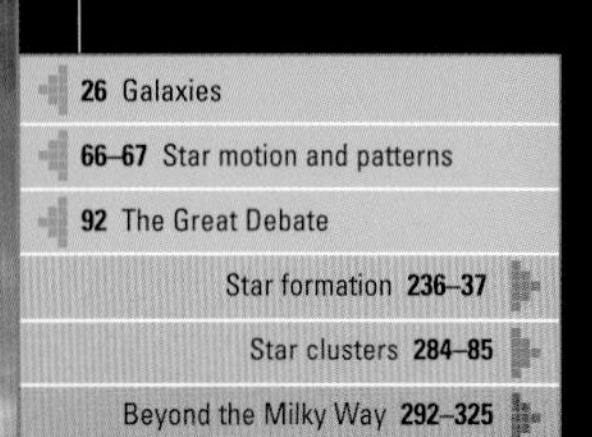
26 Galaxies
66–67 Star motion and patterns
92 The Great Debate
Star formation 236–37
Star clusters 284–85
Beyond the Milky Way 292–325

**THE SUN IS ONE STAR** of around 100 billion that make up the Milky Way, a relatively large spiral galaxy (see p.294) that started to form around 13.5 billion years ago. From our position inside the Milky Way, it appears as a bright band of stars stretching across the night sky.

**BAND OF STARS**
As we look out along the disc of the Milky Way from our position within, we see a bright band of thousands of stars that has captured humankind's imagination throughout history.

## THE GEOGRAPHY OF THE MILKY WAY

At the very centre of the Milky Way lies a black hole with a mass of about 3 million solar masses. This core or nucleus of the galaxy is surrounded by a bulge of stars that grows denser closer to the centre. This forms an ellipsoid of about 15,000 by 6,000 light-years, the longest dimension lying along the plane of the Milky Way. Lying in the plane is the disc containing most of the Galaxy's stellar materials. Young stars etch out a spiral pattern, and it is thought that they radiate out from a bar. Surrounding the bulge and disc is a spherical halo in which lie some 200 globular clusters, and this in turn may be surrounded by a dark halo, the corona.

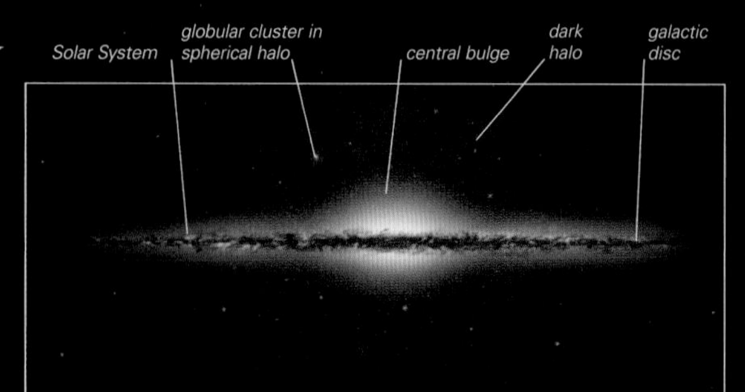


**MILKY WAY GALAXY**
The Milky Way has a diameter of about 100,000 light-years and a thickness of about 2,000 light-years. The Sun lies about 25,000 light-years from the centre.

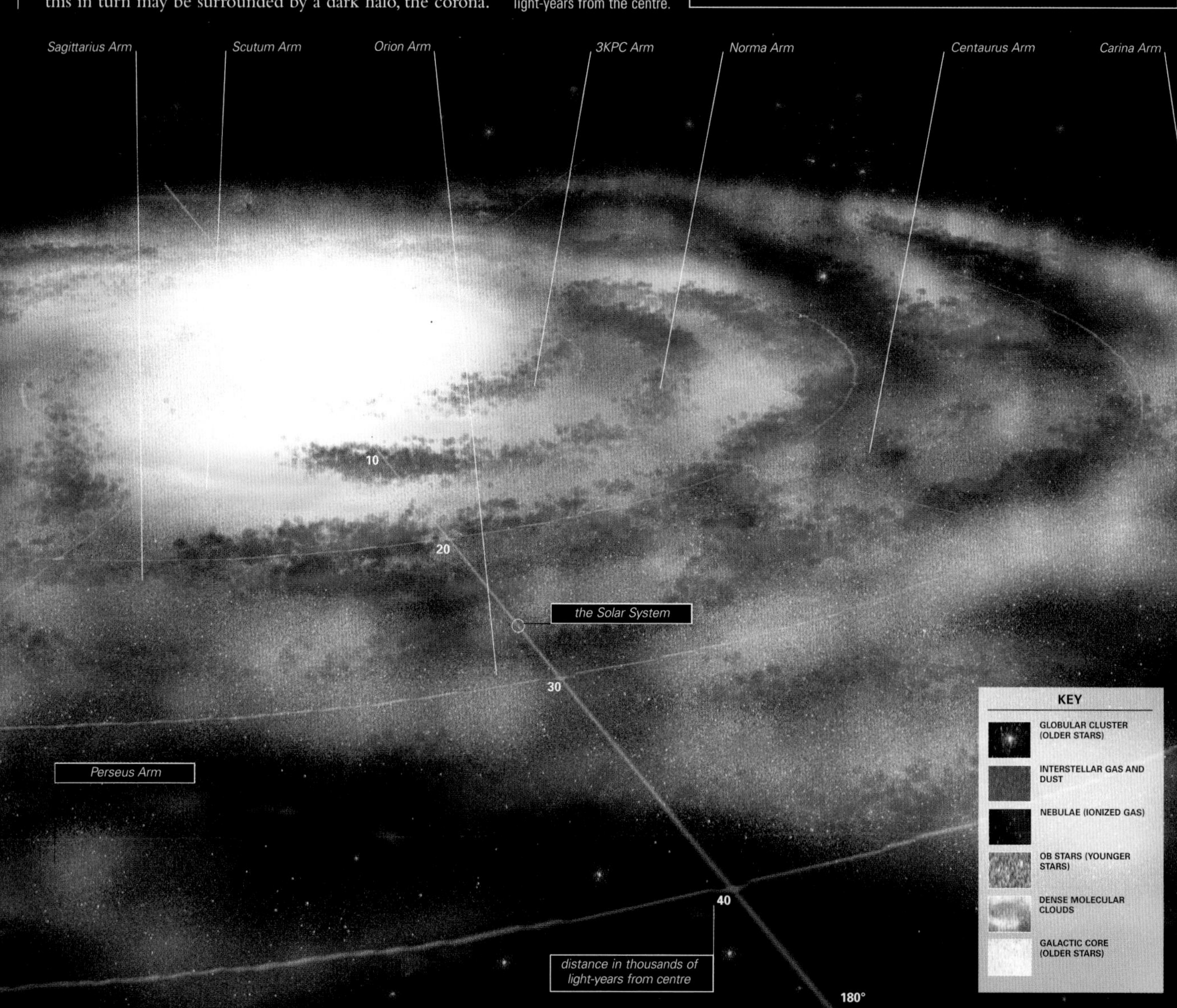

# THE SPIRAL ARMS

Seen "face-on", the Milky Way would look like a huge Catherine wheel, with the majority of its light coming from the arms spiralling out from the central bulge. In fact, the material in the spiral arms is generally only slightly denser than the matter in the rest of the disc. It is only because the stars that lie within them are younger, and therefore brighter, that the pattern in spiral galaxies shows up. Two mechanisms are thought to create the Milky Way's spiral structure. Density waves, probably caused by gravitational attraction from other galaxies, ripple out through the disc, creating waves of slightly denser material and triggering star formation (see pp.236–37). By the time the stars have become bright enough to etch out the spiral pattern, the density waves have moved on through the disc, starting more episodes of star formation and leaving the young stars to age and fade. High-mass stars eventually explode as supernovae, sending out blast waves that also pass through the star-making material, triggering further star formation.

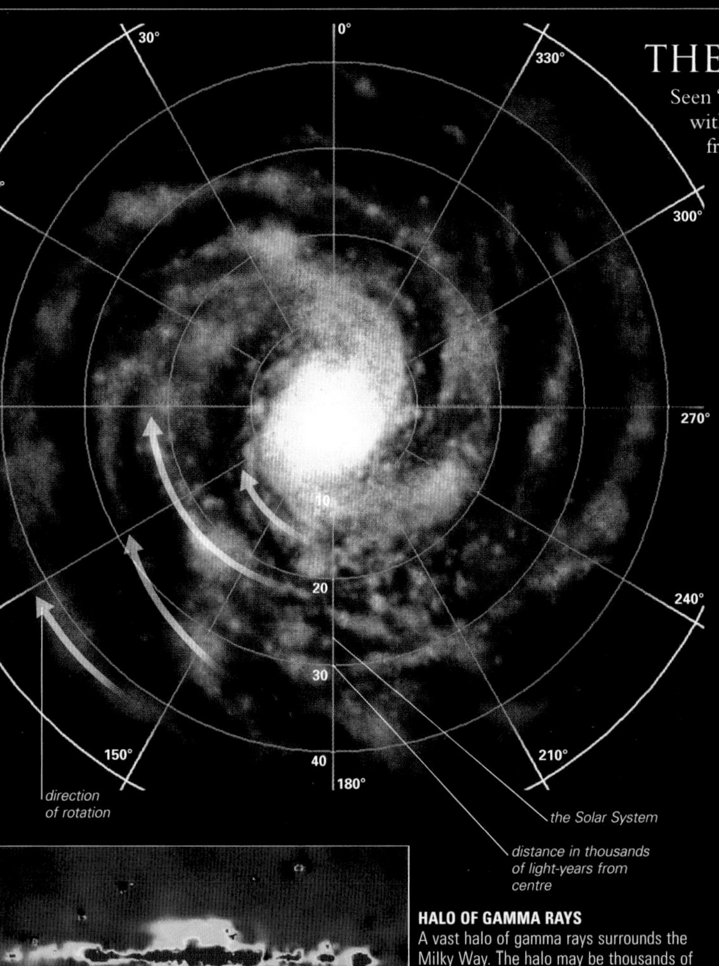


**SPINNING GALAXY**
The galaxy rotates differentially – the closer objects are to the centre, the less time they take to complete an orbit. The Sun travels around the galactic centre at about 800,000kph (500,000mph), taking around 225 million years to make one orbit.

**HALO OF GAMMA RAYS**
A vast halo of gamma rays surrounds the Milky Way. The halo may be thousands of light-years thick and might help to define the edges of the Milky Way.

MYTHS AND STORIES

## HEAVENLY MILK

There are many myths involving the formation of the Milky Way. In Greek mythology, Hercules was the illegitimate son of Zeus and a mortal woman, Alcmene. It was said that when Zeus's wife, while suckling Hercules, heard he was the son of Alcmene, she pulled her breast away and her milk flowed among the stars.

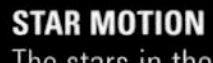

**MYTH IN ART**
*The Origin of the Milky Way* (c.1575) by Jacopo Tintoretto was inspired by the Greek myth.

# STELLAR POPULATIONS

Stars are broadly classified into two groups, called populations, based on age and chemical content. Population I consists of the youngest stars, which tend to be richer in heavy elements. These elements are primarily produced by stars, and Population I stars are created from materials shed by existing stars. In the Milky Way, the majority of Population I stars lie in the galactic disc, where there is an abundance of star-making material. Population II stars are older, metal-poor stars, existing primarily in the halo, but also in the bulge. Most are found within globular clusters, where all star-making materials have been used up and no new star formation is taking place.

**STAR MOTION**
The stars in the bulge have the highest orbital rates. They can travel hundreds of light-years above and below the plane of the Milky Way. Within the disc, stars stay mainly in the plane of the galaxy as they orbit the galactic centre. Stars in the halo plunge through the disc, reaching distances many hundreds of light-years above and below it.

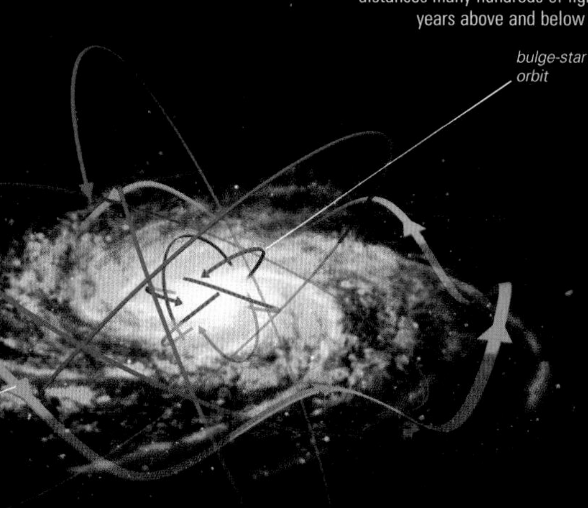


**MAPPING THE MILKY WAY**
The Milky Way's structure is defined by its major arms, each named after the constellation in which it is most prominent – the brightest arm is that in Sagittarius, beyond which lies the galactic nucleus. The Solar System lies near the inner edge of the Orion Arm. All the arms lie in a plane defined by the galactic disc. The nucleus forms a bulge at its centre, and globular clusters orbit above and below it in the halo region.

# THE INTERSTELLAR MEDIUM

The interstellar medium, permeating the space between the stars, consists mainly of hydrogen in various states, together with dust grains. It constitutes about ten per cent of the mass of the Milky Way and is concentrated in the galactic disc. It is not distributed uniformly: there are clouds of denser material, where star formation takes place, and regions where material has been shed by stars, interspersed with areas of very low density. Within the interstellar medium there is a wide range of temperatures. In the cooler regions, at around -260°C (-440°F), hydrogen exists as clouds of molecules. These cold molecular clouds contain molecules other than hydrogen, and star formation occurs where such clouds collapse. There are also clouds of neutral hydrogen (HI regions) with temperatures ranging from -170°C (-280°F) to 730°C (1,340°F), and areas of ionized hydrogen heated by stars (HII regions) with temperatures around 10,000°C (18,000°F). Dust grains contribute about one per cent of the galactic mass and are found throughout the medium. They are mostly small, solid grains, 0.01 to 0.1 micrometres in diameter, consisting of carbon, silicates (compounds of silicon and oxygen), or iron, with mantles of water and ammonia ice or, in the cooler clouds, possibly solid carbon dioxide.

**NONUNIFORM MEDIUM**
As this image of the Cygnus Loop supernova remnant (see p.265) shows, material in the interstellar medium is very uneven. The blast wave from the supernova explosion is still expanding through the interstellar matter. Where it hits denser areas and slows down, atoms in the medium become excited and emit optical and ultraviolet light.

**INVISIBLE COSMIC RAYS**
Cosmic rays travel throughout the Milky Way. These are highly energetic particles that spiral along magnetic field lines. Cosmic rays are primarily ions and electrons and are an important part of the interstellar medium, producing a pressure comparable to that of the interstellar gas.

**STARS**
Stars are an important factor in the composition of the interstellar medium as they enrich the medium with heavy, metallic elements. A supernova explosion, the death of a massive star (see p.262), is the only mechanism that produces elements heavier than iron.

**DARK NEBULAE**
Dark nebulae are cool clouds composed of dust and the molecular form of hydrogen. They are only observed optically when silhouetted against a brighter background as they absorb light and re-radiate the energy in infrared wavelengths. Stars are formed when dark nebulae collapse.

**MAGNETIC FIELDS**
Galactic magnetic fields are weak fields that appear to lie in the plane of the Milky Way, increasing in strength towards the centre. They are aligned with the spiral arms, but are distorted locally by events such as the collapse of molecular clouds and supernovae.

**BETWEEN THE STARS**
Contrary to early popular belief, the space between stars is not empty. The interstellar medium is fundamental in the process of star formation and galaxy evolution. Temperature defines the material's appearance and the processes occurring within it.

**DUST CLOUDS**
Young stars are often surrounded by massive discs of dust. These discs are believed to be the material from which solar systems are formed. There are often high levels of dust around stars in the later stages of their lives as they lose material to the interstellar medium.

**REFLECTION NEBULAE**
Material surrounding young stars contains dust grains that scatter starlight. In these nebulae, the density of the dust is sufficient to produce a noticeable optical effect. The nebulae appear blue because the shorter-wavelength, bluer light is scattered more efficiently.

**EMISSION NEBULAE**
When the interstellar medium is heated by stars, the hydrogen is ionized, producing a so-called HII region. The electrons freed by the ionization process are continually absorbed and re-emitted, producing the red colouring observed in emission nebulae.

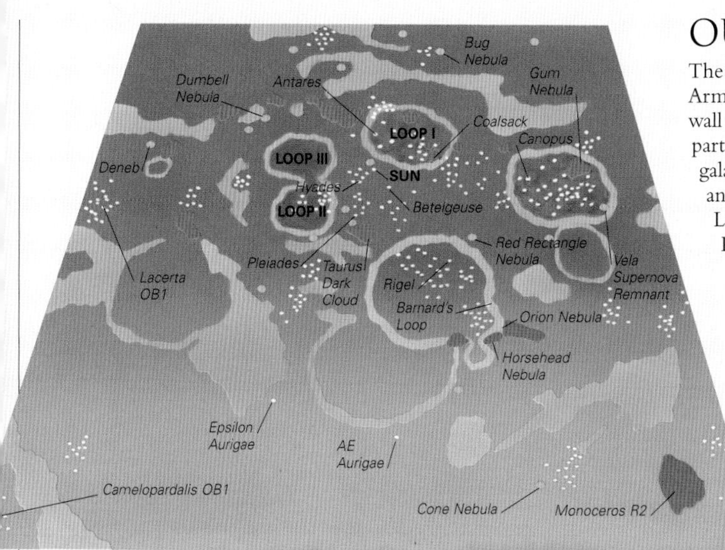


**REGIONAL MAP**
This schematic representation of the Solar System's local neighbourhood maps out a section of the Milky Way's Orion Arm about 5,000 light-years across. The Sun is located just above centre. Hydrogen gas clouds are marked in brown, molecular clouds in red, and interstellar bubbles are coloured green. Nebulae are shown in pink, while star clusters and giant stars are picked out in white.

# OUR LOCAL NEIGHBOURHOOD

The Sun lies in one of the less-dense regions of the Milky Way's Orion Arm. It sits in a "bubble" of hot, ionized hydrogen gas bounded by a wall of colder and denser neutral hydrogen gas. The Local Bubble is part of a tube-like chimney that extends through the disc into the galactic halo. The largest local coherent structure, detected by radio- and X-rays, is known as Loop I. This is believed to be part of the Local Bubble impacting into a molecular cloud known as the Aquila Rift. Two other expanding bubbles, Loops II and III, lie nearby. The Sun is travelling through material flowing out from the young stars known as the Scorpius–Centaurus Association, towards the Local Interstellar Cloud, a mass of dense interstellar gas.

to the Galactic Centre
Scorpius–Centaurus Association
Gum Nebula
Vela Supernova Remnant
Aquila Rift
direction of Sun's movement
Sun
Local Bubble
Orion Shelf
Orion Association

**LOCAL BUBBLE**
The Sun moves within the boundaries of the Local Bubble (shown in black). It is passing through strong stellar winds (shown in blue) thrown out by the Scorpius–Centaurus Association of young stars. High-density molecular clouds are highlighted in red.

# THE GALACTIC CENTRE

Dense layers of dust and gas obscure the centre of the Milky Way from us in optical wavelengths. However, the brightest radio source in the sky is located towards the Galactic Centre in the constellation of Sagittarius. This source, known as Sagittarius A consists of two parts. Sagittarius A East is believed to be a bubble of ionized gas, possibly a supernova remnant. Sagittarius A West is a cloud of hot gas, and embedded within it is a very strong and compact radio source, called Sagittarius A★ (Sgr A★). Sgr A★ appears to have no orbital motion and therefore probably lies at the very centre of the Milky Way. It has a radius of less than 2.2 billion km (1.4 billion miles) – smaller than that of Saturn's orbit – and orbital motions of the gas clouds around it indicate that it surrounds a supermassive black hole of about 3 million solar masses. Centred on Sgr A★ is a three-pronged mini-spiral of hot gas, about 10 light-years in diameter, and surrounding this is a disc of cooler gas and dust called the Circumnuclear Disc.

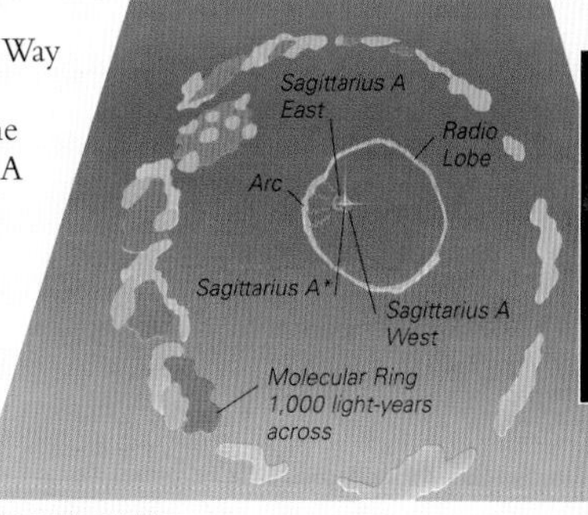


**GALACTIC CENTRE**
Surrounding Sagittarius A, the Radio Lobe is a region of magnetized gas including an arc of twisted gas filaments. Farther out, the expanding Molecular Ring consists of a series of huge molecular clouds (red), and an association of hydrogen clouds (brown) and nebulae (pink). The two smaller gas discs around Sagittarius A cannot be seen at this scale.

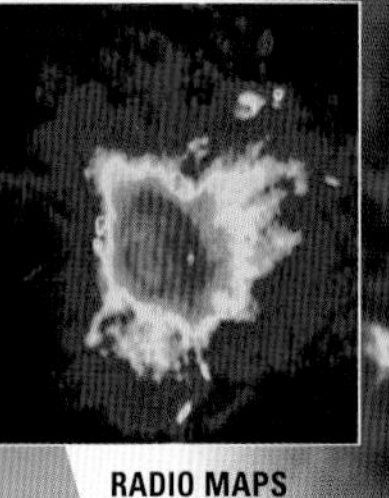

**RADIO MAPS**
Radio maps of Sagittarius A show a spiral pattern of hot, ionized gas that appears to be falling into the very centre of the Milky Way. Situated at the middle of the maps is the point source Sagittarius A*, thought to be a supermassive black hole at the very heart of the Milky Way.

## J.C. KAPTEYN

The Dutch astronomer, Jacobus Cornelius Kapteyn (1851–1922) was fascinated by the structure of the Milky Way. Studying at the University of Groningen, he used photography to plot star densities. He arrived at a lens-shaped galaxy with the Sun near its centre. Although his positioning of the Sun was incorrect, many subsequent studies of the structure of the Milky Way stemmed from his work.

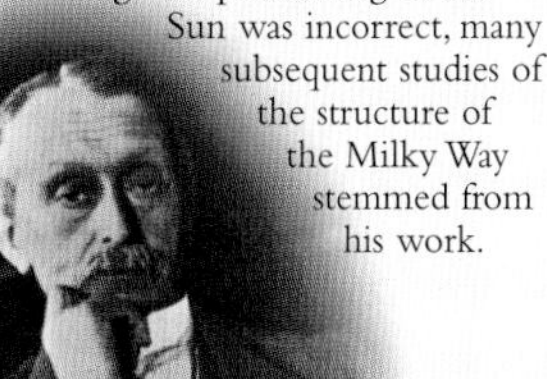

**GLOBULAR CLUSTER**
Like bees around a honey pot, the stars of a globular cluster swarm in a compact sphere. Containing up to a million (mostly Population II) stars, most of these clusters are found in the Milky Way's halo.

# THE EDGES OF THE MILKY WAY

Surrounding the disc and central bulge of the Milky Way is the spherical halo, stretching out to a diameter of more than 100,000 light-years. Compared to the density of the disc and the bulge, the density of the halo is very low, and it decreases as it extends away from the disc. Throughout the halo are about 200 globular clusters (see pp.284–85), spherical concentrations of older, Population II stars (see p.227). Individual Population II stars also exist in the halo. These halo stars orbit the galactic centre in paths that take them far from the galactic disc, and because they do not follow the motion of the majority of the stars in the disc, their relative motion to the Sun is high. For this reason, they are sometimes called high-velocity stars. Calculations of the mass of the Milky Way suggest that 90 per cent consists of mysterious dark matter (see p.27). Some of this may be composed of objects with low luminosities, such as brown dwarfs and black holes, but most is believed to be composed of exotic particles, the nature of which have yet to be discovered. The halo extends into the corona which reaches out to encompass the Magellanic Clouds (see pp.300–301), the Milky Way's nearby neighbours in space.

# STARS

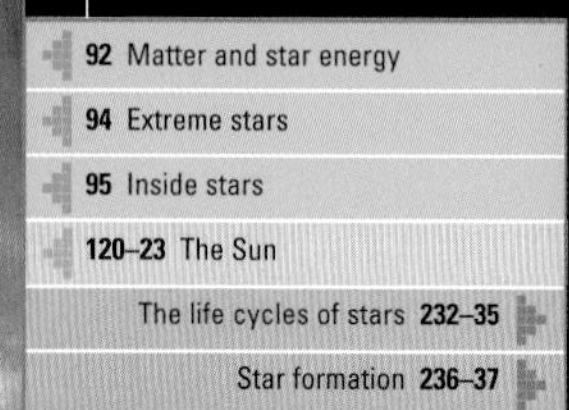
92 Matter and star energy
94 Extreme stars
95 Inside stars
120–23 The Sun
The life cycles of stars 232–35
Star formation 236–37

**STARS ARE MASSIVE** gaseous bodies that generate energy by nuclear reactions and shine because of this energy source. The mass of a star determines its properties – such as luminosity, temperature, and size – and its evolution over time. Throughout its life, a star achieves equilibrium by balancing its internal pressure against gravity.

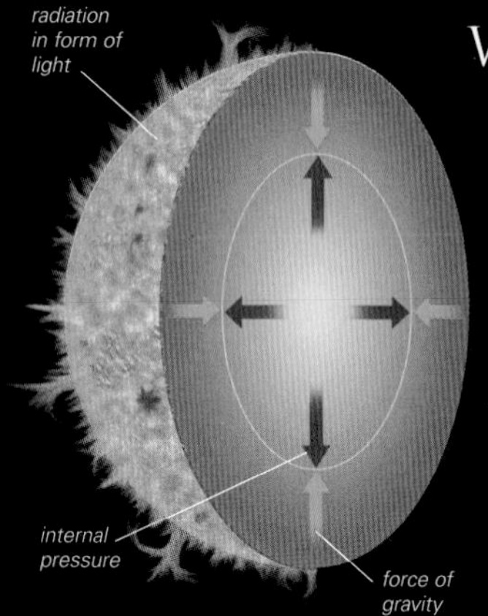


**PRESSURE BALANCE**
The state and behaviour of any star, at any stage in its evolution, is dictated by the balance between its internal pressure and its gravitational force.

## WHAT IS A STAR?

A collapsing cloud of interstellar matter becomes a star when the pressure and temperature at its centre become so high that nuclear reactions start (see pp.236–37). A star converts the hydrogen in its core into helium, releasing energy that escapes through the star's body and radiates out into space. The pressure of the escaping energy would blow the star apart if it were not for the force of gravity acting in opposition. When these forces are in equilibrium, the star is stable, but a shift in the balance will change the star's state. Stars fall within a relatively narrow mass range, as nuclear reactions cannot be sustained below about 0.08 solar masses, and in excess of about 100 solar masses stars become unstable. A star's life cycle, as well as its potential age, is directly linked to its mass. High-mass stars burn their fuel at higher rates and live much shorter lives than low-mass stars.

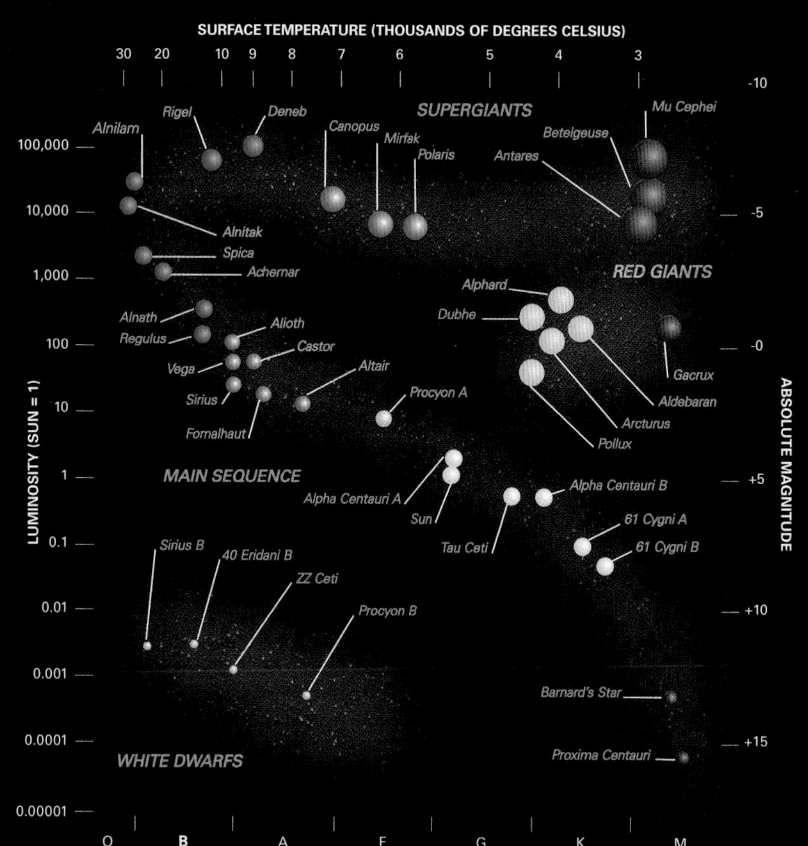


## THE H–R DIAGRAM

Named after the Danish and American astronomers Ejnar Hertzsprung and Henry Russell, the Hertzsprung–Russell (H–R) diagram graphically illustrates the relationship between the luminosity, surface temperature, and radius of stars. The astronomers' independent studies had revealed that a star's colour and spectral type are indications of its temperature. When the temperature of stars was plotted against their luminosity, it was noticed that stars did not fall randomly, but tended to be grouped. Most stars lie on the main sequence, a curved diagonal band stretching across the diagram. Star radius increases diagonally from bottom left to top right. Protostars evolve onto the main sequence as they reduce in radius and increase in temperature. On the main sequence, stars remain at their most stable before evolving into red giants or supergiants, moving to the right of the diagram as their radius increases and their temperature falls. White dwarfs are at the bottom left with small radii and high temperatures.

**IMPORTANT DIAGRAM**
The H–R diagram is the most important diagram in astronomy. It illustrates the state of a star throughout its life. Distinct groupings represent different stellar stages, and few stars are found outside these groups, as they spend little time migrating.

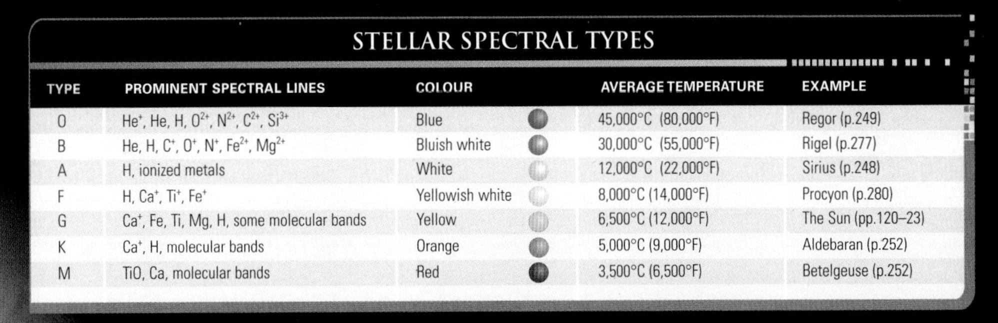

## STELLAR SPECTRAL TYPES

| TYPE | PROMINENT SPECTRAL LINES | COLOUR | AVERAGE TEMPERATURE | EXAMPLE |
|---|---|---|---|---|
| O | $He^{+}$, He, H, $O^{2+}$, $N^{2+}$, $C^{2+}$, $Si^{3+}$ | Blue | 45,000°C (80,000°F) | Regor (p.249) |
| B | He, H, $C^{+}$, $O^{+}$, $N^{+}$, $Fe^{2+}$, $Mg^{2+}$ | Bluish white | 30,000°C (55,000°F) | Rigel (p.277) |
| A | H, ionized metals | White | 12,000°C (22,000°F) | Sirius (p.248) |
| F | H, $Ca^{+}$, $Ti^{+}$, $Fe^{+}$ | Yellowish white | 8,000°C (14,000°F) | Procyon (p.280) |
| G | $Ca^{+}$, Fe, Ti, Mg, H, some molecular bands | Yellow | 6,500°C (12,000°F) | The Sun (pp.120–23) |
| K | $Ca^{+}$, H, molecular bands | Orange | 5,000°C (9,000°F) | Aldebaran (p.252) |
| M | TiO, Ca, molecular bands | Red | 3,500°C (6,500°F) | Betelgeuse (p.252) |

# STELLAR CLASSIFICATION

Stars are classified by group, according to the characteristics of their spectra. If the light from a star is split into a spectrum, dark absorption and bright emission lines are seen (see p.33). The positions of these lines indicate what elements exist in the photosphere of the star, and the strengths of the lines give an indication of its temperature. The classification system has seven main spectral types, running from the hottest O stars to the coolest M stars. Each spectral type is further divided into 10 subclasses denoted by a number from 0 to 9. Stars are also divided into luminosity classes, denoted by a Roman numeral, which indicates the type of star and its position on the H–R diagram. For example, class V is for main-sequence stars and class II for bright giants, while dim dwarfs are class VI. In addition to the main spectral types, there are classes for stars that show unusual properties, such as the carbon stars (C class). A small letter after the spectral class can also indicate a special property – for example, "v" means variable.

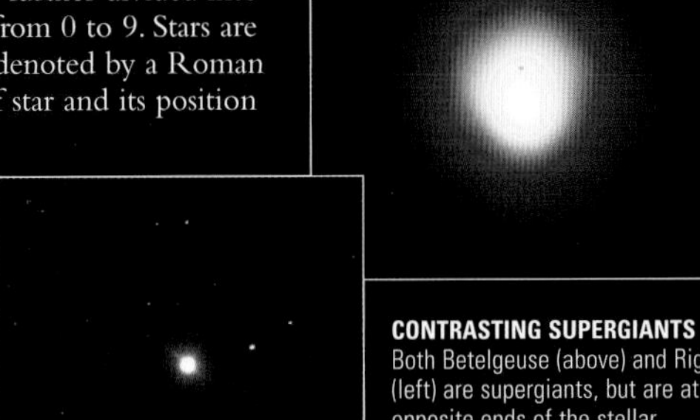

**CONTRASTING SUPERGIANTS**
Both Betelgeuse (above) and Rigel (left) are supergiants, but are at opposite ends of the stellar spectrum. Betelgeuse (see p.252) is a cool, red star, in its later stages, while Rigel is a hot, blue, relatively young star (see p.277).

**MAIN-SEQUENCE STAR**
Shown here in a false-colour image, the Sun is a yellowish main-sequence star with a surface temperature of 5,500°C (9,900°F) and spectral type G2, class V.

# LUMINOSITY

The luminosity of a star is its brightness, defined as the total energy it radiates per second. It can be calculated over all wavelengths – the bolometric luminosity – or at particular wavelengths. Measuring the brightness of a star as it appears in the night sky gives its apparent magnitude, but this does not take account of its distance from Earth. Stars that are located at vastly different distances from Earth can have the same apparent magnitudes if the farther star is sufficiently bright (see p.67). Once a star's distance is known, its absolute magnitude can be determined. This is its intrinsic brightness, and from this its luminosity can be determined. Stellar luminosities are generally expressed as factors of the Sun's luminosity. There is a very large range of stellar luminosities, from less than one ten-thousandth to about a million times that of the Sun. If stars are of the same chemical composition, their luminosities are dependent on their mass. Apart from highly evolved stars, they generally obey a consistent mass–luminosity relation, which means that if a star's luminosity is known, its mass can be determined.

**DENEB AND VEGA**
Although Deneb (bottom) and its neighbour Vega (top) are similar in apparent brightness, Deneb is about 300 times more distant. If Deneb was moved to Vega's distance of only 25 light-years from Earth, it would appear to be as bright as the crescent moon.

## CECILIA PAYNE-GAPOSCHKIN

Born in America but raised in England, Cecilia Helena Payne (1900–79) married fellow astronomer Sergei Gaposchkin. Initially studying at Cambridge University, England, Payne-Gaposchkin was one of the first astronomy graduates to enter Harvard College Observatory, USA. She studied the spectra of stars and suggested in her doctoral thesis that the different strengths of absorption lines in stellar spectra were a result of temperature differences, rather than chemical content. She also suggested that hydrogen was the most abundant element in stars. Her ideas were initially dismissed, but finally accepted in 1929.

**HARVARD PROFESSOR**
Cecilia Payne-Gaposchkin was the first woman to become a full professor at Harvard.

# THE LIFE CYCLES OF STARS


230–31 Stars
Star formation 236–37
Main-sequence stars 246–47
Old stars 250–51
Stellar endpoints 262–63
Extra-solar planets 290–91


STARS FORM WHEN clouds of interstellar gas collapse under the influence of gravity (see pp.236–37). During their lifetimes, stars pass through a series of stages, with the sequence and timing depending crucially on the mass of the star. As a star passes through these stages, different elements are created, again depending on the star's mass. When stars have completed their development, they shed their material back into the interstellar medium, enriching the matter from which future generations of stars will form.

**LIFE STAGES**
The environs of the nebula NGC 3603 display most stellar life stages, from "pregnant" dark nebulae and pillars of hydrogen, to a cluster of young stars, and a red star nearing its end.

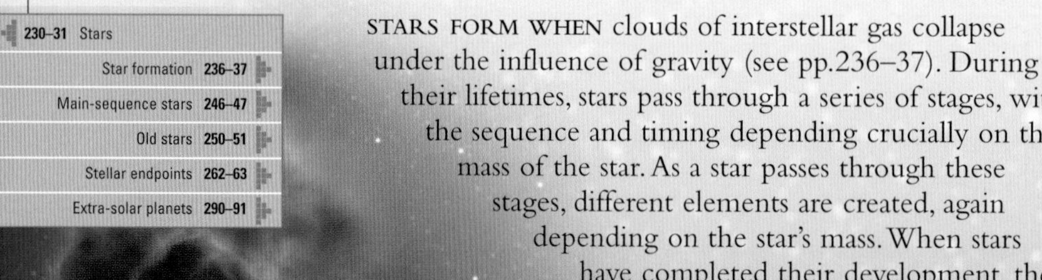


**DENSE CLOUDS START TO COLLAPSE**
Stars form from cold interstellar clouds. The colder the cloud, the less resistant it is to gravitational collapse. Clouds are formed mostly of hydrogen. At low temperatures, hydrogen atoms combine to form molecules (molecular hydrogen).

**PROTOSTARS BEGIN TO FORM**
If the cloud is over a certain mass, and it experiences a gravitational tug, it will begin to collapse. As it does, it will fragment into smaller parts of differing size and mass. These fragmented cloud sections become protostars.

**PRESSURE AND TEMPERATURE RISE**
The protostar continues to collapse, and the central temperature and pressure build up. The temperature and pressure levels will depend on the initial mass of the fragment – the higher the mass, the higher the temperature and pressure.

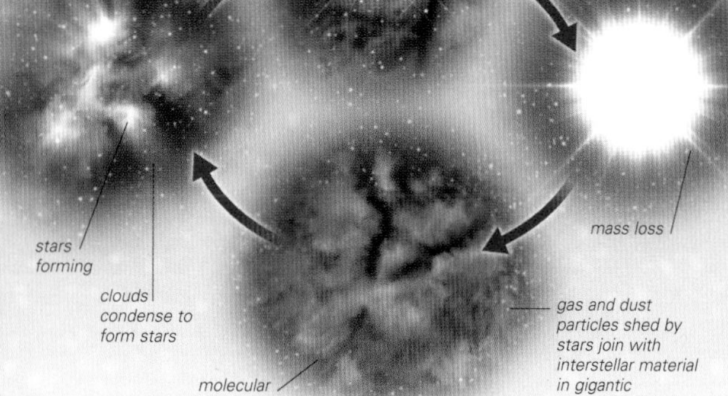


**ON-GOING CYCLE**
Stars form from material shed by previous generations of stars, and the death of massive stars can trigger the birth of others.

## STAR-MAKING RECIPE

The basic ingredients of stars are found in cold clouds made mostly of hydrogen molecules. The early stages of star formation are initiated by gravity, which can be exerted by the tug of a passing object, a supernova shock wave, or the compression of one of the Milky Way's density waves. If the cloud has sufficient mass, it will collapse into a protostar, which contracts until nuclear reactions start in its core. At this point a star is born. During its lifetime, a star will convert hydrogen to helium and a series of heavier materials, depending on its mass. These materials are gradually lost to the interstellar medium, until the star has used up most of its fuel and begins to collapse. For a high-mass star, this will result in a supernova that scatters much of the remaining material into space.

**BROWN DWARF**
In protostars less than ( solar masses, the press and temperature at the do not get high enough nuclear reactions to beg These protostars becom brown dwarfs.

# STELLAR EVOLUTION

If they are of a sufficient mass, new stars will go onto the main sequence, where they will remain for most of their lives. When the hydrogen fuel in their cores is exhausted, they will evolve off the main sequence to become red giants or supergiants. Mass dictates what path stars will follow in their maturity. When a star expands as it burns fuel in its atmosphere or collapses after using up its fuel, it crosses a region to the right of the main sequence on the Hertzsprung–Russell (H–R) diagram (see p.230) known as an instability strip. The more massive the star, the more times it will expand and contract. High-mass stars explode as supernovae in the supergiant region of the H–R diagram, while low-mass stars cross back over the main-sequence band as they collapse to form white dwarfs. Being small and hot, white dwarfs appear in the bottom left of the H–R diagram. As they cool they move to the right, eventually cooling to become black dwarfs. Neutron stars and black holes do not appear on the H–R diagram as they do not fit the mass–luminosity relationship that it represents.

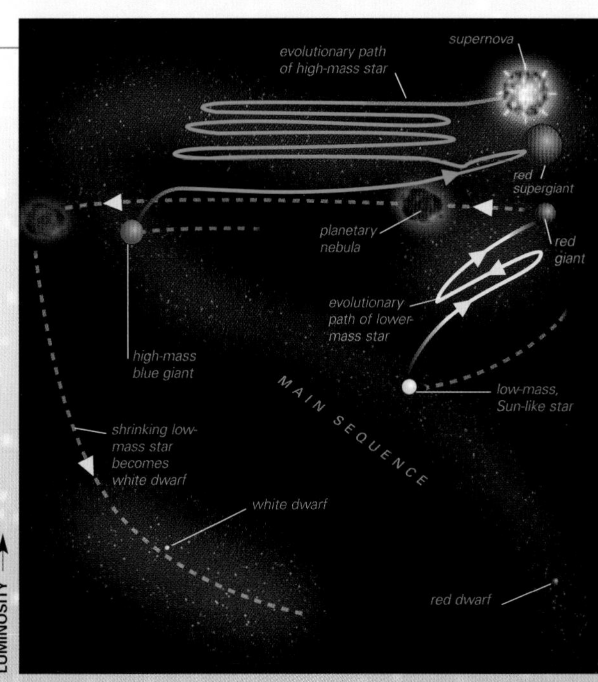


**STELLAR MATURITY**
The paths of mature stars on their journey towards death can be traced on the H–R diagram. Stars expand off to the right as they get larger and cooler to become red giants or supergiants. They travel back leftwards as they collapse, after burning fuel in their atmospheres.

strong stellar winds

**STELLAR ADOLESCENCE**
The gas that contracts to make a protostar starts to rotate slowly and speeds up as it is pulled inwards, creating a disc of stellar material. Before joining the main sequence, the protostar exhibits unstable behaviour such as rapid rotation and strong winds.

**STAR REACHES MAIN SEQUENCE**
For protostars with a mass of more than about 0.08 solar masses, the pressure and temperature within become high enough for nuclear reactions to start. The pressure balances gravity, and the protostar becomes a star.

circumstellar disc

**FORMATION OF ORBITING PLANETS**
Once a star is on the main sequence and stable, any disc of remaining material will start to cool. As it cools, elements condense out and begin to stick together. The larger clumps attract the smaller ones, until conglomerations are planet-sized.

# FORMATION OF A PLANETARY SYSTEM

Most young stars, unless they are in a close binary system, are surrounded by the remnants of the material from which they have formed. Rotation and stellar winds often shape the material into a flattened disc about the equatorial radius. Initially, the disc of material is hot, but as the star settles down onto the main sequence, it begins to cool. As it cools, different elements condense out, depending on the disc's temperature. Elements can exist in different states throughout the disc. Moving out from the star, temperatures fall, so water, for example, will exist as ice far away from the star and steam close to the star. Tiny condensing particles gradually stick together and grow larger. The ones that grow fastest will gravitationally attract others, becoming larger still, though in the dynamic early stages they may be broken back into pieces by collisions with other growing particles. Eventually, as the disc cools down, it becomes a calmer environment, and some particles will grow large enough to be classed as planetesimals – embryonic planets. Remnants of the original disc that do not form planets become asteroids or comets, depending on their distance from the parent star. Atmospheres are formed by gas attracted from the circumstellar disc, from gases erupting from the planets, or from bombarding comets.

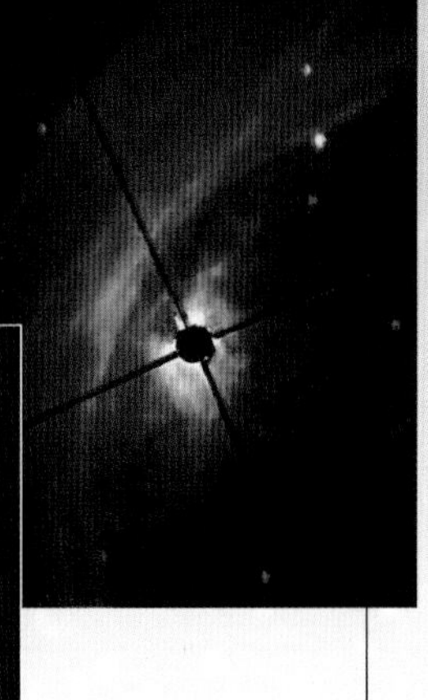

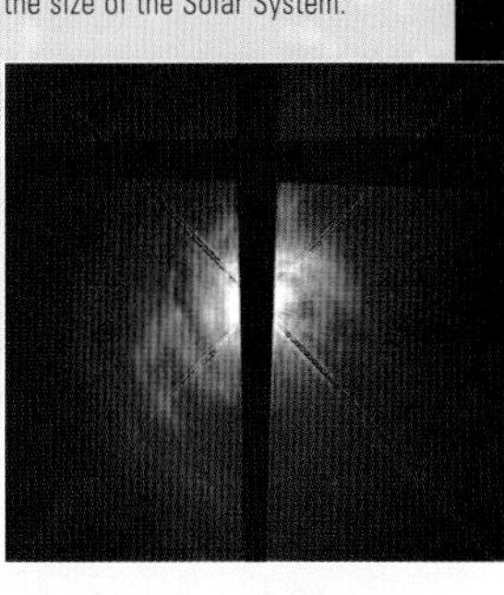

**CIRCUMSTELLAR DISC**
Within the circumstellar disc of AB Aurigae, knots of material may be in the early stages of planet formation. This swirling disc of stellar material is about 30 times the size of the Solar System.

# FROM MATURITY TO OLD AGE

When a star has finished burning hydrogen in its core, it will start burning its outer layers in a series of concentric shells. The star will expand as the source of heat moves outwards, and its outer layers cool. Stars with very low mass will eventually fade and cool; Sun-like stars will evolve into red giants; and high-mass stars will become supergiants. Once a star has used up all its available nuclear fuel, it will deflate, because there is no longer any power source to replace the energy lost from its surface. As it collapses, if it has enough mass, its helium core starts to burn and change into carbon. Once the fuel in its core is used up again, helium-shell burning begins in the star's atmosphere and the star expands. In very massive stars, this process is repeated until iron is produced. When a Sun-like star has used up all of its fuel, it will lose its outer atmosphere in a spectacular planetary nebula and collapse to become a white dwarf. A high-mass star will explode as a supernova and leave behind a neutron star or black hole.

*star expands as hydrogen-shell burning occurs*

*star starts to collapse as hydrogen is used up*

**LOW-MASS STAR**

Once a star with a mass less than half that of the Sun has used up the hydrogen in its core, it will convert the hydrogen in its atmosphere to helium and collapse, just as in higher-mass stars. However, low-mass stars do not have enough mass for the temperature and pressure at its core to get high enough for helium burning to occur. These stars will just gradually fade as they cool.

**STAR NOW ON THE MAIN SEQUENCE**

Stars spend the greatest proportion of their lives on the main sequence. The more massive the star, the shorter the period of time it will spend on the main sequence, as larger stars burn their fuel at a faster rate than smaller ones.

**SUN-LIKE STAR**

When a Sun-like star exhausts the hydrogen in its core, hydrogen-shell burning begins and it becomes a red giant, often losing its outer layers to produce a planetary nebula. It eventually collapses, and the temperature and pressure at its core initiate helium-core burning. The star again expands as helium-shell burning occurs, before finally collapsing to become a white dwarf that gradually fades to black.

*star becomes a red giant as hydrogen-shell burning starts*

**MOSTLY MAIN SEQUENCE STARS**
About 90 per cent of the visible stars in a typical view of the night sky are on the main sequence. This corresponds with the fact that most stars spend 90 per cent of their life on the main sequence.

**HIGH-MASS STAR**

The higher the mass of the star, the more times it will expand and contract, as its mass dictates the temperature of the core each time it contracts. Different elements are produced at each stage. If the star is massive enough, an iron core is formed, but elements heavier than iron cannot be formed within stellar cores. They are formed in supernova explosions that leave behind neutron stars or black holes.

*supergiant star produces heavier elements through nuclear reactions*

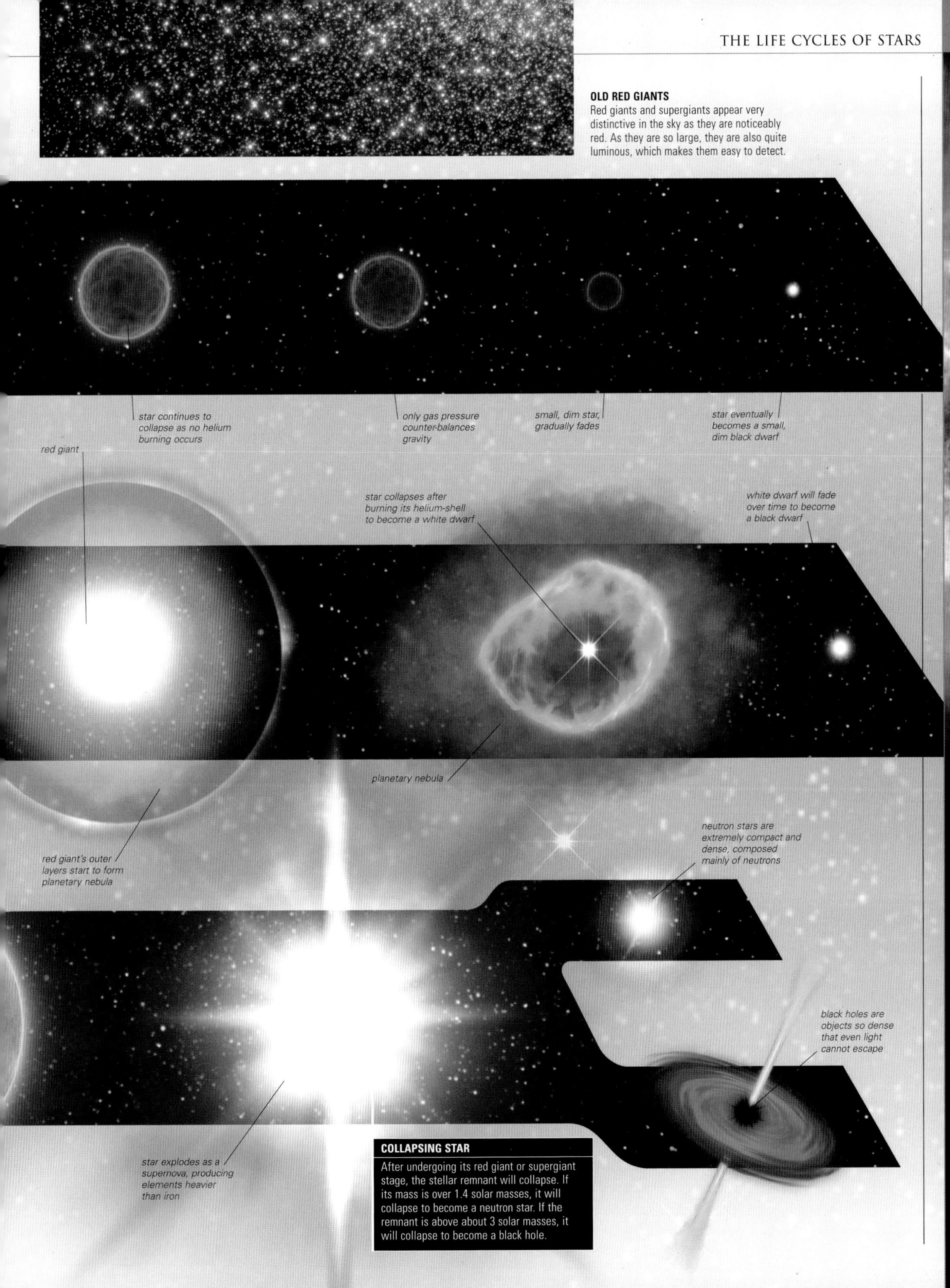
star continues to collapse as no helium burning occurs
only gas pressure counter-balances gravity
small, dim star, gradually fades
star eventually becomes a small, dim black dwarf
red giant
star collapses after burning its helium-shell to become a white dwarf
white dwarf will fade over time to become a black dwarf
planetary nebula
red giant's outer layers start to form planetary nebula
neutron stars are extremely compact and dense, composed mainly of neutrons
black holes are objects so dense that even light cannot escape
star explodes as a supernova, producing elements heavier than iron

**OLD RED GIANTS**
Red giants and supergiants appear very distinctive in the sky as they are noticeably red. As they are so large, they are also quite luminous, which makes them easy to detect.

**COLLAPSING STAR**
After undergoing its red giant or supergiant stage, the stellar remnant will collapse. If its mass is over 1.4 solar masses, it will collapse to become a neutron star. If the remnant is above about 3 solar masses, it will collapse to become a black hole.

# STAR FORMATION

24–27 Celestial objects
92 Matter and star energy
228 The interstellar medium
230–31 Stars
232–35 The life cycles of stars
Star clusters 284–85

STARS ARE FORMED by the gravitational collapse of cool, dense interstellar clouds. These clouds are composed mainly of molecular hydrogen (see p.228). A cloud has to be of a certain mass for gravitational collapse to occur, and a trigger is needed for the collapse to start, as the clouds are held up by their own internal pressure. Larger clouds fragment as they collapse, forming sibling protostars that initially lie close together – some so close they are gravitationally bound. The material heats up as it collapses until, in some clouds, the temperature and pressure at their centres become so great that nuclear fusion begins and a star is born.

**STAR-FORMING REGION**
The young stars in the star cluster IC 1590 (centre right) are heating up the nebula NGC 281 behind. The lanes of dust and small Bok globules are future regions of possible star formation.

## STELLAR NURSERIES

As well as being among the most beautiful objects in the Universe, star-forming nebulae contain a combination of raw materials that makes star birth possible. These clouds of hydrogen molecules, helium, and dust can be massive systems, hundreds of light-years across or smaller individual clouds, known as Bok globules. Although they may lie undisturbed for millions of years, disturbances can trigger these nebulae to collapse and fragment into smaller clouds from which stars are formed. Remnants from the star-forming nebulae will surround the stars, and the stellar winds produced by the new stars can, in turn, cause these remnants to collapse. If the clouds are part of a larger complex, this can become a great stellar nursery. Massive stars have relatively short lives, and they can be born, live, and die as a supernova while their less-massive siblings are still forming. The shock wave from the supernova may plough through nearby interstellar matter, triggering yet more star birth.

**BOK GLOBULE**
Small, cool clouds of dust and gas, known as Bok globules, are the origins of some of the Milky Way's lower-mass stars.

*Bok globule*

**STELLAR EGGS**
Within the evaporating gaseous globules (EGGS) of the Eagle Nebula, interstellar material is collapsing to form stars.

**FORMATION IN ACTION**
Within the nebula NGC 2467 lie stars at various stages of formation. At the lower left lies a very young star that is breaking free of its surrounding birth cocoon of gas. On the far right, a wall of bright gas glows as it is evaporated by the energy of many newly formed hot stars. Dark lanes of dust at the centre hide parts of the nebula that are probably forming new stars.

# TRIGGERS TO STAR FORMATION

Clouds of interstellar material need a trigger to start them collapsing, as they are held up by their own pressure and that of internal magnetic fields. Such a trigger might be as simple as the gravitational tug from a passing star, or it might be a shock wave caused by the blast from a supernova or the collision of two or more galaxies. In spiral galaxies such as the Milky Way, density waves move through the dust and gas in the galactic disc (see p.227). As the waves pass, they temporarily increase the local density of interstellar material, causing it to collapse. Once the waves have passed, their shape can be picked out by the trails of bright young stars.

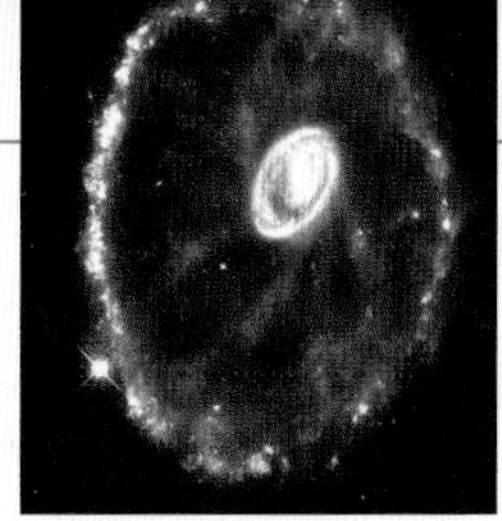

**GALACTIC COLLISIONS**
A ring of stars is created when two galaxies collide. Here, shock waves have rippled out, triggering star formation in the interstellar material.

**FROM OLD TO NEW**
Shock waves and material from a supernova blast spread out through the interstellar medium, triggering new star formation.

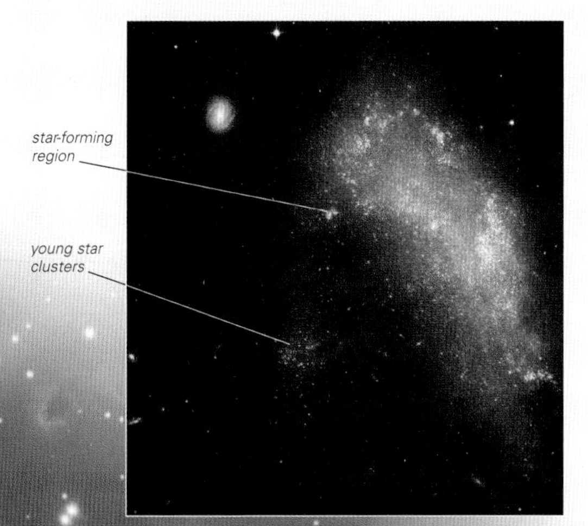


**VIOLENT STAR FORMATION**
Young star clusters (blue) and star-forming regions (pink) abound in NGC 1427A. As the galaxy's gas collides with the intergalactic medium through which the galaxy is travelling, the resulting pressure triggers violent but stunning star-cluster formation.

# STAR CLUSTERS

When they have formed from the fragmentation of a single collapsing molecular cloud, young stars are often clustered together. Many stars are formed so close to their neighbours that they are gravitationally bound, and some are even close enough to transfer material. It is unusual for a star not to be in a multiple system such as a binary pair (see pp.270–71), and in this respect, the Sun is uncommon. Stars within a cluster will usually have a similar chemical composition, although, as successive generations of stars may be produced by a single nebula, clusters may contain stars of different ages (see pp.284–85). Remnants of dust and gas from the initial cloud will linger, and the dust grains often reflect the starlight, predominantly in the shorter blue wavelengths. Thus, young star clusters are often surrounded by distinctive blue reflection nebulae. Young stars are hot and bright, and any nearby interstellar material will be heated by new stars' heat, producing red emission nebulae. Stars' individual motions will eventually cause a young star cluster to dissipate, though multiple stellar systems may remain gravitationally bound and may move through a galaxy together.

# TOWARDS THE MAIN SEQUENCE

As collapsing fragments of nebulae continue to shrink, their matter coalesces and contracts to form protostars. These stellar fledglings release a great deal of energy as they continue to collapse under their own gravity. However, they are not easily seen as they are generally surrounded by the remnants of the cloud from which they have formed. The heat and pressure generated within protostars acts against the gravity of their mass, opposing the collapse. Eventually, matter at the centres of the protostars gets so hot and dense that nuclear fusion starts and a star is born. At this stage, stars are very unstable. They lose mass by expelling strong stellar winds, which are often directed in two opposing jets channelled by a disc of dust and gas that forms around their equators. Gradually, the balance between gravity and pressure begins to equalize and the stars settle down on to the main sequence (see pp.232–35).

## J.L.E. DREYER

Danish-Irish astronomer, John Louis Emil Dreyer (1852–1926) compiled the New General Catalogue of Nebulae and Clusters of Stars, from which nebulae and galaxies get their NGC number. At the time of compilation, it was not known if all the nebulous objects were within the Milky Way. Dreyer studied the proper motions of many and concluded the "spiral nebulae", now known to be spiral galaxies, were likely to be more distant objects.

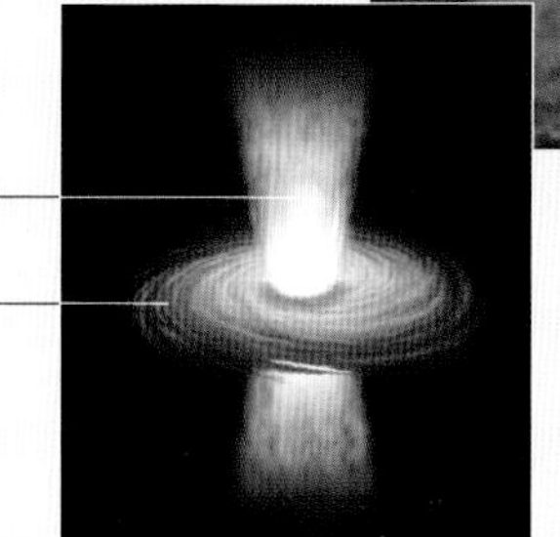


**ADOLESCENT STAR**
T Tauri (above) is the prototype of a type of adolescent star that is still undergoing gravitational contractions. These stars have extensive accretion discs and violent stellar winds coming from their poles (left).

# STAR-FORMING NEBULAE

Star formation can be seen throughout the Milky Way, but it is principally evident in the spiral arms and towards the Galactic Centre, where there is an abundance of star-making ingredients: dust and gas. In these regions, the interstellar matter is dense enough for molecular clouds to exist. These clouds are cold and appear as dark nebulae that are visible only when framed against a brighter background. When stars are born, these clouds are illuminated from within to become emission nebulae, some of the most beautiful sights in the Milky Way.

**STELLAR NURSERY**
Bright young stars within the Omega Nebula, M17, light up the nebula from which they were born.

**BINARY FORMATION**
Jets from BHR 71's newly forming binary star system have created the filamentary structure seen in this composite image made from four separate images.

DARK NEBULA

## BHR 71

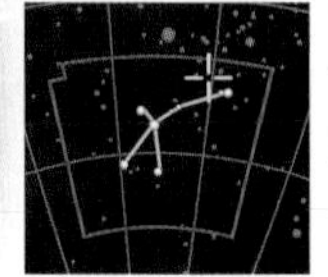

**CATALOGUE NUMBER**
BHR 71

**DISTANCE FROM SUN**
600 light-years

**MUSCA**

The small dark nebula BHR 71 is called a Bok globule (see p.236) and has a diameter of about one light-year. Within the dark molecular cloud are two sources of infrared and radio rays believed to be very close embryonic stars: HH 320 and HH 321, both losing vast amounts of material as they collapse. HH 320 has the strongest outflow, and it is probably surrounded by a massive disc of previously ejected stellar material. Although not optically visible, HH 320 has ten times the luminosity of the Sun. BHR 71 and its protostars offer a rare opportunity for the study of star-formation processes.

DARK NEBULA

## Horsehead Nebula

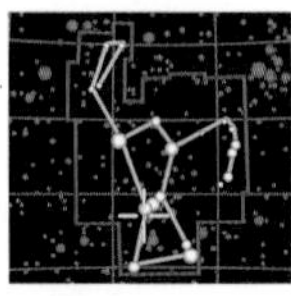

**CATALOGUE NUMBER**
Barnard 33

**DISTANCE FROM SUN**
1,500 light-years

**ORION**

One of the most beautiful and well-known astronomical sights, the Horsehead Nebula can be located in the night sky just south of the bright star Zeta (ζ) Orionis, the left star of the three in Orion's belt (see pp.374–75). The nebula is an extremely dense, cold, dark cloud of gas and dust, silhouetted against the bright, active nebula IC 434. It is about 16 light-years across and has a total mass about 300 times that of the Sun. The Horsehead shape is sculpted out of dense interstellar material by the radiation from the hot young star Sigma (σ) Orionis. Within the dark cloud, from which the Horsehead rears, is a scattering of young stars in the process of forming. The streaks that extend through the bright area above the Horsehead are probably caused by magnetic fields within the nebula.

**DARK KNIGHT**
One of the most photographed objects in the night sky, this dark nebula resembles the head of a sea horse or a knight on a chessboard. Its unusual shape was first discovered on a photographic plate in 1888.

**NEW STARS**
At the top of this image are the Trapezium stars forming within the Orion Nebula. Also visible, towards the bottom left-hand corner, is a line of shock waves created by material outflowing from the embryonic stars at speeds of 720,000kph (450,000mph).

EMISSION NEBULA

## Orion Nebula

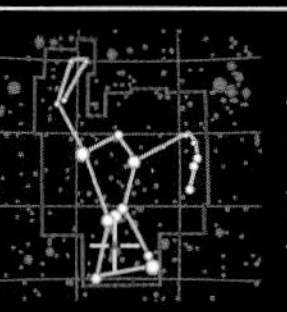

ORION

**CATALOGUE NUMBERS**
M42, NGC 1976

**DISTANCE FROM SUN**
1,500 light-years

**MAGNITUDE** 4

The most famous and the brightest nebula in the night sky, the Orion Nebula is easily visible with the naked eye as a diffuse, reddish patch below Orion's belt (see pp. 374–75). It is also the nearest emission nebula to Earth and has been extensively studied. The nebula spans about 30 light-years and has an apparent diameter four times that of the full Moon. However, it is a small part of a much larger molecular cloud system known as OMC-1, which has a diameter of several hundred light-years. The Orion Nebula sits at the edge of OMC-1, which stretches as far as the Horsehead Nebula (opposite).
The nebula glows with the ultraviolet radiation of the new stars forming within it. Many of these stars have been shown to have protoplanetary discs surrounding them. The principal stars whose radiation is ionizing the cloud of dust and gas belong to the Trapezium star cluster (see p.375), located at the heart of the nebula. At about 30,000 years old, the Trapezium is one of the youngest clusters known. It is a quadruple star system consisting of hot OB stars (see pp.230–31). In 1967, an extended dusty region was discovered directly behind the Orion Nebula. Known as the Kleinmann–Low Nebula, it has strong sources of infrared radiation embedded within it. These sources are believed to be protostars and newly formed stars.

EXPLORING SPACE

### FIRST PHOTOGRAPH

A pioneer of astrophotography, the American scientist Henry Draper (1837–82) took the first photograph of a nebula in September 1870 after he turned his camera to the Orion Nebula, the brightest one in the sky. Although his photograph was relatively crude, 12 years later he used a 28cm (11in) photographic refractor to obtain a much-improved image. The Orion Nebula has since been photographed probably more times than any other nebula.

**THE GREAT ORION NEBULA**
The Orion Nebula is separated from a smaller emission nebula known as M43 (at the top of this image) by a dark lane of dust popularly called the Fish's Mouth.

DARK NEBULA

## Cone Nebula

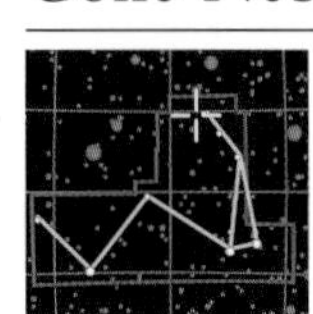

**MONOCEROS**

**CATALOGUE NUMBER**
NGC 2264

**DISTANCE FROM SUN**
2,500 light-years

**MAGNITUDE** 3.9

Discovered by William Herschel (see p.90) in 1785, the Cone Nebula is a dark nebula located at the edge of an immense, turbulent star-forming region. This conical pillar of dust and gas is more than 7 light-years long and at its "top" is 2.5 light-years across. The Cone Nebula is closely associated with the star cluster NGC 2264, commonly known as the Christmas Tree Cluster. This cluster, which spans a distance of 50 light-years, is made up of at least 250 stars, and it is the light from some of its newborn stars that allows the Cone Nebula to be seen in silhouette. The Cone Nebula is located at the top of the Christmas Tree Cluster, pointing downwards to the bottom of the tree. At the opposite end, the 5th-magnitude star S Mon marks the left of the base of the tree (see below left). Jets of stellar material thrown out by newly forming stars have been detected within the star cluster. These Herbig Haro objects also help to shape the material in the surrounding nebula. One explanation for the shape of the Cone Nebula suggests that it was formed by stellar wind particles from an energetic source blowing past a Bok globule at the head of the cone. Buried in the dust and gas near the top of the Cone is a massive star known as NGC 2264 IRS, which is surrounded by six smaller Sun-like stars. It is thought that the outflow of stellar material during the early years of this massive star triggered the formation of the surrounding six and also helped to sculpt the shape of the Cone Nebula itself. None of these stars are visible with optical telescopes. Infrared observations have revealed further embryonic stars embedded in the nebulosity (above), making this one of the most active star-forming regions in this area of the Milky Way.

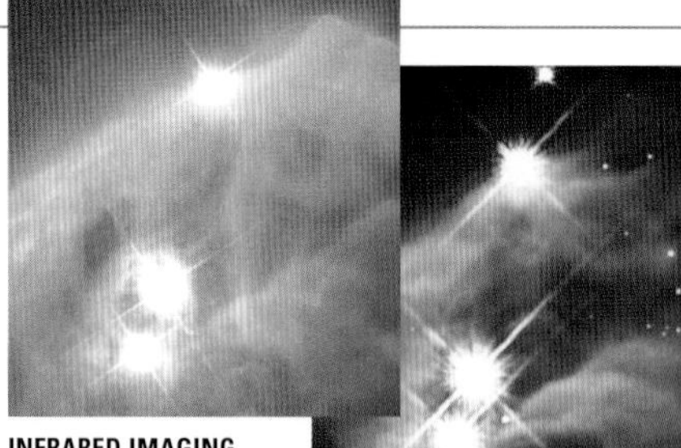

**INFRARED IMAGING**
Unseen in an optical image (left), a remarkable infrared view of the tip of the Cone Nebula (right) reveals, to the right of the image, a clutch of faint newborn stars.

**CHRISTMAS TREE CLUSTER**
The stars of the open cluster NGC 2264 can be seen in this image resembling an upside-down Christmas tree, with the Core Nebula (boxed) at the apex of the tree.

**TOWER OF RESISTANCE**
Born in immense clouds of dust and gas, the great tower of the Cone Nebula is a slightly denser region of material that has resisted erosion by radiation from its neighbouring stars.

EMISSION NEBULA

## IC 1396

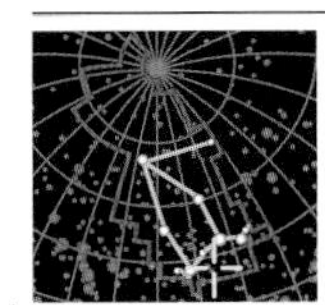

CEPHEUS

**CATALOGUE NUMBER**
IC 1396

**DISTANCE FROM SUN**
3,000 light-years

Occupying an area hundreds of light years across, the IC 1396 complex contains one of the largest emission nebulae close enough to be observed in detail. It has an apparent diameter in the night sky ten times that of the full Moon. The mass of the nebula is estimated to be an immense 12,000 times the mass of the Sun, mainly consisting of hydrogen and helium in various forms. HD 206267, a massive, young blue star at the centre of the region, produces most of the radiation that illuminates the nebula's interstellar material. Observations have shown that ionized clouds form a rough ring around this star at distances between 80 and 130 light-years. These clouds are the remains of the molecular cloud that originally gave birth to HD 206267 and its siblings, which compose the star cluster known as Tr37. Tracts of cool, dark material lie farther away from HD 206267. Among the most dramatic of these structures is one commonly known as the Elephant's Trunk Nebula. Research suggests that some of this material has been blown away from the star by strong stellar winds, causing the material to form elongated structures such as the Elephant's Trunk. Some of these structures stretch radially away from HD 206267 for up to 20 light-years. Within IC 1396 lies Mu (μ) Cephei, also known as Herschel's Garnet Star. One of the largest and brightest stars known, Mu Cephei is a red supergiant emitting 350,000 times the power of the Sun.

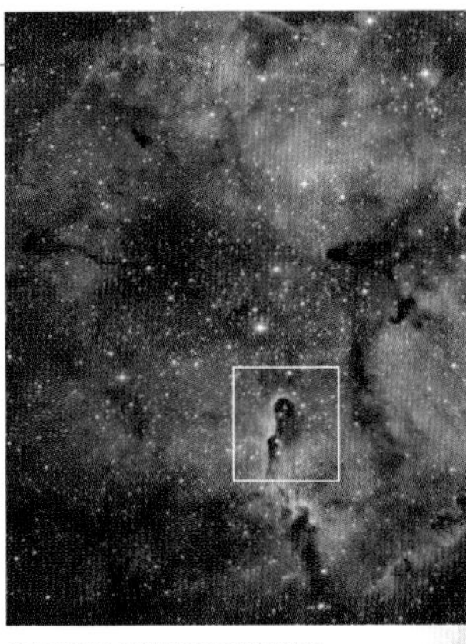

**GIGANTIC STELLAR NURSERY**
The immense IC 1396 complex of emission nebulae, dark nebulae, and a young star cluster is shown here in a composite image. Mu Cephei is located at the centre, and the Elephant's Trunk nebula is boxed.

**ELEPHANT'S TRUNK**
The Elephant's Trunk Nebula is sculpted from a huge cloud of interstellar material in which star formation may take place in the future.

EMISSION NEBULA

## DR6

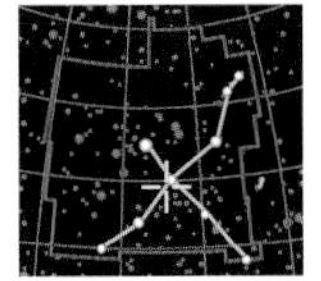

CYGNUS

**CATALOGUE NUMBER**
DR6

**DISTANCE FROM SUN**
4,000 light-years

Strong stellar winds from about 10 young stars at the centre of this unusual nebula have created cavities within its interstellar material, making it resemble a human skull. The nebula has a diameter of about 15 light-years, and the "nose", where the stars that have sculpted the nebula are located, is about 3.5 light-years across. The central group of stars is very young, having formed less than 100,000 years ago. The picture below is a composite of four infrared images.

**HOLLOW SKULL**

**TWISTS OF GAS**
Creative chaos is revealed within the vast Lagoon Nebula, as radiation and strong winds from forming stars interact with surrounding clouds of interstellar dust and gas.

**DARK GLOBULES**
One of the key features of the Lagoon Nebula is the presence of a large number of dark, comet-shaped clouds of collapsing dust and gas called Bok globules, where future stars may be born.

EMISSION NEBULA

## Lagoon Nebula

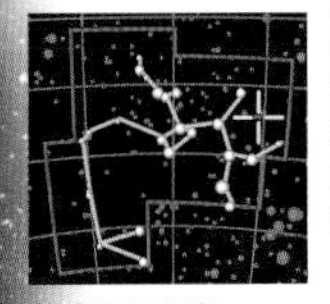

SAGITTARIUS

**CATALOGUE NUMBERS**
M8, NGC 6523

**DISTANCE FROM SUN**
5,200 light-years

**MAGNITUDE** 6

The Lagoon Nebula is a productive star-forming region situated within rich, conspicuous fields of interstellar matter. Covering an apparent diameter of more than three full Moons, the Lagoon Nebula is so large and luminous that it is visible to the naked eye. The region contains young star clusters, distinctive Bok globules (see p.236), and very energetic star-forming regions. There are also many examples of twisted-rope structures thought to have been created by hot stellar winds colliding with cooler dust clouds. The bright centre of the Lagoon Nebula is illuminated by the energy of several very hot young stars, including the 6th-magnitude 9 Sagittarii and the 9th-magnitude Herschel 36. Also found in the brightest region is the famous Hourglass Nebula (see p.259). The open cluster NGC 6530 (to the left of centre in the main image) contains 50 to 100 stars that are only a few million years old. Clearly visible across the Lagoon Nebula are dark Bok globules.

EMISSION NEBULA

## Eagle Nebula

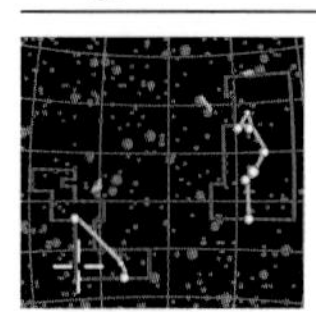

SERPENS

**CATALOGUE NUMBER** IC 4703

**DISTANCE FROM SUN** 7,000 light-years

**MAGNITUDE** 6

Observations of the Eagle Nebula have introduced new ideas to the theory of star formation. Lying in one of the dense spiral arms of the Milky Way, this is an immense stellar nursery where young stars flourish, new stars are being created, and the material and triggers exist for future star formation. In optical wavelengths, this region is dominated by the light from the bright young star cluster M16. This cluster was discovered by the Swiss astronomer Philippe Loys de Chéseaux in around 1745, but it was nearly 20 years later that the surrounding nebula, from which the star cluster had formed, was discovered by Charles Messier (see p.69). The star cluster itself is only about 5 million years old and has a diameter of about 15 light-years. The Eagle Nebula is much larger than the star cluster, with a diameter of about 70 light-years.

In 1995, the Hubble Space Telescope imaged features within the nebula that are commonly known as the Pillars of Creation (see panel, below). These famous pillars are towers of dense material that have resisted evaporation by radiation from local young stars. However, the stars' ultraviolet radiation is gradually boiling their surfaces away, through a process called photo-evaporation. As the towers themselves are not of a consistent density, the continuing photo-evaporation has caused some of the smaller nodules, known as evaporating gaseous globules (EGGs), to become detached from the main gas towers. At this point, these dense stellar nurseries cease to accrue more material, and any embryonic star within has its upper mass limit fixed. It is thought that this method of star formation inhibits the formation of accretion discs around the stars, which are believed to be the material from which planets are formed. These detailed images of the Pillars of Creation were the first to suggest this method of star creation. The Eagle Nebula also contains many Bok globules, regions where future star formation is probably occurring.

**HUGE STELLAR NURSERY**
This wide-field image shows the immensity of the Eagle Nebula, with the three Pillars of Creation located near the centre. This huge cloud of gas lies in the galaxy's Sagittarius–Carina arm, towards the Galactic Centre.

EXPLORING SPACE

### THE PILLARS OF CREATION

This image, taken by the Hubble Space Telescope in 1995, has become one of the most famous and iconic astronomical images. Revealing, for the first time in dramatic detail, a previously unsuspected method of star formation, it captured the public's imagination and inspired a new interest in astronomy. The image's aesthetic appeal and the sense of wonder it inspires have led to it being displayed on posters, in magazines, and even on stamps.

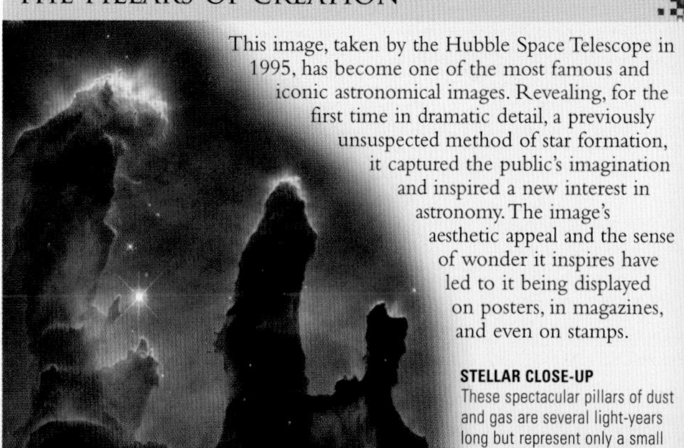

**STELLAR CLOSE-UP**
These spectacular pillars of dust and gas are several light-years long but represent only a small section of the Eagle Nebula.

**TWISTED PILLARS**
The three Pillars of Creation are shown twisting through a rich star field in this composite infrared image. Not all these stars lie in the Eagle Nebula, as some lie far behind and others lie in front.

EMISSION NEBULA

## IC 2944

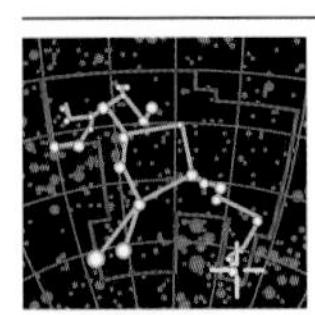

CATALOGUE NUMBER
IC 2944

DISTANCE FROM SUN
5,900 light-years

MAGNITUDE 4.5

CENTAURUS

Between the constellations Crux and Centaurus lies the bright, busy star-forming nebula IC 2944. This nebula is made up of dust and gas that is illuminated by a loose cluster of massive young stars. IC 2944 is perhaps best known for the many Bok globules that are viewed in silhouette against its backdrop. Bok globules are thought to be cool, opaque regions of molecular material that will eventually collapse to form stars. However, studies of the globules in IC 2944 have revealed that the material of which they are composed is in constant motion. This may be caused by radiation from the loose cluster of massive young stars embedded in IC 2944. The stars' ultraviolet radiation is gradually eroding the globules, and it is possible that this could prevent them from collapsing to form stars. In addition to radiation, the stars also emit strong stellar winds that send out material at high velocities, causing heating and erosion of interstellar material. The largest Bok globule in IC 2944 (below) is about 1.4 light-years across, with a mass about 15 times that of the Sun.

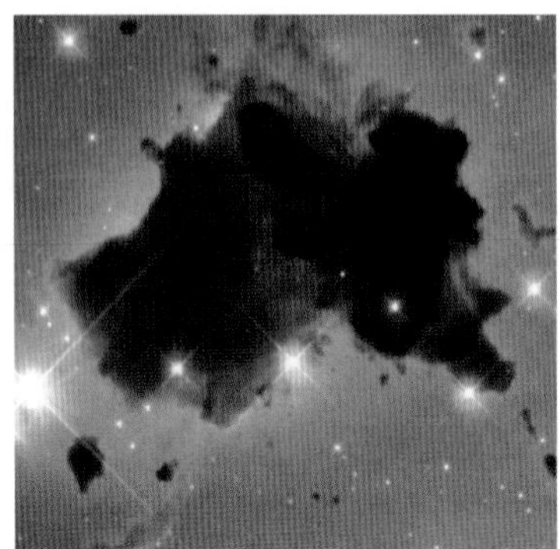

**THACKERAY'S GLOBULES**
The Bok globules in IC 2944 were first observed in 1950 by the South African astronomer A.D. Thackeray. This globule has recently been shown to be two overlapping clouds.

EMISSION NEBULA

## DR 21

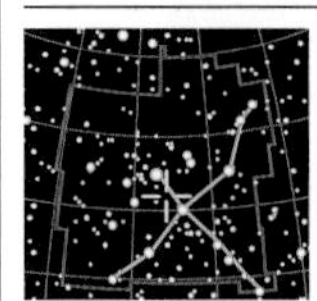

CATALOGUE NUMBER
DR 21

DISTANCE FROM SUN
6,000 light-years

CYGNUS

The birth of some of the Milky Way's most massive stars has been discovered within DR 21, a giant molecular cloud spanning about 80 light-years. Infrared images have revealed an energetic group of newborn stars tearing apart the gas and dust around them. One star alone is 100,000 times as bright as the Sun. This star is ejecting hot stellar material into the surrounding molecular cloud, suggesting it may have a planet-forming disc around it.

**GIGANTIC EMBRYOS**
This infrared image reveals a clutch of gigantic newborn stars, shown here in green. In optical light, the surrounding molecular cloud is opaque.

EMISSION NEBULA

## Trifid Nebula

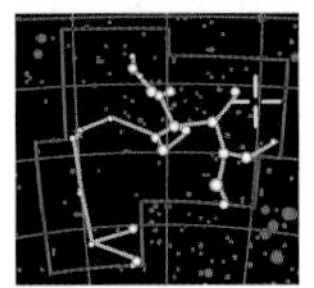

CATALOGUE NUMBER
M20

DISTANCE FROM SUN
7,600 light-years

MAGNITUDE 6.3

SAGITTARIUS

This emission nebula is one of the youngest yet discovered. It was first called the Trifid Nebula by the English astronomer John Herschel because of its three-lobed appearance when seen through his 18th-century telescope. The nebula is a region of interstellar dust and gas being illuminated by stars forming within it. It spans a distance of around 50 light-years. The young star cluster at its centre, NGC 6514, was formed only about 100,000 years ago. The Trifid's lobes, the brightest of which is actually a multiple system, are created by dark filaments lying in and around the bright nebula. The whole area is surrounded by a blue reflection nebula, particularly conspicuous in the upper part, where dust particles disperse light.

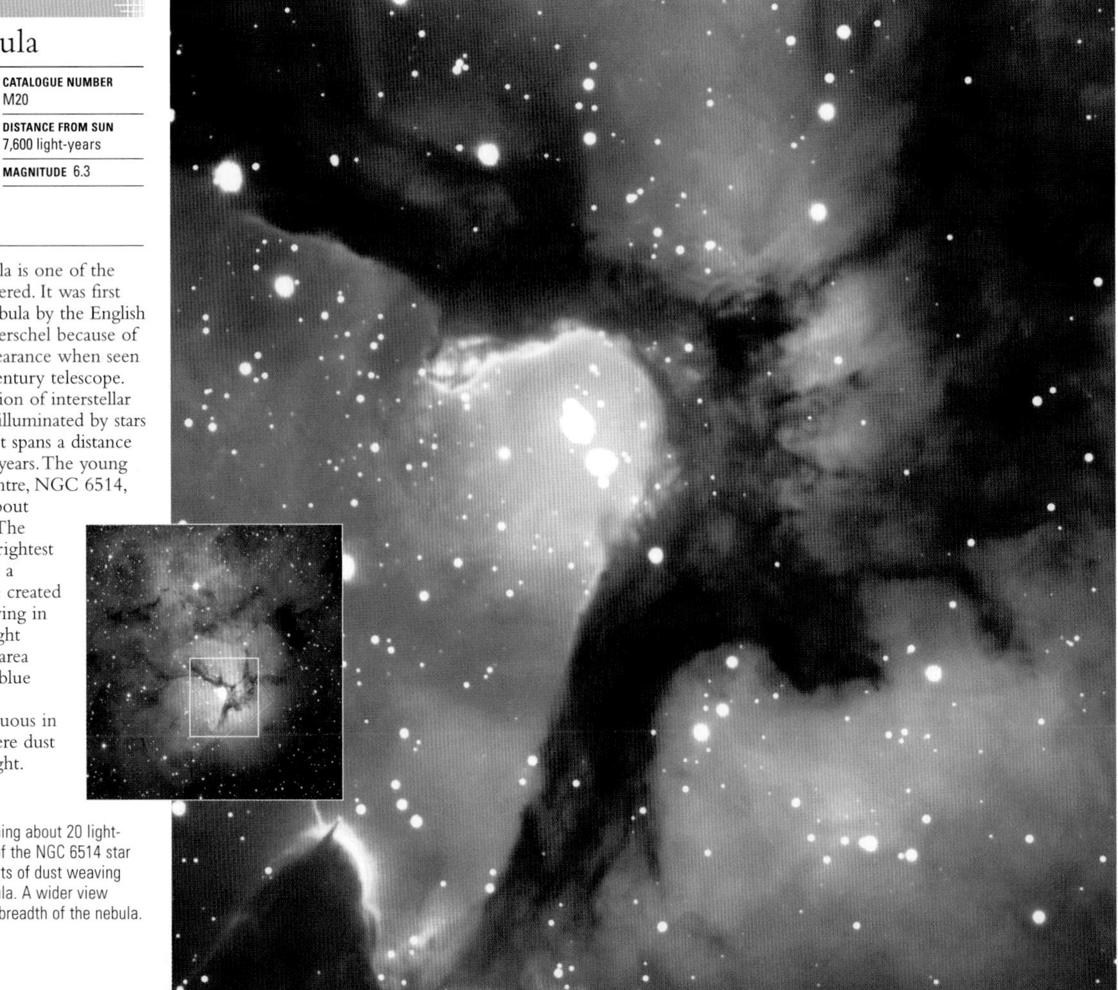

**HEART OF THE TRIFID**
The main image, spanning about 20 light-years, reveals details of the NGC 6514 star cluster and the filaments of dust weaving through the Trifid Nebula. A wider view (above) shows the full breadth of the nebula.

EMISSION NEBULA

## Carina Nebula

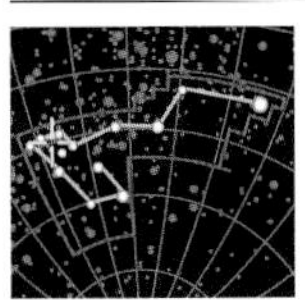

CARINA

**CATALOGUE NUMBER** NGC 3372

**DISTANCE FROM SUN** 8,000 light-years

**MAGNITUDE** 1

Also known as the Eta (η) Carina Nebula, this is one of the largest and brightest nebulae to be discovered. It has a diameter of more than 200 light-years, stretching up to 300 light-years if its fainter outer filaments are included. Within its heart, and heating up its dust and gas, is an interesting zoo of young stars. These include examples of the most massive stars known, with a spectral type of O3 (see pp.230–31). This type of star was first discovered in the Carina Nebula, and the nebula remains the closest location of O3 stars to Earth. Also within the Carina Nebula are three Wolf–Rayet stars with spectral type WN (see pp.250–51). These stars are believed to be evolved O3 stars with very large rates of mass ejection. One of the best-known features within the Carina Nebula is the blue supergiant star Eta (η) Carinae (see p.258), embedded within part of the nebula known as the Keyhole Nebula. Recent observations made with infrared telescopes reveal that portions of the Carina Nebula are moving at very high speeds – up to 828,000kph (522,000mph) – in varying directions. Collisions of interstellar clouds at these speeds heat material to such high temperatures that it emits high-energy X-rays, and the entire Carina Nebula is a source of extended X-ray emission. The movement of these clouds of material is thought to be due to the strong stellar winds emitted by the massive stars within, bombarding the surrounding material and accelerating it to its high velocities.

**PROBING THE NEBULA**
An infrared image reveals the stars lying within the nebula's dense dust and gas. The open clusters Trumpler 14 and Trumpler 16 are visible to the left and top of the image.

**DETACHED CLOUD**
This unusually shaped cloud of dust and gas has been separated from the main nebula by radiation and hot jets of material emitted by stars forming locally. It spans a distance of about two light-years.

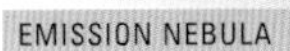

EMISSION NEBULA

## RCW 49

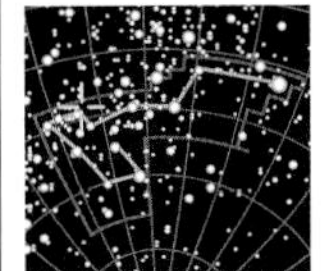

CARINA

**CATALOGUE NUMBERS** RCW 49, GUM 29

**DISTANCE FROM SUN** 14,000 light-years

One of the most productive regions of star formation to have been found in the Milky Way, RCW 49 spans a distance of about 350 light-years. It is thought that over 2,200 stars reside within RCW 49, but because of the nebula's dense areas of dust and gas, the stars are hidden from view at optical wavelengths of light. However, the infrared telescope onboard the Spitzer spacecraft (see panel, right) has recently revealed the presence of up to 300 newly formed stars. Stars have been observed at every stage of their early evolution in this area, making it a remarkable source of data for studying star formation and development. One surprising preliminary observation suggests that most of the stars have accretion discs around them. This is a far higher ratio than would usually be expected. Detailed observations of two of the discs reveal that they are composed of exactly what is required in a planet-forming system. These are the farthest and faintest potential planet-forming discs ever observed. This discovery supports the theory that planet-forming discs are a natural part of a star's evolution. It also suggests that solar systems like our own are probably not rare in the Milky Way (see pp.290–91).

**COSMIC CONSTRUCTION**
This false-colour image, composed of four separate images taken in different infrared wavelengths, reveals more than 300 newborn stars scattered throughout the RCW 49 nebula. The oldest stars of the nebula appear in the centre in blue, gas filaments appear in green, and dusty tendrils are shown in pink.

EXPLORING SPACE

### SPITZER TELESCOPE

Launched in August 2003, the Spitzer telescope is the largest infrared telescope ever put into orbit. It has been very successful in probing the dense dust and gas that lies in the interstellar medium and has revealed features and details within star-forming clouds that have never been seen before. As Spitzer observes in infrared, its instruments are cooled almost to absolute zero, to ensure that their own heat does not interfere with the observations. A solar shield protects the telescope from the Sun.

**INSIDE SPITZER**
The Spitzer craft has an 85cm (34in) telescope and three super-cooled processing instruments.

# MAIN-SEQUENCE STARS

92 Matter and star energy
230–31 Stars
232–35 The life cycles of stars
237 Towards the main sequence
Old stars 250–51
Stellar end points 262–63

**MAIN-SEQUENCE STARS** are those that convert hydrogen into helium in their cores by nuclear reactions. Stars spend a high proportion of their lives on the main sequence, during which time they are very stable. The higher the mass of the star, the less time it spends on the main sequence, as nuclear reactions occur faster in higher-mass stars.

**STAR FLARES**
The Sun's photosphere radiates huge amounts of energy as solar flares contribute to the solar wind.

## STAR ENERGY

The cores of main-sequence stars initially consist mainly of hydrogen. When the temperature and pressure become high enough, the hydrogen is converted into helium by nuclear reactions. For stars of less than about 1.5 solar masses, this is done by means of a process called the proton–proton chain reaction (the pp chain). For stars of more than about 1.5 solar masses and with core temperatures of more than about 20 million °C (36 million °F), carbon, nitrogen, and oxygen are used as catalysts in a process called the carbon cycle (CNO cycle). When hydrogen is converted to helium, a tiny amount of energy is released as gamma rays, which gradually permeate their way out through the photosphere (the Sun's visible surface). The huge amounts of energy radiated by main-sequence stars are due to the immense masses of hydrogen they contain. In the core of the Sun, 600 million tonnes of hydrogen are converted into helium every second.

**MASSIVE STAR**
Achernar, or Alpha (α) Eridani, the ninth-brightest star in the sky, is a blue main-sequence star of about six to eight solar masses. Main-sequence stars of this size convert hydrogen to helium through a process called the carbon cycle.

## STELLAR STRUCTURE

Energy, in the form of gamma rays, is released in the nuclear reactions occurring within stellar cores. This energy can be transported outwards by two processes: convection and radiation. In convection, hot material rises to cooler zones, expanding and cooling, then sinks back to hotter levels, just like water being boiled in a saucepan. In the radiation process, photons are continually absorbed and re-emitted. They can be emitted in any direction, so sometimes travel back into the central core. They follow a path termed a "random walk", but gradually diffuse outwards, losing energy as they do so. Their energy matches the temperature of the surrounding material, so they start as gamma rays, but at the Sun's surface, the photosphere, they appear in the visible part of the electromagnetic spectrum.

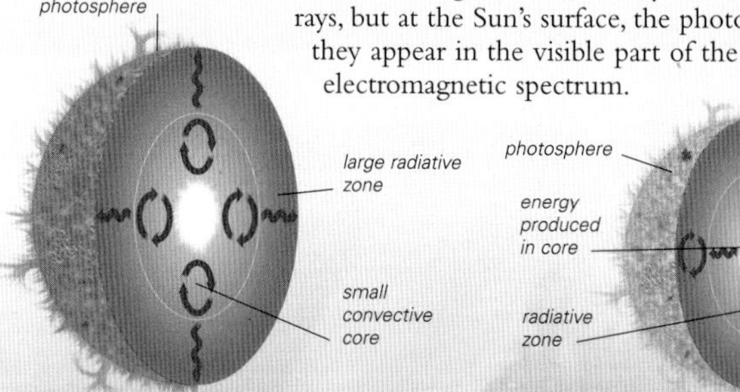


**HIGH-MASS STAR**
Stars with a mass greater than 1.5 solar masses produce energy through the CNO cycle. They have convective cores and a large radiative zone reaching to the photosphere.

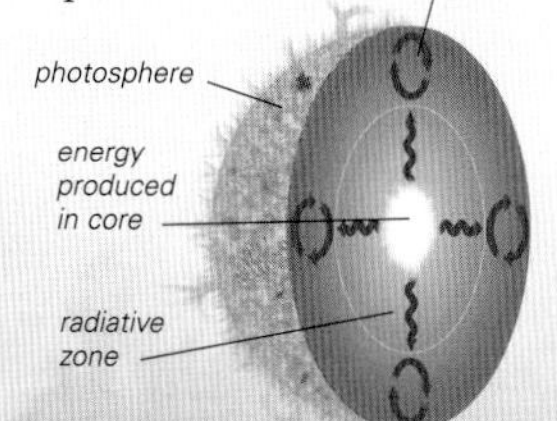


**LOW-MASS STAR**
In stars with a mass smaller than 1.5 solar masses, the pp chain dominates, and a large, inner radiative zone reaches out to a smaller convection zone near the star's photosphere.

**ERUPTIVE SURFACE**
Main-sequence stars, such as the Sun, appear smooth in optical light, but in reality their photospheres are extremely turbulent with huge prominences of material constrained by magnetic fields.

# ROTATION AND MAGNETISM

The pressures and temperatures within stars mean they are composed of plasma (see p.30). Within this ionized matter, negatively charged electrons travel free from the positively charged ions. This has a profound effect on magnetic fields, as charged particles do not cross magnetic field lines easily. Magnetic field lines can dictate the movement of stellar material, but the movement of the plasma can also affect magnetic fields. All stars rotate, and some spin so fast they bulge out at the equator and are very flattened at the poles. As stars rotate, magnetic-field lines are carried around by the plasma. This "winds up" the magnetic field and creates pockets of intense magnetic flux where field lines are brought close together. The movement of stellar material and the transfer of heat is restricted in these areas, so they are appreciably cooler than the surrounding material. As they are cooler, they appear dark against the rest of the photosphere. Dark star spots on the surface of stars are areas of intense activity, because the build-up of heat around them can suddenly be released as flares.

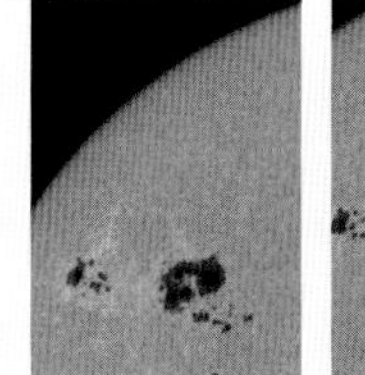
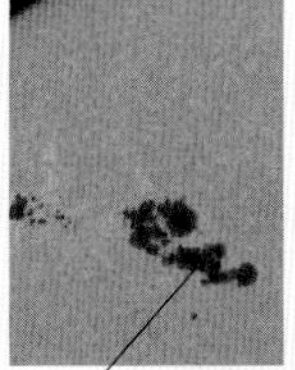
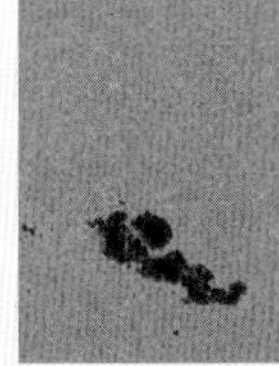
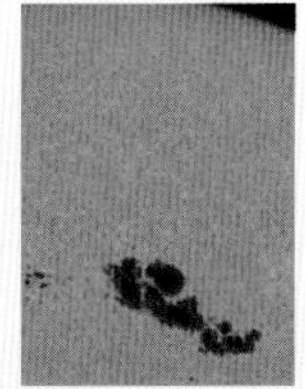
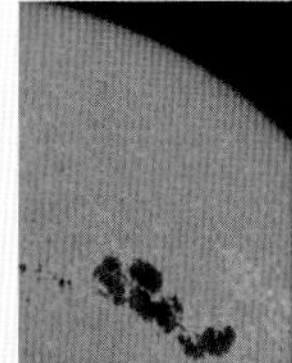
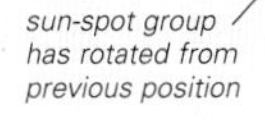
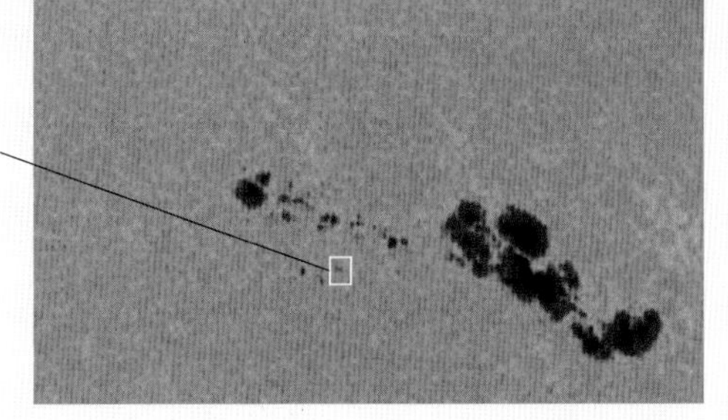


**SOLAR ROTATION**
As the Sun rotates, sun-spot groups are observed travelling across its disc. Main-sequence stars rotate differentially, with material at the equator rotating faster than that at the poles. On the Sun, sunspots nearer the equator travel across the solar disc more rapidly.

## ARTHUR EDDINGTON

British astronomer Arthur Stanley Eddington (1882–1944) studied the internal structure of stars and derived a mass–luminosity relationship for main-sequence stars. In 1926, he published *The Internal Constitution of Stars*, in which he suggested that nuclear reactions were the power source of stars. While working at the Royal Greenwich Observatory, Eddington led two expeditions to view total solar eclipses and in 1919 provided evidence for the theory of general relativity. Eddington also calculated the abundance of hydrogen within stars and developed a model for Cepheid variable pulsation (see p.278). He became Plumian professor of astronomy at Cambridge in 1913 and director of the Cambridge Observatory in 1914. He was knighted in 1930.

# THE MAIN SEQUENCE

A star enters the main sequence when it starts to burn hydrogen in its core. As soon as the nuclear reactions instigating this process begin, it is said to be at age zero on the main sequence. A star's life on the main sequence is very stable, with the pressure from the nuclear reactions in its core being balanced by its gravity trying to compress all of its mass into the centre. A star will spend most of its life on the main sequence, and consequently about 90 per cent of the stars observed in the sky are main-sequence stars. A star's time on the main sequence is dependent on its mass. The more massive the star, the hotter and denser its core and the faster it will convert hydrogen into helium. The Sun is a relatively small main-sequence star and will be on the main sequence for about ten billion years. A ten-solar-mass star will be on the main sequence for only ten million years. While on the main sequence, a star will conform to the mass–luminosity relation, which means that the absolute magnitude or luminosity of a star will give an indication of its mass. As it converts hydrogen into helium, a star's chemical composition and internal structure will change and it will move slightly to the right of its zero-age position on the H–R diagram (below). As soon as the hydrogen in the core is depleted, and hydrogen burning in the atmosphere begins, the star leaves the main sequence (see p.234).

**DIAGONAL PATH**
The main sequence is a diagonal curving path of stars on the Hertzsprung–Russell diagram, a simplified version of which is shown here (see also p.230). The curve runs from bottom right (low mass and cool) to top left (massive and hot). Each star has a "zero-age" position (a point on the curve indicating its mass and temperature). It hardly strays at all from this position during its time on the main sequence.

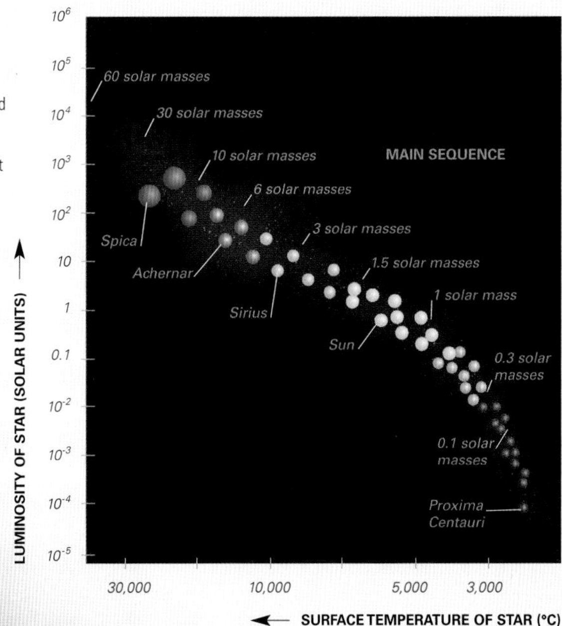

# MAIN-SEQUENCE STARS

During a star's life, it passes through many phases, but most of its time will be spent on the main sequence. This means that the chances of seeing any star are greatest during its main-sequence life time. In fact, about 90 per cent of all observed stars are on the main sequence. Although main-sequence stars are spread throughout the Milky Way, they appear predominantly in its plane and central bulge.

**PROMINENT STARS**
Known as the Pointers, Alpha and Beta Centauri are prominent main-sequence stars guiding the way to the Southern Cross.

ORANGE-RED STAR

## Proxima Centauri

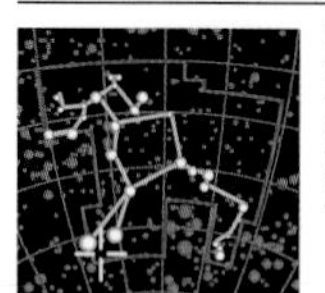

CENTAURUS

**DISTANCE FROM SUN** 4.2 light-years

**MAGNITUDE** 11.05

**SPECTRAL TYPE** M

The closest star to the Sun, Proxima Centauri is thought to be a member of the Alpha Centauri system (right), orbiting the binary system at a distance 10,000 times the distance of the Earth from the Sun. Its orbital period is at least one million years, prompting some astronomers to question whether Proxima is gravitationally bound to Alpha Centauri at all. Proxima is a flare star, a cool red dwarf that undergoes outbursts of energy, when it brightens by about one magnitude (see pp.278–79). Even when in eruption, it is very faint – 18,000 times dimmer than the Sun – but it is an intense source of low-energy X-rays and high-energy ultraviolet rays. With a low luminosity and small size, it was not discovered until 1915. It has only about a tenth the mass of the Sun, and is a good example of a main-sequence star nearing the end of its life.

**POSSIBLE PLANET**
Small variations in Proxima Centauri's movement across the sky have suggested that it may be orbited by a planet with a mass 80 per cent that of Jupiter.

YELLOW AND ORANGE STARS

## Alpha Centauri

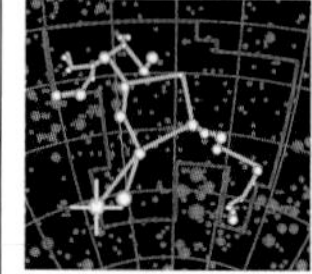

CENTAURUS

**DISTANCE FROM SUN** 4.4 light-years

**MAGNITUDES** -0.1 and 1.33

**SPECTRAL TYPES** G and K

The two stars of Alpha Centauri – also known as Rigil Kentaurus – orbit each other every 79.9 years. They are very close, and, in some images (below), are distinguishable only by seeing two sets of diffraction spikes. Alpha Centauri A is the brighter and more massive, at 1.57 times the luminosity and 1.1 times the mass of the Sun. Alpha Centauri B is both less massive and less luminous than the Sun.

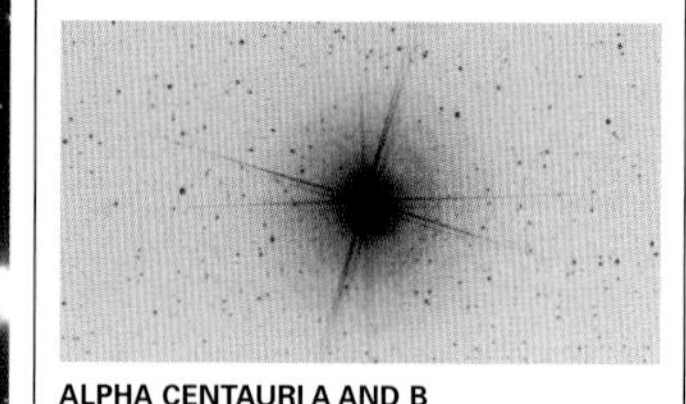

ALPHA CENTAURI A AND B

WHITE STAR

## Sirius A

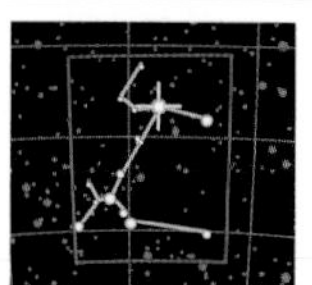

CANIS MAJOR

**DISTANCE FROM SUN** 8.6 light-years

**MAGNITUDE** -1.46

**SPECTRAL TYPE** A

The brightest star in the night sky, Sirius is the ninth closest star to Earth. It is a binary star, with Sirius A being a main-sequence star and its companion a white dwarf. Sirius A has twice the mass of the Sun and is 23 times as luminous. Recent observations suggest that it may have a stellar wind – the first spectral type A star to show evidence of one.

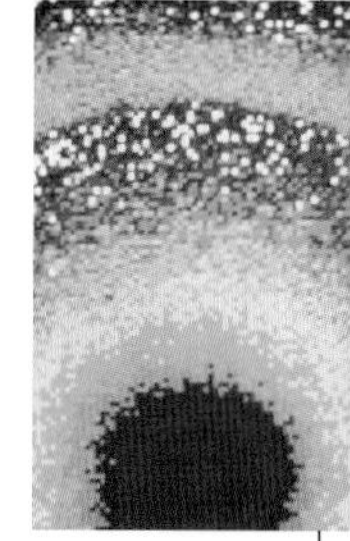

**SCORCHING STAR**
A false-colour image shows the diffraction pattern of Sirius, the brightest star in the sky. Its name is from the Greek for "scorching".

ORANGE STAR

## 61 Cygni

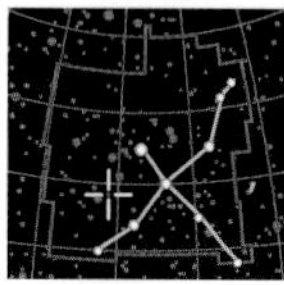

CYGNUS

**DISTANCE FROM SUN** 11.4 light-years

**MAGNITUDES** 5.2 and 6.1

**SPECTRAL TYPE** K

61 Cygni is a binary system of two main-sequence stars that orbit each other every 653 years. It is believed that 61 Cygni has at least one massive planet and possibly as many as three. In 1838, German astronomer Friedrich Bessel became the first to measure the distance of a star from Earth accurately, when he calculated 61 Cygni's annual parallax (see p.66). He chose 61 Cygni because, at that time, it was the star with the largest known proper motion.

FAST STAR

WHITE STAR

## Altair

AQUILA

**DISTANCE FROM SUN** 16.8 light-years

**MAGNITUDE** 0.77

**SPECTRAL TYPE** A

One of the three stars of the Summer Triangle, Altair is the 12th-brightest star in the sky. With a diameter about 1.6 times that of the Sun, it rotates once every 6.5 hours. This puts its equatorial spin rate at about 900,000kph (559,000mph), which causes distortion of its overall shape. This distortion is such that the star becomes wider at the equator and flattened at the poles, and estimates have suggested that its equatorial diameter is as much as double its polar diameter. It has a surface temperature of about 9,500°C (17,000°F) and a high rate of proper motion through the Milky Way.

**DUSTY BACKDROP**
In this optical image, Altair (boxed), the brightest star in Aquila, the Eagle, shines out against the dusty backdrop of the Milky Way.

WHITE STAR

## Fomalhaut

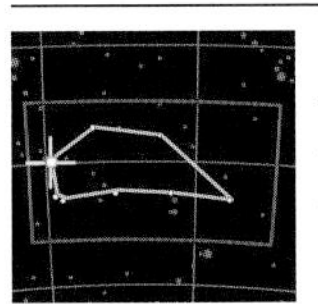

PISCIS AUSTRINUS

**DISTANCE FROM SUN** 25.1 light-years

**MAGNITUDE** 1.16

**SPECTRAL TYPE** A

The brightest star in Piscis Austrinus, Fomalhaut is the 18th-brightest star in the sky. It has a surface temperature of about 8,500°C (15,000°F), with a luminosity 16 times that of the Sun. In 1983 the infrared telescope IRAS revealed that it was a source of greater infrared radiation than expected. Further observations revealed that the infrared is being emitted by a cool disc of icy dust particles that surrounds Fomalhaut. The disc has a diameter at least twice that of the Solar System and is believed to be made of material from which planets are forming. Recently a hole was discovered within the disc, which may be consistent with embryonic planets removing the material as they form.

**DUSTY DISC**
The dusty disc that surrounds Fomalhaut is possibly the precursor of a solar system. Here, Fomalhaut's position is marked with a black star symbol, and false-colours show the intensity of radio emissions from the surrounding disc.

**DISTINCTIVE STAR**
Fomalhaut, the "mouth of the fish", is the most distinctive star in the constellation Piscis Austrinus.

WHITE STAR

## Vega

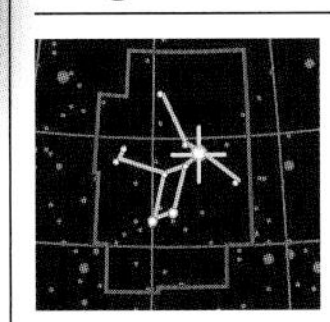

LYRA

**DISTANCE FROM SUN** 25.3 light-years

**MAGNITUDE** 0.03

**SPECTRAL TYPE** A

Also known as Alpha (α) Lyrae, Vega is the fifth-brightest star in the sky. Along with Altair (opposite) and Deneb, it makes up the Summer Triangle. Vega has a mass of about 2.5 solar masses, a luminosity 54 times that of the Sun, and a surface temperature of about 9,300°C (16,500°F). Around 12,000 years ago, it was the north Pole Star, and it will be so again in about 14,000 years' time. In 1983, the infrared satellite IRAS revealed that it is surrounded by a disc of dusty material that is possibly the precursor to a planetary system. Vega is the ultimate "standard" star, used to calibrate the spectral range and apparent magnitude of stars in optical astronomy (see p.231).

**BRIGHT BEACON**
The brightest star in the northern summer sky, Vega takes its name from an Arabic word meaning "swooping eagle".

YELLOW-WHITE STAR

## Porrima

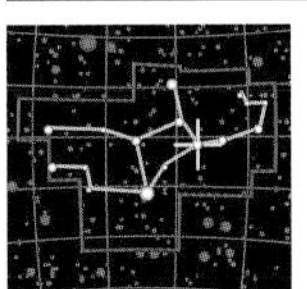

VIRGO

**DISTANCE FROM SUN** 38 light-years

**MAGNITUDE** 0.36

**SPECTRAL TYPE** F

**THE PORRIMA PAIR**

Porrima, also known as Gamma (γ) Virginis, is a binary system made up of two almost identical stars, both about 1.5 times the mass of the Sun. Their surface temperatures are around 7,000°C (13,000°F) and they appear creamy white in amateur telescopes. Their luminosities are each about four times that of the Sun. They orbit each other in a very elliptical path that takes around 170 years to complete.

BLUE-WHITE STAR

## Regulus

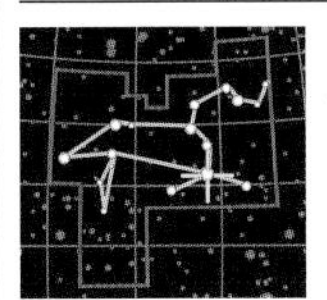

LEO

**DISTANCE FROM SUN** 78 light-years

**MAGNITUDE** 1.35

**SPECTRAL TYPE** B

The brightest star in the constellation Leo, Regulus just makes it into the top 25 brightest stars as seen from Earth. *Regulus* is a Latin word meaning "little king". The star is situated at the base of the distinctive sickle asterism (shaped like a reversed question mark) in the constellation. It lies very close to the ecliptic (see pp.58–61) and is often occulted by the Moon (right). Regulus is a triple system. The brightest component is a blue-white main-sequence star about 3.5 times the mass of the Sun and with a diameter also around 3.5 times that of the Sun. It has a surface temperature of about 12,000°C (22,000°F) and shines at about 140 times the brightness of the Sun. It is also an emitter of high levels of ultraviolet radiation. Regulus has a companion binary star, composed of an orange dwarf and a red dwarf separated by about 14 billion km (9 billion miles). These dwarf components orbit each other over a period of about 1,000 years, and they in turn orbit the main star once every 130,000 years.

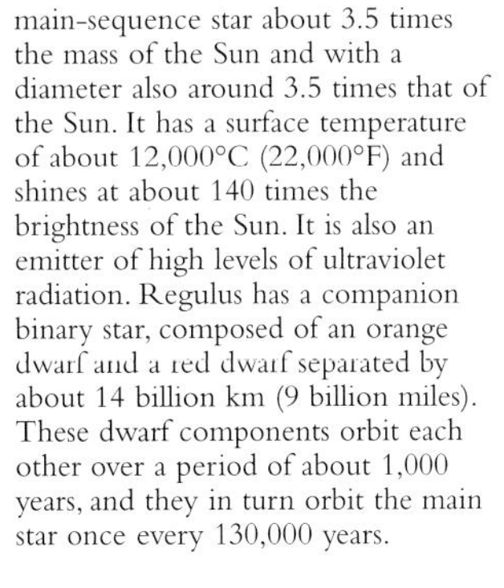

**REGULUS OCCULTED**
Poised at the top-left curve of the Moon, Regulus is about to be occulted as the Moon passes in front of it. Occultations can help astronomers to determine the diameters of large stars and ascertain whether they are binary systems. Occultations by the Moon can also reveal details about the Moon's surface features.

BLUE STAR

## Regor

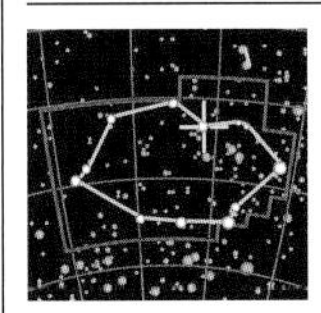

VELA

**DISTANCE FROM SUN** 840 light-years

**MAGNITUDE** 1.78

**SPECTRAL TYPES** O and WR

Unusually, this blue star has been very recently renamed. Its new name, spelling "Roger" backwards, honours the astronaut Roger Chaffee, who died in a fire during a routine test onboard the Apollo 1 spacecraft in 1967 (see p.103). Regor, or Gamma (γ) Velorum, is a complex star system dominated by a blue subgiant poised to evolve off the main sequence. Its evolution has been affected by being in a very close binary orbit with a star that is now a Wolf–Rayet star. They lie as close as the Earth does to the Sun and orbit each other every 78.5 days. The Wolf–Rayet star is now the less massive component of the close binary, but probably started as the more massive and evolved much more rapidly. The subgiant has around 30 times the mass of the Sun, with a surface temperature of 35,000°C (60,000°F) and a luminosity around 200,000 times that of the Sun. There are also two other components to the system, lying much further away, one of which is a hot B-type star (see pp. 230–31) at a distance of about 0.16 light-years.

# OLD STARS

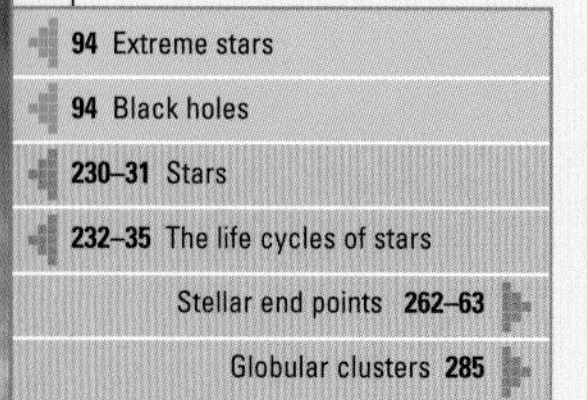

94 Extreme stars
94 Black holes
230–31 Stars
232–35 The life cycles of stars
Stellar end points 262–63
Globular clusters 285


**OLD STARS INCLUDE** low-mass main-sequence stars that came into existence billions of years ago and also some high-mass stars that will explode as supernovae after existing for less than a million years. Some of the most beautiful sights in the Milky Way are old stars undergoing their death throes.

## RED GIANTS

**EVOLVED STARS**
It is easy to pick out the evolved red giant stars in this image of the ancient star cluster NGC 2266.

When a star has depleted the hydrogen in its core, it will start to burn the hydrogen in a shell surrounding the core. This shell gradually moves outwards through the atmosphere of the star as fuel is used up. The expanding source of radiation heats the outer atmosphere, which expands, and then cools. The result is a large star with a relatively low surface temperature. It remains luminous because of its huge size, though some red giants are hidden from view by extensive dust clouds. Red giants have surface temperatures of 2,000–4,000°C (3,600–7,200°F) and radii 10–100 times that of the Sun. As these stars are so large, gravity does not have much effect on their outer layers and they can lose a great deal of mass to the interstellar medium, either by stellar winds or in the form of planetary nebulae. Red giants are often variable stars, as their outer layers pulsate, causing changes in luminosity (see p.278).

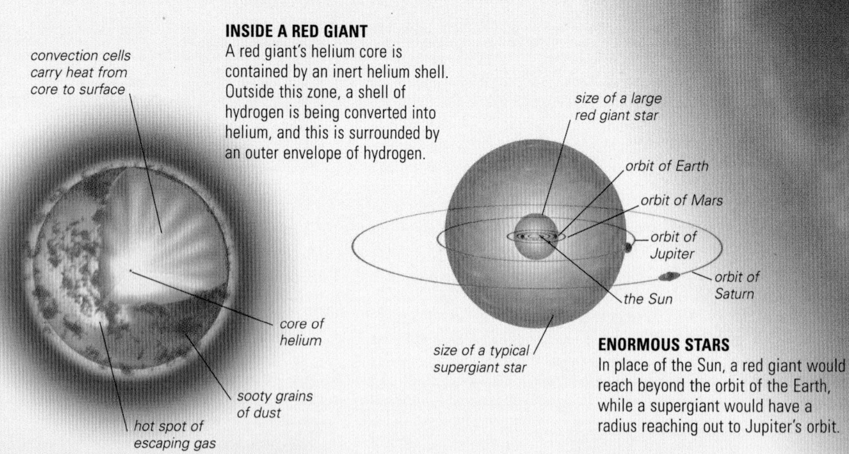


**INSIDE A RED GIANT**
A red giant's helium core is contained by an inert helium shell. Outside this zone, a shell of hydrogen is being converted into helium, and this is surrounded by an outer envelope of hydrogen.

**ENORMOUS STARS**
In place of the Sun, a red giant would reach beyond the orbit of the Earth, while a supergiant would have a radius reaching out to Jupiter's orbit.

## SUPERGIANTS

Stars of very high mass expand to become even larger than red giants. Red supergiants can have radii several hundred times that of the Sun. Just like red giants, they undergo hydrogen-shell burning (see p.234) and leave the main sequence (see p.230). When they have finished hydrogen-shell burning, they collapse and the helium core reaches a high enough temperature for the helium to be converted into carbon and oxygen. Helium-core burning is briefer than hydrogen burning, and when the helium core is depleted, helium-shell burning begins. If massive enough, further nuclear burning will occur, producing elements with an atomic mass up to that of iron. Near the end of the supergiant phase, a high-mass star will develop several layers of increasingly heavy elements. Eventually, supergiants die as supernovae.

**GARNET STAR**
One of the largest stars visible in the night sky, Mu Cephei or the Garnet Star is a red supergiant with a radius greater than that of Jupiter's orbit.

# HELIUM FLASHING

Once hydrogen burning has produced a core of helium, if its temperature reaches higher than about 100 million °C (180 million °F), the helium will be fused together to form carbon. In stars of around two to three solar masses, helium burning can start in an explosive process called a helium flash. As the core collapses after hydrogen burning, it temporarily arrives at a dormant or "degenerate" state as the collapse is halted by the pressure between the helium's electrons. The temperature continues to rise, but the dormant core does not change in pressure, so does not expand and cool. The rising temperature causes the helium to burn at an increasing rate, causing a "flash" that rids the core of the degenerate electrons. In higher-mass stars, the temperature rises high enough for helium fusion to begin before the core becomes degenerate.

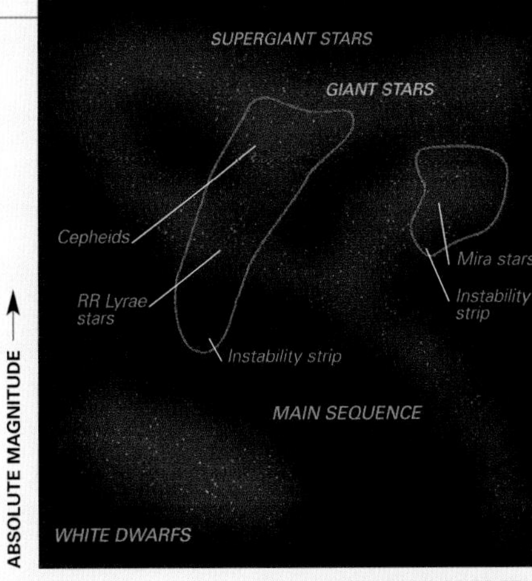


**INSTABILITY STRIPS**
Many red giants and supergiants are pulsating variable stars that appear in regions of the H–R diagram (right, see also p.230) called instability strips. Three types of variable star are shown here.

# WOLF–RAYET STARS

Massive stars, of about ten solar masses, that have strong, broad emission lines in their spectra (see p.33), but few absorption lines, are named Wolf–Rayet stars, after Charles Wolf and Georges Rayet (see p.260), who discovered them in 1867. They are hot, luminous stars whose strong stellar winds have blown away their outer atmospheres, revealing the stars' inner layers. They are broadly classified as WN, WC, and WO stars, depending on their spectra. The emission lines of WN stars are dominated by hydrogen and nitrogen, those of WC stars by carbon and helium, and those of WO by oxygen as well as carbon and helium. More than half of the known Wolf–Rayet stars are members of binary systems (see pp.270–71) with O or B stars as companions. It is believed that the Wolf-Rayet star was originally the more massive partner but lost its outer envelope to the companion star.

**GREAT ILLUMINATION**
A Wolf–Rayet star illuminates the heart of N44C, a nebula of glowing hydrogen gas surrounding young stars in the Large Magellanic Cloud.

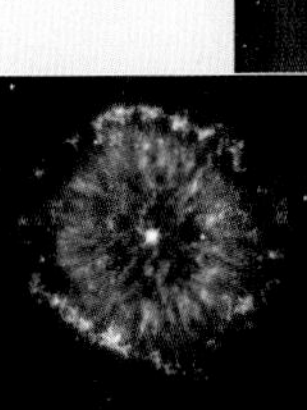

**STRONG WINDS**
The planetary nebula NGC 6751 may have a Wolf–Rayet star at its centre. Its strong winds created the elaborate filaments.

# PLANETARY NEBULAE

Planetary nebulae are heated halos of material shed by dying stars. They were termed planetary nebulae by William Herschel (see p.90) in 1785 because of their disc-like appearance through 18th-century telescopes. Planetary nebulae include some of the most stunning sights in the Universe, contorted into various shapes by magnetic fields and the orbital motion of binary systems (see p.270–71). They are composed of low-density gas thrown off by low-mass stars in the red-giant phase of their lives, and this gas is heated by the ultraviolet radiation given off by the hot inner cores of the dying stars. This stage of a star's life is relatively short. Eventually, the planetary nebula will disperse back into the interstellar medium, enriching the material there with the elements that have been produced by its parent star. These elements include hydrogen, nitrogen, and oxygen. At one time, the oxygen identified in the emission spectra of planetary nebulae (see p.33) was regarded as a new element called nebulium. It was later realised that "forbidden" emission lines of oxygen were present – forbidden because under usual conditions on Earth, they are very unlikely to occur. The central stars of planetary nebulae are among the hottest stars known. They are the contracting cores of red giants evolving into white dwarfs. Some planetary nebulae have been observed surrounding the resulting white dwarf. Current studies of planetary nebulae are revealing new facts about the late evolution of red giants and the manner of mass loss from these ageing stars.

**BUTTERFLY NEBULA**
The Hubble 5 nebula is a prime example of a "butterfly" or bi-polar nebula, created by the funnelling of expanding gas.

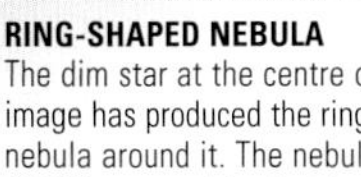

**RING-SHAPED NEBULA**
The dim star at the centre of this image has produced the ring-shaped nebula around it. The nebula (NGC 3132) is crossed by dust lanes and surrounded by a cooler gas shell.

**UNUSUAL NEBULA**
The Saturn Nebula was shaped by early ejected material confining subsequent stellar winds into jets.

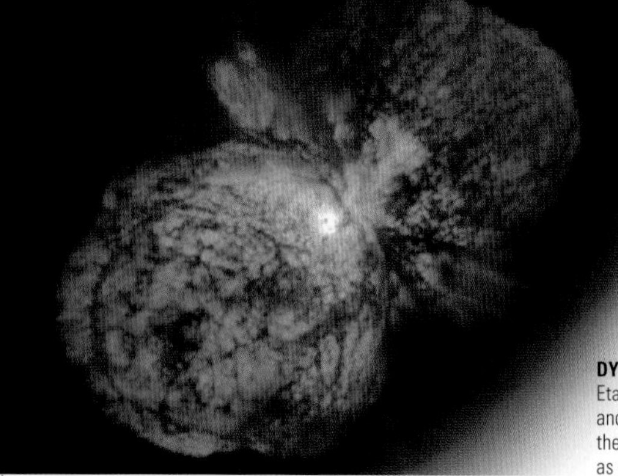

**DYING STAR**
Eta Carinae is a large, extremely old, and unstable star ejecting material into the interstellar medium. It could explode as a supernova at any time.

# OLD STARS

Some of the most visible and familiar bodies in the sky are stars that are approaching the ends of their lives or are experiencing their final death throes. In Wolf–Rayet stars and planetary nebulae, these old stars also present some of the most dramatic events and most beautiful sights in the Universe. Although different types of old stars exist throughout the Milky Way, the oldest are situated far out in the galactic halo (see pp.226–29) or within the globular clusters (see pp.284–89). Some of these stars are nearly as old as the Universe itself.

RED GIANT

## Aldebaran

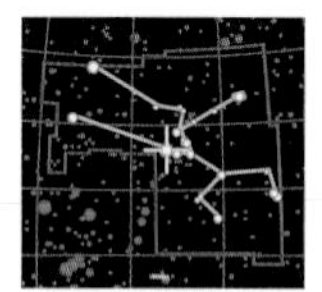

**TAURUS**

**DISTANCE FROM SUN** 65 light-years

**MAGNITUDE** 0.85

**SPECTRAL TYPE** K5

Also known as Alpha (α) Tauri, Aldebaran is the brightest star in the constellation Taurus and the 13th-brightest star in the sky. Its surface temperature of only 3,727°C (6,740°F) makes it glow a dull red that can easily be seen by the naked eye. Aldebaran's diameter is about 45 times that of the Sun, and, in place of the Sun, it would extend halfway to the orbit of Mercury. The star appears to be part of the Hyades cluster (see p.286), but this is a line-of-sight effect, with Aldebaran lying about 40 light-years closer to the Sun. This elderly star is a slow rotator, taking two years for each rotation, and an irregular variable, pulsating erratically. It has at least two faint stellar companions. Its name is derived from the Arabic *Al Dabaran*, meaning "the Follower", because it rises after the prominent Pleiades star cluster and pursues it across the sky. Aldebaran was one of the Royal Stars or Guardians of the Sky of ancient Persian astronomers and marked the coming of spring.

**BULL'S EYE**
The red tinge of Aldebaran makes it very distinctive against the whiter stars of the Hyades cluster. It is often depicted as the eye of the bull in the constellation Taurus.

RED SUPERGIANT

## Betelgeuse

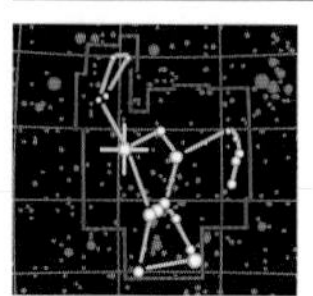

**ORION**

**DISTANCE FROM SUN** 500 light-years

**MAGNITUDE** 0.5

**SPECTRAL TYPE** M2

The right shoulder of the hunter, Orion, is marked by this distinctive, bright red star. Betelgeuse, or Alpha (α) Orionis, is a massive supergiant and the first star after the Sun to have its size reliably determined. Its diameter is more than twice that of the orbit of Mars, or about 500 times that of the Sun, and because of its huge size it is about 14,000 times brighter. Betelgeuse is the 10th-brightest star in the sky, although as it pulsates its brightness varies over a period of about six years. It is a strong emitter of infrared radiation, which is produced by three concentric shells of material ejected by the star over its lifetime. It is slowly using up its remaining fuel and one day will probably explode as a supernova.

**CLOSE SUPERGIANT**
Due to its enormous size and proximity to the Solar System, Betelgeuse was the first star after the Sun to have its surface imaged (above). It has a distinctive red colour when seen against other stars in Orion (left).

RED SUPERGIANT

## Antares

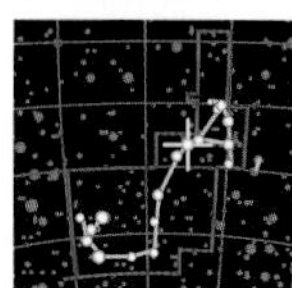

**SCORPIUS**

**DISTANCE FROM SUN** 520 light-years

**MAGNITUDE** 0.96

**SPECTRAL TYPE** M1.5

Antares or Alpha (α) Scorpii is the 15th-brightest star in the sky. Estimates of its diameter range from 280 to 700 times that of the Sun. It is about 15 times more massive than the Sun and shines 10,000 times brighter. This elderly star pulsates irregularly and has a binary companion that orbits in a period of about 1,000 years. This companion lies close enough to be affected by Antares's stellar wind and is a hot radio source. When viewed through an optical telescope, this blue companion looks green because of the colour contrast with red Antares.

**A RIVAL OF MARS**
The glowing Antares (bottom right) looks a lot like Mars, the red planet. Its name derives from the Greek for "rival of Mars" (or *anti Ares*).

RED GIANT

## TT Cygni

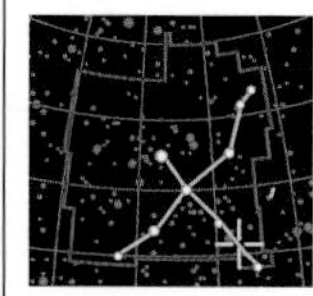

**CYGNUS**

**DISTANCE FROM SUN** 1,500 light-years

**MAGNITUDE** 7.55

**SPECTRAL TYPE** G

With a high ratio of carbon to oxygen in its surface layers, TT Cygni is known as a carbon star. The carbon, produced during helium burning, has been dredged up from inside the star. An outer shell, about half a light-year across, was emitted about 6,000 years before the star was as it appears to us now.

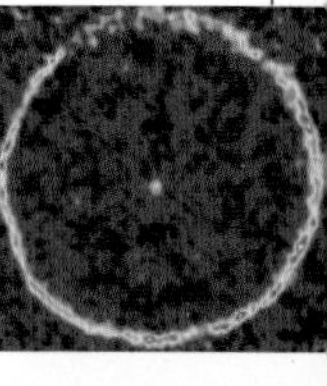

**CARBON RING**
This false-colour image shows a shell of carbon monoxide surrounding the carbon star TT Cygni.

PLANETARY NEBULA

## Helix Nebula

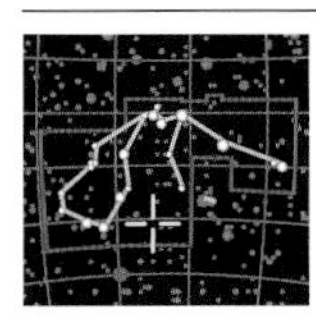

**CATALOGUE NUMBER** NGC 7293

**DISTANCE FROM SUN** Up to 650 light-years

**MAGNITUDE** 6.5

**AQUARIUS**

The Helix Nebula is the closest planetary nebula to the Sun, but its actual distance is uncertain, and estimates vary from 85 to 650 light-years. It is called the Helix Nebula because, from Earth, the outer gases of the star expelled into space give the impression that we are looking down the length of a helix. One of the largest known planetary nebulae, its main rings are about 1.5 light-years in diameter and span an apparent distance of more than half the width of the full Moon. Its outer halo extends up to twice this distance. The dying star at the centre of the nebula is destined to become a white dwarf, and as it continues to use up all its energy it will continue to expel material into the interstellar medium. The Helix Nebula presents an impressive example of the final stage that stars like our Sun will experience before collapsing for the last time. It was first discovered by the German astronomer Karl Ludwig Harding in around 1824, and its size and proximity mean that it has been extensively observed and imaged.

Detailed images made of the inner edge of the ring of material surrounding the central star have shown "droplets" of cooler gas, twice the diameter of our solar system, radiating outwards for billions of kilometres. These were probably formed when a fast-moving shell of gas, expelled by the dying star, collided with slower-moving material thrown off thousands of years before.

**GLOWING HALO**
Rings of expelled material glow red in the light produced by nitrogen and hydrogen atoms when they are energized by ultraviolet radiation.

**COMET-LIKE KNOTS**
Resembling comets, these tadpole-shaped gaseous knots are several thousand million miles across. They lie like spokes in a wheel along the inner edge of the ring of ejected gas surrounding the central star.

PLANETARY NEBULA

## Ring Nebula

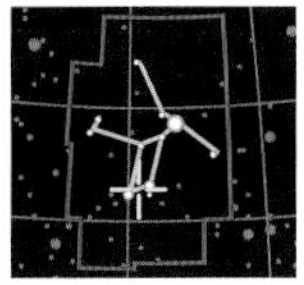

**CATALOGUE NUMBER** M57

**DISTANCE FROM SUN** 2,000 light-years

**MAGNITUDE** 8.8

**LYRA**

One of the best known planetary nebulae, the Ring Nebula was discovered in 1779 by French astronomer Antoine Darquier de Pellepoix. When seen through a small telescope, it appears larger than the planet Jupiter. Its central star, a planet-sized white dwarf of only about 15th magnitude, was not discovered until 1800, when it was found by the German astronomer Friedrich von Hahn. There has been a great deal of discussion about the true shape of the Ring Nebula. Although it appears like a flattened ring, some astronomers believe the stellar material has been expelled in a spherical shell that only looks like a ring because we view it through a thicker layer at its edges. Others believe it is a torus (shaped like a ring doughnut), which would look similar to the Dumbbell Nebula if viewed side-on, or that it is cylindrical or tube-like. The nebula appears to be about one light-year in diameter, but it has an outer halo of material that extends for more than two light-years. This is possibly a remnant of the central star's stellar winds before the nebula itself was ejected. The nebula is lit by fluorescence caused by the large amount of ultraviolet radiation emitted by the central star. The rate of the ring's expansion indicates that the nebula started to form about 20,000 years before it was as it appears to us now.

**TRUE COLOURS**
An optical view shows the Ring Nebula in its true colours. Blue indicates very hot helium, green represents ionized oxygen, and red is ionized nitrogen. The star that produced the nebula, now a white dwarf, is visible at the centre.

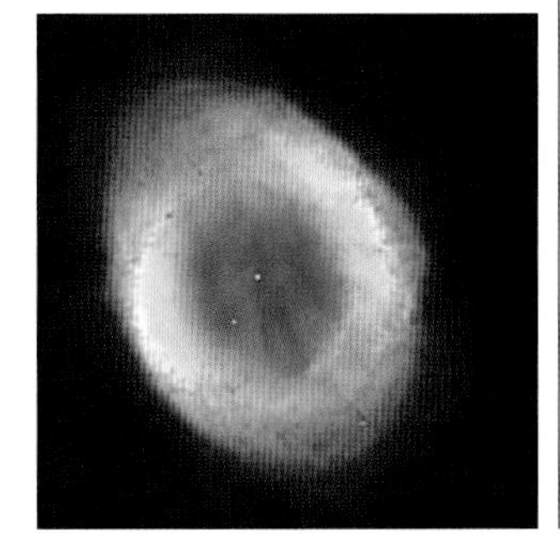

PLANETARY NEBULA

## Twin Jet Nebula

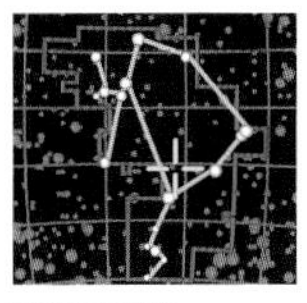

**CATALOGUE NUMBER** M2–9

**DISTANCE FROM SUN** 2,100 light-years

**MAGNITUDE** 14.7

**OPHIUCHUS**

The Twin Jet Nebula is one of the most striking examples of a butterfly or bipolar planetary nebula. It is believed that the star at the centre of this nebula is actually an extremely close binary that has affected the shape of the resulting planetary nebula. The gravitational interaction between the stars has pulled stellar material around them into a dense disc with a diameter about 10 times that of Pluto's orbit. About 1,200 years before this happened, one of the stars had an outburst, ejecting material in a strong stellar wind. This rammed into the disc, which acted like a nozzle, deflecting the material in perpendicular directions, forming the two lobes stretching out into space. This is very similar to the process that takes place in jet propulsion engines. Studies have suggested that the nebula's size has increased steadily with time and that the material is flowing outwards at up to 720,000kph (450,000mph).

**EXHAUST JETS**
This false-colour image reveals apparent jets of material radiating outwards. Neutral oxygen is shown in red, ionized nitrogen in green, and ionized oxygen in blue.

PLANETARY NEBULA

## Red Rectangle Nebula

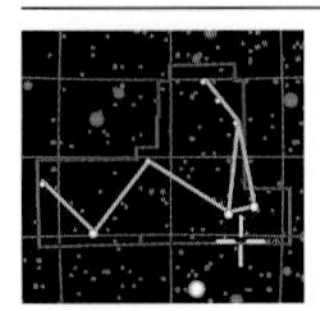

MONOCEROS

**CATALOGUE NUMBER** HD 44179

**DISTANCE FROM SUN** 2,300 light-years

**MAGNITUDE** 9.02

Nature does not often create rectangles, so astronomers were surprised to observe this planetary nebula's unusual shape. The shape of the Red Rectangle nebula is created by a pair of stars orbiting so close to each other that they experience gravitational interactions. This close binary star has created a dense disc of material around itself, which has restricted the direction of further outflows. This has caused subsequently ejected material to be expelled in expanding cone shapes perpendicular to the disc. Our view of the Red Rectangle is from the side, at right angles to these cones.

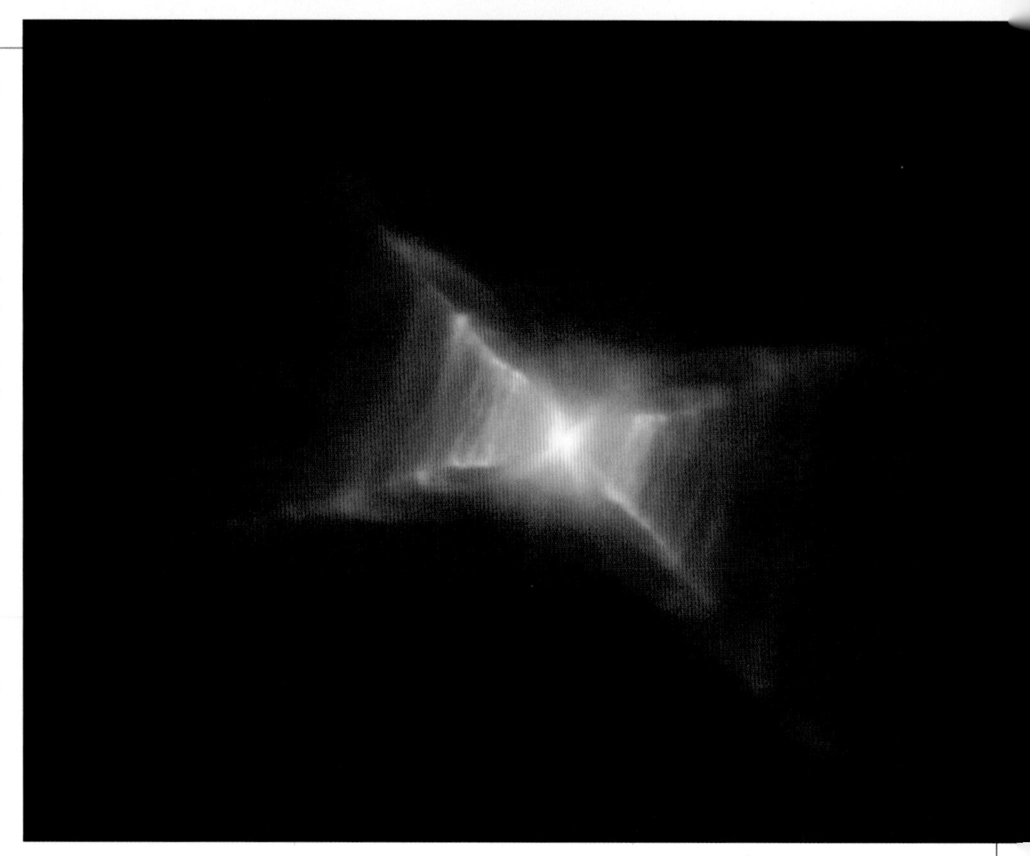

**COMPLEX STRUCTURE**
One of the most unusual celestial bodies in the Milky Way, the Red Rectangle Nebula has a distinctive shape that reflects an extremely complex inner structure.

PLANETARY NEBULA

## Cat's Eye Nebula

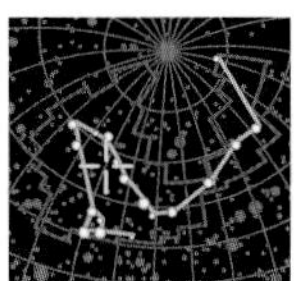

DRACO

**CATALOGUE NUMBER** NGC 6543

**DISTANCE FROM SUN** 3,000 light-years

**MAGNITUDE** 9.8

The Cat's Eye Nebula is one of the most complex of all planetary nebulae. It is thought that its intricate structures may be produced by either the interactions of a close binary system or by the recurring magnetic activity of a solitary central star. At 3,000 light-years away, it is too far even for the Hubble Space Telescope to resolve its central star. The "eye" of the nebula is estimated to be more than half a light-year in diameter, with a much larger outer halo stretching into the interstellar medium. Although models of planetary nebulae once assumed a continuous outflow of stellar material, this nebula contains concentric rings that are the edges of bubbles of stellar material ejected at intervals. Eleven of these bubbles have been identified, possibly ejected at intervals of 1,500 years. The Cat's Eye also contains jets of high-speed gas, as well as bow waves created when the gas has slammed into slower-moving, previously ejected material.

**WAVES AND SYMMETRIES**
A composite picture (above) shows emission from nitrogen atoms as red and oxygen atoms as green and blue shades, thus revealing successive waves of expelled stellar material. The nebula's symmetrical properties are further revealed by a false-colour image processed to highlight its ring structure (right).

PLANETARY NEBULA

## Egg Nebula

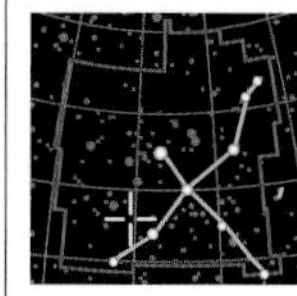

CYGNUS

**CATALOGUE NUMBER** CRL 2688

**DISTANCE FROM SUN** 3,000 light-years

**MAGNITUDE** 14

The Egg Nebula's central star, which was a red giant until a few hundred years ago, is hidden by a dense cocoon of dust (visible in the image below as the dark band of material across the middle of the nebula). The material shed by the dying star is expanding at the rate of 72,000kph (45,000mph). Distinct arcs of material suggest a varying density throughout the nebula. The light from the central star shines like searchlights through the thinner parts of its cocoon and reflects off dust particles in the outer layers of the nebula.

**BRIGHT SEARCHLIGHTS**

PLANETARY NEBULA

## Ant Nebula

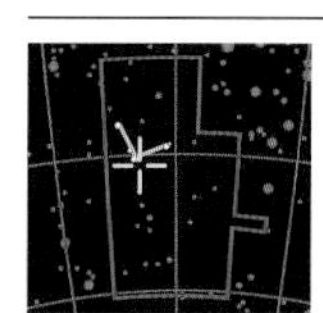

**CATALOGUE NUMBER** Menzel 3

**DISTANCE FROM SUN** 4,500 light-years

**MAGNITUDE** 13.8

**NORMA**

There are two main theories about what has caused the unusual shape of this planetary nebula. Either the central star is a close binary, its interacting gravitational forces shaping the outflowing gas, or it is a single spinning star whose magnetic field is directing the material it has ejected. The expelled stellar material is travelling at around 3.6 million kph (2.25 million mph) and impacting into the surrounding slower-moving medium; the lobes of the nebula stretch to a distance of more than 1.5 light-years. Observations of the Ant Nebula may reveal the future of our own star, as its central star appears to be very similar to the Sun.

**HEAD AND THORAX**
Even through a small telescope, this planetary nebula resembles the head and thorax of a common garden ant.

PLANETARY NEBULA

## Crescent Nebula

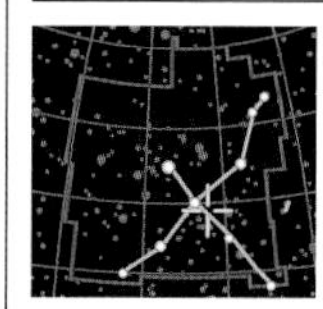

**CATALOGUE NUMBER** NGC 6888

**DISTANCE FROM SUN** 4,700 light-years

**MAGNITUDE** 7.44

**CYGNUS**

The central star of the Crescent Nebula is a Wolf–Rayet star. Only about 4.5 million years after its formation (one-thousandth the age of the Sun), this massive star expanded to become a red giant and ejected its outer layers at about 35,000kph (22,000mph). Two hundred thousand years later, the intense radiation from the exposed, hot inner layer of the star began pushing gas away at speeds in excess of 4.5 million kph (2.8 million mph). This strong stellar wind expelled material equivalent to the Sun's mass every 10,000 years, forming a series of dense, concentric shells that are visible today. Typical of emission nebulae, the radiation from the hot central star excites the stellar material, principally hydrogen, causing it to shine in the red part of the spectrum. It is thought that the nebula's central star will probably explode as a supernova in about 100,000 years.

**GASEOUS COCOON**
This composite image of the Crescent Nebula shows a compact semicircle of dense material surrounding a pre-supernova star (centre). The Crescent spans a distance of about three light-years.

WOLF–RAYET STAR

## WR 104

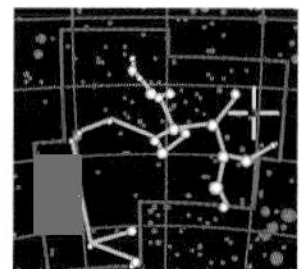

**DISTANCE FROM SUN** 4,800 light-years

**MAGNITUDE** 13.54

**SPECTRAL TYPE** WCvar+

**SAGITTARIUS**

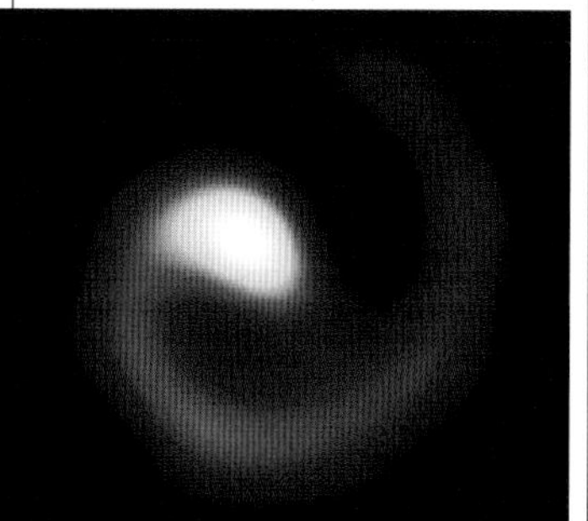

**STELLAR SPIRAL**

Like water from a cosmic lawn sprinkler, dust streaming from this rotating star system creates a pinwheel pattern. As Wolf–Rayet stars are so hot that any dust they emit is usually vaporized, it is surprising that WR 104 has dust streaming away from it in this obvious spiral pattern. One theory is that this is a binary system, with each star emitting a strong stellar wind. Where these winds meet, there is a "shock front" that compresses the outflowing material, creating a denser, slightly cooler environment in which dust can exist. The orbital motion of the two stars then causes the spiral shape.

PLANETARY NEBULA

## Eskimo Nebula

**CATALOGUE NUMBER** NGC 2392

**DISTANCE FROM SUN** 5,000 light-years

**MAGNITUDE** 10.11

**GEMINI**

The German-born astronomer William Herschel (see p.90) discovered the Eskimo Nebula in 1787, and it has since become a much-loved sight for amateur astronomers. Even through small telescopes, this nebula's form, suggesting a face ringed by a fur parka hood, is clearly visible.

Hubble Space Telescope images reveal a complex structure, featuring an inner nebula and an outer halo. The inner nebula consists of material ejected from the central star in two elliptical lobes around 10,000 years before the star was as we now see it. Each lobe is about one light-year long and about half a light-year wide, and contains filaments of dense matter. Astronomers think that a ring of dense material around the star's equator, ejected during its red-giant phase, helped create the nebula's "face". The surrounding "hood" contains unusual orange filaments, each about one light-year long, streaming away from the central star at up to 120,000kph (75,000mph). One explanation for these is that they were created when a fast-moving outflow from the central star impacted into slower-moving, previously ejected material.

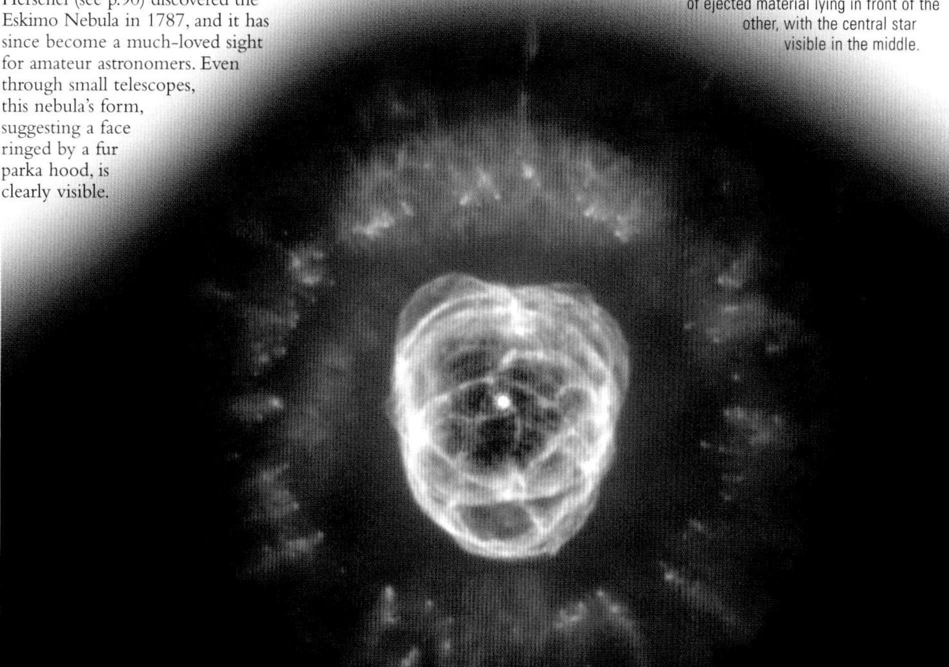

**HOODED NEBULA**
In the centre of this image, the apparent "face" of the Eskimo consists of one bubble of ejected material lying in front of the other, with the central star visible in the middle.

PLANETARY NEBULA

# Bug Nebula

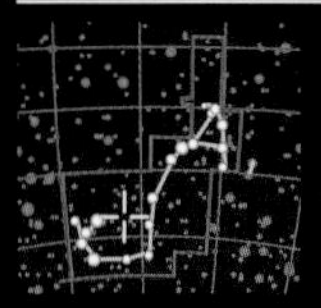

**CATALOGUE NUMBER**
NGC 6302

**DISTANCE FROM SUN**
4,000 light-years

**MAGNITUDE** 7.1

**SCORPIUS**

First discovered in 1826 by Scottish astronomer James Dunlop, then rediscovered in the late 19th century by the great American astronomer E.E. Barnard, the Bug Nebula is one of the brightest planetary nebulae. The central star is thought to have an extremely high temperature, and its intense ultraviolet radiation lights up the surrounding stellar material. The star itself is not visible at optical wavelengths as it is hidden by a blanket of dust. It is believed that the central star ejected a ring of dark material about 10,000 years before it was as we see it now, but astronomers cannot explain why it was not destroyed long ago by the star's ultraviolet emissions. The composition of the surrounding material is also surprising, as it contains carbonates, which usually form when carbon dioxide dissolves in liquid water. Although ice exists in the nebula, along with hydrocarbons and iron, there is no evidence of liquid water.

**GOLDEN NEBULA**
A dramatic close-up of the central area of the Bug Nebula reveals a brilliant cocoon of dust, gases, and ice surrounding one of the Milky Way's hottest stars during the final stages of its life.

**COLOURFUL BUG**
This butterfly-shaped planetary nebula is here imaged through different colour filters. The blue colour indicates the presence of hydrogen-alpha, while the red indicates the presence of nitrogen in the cooler outer layers.

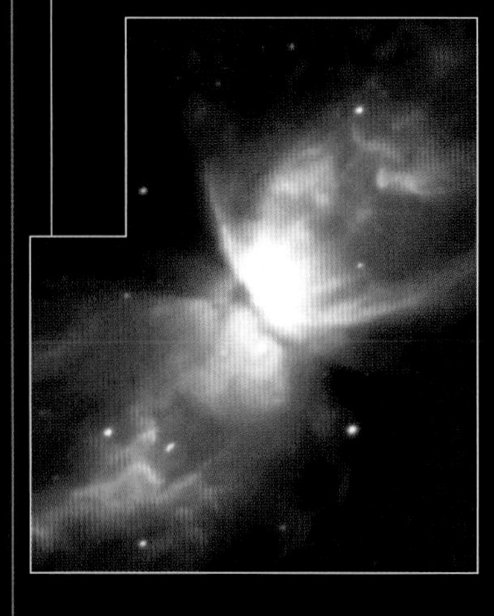

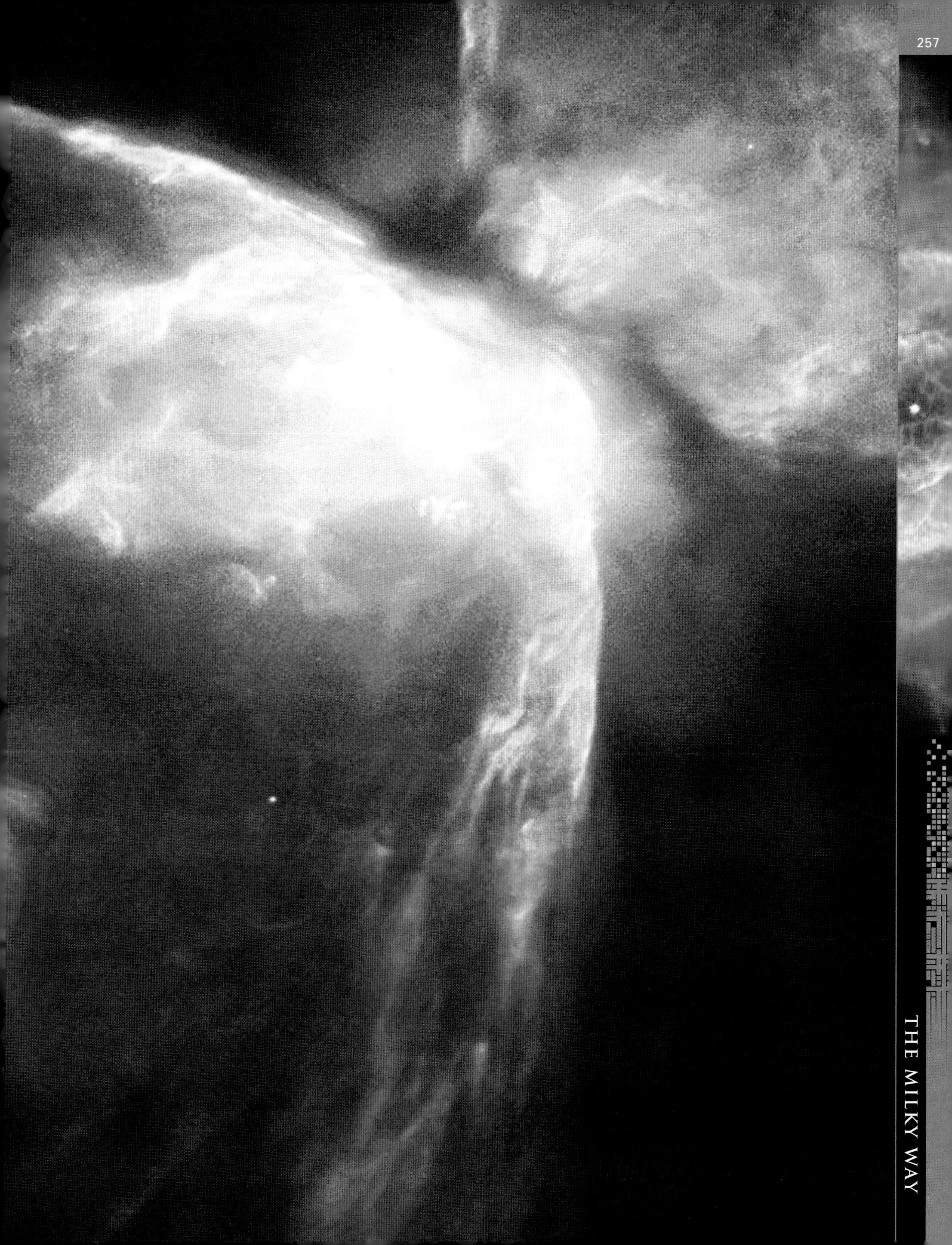

PLANETARY NEBULA

## Calabash Nebula

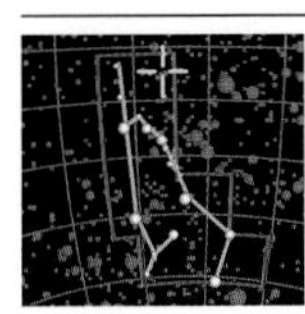

**CATALOGUE NUMBER** OH231.8+4.2

**DISTANCE FROM SUN** 5,000 light-years

**MAGNITUDE** 9.47

**PUPPIS**

One of the most dynamic planetary nebulae, the Calabash Nebula's central star is expelling gas at a speed of 700,000kph (435,000mph). The fast-moving material is being channelled into streamers on one side and into a jet on the other. The jet of material appears to be striking denser, slower-moving material, creating shock waves. Radio observations have revealed an unusually large amount of sulphur in the gas around the star, which may have been produced by the shock waves. This planetary nebula is in the earliest stages of formation and has offered astronomers the chance to observe the kind of processes that led to the creation of more established planetary nebulae elsewhere in the Milky Way.

**ROTTEN EGG NEBULA**
The Calabash Nebula is popularly called the Rotten Egg Nebula because it contains a lot of sulphur, which smells of rotten eggs. The outflows of expelled gas show up bright yellow-orange in the centre of this picture.

PLANETARY NEBULA

## Gomez's Hamburger Nebula

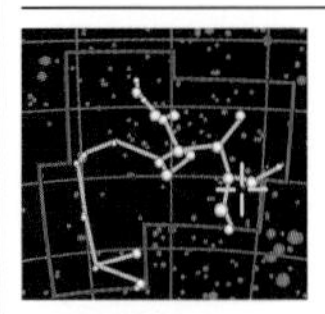

**CATALOGUE NUMBER** IRAS 18059-3211

**DISTANCE FROM SUN** 6,500 light-years

**MAGNITUDE** 14.4

**SAGITTARIUS**

Discovered in 1985 by the Chilean astronomer Arturo Gomez at the Cerro Tololo Inter-American Observatory in Chile, this dramatic, hamburger-shaped object is a planetary nebula in the making. The central star, obscured by a dark band of dust, is a red giant throwing off its outer layers. Eventually the star's hot core will be exposed and its ultraviolet radiation will heat up the clouds of dust and gas surrounding it, giving us a fully fledged planetary nebula. It is rare to see nebulae at this early stage of evolution, as this process does not last long. In less than 1,000 years from its presently observed state, the central star will be hot enough to vaporize the dust surrounding it. This nebula is only a small fraction of a light-year across but it will expand as the star continues to eject material.

**CELESTIAL SANDWICH**
The two "buns"of Gomez's Hamburger are dust clouds illuminated by the central star. The "meat" of the hamburger is a thick disc of dust surrounding this red giant and obscuring it from our view.

BLUE SUPERGIANT

## Eta Carinae

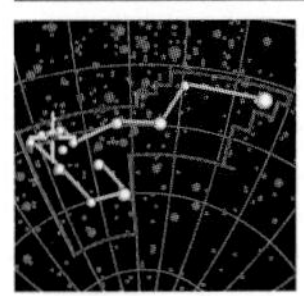

**DISTANCE FROM SUN** 8,000 light-years

**MAGNITUDE** 6

**SPECTRAL TYPE** B0

**CARINA**

With a mass more than 100 times that of the Sun, this star, which is embedded in an impressive dumbbell of stellar material, is one of the most massive known. Eta Carinae is classified as an eruptive variable star (see pp.278–79), and it experiences two types of irregular eruptions. The first involves a brightening of one to two magnitudes (see pp.230–31) lasting a few years; the second features a briefer, giant eruption that produces a significant increase in total luminosity and the ejection of more than a solar mass of material. Since it was first catalogued by the English astronomer Edmond Halley in 1677, Eta Carinae has varied in brightness from eighth magnitude to a magnitude as bright as -1. It is currently around sixth magnitude. In 1841, when it reached a magnitude rivalling that of Sirius, it underwent a giant outburst that produced the two distinctive lobes of outflowing material. These lobes are moving outwards at a rate of about 2 million kph (1.2 million mph). This highly unstable star survived that outburst, but will probably eventually erupt as a supernova.

**HOMUNCULUS NEBULA**
This false-colour optical image shows the Homunculus Nebula surrounding Eta Carinae, which lies at the very centre of this image.

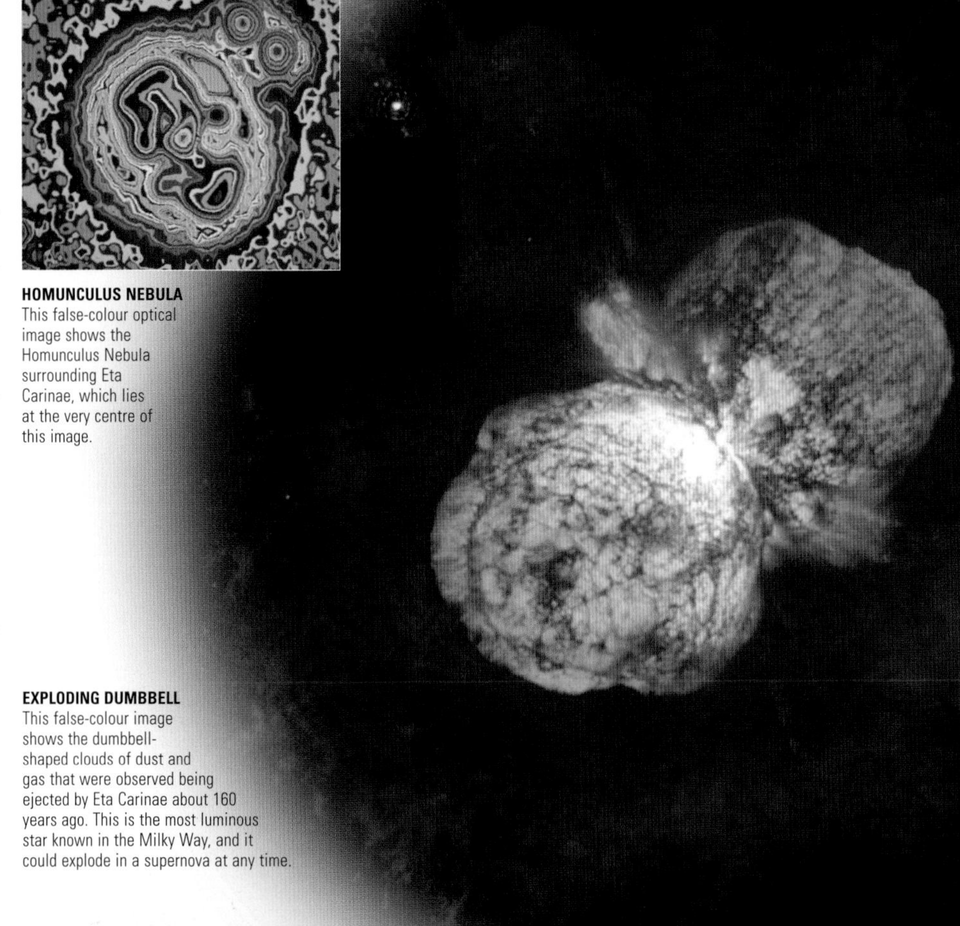

**EXPLODING DUMBBELL**
This false-colour image shows the dumbbell-shaped clouds of dust and gas that were observed being ejected by Eta Carinae about 160 years ago. This is the most luminous star known in the Milky Way, and it could explode in a supernova at any time.

PLANETARY NEBULA

## Hourglass Nebula

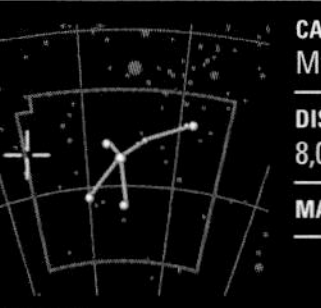

MUSCA

**CATALOGUE NUMBER** MyCn18

**DISTANCE FROM SUN** 8,000 light-years

**MAGNITUDE** 11.8

The distinctive shape of the stunning Hourglass Nebula has fired much debate over its formation among astronomers. One suggestion is that as the ageing, intermediate-mass star started to expand into a red giant, the escaping gas and dust accumulated first as a belt around the star's equator. As the volume of escaping gas continued to grow, the belt constricted the star's midsection, forcing the increasingly fast-moving gas into an hourglass shape. Other astronomers argue that the central star has a massive, heavy-element core that produces a strong magnetic field. In this scenario, the shape is a result of the ejected material being constrained by the magnetic field. Yet another suggestion is that the central star is in fact a binary and one of the pair is a white dwarf. A disc of dense material is produced around its middle by the gravitational interactions between the two components, which pinches in the "waist" of the expanding nebula. However, other features of the Hourglass Nebula have so far defied explanation. Astronomers have observed a second hourglass-shaped nebula within the larger one, but, unusually, neither is positioned symmetrically around the central star. Two rings of material seen around the "eye" of the hourglass, perpendicular to one another, are the subjects of continuing studies.

EXPLORING SPACE

### NEBULA IN ACTION

The beautiful images of the Hourglass Nebula captured by the Hubble Space Telescope have revealed details within planetary nebulae that have revolutionized the study of these elusive but beautiful objects, especially as regards the creation of non-spherical planetary nebulae. These fascinating nebulae are observed in many varied shapes, and an equally large number of hypotheses have been suggested to account for them. The life of a planetary nebula is a mere blink of an eye when compared to the lifetime of a star, but it is a very important stage. When a star is evolving off the main sequence it loses huge quantities of its material and thus enriches the interstellar medium in elements heavier than helium, which can then be recycled to form other celestial objects.

**GAS SHELLS**
This revealing picture of the Hourglass Nebula is a composite of three images taken in different wavelengths. The colorful gas rings are nitrogen (red), hydrogen (green), and oxygen (blue).

WOLF–RAYET STAR

## HD 56925

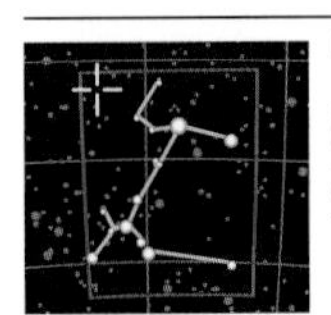

CANIS MAJOR

**DISTANCE FROM SUN** 15,000 light-years

**MAGNITUDE** 11.4

**SPECTRAL TYPE** WN5

The emission nebula NGC 2359, which has a diameter of around 30 light-years, has been produced by an extremely hot Wolf–Rayet star, visible at its centre. This star, designated HD 56925, has a surface temperature of between 30,000°C (54,000°F) and 50,000°C (90,000°F) – six to ten times as hot as the Sun. It is also highly unstable, ejecting stellar material into the interstellar medium at speeds approaching 7.2 million kph (4.5 million mph). Even though it is a massive star of around 10 solar masses, it is losing about the equivalent of the mass of the Sun every thousand years. With this level of mass loss, Wolf–Rayet stars like HD 56925 are unable to exist in this stage of their life for long, and are therefore rarely observed: only about 220 such stars are known in the Milky Way. Material from the star has been ejected in an even, spherical manner, producing a bubble of material. This bubble has been further shaped by interactions with the surrounding interstellar medium. HD 56925 is unusual because it lies at the edge of a dense, warm molecular cloud, and the asymmetrical shape of the outer parts of the surrounding nebula is due to "bow shocks", produced when fast stellar winds hit denser, static material.

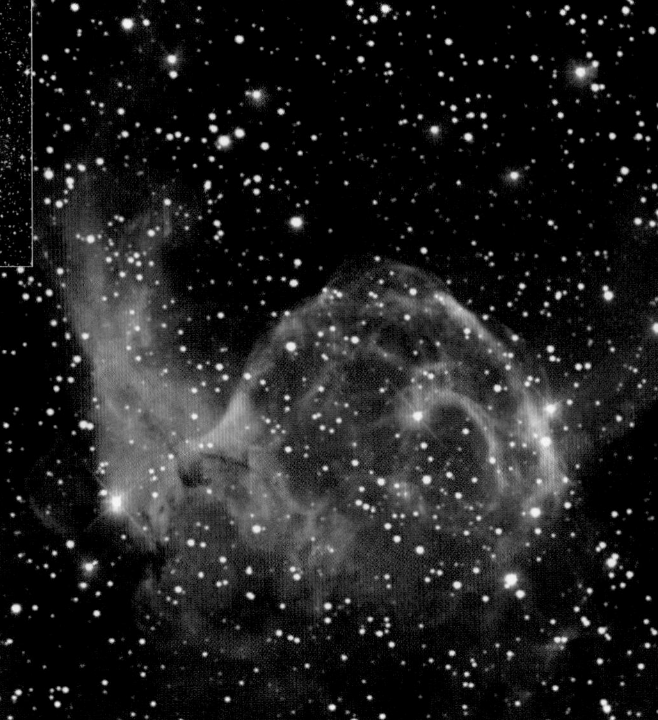

**THOR'S HELMET**
The popular name for the nebula surrounding HD 56925 is Thor's Helmet, because it looks like a helmet with wings (above). The nebulae surrounding Wolf–Rayet stars are sometimes called bubble nebulae, and HD 56925 lies at the centre of the nebula's main bubble of hot gas (the star is above and to the right of centre in the image to the right).

WOLF–RAYET STAR

## WR 124

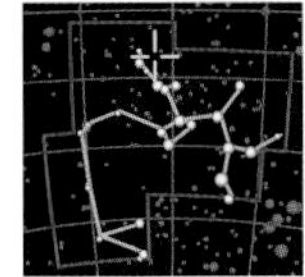

SAGITTARIUS

**DISTANCE FROM SUN** 15,000 light-years

**MAGNITUDE** 11.04

**SPECTRAL TYPE** WN

With a surface temperature of around 50,000°C (90,000°F), WR 124 is one of the hottest known Wolf–Rayet stars. This massive, unstable star is blowing itself apart – its material is travelling at up to 150,000kph (90,000mph). The observed state of M1-67, the relatively young nebula surrounding WR 124, is only 10,000 years old and it contains clumps of material with masses about 30 times that of the Earth and diameters of 150 billion km (90 billion miles).

**STELLAR FIREBALL**
WR 124 can be seen as a glowing body at the centre of a huge, chaotic fireball. The fiery nebula surrounding the star consists of vast arcs of glowing gas violently expanding outwards into space.

### CHARLES WOLF AND GEORGES RAYET

**GEORGES RAYET**

The French astronomers, Charles Wolf (1827–1918) and Georges Rayet (1839–1906) co-discovered the type of unusual, hot stars that now bear their name. In 1867, they used the Paris Observatory's 40cm (16in) Foucault telescope to discover three stars whose spectra were dominated by broad emission lines rather than the usual narrow absorption lines (see pp.250–51). Today, in total around 300 Wolf–Rayet stars are known. Rayet later became Director of the Floriac Bordeaux Observatory.

PLANETARY NEBULA

## Stingray Nebula

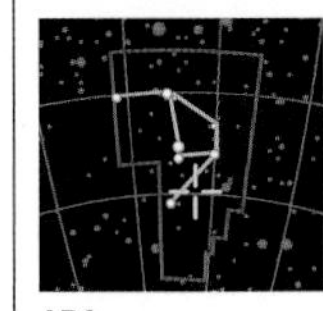

ARA

**CATALOGUE NUMBER** Hen-1357

**DISTANCE FROM SUN** 18,000 light-years

**MAGNITUDE** 10.75

The Stingray Nebula is the youngest known planetary nebula. Observations made in the 1970s revealed that the dying star at the centre of the nebula was not hot enough to cause the surrounding gases to glow. By the 1990s, further observations had shown that the central star had rapidly heated up as it entered the final stages of its life, causing the nebula to shine. This afforded astronomers a remarkable opportunity to observe the star in an exceedingly brief phase of its evolution. Because of its young age, the Stingray Nebula is one-tenth the size of most planetary nebulae, with a diameter only about 130 times that of the Solar System. A ring of ionized oxygen surrounds the central star, and bubbles of gas billow out in opposite directions above and below the ring. Material travelling at speed outwards from the central star has opened holes in the ends of the bubbles, allowing streams of gas to escape in opposite directions. On the outer edges of the nebula, the central star's winds crash into the walls of the gas bubbles, generating shock waves and heat that cause the gas to glow brightly.

**GRACEFUL SYMMETRY**
The graceful, symmetrical shape of this very young planetary nebula gives it its popular name. In this enhanced true-colour image, the Stingray Nebula's central star has a companion star just visible above it to the left.

RED SUPERGIANT

## V838 Monocerotis

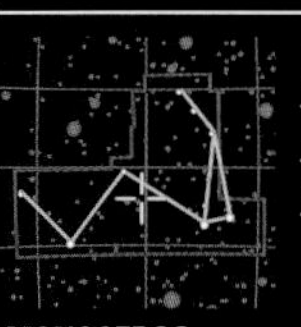

MONOCEROS

**DISTANCE FROM SUN** 20,000 light-years

**MAGNITUDE** 10

**SPECTRAL TYPE** K

Discovered on 6 January 2002 by an amateur astronomer, V838 Monocerotis is one of the most interesting stars. Its precise nature is not yet fully understood, but astronomers believe its recent evolution has moved it off the main sequence to become a red supergiant. While this phase would usually take hundreds or thousands of years, here it has happened in a matter of months.

Its first viewed outburst, in January 2002, was followed a month later by a second in which it brightened from magnitude 15.6 to 6.7 in a single day – an increase of several thousand times. Finally, in March 2002 it brightened from magnitude 9 to 7.5 over just a few days. The energy emitted in the outbursts caused previously ejected shells of material to brighten and become visible.

**LIGHT ECHOES**
Light echoes from recent outbursts illuminate the ghostly shells of ejected material around the enigmatic star V838 Monocerotis (seen glowing red).

BLUE SUPERGIANT

## Sher 25

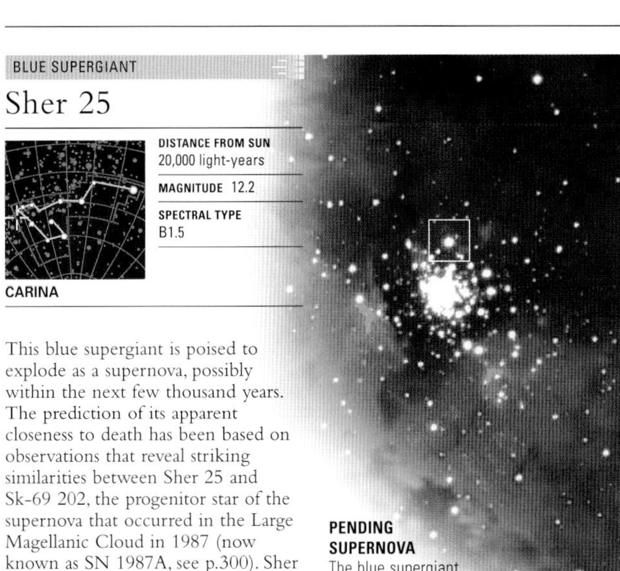

CARINA

**DISTANCE FROM SUN** 20,000 light-years

**MAGNITUDE** 12.2

**SPECTRAL TYPE** B1.5

This blue supergiant is poised to explode as a supernova, possibly within the next few thousand years. The prediction of its apparent closeness to death has been based on observations that reveal striking similarities between Sher 25 and Sk-69 202, the progenitor star of the supernova that occurred in the Large Magellanic Cloud in 1987 (now known as SN 1987A, see p.300). Sher 25 lies at the centre of a clumpy ring of ejected material, and additional material from the star is escaping perpendicular to this ring. This has caused the ejected stellar material to form an hourglass-shaped nebula with Sher 25 lying at its middle. The ring and nebula are similar to those observed around Sk-69 202 before that blue supergiant exploded. Spectroscopy reveals that the nebula surrounding Sher 25 is rich in nitrogen, indicating that it has passed through a red supergiant phase, again displaying an evolutionary path similar to that of the star Sk-69 202.

**PENDING SUPERNOVA**
The blue supergiant shown boxed in this image is likely to explode as a supernova. The open cluster of bright white stars and the surrounding red nebula are known as NGC 3603.

BLUE VARIABLE

## Pistol Star

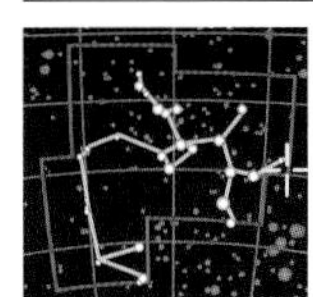

SAGITTARIUS

**DISTANCE FROM SUN** 25,000 light-years

**SPECTRAL TYPE** LBV

The most luminous star ever discovered is located at the centre of the Pistol Nebula and is known as a luminous blue variable. The Pistol Star emits around 10 million times more light than the Sun, unleashing as much energy in six seconds as the Sun does in one year. It is also one of the most massive stars known, weighing in at 100 times the mass of the Sun. When it originally formed, it may have been up to 200 times the mass of the Sun, but it has ejected at least 10 solar masses of material in giant eruptions. These occurred about 4,000 and 6,000 years before its presently seen state. In the Sun's position, the star would fill the diameter of the Earth's orbit. Despite its size and luminosity, the star is obscured at visible wavelengths by the ejected material that has formed the pistol-shaped nebula surrounding it.

**VAST NEBULA**
Seen in infrared light, the Pistol Nebula glows brightly. The nebula is four light-years across and would nearly span the distance from the Sun to Proxima Centauri, the closest star to the Solar System.

# STELLAR END POINTS

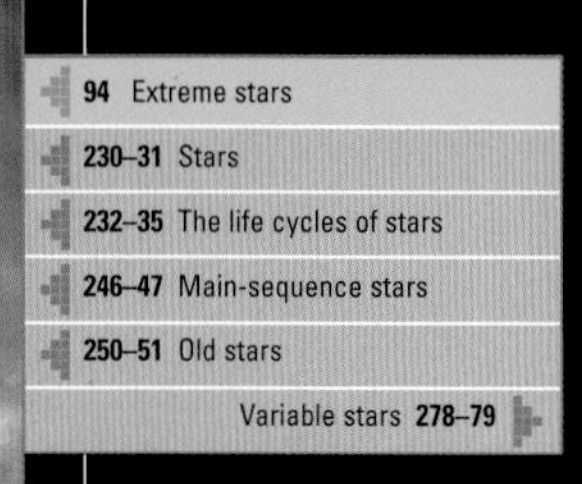
94 Extreme stars
230–31 Stars
232–35 The life cycles of stars
246–47 Main-sequence stars
250–51 Old stars
Variable stars 278–79

THE FORM A STAR TAKES in the ultimate stage of its life is called a stellar end point. Such end points include some of the most exotic objects in the Milky Way. The fate of a star is dictated by its mass, with lower-mass stars becoming white dwarfs, and the highest-mass stars becoming black holes, from which not even light can escape. Between these are neutron stars, including spinning pulsars.

**NEW WHITE DWARF**
This artist's impression depicts the white-dwarf star H1504+65 from a distance similar to that between the Earth and the Sun. This is one of the youngest white dwarfs ever discovered and the hottest ever detected.

## WHITE DWARFS

Once a star has used up all of its fuel through nuclear fusion, the stellar remnant will collapse, as it cannot maintain enough internal pressure to counteract its gravity. Stars of less than about eight solar masses will lose up to 90 per cent of their material in stellar winds and by creating planetary nebulae (see p.251). If the remnants of these stars have less than 1.4 solar masses (the Chandrasekhar limit), they will become white dwarfs. White dwarfs are supported by what is known as electron degeneracy pressure, created by the repulsion between electrons in their core material. More massive stars collapse to the smallest diameters and highest densities. The first white dwarf to be discovered, Sirius B (see p.264), has a mass similar to that of the Sun but a radius only twice that of the Earth. Although they have surface temperatures of around 100,000°C (180,000°F) at first, white dwarfs fade over periods of hundreds of millions of years, eventually becoming cold black dwarfs.

**MORGUE OF STARS**
Spanning a distance of 900 light-years, this mosaic of X-ray images of the centre of the Milky Way reveals hundreds of white-dwarf stars, neutron stars, and black holes. They are all embedded in a hot, incandescent fog of interstellar gas. The supermassive black hole at the centre of the galaxy is located inside the central bright white patch.

## SUPERNOVAE

Massive stars die spectacularly, blasting their outer layers off into space in type II supernovae explosions. A type I supernova is a type of variable star (see p.279). When a star of more than about ten solar masses reaches the end of its hydrogen-burning stage, it will eventually produce an iron core. Initially this core is held up by its internal pressure, but when it reaches a mass greater than 1.4 solar masses (the Chandrasekhar limit), it starts to collapse, forming an extremely dense core almost entirely made of neutrons. Supernova detonation occurs when the outer layers of the star, which have continued to implode, impact on the rigid core and rebound back into space at speeds of up to 70 million kph (45 million mph). This releases massive amounts of energy, creating a great rise in luminosity that may last for several months, before fading. A supernova remnant consisting of the debris will become a nebula.

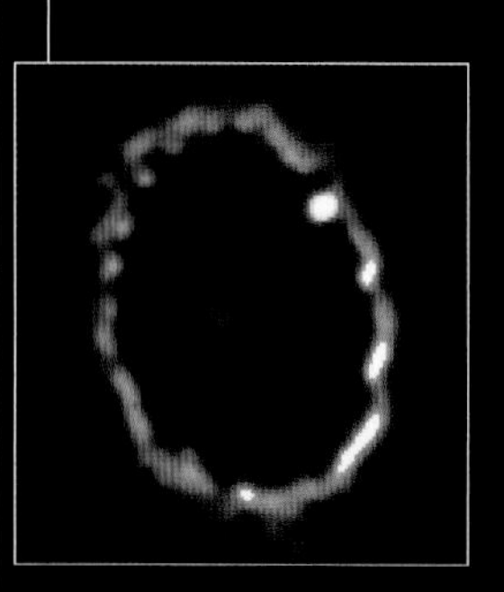

**DEATH RING**
The envelope of Supernova 1987A is still expanding outwards at very high velocities, slamming into interstellar material and creating this ring of glowing gas.

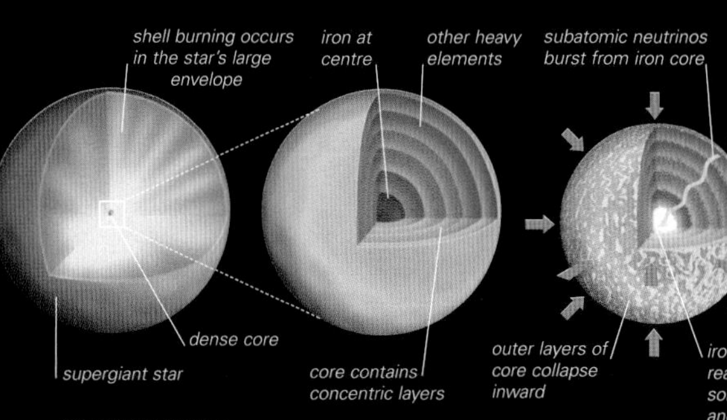


**COLLAPSING STAR**
As a massive star collapses, elements heavier than helium are produced in a series of shell-burning layers. Elements heavier than iron cannot be produced in this way, and an iron core may collapse to produce a neutron star.

# NEUTRON STARS

Neutron stars are one of the by-products of type II supernovae explosions. During an explosion, the outer layers of a star are blown off, leaving an extremely dense, compact star, consisting predominantly of neutrons with a smaller amount of electrons and protons. Neutron stars have a mass between 0.1 and 3 solar masses. Beyond this limit, a star will collapse further to become a black hole (below). As the neutron star forms, the magnetic field of the parent star becomes concentrated and grows in strength. Similarly, the original rotation of the star increases in speed as the star collapses. Neutron stars are characterized by their strong magnetic fields and rapid rotation. Over time, their rotation slows as they lose energy. However, some neutron stars show a temporary rise in rotation rate, possibly due to tremors, known as starquakes, in their thin, crystalline outer crusts. Neutron stars that emit directed pulses of radiation at regular intervals are known as pulsars (below).

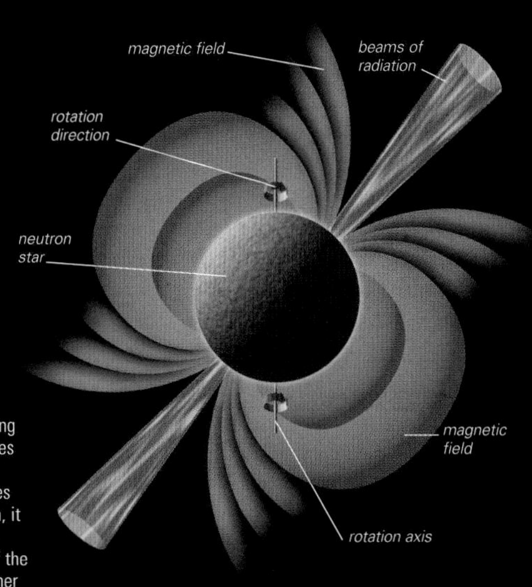


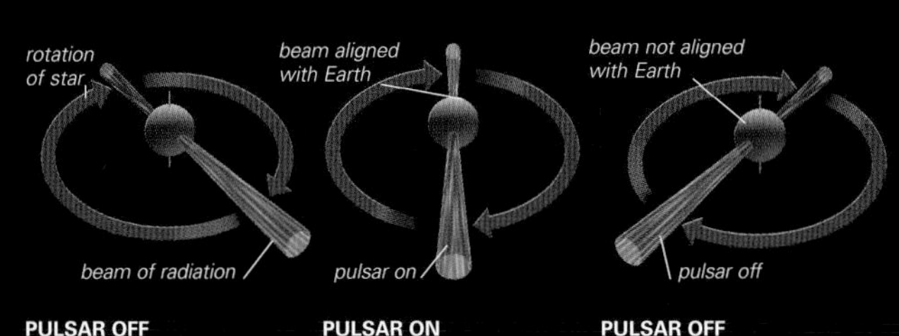


**HOW PULSARS WORK**
Charged particles spiral along the star's magnetic-field lines and produce a beam of radiation. If the beam passes across the field of the Earth, it can be detected as a pulse. Depending on the energy of the radiation, this can be in either the radio or X-ray part of the electromagnetic spectrum.

# BLACK HOLES

If the remnant of a supernova explosion is greater than about three solar masses, there is no mechanism that can stop it collapsing. It becomes so small and dense that its resulting gravitational pull is great enough to stop even radiation, including visible light, from escaping. Stellar-mass black holes, as such objects are known, can be detected only by the effect they have on objects around them. Light from far-off objects can be bent around a black hole as it acts as a gravitational lens, while the movement of nearby objects can be affected by a black hole's strong gravitational field (see pp.40–41). If a stellar-mass black hole is a member of a close binary system (see pp.270–71), the material from its companion star will be pulled towards it by its immense gravity. Matter will not fall directly onto the black hole, due to its rotational motion. Instead it will first be pulled into a accretion disc around the black hole. Matter impacts onto this disc, creating hot spots that can be detected by the radiation they emit. As matter in the disc gradually spirals into the black hole, friction will heat up the gas and radiation is emitted, predominantly in the X-ray part of the electromagnetic spectrum.

**BLACK HOLE**
Here, the gas from a companion star is drawn into a black hole via an accretion disc. When the gas crosses a limit called the event horizon, the gravitational field has become so strong that light cannot escape, and it disappears from view.

**NEUTRON STAR**
The gas drawn from a companion star approaches a neutron star in the same manner. However, when the gas strikes the solid surface of the neutron star, light is emitted and the star glows.

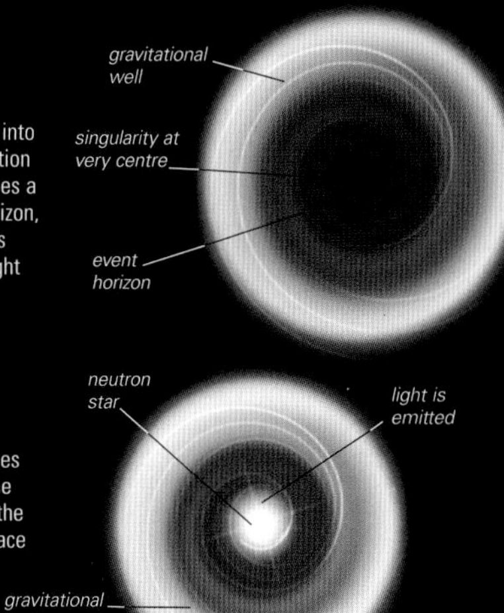

# STELLAR END POINTS

Stars end their lives in a variety of ways, but many are difficult or impossible to observe. It is thought that unobserved dead stars contribute significantly to the Milky Way's mysterious missing mass (see pp.226–29). Often, black holes and small white dwarfs can be observed only by the effect they have on nearby objects, and neutron stars are visible only in gamma-ray wavelengths. However, some stellar end points and their remnants, such as supernovae, are among the Galaxy's most spectacular sights.

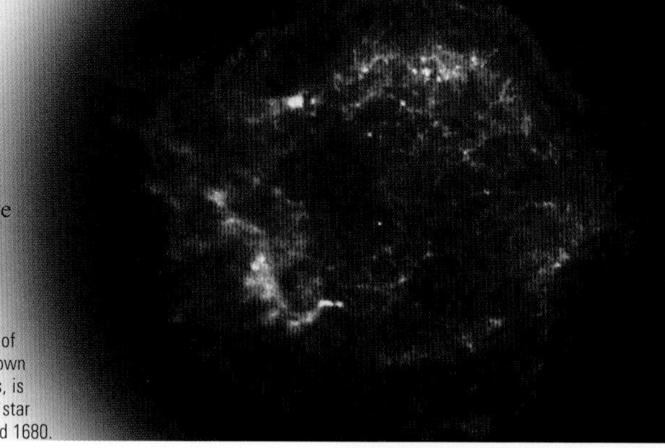

**STAR REMNANT**
A rapidly expanding shell of hot gas, Cassiopeia A, shown here in X-ray wavelengths, is the remnant of a massive star that died unnoticed around 1680.

WHITE DWARF

## Sirius B

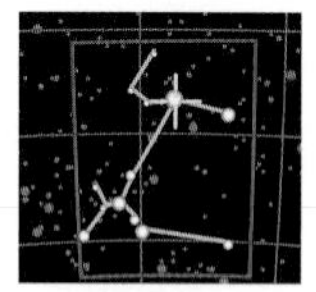

**CANIS MAJOR**

**CATALOGUE NUMBER** HD 48915 B

**DISTANCE FROM SUN** 8.6 light-years

**MAGNITUDE** 8.5

This was the first white dwarf to be discovered. First observed in 1862, it was found to be a stellar remnant when its spectrum was analysed in 1915. Although Sirius A, its companion, is the brightest star in the sky, Sirius B appears brighter in X-ray images (such as the one below). Sirius B's diameter is only 90 per cent that of Earth's but, as its mass is equal to that of the Sun, its gravity is 400,000 times that on Earth.

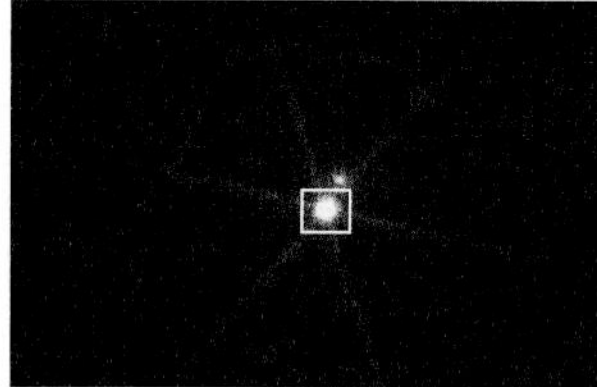

**CLOSE COMPANIONS**

NEUTRON STAR

## RX J1856.5-3754

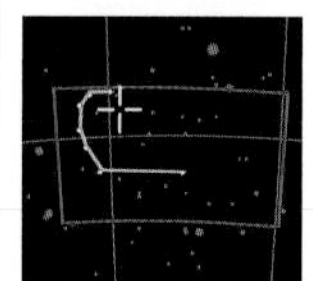

**CORONA AUSTRALIS**

**CATALOGUE NUMBER** 1ES 1853-37.9

**DISTANCE FROM SUN** 200–400 light-years

**MAGNITUDE** 26

This lone star is the closest known neutron star to Earth. Discussions are ongoing as to its true distance, but estimates vary from 200 to 400 light-years. There is also much speculation about its age. Some astronomers believe it is an old neutron star emitting X-rays because it is accreting material onto its surface from the surrounding interstellar medium. Others believe it is a young neutron star, emitting X-rays as it cools. It is possible that it formed about 1 million years ago, when a massive star in a close binary system exploded. It is travelling through the interstellar medium at about 390,000kph (240,000mph). RX J1856.5-3754 is moving away from a group of young stars in the constellation of Scorpius. Also moving away from this group of stars is the ultra-hot blue star now known as Zeta (ζ) Ophiuchi. It is possible that RX J1856.5-3754 is the remnant of Zeta Ophiuchi's original binary companion. As the closest neutron star, it is being extensively studied, but its diminutive size makes it difficult for astronomers to obtain conclusive results. Estimates of the diameter of RX J1856.5-3754 vary from 10km (6 miles) to about 30km (20 miles). This puts it very close to the theoretical limit of how small a neutron star can be, challenging some models of their internal structure. Its X-ray emissions suggest it has a surface temperature of around 600,000°C (1,000,000°F). Its visual magnitude of only 26 means that this star is 100 million times fainter than an object on the limit of naked-eye visibility.

**RARE VIEWS**
Taken in 1997, a Hubble image (above), offered astronomers an unusual glimpse of a neutron star in visible light. The star's movement through the interstellar medium has produced a cone-shaped nebula, visible in a later image (below).

NEUTRON STAR

## Geminga Pulsar

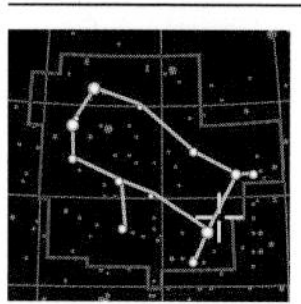

**GEMINI**

**CATALOGUE NUMBER** SN 437

**DISTANCE FROM SUN** 500 light-years

**MAGNITUDE** 25.5

Discovered in 1972, the Geminga Pulsar, a pulsating neutron star, is the second-brightest source of high-energy gamma rays known in the Milky Way. Its name is a contraction of "Gemini gamma-ray source"; it is also an expression, in the Milanese dialect, meaning "It's not there", because only recently has this object been observed in wavelengths other than gamma rays. Variations in the pulsar's period of luminosity (see pp.276–77) have suggested that it may have a companion planet, but they may also be due to irregularities in the star's rotation. Geminga is believed to be the remnant of a supernova that took place about 300,000 years earlier in the star's life. It is travelling through space at almost 25,000kph (15,000mph), at the head of a shock wave 3.2 billion km (2 billion miles) long.

**GAMMA RAY SOURCE**
The Geminga Pulsar shines bright in an image taken through a gamma-ray telescope. Gamma-ray photons are blocked from the Earth's surface by the atmosphere.

WHITE DWARF

## NGC 2440 nucleus

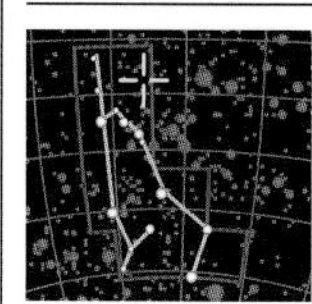

**PUPPIS**

**CATALOGUE NUMBER** HD 62166

**DISTANCE FROM SUN** 3,600 light-years

**MAGNITUDE** 11

The central star of the planetary nebula NGC 2440 has one of the highest surface temperatures of all known white dwarfs. This stellar remnant has a surface temperature of around 200,000°C (360,000°F) – 40 times hotter than that of the Sun. This also makes it intrinsically very bright, with a luminosity more than 250 times that of the Sun. The complex structure of the surrounding nebula has led some astronomers to believe that there have been periodic ejections of material from the dying central star. The structure of the nebula also suggests that the material was ejected in various directions during each episode.

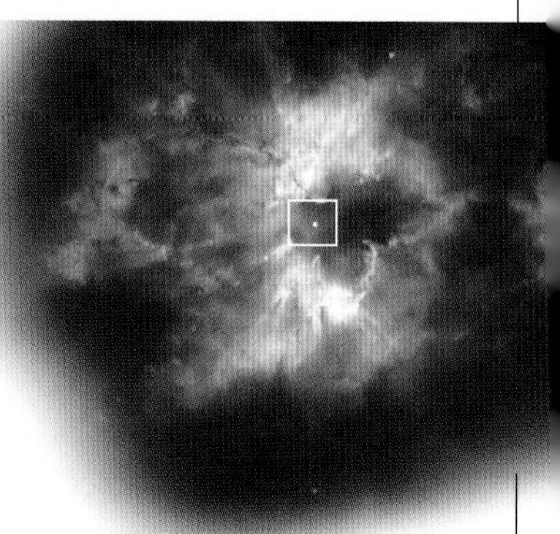

**INNER LIGHT**
Energy from the extremely hot surface of NGC 2440's central white dwarf makes this beautiful and delicate-looking planetary nebula fluoresce.

SUPERNOVA REMNANT

# The Cygnus Loop

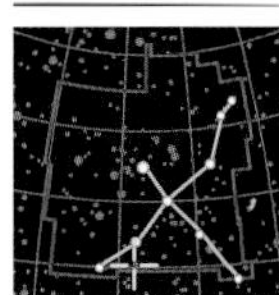

CYGNUS

| | |
|---|---|
| CATALOGUE NUMBER | NGC 6960/95 |
| DISTANCE FROM SUN | 2,600 light-years |
| MAGNITUDE | 11 |

The Cygnus Loop is the remnant of a dying star that blew itself up in a supernova. Estimates of how long ago in the star's lifetime this event occurred vary from 5,000 to 15,000 years. The most prominent parts of the nebula seen in visible light are often called the Veil Nebula, and, because it is so large, the Cygnus Loop has been catalogued using many different reference numbers. The supernova remnant is some 80 light-years long and sprawls 3.5 degrees across the sky – about seven full Moons across. It shines in the light generated by shock waves produced as stellar material from the supernova hits material in the interstellar medium. Observations of this stellar laboratory have revealed an inconsistent composition and structure of the interstellar medium as well as that of the supernova remnant.

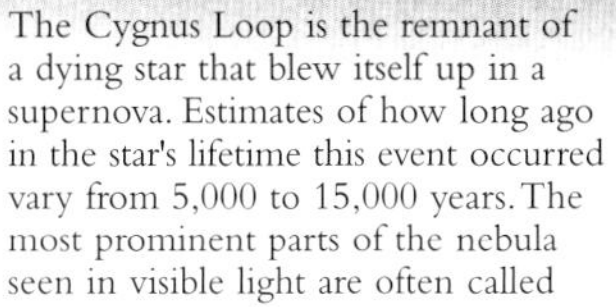

**GLOWING FILAMENTS**
Filaments of shocked interstellar gas glow in the light emitted by excited hydrogen atoms. This side-on view shows a small portion of the Cygnus Loop moving upwards at about 612,000kph (380,000mph).

**COLOURFUL GASES**
This composite image of a section of the Cygnus Loop reveals the presence of different kinds of atoms excited by shock waves: oxygen (blue), sulphur (red), and hydrogen (green).

SUPERNOVA REMNANT

# Vela Supernova

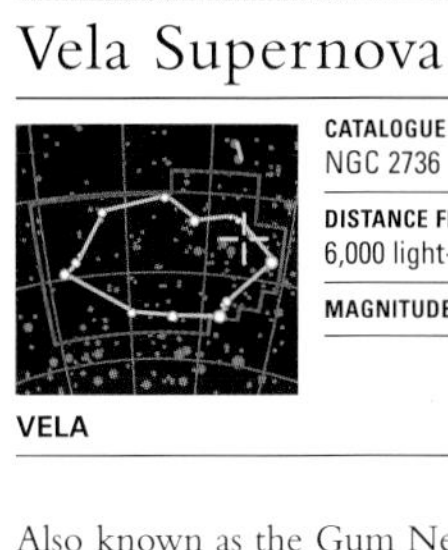

VELA

| | |
|---|---|
| CATALOGUE NUMBER | NGC 2736 |
| DISTANCE FROM SUN | 6,000 light-years |
| MAGNITUDE | 12 |

Also known as the Gum Nebula, the Vela Supernova is the brightest object in the sky at gamma-ray wavelengths. It is estimated that the star that produced it exploded between 5,000 and 11,000 years previously, and that its final explosion would have rivalled the Moon as the brightest object in the night sky. The star that died has become a pulsar, a rapidly spinning neutron star, which rotates about 11 times each second. The Vela Pulsar is about 19km (12 miles) in diameter and was only the second pulsar to be discovered optically – the optical flashes being observed in 1977. As with other pulsars, the rotation rate of the Vela Pulsar is gradually slowing down. Since 1967, it has suffered several brief glitches where its rotation rate has temporarily increased before continuing to slow.

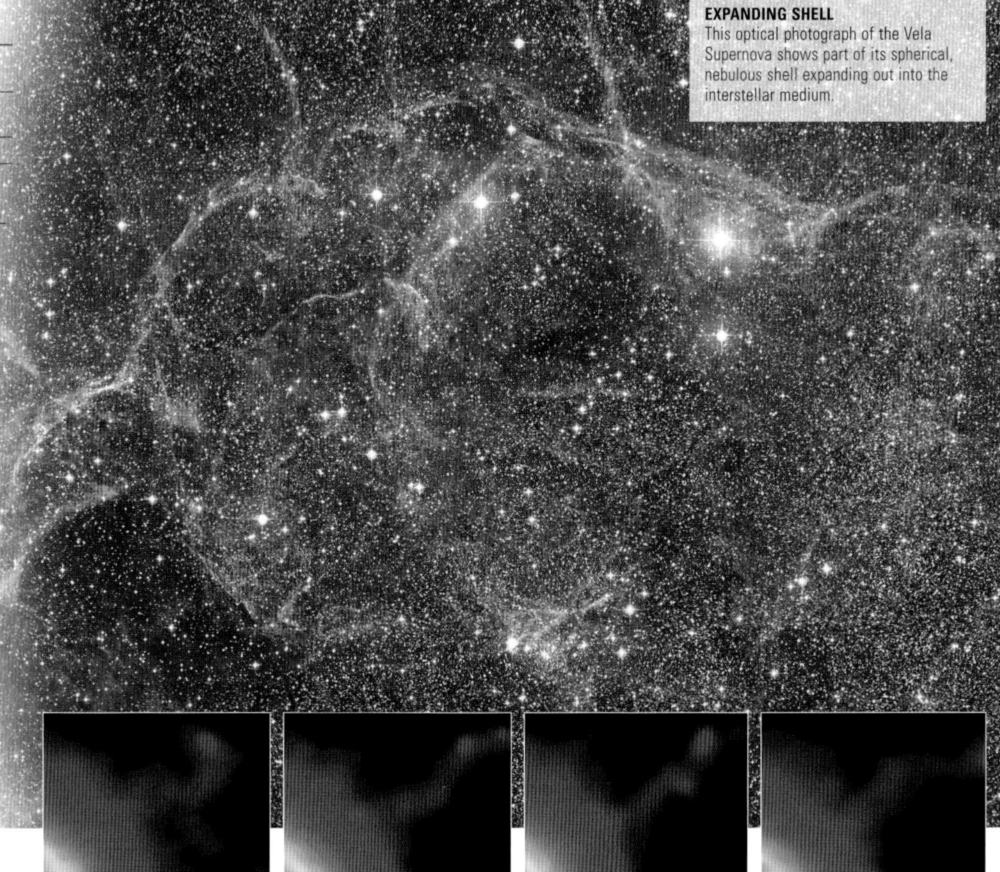

**EXPANDING SHELL**
This optical photograph of the Vela Supernova shows part of its spherical, nebulous shell expanding out into the interstellar medium.

**DYNAMIC JET**
This series of false-colour X-ray images reveals a flailing jet of high-energy particles, half a light-year long, emitted by the Vela Pulsar. These images are part of a series of 13 images made over a period of two and a half years.

30 NOVEMBER 2000 | 11 DECEMBER 2001 | 29 DECEMBER 2001 | 3 APRIL 2002

**CONSPICUOUS REMNANT**
The Crab Nebula is said to take its name from its shape in the drawings of William Parsons (see p.305). First photographed in 1862, the nebula is seen here in a Hubble composite image.

SUPERNOVA REMNANT

## Crab Nebula

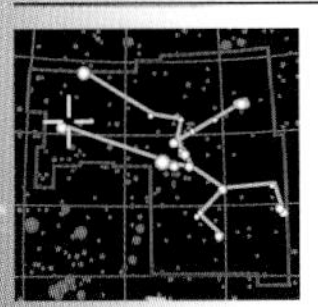

TAURUS

**CATALOGUE NUMBERS** M1, NGC 1952

**DISTANCE FROM SUN** 6,500 light-years

**MAGNITUDE** 8.4

In the summer of 1054, during the Sung dynasty, Chinese astronomers recorded that a star, in the present-day constellation Taurus, had suddenly become as bright as the full Moon. They described it as a reddish-white "guest star", and observed it over a period of two years as it slowly faded. Their records show it was visible in daylight for more than three weeks. They had witnessed a supernova, and the stellar material flung off in this cataclysmic explosion now shines as the wispy filaments of the Crab Nebula. This nebula is the very first object, and the only supernova remnant, to be listed by Charles Messier (see p.69) in his famous catalogue. The nebula is easily visible in binoculars and small telescopes. It spans a distance of about 10 light-years with a magnitude of between 8 and 9.

The remains of the original star have become a spinning neutron star, a pulsar, rotating at about 30 times per second.

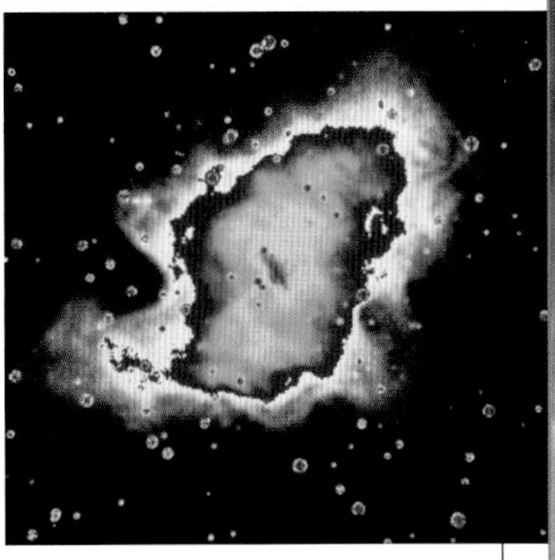

**FALSE-COLOUR MAP**
This false-colour optical image maps the intensity of light emitted from the Crab Nebula. The brightest regions are shown in red, followed by yellow, green, then blue, to the coolest regions represented in grey.

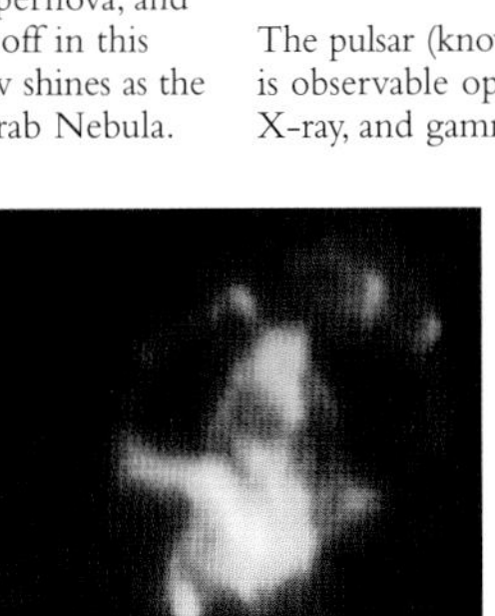

**RADIO MAP**
A false-colour radio map of the Crab Nebula shows the glowing emission of electrons spiralling in the central pulsar's strong magnetic fields. These are created by the pulsar rotating about 30 times per second.

The pulsar (known as PSR 0531 +21) is observable optically and in radio, X-ray, and gamma-ray wavelengths as the beams it generates happen to be directed towards Earth during part of its revolution. It was discovered in 1967, but had been known previously as a powerful emitter of radio waves and X-rays. It was the first pulsar to be identified optically and is of 16th magnitude. It is estimated to have a diameter of only about 10km (6 miles) but a mass greater than the Sun's. Its energy output is more than 750,000 times that of the Sun. Its rotation is decreasing by about 36.4 nanoseconds every day, which means that over 2,500 years from its presently observed state its rotation period will have doubled (see pp.278–79). The loss of rotational energy is being translated into energy, which is heating the surrounding Crab Nebula.

As the most easily observable supernova remnant, the Crab Nebula has been extensively studied. Detailed observations show that the material within the central portion of the nebula changes within a timescale of only a few weeks. Wisp-like features, each about a light-year across, have been observed streaming away from the pulsar at half the speed of light. These are created by an equatorial wind emitted by the pulsar (see left). They brighten and then fade as they move away from the pulsar and expand out into the main body of the nebula. The most dynamic feature within the centre is the point where one of the polar jets from the pulsar cannons into the surrounding previously ejected material, forming a shock front. The shape and position of this feature has been observed to change over very short timescales.

**PULSAR CLOSE-UP**
Detailed images of the central region of the Crab Nebula show its pulsar as one of a pair of bright stars. The jets from the pulsar (left) eject material in opposite directions, to form a halo of excited electrons, while a knot of dense matter lies adjacent to the pulsar.

NEUTRON STAR

## PSR B1620-26

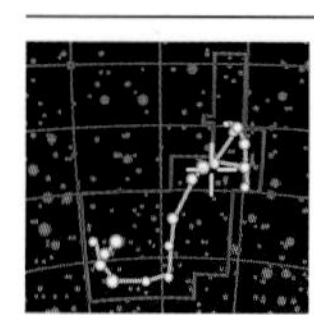

**SCORPIUS**

**CATALOGUE NUMBER** PSR B1620-26

**DISTANCE FROM SUN** 7,000 light-years

**MAGNITUDE** 21.3

Situated in the globular cluster M4, the pulsar PSR B1620-26 rotates more than 90 times per second and has a mass of about 1.3 solar masses. It has a white-dwarf companion (boxed in the image below). A third companion is thought to be a planet twice the mass of Jupiter (see p.291). This planet is named Methuselah, as it may be up to 13 billion years old.

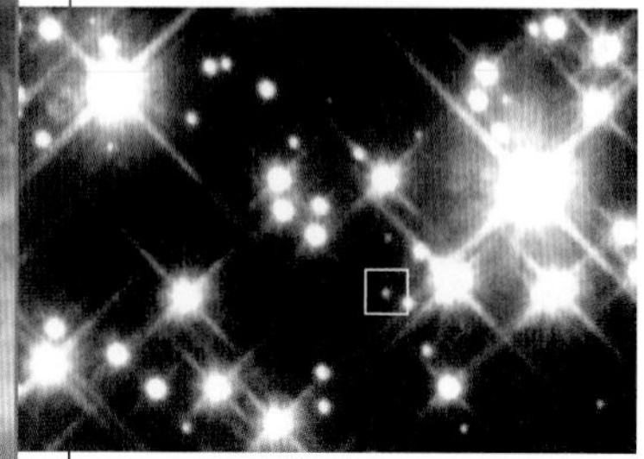

**WHITE-DWARF COMPANION**

BLACK HOLE

## GRO J1655-40

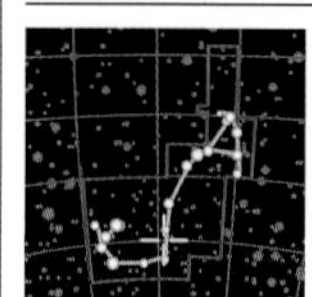

**SCORPIUS**

**CATALOGUE NUMBER** V* V1033 Sco

**DISTANCE FROM SUN** 6,000–9,000 light-years

**MAGNITUDE** 17

Discovered in 1994, as a source of unusual X-ray emissions, this black hole produces outbursts in which jets of material are ejected at speeds close to the speed of light. In addition to this, the gas surrounding GRO J1655-40 displays an unusual flicker (at a rate of 450 times per second) that can be explained as a rapidly rotating black hole. This is only the second object of this type to have been found in the Milky Way. It has been suggested that a subgiant star is orbiting the black hole, which is six to seven times the mass of the Sun. Their orbits are thought to be inclined at 70 degrees to each other, causing partial eclipses. Mass has been pulled off the subgiant star by the gravitational interaction from the black hole and formed a disc of material around the system. This system has been dubbed a mini-quasar because of its similarity to active galactic nuclei (AGNs) (see pp.298–99).

BLACK HOLE

## Cygnus X-1

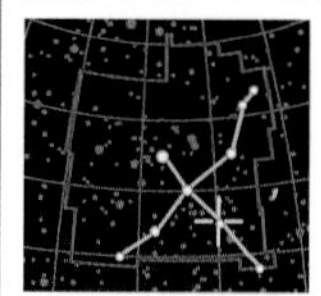

**CYGNUS**

**CATALOGUE NUMBER** HDE 226868

**DISTANCE FROM SUN** 8,200 light-years

**MAGNITUDE** 8.95

This X-ray source was one of the first to be discovered, and is one of the strongest X-ray sources in the sky. The X-ray emissions from Cygnus X-1 flicker at a rate of 1,000 times per second. In 1971, astronomers observed a radio source at the same position in the sky and also identified an optical object, the blue supergiant star HDE 226868. This star has a mass of 20–30 solar masses and is visible through binoculars. It is in a 5.6-day orbit with Cygnus X-1, which has a mass of about six solar masses. Further observations have shown that the black hole is slowly pulling material from its companion supergiant and increasing its own mass. Cygnus X-1 was the first object to be identified as a stellar-mass black hole.

**ELUSIVE BLACK HOLE**
Cygnus X-1 is located close to the red emission nebula Sh2-101, within the rich Cygnus Star Cloud (below). A negative optical image helps to pinpoint its companion, HDE 226868.

SUPERNOVA

## Tycho's Supernova

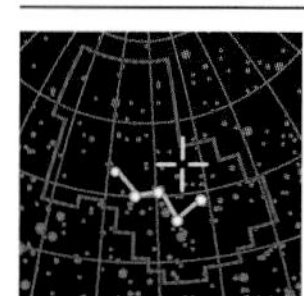

**CASSIOPEIA**

**CATALOGUE NUMBER** SN 1572

**DISTANCE FROM SUN** 7,500 light-years

**MAXIMUM MAGNITUDE** -3.5

In 1572, Tycho Brahe (see panel, below) observed a supernova in the constellation Cassiopeia and recorded its brightness changes in exceptional detail. It brightened to around -3.5 – as bright in the sky as Venus – before fading over a period of about six months. This brilliant new object was to help astronomers reject the idea that the heavens were immutable. The remnant from this supernova is still expanding and has a current diameter estimated at nearly 20 light-years. Its stellar material is estimated to be travelling at 21.5–27 million kph (14.5–18 million mph), which is the highest expansion rate observed for any supernova remnant. No strong central point source is detected in the remnant, which suggests that Tycho was a Type Ia supernova. The model for this type of supernova is the destruction of a white dwarf when in-falling matter from a companion star increases its mass beyond the Chandrasekhar limit (see pp.262–63). This concurs with the recent discovery of what astronomers think is the burned-out star from the heart of the supernova. The star was discovered because it is moving at three times the speed of other objects in the region. At the edge of the remnant is a shock wave heating the stellar material to 20 million °C (36 million °F); the interior gas is much cooler, at 10 million °C (18 million °F).

### TYCHO BRAHE

The leading astronomer of his day, Tycho Brahe (1546–1601) founded a great observatory at Uraniborg, Denmark, and spent years making detailed observations of planetary movements and the positions of the stars. Johannes Kepler became his assistant, and Tycho's work was to give the empirical basis for Kepler's laws of planetary motion (see p.87).

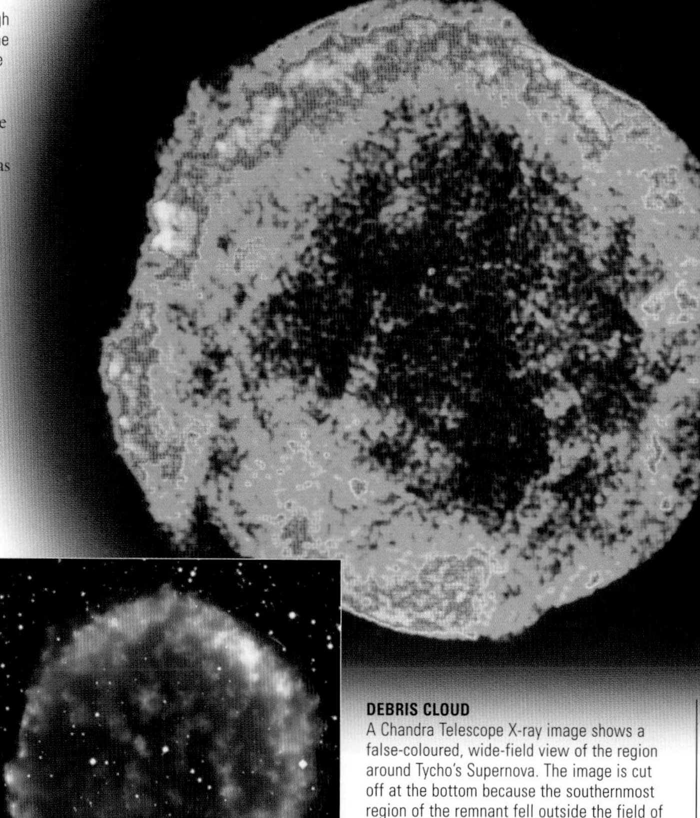

**RADIO ENERGY**
A radio image of Tycho's Supernova shows areas of low (red), medium (green), and high (blue) energy. A shock wave produced by the expanding debris is shown by the pale blue circular arcs on the outer rim.

**DEBRIS CLOUD**
A Chandra Telescope X-ray image shows a false-coloured, wide-field view of the region around Tycho's Supernova. The image is cut off at the bottom because the southernmost region of the remnant fell outside the field of view of the Chandra camera.

SUPERNOVA

## Kepler's Star

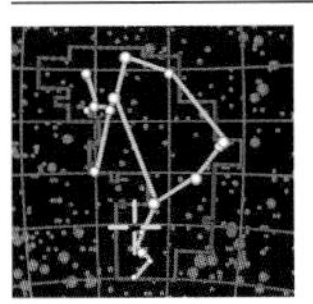

OPHIUCHUS

CATALOGUE NUMBER
SN 1604

DISTANCE FROM SUN
13,000 light-years

MAXIMUM MAGNITUDE
-2.5

**VISIBLE WISPS**
In this optical image, the supernova remnant appears as a faint ring of gas filaments. Having been expelled by the original explosion, this stellar material becomes heated and glows as it ploughs through the interstellar medium.

The last supernova explosion in the Milky Way to be observed is named after Johannes Kepler, who witnessed it in October 1604. This previously unremarkable star reached a magnitude of -2.5 and remained visible to the naked eye for more than a year. Its position is now marked by a strong radio source and, in optical light, by a wispy supernova remnant, generally known as Kepler's Star. Observations have revealed that the supernova remnant has a diameter of about 14 light-years and that the material within it is expanding at 7.2 million kph (4.5 million mph). Kepler's Star has been imaged by three of NASA's great observatories: the Hubble Space Telescope, the Spitzer Space Telescope, and the Chandra X-ray Observatory.

A combination of these images (right) has highlighted the remnant's distinct features. It shows an expanding bubble of iron-rich material surrounded by a shock wave, created as ejected material slams into the interstellar medium. This shock wave, shown in yellow, can also be seen optically (above). The red colour is produced by microscopic dust particles, which have been heated by the shock wave. The blue and green regions represent locations of hot gas: blue indicates high-energy X-rays and the highest temperatures; green represents lower-energy X-rays.

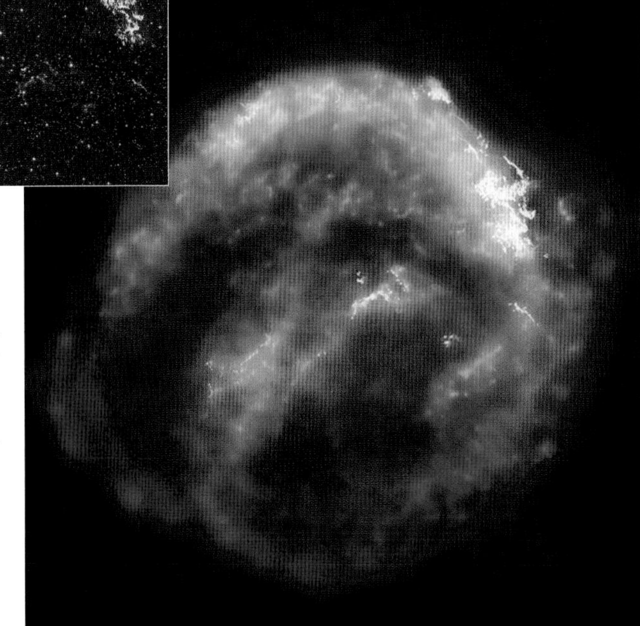

**COMBINED IMAGE**
A composite picture made using images from three separate telescopes offers a view ranging from X-ray through to infrared.

SUPERNOVA

## Cassiopeia A

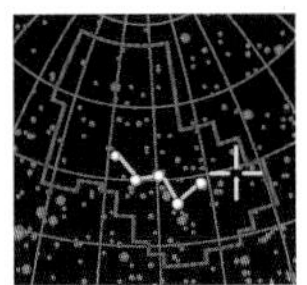

CASSIOPEIA

CATALOGUE NUMBER
SN 1680

DISTANCE FROM SUN
10,000 light-years

MAXIMUM MAGNITUDE 6

An intense radio source, Cassiopeia A is the remnant of a supernova explosion that occurred in the middle of the 17th century. The fact that no reports of the original explosion have been found suggests it may have been of unusually low optical luminosity. Today, Cassiopeia A is the strongest discrete low-frequency radio source in the sky (after the Sun). The radio waves are produced by electrons spiralling in a strong magnetic field. Cassiopeia A is about 10 light-years in diameter and is expanding at a rate of about 8 million kph (5 million mph).

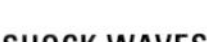

**COLOUR-CODED IMAGE**
This Hubble Space Telescope image of Cassiopeia A's cooling filaments and knots has been colour-coded to help astronomers understand the chemical processes involved in the recycling of stellar material.

**SHOCK WAVES**
This false-colour X-ray image clearly shows (in green) the edges of Cassiopeia A's expanding shock wave. The tiny white dot at the centre is the neutron star created by the supernova explosion.

BLACK HOLE

## MACHO 96

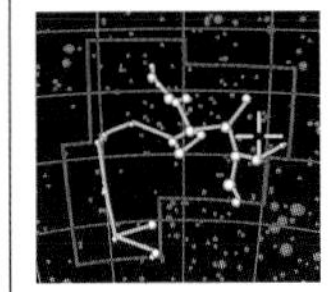

SAGITTARIUS

CATALOGUE NUMBER
MACHO 96

DISTANCE FROM SUN
Up to 100,000 light-years

Although we cannot see black holes, we can detect their presence by measuring their effects on objects around them. The existence of the black hole named MACHO 96 is inferred from the observed brightening of a star lying beyond the black hole caused by a process called lensing (see p.317). Through this process, the black hole's mass bends the light from the star in the same way as a lens does. The distant star is temporarily magnified, and we see a brief and subtle brightening in the star's output. The dark lensing object MACHO 96 has been calculated to be a six-solar-mass black hole that is moving independently among other stars. The chances of observing such a lensing event are estimated to be extremely slim. Therefore astronomers monitor millions of stars every night, using computers to analyse the brightness of the stellar images captured by advanced camera systems. So far, fewer than 20 events have been detected looking towards the Large Magellanic Cloud, a nearby galaxy (see pp.300-01). MACHO 96 was initially detected by the MACHO Alert System in 1996 and subsequently monitored by the Global Microlensing Alert Network. However, it was only by studying images taken by the Hubble Space Telescope that astronomers could identify the lensed star and determine its true brightness (see below). Observations have suggested that the distant star may be a close binary system, but astronomers are still debating whether the lensing object lies in the Milky Way's Galactic Halo or in the Large Magellanic Cloud.

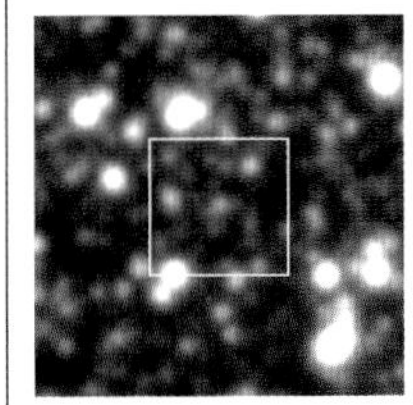

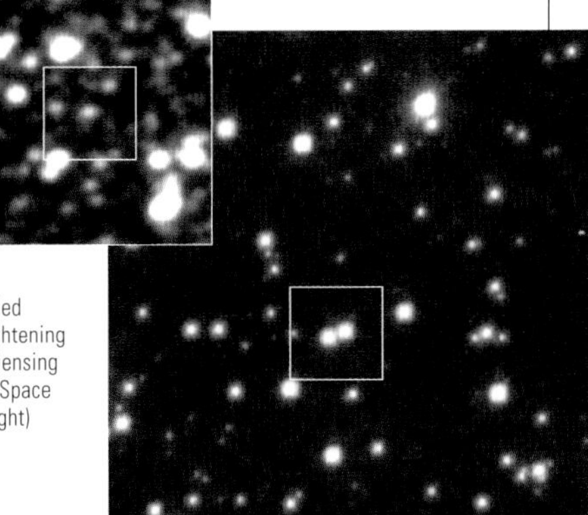

**PASSING BLACK HOLE**
Two ground-based images of a crowded star field (above) show the slight brightening of a star caused by the gravitational lensing of the passing MACHO 96. A Hubble Space Telescope image of the same area (right) resolves the star and allows its true brightness to be determined.

# MULTIPLE STARS

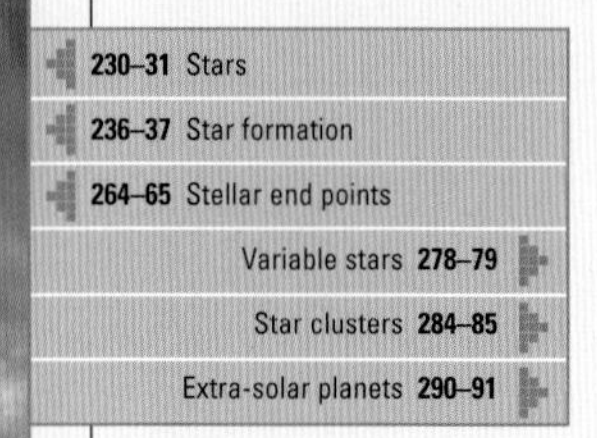


A MULTIPLE STAR IS A SYSTEM of two or more stars bound together by gravity. Systems with two stars are called binary or double stars. Although at first sight only a few stars appear to be multiple, it is estimated that they may account for over 60 per cent of stars in the Milky Way. Binary stars orbit each other at a great variety of distances, with orbital periods ranging from a few hours to millions of years. Multiple stars allow astronomers to determine stellar masses and diameters and give them insights into stellar evolution.

## BINARIES AND BEYOND

Although there are many millions of multiple systems within the Milky Way, not all of them consist of just two stars in mutual orbit. What may appear to be a double or binary star can often reveal itself to be a more complex system of three or more stars. A simple binary system consists of two stars orbiting each other. If the stars are of similar mass, they orbit around a common centre of gravity, located between them. If one of the stars is much more massive than the other, the common centre of gravity may be located inside the massive star. The more massive star then merely exhibits a wobble, while the secondary star appears to take on all the orbital motion. However, multiple systems may have a greater orbital complexity, with multiple centres of gravity. For example, a quadruple system may have two pairs of stars orbiting each other, while the individual stars within the pair are also in mutual orbit.

**MULTIPLE STAR SYSTEM**
Albireo is a striking double star made up of a golden giant star and a cool blue dwarf. Albireo's stars have a relatively long orbital period of about 7,300 years.

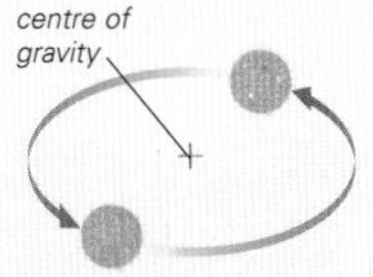


**EQUAL MASS**
In binaries with stars of equal mass, the common centre of gravity lies midway between the stars.

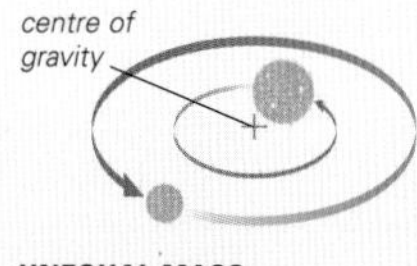


**UNEQUAL MASS**
If one star in a binary system is more massive, the centre of gravity lies closer to the higher-mass star.

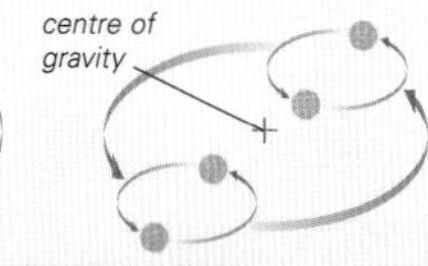


**DOUBLE BINARY**
In a double binary system, each star orbits its companion, and the two pairs orbit a single centre of gravity.

## DETECTING BINARIES

Astronomers detect binary stars in a variety of ways. Line-of-sight binaries consist of stars that appear in the sky to be related but, in fact, are not physically associated. These are usually identified by determining the true distances to the individual stars. Visual binaries are detected when the naked eye or magnification splits the stars and shows each one separately. Measurements of each star's position over time allow astronomers to compute their orbit. Although they cannot be separated with a telescope, astrometric binaries are detected when an unseen companion causes a star to wobble periodically through its gravitational influence. Spectroscopy can also be used to identify binary stars, when a star's spectrum appears doubled up and actually consists of the combined spectrums of two stars orbiting each other. These systems are known as spectroscopic binaries. The apparent magnitude of a binary star may show periodic fluctuations, caused by the stars eclipsing each other. Such stars are known as eclipsing binaries.

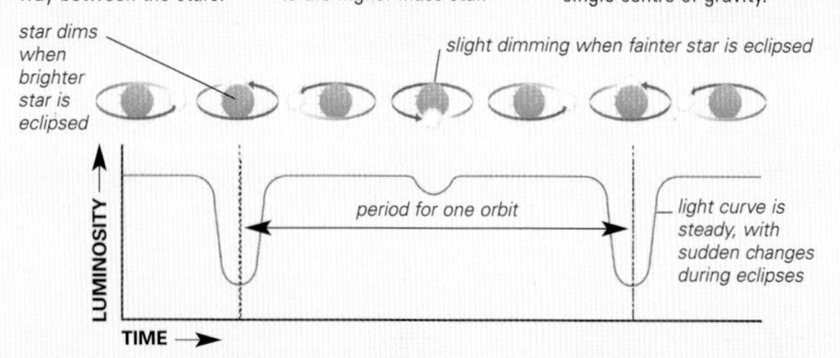


**ECLIPSING BINARIES**
Eclipsing binary stars are detected by variations in a star's magnitude. These variations occur when stars periodically pass in front of each other during orbit.

## CO-EVOLUTION

Like all stars, those within a multiple system evolve. A binary system can start out as two main-sequence stars with a mutual, regular orbit and predictable eclipses. However, over millions of years, the stars progress through their evolutionary stages, which may result in a binary system with two stars of completely different characteristics. One example is the Sirius system (see p.248). The evolution of one star within a system can change the behaviour of the whole system. For example, should a star expand and become a red giant, the expansion can bring the evolving star to interact with its companion star. This leads to mass transfer, and if the companion has itself evolved into a white dwarf, the result can be a cataclysmic explosion (see p.279). Stellar evolution can thus convert a stable binary system into a scene of immense violence.

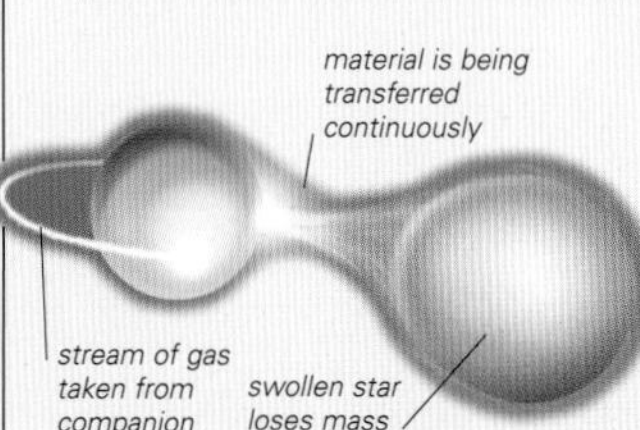


**INTERACTING BINARIES**
The stars in some binary systems are located so close together that material can pass between them. Here, one of the stars has swollen and is spilling gas onto the other.

## EXTREME BINARIES

Many binary systems exhibit perfectly regular behaviour, with the stars orbiting each other for millions of years with no dramatic changes. However, other binary systems, particularly those that have undergone evolutionary changes, may exhibit much more extreme behaviour. One example is a contact binary system, in which the two stars are touching each other. In this case, a massive star transfers material to the secondary star at a faster rate than the secondary star can absorb. This results in the material forming a common envelope that surrounds both stars. The envelope then creates frictional drag, causing the stars' orbital periods to change. In this way, a binary system with a wide separation and an orbital period of about a decade may be converted into a rapid system with the stars orbiting in a matter of hours. Other binary systems seem to operate at the extremes of physics. The discovery in 1974 of a binary pulsar system opened up a new field of observation in gravitational physics. A strong source of gravitational waves, binary pulsars are very regular and precise systems.

**HUB OF STARS**
One of the most famous multiple star systems, Theta (θ) Orionis, or the Trapezium (top left of image), is the middle star in the sword of Orion (see pp.374–75). Its four brightest stars are easily separated with a telescope, but it is made up of a total of at least 10 stars.

# MULTIPLE STARS

Most of the stars in the Milky Way are members of either binary or multiple systems – single stars like the Sun are more unusual. These systems vary from distant pairs in slow, centuries-long orbits around a common centre of mass to tightly bound groups that orbit each other in days and may even distort each other's shape. Most multiples are so close together that we know about them only from their spectra. They also vary widely in size and colour – stars of any age and type can be members of a multiple star system.

**TRAPEZIUM**
The multiple star known as Theta Orionis, or the Trapezium, is a system containing at least ten individual stars.

TRIPLE STAR

## Omicron Eridani

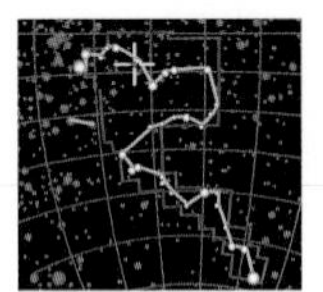

**ERIDANUS**

**DISTANCE FROM SUN** 16 light-years
**MAGNITUDE** 9.5
**SPECTRAL TYPE** DA

Originally Omicron (ο) Eridani was classed as a double star, Omicron-1 Eridani and Omicron-2 Eridani. Nineteenth-century observations revealed that the system is actually three stars, now called 40 Eridani A, B, and C. A is a main sequence orange-red dwarf, and C is a faint red dwarf. However it is 40 Eridani B that is the gem. This young white dwarf is the brightest white dwarf visible through a small telescope.

**TRIPLE SYSTEM**

SEXTUPLE SYSTEM

## Castor

**GEMINI**

**DISTANCE FROM SUN** 52 light-years
**MAGNITUDE** 1.6
**SPECTRAL TYPE** A2

Easily visible to the naked eye, Castor appears to be an ordinary A-type star. However, a telescope reveals that Castor is in fact a pair of bright A-type stars, Castor A and Castor B, with a fainter third companion, Castor C. Spectrographic analysis shows that both the A and B components of Castor are themselves double stars.

Castor A consists of two stars in a very close 9.2-day orbit, while Castor B's components orbit each other in a rapid 2.9 days. The faint Castor C star is also a double – a pair of faint red-dwarf stars orbiting each other with a period of only 20 hours. Castor is therefore a sextuplet star, a double-double-double.

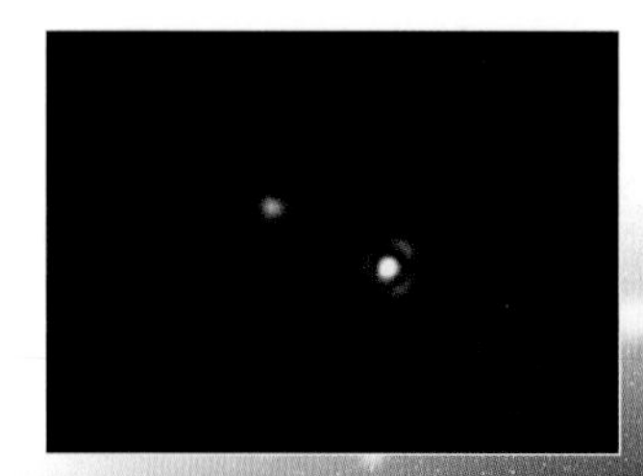

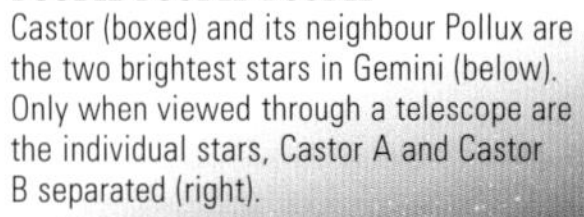

**DOUBLE-DOUBLE-DOUBLE**
Castor (boxed) and its neighbour Pollux are the two brightest stars in Gemini (below). Only when viewed through a telescope are the individual stars, Castor A and Castor B separated (right).

QUADRUPLE STAR

## Mizar and Alcor

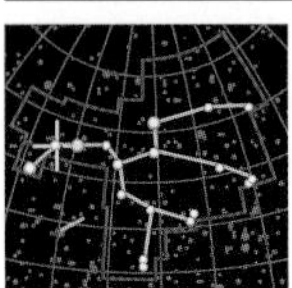

**URSA MAJOR**

**DISTANCE FROM SUN** 81 light-years
**MAGNITUDE** 2
**SPECTRAL TYPE** A2

Although Mizar and Alcor are a famous naked-eye double, easily visible in the handle of the Plough and known since ancient times as the horse and rider, it is still unknown whether or not they are a genuine double. Mizar itself is a double star – the first double star to be discovered. Spectrography reveals that Mizar however is a double-double star, that is, two double stars in orbit around each other.

**FAMOUS DOUBLE**

QUADRUPLE STAR

## Algol

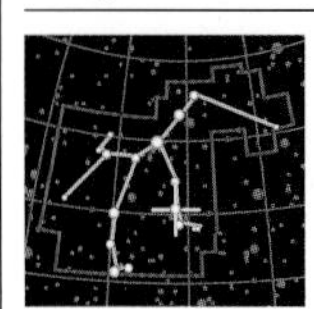

**PERSEUS**

**DISTANCE FROM SUN** 93 light-years
**MAGNITUDE** 2.1
**SPECTRAL TYPE** B8

Algol, or Beta (β) Persei, appears to the naked eye as a single star. However, exactly every 2.867 days, the star's brightness drops by 70 per cent for a few hours – a variation that was discovered as early as 1667. This variation is caused by Algol being eclipsed by a faint giant star Algol B, which is larger than the bright primary Algol A.

**ECLIPSING BINARY**

QUADRUPLE STAR

## Epsilon Lyrae

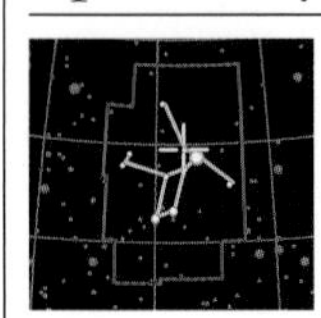

**LYRA**

**DISTANCE FROM SUN** 160 light-years
**MAGNITUDE** 3.9
**SPECTRAL TYPE** A4

Epsilon (ε) Lyrae is visible as a double star on a clear, dark night, but closer observation reveals that, in fact, each star is itself a double. Unlike other double-double systems, Epsilon Lyrae is within reach of amateur astronomers – its four component stars can each be seen through a telescope, and spectroscopy is not needed to detect their presence (see p.270). The two bright stars visible to the naked eye, Epsilon-1 and Epsilon-2, are widely separated, with an orbital period of millions of years. The components of each pair orbit much more closely, with a period of about 1,000 years. Epsilon-1 and Epsilon-2 are so far apart they are hardly bound by gravity at all, and eventually Epsilon Lyrae will become two isolated star systems.

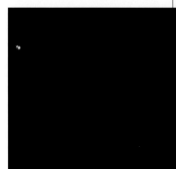

**ISOLATED PAIRS**
This double-double system is easily separated into its four components through a telescope. Although the stars in each pair are strongly bound to one another, the link between the pairs is tenuous.

DOUBLE STAR

## Zeta Boötis

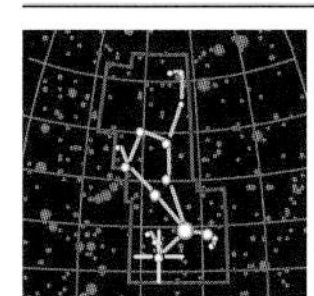
BOOTES

**DISTANCE FROM SUN** 180 light-years
**MAGNITUDE** 3.8
**SPECTRAL TYPE** A3

Zeta (ζ) Boötis would appear to be a standard double star – two A-type stars orbiting each other with a period of about 123 years. However, anomalies in calculations of its mass have suggested that there is something strange about the Zeta Boötis system. The answer lies in a highly elongated orbit, in which the stars range from 210–9,500 million km (130–5,900 million miles) apart. At their closest, they are almost as close as the Sun and Earth, and no telescope can visually split them. The Zeta Boötis system is about 40 times as luminous as the Sun, with about four times its mass, and has a temperature of about 8,700°C (15,700°F).

**ENHANCED IMAGES**
When the components of Zeta Boötis are at their farthest apart, image-processing software can be used to separate them, and even split their spectra.

DOUBLE STAR

## Izar

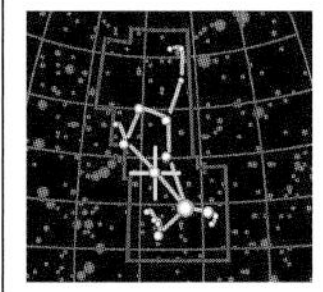
BOOTES

**DISTANCE FROM SUN** 210 light-years
**MAGNITUDE** 2.4
**SPECTRAL TYPE** A0

Izar, or Epsilon (ε) Boötis, is one of the best double stars in the sky. Its stars exhibit a striking colour contrast – an orange giant close to a white dwarf – and it was given the name Pulcherrima, "most beautiful", by its discoverer, German-born Friedrich Struve. The dwarf star is about twice the size of the Sun, while the orange giant is about 34 times the size. The dwarf and giant orbit each other with a period of more than 1,000 years. This double star is not particularly astronomically unusual, but is well known to amateur astronomers for its visual splendour.

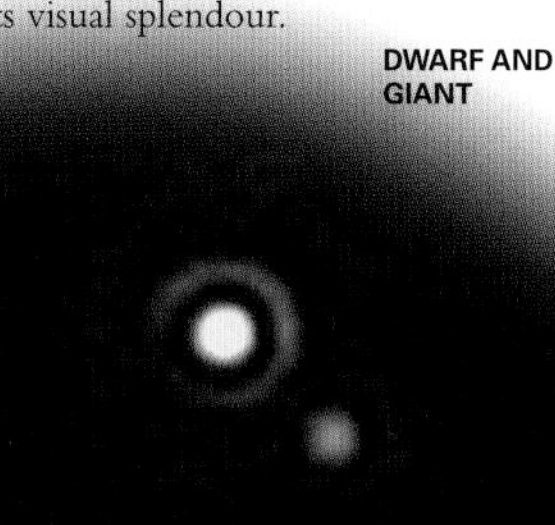
DWARF AND GIANT

QUADRUPLE STAR

## Almach

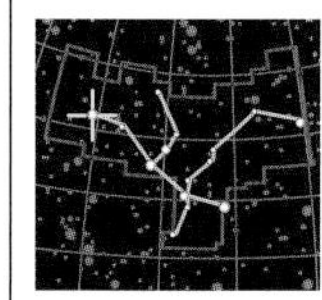
ANDROMEDA

**DISTANCE FROM SUN** 355 light-years
**MAGNITUDE** 2.3
**SPECTRAL TYPE** K3

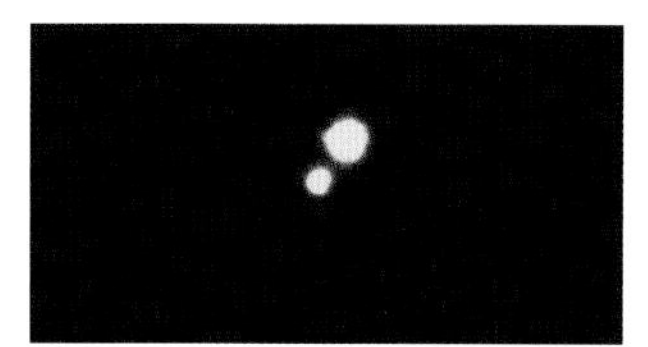
ALMACH

Almach, or Gamma (γ) Andromedae, is well known to amateur astronomers as being a fine example of a double star with contrasting colours. The brighter star is yellow-orange, and the fainter star is blue, and through a telescope the two colours enhance each other. The brighter star is a giant K-type star, while the fainter star is itself a double star, consisting of two hot, white main-sequence stars in a mutual orbit, with a period of about 60 years. It is difficult to split these two stars visually, but spectroscopic analysis reveals that one of them is also a double star in turn, making Almach a quadruple system.

DOUBLE STAR

## M40

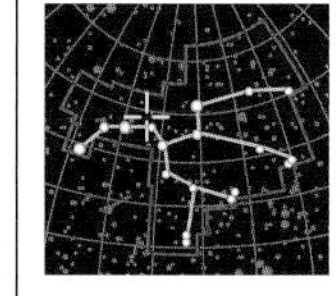
URSA MAJOR

**DISTANCE FROM SUN** 1900 and 550 light-years
**MAGNITUDE** 8.4
**SPECTRAL TYPE** G0

OPTICAL PAIR

Some multiple stars are famous for their beauty, others for the dramatic astrophysics played out within the system. In the case of M40, neither applies. When compiling his well-known catalogue of star clusters and nebulae, Charles Messier (see p.69) observed two stars close to each other in the night sky and erroneously included them. The two stars are nothing more than an optical double – that is, they happen to lie on the same line of sight. Modern distance measurements have shown that they are not truly associated with each other. M40 is therefore a double that achieves fame through error.

QUADRUPLE STAR

## Alcyone

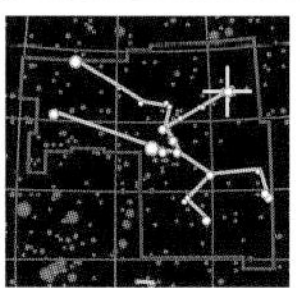
TAURUS

**DISTANCE FROM SUN** 368 light-years
**MAGNITUDE** 2.9
**SPECTRAL TYPE** B7

Alcyone, one of the sisters of the Pleiades (see p.287), is a bright giant star of spectral type B that shines about 1,400 times brighter than the Sun. Orbiting around Alcyone are three stars forming a compact system: 24 Tau and V647 Tau are A-type stars, while HD 23608 is a G-type star. The system of three stars orbits Alcyone at a distance of a few billion kilometres. Alcyone is unusual in that it rotates at high speed. This has caused it to throw gas off at its equator, which forms a light-emitting disc.

**SEASONAL SIGNAL**
Alcyone is the brightest star in the Pleiades Cluster (see p.287). Its appearance over the eastern horizon signals the start of autumn in the northern hemisphere.

TRIPLE STAR

## Albireo

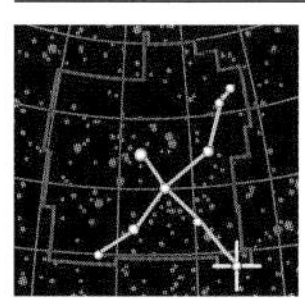
CYGNUS

**DISTANCE FROM SUN** 385 light-years
**MAGNITUDE** 3.1
**SPECTRAL TYPE** K3

Albireo, or Beta (β) Cygni, is regarded as one of the finest multiple star systems in the night sky. It consists of a golden-coloured giant star and a cooler blue-dwarf star. Star colours are often subtle, but when different coloured stars appear in the same field of view, as in this case, the colours often enhance each other. The separation between the two stars is relatively large and Albireo is easily separated through a small telescope. The blue dwarf is a rapidly rotating star that is throwing off atmospheric material from its equator and forming a surrounding disc of gas and dust. The large separation between the two visible stars of Albireo means that the orbital period is relatively long – about 7,300 years – and there is some doubt as to whether these two stars form a true double star (see p.270). The brighter component of Albireo is also a double star, consisting of a bright giant star and a dimmer dwarf star, but they are too close to each other to be separated with a telescope. As there is no evidence to suggest that the blue dwarf is a multiple star, Albireo is classified as a triple star system.

**COMPLEMENTARY COLOURS**
One of the most beautiful star systems in the night sky, Albireo's stars display complementary colours that astronomers have called topaz and sapphire.

**NORTH POLE STAR**
Polaris may appear motionless, but a long-exposure photograph reveals it is slightly offset from the celestial pole. Polaris's movement is marked by the bright arc just left of centre.

DOUBLE STAR

# Polaris

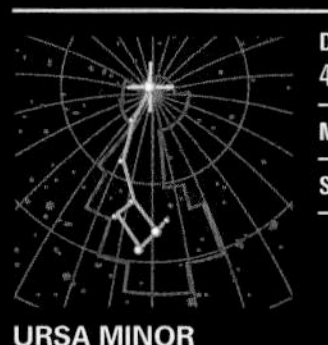

URSA MINOR

**DISTANCE FROM SUN** 430 light-years

**MAGNITUDE** 2

**SPECTRAL TYPE** F7

Polaris is famous as the current north Pole Star, and consequently is known to every observer of the northern sky (see panel, below). However, it is also an interesting system in terms of its component stars. Polaris is a double star, consisting of Polaris A, a supergiant, and Polaris B, a main-sequence star. The two stars can be separated through a modest amateur telescope, and Polaris B was first detected by William Herschel (see p.90) in 1780. The distance between them has been estimated at more than 300 billion km (190 billion miles), Polaris A is more than 1,800 times more luminous than the Sun, and is also a Cepheid variable with a period of just under four days (see p.278). The radial velocity, or line-of-sight motion, of Polaris has been accurately measured (see p.66), and found to vary regularly with a period of 30.5 years. This indicates that Polaris is also an astrometric binary – that is, the presence of an unseen companion is detected by the movement it induces in the primary star (see p.270). The companion, which was seen for the first time in a Hubble Space Telescope image in 2005, orbits Polaris with a 30.5-year period, but it is so faint that it has no effect on Polaris's spectrum.

**DISTINCTIVE STAR**
One of the best-known stars in the northern sky, Polaris lies just away from the celestial pole, in the tail of Ursa Minor, the Little Bear. This telescope view reveals its faint companion, Polaris B, but a second smaller companion is not visible.

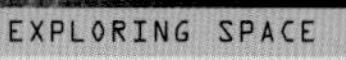

EXPLORING SPACE

## CELESTIAL SIGN POST

Polaris has long been regarded as the most important star in the northern sky. As it is located almost directly overhead at the north pole, it has long been used, just like a compass, to locate north (see p.83). By calculating the relative angle of Polaris above the horizon, travellers by land and sea have also used Polaris to establish approximate latitudinal positions on the Earth's surface. The status accorded to Polaris by disparate cultures is reflected in their myths. In Norse mythology, Polaris was the jewel on the head of the spike that the gods stuck through the universe. The Mongols called Polaris the Golden Peg that held the world together. In ancient China, Polaris was known as Tou Mu, the goddess of the North Star.

**IN THE LITTLE BEAR'S TAIL**
In Arabic mythology, Polaris was an evil star who killed the great warrior of the sky. The dead warrior was said to lie in the tail of the little bear, a constellation that also represented a funeral bier.

DOUBLE STAR

## 15 Monocerotis

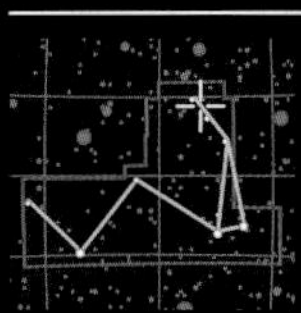

MONOCEROS

**DISTANCE FROM SUN** 1,020 light-years

**MAGNITUDE** 4.7

**SPECTRAL TYPE** O7

15 Monocerotis (15 Mon), also known as S Monocerotis, is an O-type binary system located within the open cluster NGC 2264. It is a blue supergiant star – young, massive, and about 8,500 times more luminous than the Sun. It is also a variable star, exhibiting a small (0.4 magnitude) change in brightness. 15 Mon is responsible for illuminating the Cone Nebula (see p.240), and consequently is an easy target for amateur astronomers. 15 Mon is an astrometric and spectroscopic binary – that is, its companion star is detected through observations of the motion of 15 Mon, and also through spectroscopic analysis of 15 Mon's starlight (see p.270). The companion orbits 15 Mon with a period of 24 years, and recent studies using the Hubble Space Telescope show that the closest approach between the stars occurred in 1996. It has been suggested that 15 Mon is a multiple system, with 3 other bright giants nearby. However, there is no evidence that the other giants are associated with 15 Mon.

**BRIGHTEST STAR**
The brightest star in the open star cluster NGC 2264, 15 Monocerotis sits in close visual proximity to the Cone Nebula (see p.240).

**BRILLIANT ILLUMINATION**
Even the most powerful telescopes cannot separate the two stars of 15 Monocerotis visually. The brilliant blue star lights up the emission nebula that surrounds it.

TRIPLE STAR

## Beta Monocerotis

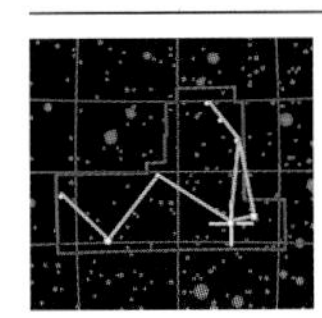

MONOCEROS

| | |
|---|---|
| DISTANCE FROM SUN | 700 light-years |
| MAGNITUDE | 5.4 |
| SPECTRAL TYPE | B2 |

Beta (β) Monocerotis is a triple star system, with components A, B, and C. The BC pair orbits each other with a period of about 4,000 years, and A orbits the BC pair with a period of about 14,000 years. The system is unusual because the three stars are so similar. All are hot, blue-white B-type stars, each more than 1,000 times as luminous and six times as massive as the Sun. All three stars also exhibit the same rotation speed and have circumstellar discs.

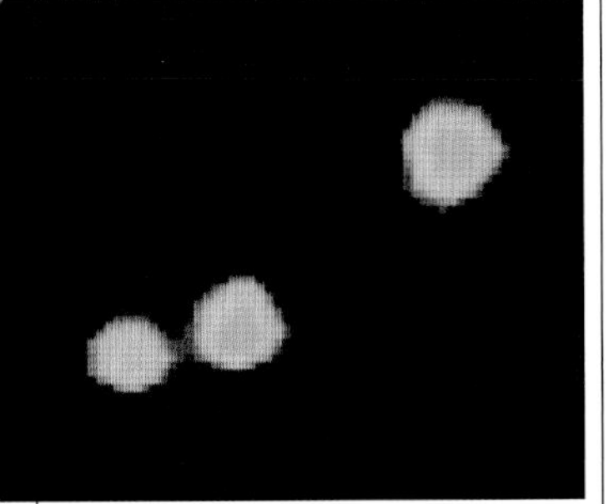

COMPUTER-ENHANCED OPTICAL IMAGE

TRIPLE STAR

## Rigel

ORION

| | |
|---|---|
| DISTANCE FROM SUN | 770 light-years |
| MAGNITUDE | 0.1 |
| SPECTRAL TYPE | B8 |

Rigel is a blue supergiant star shining 40,000 times more brightly than the Sun and has a faint close companion, Rigel B. The luminosity of Rigel makes observation of the companion difficult. Rigel B has been discovered to be a double star itself. It consists of two faint B-type main-sequence stars, Rigel B and Rigel C, separated by about 4 billion km (2.5 billion miles) and orbiting each other in an almost circular orbit. By contrast, the separation between the bright supergiant and the BC pair is over 300 billion km (190 billion miles).

**BRILLIANT GIANT**
Rigel is the brightest star in the constellation Orion and the 7th-brightest star in the night sky. Rigel B and C, its companion stars, are obscured by Rigel's great luminosity.

DOUBLE STAR

## Beta Lyrae

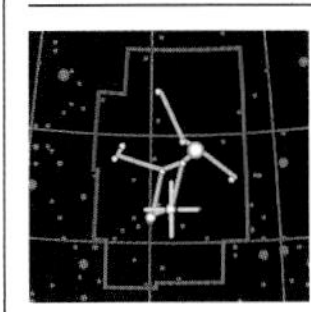

LYRA

| | |
|---|---|
| DISTANCE FROM SUN | 880 light-years |
| MAGNITUDE | 3.5 |
| SPECTRAL TYPE | B7 |

Beta (β) Lyrae, or Sheliak, is the prototype of a class of eclipsing binary stars known as Beta Lyrae stars or EB variables (see p.270). The brightness of the system varies by about one magnitude every 12 days 22 hours and is easily visible to the naked eye. Beta Lyrae's component stars are contact binaries, and are so close together that they are greatly distorted by their mutual attraction. Material pouring out of the stars is forming a thick accretion disc.

CLOSE BINARY

QUINTUPLE STAR

## Sigma Orionis

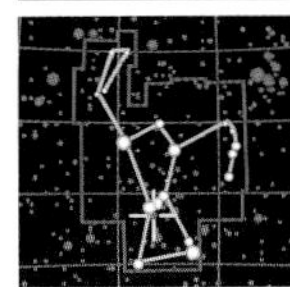

ORION

| | |
|---|---|
| DISTANCE FROM SUN | 1,150 light-years |
| MAGNITUDE | 3.8 |
| SPECTRAL TYPE | O9 |

Sigma (σ) Orionis is a quintuple system, containing four bright, easily visible stars and one fainter component, with the brightest being a close double. The two main stars, A and B, are more than 30,000 times as luminous as the Sun and have a combined mass more than 30 times greater than the Sun. The AB pair is one of the more massive binary systems in the Milky Way. It is in a stable orbit, but the C, D, and E stars are not, and gravitational forces may well throw them out of the system in the future.

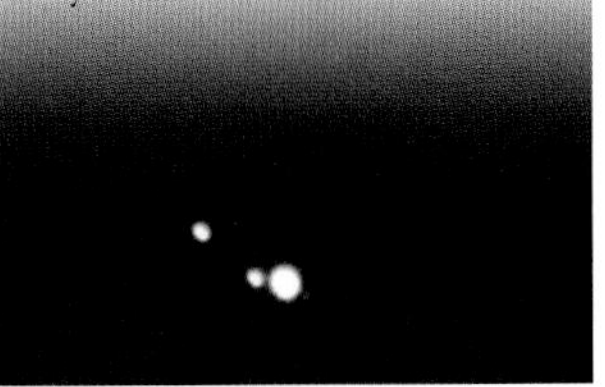

SIGMA ORIONIS'S FOUR BRIGHT STARS

QUADRUPLE STAR

## Theta Orionis

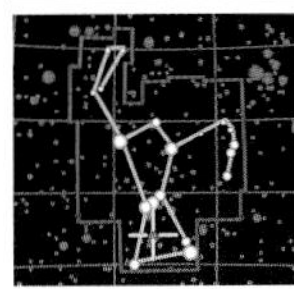

ORION

| | |
|---|---|
| DISTANCE FROM SUN | 1,800 light-years |
| MAGNITUDE | 4.7 |
| SPECTRAL TYPE | B |

Theta (θ) Orionis, perhaps better known as the Trapezium, appears to the naked eye to be a single star, but is revealed by any telescope to be a quadruple system. Theta Orionis provides much of the ultraviolet radiation that illuminates the Orion Nebula (see p.239). All four stars are hot O- and B-type stars, the largest being Theta-1 C, with 40 times the mass of the Sun, about 200,000 times its luminosity, and a temperature of 40,000°C (72,000°F). Theta-1 C is the hottest star visible to the naked eye. Theta-1 A is an eclipsing double star with an additional companion; Theta-1 D is a double star; and Theta-1 B is an eclipsing binary star, with a companion double (making it quadruple in itself). Although known as a quadruple star, Theta Orionis in fact consists of at least ten stars.

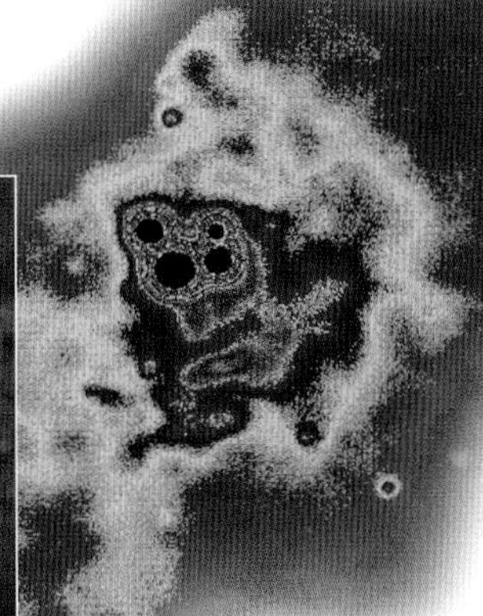

**THE TRAPEZIUM GROUP**
The stars of Theta Orionis light up the centre of the Orion Nebula. A false-colour image (below) helps to define the the four main stars in the system.

DOUBLE STAR

## Epsilon Aurigae

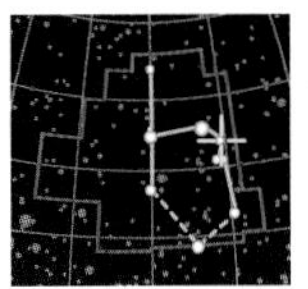

AURIGA

| | |
|---|---|
| DISTANCE FROM SUN | 2,040 light-years |
| MAGNITUDE | 3 |
| SPECTRAL TYPE | A8 |

The hot supergiant Epsilon (ε) Aurigae, or Almaaz, is an eclipsing binary star. Unusually, its eclipse lasts for two years, suggesting that the system is huge. The supergiant is being eclipsed by something far bigger than itself, but exactly what is uncertain. One theory is that Epsilon Aurigae's companion is an unseen star surrounded by a huge, dusty ring, and the bright star shines through this ring during an eclipse.

**DISTANT BINARY**
Dwarfed in this image by its celestial neighbour Capella, Epsilon Aurigae is in fact some 2,000 light-years more distant.

# VARIABLE STARS

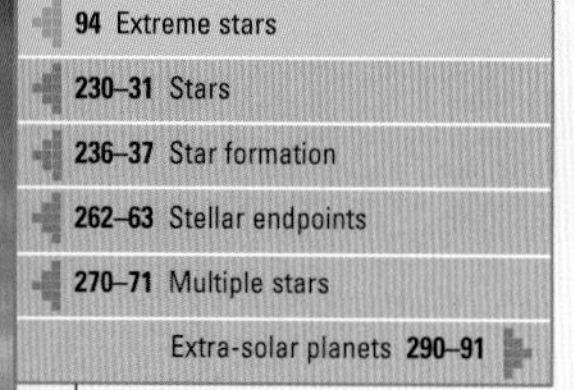

94 Extreme stars
230–31 Stars
236–37 Star formation
262–63 Stellar endpoints
270–71 Multiple stars
Extra-solar planets 290–91


ALTHOUGH AT FIRST SIGHT the stars in the night sky seem to be unchanging, many thousands of stars change their brightness over periods ranging from a few days to decades. True, or intrinsic, variable stars vary in brightness due to physical changes within the star. Others, such as eclipsing binaries (see p.270), only appear to vary because they have orbiting companions.

**PROTOTYPE**
Mira is one of the most famous variable stars in the Milky Way (see p.281). It is a long-period, pulsating star that has given its name to one of the main types of variable star.

## PULSATING VARIABLES

Pulsating variable stars are intrinsic variables that undergo repetitive expansion and contraction of their outer layers. A pulsating star is constantly trying to reach equilibrium between the inward gravitational force and the outward radiation and gas pressure. This causes the star's brightness to vary. In many types of pulsating stars, including Cepheids (see p.282), a star's period of variation is related to its luminosity. Knowledge of the star's luminosity, coupled with its apparent magnitude, enables astronomers to calculate a star's distance. Pulsating variables are therefore a useful tool for determining distances to far-away objects such as other galaxies.

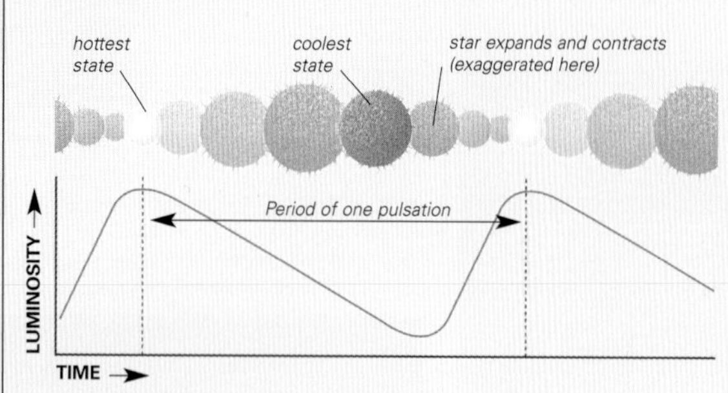


**LIGHT CURVE**
The light curve of a Cepheid variable shows the regular variation in luminosity during a period of one pulsation.

## NOVAE

A nova is a binary system, consisting of a giant star that is being orbited by a smaller white dwarf. The giant star has grown so large that its outer material is no longer gravitationally bound to the star and instead falls onto the white-dwarf companion. Eventually, this gain in material triggers a thermonuclear explosion on the surface of the white dwarf, which brightens it by many magnitudes, increasing its energy output by a factor of a million or more. The surface gases of the white dwarf are in a "degenerate" state, and they do not obey the normal gas laws. Usually a gas explosion will cause the gas to expand, thereby reducing the explosion and finishing it. However, the degenerate gases of a white dwarf do not expand, and the explosion turns into a runaway event that does not finish until the fuel is exhausted. Prior to this, the binary system would be invisible to the naked eye, and the nova outburst would then bring the system into visibility as a "new" – in Latin, *nova* – star.

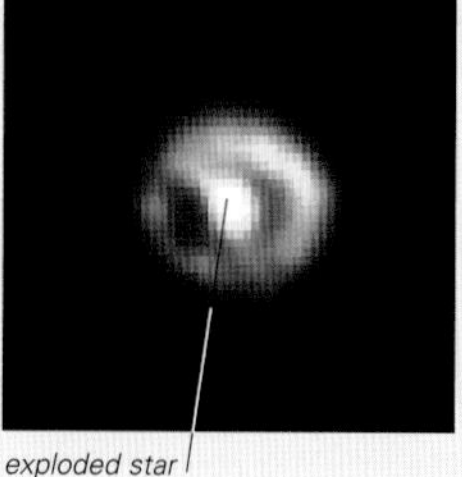


hot bubble of gas

**CATACLYSMIC BINARY**
The most widely studied nova, Nova Cygni 1992, was witnessed exploding in 1992 (see p.283). Its magnitude rose by such a degree that at its brightest the nova was visible to the naked eye.

**LIGHT ECHOES**
Light echoes from an expanding dusty cloud have been produced by the pulsating red supergiant V838 Monocerotis (see p.261). Initially thought to be a nova, astronomers think that it might, in fact, be a new type of eruptive star.

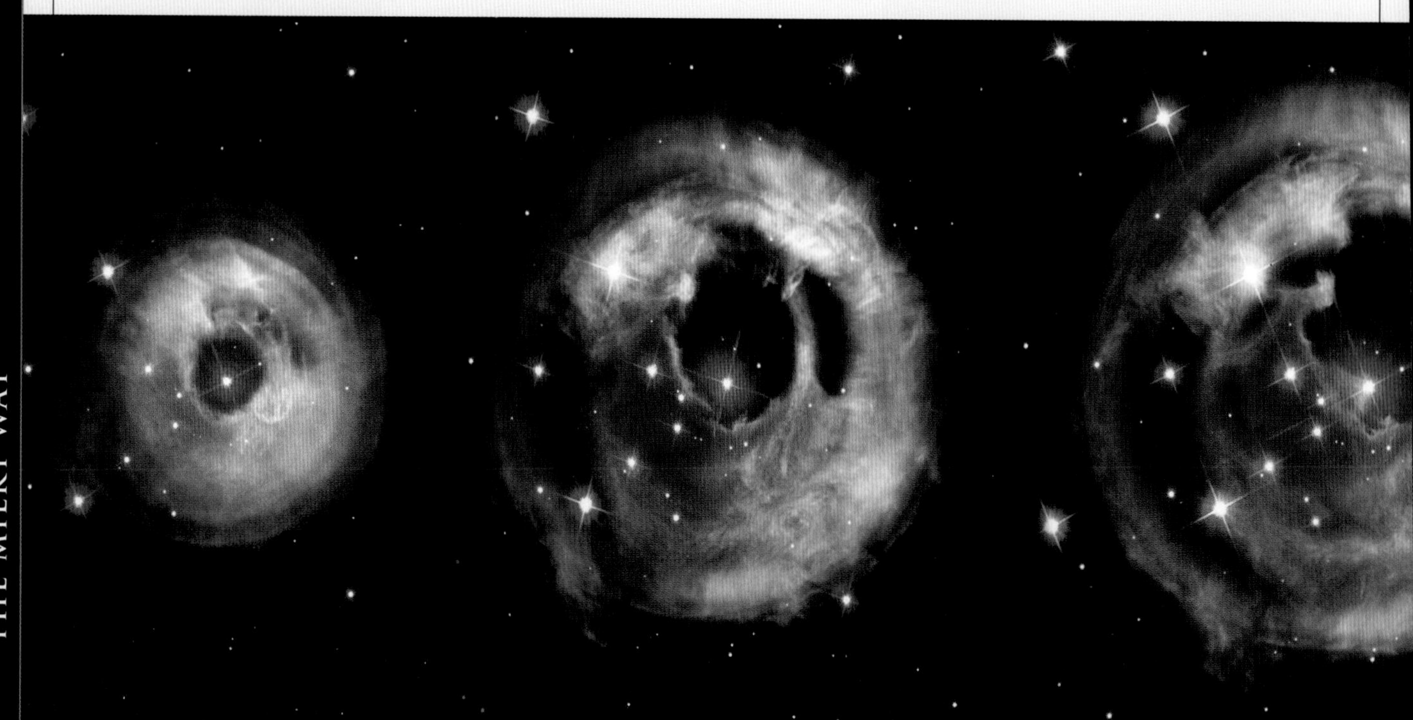

# TYPE I SUPERNOVAE

As in a nova (opposite), the source of a type I supernova is a binary system consisting of a giant star and a white dwarf. In type I supernovae, rather than triggering a nova, the material transfer onto the white dwarf continues to increase the mass of the star until it collapses and then explodes, destroying itself. The class of type I supernovae is subdivided depending on which chemical elements are present in the supernova's spectrum. In type Ia supernovae, the core of the white dwarf reaches a critical density, triggering the fusion of carbon and oxygen. This fusion is unconstrained and results in a massive explosion, with an associated leap in luminosity and the ejection of matter into interstellar space. According to theory, all type Ia supernovae have identical luminosities. This means that the distance to a supernova of this type can be determined by comparing its intrinsic luminosity with its apparent brightness.

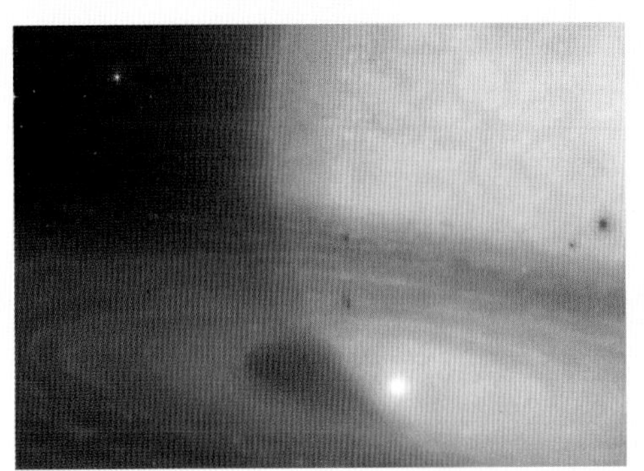

**POWERFUL SUPERNOVA**
This white-dwarf star pulls gas from a larger companion. Its mass rises until it can no longer support itself and it collapses in a huge explosion.

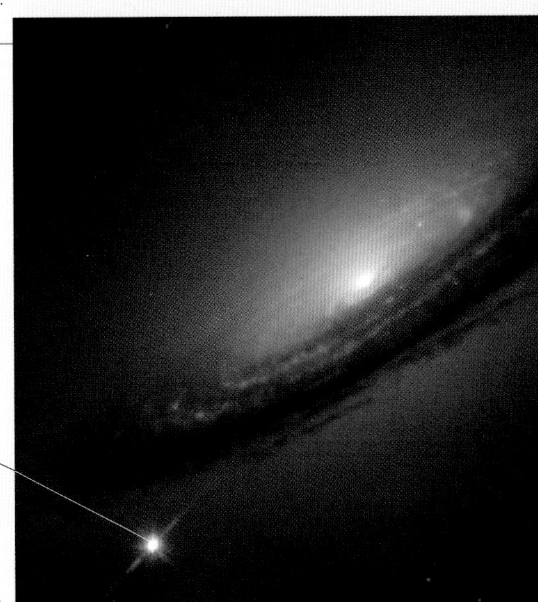


**DISTANT SUPERNOVA**
Like other type Ia supernovae, 1994D, seen in the outskirts of the distant galaxy NGC 4526, has an intrinsic brightness that allows its distance to be known.

# BIZARRE VARIABLES

Many variable stars exhibit magnitude variations that are regular, and are easily explained by eclipsing or by a pulsation mechanism occurring in a star's outer layers. However, there are other variable stars that seem to defy explanation. One example is Epsilon (ε) Aurigae or Almaaz, a supergiant with eclipses that last for two years, far longer than expected for a normal eclipsing system. As Almaaz is itself huge, whatever is eclipsing it must be even larger, but in the absence of decisive observations astronomers can only theorize. One theory is that there is an unseen companion star or stars surrounded by a large dust ring, and it is the extended dust ring that eclipses Almaaz. The answer will most likely not be available until 2010, when the next eclipse is due and new instruments will be able to study the system. Another bizarre variable is R Coronae Borealis (R CrB). This star can suddenly drop eight magnitudes, a large range that cannot be explained by physical changes within the star's structure (see p.283). The variation cannot be due to an eclipse, as the drop in magnitude is irregular and not periodic. Some astronomers have suggested that an orbiting dust cloud is responsible, but the more popular theory is that R CrB is ejecting material from its surface, which blocks the light from the star before being blown away. Although the majority of variable stars are well understood, even to the extent that they can be used as reliable distance indicators, there are many individual stars that require further study before they reveal their secrets.

**MYSTERIOUS STAR**
One of the strangest stars known to astronomers, Epsilon Aurigae is a supergiant star that is being eclipsed by something even bigger than itself. One theory is that it is being orbited by a large dusty disc surrounding a companion.

# VARIABLE STARS

More than 30,000 variable stars are known within the Milky Way, and it is likely that there are many thousands more waiting to be discovered. Variable star research is a fundamental and vital branch of astronomy, as it provides information about stellar masses, temperatures, structure, and evolution. Variable stars often have periods ranging from years to decades, and professional astronomers do not have the resources to continuously monitor such stars. Consequently amateur astronomers play a key role within this field, submitting thousands of observations into an international database.

**IRREGULAR VARIABLE**
The brightness of the variable star, Gamma (γ) Cassiopeiae, changes irregularly and unpredictably by up to two magnitudes.

ROTATING VARIABLE

## Procyon

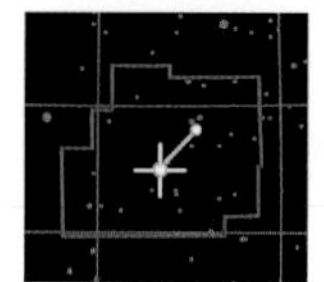

CANIS MINOR

**DISTANCE FROM SUN** 11.4 light-years
**MAGNITUDE** 0.34
**SPECTRAL TYPE** F5

Procyon is only about seven times as luminous as the Sun, but appears bright in the sky due to its proximity to Earth. Procyon has a companion, Procyon B, a white-dwarf star about the same size as Earth. Historical records suggest that Procyon may have changed its colour over the millennia. Procyon also shows small changes in magnitude, caused by surface features, such as star spots, passing in and out of view, as the star rotates. This type of variation classifies Procyon as a BY Draconis-type variable. In addition to surface changes, the tiny, brighter, companion also increases the apparent brightness of Procyon when it passes in front of the star as seen from Earth.

**CONSPICUOUS VARIABLE**
Seven times more luminous than the Sun, Procyon is the eighth-brightest star in the night sky.

ERUPTIVE VARIABLE

## U Geminorum

GEMINI

**DISTANCE FROM SUN** 250 light-years
**MAGNITUDE** 8.8
**SPECTRAL TYPE** B

The prototype cataclysmic variable star, U Geminorum, is a close binary system, consisting of a red main-sequence star orbiting and eclipsing a white dwarf and its accretion disc. Material falls from the main-sequence star onto the disc, causing localized heating and rapid increases in brightness of three to five magnitudes.

**PROTOTYPE**
U Geminorum lends its name to a type of irregular variable star that displays sudden increases in brightness.

ECLIPSING BINARY

## Lambda Tauri

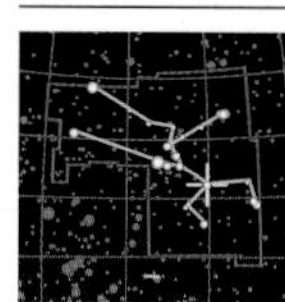

TAURUS

**DISTANCE FROM SUN** 370 light-years
**MAGNITUDE** 3.4
**SPECTRAL TYPE** B3

Lambda (λ) Tauri is an Algol-type eclipsing binary (see p.270). The primary eclipse occurs every 3.95 days, during which the brightness drops by half a magnitude – noticeable to the naked eye. The two stars involved are a bright spectral-type-B3 dwarf and a giant of spectral type A4. The eclipses are partial eclipses, as only a part of each star is hidden by the other as it orbits. The stars are very close to each other, separated by only about 15 million km (9 million miles). Such proximity leads to tidal distortions in the stars, and perhaps mass exchange, leading to magnitude variations even when they are not eclipsing.

ECLIPSING BINARY

## Eta Geminorum

GEMINI

**DISTANCE FROM SUN** 349 light-years
**MAGNITUDE** 3.3
**SPECTRAL TYPE** M3

Commonly known as Propus, Eta (η) Geminorum is a red giant star, and its red colouring is very apparent through binoculars. It is a semi-regular variable star that has a 0.6-magnitude variation – ranging between magnitudes 3.3 and 3.9 – over 234 days. Propus is also a spectroscopic eclipsing binary, having a cool spectral-type-B companion star orbiting it with a period of 8.2 years and at a distance of about 1 billion km (625 million miles). Propus is therefore eclipsed every 8.2 years and is a target for amateur variable-star observers. A second star orbits at a greater distance, with a period of 700 years, but with no eclipsing. Although Propus is a cool star, with a temperature of about 3,600°C (6,500°F), it is more than 2,000 times as luminous as the Sun. Its temperature and luminosity suggest that it is 130 times larger than the Sun. This agrees with optical measurements, but there is uncertainty – the star has different sizes when observed at different wavelengths, due to dark bands of titanium oxide in its spectrum. This uncertainty in measuring Propus's size is typical of large, cool stars. Propus is evolving – observations by amateur astronomers show that its average brightness has increased by 0.1 magnitude over the last decade. It has a dead helium core and is slowly entering a new phase: it is destined to become a Mira variable (see opposite).

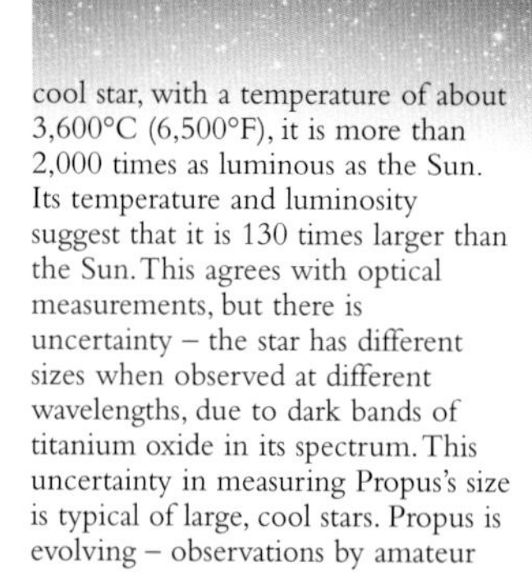

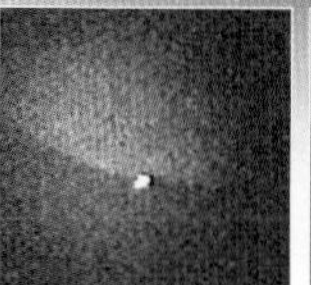

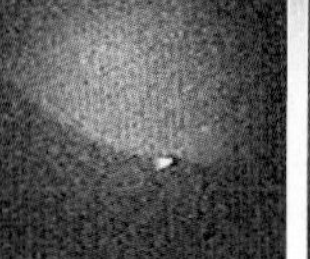

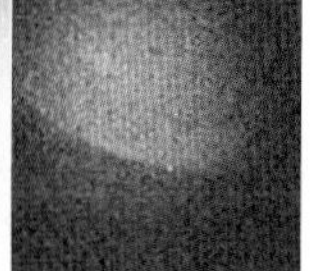

**ETA GEMINORUM OCCULTED**
In an event that takes less than one-thirtieth of a second (above), Propus is occulted by the moon (see p.65). In an optical image (top) Propus is pictured alongside the much more distant supernova remnant IC 443.

ECLIPSING BINARY

## Alpha Herculis

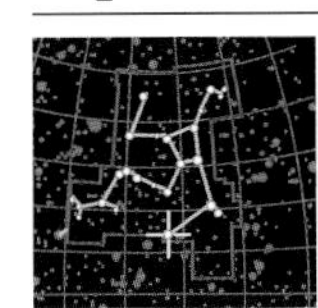

HERCULES

**DISTANCE FROM SUN** 382 light-years

**MAGNITUDE** 3

**SPECTRAL TYPE** M5

Alpha (α) Herculis, or Ras Algethi, is a cool red supergiant star that varies in brightness by almost one magnitude over a period of about 128 days. It is a complex star system with a much smaller companion that is itself a binary, consisting of a giant and a Sun-like star. There is a strong wind of stellar material blowing from the star, which reaches and engulfs its companions. Alpha Herculis is wider than the orbit of Mars. The outer atmosphere of the supergiant is slowly being removed, and the star will eventually become a white dwarf.

**GREAT CONTRAST** Although they are not particularly bright, the great contrast in size and colour of the stars that make up Alpha Herculis allow them to be separated easily through a telescope.

PULSATING VARIABLE

## Mira

CETUS

**DISTANCE FROM SUN** 418 light-years

**MAGNITUDE** 3

**SPECTRAL TYPE** M7

Omicron (ο) Ceti, better known as Mira, is among the best known of all variable stars. At its brightest, it reaches second magnitude, and at its faintest it drops to tenth – far too faint for the naked eye. It undergoes this variation with a period of 330 days. Therefore an observer can find Mira when it is at its brightest and over a period of time watch it completely disappear. Although Mira is one of the coolest stars visible in the sky, with a temperature of just 2,000°C (3,600°F), it is at least 15,000 times more luminous than the Sun. Internal changes in the star have left it so distended that the Hubble Space Telescope has revealed that it is not perfectly spherical (right). The variation in Mira's magnitude is caused by pulsations that cause temperature changes and therefore changes in the star's luminosity. Furthermore, Mira is shedding material from its outer layers in the form of a stellar wind. In the future, Mira will lose its outer structure and be left as a small white dwarf. In this way, Mira represents the future of the Sun.

EXPLORING SPACE

### WONDERFUL MIRA

When Dutch astronomer David Fabricius discovered Mira in 1596, it was the first long-period variable star to be recognized. In 1642 Johannes Hevelius named the star Mira, meaning "wonderful". It has become the most famous long-period pulsating variable in the sky, and one of the most popular stars for amateur astronomers. The American Association of Variable Star Observers has received more than 50,000 observations of Mira by over 1,600 observers.

**THE ORIGINAL MIRA** Easily recognized in the night sky, Mira lends its name to a type of long-period variable, of which thousands are known.

**MIRA AND COMPANION** In 2003, scientists used the Hubble Space Telescope to separate Mira (right) from its hot companion star (left) for the first time.

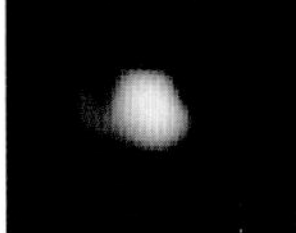

**DISTORTED SHAPE** Enhancement of Hubble's Mira images reveals the star's asymmetrical atmosphere in visible (left) and ultraviolet light (right).

IRREGULAR VARIABLE

## Gamma Cassiopeiae

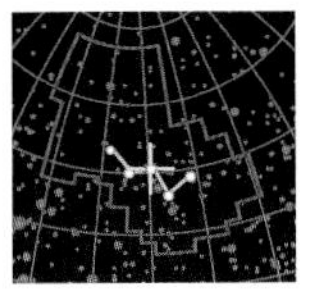

CASSIOPEIA

**DISTANCE FROM SUN** 613 light-years

**MAGNITUDE** 2.4

**SPECTRAL TYPE** B0

A hot blue star with a surface temperature of 25,000°C (45,000°F), Gamma (γ) Cassiopeiae is about 70,000 times more luminous than the Sun. It is a variable star with unpredictable changes in magnitude. Astronomers have observed it as bright as 1st magnitude and as faint as 3rd magnitude. It may have been fainter in ancient times, which might explain its lack of a common name. Gamma Cassiopeiae is a Be star (see panel, right), rotating at more than 1 million kph (625,000mph) at its equator and shedding material from its surface. The ejected material forms a surrounding disc, and it is the disc that makes varying and unpredictable emissions. Gamma Cassiopeiae may also be transferring material to an undiscovered dense companion star.

**NAMELESS STAR** Pictured here with the red-coloured emission nebula IC 63, Gamma Cassiopeiae is among the most prominent stars in the sky that carries no common name.

EXPLORING SPACE

### THE FIRST "BE" STAR

When in 1866 Father Angelo Secchi, director of the Vatican Observatory and scientific advisor to Pope Pius IX, studied the spectrum of Gamma Cassiopeiae, he discovered that the star emitted light at particular wavelengths associated with hydrogen emission (see p.33). He is therefore credited with the discovery of the first Be star – a star of spectral type B but with "e" for emission. Be stars are characterized by their high rotation speeds, high surface temperatures, and strong stellar winds focused into equatorial discs.

**CENTRAL STAR** Gamma Cassiopeiae, the brightest star in this image, is the central star in the distinctive "W" of Cassiopeia (see p.341).

PULSATING VARIABLE

## W Virginis

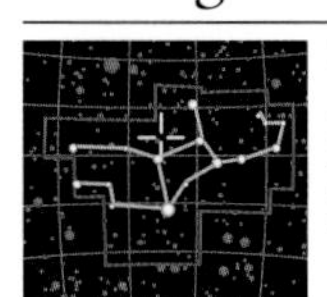

**DISTANCE FROM SUN** 10,000 light-years

**MAGNITUDE** 9.6

**SPECTRAL TYPE** F0

**VIRGO**

W Virginis lends its name to a class of variable stars that are similar to Cepheid variables (see p.278) and are also known as Population II Cepheids. W Virginis is a pulsating yellow giant star. The outer layers of its atmosphere expand and contract with a period of 17.27 days. The period has lengthened over the last 100 years of observation. The pulsation causes a 1.2-magnitude variation, as the star doubles its size during the cycle. As a Population II star (see p.227), W Virginis is among the oldest stars in the Milky Way.

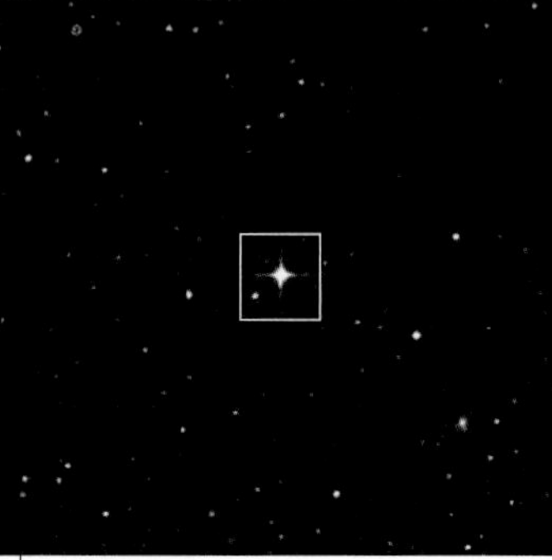

**W VIRGINIS**
W Virginis is located high above the galactic plane in the diffuse halo of old stars that surrounds the Milky Way (see p.226–29). Like other W Virginis variables, it is an old Population II star, on average lower in mass and magnitude than a Cepheid variable.

PULSATING VARIABLE

## RR Lyrae

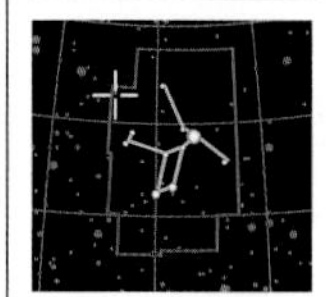

**DISTANCE FROM SUN** 744 light-years

**MAGNITUDE** 7.1

**SPECTRAL TYPE** F5

**LYRA**

RR Lyrae is the brightest member of the class of variables that takes its name. RR Lyrae stars are similar to Cepheid variables (see p.278), but are less luminous and tend to have shorter periods – ranging from about 5 hours to just over a day. RR Lyrae's period is 0.567 days, and its magnitude varies between 7.06 and 8.12. By comparing the luminosity of RR Lyrae variables with their apparent magnitude, a good distance determination can be made. In this way, RR Lyrae variable stars are important tools for calculating astronomical distances.

**BRIGHT VARIABLE**
RR Lyrae has an average luminosity 40 times that of the sun and a surface temperature of about 6,700°C (12,000°F). RR Lyrae stars are often found in globular clusters, and they are sometimes referred to as cluster variables.

PULSATING VARIABLE

## Delta Cephei

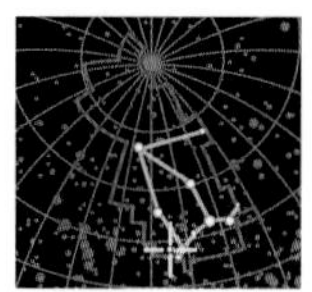

**DISTANCE FROM SUN** 982 light-years

**MAGNITUDE** 4

**SPECTRAL TYPE** F5

**CEPHEUS**

Delta (δ) Cephei is the prototype of the Cepheid class of variable stars (see p.278), and to astronomers it is one of the most famous stars in the sky. Its magnitude variation, from 3.48 to 4.37, is visible to the naked eye, and its short period of 5 days, 8 hours, and 37.5 minutes makes it a popular target for amateur observers. Its position in the sky makes it easy to find, and it is close to two comparison stars with magnitudes at the ends of Delta Cephei's range. Delta Cephei is a supergiant with a spectral type that varies between F5 and G2.

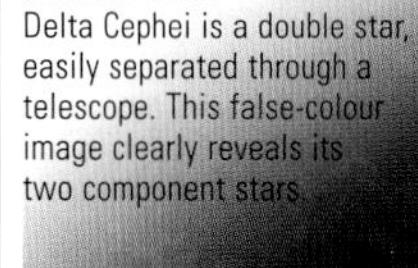

**DOUBLE STAR**
Delta Cephei is a double star, easily separated through a telescope. This false-colour image clearly reveals its two component stars.

EXPLORING SPACE

### THE CEPHEID PROTOTYPE

In 1921, Henrietta Leavitt (1868–1921), an astronomer based at the Harvard Observatory, discovered a strong link between the period and luminosity of a group of stars later known as Cepheid variables (see p.278), of which Delta Cephei was the prototype. This correlation provided astronomers with a new way of measuring distances in space. In 1923, Edwin Hubble used it to prove that the Andromeda Galaxy is situated outside the Milky Way. Since then, Cepheids have provided more useful information about the Universe than any other star type.

**HENRIETTA LEAVITT**

PULSATING VARIABLE

## Zeta Geminorum

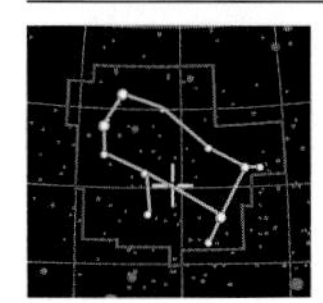

**DISTANCE FROM SUN** 1,168 light-years

**MAGNITUDE** 4

**SPECTRAL TYPE** G0

**GEMINI**

Also known as Mekbuda, Zeta (ζ) Geminorum is a yellow supergiant, about 3,000 times as luminous as the Sun. It is one of the easiest Cepheid variable stars in the night sky to observe (see p.278). Zeta Geminorum, like all Cepheids, is unstable and pulsates, changing its temperature, size, and spectral type. It has a period of 10.2 days and a magnitude that varies from 3.6 to 4.2. Its period is shortening at the rate of about three seconds per year. Zeta Geminorum is also a binary star, with a faint, magnitude-10.5 companion.

PULSATING VARIABLE

## Eta Aquilae

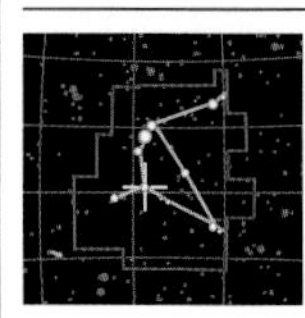

**DISTANCE FROM SUN** 1,173 light-years

**MAGNITUDE** 3.9

**SPECTRAL TYPE** F6

**AQUILA**

Eta (η) Aquilae is a yellow supergiant star, with a luminosity 3,000 times that of the Sun. It is one of the brightest Cepheid variables in the night sky (see p.278), and also one of the first to be discovered. Eta Aquilae varies in magnitude from 3.5 to 4.3 over a period of 7.176 days. Such a brightness variation is easily detectable with the naked eye. The magnitude range, by coincidence, is the same as the prototype of the class, Delta Cephei (above). Over this period, Eta Aquilae also varies in spectral type between G2 and F6.

NOVA

## T Coronae Borealis

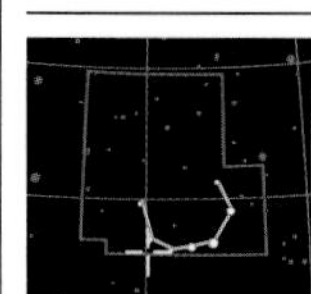

**DISTANCE FROM SUN** 2,025 light-years

**MAGNITUDE** 11

**SPECTRAL TYPE** M3

**CORONA BOREALIS**

T Coronae Borealis (T CrB), also known as the Blaze Star, is a recurrent nova (see p.278). It has displayed two major outbursts, one witnessed in 1866, the other in 1946. T CrB's usual apparent magnitude is 10.8, but during outbursts it has reached 2nd or 3rd magnitude. T CrB is a spectroscopic binary, consisting of a red giant of spectral type M3, and a smaller blue-white dwarf. T CrB is usually about 50 times as luminous as the Sun, but during outbursts it becomes more than 200,000 times as luminous. In between the outbursts, stellar dust and gas from the outer layers of the red giant are drawn onto the white dwarf. Eventually the total mass of the white dwarf reaches a critical level, causing the outer layers of the white dwarf to explode violently. After the explosion, the two stars return to normality, to repeat the process many years later.

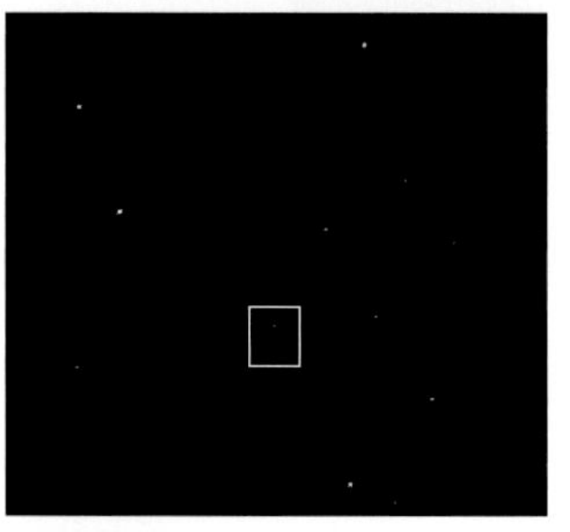

**BLAZE STAR**
Although T Coronae Borealis cannot usually be seen without a telescope, during eruptions it has "blazed" bright enough to be seen with the naked eye.

PULSATING VARIABLE

## Mu Cephei

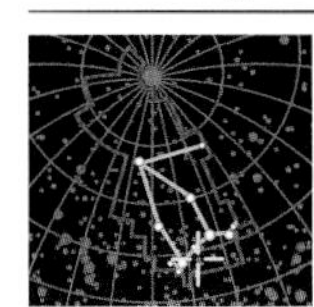

CEPHEUS

**DISTANCE FROM SUN** 5,258 light-years

**MAGNITUDE** 4

**SPECTRAL TYPE** M2

Mu (μ) Cephei, or the Garnet Star, is also known as Herschel's Garnet Star, after the pioneering German-born astronomer William Herschel (see p.90), who first described its distinctive red colour and noted its resemblance to the precious stone garnet. Mu Cephei is one of the most luminous stars in the Milky Way, outshining the Sun by a factor of more than 200,000. A red supergiant, it is also one of the largest stars that can be seen with the naked eye. Its great size means that if placed in the Sun's position at the centre of the Solar System, its outer layers would fall between Jupiter and Saturn. As with most large supergiants, Mu Cephei is an unstable star, expanding and contracting in diameter with a corresponding variation in magnitude. It is classed as a semi-regular supergiant variable with a spectral type of M2 and a magnitude varying between 3.43 and 5.1. It has two periods of variation (730 and 4,400 days) overlaid on one another. The pulsations of Mu Cephei, caused by internal absorption and release of energy, have thrown off the outer layers of the star's atmosphere, creating concentric shells of dust and gas around it. Observations have also shown that Mu Cephei is surrounded by a sphere of water vapour. Mu Cephei probably started its life as a star of around 20 solar masses. Typically for such a high-mass star, it has evolved very rapidly, and we are seeing it as it hurtles headlong towards the end of its short life. One day soon (on an astronomical timescale), Mu Cephei will erupt in a cataclysmic supernova, after which only the core will remain, ending its days as a neutron star or black hole.

**VARIABLE JEWEL**
Known for its distinctive colour, Mu Cephei, or the Garnet Star, is the bright reddish-orange star at the top left of the image. It is pictured above the red emission nebula IC 1396 (see p.241).

NOVA

## RS Ophiuchi

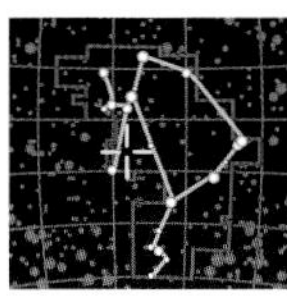

OPHIUCHUS

**DISTANCE FROM SUN** 2,000–5,000 light-years

**MAGNITUDE** 12.5

**SPECTRAL TYPE** M2

RS Ophiuchi is a recurrent nova (see p.278), having been witnessed erupting in 1898, 1933, 1958, 1967, and, most recently, in 1985. During its periods of normality, it shines at magnitude 12.5, but during outbursts it has reached 4th magnitude. While RS Ophiuchi is usually invisible to the naked eye, during outbursts it can be seen in the night sky without a telescope. RS Ophiuchi is classed as a cataclysmic variable – a binary system consisting of a giant star shedding material and a dwarf companion receiving the material. Eventually a thermonuclear explosion is triggered on the surface of the dwarf, resulting in the ejection of a shell and an increase in brightness. RS Ophiuchi is constantly monitored by amateur astronomers, and the American Association of Variable Star Observers has more than 30,000 observations in its database.

IRREGULAR VARIABLE

## R Coronae Borealis

CORONA BOREALIS

**DISTANCE FROM SUN** 6,037 light-years

**MAGNITUDE** 5.9

**SPECTRAL TYPE** G0

The prototype of a class of variable stars, R Corona Borealis (R CrB) drops in magnitude from 5.9 to 14.4 at irregular intervals. There are two theories for this variation. One is that an orbiting dust cloud obscures R CrB when it passes in front of the star. The other is that R CrB ejects material, which obscures the light the star emits, before being blown away.

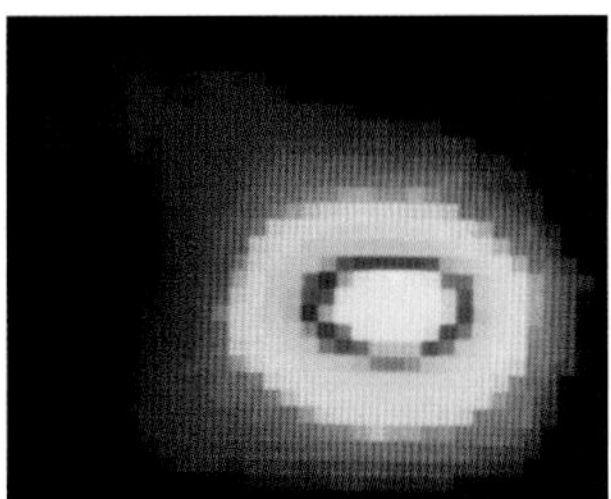

FALSE-COLOUR INFRARED IMAGE

NOVA

## Nova Cygni 1992

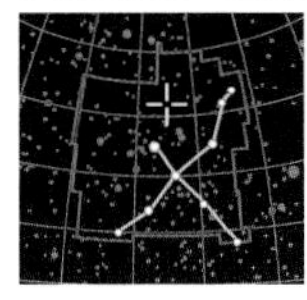

CYGNUS

**DISTANCE FROM SUN** 10,430 light-years

**MAGNITUDE** 4.3

**SPECTRAL TYPE** Q

Nova Cygni 1992, a cataclysmic binary, was discovered on the night of 18–19 February 1992, shining with a magnitude of 7.2 at a location where there should have been no such star. The discoverer, Peter Collins (see p.74), alerted astronomical authorities, and soon a whole range of instruments, both ground-based and space-borne (including the Voyager spacecraft), were observing it at a variety of wavelengths. Over the next few days, the nova continued to brighten to magnitude 4.3, making Nova Cygni 1992 not only the first nova to be observed so extensively, but also the first to be thoroughly observed before it had reached its peak. The nova eruption was the result of material falling from one star onto a white dwarf, triggering an explosion and the ejection of a shell of material. The Hubble Space Telescope observed the system in 1993, detecting the ring thrown out by the binary system and also an unusual bar-like structure across the middle of the ring, the origin of which is unknown.

**BRIGHT NOVA**
Nova Cygni 1992 was one of the brightest nova to be witnessed erupting in recent history. Targeted by some of the most powerful telescopes in the world, it could be seen, at its brightest, with the naked eye.

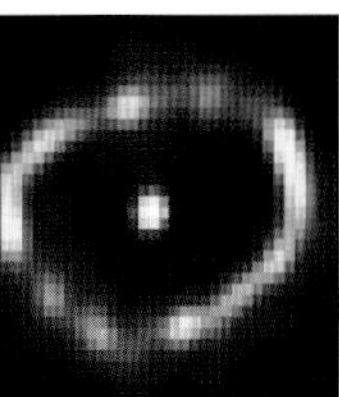

**HOT BUBBLE**
This Hubble Space Telescope photograph reveals the irregularly shaped bubble of hot stellar material blasted into space by the eruption of Nova Cygni 1992.

# STAR CLUSTERS

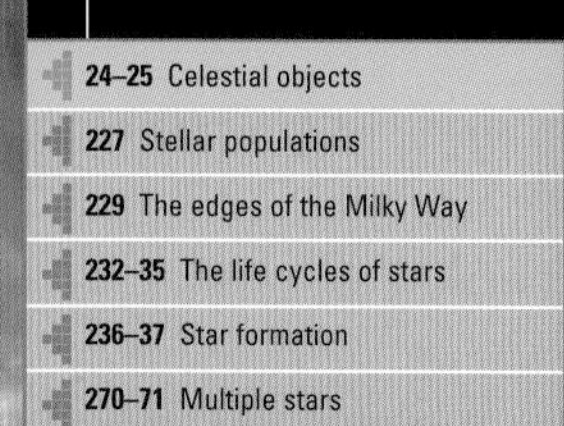


24–25 Celestial objects
227 Stellar populations
229 The edges of the Milky Way
232–35 The life cycles of stars
236–37 Star formation
270–71 Multiple stars


ALTHOUGH THE STARS in our night sky appear to live out their lives in isolation, many millions of stars reside in groups called open and globular clusters. Open clusters are young and often the site of new star creation, whereas globulars are ancient, dense cities of stars, some of which contain as many stars as a small galaxy.

## OPEN STAR CLUSTERS

Open clusters are made up of "sibling" stars of similar age formed from the same nebulous clouds of interstellar gas and dust. This often results in stars within an open cluster having the same chemical composition. However, an open cluster's stars can exhibit a wide range of masses, due to variations within the original nebula and other influences during their formation. Open clusters reside within the galactic plane, and often remain associated with the nebulous clouds from which they were produced. Open clusters do not hold on to their stars for long – as they orbit the centre of the galaxy, they lose their members over a period of hundreds of millions of years. More than 1,100 open clusters have been discovered within the Milky Way, representing perhaps only 1 per cent of the total population.

**YOUNG OPEN CLUSTER**
Spanning an area in the sky larger than the full Moon, M39 is a large but sparsely populated open cluster. It contains about 30 loosely bound stars, each around 300 million years old, and therefore much younger than the Sun.

**LARGEST CLUSTER**
The largest globular cluster in the Milky Way, Omega (ω) Centauri probably contains more than 10 million stars. This makes it larger than some small galaxies.

# GLOBULAR CLUSTERS

A globular cluster is a massive group of stars bound together by gravity within a spherical volume. Globular clusters can contain between 10,000 and several million stars, all within an area often less than 200 light-years across. As in open clusters, the stars within a globular cluster all have the same origin, and thus similar ages and chemical compositions. Spectroscopic studies of the starlight from globulars reveal that their stars are very old – older than most of the stars currently within the disc of the Milky Way. Analysis of their properties also reveals that they are about the same age, implying that they all formed together, over a short period of time. Estimates of their ages vary, but they are thought to be over 10 billion years old. More than 150 globular clusters have been discovered in the Milky Way. Although a few are found in its central bulge, most are located in the halo. The chemistry of globular clusters shows that they represent the remnants of the early stages of the formation of the Milky Way, and perhaps formed even before the Milky Way had a disc. Four globulars may have originally been part of a dwarf galaxy that has been absorbed into the Milky Way. Globular clusters are made up of Population II stars (see p.227), which have their own independent orbits. These orbits are highly elliptical, and can take the globulars out to distances of hundreds of thousands of light-years from the centre of the Milky Way. Globular clusters are not unique to the Milky Way, and some galaxies have more globular clusters than our own.

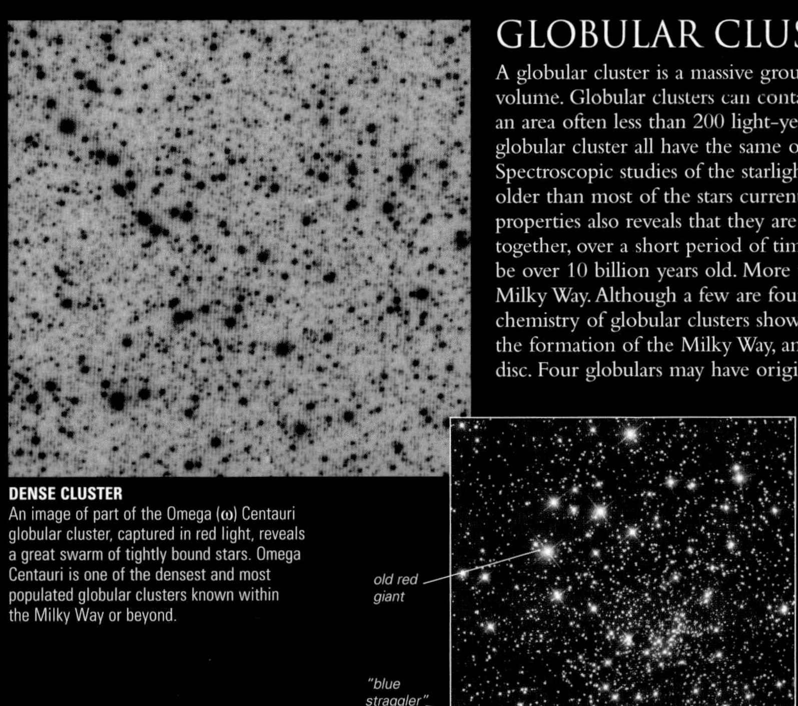


**DENSE CLUSTER**
An image of part of the Omega (ω) Centauri globular cluster, captured in red light, reveals a great swarm of tightly bound stars. Omega Centauri is one of the densest and most populated globular clusters known within the Milky Way or beyond.

**BLUE STRAGGLERS**
In the central region of the globular cluster NGC 6397, among its old red stars, are seen a few young blue stars. These stars, called "blue stragglers", are thought to have been created by densely packed stars colliding.

# CLUSTER EVOLUTION

Star clusters, whether open or globular, are not static through time. Over millions of years, the clusters change physically and the stars within them age and die. However, there are major differences between the evolution of globular clusters and open clusters. An open cluster starts its life with a set of stars of similar chemical composition and age. Over hundreds of millions of years, it loses its members, either due to death of the stars or losing them to the gravitational tugs of other stars within the Milky Way. However, an open cluster can continue to manufacture stars from the original nebulous cloud from which it formed. Because of this, open clusters often contain stars of different ages at various stages of evolution. A globular cluster is more tightly bound, and less likely to lose its stars. It also spends most of its time away from the disc of the Galaxy, avoiding interactions. In this way its structure is preserved for thousands of millions of years – far longer than open clusters. Similarly, once a globular cluster has formed, the original gas and dust is ejected, and the cluster is then unable to form new stars. As the stars within a globular cluster age and die, so the cluster itself ages and dies.

**CLUSTER DISTRIBUTION**
The difference in the distributions of open and globular clusters within the Milky Way reflects their differences in age and orbit. Open clusters, formed from relatively young, Population I stars are located within the Milky Way's rotating disc. Globular clusters, made up of Population II stars, have independent orbits mostly located out in the Milky Way's halo.

**EVOLVED CLUSTER**
At about 1 billion years old, NGC 2266 is a relatively old and well-evolved open cluster. Many of its stars, clearly seen here, have reached the red-giant stage of their life cycle, while young blue stars are also present.

# STAR CLUSTERS

More than 1,000 open clusters have been catalogued in the Milky Way. About half contain fewer than 100 stars, but the largest have more than 1,000. Open clusters are asymmetrical and range in size from 5 to 75 light-years across. By contrast, globular clusters contain up to 1,000,000 stars, spread symmetrically across several hundred light-years. Only about 150 globular clusters are known in the Milky Way and, unlike open clusters, which are found mainly in the galaxy's spiral arms, most are scattered around the periphery.

**MASSIVE CLUSTER**
Omega Centauri is a prime example of a globular star cluster. It contains more than 10 million old stars and has a mass of 5 million solar masses.

OPEN CLUSTER

## Hyades

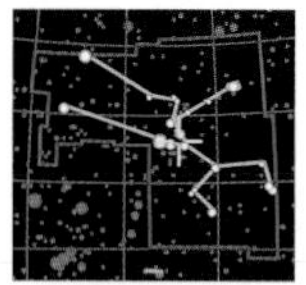

**TAURUS**

**CATALOGUE NUMBER** MEL 25

**DISTANCE FROM SUN** 150 light-years

**MAGNITUDE** 0.5

The Hyades cluster is one of the closest open clusters to Earth and has been recognized since ancient times. The brightest of the cluster's 200 stars form a V-shape in the sky, clearly visible to the naked eye. The cluster's central group is about 10 light-years in diameter, and its outlying members span up to 80 light-years. Most of the stars in this cluster are of spectral classes G and K (see pp.230–31) and are average in size, with temperatures comparable to that of the Sun. The brightest star in the field of the Hyades, the red giant Aldebaran (see p.252), is not a member of the cluster and is much closer to Earth. The cluster's stars all move in a common direction, towards a point east of Betelgeuse in Orion (see p.252). Studies of the movement of the stars of the Hyades show that they have a common origin with the Beehive Cluster (see below). The Hyades cluster is thought to be about 790 million years old, and this age matches that of the Beehive Cluster. The parallel movement of stars in the Hyades has allowed their distance to be measured, using the moving cluster method for stellar distances (see pp.230–31).

**PROMINENT CLUSTER**
First recorded by Homer in about 750 BC, the Hyades is one of the few star clusters visible to the naked eye. Aldebaran, the bright red giant in this image, is not part of the cluster, but is 90 light-years closer to Earth.

OPEN CLUSTER

## Butterfly Cluster

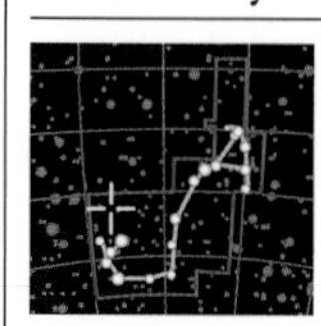

**SCORPIUS**

**CATALOGUE NUMBERS** M6, NGC 6405

**DISTANCE FROM SUN** 2,000 light-years

**MAGNITUDE** 5.3

The Butterfly Cluster, located towards the centre of the Milky Way, is about 12 light-years across and has an estimated age of 100 million years. In the night sky, the cluster occupies an area the size of the full Moon, and, to some, it resembles the shape of a butterfly. The cluster is made up of about 80 stars, most of them very hot, blue main-sequence stars with spectral types B4 and B5 (see pp.230–31). The brightest star in the cluster, BM Scorpii, is an orange supergiant star that is also a semi-regular variable (see pp.278–79). At its brightest, this star is visible to the naked eye; at its faintest, binoculars are needed. The Butterfly Cluster displays a striking contrast between the blue main-sequence stars and the orange supergiant.

**SKY SPECTACLE**
The Butterfly Cluster is one of the largest and brightest open star clusters in the Milky Way. It can best be seen with binoculars in a dark sky and can be located within the constellation Scorpius.

OPEN CLUSTER

## Beehive Cluster

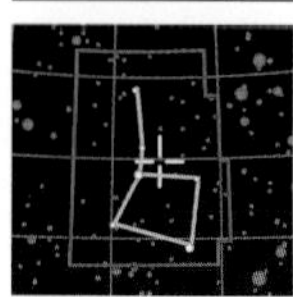

**CANCER**

**CATALOGUE NUMBER** M44

**DISTANCE FROM SUN** 577 light-years

**MAGNITUDE** 3.7

The Beehive Cluster, also known as Praesepe, is easily visible to the naked eye. The cluster contains over 350 stars, spread across 10 light-years, but most of them can be seen only with a large telescope. It is thought to be about 730 million years old. Age, distance, and motion measurements suggest that the Beehive Cluster most likely originated in the same star-forming nebula as the Hyades (above).

**CELESTIAL BEEHIVE**

OPEN CLUSTER

## M93

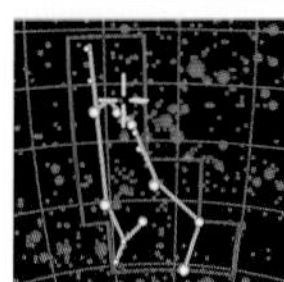

**PUPPIS**

**CATALOGUE NUMBERS** M93, NGC 2447

**DISTANCE FROM SUN** 3,600 light-years

**MAGNITUDE** 6

M93 is a bright open cluster and, at about 25 light-years across, is relatively small. It lies in the southern sky, close to the galactic equator. The cluster consists of about 80 stars, but only a few of the stars, blue giants of spectral type B9 (see pp.230–31), provide most of the cluster's light. At about 100 million years old, M93 is young in astronomical terms.

**SOUTHERN CLUSTER**

OPEN CLUSTER

## M52

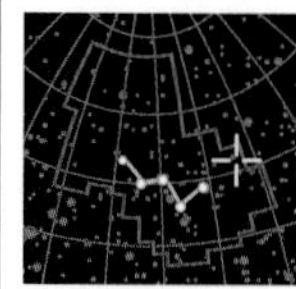

**CASSIOPEIA**

**CATALOGUE NUMBERS** M52, NGC 7654

**DISTANCE FROM SUN** 3,000–7,000 light-years

**MAGNITUDE** 7.5

An open cluster of about 200 stars, M52 lies against a rich Milky Way background. It was first catalogued in 1774 by Charles Messier (see p.69). The distance to the cluster is uncertain, with estimates ranging from 3,000 to 7,000 light-years. The uncertainty is due to high interstellar absorption that affects the light from the cluster during its journey to Earth. The uncertain distance also means that the cluster's size is unknown, but mid-range estimates give a size of about 20 light-years across. The age of the cluster is calculated to be about 35 million years. The brightest stars in M52 have magnitudes of only 7.7 and 8.2, and with an overall magnitude of 7.5 the cluster is too faint to be seen with the naked eye. However, through binoculars the cluster can be viewed as a faint nebulous patch, while a small telescope reveals a rich, compressed cluster of stars.

**CLUSTER AND NEBULA**
This image, stretching more than twice the diameter of the full Moon, captures the open cluster M52 (top left) and the glowing Bubble Nebula (bottom right).

OPEN CLUSTER

# Pleiades

TAURUS

| | |
|---|---|
| CATALOGUE NUMBER | NGC 1435 |
| DISTANCE FROM SUN | 380 light-years |
| MAGNITUDE | 4.17 |

The Pleiades, also known as the Seven Sisters, is the best-known open cluster in the sky, and has been recognized since ancient times (see panel, right). The cluster is easily visible to the naked eye, but although most people can see seven stars in the Pleiades, the seventh can often be elusive. Nine stars can be seen on a very dark and clear night. The nine brightest stars are known as the father, Atlas, the mother, Pleione, and the sisters Alcyone, Maia, Asterope, Taygeta, Celaeno, Merope, and Electra. Small telescopes and binoculars reveal many more stars, and larger telescopes show that the cluster, in fact, contains hundreds of stars. The Pleiades is about 100 million years old and will remain a cluster for only another 250 million years or so, by which time it will have broken up into separate isolated stars. The stars of the Pleiades are blue giants of spectral class B (see pp.230–31) and are hotter and more luminous than the Sun. Long-exposure photography reveals that the Pleiades stars are embedded in clouds of interstellar dust. The clouds are illuminated by radiation from the stars, and they glow as reflection nebulae (see p.228). Although most gas and dust surrounding star clusters represents the material that gave birth to the stars, here the clouds are merely moving through the cluster. The clouds are travelling relative to the Pleiades at 40,000kph (25,000mph), and will eventually pass through the cluster and travel into deep space, where they will once again become dark and invisible.

**CLUSTERS IN TAURUS**
The two best-known clusters, the Pleiades (boxed) and the Hyades (opposite), both lie in Taurus. However, the Pleiades is more than 200 light-years more distant.

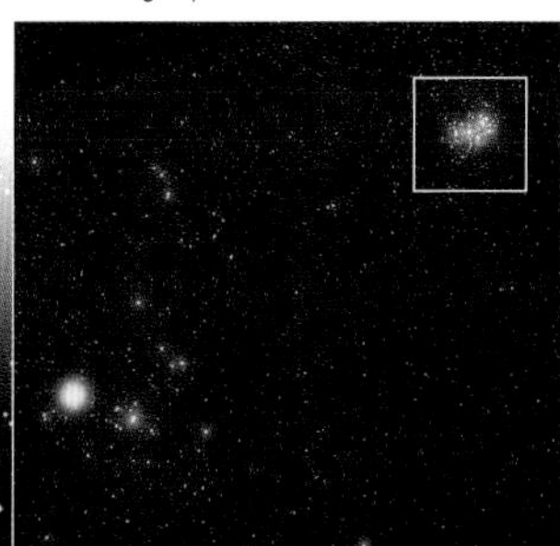

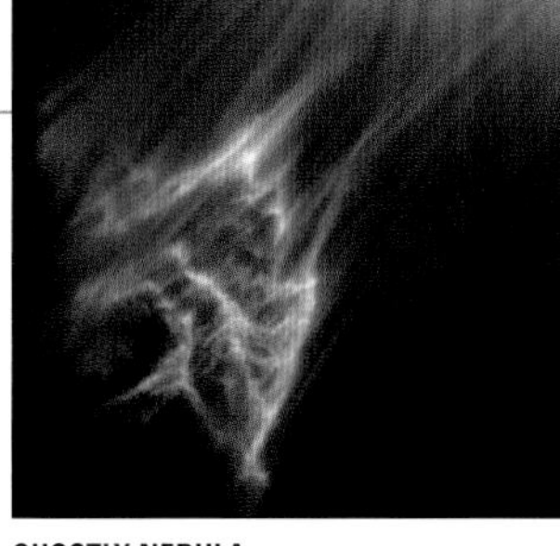

**GHOSTLY NEBULA**
This haunting image shows an interstellar cloud caught in the process of destruction by strong radiation from the star Merope in the Pleiades. The cloud is called IC 349 or Barnard's Merope Nebula.

**GLOWING NEBULOSITY**
The stars of the Pleiades are surrounded by clouds of dusty material that are reflecting the blue light of the stars. However, the stars were not produced from this material, which seems simply to be passing by.

EXPLORING SPACE

## BRONZE AGE CLUSTER

The Nebra Disc is perhaps the oldest semi-realistic depiction of the night sky. It was discovered in 1999, near the German town of Nebra, and other artefacts found nearby have allowed it to be dated to about 1600 BC. The disc depicts a crescent Moon, the full Moon, randomly placed stars, and a star cluster likely to be the Pleiades. Although its authenticity remains uncertain, the Nebra Disc may be proof that European Bronze Age cultures had a more sophisticated appreciation of the night sky than had previously been accepted.

**ANCIENT PLEIADES**
The cluster of seven gold dots (above and right of centre) has been interpreted as the Pleiades cluster as it appeared 3,600 years ago.

GLOBULAR CLUSTER

## M4

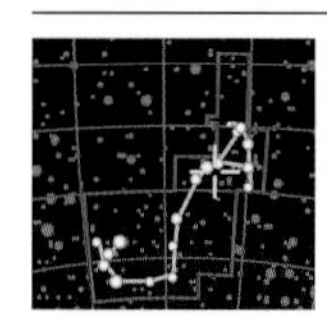

**CATALOGUE NUMBERS** M4, NGC 6121

**DISTANCE FROM SUN** 7,000 light-years

**MAGNITUDE** 7.1

**SCORPIUS**

M4 is one of the closest globular clusters to Earth and can be seen by the naked eye on a dark, clear night. The cluster has a diameter of about 70 light-years and contains more than 100,000 stars, but about half the cluster's mass resides within eight light-years of its centre. The Hubble Space Telescope has revealed a planet within M4, with about twice the mass of Jupiter, orbiting a white dwarf star. The planet is estimated to be 13 billion years old.

**DENSE CENTRE**

OPEN CLUSTER

## Jewel Box

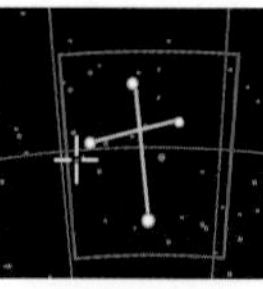

**CATALOGUE NUMBER** NGC 4755

**DISTANCE FROM SUN** 8,150 light-years

**MAGNITUDE** 4.2

**CRUX**

**GLITTERING JEWELS**

The Jewel Box, also known as the Kappa Crucis cluster, is an open cluster of about 100 stars and is about 20 light-years across. At less than 10 million years old, it is one of the youngest open clusters known. The three brightest stars are blue giants, while the fourth-brightest star is a red supergiant. The different colours are very apparent in photographs of the cluster, hence its popular name. Lying within the constellation Crux, the Jewel Box is visible only to observers in the southern hemisphere.

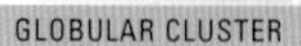

GLOBULAR CLUSTER

## 47 Tucanae

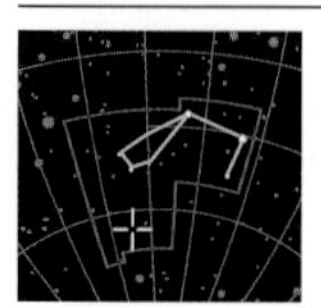

**CATALOGUE NUMBER** NGC 104

**DISTANCE FROM SUN** 13,400 light-years

**MAGNITUDE** 4.9

**TUCANA**

47 Tucanae is so named as it was originally catalogued as a star – the 47th in order of right ascension in the constellation Tucana. In reality, it is the second-largest and second-brightest globular cluster in the sky, containing several million stars – enough to make a small galaxy. These stars are spread over an area about 120 light-years across, and the cluster's central region is so crowded there is a high rate of stellar collisions. As a globular cluster ages, the stars within it also age, but 47 Tucanae is home to a number of blue stragglers – stars that are too blue and too massive to still be there if they were original members of the cluster. Astronomers have determined that it is the stellar collisions within the cluster that cause the formation of these blue stragglers.

**SOUTHERN SPECTACLE**

An optical image (top) captures 47 Tucanae and the Small Magellanic Cloud, a satellite galaxy of the Milky Way (see p.301). A close-up of 47 Tucanae (boxed) reveals one of the most spectacular globular clusters in the sky.

GLOBULAR CLUSTER

## Omega Centauri

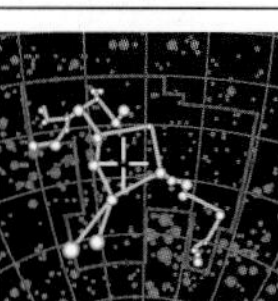

**CATALOGUE NUMBER** NGC 5139

**DISTANCE FROM SUN** 17,000 light-years

**MAGNITUDE** 5.33

**CENTAURUS**

Omega Centauri is the largest globular cluster in the Milky Way – up to ten times as massive as other globular clusters. Containing more than 10 million stars and having a width of 150 light-years, Omega Centauri is as massive as some small galaxies. To the naked eye, it appears as a fuzzy star, but a small telescope starts to resolve its individual stars. Studies of the cluster's stellar population have revealed that Omega Centauri is one of the oldest objects in the Milky Way – almost as old as the Universe itself – and that there have been several episodes of star formation within the cluster. This is unusual for a globular cluster, and one explanation for this is that Omega Centauri may once have been a dwarf galaxy that collided with our own. It would have had about 1,000 times its current mass, but the Milky Way would have ripped it apart, leaving Omega Centauri as the remnant core.

**GIGANTIC GLOBULAR**

Easily the biggest of all known globular clusters in the Milky Way, Omega Centauri has a mass of more than 5 million solar masses. The stars in this globular cluster are generally older, redder, and less massive than the Sun.

GLOBULAR CLUSTER

## NGC 3201

**CATALOGUE NUMBER** NGC 3201

**DISTANCE FROM SUN** 15,000 light-years

**MAGNITUDE** 8.2

**VELA**

The globular cluster NGC 3201 contains many bright red giant stars, which give the cluster an overall reddish appearance. The cluster lies close to the galactic plane, and so its appearance is further reddened by interstellar absorption. With a visual magnitude of only 8.2, the cluster is too faint to be seen with the naked eye. NGC 3201 is less condensed than most globular clusters, and several observers have suggested that some of the stars appear in short, curved rays, like jets of water from a fountain.

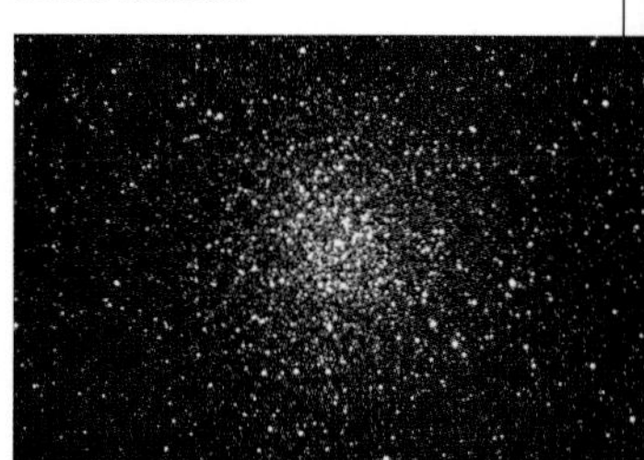

**RED-TINGED CLUSTER**

GLOBULAR CLUSTER

## M12

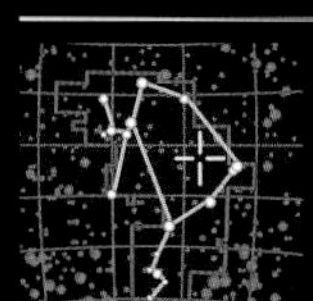

**CATALOGUE NUMBERS** M12, NGC 6128

**DISTANCE FROM SUN** 16,000 –18,000 light-years

**MAGNITUDE** 7.7

**OPHIUCHUS**

Discovered by Charles Messier (see p.69) in 1764, M12 was one of the first globular clusters to be recognized. M12 is at the very limit of naked-eye visibility and therefore best viewed with a telescope. The cluster contains many bright stars and is condensed towards the centre. Its stars are spread across a distance of about 70 light-years, making it less compact than most. Because of this, M12 was originally regarded as an intermediate form of cluster, between open clusters and globular clusters, before the two types were recognized as being fundamentally different.

**EARLY DISCOVERY**

GLOBULAR CLUSTER

## NGC 4833

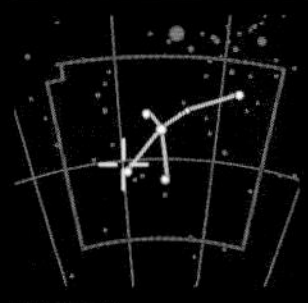

**CATALOGUE NUMBER** NGC 4833

**DISTANCE FROM SUN** 17,000 light-years

**MAGNITUDE** 7.8

**MUSCA**

NGC 4833 is a small globular cluster in the southern constellation Musca and therefore is not visible to most observers in the northern hemisphere. It was discovered by Nicolas Louis de Lacaille (see p.406) during his 1751–52 journey to South Africa. Although NGC 4833 is too faint to see with the the naked eye, it is easily visible through a small telescope. However, because the cluster is rich and compact, even a moderate amateur telescope fails to resolve its stars fully. The centre of the cluster is only slightly more dense than its surroundings, and consequently the cluster lacks the gravitational pull needed to hold on to its stars, and many have already left the cluster. NGC 4833 is located below of the galactic plane behind a dusty region. The dust absorbs light from the cluster and causes its starlight to redden. Because of this reddening, astronomers studying this globular cluster have had to correct the apparent magnitudes of the various stars being studied. The technique used is applied to all globulars lying near the galactic plane. The cluster contains at least 13 confirmed RR Lyrae variable stars (see pp.278–79), which have helped astronomers to estimate the cluster's age at about 13 billion years.

**COMPACT CLUSTER**
NGC 4833 was first recorded by Nicolas Louis de Lacaille in 1752 as resembling a comet. However, with modern, high-powered telescopes it is seen as a well-resolved and compact cluster, with a scattering of outlying stars.

**DISTANT GLOBULAR**

GLOBULAR CLUSTER

## M14

**CATALOGUE NUMBERS** M14, NGC 6402

**DISTANCE FROM SUN** 23,000–30,000 light-years

**MAGNITUDE** 8.3

**OPHIUCHUS**

The globular cluster known as M14 has a diameter of about 100 light-years and contains several hundred thousand stars. Because of its considerable distance, it is too faint to be seen with the naked eye, and, although binoculars or a small telescope will reveal the cluster, a larger instrument is needed to resolve individual stars. Many amateur observers mistakenly identify this object as an elliptical galaxy. In 1938, M14 was home to the first nova photographed in a globular cluster. However, subsequent searches with some of the world's most powerful telescopes have failed to find either the nova star or any of its remnants.

GLOBULAR CLUSTER

## M107

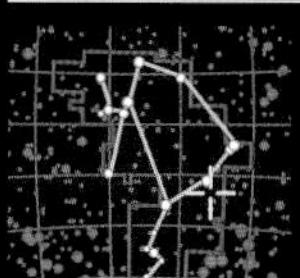

**CATALOGUE NUMBERS** M107, NGC 6171

**DISTANCE FROM SUN** 27,000 light-years

**MAGNITUDE** 8.9

**OPHIUCHUS**

A relatively "open" globular cluster lying close to the galactic plane, M107 is too faint to be seen with the unaided eye. Observations through large telescopes have revealed that the cluster contains dark regions of interstellar dust that obscure some of its stars. This is quite unusual in globular clusters. M107 spans a distance of about 50 light-years.

**LOOSE CLUSTER**

GLOBULAR CLUSTER

## M68

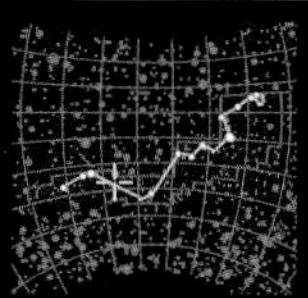

**CATALOGUE NUMBERS** M68, NGC 4590

**DISTANCE FROM SUN** 33,000–44,000 light-years

**MAGNITUDE** 9.7

**HYDRA**

M68 is a globular cluster that is visible only through telescopes. It appears as a small patch when viewed with binoculars, but small telescopes can reveal its constituent stars and its densely populated centre. The cluster has a diameter of about 105 light-years, and its orbit around the centre of the Milky Way means that it is approaching the Solar System at about 400,000kph (250,000mph). Although many variable stars (see pp.278–79) have been detected within the cluster – more than 40 to date, including RR Lyrae stars – the distance to M68 is still uncertain.

**DENSE BALL**

GLOBULAR CLUSTER

## M15

**CATALOGUE NUMBERS** M15, NGC 7078

**DISTANCE FROM SUN** 35,000–45,000 light-years

**MAGNITUDE** 6.4

**PEGASUS**

At the limit of naked-eye visibility, M15 is one of the densest globular clusters in the Milky Way. The cluster has a diameter of about 175 light-years, but, as the centre of the cluster has collapsed in on itself, half of its mass is located within its one-light-year-wide superdense core. M15 also contains nine pulsars, remnants of ancient supernova explosions (see pp.262–63) from the time when the cluster was young. Unusually, two of these pulsating neutron stars form a contact binary pair (see p.270).

**PACKED CORE**
At its core, this globular cluster has the highest concentration of stars in the Milky Way outside the galactic centre.

**TRUE COLOURS**
The brightest stars in M15 are red giants, with surface temperatures lower than the Sun's. Most of the fainter stars are hotter, giving them a bluish-white tint.

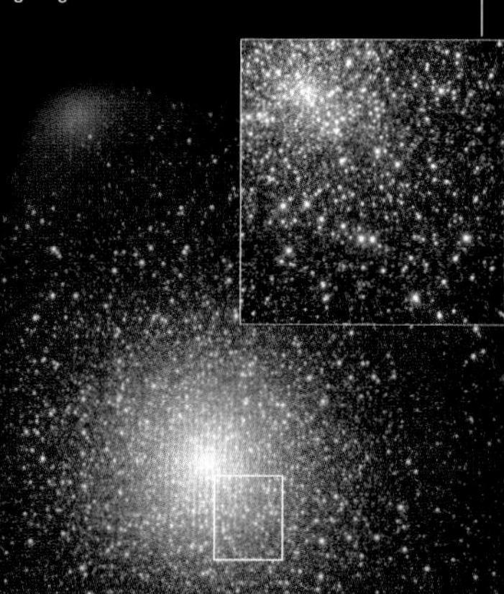

# EXTRA-SOLAR PLANETS

25 Stars and brown dwarfs
52–53 Life in the Universe
96 Extra-solar planets
233 Formation of a planetary system
236–37 Star formation
270–71 Multiple stars

THE SUN IS NOT the only star with a planetary system. More than 200 planets have so far been found orbiting other stars, and more are found almost every month. Extra-solar planets have been detected around stars of a range of types and ages, suggesting that planet formation is a robust process and that planetary systems might be commonplace.

**FLYING SAUCER DISC**
A young star near Rho (ρ) Ophiuchi shines out from within a dust disc that might contain planets.

## PLANET-FORMING DISCS

The first evidence on the trail leading to extra-solar planets was the discovery of flattened discs of material around young stars. This fits the standard theory of solar-system formation, with planets forming from a disc of dust and gas rotating around a star. Some circumstellar discs are symmetrical, suggesting that they are in the early stages, before planet formation. Others are distorted or have gaps, suggesting that planets have formed and are disturbing the material in the disc. Astronomers can infer information about the structure of such discs by methods such as spectroscopy. For example, spectral analysis of CoKu Tau 4, a young star in the Taurus Molecular Cloud, suggests that the hotter inner parts of its dust disc are missing. This is an indication that a giant planet has already formed close to the star. If so, it has formed surprisingly quickly – the star is about 1 million years old, but astronomers have not expected to see planets emerge around stars younger than about 4 million years old. Dusty discs are also found around mature stars. Vega (see p.249) is surrounded by an extensive dust disc fully revealed only at infrared wavelengths. This fine dust is thought to be the debris from a large and relatively recent collision between Pluto-sized bodies orbiting the star at a distance of 13 billion km (8 billion miles).

**DEBRIS DISC**
A debris disc surrounds the red dwarf AU Microscopii. It is thought to have been caused by the collision of asteroids and comets, although planets may also exist within it.

**DENSE DISC**
The disc around the Sun-like yellow dwarf HD 107146 lies face-on to Earth. The Sun is believed to have a similar debris disc beyond Neptune, the Kuiper Belt (see p.206), but that of HD 107146 is 10 times thicker and contains 1,000 to 10,000 times more material.

**PLANETARY DISC**
In this image of the planet-forming disc surrounding Beta (β) Pictoris, the star itself is obscured, but its light is reflected on the disc and has been colour-enhanced. Distortions in the disc suggest the presence of planets orbiting the star.

# NEW SOLAR SYSTEMS

The first extra-solar planets to be found around Sun-like stars appeared to be very unusual: massive gas giants that had short orbital periods very close to their host stars or had been flung into highly elliptical orbits. Astronomers were surprised, since their theories suggested that planet-forming discs generally ended up looking similar to our own solar system, with small rocky planets close in and gas giants farther out. It is unlikely that gas giants can form so close to their host stars, since any gas would be dissipated by the star's heat and radiation pressure. They might have initially followed the standard model for planet formation, but then some mechanism might have caused the outer gas giants to migrate inwards. There seems little room for Earth-like planets in Upsilon (υ) Andromedae (see below) and similar systems. Smaller planets would most likely be flung out of orbit as the gas giants migrated in. However, the method used to detect almost all these new planets is biased towards massive planets with short orbital periods, because they have a larger effect on the motion of the host star (see panel, right). There may be many more, less violent, Sun-like systems waiting to be found.

EXPLORING SPACE

## DETECTING EXTRA-SOLAR PLANETS

Planets are much smaller and much dimmer than their parent stars, so observing them directly from a distance presents many challenges. Astronomers use indirect methods to detect extra-solar planets, mostly relying on the gravitational pull of the planet on its parent star. This gives the star a "wobble", which may be measured by looking at the red shift and blue shift of its light during its orbit. Advanced telescopes will allow the direct imaging of extra-solar planets over the next few years.

**POSSIBLE PLANET**
Observations suggest that a red object seen next to the brown dwarf 2M1207 may be the first extra-solar planet imaged from Earth.

**ELLIPTICAL ORBITS**
Three gas giants orbit the star Upsilon Andromedae. The inner planet orbits every four days at a distance of 7.5 million km (4.7 million miles) – well within the orbit of Mercury. The outer planet orbits every 3.5 years, closer in than Jupiter and with a highly elliptical orbit.

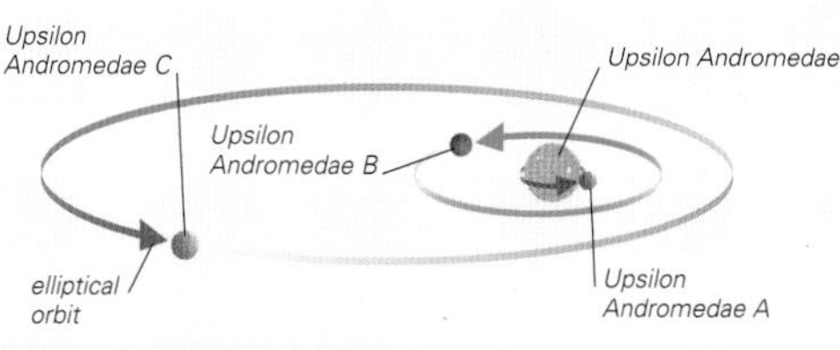


**UPSILON ANDROMEDAE PLANETARY SYSTEM**

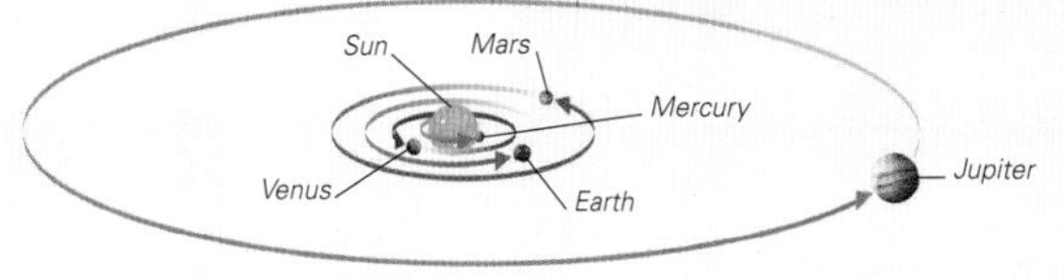


**THE SOLAR SYSTEM (INNER PLANETS AND JUPITER)**

ejected planet

**EJECTED PLANET**
The object TMR-1C, either a gas giant or a brown dwarf, appears to have been flung away from a nearby pair of binary stars. Travelling at about 35,000kph (22,000mph) and leaving a trail of material, it is destined to wander interstellar space.

# STRANGE WORLDS

The gas giants orbiting very close to their host stars are hellish places, with typical surface temperatures of 1,100°C (2,000°F). They are slowly evaporating in the heat and have been called "roasters" or "hot Jupiters". Eccentric giants are found further out in highly elliptical orbits. The planet system 47 Ursae Majoris looks a little more familiar, with two relatively distant gas giants in stable, almost circular orbits. A new class of extra-solar planet was discovered in 2004: a planet 14 times heavier than Earth orbiting the star Mu (μ) Arae. This could be a gaseous planet the size of Neptune, or, if made of rock, it might be a "super Earth" three times the size of Earth. The first extra-solar planets were found in 1991, orbiting the pulsar PSR1257. Although Earth-sized, they are not Earth-like. They are probably completely barren, bathed in the deathly glow from the pulsar, a very intense source of X-ray and gamma radiation.

**BROWN DWARF TWA 5B**
Brown dwarfs are failed stars that can be confused with large planets. The smallest-known brown dwarf, TWA-5B, has a mass which may be as low as 15 times that of Jupiter.

# LOOKING FOR EARTHS

If there is life elsewhere in the Universe, it seems reasonable to expect to find it on a world similar to Earth: a rocky planet orbiting a main-sequence star. Evidence suggests that about 10 per cent of nearby Sun-like stars have planets. Current techniques can be used to estimate the orbital parameters of gas giants around these stars and thereby identify those with stable, distant, circular orbits. Within such systems – perhaps one-quarter of the total – there might lie an inner zone where rocky terrestrial planets have formed. These systems will be first on the list for examination by a new generation of dedicated planet-hunting telescopes due to operate between 2010 and 2020. These telescopes will be sensitive enough to capture the light from terrestrial planets and analyse it for telltale signs of life, such as oxygen and methane. If these chemical signatures are detected, we will have found another home for life in the Universe – a true Earth-twin.

**HABITABLE ZONE**
For life to develop on a planet it must lie within the host star's "habitable zone", where liquid water can permanently exist on the planet's surface. This zone's extent depends on the star's mass and luminosity.

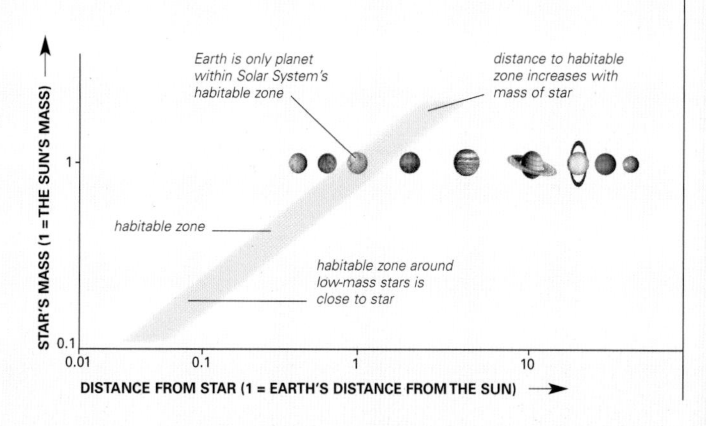

# BEYOND THE MILKY WAY

*"The history of astronomy is a history of receding horizons."*

Edwin Hubble

OUTSIDE THE BOUNDS of the Milky Way stretch vast gulfs of space, the realm of the galaxies. The closest are on our own galactic doorstep – there is even a small galaxy currently in collision with the Milky Way. The farthest lie billions of light-years away, at the edge of the visible Universe – their light has been travelling towards Earth for most of time. Galaxies range from great wheeling discs of matter to giant, diffuse globes of billions of stars and from starless clouds of gas to brilliant furnaces lit up by star formation. They are also violent – despite their stately motion over millions and billions of years, collisions are frequent and spectacular. Collisions disrupt galaxies, sending material spiralling into the supermassive black holes at their centres, fuelling activity that may outshine ordinary galaxies many times over. Galaxies influence their surroundings and form constantly evolving clusters and superclusters. At the largest scale, it is these galaxy superclusters that define the structure of the Universe itself.

**COSMIC RING**
A circlet of brilliant star-forming regions, 300 million light-years from Earth, surrounds the yellow hub of what was once a normal spiral galaxy. This ring galaxy, AM 0644-741, is probably the result of a cosmic collision with a smaller galaxy.

# TYPES OF GALAXY

24–27 Celestial objects
32–35 Radiation
36–37 Gravity, motion, and orbits
Galaxy evolution 298–99
Galaxy clusters 316–17

THROUGHOUT THE UNIVERSE, galaxies exist in enormous diversity. These vast wheels, globes, and clouds of material vary hugely in size and mass – the smallest contain just a few million stars, the largest around a million million. Some are just a few thousand light-years across, others can be a hundred times that size. Some contain only old red and yellow stars, while others are blazing star factories, full of young blue and white stars, gas, and dust. The features of galaxies are clues to their history and evolution, but astronomers have only recently begun to put the entire story together – and there are still many gaps in their knowledge.

**EDGE-ON SPIRALS**
NGC 4013 is a spiral galaxy that happens to lie edge-on to Earth. Such edge-on views reveal the thinness and flatness of spiral galaxies. This Hubble image displays the dense dust within the disc, and shows how few stars lie above or below the disc.

## THE VARIETY OF GALAXIES

Galaxies can be classified by their shape, size, and colour. At the most basic level, they are divided by shape into spiral, elliptical, and irregular galaxies. Edwin Hubble (see p.43), devised a more precise classification, still used today, that subdivides these galaxy shapes. Hubble classed spiral galaxies as types Sa to Sd – an Sa galaxy has tightly wound spiral arms, an Sd very loose arms. Spirals with a bar across their centre are classed as SBa to SBd. Hubble classed elliptical galaxies as E0 to E7 according to their shape in the sky – circular galaxies are E0, and elongated ellipses E7. Elliptical galaxies appear as two-dimensional ellipses, but in reality they are three-dimensional ellipsoids ranging from roughly ball-shaped star clouds to cigar shapes. So Hubble's classification does not reflect their true geometry, since an E0 galaxy could be a cigar shape viewed end-on from Earth. Hubble also recognized an intermediate type of galaxy – the lenticular (type S0), with a spiral-like disc, a hub of old yellow stars, but no spiral arms. Finally, irregular galaxies (type Irr) are usually small, rich in gas, dust, and young stars, but have few signs of structure.

**NUMBERING ELLIPTICALS**
The class of an elliptical galaxy is found by dividing the difference between its long and short axes by the long-axis length and then multiplying by ten, making this galaxy, M110, an E6.

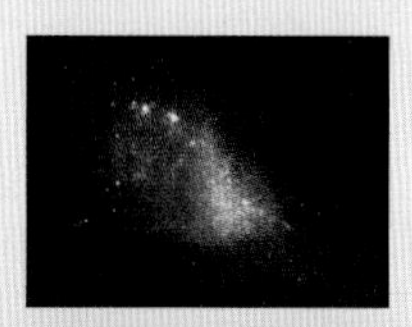

**IRREGULAR GALAXY**
Clouds of stars that lack clear disc- or ellipse-like structure are called irregular galaxies. The Small Magellanic Cloud is one such irregular galaxy.

**ELLIPTICAL GALAXY**
Balls of stars, from perfect spheres, through egg shapes (such as M59, pictured here) to cigar-shaped ellipsoids, are called elliptical galaxies.

**SPIRAL GALAXY**
Vast, rotating discs of stars, dust, and gas are classed as spiral galaxies. Spirals have a ball-shaped nucleus inside a disc with spiral arms. M33 is a nearby spiral galaxy.

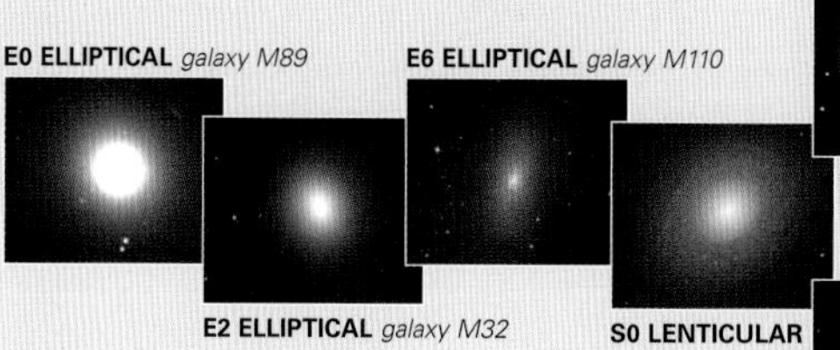

**E0 ELLIPTICAL** *galaxy M89*

**E2 ELLIPTICAL** *galaxy M32*

**E6 ELLIPTICAL** *galaxy M110*

**S0 LENTICULAR** *galaxy NGC 2755*

**Sa SPIRAL** *galaxy NGC 7217*

**Sb SPIRAL** *galaxy NGC 4622*

**Sc SPIRAL** *the whirlpool galaxy (M51)*

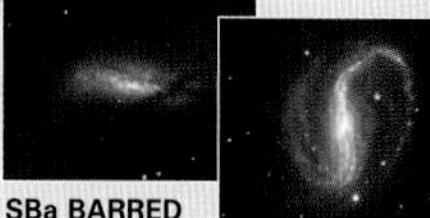

**SBa BARRED SPIRAL** *galaxy NGC 660*

**SBb BARRED SPIRAL** *galaxy NGC 7479*

**SBc BARRED SPIRAL** *galaxy NGC 1300*

**HUBBLE'S CLASSIFICATION**
Hubble arranged his galaxy types in a fork shape, with ellipticals along the handle, and spirals and barred spirals as prongs. This excludes irregular galaxies. He thought his scheme indicated the evolution of galaxies – today astronomers know it is not so simple.

## SPIRAL GALAXIES

Some 25–30 per cent of galaxies in the nearby Universe are spirals. In each one, a flattened disc of gas- and dust-rich material orbits a spherical nucleus, or hub, of old red and yellow stars, which is often distorted into a bar. Stars occur throughout the disc, but the brightest clusters of young blue and white stars are found only in the spiral arms. The space between the arms often looks empty viewed from Earth, but it is also full of stars. Above and below the disc is a spherical "halo" region, where globular clusters (see p.285) and stray stars orbit. Spiral galaxies rotate slowly – typically once every few hundred million years – but they do not behave like a solid object. Stars orbiting farther away take longer to complete an orbit than those close to the core. The resulting "differential rotation" is the key to understanding the spiral arms.

**BARRED SPIRAL**
Galaxy M83 is typical of the majority of spiral galaxies in that it has a straight bar either side of the nucleus.

**ORBITS IN SPIRALS**
Stars in the disc of a spiral galaxy follow elliptical, nearly circular orbits in a single plane. Those in the hub have wildly irregular orbits at a multitude of angles.

**FLOCCULENT SPIRAL**
The spiral galaxy NGC 4414 is flocculent, with bright stars clumped throughout the disc. Its star formation seems to be caused by local collapses of material rather than a large-scale density wave.

# SPIRAL ARMS

The continued presence of spiral arms in most disc-shaped galaxies was once a mystery. If the arms orbit more quickly near the nucleus, then, during a galaxy's multi-billion-year lifetime, they would become tightly wrapped around the core. It now seems that the arms are in fact rotating regions of star formation, not rotating chains of stars themselves. The arms arise from a "density wave" – a zone that rotates far more slowly than the galaxy itself. The density wave is like a traffic jam – stars and other material slow down as they move into it and accelerate as they move out, but the jam itself advances only slowly. The increased density helps to trigger the collapse of gas clouds and the start of star formation. The strength of the density wave varies between spirals. If the wave is strong, the result is a neat, "grand design" spiral with two clearly defined arms. If it is weak or non-existent, disc stars will tend to form in localized regions, creating the more clumpy "flocculent" spirals.

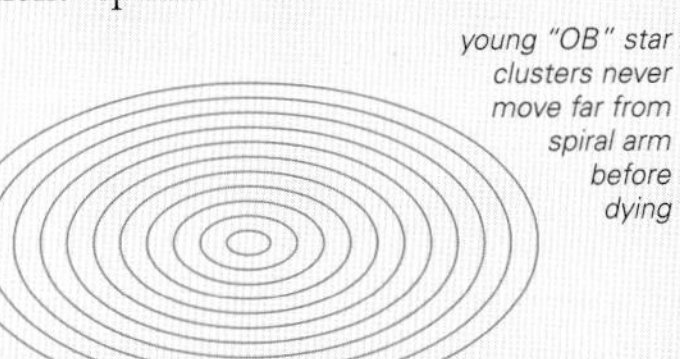

**PERFECT GALAXY**
In this diagram of an ideal galaxy, objects follow neatly aligned elliptical orbits around the nucleus. They travel fastest when close to the nucleus and slowest when farthest away.

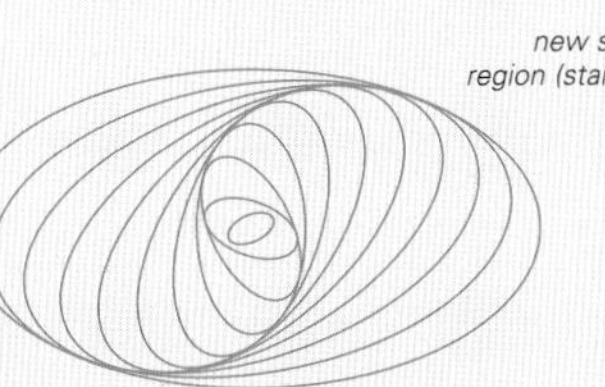

**SPIRAL REALITY**
In a real galaxy, the orbits do not line up neatly. The variety of alignments, coupled with the slower movement when farther from the nucleus, creates spiral zones in which objects are moving more slowly and so become bunched together.

open clusters of longer-lived stars move out from spiral arm
young "OB" star clusters never move far from spiral arm before dying
new stars ignite in HII region (star-forming nebula)
molecular cloud is compressed
density wave causes material to build up
sparse stars orbit faster than the spiral arm and move into arm from behind

**DETAIL OF A SPIRAL ARM**
As material orbiting in a galaxy's disc approaches the denser region marked by the spiral arm, it packs together. Dark molecular clouds form, some of which turn into star-forming nebulae (see pp.236–37). New stars of all kinds ignite here, but the brightest ones soon die, so they always mark the spiral arms.

# ELLIPTICAL GALAXIES

Elliptical galaxies show little structure other than a simple ball shape. They span the range from the largest to the smallest galaxies. At one end, dwarf ellipticals are relatively tiny clusters of a few million stars, often very loosely distributed, appearing faint and diffuse. Such galaxies are scattered in the space between larger galaxies and must contain significant amounts of invisible material simply to hold them together. Some of this could be in a central black hole, but much of it seems to be mysterious "dark matter" (see p.27) scattered through the whole of the galaxy. At the other extreme lie the giant ellipticals – galaxies only found near the centres of large galaxy clusters and often containing many hundreds of billions of stars. Some giant ellipticals, called cD galaxies, have large outer envelopes of stars and even multiple concentrations of stars at their centres, suggesting they may have formed from the merging of smaller ellipticals. Almost all the stars in elliptical galaxies are yellow and red, and there is rarely any sign of star-forming gas and dust. The dominance of old, long-lived stars implies that any star formation in these galaxies has long since ended. Each star orbits the galaxy's dense core in its own path. The chances of collision are very remote, because stars are so small relative to the distances between them. With no gas and dust clouds to interact with, there is nothing else to flatten the stars into a single plane of rotation. Ellipticals are described according to their degree of elongation – how much they deviate from a perfect sphere (see p.294) – but the largest galaxies are always very close to perfect spheres.

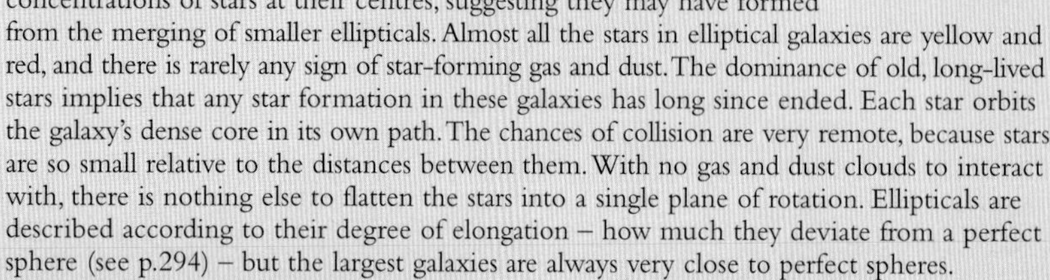

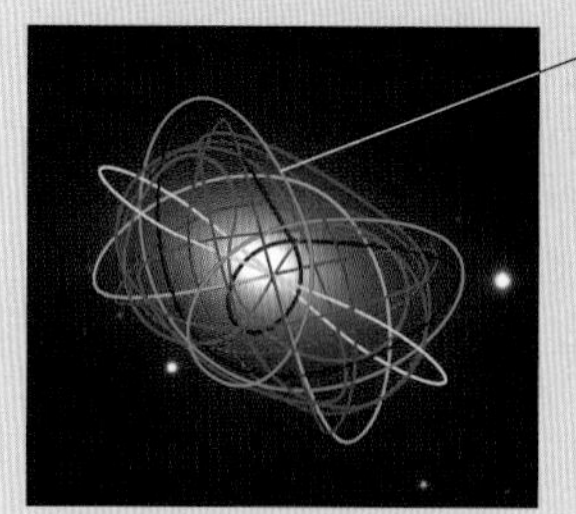


**ORBITS IN ELLIPTICAL GALAXIES**
The orbits of stars in an elliptical galaxy vary wildly, from circles to very long ellipses, and are not confined to any specific direction.

**GIANT ELLIPTICAL**
M87 is the giant elliptical at the heart of the nearby Virgo cluster. It is a type E1 or E0, almost perfectly spherical and containing roughly a trillion stars. At lower right, three smaller galaxies can be seen.

**INTERMEDIATE GALAXY**
M49 in the Virgo galaxy cluster is a large elliptical of type E4. With a diameter of about 160,000 light-years, it is classed by some astronomers as a giant elliptical, although its mass is much less than that of the true giants.

**DWARF ELLIPTICAL**
The Leo I galaxy is a nearby dwarf elliptical, and one of the few we can study closely. With so few stars, there must be a large amount of dark matter holding the galaxy together with its gravity.

# LENTICULAR GALAXIES

At first glance, lenticular galaxies appear to be relatives of ellipticals – they are dominated by a roughly spherical nucleus of old red and yellow stars. However, around this nucleus, these galaxies also have a disc of stars and gas. This links them to spiral galaxies, and they are similar in overall size and general shape, although the nucleus is often considerably bigger than it would be in a spiral of similar size. The overall shape is often described as that of a lens, which is the root of the name "lenticular". The key difference between lenticulars and spirals is that lenticulars have no spiral arms and little sign of star-forming activity in their discs. Without the bright blue star clusters that illuminate the discs of spirals, lenticulars are sometimes hard to tell apart from ellipticals. Those that are face-on may be indistinguishable from ellipticals and misclassified. An edge-on spiral galaxy with a large nucleus can equally be misclassified as lenticular, because at oblique angles spiral structure is often invisible. Astronomers are uncertain how lenticular galaxies form, but they could be spiral galaxies that have lost most of their dust and gas.

**DUSTY LENTICULAR**
Lying 25 million light-years away, galaxy NGC 2787 is one of the closest lenticular galaxies. Dust lanes can be seen silhouetted against the nucleus, marking the plane of its disc.

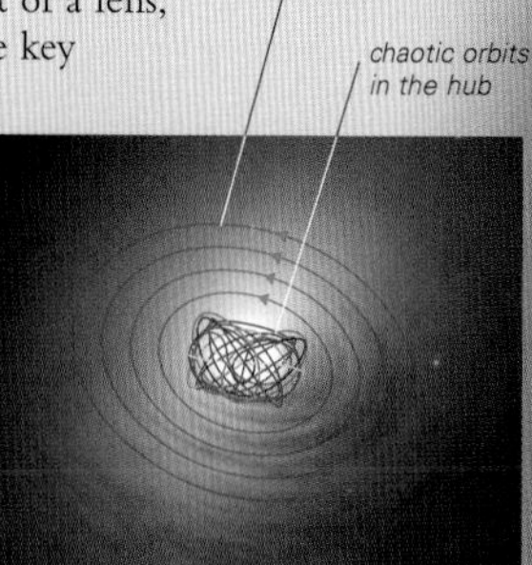


**ORBITS IN LENTICULAR GALAXIES**
Stars in the nucleus of a lenticular galaxy follow orbits with no specific plane, similar to those in an elliptical galaxy or a spiral nucleus. Gas and dust in the disc orbits in a more orderly plane.

**IRREGULAR DWARF**
NGC 4449 is an irregular galaxy 12 million light-years away. It lacks structure, but includes chains of young blue stars, pink regions of star formation, and dark clouds.

EXPLORING SPACE

## GALAXIES AT DIFFERENT WAVELENGTHS

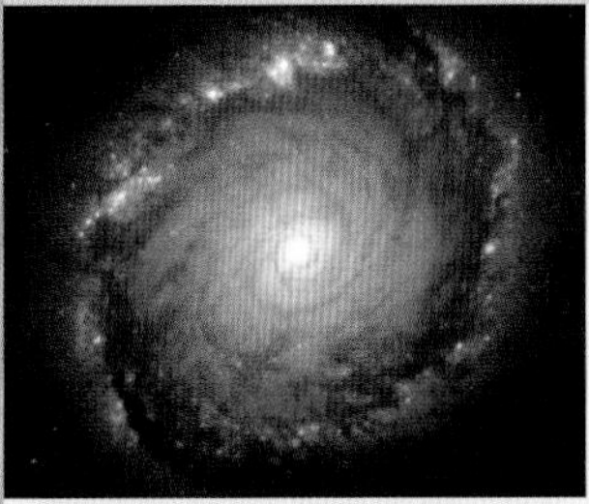

COMBINED IMAGE OF NGC 1512

Radiation of different wavelengths can reveal hidden structures within galaxies. The hottest stars appear brightest in ultraviolet, while cool, diffuse gas may be visible only in infrared. By overlaying images from different spectral regions, astronomers build up a full picture of a galaxy.

**FROM ULTRAVIOLET TO INFRARED**
These images of galaxy NGC 1512 increase in wavelength from left to right. Each wavelength is represented by a false colour.

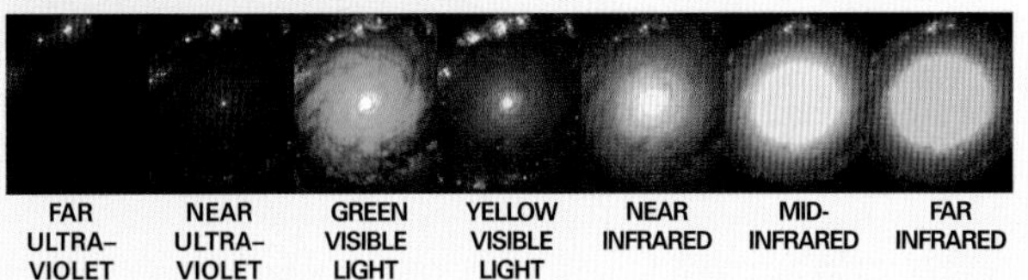


# IRREGULAR GALAXIES

Not all galaxies fit into the scheme of spirals, ellipticals, and lenticulars. Some of these misfit galaxies are colliding with companions or being pulled out of shape by a neighbour's gravity. These are usually classed under the catch-all term "peculiar" or "Pec". Many more are true irregulars (type Irr). These galaxies typically contain a lot of gas, dust, and hot blue stars. In fact, many irregulars are "starburst" galaxies, with great waves of star formation sweeping through them. Irregulars frequently have vast, pink hydrogen emission nebulae where star formation is taking place. Some irregulars show signs of structure – central bars and sometimes the beginnings of spiral arms. The Milky Way's brightest companion galaxies, the Large and Small Magellanic Clouds (see pp.300–301), are typical irregular galaxies.

**IRREGULAR STARBURST**
M82 is an irregular starburst galaxy crossed with dark dust lanes. It is undergoing an intense period of star birth.

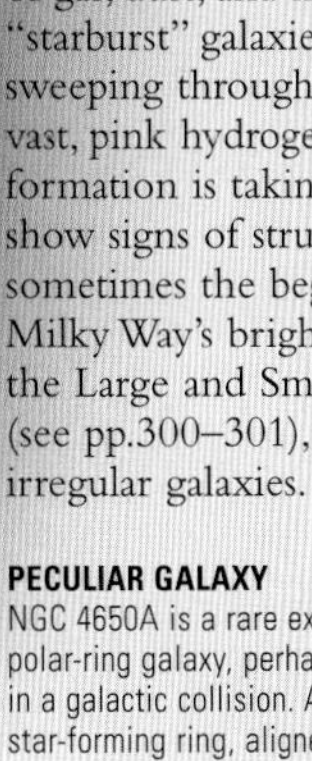

**PECULIAR GALAXY**
NGC 4650A is a rare example of a polar-ring galaxy, perhaps created in a galactic collision. A blue-white star-forming ring, aligned with the poles, extends from the nucleus.

EXPLORING SPACE

## ASTRONOMY FROM THE SOUTH POLE

Some of the best Earth-based observations of galaxies come from an automated observatory at the South Pole. The AASTO project takes advantage of the dryness on the Antarctic Plateau – the driest place on Earth. With no water vapour in the atmosphere, near-infrared light is not absorbed, so it reaches the ground unhindered.

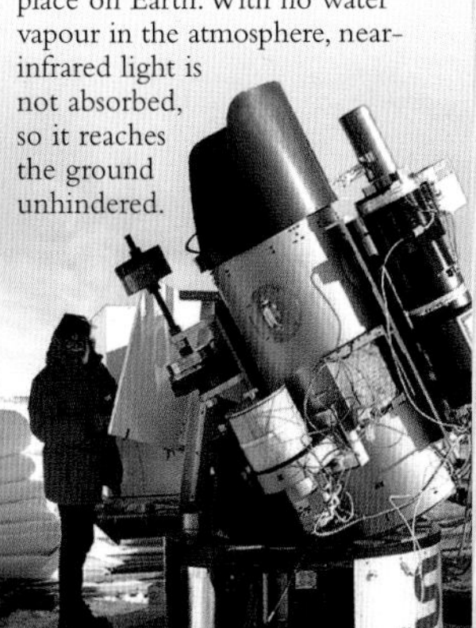

# CENTRAL BLACK HOLES

Many, if not all, galaxies have a dark region within their nucleus that seems strange by contrast with the outer parts. The fast orbits of stars near galactic nuclei suggest an enormous concentration of mass in a tiny volume at the centre of most spiral and elliptical galaxies – often billions of Suns' worth of material in a space little larger than the Solar System. The only object that can reach such a density is a black hole (see p.26). Despite the tremendous gravity of this "supermassive" black hole, in most nearby galaxies the material has long since settled into steady orbits around it. With no material to absorb, the black hole remains dormant. When a gas cloud or other object comes too close, however, the black hole may awake, pulling in the stray material and heating it, producing radiation. The black hole may generate any type of radiation from low-energy radio waves to high-energy X-rays. In extreme cases called "active galaxies" (see pp.310–11), the radiation from the nucleus is the galaxy's dominant feature.

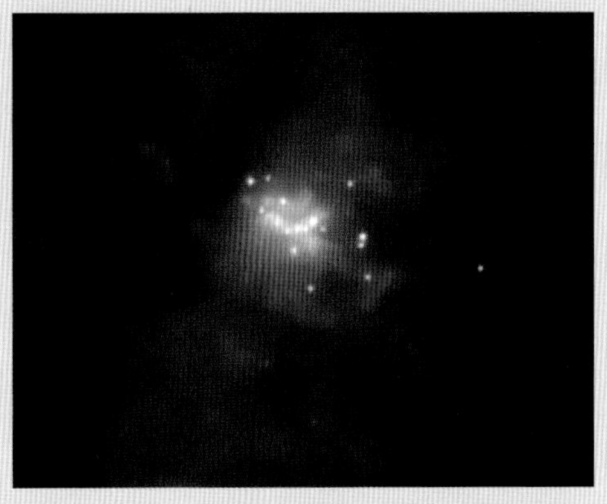

**HIDDEN SUPERMASSIVE BLACK HOLE**
An X-ray image of galaxy M82 shows glowing hot gas and intense point sources of X-rays. These are probably stellar-mass black holes surrounding a central supermassive black hole.

# GALAXY EVOLUTION

22–23 The scale of the Universe
50–51 Out of the darkness
294–97 Types of galaxy
Active galaxies 310–11
Galaxy clusters 316–17
Galaxy superclusters 324–25

THE FORMATION OF GALAXIES has puzzled astronomers for decades. Today, telescopes are used to study galaxies billions of light-years away to uncover their early life. Light reaching Earth from these objects left the galaxies when the Universe was young. Such distant views show a multitude of irregular galaxies, violent collisions, and intense activity. Astronomers are now settling on a new model of the lives of galaxies.

## THE DISTRIBUTION OF GALAXIES

Because astronomers see only a "snapshot" from the evolution of any single galaxy, they have to build up the life stories of galaxies by looking at many objects to detect patterns that might link galaxies across space and time. Spatially, large elliptical galaxies are found only in substantial galaxy clusters – the largest lie at the exact centre of the largest clusters. This suggests that the development of elliptical structure depends on the proximity of other galaxies. Across time, the proportion of large irregular galaxies was much higher close to the Big Bang, with more spirals and fewer ellipticals. This suggests that irregular galaxies evolve into spiral, then elliptical galaxies.

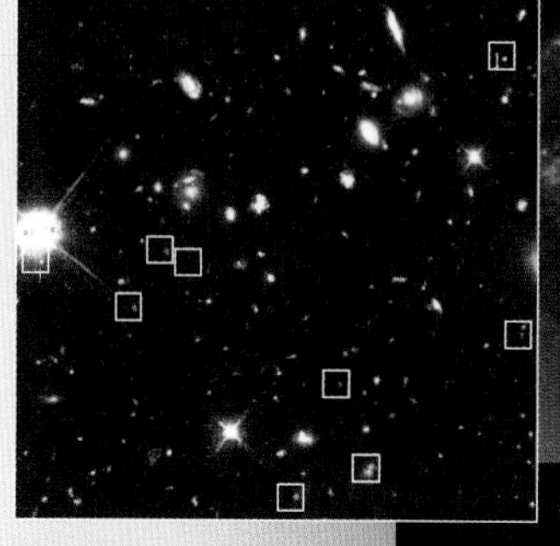

**ANCIENT IRREGULARS**
This detail from the Hubble "Deep Field" image shows a profusion of small, irregular, blue "protogalaxies" (in squares) in the very distant, hence very young, Universe.

**YOUNG GALAXY**
I Zwicky 18 is a mysterious irregular galaxy in the nearby Universe. It contains no old stars and is possibly only 500 million years old (the Milky Way is about 20 times older). It may resemble the first galaxies.

## GALAXY FORMATION

Until recently, there were two competing theories of galaxy formation: a "top-down" one, in which galaxies condensed from huge ribbons of matter, eventually forming stars; and a "bottom-up" theory, in which stars initially formed in small groups, and then merged together to form larger structures: galaxies and then clusters. Either way, the influence of dark matter was crucial. Just as today, dark matter would have vastly outweighed ordinary matter. Was this matter hot and fast-moving, or cold and relatively slow? Hot dark matter models suggest that large-scale structures would have formed first, with galaxies slowly condensing out of them. This clashes with observations that continue to find evidence of the first stars and galactic black holes very soon after the Big Bang. It now seems likely that the origin of individual galaxies lies in a mixture of rapid bottom-up development and slower formation of larger-scale structures, driven in part by the gravity of cold dark matter.

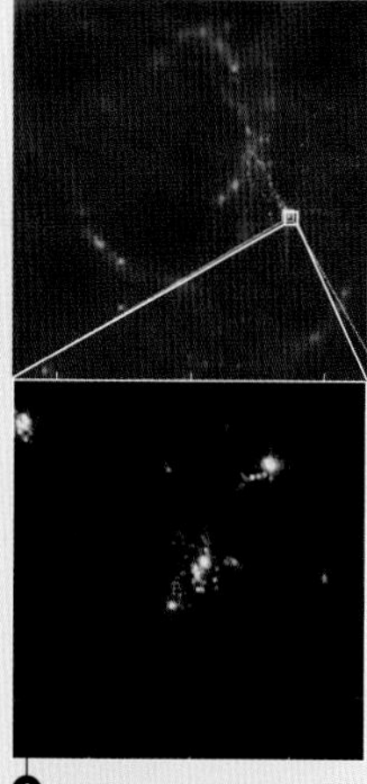

**1 1.5 BILLION YEARS OLD**
When the Universe is just one-tenth of its present age, matter is clumping into strands (top). Within these strands, protogalaxies are forming (bottom).

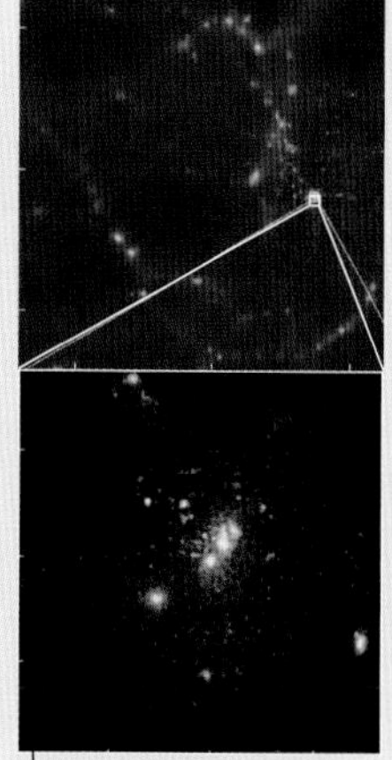

**2 2.6 BILLION YEARS OLD**
At the largest scale (top), matter strands become more defined. On a smaller scale, small, irregular protogalaxies are merging into larger irregular galaxies (bottom).

**3 5.0 BILLION YEARS OLD**
The strands of matter are becoming chains of galaxies (top), within which mature galaxies, including ellipticals (bottom), have formed from the irregular building blocks.

**4 7.4 BILLION YEARS OLD**
Ever more matter falls into the matter strands (top). Now there are mature galaxy clusters containing some giant elliptical galaxies (bottom).

**COMPUTER MODEL OF GALAXY BIRTH**
The upper row of pictures in this computer simulation shows the formation of large-scale chains and clusters of galaxies, while the lower row shows the formation of a giant elliptical galaxy by collisions and mergers.

**GRAZING ENCOUNTER**
The galaxies NGC 2207 and IC 2163 have become distorted by a close encounter. NGC 2207, on the left, has torn a long streamer of stars from the smaller galaxy. Violent events such as this are a key influence on the evolution of galaxies.

# COLLIDING GALAXIES

Relative to their size, galaxies are packed much more tightly than are stars, so violent encounters between them are common. During collisions, the interacting galaxies' gas clouds are forced together, triggering bursts of star formation. The clouds can be heated so much that they "boil" away from the galaxy in a process called pressure stripping. Stars themselves rarely collide. They are so small relative to their spacing that they pass like the members of marching bands. Whole spiral arms, however, can be flung free of the galaxy by gravity. Computer models of galaxy collisions have recently been combined with observations of the early Universe, during which collisions were even more frequent. This research suggests that galaxy interactions are responsible for changing galaxies from one type to another over billions of years.

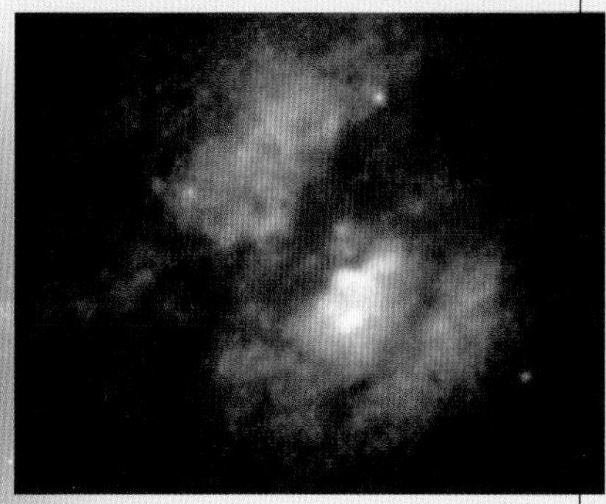

**HIDDEN ACTIVITY**
The peculiar double-lobed galaxy Arp 220 is host to a burst of star formation triggered by an intergalactic collision, only visible in infrared. Many more such luminous infrared galaxies are now coming to light, suggesting that galaxy collisions may be even more frequent than was previously thought.

# EVOLUTION BY MERGING

The most popular model of galaxy evolution today suggests galaxies change through a series of collisions and interactions. A key element of the theory is that the free hydrogen in intergalactic space is steadily diminishing as it is absorbed into galaxies. In the early, gas-rich era, the first galaxies to form were small irregulars or ellipticals. As these pulled in more material and sometimes merged, they developed more internal structure, becoming the first spiral galaxies. When spirals merge, the gas from their discs is stripped away and their stars are thrown into chaotic orbits – they become elliptical galaxies. If there is enough intergalactic hydrogen, it falls into the elliptical galaxy, forming a new disc of gassy and dusty material that may eventually develop its own spiral arms. By this time, the only stars remaining from the original spiral are the old red and yellow stars, which explains the dominance of old red and yellow stars in the hubs of old spirals. In regions where many such events have happened (for instance, within old, dense galaxy clusters), the available free gas is reduced until eventually the galaxies are all ellipticals, merging occasionally to form ever-larger systems such as cD galaxies (see p.316).

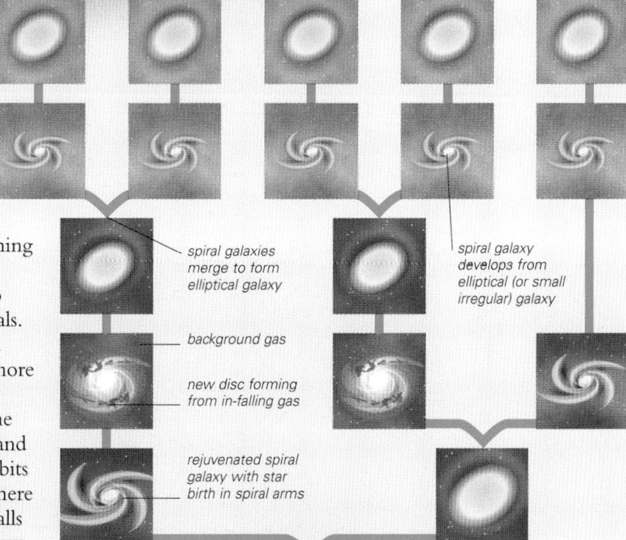


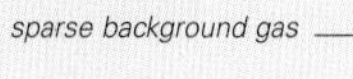


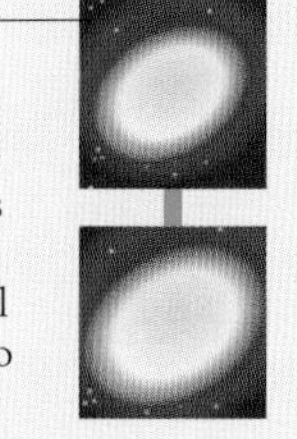

**THE MERGER MODEL**
According to recent thinking, galaxy mergers tend to form ellipticals, but in-falling gas can rejuvenate them, forming new discs and star-forming spiral arms. The way a galaxy develops depends on the time between mergers, and the amount of background gas. There is a general trend towards large, gas-poor elliptical galaxies.

SPIRAL GALAXY (NGC 3370)

DUSTY ELLIPTICAL (NGC 1316)

ELLIPTICAL GALAXY (M87)

# GALAXIES

Astronomers are drawn naturally to the brightest, the most beautiful, and the most intriguing galaxies. However, of the 100 billion galaxies in the observable Universe, only a minority are spectacular spirals and giant ellipticals. Astronomers are beginning to understand that most galaxies are relatively small and faint – diffuse balls and irregular clouds of stars. The faintest and commonest galaxies are dwarf ellipticals, which are like oversized globular star clusters of only a few million stars. These feeble galaxies are visible only if they lie nearby in intergalactic terms. The most brilliant are the giant ellipticals, which can be 20 times as luminous as the Milky Way.

**BIG AND BRIGHT**
Spirals such as Bode's Galaxy, M81, may be the most attractive type of galaxy, but they are far from the most common. Making up less than 30 per cent of all galaxies, they are outnumbered by smaller, fainter galaxies.

DWARF ELLIPTICAL GALAXY

## SagDEG

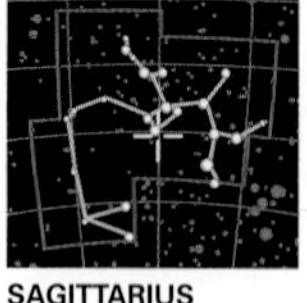

**SAGITTARIUS**

**CATALOGUE NUMBER** None
**DISTANCE** 88,000 light-years
**DIAMETER** 10,000 light-years
**MAGNITUDE** 7.6 for M54 star cluster in SagDEG

The Sagittarius Dwarf Elliptical Galaxy, often called SagDEG, was until recently our closest known galactic neighbour. It was not found until 1994 and was supplanted only by the discovery of the even closer Canis Major Dwarf in 2004. SagDEG remained hidden for so long because, like all dwarf ellipticals, it is a very faint scattering of stars. It is also well disguised by its position behind the great Sagittarius star clouds that mark our galaxy's centre. SagDEG is small and obscure, but it has at least four orbiting globular clusters, which are brighter and more obvious. One of these, M54, was discovered by Charles Messier more than 200 years before the parent galaxy was found.

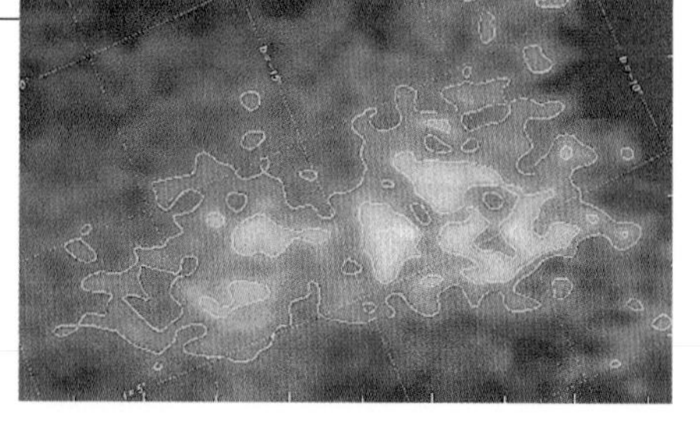

**STAR DENSITY**
SagDEG's existence came to light only when a survey of Sagittarius found regions of increased star density – the bright patches in this image.

SagDEG's existence so close to our galaxy is a puzzle. It orbits the Milky Way in less than a billion years and so must have gone through several close encounters that should have ripped it apart and scattered its stars through the galactic halo. It has survived only due to a large amount of dark matter, producing more gravity than SagDEG's visible stars.

IRREGULAR GALAXY

## Large Magellanic Cloud

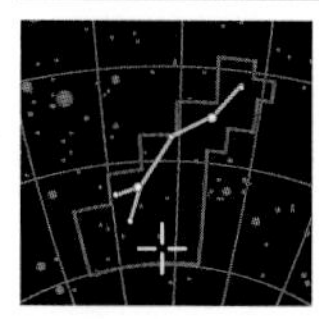

**DORADO**

**CATALOGUE NUMBER** None
**DISTANCE** 179,000 light-years
**DIAMETER** 20,000 light-years
**MAGNITUDE** 0.1

The Large Magellanic Cloud (LMC) bears the name of 16th-century explorer Ferdinand Magellan (see panel, opposite). However, cultures native to the southern hemisphere have recognized its existence since prehistoric times. Like its smaller counterpart, the Small Magellanic Cloud, the LMC appears from Earth to be a distinctive, isolated region of the Milky Way, some 10 degrees across, with its own areas of nebulosity and star clusters.

**RADIO MAP**
This false-colour radio image of the LMC is centred on the Tarantula Nebula. It shows intense radiation as red and black, indicating ionized hydrogen and star formation.

**CLOUD DETAILS**
In this image, the LMC's main "bar" appears as the yellowish star cloud on the left, outshone by the pink star-birth regions of the Tarantula Nebula.

The LMC is in fact an irregular galaxy, orbiting the Milky Way roughly once every 1.5 billion years on a path that brought it to within 120,000 light-years of our galaxy at its closest approach around 250 million years ago. Although the LMC is irregular and is being distorted by the gravity of the Milky Way, it shows some signs of basic structure. Many of its stars are concentrated in a central bar-like nucleus, curved at one end. Some astronomers have likened the LMC to a barred spiral with just one arm.

Like all irregular galaxies, the LMC is rich in gas, dust, and young stars, including some of the largest known regions of star birth. One such region is the magnificent Tarantula Nebula, also known as 30 Doradus. It is so brilliant that, if transported to the location of the Orion Nebula (see p.239) – only 1,500 light-years away in the Milky Way – it would be bright enough to cast shadows on Earth at night.

In recent times, the LMC was host to the only bright supernova since the invention of the telescope. Supernova 1987A (see p.262) was observed by astronomers around the world both during and after its explosion, and it has taught astronomers a lot about the final stages of the stellar life cycle.

IRREGULAR GALAXY

## Small Magellanic Cloud

**CATALOGUE NUMBER** NGC 292

**DISTANCE** 210,000 light-years

**DIAMETER** 10,000 light-years

**MAGNITUDE** 2.3

**TUCANA**

Like the Large Magellanic Cloud, the Small Magellanic Cloud (SMC) is an irregular galaxy in orbit around the Milky Way. It was in the SMC that Henrietta Leavitt discovered the Cepheid variable stars that were to unlock the secrets of the galactic distance scale (see pp.278, 340). Thanks to her discovery, astronomers know that the SMC is both more distant and genuinely smaller than the LMC, with around one-tenth of the larger cloud's mass. Like the LMC, the small cloud is also undergoing intense star formation. Some astronomers argue that the SMC also shows signs of a central bar-like structure, but the case is far from proven. It has one known globular cluster in orbit, but the SMC lies deceptively close in the sky to one of the Milky Way's largest globulars – 47 Tucanae.

Both the Magellanic Clouds are ultimately doomed to be torn to shreds and absorbed into our own galaxy. They have survived several close passes of the Milky Way, but now share their orbit with a trail of gas, dust, and stars torn away during previous encounters. This "Magellanic Stream" has allowed astronomers to trace and refine their models for the orbits of the clouds.

**CLOUD OF STARS**
The SMC forms a distinctive wedge-shaped cloud in southern skies. The pinkish areas in this optical photograph show the galaxy's major star-forming regions.

EXPLORING SPACE

### MAGELLAN'S DISCOVERY

The southernmost sky was not visible to Europeans until they visited the southern hemisphere. The Portuguese explorer Ferdinand Magellan was among the first to do so during his round-the-world voyage of 1519–21. He was the first European to record two isolated patches of the Milky Way, which were later named after him.

**FERDINAND MAGELLAN**

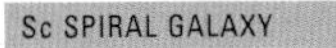

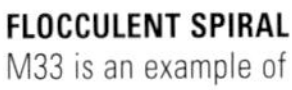

**SUPERNOVA ECHO**
This unique image shows reflected light from the LMC's supernova of 1987. The rings are created by light from the explosion bouncing off sheets of dust close to the supernova and being deflected onto a path towards Earth. The reflected light reaches us years after light that took the direct route. The "light echo" was made visible only by subtracting an image of the region before the supernova from a recent picture. Unchanged stars therefore appear black.

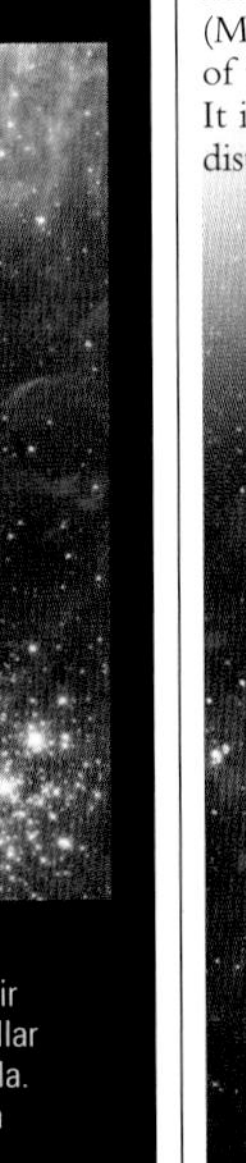

**TARANTULA NEBULA**
Massive stars run through their entire life cycle inside the stellar nursery of the Tarantula Nebula. This image shows a new open cluster, Hodge 301, whose biggest stars have already gone supernova. As the shock waves spread across space, they ripple the nearby gas clouds, triggering further star formation.

Sc SPIRAL GALAXY

## Triangulum Galaxy

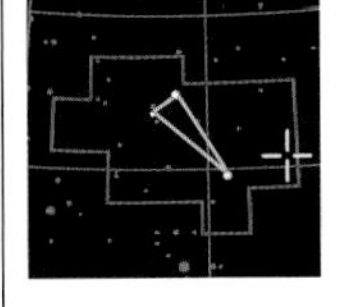

**CATALOGUE NUMBERS** M33, NGC 598

**DISTANCE** 3 million light-years

**DIAMETER** 50,000 light-years

**MAGNITUDE** 5.7

**TRIANGULUM**

After the Andromeda Galaxy and the Milky Way, the Triangulum Galaxy (M33) is the third major member of the Local Group of galaxies. It is slightly more distant than the larger and brighter Andromeda Galaxy (M31), and the two lie close to each other in the sky. M33 is affected by its larger neighbour's gravity, and it may even be in a long, slow orbit around the giant Andromeda spiral.

Seen from Earth, M33 is fainter and more diffuse than M31 – partly because it is closer to face-on than edge-on, and partly because it really is less spectacular. However, the Triangulum Galaxy is more typical of spiral galaxies than its unusually bright companions. As with several Local Group galaxies, M33 is large and bright enough in the sky for its features to be catalogued, and several of them have NGC numbers. Most prominent is the star-forming region NGC 604, the largest emission nebula known. At 1,500 light-years across, it dwarfs anything in our own galaxy.

**FLOCCULENT SPIRAL**
M33 is an example of a flocculent spiral – a galaxy with arms that divide like split ends and separate into patches. The clumpy star clouds are thought to form due to localized changes in density.

**NEBULA NGC 604**
This emission nebula's gas glows as it is excited by ultraviolet light from a central star cluster. The stars are so massive and bright that they emit most of their light in ultraviolet, and so are not prominent in visible-light photographs such as this.

Sb SPIRAL GALAXY

## Andromeda Galaxy

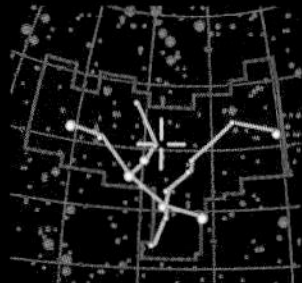


**CATALOGUE NUMBERS** M31, NGC 224

**DISTANCE** 2.5 million light-years

**DIAMETER** 250,000 light-years

**MAGNITUDE** 3.4

The Andromeda Galaxy (M31) is the closest major galaxy to the Milky Way and the largest member of the Local Group of galaxies. Its disc is twice as wide as our galaxy's.

M31's brightness and size mean it has been studied for longer than any other galaxy. First identified as a "little cloud" by Persian astronomer Al-Sufi (see p.405) in the tenth century, it was for centuries assumed to be a nebula, at a similar distance to other objects in the sky. Improved telescopes revealed that this "nebula", like many others, had a spiral structure. Some astronomers thought that M31 and other "spiral nebulae" might be solar systems in the process of formation, while others guessed rightly that they were independent systems of many stars. It was in the early 20th century that Edwin Hubble (see p.43) revealed the true nature of M31, at a stroke hugely increasing estimates of the size of the Universe (see panel, opposite). Astronomers now know that M31, like the Milky Way, is a huge galaxy attended by a cluster of smaller orbiting galaxies, which occasionally fall inwards under M31's gravity and are torn apart.

Despite being intensively studied, the Andromeda Galaxy still holds many mysteries, and it may not be as typical a spiral galaxy as it appears. For example, despite its huge size, it appears to be less massive than the Milky Way, with a sparse halo of dark matter. Despite this, astrophysicists calculate that M31's central black hole has the mass of 30 million Suns, almost ten times more than the Milky Way's central black hole. The huge mass of M31's black hole is surprising, because a galaxy's black hole is thought generally to reflect the mass of its parent galaxy. Furthermore, studies at different wavelengths have revealed disruption in the galaxy's disc, possibly caused by an encounter with one of its satellite galaxies in the past few million years.

M31 and the Milky Way are moving towards each other, and they should collide and begin to coalesce in around 5 billion years' time.

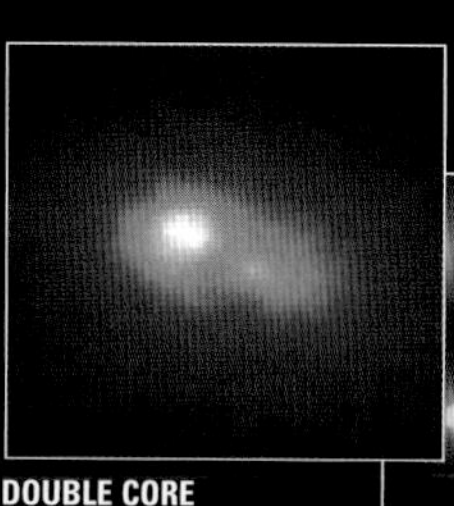

**DOUBLE CORE**
Analysis of optical images (above) reveals that M31 has two cores. The galaxy may have collided with a smaller galaxy millions of years ago. The double core could be evidence of the smaller galaxy.

**CENTRAL BLACK HOLE**
This X-ray image shows M31's central black hole as a blue dot – it is cool and inactive compared to the galaxy's other X-ray sources (yellow dots).

**GALACTIC NEIGHBOURS**
Dark dust lanes are silhouetted against glowing gas and stars in this view of the Andromeda Galaxy and its two close companions, the dwarf elliptical galaxies M32 (upper left) and M110 (bottom).

EXPLORING SPACE

## INTERGALACTIC DISTANCE

The study of M31 played a key role in the discovery that galaxies exist beyond our own. Although the spectra of galaxies suggested they shone with the light of countless stars, no one could measure their immense distance. In 1923, Edwin Hubble (see p.43) proved that M31 lay outside our galaxy. He found the true distance of M31 by calculating the luminosities of its Cepheid variable stars (see pp.278–79), and relating their true brightness to their apparent magnitude.

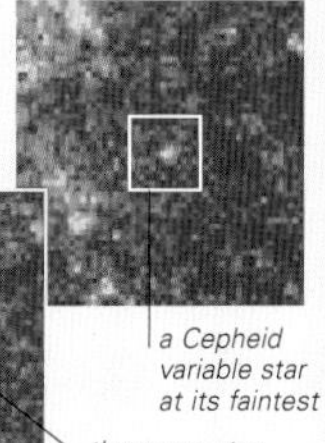

Sb SPIRAL GALAXY

## Bode's Galaxy

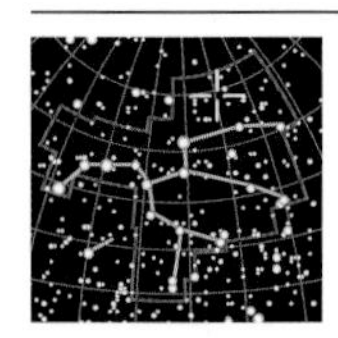

URSA MAJOR

**CATALOGUE NUMBERS** M81, NGC 3031

**DISTANCE** 10.5 million light-years

**DIAMETER** 95,000 light-years

**MAGNITUDE** 6.9

**X-RAY SOURCES**
A Chandra X-ray image shows a strong X-ray source at the galaxy's core, surrounded by smaller sources, probably X-ray binary stars.

Bode's Galaxy, also known as M81, is one of the brightest spiral galaxies visible from the northern hemisphere. It is the dominant member of a galaxy group lying near to the Local Group. The galaxy is named after Johann Elert Bode, a German astronomer who found it in 1774.

Bode's Galaxy has had a close encounter with M82, the Cigar Galaxy (see below), in the past few tens of millions of years. The near miss created tidal forces that enhanced the density waves (see p.295) in M81. The rate of star birth around the density waves increased, highlighting the spiral arms. A long, straight dust lane along one side of the core could also have been created in the encounter.

By measuring the Doppler shifts of light from either side of the core, astronomers have found that the outer regions rotate more slowly than in most galaxies. This suggests that M81 has little of the dark matter that creates higher rotation rates in other galaxies.

**CLUSTERS REVEALED**
This combined visible and ultraviolet image shows the hottest and brightest star clusters (blue and white blobs), lying in the core and spiral arms.

**M81'S COLOURS**
An enhanced image emphasizes the colour difference between the old red and yellow stars of Bode's Galaxy's core and the young blue stars of the disc and spiral arms.

IRREGULAR DISC GALAXY

## Cigar Galaxy

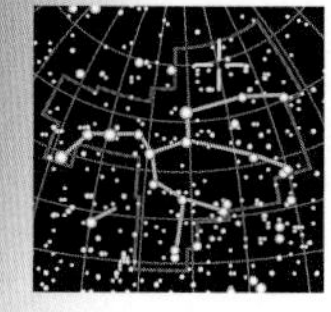

URSA MAJOR

**CATALOGUE NUMBERS** M82, NGC 3034

**DISTANCE** 12 million light-years

**DIAMETER** 40,000 light-years

**MAGNITUDE** 8.9

The brightest and most spectacular example of a "starburst galaxy", the Cigar Galaxy (M82) is an irregularly shaped cloud of stars that looks like a cigar from Earth. It is undergoing a period of intense star birth as a result of a close encounter with Bode's Galaxy (M81). The near miss has disrupted the galaxy's centre, creating the dark dust lanes that obscure much of the core and triggering the creation of many massive, brilliant star clusters in an area a few thousand light-years across. At infrared wavelengths, M82 is the brightest galaxy in the sky, and it is also a strong radio source. The infrared light comes from disturbed gas and dust around the core.

**GAS STREAMERS**
M82's most spectacular features can be observed only at the radio wavelength emitted by ionized hydrogen, here represented as magenta. This wavelength reveals a huge envelope of gas above and below the core, blown out in long streamers by fierce radiation from the central star clusters.

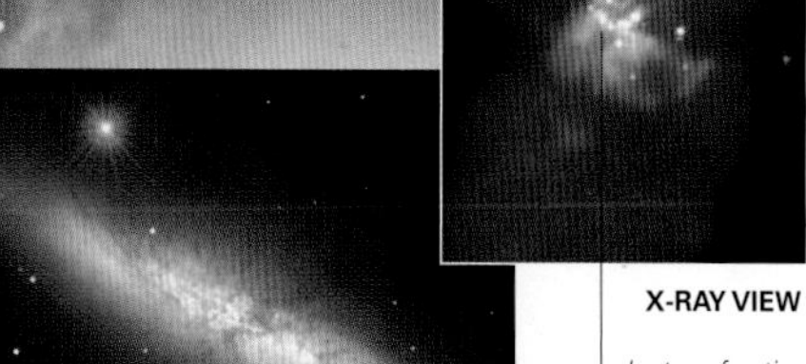

**X-RAY VIEW**

*cluster of active black holes*

**OPTICAL IMAGE**

**STARBURST GALAXY**
The intense activity in M82's core is luminous at optical and X-ray wavelengths. The young stars illuminate the nebulae with visible light, while those that have rapidly completed their life cycle form active black holes, emitting X-rays.

Sb SPIRAL GALAXY

## Black Eye Galaxy

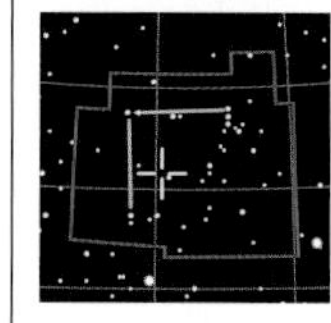

COMA BERENICES

**CATALOGUE NUMBERS** M64, NGC 4826

**DISTANCE** 19 million light-years

**DIAMETER** 51,000 light-years

**MAGNITUDE** 8.5

This distinctive galaxy has a dark dust lane, running in front of its core, from which it gets its name. The dust lane is unusual because it arcs above the galaxy's core in an orbit of its own. Because it has not yet settled into the plane of the galaxy's rotation, it must have a recent origin and probably dates from the galaxy's absorption of a smaller galaxy that strayed too close. Another bizarre feature of the Black Eye Galaxy is that its outer regions are rotating in the opposite direction to the inner regions. This could be another effect of the collision.

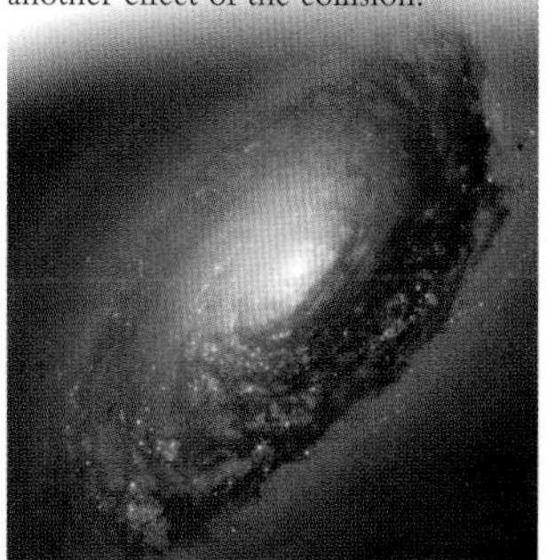

**M64'S CENTRAL REGION AND DUST LANE**

Sc SPIRAL AND IRREGULAR GALAXIES

# Whirlpool Galaxy

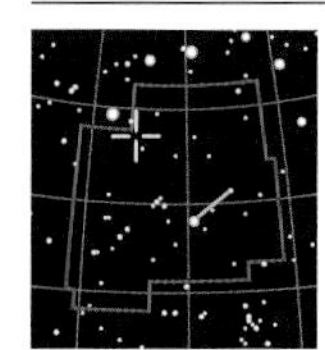

CANES VENATICI

**CATALOGUE NUMBERS** M51, NGC 5194, NGC 5195

**DISTANCE** 31 million light-years

**DIAMETER** 100,000 light-years

**MAGNITUDE** 8.4

Discovered by Charles Messier (see p.69) in 1773, the Whirlpool Galaxy is now known to be a pair of galaxies that is interacting – the brightest and clearest example of such a pair visible from Earth. The individual components are a spiral galaxy viewed face-on (NGC 5194) and a smaller irregular galaxy (NGC 5195). In visible light, the connection between them cannot be seen, but images at other wavelengths reveal an envelope of gas connecting the two. One effect of the interaction is to enhance the density wave in the larger galaxy, triggering increased star formation and making its spiral arms stand out very clearly. The Whirlpool was in fact the first "nebula" in which spiral structure was recognized, by William Parsons (see panel, right).

The interaction has also triggered increased activity in the cores of both of the galaxies – NGC 5195 is undergoing a burst of star formation, which explains its unusual brightness, while NGC 5194's core is also much brighter than expected. It is even classified by some astronomers as an active Seyfert Galaxy (see p.310).

The Whirlpool is very bright despite its distance, indicating that it is large and luminous – it is similar in size to the Milky Way, but brighter overall because of the large young star clusters in its spiral arms. It is thought to be the dominant member of a small group of galaxies, called simply the M51 group, which also includes the galaxy M63.

**THE MISSING LINK**
Combining a visible-light image of M51 (green) with one in hydrogen-alpha radio waves (blue) reveals the trail of gas linking the two galaxies.

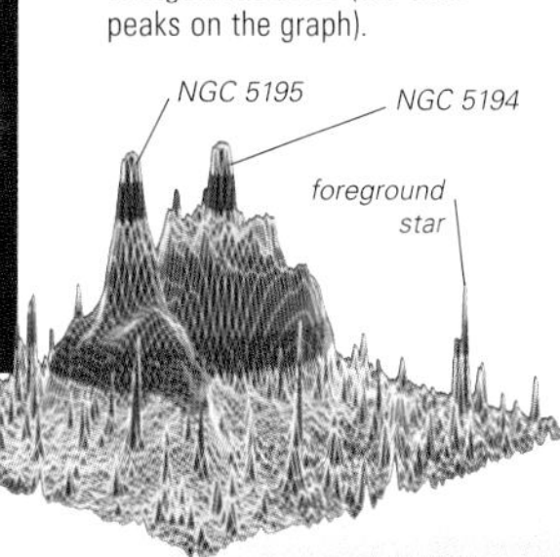


**LIGHT INTENSITY**
Plotting the intensity of light from different regions of M51 reveals the brightness of the two galactic cores (the twin peaks on the graph).

## WILLIAM PARSONS

William Parsons (1800–67) was an Irish nobleman who used his great wealth to build the largest telescope of his time and made the first detailed studies of nebulae. In 1845, he made detailed drawings and noticed the spiral structure of some "nebulae", as galaxies were thought to be at the time. This was an important step to discovering that galaxies were not nebulae but separate star systems.

**PARSONS'S SKETCH OF M51**

**LUMINOUS WHIRLPOOL**
This Hubble image combines data from different filters to reveal detail in M51, such as dark dust behind each spiral arm and bright pink regions of star birth.

Sc SPIRAL GALAXY

## Pinwheel Galaxy

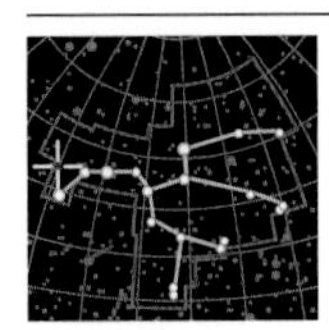

URSA MAJOR

**CATALOGUE NUMBERS**
M101, NGC 5457

**DISTANCE**
27 million light-years

**DIAMETER**
170,000 light-years

**MAGNITUDE** 7.9

Catalogued by Charles Messier (see p.69) as M101, the Pinwheel Galaxy is a bright, nearby spiral galaxy, but one that reveals its nature only when studied with powerful telescopes or seen on long-exposure photographs. Because it lies face-on to Earth, most of the Pinwheel's light is spread out across its disc, and a casual glance reveals only the bright central core. Detailed photographs show that M101 has an extensive, though rather lopsided, spiral-arm system, giving the appearance that the core is offset from the galaxy's true centre. M101 is one of the largest spirals known – its visible diameter is more than twice that of our own galaxy. Its large angular size in the sky (larger than the size of the full Moon) makes it one of the few galaxies whose individual regions can be isolated for study.

**RELATIVE RED SHIFT**
This computer image shows the red shift and blue shift of objects within M101, revealing its rotation. Yellow and red regions are moving away, green and blue parts are approaching.

**ASYMMETRICAL DISC**
M101's lopsided shape is thought to be caused by uneven distribution of mass in the disc affecting the orbit of its stars.

**DUST LANE**
The thick dust lane around the Sombrero Galaxy is silhouetted against its bright disc in this Hubble Space Telescope image.

Sa SPIRAL GALAXY

## Sombrero Galaxy

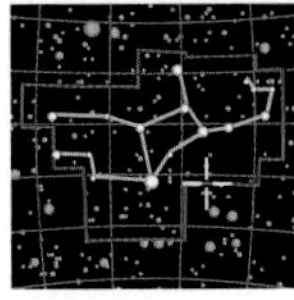

VIRGO

**CATALOGUE NUMBERS**
M104, NGC 4594

**DISTANCE**
50 million light-years

**DIAMETER**
50,000 light-years

**MAGNITUDE** 8.0

The dark dust lane and bulbous core of the Sombrero Galaxy (M104) give it a likeness to the traditional Mexican hat after which it is named. From Earth, we see the Sombrero Galaxy from just six degrees above its equatorial plane – an ideal angle to provide a clear view of the core while also revealing the spiral arms. It is usually classified as an Sa or Sb spiral, although its core is unusually large and bright. Another odd feature is the dense swarm of globular star clusters orbiting the galaxy. More than 2,000 have been counted – ten times more than orbit the Milky Way.

In the galaxy's core is a disc of bright material tilted relative to the galaxy's plane. It is probably the accretion disc of a central supermassive black hole. X-ray emission from the region suggests some material is still being absorbed by the hole.

M104 was a late addition to Messier's catalogue of celestial objects. He added it by hand to his copy of the catalogue after discovering it in 1781. Several other astronomers also found it independently. One of these was William Herschel (see p.90), who was the first to note the dark dust lanes that are M104's most distinctive feature. More recently, the Sombrero provided some of the first evidence for objects lying far beyond our own galaxy (see panel, below).

**LIGHT CONTOURS**
Computer image manipulation reveals variations in the intensity of light within the Sombrero Galaxy.

### VESTO SLIPHER

US astronomer Vesto Slipher (1875–1969) was one of the first to suggest that the Universe is bigger than our galaxy. In 1912, at Lowell Observatory in Flagstaff, Arizona, he identified red-shifted lines in M104's spectrum. The lines told him the galaxy was receding at 3.6 million kph (2.25 million mph) – too fast for it to reside within the Milky Way.

S0 LENTICULAR GALAXY

## Spindle Galaxy

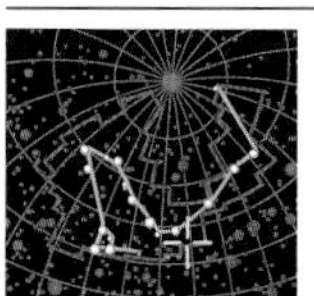

DRACO

**CATALOGUE NUMBERS** M102 (not confirmed), NGC 5866

**DISTANCE** 40 million light-years

**DIAMETER** 60,000 light-years

**MAGNITUDE** 9.9

The Spindle (NGC 5866) is an attractive galaxy orientated edge-on to observers on Earth. It is usually classified as a lenticular galaxy – a disc of stars, gas, and dust with a typical bulging core, but with no sign of true spiral arms. However, spiral structure is hard to detect in an edge-on galaxy.

The Spindle Galaxy is the major member of the NGC 5866 Group, a small cluster of galaxies. Astronomers have measured the way these galaxies move and have found that the Spindle must contain an enormous mass of material – up to 1 billion solar masses, or 30 to 50 per cent more than the Milky Way.

The Spindle Galaxy could be the mysterious entry number 102 in Charles Messier's catalogue of astronomical features. Messier included the object at first without a location, then later gave coordinates that did not match any feasible object. Some believe that Messier had listed the Pinwheel Galaxy, M101, twice. More likely, however, is that M102 was the Spindle, and he added 5 degrees to his measurements in error.

**MASSIVE SPINDLE**
From Earth we see the Spindle Galaxy edge-on, giving it a cigar-shaped appearance with a fine silhouetted dust lane.

E2 ELLIPTICAL GALAXY

## M60

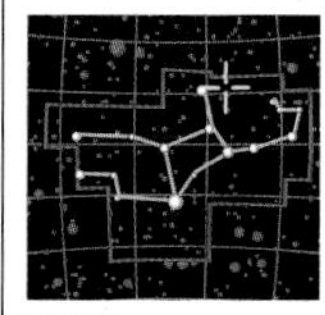

VIRGO

**CATALOGUE NUMBERS** M60, NGC 4649

**DISTANCE** 58 million light-years

**DIAMETER** 120,000 light-years

**MAGNITUDE** 8.8

M60 is one of several giant elliptical galaxies in the Virgo galaxy cluster (see p.319), the central cluster in our own Local Supercluster of galaxies. The galaxy and its neighbour, M59, were discovered in 1779 by German astronomer Johann Köhler, who was observing a comet that passed close by. Charles Messier (see p.69) found them a few nights later, and added them to his catalogue of objects that might confuse comet hunters.

M60 is similar in diameter to many spiral galaxies but, as an E2 elliptical, it is very nearly spherical, containing a much larger volume. It probably has a mass of several trillion suns, and is orbited by thousands of globular clusters. Using the Hubble Space Telescope to measure the motions of M60's stars, astronomers have discovered that a black hole of 2 billion solar masses lies at the galaxy's heart.

**CLOSE NEIGHBOURS**
M60 lies very close to the spiral M59 (upper right), and the two galaxies are thought to be interacting. In 1 billion years' time, M60 may even swallow M59 entirely.

DISRUPTED SPIRAL GALAXIES

## Antennae Galaxies

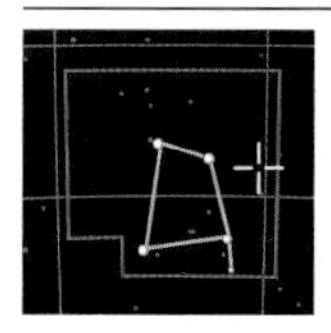

CORVUS

**CATALOGUE NUMBERS** NGC 4038, NGC 4039

**DISTANCE** 63 million light-years

**DIAMETER** 360,000 light-years (total)

**MAGNITUDE** 10.5

The Antennae Galaxies, NGC 4038 and 4039, are among the sky's most spectacular interacting galaxies. Seen from Earth, they appear as a central bright double-knot of material, with two long streamers of stars stretching in opposite directions, resembling an insect's antennae. However, powerful telescopes reveal that each streamer is in fact a spiral arm, uncurled from its parent galaxy by the tremendous gravitational forces of an intergalactic collision that began around 700 million years ago and continues today.

**THE BIGGER PICTURE**
A wide-field view of the Antennae taken from Earth reveals both the bright, distorted cores and the long, faint streamers formed by the disrupted spiral arms.

The Antennae have been studied for what they can tell us about galaxy collisions. Detailed images of the central region show that it is lit by hundreds of bright, intense star clusters. These are thought to be forming as gas clouds in the galaxies become compressed by the collision, triggering starbursts (see the Cigar Galaxy, p.304). Astronomers can use the clusters' redness to estimate their age – older clusters emit redder light because the brighter blue stars are the most massive and therefore the first to die.

**COLLIDING CORES**
Turbulent clouds of dust and brilliant star clusters are revealed in this Hubble Space Telescope view.

**STAR NURSERIES**
This enhanced optical image highlights, in patches of red and yellow, the Antennae's most intense star-birth regions. These patches are mostly in the compressed dust cloud between the two galactic cores.

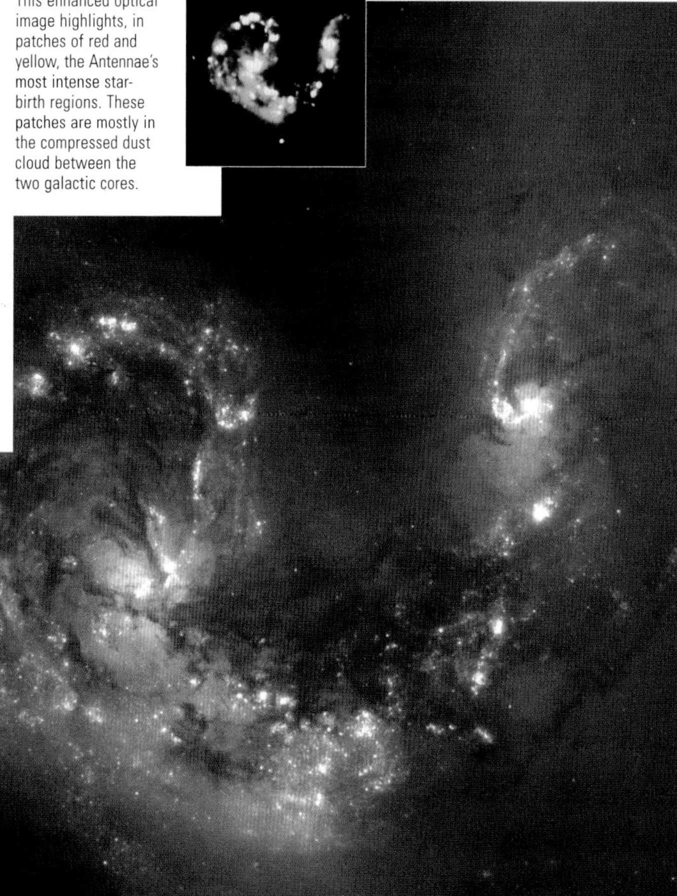

DISRUPTED SPIRAL GALAXY

## ESO 510-G13

**CATALOGUE NUMBER** ESO 510-G13

**DISTANCE** 150 million light-years

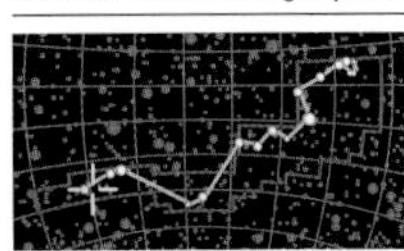

**DIAMETER** 105,000 light-years

**MAGNITUDE** 13.3

**HYDRA**

Despite being referred to only by a number rather than a name (its long designation comes from the European Southern Observatory's catalogue), ESO 510-G13 is one of the most intriguing galaxies in the sky. It is an edge-on spiral with a clear dust lane marking its central plane. The dust lane has an obvious twist.

The most obvious explanation for the kink is that ESO 510-G13 has had a close encounter or collision with another galaxy in its recent past. Some astronomers have suggested that the collision is still going on, and the dust lane is the "ghost" of a galaxy that ESO 510-G13 has swallowed – as seen in the active galaxy Centaurus A (see p.312). Alternatively, the disc might have been warped by the gravity of a nearby galaxy. The galaxy responsible might be a small neighbour or a more distant but larger member of the same group. As their techniques and instruments improve, astronomers are finding this kind of distortion is common in spirals – although it often shows up more in the distribution of gas than in the stars, so it is usually most obvious at radio wavelengths. Our near neighbour M31 (see pp.302–303) has such a distortion, and the Milky Way seems to have one, too – perhaps caused by interaction with its own family of smaller neighbours.

**WARPED DISC**
The bright core of ESO 510-G13 silhouettes the galaxy's warped dust lane in this image. The blue glow on the right is a huge area of bright young stars – evidence perhaps of a collision in the galaxy's recent history.

SB0 BARRED SPIRAL GALAXY

## NGC 6782

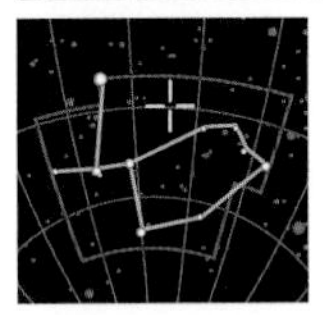

**CATALOGUE NUMBER** NGC 6782

**DISTANCE** 183 million light-years

**DIAMETER** 82,000 light-years

**MAGNITUDE** 12.7

**PAVO**

The Hubble Space Telescope imaged the apparently normal barred spiral galaxy NGC 6782 in 2001. Using ultraviolet detectors, it studied the pattern of the galaxy's hottest material. The image (see below) showed, in pale blue, two rings of stars so brilliant and hot that they emit most of their light as ultraviolet. The inner ring lies in the galaxy's bar and could have been ignited by tidal forces between the bar and the rest of the galaxy. The outer star ring is at the galaxy's edge.

**ULTRAVIOLET STAR RINGS**

DISRUPTED SPIRAL GALAXIES

## The Mice

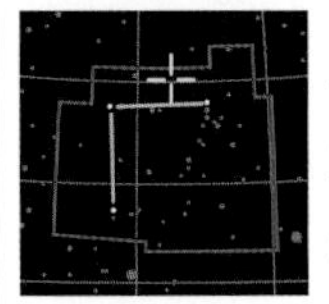

**CATALOGUE NUMBER** NGC 4676

**DISTANCE** 300 million light-years

**DIAMETER** 300,000 light-years

**MAGNITUDE** 14.7

**COMA BERENICES**

The object classified as NGC 4676 is in fact a pair of colliding galaxies – known as the Mice because they appear to have white bodies and long, narrow tails. As with the Antennae Galaxies, the long streamers are the result of the spiral arms "unwinding" during the collision – though in this case one of the arms lies edge-on to us and so appears to be long and straight, despite being strongly curved away from us. Knots of bright blue stars in the streamers and the main bodies of the galaxies show where bursts of star formation are taking place. Computer simulations of the collision (see panel, right) suggest that the galaxies are now separating after a closest approach 160 million years ago.

**HIDDEN EXTENT**
Image processing allows astronomers to amplify faint light from the outlying parts of the Mice, revealing their true shape and extent.

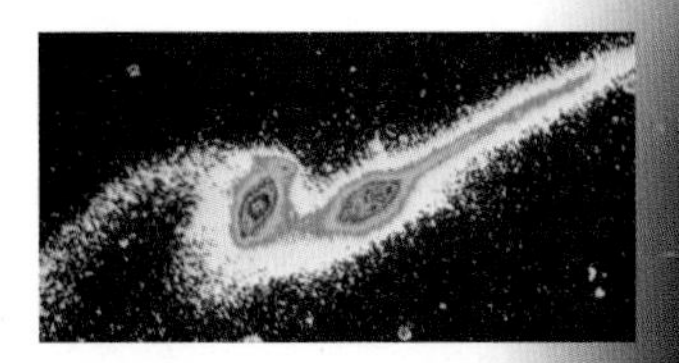

EXPLORING SPACE

## SIMULATING GALAXY COLLISIONS

The great challenge for astronomers studying colliding galaxies is that they can only ever see one stage in a story that unfolds over millions of years. Fortunately, today's supercomputers can help to speed things up. By building "model" galaxies with simplified star clouds, gas, dust, and dark matter, then smashing them into each other in a computer, astronomers can measure how gravity affects the fate of the galaxies.

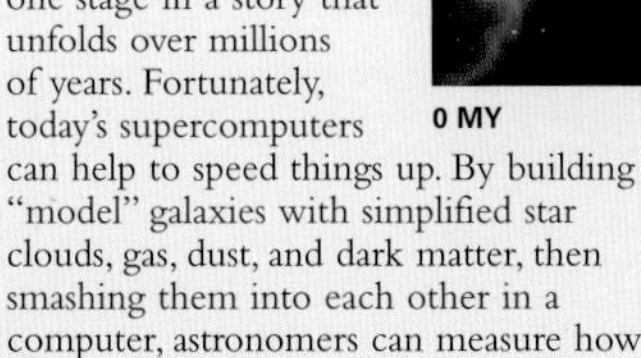


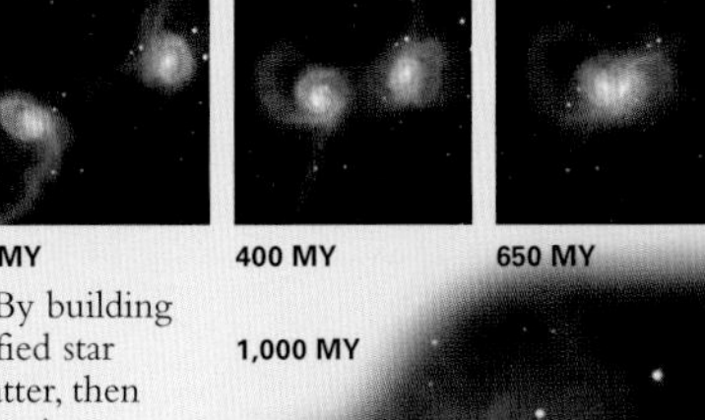


**SPIRAL COLLISION SIMULATION**
This computer simulation shows two spiral galaxies interacting and merging to form a large, irregular galaxy. Time is measured in millions of years (My).

**DESTINED TO UNITE**
Although currently moving apart from a close encounter, the Mice are gravitationally locked together and doomed eventually to merge, perhaps resulting in the formation of a new giant elliptical galaxy.

DISRUPTED SPIRAL GALAXY

## Cartwheel Galaxy

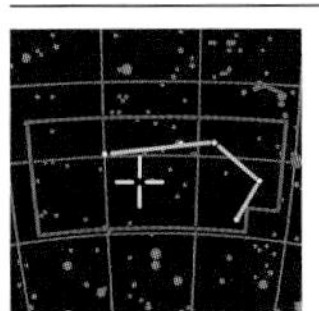

SCULPTOR

**CATALOGUE NUMBER** ESO 350-G40

**DISTANCE** 500 million light-years

**DIAMETER** 150,000 light-years

**MAGNITUDE** 19.3

If the Cartwheel Galaxy looks unusual, it's because it is the victim of an intergalactic "hit-and-run". The Cartwheel was once a normal spiral galaxy. As we see the galaxy, it is recovering from a head-on collision with a smaller runaway galaxy many millions of years earlier in its history. Such events are rare in the cosmos – galactic collisions usually involve grazing encounters or a slow dance towards an eventual merger. The Cartwheel shows what happens when two galaxies pass through each other at high speed while orientated at right angles to each other. The rotating density wave that is normally responsible for the spiral arms was disrupted in this case, resulting in the disappearance of the spiral structure. Meanwhile a shock wave spread to the outer edge of the galaxy, creating a ring of vigorous star formation. An inward-travelling shock wave is probably responsible for the core's unusual "bull's-eye" appearance.

For years, most astronomers suspected that one of the Cartwheel's two immediate neighbours was responsible for the collision. Both showed signs of being the culprit – a nearby small, blue galaxy has a disrupted shape and vigorous star formation, while a yellow galaxy could have been stripped of its star-forming gas in the encounter. However, recent radio observations have shown a telltale stream of gas leading from the Cartwheel towards another small galaxy, a quarter of a million light-years away.

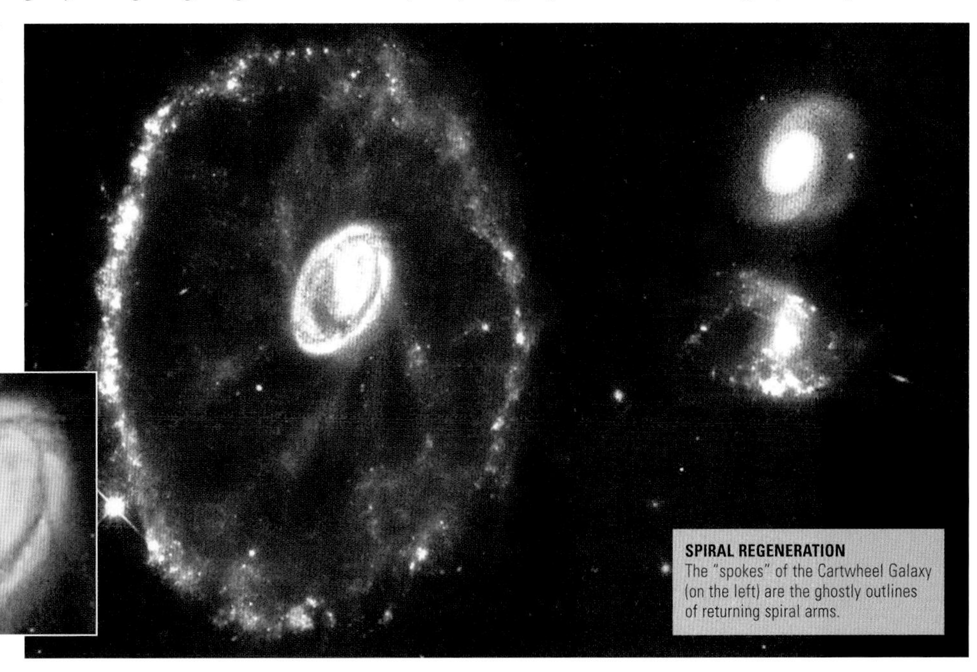

**CLOUDS IN THE CORE**
So-called "comet clouds", each a thousand light-years long, are found in the Cartwheel's core. They are thought to arise as hot, fast-moving gas set in motion by the collision ploughs through denser, slower-moving matter.

**SPIRAL REGENERATION**
The "spokes" of the Cartwheel Galaxy (on the left) are the ghostly outlines of returning spiral arms.

RING GALAXY

## Hoag's Object

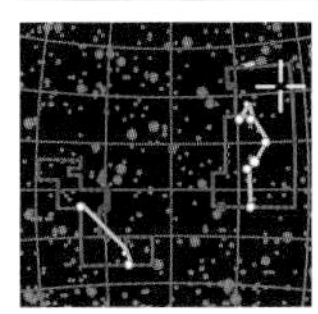

SERPENS

**CATALOGUE NUMBER** PGC 54559

**DISTANCE** 500 million light-years

**DIAMETER** 120,000 light-years

**MAGNITUDE** 15.0

Hoag's Object is one of the most bizarre galaxies in the sky. Although its ring structure suggests parallels to the Cartwheel Galaxy (a spiral disrupted by a head-on collision, see above), there are no nearby galaxies that could have caused an impact. One of two theories might account for the shape of Hoag's Object and that of similar ring galaxies. The galaxies may be members of an unusual class of spiral in which the two arms develop into a circle. Alternatively, they may be former elliptical galaxies that have each swallowed another galaxy, creating a surrounding ring of star-forming material.

**SEE-THROUGH GALAXY**
The gap between Hoag's Object's core and its ring is truly transparent – a background galaxy can be seen through it near the top of this image. However, the gap could still contain large numbers of faint stars.

LOW-SURFACE-BRIGHTNESS GALAXY

## Malin 1

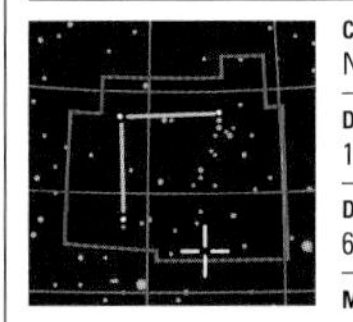

COMA BERENICES

**CATALOGUE NUMBER** None

**DISTANCE** 1 billion light-years

**DIAMETER** 600,000 light-years

**MAGNITUDE** 25.7

Despite its dull appearance, Malin 1 is an extremely important galaxy. Discovered by accident in 1987, it is an enormous but faint spiral that is for some reason poor at forming stars. It seems that such low-surface-brightness galaxies could account for up to half the galaxies in the Universe, though Malin 1 is one of the largest of the type.

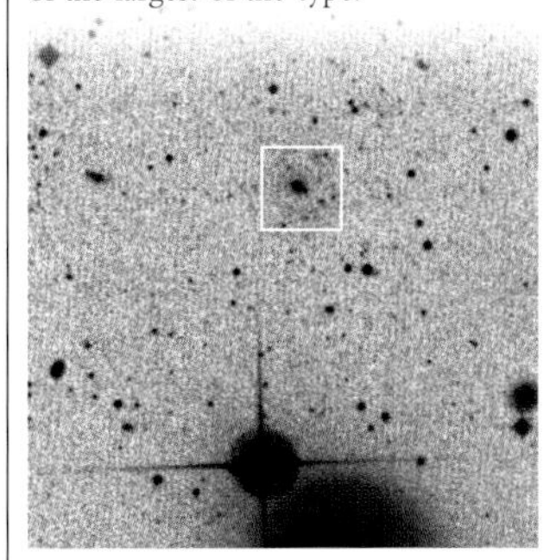

**MALIN 1 IN A NEGATIVE IMAGE**

# ACTIVE GALAXIES

material blasted from the nucleus expands into a lobe as it is slowed by the intergalactic medium


28–31 Matter
32–35 Radiation
38–41 Space and time
226–29 The Milky Way
294–97 Types of galaxy
298–99 Galaxy evolution


MANY GALAXIES ACROSS THE UNIVERSE show surprising features that mark them as out of the ordinary. Although there are several types of these strange galaxies, their unusual behaviour can always be traced back to powerful activity in their nucleus – it seems that there is an underlying similarity between them, and for this reason they are often studied together under the term "active galaxies".

## WHAT ARE ACTIVE GALAXIES?

Astronomers think that the features of active galaxies are linked to their central giant black holes. Most, if not all, galaxies have black holes with the mass of many millions of suns, known as supermassive black holes, at their nuclei (see p.297), but most such black holes are dormant – all material in these galaxies is in a stable orbit around the black hole. In active galaxies, matter is still falling inwards, and as it falls it is heated by intense gravity, generating a brilliant blast of radiation. As the black hole "engine" pulls matter in, the superheated material forms a spiralling accretion disc. The hot disc emits X-rays and other fierce, high-energy radiation. Around the outer edge of the disc, a dense torus (doughnut-shape) of dust and gas forms. The intense magnetic field surrounding the black hole also catches some of the infalling material, firing it out as two narrow beams at the poles, at right-angles to the plane of the accretion disc. These jets shine with radio-wavelength radiation, due to the synchrotron mechanism (right).

jet of particles shooting from black hole's magnetic pole

star being ripped apart by intense gravity

location of black hole

spinning accretion disc of heated gas

torus of dust, typically 10 light-years across

jet expands into lobe thousands of light-years long

**BLACK-HOLE ENGINE**
The black hole of an active galactic nucleus is surrounded by a bright accretion disc and an outer dust cloud. Jets of material flow outwards from the black hole's poles.

electron

magnetic field line

photon of radio-wavelength radiation

**SYNCHROTRON RADIATION**
As electrons from the black-hole jets move through the black hole's magnetic field, they are forced into spiral paths, releasing synchrotron radiation – a type of EM radiation that is most intense at long radio wavelengths.

## ACTIVE TYPES

Astronomers distinguish between four major types of active galaxy. Each displays its own set of active features, and in each case these features are evidence of the violent activity at the nucleus. Radio galaxies are the most intense natural sources of radio waves in the sky. The emissions typically come from two huge lobes on either side of an apparently innocuous parent galaxy (and often linked to it by narrow jets). Seyfert galaxies are relatively normal spirals with a compact, luminous nucleus that may vary in brightness over just a few days. Quasars appear as starlike points of light that show similar but more extreme variability. Red-shifted lines in their spectra reveal that they are extremely distant galaxies – powerful modern telescopes can resolve them as galaxies with incredibly brilliant cores. They are more powerful and more distant cousins of the Seyfert galaxies. Finally, blazars (also known as BL Lacertae objects) are starlike variable points similar to quasars, but with no significant lines in their spectra. The standard model of the black-hole engine (above) can explain the major features of each type – how the galaxy appears depends on the intensity of its activity, and the angle at which we see it.

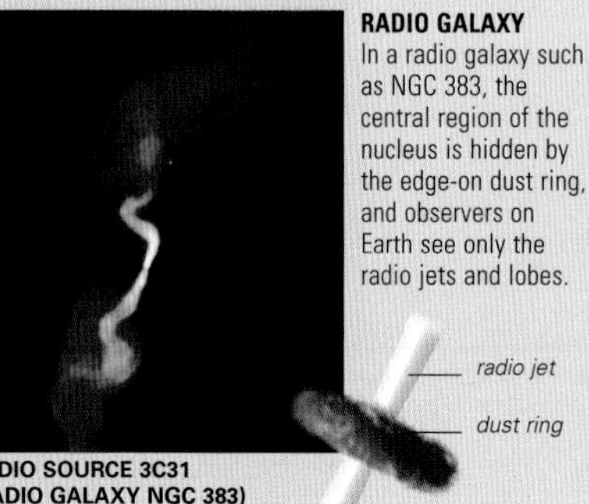
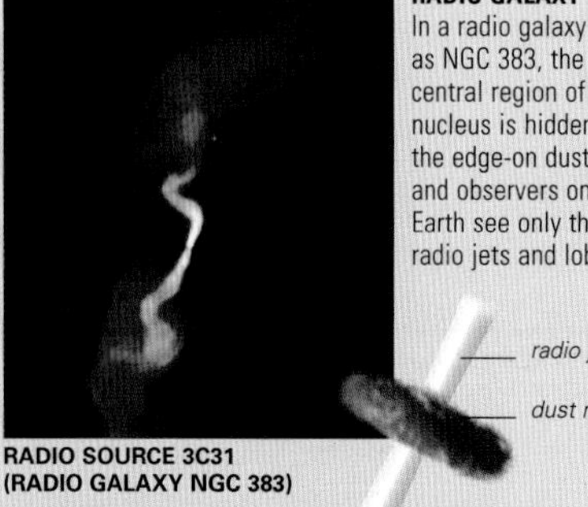


**RADIO SOURCE 3C31 (RADIO GALAXY NGC 383)**

**RADIO GALAXY**
In a radio galaxy such as NGC 383, the central region of the nucleus is hidden by the edge-on dust ring, and observers on Earth see only the radio jets and lobes.

**QUASAR PG 0052+251**

**QUASAR**
In quasars, Earth-bound observers can see over the dust ring, and brilliant light from the nucleus and disc drowns out the light of the surrounding galaxy.

**BLAZAR 3C 279**

**BLAZAR**
Blazars are active galaxies aligned so that observers on Earth look straight down the black-hole jet onto the nucleus. The galaxy is hidden by the brilliant light, but radio lobes can sometimes be detected, as in blazar 3C 279.

**SEYFERT GALAXY M106**

**SEYFERT GALAXY**
In Seyfert galaxies such as M106, the nucleus and accretion disc are exposed to our view, as in a quasar, but the activity is weak.

# THE HISTORY OF ACTIVE GALAXIES

The distribution of different types of active galaxies in the Universe provides clues about how they evolve. Quasars and blazars are never seen close to Earth. They are always faint and distant, with red shifts indicating that they lie billions of light-years from Earth – we are seeing them as they were in much earlier times. Radio and Seyfert galaxies, in contrast, are scattered throughout the nearby Universe, and radio jets are linked to both spiral and elliptical galaxies. So what happened to the quasars and blazars? It seems likely that they represent a brief phase in a galaxy's evolution, soon after its birth. At this time, material in the central regions would have had chaotic orbits, and the central black hole engine would have been fuelled by a continuous supply of infalling stars, gas and dust. As the black hole swept up the available matter, objects with stable orbits at a safe distance remained. Starved of fuel, the engine would have petered out, and the quasar became dormant – a normal galaxy such as the Milky Way. Today, such galaxies can become active again if they are involved in collisions that send new material falling in towards the black hole. Many nearby radio and Seyfert galaxies show evidence of recent collisions or close encounters, and some of these galaxies are close enough for infrared telescopes to image the dust rings around their cores directly (see p.313). However, levels of recent activity are restrained – even the most spectacular radio galaxies generate little energy compared to quasars, while Seyferts are the feeblest type of active galaxy.

**NUDGED BACK INTO LIFE**
Optical images of Centaurus A clearly show the dark dust lane of a spiral colliding with this elliptical galaxy. The overlaid radio map shows the burst of activity – the jets and plumes – triggered by this event.

*false-colour radio image of jet of particles*

*dust lane (optical image)*

*optical view of galaxy's elliptical arrangement of stars*

*false-colour radio image of galaxy's lobe*

*disc of spiral galaxy*

*jet of particles emitting radio waves*

*active nucleus of galaxy, containing an active black hole surrounded by a bright accretion disc and a dust ring*

**ACTIVE GALAXY**
This idealized active galaxy is a spiral with a bright nucleus, which hides an active black hole. From the black hole's poles blast two jets of particles, leaving at close to light speed, only slowing and billowing out into lobes many thousands of light-years away, as the particles hit the intergalactic medium.

EXPLORING SPACE

## SUPERLUMINAL JETS

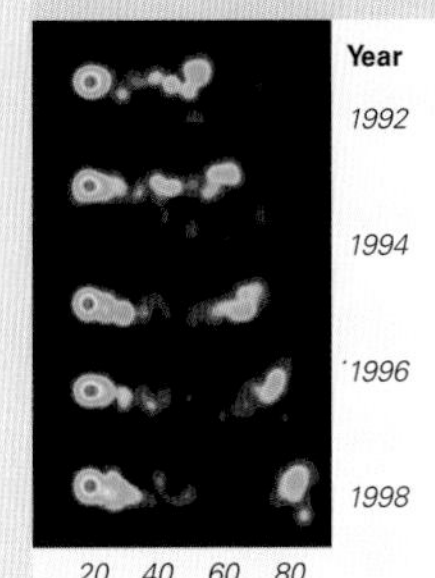


Some quasars and blazars appear to defy the laws of physics. Image sequences, taken years apart, show jets of material blasting away from the nucleus, apparently travelling faster than the speed of light. This apparent motion is called "superluminal". In reality, it is an illusion, created when jets travelling at very high speeds, of up to 99 per cent of the speed of light, happen to be pointing almost directly towards us.

**TIME-LAPSE SEQUENCE**
These images show jet emissions from blazar 3C 279, taken at intervals of almost two years, and showing motion apparently five times the speed of light.

# IS THE MILKY WAY ACTIVE?

The Milky Way galaxy, like any galaxy with a central black hole, has the potential to be active, and there is intriguing evidence that it might have burst into activity in the recent past. In 1997, scientists discovered a huge cloud of gamma-ray emission above the galactic centre. The radiation has a distinctive frequency, suggesting it is the result of electrons encountering positrons – their antimatter equivalent (see p.31) – and annihilating in a burst of energy. The positrons might have been generated by activity at the core – perhaps an infall of matter into the black hole – and are now meeting scattered electrons in the outer galaxy and mutually annihilating to produce the distinctive glow. Since the clouds lie just 3,000 light-years from the galactic centre, the activity must have occurred recently.

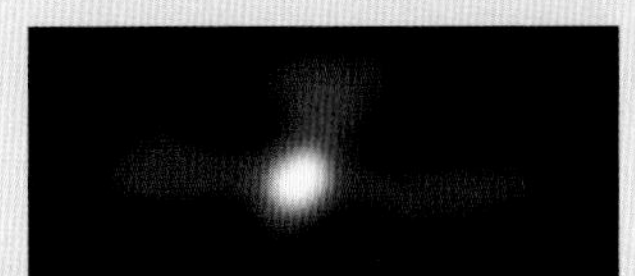

**ANTIMATTER FOUNTAIN**
This gamma-ray image traces positrons (antielectrons) around the Milky Way. The horizontal feature is the plane of the Galaxy, with the fountain above it.

**ACTIVITY AT THE CORE**
This Chandra X-ray image of the Milky Way's nucleus shows brilliant high-mass stars and a supernova remnant around the centre. Material from this crowded region may occasionally feed the black hole.

# ACTIVE GALAXIES

There are no simple rules governing the appearance of active galaxies. Some have a disrupted structure, seen either in visible light or at other wavelengths, while others appear normal at first, but radiate unusually large amounts of energy at certain wavelengths. In fact, the majority of galaxies show activity of one kind or another. However, a smaller proportion of galaxies have particularly active nuclei, powered by matter spiralling into their central black hole. These include Seyfert galaxies, radio galaxies, quasars, and blazars. The vast majority of known active galaxies are distant quasars. Objects lying nearer to the Milky Way, although less spectacularly violent, are at least close enough for astronomers to study in detail.

**JET FROM AN ACTIVE GALAXY**
Pictured in radio waves and false colours, this jet of particles blasted from the core of the galaxy M87 is a typical feature of active galaxies with black-hole engines.

TYPE-II SEYFERT GALAXY

## Circinus Galaxy

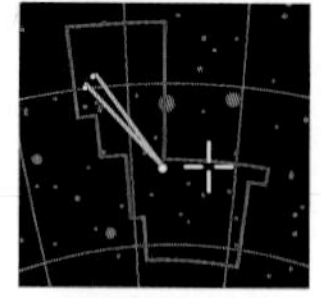

CIRCINUS

**CATALOGUE NUMBER** ESO 97-G13
**SHAPE** Sb spiral
**DISTANCE** 13 million light-years
**DIAMETER** 37,000 light-years
**MAGNITUDE** 11.0

Although the spiral galaxy in Circinus is probably the nearest active galaxy to Earth, it went undiscovered until just a few decades ago. It remained hidden for so long partly because it lies just 4 degrees below the plane of the Milky Way and is obscured by star clouds. The full extent of the Circinus Galaxy's extraordinary nature was revealed only when it was observed by the Hubble Space Telescope in 1999. The galaxy is a Seyfert (see p.310) – a spiral with an unusually bright, compact region at its core, thought to result from material slowly drifting onto a massive central black hole. Hubble's infrared camera revealed how the galaxy's gas is concentrated in a central ring, just 250 light-years in diameter, around the black hole. Also apparent is a loose outer ring in the plane of the galaxy, around 1,300 light-years across, where great bursts of star formation are occurring. Finally, Hubble showed a cone-shaped cloud billowing above the plane of the galaxy. This is matter ejected by the magnetic fields of the black hole and glows as it is heated by the ultraviolet radiation from the nucleus.

**CONE OF MATTER**
The pinkish-white region near the core of the Circinus Galaxy shows where matter is being flung out, in a cone shape, from the central black hole into the gas cloud above the galaxy.

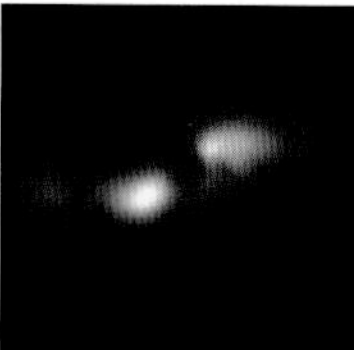

RADIO CONTINUUM

RADIO (21CM WAVELENGTH)

OPTICAL WAVELENGTHS

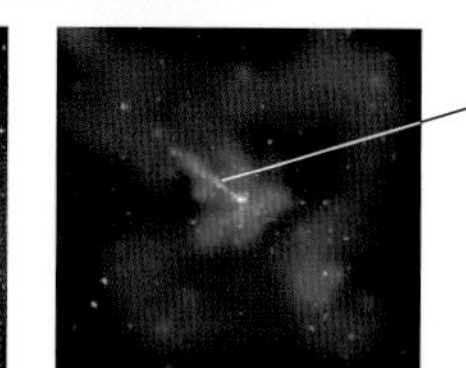


X-RAY WAVELENGTHS

**COMPOSITE PICTURE**
Astronomers have captured images of Centaurus A in radio, optical, and X-ray radiation, combining them to make the composite at far left.

**DUSTY DISC**
This Hubble Space Telescope close-up of Centaurus A (right) reveals dark interstellar dust, glowing orange gas clouds, and brilliant blue star clusters formed in the collision between two galaxies.

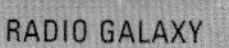

RADIO GALAXY

## Centaurus A

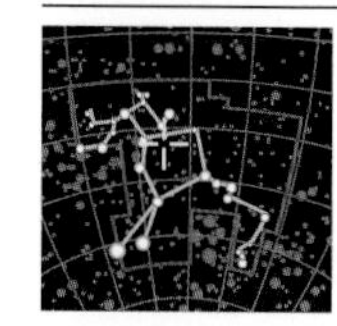

CENTAURUS

**CATALOGUE NUMBER** NGC 5128
**SHAPE** Peculiar elliptical
**DISTANCE** 15 million light-years
**DIAMETER** 80,000 light-years
**MAGNITUDE** 7.0

A ball of old yellow stars, NGC 5128 shows some features typical of an elliptical galaxy, but its most striking aspect is the dark dust lane that cuts across it, bisecting the uniform glow of stars with a ragged silhouette. What is more, the galaxy is at the centre of a pair of vast radio lobes, 1 million light-years across. The name of this radio source, Centaurus A, is now the most widely used name for the galaxy itself. Astronomers have studied Centaurus A in detail at a range of wavelengths. The Hubble Space Telescope looked through the dust lanes with its infrared camera and found a huge accretion disc at the centre – a sure sign of an active black hole pulling in matter at Centaurus A's core.

It is now generally agreed that NGC 5128 is an elliptical galaxy absorbing a spiral. The ghost of the spiral is shown by the dust lane and by the bright star clusters that stud it – perhaps generated by shock waves as the two galaxies merge.

RADIO GALAXY

## M87

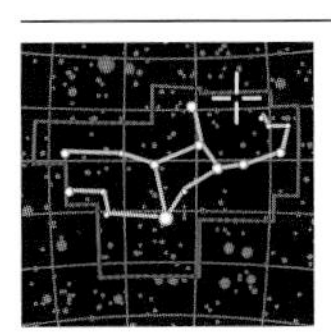

VIRGO

| | |
|---|---|
| **CATALOGUE NUMBERS** | M87, NGC 4486 |
| **SHAPE** | E1 giant elliptical |
| **DISTANCE** | 60 million light-years |
| **DIAMETER** | 120,000 light-years |
| **MAGNITUDE** | 8.6 |

**PARTICLE JET**
The blue glow of M87's jet results from synchrotron radiation – light emitted by electrons spinning through an intense magnetic field.

Lying at the heart of the Virgo galaxy cluster (see p.319), M87 is the closest example of a giant elliptical galaxy – a class of galaxy often found at the cores old galaxy clusters. This huge ball of stars seems to have a diameter roughly equivalent to that of the Milky Way, but, because its stars are distributed across its spherical structure, it contains many more stars – probably several trillion. Long-exposure photographs have revealed that the galaxy also has an extensive halo of more loosely scattered stars, extending well beyond the central region in a more elongated shape. The galaxy also has an unrivalled collection of globular star clusters in orbit – some astronomers estimate as many as 15,000 such groups.

What is more, M87 is an active galaxy – its location coincides with the Virgo A radio source, and with a strong source of X-rays. There is even a sign of this activity that is visible at optical wavelengths, in the form of a long, narrow jet of material being blasted from its interior.

**BALL OF STARS**
M87's full extent is shown in this wider image. The elliptical galaxy is remarkably uniform, and the star-like specks around it are just a few of its many thousands of globular clusters.

TYPE-II SEYFERT GALAXY

## Fried Egg Galaxy

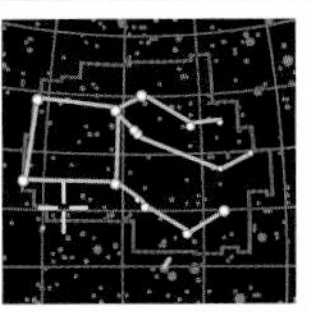

PEGASUS

| | |
|---|---|
| **CATALOGUE NUMBER** | NGC 7742 |
| **SHAPE** | Sb spiral |
| **DISTANCE** | 72 million light-years |
| **DIAMETER** | 36,000 light-years |
| **MAGNITUDE** | 11.6 |

The small spiral galaxy NGC 7742 resembles a fried egg because of the intense yellow glow from its core. The core is much brighter than is usual for a galaxy of this size, because this is a Seyfert galaxy, with a moderately active core. Seyferts emit radiation across a broad band of wavelengths – NGC 7742 is a Type-II – a galaxy that is brightest in infrared light.

**CELESTIAL EGG**

## CARL SEYFERT

US astronomer Carl Seyfert (1911–60) was the son of a pharmacist from Cleveland, Ohio. He studied at Harvard and went on to work at McDonald Observatory, then at Mount Wilson in California. It was here that he first identified the class of galaxies with unusually bright nuclei that bear his name (see p.310). In 1951, he also discovered Seyfert's Sextet, an interesting, compact cluster of galaxies (see p.319).

**SEYFERT'S OBSERVATORY**
At Nashville, Seyfert found time to give public lectures as well as raising support and supervising the construction of the Arthur J. Dyer Observatory (above).

RADIO GALAXY

## NGC 4261

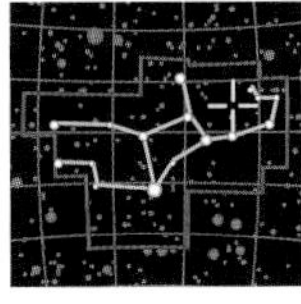

VIRGO

| | |
|---|---|
| **CATALOGUE NUMBER** | NGC 4261 |
| **SHAPE** | E1 elliptical |
| **DISTANCE** | 100 million light-years |
| **DIAMETER** | 60,000 light-years |
| **MAGNITUDE** | 10.3 |

**RADIATING PLUMES**
Combining optical and radio images of NGC 4261 reveals its full extent. The visible part of the galaxy is the white blob in the centre, while the orange plumes mark the radio-emitting regions.

The elliptical galaxy NGC 4261 lies at the centre of two great lobes of radio emission measuring 150,000 light-years from tip to tip. In many ways a typical radio galaxy, it is also one of the few active elliptical galaxies to have revealed its internal structure to astronomers. Infrared images from the Hubble Space Telescope pierced the obscuring clouds of stars to reveal an unexpectedly dense disc of dusty material, apparently spiralling onto the galaxy's central black hole.

Most elliptical galaxies are thought to be relatively dust-free, so where did the material in NGC 4261 come from? The most likely answer is that the elliptical galaxy has merged with a spiral in its relatively recent history. The spiral's individual stars have now become indistinguishable from the stars that were originally part of the elliptical galaxy, but the ghostly outline of the galaxy's gas and dust remains.

**DUST WHIRLPOOL**
The Hubble Space Telescope's close-up image of the core reveals a dusty spiral of matter within a ring of glowing outer clouds. A distinct cone shows where matter is being flung off from the active galactic nucleus into the radio lobes.

TYPE-I SEYFERT GALAXY

## NGC 5548

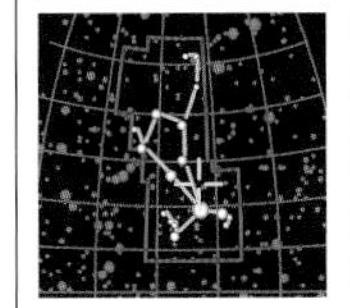

BOÖTES

| | |
|---|---|
| **CATALOGUE NUMBER** | NGC 5548 |
| **SHAPE** | Sb spiral |
| **DISTANCE** | 220 million light-years |
| **DIAMETER** | 100,000 light-years |
| **MAGNITUDE** | 10.5 |

NGC 5548 is a Type-I Seyfert galaxy – that is, a Seyfert that emits more ultraviolet and X-ray radiation than visible light. Like all Seyferts, it has a bright, compact core, but, unlike the Fried Egg Galaxy (see above), its core is an intense blue-white. Using the Chandra X-ray telescope, astronomers have detected an envelope of warm gas expanding around the core. The gas eventually forms two lobes of weak radio emission around the galaxy.

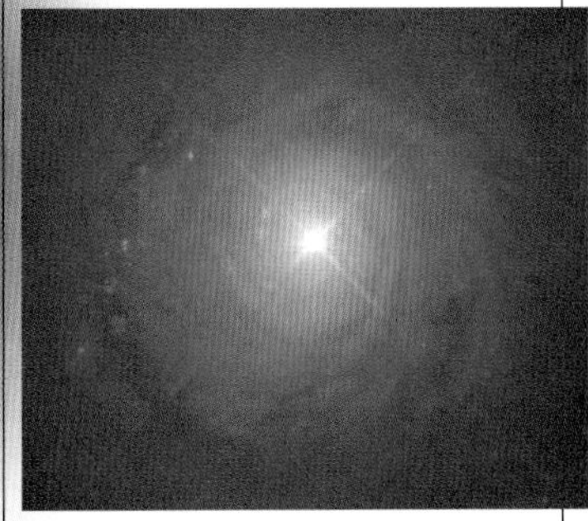

**HUBBLE IMAGE OF NGC 5548**

RADIO GALAXY

# NGC 1275

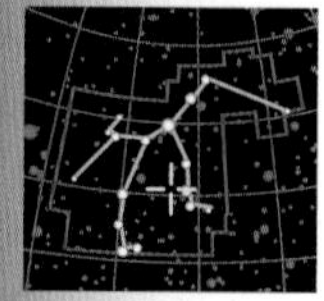


**CATALOGUE NUMBER** NGC 1275

**SHAPE** Elliptical and distorted spiral

**DISTANCE** 235 million light-years

**DIAMETER** 70,000 light-years

**MAGNITUDE** 11.6

Despite being catalogued as a Seyfert galaxy by Carl Seyfert himself (see p.313), NGC 1275 has remained a mystery. Recent observations have shown that there are two objects – one in front of the other. A ghostly spiral galaxy, revealed by its bright blue star clusters, is responsible for the dust lanes that cross the bright central region, but this brighter region is in fact a separate galaxy. Despite its Seyfert-like core, it is an elliptical, not a spiral. This galactic giant lies at the heart of the Perseus galaxy cluster, and the foreground spiral is racing towards it at 10.8 million kph (6.7 million mph), its structure already disrupted by the elliptical's gravity. Adding to the complexity, the elliptical galaxy is also a radio source, and some astronomers have argued that it shows blazar-like activity (see BL Lacertae, opposite). Whatever the details, NGC 1275 displays many of the typical features of an active galactic nucleus.

**COLLISION DEBRIS**
Young, blue star clusters in the dark dust of NGC 1275 show that the galactic collision triggered star birth.

**CLUSTERS IN THE NUCLEUS**
The core of NGC 1275 offers clues to the origin of globular clusters – numerous globular-like clusters are found here, but they are composed of young blue, rather than old yellow stars.

RADIO GALAXY

# Cygnus A

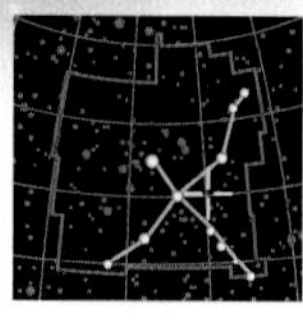


**CATALOGUE NUMBER** 3C 405

**SHAPE** Pec (peculiar)

**DISTANCE** 600 million light-years

**DIAMETER** 120,000 light-years (excluding radio lobes)

**MAGNITUDE** 15.0

The most spectacular and powerful radio galaxy in the nearby Universe, Cygnus A was discovered as soon as radio telescopes began operating in the 1950s. It features two huge lobes of material emitting radio waves. The lobes are visibly linked to their origin at the heart of a faint, central, elliptical galaxy by two long narrow jets. From lobe to lobe, the entire structure extends over half a million light-years.

Despite its prominence in the radio sky, mysteries still surround Cygnus A, largely because of its great remoteness. Early observations led astronomers to believe the central galaxy was in fact a pair of colliding galaxies. Hubble Space Telescope images suggested a resemblance to NGC 5128, the Centaurus A galaxy (see p.312), which is thought to be an elliptical galaxy that has recently swallowed a spiral. Recent detection of a large cloud of red-shifted gas moving through the Cygnus A galaxy suggests that a collision may indeed be the root cause of the activity.

Astronomers have also argued about the origin of the "hot spots", where the radio lobes glow brightest at either end. Studies by the Chandra X-ray telescope have shown that Cygnus A lies at the centre of a cloud of hot but sparse gas. The jets have blown out a rugby-ball-shaped cavity in the gas so vast that it dwarfs the central galaxy. Tendrils of gas, which are emitting X-rays and radio waves, are also falling back down through the cavity onto the poles of the galaxy, drawn by its gravitational pull. The hot spots are apparently created where the outward blast of the jets collides with the hot gas falling inwards.

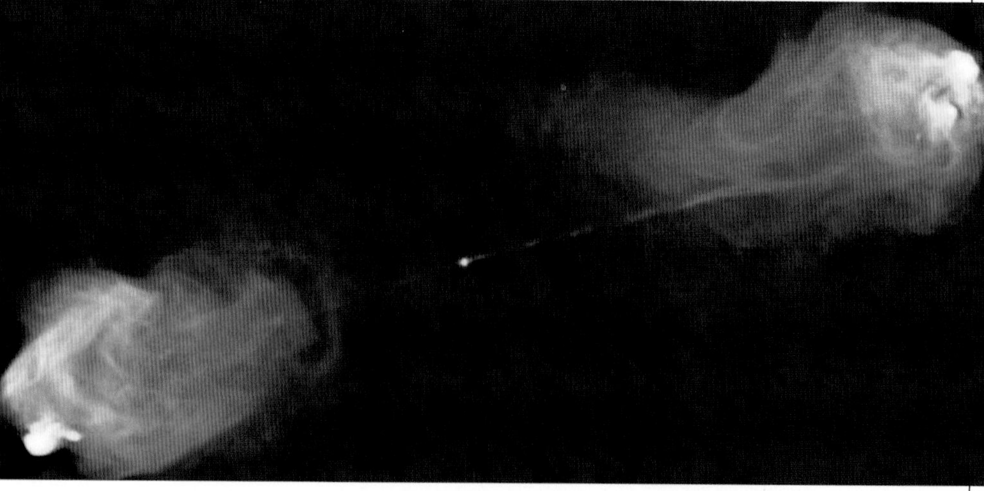

**LOBES EMITTING RADIO WAVES**
This radio map of Cygnus A shows the galaxy's extremely narrow jets blasting from its core, the hot spots at the end its radio lobes, and the tendrils of hot gas falling back towards the central galaxy.

BLAZAR (BL LAC OBJECT)

## BL Lacertae

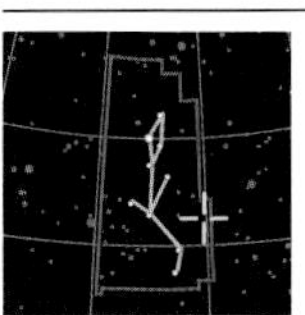

LACERTA

**CATALOGUE NUMBER** BL Lac

**SHAPE** Elliptical

**DISTANCE** 1 billion light-years

**DIAMETER** Unknown

**MAGNITUDE** 12.4–17.2

BL Lacertae (BL Lac for short) was first catalogued as an irregular variable star by German astronomer Cuno Hoffmeister in the 1920s. Since then, astronomers' understanding of the object has changed. For a variable star, it was very mysterious, showing rapid but completely unpredictable variations. At the same time, it displayed a totally featureless spectrum – it had neither the dark absorption lines seen in stars, nor the bright emission lines found in galaxies (see p.33). It was not until 1969, when BL Lac was found to be a strong radio source, that astronomers realized it might be a new type of active galaxy. Today it is seen as the founder member of a class of active galaxies called blazars or BL Lac objects. Blazars show many similarities to quasars but also some differences, most notably their featureless spectra.

The mystery of BL Lac was solved in the 1970s, when two astronomers blocked out or "occulted" BL Lac's bright core to study its surroundings. This revealed that it was embedded in a faint elliptical galaxy, whose light was normally drowned out. Red-shifted lines in the spectrum of this galaxy confirmed BL Lac's great distance (see p.42). Today, blazars are accepted as rare cases in which Earth's position happens to align directly with the jet of material blasting out of an active galactic nucleus, with no obscuring material in the way.

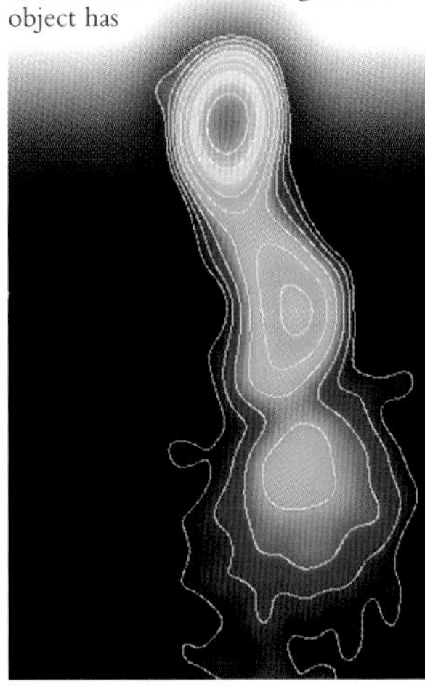

**MAP OF A BLAZAR**
This radio map of BL Lacertae shows the intensity of radiation (contour lines) and also its polarization (colour) – an indication of magnetic field strength. The red object at the top is the galaxy's nucleus, while the lower regions are parts of a radio jet.

QUASAR

## PKS 2349

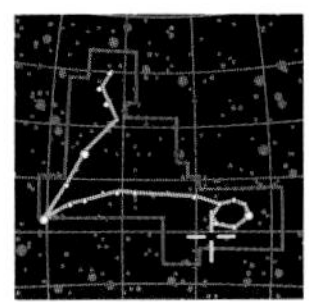

PISCES

**CATALOGUE NUMBER** PKS 2349

**SHAPE** Disrupted

**DISTANCE** 1.5 billion light-years

**DIAMETER** Unknown

**MAGNITUDE** 15.3

The Hubble Space Telescope offered astronomers an unprecedented chance to study quasars in detail during the 1990s. One of their most intriguing subjects was the otherwise undistinguished quasar PKS 2349 (referred to by its designation in the catalogue of the Australian Parkes radio telescope). For the first time, astronomers were able to see the faint host galaxies surrounding quasars, as well as other galaxies close to the quasars. The images showed that in many cases quasars do not just sit at the centres of their host galaxies, but are involved in violent interactions with neighbouring galaxies and other quasars. PKS 2349 was referred to as a "smoking gun" because it showed these interactions so clearly. The quasar is surrounded by a ring of faint material that may mark the outline of its host galaxy – though, if so, the quasar itself is remarkably "off-centre". A small companion galaxy, about the size of the Large Magellanic Cloud (see p.300), also lies nearby and seems doomed to collide with the quasar itself.

**QUASAR CLOSE-UP**
In Hubble's image of PKS 2349, the quasar is the bright central object, the companion galaxy is the smaller bright region above it, and the supposed host galaxy is the fainter ring extending from the quasar.

QUASAR

## 3C 273

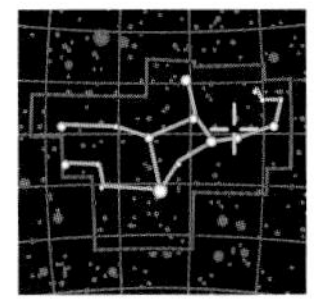

VIRGO

**CATALOGUE NUMBERS** 3C 273, PKS 1226+02

**SHAPE** E4 elliptical

**DISTANCE** 2.1 billion light-years

**DIAMETER** 160,000 light-years (excluding jet)

**MAGNITUDE** 12.8

The brightest quasar in the sky, 3C 273 was the second to be discovered. The existence of this radio source was already known when, in 1963, Australian astronomer Cyril Hazard used an occultation by the Moon (see p.65) to precisely establish its position, linking the radio source to what appeared to be an irregular variable star. The star's spectrum had a forest of unidentifiable dark emission lines (see p.33). Astronomers finally realized that the lines could have been formed by hydrogen, oxygen, and magnesium if the light was heavily red-shifted and its source was racing away from us at 16 per cent of the speed of light, or 173 million kph (107 million mph). We now know that the object is not a star, but a distant active galaxy.

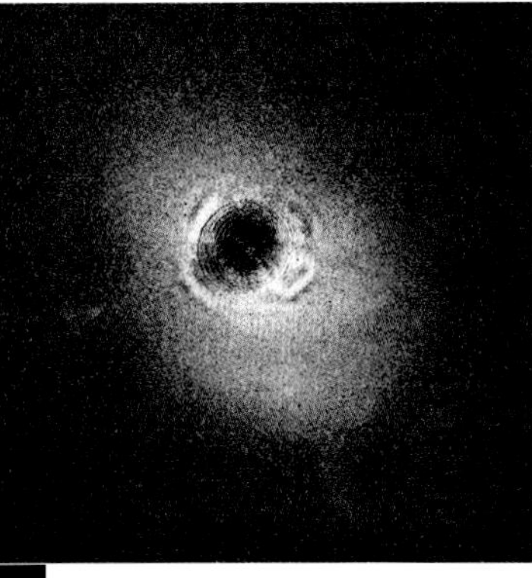

**HOST GALAXY**
By blocking the light from 3C 273's nucleus, the Hubble Space Telescope was able to photograph detail (right) in the fainter surrounding galaxy.

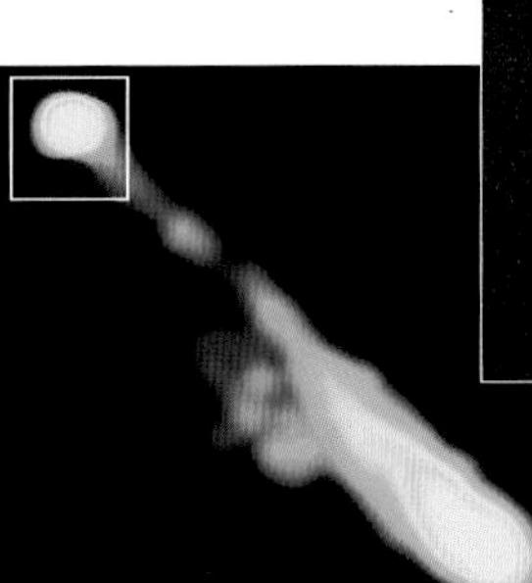

**RADIO JET**
This long-wavelength radio image shows a jet shooting out 1.2 million light-years from 3C 273's core (top left of the image). The end of the jet is as bright as the core.

QUASAR

## 3C 48

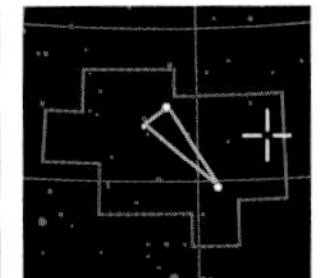

TRIANGULUM

**CATALOGUE NUMBERS** 3C 48, PKS 0134+029

**SHAPE** SB interacting

**DISTANCE** 2.8 billion light-years

**DIAMETER** 100,000 light-years

**MAGNITUDE** 16.2

The radio source 3C 48 has a unique place in the history of the study of active galaxies. It was detected in the 1950s, and in 1960 Allan Sandage (see panel, below) confirmed that it coincided with a faint, blue, star-like object. The object's spectrum revealed strange emission lines (see p.33) that could not have been emitted by any known element. Studies of similar lines in the optical counterpart of 3C 273 (left) suggested that the lines of 3C 48 were hydrogen lines with a huge red shift, suggesting the object was extremely distant and receding at great speed. 3C 48 was therefore the first quasi-stellar object, or quasar, to be discovered.

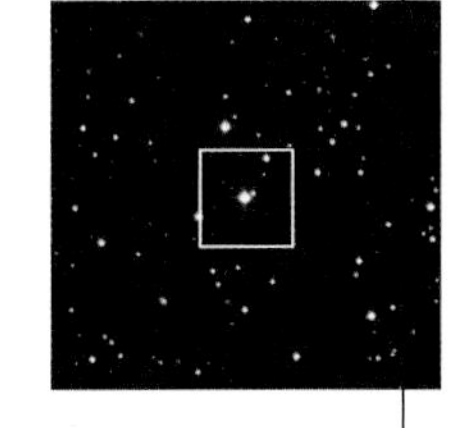

**FIRST QUASAR**
At first, 3C 48 is indistinguishable from foreground stars. It was only its unpredictable variability and radio emission that marked it out as something special.

### ALLAN SANDAGE

Beginning his astronomical career as a student under Edwin Hubble himself (see p.43), Allan Sandage (b.1926) has had a great influence on our understanding of the Universe's evolution. Sandage's studies have focused on detecting Cepheid variable stars in distant galaxies, for use in measuring cosmological expansion. His many quasar discoveries were a natural offshoot from his studies of deep space.

# GALAXY CLUSTERS

22–23 The scale of the Universe
24–27 Celestial objects
36–37 Gravity, motion, and orbits
38–41 Space and time
42–43 Expanding space
294–97 Types of galaxy

GALAXIES ARE NATURALLY GREGARIOUS. Pulled together by their enormous gravity, they cluster tightly, sometimes orbiting one another, often colliding. As galaxies slowly move within a cluster, the cluster's structure changes. The evolution of clusters can tell astronomers about dark matter, and clusters can even be used as cosmic "lenses" to peer back into the early Universe.

**DENSE CLUSTER**
The massive galaxy cluster Abell 1689 lies 2.2 billion light-years away. The yellow elliptical galaxies are surrounded by arcs of light, which are images of more distant galaxies distorted by the cluster's gravitational lensing.

## TYPES OF CLUSTERS

Some galaxy clusters are sparse, loose collections of galaxies. The smallest clusters are usually termed "groups". The Local Group (see p.318), of which the Milky Way is a member, is one such cluster. Other clusters, such as the nearby Virgo Cluster (see p.319), are denser, containing many hundreds of galaxies in a chaotic distribution. Yet other clusters, such as the Coma Cluster (see p.320), are even more dense, with galaxies settled into a neat, spherical pattern around a centre dominated by giant elliptical galaxies. Although clusters differ in density, the volume of space they occupy is generally the same – a few million light-years across. Not all galaxies exist in clusters – there are more isolated "field galaxies" than there are cluster galaxies. Some galaxy types do not exist outside clusters, however. Giant ellipticals (see p.296) always lie near the centre of large clusters, as do vast, diffuse cD galaxies (below right). The most numerous cluster components may be invisible, including faint, diffuse dwarf elliptical galaxies and proposed "dark galaxies". A dark galaxy would consist of hydrogen gas and material too thin to condense and ignite stars. The first such galaxy may have been found, in the Virgo Cluster, in early 2005.

*Andromeda Galaxy (M31)*

*Milky Way*

**SPARSE CLUSTER**
This sparse cluster, or group, of galaxies is in fact the Local Group, containing the Milky Way and its galactic neighbours. Most of the galaxies are orbiting either the Milky Way or the Andromeda Galaxy (M31).

*dense core of cluster containing many large galaxies*

**IDEAL DENSE CLUSTER**
A dense cluster occupies the same volume as a sparse cluster such as the Local Group, but the galaxies are mainly elliptical and have a roughly spherical distribution around the cluster's centre.

**DWARF ELLIPTICAL**
Most galaxies in the Local Group, including the Sculptor Dwarf, are dwarf ellipticals. They are invisible in distant clusters, but must be present.

**cD GALAXY**
cD galaxies are similar to giant ellipticals but have extensive, sparse outer haloes of stars. They sometimes have hints of multiple cores, suggesting the merger of several smaller ellipticals. NGC 4889 (left) is a cD galaxy at the heart of the dense Coma Cluster.

**ABELL 2029**
This visible-light image of Abell 2029 shows that it is an old, regular, spherical cluster full of elliptical galaxies.

**INTERGALACTIC GAS**
An X-ray image of cluster Abell 2029 shows the hot gas cloud around its centre. If not for the gravity of the cluster's dark matter, this gas would escape.

# THE INTERGALACTIC MEDIUM

Astronomers can estimate the overall mass of a galaxy cluster from the way in which its galaxies are moving, but also through the phenomenon of gravitational lensing – an effect of general relativity (see pp.40–41). When a compact cluster lies in front of more distant galaxies, its mass bends the light passing close to it and deflects distorted images of the distant galaxies towards Earth. By measuring the strength of this effect, it is possible to measure the mass of the cluster and model how it is distributed. Galaxy clusters contain far more mass than the visible galaxies can account for and most of it is in the matter that permeates the space between galaxies. This intergalactic medium is distributed around the cluster's centre, rather than around the galaxies.

X-ray satellites such as Chandra have revealed the nature of part of this material – large galaxy clusters often contain huge clouds of sparse, hot gas, glowing at X-ray wavelengths. Most is hydrogen, but heavier elements are present. It is thought to originate in the cluster galaxies, and to be stripped away during encounters and collisions. Most of a cluster's mass is not gas, however, but dark matter.

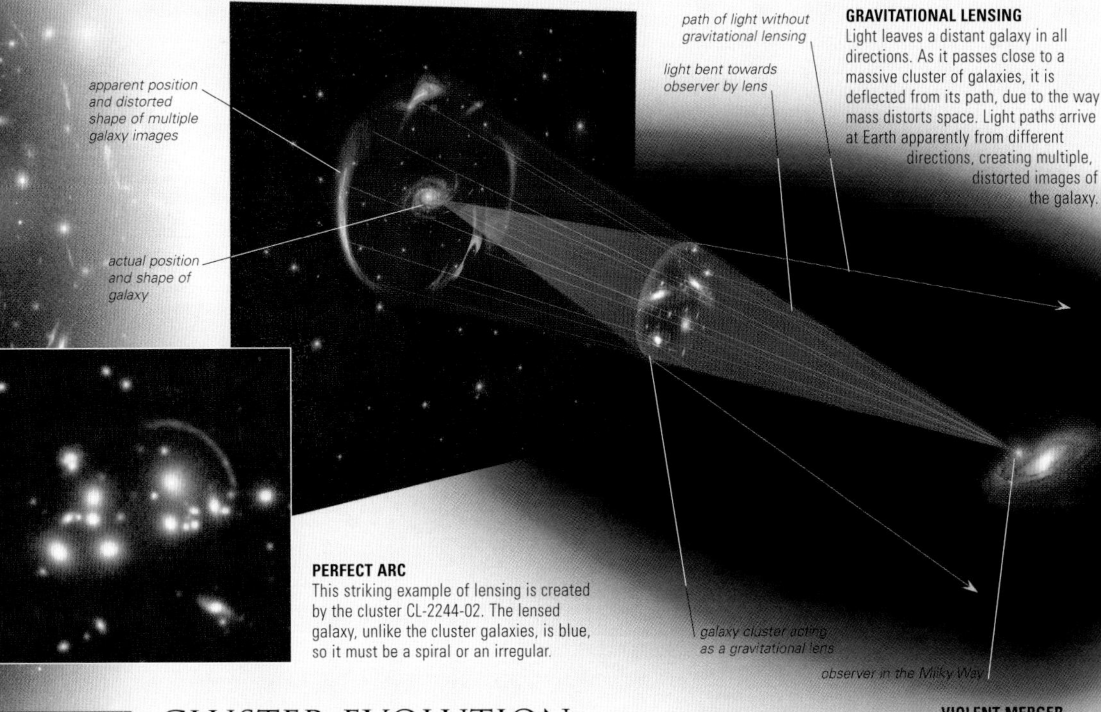


**GRAVITATIONAL LENSING**
Light leaves a distant galaxy in all directions. As it passes close to a massive cluster of galaxies, it is deflected from its path, due to the way mass distorts space. Light paths arrive at Earth apparently from different directions, creating multiple, distorted images of the galaxy.

**PERFECT ARC**
This striking example of lensing is created by the cluster CL-2244-02. The lensed galaxy, unlike the cluster galaxies, is blue, so it must be a spiral or an irregular.

# CLUSTER EVOLUTION

Astronomers have built a picture of cluster development that complements their models of galaxy evolution (see p.299). According to this thinking, galaxy clusters start as loose collections of gas-rich spirals, irregulars, and small ellipticals. Because of their proximity and huge gravity, the spirals tend to merge, regenerating as spirals or forming ellipticals. Each interaction drives off more of the galaxies' free gas into the intergalactic medium. The high temperature and speed of atoms in this medium prevents their recapture by the cluster's galaxies. At this stage, the cluster is irregular, or "unrelaxed", and the pattern of galaxies and intergalactic gas is irregular and chaotic. However, as galaxies swing round each other, their random motions are eliminated and they settle into a stable, spherical distribution around the cluster's centre. Eventually even the largest elliptical galaxies begin to merge, forming giant ellipticals and cD galaxies. The hot gas, freed from ties to individual galaxies, sinks into the centre of the cluster, where it lies evenly around the cluster's major elliptical galaxies. What remains is an old, spherical, "relaxed" cluster full of ellipticals.

**IRREGULAR AND RELAXED CLUSTERS**
The central regions of the Virgo Cluster (above) and the Coma Cluster (below) show the difference between an irregular and a more spherical (relaxed) pattern of galaxies.

**VIOLENT MERGER**
An optical image (left) of cluster Abell 400 shows two galaxies merging at its centre to form a giant elliptical. Radio images (below) reveal that they are both active radio galaxies. Such an event is typical of those that shape galaxy clusters.

# GALAXY CLUSTERS

The shape and size of galaxy clusters are thought to be linked to their evolution. Clusters range from small groups comprising young, gas-rich irregular and spiral galaxies, to highly evolved clusters dominated by giant ellipticals, with a central cloud of gas so hot that it emits X-rays. Astronomers can study details in nearby clusters that are too faint to see in distant clusters. Earth's neighbouring clusters do not offer a spectacle to stargazers, however, because clusters are so vast that their members are widely scattered across the sky. To appreciate clusters in a single picture, it is necessary to peer tens of millions of light-years into deep space.

**STEPHAN'S QUINTET**
This elegant group of five galaxies shows that clusters are constantly changing – two of its spiral galaxies are colliding, while a third is being distorted by their gravity and is doomed to collide with them one day.

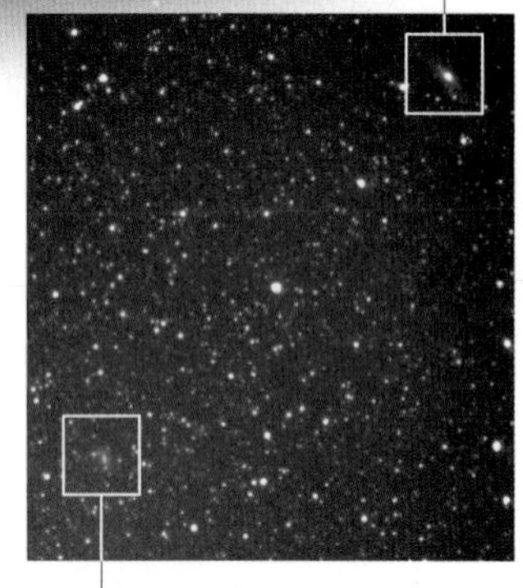


**LOCAL GROUP MEMBERS**
Since Earth is in the midst of the Local Group, the galaxies are scattered around the sky. However, two large members, M33 and M31, are near enough in the sky to appear in the same frame.

IRREGULAR CLUSTER

## Local Group

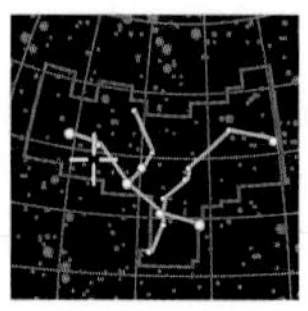

**DISTANCE**
0–5 million light-years

**NUMBER OF GALAXIES** 46

**BRIGHTEST MEMBERS**
Milky Way;
M31 (magnitude 3.5)

**ANDROMEDA AND TRIANGULUM**

The Local Group is the small galaxy cluster of which the Milky Way is a member. From Earth, its members appear dispersed throughout the sky, but some of its galaxies are grouped in the constellations of Andromeda and Triangulum. In space, the core of the group comprises about 30 members in a region just over 3 million light-years across. It is dominated by the Andromeda Galaxy (M31; see pp.302–303), and the Milky Way. Most of the smaller galaxies orbit close to one or the other of these large spirals. The third large spiral in the group, M33 (see p.301), may also be trapped in a long orbit around M31.

Outnumbering these spirals is a host of dwarf elliptical and irregular galaxies. Examples include SagDEG and the two Magellanic clouds (see pp.300–301), as well as M110 and M32, both ellipticals orbiting the M31 spiral. The Local Group appears to be relatively young. Its major galaxies are all spirals, and there is little matter in the space between galaxies – most of the cluster's gas is still trapped in the spirals. It is in an early state of cluster evolution. The Milky Way is currently colliding with the Magellanic Clouds, and is heading inexorably towards an ultimate merger with M31.

**BARNARD'S GALAXY**
This small, irregular galaxy (right), catalogued as NGC 6822, lies 1.7 million light-years away within the Local Group. It is rich in gas and dust, with many pinkish star-birth regions.

**FORNAX DWARF GALAXY**
This dwarf spheroidal galaxy (left) has no obvious nucleus. Such faint and diffuse galaxies are easily missed in more distant galaxy clusters, but they are probably the most numerous.

**THE MILKY WAY GALAXY**
A major member of the Local group is the Milky Way galaxy. Earth is within the galaxy's disc, so our view is edge-on and stretched acoss the sky.

IRREGULAR CLUSTER

## Sculptor Group

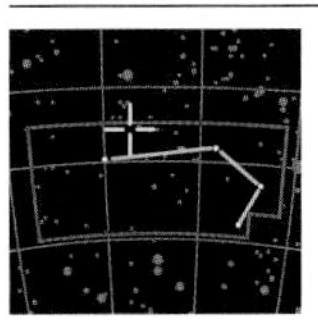

SCULPTOR

ALTERNATIVE NAME South Polar Group

DISTANCE 9 million light-years to centre

NUMBER OF GALAXIES 19 (6 major)

BRIGHTEST MEMBER NGC 253 (8.2)

Lying just beyond the gravitational boundaries of the Local Group, the Sculptor Group is similar in size to the Local Group. It is also a young cluster of irregular and spiral galaxies, with no major ellipticals. It is possible that this group, the Local Group, and another group called Maffei 1 were once part of the same larger cluster.

The nearest member to Earth is NGC 55, an irregular galaxy that, like the Large Magellanic Cloud (see p.300), shows enough structure for some astronomers to consider it a single-armed spiral. The dominant galaxy, however, is NGC 253. This large spiral is the same size as the Milky Way and more than twice the size of any other galaxy in the group.

**GALAXY NGC 253**
This large spiral dominates the Sculptor Group in this wide-field image. Most of the other galaxies are too faint to be seen without powerful telescopes.

**STARBURST GALAXY**
NGC 253 is a spiral starburst galaxy – a galaxy undergoing a surge of star formation. The surge may have been triggered by a series of supernovae.

IRREGULAR CLUSTER

## Virgo Cluster

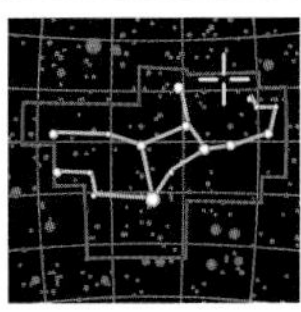

VIRGO

ALTERNATIVE NAME Virgo I Cluster

DISTANCE 52 million light-years to centre

NUMBER OF GALAXIES 2000 (160 major)

BRIGHTEST MEMBER M49 (9.3)

The Virgo Cluster is the nearest galaxy cluster worthy of the name; it is a dense collection of galaxies at the heart of the larger supercluster to which the Local Group also belongs. The contrast with smaller galaxy "groups" is striking – the Virgo Cluster contains around 160 major spiral and elliptical galaxies crammed into a volume little larger than that of the Local Group, along with more than 2,000 smaller galaxies. At its heart lie the giant ellipticals M87 (see p.313), M84, and M86, which are thought to have formed from the collisions of spirals over billions of years. Each giant elliptical seems to be at the centre of its own subgroup of galaxies – the cluster has not yet settled to become uniform. The cluster's gravity influences a huge region, extending as far as the Local Group and beyond – the Milky Way and its neighbours are falling towards the Virgo Cluster at 1.4 million kph (900,000 mph).

**CENTRE OF THE CLUSTER**
The Virgo Cluster's core has a high density of large galaxies. The two bright galaxies on the right are the ellipticals M84 and M86.

EXPLORING SPACE

### X-RAY IMAGING AND CLUSTER GAS

Many galaxy clusters are strong sources of X-rays, and orbiting X-ray telescopes can reveal features that remain hidden in visible-light images. While some X-ray sources are located at the centres of the cluster galaxies, the majority of radiation often comes from diffuse gas clouds, independent of the individual galaxies. The process that strips gas out of the cluster galaxies (see p.317) also heats it to generate the X-rays. The distribution of gas offers clues to a cluster's age and history.

**FORNAX IN X-RAYS**
This image of the Fornax cluster shows X-ray-emitting gas in blue. Both central galaxies have trailing plumes of gas, suggesting the entire cluster is moving through sparser clouds.

REGULAR CLUSTER

## Fornax Cluster

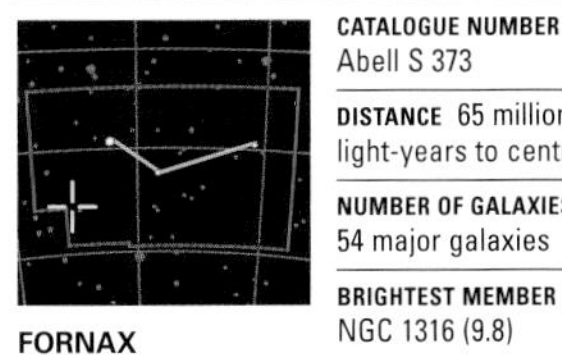

FORNAX

CATALOGUE NUMBER Abell S 373

DISTANCE 65 million light-years to centre

NUMBER OF GALAXIES 54 major galaxies

BRIGHTEST MEMBER NGC 1316 (9.8)

Fornax is home to a relatively nearby galaxy cluster, centred at around the same distance as the Virgo Cluster. However, the Fornax Cluster is at a later stage of evolution than the younger Virgo group. Here, spiral galaxies are rare – the cluster's major galaxies are mostly ellipticals, distributed evenly around the giant elliptical NGC 1399. Dwarf galaxies lying between the major ones are also mostly small ellipticals, suggesting that the cluster formed long ago and that interactions between its galaxies have had time to strip away most of their star-forming gas (see p.317). This account of the cluster's evolution has recently been confirmed by the orbiting Chandra X-ray observatory (see panel, left).

**GALAXY NGC 1365**
One of the Fornax Cluster's few spirals, NGC 1365 has a dust bar through its core.

**CLUSTER CORE**
In the Fornax Cluster's central region lie NGC 1399 (upper left of centre) and NGC 1365 (bottom right). As a rule, elliptical galaxies predominate.

COMPACT GROUP

## Seyfert's Sextet

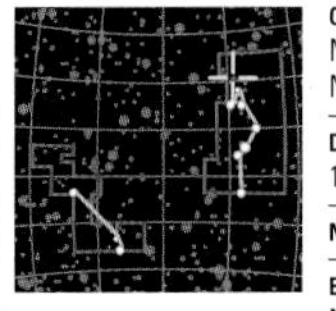

SERPENS

CATALOGUE NUMBERS NGC 6027 and NGC 6027A–C

DISTANCE 190 million light-years

NUMBER OF GALAXIES 4

BRIGHTEST MEMBER NGC 6027 (14.7)

Seyfert's Sextet actually contains just four members – each a misshapen spiral galaxy locked to the others in a gravitational waltz within a region of space no larger than the Milky Way. The sextet, as seen from Earth, is completed by a small face-on spiral that happens to lie in the background, and by a distorted star cloud (at lower right in the image below).

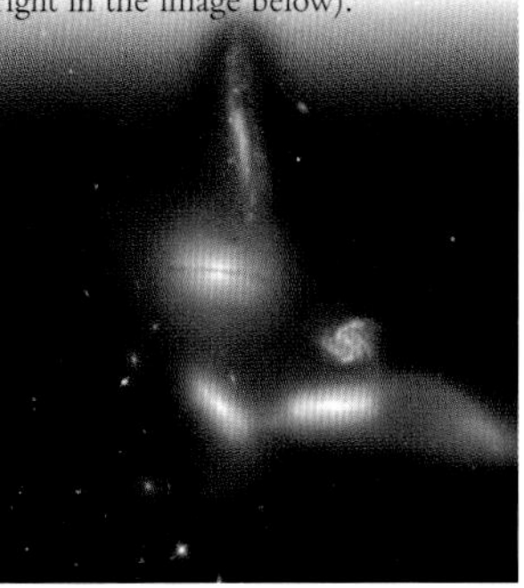

**QUARTET PLUS TWO**

REGULAR CLUSTER

## Hydra Cluster

**CATALOGUE NUMBER** Abell 1060

**DISTANCE** 160 million light-years

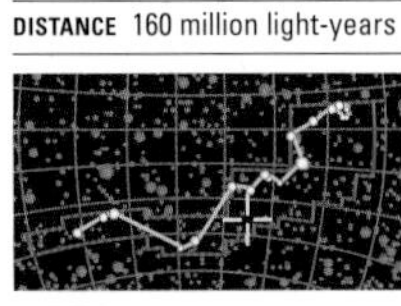

HYDRA

**NUMBER OF GALAXIES** 1,000+

**BRIGHTEST MEMBER** NGC 3311 (11.6)

The Hydra Cluster is similar in size to the huge Virgo Cluster (see p.319). It is the closest example of a "relaxed" cluster (see p.317) of mainly elliptical galaxies in a spherical distribution. Its hot X-ray gas also forms a spherical cloud around the core. The cluster is centred on two giant elliptical galaxies and an edge-on spiral, each 150,000 light-years across. These galaxies are interacting – the ellipticals' gravity has warped the spiral, while both ellipticals have distorted outer haloes. The cluster is the major member of the Hydra Supercluster, which adjoins the Local Supercluster (see p.324).

**HEART OF THE HYDRA CLUSTER**
In this image, the central giant ellipticals NGC 3309 and 3311 lie below the large, blue spiral NGC 3312. The two bright objects on either side are foreground stars.

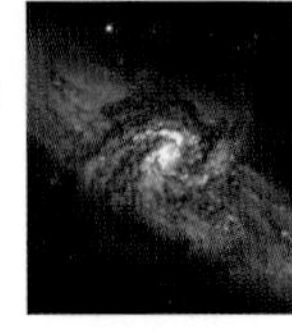

**SPIRAL SILHOUETTE**
NGC 3314, an unusual case of one spiral galaxy silhouetted against another, is one of Hydra's most beautiful objects.

COMPACT GROUP

## Stephan's Quintet

PEGASUS

**CATALOGUE NUMBER** Hickson 92

**DISTANCE** 340 million light-years (NGC 7320: 41 million light-years)

**NUMBER OF GALAXIES** 4/5

**BRIGHTEST MEMBER** NGC 7320 (13.6)

First observed by French astronomer E. M. Stephan at the University of Marseilles in 1877, Stephan's Quintet appears to be a remarkably compact cluster of five galaxies. The galaxies are a mixture of spirals, barred spirals, and ellipticals and show clear signs of disruption from interactions. The largest galaxy as seen from Earth, NGC 7320, is probably a foreground object lying in front of a quartet of interacting galaxies. The spectral red shift (see p.33) of NGC 7320 is much smaller than those of the other four galaxies, and instead matches that of several other galaxies close to it in the sky. Since it also appears physically different from the quartet, it seems likely that NGC 7320 is much closer and the unusual red shift is a normal result of the expansion of space (see p.42). However, a few astronomers claim that trails of material link NGC 7320 to other Quintet galaxies. If this is the case, then the red shift suggests that the galaxy is moving very fast relative to its neighbours and towards Earth, therefore reducing its overall speed of recession and its red shift. Or perhaps the red shift does not originate from its motion at all. These competing theories have turned Stephan's Quintet into a battleground for the small minority of astronomers who think that red shifts are not all caused by the expansion of the space, and that Hubble's Law (see p.42) does not always apply.

**FOUR OR FIVE?**
The quintet consists of a quartet of yellow galaxies beside the white spiral NGC 7320. The contrasting appearance of NGC 7320 suggests it lies in front of the other galaxies.

**QUINTET CLOSE-UP**
This detailed Hubble Space Telescope view of Stephan's Quintet shows chains of stars linking several of its interacting galaxies.

REGULAR CLUSTER

## Coma Cluster

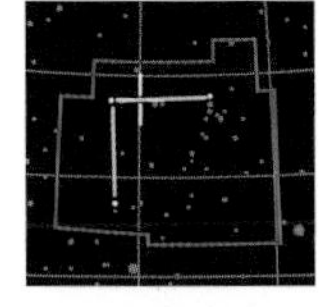

COMA BERENICES

**CATALOGUE NUMBER** Abell 1656

**DISTANCE** 300 million light-years

**NUMBER OF GALAXIES** 3,000+

**BRIGHTEST MEMBER** NGC 4889 (13.2)

Although it lies near the Virgo Cluster in the sky (see p.319), the Coma Cluster is much farther away. First recognized by William Herschel (see p.90) as a concentration of "fine nebulae" in 1785, this is one of the nearest highly evolved or "relaxed" galaxy clusters (see p.317). It is very dense, with over 3,000 galaxies, and is dominated by elliptical and lenticular galaxies. Because it is near the north galactic pole (and therefore free of the dense star fields of the Milky Way), it is well studied. Swiss-American astronomer Fritz Zwicky used Coma when he made the first measurements of galaxy movements within a cluster in the 1930s. He found the cluster contained many times more mass than its visible galaxies suggested – an idea that was not accepted until the 1970s. Overall, the cluster is moving away at 25 million kph (16 million mph). At the cluster's centre lie the giant elliptical NGC 4889 and the lenticular galaxy NGC 4874. Most of the spirals and irregulars are in the outer regions. X-ray images show two distinct patches of cluster gas, suggesting that the cluster is absorbing a smaller cluster of galaxies. Like the Virgo and Hydra clusters, Coma forms the core of its own galaxy supercluster.

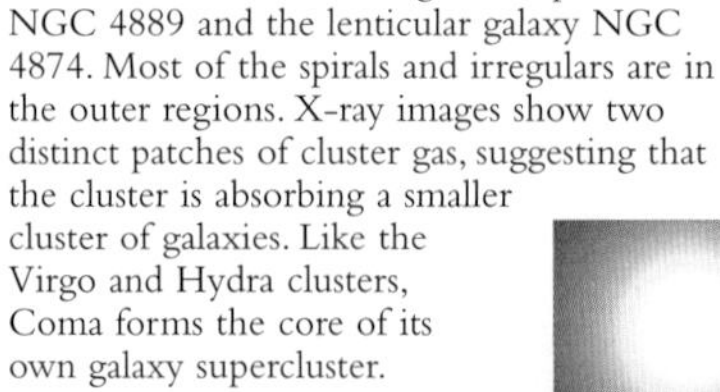

**COMA ELLIPTICAL**
This image is dominated by the Coma Cluster elliptical NGC 4881 and a nearby spiral. The other galaxies are far more distant.

IRREGULAR CLUSTER

## Hercules Cluster

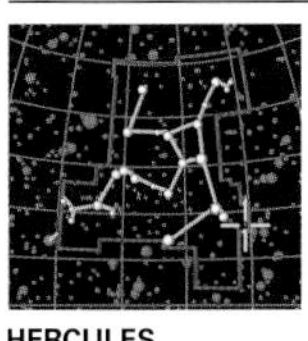

HERCULES

| | |
|---|---|
| CATALOGUE NUMBER | Abell 2151 |
| DISTANCE | 500 million light-years |
| NUMBER OF GALAXIES | 100+ |
| BRIGHTEST MEMBER | NGC 6041A (14.4) |

The small Hercules Cluster is dominated by spiral and irregular galaxies, suggesting that it is in an early stage of development. In keeping with the best models of such clusters' formation (see p.317), it shows little sign of structure. Within the cluster, several pairs or groups of galaxies seem to be merging or interacting – encounters that will transform them into different kinds of galaxies and reduce their random movements until they become more evenly distributed. The most prominent of these mergers is NGC 6050, a pair of interlocking spiral galaxies near the cluster's centre that may eventually form the core of a giant elliptical, such as those found in more evolved clusters.

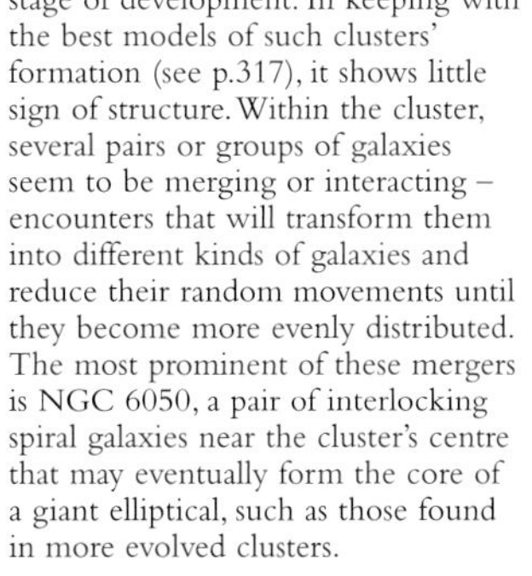

**HERCULES FIELD**
This wide-field view captures most of the bright galaxies in Hercules and shows their irregular, "unrelaxed" distribution.

REGULAR CLUSTER

## Abell 1689

VIRGO

| | |
|---|---|
| CATALOGUE NUMBER | Abell 1689 |
| DISTANCE | 2.2 billion light-years |
| NUMBER OF GALAXIES | 3,000+ |
| BRIGHTEST MEMBER | Unnamed galaxy (17.0) |

Abell 1689 is one of the densest galaxy clusters known, with thousands of galaxies packed into a volume of space only 2 million light-years across. Its ball shape makes it a fine gravitational lens, bending the images of distant galaxies into arcs. By noting the lensing power throughout the cluster, astronomers have worked out the distribution of the cluster's dark matter.

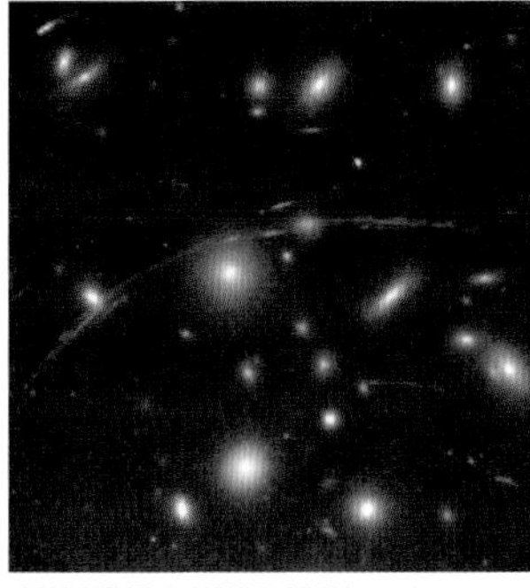

LENSING IN CLUSTER ABELL 1689

### GEORGE ABELL

George Abell (1927–1983) was a career astronomer and popularizer of science who carried out the first, and most influential, survey of galaxy clusters. After working on the Palomar Sky Survey during the 1940s and 1950s, using the powerful Palomar Schmidt telescope, he turned his attention to analysing the results, developing methods for distinguishing galaxy clusters from isolated field galaxies, and classifying clusters into types.

REGULAR CLUSTER

## Abell 2065

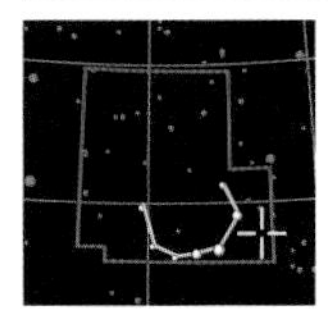

CORONA BOREALIS

| | |
|---|---|
| CATALOGUE NUMBER | Abell 2065 |
| DISTANCE | 1 billion light-years |
| NUMBER OF GALAXIES | 1,000+ |
| BRIGHTEST MEMBER | PGC 54876 (16.0) |

Abell 2065, also known as the Corona Borealis Cluster, contains 400 or more large galaxies. A highly evolved cluster like the Coma Cluster (opposite), it emits X-rays from a diffuse cloud of hot gas. However, X-ray observations have found two distinct X-ray cores, suggesting that Abell 2065 may be two already ancient clusters merging together. The cluster lies at the centre of the Corona Borealis Supercluster.

THE CORONA BOREALIS CLUSTER

IRREGULAR CLUSTER

## Abell 2125

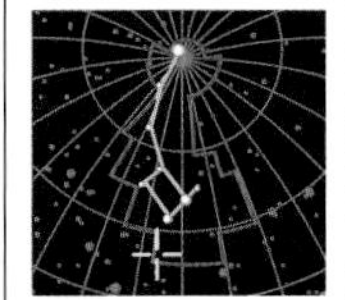

URSA MINOR

| | |
|---|---|
| CATALOGUE NUMBER | Abell 2125 |
| DISTANCE | 3 billion light-years |
| NUMBER OF GALAXIES | 1,000+ |
| BRIGHTEST MEMBER | Magnitude 17.0 |

Abell 2125 has been the subject of intense scrutiny from the orbiting Chandra X-ray Observatory. The cluster lies close enough to Earth to see detail, but so far away that images reaching Earth show an early and still active phase of its evolution, 3 billion years ago. Abell 2125 is therefore ideal for testing ideas on cluster formation. X-ray images reveal what optical ones cannot – that the cluster is forming from the merger of several smaller clusters. The most intense cloud of X-ray emitting gas shows "clumpiness", which indicates it has recently come together. Spectra reveal that the cloud is enriched with heavy elements such as iron, and close-up images show gas actively being stripped away from galaxies such as C153. With it, the gas carries atoms of heavy metals created in supernova explosions, distributing them through the intergalactic medium. A fainter cloud of almost equal size, enveloping hundreds more galaxies, has remarkably few heavy elements, suggesting that the gas-stripping process becomes more powerful and thorough over time, and that the cloud is much younger than its fainter neighbour.

Since X-ray evidence shows so much activity within the cluster, astronomers have also imaged it at other wavelengths. Infrared telescopes, for example, have revealed enormous bursts of star formation going on in galaxies far from the cluster centre. One possible explanation is that, even at distances of up to 1 million light-years, the tidal forces from the centre of a large cluster are enough to disrupt nearby galaxies and trigger starbursts.

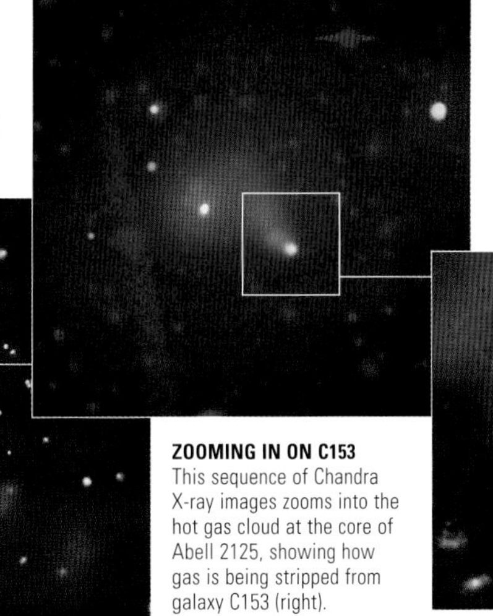

**ZOOMING IN ON C153**
This sequence of Chandra X-ray images zooms into the hot gas cloud at the core of Abell 2125, showing how gas is being stripped from galaxy C153 (right).

REGULAR CLUSTER

## Abell 2218

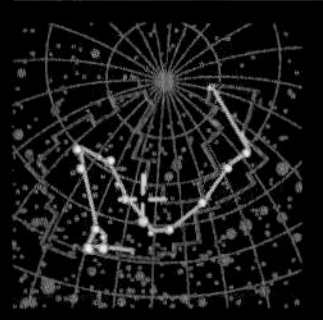

DRACO

**CATALOGUE NUMBER**
Abell 2218

**DISTANCE**
2 billion light-years

**NUMBER OF GALAXIES**
250 or more

**BRIGHTEST MEMBER**
Unnamed galaxy (17.0)

Abell 2218 is a spectacular example of a highly evolved and extremely dense galaxy cluster. It contains more than 250 mostly elliptical galaxies in a volume of space roughly 1 million light-years across.

The cluster has taught astronomers much about galaxy clusters, and about galaxies themselves. The cluster's density is so great that it affects the shape of the surrounding space, as predicted by Einstein's theory of general relativity (see p.40). Many more distant galaxies lie directly behind the cluster, and as light rays from these objects pass close to Abell 2218 their paths are deflected and focused towards Earth, in the same way that a magnifying lens focuses sunlight. This gravitational lensing (see p.317) brightens the images of galaxies that would otherwise be too far away to detect. It results in a series of distorted images of distant galaxies ringing the centre of Abell 2218.

The galaxies beyond Abell 2218 lie much farther away, and therefore, their images come from a much earlier time. Most of the lensed galaxies are blue-white, suggesting they are young irregulars and spirals very different from Abell 2218's own aged ellipticals. Some of the lensed galaxies align with X-ray sources, suggesting they are active galaxies. Recent studies yielded images of a galaxy so far beyond Abell 2218 that all its light has been red-

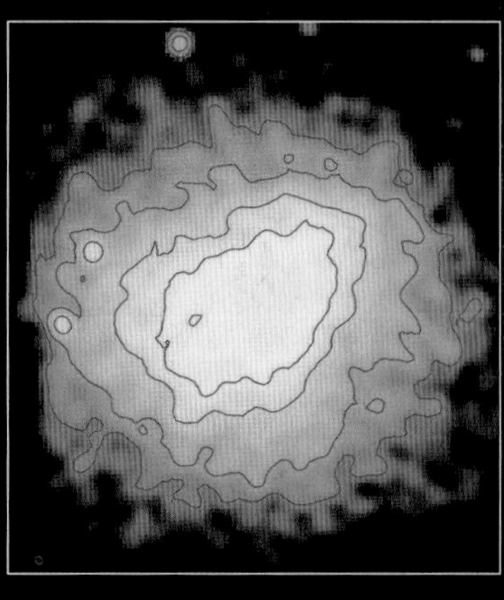

**HOLE IN THE COSMIC BACKGROUND**
In this composite image of Abell 2218, yellow and red depict the X-ray-emitting gas around its core. The gas scatters the cosmic microwave background radiation, creating a hole, outlined here by contours.

**DISTORTED BY GRAVITY**
Most of the bright objects in this image are galaxies in the Abell 2218 cluster. The arcs are much more remote galaxies, their images distorted by Abell 2218's gravity.

shifted into the infrared part of the spectrum. At the time, it was the most distant galaxy known, at 13 billion light-years from Earth. It must have formed shortly after the first stars, in the aftermath of the Big Bang.

Gravitational lensing can also reveal hidden properties of Abell 2218 itself. Because the strength of lensing depends on the cluster's density, it offers a measure of the distribution of all matter in the cluster – including the dark matter. Abell 2218 is one of the few galaxy clusters in which the pattern of visible matter (galaxies and X-ray-emitting gas) and the calculated distribution of dark matter do not match, suggesting the cluster is not as uniform as it appears in visible light.

Astronomers have now begun to use Abell 2218 to probe the origins of the Universe. A phenomenon called the Sunyaev–Zel'dovich effect (see caption, opposite) creates holes and ripples in the cosmic microwave background radiation shining through the cluster. This happens because gas around Abell 2218's core scatters photons of microwave radiation, just as Earth's atmosphere scatters light. The strength of these ripples can be used to estimate the true diameter of the cluster's core, and therefore its distance from Earth, independently of its red shift. The red shift and distance can then be used together to find the expansion rate of the Universe (see p.42).

EXPLORING SPACE

## MAPPING THE MISSING MASS

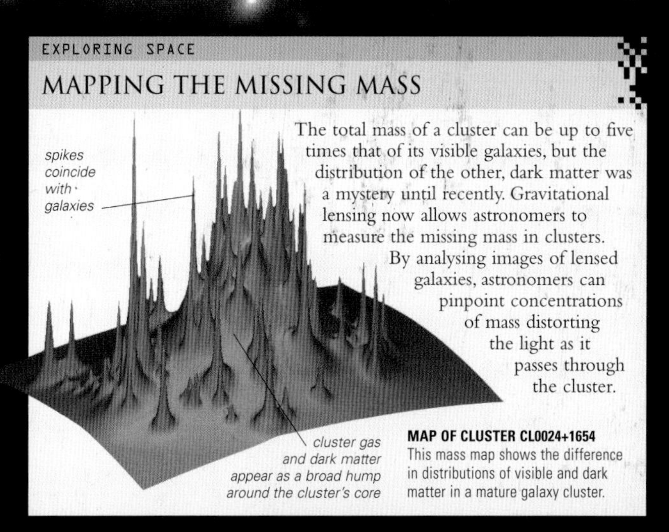
spikes coincide with galaxies
cluster gas and dark matter appear as a broad hump around the cluster's core

The total mass of a cluster can be up to five times that of its visible galaxies, but the distribution of the other, dark matter was a mystery until recently. Gravitational lensing now allows astronomers to measure the missing mass in clusters. By analysing images of lensed galaxies, astronomers can pinpoint concentrations of mass distorting the light as it passes through the cluster.

**MAP OF CLUSTER CL0024+1654**
This mass map shows the difference in distributions of visible and dark matter in a mature galaxy cluster.

# GALAXY SUPERCLUSTERS


22–23 The scale of the Universe
24–27 Celestial objects
28–31 Matter
32–35 Radiation
316–17 Galaxy clusters


THE LARGEST SCALE OF STRUCTURE in the Universe is that of galaxy superclusters – collections of neighbouring galaxy clusters bunched together in chains and sheets that stretch across the cosmos. This structure is an echo of the Big Bang itself. By studying the Universe at these enormous scales, astronomers can learn about our place within it and the way in which it formed.

## GALAXY SUPERCLUSTERS

Just as galaxies are bound by gravity into clusters, galaxy clusters are linked to form even larger structures called superclusters. A chain of galaxy clusters links the Local Group of galaxies, containing the Milky Way, to the Virgo Cluster, 52 million light-years away. This is the heart of the Local Supercluster, or Virgo Supercluster. As in other superclusters, the galaxy clusters are not discrete, but merge at their edges. The entire supercluster stretches over 200 million light-years, itself merging with others at the edges. Within a supercluster, each cluster is an evolving, collapsing knot of gravity. Even on this vast scale, the gravity of superclusters resists the background cosmological expansion of space (see pp.42–43). The Virgo Cluster's gravity is counteracting cosmic expansion, pulling other clusters towards it at 5.4 million kph (3.4 million mph).

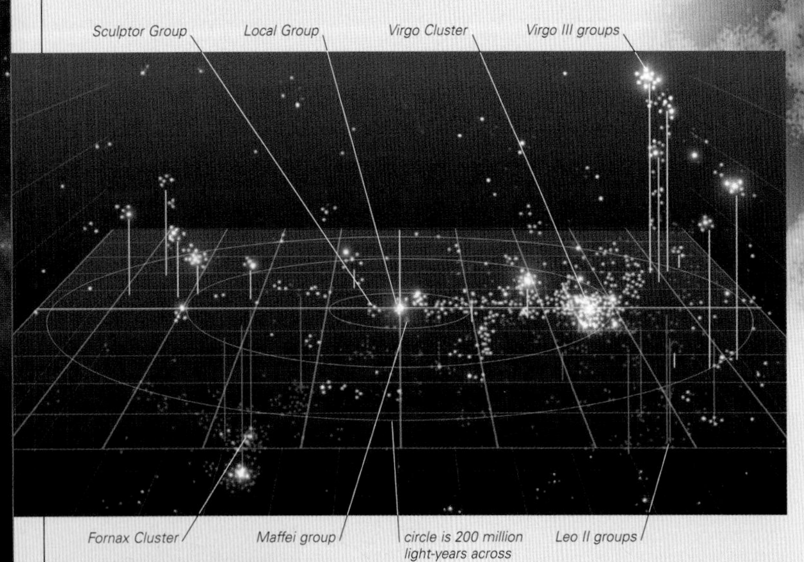


**THE LOCAL SUPERCLUSTER**
This map of the Virgo Supercluster is centred on the Local Group and shows groups and clusters of galaxies linked into a curved, branched chain. Each point denotes a major galaxy – there are thousands of smaller ones not pictured.

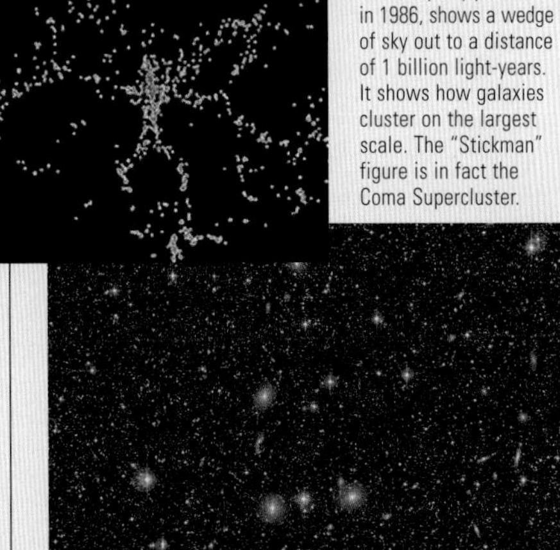

**PLOT OF GALAXIES**
This deep-sky plot, made in 1986, shows a wedge of sky out to a distance of 1 billion light-years. It shows how galaxies cluster on the largest scale. The "Stickman" figure is in fact the Coma Supercluster.

**THE GREAT ATTRACTOR**
This view of the sky in Centaurus looks towards the "Great Attractor", which is probably a massive supercluster centred on the Norma Cluster but hidden by the southern Milky Way.

## THE NEARBY UNIVERSE

Astronomers map the Universe by measuring red shifts in the spectral lines of millions of galaxies (see p.33). If the red shifts are due to galaxies moving apart as space expands (see p.42), they can be used as a measure of distance. Astronomers often use red-shift values instead of light-years for vast distances. They adjust the red-shift distance estimates if they know of large-scale motions of superclusters and clusters not due to cosmological expansion. The entire Virgo Supercluster, for example, is falling towards a mysterious concentration of mass called the Great Attractor at 2.2 million kph (1.4 million mph). Maps of the Universe out to a few hundred million light-years reveal that galaxies cluster into strands, and holes appear between the strands. As the maps reach further, these features turn out to be repeated across the entire cosmos, appearing the same in all directions.

EXPLORING SPACE

### MAPPING THE UNIVERSE

A project, currently underway, is charting the large-scale structure of the Universe to unprecedented depth. The Sloan Digital Sky Survey uses a telescope (below) with a wide field of view ideal for imaging large areas of the sky at once. In total, it will measure the positions and red shifts of one million nearby galaxies and 100,000 quasars, building up a new map of the cosmos.

# FILAMENTS AND VOIDS

At the largest scales measured, the Universe reveals a clear general pattern. Galaxy superclusters join to form string-like "filaments" or flat "sheets", surrounding enormous and apparently empty regions called "voids". While structure up to the level of clusters can be explained by the action of gravity since the Big Bang, the present age of the Universe (13.7 billion years) is not nearly long enough for gravity alone to have organised the Universe on the scale of filaments and voids. Studies of the cosmic background radiation (CMBR; see p.50) suggest that filaments and voids are an expanded echo of features from the earliest times. The CMBR shows ripples in the temperature and matter distribution of the Universe when it was just 300,000 years old – irregularities that were seeded in the Big Bang itself. The ripples of matter created "rivers" of gravity into which matter fell, forming filaments when the Universe was still very small. Superclusters condensed out of the densest parts of the filaments, while clusters and galaxies formed within them from the "bottom up" (see p.298).

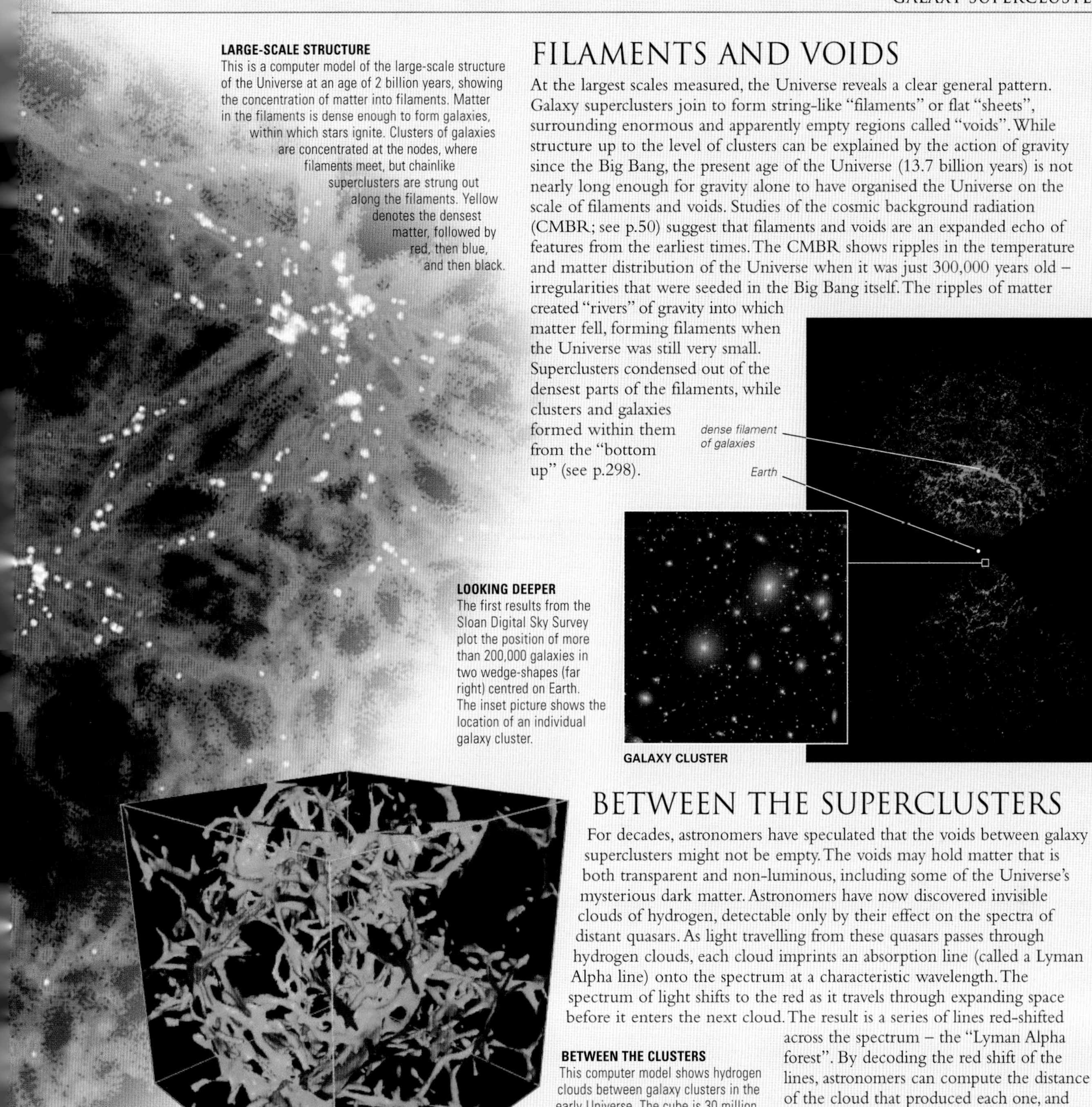


**LARGE-SCALE STRUCTURE**
This is a computer model of the large-scale structure of the Universe at an age of 2 billion years, showing the concentration of matter into filaments. Matter in the filaments is dense enough to form galaxies, within which stars ignite. Clusters of galaxies are concentrated at the nodes, where filaments meet, but chainlike superclusters are strung out along the filaments. Yellow denotes the densest matter, followed by red, then blue, and then black.

**LOOKING DEEPER**
The first results from the Sloan Digital Sky Survey plot the position of more than 200,000 galaxies in two wedge-shapes (far right) centred on Earth. The inset picture shows the location of an individual galaxy cluster.

**BETWEEN THE CLUSTERS**
This computer model shows hydrogen clouds between galaxy clusters in the early Universe. The cube is 30 million light-years along each side. The clouds are thought to be more sparsely distributed today.

# BETWEEN THE SUPERCLUSTERS

For decades, astronomers have speculated that the voids between galaxy superclusters might not be empty. The voids may hold matter that is both transparent and non-luminous, including some of the Universe's mysterious dark matter. Astronomers have now discovered invisible clouds of hydrogen, detectable only by their effect on the spectra of distant quasars. As light travelling from these quasars passes through hydrogen clouds, each cloud imprints an absorption line (called a Lyman Alpha line) onto the spectrum at a characteristic wavelength. The spectrum of light shifts to the red as it travels through expanding space before it enters the next cloud. The result is a series of lines red-shifted across the spectrum – the "Lyman Alpha forest". By decoding the red shift of the lines, astronomers can compute the distance of the cloud that produced each one, and perhaps map out the distribution of clouds between the galaxy clusters.

QUASAR
INTERGALACTIC CLOUD
photons
INTERGALACTIC CLOUD
EARTH
intensity
wavelength
peak radiation from quasar
red-shifted peak
absorption by cloud introduces line
red-shifted line
absorption by cloud introduces another line
red-shifted lines building up into a forest

**LYMAN ALPHA FOREST**
Light travelling from a distant quasar passes through a series of hydrogen clouds. Each superimposes an absorption line onto the quasar's spectrum, but red shifts mean the lines do not overlap. The result is a series of red-shifted lines called a Lyman Alpha forest.

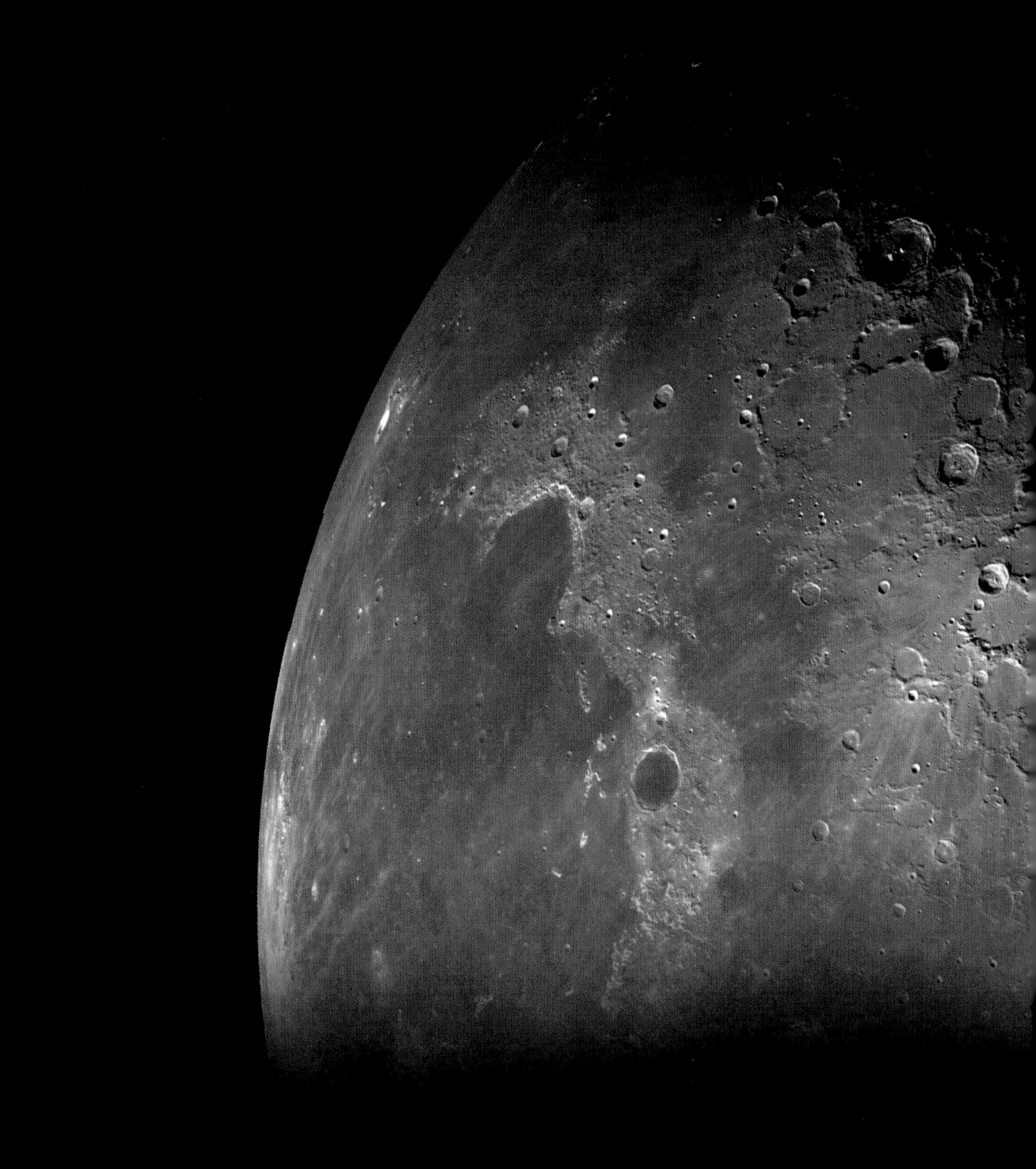

# THE NIGHT SKY

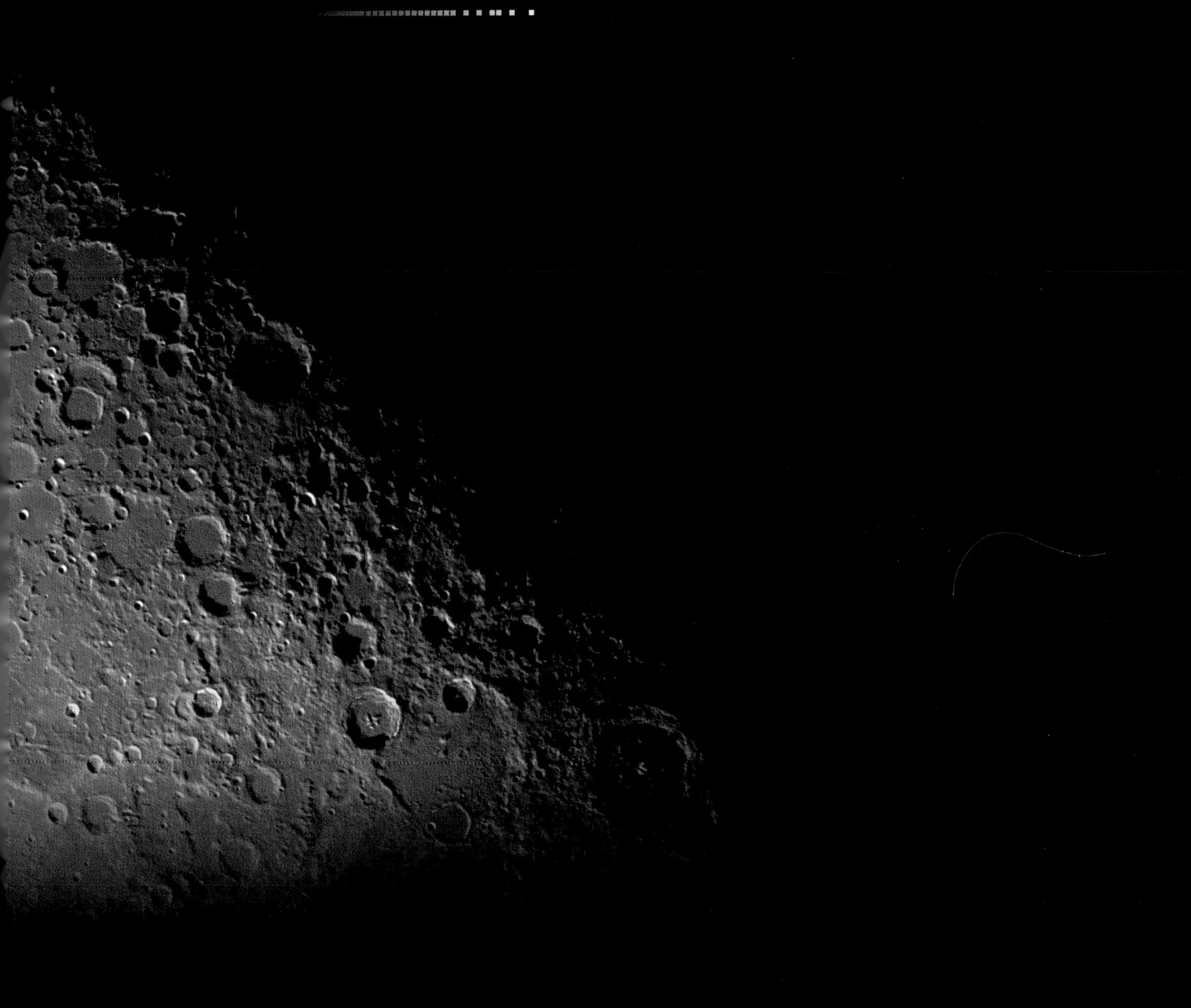

*"Why did not somebody teach me the constellations, and make me at home in the starry heavens, which are always overhead, and which I don't half know to this day?"*

Thomas Carlyle

THE HUMAN EYE HAS ALWAYS seen patterns among the stars. Ancient peoples traced the figures of gods, heroes, and mythical animals onto the skies and used the relationship between these constellations to illustrate their myths and legends. In most cases, stars within a constellation lie in the same region of sky merely by chance, however, and are not related. Despite the apparent permanence of the skies, these patterns are not fixed, because all the stars are moving relative to the Earth. Over time, the shape of all the constellations will change, and hundreds of thousands of years from now, they will be unrecognizable. Future generations will need to invent constellations of their own. But for now, 88 constellations fill our sky, interlocking like pieces of an immense jigsaw puzzle. Some are large, others small, some richly stocked with objects of note, others faint and seemingly barren. All are featured in the following pages.

**PATTERNS IN THE SKY**
As darkness falls, a stargazer scans the sky with binoculars. The familiar shape of the Plough looms overhead, part of the constellation Ursa Major, the Great Bear. The north pole star, Polaris, can be seen high up on the right.

# THE CONSTELLATIONS

# THE HISTORY OF CONSTELLATIONS

66–67 Star motion and patterns
72–73 Naked-eye astronomy
82–83 Ancient astronomy
85 Arabic astronomy

THE FIRST CONSTELLATIONS were patterns of stars that ancient peoples employed for navigation, timekeeping, and storytelling. Recently, the pictorial aspect of constellations has become less significant, and they have become simply delineated regions of the sky, although the attraction of the myths and legends remains.

## EARLY CONSTELLATION LORE

The constellation system used today stems from patterns recognized by ancient Greek and Roman civilization. The earliest surviving account of ancient Greek constellations comes from the poet Aratus of Soli (*c.*315–*c.*245 BC). His poem, the *Phaenomena*, written around 275 BC, describes the sky in storybook fashion and identifies 47 constellations. It is based on a lost book of the same name by the Greek astronomer Eudoxus (*c.*390–*c.*340 BC). Eudoxus reputedly introduced the constellations to the Greeks after learning them from priests in Egypt. These constellations had been adopted from Babylonian culture; they were originally created by the Sumerians around 2,000 BC. However, the Greeks attached their own myths to the constellations detailed by Eudoxus, and Aratus's storybook of the stars proved immensely popular. Sometime in the 2nd century AD, it was joined by a more elaborate work of constellation lore called *Poetic Astronomy*, written by the Roman author Hyginus. Many editions of both these works were produced and translated over the centuries.

**ANTICANIS**
This page from a 9th-century edition of the star myths of Hyginus shows the constellation Canis Minor, here termed Anticanis. Hyginus's words, in Latin, form the shape of the dog's body.

## FILLING THE HEAVENLY SPHERE

The oldest surviving star catalogue dates from the 2nd century AD and is contained in a book called the *Almagest*, written by the Greek astronomer and geographer Ptolemy (see p.85). It records the positions and brightnesses of one thousand stars, arranged into 48 constellations, based on an earlier catalogue by Hipparchus of Nicaea (*c.*190–*c.*120 BC). In the 10th century AD, an Arab astronomer, al-Sufi (see p.405), updated the *Almagest* in his *Book of Fixed Stars*, which included Arabic names for many stars. These Arabic names are still used today, although often in corrupted form.

No more constellations were introduced until the end of the 16th century, when Dutch explorers sailed to the East Indies. From there, they could observe the southern sky that was below the European horizon. Two navigators, Pieter Dirkszoon Keyser and Frederick de Houtman (see p.400), catalogued nearly 200 new southern stars, from which they and their mentor, Petrus Plancius (see p.342), a leading Dutch cartographer, created 12 new constellations. Plancius also created other northern constellations, forming them between those listed by Ptolemy. Nearly a century later, Johannes Hevelius (see p.368), a Polish astronomer, filled the remaining gaps in the northern sky, and in the mid-18th century, the French astronomer Nicolas Louis de Lacaille (see p.406) introduced another 14 constellations in the southern sky.

**POCKET GLOBE**
This pocket globe from the National Maritime Museum, England, positions the Earth within a shell that represents the surrounding celestial sphere. On the inside of the open shell are the constellations, painted as mirror images.

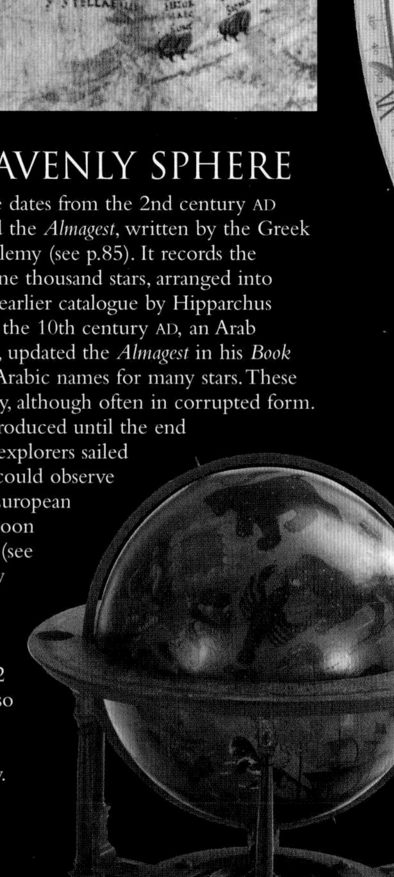

**GLOBAL COVERAGE**
This beautiful celestial globe was made around 1625 by Arnold van Langren, a celebrated Dutch globe maker. As with all celestial globes, the figures are shown reversed in comparison to their appearance in the sky.

HEAVENLY PICTURE BOOK
Ancient people imagined gods, heroes, and beasts among the stars, and these figures were depicted on star charts until the 19th century. These charts, from John Flamsteed's *Atlas Coelestis* (1729), show those 48 constellations known to the ancient Greeks depicted on the northern and southern halves of the sky.

# STAR CHARTS AND ATLASES

The first printed star chart was produced in 1515 by the great German artist Albrecht Dürer. Like a celestial globe, Dürer's chart depicted the constellations in reverse, showing the sky as it would be seen from an imaginary position outside the celestial sphere, but before long, charts were being made that could be compared directly with the sky. The finest early star atlas was *Uranometria*, produced in 1603 by the German astronomer Johann Bayer (see panel, below). This atlas remains one of the most beautiful examples of the celestial cartographer's art. Shortly after its publication, astronomy was revolutionized by the invention of the telescope. The first major star catalogue and atlas of this new era was produced by England's first Astronomer Royal, John Flamsteed (1646-1719). *Atlas Coelestis* shows the Ptolemaic constellations visible from Greenwich, England, based on Flamsteed's own painstaking observations. The pinnacle of celestial mapping came in 1801 when Johann Bode, a German astronomer, published an atlas called *Uranographia*. Covering the entire sky, this atlas depicted over 100 constellations, some invented by Bode himself. Finally, in 1922, a list of 88 constellations was agreed upon by the International Astronomical Union, astronomy's governing body, which also defined the boundaries of each constellation. On modern star charts, the only sign of the traditional pictorial charts are the few lines that link the main stars, suggesting the overall shape of each constellation.

**SKETCHY FIGURES**
Leo, the Lion, an easily recognizable constellation of the zodiac, is here depicted on the *Atlas Coelestis*, by English astronomer John Flamsteed, published in 1729.

## JOHANN BAYER

Johann Bayer (1572–1625) was a German lawyer and amateur astronomer who produced the first major printed star atlas, called *Uranometria*, in 1603. As well as Ptolemy's 48 constellations, it included 12 recently introduced southern constellations. It was the first atlas to include the entire sky. On his charts, Bayer introduced the convention of labelling the brightest stars in each constellation with Greek letters. These are now known as Bayer letters (see p.68).

**BAYER'S BEASTS**
Cygnus, the swan, is seen flying along the Milky Way on this hand-coloured chart from Bayer's *Uranometria*.

# MAPPING THE SKY

The following pages divide the celestial sphere into six parts – two polar regions and four equatorial regions – which show the location of the 88 constellations. Each constellation is then profiled in the following section. Each entry places the constellation and its main features into the context of the rest of the sky.

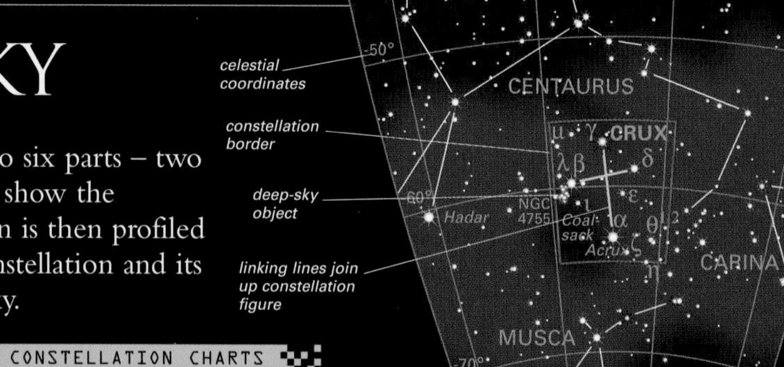


## VISIBILITY MAPS

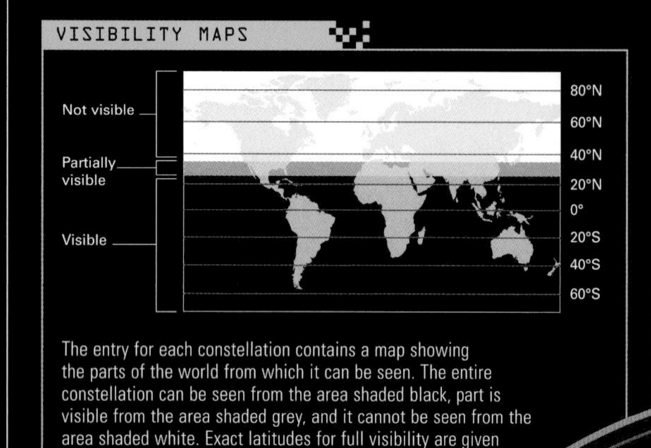


The entry for each constellation contains a map showing the parts of the world from which it can be seen. The entire constellation can be seen from the area shaded black, part is visible from the area shaded grey, and it cannot be seen from the area shaded white. Exact latitudes for full visibility are given in the accompanying dataset.

## CONSTELLATION CHARTS

Each of the 88 constellation entries has its own chart, centred around the constellation area. These charts show all stars brighter than magnitude 6.5. Within the constellation borders, every star brighter than magnitude 5 is labelled. Deep-sky objects are represented by an icon.

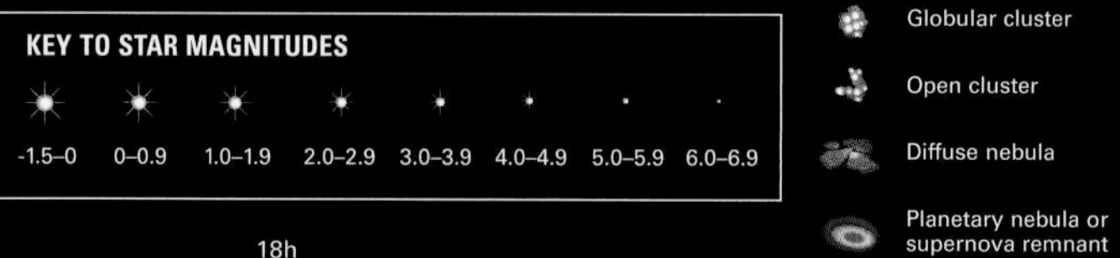


**DEEP-SKY OBJECTS**

- Galaxy
- Globular cluster
- Open cluster
- Diffuse nebula
- Planetary nebula or supernova remnant
- Black hole or X-ray binary

## THE NORTH POLAR SKY

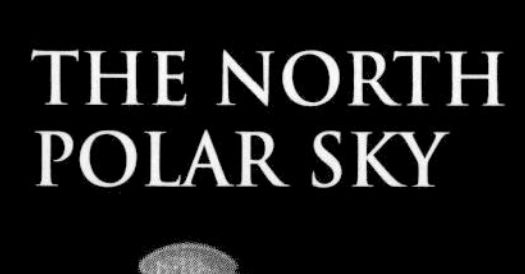

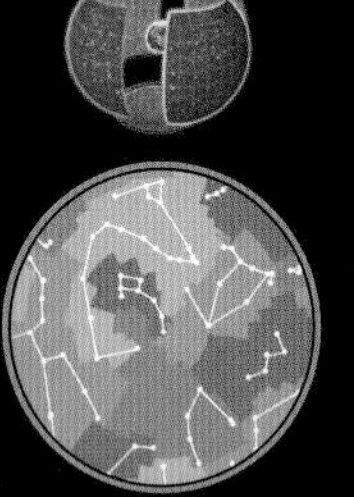

Almost in the centre of this chart is the star Polaris, in Ursa Minor, which lies less than 1° from the north celestial pole. For observers in the northern hemisphere, the stars around the pole never set – they are circumpolar. The viewer's latitude will determine how much of the sky is circumpolar: the farther north, the larger the circumpolar area. This chart shows the sky from declinations 90° to 50°.

DRACO · BOOTES · CYGNUS · CEPHEUS · LACERTA · URSA MINOR · URSA MAJOR · CASSIOPEIA · PERSEUS · CAMELOPARDALIS · LYNX · AURIGA

17h · 18h · 19h · 21h · 22h · 23h · 0h · 1h · 2h · 3h · 4h · 5h · 6h · 7h · 90° · 80° · 70° · 60° · 50°

### STAR MAGNITUDES

-1 · 0 · 1 · 2 · 3 · 4 · Variable star

Star magnitudes shown here are for the equatorial and polar sky charts

## THE GREEK ALPHABET

On most star charts, bright stars are identified by Greek letters according to a system invented by Johann Bayer (see p.331).

| | | | | | | | |
|---|---|---|---|---|---|---|---|
| Alpha | α | Eta | η | Nu | ν | Tau | τ |
| Beta | β | Theta | θ | Xi | ξ | Upsilon | υ |
| Gamma | γ | Iota | ι | Omicron | ο | Phi | φ |
| Delta | δ | Kappa | κ | Pi | π | Chi | χ |
| Epsilon | ε | Lambda | λ | Rho | ρ | Psi | ψ |
| Zeta | ζ | Mu | μ | Sigma | σ | Omega | ω |

## VISIBILITY ICONS

Beside every photograph is an icon indicating the kind of view it illustrates. Some photographs show the star or deep-sky object as it can be seen by the naked eye, through binoculars, or through amateur telescopes. Others are the result of CCD photography or show the view through professional observing equipment.

- Naked eye
- Binoculars
- Telescope (amateur)
- CCD
- Professional equipment

## ALPHABETICAL INDEX OF THE 88 CONSTELLATIONS

The constellation entries are ordered by their position on the celestial sphere, beginning with Ursa Minor in the north and spiralling south in a clockwise direction, before finishing with Octans. This alphabetical list provides an alternative way of locating constellation entries.


Andromeda p.352
Antlia p.380
Apus p.407
Aquarius p.371
Aquila p.367
Ara p.399
Aries p.355
Auriga p.343
Boötes p.347
Caelum p.389
Camelopardalis p.342
Cancer p.359
Canes Venatici p.346
Canis Major p.376
Canis Minor p.376
Capricornus p.387
Carina p.395
Cassiopeia p.341
Centaurus p.382
Cepheus p.340
Cetus p.373
Chamaeleon p.407
Circinus p.397
Columba p.392
Coma Berenices p.360
Corona Australis p.399
Corona Borealis p.363
Corvus p.381
Crater p.381
Crux p.396
Cygnus p.350
Delphinus p.369
Dorado p.405
Draco p.339
Equuleus p.369
Eridanus p.390
Fornax p.389
Gemini p.358
Grus p.401
Hercules p.348
Horologium p.403
Hydra p.378
Hydrus p.403
Indus p.400
Lacerta p.353
Leo p.361
Leo Minor p.360
Lepus p.391
Libra p.363
Lupus p.383
Lynx p.343
Lyra p.349
Mensa p.406
Microscopium p.387
Monoceros p.377
Musca p.397
Norma p.398
Octans p.409
Ophiuchus p.365
Orion p.374
Pavo p.408
Pegasus p.370
Perseus p.354
Phoenix p.401
Pictor p.404
Pisces p.372
Piscis Austrinus p.388
Puppis p.393
Pyxis p.392
Reticulum p.404
Sagitta p.366
Sagittarius p.384
Scorpius p.386
Sculptor p.388
Scutum p.366
Serpens (Caput and Cauda) p.364
Sextans p.380
Taurus p.356
Telescopium p.400
Triangulum p.353
Triangulum Australe p.398
Tucana p.402
Ursa Major p.344
Ursa Minor p.338
Vela p.394
Virgo p.362
Volans p.406
Vulpecula p.368


# THE SOUTH POLAR SKY

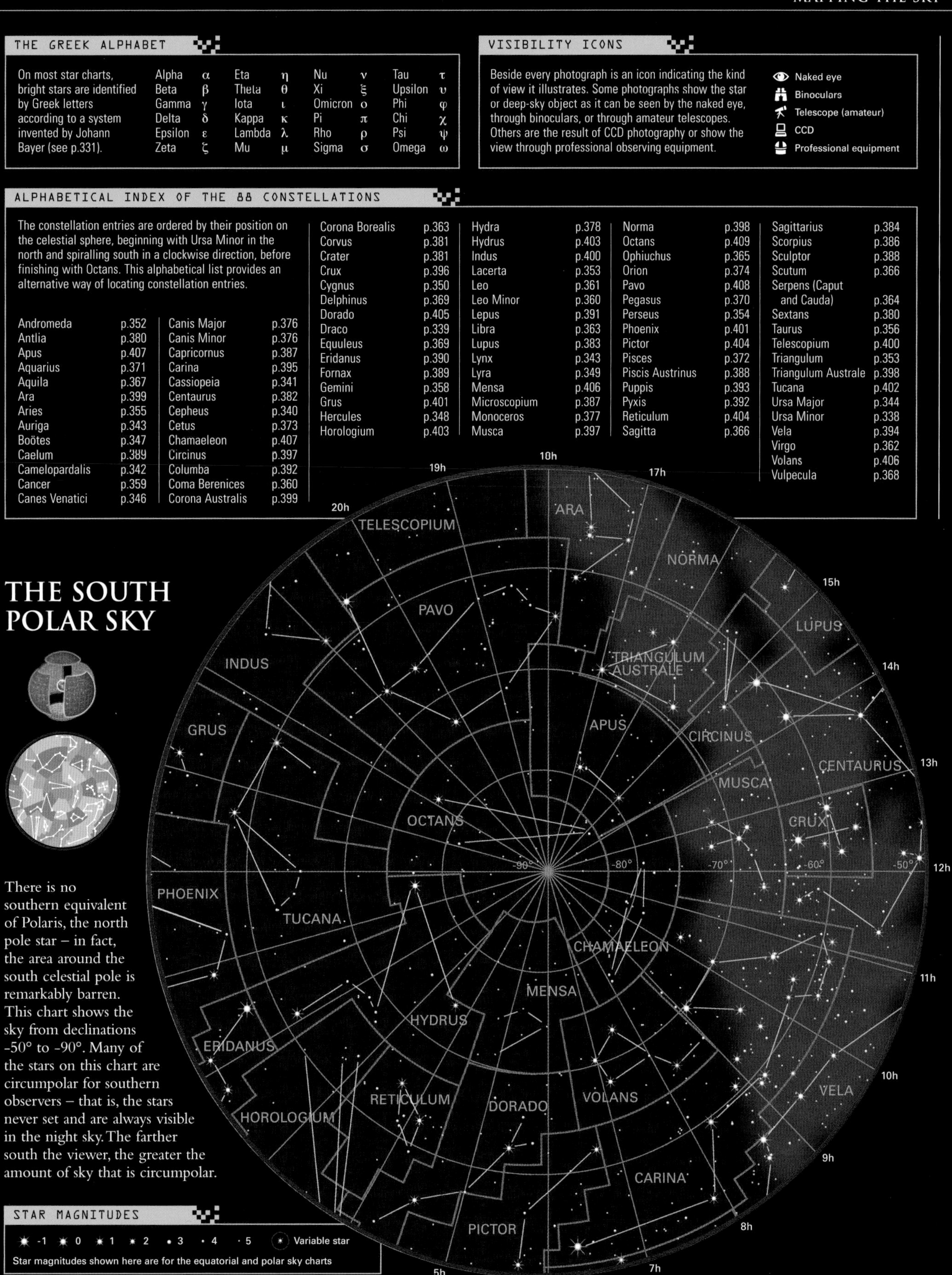


There is no southern equivalent of Polaris, the north pole star – in fact, the area around the south celestial pole is remarkably barren. This chart shows the sky from declinations -50° to -90°. Many of the stars on this chart are circumpolar for southern observers – that is, the stars never set and are always visible in the night sky. The farther south the viewer, the greater the amount of sky that is circumpolar.

## STAR MAGNITUDES

-1 0 1 2 3 4 5 Variable star

Star magnitudes shown here are for the equatorial and polar sky charts

# EQUATORIAL SKY CHART 1

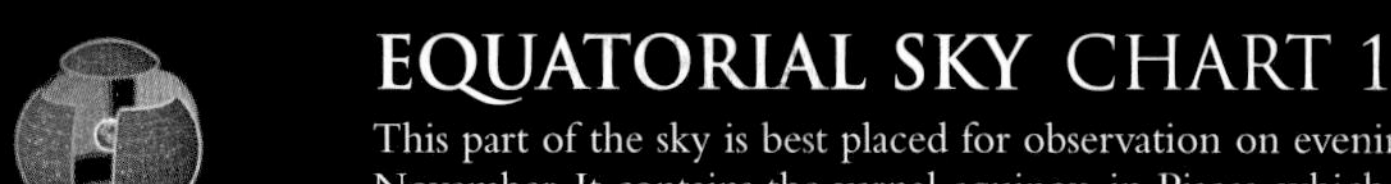

This part of the sky is best placed for observation on evenings in September, October, and November. It contains the vernal equinox, in Pisces, which is the point at which the Sun's path, the ecliptic, crosses the celestial equator into the northern half of the sky. The Sun reaches this point in late March each year. The 0h line of right ascension also passes through this point; this is the celestial equivalent of 0° longitude (the prime meridian) on Earth. The most distinctive feature in this region of the night sky is the great Square of Pegasus – although one star in the square actually belongs to neighbouring Andromeda.

# EQUATORIAL SKY CHART 2

This area of sky is best placed for observation on evenings in June, July, and August. It contains the point where the Sun reaches its most southerly declination each year, in Sagittarius. This happens around 21 December, which is the longest day in the southern hemisphere and the shortest day in the northern. Rich Milky Way star fields cross this region of sky, from Cygnus in the north to Sagittarius and Scorpius in the south. Hercules and Ophiuchus, both representing mythical giants, stand head to head in the north. Notable star patterns in the south are the Teapot asterism in Sagittarius and the curving tail of Scorpius, the Scorpion.

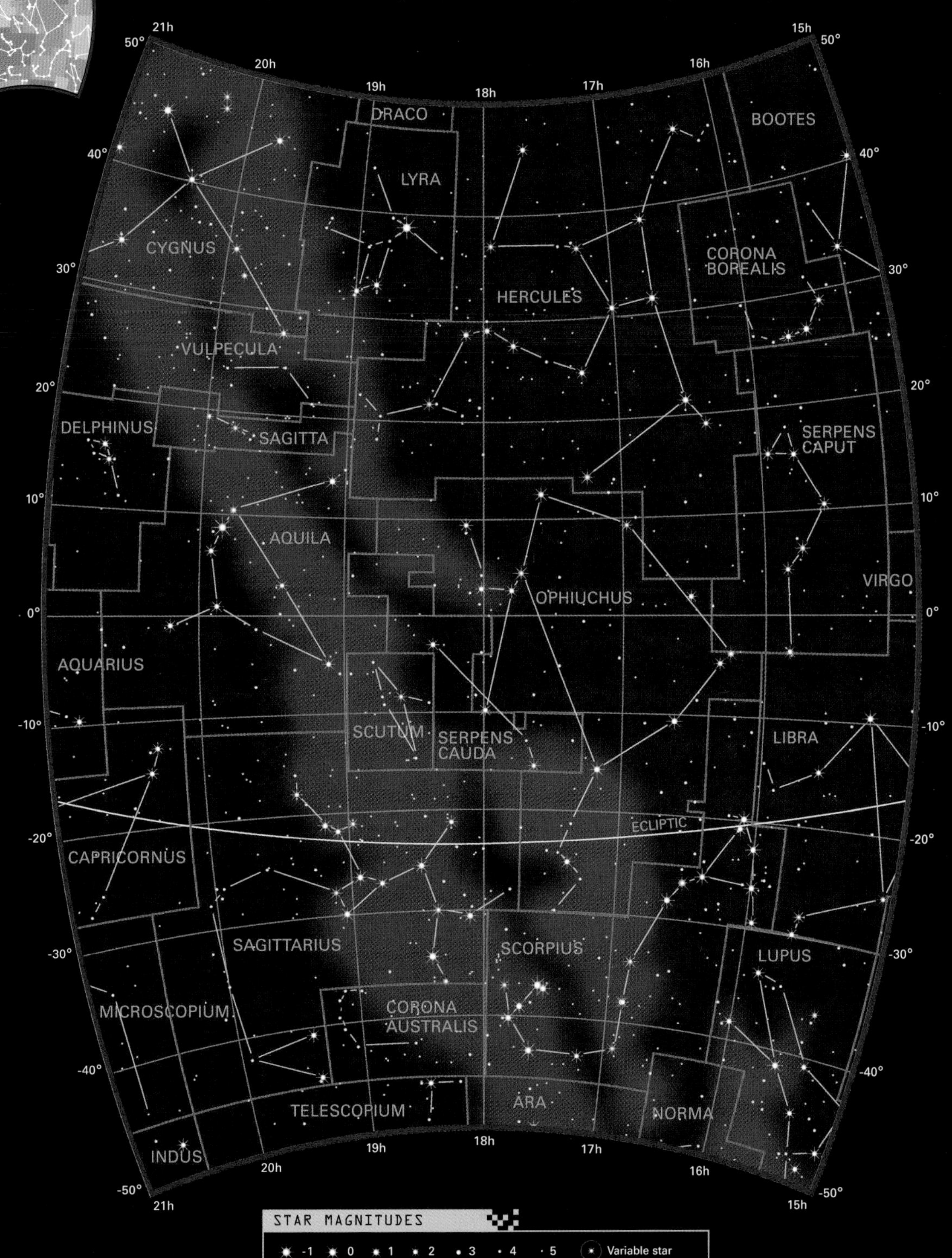


STAR MAGNITUDES

-1 0 1 2 3 4 5 Variable star

Star magnitudes shown here are for the equatorial and polar sky charts

# EQUATORIAL SKY CHART 3

This region is best placed for observation on evenings in March, April, and May. It contains the point at which the Sun moves across the celestial equator into the southern hemisphere each year. This point lies in Virgo, and the Sun reaches it around 21 September. In the northern constellation Boötes lies Arcturus, a notably orange-coloured star whose visibility marks the arrival of northern spring. South of it is the zodiacal constellation of Virgo, whose brightest star is the blue-white Spica. Adjoining Virgo is Leo, one of the few constellations that genuinely resembles the animal it is said to represent – in this case, a crouching lion.

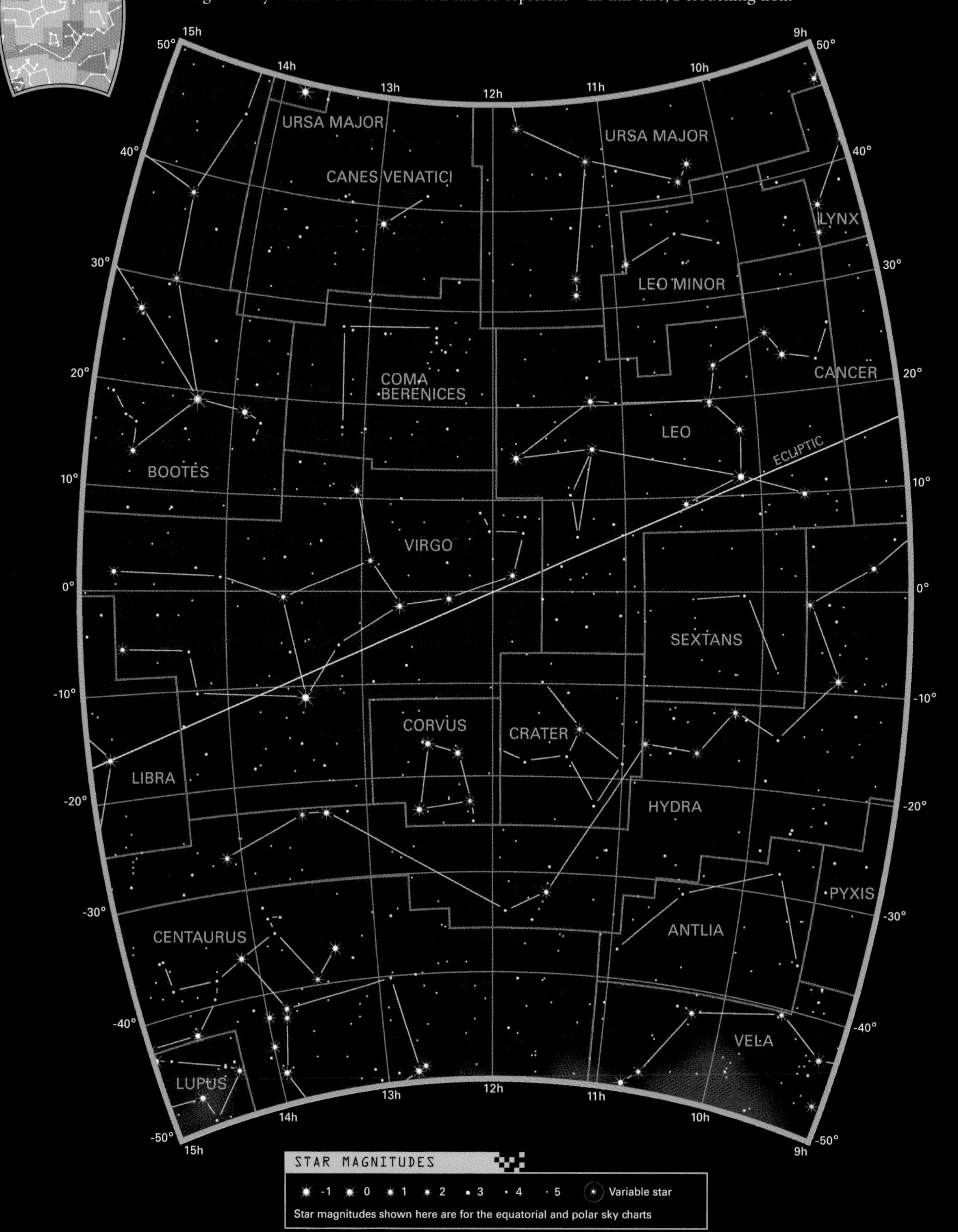

# EQUATORIAL SKY CHART 4

This region is best placed for observation on December, January, and February evenings. It contains the point at which the Sun is farthest north of the celestial equator, on the border of Taurus with Gemini. This occurs around 21 June, when days are longest in the northern hemisphere and shortest in the southern. Glittering stars and magnificent constellations abound in this region of sky, including the brightest star of all, Sirius in Canis Major. A distinctive line of three stars marks the belt of Orion, while in Taurus the bright star Aldebaran glints like the eye of the bull, along with the Hyades and Pleiades star clusters.

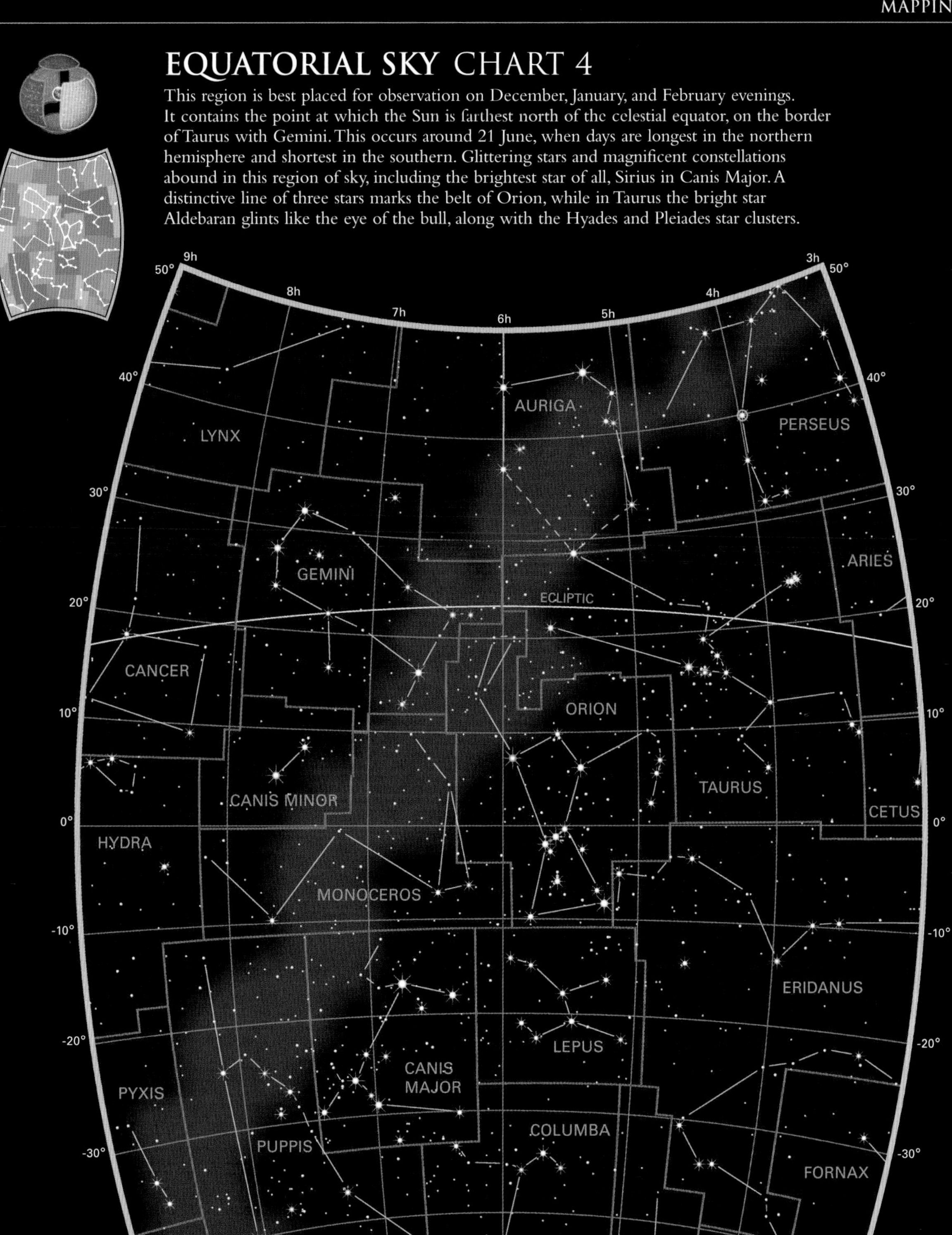


STAR MAGNITUDES

-1 0 1 2 3 4 5 Variable star

Star magnitudes shown here are for the equatorial and polar sky charts

**LONG-TAILED BEAR**
The tail of the Little Bear curves away from the north Pole Star, Polaris (upper left). Unlike real bears, the celestial bears Ursa Minor and Ursa Major both have long tails.

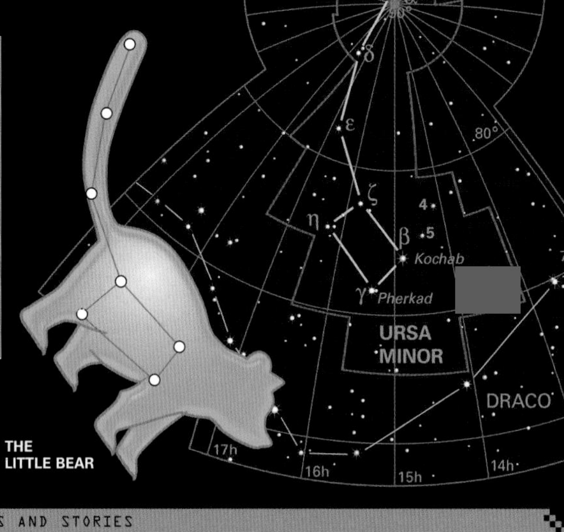


THE LITTLE BEAR

# Ursa Minor

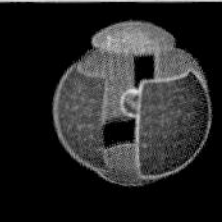

**SIZE RANKING** 56

**BRIGHTEST STAR** Polaris (α) 2.0

**GENITIVE** Ursae Minoris

**ABBREVIATION** UMi

**HIGHEST IN SKY AT 10PM** May–July

**FULLY VISIBLE** 90°N–0°

Ursa Minor is an ancient Greek constellation, which is said to represent Ida, one of the nymphs who nursed the god Zeus when he was an infant (see panel, right). Ursa Minor contains the north celestial pole and also its nearest naked-eye star, Polaris or Alpha (α) Ursae Minoris (see pp.274–75), which is currently less than one degree from the north celestial pole. The distance between them is steadily decreasing due to precession (see p.60). They will come closest around 2100, when the separation will be about 0.5 degrees.

The main stars of Ursa Minor form a shape known as the Little Dipper, reminiscent of the larger and brighter Big Dipper in Ursa Major, although its handle curves in the opposite direction. The two brightest stars in the bowl of the Little Dipper, Gamma (γ) and Eta (η) Ursae Minoris, are popularly known as the Guardians of the Pole.

**THE NORTH POLE STAR**
Seen through a small telescope, Polaris appears to have a faint companion (right), but this background star is unrelated. Its true companion is seen here just below Polaris.

## SPECIFIC FEATURES

Polaris, the north Pole Star, is a creamy white supergiant and a Cepheid variable (see p.278), but its brightness changes are too slight to be noticeable to the naked eye. With a telescope, an unrelated 8th-magnitude star can be seen nearby.

Two stars in the bowl of the Little Dipper – Gamma and Eta Ursae Minoris – are both wide doubles. Gamma is the brighter of the two, at magnitude 3.0, and its 5th-magnitude companion, 11 Ursae Minoris, can be seen with the naked eye or binoculars. Eta – at magnitude 5.0 – can also be seen with the naked eye. It has a partner of magnitude 5.5, 19 Ursae Minoris; both stars are easily visible with binoculars. Each of the component stars in both Gamma and Eta lie at different distances from the Earth and, hence, are unrelated.

MYTHS AND STORIES

## NURSING NYMPHS

According to Greek mythology, at his birth, the infant Zeus was hidden from his murderous father, Cronus, and taken to a cave on the island of Crete, where he was nursed by two nymphs, usually named as Adrastea and Ida. In gratitude, Zeus later placed the nymphs in the sky as the Great Bear and the Little Bear, respectively.

**THE PROTECTED CHILD**
The infant Zeus is cared for by nymphs and shepherds, in the *Feeding of Jupiter* by the French artist Nicolas Poussin.

THE DRAGON

# Draco

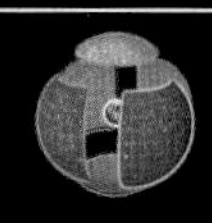

**SIZE RANKING** 8

**BRIGHTEST STAR** Etamin (γ) 2.2

**GENITIVE** Draconis

**ABBREVIATION** Dra

**HIGHEST IN SKY AT 10PM** April–August

**FULLY VISIBLE** 90°N–4°S

One of the ancient Greek constellations, Draco represents the dragon of Greek myth that was slain by Hercules (see panel, below). This large constellation winds for nearly 180 degrees around the north celestial pole. Despite its size, Draco is not particularly easy to identify, apart from a lozenge shape marking the head. This is formed by four stars, including the constellation's brightest member, Gamma (γ) Draconis, popularly known as Etamin or Eltanin.

## SPECIFIC FEATURES

Double and multiple stars are a particular feature of Draco. Nu (ν) Draconis, the faintest of the four stars in the dragon's head, is a readily identifiable pair. It consists of identical white components of 5th magnitude and is considered to be among the finest doubles visible with binoculars. Psi (ψ) Draconis is a somewhat closer pair, with components of 5th and 6th magnitudes, and requires a small telescope to be divided. More challenging to discern is Mu (μ) Draconis, with its two 6th-magnitude stars, which requires a telescope with high magnification to be seen as double.

The wide pair of stars 16 and 17 Draconis is easily spotted with binoculars, and the brighter of the two – 17 Draconis – can be further divided with a small telescope with high magnification, turning this into a triple star. A similar triple is 39 Draconis; when viewed with a small telescope with low magnification, it appears a double but at higher magnification the brighter star divides into a closer pair with components of magnitudes 5.0 and 8.0. Two more doubles that can readily be seen with a small telescope are Omicron (ο) Draconis, with stars of 5th and 8th magnitudes, and 40 and 41 Draconis, which are both 6th-magnitude orange dwarfs.

In central Draco lies a planetary nebula made famous by a striking Hubble Space Telescope image: NGC 6543, or the Cat's Eye Nebula (see p.254). Processed in false colour, the Hubble picture shows the nebula as red, but when seen through a small telescope it appears blue-green, as do all planetary nebulae.

**BEAR AND DRAGON**
The long body of Draco curls around the stars of Ursa Minor, the Little Bear. The head of the dragon is easily identifiable.

**THE CAT'S EYE NEBULA**
This amateur CCD image of NGC 6543 shows some of the colour and structure captured by the Hubble Space Telescope, but visually the nebula appears as a blue-green ellipse.

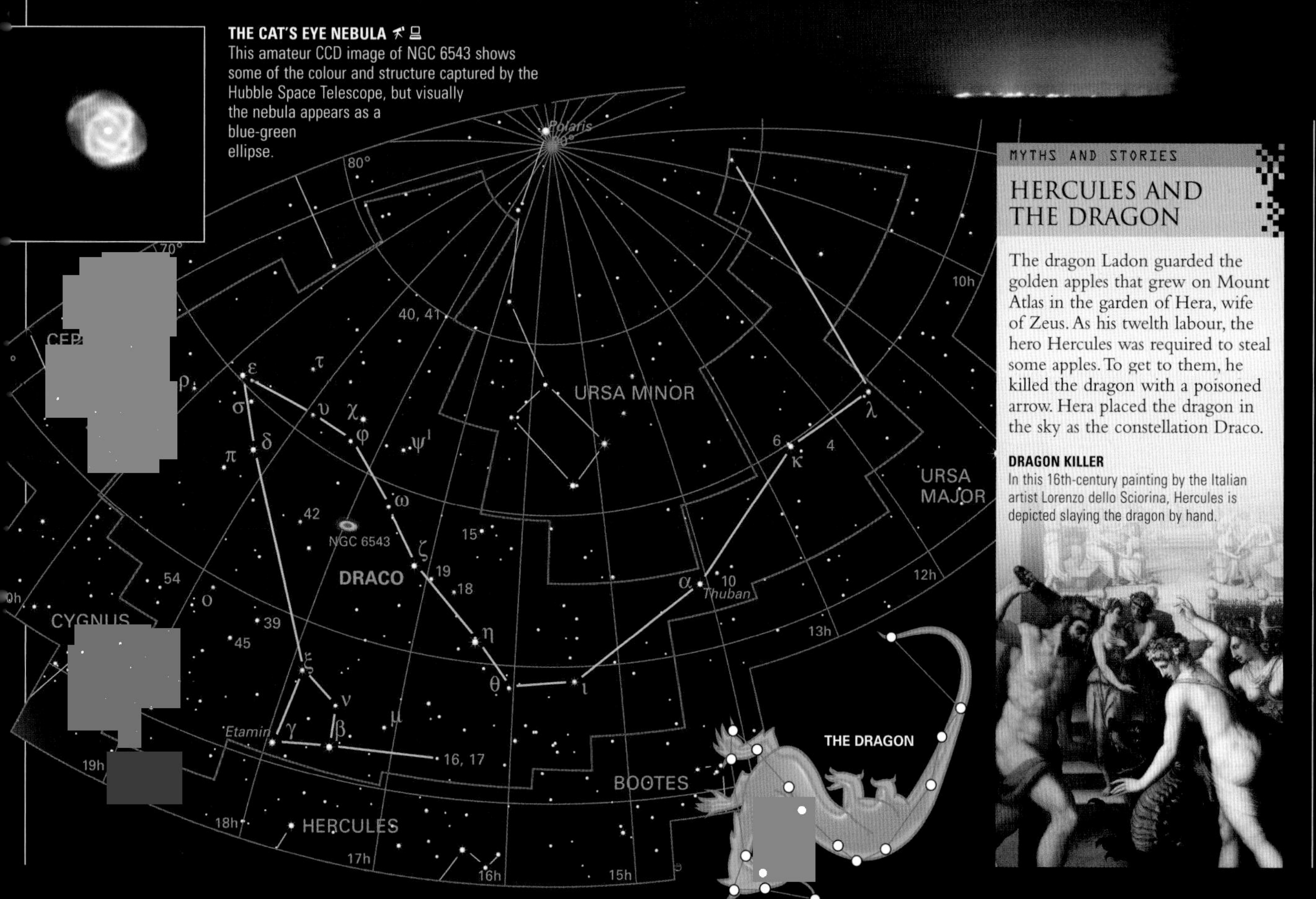


MYTHS AND STORIES

## HERCULES AND THE DRAGON

The dragon Ladon guarded the golden apples that grew on Mount Atlas in the garden of Hera, wife of Zeus. As his twelfth labour, the hero Hercules was required to steal some apples. To get to them, he killed the dragon with a poisoned arrow. Hera placed the dragon in the sky as the constellation Draco.

**DRAGON KILLER**
In this 16th-century painting by the Italian artist Lorenzo dello Sciorina, Hercules is depicted slaying the dragon by hand.

CEPHEUS

# Cepheus

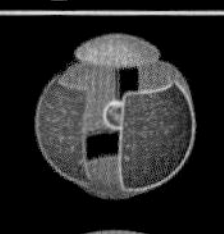

**SIZE RANKING** 27

**BRIGHTEST STAR** Alpha (α) 2.5

**GENITIVE** Cephei

**ABBREVIATION** Cep

**HIGHEST IN SKY AT 10PM** September–October

**FULLY VISIBLE** 90°N–1°S

Cepheus lies in the far northern sky between Cassiopeia and Draco. Its main stars form a distorted tower or steeple shape, yet this ancient Greek constellation in fact represents the mythical King Cepheus of Ethiopia, who was the husband of Queen Cassiopeia and the father of Andromeda. Cepheus is not a particularly prominent constellation.

## SPECIFIC FEATURES

The constellation's most celebrated star is Delta (δ) Cephei (see p.282), from which all Cepheid variables take their name. Just under 1,000 light-years away, this yellow-coloured supergiant varies between magnitudes 3.5 and 4.4 every five days nine hours.

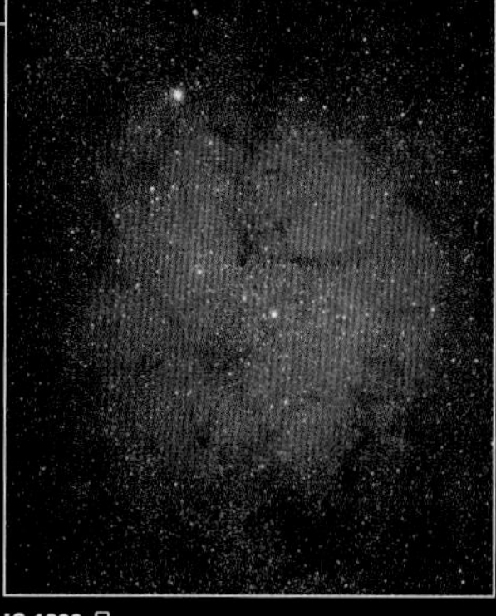

**IC 1396** The Garnet Star or Mu Cephei (top left) lies on the edge of the large but faint nebula IC 1396. The nebula is centred on the 6th-magnitude multiple star Struve 2816.

These changes can be followed with the naked eye. Delta (δ) Cephei is also a double star; its 6th-magnitude, blue-white companion is visible through a small telescope.

A significant variable star of a different kind is Mu (μ) Cephei, which is a red supergiant that ranges anywhere between magnitudes 3.4 and 5.1 every two years or so. This supergiant is also known as the Garnet Star on account of its strong red coloration.

Non-variable stars near Delta (δ) and Mu (μ) Cephei can be used to gauge the magnitude of these two variable stars at any given time. For example, they can be compared to Zeta (ζ) at magnitude 3.4, Epsilon (ε) at magnitude 4.2, or Lambda (λ) Cephei at magnitude 5.1 (see chart, below).

CEPHEUS

**THE KING** Shaped like a bishop's mitre, Cepheus is not easy to pick out in the sky. He is flanked by his prominent wife, Cassiopeia, and Draco, the dragon.

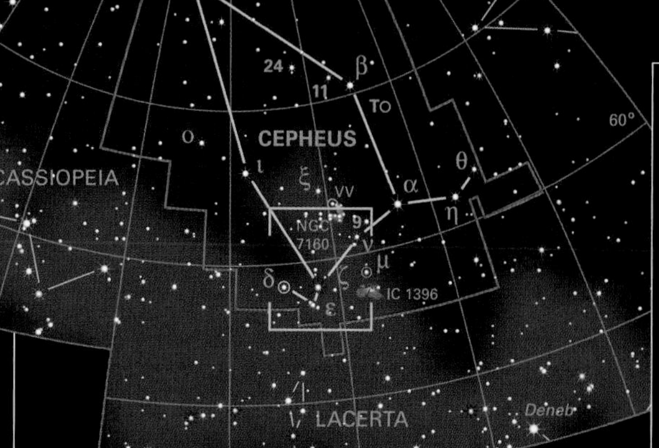


**DELTA (δ) AND MU (μ) CEPHEI**

22h 30m · 22h 00m · NGC 7160 · 19 · 12 · 14 · 13 · λ · ζ · ε · δ · μ · ν · 60° · 58° · 56°

**MAGNITUDE KEY**

- 0.0–0.9
- 1.0–1.9
- 2.0–2.9
- 3.0–3.9
- 4.0–4.9
- 5.0–5.9
- 6.0–6.9

## HENRIETTA LEAVITT

Henrietta Swan Leavitt (1868–1921) worked at Harvard College Observatory in the early 20th century. Her study of variable stars in the Small Magellanic Cloud led to the period-luminosity law. This law links the variation period of a Cepheid variable to its intrinsic brightness, which in turn can indicate distance. Her law remains fundamental to our knowledge of the scale of the Universe.

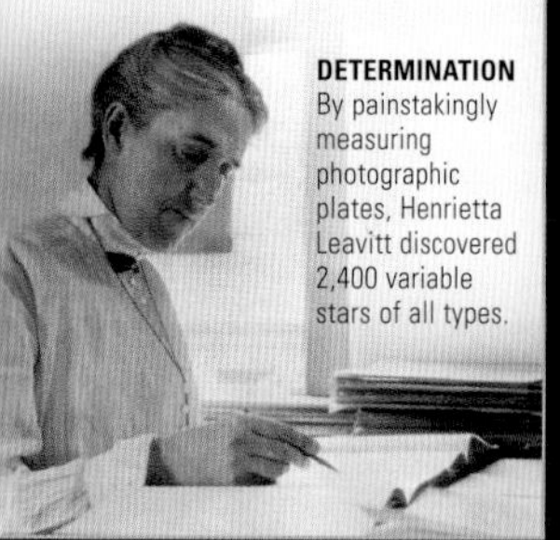

**DETERMINATION** By painstakingly measuring photographic plates, Henrietta Leavitt discovered 2,400 variable stars of all types.

CASSIOPEIA

# Cassiopeia

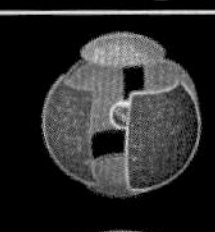

**SIZE RANKING** 25

**BRIGHTEST STARS** Shedir ($\alpha$) 2.2, Gamma ($\gamma$) 2.2

**GENITIVE** Cassiopeiae

**ABBREVIATION** Cas

**HIGHEST IN SKY AT 10PM** October–December

**FULLY VISIBLE** 90°N–12°S

This distinctive constellation of the northern sky is found within the Milky Way between Perseus and Cepheus and north of Andromeda. The large W shape formed by its five main stars is easily recognizable. It is an ancient Greek constellation, representing the mythical Queen Cassiopeia of Ethiopia.

## SPECIFIC FEATURES

Gamma ($\gamma$) Cassiopeiae (see p.281) is a hot, rapidly rotating star that occasionally throws off rings of gas from its equator, which causes unpredictable changes in its brightness. It has ranged between magnitudes 3.0 and 1.6, but it ...y lies at magnitude 2.2, ...makes it the equal-brightest ...he constellation.

A variable with a more predictable cycle is Rho ($\rho$) Cassiopeiae, an intensely luminous, yellow-white supergiant that fluctuates between 4th and 6th magnitudes every 10 or 11 months. It is estimated that it lies more than 10,000 light-years away, which is exceptionally distant for a naked-eye star.

Eta ($\eta$) Cassiopeiae is an attractive stellar pair consisting of a yellow and a red star. Its components are of magnitudes 3.5 and 7.5 and can be seen through a small telescope. This pair forms a true binary; the fainter companion orbits the brighter star every 480 years.

Cassiopeia contains a number of open clusters within range of small instruments. Chief among them is M52 (see p.286), near the border with Cepheus. It is visible through binoculars as a somewhat elongated patch of light, and its individual stars – including a bright orange giant at one edge – can be seen through a small telescope. M103 is a small, elongated group, best viewed through a small telescope. Nearby is a larger cluster, NGC 663, which is more suitable for binocular observation. NGC 457 is a looser star cluster, containing the 5th-magnitude star Phi ($\varphi$) Cassiopeiae. This cluster's appearance has been likened to an owl – its two brightest stars mark the owl's eyes.

**M103**
M103's main feature is a chain of three stars like a mini Orion's belt. The northernmost member of the line (top) is not a true member of the cluster but lies closer to Earth.

**M52**
Through binoculars, this cluster appears as a misty patch about one-third the diameter of the full Moon. A telescope is needed to resolve its individual stars.

MYTHS AND STORIES

## THE VAIN QUEEN

Wife of Cepheus and mother of Andromeda, Queen Cassiopeia was notoriously vain. She enraged the Nereids, daughters of Poseidon, by boasting she was more beautiful. In punishment, Poseidon sent a sea-monster to ravage her kingdom, which eventually led to the rescue of Andromeda by Perseus (see p.352).

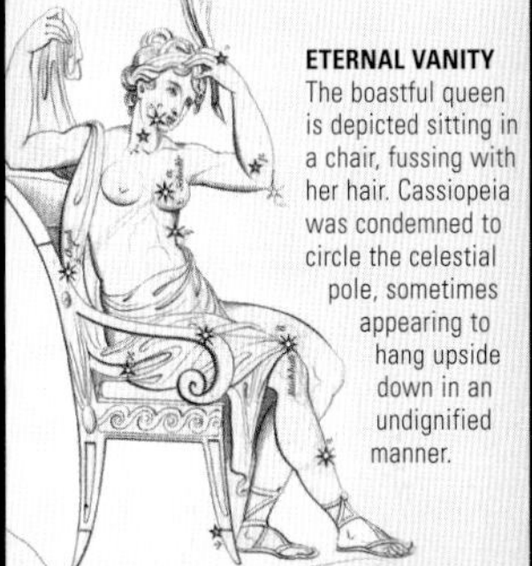

**ETERNAL VANITY**
The boastful queen is depicted sitting in a chair, fussing with her hair. Cassiopeia was condemned to circle the celestial pole, sometimes appearing to hang upside down in an undignified manner.

**POLAR POINTER**
The distinctive W shape formed by the main stars of Cassiopeia is easy to locate in the sky. The centre of the W points towards the north celestial pole.

THE GIRAFFE

# Camelopardalis

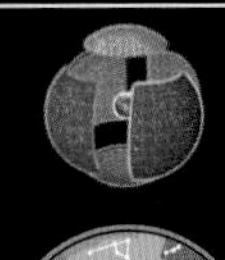

**SIZE RANKING** 18

**BRIGHTEST STAR** Beta (β) 4.0

**GENITIVE** Camelopardalis

**ABBREVIATION** Cam

**HIGHEST IN SKY AT 10PM** December–May

**FULLY VISIBLE** 90°N–3°S

This dim constellation of the far northern sky, representing a giraffe, was introduced in the early 17th century on a celestial globe created by the Dutch astronomer Petrus Plancius (see panel, below). The giraffe's long neck can be visualized as stretching around the north celestial pole towards Ursa Minor and Draco.

## SPECIFIC FEATURES

The brightest star in the constellation, Beta (β) Camelopardalis, is a double star whose fainter companion can be seen with a small telescope or even powerful binoculars. South of Beta (β) is 11 and 12 Camelopardalis, a wide double star with components of 5th and 6th magnitudes.

Within the giraffe's hindquarters is NGC 1502, a small open star cluster visible through binoculars or a small telescope. Binoculars also show a long chain of faint stars called Kemble's Cascade, which lead away from NGC 1502 towards Cassiopeia. This star feature is named after Lucian Kemble, a Canadian amateur astronomer who first drew attention to it in the late 1970s. None of the stars, however, are actually related.

NGC 2403 is a 9th-magnitude spiral galaxy that looks like a comet when seen through a small telescope. It is one of the brightest and closest galaxies to the Earth, outside the Local Group.

**KEMBLE'S CASCADE**
In an area five times the diameter of the full Moon, the stars of Kemble's Cascade seem to tumble down the sky. The small star cluster NGC 1502 can be seen in the lower left of the picture.

THE GIRAFFE

URSA MINOR
DRACO
Polaris
CAMELOPARDALIS
70°
γ
α
NGC 2403
NGC 1502
β
11,12
7
URSA MAJOR
LYNX
8h
AURIGA
Capella
6h
4h

**NGC 2403**
Colour images of this galaxy reveal the pink glow of large emission nebulae in its spiral arms. It is about 11 million light-years away.

**PARTIAL VIEW**
It can be difficult to relate the figure of a giraffe to the stars of Camelopardalis. Here, the stars of the giraffe's legs are shown. The animal's long neck would stretch off the top of the picture.

## PETRUS PLANCIUS

This Dutch church minister was also an expert geographer and astronomer. Petrus Plancius (1552–1622) taught the navigators on the first Dutch sea voyages to the East Indies how to measure star positions. In turn, they produced for him a catalogue of the southern stars divided into 12 new constellations, which Plancius depicted on his celestial globes. He also invented several constellations, such as Columba, Camelopardalis, and Monoceros, using some of the fainter stars visible from Europe.

THE CHARIOTEER

# Auriga

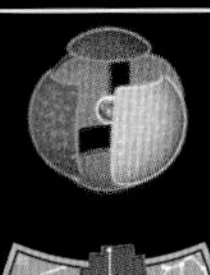

**SIZE RANKING** 21

**BRIGHTEST STAR** Capella (α) 0.1

**GENITIVE** Aurigae

**ABBREVIATION** Aur

**HIGHEST IN SKY AT 10PM** December–February

**FULLY VISIBLE** 90°N–34°S

Auriga is easily identified in the northern sky by the presence of Capella (α), the most northerly first-magnitude star. Auriga lies in the Milky Way between Gemini and Perseus, to the north of Orion. The constellation represents a charioteer.

## SPECIFIC FEATURES

Auriga's outstanding feature is a chain of three large and bright open star clusters. All three will just fit within the same field of view in wide-angle binoculars. Of the trio, M38's stars are the most scattered and, when viewed with a small telescope, seem to form chains. The middle cluster is M36, the smallest cluster but also the easiest to spot, while M37 is the largest and contains the most stars, but these are faint. All three clusters lie about 4,000 light-years away.

The star-forming nebula IC 405 is located nearby. Bright light from 6th-magnitude AE Aurigae near its centre lights up the surrounding gases.

Auriga also contains two extraordinary eclipsing binaries of long period. One is Zeta (ζ) Aurigae, which is an orange giant orbited by a smaller blue star that eclipses it every 2.7 years. This causes a 30 per cent decrease in brightness for six weeks, from magnitude 3.7 to 4.0. More remarkable, however, is Epsilon (ε) Aurigae (see p.277). This intensely luminous supergiant is orbited by a mysterious dark partner that eclipses it every 27 years – the longest interval of any eclipsing binary. During the eclipse, Epsilon's brightness is halved, from magnitude 3.0 to 3.8, and it remains dimmed for more than a year. Astronomers think that its companion is a close binary star enveloped in a disc of dust. The next eclipse is due at the end of 2009.

**SHARED STAR**
Neighbouring Beta (β) Tauri completes the charioteer figure. Auriga is usually identified as a king of Athens, Erichthonius.

**THE FLAMING STAR NEBULA**
AE Aurigae is a hot, massive star of magnitude 6 that lights up the surrounding cloud of gas and dust that is the Flaming Star Nebula, IC 405.

THE LYNX

# Lynx

**SIZE RANKING** 28

**BRIGHTEST STAR** Alpha (α) 3.1

**GENITIVE** Lyncis

**ABBREVIATION** Lyn

**HIGHEST IN SKY AT 10PM** February–March

**FULLY VISIBLE** 90°N–28°S

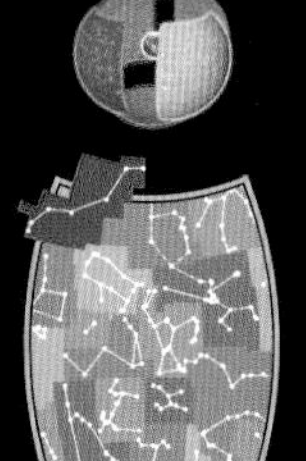

Lynx is a fair-sized but faint constellation in the northern sky. It was introduced in the late 17th century by Johannes Hevelius (see p.368), who wanted to fill the gap between Ursa Major and Auriga. Hevelius is reputed to have named it Lynx because only the lynx-eyed would be able to see it – Hevelius himself had very sharp eyesight. The animal he drew on his star chart, however, looked little like a real lynx.

## SPECIFIC FEATURES

Lynx contains many interesting double and multiple stars. For example, 12 Lyncis appears double with a small telescope, but with a telescope of 75mm (3in) or larger aperture the brighter star divides into two components of 5th and 6th magnitudes, which have an orbital period of about 700 years.

An easier triple to identify is 19 Lyncis. This consists of two stars of 6th and 7th magnitudes and a wider 8th-magnitude companion, all visible through a small telescope. A more challenging double star is 38 Lyncis, with components of 4th and 6th magnitudes. A telescope of 75mm (3in) aperture is required to separate the individual stars.

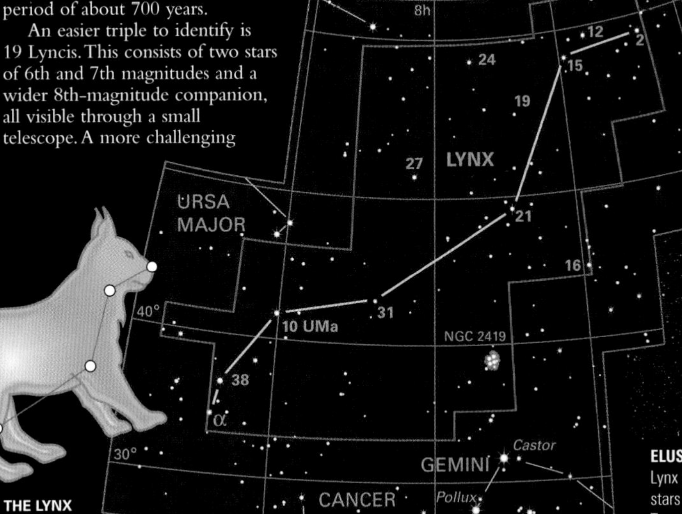


**ELUSIVE FELINE**
Lynx consists of nothing more than a few faint stars zigzagging between Ursa Major and Auriga. To spot it, keen eyesight or binoculars are required.

THE GREAT BEAR

# Ursa Major

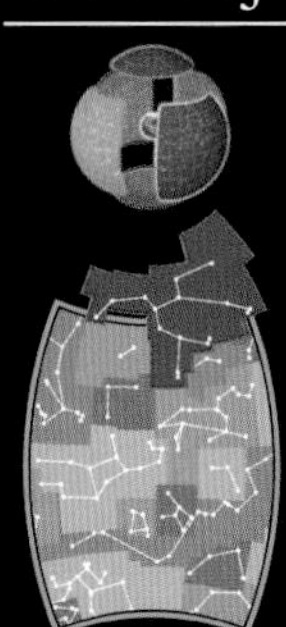

**SIZE RANKING** 3

**BRIGHTEST STARS** Alpha (α) 1.8, Epsilon (ε) 1.8.

**GENITIVE** Ursae Majoris

**ABBREVIATION** UMa

**HIGHEST IN SKY AT 10PM** February–May

**FULLY VISIBLE** 90°N–16°S

Ursa Major is one of the best-known constellations and a prominent feature of the northern sky. Seven of its stars form the familiar shape of the Plough, also known as the Big Dipper. But as a whole, Ursa Major is much larger than this; it is the third-largest constellation in the sky. The two stars in the Plough's bowl farthest from the handle, Dubhe (α) and Merak (β), point towards the north Pole Star, Polaris, while the curved handle of the Plough points towards the bright star Arcturus in the adjoining constellation of Boötes.

## SPECIFIC FEATURES

The Plough is one of the most famous patterns in the sky. Its shape is formed by the stars Dubhe (α), Merak (β), Phad (γ), Delta (δ) Ursae Majoris, Alioth (ε), Mizar (ζ) (see p.272), and Alkaid (η). With the exception of Dubhe and Alioth, these stars travel through space in the same direction, and they form what is known as a moving cluster.

Mizar (ζ), the second star in the Plough's handle, is next to Alcor (see p.272), an eighth, fainter star in the Plough, which can be seen with good eyesight. A small telescope reveals that Mizar also has a closer 4th-magnitude companion.

In southern Ursa Major lies a more difficult double star, Xi (ξ) Ursae Majoris, which needs a telescope with an aperture of 75mm (3in) to divide it. This pair, with components of 4th and 5th magnitudes, form a true binary relationship, orbiting every 60 years, which is quick by the standards of visual binary stars.

One of the easiest galaxies to identify with binoculars is M81, which is in northern Ursa Major, and is also known as Bode's Galaxy (see p.304). This spiral galaxy is at an angle and can be seen on clear, dark nights as a slightly elongated patch of light. A telescope is needed to spot the rather more elongated shape of the smaller and fainter Cigar Galaxy (see p.304), or M82, which is found one diameter of the Moon away from Bode's Galaxy. This unusual-looking object is now thought to be a spiral galaxy, seen edge-on, mottled with dust clouds and undergoing a burst of star formation following an encounter with M81.

Another major spiral galaxy in this constellation is the Pinwheel Galaxy, M101 (p.306), which lies near the end of the Plough's handle. Although larger than Bode's Galaxy, it is fainter and thus more difficult to see. An even greater challenge to find and identify, however, is the Owl Nebula, or M97, which is located under the bowl of the Plough. This planetary nebula is one of the faintest objects in Charles Messier's catalogue, and a telescope of around 75mm (3in) aperture is needed to make out its grey-green disc, which is three times larger than that of Jupiter. A telescope with an even larger aperture reveals the two dark patches, like an owl's eyes, that give rise to its popular name.

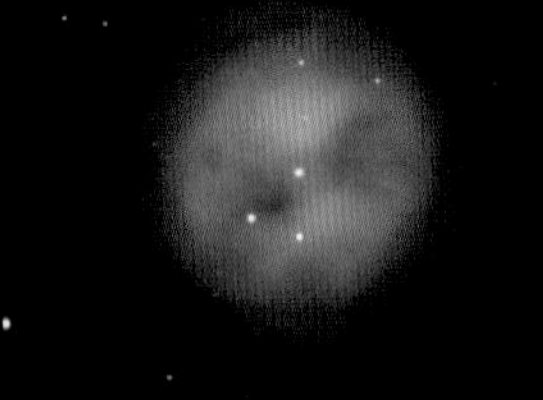

**THE OWL NEBULA**
The dark, owl-like eyes of the faint planetary nebula M97 are visible only through large telescopes or on photographs and CCD images such as this.

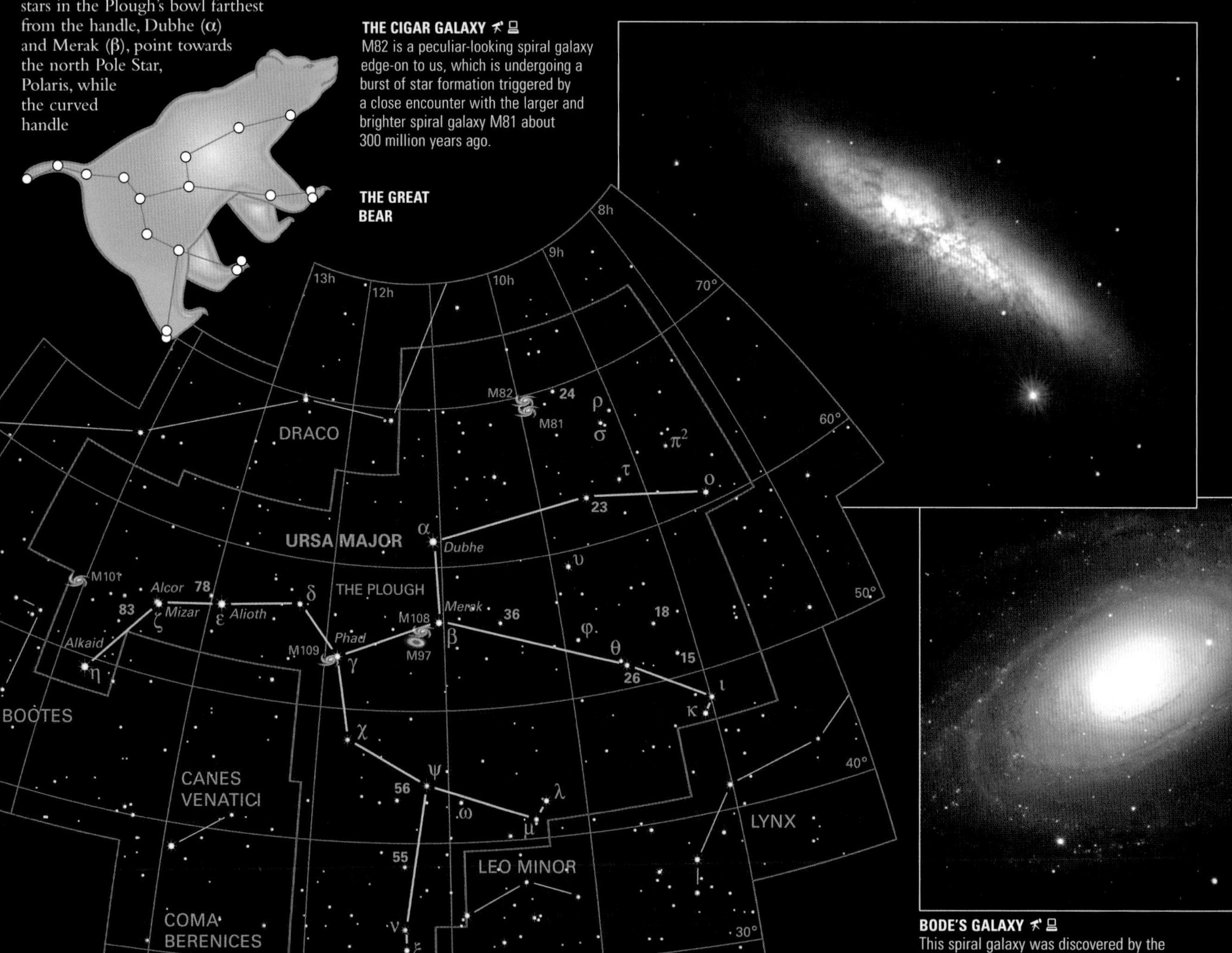


**THE CIGAR GALAXY**
M82 is a peculiar-looking spiral galaxy edge-on to us, which is undergoing a burst of star formation triggered by a close encounter with the larger and brighter spiral galaxy M81 about 300 million years ago.

**BODE'S GALAXY**
This spiral galaxy was discovered by the German astronomer Johann Elert Bode on 31 December 1774. Located approximately 11 million light-years away, M81 is nevertheless one of the brightest and most visible galaxies in the sky.

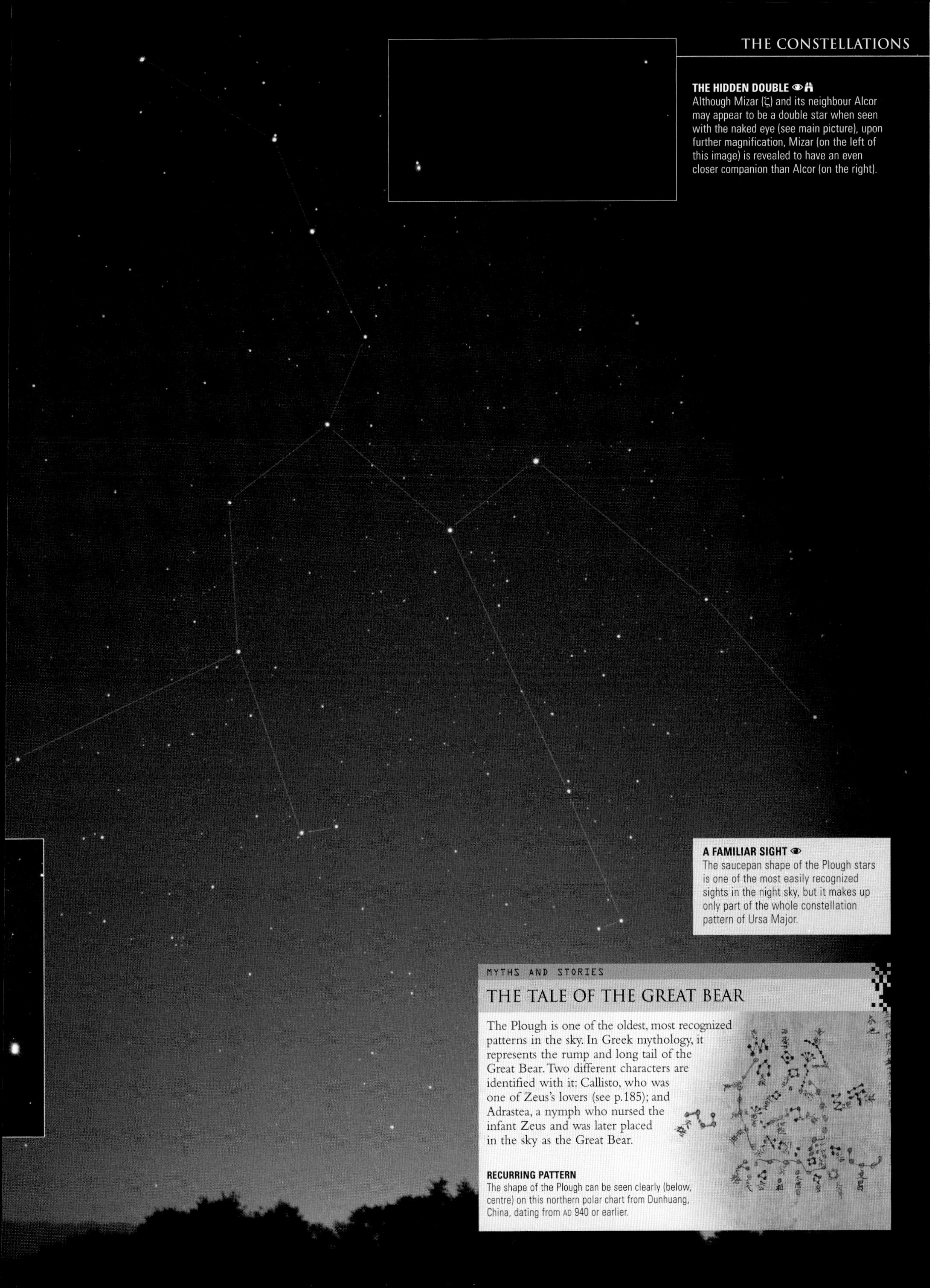

**THE HIDDEN DOUBLE**
Although Mizar (ζ) and its neighbour Alcor may appear to be a double star when seen with the naked eye (see main picture), upon further magnification, Mizar (on the left of this image) is revealed to have an even closer companion than Alcor (on the right).

**A FAMILIAR SIGHT**
The saucepan shape of the Plough stars is one of the most easily recognized sights in the night sky, but it makes up only part of the whole constellation pattern of Ursa Major.

MYTHS AND STORIES

## THE TALE OF THE GREAT BEAR

The Plough is one of the oldest, most recognized patterns in the sky. In Greek mythology, it represents the rump and long tail of the Great Bear. Two different characters are identified with it: Callisto, who was one of Zeus's lovers (see p.185); and Adrastea, a nymph who nursed the infant Zeus and was later placed in the sky as the Great Bear.

**RECURRING PATTERN**
The shape of the Plough can be seen clearly (below, centre) on this northern polar chart from Dunhuang, China, dating from AD 940 or earlier.

THE HUNTING DOGS

# Canes Venatici

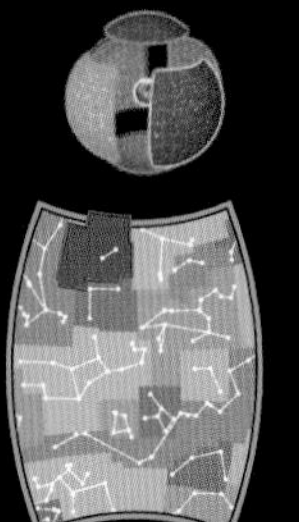

**SIZE RANKING** 38

**BRIGHTEST STAR** Cor Caroli (α) 2.9

**GENITIVE** Canum Venaticorum

**ABBREVIATION** CVn

**HIGHEST IN SKY AT 10PM** April–May

**FULLY VISIBLE** 90°N–37°S

Canes Venatici lies in the northern sky between Boötes and Ursa Major. This constellation represents two dogs held on a leash by the herdsman Boötes. It was formed by Johannes Hevelius (see p.368) at the end of the 17th century from stars that had previously been part of Ursa Major.

## SPECIFIC FEATURES

The constellation's brightest star, Alpha (α) Canum Venaticorum, is known as Cor Caroli, meaning Charles's Heart, in commemoration of King Charles I of England. This wide double star, with components of magnitudes 2.9 and 5.6, is easily separated with a small telescope. The brighter star is slightly variable, by about one-tenth of a magnitude, which is too small to be noticeable to the naked eye. Larger variation is found in Gamma (γ) Canum Venaticorum, a deep red supergiant popularly known as La Superba. It fluctuates between magnitudes 5.0 and 6.5 every 160 days or so.

**THE SUNFLOWER GALAXY**
M63 is a spiral galaxy, with patchy outer arms, that is seen at an angle from Earth. The arms give rise to comparisons with the appearance of a sunflower. The star to its right in this photograph is of 9th magnitude.

**GLOBULAR CLUSTER M3**
This cluster is one of the biggest and brightest globular clusters in the northern sky. A telescope with 100mm (4in) aperture is needed to resolve its individual stars.

**THE WHIRLPOOL GALAXY**
The core of this beautiful spiral galaxy (also known as M51) appears as a point of light in a small telescope, as does its companion galaxy NGC 5195 (top) at the end of one arm.

Canes Venatici also contains some fine galaxies, such as the Whirlpool Galaxy (see p.305), or M51, which is found seven diameters of the full Moon from the star at the end of the handle of the Plough (in Ursa Major). The Whirlpool Galaxy was the first galaxy in which spiral form was detected – the observation being made in 1845 by William Parsons (see p.305) in Ireland. The galaxy appears as a round patch of light through binoculars, but a moderate-sized telescope is needed to make out the spiral arms. At the end of one of the arms lies a smaller galaxy, NGC 5195, which is passing close to M51.

Two spiral galaxies worth looking for through a small telescope are the Sunflower Galaxy (M63) and M94.

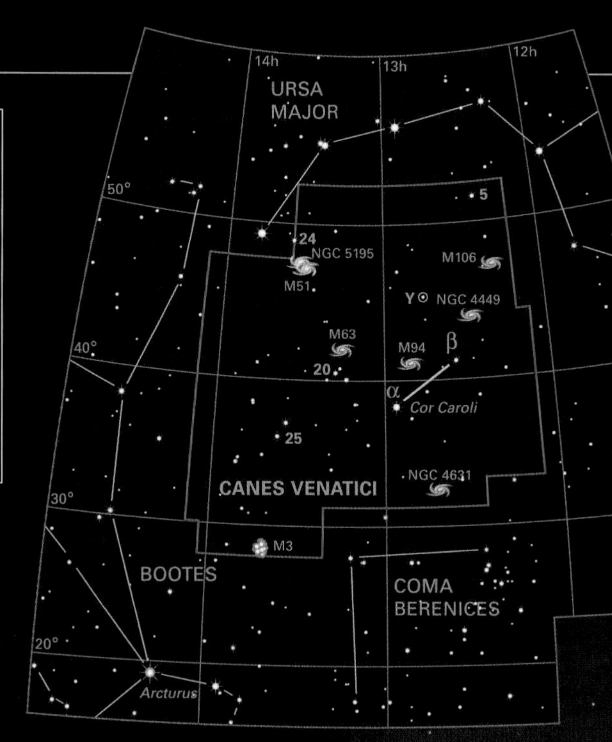


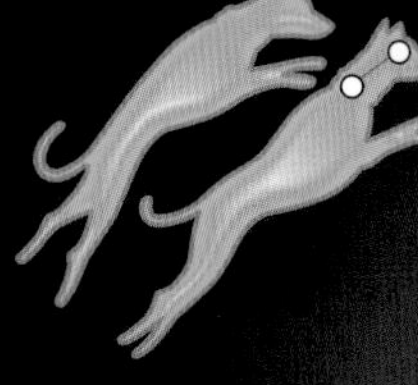

THE HUNTING DOGS

**TWO BRIGHT STARS**
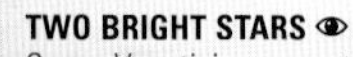
Canes Venatici represents a pair of hounds, but the unaided eye can see little more than the constellation's brightest stars, Cor Caroli and Beta Canum Venaticorum.

THE HERDSMAN

# Boötes

**SIZE RANKING** 13

**BRIGHTEST STAR** Arcturus (α) -0.1

**GENITIVE** Boötis

**ABBREVIATION** Boo

**HIGHEST IN SKY AT 10PM** May–June

**FULLY VISIBLE** 90°N–35°S

The Greek constellation Boötes contains the brightest star north of the celestial equator, Arcturus – Alpha (α) Boötis – which is also the fourth-brightest star in the entire sky. This large and conspicuous constellation extends from Draco and the handle of the Plough (in Ursa Major) to Virgo. Faint stars in the northern part of Boötes once formed the now-defunct constellation of Quadrans Muralis, which gave its name to the Quadrantid meteor shower that radiates from this area every January.

## SPECIFIC FEATURES

Arcturus is classified as a red giant, but as with most supposedly "red" stars, it actually looks orange to the unaided eye. Its colouring becomes stronger when viewed through binoculars. In billions of years' time, our Sun will swell into a red giant similar to this star.

Boötes is noted for its double stars, the most celebrated of which is Izar (see p.273), or Epsilon (ε) Boötis, at the heart of the constellation. To the naked eye, it appears of magnitude 2.4, but high magnification on a telescope of at least 75mm (3in) aperture reveals a close, 5th-magnitude companion that is blue-green in colour, providing one of the most beautiful contrasts of all double stars.

Much easier to divide with any small telescope are Kappa (κ) and Xi (ξ) Boötis. Kappa's stars, with components of 5th and 7th magnitudes, are unrelated but Xi, with stars also of 5th and 7th magnitudes, is a true binary with an orbital period of 150 years and has warm yellow-orange hues.

Easiest of all are the doubles Mu (μ) Boötis, with components of 4th and 6th magnitudes, and Nu (ν) Boötis, with two 5th-magnitude components – both are widely spaced enough to divide with binoculars.

THE HERDSMAN

MYTHS AND STORIES

## THE BEAR KEEPER

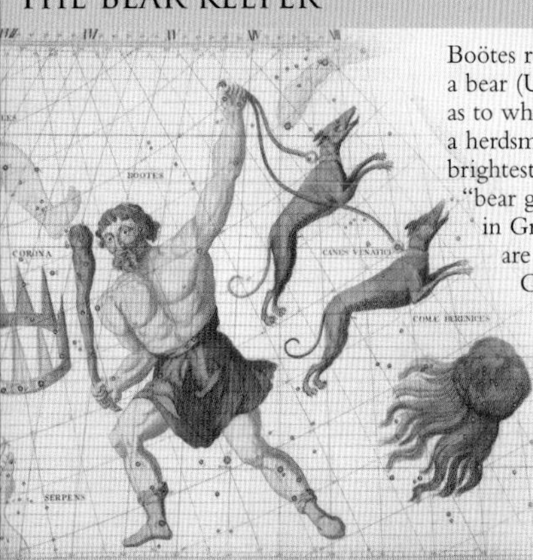

Boötes represents a man herding a bear (Ursa Major). Myths differ as to whether he is a hunter or a herdsman, as the constellation's brightest star, Arcturus, means "bear guard" or "bear keeper" in Greek. The man's two dogs are represented by adjoining Canes Venatici. In Greek myth, Boötes was identified with Arcas, son of Zeus and Callisto.

**ADJACENT STARS**
Boötes is depicted here leading the two hunting dogs, on an 18th-century star chart by Sir James Thornhill.

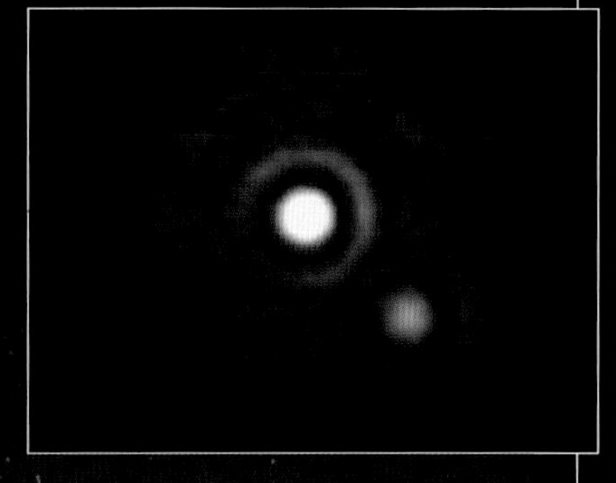

**DOUBLE STAR IZAR**
Epsilon (ε) Boötis, which is also known as Izar or Pulcherrima, is a challenging double star consisting of a bright orange star with a fainter blue-green companion star.

16h 15h 14h 50° 20° HERCULES BOOTES URSA MAJOR CORONA BOREALIS COMA BERENICES SERPENS CAPUT Izar Arcturus

**KITE-SHAPED CONSTELLATION**
Boötes, containing the bright star Arcturus, stands aloft in spring skies in the northern hemisphere. The crown of Corona Borealis can be seen to its left.

HERCULES

# Hercules

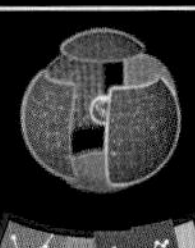

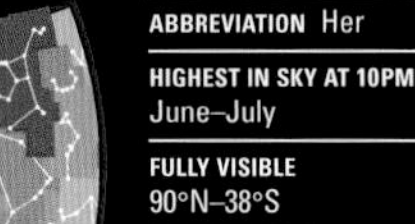

**SIZE RANKING** 5

**BRIGHTEST STAR** Kornephoros (β) 2.8

**GENITIVE** Herculis

**ABBREVIATION** Her

**HIGHEST IN SKY AT 10PM** June–July

**FULLY VISIBLE** 90°N–38°S

This large but not particularly prominent constellation of the northern sky represents Hercules, the strong man of Greek myth. In the sky, Hercules is depicted clothed in a lion's pelt, brandishing a club and the severed head of the watchdog Cerberus, and kneeling with one foot on the head of the celestial dragon, Draco – the tools and conquests of some of his 12 labours.

HERCULES

The most distinctive feature of this constellation is a quadrilateral of stars called the Keystone, which is composed of Epsilon (ε), Zeta (ζ), Eta (η), and Pi (π) Herculis.

## SPECIFIC FEATURES

Alpha (α) Herculis, (see p.281), or Rasalgethi, is actually the second-brightest star in Hercules. It fluctuates between 3rd and 4th magnitudes. As with most such erratic variables, Rasalgethi is a bloated red giant that pulsates in size, causing the brightness changes. A small telescope brings a 5th-magnitude blue-green companion star into view.

On one side of the Keystone lies M13, which is regarded as the finest globular cluster of northern skies. Under ideal conditions, M13 can be glimpsed with the naked eye, and through binoculars it appears like a hazy star half the width of the full Moon. Slightly farther away from the Keystone is a second globular cluster – M92. This often overlooked cluster is smaller and fainter than M13, and when seen through binoculars it can easily be mistaken for an ordinary star.

Several readily seen double stars are to be found in Hercules, including Kappa (κ) Herculis, with components of 5th and 6th magnitudes, and 100 Herculis, with its two 6th-magnitude stars. Positioned closer together, and hence requiring higher magnification, are 95 Herculis, with two 5th-magnitude components, and Rho (ρ) Herculis, with components of 5th and 6th magnitudes.

**UPSIDE DOWN** In the sky, Hercules is positioned with his feet towards the pole (top left in this picture) and his head pointing south.

**GLOBULAR CLUSTER M13** Through binoculars, this cluster appears as a rounded patch of light. It breaks up into countless starry points when viewed through a small telescope.

**THE HERCULES GALAXY CLUSTER** Every fuzzy object in this picture is a faint galaxy in the cluster Abell 2151, some 500 million light-years away.

THE LYRE

# Lyra

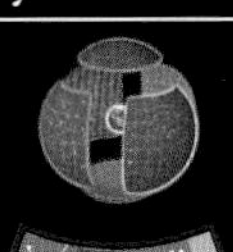

**SIZE RANKING** 52

**BRIGHTEST STAR** Vega (α) 0.0

**GENITIVE** Lyrae

**ABBREVIATION** Lyr

**HIGHEST IN SKY AT 10PM** July–August

**FULLY VISIBLE** 90°N–42°S

Lyra lies on the edge of the Milky Way next to Cygnus and is a compact constellation of the northern sky. It includes Vega, or Alpha (α) Lyrae (see p.249), which is the fifth-brightest star in the sky and one of the so-called Summer Triangle of stars – the other two being Deneb (in Cygnus) and Altair (in Aquila). The Lyrid meteors radiate from a point near Vega around 21–22 April every year. Lyra represents the stringed instrument played by Orpheus (see panel, below).

## SPECIFIC FEATURES

Vega dazzles at magnitude 0.0, appearing somewhat blue-white in colour to the unaided eye. It is the standard star against which astronomers compare the colour and brightness of all other stars.

The finest quadruple star in the sky – Epsilon (ε) Lyrae (see p.272) – is found three diameters of the full Moon from Vega. Binoculars easily show it as a neat pair of 5th-magnitude white stars, but each of these has a closer companion that is brought into view with a telescope of 60–75mm (2.5–3in) aperture and high magnification. All four stars are linked by gravity and are in long-term orbit around each other.

Two other double stars near Vega that are easy to identify with binoculars are Zeta (ζ) and Delta (δ) Lyrae, each with components of 4th and 6th magnitudes. Beta (β) Lyrae is another double star, easily resolved by a small telescope into its cream and blue components. The brighter star (the cream one) is an eclipsing binary that fluctuates between magnitudes 3.3 and 4.4 every 12.9 days. Many years of study have established that Beta's two stars are so close that gas from the larger of the pair falls towards the smaller companion, and some of it spirals off into space. Almost midway between Beta and Gamma (γ) Lyrae lies the most photographed of Lyra's celestial treasures, the Ring Nebula (see p.253), or M57. This planetary nebula is shaped like a smoke ring, and appears through a small telescope as a disc larger than that of Jupiter. Larger apertures are needed to make out the central hole. Studies with the Hubble Space Telescope have revealed that the "ring" is in fact a cylinder of gas thrown off from the central star, oriented almost end-on to the Earth.

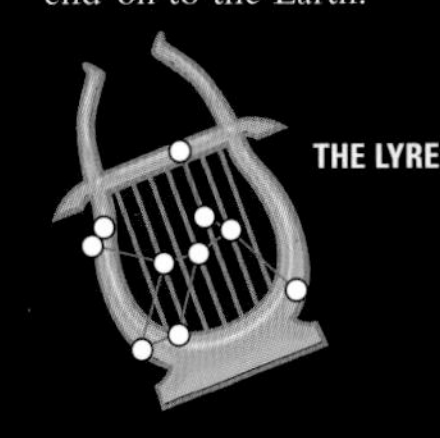
THE LYRE

**THE RING NEBULA**
One of the most famous planetary nebulae in the whole sky, the Ring Nebula, or M57, consists of hot gas shed from a central star. Its beautiful colours are revealed only on photographs such as this.

MYTHS AND STORIES

## ORPHEUS

Heartbroken Orpheus descended into the Underworld to retrieve his wife, Eurydice, who had been killed by a snake. His songs charmed Hades, god of the Underworld, who agreed to release Eurydice provided Orpheus did not look back as he led her to the surface. At the last minute, Orpheus glanced behind him, and Eurydice faded away. Orpheus then roamed the Earth, disconsolately playing his lyre.

**ENTRANCED**
Orpheus was said to have charmed even the rocks and streams with his music. In this 19th-century painting, he tames the wild animals with his songs.

**STRINGED INSTRUMENT**
Lyra, dominated by dazzling Vega, represents the harp played by Orpheus, the musician of Greek myth. Arab astronomers visualized the constellation as an eagle or vulture.

THE SWAN

# Cygnus

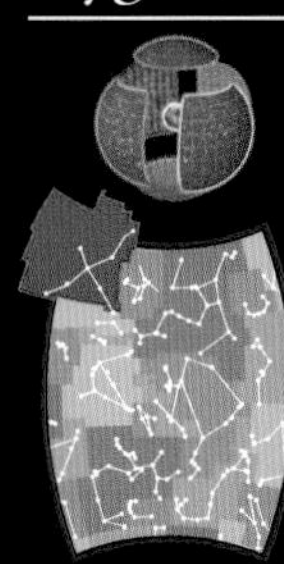

**SIZE RANKING** 16

**BRIGHTEST STAR** Deneb (α) 1.2

**GENITIVE** Cygni

**ABBREVIATION** Cyg

**HIGHEST IN SKY AT 10PM** August–September

**FULLY VISIBLE** 90°N–28°S

Situated in a rich area of the Milky Way, Cygnus is one of the most prominent constellations of the northern sky and contains numerous objects of interest. The relatively large constellation depicts a swan in flight, but its main stars are arranged in the shape of a giant cross, hence its alternative popular name of the Northern Cross.

## SPECIFIC FEATURES

Cygnus's brightest star, Deneb – Alpha (α) Cygni – lies in the tail of the swan, or at the top of the cross, depending on how the constellation is visualized. Deneb is an immensely luminous supergiant star located more than 3,000 light-years away, making it the most distant 1st-magnitude star. It forms one corner of the northern Summer Triangle – a familiar sight in the skies of northern summers and southern winters – which is completed by Vega (in Lyra) and Altair (in Aquila).

The beak of the swan (or the foot of the cross) is marked by a double star, Beta (β) Cygni, known as Albireo. Its two stars are sufficiently far apart that they can be seen separately with ordinary binoculars, if steadily mounted, and they are easy targets for a small telescope. The brighter star, of magnitude 3.1, is orange, and the fainter star, magnitude 5.1, is blue-green.

A similar colour difference is evident between Omicron-1 (ο¹) Cygni, a 4th-magnitude orange star, and its wide 5th-magnitude companion, 30 Cygni, which has a noticeable bluish colour when seen through binoculars. A 7th-magnitude star, again bluish, and even closer to Omicron-1, can also be seen with binoculars or a small telescope. Another pair of stars that is easy

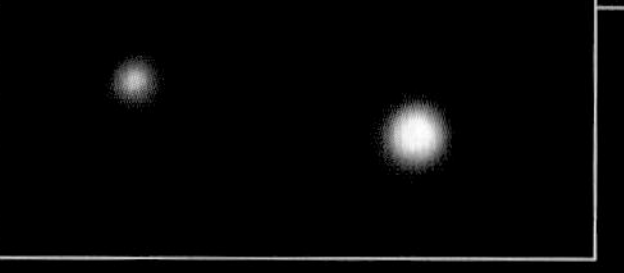

**ALBIREO**
Beta (β) Cygni, also known as Albireo, marks the beak of the swan. This double star, with its strikingly contrasting colours, is easily separated with a small telescope.

THE SWAN

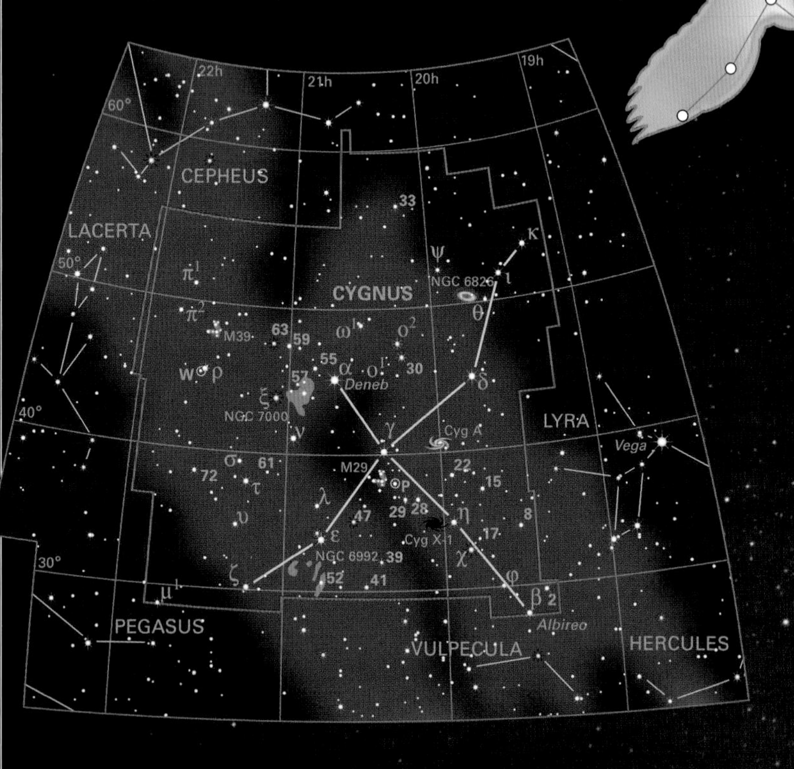


**POISED IN FLIGHT**
Among the stars of Cygnus, it is comparatively easy to visualize a swan, with its wings outstretched, as it flies along the Milky Way.

to spot with a small telescope is 61 Cygni (see p.248), which consists of two orange dwarfs of 5th and 6th magnitudes that orbit each other every 650 years. A large open star cluster, M39, covers an area of sky of similar size to the full Moon near the constellation's border with Lacerta.

On clear nights, the Milky Way appears as a hazy band of light running through Cygnus, divided in two by an intervening cloud of dust known as the Cygnus Rift or the Northern Coalsack. The rift continues, via Aquila, into Ophiuchus.

Two large and remarkable nebulae are found in Cygnus, although neither is easy to identify. The glowing gas cloud of the North America Nebula (NGC 7000), near Deneb, can be glimpsed through binoculars on clear, dark nights, but its full majesty becomes apparent only on long-exposure photographs or CCD images. The Veil Nebula is a diffuse nebula found in the wing of the swan. Again, it is best seen on photographs, although the brightest part – NGC 6992 – can just be made out with binoculars or a small telescope and becomes more prominent with the addition of filters to the telescope. Considerably smaller, but much easier to spot, is the Blinking Planetary (NGC 6826) in the other wing of the swan, with a blue-green disc similar in size to that of Jupiter. It is popularly known as the Blinking Planetary because of an odd optical effect in which, as the observer looks alternately directly at it and off to one side, it appears to blink on and off.

Two objects of considerable astrophysical interest in Cygnus are beyond the reach of amateur observers. Cygnus A (see p.314) is a powerful radio source, the result of two galaxies in collision millions of light-years away. Cygnus X-1 (see p.268), near Eta (η) Cygni, is an intense X-ray source, thought to be a black hole orbiting a 9th-magnitude blue supergiant in our galaxy.

MYTHS AND STORIES

## LEDA AND THE SWAN

The swan represents the disguise adopted by Zeus for an illicit love tryst. The object of his desire is sometimes said to have been a nymph called Nemesis or, in a more popular version, Queen Leda of Sparta. After her union with Zeus, Leda is said to have given birth to either one or two eggs, according to different versions of the story, from which hatched Castor, Pollux, and their sister Helen of Troy. Pollux and Helen were reputedly the offspring of Zeus, but Castor was the son of Leda's husband, King Tyndareus.

**FAMILY GROUPING**
Queen Leda, the twins Castor and Pollux, and the swan are captured in this painting after the original by Leonardo da Vinci.

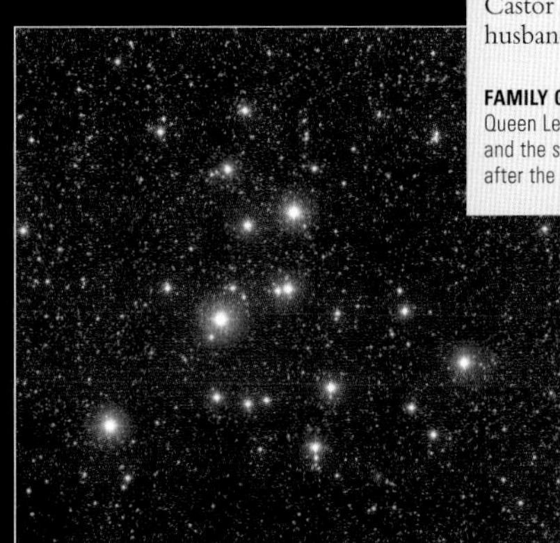

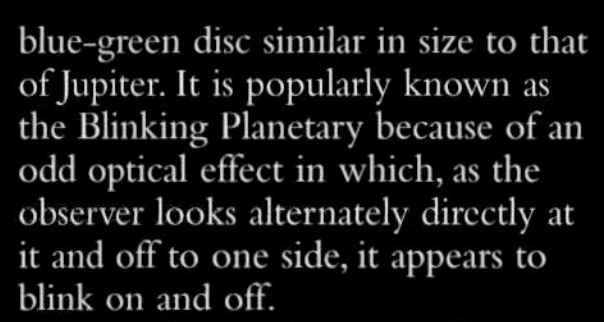

**OPEN CLUSTER M39**
M39 is the larger and brighter of the two Messier clusters in Cygnus and contains around 30 members arranged in a triangular shape, with a double star near the centre. It lies 900 light-years away and is easily spotted with binoculars. Under good conditions, M39 is visible to the naked eye.

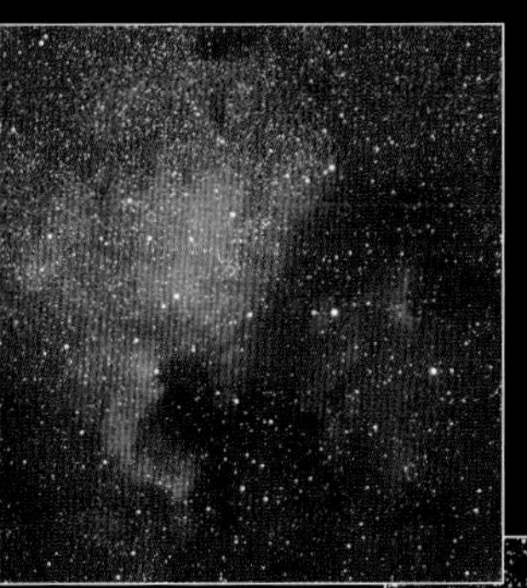

**NORTH AMERICA NEBULA**
In the tail of the swan lies NGC 7000, which is popularly known as the North America Nebula, on account of its similarity in shape to that continent.

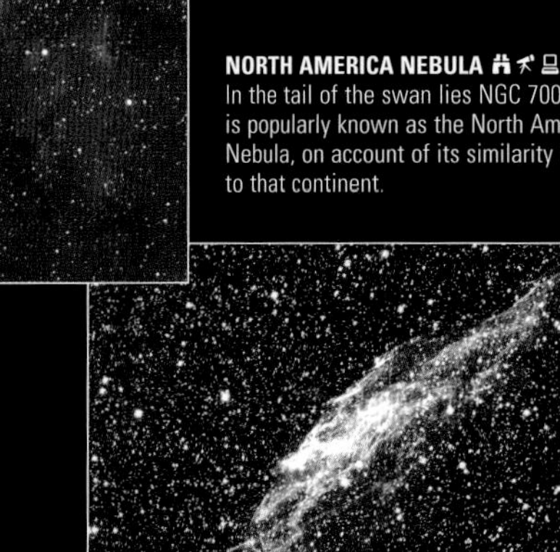

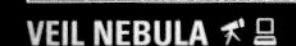

**VEIL NEBULA**
Splashed across an area wider than six full Moons is the Veil Nebula, a loop of gas that is the remains of a star that exploded as a supernova thousands of years ago.

THE NIGHT SKY

ANDROMEDA

# Andromeda

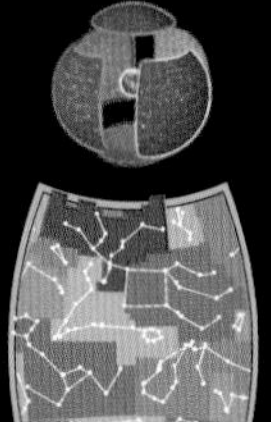

**SIZE RANKING** 19

**BRIGHTEST STARS**
Alpheratz (α) 2.1,
Mirach (β) 2.1

**GENITIVE**
Andromedae

**ABBREVIATION** And

**HIGHEST IN SKY AT 10PM**
October–November

**FULLY VISIBLE**
90°N–37°S

This celebrated constellation of the northern skies depicts the daughter of the mythical Queen Cassiopeia, who is represented by a neighbouring constellation. The head of the princess is marked by Alpheratz (or Sirrah ) – Alpha (α) Andromedae – which is the star at the nearest corner of the Square of Pegasus, in another adjacent constellation. Long ago, Alpheratz was regarded as being shared with the constellation Pegasus, where it marked the navel of the horse. The star's two names – Alpheratz and Sirrah – are both derived from an Arabic term that means "the horse's navel".

## SPECIFIC FEATURES

On a clear night, the farthest it is possible to see with the naked eye is about 2.5 million light-years, which is the distance to the Andromeda Galaxy (see pp.302–303), a huge spiral of stars similar to our own galaxy. Also known as M31, this galaxy spans several diameters of the full Moon and lies high in the mid-northern sky on autumn evenings. The naked eye sees it as a faint patch; it looks elongated, rather than spiral, because it is tilted at a steep angle towards the Earth. When looking at M31 through a telescope, low magnification must be used to give the widest field of view and to concentrate the light. The small companion galaxies, M32 and M110, are difficult to see through a small telescope.

Gamma (γ) Andromedae, known also as Almaak or Almach (see p.273), is a double star of contrasting colours. It consists of an orange giant star of magnitude 2.3 and a fainter blue companion, and it is easily seen through a small telescope.

The open star cluster NGC 752 spreads over an area larger than the full Moon and can be identified with binoculars, but a small telescope is needed to resolve its individual stars of 9th magnitude and fainter.

NGC 7662, which is popularly known as the Blue Snowball, is one of the easiest planetary nebulae to identify, and it can be found through a small telescope.

**THE ANDROMEDA GALAXY**
Only the inner parts of M31 are bright enough to be seen with small instruments. CCD images such as this bring out the full extent of the spiral arms. Below M31 on this image lies M110, while M32 is on its upper rim.

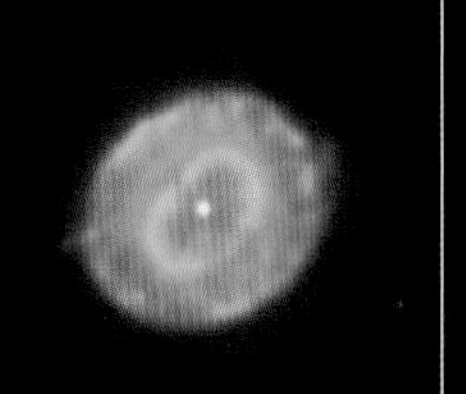

**THE BLUE SNOWBALL**
When seen through a small telescope, NGC 7662 appears as a bluish disc. Its structure is brought out only on CCD images such as this one.

MYTHS AND STORIES

## HEROIC RESCUE

According to Greek mythology, Andromeda was chained to a rock on the seashore and offered as a sacrifice to a sea monster in atonement for the boastfulness of her mother, Queen Cassiopeia. The Greek hero Perseus, flying home after slaying Medusa, the Gorgon, noticed the maiden's plight. He responded by swooping down in his winged sandals and killing the sea monster. He then whisked Andromeda to safety and married her.

**DAMSEL IN DISTRESS**
The Flemish artist Rubens added the flying horse Pegasus to his 17th-century depiction of Andromeda's dramatic rescue by Perseus from captivity on the rock.

**HEAD TO TOE**
Andromeda is one of the original Greek constellations. Its brightest stars represent the princess's head (α), her pelvis (β), and her left foot (γ).

THE LIZARD

# Lacerta

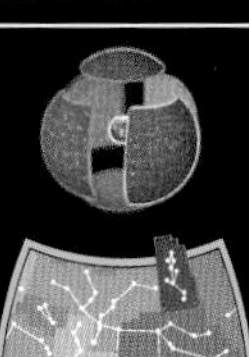

**SIZE RANKING** 68

**BRIGHTEST STAR** Alpha (α) 3.8

**GENITIVE** Lacertae

**ABBREVIATION** Lac

**HIGHEST IN SKY AT 10PM** September–October

**FULLY VISIBLE** 90°N–33°S

Lacerta consists of a zigzag of faint stars in the northern sky, squeezed between Andromeda and Cygnus like a lizard between rocks. It is one of the seven constellations invented by Johannes Hevelius (see p.368) during the late 17th century.

This constellation contains no objects of note for amateur astronomers, although BL Lacertae (see p.315), which was once thought to be a peculiar 14th-magnitude variable star, has given its name to a class of galaxies with active nuclei called BL Lac objects or "blazars".

A BL Lac object is a type of quasar that shoots jets of gas from its centre directly towards the Earth. Because we see these jets of gas head-on, these BL Lac objects tend to look star-like.

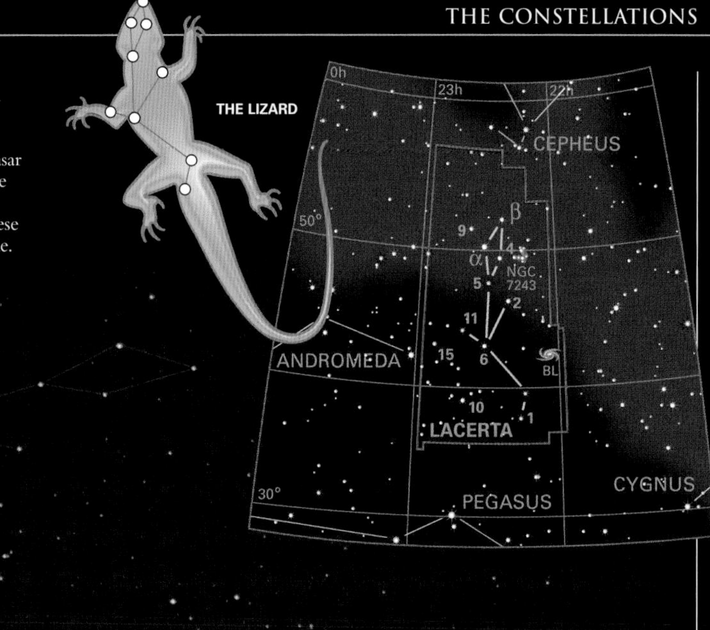


THE TRIANGLE

# Triangulum

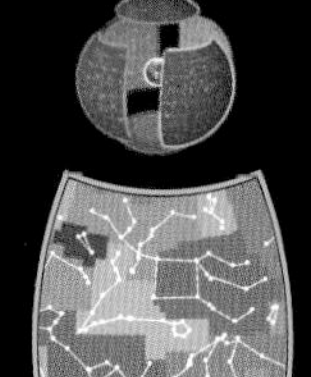

**SIZE RANKING** 78

**BRIGHTEST STAR** Beta (β) 3.0

**GENITIVE** Trianguli

**ABBREVIATION** Tri

**HIGHEST IN SKY AT 10PM** November–December

**FULLY VISIBLE** 90°N–52°S

This small northern constellation is to be found lying between Andromeda and Aries. It consists of little more than a triangle of three stars. Triangulum is one of the constellations known to the ancient Greeks, who visualized it as the Nile delta or the island of Sicily.

SPECIFIC FEATURES

Triangulum contains the third-largest member of our Local Group of galaxies, M33 or the Triangulum Galaxy (see p.301). In physical terms, M33 is about one-third the size of the Andromeda Galaxy, or M31 (see pp.302–303), and is much fainter.

The spiral galaxy M33 appears as a large pale patch of sky. It is of a similar size to the full Moon, when viewed through binoculars or a small telescope on a dark, clear night. To see the spiral arms, a large telescope is needed. M33 looks like a starfish on long-exposure photographs.

There is little else of note in the constellation apart from 6 Trianguli. This yellow star has a magnitude of 5.2 and has a 7th-magnitude companion which can be detected through a small telescope.

**TRIANGULUM AND MARS** This image of the three stars that make up the shape of Triangulum also includes the planet Mars, passing through neighbouring Pisces.

**M33** The clouds of pinkish gas in the arms of M33 show up in CCD images of this spiral galaxy in the Local Group. It is presented almost face-on to the Earth.

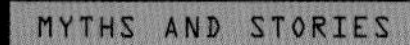

THE VICTORIOUS HERO

# Perseus

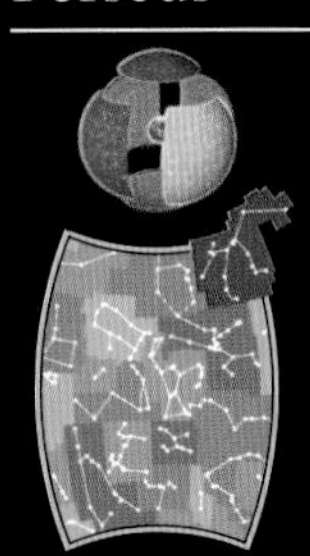

**SIZE RANKING** 24

**BRIGHTEST STAR**
Mirphak (α) 1.8

**GENITIVE** Persei

**ABBREVIATION** Per

**HIGHEST IN SKY AT 10PM**
November–December

**FULLY VISIBLE**
90°N–31°S

Perseus is a prominent northern constellation lying in the Milky Way between Cassiopeia and Auriga. It is an original Greek constellation and represents Perseus, who was sent to slay Medusa, the Gorgon. In the sky, Perseus is depicted with his left hand holding the Gorgon's head, which is marked by Algol – Beta (β) Persei – a famous variable star (see p.272). His right hand brandishes his sword, marked by the twin clusters NGC 869 and NGC 884.

## SPECIFIC FEATURES

The constellation's brightest member – Mirphak, or Alpha (α) Persei – is of magnitude 1.8. It lies at the centre of a group of stars known as the Alpha Persei Cluster or Melotte 20. Scattered over an area of sky that is several times the diameter of the full Moon, the cluster is an excellent sight through binoculars.

Algol is an eclipsing binary consisting of two stars in close orbit, one much hotter and brighter than the other. Together they shine at magnitude 2.1, but every 69 hours the fainter star eclipses its companion. Over a period of five hours, the combined light of the pair drops to just one-third its normal value, a change that is readily noticeable to the naked eye. Algol's brightness returns to normal after another five hours. Predictions of Algol's eclipses can be found in astronomical annuals and magazines.

Rho (ρ) Persei is a variable of a different kind: it is a red giant that fluctuates by about 50 per cent in brightness every seven weeks or so.

Popularly termed the Double Cluster, the twin open clusters NGC 869 and NGC 884 are one of the showpieces of the northern sky. Each cluster contains hundreds of stars of 7th magnitude and fainter, and covers an area of sky similar to that of the full Moon. They lie more than 7,000 light-years away in the Perseus spiral arm of our galaxy. Both clusters are noticeable to the naked eye as a brighter patch in the Milky Way near the border with Cassiopeia and can be seen well through binoculars or a small telescope.

M34 is a scattered open cluster of several dozen stars near the border with Andromeda. It covers a similar apparent area to the full Moon and is easy to spot through binoculars.

**ALPHA PERSEI CLUSTER**
Mirphak and its surrounding cluster lies above centre. The Pleiades Cluster is lower right, and Capella, in Auriga, is lower left.

MYTHS AND STORIES

## MEDUSA

Perseus, the son of Zeus and Danaë, was sent to bring back the head of Medusa, the Gorgon, whose evil gaze turned everything to stone. He was given a bronze shield by the goddess Athene, a sword of diamond by Hephaestus, and winged sandals by Hermes. Looking only at Medusa's reflection in his shield, Perseus managed to decapitate the Gorgon.

**SUCCESSFUL MISSION**
Perseus proudly displays the severed head of Medusa, the Gorgon, in this neoclassical sculpture by Antonio Canova.

**DOUBLE CLUSTER**
Of these two star clusters, NGC 869 (left) appears to be more densely packed. NGC 884 (right) contains some red giant stars, which its neighbour lacks.

THE RAM

# Aries

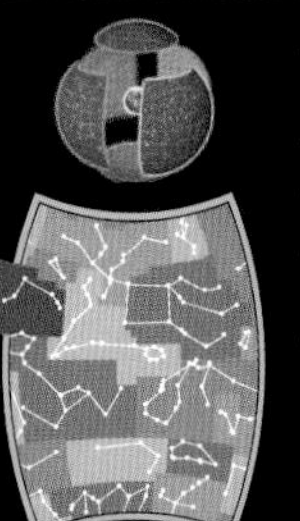

**SIZE RANKING** 39

**BRIGHTEST STAR** Hamal (α) 2.0

**GENITIVE** Arietis

**ABBREVIATION** Ari

**HIGHEST IN SKY AT 10PM** November–December

**FULLY VISIBLE** 90°N–58°S

This not particularly conspicuous constellation of the zodiac is found between Pisces and Taurus. Its most recognizable features are three stars near the border with Pisces: Alpha (α), Beta (β), and Gamma (γ) Arietis, of 2nd, 3rd, and 4th magnitudes.

Aries depicts the golden-fleeced ram of Greek legend (see panel, below). Over 2,000 years ago, the vernal equinox – the point at which the ecliptic crosses the celestial equator – lay near the border of Aries and Pisces. The effect of precession (see p.60) has now moved the vernal equinox almost into Aquarius, but it is still called the first point of Aries.

## SPECIFIC FEATURES

Gamma was one of the first stars discovered to be double, and it was found by the English scientist Robert Hooke in 1664, when telescopes were still quite crude and it was not realized that double stars are numerous. To the naked eye, it appears of 4th magnitude, but when viewed through a small telescope it consists of nearly identical white stars of magnitudes 4.6 and 4.7.

Lambda (λ) Arietis, of 5th magnitude, has a companion of 7th magnitude that can be seen through large binoculars. Pi (π) Arietis, also of 5th magnitude, has a very close companion of 8th magnitude.

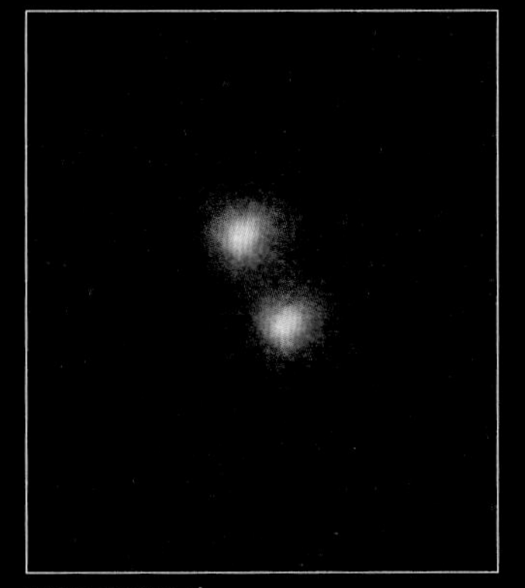

**EASY DOUBLE** Gamma (γ) Arietis is readily separable by a small telescope to reveal a pair of white stars, each of 5th magnitude.

**LEGENDARY RAM** From a crooked line formed by three faint stars, ancient astronomers visualized the figure of a crouching ram, with its head turned back over its shoulder.

MYTHS AND STORIES

## THE GOLDEN FLEECE

Aries represents the ram whose golden fleece hung on a tree in Colchis on the Black Sea. Jason and the Argonauts undertook an epic voyage to bring this fleece back to Greece. Jason was aided in his task by Medea, who had fallen in love with him. She was the daughter of King Aeetes, who owned the fleece. Medea bewitched the serpent guarding the fleece so that Jason could steal it. Taking Medea and the fleece with him, Jason then sailed away in the *Argo*.

**GOLDEN MOMENT** Watched by an admiring Medea, Jason removes the glittering fleece from the oak tree on which it hung at Colchis, in this illustration by L. du Bois-Reymond.

THE BULL

# Taurus

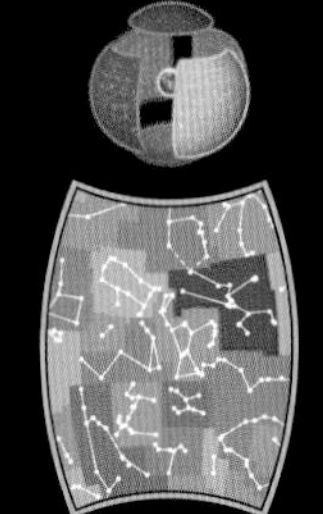

**SIZE RANKING** 17

**BRIGHTEST STAR** Aldebaran (α) 0.85

**GENITIVE** Tauri

**ABBREVIATION** Tau

**HIGHEST IN SKY AT 10PM** December–January

**FULLY VISIBLE** 88°N–58°S

Taurus is a large and prominent northern constellation of the zodiac, and it contains a wealth of objects including the Pleiades and Hyades star clusters (see p.287 and p.286 respectively) and M1, the Crab Nebula (see pp.266–67). Its stars represent the head and forequarters of a mythical Greek bull. The Hyades cluster is centred on the bull's face, while the constellation's brightest star, Aldebaran – Alpha (α) Tauri (see p.252) – is its glinting eye. Alnath (or Elnath) – Beta (β) Tauri – and Zeta (ζ) Tauri mark the tips of the bull's long horns. Each November, the Taurid meteors appear to radiate from a point south of the Pleiades.

## SPECIFIC FEATURES

Aldebaran is a red giant whose colour is clearly apparent to the naked eye. As with many red giants, it is slightly variable in brightness but the amount is only about one-tenth of a magnitude either side of its average value of 0.85 and is barely noticeable. Although Aldebaran appears to be part of the Hyades cluster, it lies 65 light-years away – less than half the cluster's distance – and is superimposed only by chance.

The main stars of the Hyades are arranged in a V-shape that is the width of over ten diameters of the full Moon. More than a dozen stars are visible with the unaided eye, and dozens more come into view through binoculars. At 150 light-years away, the Hyades is the nearest major star cluster to the Earth. On one arm of the Hyades' V-shape is a wide double star, Theta (θ) Tauri. At magnitude 3.4, the brighter of the pair, Theta-1 ($\theta^1$), is also the brightest member of the Hyades. Another double star that is easy to spot is Sigma (σ) Tauri, which has two 5th-magnitude components, near Aldebaran. The apex of the Hyades cluster points towards Lambda (λ) Tauri, an eclipsing binary of the same type as Algol (in Perseus). It varies between magnitudes 3.4 and 3.9 in a cycle lasting just under four days.

An even brighter star cluster is the Pleiades, which hovers over the bull's shoulders. Although popularly known as the Seven Sisters, after a group of mythical Greek nymphs (see panel, opposite), the Pleiades in fact contains nine named stars – the seven sisters themselves and their parents, Atlas and Pleione. The brightest member is Alcyone (see p.273), which is of magnitude 2.9 and lies near the centre of the cluster. The Pleiades covers an area of sky three times the width of the full Moon. On long-exposure photographs of the Pleiades, a surrounding haze is visible. This was once thought to be left-over gas and dust from the stars' formation, but it is now recognized as an unrelated cloud into which the cluster has drifted.

The first object on Charles Messier's list of comet-like objects (see p.69), M1 is the remains of a star that exploded as a supernova in AD 1054. It was given its popular name, the Crab Nebula, by the Irish astronomer William Parsons (see p.305) in 1844, because he thought the filaments of gas that protruded from the supernova remnant resembled the legs of a crab. The Crab Nebula is found two diameters of the full Moon away from Zeta Tauri. Through a small telescope, it appears as a faint elliptical glow several times larger than the disc of Jupiter. Large apertures are needed to make out the level of detail seen by Parsons.

**THE CRAB NEBULA**

This supernova reveals the beauty of a massive star's violent death throes. Convoluted filaments of gas expand away from the site of the supernova explosion, which was seen from Earth in AD 1054.

**HYADES AND PLEIADES**

The Hyades (lower left) is the larger of these two dazzling star clusters; the Pleiades (upper right) is a tighter bunch that appears hazy at first glance – good viewing conditions are needed to see all nine named stars with the naked eye.

MYTHS AND STORIES

## THE LOST PLEIAD

The popular name for the Pleiades is the Seven Sisters, although only six stars are easily visible to the naked eye. Two myths have arisen to explain the "missing" Pleiad. One myth says that the star that shines least brightly is Merope, the only one of the seven sisters to marry a mortal. Another story says that it is Electra, who could not bear to stay and watch the fall of Troy, the city founded by her brother. The names of the stars in the cluster do not follow either of these legends, however, for the faintest named member is actually Asterope.

**WANDERING STAR**
This 19th-century painting, *The Lost Pleiad*, depicts the separation of one of the Pleiades from her sisters.

**RAGING BULL**
Taurus, the celestial bull, thrusts his star-tipped horns into the night air. The bull is said to represent a disguise adopted by Zeus in a Greek myth. The bright reddish "star" seen here on the bull's back, below the Pleiades, is actually the planet Mars.

THE TWINS

# Gemini

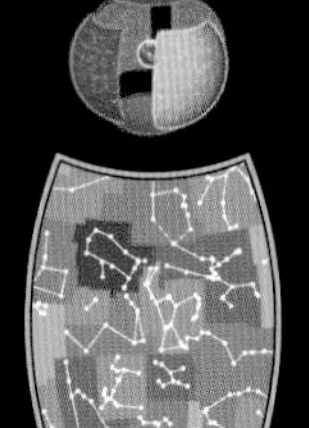

**SIZE RANKING** 30

**BRIGHTEST STAR** Pollux (β) 1.2

**GENITIVE** Geminorum

**ABBREVIATION** Gem

**HIGHEST IN SKY AT 10PM** January–February

**FULLY VISIBLE** 90°N–55°S

This prominent zodiacal constellation represents the mythical twins Castor and Pollux, who were the sons of Queen Leda of Sparta and the brothers of Helen of Troy (see Leda and the Swan, p.351). The constellation is easily identifiable within the northern sky because of its two brightest stars, which are named after the twins. Even though it is labelled Beta (β) Geminorum, Pollux is brighter than Castor, or Alpha (α) Geminorum (see p.272). The two stars mark the heads of the twins, while their feet lie bathed in the Milky Way. In mid-December each year, the Geminid meteors radiate from a point in Gemini near Castor.

## SPECIFIC FEATURES

Castor is a remarkable multiple star. To the naked eye, it appears as a single entity of magnitude 1.6, but through a small telescope with suitably high magnification, it divides into a sparkling blue-white duo of 2nd and 3rd magnitudes. The two stars form a genuine binary, with an orbital period of 470 years, which also has a 9th-magnitude red dwarf companion. Although these three stars cannot be divided further visually, each is a spectroscopic binary, bringing the total number of stars in the Castor system to six.

Although Castor and Pollux are named after twins, the stars themselves are far from identical. Being an orange giant, Pollux is noticeably warmer-toned than Castor. It is also closer to the Earth, lying only 34 light-years away, compared to Castor's 52 light-years.

The open star cluster M35 lies at the feet of the twins. Under clear skies, this cluster can be glimpsed with the naked eye, but it is more easily found with binoculars, through which it appears as an elongated, elliptical patch of starlight spanning the same apparent width as the full Moon. When viewed through a small telescope, its individual stars seem to form chains or curved lines.

Two variable stars of note in Gemini are Zeta (ζ) Geminorum (see p.282), which is a Cepheid variable that ranges between magnitudes 3.6 and 4.2 every 10.2 days, and Eta (η) Geminorum (see p.280), which is a red giant whose brightness can vary anywhere between magnitudes 3.1 and 3.9. This constellation also contains the Eskimo Nebula, or NGC 2392 (see p.255), a planetary nebula with a bluish disc similar in size to that of the globe of Saturn and visible through a small telescope. Larger telescope apertures are needed to reveal the nebula's surrounding fringe of gas, reminiscent of an Eskimo's parka, that gives NGC 2392 its popular name. An alternative name for this nebula is the Clown-face Nebula.

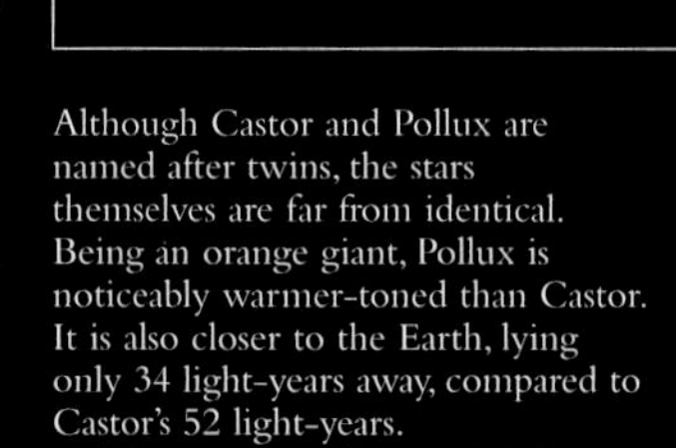

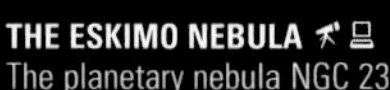

**THE ESKIMO NEBULA**
The planetary nebula NGC 2392 is so-called because it is surrounded by a fringe of gas that resembles the fur-lined hood of an Eskimo's parka.

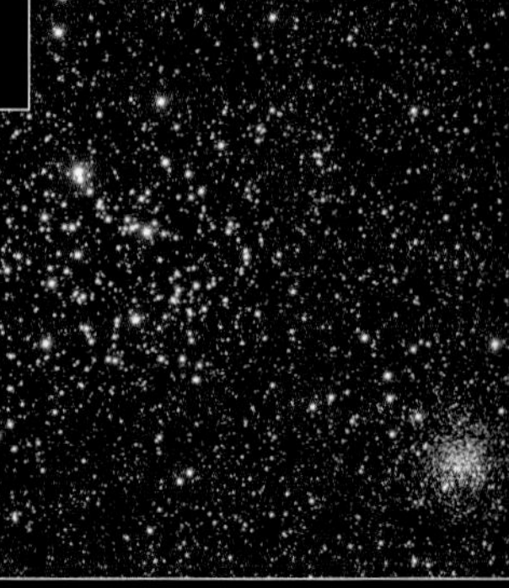

**LARGE AND SMALL CLUSTER**
The large star cluster M35 is visible through binoculars; larger telescopes reveal a fainter and more distant cluster, NGC 2158 (bottom right), in the same field of view.

**CELESTIAL TWINS**
Castor and Pollux, the twins of the Greek myth, stand side by side in the sky between Taurus and Cancer. The bright "star" in the middle of Gemini in this picture is actually the planet Saturn.

THE CRAB

# Cancer

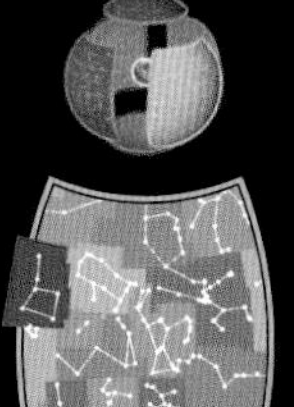

**SIZE RANKING** 31

**BRIGHTEST STAR** Beta (β) 3.5

**GENITIVE** Cancri

**ABBREVIATION** Cnc

**HIGHEST IN SKY AT 10PM** February–March

**FULLY VISIBLE** 90°N–57°S

Cancer is the faintest of the 12 zodiacal constellations, lying in the northern sky between Gemini and Leo, and it represents the crab of Greek mythology (see panel, right). Cancer includes the major open star cluster M44 (see p.286), which is alternatively known as the Beehive Cluster, the Manger Cluster, or Praesepe – which is the Latin for both "hive" and "manger". It also includes the stars Gamma (γ) and Delta (δ) Cancri, which represent two donkeys feeding at the manger. These t stars are sometimes known as A llus Borealis and Asellus Australis, t northern and southern asses.

## SPECIFIC FEATURES

Iota (ι) Cancri is a 4th-magnitude yellow giant with a nicely contrasting 7th-magnitude blue-white companion. The companion is just detectable through 10 x 50 binoculars, and it is easy to identify through a small telescope. Another double star that can be seen through a small telescope is Zeta (ζ) Cancri. Its components, of 5th and 6th magnitude, form a binary star with an orbital period of more than 1,000 years.

The Beehive Cluster (M44) is a large open cluster at the the heart of Cancer, located between Gamma (γ) and Delta (δ) Cancri. The ancient Greeks could see the cluster as a misty spot with the unaided eye, but under modern urban skies it is unlikely to be visible without binoculars. This cluster consists of a scattering of stars of 6th magnitude and fainter. It appears to cover an area more than three times wider than the diameter of the full Moon, and although it can be seen through binoculars, it is too wide to fit in the field of view of most telescopes.

The Beehive Cluster's glory overshadows another open cluster, M67, which is smaller and denser yet still the width of the full Moon in the sky. It lies about 2,600 light-years away – more distant than the Beehive Cluster, which is 520 light-years away. M67 can be found with binoculars, but a telescope is needed to resolve individual stars. At an estimated age of around 5 billion years, it is one of the oldest open clusters known – it is also of similar age to Earth.

**M67**
Inferior to M44, but still worthy of note, M67 can be found with binoculars in the region of Cancer south of the ecliptic.

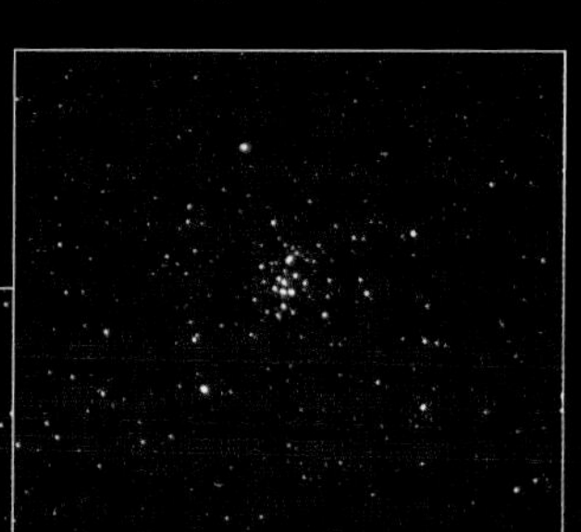

**THE BEEHIVE CLUSTER**
Also known as the Manger Cluster, M44 is an open cluster located between the two asses feeding from the manger, Gamma (γ) (centre, top) and Delta (δ) Cancri (centre bottom).

MYTHS AND STORIES

## A SMALL VICTORY

According to the Greek story, a crab attacked Hercules during his fight with the multi-headed Hydra but was crushed underfoot during the struggle. Such a minor role befits this faint constellation.

**SCUTTLING AWAY**
A small crab can be seen in the foreground of this 18th-century engraving, *Hercules Fights the Lernean Hydra.*

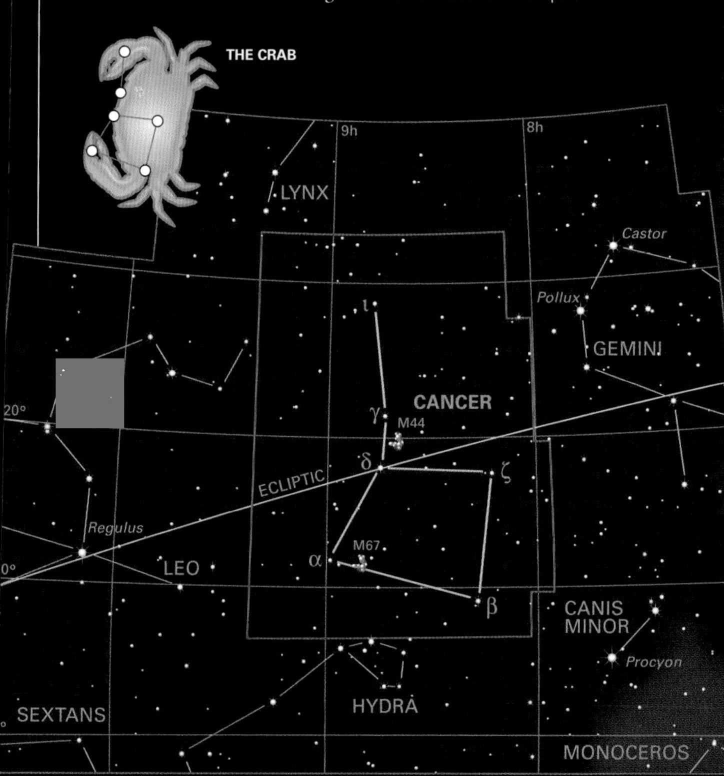


**HIDDEN CRAB**
Cancer is the faintest constellation in the zodiac, but it contains a major star cluster, M44, which is just visible in this photograph as a hazy patch near the centre of the constellation.

THE LITTLE LION

## Leo Minor

**SIZE RANKING** 64

**BRIGHTEST STAR** 46 Leonis Minoris 3.8

**GENITIVE** Leonis Minoris

**ABBREVIATION** LMi

**HIGHEST IN SKY AT 10PM** March–April

**FULLY VISIBLE** 90°N–48°S

This small, insignificant constellation, adjacent to Leo in the northern sky, represents a lion cub, although this is not suggested by the pattern of its stars. It was introduced in the 17th century by the Polish astronomer Johannes Hevelius (see p.368).

SPECIFIC FEATURES

Unusually, this constellation has no star labelled Alpha. This is due to an error by the 19th-century English astronomer Francis Baily, who assigned the Greek letters to the constellation's stars. When doing so, he overlooked assigning a Bayer letter to the brightest star, 46 Leonis Minoris, which should have been recorded as Alpha (α), although he did label the second-brightest star as Beta (β) Leonis Minoris.

Although Leo Minor contains no objects of interest for users of binoculars or a small telescope, Beta (β) is a close double star that can be separated by a telescope with very large aperture. It has a magnitude of 4.2, and its component stars orbit each other every 37 years.

**THE LION CUB**
Having located the distinctive shape of the Sickle in Leo (top, right), look north of it to find the faint stars of Leo Minor.

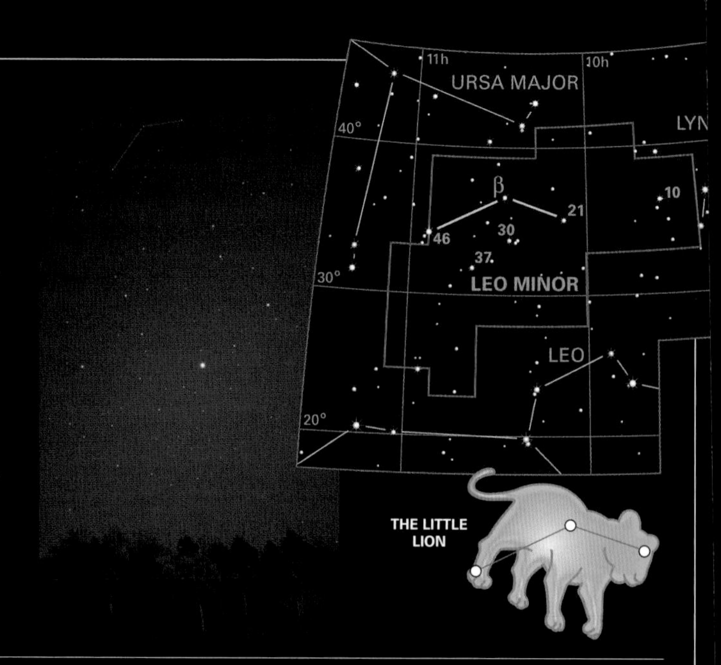


BERENICE'S HAIR

## Coma Berenices

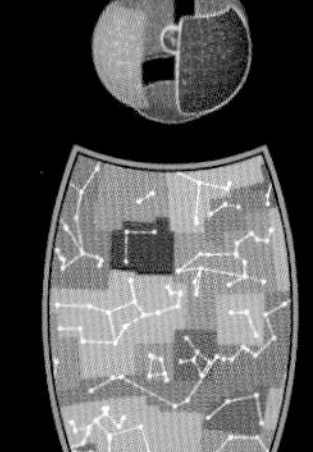

**SIZE RANKING** 42

**BRIGHTEST STAR** Beta (β) 4.2

**GENITIVE** Comae Berenices

**ABBREVIATION** Com

**HIGHEST IN SKY AT 10PM** April–May

**FULLY VISIBLE** 90°N–56°S

Coma Berenices represents the flowing locks of Queen Berenice of Egypt, which she cut off as a tribute to the gods after the safe return of her husband, Ptolemy III, from battle during the 3rd century BC. It is a faint but interesting northern constellation, lying between Leo and Boötes. In the mid-16th century, it was named as a constellation by the Dutch cartographer Gerardus Mercator. Before then, its stars were regarded as forming the tail of Leo.

SPECIFIC FEATURES

The Coma Star Cluster, also known as Melotte 111, is the constellation's main feature. It comprises several dozen faint stars, which fan out distinctively for several diameters of the Moon southwards from Gamma (γ) Comae Berenices. This open cluster, which is seen to best advantage through binoculars, has been imagined as both the bushy tip of a lion's tail and a lock of Berenice's hair.

Coma Berenices contains numerous galaxies in its southern half. Most of these are members of the Virgo Cluster, such as M85, M88, M99, and M100, but two notable exceptions, M64 (see p.304) and NGC 4565, are closer to the Earth.

Popularly known as the Black Eye Galaxy, M64 is a spiral galaxy tilted at an angle to the Earth, which is seen as an elliptical patch of light through a small telescope; it is best detected with a telescope with an aperture of 150mm (6in) or more. A dust cloud near the galaxy's nucleus creates the "black eye" effect.

NGC 4565, another spiral galaxy, lies edge-on to the Earth and is more difficult to spot. It appears long and thin when viewed through a telescope with 100mm (4in) aperture, and a lane of dark dust is revealed in long-exposure photographs.

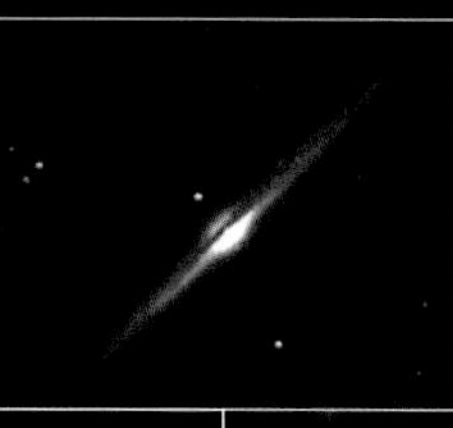

**NGC 4565**
Seen edge-on, this spiral galaxy displays a lane of dark dust along its spiral arms when viewed through larger apertures.

**THE BLACK EYE GALAXY**
The spiral galaxy M64 sports a large, dark dust cloud near its core, giving it the appearance of a blackened eye.

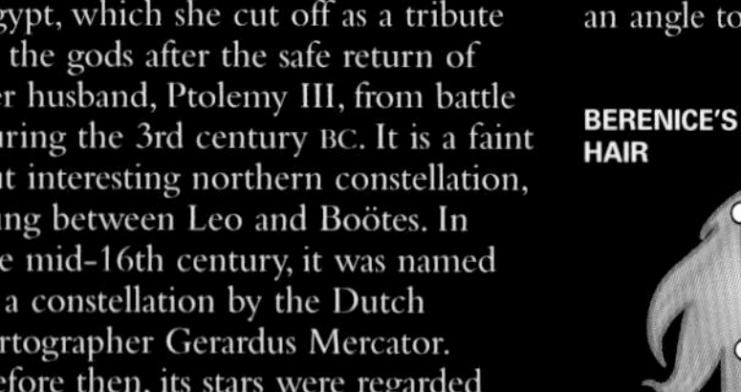


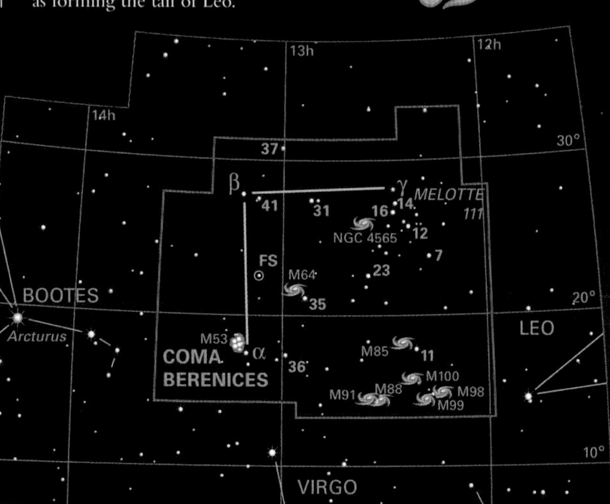


**MANE OF HAIR**
The distinctive splay of the Coma Star Cluster marks out Coma Berenices in the night sky. Leo's hindquarters can be seen closer to the horizon.

THE LION

# Leo

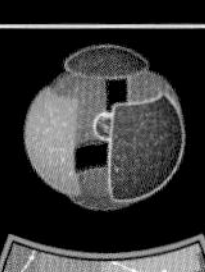

**SIZE RANKING** 12

**BRIGHTEST STAR** Regulus (α) 1.4

**GENITIVE** Leonis

**ABBREVIATION** Leo

**HIGHEST IN SKY AT 10PM** March–April

**FULLY VISIBLE** 82°N–57°S

The outline stars of Leo really do bear a marked resemblance to a crouching lion, in this large constellation of the zodiac, located just north of the celestial equator. It is one of the easiest constellations to recognize. The pattern of six stars that marks the lion's head and chest is known as the Sickle and is shaped like a reversed question mark or a hook. The Leonid meteors radiate from the region of the Sickle every November (see pp.220–21).

## SPECIFIC FEATURES

Regulus – Alpha (α) Leonis (see p.249) – lies at the foot of the Sickle. It is the faintest of the first-magnitude stars, at magnitude 1.4, and its wide companion is of 8th magnitude.

The double star Algieba, or Gamma (γ) Leonis, consists of components of magnitudes 2.2 and 3.5. Both stars are orange giants, and they orbit each other every 600 years or so. A nearby star – 40 Leonis – is unrelated.

Zeta (ζ) Leonis is a wide triple star, consisting of a 3rd-magnitude star with a 6th-magnitude companion to both the north and south, which can be seen with binoculars. All three stars are at different distances from Earth and, hence, they are unrelated.

A pair of spiral galaxies, M65 and M66, can be glimpsed with a small telescope beneath the hind quarters of Leo. A fainter pair of spirals, M95 and M96, lie under the lion's body, as does an elliptical galaxy, M105, about one degree away.

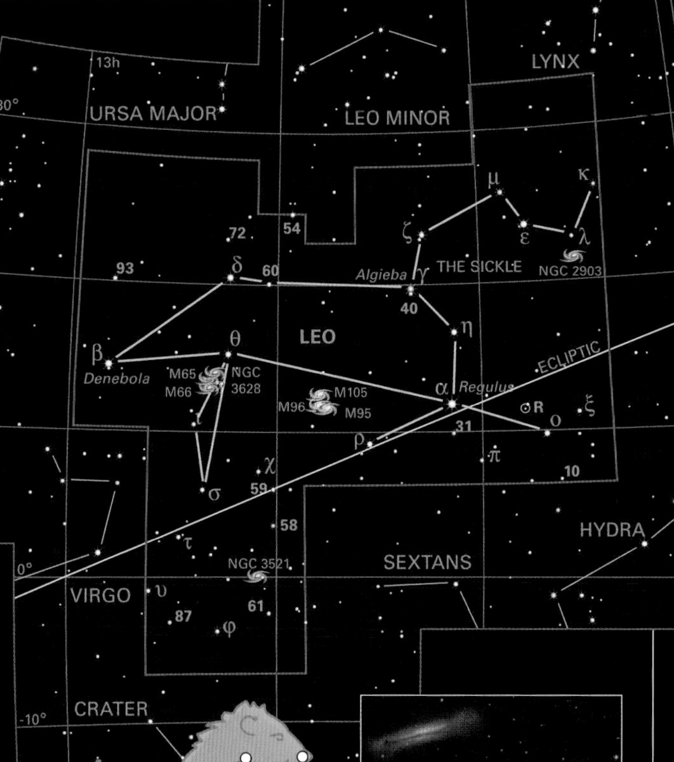


THE LION

**LEO TRIPLET**
A trio of galaxies lies near Theta (θ) Leonis: M65 (lower right); M66 (lower left); and the edge-on spiral NGC 3628 (top). Although NGC 3628 appears the largest on photographs, it is less bright than the others and is difficult to see through small telescopes.

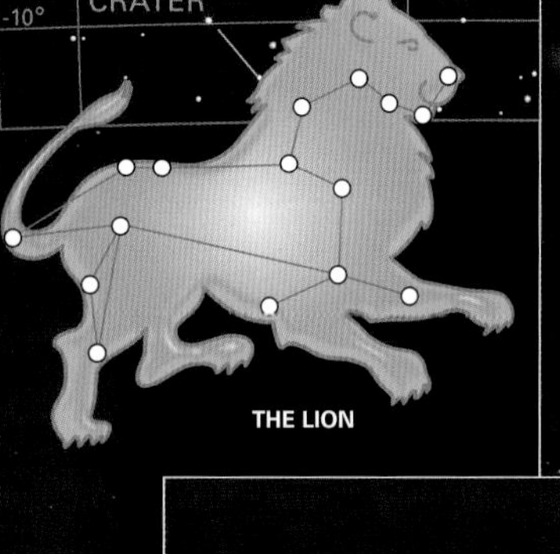

**ALGIEBA**
This beautiful pair of golden-coloured orange giants is clearly visible through small telescopes.

**THE BIG CAT**
The crouching lion is a distinctive sight in the night sky. The pattern of its stars is disturbed here by the presence of Jupiter under the lion's body.

MYTHS AND STORIES

## HERCULES AND THE LION

Leo represents the mythical lion that lived in a cave near the Greek town of Nemea, terrorizing the area and emerging to attack and devour local inhabitants. As the first of the 12 labours in his quest for immortality, Hercules was sent by his cousin Eurystheus to kill the lion. Finding that the creature's hide was impervious to his arrows, Hercules instead wrestled and strangled the beast. He then used the lion's own razor-sharp claws to cut off its pelt, which he wore victoriously as a cloak.

**THE HERO AND THE BEAST**
Hercules grapples with the Nemean Lion in a sculpture by the 16th-century Flemish artist Jean de Boulogne, or Giambologna.

THE VIRGIN

# Virgo

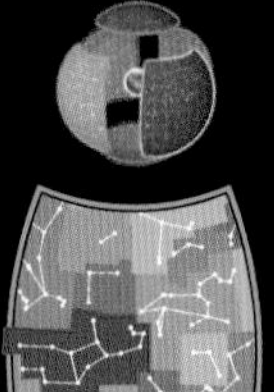

**SIZE RANKING** 2

**BRIGHTEST STAR** Spica (α) 1.0

**GENITIVE** Virginis

**ABBREVIATION** Vir

**HIGHEST IN SKY AT 10PM** April–June

**FULLY VISIBLE** 67°N–75°S

Virgo straddles the celestial equator, between Leo and Libra. It is the largest constellation of the zodiac, and the second-largest overall. The constellation depicts a Greek virgin goddess (see panel, right). Virgo contains the Virgo Cluster (see p.319), the nearest large cluster of galaxies to Earth, which is some 50 million light-years away and which extends over the border of Virgo into Coma Berenices. The Sun is in Virgo during the September equinox each year.

## SPECIFIC FEATURES

Porrima, or Gamma (γ) Virginis (see p.249), is a binary star in which the effects of orbital motion will be seen in forthcoming years. Having been closest together in 2005, when a telescope with a 250mm (10in) aperture was needed to separate Porrima's components, the two stars are now moving further apart and hence are coming within range of a smaller telescope. They will become divisible with a telescope of 100mm (4in) aperture by 2010 and of 60mm (2.4in) after 2012. For the rest of the 21st century, it will be possible to split Porrima with a small-aperture telescope. Both of Porrima's stars are of magnitude 3.5, and their orbital period is 169 years.

In the upper part of Virgo's body lie the numerous galaxies of the Virgo Cluster. None is easy to see with a small instrument. The brightest members are giant ellipticals, notably M49, M60 (see p.307), M84, M86, and M87 (see p.313). M87 is a strong radio and X-ray source also known as Virgo A. Long-exposure photographs show it is ejecting a jet of gas, like certain quasars.

The Sombrero Galaxy (see p.306), or M104, is Virgo's best-known galaxy. This spiral is about two-thirds as far away as the Virgo Cluster. It is oriented almost edge-on to the Earth, so that a dark lane of dust in the galaxy's plane crosses its central bulge. The bulge may be all that can be seen through a small telescope; the dust lane is only revealed when seen through a large-aperture telescope or on long-exposure photographs.

The brightest quasar in the sky, 3C 273 (see p.315), also lies in the bowl of Virgo. However, it is much more distant than the Virgo Cluster. Through most telescopes, it appears as nothing more than a 13th-magnitude star. Only professional equipment will reveal it as the centre of an active galaxy, which is some 2,000 million light-years away from Earth.

**THE SOMBRERO GALAXY**
The Sombrero Galaxy (M104) is a spiral galaxy with a large central bulge, seen almost edge-on, and resembling a Mexican hat. It lies about 30 million light-years away.

**M87**
Through a small telescope, the giant elliptical galaxy M87 appears as a rounded glow, but photographs and CCD images reveal the jet of gas that is being expelled from its highly active nucleus. Here, the jet is just visible near the top right of the core.

MYTHS AND STORIES

## THE VIRGIN GODDESS

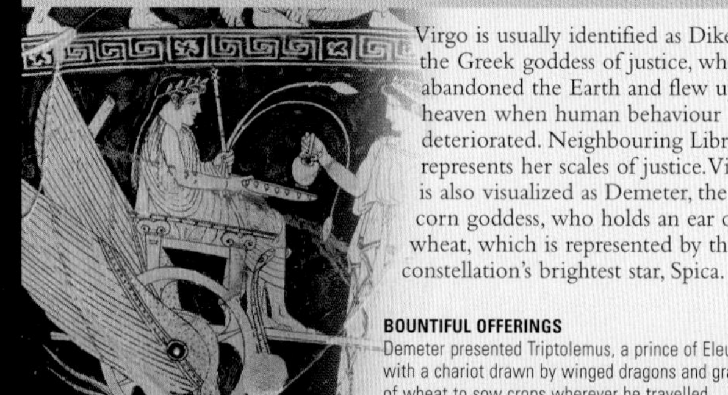

Virgo is usually identified as Dike, the Greek goddess of justice, who abandoned the Earth and flew up to heaven when human behaviour deteriorated. Neighbouring Libra represents her scales of justice. Virgo is also visualized as Demeter, the corn goddess, who holds an ear of wheat, which is represented by the constellation's brightest star, Spica.

**BOUNTIFUL OFFERINGS**
Demeter presented Triptolemus, a prince of Eleusis, with a chariot drawn by winged dragons and grains of wheat to sow crops wherever he travelled.

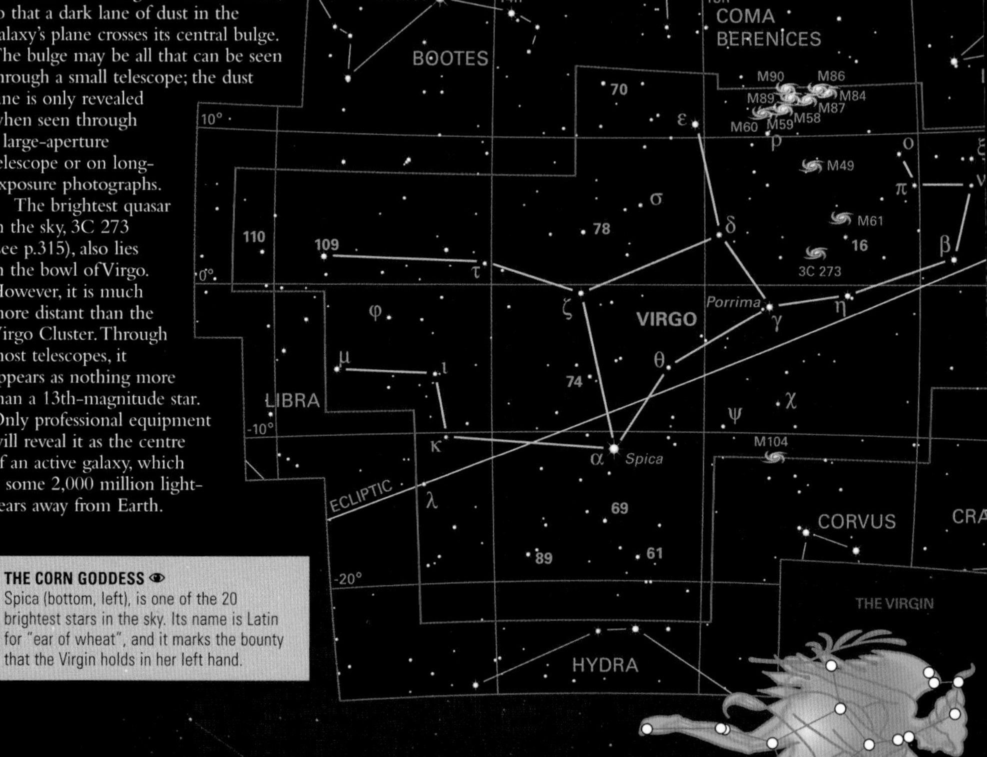


**THE CORN GODDESS**
Spica (bottom, left), is one of the 20 brightest stars in the sky. Its name is Latin for "ear of wheat", and it marks the bounty that the Virgin holds in her left hand.

THE SCALES

# Libra

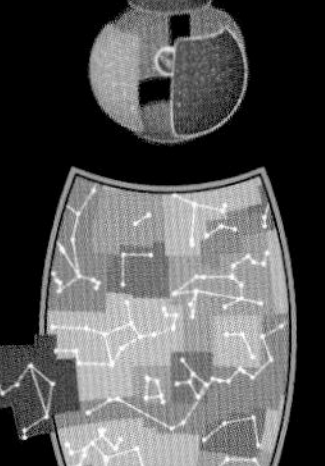

SIZE RANKING 29

BRIGHTEST STAR
Beta (β) 2.0

GENITIVE Librae

ABBREVIATION Lib

HIGHEST IN SKY AT 10PM
May–June

FULLY VISIBLE
60°N–90°S

This constellation of the zodiac lies just south of the celestial equator between Virgo and Scorpius. Originally, the ancient Greeks visualized the constellation as the claws of the neighbouring Scorpius, which is why Libra's brightest stars have names that mean the northern claw and the southern claw. Libra's present-day identification as Virgo's scales of justice became more common in Roman times.

SPECIFIC FEATURES

Zubenelgenubi (Arabic for "the southern claw") or Alpha (α) Librae is a wide double star of 3rd and 5th magnitudes and is easily divisible with binoculars or even sharp, unaided eyesight. To the north of this pair is the constellation's brightest star, Zubeneschamali ("the northern claw") or Beta (β) Librae, which shows a greenish tinge when viewed through binoculars or a telescope. This highly unusual colouring is due, presumably, to the chemical composition of Zubeneschamali's outer layers.

In the heart of the constellation lies Iota (ι) Librae, a double with stars of 5th and 6th magnitudes which can be viewed through binoculars. A small telescope will reveal the closer 9th-magnitude companion of the brighter star. Mu (μ) Librae, with components of 6th and 7th magnitude, is a more difficult pair to separate; a telescope with 75mm (3in) aperture is needed.

Delta (δ) Librae is an eclipsing variable. Every two days eight hours, it rises and falls between 5th and 6th magnitudes. This change can be readily followed with binoculars.

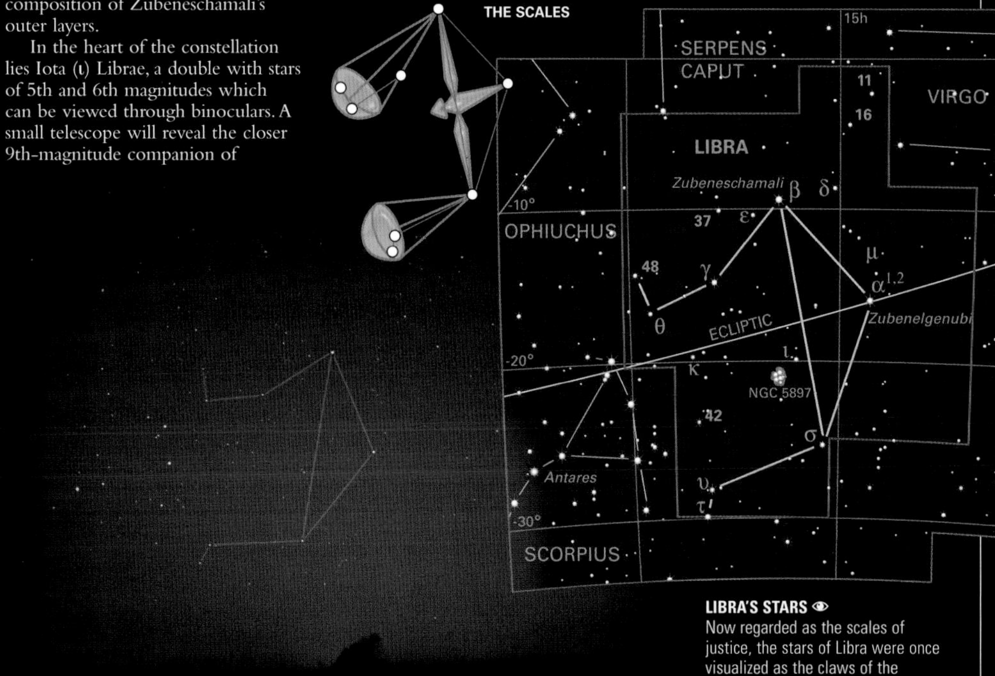


**LIBRA'S STARS**
Now regarded as the scales of justice, the stars of Libra were once visualized as the claws of the adjacent scorpion, Scorpius.

THE NORTHERN CROWN

# Corona Borealis

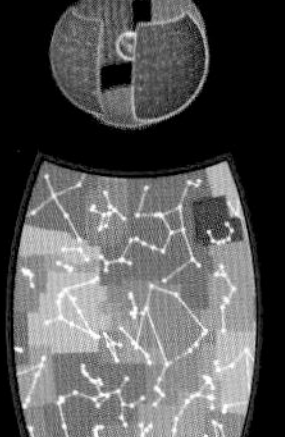

SIZE RANKING 73

BRIGHTEST STAR
Alphekka or Gemma (α) 2.2

GENITIVE
Coronae Borealis

ABBREVIATION CrB

HIGHEST IN SKY AT 10PM
June

FULLY VISIBLE
90°N–50°S

Corona Borealis is a small but distinctive constellation in the northern sky, between Boötes and Hercules, consisting of a horseshoe shape of seven stars. It is one of the original Greek constellations and represents the crown worn by Princess Ariadne (see panel, right).

SPECIFIC FEATURES

The arc of the northern crown contains the remarkable variable star R Coronae Borealis (see p.283), a yellow supergiant normally of 6th magnitude, which shows sudden dips in brightness. These fades, which are due to a build-up of sooty particles in its atmosphere, occur every few years and can last for months.

Corona Borealis has three double stars of note for small-instrument users, although none is particularly bright. Nu (ν) Coronae Borealis is a pair of 5th-magnitude red giants divisible with binoculars. Zeta (ζ) Coronae Borealis is a blue-white pair, with components of 5th and 6th magnitudes – an attractive sight when seen through a small telescope – while Sigma (σ) Coronae Borealis is a yellow pair with components of 6th and 7th magnitudes, which can also be split with a small telescope.

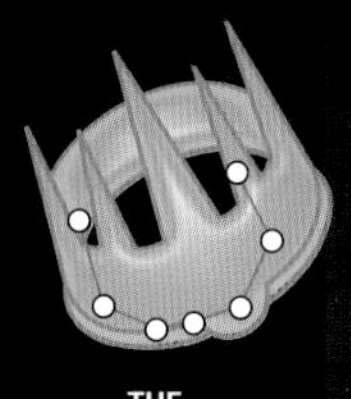


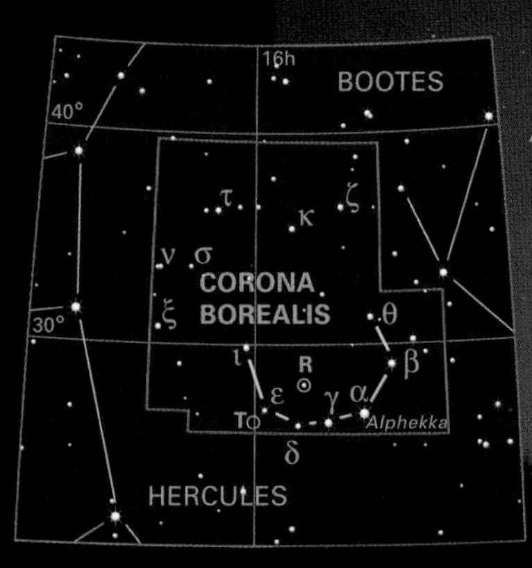


**CROWN OF STARS**
Like a celestial tiara, the seven main stars of Corona Borealis form a distinctive arc between Boötes and Hercules.

MYTHS AND STORIES

## PRINCESS ARIADNE

Ariadne, daughter of King Minos of Crete, helped Theseus slay the Minotaur, a gruesome creature that was half-bull, half-human. Theseus sailed off with Ariadne to the island of Naxos, where he then abandoned her. The god Dionysus looked down on the princess and was overcome. At their wedding, Ariadne wore a jewel-studded crown, which Dionysus threw into the sky, where the crown's jewels were changed into stars.

**CROWNING GLORY**
Dionysus, known as Bacchus by the Romans, holds Ariadne's jewelled crown, in this painting by the 17th-century French artist Eustache Le Sueur.

THE SERPENT

# Serpens

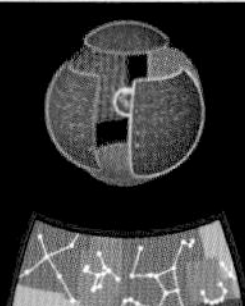

**SIZE RANKING** 23

**BRIGHTEST STAR**
Unukalhai (α) 2.6

**GENITIVE**
Serpentis

**ABBREVIATION** Ser

**HIGHEST IN SKY AT 10PM**
June–August

**FULLY VISIBLE**
74°N–64°S

Although counted as a single constellation, Serpens is in fact split into two separate areas, and is thus unique. It is one of the original 48 Greek constellations and straddles the celestial equator. Serpens represents a huge snake coiled around Ophiuchus, who holds the head (Serpens Caput) in his left hand and the tail (Serpens Cauda) in his right. In Greek mythology, snakes were a symbol of rebirth, because of the fact that they shed their skins. Ophiuchus represents the great healer Asclepius, who was reputedly able to revive the dead (see panel, opposite).

## SPECIFIC FEATURES

The Eagle Nebula (see pp.242–43) in Serpens Cauda was made world-famous by a spectacular Hubble Space Telescope picture of dark columns of dust embedded within its glowing gas. Unfortunately, the dust columns show up only through a telescope of large aperture and on long-exposure photographs such as those from the Hubble Space Telescope.

The Eagle Nebula contains a star cluster, M16, which can be spotted readily through binoculars or a small telescope. It appears as a hazy patch covering an area of sky that is similar in size to the full Moon. Another open cluster that is visible through binoculars is IC 4756, which appears about twice the size of M16. It is situated in Serpens Cauda near the tip of the serpent's tail.

Close to the border with Virgo lies M5, which is about 25,000 light-years away. Its condensed centre appears as a faint area about half the size of the full Moon, when viewed with binoculars, while the curving chains of stars in its outskirts are revealed only through a telescope with an aperture of 100mm (4in) or more.

Delta (δ) Serpentis, near the serpent's head, is a binary with components of 4th and 5th magnitudes. It is divisible using high powers of magnification on a small telescope.

Theta (θ) Serpentis, near the serpent's tail, is a pair of white stars that are easily split through a small telescope. This wide double star has components of magnitude 4.6 and 5.0.

**M5**
This is one of the finest globular clusters in northern skies. M5 is noticeably elliptical in shape when viewed through a telescope.

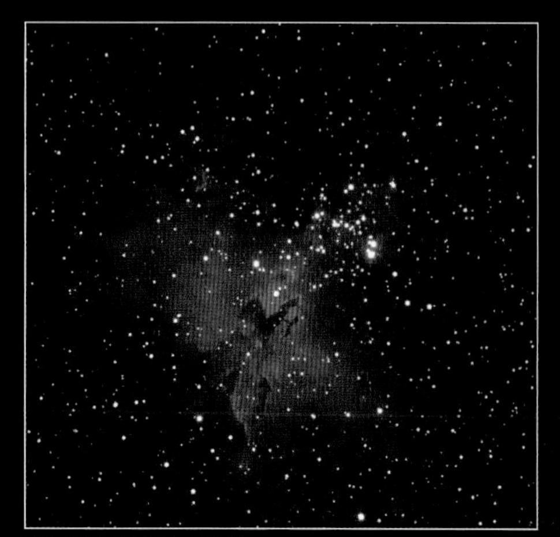

**THE EAGLE NEBULA**
This image was captured by a professional four-metre telescope. It can only be seen well with telescopes of large aperture.

**SERPENTINE STARS**
The upper part of the snake (above, right) contains Unukalhai (α), which derives its name from the Arabic for "the serpent's neck".

THE SERPENT HOLDER

# Ophiuchus

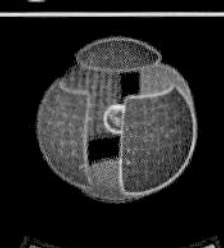

**SIZE RANKING** 11

**BRIGHTEST STAR**
Rasalhague (α) 2.1

**GENITIVE**
Ophiuchi

**ABBREVIATION** Oph

**HIGHEST IN SKY AT 10PM**
June–July

**FULLY VISIBLE**
59°N–75°S

This large constellation straddling the celestial equator depicts a man holding a snake. The head of Ophiuchus adjoins Hercules in the north while his feet rest on Scorpius in the south. The Sun passes through Ophiuchus in the first half of December, but despite this the constellation is not regarded as a true member of the zodiac.

Ophiuchus was the site of the last supernova explosion seen in our Galaxy, which appeared in 1604. It far outshone all other stars and is known as Kepler's Star (see p.269) after Johannes Kepler who wrote about it in *De stella nova* (see p.64).

## SPECIFIC FEATURES

Lying on the edge of the Milky Way in the direction of the centre of our Galaxy, Ophiuchus contains numerous star clusters. Messier catalogued seven globular clusters, although none is particularly prominent. M10 and M12 (see p.289) are both near the centre of the constellation and detectable through binoculars on a clear night. Better sights for binoculars are two large and scattered open clusters, NGC 6633 and IC 4665.

An outstanding multiple star is Rho (ρ) Ophiuchi, lying near Antares (in neighbouring Scorpius). This 5th-magnitude star has a 7th-magnitude companion either side of it, and these are best viewed through binoculars. Another 6th-magnitude companion that is much closer to the central star can be identified through a small telescope using high magnification. The complex nebulosity in this area, including around Antares, is revealed only in long-exposure photographs.

The beautiful double star 70 Ophiuchi consists of yellow and orange dwarfs, with components of 4th and 6th magnitudes, while the double star 36 Ophiuchi is a pair of orange dwarfs with components of 5th magnitude.

Barnard's Star is the most celebrated star in Ophiuchus and is the second-closest star to the Sun. Even though this red dwarf is a mere 5.9 light-years away, its light output is so feeble that it appears as only magnitude 9.5, and it is too faint to see without a telescope. Barnard's Star is moving so quickly relative to the background stars that its change in position is noticeable over a matter of only a few years (see chart, right).

**THE SERPENT HOLDER**

**BARNARD'S STAR MOVEMENT**

**MAGNITUDE KEY**

0.0–0.9
1.0–1.9
2.0–2.9
3.0–3.9
4.0–4.9
5.0–5.9
6.0–6.9

**INTRICATE NEBULOSITY**
Complex nebulosity extends from the area around Rho (ρ) Ophiuchi (at the top of the image below), southwards to Antares (bottom).

**M10**
The large globular cluster M10 is some 14,000 light-years away. Like its neighbour M12, it is detectable through binoculars on a clear night.

**SNAKE MAN**
Ophiuchus represents a man encoiled by a huge snake, the constellation Serpens. The ecliptic runs through Ophiuchus, and planets can be seen within its borders.

MYTHS AND STORIES

## ASCLEPIUS

Ophiuchus is identified with Asclepius, the Greek god of medicine who reputedly had the power to revive the dead. Hades, god of the Underworld, feared that this ability endangered his trade in dead souls and asked Zeus to strike Asclepius down. Zeus then placed the great healer among the stars.

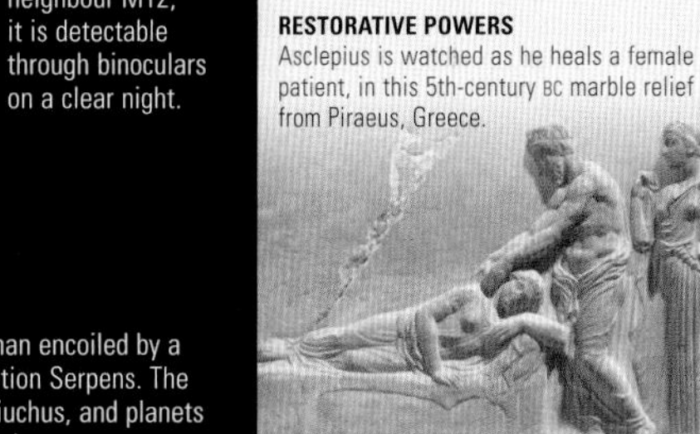

**RESTORATIVE POWERS**
Asclepius is watched as he heals a female patient, in this 5th-century BC marble relief from Piraeus, Greece.

THE SHIELD

# Scutum

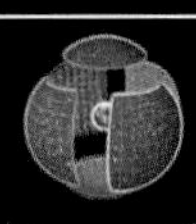

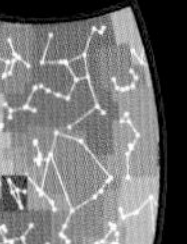

**SIZE RANKING** 84

**BRIGHTEST STAR** Alpha (α) 3.8

**GENITIVE** Scuti

**ABBREVIATION** Sct

**HIGHEST IN SKY AT 10PM** July–August

**FULLY VISIBLE** 74°N–90°S

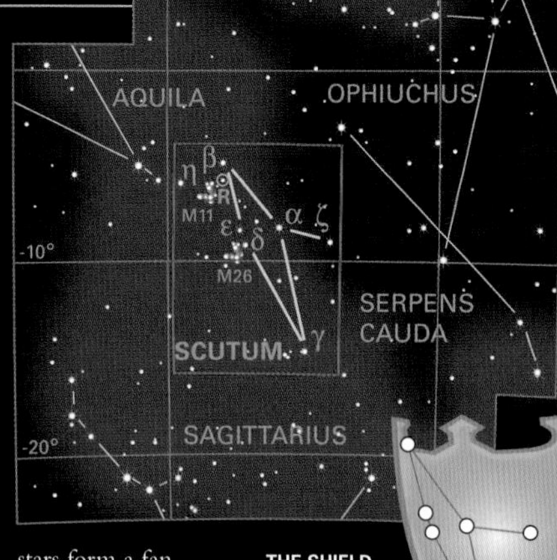


THE SHIELD

This minor constellation is situated in a rich area of the Milky Way, between Aquila and Sagittarius, south of the celestial equator. It was introduced by Johannes Hevelius (see p.368) in the late 17th century. He gave it the name *Scutum Sobiescianum*, meaning Sobieski's Shield, to honour his patron, King John Sobieski of Poland.

SPECIFIC FEATURES

Delta (δ) Scuti is the prototype of a class of variable star that pulsates in size every few hours, changing brightness by only a few tenths of a magnitude. Delta itself varies between magnitude 4.6 and 4.8 in less than five hours, but the change is only detectable with sensitive instruments. Far more obvious is R Scuti, an orange supergiant that rises and falls between magnitudes 4.2 and 8.6 in a 20-week cycle.

Near R Scuti is the beautiful Wild Duck Cluster (M11), which appears as a smudgy glow half the apparent width of the full Moon when viewed through binoculars. This open cluster gained its popular name because its stars form a fan shape, like a flock of ducks in flight, when seen through a small telescope. Near the apex of the fan is an 8th-magnitude red giant. The Wild Duck Cluster is in an area of the constellation that is known as the Scutum Star Cloud. This rich star field is located just south of Beta (β) Scuti.

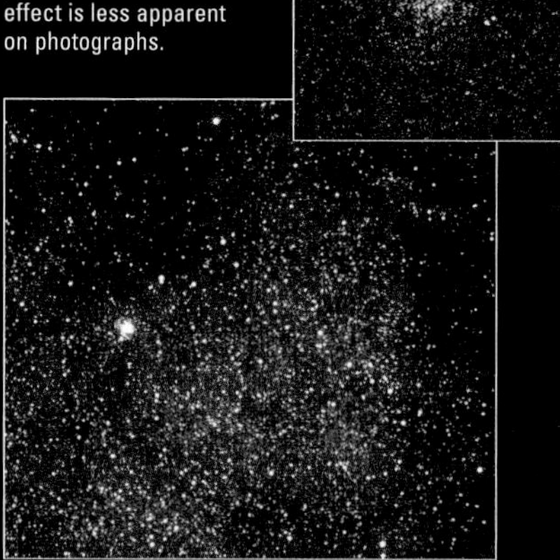

**WILD DUCKS** Seen through a small telescope, M11 looks like the V-shaped flight pattern of wildfowl. This effect is less apparent on photographs.

**SCUTUM STAR CLOUD** One of the brightest parts of the Milky Way lies in Scutum and is known as the Scutum Star Cloud. The bright spot at centre left is the Wild Duck Cluster.

**SOBIESKI'S SHIELD** Scutum has no bright stars of its own, but it lies in an area of the Milky Way, between Aquila and Sagittarius, that is particularly rich with stars.

THE ARROW

# Sagitta

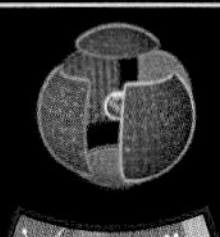

**SIZE RANKING** 86

**BRIGHTEST STAR** Gamma (γ) 3.5

**GENITIVE** Sagittae

**ABBREVIATION** Sge

**HIGHEST IN SKY AT 10PM** August

**FULLY VISIBLE** 90°N–69°S

Sagitta was known to the ancient Greeks, who believed it represented an arrow shot by either Apollo, Hercules, or Eros. It is the third-smallest constellation, lying in the Milky Way between Vulpecula and Aquila in the northern sky. It is faint and easily overlooked.

SPECIFIC FEATURES

There is little of note in Sagitta for users of small instruments. Zeta (ζ) Sagittae is a 5th-magnitude star with a 9th-magnitude companion that is visible in a small telescope, but it is not a particularly impressive double. S Sagittae is a Cepheid variable that halves in brightness every 8.4 days before recovering again, as it swings between magnitudes 5.2 and 6.0.

Midway along the shaft of the arrow is M71, a modest globular cluster detectable with binoculars but better seen through a telescope. M71 lacks the central condensation typical of most globulars and instead looks more like a dense open cluster.

WZ Sagittae is a dwarf nova variable (see Novae, p.278). It is rarely in outburst.

**ARROW IN FLIGHT** The small arrow Sagitta flies over the stars of Aquila, the eagle, and towards Delphinus, the dolphin.

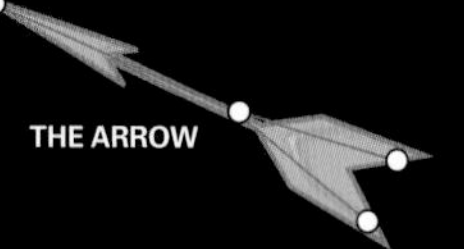

THE EAGLE

# Aquila

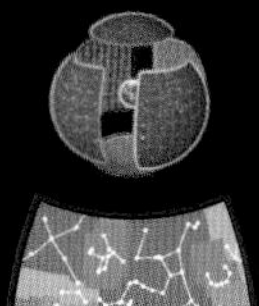

**SIZE RANKING** 22

**BRIGHTEST STAR** Altair (α) 0.8

**GENITIVE** Aquilae

**ABBREVIATION** Aql

**HIGHEST IN SKY AT 10PM** July–August

**FULLY VISIBLE** 78°N–71°S

Aquila depicts an eagle in flight (see panel, right). It lies on the celestial equator in a rich area of the Milky Way near Cygnus, Scutum, and Sagittarius, yet there are no deep-sky objects of particular note within it. Aquila's brightest star, Altair or Alpha (α) Aquilae (see p.248), forms one corner of the northern Summer Triangle of stars, completed by Vega (in Lyra) and Deneb (in Cygnus).

Altair is flanked by 4th-magnitude Alshain, or Beta (β) Aquilae, and 3rd-magnitude Tarazed, or Gamma (γ) Aquilae, which form a distinctive trio.

## SPECIFIC FEATURES

Aquila's main feature of interest is Eta (η) Aquilae (see p.278), which is one of the brightest Cepheid variables. Eta ranges between magnitudes 3.5 and 4.4 on a cycle of 7.2 days. As with all members of this class, it is a brilliant supergiant. Its distance is estimated at 1,200 light-years.

The constellation also has a couple of faint double stars that can readily be split with a small telescope: 15 Aquilae, with stars of 5th and 7th magnitudes; and 57 Aquilae, with two 6th-magnitude components.

MYTHS AND STORIES

## WINGED CARRIERS

The eagle has at least two identifications in Greek mythology. It was the bird that carried the thunderbolts for the god Zeus, and in one myth Zeus sent an eagle, or took the form of an eagle, to carry the shepherd boy Ganymede up to Mount Olympus, where he was made a servant of the gods. Zeus had spied the boy tending sheep in a field and had become infatuated with him. Ganymede is represented by neighbouring Aquarius.

**ON EAGLE'S WINGS**
The beautiful youth Ganymede is carried aloft by an eagle in Peter Paul Rubens's 17th-century painting *The Abduction of Ganymede*.

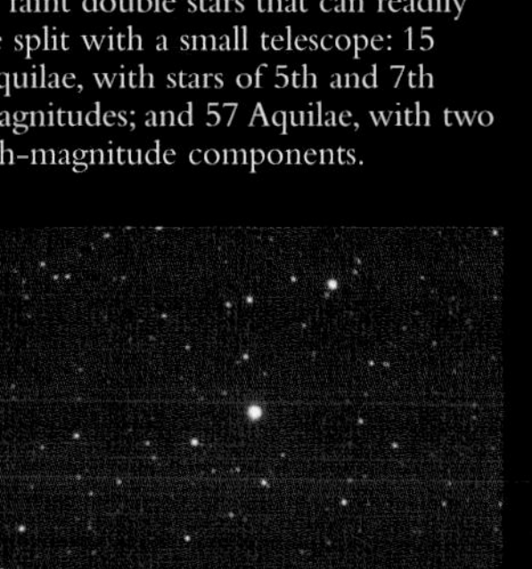

**STELLAR TRIO**
Altair, the constellation's brightest star, is flanked by 3rd-magnitude Tarazed (top), which has a noticeably orange colour, and 4th-magnitude Alshain (bottom), forming an attractive stellar trio.

**THE HOOK**
This easily recognizable group of stars in southern Aquila includes Lambda (λ) Aquilae (centre left) and branches into neighbouring Scutum.

THE EAGLE

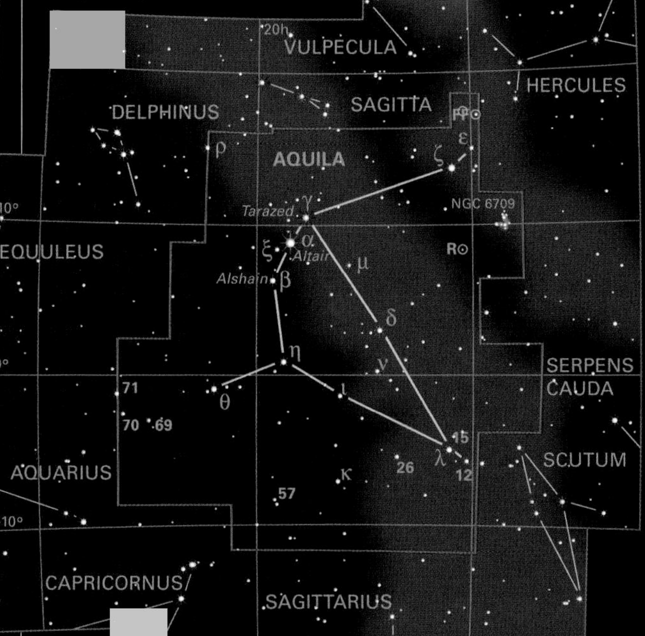


**SWOOPING ACROSS THE SKIES**
The eagle swoops across the evening skies in the second half of the year. Its main star, Altair, is the most southerly of those that form the northern Summer Triangle. Aquila points towards the stars of Capricornus.

THE FOX

# Vulpecula

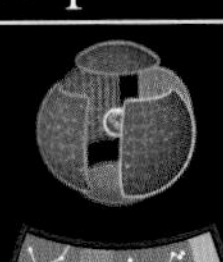

**SIZE RANKING** 55

**BRIGHTEST STAR** Alpha (α) 4.4

**GENITIVE** Vulpeculae

**ABBREVIATION** Vul

**HIGHEST IN SKY AT 10PM** August–September

**FULLY VISIBLE** 90°N–61°S

This small, faint northern constellation lies in the Milky Way south of Cygnus. When it was first introduced in the late 17th century by the Polish astronomer Johannes Hevelius (see panel, below), it was named *Vulpecula cum Anser* (the fox with the goose). Its name has since been simplified to Vulpecula. Despite its relative obscurity, it contains two unmissable objects for binocular users.

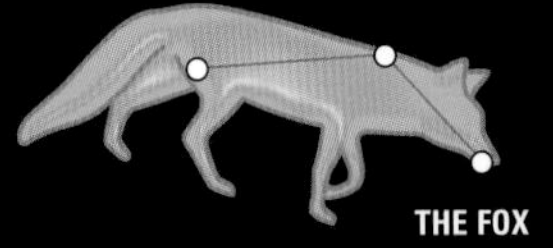

THE FOX

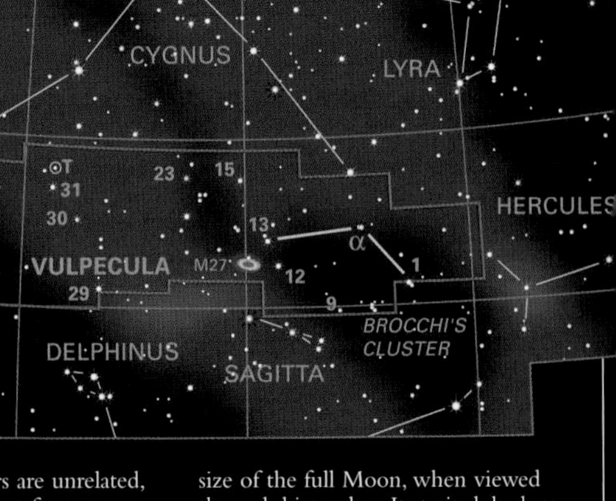


## SPECIFIC FEATURES

The brightest star in the constellation, Alpha (α) Vulpeculae, is a 4th-magnitude red giant with a 6th-magnitude orange star nearby, which is visible with binoculars. The two lie at different distances and are unrelated.

Brocchi's Cluster is one of the binocular treasures of the sky. This grouping of ten stars, with components ranging from 5th to 7th magnitude, is better known as the Coathanger because of its shape: a line of six stars forms the bar of the hanger while the remaining four are the hook. All the stars are unrelated, however, and so do not form a true cluster. The Coathanger's shape is therefore the delightful product of a chance alignment.

Popularly known as the Dumbbell Nebula, M27 is the easiest planetary nebula to spot in the sky. It appears as a rounded patch, about one-third the size of the full Moon, when viewed through binoculars. Its twin-lobed or hourglass shape is revealed only with larger instruments and on long-exposure photographs. It is about 1,000 light-years away. CCD images and photographs show a variety of colours, but visually the Dumbbell appears grey-green.

**THE DUMBBELL NEBULA**
Reputedly the easiest planetary nebula to spot, M27 can be found with binoculars on dark nights. A telescope is needed to make out the twin lobes that give rise to its popular name.

**THE COATHANGER**

Perhaps the most charming of all star clusters is Brocchi's Cluster, also known as the Coathanger. This group of stars, easily visible through binoculars, appears to mark out the shape of a simple coathanger.

## JOHANNES HEVELIUS

Johannes Hevelius (1611–87) was born and worked in the town of Danzig, Germany (now Gdansk, Poland), where he established an observatory equipped with the finest instruments of his time. Among his legacies was a star catalogue and atlas, published posthumously by his assistant and second wife, Elizabeth, introducing seven new constellations and filling the gaps in the northern skies.

**JOINT EFFORT**
Johannes Hevelius and his wife Elizabeth measured star positions with a large sextant. This instrument is commemorated in one of the constellations Hevelius invented, Sextans.

**FOX IN THE MILKY WAY**
Vulpecula is a shapeless constellation sandwiched between the more easily recognizable pattern of Sagitta, the arrow, at left of this picture, and the head of the swan, Cygnus.

THE DOLPHIN

# Delphinus

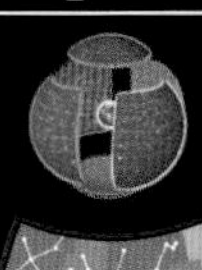

**SIZE RANKING** 69

**BRIGHTEST STAR**
Rotanev (β) 3.6

**GENITIVE** Delphini

**ABBREVIATION** Del

**HIGHEST IN SKY AT 10PM**
August–September

**FULLY VISIBLE**
90°N–69°S

This small but distinctive constellation is situated between Aquila and Pegasus. According to Greek myth, Delphinus represents the dolphin that saved the poet and musician Arion from drowning after he leapt into the sea to escape robbers onboard a ship. Alternatively, the constellation is said to depict one of the dolphins sent by Poseidon to bring the sea nymph Amphitrite to him to marry. It is one of the constellations listed by the astronomer Ptolemy (see p.85).

The whole constellation was once popularly known as Job's Coffin, presumably because of the box-like shape of its area, although sometimes this name is restricted to the diamond asterism formed by the four brightest stars: Sualocin (α), Rotanev (β), and Gamma (γ) and Delta (δ) Delphini. Who applied the name Job's Coffin, and when, is not known.

**GAMMA DELPHINI**
Gamma (γ) Delphini is an attractive double star. Although both the component stars are usually described as yellow, some observers see the fainter star as bluish.

## SPECIFIC FEATURES

Gamma (γ) Delphini is normally described as an attractive orange-yellow double star. Its components are of 4th and 5th magnitudes, and they are easily separated by a small telescope.

The fainter and closer double star Struve 2725, which has components of 7th and 8th magnitudes, can also be seen through a small telescope and is visible in the same field of view as Gamma (γ) Delphini.

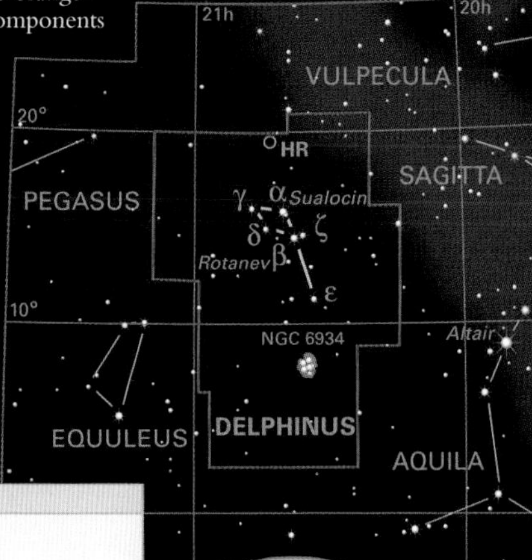


THE DOLPHIN

**THE PLAYFUL DOLPHIN**
The kite-shaped Delphinus, on the edge of the Milky Way near Cygnus, brings to mind a dolphin jumping from ocean waters.

## NICCOLÒ CACCIATORE

Alpha (α) and Beta (β) Delphini bear the unusual names Sualocin and Rotanev. When reversed, these names spell Nicolaus Venator. This is the Latinized name of Niccolò Cacciatore (1780–1841), an Italian astronomer who was assistant to Giuseppe Piazzi, the director of the Palermo Observatory, Sicily. Cacciatore defied convention by surreptitiously naming two stars after himself in the Palermo star catalogue of 1814. No one realized what he had done until much later, by which time the star names had become established.

THE FOAL

# Equuleus

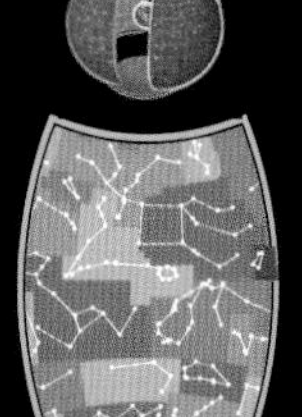

**SIZE RANKING** 87

**BRIGHTEST STAR**
Alpha (α) 3.9

**GENITIVE** Equulei

**ABBREVIATION** Equ

**HIGHEST IN SKY AT 10PM**
September

**FULLY VISIBLE**
90°N–77°S

The second-smallest constellation in the sky represents the head of a young horse, or foal, and lies next to the larger celestial horse, Pegasus. No myths or legends are associated with Equuleus, which is thought to have been added to the sky by the Greek astronomer Ptolemy (see p.85) in his 2nd-century AD compendium of the original Greek constellations.

## SPECIFIC FEATURES

Gamma (γ) Equulei is a wide double star, with components of 5th and 6th magnitudes and is easily separated with binoculars. Its two stars are unrelated. The 5th-magnitude double star 1 Equulei – labelled as Epsilon (ε) Equulei on some maps – has a 7th-magnitude companion, which can be seen through a small telescope, and a fainter true companion, which can be seen only through instruments with larger apertures. Other than these two double stars, there is nothing of note in Equuleus for users of binoculars or small telescopes.

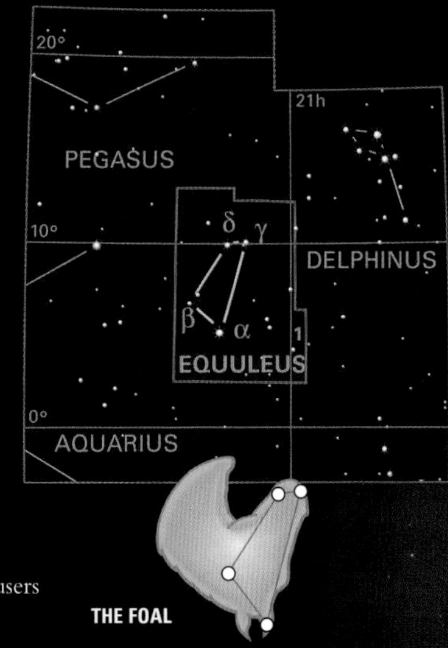


THE FOAL

**THE HORSE'S HEAD**
Equuleus consists of a small area of faint stars wedged between Pegasus and Delphinus and is easily overlooked.

THE WINGED HORSE

# Pegasus

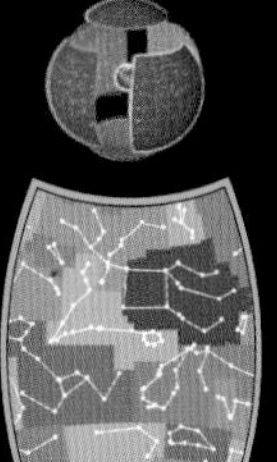

**SIZE RANKING** 7

**BRIGHTEST STARS**
Beta (β) 2.4,
Epsilon (ε) 2.4

**GENITIVE** Pegasi

**ABBREVIATION** Peg

**HIGHEST IN SKY AT 10PM**
September–October

**FULLY VISIBLE**
90°N–53°S

Pegasus lies north of the zodiacal constellations Aquarius and Pisces, in low northern declinations, and it adjoins Andromeda. It was one of the original 48 Greek constellations. Pegasus represents the flying horse ridden by the hero Bellerophon, although he is sometimes wrongly identified as the steed of Perseus (see panel, below). Although only the forequarters of the horse are indicated by stars, the constellation is still the seventh-largest in the sky.

## SPECIFIC FEATURES

The Great Square of Pegasus is formed by the stars Alpha (α), Beta (β), and Gamma (γ) Pegasi, plus Alpha (α) Andromedae. Long ago, the fourth star of the Square was also known as Delta (δ) Pegasi and was shared with Andromeda, but now it is exclusively Andromeda's. A line of more than 30 full Moons would fit into the Square, yet for such a large area it is surprisingly devoid of stars – its brightest star being Upsilon (υ) Pegasi, of magnitude 4.4. Therefore, when the Great Square is viewed through polluted skies, it may seem completely empty.

The constellation's two brightest stars are Beta, a red giant that varies between magnitudes 2.3 and 2.7, and Epsilon (ε) Pegasi, a yellow star of magnitude 2.4 with a wide 8th-magnitude companion that can be seen through a small telescope.

Not far from Epsilon lies the globular cluster M15 (see p.289), which is one of the finest such objects in northern skies. It is just at the limit of naked-eye visibility when viewed through clear skies.

Just outside the Great Square of Pegasus is a 5th-magnitude star called 51 Pegasi, which was the first star beyond the Sun confirmed to have a planet in orbit around it. This planet was discovered in 1995, and its mass is about half that of Jupiter.

**THE GREAT SQUARE**
The most distinctive feature of this constellation is the Great Square of Pegasus, which forms the horse's body.

**M15**
A telescope with 150mm (6in) aperture resolves this globular cluster into individual stars. It is over 30,000 light-years away.

THE WINGED HORSE

MYTHS AND STORIES

## BELLEROPHON AND PEGASUS

Pegasus the winged horse was born from the body of Medusa, the Gorgon, when she was decapitated by Perseus. He flew to Mount Helicon, home of the Muses, where he stamped on the ground and brought forth a spring called Hippocrene, the horse's fountain. With the aid of a golden bridle from Athena, the hero Bellerophon tamed Pegasus, and rode the horse on his successful mission to kill the fire-breathing monster Chimaera. Bellerophon later attempted to ride Pegasus up to Olympus, to join the gods, but he fell off. The horse arrived safely.

**TAKING FLIGHT**
Pegasus beats his wings, as though attempting to ascend to the skies, in this statue in Powerscourt Gardens, Dublin, Ireland.

THE WATER CARRIER

# Aquarius

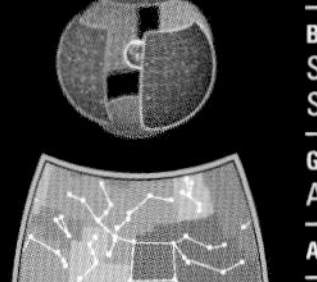

**SIZE RANKING** 10

**BRIGHTEST STARS**
Sadalmelik (α) 2.9,
Sadalsuud (β) 2.9

**GENITIVE**
Aquarii

**ABBREVIATION** Aqr

**HIGHEST IN SKY AT 10PM**
June–July

**FULLY VISIBLE**
65°N–86°S

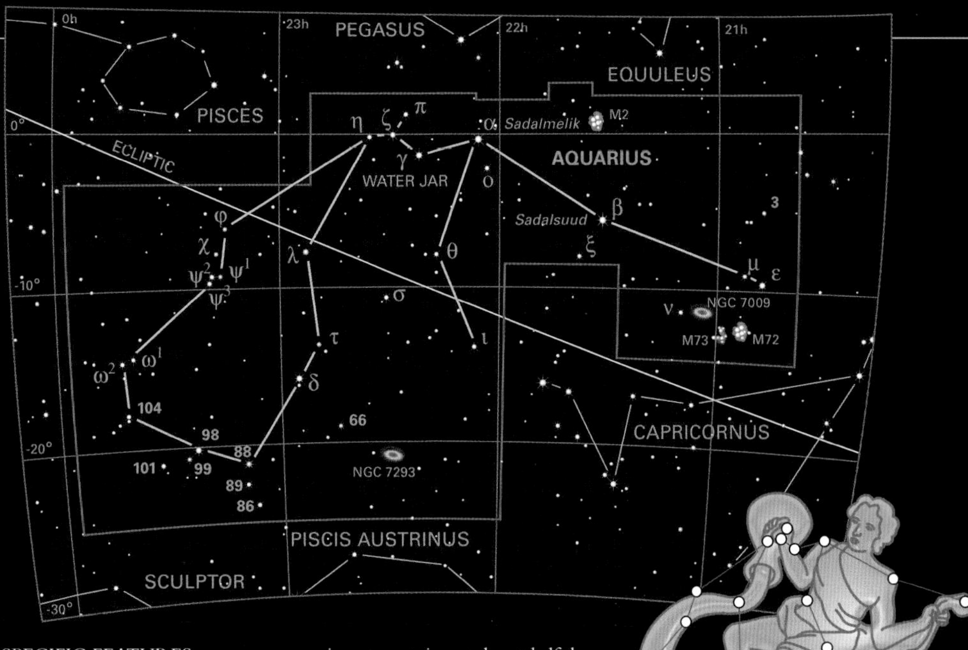


This large constellation of the zodiac is visualized as a youth (or, sometimes, an older man) pouring water from a jar. It is is found between Capricornus and Pisces, near the celestial equator. The stars Gamma (γ), Zeta (ζ), Eta (η), and Pi (π) Aquarii form a Y-shaped grouping that makes up the Water Jar, from which a stream of stars represents water flowing towards Piscis Austrinus. In early May each year, the Eta Aquarid meteor shower radiates from the area of the water jar.

In Greek myths and stories, Aquarius represents Ganymede – a beautiful shepherd boy to whom the god Zeus took a fancy. Zeus dispatched his eagle (or, in some stories, turned himself into an eagle) to carry Ganymede up to Mount Olympus, where he became a waiter to the gods. The Eagle is represented by neighbouring Aquila.

## SPECIFIC FEATURES

Zeta – the star at the centre of the Water Jar group – is a close binary of 4th-magnitude stars just at the limit of resolution with a telescope of 60mm (2.4in) aperture. Located near the border with Equuleus, the globular cluster M2 appears as a fuzzy star when viewed through binoculars or a small telescope.

Aquarius contains two of the best-known planetary nebulae in the sky. The Helix Nebula (NGC 7293; see p.253) is reckoned to be the closest planetary nebula to the Earth, being some 300 light-years away. It is therefore one of the largest nebulae in apparent size, at almost half the diameter of the full Moon. However, because its light is spread over such a large area, the Helix Nebula can be identified only when skies are clear and dark. Visually, this nebula appears as a pale grey patch, showing none of the beautiful colours captured on photographs.

The second planetary nebula – the Saturn Nebula (NGC 7009) – is easier to spot, appearing to be of similar size to the disc of Saturn when viewed with a small telescope. Its faint extensions on either side, rather like the rings of Saturn, give rise to the object's popular name.

**THE WATER CARRIER**

**THE HELIX NEBULA**

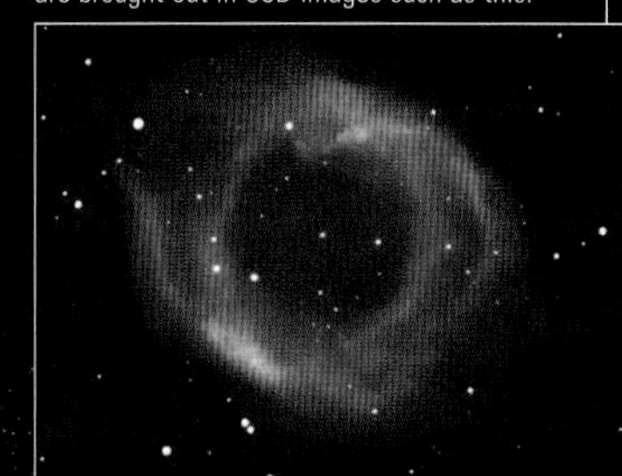

NGC 7293 is visible as a pale rounded patch through binoculars under dark skies, but its detailed structure and approximate colours are brought out in CCD images such as this.

**THE SATURN NEBULA**

NGC 7009's resemblance to the ringed planet Saturn is most evident when it is viewed through a large telescope or on a CCD image.

**POURING WATER**

The cascade of stars that represent the flow of water from Aquarius's jar is to the left of this image. The distinctive Water Jar is centre, top.

THE FISHES

# Pisces

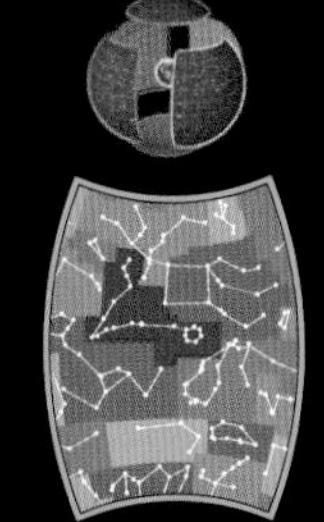

**SIZE RANKING** 14

**BRIGHTEST STAR**
Eta (η) 3.6

**GENITIVE**
Piscium

**ABBREVIATION** Psc

**HIGHEST IN SKY AT 10PM**
October–November

**FULLY VISIBLE**
83°N–56°S

This zodiacal constellation represents two mythical fishes (see panel, right). Its main claim to fame is that it contains the vernal equinox, which is the point where the Sun crosses the celestial equator into the northern hemisphere each year in March – on star maps, this is where 0h right ascension intersects 0° declination. Because of the slow wobble of the Earth, known as precession (see p.60), the point of the vernal equinox is gradually moving along the celestial equator and will enter Aquarius in about AD 2600.

## SPECIFIC FEATURES

The most distinctive feature of Pisces is the ring of seven stars lying south of the Great Square of Pegasus. Known as the Circlet, this ring marks the body of one of the fishes. It includes TX Piscium (also known as 19 Piscium), a deep-orange-coloured red giant that fluctuates irregularly between magnitudes 4.8 and 5.2.

Alrescha (α) is a close pair of stars of 4th and 5th magnitudes that can be separated with a telescope with an aperture of 100mm (4in). These two stars form a true binary with an orbital period of more than 900 years. Zeta (ζ) and Psi-1 ($\psi^1$) Piscium are two more doubles that can be divided with a small telescope.

A beautiful face-on spiral galaxy, M74, lies just over two diameters of the Moon from the constellation's brightest star, Eta (η) Piscium. It appears as a round, bright glow through a small telescope; the spiral arms only show up well through a telescope with larger aperture and on long-exposure photographs.

**THE CIRCLET**
The body of the southerly fish is marked by a ring of stars called the Circlet. One of the stars, TX Piscium, is a red giant of variable brightness, which appears noticeably orange through binoculars.

**DIVERGENCE**
Pisces represents a pair of fishes tied together by their tails with ribbon. The point where the two ribbons are knotted together is marked by the star Alpha (α) Piscium.

MYTHS AND STORIES

## EROS AND APHRODITE

Ancient Greek myths concerning the origins of the constellation of Pisces are rather vague. In one myth, Aphrodite and her son Eros transformed themselves into fishes and plunged into the Euphrates to escape the fearsome monster Typhon. In another version of the same story, two fishes swam up and carried Aphrodite and Eros to safety on their backs.

**RESCUE AT SEA**
In this 17th-century painting by the Flemish artist Jacob Jordaens, Aphrodite and Eros are carried away on the back of a fish.

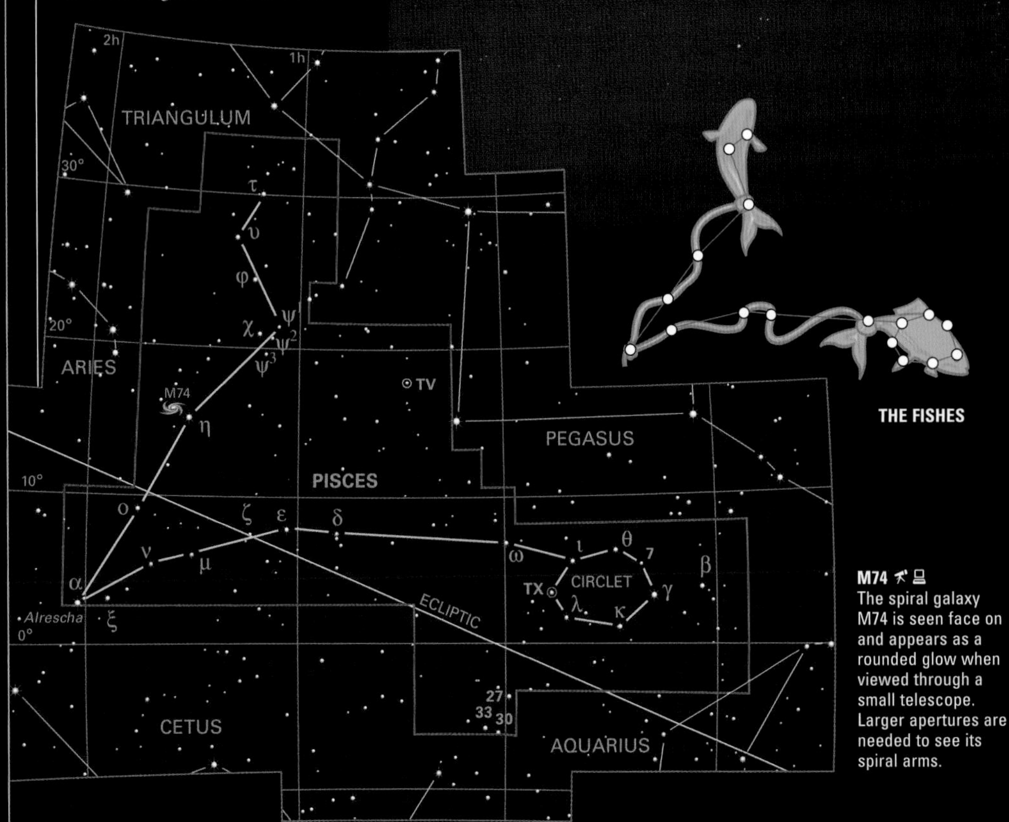


THE FISHES

**M74**
The spiral galaxy M74 is seen face on and appears as a rounded glow when viewed through a small telescope. Larger apertures are needed to see its spiral arms.

THE SEA MONSTER

# Cetus

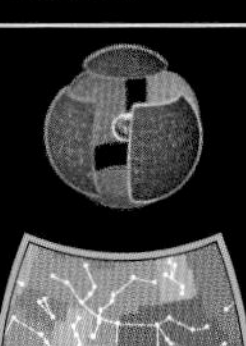

**SIZE RANKING** 4

**BRIGHTEST STAR**
Diphda or Deneb Kaitos (β) 2.0

**GENITIVE** Ceti

**ABBREVIATION** Cet

**HIGHEST IN SKY AT 10PM**
October–December

**FULLY VISIBLE**
65°N–79°S

Cetus is represented on old star charts as an unlikely looking, almost comical, hybrid sea monster, although the figure is also sometimes referred to as a whale. It is one of the original 48 Greek constellations listed by Ptolemy in his *Almagest*. It is a large but not very obvious constellation found in the equatorial region of the sky, and it lies south of the zodiacal constellations Pisces and Aries. Cetus is home to the celebrated variable star, Mira (o) (see p.281), as well as a peculiar spiral galaxy, M77.

## SPECIFIC FEATURES

Menkar (α) is the second-brightest star in the constellation. It forms part of the loop of stars that mark the sea monster's head, and it has a wide and unrelated 6th-magnitude companion that is visible through binoculars. Positioned near the neck of the sea monster is Gamma (γ) Ceti, a close double star that is more challenging to divide than Menkar. High magnification on a telescope is required to see the two component stars, of 4th and 7th magnitudes.

Mira (o) is the prototype of a common type of red giant that pulsates in size over months or years. Mira can reach magnitude 2 at its brightest – although magnitude 3 is more usual – while at its faintest it drops to magnitude 10. Hence, depending on how much it has swollen or contracted within its 11-month cycle, Mira can be either a naked-eye star or one that is visible only with a telescope.

**M77**
Because it is a Seyfert galaxy, the spiral galaxy M77 looks like a fuzzy star through smaller telescopes – only its extremely bright core can be seen.

Tau (τ) Ceti is 11.9 light-years away. Its temperature and brightness make it the most Sun-like of all Earth's nearby stars. Tau is, however, surrounded by a swarm of asteroids and comets, which would subject any local planets to devastating bombardments. Thus the prospects for life in its vicinity seem rather slim.

M77 is found near Delta (δ) Ceti. This spiral galaxy is the brightest example of a Seyfert galaxy (see Types of Active Galaxies, p.310). Related to quasars, Seyfert galaxies are a class of galaxies that have extremely bright centres. M77 is oriented face-on towards the Earth, although only its core is visible through a small telescope, and it looks only like a small, round patch. M77 lies just under 50 million light-years away.

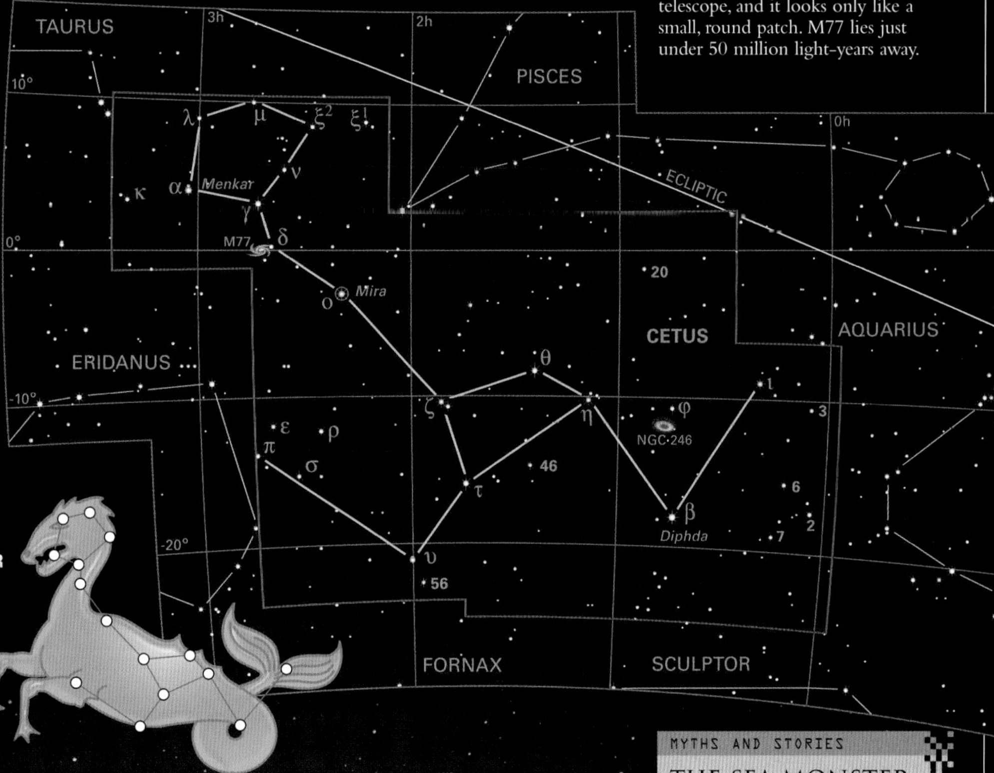


THE SEA MONSTER

**LURCHING MONSTER**
Cetus is large but not particularly prominent. Its most celebrated star is the variable red giant Mira (o), which for much of the time is too faint to be seen with the naked eye.

MYTHS AND STORIES

## THE SEA MONSTER

Cetus was the sea monster sent to devour the princess Andromeda in the famous Greek myth (see p.352). On his return from killing Medusa the Gorgon, Perseus spied Andromeda's plight and swooped down on the sea monster as it attacked, stabbing it repeatedly with his sword in a fury of blood and foam, and leaving its waterlogged corpse on the beach for the local people to pillage.

**MYTHICAL MONSTER**
Old star charts depict Cetus with enormous jaws and a coiled tail, its flippers dipped in the neighbouring constellation, the river Eridanus.

THE HUNTER

# Orion

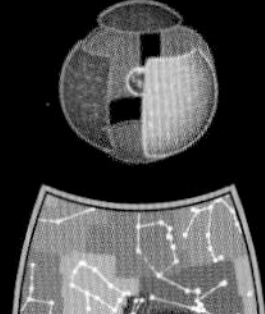

**SIZE RANKING** 26

**BRIGHTEST STARS** Rigel (β) 0.2, Betelgeuse (α) 0.5

**GENITIVE** Orionis

**ABBREVIATION** Ori

**HIGHEST IN SKY AT 10PM** December–January

**FULLY VISIBLE** 79°N–67°S

Orion is one of the most glorious constellations in the sky, representing a giant hunter or warrior followed by his dogs, Canis Major and Canis Minor (see panel, below). Its most distinctive feature is Orion's belt, formed by a line of three 2nd-magnitude stars almost exactly on the celestial equator. A complex of stars and nebulosity represents the sword that hangs from Orion's belt and contains the great star-forming region of M42, the Orion Nebula (p.239). In October each year, the Orionid meteors seem to radiate from a point near Orion's border with Gemini.

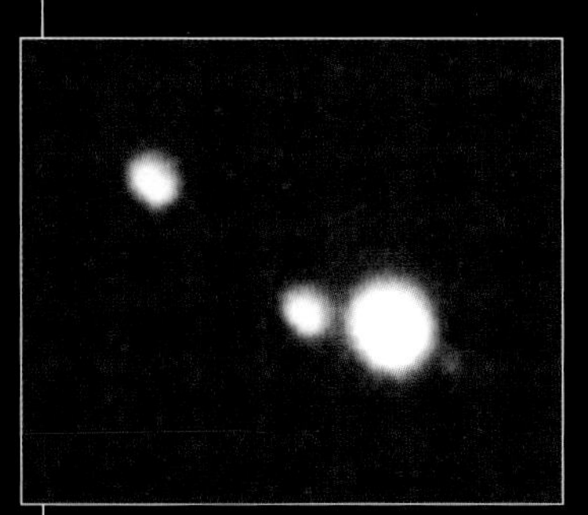

**MULTIPLE COMPANIONS**
Sigma (σ) Orionis is a remarkable multiple star with three fainter companions – two on one side and an even fainter one on the opposite side – appearing rather like a planet orbited by moons.

## SPECIFIC FEATURES

Marking one shoulder of Orion is Betelgeuse – Alpha (α) Orionis (see p.252) – a red supergiant hundreds of times larger than the Sun. Betelgeuse varies irregularly in brightness between magnitudes 0.0 and 1.3, but it averages around magnitude 0.5. It is about 430 light-years away, and it is closer to the Earth than any of the other bright stars in Orion.

Betelgeuse contrasts noticeably in colour with Rigel – Beta (β) Orionis – an even more luminous blue supergiant, which marks one of Orion's feet. Apart from the rare times when Betelgeuse is at its maximum magnitude, Rigel is the brightest star in the constellation. Rigel lies nearly 800 light-years from Earth – almost twice as far away as Betelgeuse. Its 7th-magnitude companion can be picked out from its surrounding glare using a small telescope. Two other easily seen double stars are in Orion's belt. Delta (δ) Orionis has a 7th-magnitude companion, which is visible through a small telescope or binoculars. It is a greater challenge to reveal the close 4th-magnitude companion of Zeta (ζ) Orionis – this requires a telescope with an aperture of at least 75mm (3in).

**BRIGHT HUNTER**
Orion, the hunter, is one of the most magnificent and easily recognizable constellations. A line of three stars makes up his belt, while an area of star clusters and nebulae forms his sword.

MYTHS AND STORIES

## THE GREAT HUNTER

In Greek mythology, Orion was a tall and handsome man and the son of Poseidon, god of the sea. The Greek poet Homer, in his *Odyssey*, described Orion as a great hunter, who brandished a club of bronze. Despite his hunting prowess, Orion was killed by a mere scorpion, some say in retribution for his boastfulness. In the sky, Orion is placed opposite the constellation of Scorpius and, each night, the hunter flees below the horizon as the scorpion rises.

**HUNTER AND WARRIOR**
This depiction of Orion is from an ancient manuscript based on the *Book of Fixed Stars*, which was written by the Arabic astronomer al-Sufi around AD 964.

The real treasures of this constellation lie in the area around Orion's sword. NGC 1981, for example, appears as a large, scattered cluster of stars through binoculars; its brightest stars are of 6th magnitude. NGC 1977 is an elongated patch of nebulosity surrounding the stars 42 and 45 Orionis. Nearby is the Orion Nebula, an enormous star-forming cloud of gas, 1,500 light-years away, covering an area of sky wider than two diameters of the Moon. Its glowing gas appears multi-coloured on photographs and CCD images, yet visually it looks only grey-green because the eye is not sensitive to colours in faint objects. On clear nights, it appears to the naked eye as a hazy patch of light, and is obvious through any form of optical aid. An extension of the Orion Nebula bears a separate number, M43, but both are part of the same cloud. At the centre of M42 lies a multiple star, Theta-1 ($\theta^1$) Orionis (see p.277), better known as the Trapezium because it appears as a group of four stars of 5th to 8th magnitude when seen through a small telescope. To one side of the nebula lies Theta-2 ($\theta^2$) Orionis, a double star with components of 5th and 6th magnitudes that can be separated with binoculars. At the tip of Orion's sword lies Iota (ι) Orionis, a double, with components of 3rd and 7th magnitudes, divisible with a small telescope. Struve 747 is a wider double star nearby, with components of 5th and 6th magnitudes.

Even more impressive is the multiple star Sigma (σ) Orionis (p.277). A small telescope shows that the main 4th-magnitude star has two 7th-magnitude companions on one side and a closer 9th-magnitude component on the other.

Extending from the belt star Zeta (ζ) Orionis is a strip of bright nebulosity, IC 434, against which is silhouetted the Horsehead Nebula (see p.238). This is probably the best-known dark nebula in the sky, and it shows well on photographs. To see it visually requires a large telescope and a dark viewing site.

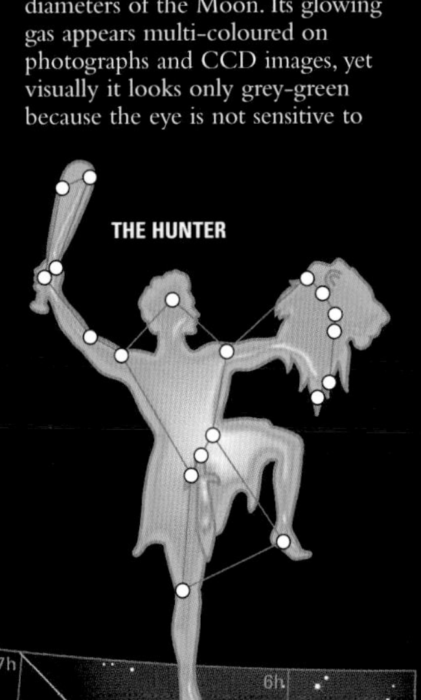


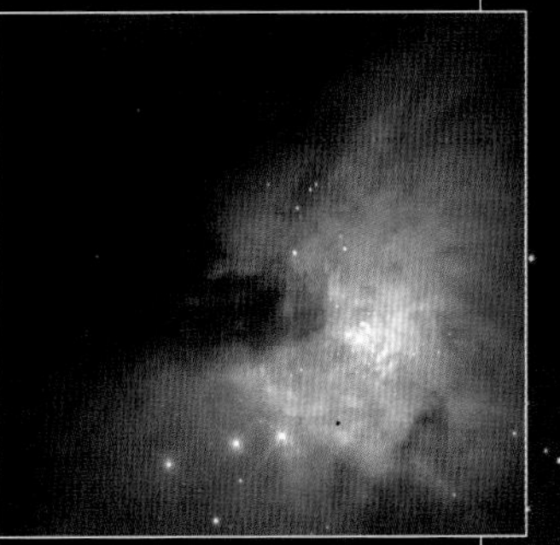

**THE TRAPEZIUM**
At the heart of the Orion Nebula lies a multiple star called the Trapezium ($\theta^1$) (centre, right). Its four stars are visible through small telescopes but with larger apertures, two additional stars can also be seen.

**THE ORION NEBULA**
To the naked eye and through binoculars, the Orion Nebula (M42) appears only as a misty patch, south of Orion's belt, its heart lit up by new-born stars. Its full beauty and its pinkish colour become apparent only on photographs and CCD images such as this.

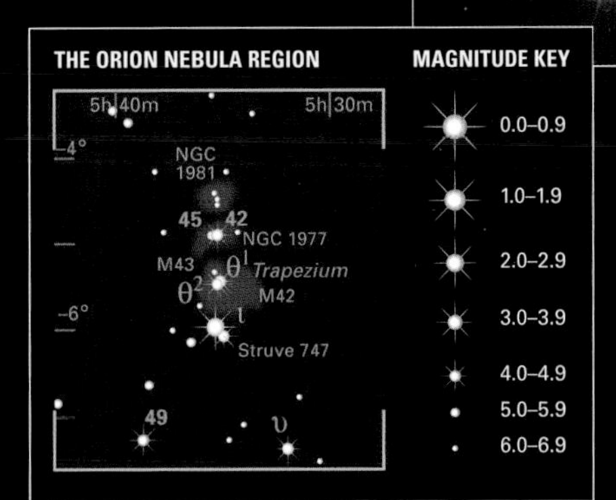


**THE HORSEHEAD NEBULA**
Looking like a knight in a celestial chess game, the Horsehead Nebula is a curiously shaped dark dust cloud silhouetted against IC 434, a backdrop of glowing hydrogen. It lies to the south of Zeta (ζ) Orionis (centre left) in Orion's belt.

THE GREATER DOG

## Canis Major

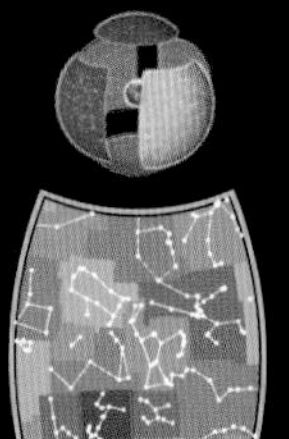

**SIZE RANKING** 43

**BRIGHTEST STARS** Sirius (α) -1.4, Adhara (ε) 1.5

**GENITIVE** Canis Majoris

**ABBREVIATION** CMa

**HIGHEST IN SKY AT 10PM** January–February

**FULLY VISIBLE** 56°N–90°S

This southern constellation contains the brightest star in the entire sky: Sirius, or Alpha (α) Canis Majoris (see p.248). It forms a triangle with Procyon (in Canis Minor) and Betelgeuse (in Orion). Canis Major was known to the ancient Greeks as one of the two dogs following Orion, the hunter (see panel, below).

### SPECIFIC FEATURES

Sirius is a more powerful star than the Sun, giving out about 20 times as much light, and it is among the closest stars to the Earth, being 8.6 light-years away. In combination, these factors give Sirius an apparent brightness twice that of the second-brightest star, Canopus (in Carina). Sirius is accompanied by a faint white dwarf, Sirius B (see p.264), which orbits it every 50 years.

M41 is a large open cluster that is visible as a hazy patch to the naked eye. Its stars, which are scattered over an area about the size of the full Moon, are revealed with binoculars, while telescopes show chains of stars radiating from its centre.

Around Tau (τ) Canis Majoris is NGC 2362, which is best viewed with a telescope. Also nearby is UW Canis Majoris, an eclipsing binary.

**NGC 2362** The brightest member of this neat cluster of stars is the 4th-magnitude blue supergiant Tau (τ) Canis Majoris, which is almost at its centre.

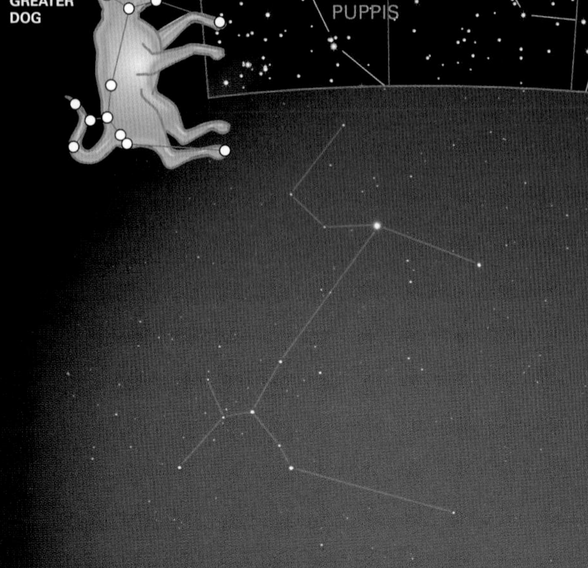


**ORION'S HUNTING DOG** The great dog stands on its hind legs in the sky, holding brilliant Sirius in its jaws like a sparkling ball.

MYTHS AND STORIES

### LAELAPS

This mythical dog was so swift that no prey could escape it, except for the Teumessian Fox, which was destined never to be caught. Laelaps was sent off in pursuit of the fox, which was creating havoc near the town of Thebes, north of Athens, but it was an unending chase. Zeus ended the pursuit by turning them both to stone, and placed the dog in the sky as Canis Major – but without the fox.

**FACING LEFT** In common with many other older depictions of the constellations, Canis Major is shown here as a mirror image.

THE LITTLE DOG

## Canis Minor

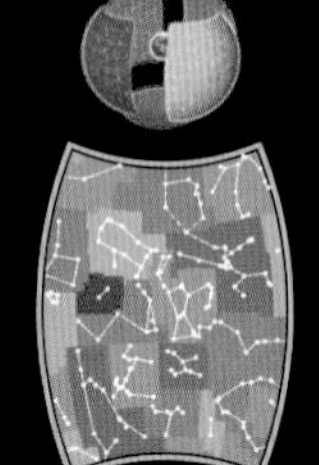

**SIZE RANKING** 71

**BRIGHTEST STAR** Procyon (α) 0.4

**GENITIVE** Canis Minoris

**ABBREVIATION** CMi

**HIGHEST IN SKY AT 10PM** February

**FULLY VISIBLE** 89°N–77°S

Canis Minor is one of the original Greek constellations and lies virtually on the celestial equator. It is usually identified as the smaller of Orion's two hunting dogs.

The constellation is easily identified by its brightest star – Procyon, or Alpha (α) Canis Minoris (see p.280) – which forms a large sparkling triangle with two other 1st-magnitude stars: Betelgeuse (in Orion) and Sirius (in Canis Major). Other than that, the constellation contains little of particular note to small telescope users.

### SPECIFIC FEATURES

Procyon is the eighth-brightest star in the sky. It is somewhat cooler and fainter than the other dog star, Sirius, and also more distant, lying 11.4 light-years away. It has a white dwarf partner, Procyon B, which is visible only with a very large telescope.

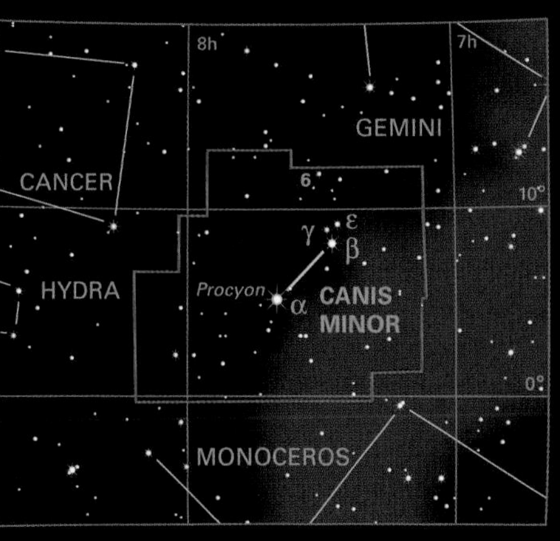

**LONE STAR** Unlike the distinctive constellation of the Greater Dog, Canis Minor consists of little more than its brightest star, Procyon.

THE UNICORN

# Monoceros

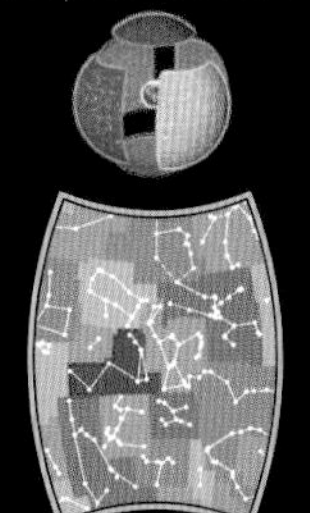

**SIZE RANKING** 35

**BRIGHTEST STAR**
Alpha (α) 3.9

**GENITIVE**
Monocerotis

**ABBREVIATION** Mon

**HIGHEST IN SKY AT 10PM**
January–February

**FULLY VISIBLE**
78°N–78°S

Monoceros is often overlooked, because it is overshadowed by neighbouring Orion, Gemini, and Canis Major. It is easy to locate, however, as it is situated on the celestial equator in the middle of the large triangle formed by the brilliant 1st-magnitude stars Betelgeuse (in Orion), Procyon (in Canis Minor), and Sirius (in Canis Major).

Although none of the stars of Monoceros is bright, the Milky Way passes through it and it contains many deep-sky objects of interest.

The constellation was introduced in the early 17th century by the Dutch astronomer and cartographer Petrus Plancius and depicts the unicorn, a mythical animal with religious symbolism.

## SPECIFIC FEATURES

Beta (β) Monocerotis (see p.277) is regarded as one of the finest triple stars in the sky for small instruments. It is readily separated to show an arc of three 5th-magnitude stars.

The double star Epsilon (ε) Monocerotis is labelled 8 Monocerotis on some charts. Its components, of 4th and 7th magnitudes, are easily spotted through a small telescope.

Prime among Monoceros's most celebrated clusters and nebulae is NGC 2244, an elongated group of stars of 6th magnitude and fainter. Surrounding the cluster is a glorious nebula known as the Rosette Nebula, although it is faint and seen well only on CCD images and photographs.

NGC 2264 is another combination of open cluster and nebula. This triangular group can be seen through binoculars or a small telescope. Its brightest member is 5th-magnitude S Monocerotis – an intensely hot and luminous star that is slightly variable. CCD images and photographs show a surrounding nebulosity into which encroaches a dark wedge called the Cone Nebula (see p.240).

M50 is an open cluster about half the apparent size of the full Moon. It is visible through binoculars but a telescope is needed to resolve its individual stars.

NGC 2232 is larger and more scattered, and its brightest stars are visible through binoculars.

**THE CONE NEBULA**
This tapering region of dark gas and dust intrudes into brighter nebulosity at the southern end of the star cluster NGC 2264. The Cone Nebula is visible only on images taken with a large telescope, as here.

**THE ROSETTE NEBULA**
The flower-like form of the Rosette Nebula glows like a pink carnation in this CCD image. At its centre is the star cluster NGC 2244, which can readily be identified through binoculars.

**FRAMED BEAST**
Monoceros occupies the space within the bright triangle of stars formed by Sirius (seen here at upper right), Betelgeuse (upper left), and Procyon (bottom centre).

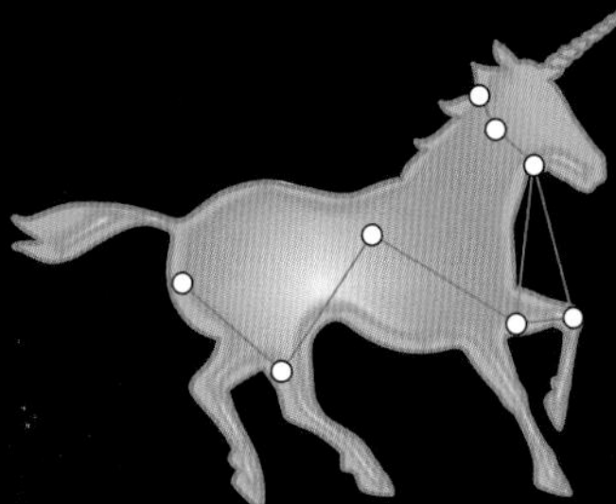


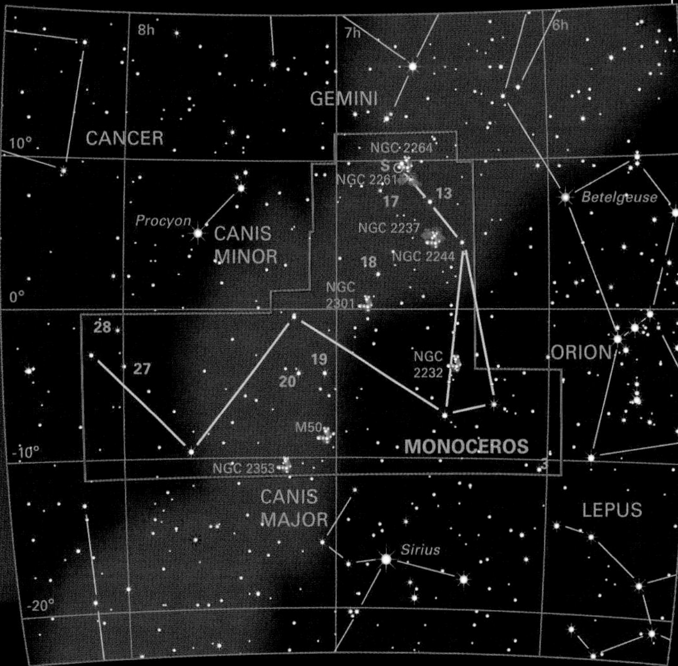

THE WATER SNAKE

# Hydra

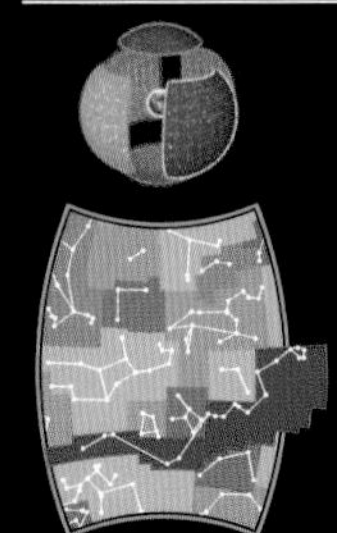

**SIZE RANKING** 1

**BRIGHTEST STAR** Alphard (α) 2.0

**GENITIVE** Hydrae

**ABBREVIATION** Hya

**HIGHEST IN SKY AT 10PM** February–June

**FULLY VISIBLE** 54°N–83°S

Hydra depicts the multi-headed monster that fought and was killed by Hercules in the second of his labours (see panel, right). During the struggle, a crab joined forces with the Hydra but was crushed underfoot by Hercules; it was later commemorated as the constellation Cancer. Although the Hydra had nine heads, it is represented in the sky with a single head – presumably its immortal one.

The constellation is the largest of all 88 and stretches for more than a quarter of the way around the sky from its head, south of Cancer and just north of the celestial equator, to its tail in the southern hemisphere between Libra and Centaurus. Despite its size, there is little to mark out this constellation other than a group of six stars of modest brightness, which forms the head of the water snake.

## SPECIFIC FEATURES

Hydra's brightest star is 2nd-magnitude Alphard, or Alpha (α) Hydrae. Alphard means "the solitary one", and this name reflects its position in an otherwise blank area of sky. This orange-coloured giant is in fact the only star in the constellation brighter than magnitude 3.0. It is about 175 light-years away. Epsilon (ε) Hydrae is a close binary star with components of contrasting colours that can be divided with a telescope with an aperture of at least 75mm (3in) and high magnification. The yellow and blue component stars are of magnitude 3.4 and 6.7 and have an orbital period of nearly 1,000 years.

M48 is an open star cluster on the border with Monoceros. It lies nearly 2,000 light-years away. M48 is larger than the full Moon and it is seen well through binoculars or a small telescope. It contrasts with the globular cluster M68 (see p.289), which resembles a fuzzy star when viewed through binoculars or a small telescope.

M83 is a spiral galaxy, towards the Hydra's tail, that lies about 15 million light-years away. Through a small telescope, it appears as an elongated glow, but a telescope of larger apertures will reveal its spiral structure and its noticeable central "bar", which may be similar to the bar that is thought to lie across the centre of the Milky Way Galaxy.

The planetary nebula known as the Ghost of Jupiter, or NGC 3242, is to be found near the star Mu (μ) Hydrae, in the central part of Hydra's body.

MYTHS AND STORIES

## HERCULES AND THE HYDRA

The Hydra was a serpent with nine heads, one of them immortal, which lived in a swamp near the town of Lerna, emerging to ravage crops and cattle. As the second of his labours, Hercules was sent to kill the monster. He flushed it from its lair with flaming arrows and cut off each head in turn, ending with the immortal head, which he buried under a rock.

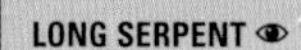

**DEADLY BLOWS**
Hercules battles with the Hydra in this sculpture by François-Joseph Bosio (1768–1845), which is exhibited in the Tuileries gardens, Paris.

**LONG SERPENT**
The Hydra's head, at the right in this photograph, lies south of Cancer (and, in this view, the disc of Jupiter), while the tip of its tail lies far to the left.

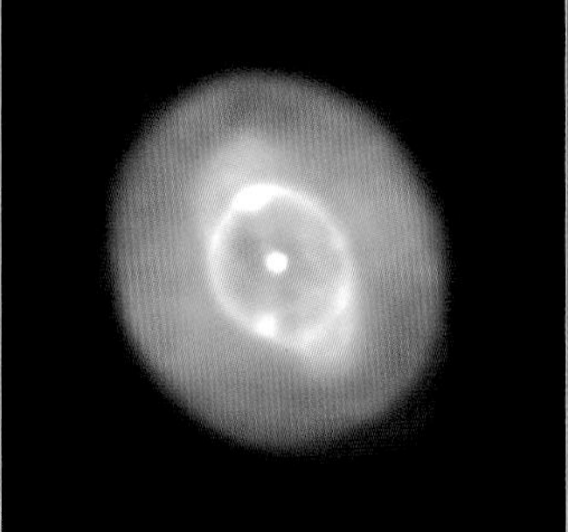

**THE GHOST OF JUPITER**
When viewed through a small telescope, the planetary nebula NGC 3242 appears as an ethereal, blue-green, elliptical glow about the size of the planet Jupiter, hence its popular name – the Ghost of Jupiter.

**M83**
This magnificent face-on spiral galaxy is to be found lying on the border of Hydra and Centaurus. M83 has a central "bar" of stars and gas, and it is sometimes known as the Southern Pinwheel.

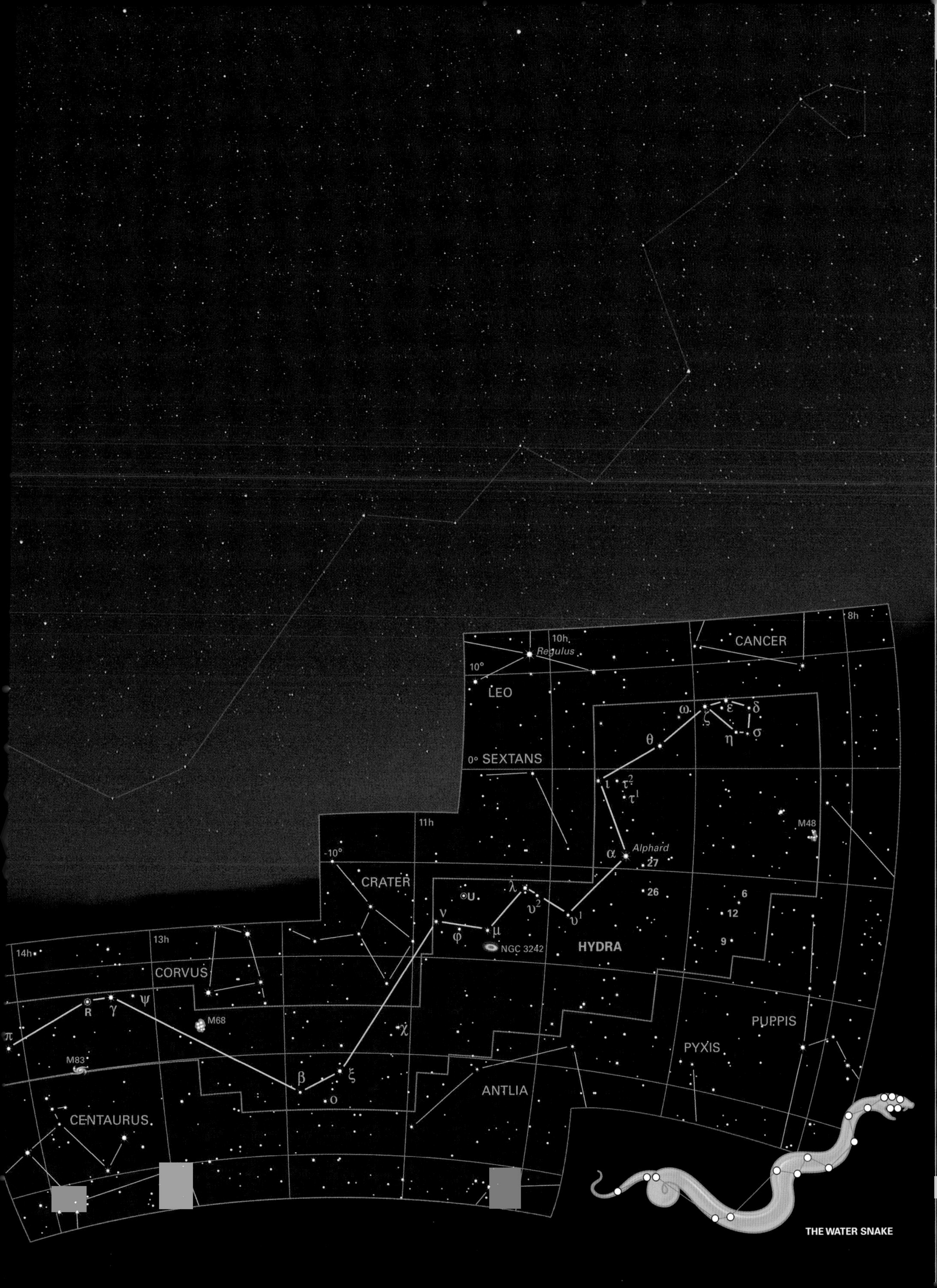

CANCER
LEO
Regulus
SEXTANS
CRATER
Alphard
NGC 3242
HYDRA
CORVUS
M68
M83
M48
PUPPIS
PYXIS
ANTLIA
CENTAURUS

THE WATER SNAKE

THE AIR PUMP

# Antlia

**SIZE RANKING** 62

**BRIGHTEST STAR** Alpha (α) 4.3

**GENITIVE** Antliae

**ABBREVIATION** Ant

**HIGHEST IN SKY AT 10PM** March–April

**FULLY VISIBLE** 49°N–90°S

This constellation was one of those introduced in the mid-18th century by the French astronomer Nicolas Louis de Lacaille (see p.406) to commemorate scientific and technical inventions – in this case, an air pump designed by the French physicist Denis Papin for his experiments on gases.

## SPECIFIC FEATURES

Zeta (ζ) Antliae appears as a wide pair of 6th-magnitude stars when viewed through binoculars. The brighter of the pair has a companion of 7th magnitude.

NGC 2997 is an elegant spiral galaxy inclined at about 45 degrees to our line of sight. Unfortunately, it is just too faint to be well seen through a small telescope, although it can be captured beautifully on photographs and CCD images. NGC 2997 is about 35 million light-years away.

**NGC 2997** This classic spiral galaxy reveals pink clouds of hydrogen gas dotted along its spiral arms in CCD images.

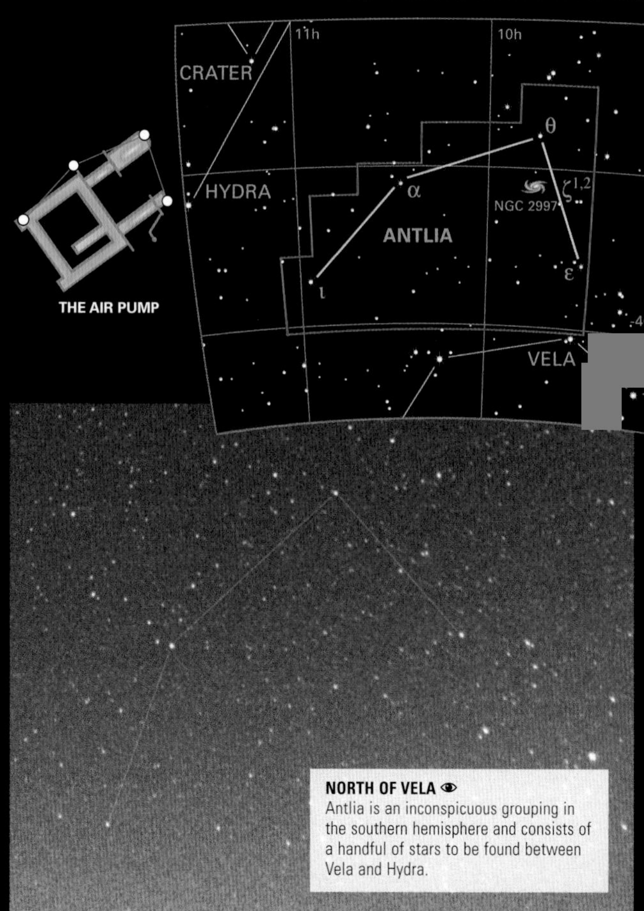


**NORTH OF VELA** Antlia is an inconspicuous grouping in the southern hemisphere and consists of a handful of stars to be found between Vela and Hydra.

THE SEXTANT

# Sextans

**SIZE RANKING** 47

**BRIGHTEST STAR** Alpha (α) 4.5

**GENITIVE** Sextantis

**ABBREVIATION** Sex

**HIGHEST IN SKY AT 10PM** March–April

**FULLY VISIBLE** 78°N–83°S

Representing a sextant used for taking star positions in the days before telescopes were invented, Sextans was introduced in the late 17th century by the Polish astronomer Johannes Hevelius (see p.368), who used such a device when cataloguing the stars.

## SPECIFIC FEATURES

Two unrelated stars of 6th magnitude – 17 and 18 Sextantis – form a line-of-sight double star, which shows neatly through binoculars.

In the same part of the constellation lies NGC 3115, which is popularly named the Spindle Galaxy because of its highly elongated shape. This lenticular galaxy is detectable through a small telescope.

**HEVELIUS'S SEXTANT** Sextans is difficult to pick out with the naked eye because it is a faint and unremarkable constellation. It lies on the celestial equator south of Leo.

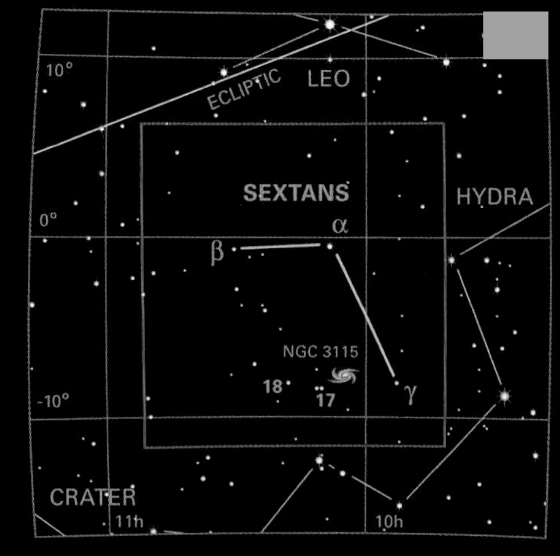


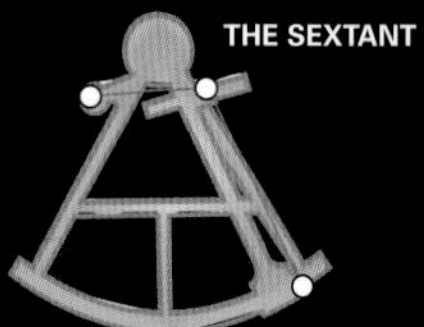

**THE SPINDLE GALAXY** NGC 3115 is a lenticular galaxy seen edge-on from Earth, so it appears highly elliptical in shape when viewed through a telescope. It is just over 30 million light-years from us.

THE CUP

# Crater

**SIZE RANKING** 53

**BRIGHTEST STAR** Delta (δ) 3.6

**GENITIVE** Crateris

**ABBREVIATION** Crt

**HIGHEST IN SKY AT 10PM** April

**FULLY VISIBLE** 65°N–90°S

Crater is a faint constellation representing a cup. Although larger than Corvus, to which it is linked in Greek myth (see panel, right), Crater contains no objects that might be of interest to users of small telescopes.

This area once contained two other constellations that have since been dropped by astronomers. In the late 18th century, a French astronomer, J.J. Lalande, introduced Felis, the cat, between Hydra and Antlia, while others introduced Noctua, the night owl, on the tail of Hydra (see panel illustration, right).

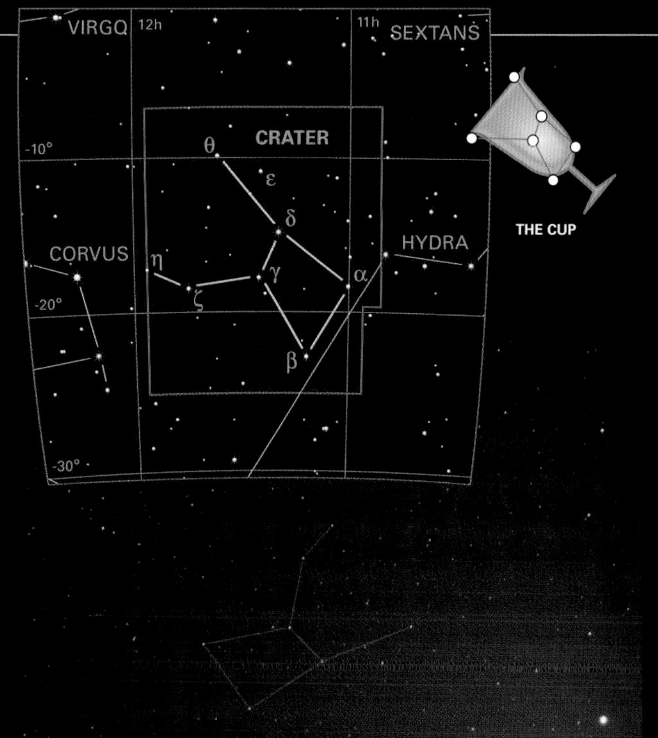


MYTHS AND STORIES

## CROW AND CUP

Crater and Corvus feature together in a Greek myth in which the god Apollo sent the crow (Corvus) to fetch water in a cup (Crater). On the way, the greedy crow stopped to eat figs. As an alibi, the crow snatched up a water snake (Hydra) and blamed it for delaying him but Apollo saw through the deception and banished the trio to the skies.

**HISTORIC DEPICTION**
The area around Crater as shown in *Urania's Mirror*, a set of 19th-century constellation cards.

**CELESTIAL VESSEL**
Crater is to be found lying next to Corvus on the back of Hydra, the water snake. This undistinguished constellation is also known as the goblet or the chalice.

THE CROW

# Corvus

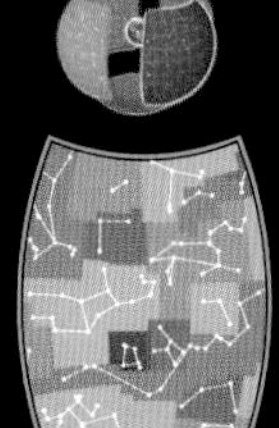

**SIZE RANKING** 70

**BRIGHTEST STAR** Gamma (γ) 2.6

**GENITIVE** Corvi

**ABBREVIATION** Crv

**HIGHEST IN SKY AT 10PM** April–May

**FULLY VISIBLE** 65°N–90°S

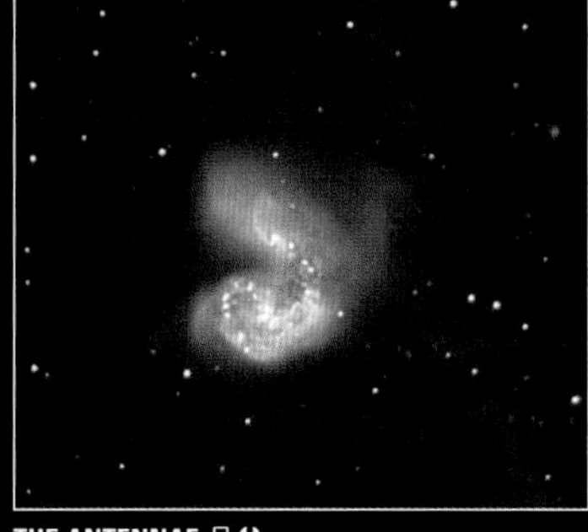

**THE ANTENNAE**
As NGC 4038 and 4039 sweep past each other, gravity draws out long streams of dust and gas from them. The streams extend off the top and bottom of this picture.

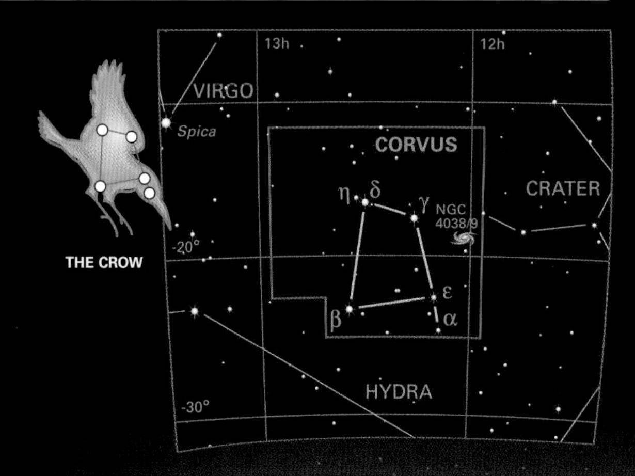

The four brightest stars of Corvus – Beta (β), Gamma (γ), Delta (δ), and Epsilon (ε) Corvi – form a distinctive keystone shape in this small constellation south of Virgo. Oddly, the star labelled Alpha (α) Corvi, at magnitude 4.0, is significantly fainter than all of these. Corvus is one of the original 48 Greek constellations and represents a crow, the sacred bird of the Greek god Apollo.

SPECIFIC FEATURES
Delta is a double star with components of 3rd and 9th magnitudes. It is divisible through a small telescope.

Corvus also boasts a remarkable pair of interacting galaxies: NGC 4038 and 4039. At 10th magnitude, they are too faint to be seen through a small telescope, but photographs reveal this as a graphic example of a galactic collision. When the galaxies passed by each other, gravity pulled out stars and gas to create a shape like an insect's feelers, hence their popular name, the Antennae (see p.307).

**PECKING BIRD**
Corvus, the crow, is linked in legend with neighbouring Crater, the cup. The crow is visualized as pecking at Hydra, the water snake, on whose back it stands.

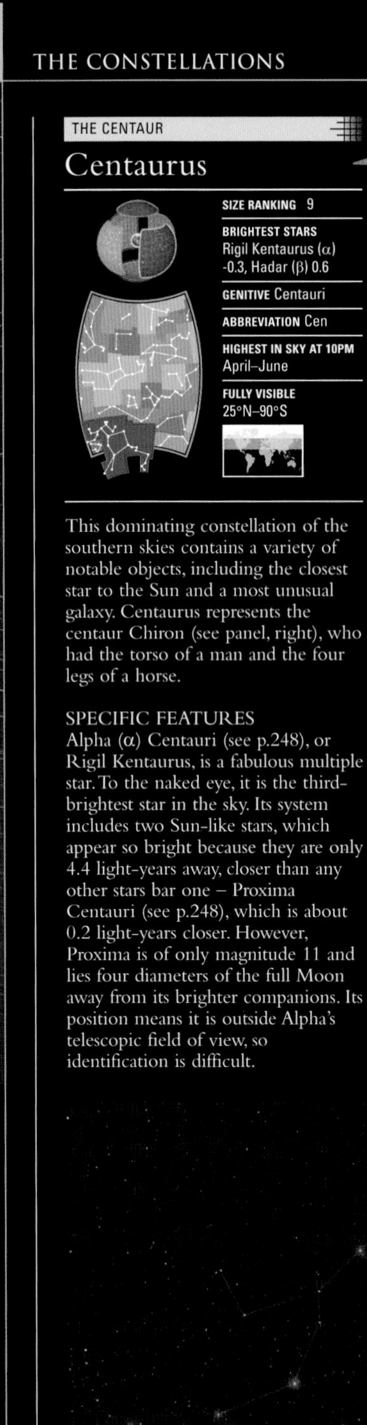

THE CENTAUR

# Centaurus

**SIZE RANKING** 9

**BRIGHTEST STARS** Rigil Kentaurus (α) -0.3, Hadar (β) 0.6

**GENITIVE** Centauri

**ABBREVIATION** Cen

**HIGHEST IN SKY AT 10PM** April–June

**FULLY VISIBLE** 25°N–90°S

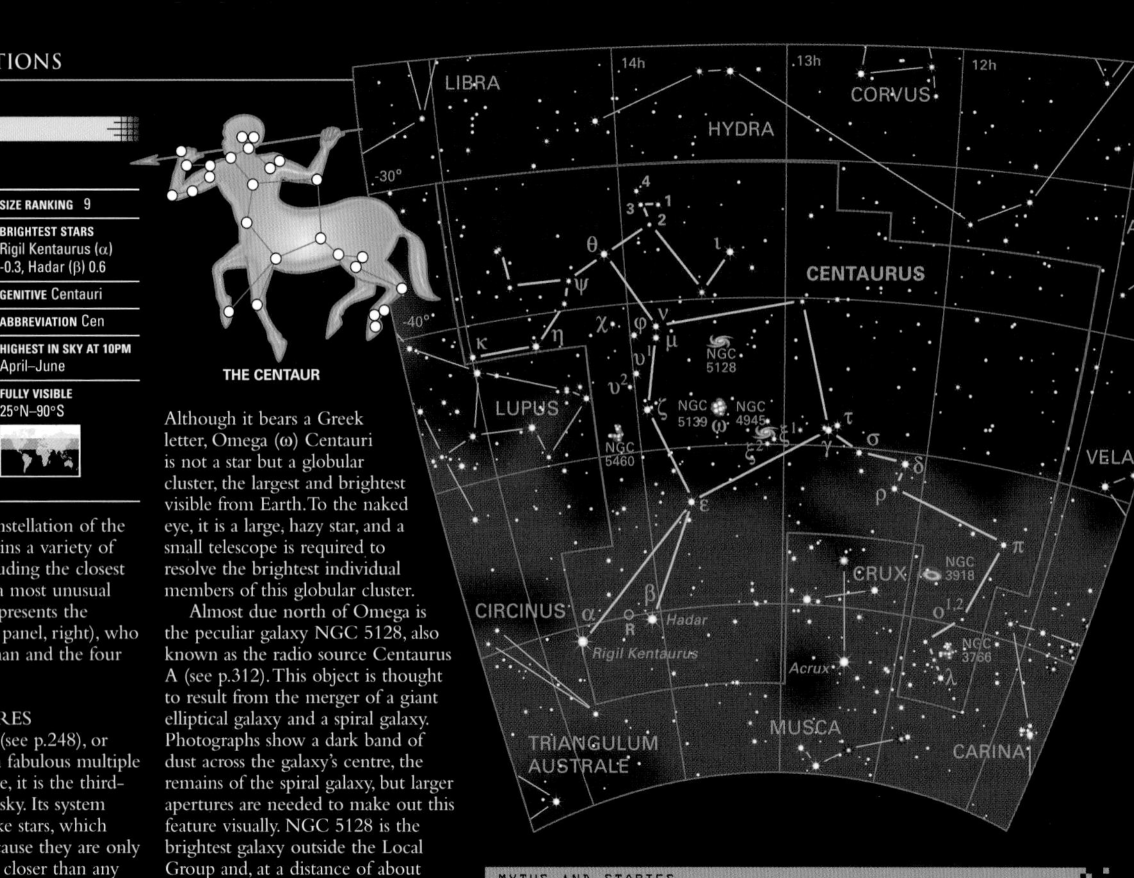


This dominating constellation of the southern skies contains a variety of notable objects, including the closest star to the Sun and a most unusual galaxy. Centaurus represents the centaur Chiron (see panel, right), who had the torso of a man and the four legs of a horse.

## SPECIFIC FEATURES

Alpha (α) Centauri (see p.248), or Rigil Kentaurus, is a fabulous multiple star. To the naked eye, it is the third-brightest star in the sky. Its system includes two Sun-like stars, which appear so bright because they are only 4.4 light-years away, closer than any other stars bar one – Proxima Centauri (see p.248), which is about 0.2 light-years closer. However, Proxima is of only magnitude 11 and lies four diameters of the full Moon away from its brighter companions. Its position means it is outside Alpha's telescopic field of view, so identification is difficult.

Although it bears a Greek letter, Omega (ω) Centauri is not a star but a globular cluster, the largest and brightest visible from Earth. To the naked eye, it is a large, hazy star, and a small telescope is required to resolve the brightest individual members of this globular cluster.

Almost due north of Omega is the peculiar galaxy NGC 5128, also known as the radio source Centaurus A (see p.312). This object is thought to result from the merger of a giant elliptical galaxy and a spiral galaxy. Photographs show a dark band of dust across the galaxy's centre, the remains of the spiral galaxy, but larger apertures are needed to make out this feature visually. NGC 5128 is the brightest galaxy outside the Local Group and, at a distance of about 12 million light-years, is the closest peculiar galaxy to us.

The planetary nebula NGC 3918, or the Blue Planetary, is easily identified through a small telescope. It appears like a larger version of the disc of Uranus. Also in Centaurus are two interesting open star clusters NGC 3766 and NGC 5460.

MYTHS AND STORIES

## CHIRON

The wise and scholarly centaur Chiron was the offspring of Cronus, king of the Titans, and the sea nymph Philyra. He lived in a cave, from where he taught hunting, medicine, and music to the offspring of the gods. His most successful pupil was Asclepius, son of Apollo, who became the greatest healer of the ancient world. Chiron was immortalized among the stars after Heracles accidentally shot him with a poisoned arrow.

**TEACHER OF THE GODS**
This Roman fresco shows Chiron teaching Achilles, his foster-son, to play the lyre. They are in Chiron's cave on Mount Pelion.

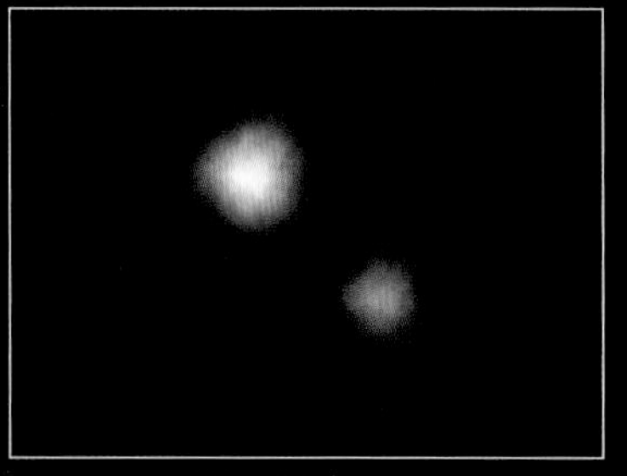

**ALPHA CENTAURI**
The two yellow stars, of magnitudes 0.0 and 1.4, of this beautiful double star orbit each other every 80 years. They are easily separated through a small telescope.

**CELESTIAL CENTAUR**
The brilliant stellar pairing of Alpha (α) and (β) Beta Centauri guides the eye to Centaurus, the celestial centaur. The familiar pattern of Crux, the Southern Cross, lies beneath the centaur's body.

THE WOLF

# Lupus

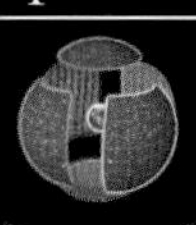

**SIZE RANKING** 46

**BRIGHTEST STAR** Alpha (α) 2.3

**GENITIVE** Lupi

**ABBREVIATION** Lup

**HIGHEST IN SKY AT 10PM** May–June

**FULLY VISIBLE** 34°N–90°S

Lupus is a southern constellation lying on the edge of the Milky Way between the better-known figures of Centaurus and Scorpius. It contains numerous double stars of interest to amateur observers.

It was one of the original 48 constellations familiar to the ancient Greeks, who visualized it as a wild animal speared by Centaurus (see panel, below).

## SPECIFIC FEATURES

Kappa (κ) Lupi, with components of magnitudes 3.9 and 5.7, and Xi (ξ) Lupi, with components of magnitudes 5.1 and 5.6, are two doubles that are easy to spot through a small telescope.

Pi (π) Lupi can be divided into matching 5th-magnitude components through a telescope with an aperture of 75mm (3in). Even more challenging is 4th-magnitude Mu (μ) Lupi, which has a wide 7th-magnitude companion visible through a small telescope. Its primary star, however, is a close double, needing an aperture of at least 100mm (4in) to separate.

The 3rd-magnitude Epsilon (ε) Lupi has a companion of 9th magnitude, and Eta (η) Lupi is a 3rd-magnitude star with an 8th-magnitude companion.

NGC 5822 is a rich open cluster within the Milky Way. Its brightest stars are of only 9th magnitude, so it is not particularly prominent. It lies 2,400 light-years away.

**NGC 5822**

This large open cluster in southern Lupus contains more than 100 stars of 9th magnitude and fainter. It can be seen through binoculars or a small telescope.

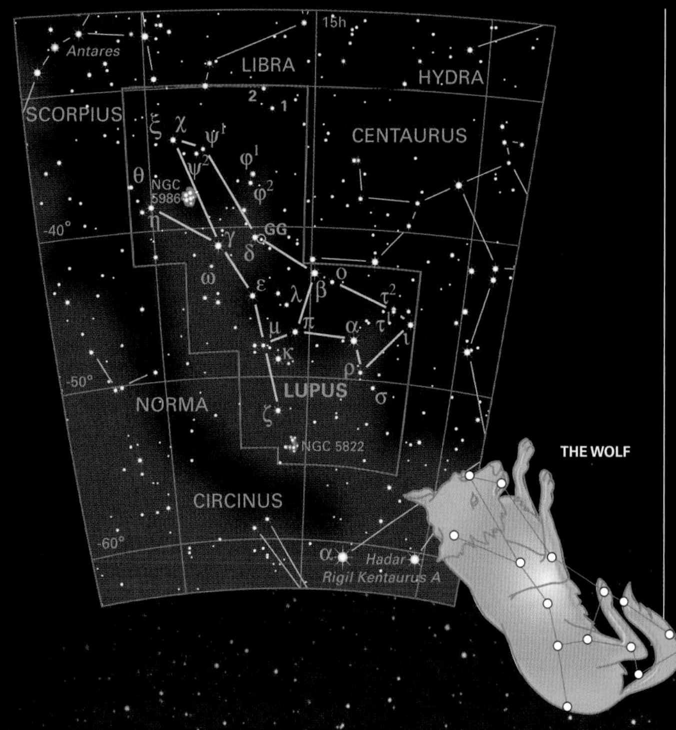


MYTHS AND STORIES

## THE LANCED BEAST

To the ancient Greeks and Romans, Lupus represented a wild animal of unspecified nature which had been speared, by neighbouring Centaurus, on a long pole called a thyrsus. In consequence, Centaurus and Lupus were often regarded as a combined figure. The identification of Lupus as a wolf seems to have become common during Renaissance times.

**IN MIRROR IMAGE**

This medieval Arabic illustration shows Centaurus holding Lupus and the thyrsus, which has become a bunch of leaves or flowers.

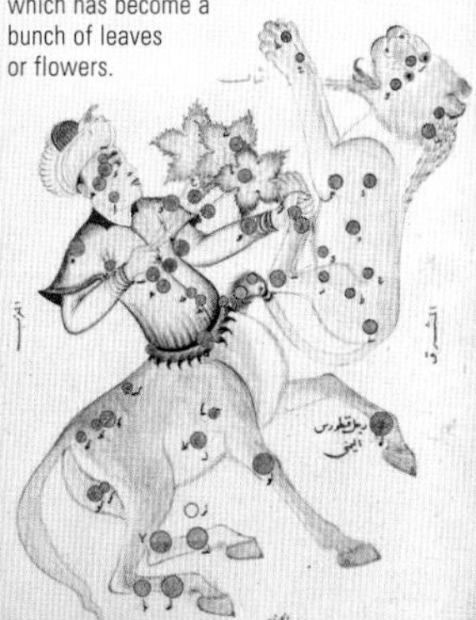

**BESTIAL OFFERING**

Here, Lupus is partly surrounded by the stars of Centaurus. In Greek myth, the centaur killed the beast and carried it to the altar, Ara.

THE ARCHER

# Sagittarius

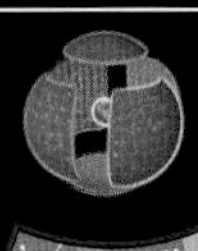

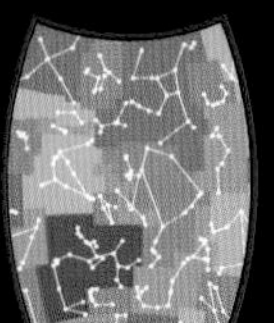

**SIZE RANKING** 15

**BRIGHTEST STAR** Epsilon (ε) 1.8

**GENITIVE** Sagittarii

**ABBREVIATION** Sgr

**HIGHEST IN SKY AT 10PM** July–August

**FULLY VISIBLE** 44°N–90°S

This prominent zodiacal constellation is found between Scorpius and Capricornus, in the southern celestial hemisphere. It includes a highly recognizable star pattern called the Teapot, with a pointed lid (λ) and large spout (γ, ε, and δ). The handle of the Teapot is sometimes also called the Milk Dipper.

The Milky Way is particularly broad and rich in Sagittarius, because the centre of our Galaxy (Sagittarius A) lies in this direction. The exact centre of the Galaxy is thought to coincide with a radio source known as Sagittarius A*, near where the borders of Sagittarius, Ophiuchus, and Scorpius meet. Sagittarius boasts more Messier objects than any other constellation – it has 15 in all.

Although old star charts depicted this constellation as a centaur, in Greek mythology Sagittarius was identified as a different type of creature, known as a satyr. He is usually said to be Crotus, son of Pan, who invented archery and went hunting on horseback. He is seen aiming his bow at neighbouring Scorpius.

## SPECIFIC FEATURES

Beta (β) Sagittarii appears to the naked eye as a pair of 4th-magnitude stars. The more northerly (and slightly brighter) of the two stars has a 7th-magnitude companion. All three stars are at different distances, and hence are unrelated.

Probably the finest object for binoculars is M8, the Lagoon Nebula (see p.241), which extends for three times the width of the full Moon. It contains the cluster NGC 6530, with stars of 7th magnitude and fainter, as well as the 6th-magnitude blue supergiant 9 Sagittarii.

The Trifid Nebula, M20 (see p.244), is so-named because it is trisected by dark lanes of dust. Visually, it is far less impressive than its photographic representation, and little more than the faint double star at its centre can be identified through a small instrument.

On the northern border of Sagittarius with Scutum lies another frequently photographed object – the Omega Nebula, M17. The loose cluster of stars within it can be detected through binoculars.

M22 is one of the finest globular clusters in the entire sky. Under good conditions, it is visible to the naked eye. Through a small telescope, it is somewhat elliptical in outline, while one with an aperture of 75mm (3in) will resolve its brightest stars.

M23 is a large open cluster visible through binoculars near the border with Ophiuchus. M25 is another binocular cluster, while M24 is a bright Milky Way star field the length of four diameters of the full Moon.

**M22**
This prominent globular cluster lies near the lid of the Teapot. Through binoculars, it appears as a woolly ball about two-thirds the apparent diameter of the Moon.

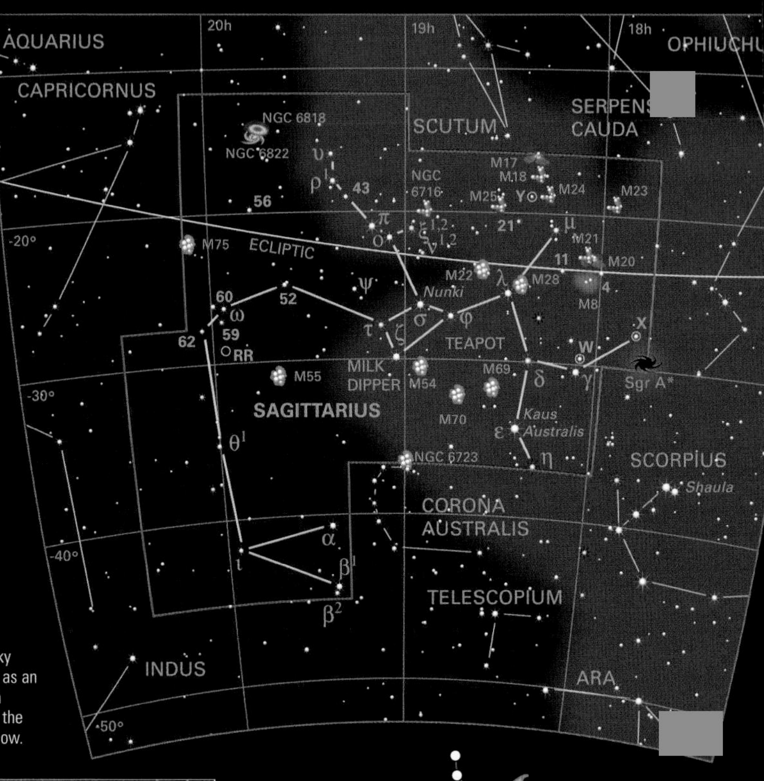


**THE LAGOON NEBULA**
One of the largest nebulae in the sky is M8, which appears in binoculars as an elongated, milky patch of light with embedded stars, including those in the cluster NGC 6530, which make it glow.

THE ARCHER

**THE TRIFID NEBULA**
The pinkish emission of the Trifid Nebula contrasts with the blue reflection nebula to its north, as revealed on long-exposure photographs and CCD images. At its heart is a faint double star, which is overexposed on this image.

**THE OMEGA NEBULA**
M17 can be glimpsed through binoculars but shows up better through a telescope. It resembles the Greek capital letter omega (Ω). However, others see it as a swan, hence its alternative name, the Swan Nebula.

**MOUNTED BOWMAN**
The stars that make up the outline of Sagittarius, the Archer, lie in front of dense Milky Way star fields towards the centre of our Galaxy. North is to the left in this photograph.

THE SCORPION

# Scorpius

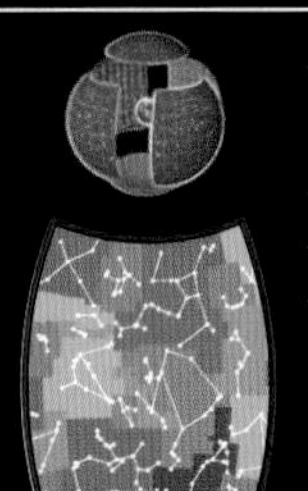

**SIZE RANKING** 33

**BRIGHTEST STAR** Antares (α) 1.0 (variable)

**ABBREVIATION** Sco

**GENITIVE** Scorpii

**HIGHEST IN SKY AT 10PM** June–July

**FULLY VISIBLE** 44°N–90°S

This beautiful and easily recognizable zodiacal constellation is situated in the southern sky. It depicts a scorpion (see panel, below), whose raised tail is marked by a curve of stars extending into a rich area of the Milky Way towards the centre of the Galaxy.

## SPECIFIC FEATURES

Antares, or Alpha (α) Scorpii (see p.252), is a red supergiant hundreds of times larger than the Sun. It fluctuates from about magnitude 0.9 to 1.2 every four to five years.

Normally, Delta (δ) Scorpii is of magnitude 2.3, but in the year 2000 it unexpectedly began to brighten by over 50 per cent. Whether it will remain at its new magnitude or return to its previous value is unknown.

Beta (β) Scorpii is a line-of-sight pair with components of 3rd and 5th magnitudes, while Omega (ω) Scorpii is an even wider unrelated pair, with stars of 4th magnitude. A small telescope easily splits Nu (ν) Scorpii into a double with components of 4th and 6th magnitudes. Mu (μ) Scorpii is another naked-eye pair, with stars of 3rd and 4th magnitudes.

More complex is Xi (ξ) Scorpii, a white and orange pair of stars of 4th and 7th magnitudes. In the same field of view a fainter and wider pair can also be seen. All four stars are gravitationally linked, making this a genuine quadruple.

The open cluster M7 is visible to the naked eye as a hazy patch. It has dozens of stars of 6th magnitude and fainter scattered over an area twice the apparent width of the full Moon. About twice as distant is M6, which is known as Butterfly Cluster (see p.286) because of its shape when viewed through binoculars. On one wing lies BM Scorpii, a variable orange giant. Near Antares, M4 (see p.288) is one of the closest globular clusters to us, at 7,000 light-years away.

Just too far south to have featured on Charles Messier's list (see p.69) is the open cluster NGC 6231. Its brightest member, 5th-magnitude Zeta (ζ) Scorpii, has a 4th-magnitude companion much closer to us.

The strongest X-ray source in the sky is Scorpius X-1. This consists of a 13th-magnitude blue star orbited by a neutron star.

**GLITTERING CLUSTERS** Two prominent star clusters, M6 and M7, adorn the tail of Scorpius in the Milky Way. M6 is at the centre of this photograph; M7 is bottom left.

THE SCORPION

**STING IN THE TAIL** This view of Scorpius has south at the top and shows the scorpion raising its curving tail as though to strike. Its heart is marked by the red star Antares.

MYTHS AND STORIES

## THE DEATH OF ORION

In Greek mythology, Scorpius was the scorpion that stung Orion to death. According to one story the scorpion was sent by Artemis, the goddess of hunting, after Orion had tried to attack her, while another account relates how Mother Earth dispatched the scorpion to humble Orion after he had boasted that he could kill any wild beast.

**MISPLACED FOOT** Like other old star charts, Jean Fortin's *Atlas Céleste* shows the foot of Ophiuchus awkwardly overlapping Scorpius.

THE SEA GOAT

# Capricornus

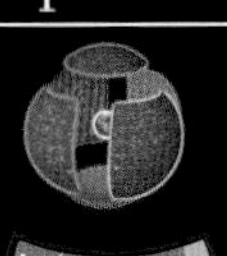

**SIZE RANKING** 40

**BRIGHTEST STAR** Deneb Algedi (δ) 2.9

**GENITIVE** Capricorni

**ABBREVIATION** Cap

**HIGHEST IN SKY AT 10PM** August–September

**FULLY VISIBLE** 62°N–90°S

This is the smallest constellation of the zodiac and not at all prominent; it is situated in the southern sky between Sagittarius and Aquarius. In Greek myth, Capricornus represents the goat-like god Pan (see panel, right), who jumped into a river and became part-fish to escape from the monster Typhon.

SPECIFIC FEATURES

Alpha (α) Capricorni is a wide pairing of unrelated 4th-magnitude stars. They can be separated through binoculars or even with good eyesight. Alpha-1 ($\alpha^1$) Capricorni is a yellow supergiant nearly 700 light-years away, while Alpha-2 ($\alpha^2$) is a yellow giant less than one-sixth that distance from the Earth.

Beta (β) Capricorni is a 3rd-magnitude yellow giant with a 6th-magnitude blue-white companion that can be seen through a small telescope or even good binoculars.

The modest globular cluster M30 is visible as a hazy patch through a small telescope.

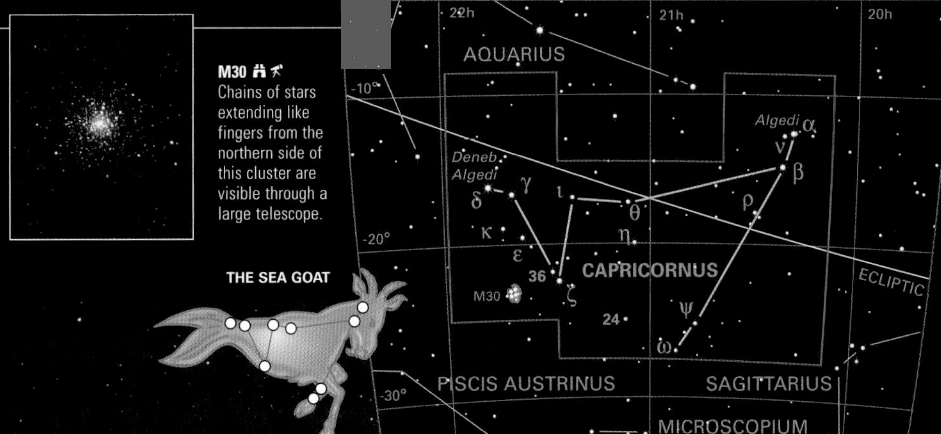


**M30** Chains of stars extending like fingers from the northern side of this cluster are visible through a large telescope.

**CAPRICORNUS AND MARS** Mars is here seen here to the left of Capricornus, whose stars form a roughly triangular shape depicting Pan as half-goat half-fish.

MYTHS AND STORIES

## PAN

This Greek god of the countryside had the hind legs and horns of a male goat but the body of a human. He created the pipes of Pan, also known as the syrinx, from reeds of different length.

**PANPIPER** This stone statue of Pan playing his reed pipes is to be found in the garden of a castle in Schwetzingen, Germany.

THE MICROSCOPE

# Microscopium

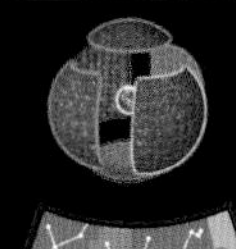

**SIZE RANKING** 66

**BRIGHTEST STARS** Gamma (γ) 4.7, Epsilon (ε) 4.7

**GENITIVE** Microscopii

**ABBREVIATION** Mic

**HIGHEST IN SKY AT 10PM** August–September

**FULLY VISIBLE** 45°N–90°S

Microscopium is a faint and obscure southern constellation to be found between Sagittarius and Piscis Austrinus. It was invented in the 18th century by the French astronomer Nicolas Louis de Lacaille (see p.406), and it represents an early design of compound microscope.

SPECIFIC FEATURES

The orange giant Alpha (α) Microscopii, of 5th magnitude, has a 10th-magnitude companion that is visible through an amateur telescope.

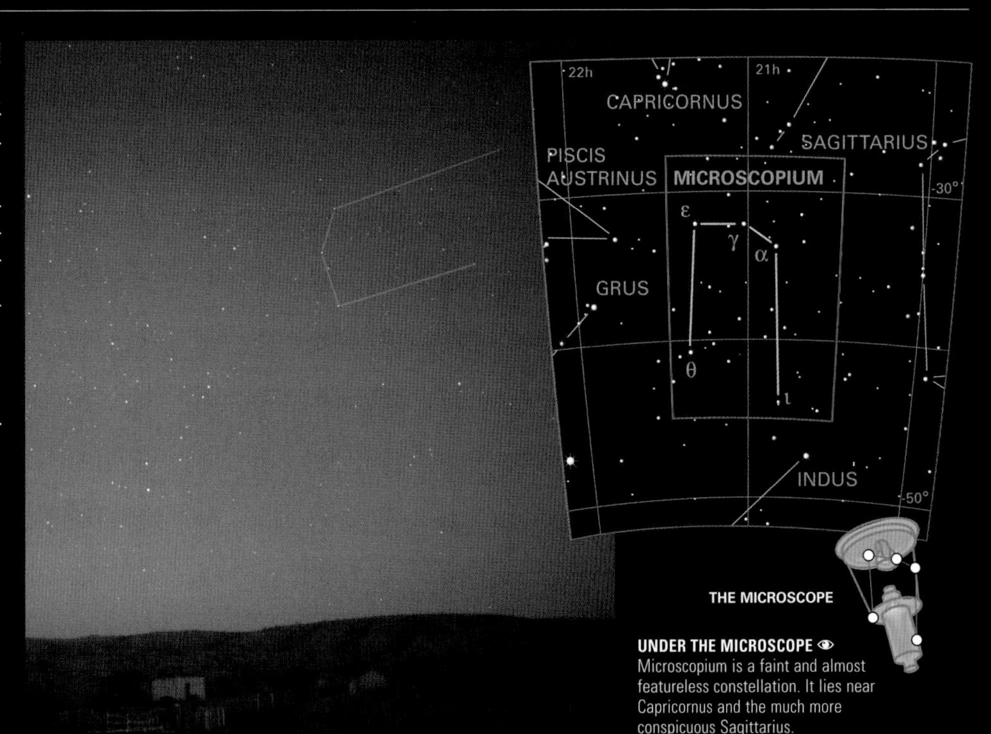


**UNDER THE MICROSCOPE** Microscopium is a faint and almost featureless constellation. It lies near Capricornus and the much more conspicuous Sagittarius.

THE SOUTHERN FISH

## Piscis Austrinus

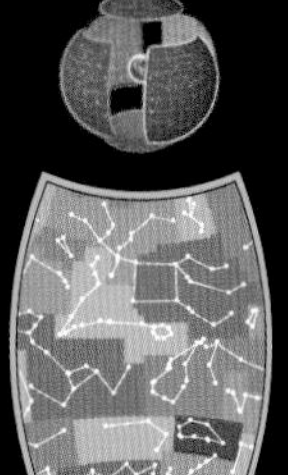

**SIZE RANKING** 60

**BRIGHTEST STAR** Fomalhaut (α) 1.2

**GENITIVE** Piscis Austrini

**ABBREVIATION** PsA

**HIGHEST IN SKY AT 10PM** September–October

**FULLY VISIBLE** 53°N–90°S

Piscis Austrinus was known to the ancient Greeks, including Ptolemy in the 2nd century AD. It depicts a fish, which was said to be the parent of the two fishes represented by the zodiacal constellation Pisces.

This constellation has also been called Piscis Australis. It is made prominent in the southern hemisphere by the presence of 1st-magnitude Fomalhaut, or Alpha (α) Piscis Austrini (see p.249). This blue-white star lies 25 light-years away.

SPECIFIC FEATURES

Beta (β) Piscis Austrini is a wide double star with components of 4th and 8th magnitudes. It is divisible through a small telescope.

More difficult to separate with a small telescope is Gamma (γ) Piscis Austrini, a closer pair of stars of 5th and 8th magnitudes.

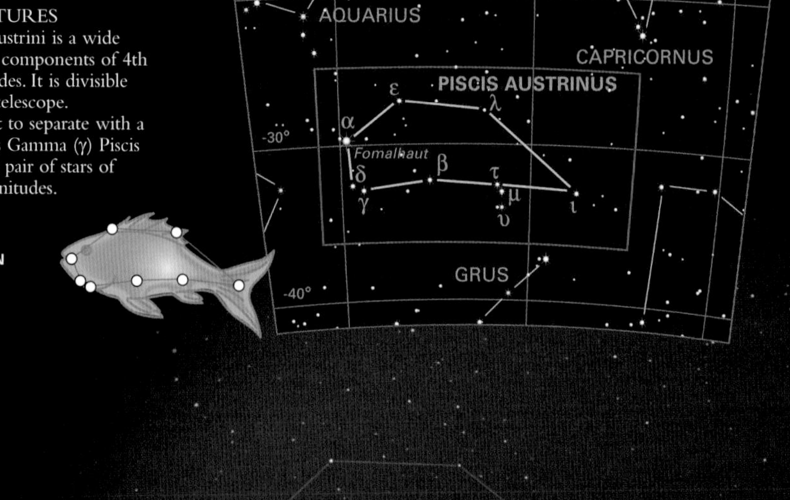


**NEVER-ENDING DRINK**

In the sky, water from the jar of the adjacent Aquarius, the Water Carrier, flows towards the mouth of the fish, marked by Fomalhaut. The star's name is an Arabic term meaning "fish's mouth".

THE SCULPTOR

## Sculptor

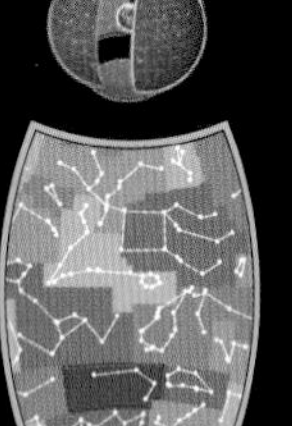

**SIZE RANKING** 36

**BRIGHTEST STAR** Alpha (α) 4.3

**GENITIVE** Sculptoris

**ABBREVIATION** Scl

**HIGHEST IN SKY AT 10PM** October–November

**FULLY VISIBLE** 50°N–90°S

This unremarkable southern constellation was introduced in the 18th century by the French astronomer Nicolas Louis de Lacaille (see p.406). He originally described it as representing a sculptor's studio, although the name has since been shortened.

Sculptor contains the south pole of our Galaxy – that is, the point 90 degrees south of the plane of the Milky Way. As a result, we can see numerous far-off galaxies in this direction, as they are unobscured by intervening stars or nebulae.

SPECIFIC FEATURES

Epsilon (ε) Sculptoris is a binary star that can be separated with a small telescope. Its components, of 5th and 9th magnitudes, have an orbital period of more than 1,000 years.

The spiral galaxy NGC 253 is seen nearly edge-on, so that it appears highly elongated. Under good sky conditions, it can be picked up through binoculars or a small telescope. Nearby lies the fainter and smaller globular cluster NGC 288. Another spiral galaxy is NGC 55, which is similar in size and shape to NGC 253.

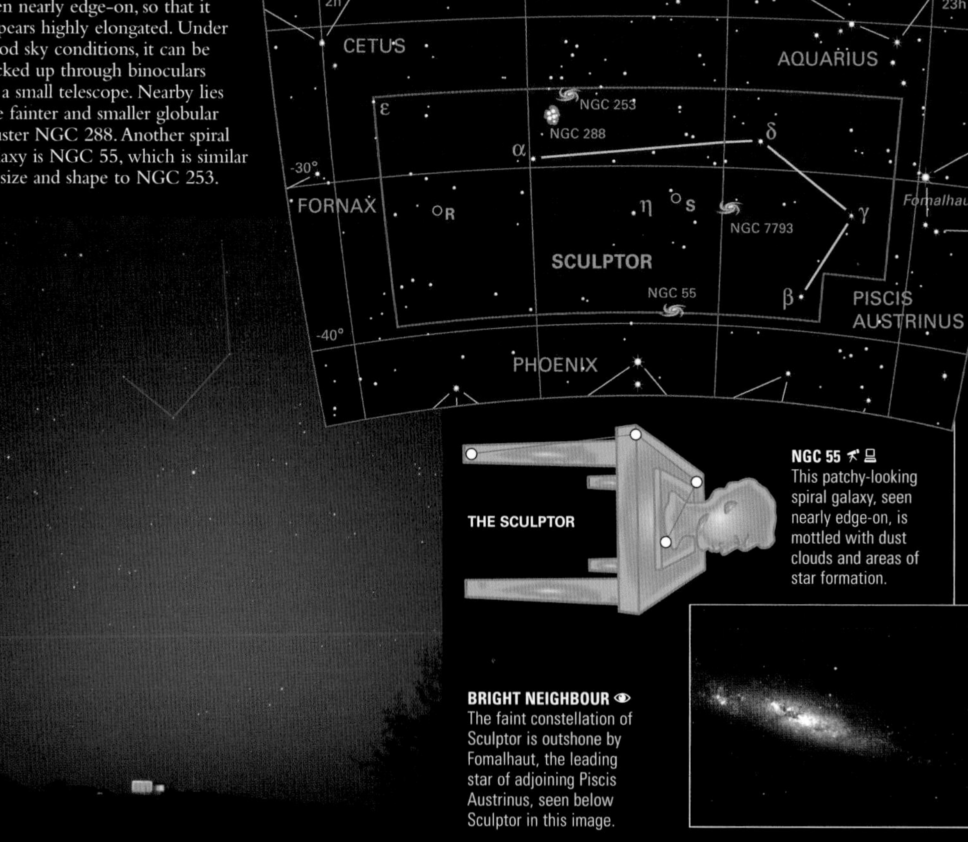


**NGC 55**

This patchy-looking spiral galaxy, seen nearly edge-on, is mottled with dust clouds and areas of star formation.

**BRIGHT NEIGHBOUR**

The faint constellation of Sculptor is outshone by Fomalhaut, the leading star of adjoining Piscis Austrinus, seen below Sculptor in this image.

THE FURNACE

# Fornax

| | |
|---|---|
| SIZE RANKING | 41 |
| BRIGHTEST STAR | Alpha (α) 3.9 |
| GENITIVE | Fornacis |
| ABBREVIATION | For |
| HIGHEST IN SKY AT 10PM | November–December |
| FULLY VISIBLE | 50°N–90°S |

A handful of faint stars makes up this undistinguished constellation of the southern sky. Fornax is situated on the edge of Eridanus and Cetus, and it represents a furnace used by chemists for distillation. It was originally known by the name Fornax Chemica, the chemical furnace, but this has since been shortened to Fornax.

## SPECIFIC FEATURES

The brightest star in the constellation, 4th-magnitude Alpha (α) Fornacis, has a yellow companion, which orbits it every 300 years. This 7th-magnitude star is visible through a small telescope.

On the border of Fornax with Eridanus lies a small cluster of galaxies known as the Fornax Cluster (see p.319). It is about 65 million light-years away, and its brightest member – the peculiar spiral NGC 1316 – is a radio source known as Fornax A. Another prominent member of the Fornax Cluster is the beautiful barred spiral galaxy NGC 1365.

**THE FORNAX CLUSTER** Most of the galaxies in this cluster in southern Fornax are ellipticals, including the 10th-magnitude NGC 1399 (left of centre in this photograph). Standing out among the elliptical galaxies is the large barred spiral NGC 1365 (bottom right).

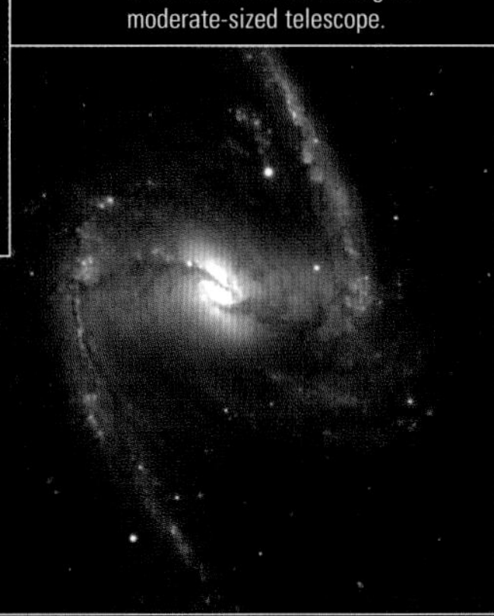

**NGC 1365** This barred spiral galaxy is the largest in the Fornax Cluster and is about as massive as the Milky Way. It can be identified through a moderate-sized telescope.

THE FURNACE

3h 2h CETUS FORNAX α ω ν -30° μ δ β NGC 1365 NGC 1316 -40° ERIDANUS PHOENIX HOROLOGIUM

**PROTECTED POSITION** Fornax is tucked into a bend in the celestial river, Eridanus. It was introduced by Nicolas Louis de Lacaille during the 18th century.

THE CHISEL

# Caelum

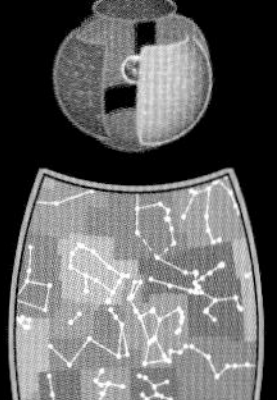

| | |
|---|---|
| SIZE RANKING | 81 |
| BRIGHTEST STAR | Alpha (α) 4.4 |
| ABBREVIATION | Cae |
| GENITIVE | Caeli |
| HIGHEST IN SKY AT 10PM | December–January |
| FULLY VISIBLE | 41°N–90°S |

Sandwiched between Eridanus and Columba is this small and faint southern constellation, which was introduced in the 18th century by the French astronomer Nicolas Louis de Lacaille (see p.406). It represents a stonemason's chisel.

## SPECIFIC FEATURES

Gamma (γ) Caeli is a double star, consisting of an orange giant of magnitude 4.6, with an 8th-magnitude companion. Because they are positioned close together, a modest-sized telescope is required in order to separate them.

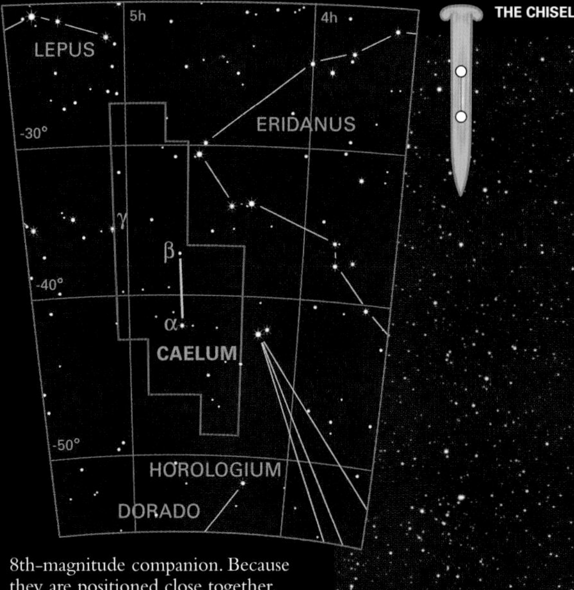


**ENGRAVED IN STONE** Beta (β) and Alpha (α) Caeli mark the shaft of the celestial chisel, which points towards the constellations Dorado and Reticulum in the south.

THE RIVER

# Eridanus

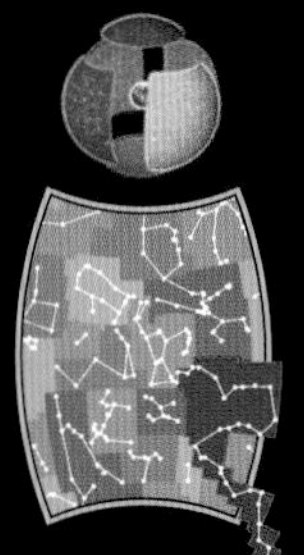

**SIZE RANKING** 6

**BRIGHTEST STAR** Achernar (α) 0.5

**GENITIVE** Eridani

**ABBREVIATION** Eri

**HIGHEST IN SKY AT 10PM** November–January

**FULLY VISIBLE** 32°N–89°S

This large constellation represents a river meandering from the foot of Taurus south to Hydrus. Its range in declination of 58 degrees is the greatest of any constellation.

The only star of any note in Eridanus is 1st-magnitude Achernar, or Alpha (α) Eridani, which lies at the southern tip of the constellation. The name Achernar is of Arabic origin and means "river's end".

Eridanus features in the story of Phaethon, son of the sun god Helios, who attempted to drive his father's chariot across the sky. He lost control and fell into the river below. This river has been identified with two real ones: the Nile in Egypt and the Po in Italy.

## SPECIFIC FEATURES

For all its size, Eridanus is short on objects of interest for a small telescope. The best is the multiple star Omicron-2 ($o^2$) Eridani (see p.272), also known as 40 Eridani, which includes both a red dwarf and a white one. To the eye, it appears as a 4th-magnitude orange star, but a small telescope reveals a 10th-magnitude companion, the white dwarf. This is the easiest white dwarf to spot with a small telescope. It forms a binary with a fainter red dwarf, although this star may require a telescope with a slightly larger aperture to be detectable.

Two double stars of note are Theta (θ) Eridani, consisting of white stars of 3rd and 4th magnitudes divisible through a small telescope, and 32 Eridani, a contrasting pair of orange and blue stars of 5th and 6th magnitudes, also within range of a small telescope.

The galaxy NGC 1300 is estimated to lie around 75 million light-years away and is too faint for viewing through a small telescope. However, it shows up beautifully on photographs.

**MULTIPLE STAR**
The primary star of Omicron-2 ($o^2$) Eridani is in the centre of this photograph, while its white-dwarf and red-dwarf companions overlap each other to the right.

**NGC 1300**
This is a classic example of a barred spiral galaxy. The length of its central bar is greater than the diameter of the Milky Way, being 150,000 light-years across.

**CELESTIAL RIVER**
Eridanus has its source next to Rigel (in Orion) and flows south to Achernar. It is fully visible to almost all of the southern hemisphere and half of the northern.

THE HARE

# Lepus

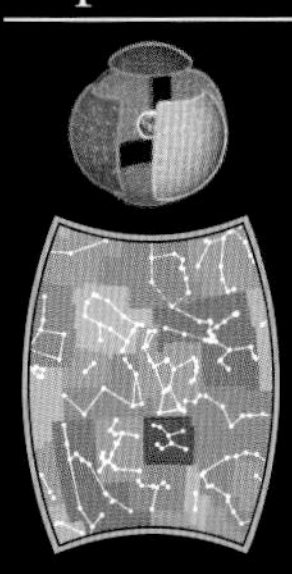

**SIZE RANKING** 51

**BRIGHTEST STAR** Arneb (α) 2.6

**GENITIVE** Leporis

**ABBREVIATION** Lep

**HIGHEST IN SKY AT 10PM** January

**FULLY VISIBLE** 62°N–90°S

Lepus is often overlooked as it is surrounded by sparkling Orion and Canis Major, yet it is worthy of note. It is one of the constellations known to the ancient Greeks.

**M79**

This somewhat sparse 8th-magnitude globular cluster, 42,000 light-years away, has long starry arms which give it the appearance of a starfish.

## SPECIFIC FEATURES

Gamma (γ) Leporis is a 4th-magnitude yellow star with a 6th-magnitude orange companion, which is visible through binoculars. Another double star is Kappa (κ) Leporis, a 4th-magnitude star with a close companion of 7th magnitude. It is difficult to separate through telescopes of small apertures.

NGC 2017 is a compact group of stars in what seems to be a chance alignment. Thus it is not a true star cluster at all.

Near the border with Eridanus lies R Leporis, an intensely red variable star of the same type as Mira (in Cetus). Its brightness ranges from 6th to 12th magnitude every 14 months or so.

The globular cluster M79 can be seen though a small telescope. In the same field of view lies Herschel 3752, a triple star with components of 5th, 7th, and 9th magnitudes.

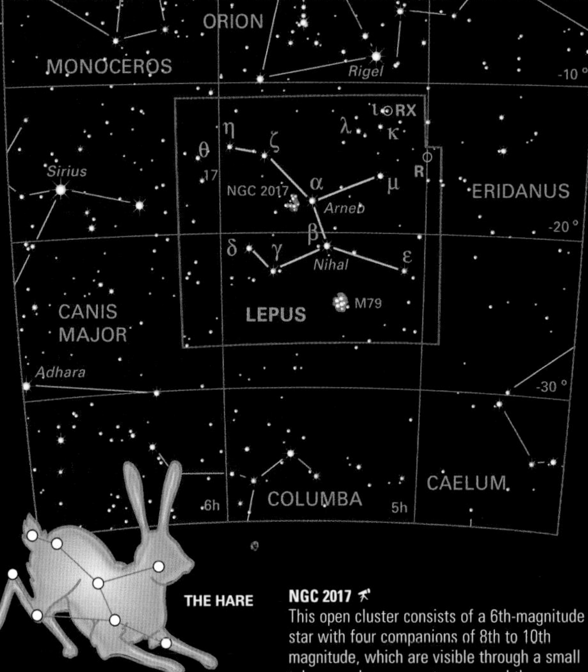


**NGC 2017**

This open cluster consists of a 6th-magnitude star with four companions of 8th to 10th magnitude, which are visible through a small telescope. Larger apertures reveal three fainter stars in the group.

**SAFE HAVEN**

Lepus, the celestial hare, crouches under the feet of Orion, like an animal trying to hide from its hunter. Orion's dogs, Canis Major and Canis Minor, lie nearby.

MYTHS AND STORIES

## A RUNNING HARE

According to Greek mythology, there were no hares on the island of Leros until one man introduced a pregnant female. Soon everyone was raising hares, but they became pests, destroying crops and reducing the population to starvation. The inhabitants eventually drove the hares out of the island and put the image of the hare among the stars as a reminder that one can have too much of a good thing.

**THE HUNTER AND HUNTED**

One of Orion's dogs chases the hare in this 15th-century Flemish illustration, which was based on the *Liber Floridus* of Lambertus, compiled during the Middle Ages.

THE DOVE

## Columba

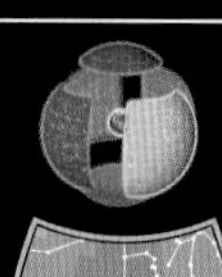

**SIZE RANKING** 54

**BRIGHTEST STAR** Phact (α) 2.7

**GENITIVE** Columbae

**ABBREVIATION** Col

**HIGHEST IN SKY AT 10PM** January

**FULLY VISIBLE** 46°N–90°S

The Dutch theologian and astronomer Petrus Plancius (see p.342) formed this southern constellation in the late 16th century from stars near Lepus and Canis Major that had not previously been allocated to any constellation. It supposedly represents Noah's dove (see panel, right).

### SPECIFIC FEATURES

Fifth-magnitude Mu (μ) Columbae is a fast-moving star apparently thrown out from the area of the Orion Nebula about 2.5 million years ago. Astronomers think that it was once a member of a binary system that was disrupted by a close encounter with another star. The other member of the former binary, moving away from Orion in the opposite direction, is 6th-magnitude AE Aurigae.

NGC 1851 is a modest globular cluster that is visible as a faint patch through a small telescope.

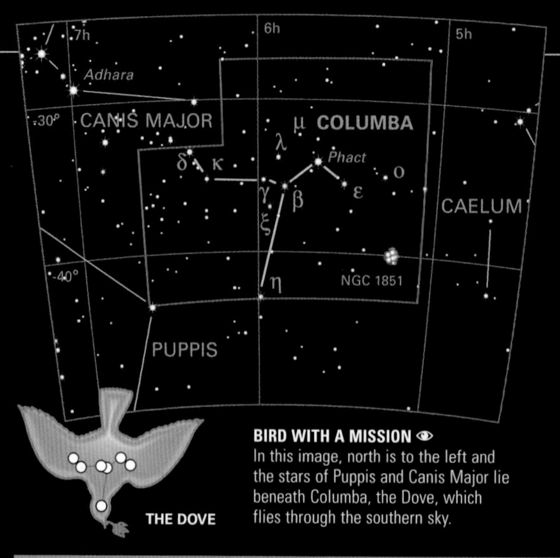


**BIRD WITH A MISSION** In this image, north is to the left and the stars of Puppis and Canis Major lie beneath Columba, the Dove, which flies through the southern sky.

MYTHS AND STORIES

### NOAH'S DOVE

In the Biblical story of the Flood, Noah loaded an ark with a male and female of every kind of animal on Earth. It then rained for 40 days and 40 nights, drowning everything except the animals aboard the ark. When the rain abated, Noah sent out a dove to find dry land. The dove came back with an olive stem in its beak – a sure sign that the waters were at last receding.

**WINGED MESSENGER** The dove returns to Noah's Ark, carrying an olive branch, in this illustration by a 10th-century Catalan monk called Emeterio.

THE COMPASS

## Pyxis

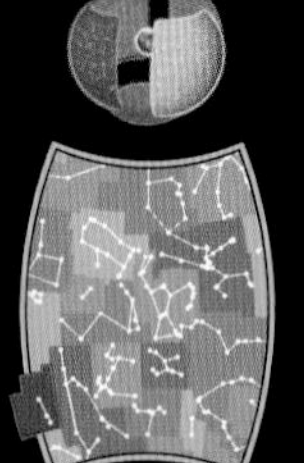

**SIZE RANKING** 65

**BRIGHTEST STAR** Alpha (α) 3.7

**GENITIVE** Pyxidis

**ABBREVIATION** Pyx

**HIGHEST IN SKY AT 10PM** February–March

**FULLY VISIBLE** 52°N–90°S

Pyxis is a faint and unremarkable southern constellation lying next to Puppis on the edge of the Milky Way. It represents a ship's magnetic compass. The constellation was introduced in the 18th century by the French astronomer Nicolas Louis de Lacaille (see p.406).

### SPECIFIC FEATURES

T Pyxidis is a recurrent nova – that is, one that has undergone several recorded outbursts. Five eruptions have been seen since 1890, the last being in 1966. During these outbursts, it has brightened to 6th or 7th magnitude. It is likely to brighten again at any time and so become visible through binoculars.

**COMPASS BEARINGS** In this image of the scattered stars of Pyxis, the Compass, north is on the left, and the stars of adjacent Puppis are to be found above Pyxis.

THE STERN

# Puppis

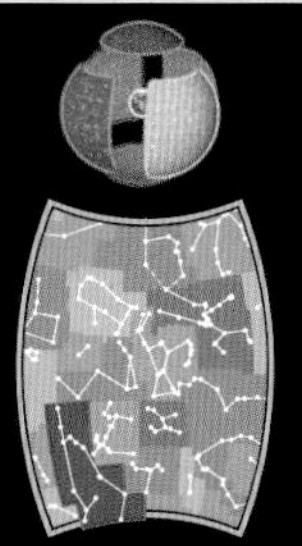

**SIZE RANKING** 20

**BRIGHTEST STAR** Naos (ζ) 2.2

**GENITIVE** Puppis

**ABBREVIATION** Pup

**HIGHEST IN SKY AT 10PM** January–February

**FULLY VISIBLE** 39°N–90°S

This rich southern constellation straddling the Milky Way was originally part of the ancient Greek constellation of Argo Navis (the ship of Jason and the Argonauts, see p.394) until it was divided into three parts in the 18th century. Puppis, representing the ship's stern, is the largest part. The stars of each section retained their original Greek letters, and in the case of Puppis the lettering now starts at Zeta (ζ) Puppis, a star that is also known as Naos.

## SPECIFIC FEATURES

Third-magnitude Xi (ξ) Puppis has a wide and unrelated 5th-magnitude companion that is visible through binoculars, while k Puppis consists of a pair of nearly identical stars with components of 5th magnitude that can be split through a small telescope.

L Puppis is a wide naked-eye and binocular pair, of which $L^2$ Puppis is a variable red giant that ranges between 3rd and 6th magnitudes every five months or so.

M46 and M47 are a pair of open clusters that together create a brighter patch in the Milky Way. Both appear of similar size to the full Moon. M46 is the richer of the two, while M47 is the closer, being about 1,500 light-years away – that is, less than one-third of the distance of its apparent neighbour. The cluster M93 lies about 3,500 light-years away.

NGC 2477 is an open cluster that looks almost like a globular cluster when seen through binoculars, while NGC 2451 is more scattered and has the 4th-magnitude orange giant c Puppis near its centre.

**NGC 2477**
This is one of the richest open clusters, containing an estimated 2,000 stars. It is about 4,000 light-years away. The star below NGC 2477 in this picture – b Puppis – is of magnitude 4.5.

**STERN OF THE *ARGO***
The stars of Puppis, representing the stern of the *Argo*, are seen here rising behind thin cloud. Sirius (in Canis Major) lies near the left edge of this picture.

**SHARP CLUSTER**
M93 is an attractive open cluster for viewing through binoculars or a small telescope. It is shaped like an arrowhead with two orange giants near its tip.

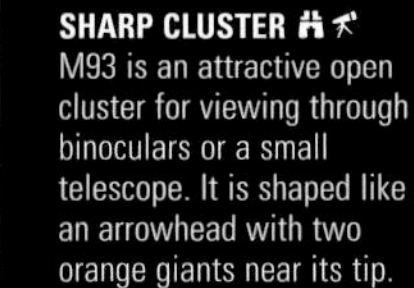

**M46 AND NEBULA**
A small planetary nebula, seen here below centre, seems to be part of M46 but in fact lies in the foreground.

THE SAILS

# Vela

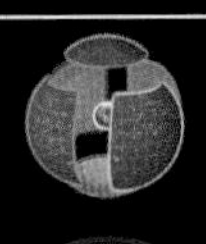

**SIZE RANKING** 32

**BRIGHTEST STAR** Regor (γ) 1.8

**GENITIVE** Velorum

**ABBREVIATION** Vel

**HIGHEST IN SKY AT 10PM** February–April

**FULLY VISIBLE** 32°N–90°S

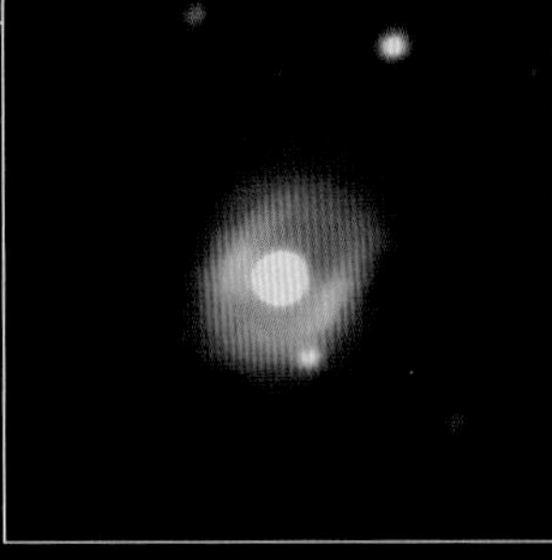

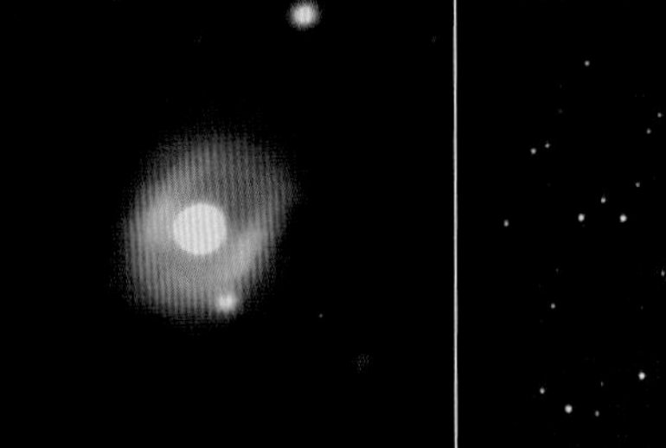

**THE EIGHT-BURST NEBULA**
The planetary nebula NGC 3132 has loops of gas that interlock like figures of eight, hence the object's popular name.

In the 18th century, the ancient Greek constellation Argo Navis (the ship of Jason and the Argonauts – see panel, below) was divided into three parts, one of which was Vela, which represents the ship's sails. Because the stars labelled Alpha (α) and Beta (β) in the former Argo Navis are now in Carina, to the south, the labelling of the stars in Vela starts with Gamma (γ) Velorum, or Regor (see p.249).

Between Gamma and Lambda (λ) Velorum are found the gaseous strands of the Vela supernova remnant (see p.265) – the supernova could have been seen from Earth around 11,000 years ago – while Delta (δ) and Kappa (κ) Velorum combine with two stars in Carina to form the False Cross (sometimes mistaken for the true Southern Cross).

SPECIFIC FEATURES
Gamma Velorum is the brightest example of a Wolf–Rayet star, a rare type of star that has lost its outer layers, thereby exposing its ultra-hot interior. A 4th-magnitude companion is visible through a small telescope or good binoculars. In addition, two wider companions, with components of 8th and 9th magnitudes, are visible through a telescope.

IC 2391 is the best star cluster in Vela for the naked eye or binoculars. It is a group of several dozen stars covering a greater area than the full Moon. To the north of it is another binocular cluster, IC 2395.

NGC 2547 is an open cluster half the size of the full Moon and can be identified through binoculars or a small telescope.

Popularly known as the Eight-Burst Nebula, NGC 3132 has complex loops that are revealed only through a large telescope or on long-exposure photographs. A small telescope will show the nebula's disc, of similar apparent size to Jupiter, and the 10th-magnitude star at its centre.

MYTHS AND STORIES

## THE ARGONAUTS

The *Argo* was a mighty 50-oared galley in which Jason and 50 of the greatest Greek heroes, called the Argonauts, sailed to Colchis, on the eastern shore of the Black Sea, on their mission to fetch the golden fleece of a ram. Their epic voyage is one of the great stories of Greek myth.

**LEGENDARY SAILING GALLEY**
The *Argo*, ship of the Argonauts, is here depicted by the Italian artist Lorenzo Costa (1459–1535).

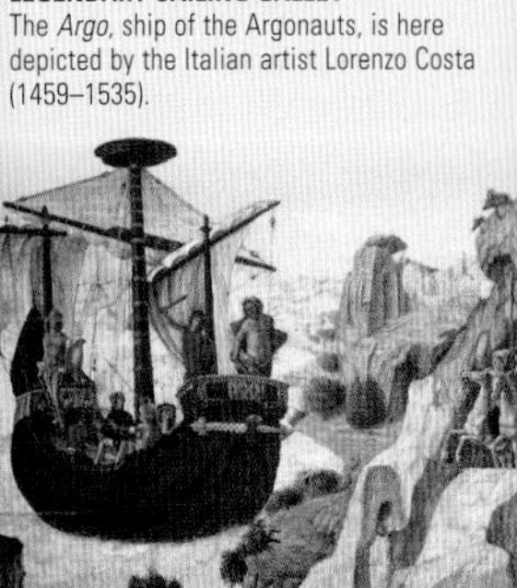

**UNDER SAIL**
Vela represents the mainsail of the *Argo*, the ship of Jason and the Argonauts, sailing through the southern sky in the quest for the golden fleece.

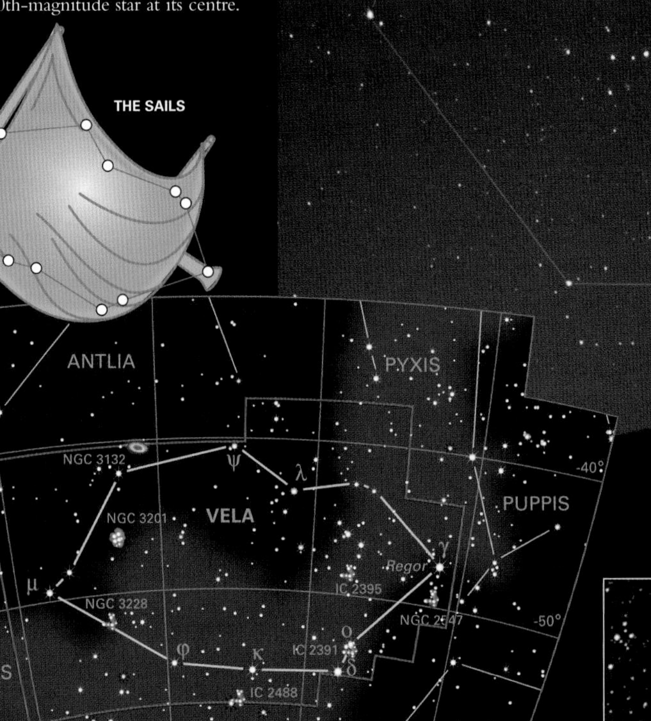


**IC 2391**
Omicron (o) Velorum, at magnitude 3.6, is the brightest member of IC 2391, a scattered cluster that lies some 500 light-years from Earth in the southern reaches of Vela.

THE KEEL

# Carina

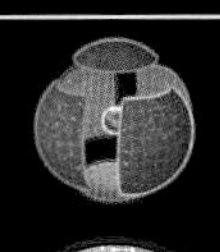

**SIZE RANKING** 34

**BRIGHTEST STAR**
Canopus (α) -0.6

**GENITIVE** Carinae

**ABBREVIATION** Car

**HIGHEST IN SKY AT 10PM**
January–April

**FULLY VISIBLE**
14°N–90°S

Carina is a major southern constellation that was originally part of the larger figure of Argo Navis, which depicted a ship, until that was split up in the 18th century. Carina represents the ship's keel.

Its most prominent star, Canopus, or Alpha (α) Carinae, is a white supergiant 310 light-years away and second in brightness only to Sirius in the entire sky. The stars Epsilon (ε) and Iota (ι) Carinae form a pseudo "southern cross", known as the False Cross, in conjunction with two stars in neighbouring Vela.

## SPECIFIC FEATURES

Splashed across the Milky Way near the border with Centaurus and Vela is the Carina Nebula, NGC 3372 (see p.245), a patch of glowing gas four diameters of the full Moon wide. It is visible to the eye and well seen through binoculars. The densest and brightest part of the nebula is around Eta (η) Carinae (see p.258), an unusual variable star that flared up during the 19th century to become temporarily the second-brightest star in the sky, although it has now subsided to around 5th magnitude. A shell of gas around Eta, which was thrown off during the outburst, is visible through a telescope, next to the Keyhole Nebula, which appears as a dark and bulbous cloud of dust silhouetted against the glowing gas of the Carina Nebula.

A glorious sight through binoculars, another treasure is IC 2602, an open cluster known as the Southern Pleiades. Twice the apparent size of the full Moon, it contains several stars visible to the naked eye – the brightest being 3rd-magnitude Theta (θ) Carinae.

Among Carina's naked-eye clusters is NGC 3532. At its widest point, this elongated group of stars is twice the width of the full Moon. NGC 3114 is about the same apparent size as the full Moon, its brightest individual members being visible through binoculars. NGC 2516 is sparser and appears cross-shaped through binoculars. Its brightest star is a 5th-magnitude red giant.

THE KEEL

**ELONGATED CLUSTER**
What appears to be the brightest member of NGC 3532, in the lower left of this photograph, is in fact an extremely luminous background star some four times farther off.

**THE CARINA NEBULA**
The brightest part of this immense cloud of glowing gas is V-shaped (shown here), while the star Eta (η) Carinae itself (below left of centre) is a peculiar variable that appears as a hazy orange ellipse.

**EVEN KEEL**
Carina represents the keel and hull of the Argonauts' ship, the *Argo*. The blade of the steering oar is marked by Canopus, Carina's brightest star.

THE SOUTHERN CROSS

# Crux

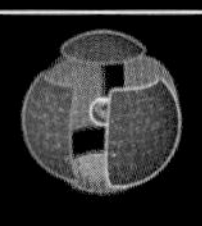

**SIZE RANKING** 88

**BRIGHTEST STARS** Acrux ($\alpha$) 0.8, Becrux (or Mimosa) ($\beta$) 1.3

**GENITIVE** Crucis

**ABBREVIATION** Cru

**HIGHEST IN SKY AT 10PM** April–May

**FULLY VISIBLE** 25°N–90°S

Crux lies in a rich area of the Milky Way. Although it is the smallest constellation, it is instantly recognizable and is squeezed between the legs of the centaur, Centaurus. The longer axis of the Southern Cross, as Crux is known, points towards the south celestial pole. Its stars were known to the ancient Greeks (see panel, below), who regarded them as part of Centaurus. They were made into a separate constellation in the 16th century.

SPECIFIC FEATURES

Alpha ($\alpha$) Crucis or Acrux is the most southerly first-magnitude star. It is a glittering double that is readily divisible through a small telescope. The two components are of magnitudes 1.3 and 1.8. A wider 5th-magnitude star can be seen through binoculars; it is not related to Acrux.

The star at the top of the cross is the 2nd-magnitude red giant Gamma ($\gamma$) Crucis or Gacrux. It has an unrelated 6th-magnitude companion visible through binoculars. Nearby, Mu ($\mu$) Crucis is a wide pair of 4th- and 5th-magnitude stars easily separated through a small telescope or even good binoculars.

One of the gems of the southern sky is the Jewel Box Cluster (see p.288), or NGC 4755, visible to the naked eye as a brighter patch within the Milky Way near Beta ($\beta$) Crucis or Becrux. Its individual stars, the brightest being of 6th magnitude, cover about one-third the width of the full Moon. They can be viewed through binoculars or a small telescope. Near the centre is a ruby-coloured su[illegible]nt that contrasts with the b[illegible]hite sparkle of the other stars, producing a resemblance to a casket of jewels, hence the popular name.

The Coalsack Nebula is to be found beside the Jewel Box. This dark cloud of dust blocks light from the stars of the Milky Way behind it. It spans the width of 12 full Moons and extends into neighbouring Centaurus and Musca, so is prominent to the naked eye and through binoculars.

THE SOUTHERN CROSS

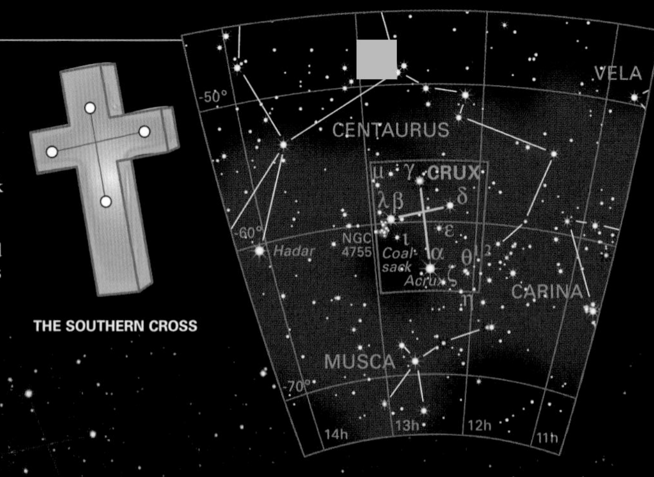


**SIGN OF THE CROSS**
Four prominent stars make up the Southern Cross, one of the most famous of all celestial patterns, which appears on the flags of several nations.

**THE COALSACK**
The Coalsack Nebula, which lies next to the stars of the Southern Cross, is a smudgy cloud of interstellar dust silhouetted against the bright background of the Milky Way.

**JEWELS OF THE SKIES**
The Jewel Box Cluster is a sparkling group of stars just north of the Coalsack Nebula although the cluster is almost ten times more distant from the Earth.

EXPLORING SPACE

## REDISCOVERING STARS

When European seafarers returned from exploring the southern latitudes in the 15th and 16th centuries, they reported stars they had never seen before. Among these explorers was Amerigo Vespucci (1454–1512), an Italian who in 1501 charted Alpha ($\alpha$) and Beta ($\beta$) Centauri and the stars of Crux. Astronomers later realized that these stars had been known to the ancient Greeks but that precession (see p.60) had subsequently carried them below the horizon in Europe.

**AMERIGO VESPUCCI**
This imaginative view of Amerigo Vespucci observing the Southern Cross with an astrolabe was painted by the 16th-century Flemish artist Joannes Stradanus (Hans van der Straet).

THE FLY

# Musca

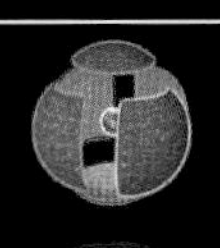

SIZE RANKING 77

BRIGHTEST STAR Alpha (α) 2.7

GENITIVE Muscae

ABBREVIATION Mus

HIGHEST IN SKY AT 10PM April–May

FULLY VISIBLE 14°N–90°S

This modest constellation is to be found in the Milky Way south of Crux and Centaurus. In fact, the southern tip of the dark Coalsack Nebula extends into it from Crux.

Musca is one of the southern constellations introduced at the end of the 16th century by the Dutch navigator-astronomers Pieter Dirkszoon Keyser and Frederick de Houtman. It represents a fly.

SPECIFIC FEATURES

Theta (θ) Muscae is a double star with components of 6th and 8th magnitude, divisible through a small telescope. The fainter component is an example of a Wolf–Rayet star – a hot star that has lost its outer layers. Musca also has a globular cluster, known as NGC 4833 (see p.289).

**NGC 4833**
This globular cluster is just visible through binoculars. Individual stars can be resolved with a telescope of 100mm (4in) aperture.

**FINDING THE FLY**
The long axis of the Southern Cross points to Musca, the fly, which lies on the edge of the Milky Way within the southern celestial hemisphere.

THE COMPASSES

# Circinus

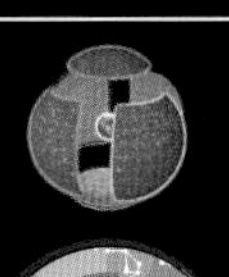

SIZE RANKING 85

BRIGHTEST STAR Alpha (α) 3.2

GENITIVE Circini

ABBREVIATION Cir

HIGHEST IN SKY AT 10PM May–June

FULLY VISIBLE 19°N–90°S

Circinus represents a pair of dividing compasses, as used by surveyors and navigators. It is one of the figures introduced in the 18th century by the French astronomer Nicolas Louis de Lacaille (see p.406).

This small southern constellation is squeezed awkwardly in between Centaurus and Triangulum Australe. It lies next to Alpha (α) Centauri, so it is not difficult to locate.

SPECIFIC FEATURES

Circinus contains little of interest for amateur astronomers. Alpha (α) Circini, however, is its one star of note. It is situated against the background of the Milky Way and is easy to identify, being a double with components of 3rd and 9th magnitudes. These are divisible through a small telescope.

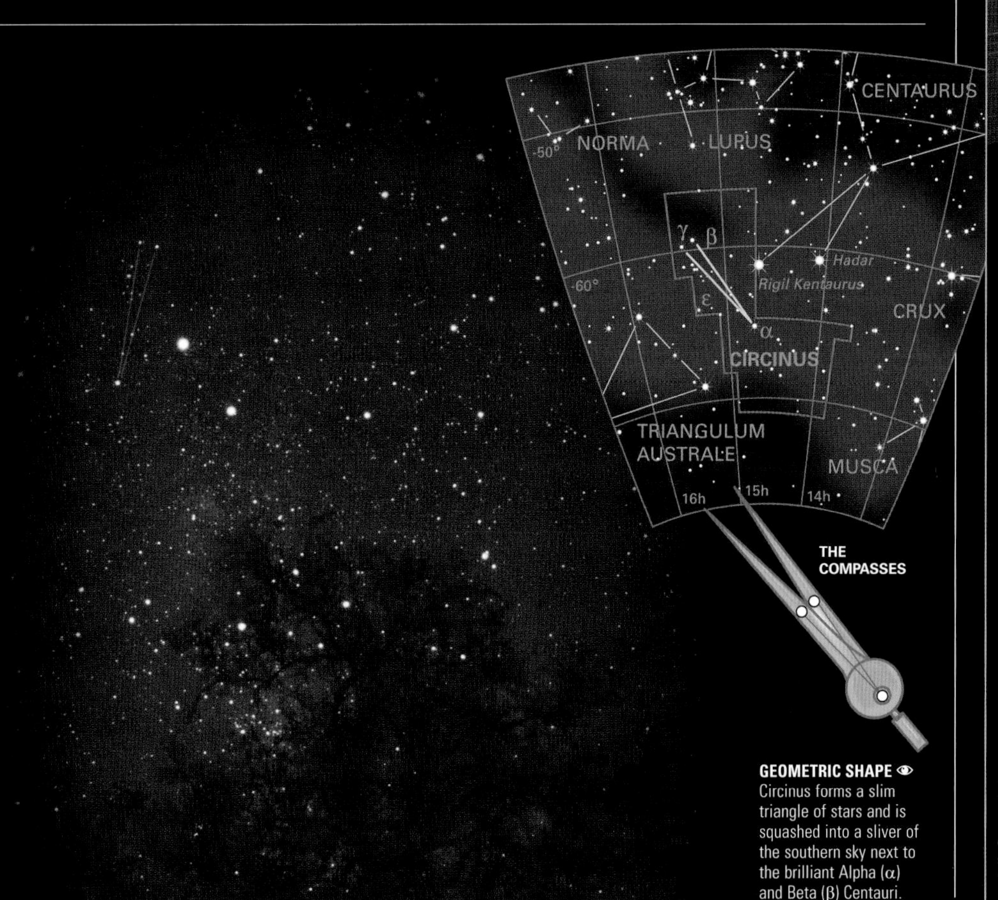


**GEOMETRIC SHAPE**
Circinus forms a slim triangle of stars and is squashed into a sliver of the southern sky next to the brilliant Alpha (α) and Beta (β) Centauri.

THE SET SQUARE

# Norma

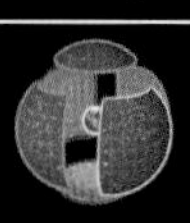

**SIZE RANKING** 74

**BRIGHTEST STAR** Gamma-2 ($\gamma^2$) 4.0

**GENITIVE** Normae

**ABBREVIATION** Nor

**HIGHEST IN SKY AT 10PM** June

**FULLY VISIBLE** 29°N–90°S

Norma was introduced in the 1750s by the Frenchman Nicolas Louis de Lacaille (see p.406), and was originally known as *Norma et Regula*, the square and ruler. It is an unremarkable southern constellation lying in the Milky Way between Lupus and the zodiacal constellation of Scorpius.

The stars that Lacaille designated Alpha (α) and Beta (β) have since been incorporated into Scorpius.

SPECIFIC FEATURES

At magnitude 4.0, Gamma-2 ($\gamma^2$) Normae is the constellation's brightest star, and it is one-half of a naked-eye double together with Gamma-1 ($\gamma^1$) Normae, of magnitude 5.0. The two stars lie at widely different distances and hence are unrelated.

Two other doubles in the constellation that are readily separated through a small telescope are Epsilon (ε) Normae, with components of 5th and 7th magnitudes, and Iota-1 ($\iota^1$) Normae, with components of 5th and 8th magnitudes.

NGC 6087 is a large open cluster that has radiating chains of stars, which are visible through binoculars. Near its centre is its brightest star, S Normae – a Cepheid variable that ranges in brightness from magnitude 6.1 to 6.8 every 9.8 days.

**NGC 6067**

This rich cluster covers an area of sky about half the apparent diameter of the full Moon. It is seen against the backdrop of the Milky Way.

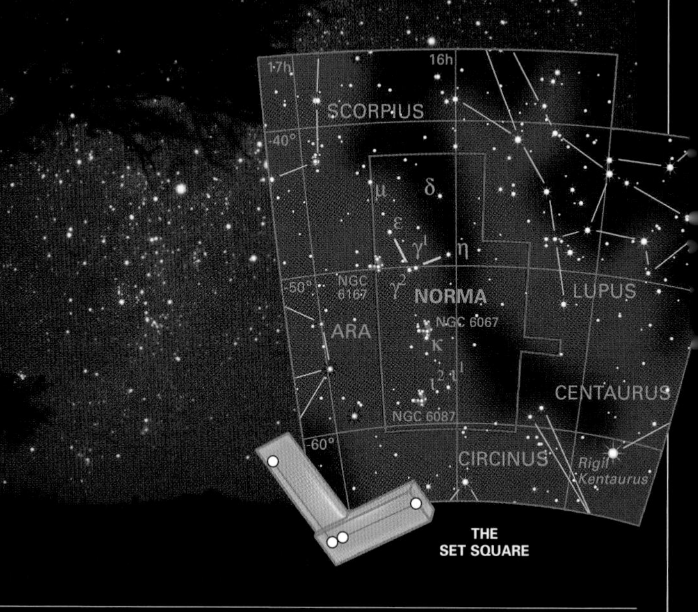


**RIGHT ANGLE**

Norma's most distinctive feature is a right-angled trio of three faint stars, which is somewhat difficult to identify among the rich Milky Way star fields.

THE SOUTHERN TRIANGLE

# Triangulum Australe

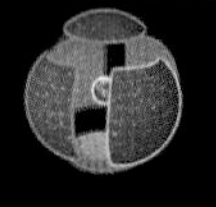

**SIZE RANKING** 83

**BRIGHTEST STAR** Alpha (α) 1.9

**GENITIVE** Trianguli Australis

**ABBREVIATION** TrA

**HIGHEST IN SKY AT 10PM** June–July

**FULLY VISIBLE** 19°N–90°S

Triangulum Australe is one of the constellations of the southern sky that was introduced at the end of the 16th century by the Dutch navigators Pieter Dirkszoon Keyser and Frederick de Houtman. It is the smallest of the 12 they identified.

Although smaller than its northern counterpart, Triangulum, this constellation contains brighter stars and so is more prominent.

SPECIFIC FEATURES

NGC 6025 lies on Triangulum Australe's northern border with Norma. It is 2,700 light-years away from the Earth. This open cluster is noticeably elongated in shape and is about one-third the apparent diameter of the full Moon. It is easily seen through binoculars.

Alpha (α) Trianguli Australis is an orange giant whose colour shows prominently through binoculars.

There is nothing else in the constellation to attract users of small telescopes.

**SOUTHERN TRIPLET**

Triangulum Australe is an easily recognized triangle of stars, lying in the Milky Way near brilliant Alpha (α) and Beta (β) Centauri, which here are visible on the right.

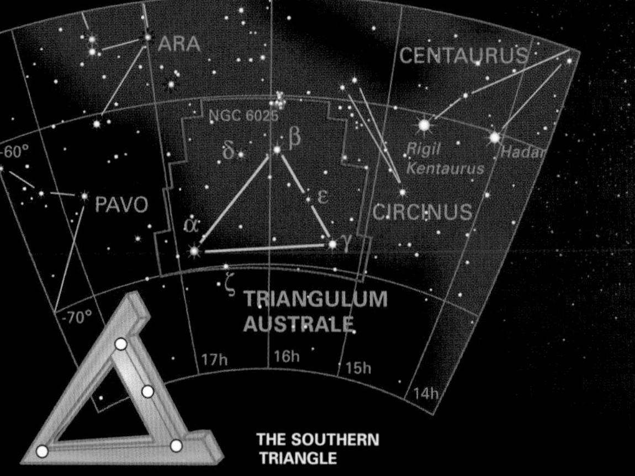

THE ALTAR

## Ara

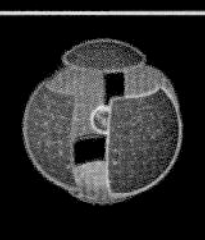

**SIZE RANKING** 63

**BRIGHTEST STARS** Alpha ($\alpha$) 2.8, Beta ($\beta$) 2.8

**GENITIVE** Arae

**ABBREVIATION** Ara

**HIGHEST IN SKY AT 10PM** June–July

**FULLY VISIBLE** 22°N–90°S

Ara was visualized by the ancient Greeks as the altar on which the gods of Olympus swore an oath of allegiance before their battle with the Titans for control of the Universe (see panel, left). This southern constellation lies within the Milky Way and is situated south of Scorpius.

MYTHS AND STORIES

### TITANOMACHIA

Titanomachia, or the Clash of the Titans, was the ten-year war for dominance of the Universe between the gods on Mount Olympus, led by Zeus, and the Titans on Mount Othrys. In gratitude for their victory, Zeus placed the altar of the gods in the sky.

**VICTORY PANEL**
Part of the battle of the gods and Titans is here depicted in the Zeus Altar of Pergamon, which was sculpted in Greece *c.*180 BC.

### SPECIFIC FEATURES

The attractive open cluster NGC 6193 consists of about 30 stars of 6th magnitude and fainter. It can be viewed through binoculars.

NGC 6397 is among the closest globular clusters to us, being around 10,000 light-years away, and is well seen through binoculars or a small telescope. Like NGC 6193, it appears relatively large – both being over half the apparent width of the full Moon.

Ara contains no stars of particular interest to users of small telescopes.

**NGC 6397**
The globular cluster NGC 6397 has a condensed centre and scattered outer regions in which chains and sprays of stars can be traced.

**INCENSE BURNER**
Ara, the celestial altar, is oriented with its top facing south. Incense burning on the altar might give off the "smoke" of the Milky Way above it.

THE SOUTHERN CROWN

## Corona Australis

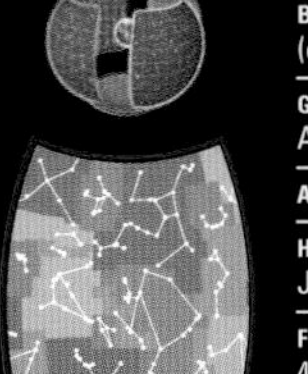

**SIZE RANKING** 80

**BRIGHTEST STARS** Alpha ($\alpha$) 4.1, Beta ($\beta$) 4.1

**GENITIVE** Coronae Australis

**ABBREVIATION** CrA

**HIGHEST IN SKY AT 10PM** July–August

**FULLY VISIBLE** 44°N–90°S

The small southern constellation of Corona Australis lies under the feet of Sagittarius. It comprises stars of 4th magnitude and fainter, and it was one of the 48 constellations recognized by the ancient Greek astronomer Ptolemy (see p.85).

### SPECIFIC FEATURES

Gamma ($\gamma$) Coronae Australis is a binary star with components of 5th magnitude. The pair orbit each other every 122 years, and they are slowly moving apart as seen from Earth. This means the components are becoming easier to view individually. Meanwhile, a 100mm (4in) aperture is needed to divide this challenging star.

Kappa ($\kappa$) Coronae Australis is an unrelated double with components of 6th magnitude, which are readily divided through a small telescope.

The modest globular cluster NGC 6541 covers about one-third the apparent diameter of the full Moon. It is visible through a small telescope or binoculars.

**SOUTHERN ARC**
Corona Australis is an attractive arc of stars that represents a crown or laurel wreath.

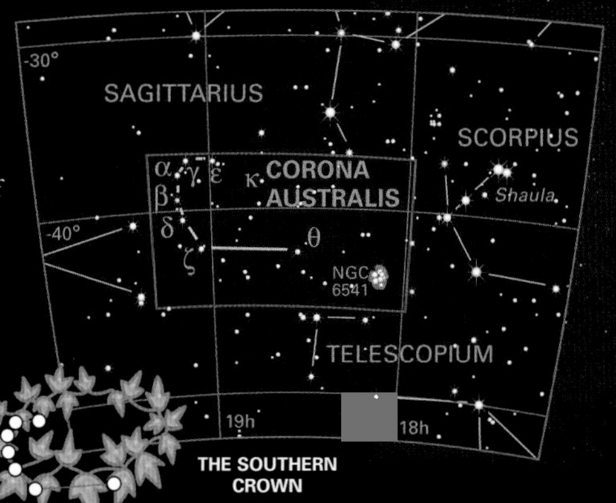

THE TELESCOPE

## Telescopium

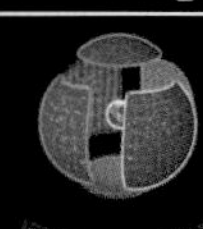
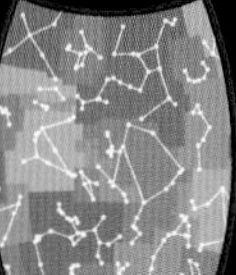

SIZE RANKING 57

BRIGHTEST STAR Alpha (α) 3.5

GENITIVE Telescopii

ABBREVIATION Tel

HIGHEST IN SKY AT 10PM July–August

FULLY VISIBLE 33°N–90°S

Telescopium is an almost entirely undistinguished southern constellation near Sagittarius and Corona Australis. It was invented by the French astronomer Nicolas Louis de Lacaille (see p.406) to commemorate the telescope. Its pattern of stars represents one of the aerial telescopes used at the Paris observatory. These were refractors with extremely long focal lengths – to reduce chromatic aberration – suspended from tall poles by ropes and pulleys.

SPECIFIC FEATURES

Delta (δ) Telescopii is an unrelated pair of stars with components of 5th-magnitude. It can be divided with binoculars or even good eyesight.

**LONG VIEW**
Telescopium depicts an early design of refracting telescope with a long tube supported by a flimsy mounting – a far cry from the massive reflectors of today.

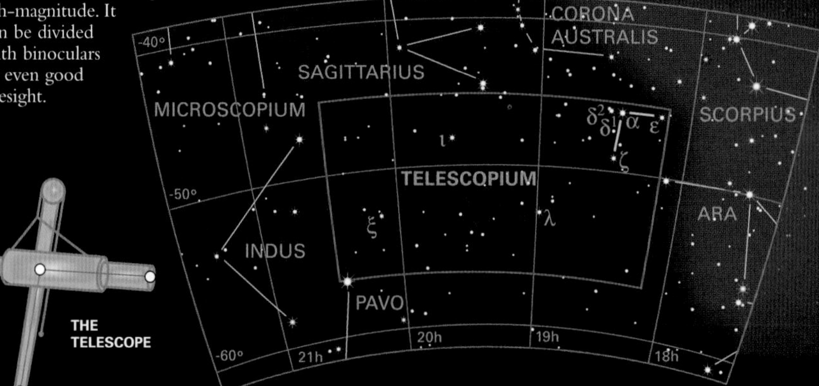


THE INDIAN

## Indus

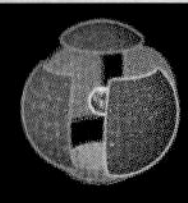

SIZE RANKING 49

BRIGHTEST STAR Alpha (α) 3.1

GENITIVE Indi

ABBREVIATION Ind

HIGHEST IN SKY AT 10PM August–October

FULLY VISIBLE 15°N–90°S

This southern constellation was introduced in the late 16th century by Pieter Dirkszoon Keyser and Frederick de Houtman (see panel, right). It represents a human figure with a spear and arrows, although it remains unclear whether this is supposed to be a native of the East Indies (as discovered by the Dutch explorers during their expeditions) or a native of the Americas.

SPECIFIC FEATURES

Fifth-magnitude Epsilon (ε) Indi is one of the closest stars to us, being 11.8 light-years away. Somewhat smaller and cooler than the Sun, it appears pale orange in colour.

Theta (θ) Indi is a 4th-magnitude star with a companion of 7th magnitude that can be identified through a small telescope.

**CONCEALED FIGURE**
Only a vivid imagination could discern the figure of a human in the constellation of Indus, which comprises a few faint stars next to the distinctive figures of Grus and Tucana.

EXPLORING SPACE

### DUTCH VOYAGES OF DISCOVERY

As well as exploring the southern oceans, Dutch traders and navigators also charted the southern sky. On the first Dutch expedition to the East Indies in 1595 were two Dutch navigator–astronomers, Pieter Dirkszoon Keyser (c.1540–96) and Frederick de Houtman (1571–1627). Keyser died during the voyage, but his celestial observations, along with those of de Houtman, were returned to the Dutch cartographer Petrus Plancius (see p.342) and formed the basis for 12 new constellations, all of which are still recognized.

**FAMILY OF EXPLORERS**
The first Dutch expedition to the East Indies consisted of four ships and was led by Cornelis de Houtman, the brother of Frederick, who was on the trip as a navigator.

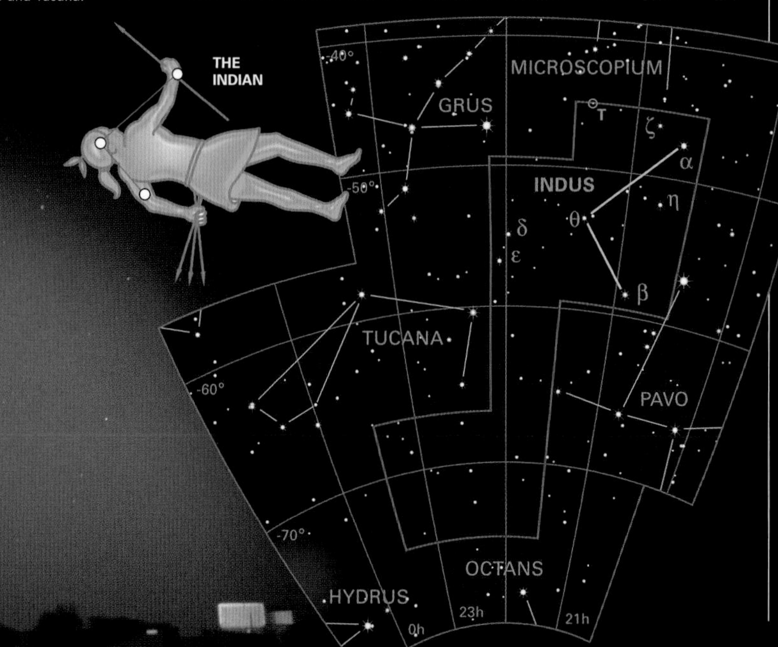

THE CRANE

# Grus

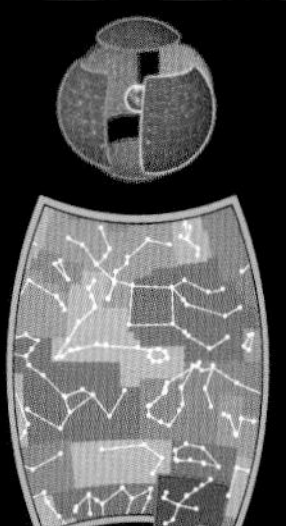

**SIZE RANKING** 45

**BRIGHTEST STAR** Alnair (α) 1.7

**GENITIVE** Gruis

**ABBREVIATION** Gru

**HIGHEST IN SKY AT 10PM** September–October

**FULLY VISIBLE** 33°N–90°S

Grus represents a long-necked wading bird – a crane – although it has also been depicted as a flamingo. It is a constellation of the southern sky and is situated between Piscis Austrinus and Tucana. Grus was introduced at the end of the 16th century by the Dutch navigator–astronomers Pieter Dirkszoon Keyser and Frederick de Houtman (see panel, opposite).

SPECIFIC FEATURES

Delta (δ) Gruis is a pair of 4th-magnitude giants, with one yellow component and one red one, while Mu (μ) Gruis is a pair of 5th-magnitude yellow giants. Both pairs are divisible with the naked eye. They appear double due to chance alignments and are not true binaries.

Beta (β) Gruis is a red giant whose brightness ranges from magnitude 2.0 to 2.3, with no set period.

**SHOWING THE WAY** Two wide doubles – Delta (δ) and Mu (μ) Gruis – appear along the extended neck of Grus, the Crane, which points to the lower right in this image.

THE PHOENIX

# Phoenix

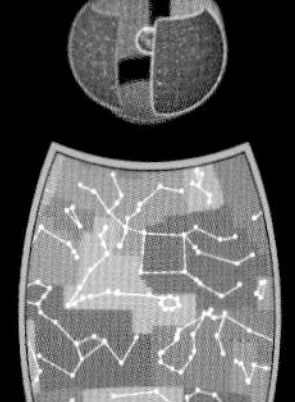

**SIZE RANKING** 37

**BRIGHTEST STAR** Ankaa (α) 2.4

**GENITIVE** Phoenicis

**ABBREVIATION** Phe

**HIGHEST IN SKY AT 10PM** October–November

**FULLY VISIBLE** 32°N–90°S

Phoenix lies at the southern end of Eridanus, next to that constellation's brightest star, Achernar. It is the largest of the 12 southern constellations introduced during the late 16th century by the Dutch navigator–astronomers Pieter Dirkszoon Keyser and Frederick de Houtman (see panel, opposite). It represents the mythical bird that was supposedly born from the ashes of its predecessor (see panel, right).

SPECIFIC FEATURES

Zeta (ζ) Phoenicis is a variable double consisting of a 4th-magnitude star with an 8th-magnitude companion. The brighter star is an eclipsing binary and varies between magnitudes 3.9 and 4.4 every 1.7 days.

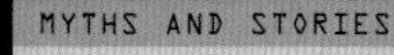

MYTHS AND STORIES

## MYTHICAL BIRD

According to legend, the phoenix was said to live for 500 years. At the end of its life span, it built a nest of cinnamon bark and incense on which it died, some say in fire. A baby phoenix was born from its ancestor's remains. The death and rebirth of the phoenix has been seen as symbolic of the daily rising and setting of the Sun.

**FUNERAL PYRE** The phoenix is consumed by fire in this 18th-century German copper engraving from *Bilderbuch für Kinder*.

**PHOENIX FALLING** The stars of Phoenix sink towards the western horizon in the morning sky, with Grus below it. North is to the right in this photograph.

THE TOUCAN

# Tucana

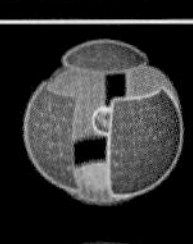

**SIZE RANKING** 48

**BRIGHTEST STAR** Alpha (α) 2.9

**GENITIVE** Tucanae

**ABBREVIATION** Tuc

**HIGHEST IN SKY AT 10PM** September–November

**FULLY VISIBLE** 14°N–90°S

This far-southern constellation is to be found at the end of the celestial river, Eridanus. It represents the large-beaked tropical bird that is native to South and Central America.

Tucana was introduced in the late 16th century by the Dutch navigator–astronomers Pieter Dirkszoon Keyser and Frederick de Houtman (see p.400).

## SPECIFIC FEATURES

Tucana contains the Small Magellanic Cloud (see p.301), the lesser of the two satellite galaxies that accompany our own galaxy. To the naked eye, it appears like a detached patch of the Milky Way and is seven times wider than the apparent diameter of the full Moon. Star fields and clusters within the Small Magellanic Cloud can be detected through binoculars or a small telescope. It is about 190,000 light-years away.

Two globular clusters lie near the Small Magellanic Cloud, although both are actually foreground objects in our galaxy and so are not associated with the Cloud. The more prominent of the two is 47 Tucanae (see p.288), which looks like a hazy 4th-magnitude star to the naked eye. It apparently covers the same area of sky as the full Moon when viewed through binoculars or a small telescope. In the entire sky, only Omega (ω) Centauri is a more impressive globular cluster than 47 Tucanae. NGC 362, the other globular cluster in Tucana, is smaller and fainter and requires binoculars or a small telescope to be seen.

Beta (β) Tucanae is a naked-eye or binocular double with stars of 4th and 5th magnitudes. The brighter component can be further separated through a telescope. Kappa (κ) Tucanae, near NGC 362, is a double star of 5th and 7th magnitudes divisible through a small telescope.

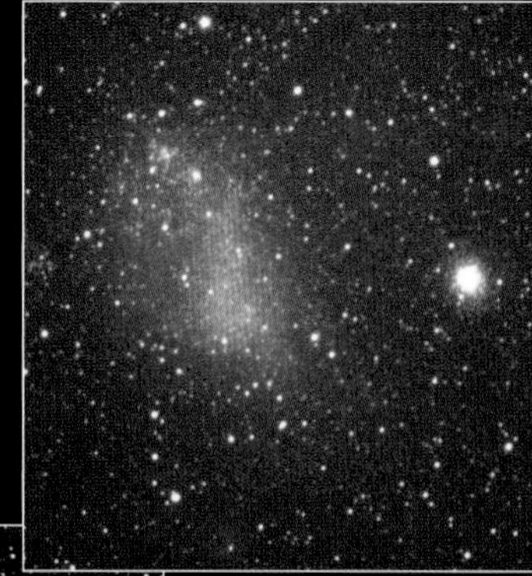

**THE SMC** This neighbouring mini-galaxy, the Small Magellanic Cloud, appears noticeably elongated. To its right in this image lies 47 Tucanae, or NGC 104, a globular cluster in our galaxy.

**47 TUCANAE** This bright globular cluster looks like a fuzzy star on wide-angle photographs like the one above right, but telescopes reveal it to be an immense swarm of stars.

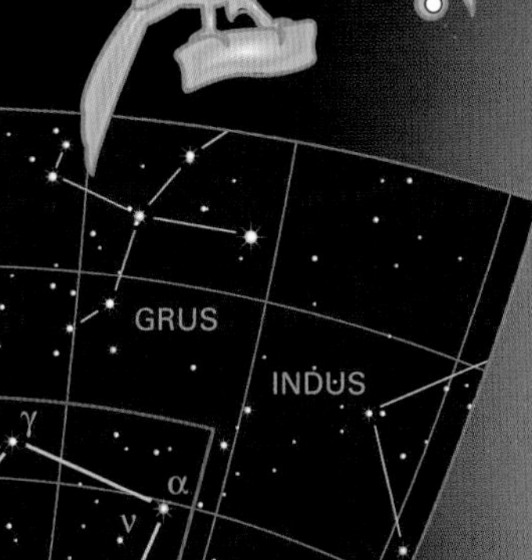


**BIRD OF THE SOUTHERN SKIES** The Toucan's huge beak points downwards as the constellation sets towards the western horizon. North is to the right in this picture.

THE LITTLE WATER SNAKE

# Hydrus

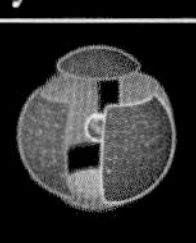

**SIZE RANKING** 61

**BRIGHTEST STAR** Beta (β) 2.8

**GENITIVE** Hydri

**ABBREVIATION** Hyi

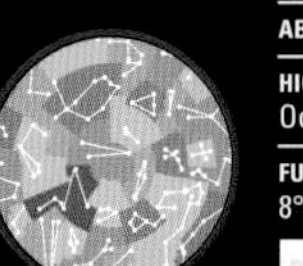

**HIGHEST IN SKY AT 10PM** October–December

**FULLY VISIBLE** 8°N–90°S

Hydrus was introduced in the late 16th century by the Dutch navigator–astronomers Pieter Dirkszoon Keyser and Frederick de Houtman (see p.400). It is a constellation of the far-southern sky and is situated between the Large Magellanic Cloud (see p.300) and the Small Magellanic Cloud (see p.301).

This constellation represents a small water snake. It should not be confused with the larger constellation Hydra, also identified as a water snake, which has been recognized since the time of the ancient Greeks.

SPECIFIC FEATURES

Pi (π) Hydri is a wide double of 6th-magnitude red giants, although they lie at different distances from us and hence are unrelated. It can be split readily through binoculars.

Pi-1 ($\pi^1$) is of magnitude 5.6 and is to be found about 740 light-years away. Pi-2 ($\pi^2$) lies much closer to us, being about 470 light-years away; it is of magnitude 5.7.

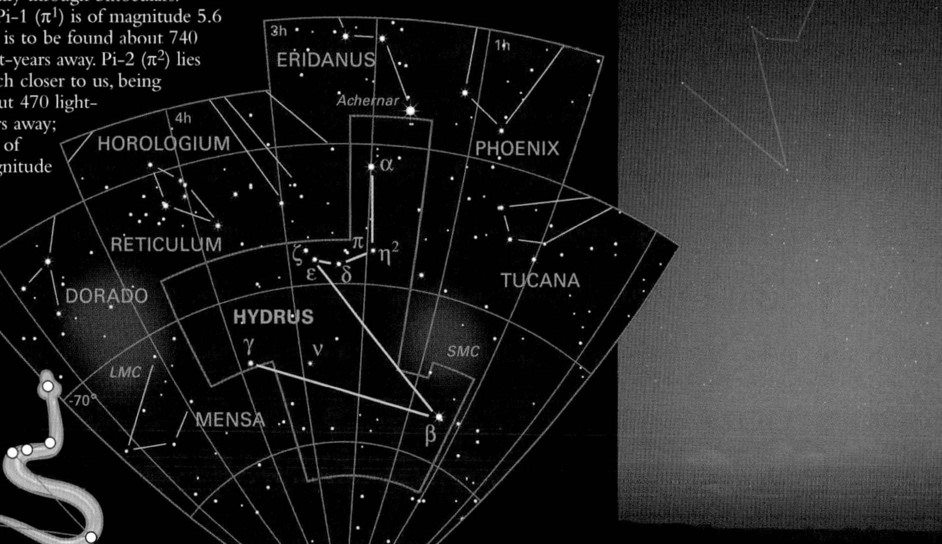


**HYDRUS AND ACHERNAR**
The sinuous little water snake winds its way across southern skies between the two Magellanic Clouds. The brightest star near it is Achernar in Eridanus (top, right.)

THE PENDULUM CLOCK

# Horologium

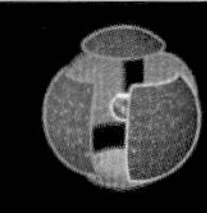

**SIZE RANKING** 58

**BRIGHTEST STAR** Alpha (α) 3.9

**GENITIVE** Horologii

**ABBREVIATION** Hor

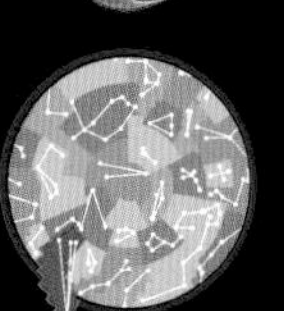

**HIGHEST IN SKY AT 10PM** November–December

**FULLY VISIBLE** 23°N–90°S

Horologium represents a pendulum clock, as used in observatories. Some depictions show its brightest star, Alpha (α) Horologii, marking the clock's pendulum (as in the illustration here), while others include it as one of the clock weights.

This faint and unremarkable constellation of the southern sky lies near the foot of Eridanus and was introduced by the French astronomer Nicolas Louis de Lacaille (see p.406).

SPECIFIC FEATURES

R Horologii is a red-giant variable star of the same type as Mira (in Cetus). It ranges between 5th and 14th magnitudes every 13 months or so.

**NGC 1261**
The best deep-sky object in Horologium for amateur instruments is NGC 1261, a compact globular cluster of 8th magnitude more than 50,000 light-years from us.

NGC 1261 is a modest globular cluster dimly detectable through a small telescope.

Arp–Madore 1 (AM1) is another globular cluster of note within the constellation Horologium. It is the most distant known globular cluster from the Sun, being nearly 400,000 light-years away. Because it is of 16th magnitude, a large telescope is needed to detect it.

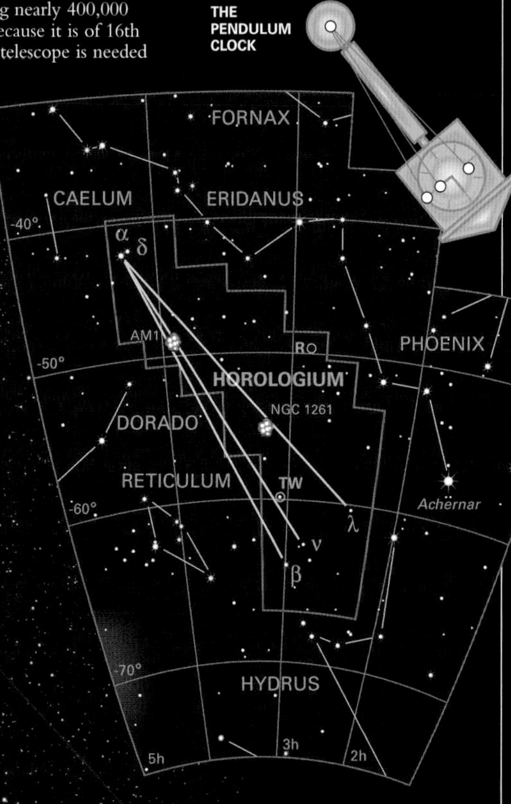


**STELLAR CLOCK**
The shape of Horologium is reminiscent of a clock with a long pendulum – unlike many of the shapeless constellations invented by de Lacaille.

THE NET

# Reticulum

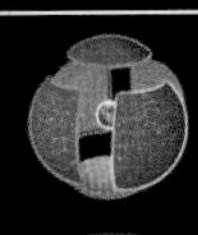

**SIZE RANKING** 82

**BRIGHTEST STAR** Alpha (α) 3.3

**GENITIVE** Reticuli

**ABBREVIATION** Ret

**HIGHEST IN SKY AT 10PM** December

**FULLY VISIBLE** 23°N–90°S

Reticulum is a small constellation in the southern sky, near the Large Magellanic Cloud (see p.300). It was introduced by the French astronomer Nicolas Louis de Lacaille (see p.406) and represents the reticule, or grid, in his eyepiece, which he used for measuring star positions.

SPECIFIC FEATURES

Zeta (ζ) Reticuli is a yellow double star. Its 5th-magnitude components can be split through binoculars.

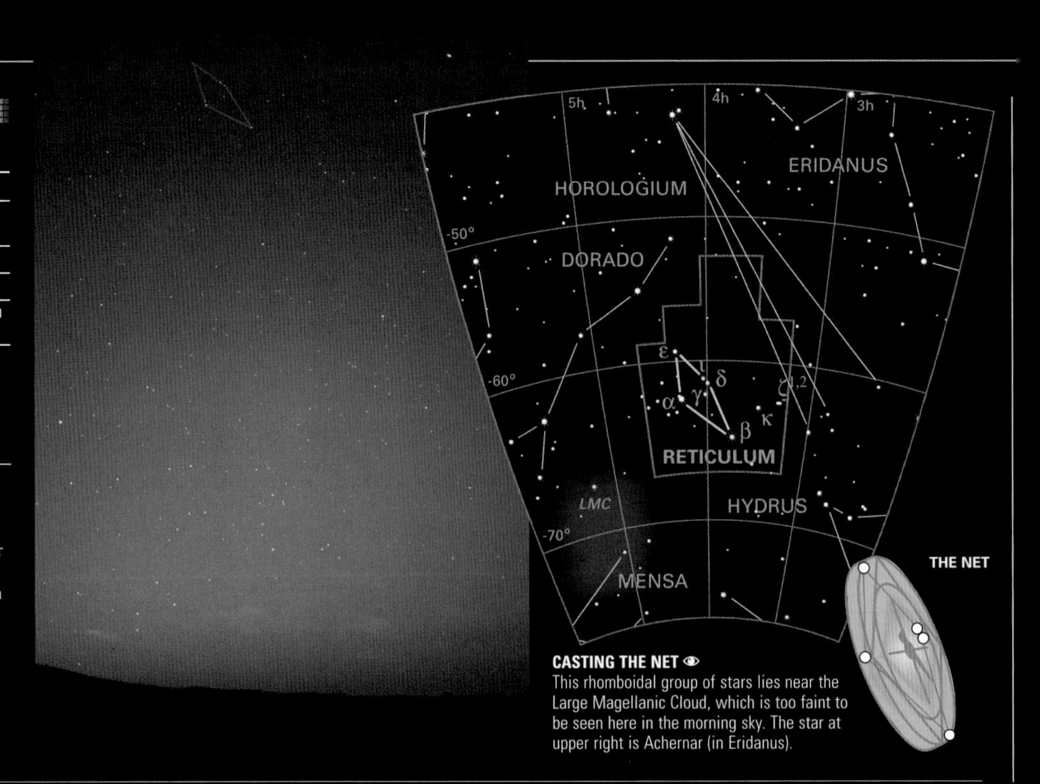


**CASTING THE NET** This rhomboidal group of stars lies near the Large Magellanic Cloud, which is too faint to be seen here in the morning sky. The star at upper right is Achernar (in Eridanus).

THE PAINTER'S EASEL

# Pictor

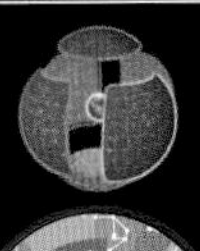

**SIZE RANKING** 59

**BRIGHTEST STAR** Alpha (α) 3.2

**GENITIVE** Pictoris

**ABBREVIATION** Pic

**HIGHEST IN SKY AT 10PM** December–February

**FULLY VISIBLE** 26°N–90°S

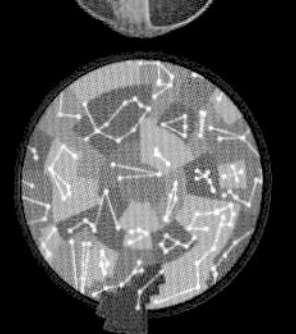

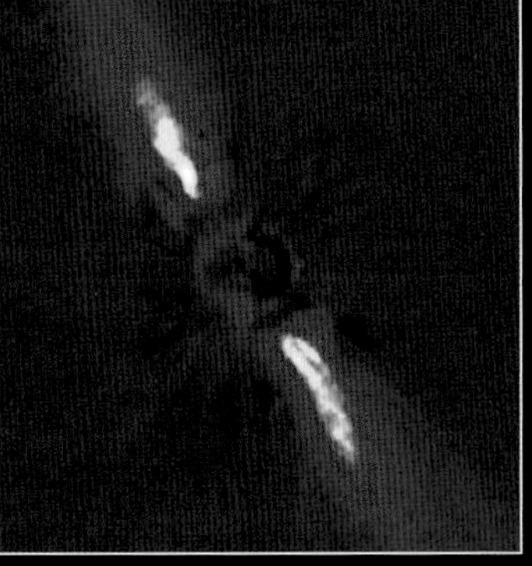

**BETA PICTORIS** The bright areas on this professional false-colour image indicate the circumstellar disc. Distortions in the shape may be due to a planetary system forming around the star.

Pictor was invented by the French astronomer Nicolas Louis de Lacaille (see p.406), who imagined it as an artist's easel, complete with palette. He originally called it *Equuleus Pictoris*, although that name has since been shortened. It is a faint constellation of the southern sky, and it is situated beside the constellations Puppis and Columba.

SPECIFIC FEATURES

Beta (β) Pictoris is 63 light-years away. It is of special interest because, in 1984, astronomers discovered a disc of dust and gas orbiting this blue-white star of magnitude 3.9. The circumstellar disc is thought to be a planetary system in the process of formation. The planets of our solar system are believed to have developed from a similar disc that existed around the Sun shortly after its formation.

Iota (ι) Pictoris is a double star with components of 6th magnitude. These are readily separated through a small telescope.

**DIVIDING LINE** Pictor consists of little more than a crooked line of stars between brilliant Canopus (in Carina), seen here on the left, and the Large Magellanic Cloud.

THE GOLDFISH

# Dorado

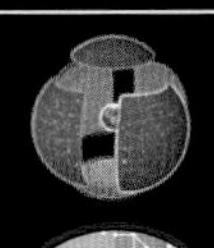

**SIZE RANKING** 72

**BRIGHTEST STAR** Alpha (α) 3.3

**GENITIVE** Doradus

**ABBREVIATION** Dor

**HIGHEST IN SKY AT 10PM** December–January

**FULLY VISIBLE** 20°N–90°S

Dorado is one of the southern constellations introduced in the late 16th century by the Dutch navigator–astronomers Pieter Dirkszoon Keyser and Frederick de Houtman (p.400). Although known as the goldfish, Dorado in fact represents the dolphinfish found in tropical waters, and not the fish common to aquaria and ponds. The constellation has also been depicted as a swordfish.

Most of the Large Magellanic Cloud (see p.300) is contained within Dorado, although this mini-galaxy also extends into Mensa. The first recorded mention of the Large Magellanic Cloud is credited to al-Sufi (see panel, below).

## SPECIFIC FEATURES

The Large Magellanic Cloud is a satellite galaxy of the Milky Way. It is situated some 170,000 light-years away from the Earth and, at first sight, looks like a detached part of the Milky Way. Its numerous star clusters and nebulous patches are brought into view through binoculars or a small telescope.

A remarkable object in the Large Magellanic Cloud is the Tarantula Nebula, or NGC 2070. It is bright enough to be visible with the naked eye and can be well seen through binoculars. A cluster of newborn stars at the heart of the Tarantula Nebula can be detected through binoculars or a small telescope, while photographs show its looping extremities, like a spider's legs, from which this large nebula of glowing gas gets its popular name.

In February 1987 a supernova flared up in the Large Magellanic Cloud. Supernova 1987A, as it was called, reached 3rd magnitude in May of that year, and this made it the brightest supernova visible from Earth since 1604. It remained visible to the naked eye for 10 months.

Beta (β) Doradus is one of the brightest Cepheid variables, ranging between magnitudes 3.5 and 4.1 every 9.8 days, while R Doradus is an erratic red giant that varies from 5th to 6th magnitude every 11 months or so.

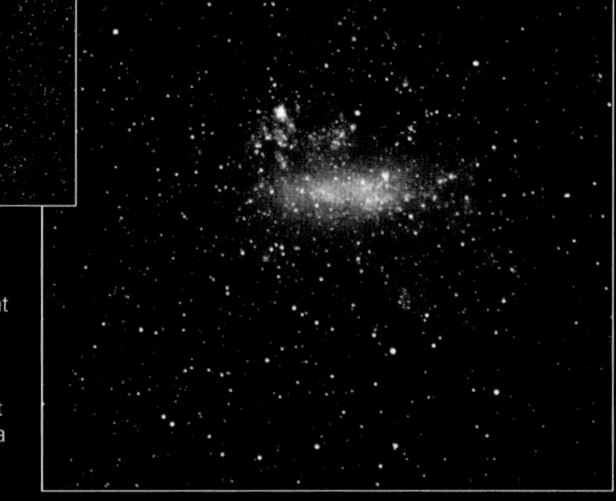

**SUPERNOVA 1987A** This supernova has faded since its dramatic flare-up in 1987. To its upper left in this image is the spider-like Tarantula Nebula.

**THE LMC** The brighter of the two mini-galaxies that accompany our own, the Large Magellanic Cloud appears elongated in shape. It includes the Tarantula Nebula (here on its upper-left edge).

**HEADING SOUTH** Dorado, the Goldfish, swims through the southern skies, apparently on its way to the south celestial pole.

## AL-SUFI

Abd al-Rahman al-Sufi (903–86), known also by his Latinized name, Azophi, was an Arabic astronomer. Around AD 964, he produced the *Book of the Fixed Stars* – an updated version of Ptolemy's *Almagest* – which introduced many star names still in use today. Later editions of the book contained Arabic illustrations of the constellations (like the one below).

**CONSTELLATION PORTRAIT**
A version of al-Sufi's *Book of the Fixed Stars* was produced in the 16th century by a Persian artist. It included this image of Boötes.

THE FLYING FISH

## Volans

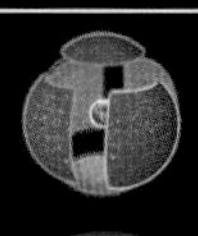

SIZE RANKING 76

BRIGHTEST STARS Beta (β) 3.8, Gamma (γ) 3.8

GENITIVE Volantis

ABBREVIATION Vol

HIGHEST IN SKY AT 10PM January–March

FULLY VISIBLE 14°N–90°S

This small and faint constellation of the southern sky between Carina and the Large Magellanic Cloud (see p.300) was introduced in the late 16th century by the Dutch navigator–astronomers Pieter Dirkszoon Keyser and Frederick de Houtman (see p.400). It represents the tropical fish that uses its outstretched fins as wings to glide through the air.

SPECIFIC FEATURES

Although it lies on the edge of the Milky Way, Volans is surprisingly bereft of deep-sky objects. It does, however, contain two good double stars, one of them being 4th-magnitude Gamma (γ) Volantis, which is jointly the brightest star in the constellation. This orange star has a yellow companion, of 6th magnitude. They form a beautiful double when viewed through a small telescope.

Epsilon (ε) Volantis is another interesting double, although it is not as colourful as Gamma. Its components, which are of 4th and 7th magnitudes, can be detected readily through a small telescope.

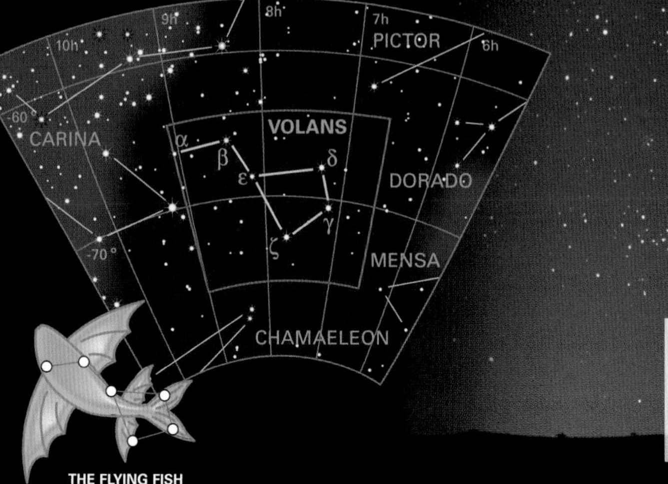


**FISH IN FLIGHT** The Flying Fish leaps into the evening sky above the eastern horizon. Beneath it here are the Milky Way and the stars of Carina and Vela, with the False Cross at left.

THE TABLE MOUNTAIN

## Mensa

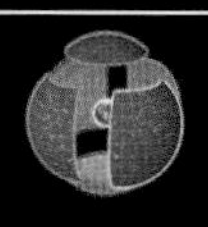

SIZE RANKING 75

BRIGHTEST STAR Alpha (α) 5.1

GENITIVE Mensae

ABBREVIATION Men

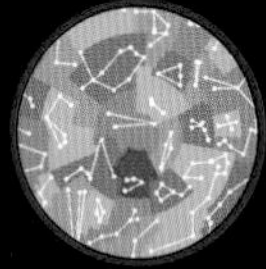

HIGHEST IN SKY AT 10PM December–February

FULLY VISIBLE 5°N–90°S

The French astronomer Nicolas Louis de Lacaille (see panel, right) introduced this constellation. It commemorates Table Mountain near the modern Cape Town, South Africa, which is close to where he set up his observatory. When viewing the wispy appearance of the Large Magellanic Cloud (see p.300) in Mensa, de Lacaille may have recalled the clouds sometimes seen over the real Table Mountain. It is the only constellation that de Lacaille did not name after a scientific or artistic tool.

Mensa is the faintest of all 88 constellations, and its brightest star, Alpha (α) Mensae, is of only 5th magnitude. Its main point of interest is that part of the Large Magellanic Cloud overlaps into it from neighbouring Dorado. Other than this Cloud, there is nothing to attract the casual observer to this small constellation of the south-polar region of the sky.

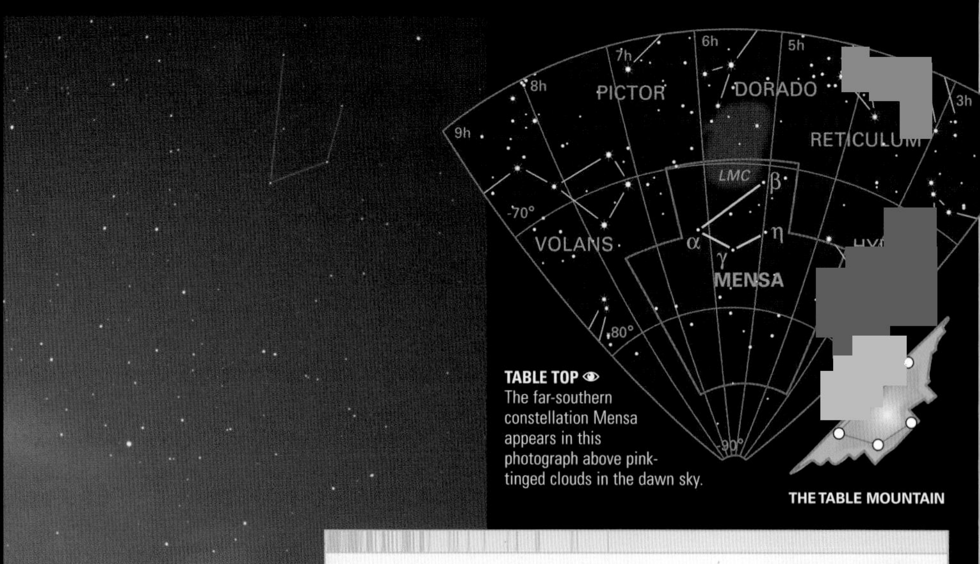


**TABLE TOP** The far-southern constellation Mensa appears in this photograph above pink-tinged clouds in the dawn sky.

### NICOLAS LOUIS DE LACAILLE

This French astronomer charted the southern skies in 1751–52 from Cape Town, South Africa. Nicolas Louis de Lacaille (1713–62) observed the positions of nearly 10,000 stars, producing a catalogue and star chart on which he introduced 14 new constellations. Most of these represented instruments of the arts and sciences.

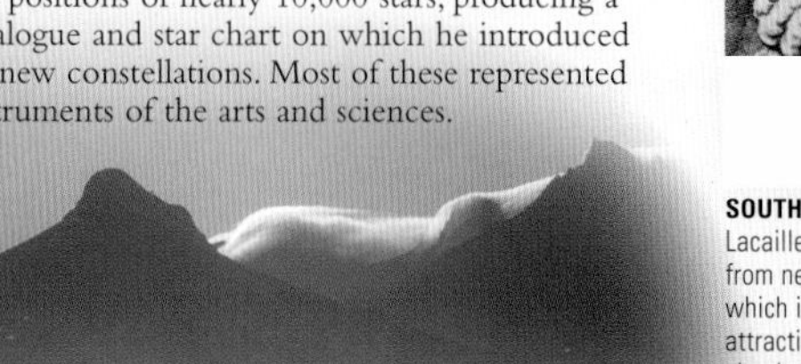

**SOUTHERN VIEWPOINT** Lacaille observed the stars from near Table Mountain, which is covered by an attractive "tablecloth" of clouds in this photograph.

THE CHAMELEON

# Chamaeleon

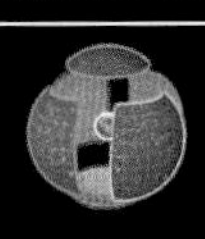

**SIZE RANKING** 79

**BRIGHTEST STARS** Alpha (α) 4.1, Gamma (γ) 4.1

**GENITIVE** Chamaeleontis

**ABBREVIATION** Cha

**HIGHEST IN SKY AT 10PM** February–May

**FULLY VISIBLE** 7°N–90°S

Chamaeleon was named after the lizard that can change its skin colour to match its surroundings. It is a small, faint constellation of the south-polar region of the sky.

The constellation was introduced at the end of the 16th century by the Dutch navigator–astronomers Pieter Dirkszoon Keyser and Frederick de Houtman (see p.400).

SPECIFIC FEATURES

Delta (δ) Chamaeleontis is a wide pair of unrelated stars of 4th and 5th magnitudes. They are easily seen through binoculars.

NGC 3195 is a planetary nebula of similar apparent size to Jupiter, but it is relatively faint and so requires a moderate-sized telescope to be seen.

THE CHAMELEON

**CAMOUFLAGE ARTIST** 👁
Chamaeleon lies close to the south celestial pole, which is to the left of it in this picture. To the north of this constellation are to be found the rich Milky Way star fields of Carina.

THE BIRD OF PARADISE

# Apus

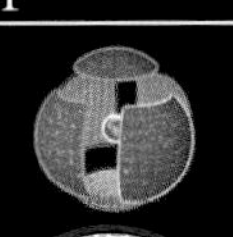

**SIZE RANKING** 67

**BRIGHTEST STAR** Alpha (α) 3.8

**GENITIVE** Apodis

**ABBREVIATION** Aps

**HIGHEST IN SKY AT 10PM** May–July

**FULLY VISIBLE** 7°N–90°S

The constellation Apus is situated in the almost featureless area around the south celestial pole. It was invented in the late 16th century by the Dutch navigator–astronomers Pieter Dirkszoon Keyser and Frederick de Houtman (see p.400).

SPECIFIC FEATURES

Delta (δ) Apodis is a wide pair of unrelated 5th-magnitude red giants, while Theta (θ) Apodis is a red giant that varies somewhat erratically between 5th and 7th magnitudes every 4 months or so.

**EXOTIC BIRD** 👁
Apus, which is south of the distinctive Triangulum Australe, represents a bird of paradise but is a disappointing tribute to such an exotic bird.

THE PEACOCK

# Pavo

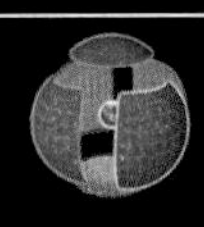

**SIZE RANKING** 44

**BRIGHTEST STAR**
Peacock (α) 1.9

**GENITIVE** Pavonis

**ABBREVIATION** Pav

**HIGHEST IN SKY AT 10PM**
July–September

**FULLY VISIBLE**
15°–90°S

Pavo is one of the far-southern constellations that were introduced at the end of the 16th century by the Dutch navigator–astronomers Pieter Dirkszoon Keyser and Frederick de Houtman (see p.400). It represents the peacock of southeast Asia, which the Dutch explorers encountered on their travels. In more recent times, its brightest star, 2nd-magnitude Alpha (α) Pavonis, was given the name Peacock.

In Greek mythology, the peacock was the sacred bird of Hera, wife of Zeus, who travelled through the air in a chariot drawn by these birds. It was Hera who placed the markings on the tail of the peacock after an episode involving Zeus and one of his illicit loves, Io. Although Zeus had disguised Io as a white cow, Hera suspected something was amiss and set the 100-eyed Argus to keep watch on the heifer. Her husband retaliated by sending his son Hermes to release Io. In order to overcome Argus, Hermes told him tales and played music on his reed pipe until the watchman's eyes closed one by one. When Argus was finally asleep, Hermes chopped off his head and set Io free. In his memory, Hera then placed the eyes of Argus on the peacock's tail.

The constellation Pavo is to be found on the edge of the Milky Way south of Sagittarius and next to another exotic bird, the toucan (the constellation Tucana).

## SPECIFIC FEATURES

Kappa (κ) Pavonis is one of the brighter Cepheid variables. Its fluctuations, between magnitudes 3.9 and 4.8 every 9.1 days, can be followed with the naked eye.

Xi (ξ) Pavonis is a double star with components of unequal brightness – 4th and 8th magnitudes. The fainter star is difficult to identify with the smallest-apertured telescopes as its brighter neighbour overwhelms it.

NGC 6752 is one of the largest and brightest globular clusters in the sky. It is just at the limit of naked-eye visibility but readily located through binoculars. It covers half the apparent width of the full Moon. A telescope with an aperture of 75mm (3in) or more will resolve its brightest individual stars.

The large spiral galaxy NGC 6744 is presented virtually face-on to the Earth. It is visible as an elliptical haze in a telescope of small to moderate aperture. NGC 6744 lies about 30 million light-years away.

**NGC 6744**
This beautiful barred spiral galaxy in Pavo is detectable through a small telescope. The Milky Way might appear like this when viewed from the outside.

**NGC 6752**
The fine globular cluster NGC 6752 remains little-known because of its far-southern declination. The bright star seen above right of it in this image is a foreground object in our galaxy.

**CELESTIAL DISPLAY**
The constellation Pavo, the Peacock, is depicted fanning its tail across the southern skies, in imitation of a real-life peacock when attracting a mate.

THE OCTANT

# Octans

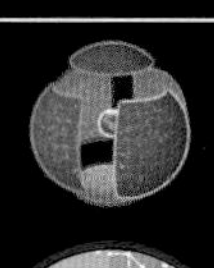

SIZE RANKING 50

BRIGHTEST STAR Nu (ν) 3.8

GENITIVE Octantis

ABBREVIATION Oct

HIGHEST IN SKY AT 10PM October

FULLY VISIBLE 0°–90°S

This constellation, which originally was also known as *Octans Nautica* or *Octans Hadleianus*, contains the south celestial pole. It was introduced in the 18th century by the French astronomer Nicolas Louis de Lacaille (see p.406).

The area of sky in which Octans lies is quite barren. Within naked-eye range, the nearest star to the south celestial pole is Sigma (σ) Octantis. It is of only magnitude 5.4 and hence far from prominent.

Because of the effect of precession (see p.60), the positions of the celestial poles are constantly changing. As a result, the south celestial pole is moving farther away from Sigma and towards the constellation of Chamaeleon. There are no bright stars in this area either, so the region of the south celestial pole will remain blank for another 1,500 years, when the pole will pass just over a degree away from 4th-magnitude Delta (δ) Chamaeleontis.

Octans represents an instrument known as an octant, which was used by navigators to help them find their position (see panel, right). It was invented by the English instrument maker John Hadley (1682–1744).

SPECIFIC FEATURES
Lambda (λ) Octantis is a double star that is divisible with a small telescope. The components are of 5th and 7th magnitudes.

**SOUTHERN STAR TRAILS**
Curving star trails, drawn out by the Earth's rotation on this long-exposure photograph, emphasize the barren nature of the area around the south celestial pole.

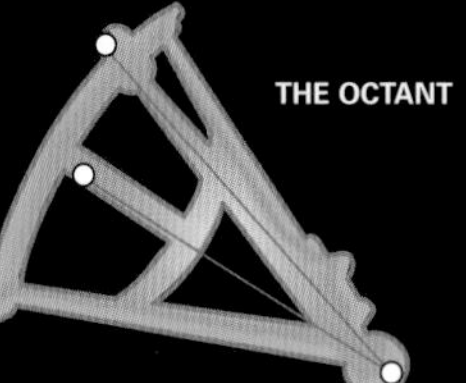

EXPLORING SPACE

## NAVIGATION

In 1731, the British mathematician John Hadley built a device called a doubly reflecting octant. The navigator sighted the horizon through a telescope and adjusted a movable arm until the reflected image of the Sun or a star overlay the direct view of the horizon. The altitude of the Sun or star could be read off a scale, from which the navigator could deduce his latitude.

**OCTANT**
This wood and brass octant is by Browning of Boston. In later designs, the arc was extended from one-eighth of a circle to one-sixth, and the octant became the modern sextant.

**AT THE POLE**
Octans comprises only a scattering of faint stars. There is no bright star to mark the southern pole, which lies centre left in this picture.

*"If the stars should appear one night in a thousand years, how would men believe and adore; and preserve for many generations the remembrance of the city of God which had been shown! But every night come out these envoys of beauty, and light the universe with their admonishing smile."*

Ralph Waldo Emerson

AS THE EARTH MAKES its year-long journey around the Sun, the night sky changes its appearance and the stars seem to move from east to west. Depending on the observer's location, some stars are circumpolar and always visible, but others are seen only at certain times of the year. For example, some stars are seen well in the evening sky in January, but are invisible six months later, when the Earth has moved around its orbit to the opposite side of the Sun. The following section tracks seasonal changes in the night sky for observers in both the northern and southern hemispheres. As well as covering the regular annual cycles of the stars and constellations, it charts the positions of the planets and provides an observer's guide to celestial events, such as meteor showers and eclipses of the Sun and Moon.

**THE LEONID METEORS**
This composite image shows the Leonid meteor shower that occurs in November each year. Also visible are the Sickle, a distinctive group of stars in the constellation Leo (top left), and the planet Jupiter (centre).

# MONTHLY SKY GUIDE

# USING THE SKY GUIDES

58–59 The celestial sphere
60–61 Celestial cycles
64–65 Planetary motion
66–67 Star motions and patterns

**THIS MONTH-BY-MONTH GUIDE** to the night sky features charts that show the whole sky as it appears from most places on the Earth's surface. It complements the **CONSTELLATIONS** section, in which detailed maps show smaller areas of sky. For each month, text, tables, and supporting charts identify good objects for observation and show the positions of the planets.

## MONTHLY HIGHLIGHTS AND PLANET LOCATORS

For each month of the year, a double-page introduction highlights different phenomena in the sky. Dates of special events, such as phases of the Moon and eclipses, are listed year-by-year in a table. The main text describes those stars, deep-sky objects, and meteor showers that feature prominently in that particular month – this text complements the whole-sky charts that follow. The introductory pages also feature a planet locator chart. This map shows the band of sky that lies either side of the ecliptic, the plane close to which the planets always appear. These charts should be used in conjunction with the extra information supplied in the Special Events table, as well as the whole-sky charts and the individual constellation entries (see pp.338–409).

**SPECIAL EVENTS**

**PHASES OF THE MOON**

| | FULL MOON | NEW MOON |
|---|---|---|
| **2011** | 19 January | 4 January |
| **2012** | 9 January | 23 January |
| **2013** | 27 January | 11 January |
| **2014** | 16 January | 1, 30 January |
| **2015** | 5 January | 20 January |
| **2016** | 24 January | 10 January |
| **2017** | 12 January | 28 January |

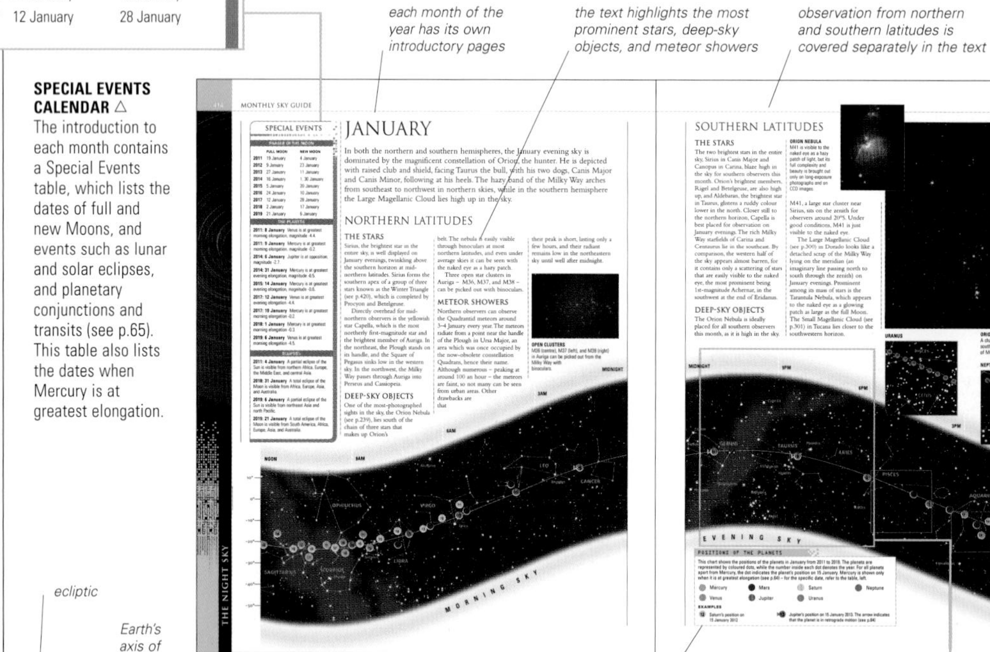
each month of the year has its own introductory pages
the text highlights the most prominent stars, deep-sky objects, and meteor showers
observation from northern and southern latitudes is covered separately in the text
photographs illustrate some of the most interesting features to be observed
JANUARY
SPECIAL EVENTS
NORTHERN LATITUDES
SOUTHERN LATITUDES
MORNING SKY
EVENING SKY
THE NIGHT SKY
NEPTUNE
AQUARIUS

**SPECIAL EVENTS CALENDAR △**
The introduction to each month contains a Special Events table, which lists the dates of full and new Moons, and events such as lunar and solar eclipses, and planetary conjunctions and transits (see p.65). This table also lists the dates when Mercury is at greatest elongation.

**△ THE OUTER PLANETS**
The two outermost planets, Uranus and Neptune, are shown on magnified insets of the main chart because they move relatively slowly through our sky.

ecliptic
Earth's axis of rotation
celestial sphere
celestial equator
position of planet shown by coloured dot
the planet locator chart shows a portion of the celestial sphere on either side of the ecliptic

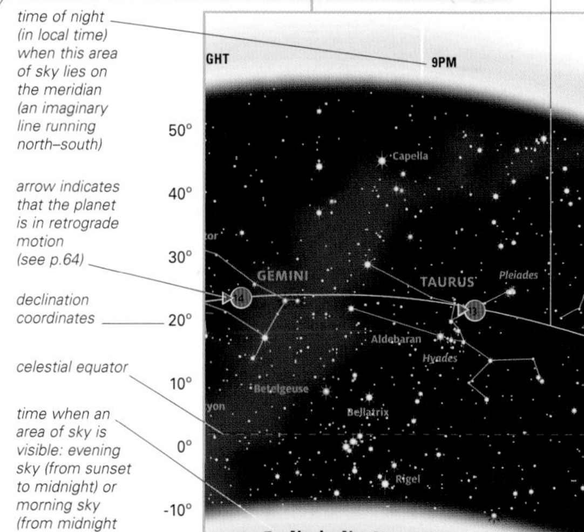
key to the coloured planet icons
time of night (in local time) when this area of sky lies on the meridian (an imaginary line running north–south)
arrow indicates that the planet is in retrograde motion (see p.64)
declination coordinates
celestial equator
time when an area of sky is visible: evening sky (from sunset to midnight) or morning sky (from midnight to sunrise)
ecliptic
GHT
9PM
6PM
50°
40°
30°
20°
10°
0°
-10°
Capella
GEMINI
TAURUS
Pleiades
ARIES
Aldebaran
Hyades
Betelgeuse
Bellatrix
Rigel
Mira
EVENING SKY

**PLANET LOCATOR CHARTS △**
These charts show the positions of the planets at 10pm local standard time on the 15th day of the month. Each planet is represented by a differently coloured dot, and the number inside the dot refers to a particular year. Each chart shows the planets' positions in relation to the 13 constellations along the ecliptic (see p.61), the area in which the planets are always found.

**THE INNER PLANETS ▷**
The six planets closest to the Sun are represented on the main body of the chart. Bands along the top and bottom indicate in local time when that area of sky is highest in the sky. However, local sunset and sunrise times will affect the darkness of the sky, and thus the visibility of the planets.

# THE WHOLE-SKY CHARTS

The introduction to each month is followed by two whole-sky charts. These show the position of the stars at 10pm local time on the 15th day of the month, for both the northern and southern hemispheres. They project the half of the celestial sphere (see pp.58–59) that would be visible to a viewer under perfect conditions – that is, without any obstruction to the horizon. Any given star rises four minutes earlier each night compared to the previous night. Thus, the night sky changes subtly from one night to the next and even more dramatically from one month to another. To use the whole-sky charts, determine the colour-coded horizon and zenith for your location (below), turn to the appropriate month, and position yourself and the whole-sky chart (right).

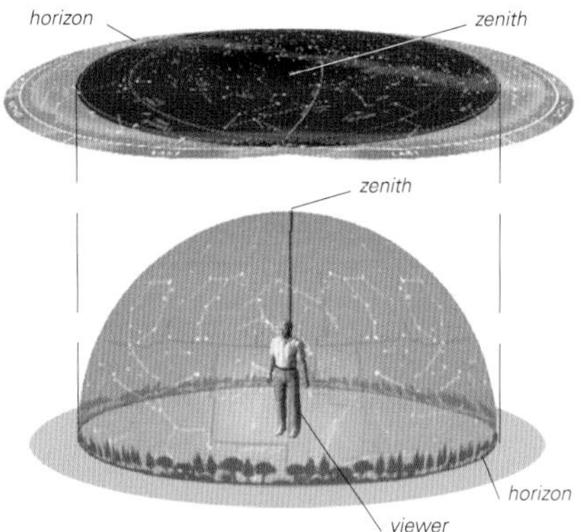


△ **CELESTIAL SPHERE**
Each whole-sky chart shows an area that equals more than half a celestial sphere because it combines three different projections of the night sky, as seen from three different latitudes on Earth. Each month, the sky charts show the night sky as it appears from 60°–20°N, on the northern hemisphere chart, and from 0°–40°S, on the southern hemisphere chart.

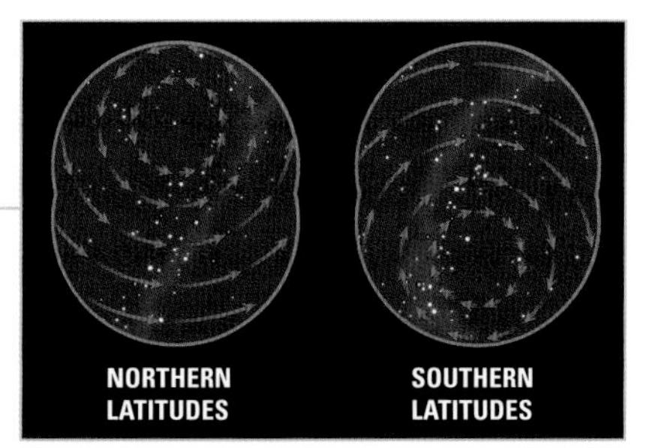


◁ **STAR-MOTION DIAGRAMS**
These diagrams show the direction in which the stars appear to move as the night progresses. Stars near the equator appear to move from east to west, while circumpolar stars circle around the celestial poles without setting.

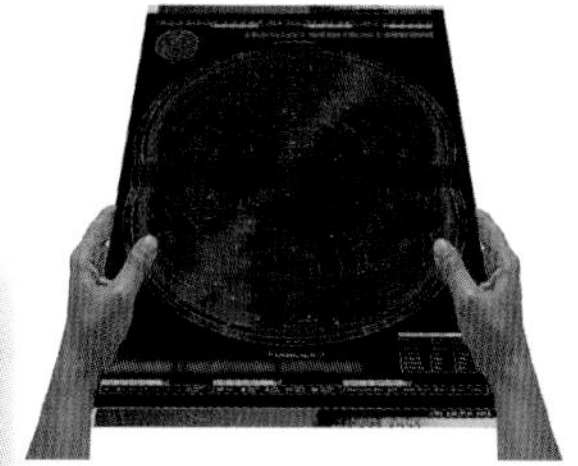

△ **ORIENTATION**
To view the sky to the north, turn northwards and hold the map flat, with the north label closest to your body. One of the colour-coded lines around the near edge of the map will relate to the horizon in front of you. To view the south, turn around and reposition the map.

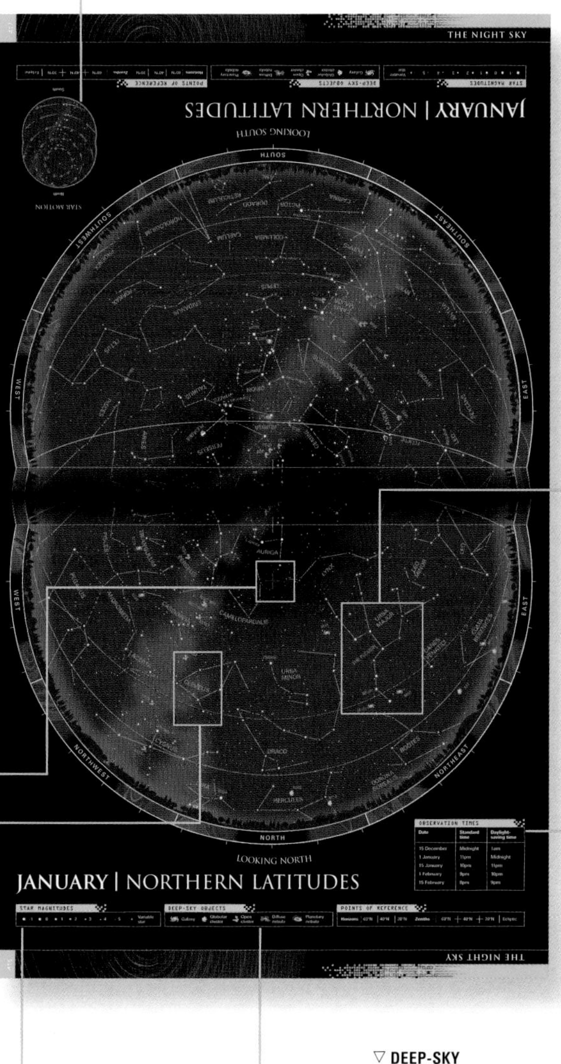


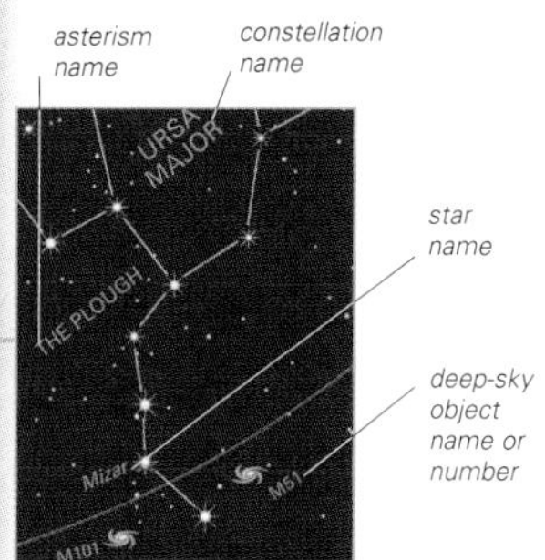


△ **MAIN FEATURES**
All 88 constellations are featured on the whole-sky charts, as well as any notable deep-sky objects within their boundaries. Well-known and easily recognizable stars, star clusters, and asterism patterns (see p.68) are also labelled.

**HORIZONS AND ZENITHS** ▷
The stars located near the centre of each chart can be seen on the zenith (the point directly overhead), while the stars near the chart's edge appear close to the horizon. Colour-coded lines and crosses are used to identify the horizon and zenith on each of the three latitude projections on each monthly chart.

OBSERVATION TIMES

| Date | Standard time | Daylight-saving time |
|---|---|---|
| 15 December | Midnight | 1am |
| 1 January | 11pm | Midnight |
| 15 January | 10pm | 11pm |
| 1 February | 9pm | 10pm |
| 15 February | 8pm | 9pm |

△ **OBSERVING TIMES**
Each chart shows the sky as it appears at 10pm local standard time, mid-month. However, this view can also be seen at other times of the month, as well as one hour later when local daylight-saving time is in use. To view the sky at a time before or after 10pm, you may need to consult a different monthly chart.

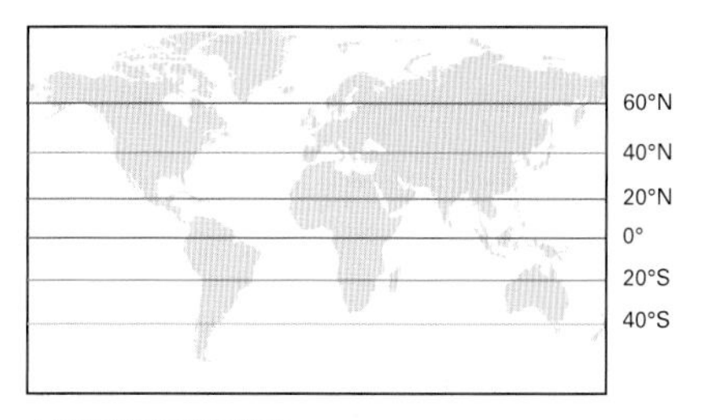


△ **LINES OF LATITUDES**
Determine the latitude line that is closest to your geographical location, and use the colour-coding on the sky charts to find the view from your location. Note that a 10° difference in latitude has little effect on the stars that can be seen.

▽ **STAR MAGNITUDE**
Stars that appear brighter than magnitude 6 are illustrated on the whole-sky charts. This key can be used to gauge their magnitude. About 25 prominent stars are also labelled with their popular names.

▽ **DEEP-SKY OBJECTS**
Icons are used to represent a selection of deep-sky objects of interest to the amateur astronomer.

DEEP-SKY OBJECTS

 Galaxy 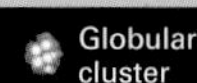 Globular cluster 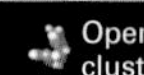 Open cluster  Diffuse nebula  Planetary nebula

STAR MAGNITUDES

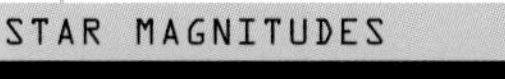
-1 0 1 2 3 4 5 Variable star

# JANUARY

In both the northern and southern hemispheres, the January evening sky is dominated by the magnificent constellation of Orion, the hunter. He is depicted with raised club and shield, facing Taurus the bull, with his two dogs, Canis Major and Canis Minor, following at his heels. The hazy band of the Milky Way arches from southeast to northwest in northern skies, while in the southern hemisphere the Large Magellanic Cloud lies high up in the sky.

### SPECIAL EVENTS

**PHASES OF THE MOON**

| | FULL MOON | NEW MOON |
|---|---|---|
| **2011** | 19 January | 4 January |
| **2012** | 9 January | 23 January |
| **2013** | 27 January | 11 January |
| **2014** | 16 January | 1, 30 January |
| **2015** | 5 January | 20 January |
| **2016** | 24 January | 10 January |
| **2017** | 12 January | 28 January |
| **2018** | 2 January | 17 January |
| **2019** | 21 January | 6 January |

**THE PLANETS**

**2011: 8 January** Venus is at greatest morning elongation, magnitude -4.4.

**2011: 9 January** Mercury is at greatest morning elongation, magnitude -0.2.

**2014: 6 January** Jupiter is at opposition, magnitude -2.7.

**2014: 31 January** Mercury is at greatest evening elongation, magnitude -0.5.

**2015: 14 January** Mercury is at greatest evening elongation, magnitude -0.6.

**2017: 12 January** Venus is at greatest evening elongation -4.4.

**2017: 19 January** Mercury is at greatest morning elongation -0.2.

**2018: 1 January** Mercury is at greatest morning elongation -0.3.

**2019: 6 January** Venus is at greatest morning elongation -4.5.

**ECLIPSES**

**2011: 4 January** A partial eclipse of the Sun is visible from northern Africa, Europe, the Middle East, and central Asia.

**2018: 31 January** A total eclipse of the Moon is visible from Africa, Europe, Asia, and Australia.

**2019: 6 January** A partial eclipse of the Sun is visible from northeast Asia and north Pacific.

**2019: 21 January** A total eclipse of the Moon is visible from South America, Africa, Europe, Asia, and Australia.

## NORTHERN LATITUDES

### THE STARS

Sirius, the brightest star in the entire sky, is well displayed on January evenings, twinkling above the southern horizon at mid-northern latitudes. Sirius forms the southern apex of a group of three stars known as the Winter Triangle (see p.420), which is completed by Procyon and Betelgeuse.

Directly overhead for mid-northern observers is the yellowish star Capella, which is the most northerly first-magnitude star and the brightest member of Auriga. In the northeast, the Plough stands on its handle, and the Square of Pegasus sinks low in the western sky. In the northwest, the Milky Way passes through Auriga into Perseus and Cassiopeia.

### DEEP-SKY OBJECTS

One of the most-photographed sights in the sky, the Orion Nebula (see p.239), lies south of the chain of three stars that makes up Orion's belt. The nebula is easily visible through binoculars at most northern latitudes, and even under average skies it can be seen with the naked eye as a hazy patch.

Three open star clusters in Auriga – M36, M37, and M38 – can be picked out with binoculars.

### METEOR SHOWERS

Northern observers can observe the Quadrantid meteors around 3–4 January every year. The meteors radiate from a point near the handle of the Plough in Ursa Major, an area which was once occupied by the now-obsolete constellation Quadrans, hence their name. Although numerous – peaking at around 100 an hour – the meteors are faint, so not many can be seen from urban areas. Other drawbacks are that their peak is short, lasting only a few hours, and their radiant remains low in the northeastern sky until well after midnight.

**OPEN CLUSTERS**
M36 (centre), M37 (left), and M38 (right) in Auriga can be picked out from the Milky Way with binoculars.

MIDNIGHT
3AM
6AM
9AM
NOON
10°
0°
-10°
-20°
-30°
-40°
-50°
Arcturus
LEO
Regulus
CANCER
OPHIUCHUS
VIRGO
Spica
Antares
LIBRA
SAGITTARIUS
SCORPIUS
Shaula
MORNING SKY

# SOUTHERN LATITUDES

## THE STARS

The two brightest stars in the entire sky, Sirius in Canis Major and Canopus in Carina, blaze high in the sky for southern observers this month. Orion's brightest members, Rigel and Betelgeuse, are also high up, and Aldebaran, the brightest star in Taurus, glistens a ruddy colour lower in the north. Closer still to the northern horizon, Capella is best placed for observation on January evenings. The rich Milky Way starfields of Carina and Centaurus lie in the southeast. By comparison, the western half of the sky appears almost barren, for it contains only a scattering of stars that are easily visible to the naked eye, the most prominent being 1st-magnitude Achernar, in the southwest at the end of Eridanus.

## DEEP-SKY OBJECTS

The Orion Nebula is ideally placed for all southern observers this month, as it is high in the sky.

**ORION NEBULA**
M41 is visible to the naked eye as a hazy patch of light, but its full complexity and beauty is brought out only on long-exposure photographs and on CCD images.

M41, a large star cluster near Sirius, sits on the zenith for observers around 20°S. Under good conditions, M41 is just visible to the naked eye.

The Large Magellanic Cloud (see p.300) in Dorado looks like a detached scrap of the Milky Way lying on the meridian (an imaginary line passing north to south through the zenith) on January evenings. Prominent among its mass of stars is the Tarantula Nebula, which appears to the naked eye as a glowing patch as large as the full Moon. The Small Magellanic Cloud (see p.301) in Tucana lies closer to the southwestern horizon.

**ORION'S BELT**
A chain of three stars forms Orion's belt, south of which can be seen the nebulosity of M42. North is to the top of this picture.

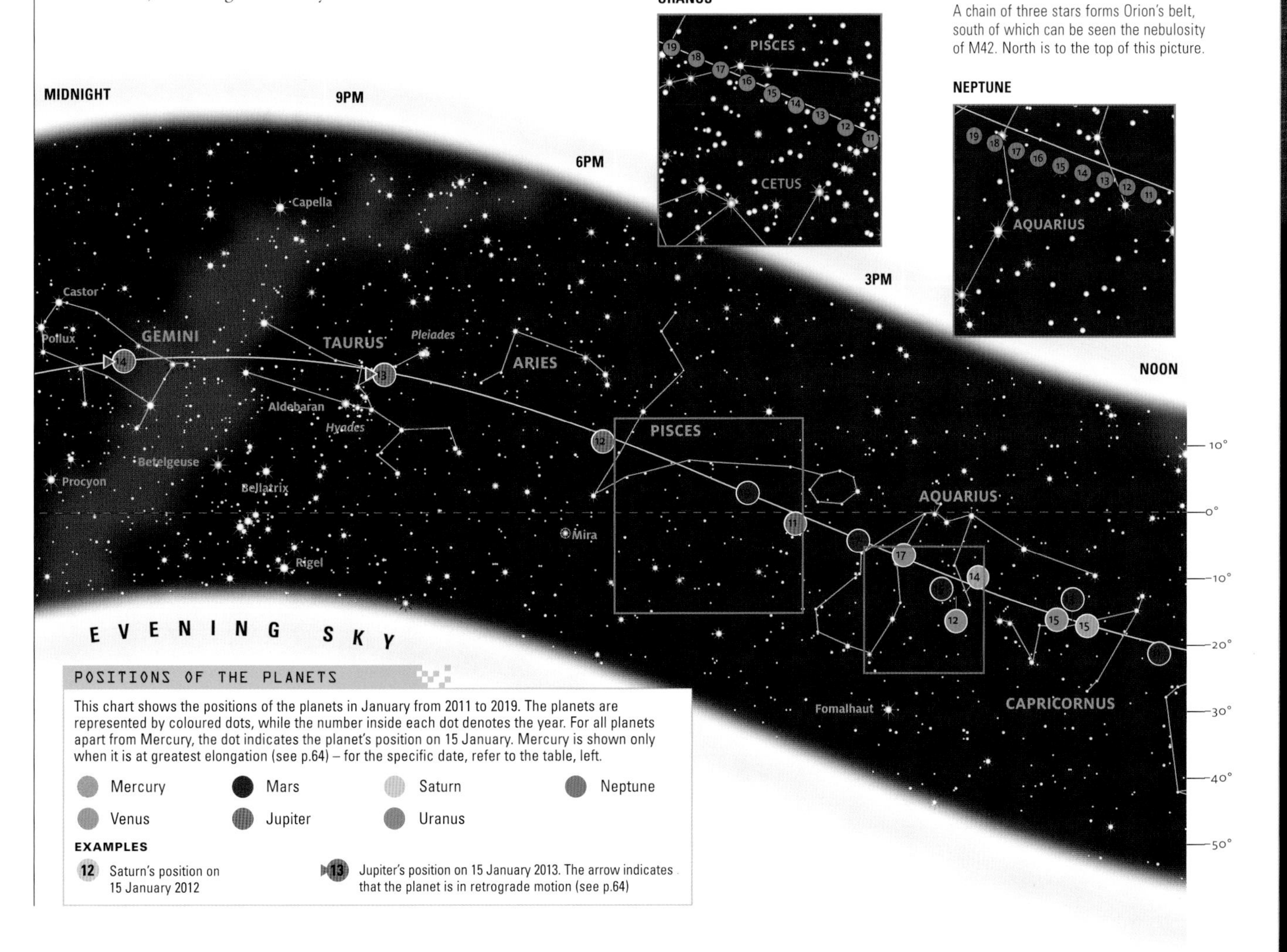


**POSITIONS OF THE PLANETS**

This chart shows the positions of the planets in January from 2011 to 2019. The planets are represented by coloured dots, while the number inside each dot denotes the year. For all planets apart from Mercury, the dot indicates the planet's position on 15 January. Mercury is shown only when it is at greatest elongation (see p.64) – for the specific date, refer to the table, left.

Mercury
Venus
Mars
Jupiter
Saturn
Uranus
Neptune

**EXAMPLES**

12 Saturn's position on 15 January 2012

13 Jupiter's position on 15 January 2013. The arrow indicates that the planet is in retrograde motion (see p.64)

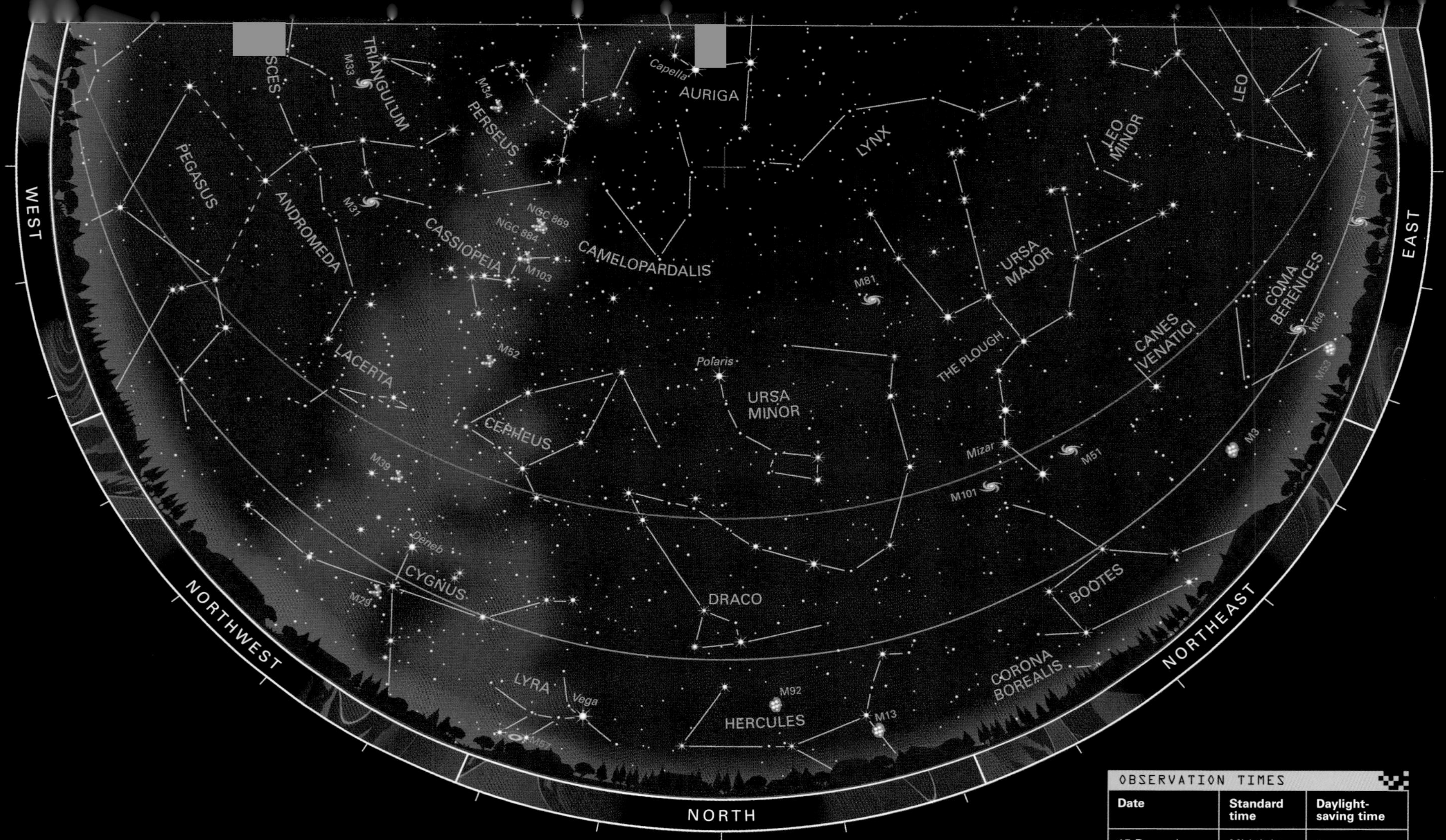

# JANUARY | NORTHERN LATITUDES

LOOKING NORTH

WEST
NORTHWEST
NORTH
NORTHEAST
EAST

PEGASUS
PISCES
TRIANGULUM
M33
ANDROMEDA
M31
PERSEUS
M34
NGC 869
NGC 884
CASSIOPEIA
M103
M52
LACERTA
CEPHEUS
M39
Deneb
CYGNUS
M29
LYRA
Vega
M57
CAMELOPARDALIS
AURIGA
Capella
LYNX
Polaris
URSA MINOR
DRACO
HERCULES
M92
M13
M81
URSA MAJOR
THE PLOUGH
Mizar
M101
M51
LEO MINOR
LEO
CANES VENATICI
COMA BERENICES
M64
M53
M3
BOOTES
CORONA BOREALIS

OBSERVATION TIMES

| Date | Standard time | Daylight-saving time |
|---|---|---|
| 15 December | Midnight | 1am |
| 1 January | 11pm | Midnight |
| 15 January | 10pm | 11pm |
| 1 February | 9pm | 10pm |
| 15 February | 8pm | 9pm |

STAR MAGNITUDES

-1 · 0 · 1 · 2 · 3 · 4 · 5 · Variable star

DEEP-SKY OBJECTS

Galaxy · Globular cluster · Open cluster · Diffuse nebula · Planetary nebula

POINTS OF REFERENCE

**Horizons** 60°N | 40°N | 20°N **Zeniths** 60°N | 40°N | 20°N | Ecliptic

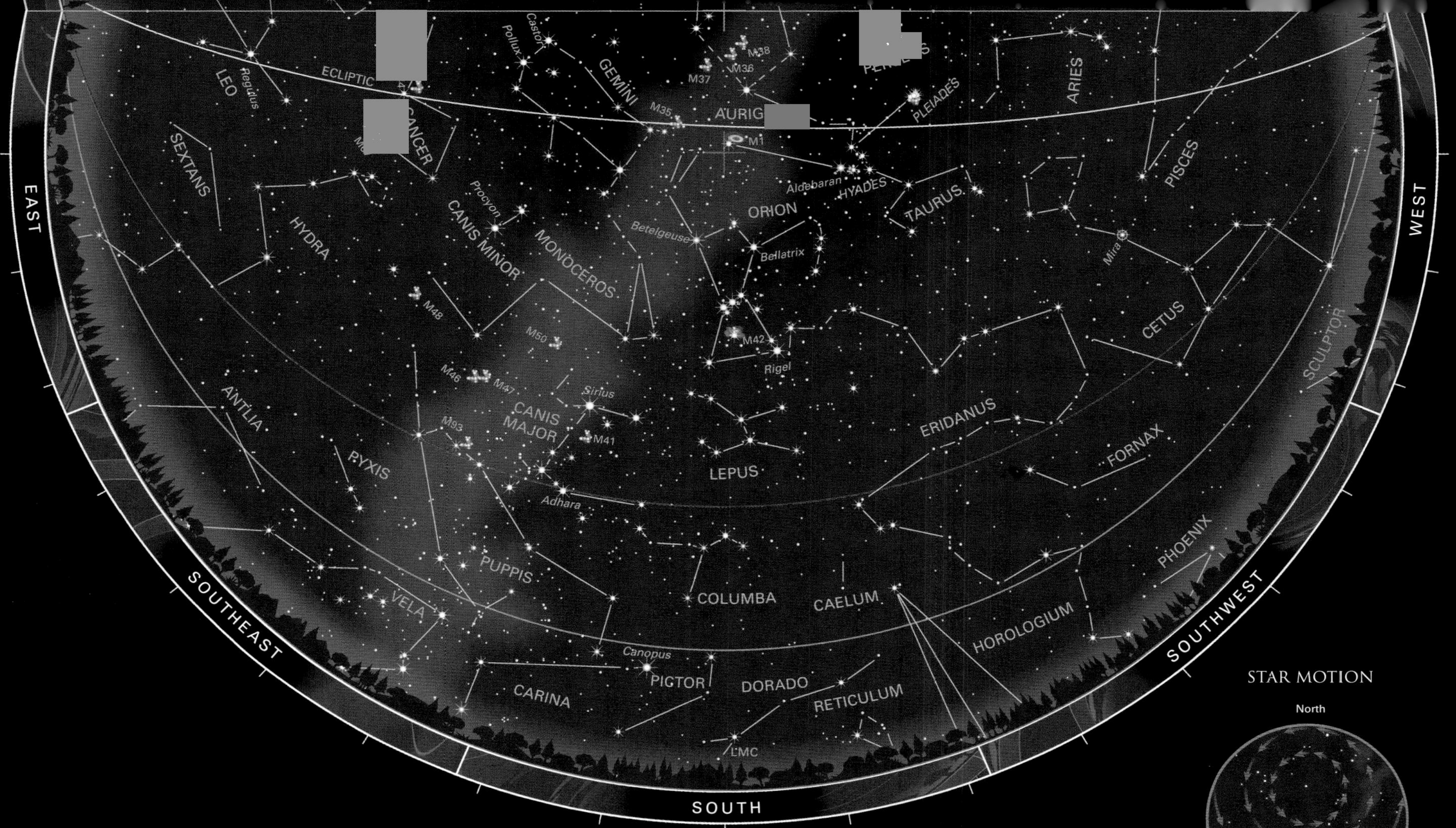

EAST
SOUTHEAST
SOUTH
SOUTHWEST
WEST
ECLIPTIC
LEO
Regulus
SEXTANS
HYDRA
CANCER
Pollux
Castor
GEMINI
M35
M37
M36
M38
AURIG
M1
PLEIADES
ARIES
PISCES
Procyon
CANIS MINOR
MONOCEROS
Betelgeuse
Bellatrix
ORION
Aldebaran
HYADES
TAURUS
Mira
CETUS
SCULPTOR
M48
M50
M42
Rigel
M46
M47
Sirius
CANIS MAJOR
M41
M93
ANTLIA
PYXIS
LEPUS
ERIDANUS
FORNAX
Adhara
PUPPIS
VELA
COLUMBA
CAELUM
PHOENIX
HOROLOGIUM
Canopus
CARINA
PICTOR
DORADO
RETICULUM
LMC

LOOKING SOUTH

# JANUARY | NORTHERN LATITUDES

STAR MOTION

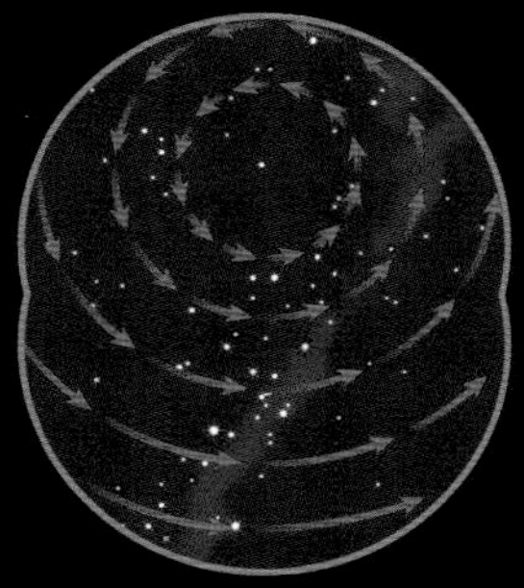

North

South

STAR MAGNITUDES

-1 0 1 2 3 4 5 Variable star

DEEP-SKY OBJECTS

Galaxy
Globular cluster
Open cluster
Diffuse nebula
Planetary nebula

POINTS OF REFERENCE

Horizons 60°N 40°N 20°N
Zeniths 60°N 40°N 20°N
Ecliptic

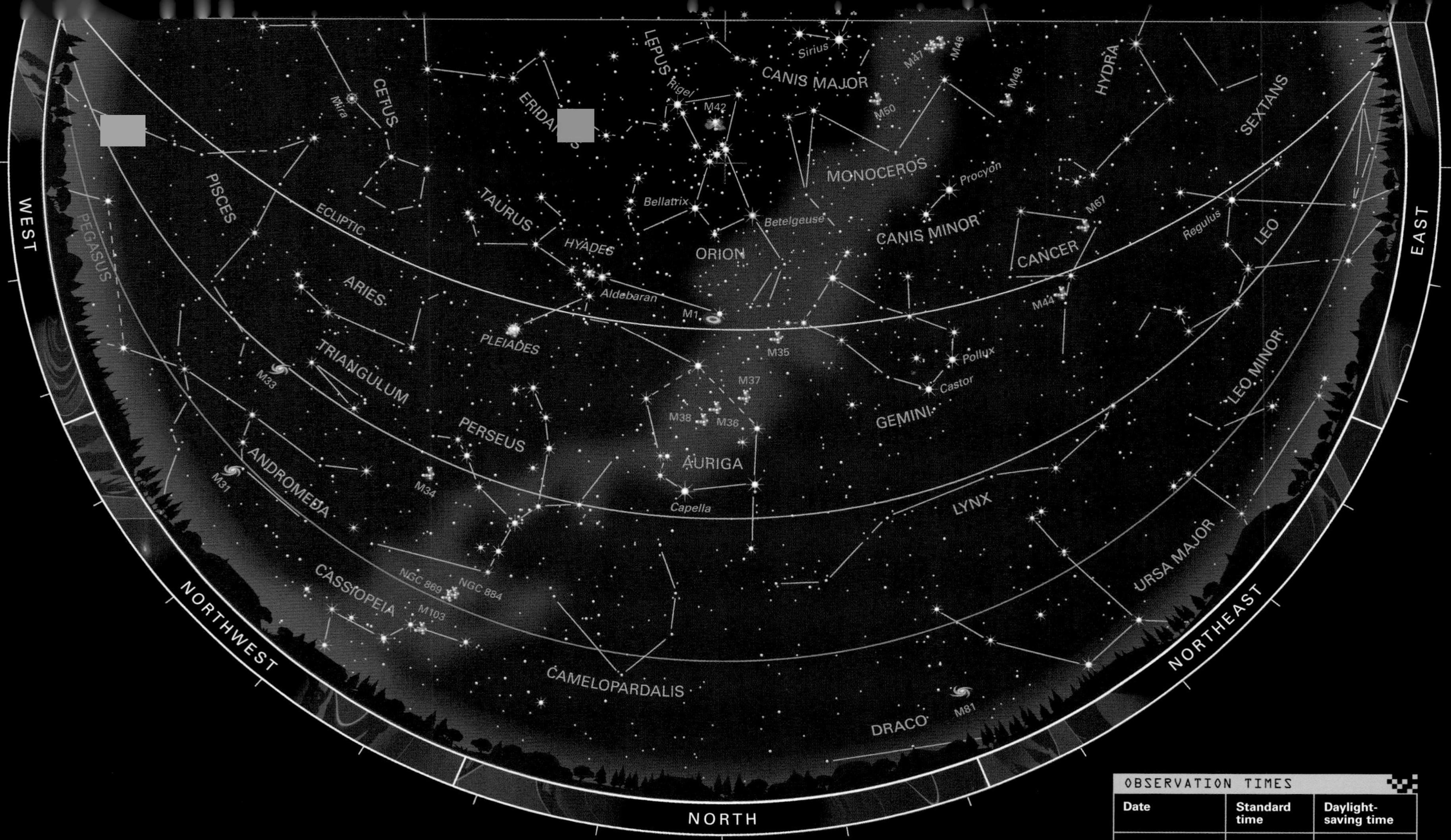

# JANUARY | SOUTHERN LATITUDES

## LOOKING NORTH

WEST
NORTHWEST
NORTH
NORTHEAST
EAST

PEGASUS
PISCES
ECLIPTIC
CETUS
Mira
ERIDANUS
LEPUS
Rigel
M42
CANIS MAJOR
Sirius
M47
M46
M48
M50
HYDRA
SEXTANS
ARIES
TAURUS
HYADES
Aldebaran
Bellatrix
ORION
Betelgeuse
MONOCEROS
Procyon
CANIS MINOR
M67
CANCER
M44
Regulus
LEO
M1
PLEIADES
TRIANGULUM
M33
M35
Pollux
Castor
GEMINI
LEO MINOR
M37
M38
M36
PERSEUS
AURIGA
Capella
ANDROMEDA
M31
M34
LYNX
URSA MAJOR
CASSIOPEIA
NGC 869
NGC 884
M103
CAMELOPARDALIS
DRACO
M81

### OBSERVATION TIMES

| Date | Standard time | Daylight-saving time |
|---|---|---|
| 15 December | Midnight | 1am |
| 1 January | 11pm | Midnight |
| 15 January | 10pm | 11pm |
| 1 February | 9pm | 10pm |
| 15 February | 8pm | 9pm |

### STAR MAGNITUDES

-1 · 0 · 1 · 2 · 3 · 4 · 5 · Variable star

### DEEP-SKY OBJECTS

Galaxy · Globular cluster · Open cluster · Diffuse nebula · Planetary nebula

### POINTS OF REFERENCE

Horizons 0° | 20°S | 40°S · Zeniths 0° | 20°S | 40°S · Ecliptic

# JANUARY | SOUTHERN LATITUDES

**STAR MAGNITUDES**
-1 · 0 · 1 · 2 · 3 · 4 · 5 · Variable star

**DEEP-SKY OBJECTS**
Galaxy · Globular cluster · Open cluster · Diffuse nebula · Planetary nebula

**POINTS OF REFERENCE**
Horizons: 0° | 20°S | 40°S · Zeniths: 0° | 20°S | 40°S · Ecliptic

# FEBRUARY

Castor and Pollux, the brightest stars in the northern zodiacal constellation of Gemini, lie close to the celestial meridian (the imaginary north–south line in the sky) on February evenings, as does Procyon in Canis Minor, which adjoins Gemini to the south. In the southern hemisphere, Carina, Puppis, and Vela – the three constellations that once formed the large ancient Greek constellation Argo Navis, ship of the Argonauts – are high in the sky.

## SPECIAL EVENTS

### PHASES OF THE MOON

| | FULL MOON | NEW MOON |
|---|---|---|
| **2011** | 18 February | 3 February |
| **2012** | 7 February | 21 February |
| **2013** | 25 February | 10 February |
| **2014** | 14 February | None |
| **2015** | 3 February | 18 February |
| **2016** | 22 February | 8 February |
| **2017** | 11 February | 26 February |
| **2018** | None | 15 February |
| **2019** | 19 February | 4 February |

### PLANETS

**2013: 8 February** Mercury and Mars are 0.5° apart in the western evening sky.

**2013: 16 February** Mercury is at greatest evening elongation, magnitude -0.5.

**2015: 7 February** Jupiter is at opposition, magnitude -2.6.

**2015: 21 February** Venus and Mars are 0.4° apart in the western evening sky.

**2015: 24 February** Mercury is at greatest morning elongation, magnitude 0.1.

**2016: 7 February** Mercury is at greatest morning elongation, magnitude 0.0.

**2019: 27 February** Mercury is at greatest evening elongation, magnitude -0.4.

### ECLIPSES

**2017: 26 February** An annular eclipse of the Sun is visible from Pacific Ocean, Chile, Argentina, Atlantic Ocean, and Africa. A partial solar eclipse is visible from southern South America, Atlantic Ocean, and Antarctica.

**2018: 15 February** A partial eclipse of the Sun is visible from southern South America and Antarctica.

## NORTHERN LATITUDES

### THE STARS

Gemini is almost overhead as seen from mid-northern latitudes in February, with the faintest of the zodiacal constellations, Cancer, close by but slightly lower in the sky. South of Gemini, the sparkling Winter Triangle formed by Sirius (in Canis Major), Betelgeuse (in Orion), and Procyon (in Canis Minor) remains prominent. Taurus, the Bull, backs away from Orion towards the western horizon, with Auriga and Perseus higher above it. Close to the northwest horizon is the W-shaped Cassiopeia. Leo, the Lion, is moving into the eastern sky, with the familiar figure of the Plough above it in the northeast.

### DEEP-SKY OBJECTS

M35, a large open star cluster at the feet of Gemini, is easily seen through binoculars. The Beehive Cluster (see p.286) – also known as M44 or Praesepe – lies nearby in Cancer. Through binoculars, the Beehive is visible as a scattering of stars three times wider than the full Moon; under ideal conditions, it can be glimpsed by the naked eye as a hazy patch – it was known to the ancient Greeks.

The Milky Way runs through Monoceros, an often-overlooked constellation framed by the Winter Triangle, which contains several open star clusters. One of the most notable of these clusters, NGC 2244, is visible through binoculars. It is located at the heart of the elusive Rosette Nebula, which is seen well only in photographs.

**NEPTUNE**

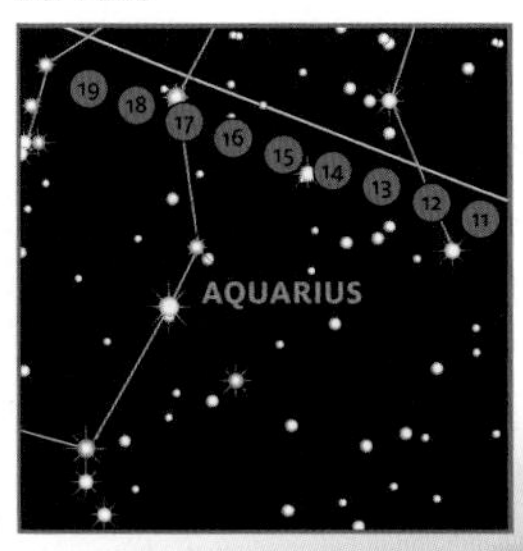

**THE WINTER TRIANGLE**
Brilliant Sirius (bottom) forms a prominent triangle in the northern winter sky with Procyon (top, left) and Betelgeuse (top, right).

MIDNIGHT
3AM
6AM
9AM
NOON
20°
10°
0°
-10°
-20°
-30°
-40°
LEO
Regulus
Arcturus
Altair
OPHIUCHUS
VIRGO
Spica
CAPRICORNUS
Antares
LIBRA
SAGITTARIUS
Shaula
SCORPIUS
MORNING SKY

# SOUTHERN LATITUDES

## THE STARS

Sirius (see p.264) and Canopus, the two brightest stars in the entire sky, remain high for southern observers throughout February, while Achernar, the 1st-magnitude star at the end of the celestial river Eridanus, sinks towards the southwestern horizon. In the southeast, Crux, the Southern Cross, enters the scene, followed by the bright stars of Centaurus. Higher up is the False Cross, which is formed by four stars in Vela and Carina and is sometimes mistaken for the true Southern Cross.

Due north lie Castor (see p.272) and Pollux in Gemini. Orion is also high in the sky, with Taurus lower in the northwest. As seen from the most southerly latitudes, Perseus has already set and Auriga is following. Meanwhile, looking northeast, the distinctive shape of Leo, the Lion, has come into view.

## DEEP-SKY OBJECTS

The Milky Way, which meanders from southeast to northwest this month, contains numerous star clusters, of which M46 and M47, adjacent in Puppis, are prominent. Both clusters are at the edge of naked-eye visibility and look superb through binoculars. Two other open clusters that can be seen excellently through binoculars are NGC 2451 and NGC 2477, also in Puppis; farther south, in Vela, IC 2391 and IC 2395 are also good examples.

Outside the boundaries of the Milky Way, the open cluster M41 is found south of Sirius, while in the north, the Beehive Cluster (see p.286), or M44, is well positioned for observation in both February and March. In Carina, another open cluster, NGC 2516, is prominent. The Large Magellanic Cloud and the Tarantula Nebula are on view, south of Canopus, in the constellation Dorado.

**FINDING THE SOUTH CELESTIAL POLE**
The south celestial pole (left) is not marked by a bright star, but it can be located by intersecting two imaginary lines. One is the extension of the long axis of Crux. The other is at right angles to the line joining Alpha (α) and Beta (β) Centauri.

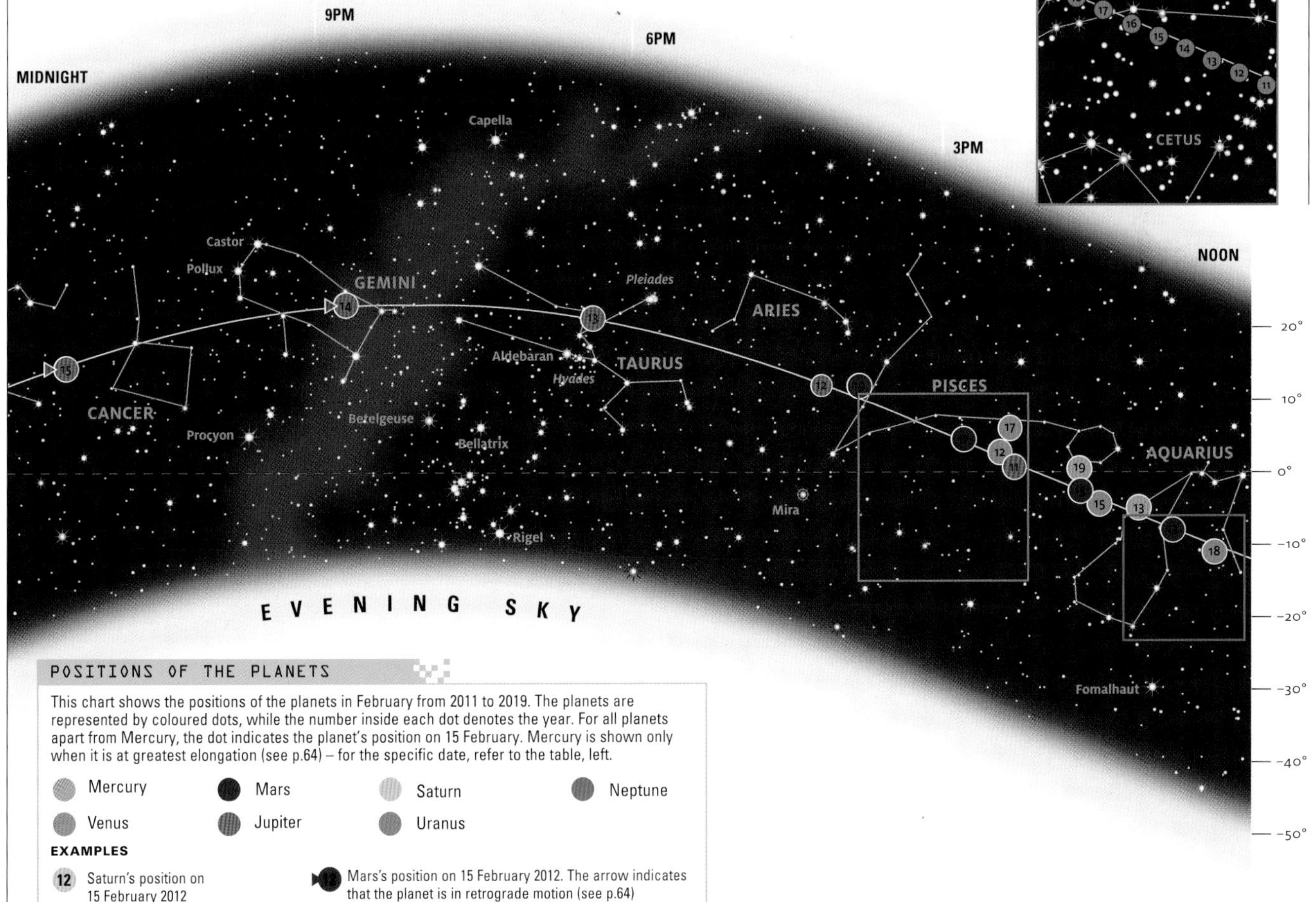


**POSITIONS OF THE PLANETS**

This chart shows the positions of the planets in February from 2011 to 2019. The planets are represented by coloured dots, while the number inside each dot denotes the year. For all planets apart from Mercury, the dot indicates the planet's position on 15 February. Mercury is shown only when it is at greatest elongation (see p.64) – for the specific date, refer to the table, left.

Mercury · Mars · Saturn · Neptune · Venus · Jupiter · Uranus

**EXAMPLES**

12 Saturn's position on 15 February 2012

Mars's position on 15 February 2012. The arrow indicates that the planet is in retrograde motion (see p.64)

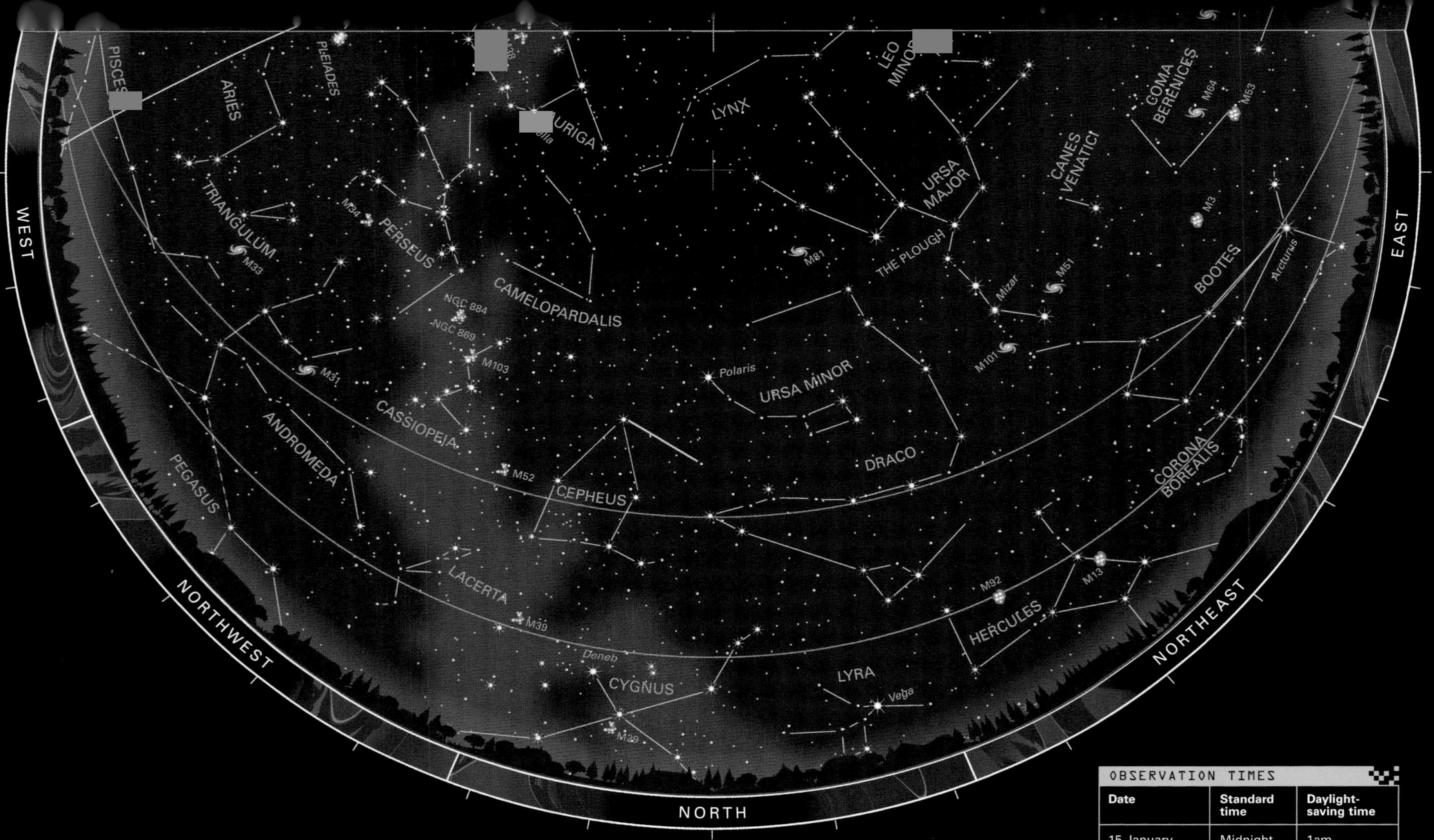

WEST
NORTHWEST
NORTH
NORTHEAST
EAST

PISCES
ARIES
PLEIADES
TRIANGULUM
M33
M34
PERSEUS
AURIGA
LYNX
LEO MINOR
COMA BERENICES
M64
M53
CANES VENATICI
URSA MAJOR
THE PLOUGH
M3
M81
M51
Mizar
BOOTES
Arcturus
CAMELOPARDALIS
NGC 884
NGC 869
M103
M31
Polaris
URSA MINOR
M101
CASSIOPEIA
ANDROMEDA
PEGASUS
M52
CEPHEUS
DRACO
CORONA BOREALIS
LACERTA
M39
M13
M92
HERCULES
Deneb
CYGNUS
LYRA
Vega
M29

LOOKING NORTH

# FEBRUARY | NORTHERN LATITUDES

| OBSERVATION TIMES | | |
|---|---|---|
| **Date** | **Standard time** | **Daylight-saving time** |
| 15 January | Midnight | 1am |
| 1 February | 11pm | Midnight |
| 15 February | 10pm | 11pm |
| 1 March | 9pm | 10pm |
| 15 March | 8pm | 9pm |

STAR MAGNITUDES
-1 · 0 · 1 · 2 · 3 · 4 · 5 · Variable star

DEEP-SKY OBJECTS
Galaxy · Globular cluster · Open cluster · Diffuse nebula · Planetary nebula

POINTS OF REFERENCE
**Horizons** 60°N | 40°N | 20°N **Zeniths** 60°N 40°N 20°N | Ecliptic

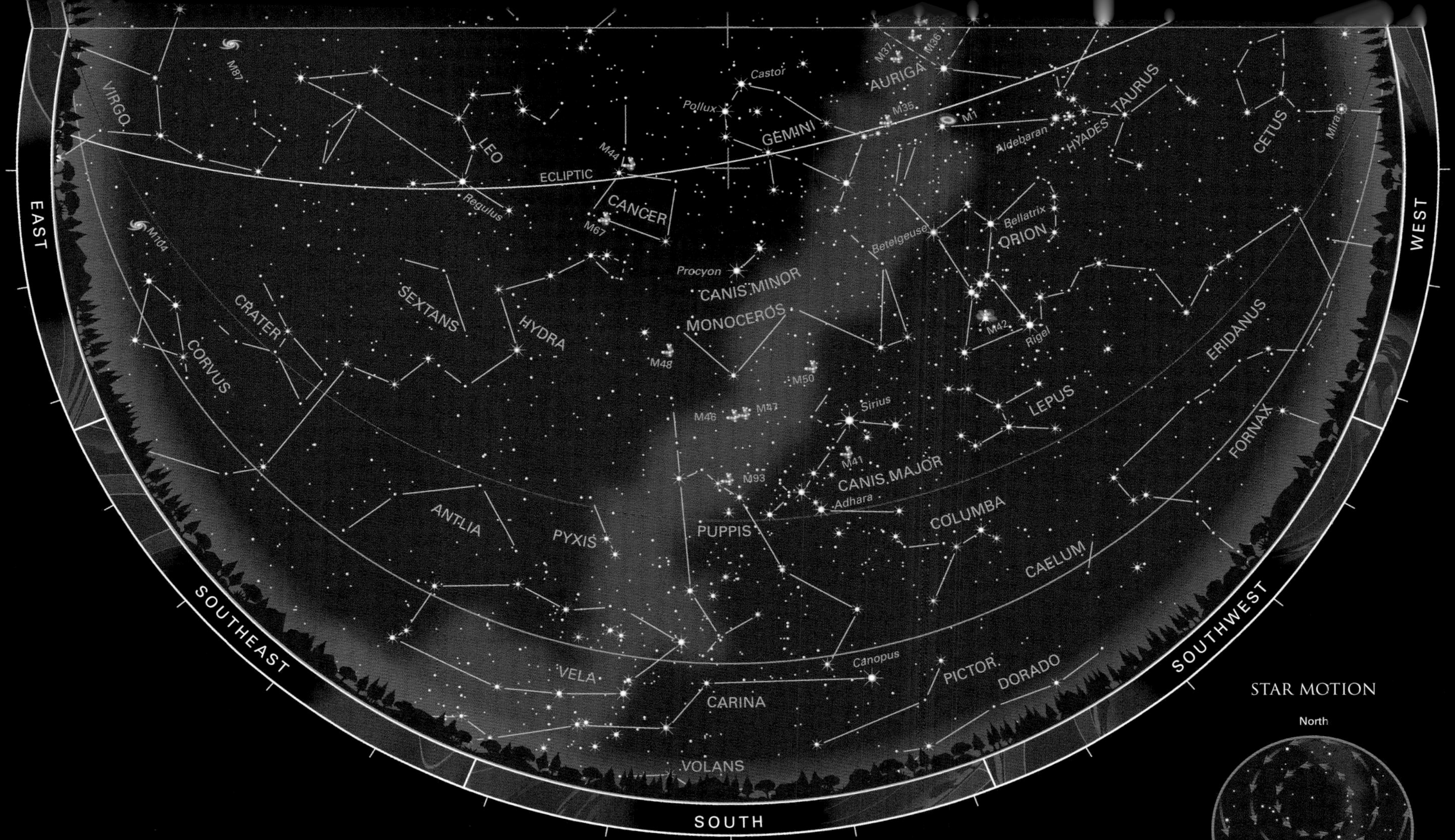

# FEBRUARY | NORTHERN LATITUDES

LOOKING SOUTH

EAST
SOUTHEAST
SOUTH
SOUTHWEST
WEST

VIRGO
M87
LEO
Regulus
ECLIPTIC
M104
CORVUS
CRATER
SEXTANS
HYDRA
ANTLIA
PYXIS
VELA
CANCER
M44
M67
Castor
Pollux
GEMINI
Procyon
CANIS MINOR
MONOCEROS
M48
M50
M46
M47
M93
PUPPIS
CARINA
VOLANS
Canopus
AURIGA
M37
M36
M35
M1
Aldebaran
HYADES
TAURUS
CETUS
Mira
Betelgeuse
Bellatrix
ORION
M42
Rigel
Sirius
M41
CANIS MAJOR
Adhara
LEPUS
COLUMBA
CAELUM
PICTOR
DORADO
ERIDANUS
FORNAX

STAR MOTION

North

South

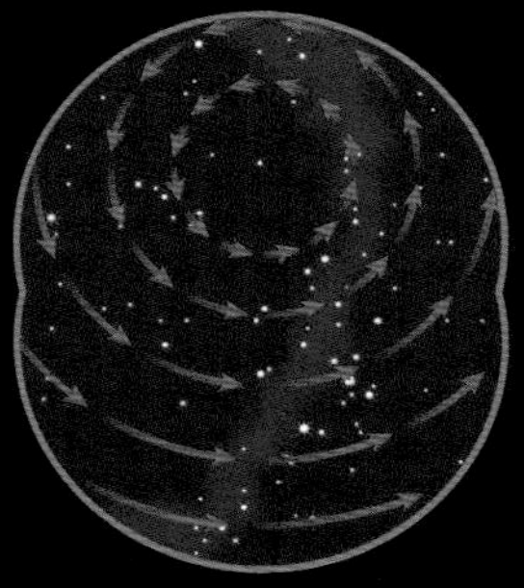

STAR MAGNITUDES

-1 0 1 2 3 4 5 Variable star

DEEP-SKY OBJECTS

Galaxy Globular cluster Open cluster Diffuse nebula Planetary nebula

POINTS OF REFERENCE

Horizons 60°N 40°N 20°N Zeniths 60°N 40°N 20°N Ecliptic

WEST
NORTHWEST
NORTH
NORTHEAST
EAST

CETUS
PISCES
Mira
ERIDANUS
LEPUS
Rigel
Sirius
M43
M47
M46
M50
M48
Bellatrix
MONOCEROS
HYDRA
SEXTANS
ECLIPTIC
VIRGO
M87
Procyon
ORION
Betelgeuse
CANIS MINOR
M67
M53
ARIES
TAURUS
HYADES
Aldebaran
CANCER
Regulus
LEO
M64
COMA BERENICES
M1
M44
PLEIADES
TRIANGULUM
M35
GEMINI
Pollux
Castor
LEO MINOR
M37
M36
M38
CANES VENATICI
LYNX
AURIGA
Capella
M34
PERSEUS
URSA MAJOR
THE PLOUGH
Mizar
CAMELOPARDALIS
M81
DRACO

LOOKING NORTH

# FEBRUARY | SOUTHERN LATITUDES

OBSERVATION TIMES

| Date | Standard time | Daylight-saving time |
|---|---|---|
| 15 January | Midnight | 1am |
| 1 February | 11pm | Midnight |
| 15 February | 10pm | 11pm |
| 1 March | 9pm | 10pm |
| 15 March | 8pm | 9pm |

STAR MAGNITUDES

-1 0 1 2 3 4 5 Variable star

DEEP-SKY OBJECTS

Galaxy Globular cluster Open cluster Diffuse nebula Planetary nebula

POINTS OF REFERENCE

Horizons 0° 20°S 40°S Zeniths 0° 20°S 40°S Ecliptic

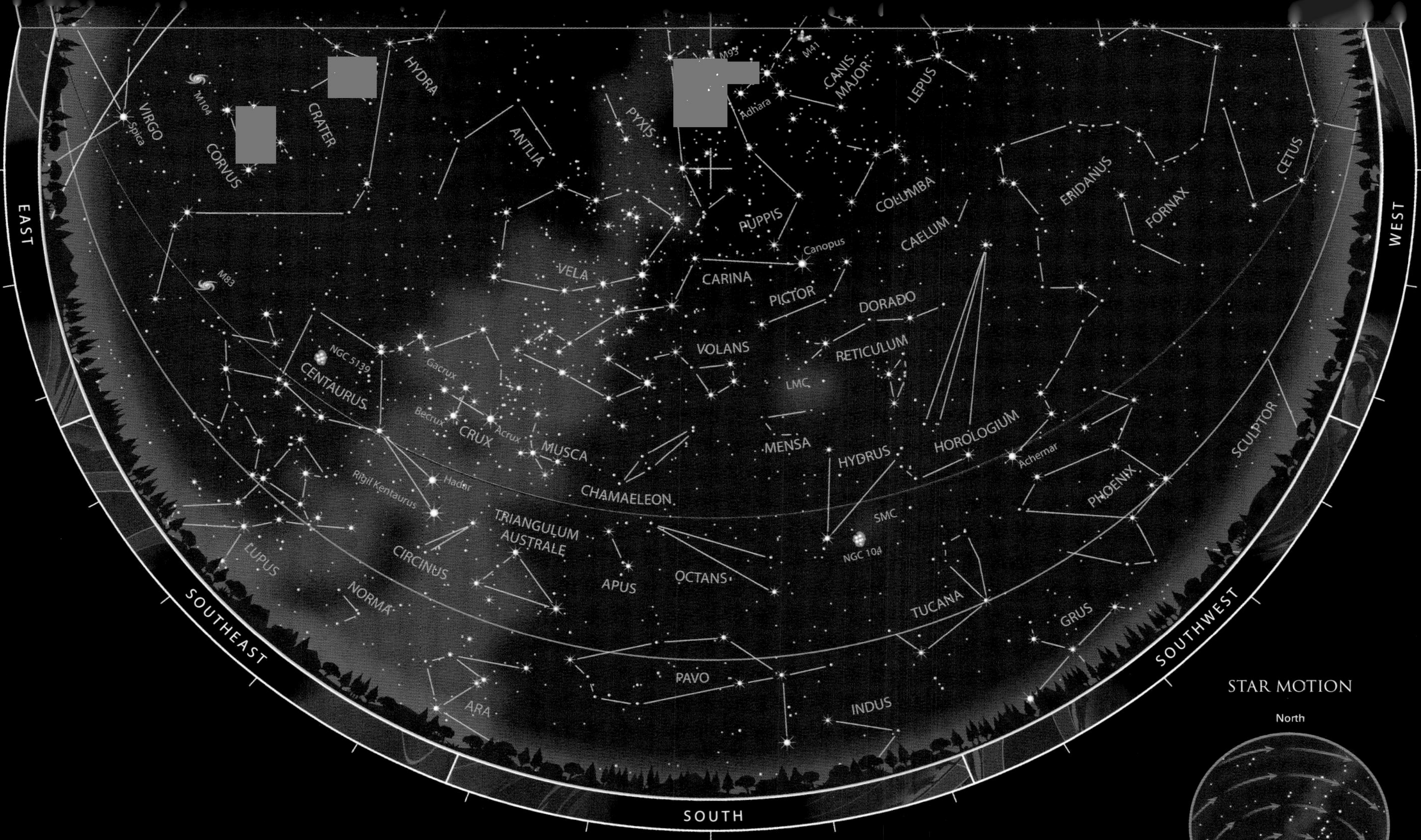

EAST

SOUTHEAST

SOUTH

SOUTHWEST

WEST

VIRGO
Spica
M104
CORVUS
CRATER
HYDRA
ANTLIA
PYXIS
M93
Adhara
M41
CANIS MAJOR
LEPUS
M83
PUPPIS
COLUMBA
ERIDANUS
CETUS
FORNAX
CAELUM
Canopus
VELA
CARINA
PICTOR
DORADO
NGC 5139
CENTAURUS
VOLANS
RETICULUM
LMC
Gacrux
Becrux
CRUX
Acrux
MUSCA
MENSA
HYDRUS
HOROLOGIUM
Achernar
SCULPTOR
Hadar
Rigil Kentaurus
CHAMAELEON
PHOENIX
SMC
NGC 104
TRIANGULUM AUSTRALE
LUPUS
CIRCINUS
APUS
OCTANS
NORMA
TUCANA
GRUS
PAVO
ARA
INDUS

STAR MOTION

North

South

LOOKING SOUTH

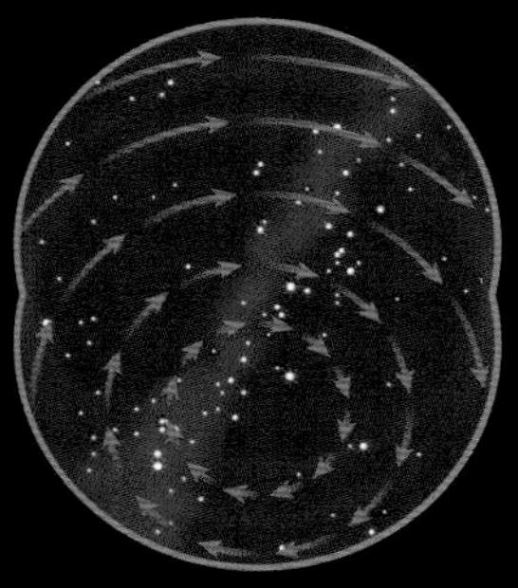

# FEBRUARY | SOUTHERN LATITUDES

STAR MAGNITUDES

-1 0 1 2 3 4 5 Variable star

DEEP-SKY OBJECTS

Galaxy | Globular cluster | Open cluster | Diffuse nebula | Planetary nebula

POINTS OF REFERENCE

Horizons 0° | 20°S | 40°S Zeniths 0° | 20°S | 40°S | Ecliptic

# MARCH

Nights grow shorter in the northern hemisphere, but longer in the southern hemisphere, as the Sun moves towards the equinox on March 20. On that date, the Sun lies exactly on the celestial equator, and all over the world day and night are of equal length. For northern observers, Orion and the other brilliant constellations of winter are departing towards the western horizon, while for southern observers the rich star fields of Carina and Centaurus are moving to centre stage.

## SPECIAL EVENTS

### PHASES OF THE MOON

| | FULL MOON | NEW MOON |
|---|---|---|
| **2011** | 19 March | 4 March |
| **2012** | 8 March | 22 March |
| **2013** | 27 March | 11 March |
| **2014** | 16 March | 1, 30 March |
| **2015** | 5 March | 20 March |
| **2016** | 23 March | 9 March |
| **2017** | 12 March | 28 March |
| **2018** | 2, 31 March | 17 March |
| **2019** | 21 March | 6 March |

### PLANETS

**2011: 23 March** Mercury is at greatest evening elongation, magnitude 0.0.

**2012: 3 March** Mars is at opposition, magnitude -1.2.

**2012: 5 March** Mercury is at greatest evening elongation, magnitude -0.3.

**2012: 27 March** Venus is at greatest evening elongation, magnitude -4.3.

**2013: 31 March** Mercury is at greatest morning elongation, magnitude 0.3.

**2014: 14 March** Mercury is at greatest morning elongation, magnitude 0.2.

**2014: 22 March** Venus is at greatest morning elongation, magnitude -4.3.

**2016: 8 March** Jupiter is at opposition, magnitude -2.5.

**2018: 15 March** Mercury is at greatest evening elongation, magnitude -0.3.

### ECLIPSES

**2015: 20 March** A total eclipse of the Sun is visible from the Faroes (between Scotland and Iceland), the Norwegian Sea, and Svalbard. A partial solar eclipse is visible from Europe, North Africa, and northwestern Asia.

**2016: 9 March** A total eclipse of the Sun is visible from Indonesia and the North Pacific. A partial solar eclipse is visible from east Asia, Australia, and Pacific Ocean.

## NORTHERN LATITUDES

### THE STARS

The distinctive sickle-shaped group of stars that makes up the head of Leo, the Lion, takes pride of place in the northern evening sky this month, with the fainter stars of Cancer to its right. Below it, in the south, lies a blank-looking area of sky occupied by the faint constellations Sextans, Crater, and Hydra. The only notable star in this area is 2nd-magnitude Alphard (in Hydra) – which appropriately means "the solitary one"– lying on the north–south meridian.

The saucepan shape of the Plough rides high in the northeast, its handle pointing down towards the bright star Arcturus, in Boötes, which is the harbinger of northern spring. Closer again to the horizon is Spica in Virgo. In the west, the stars of Gemini and Auriga remain high, with Taurus and Orion lower down. Sirius twinkles near the southwest horizon.

### DEEP-SKY OBJECTS

The beautiful spiral galaxy M81 (see p.304) in northern Ursa Major, lies near the north–south meridian on March evenings and is detectable through binoculars in clear skies. Farther south, the Beehive cluster (see p.286), or M44, in Cancer remains well positioned for observation.

**THE SICKLE OF LEO**
The stars that represent the head and neck of Leo, the Lion, form a distinctive shape like a sickle or a reversed question mark.

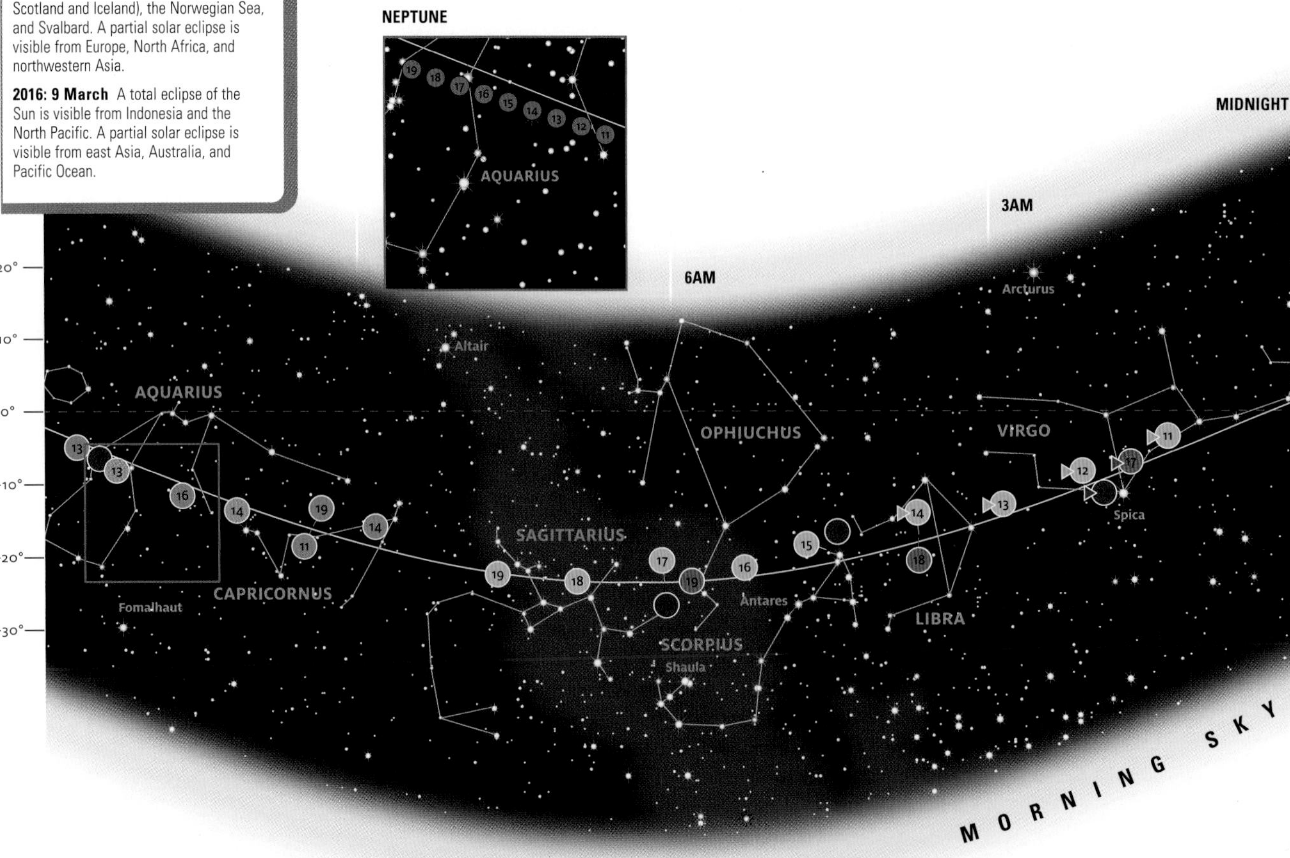

**THE FALSE CROSS**
Two stars in Vela (top left and centre right) and two in Carina (centre left and bottom right) form the False Cross in the southern sky.

# SOUTHERN LATITUDES

## THE STARS

Leo, the Lion, and its brightest star Regulus (see p.249) are high in the northern half of the sky for all southern observers, with Castor (see p.272) and Pollux in Gemini lower in the northwest. Sirius (see p.264) still sparkles high in the western sky but Orion sinks on its side towards the western horizon. Almost overhead for observers in mid-latitudes is Alphard, the brightest star in the constellation Hydra, which sprawls across an otherwise barren region of sky towards the southeast horizon.

Spica, the brightest star in Virgo, is well-placed in the east, and Canopus, in Carina, is prominent in the southwest sky. However, the main focus of attention is in the southeast, where the Southern Cross, Crux, now rides high along with brilliant Alpha ($\alpha$) and Beta ($\beta$) Centauri – Rigil Kentaurus (see p.248) and Hadar – which point towards it.

## DEEP-SKY OBJECTS

An open star cluster popularly known as the Southern Pleiades, IC 2602 lies close to the meridian on March evenings. Its brightest member, 3rd-magnitude Theta ($\theta$) Carinae, is easily visible to the naked eye, and binoculars reveal at least two dozen more members.

Four degrees to the north of the Southern Pleiades lies a large glowing region visible to the naked eye, NGC 3372, also known as the Carina Nebula (see p.245), which contains the erratic variable star Eta ($\eta$) Carinae (see p.258). Farther north, between Antlia and Vela, telescopes will pick up the planetary nebula NGC 3132, also known as the Eight-Burst Nebula. On view in the southwest sky are the Large Magellanic Cloud and the Tarantula Nebula (in Dorado).

URANUS

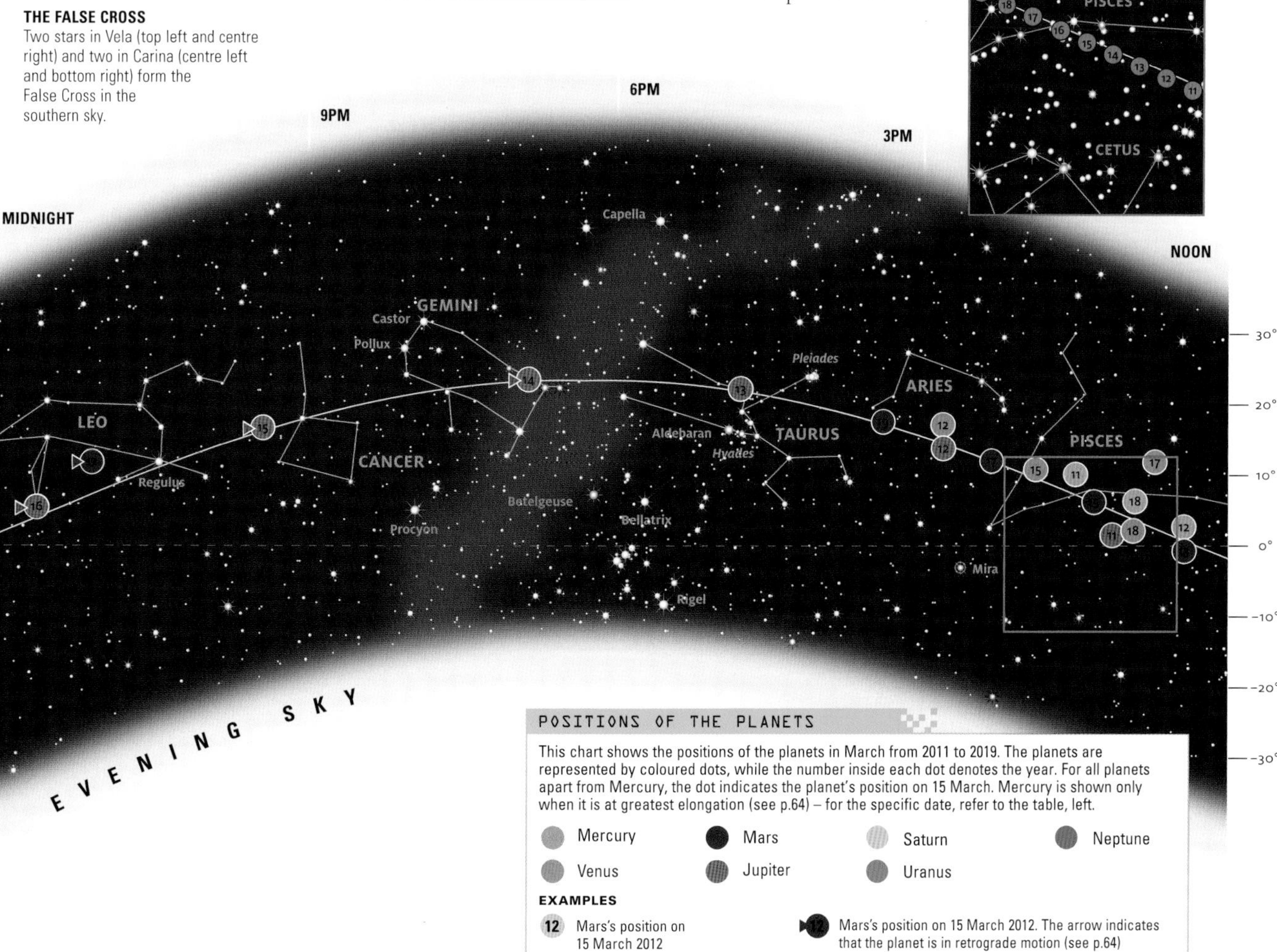
PISCES
CETUS
9PM
6PM
3PM
MIDNIGHT
NOON
Capella
GEMINI
Castor
Pollux
Pleiades
ARIES
LEO
CANCER
Regulus
Aldebaran
TAURUS
Hyades
PISCES
Betelgeuse
Bellatrix
Procyon
Mira
Rigel
30°
20°
10°
0°
–10°
–20°
–30°
EVENING SKY
POSITIONS OF THE PLANETS
This chart shows the positions of the planets in March from 2011 to 2019. The planets are represented by coloured dots, while the number inside each dot denotes the year. For all planets apart from Mercury, the dot indicates the planet's position on 15 March. Mercury is shown only when it is at greatest elongation (see p.64) – for the specific date, refer to the table, left.
Mercury
Mars
Saturn
Neptune
Venus
Jupiter
Uranus
EXAMPLES
Mars's position on 15 March 2012
Mars's position on 15 March 2012. The arrow indicates that the planet is in retrograde motion (see p.64)

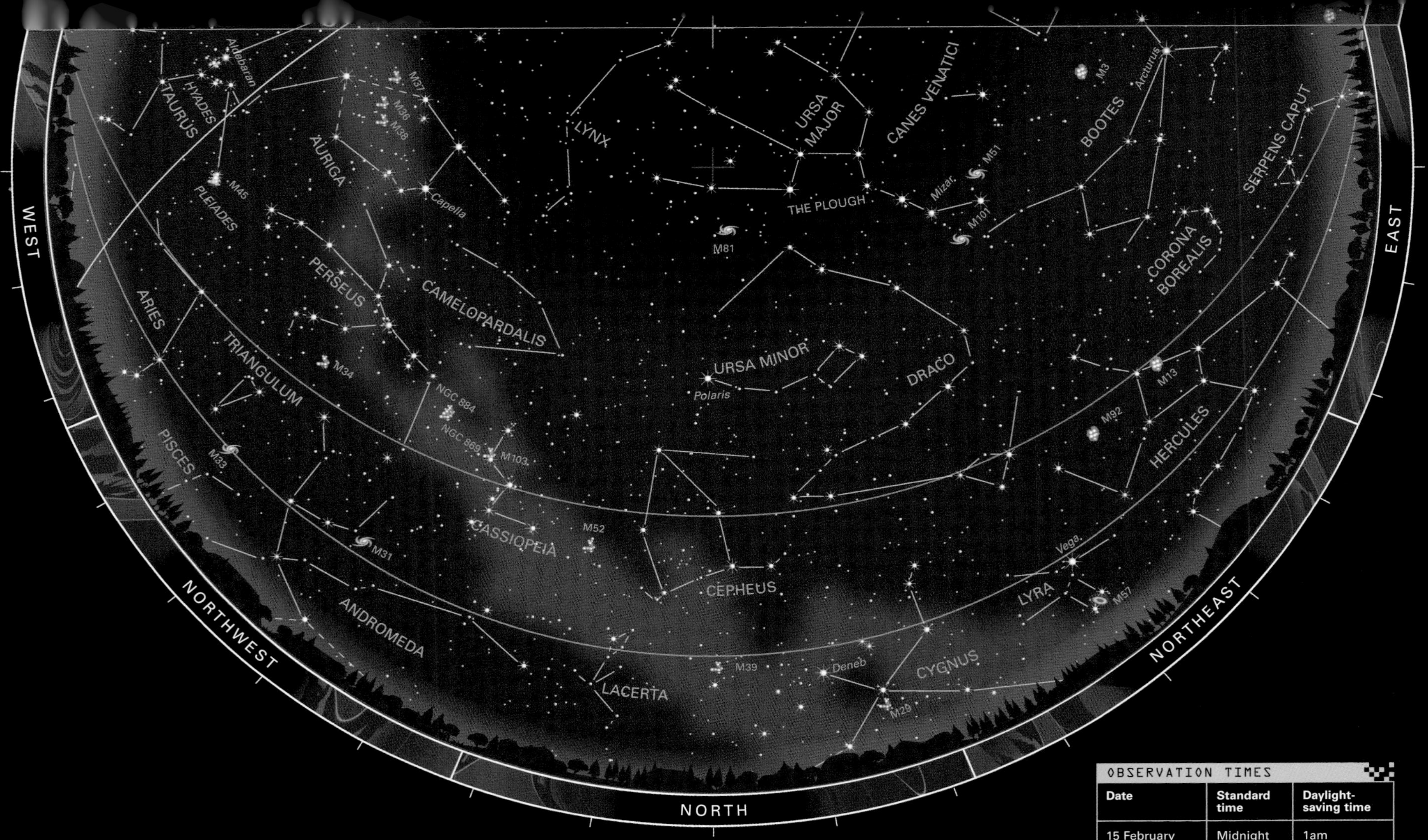

WEST
EAST
NORTHWEST
NORTH
NORTHEAST

TAURUS
HYADES
Aldebaran
PLEIADES
M45
AURIGA
M37
M36
M38
Capella
LYNX
URSA MAJOR
THE PLOUGH
CANES VENATICI
M51
Mizar
M101
M81
M3
BOOTES
Arcturus
SERPENS CAPUT
CORONA BOREALIS
ARIES
PERSEUS
CAMELOPARDALIS
TRIANGULUM
M34
NGC 884
NGC 869
M103
URSA MINOR
Polaris
DRACO
M13
M92
HERCULES
PISCES
M33
M31
ANDROMEDA
CASSIOPEIA
M52
CEPHEUS
LACERTA
M39
Deneb
M29
CYGNUS
LYRA
Vega
M57

LOOKING NORTH

# MARCH | NORTHERN LATITUDES

**OBSERVATION TIMES**

| Date | Standard time | Daylight-saving time |
|---|---|---|
| 15 February | Midnight | 1am |
| 1 March | 11pm | Midnight |
| 15 March | 10pm | 11pm |
| 1 April | 9pm | 10pm |
| 15 April | 8pm | 9pm |

**STAR MAGNITUDES**

-1 · 0 · 1 · 2 · 3 · 4 · 5 · Variable star

**DEEP-SKY OBJECTS**

Galaxy · Globular cluster · Open cluster · Diffuse nebula · Planetary nebula

**POINTS OF REFERENCE**

**Horizons** 60°N | 40°N | 20°N **Zeniths** 60°N 40°N 20°N | Ecliptic

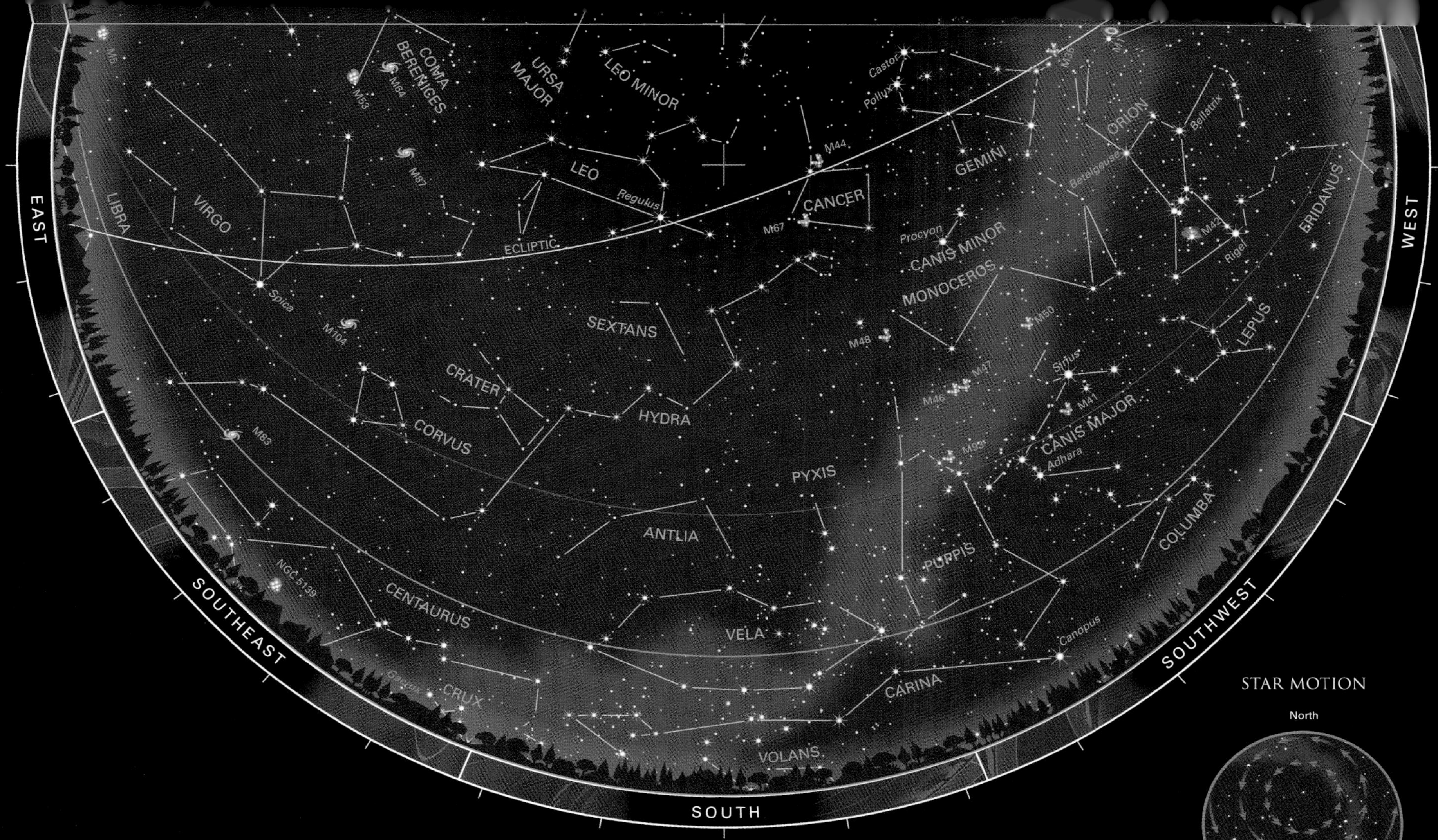

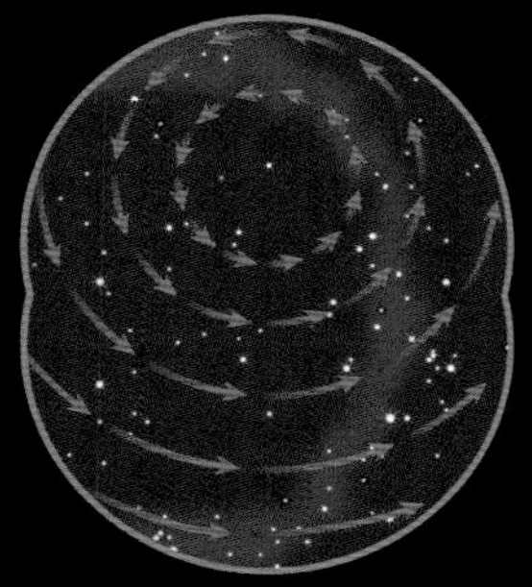

# MARCH | NORTHERN LATITUDES

LOOKING SOUTH

EAST
SOUTHEAST
SOUTH
SOUTHWEST
WEST

LIBRA
M5
VIRGO
Spica
COMA BERENICES
M53
M64
M87
M104
URSA MAJOR
LEO MINOR
LEO
Regulus
ECLIPTIC
CANCER
M44
M67
Castor
Pollux
GEMINI
M35
M1
ORION
Bellatrix
Betelgeuse
M42
Rigel
ERIDANUS
LEPUS
Procyon
CANIS MINOR
MONOCEROS
M50
M48
M47
M46
Sirius
M41
CANIS MAJOR
Adhara
M93
COLUMBA
SEXTANS
CRATER
HYDRA
CORVUS
M83
PYXIS
ANTLIA
PUPPIS
NGC 5139
CENTAURUS
VELA
Canopus
CARINA
Gacrux
CRUX
VOLANS

STAR MOTION
North
South

STAR MAGNITUDES
-1 | 0 | 1 | 2 | 3 | 4 | 5 | Variable star

DEEP-SKY OBJECTS
Galaxy | Globular cluster | Open cluster | Diffuse nebula | Planetary nebula

POINTS OF REFERENCE
**Horizons** 60°N | 40°N | 20°N **Zeniths** 60°N 40°N 20°N | Ecliptic

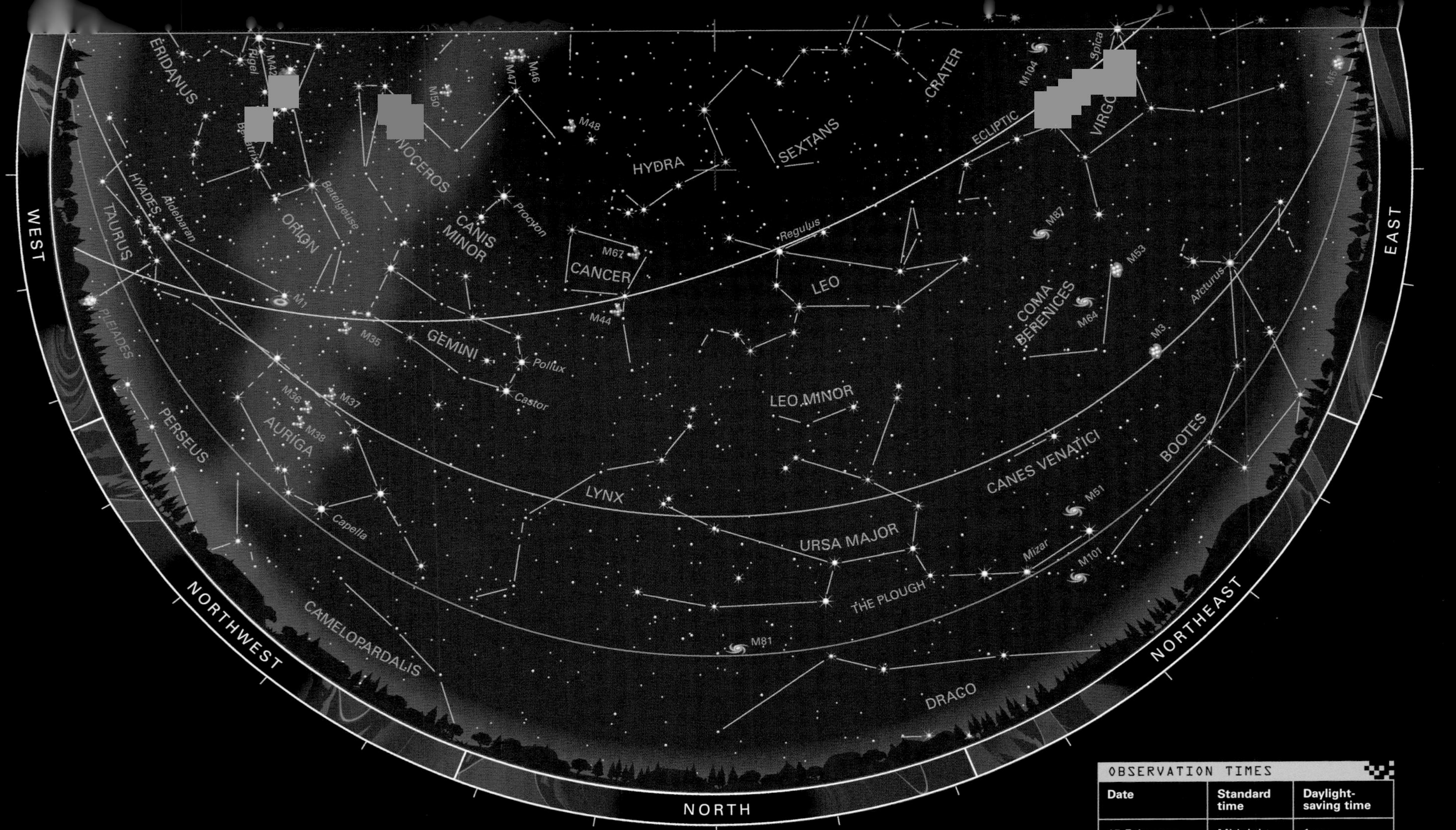

# MARCH | SOUTHERN LATITUDES

LOOKING NORTH

| OBSERVATION TIMES | | |
|---|---|---|
| Date | Standard time | Daylight-saving time |
| 15 February | Midnight | 1am |
| 1 March | 11pm | Midnight |
| 15 March | 10pm | 11pm |
| 1 April | 9pm | 10pm |
| 15 April | 8pm | 9pm |

STAR MAGNITUDES

-1 · 0 · 1 · 2 · 3 · 4 · 5 · Variable star

DEEP-SKY OBJECTS

Galaxy · Globular cluster · Open cluster · Diffuse nebula · Planetary nebula

POINTS OF REFERENCE

Horizons: 0° | 20°S | 40°S · Zeniths: 0° | 20°S | 40°S · Ecliptic

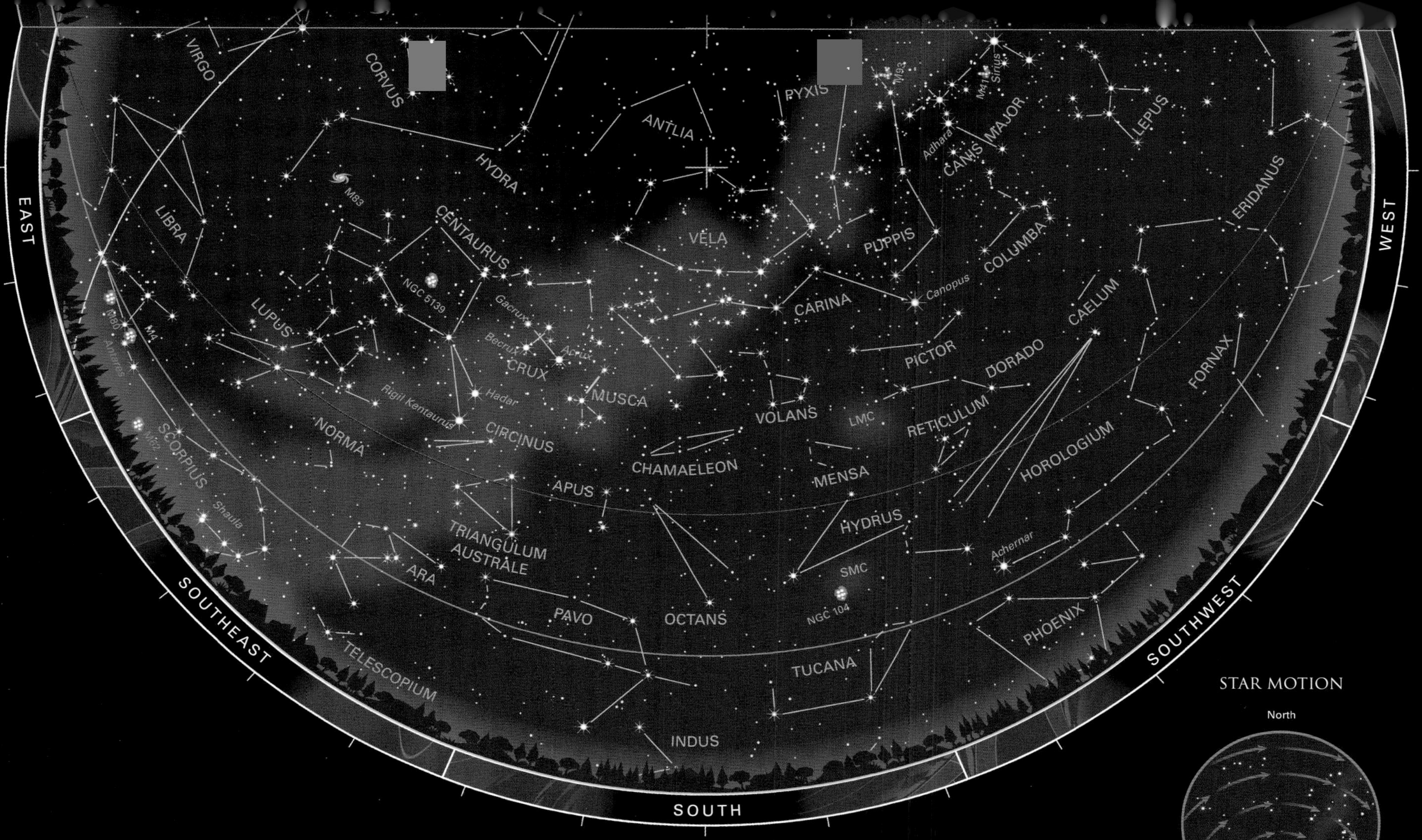

VIRGO
CORVUS
PYXIS
M93
M41
Sirius
ANTLIA
HYDRA
M83
LIBRA
CENTAURUS
VELA
PUPPIS
Adhara
CANIS MAJOR
LEPUS
COLUMBA
ERIDANUS
NGC 5139
Canopus
CARINA
CAELUM
LUPUS
M80
M4
Antares
Gacrux
Becrux
Acrux
CRUX
MUSCA
PICTOR
DORADO
FORNAX
Hadar
Rigil Kentaurus
CIRCINUS
VOLANS
LMC
RETICULUM
NORMA
M62
SCORPIUS
CHAMAELEON
MENSA
HOROLOGIUM
APUS
Shaula
TRIANGULUM AUSTRALE
HYDRUS
Achernar
ARA
SMC
NGC 104
PAVO
OCTANS
PHOENIX
TELESCOPIUM
TUCANA
INDUS
EAST
SOUTHEAST
SOUTH
SOUTHWEST
WEST

LOOKING SOUTH

# MARCH | SOUTHERN LATITUDES

STAR MOTION

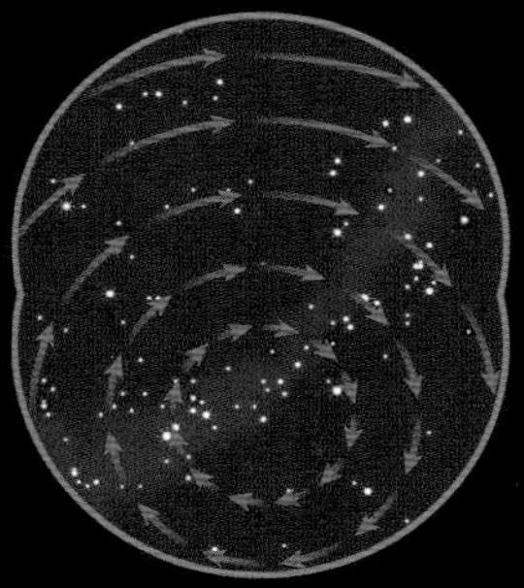

North

South

STAR MAGNITUDES

-1 0 1 2 3 4 5 Variable star

DEEP-SKY OBJECTS

Galaxy | Globular cluster | Open cluster | Diffuse nebula | Planetary nebula

POINTS OF REFERENCE

**Horizons** 0° | 20°S | 40°S **Zeniths** 0° | 20°S | 40°S | Ecliptic

# APRIL

One of the most familiar patterns in the sky, the seven stars that make up the Plough lie overhead from mid-northern latitudes, with the crouching figure of Leo, the Lion, reigning further south. In the eastern sky, the daffodil-coloured Arcturus, in Boötes, announces the arrival of spring in the north. In southern latitudes, the Southern Cross lies close to the north–south meridian, and Alpha (α) and Beta (β) Centauri – Rigil Kentaurus and Hadar – are high in the southeast.

## SPECIAL EVENTS

**PHASES OF THE MOON**

| | FULL MOON | NEW MOON |
|---|---|---|
| **2011** | 18 April | 3 April |
| **2012** | 6 April | 21 April |
| **2013** | 25 April | 10 April |
| **2014** | 15 April | 29 April |
| **2015** | 4 April | 18 April |
| **2016** | 22 April | 7 April |
| **2017** | 11 April | 26 April |
| **2018** | 30 April | 16 April |
| **2019** | 5 April | 19 April |

**PLANETS**

**2011: 4 April** Saturn is at opposition, magnitude 0.4.

**2012: 15 April** Saturn is at opposition, magnitude 0.2.

**2012: 18 April** Mercury is at greatest morning elongation, magnitude 0.5.

**2013: 28 April** Saturn is at opposition, magnitude 0.13.

**2014: 8 April** Mars is at opposition, magnitude -1.5.

**2016: 18 April** Mercury is at evening elongation, magnitude 0.2.

**2017: 1 April** Mercury is at evening elongation, magnitude -0.1.

**2018: 29 April** Mercury is at morning elongation, magnitude 0.5.

**2019: 11 April** Mercury is at morning elongation, magnitude 0.4.

**ECLIPSES**

**2014: 15 April** A total eclipse of the Moon is visible from North America, South America, and New Zealand.

**2014: 29 April** A partial solar eclipse is visible from west Australia.

**2015: 4 April** A total eclipse of the Moon is visible from western North America, east Asia, and Australia.

## NORTHERN LATITUDES

### THE STARS

On April evenings, the Plough is high in the sky. The stars in the bowl point north to Polaris (see pp.274–75), the north Pole Star, while following the curve of its handle leads to Arcturus, in Boötes, which is the brightest star north of the celestial equator. Continuing this curve leads to Spica, the brightest star in Virgo, close to the southeastern horizon. South of Leo and Virgo, the sprawling figure of Hydra occupies a large but mostly blank area of sky. By April, most of the stars of winter have disappeared in the west, although Gemini remains on view and Capella, in Auriga, twinkles in the northwest.

### DEEP-SKY OBJECTS

M81 (see p.304), the beautiful spiral galaxy in northern Ursa Major, is well placed for observation this month. A large open star cluster worthy of attention can be found in Coma Berenices and consists of a scattering of stars of 5th magnitude and fainter fanned out over an area of sky several times wider than the full Moon. Known as the Coma Star Cluster, this is best viewed through wide-angle binoculars. To its south is the Virgo Cluster (see p.319); a telescope is needed to see its numerous but faint member galaxies.

**NEPTUNE**

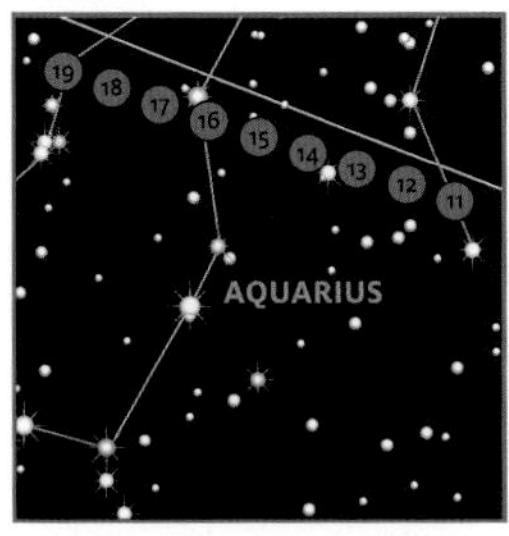

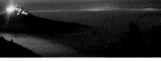

**THE PLOUGH**
The familiar shape of the Plough, or the Big Dipper, can be seen high in the sky on northern spring evenings.

### METEOR SHOWER

One of the weaker annual meteor showers, the Lyrids reaches its peak around 21–22 April, when a dozen or so meteors per hour can be seen radiating from a point near Vega (see p.249) in Lyra. Although not numerous, Lyrids are bright and fast. Rates are highest towards dawn, when Vega is highest in the sky, and they are much lower for a day or so either side of the peak.

**URANUS**

MORNING SKY

**THE CARINA NEBULA**
This huge nebula in the southern Milky Way is visible to the naked eye. Eta (η) Carinae (centre, left) is a peculiar variable star, which is surrounded by a glowing shell of gas.

# SOUTHERN LATITUDES

## THE STARS

In the southern hemisphere, Crux lies almost on the north–south meridian line, with Rigil Kentaurus (see p.248) and Hadar – Alpha (α) and Beta (β) Centauri – slightly to its lower left. Antares, in Scorpius, is rising in the southeast while Canopus, in Carina, sinks low in the southwest. Hydra's long body meanders overhead, its head adjoining Cancer in the northwest and its tail ending between Libra and Centaurus in the southeast. Spica, the brightest star in Virgo, is high in the east. Leo lies in the north with Arcturus, in Boötes, in the northeast. Observers north of latitude 40°S can see the Plough low on the northern horizon.

## DEEP-SKY OBJECTS

Next to the Southern Cross, an apparent gap in the rich stream of the Milky Way is visible to the naked eye. This is, in fact, a dark nebula, known as the Coalsack, which obscures the light of the background stars. On its edge is the Jewel Box cluster (see p.288), or NGC 4755, which looks like a hazy star to the naked eye.

On show in Carina is the cluster IC 2602 and the Carina Nebula (see p.245), or NGC 3372. To the east, among the rich star fields of Centaurus, is the globular cluster NGC 5139 or Omega (ω) Centauri, which looks like a hazy 4th-magnitude star. In the north of the sky, members of the Virgo Cluster are well placed for telescopic observation this month.

**THE COALSACK**
This dark cloud of dust (centre left), next to the Southern Cross, is silhouetted against the bright background of the Milky Way.

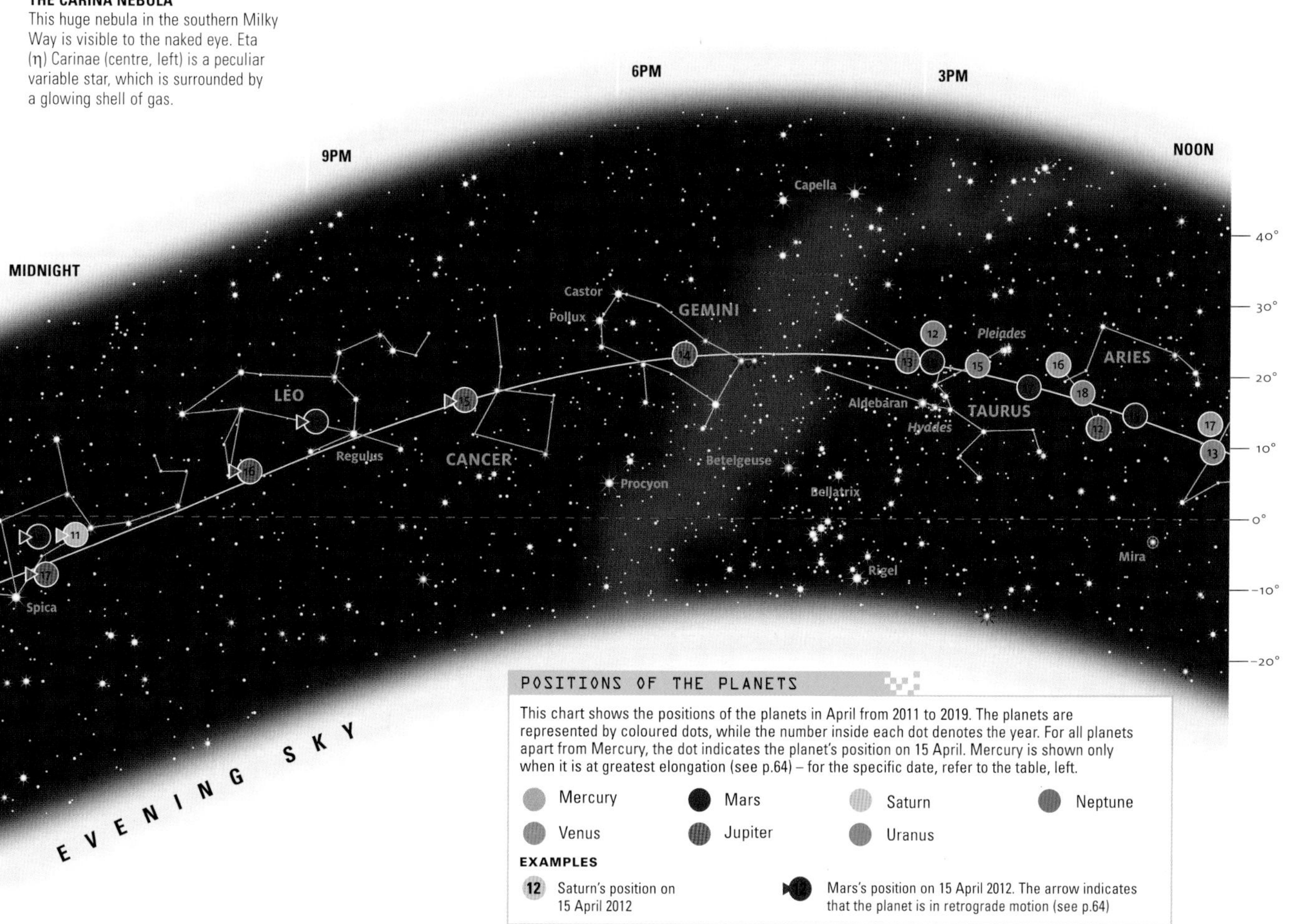


## POSITIONS OF THE PLANETS

This chart shows the positions of the planets in April from 2011 to 2019. The planets are represented by coloured dots, while the number inside each dot denotes the year. For all planets apart from Mercury, the dot indicates the planet's position on 15 April. Mercury is shown only when it is at greatest elongation (see p.64) – for the specific date, refer to the table, left.

Mercury, Mars, Saturn, Neptune, Venus, Jupiter, Uranus

**EXAMPLES**

12 Saturn's position on 15 April 2012

Mars's position on 15 April 2012. The arrow indicates that the planet is in retrograde motion (see p.64)

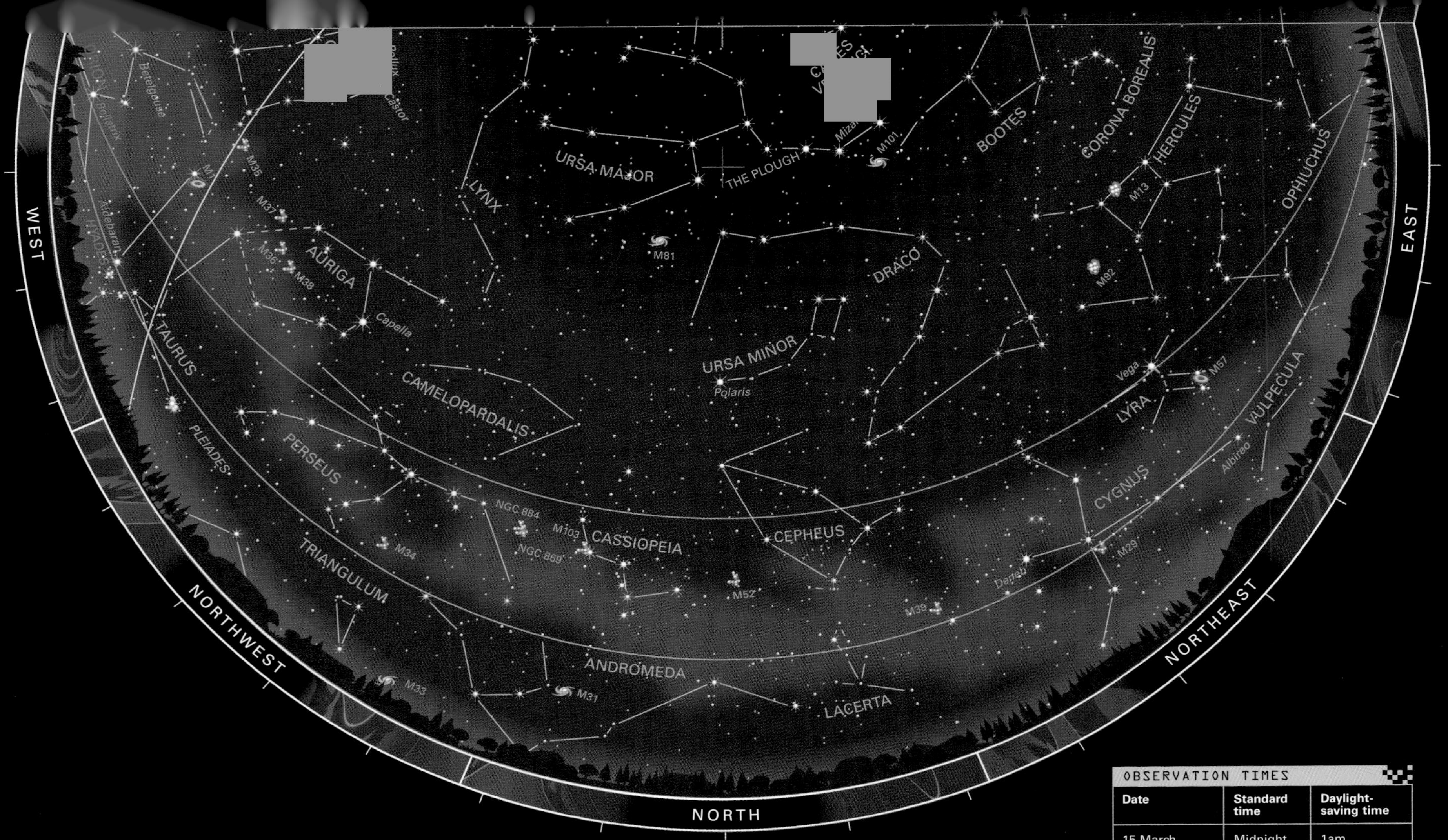

# APRIL | NORTHERN LATITUDES

LOOKING NORTH

| OBSERVATION TIMES | | |
|---|---|---|
| Date | Standard time | Daylight-saving time |
| 15 March | Midnight | 1am |
| 1 April | 11pm | Midnight |
| 15 April | 10pm | 11pm |
| 1 May | 9pm | 10pm |
| 15 May | 8pm | 9pm |

STAR MAGNITUDES

-1 · 0 · 1 · 2 · 3 · 4 · 5 · Variable star

DEEP-SKY OBJECTS

Galaxy · Globular cluster · Open cluster · Diffuse nebula · Planetary nebula

POINTS OF REFERENCE

**Horizons** 60°N · 40°N · 20°N **Zeniths** 60°N · 40°N · 20°N · Ecliptic

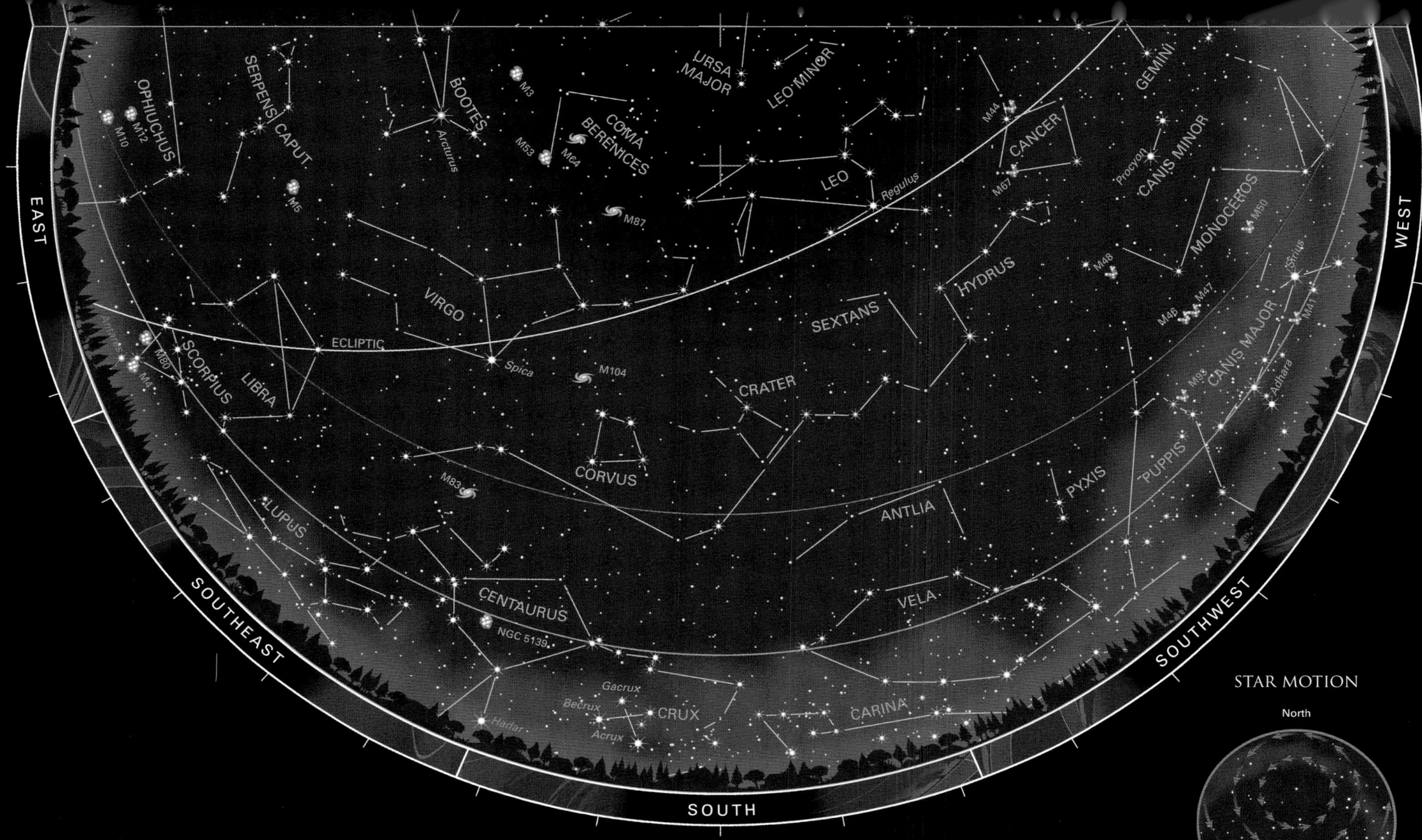

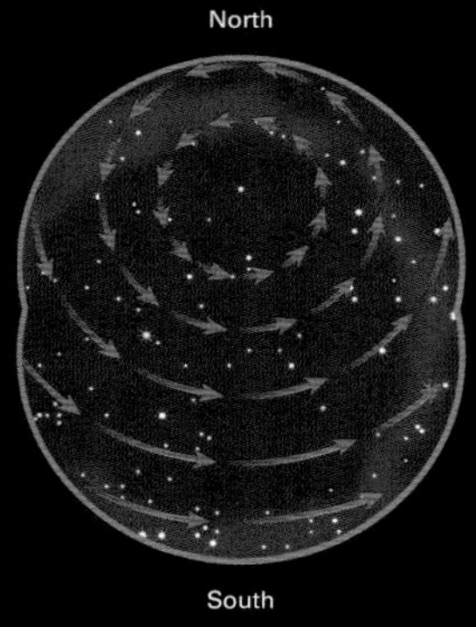

# APRIL | NORTHERN LATITUDES

LOOKING SOUTH

EAST
SOUTHEAST
SOUTH
SOUTHWEST
WEST

OPHIUCHUS
M10
M12
SERPENS CAPUT
M5
BOOTES
Arcturus
M3
M53
M64
COMA BERENICES
M87
URSA MAJOR
LEO MINOR
LEO
Regulus
M44
CANCER
M67
GEMINI
Procyon
CANIS MINOR
MONOCEROS
M50
M48
HYDRUS
SEXTANS
VIRGO
ECLIPTIC
Spica
M104
CRATER
CORVUS
SCORPIUS
LIBRA
Antares
M80
M4
LUPUS
M83
CENTAURUS
NGC 5139
Hadar
Becrux
Gacrux
Acrux
CRUX
CARINA
VELA
ANTLIA
PYXIS
PUPPIS
M46
M47
M93
CANIS MAJOR
Sirius
M41
Adhara

STAR MOTION

North

South

| STAR MAGNITUDES |
|---|
| -1 · 0 · 1 · 2 · 3 · 4 · 5 · Variable star |

| DEEP-SKY OBJECTS |
|---|
| Galaxy · Globular cluster · Open cluster · Diffuse nebula · Planetary nebula |

| POINTS OF REFERENCE |
|---|
| **Horizons** 60°N · 40°N · 20°N **Zeniths** 60°N · 40°N · 20°N · Ecliptic |

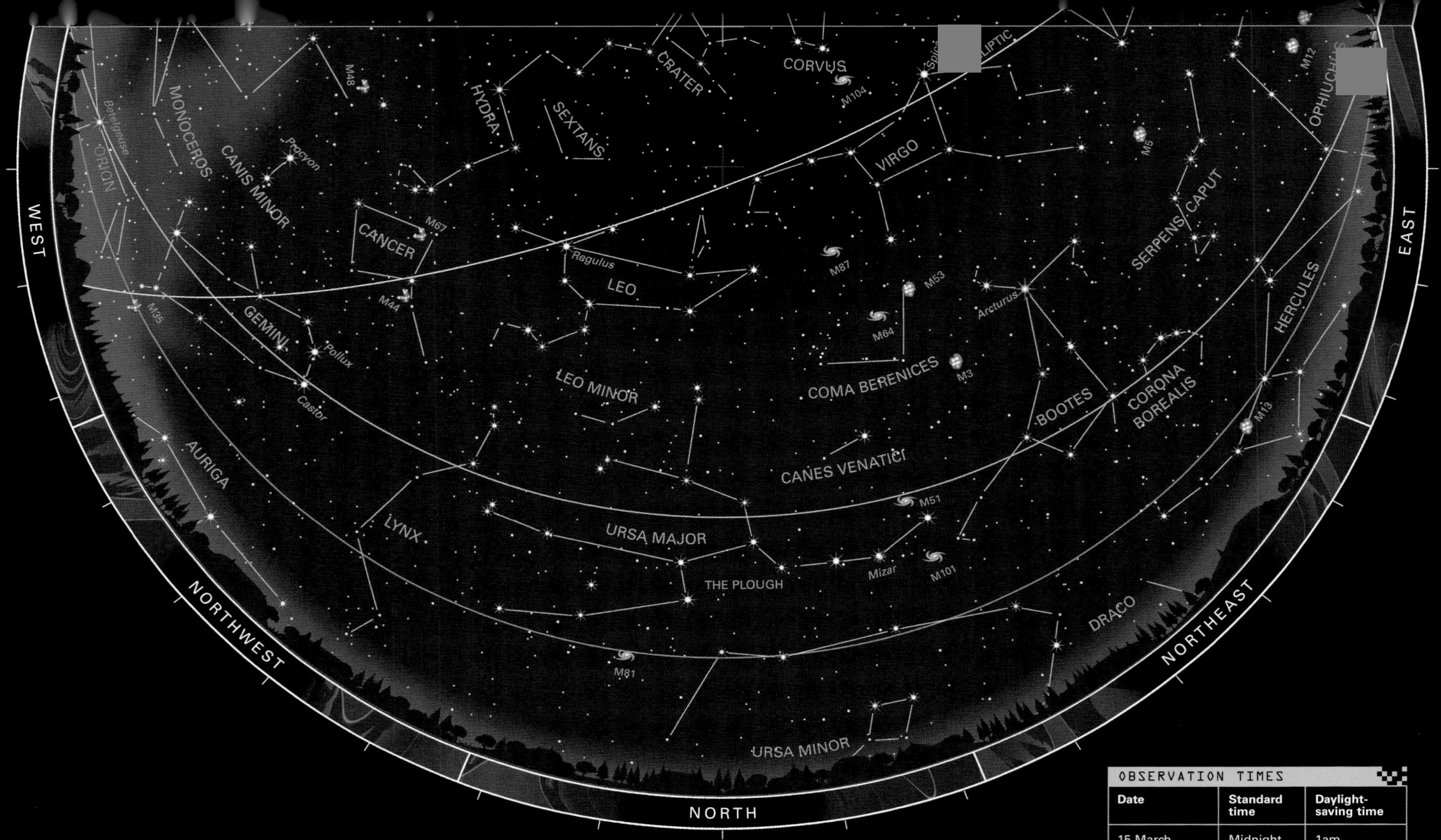

# APRIL | SOUTHERN LATITUDES

## LOOKING NORTH

| OBSERVATION TIMES | | |
|---|---|---|
| Date | Standard time | Daylight-saving time |
| 15 March | Midnight | 1am |
| 1 April | 11pm | Midnight |
| 15 April | 10pm | 11pm |
| 1 May | 9pm | 10pm |
| 15 May | 8pm | 9pm |

STAR MAGNITUDES

-1 · 0 · 1 · 2 · 3 · 4 · 5 · Variable star

DEEP-SKY OBJECTS

Galaxy · Globular cluster · Open cluster · Diffuse nebula · Planetary nebula

POINTS OF REFERENCE

Horizons: 0° · 20°S · 40°S

Zeniths: 0° · 20°S · 40°S

Ecliptic

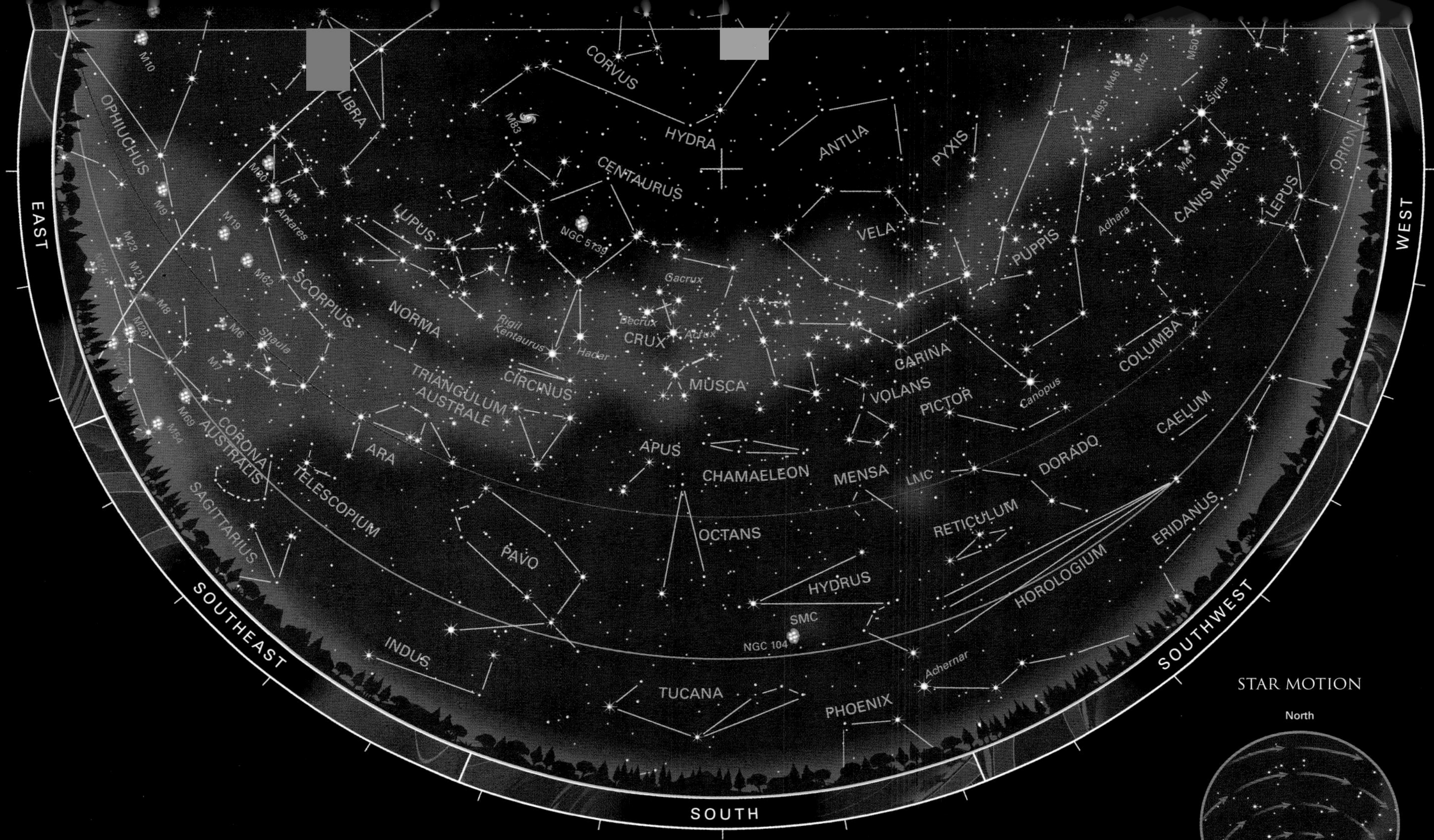

EAST
SOUTHEAST
SOUTH
SOUTHWEST
WEST

OPHIUCHUS
M10
M9
M23
M21
M8
M28
M22
M6
M7
M69
M54
LIBRA
M80
M4
Antares
M19
M62
SCORPIUS
Shaula
CORONA AUSTRALIS
SAGITTARIUS
TELESCOPIUM
ARA
LUPUS
NORMA
TRIANGULUM AUSTRALE
CORVUS
M83
HYDRA
CENTAURUS
NGC 5139
Rigil Kentaurus
Hadar
CIRCINUS
Gacrux
Becrux
Acrux
CRUX
MUSCA
APUS
PAVO
INDUS
CHAMAELEON
OCTANS
TUCANA
ANTLIA
VELA
PYXIS
CARINA
VOLANS
MENSA
HYDRUS
SMC
NGC 104
PHOENIX
Achernar
PICTOR
Canopus
LMC
RETICULUM
HOROLOGIUM
DORADO
CAELUM
ERIDANUS
COLUMBA
PUPPIS
Adhara
M93
M46
M47
M50
M41
Sirius
CANIS MAJOR
LEPUS
ORION

STAR MOTION

North

South

LOOKING SOUTH

# APRIL | SOUTHERN LATITUDES

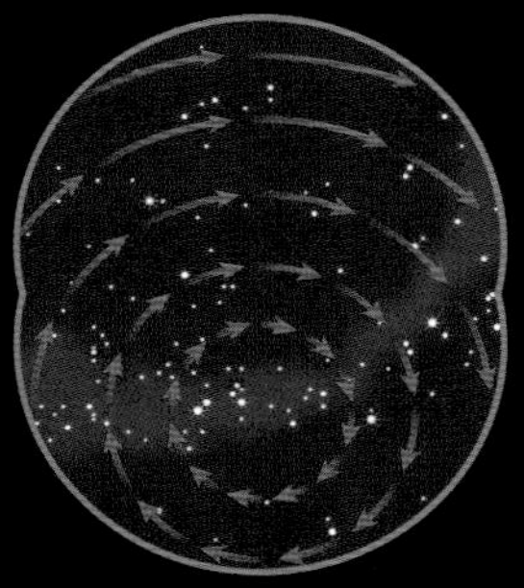

STAR MAGNITUDES

-1 · 0 · 1 · 2 · 3 · 4 · 5 · Variable star

DEEP-SKY OBJECTS

Galaxy · Globular cluster · Open cluster · Diffuse nebula · Planetary nebula

POINTS OF REFERENCE

Horizons 0° | 20°S | 40°S · Zeniths 0° | 20°S | 40°S · Ecliptic

## SPECIAL EVENTS

**PHASES OF THE MOON**

| | FULL MOON | NEW MOON |
|---|---|---|
| **2011** | 17 May | 3 May |
| **2012** | 6 May | 20 May |
| **2013** | 25 May | 10 May |
| **2014** | 14 May | 28 May |
| **2015** | 4 May | 18 May |
| **2016** | 21 May | 6 May |
| **2017** | 10 May | 25 May |
| **2018** | 29 May | 15 May |
| **2019** | 18 May | 4 May |

**PLANETS**

**2011: 7 May** Mercury is at greatest morning elongation, magnitude 0.5.

**2014: 10 May** Saturn is at opposition, magnitude 0.06.

**2014: 25 May** Mercury is at greatest evening elongation, magnitude 0.6.

**2015: 7 May** Mercury is at greatest evening elongation, magnitude 0.5.

**2015: 23 May** Saturn is at opposition, magnitude 0.02.

**2016: 23–24 May** Mars is at opposition, magnitude -2.1.

**2017: 17 May** Mercury is at greatest morning elongation, magnitude 0.6.

**2018: 9 May** Jupiter is at opposition, magnitude -2.1.

**ECLIPSES**

**2012: 20–21 May** An annular eclipse of the Sun is visible from the northern Pacific Ocean, southern Japan, and the western United States. A partial solar eclipse is visible from northeast Asia, the northern Pacific Ocean, and western North America.

**2013: 10 May** An annular eclipse of the Sun is visible from the northern Australia and into the central Pacific Ocean.

# MAY

As summer approaches, the days get longer in the northern hemisphere, restricting early evening observation, while in the southern hemisphere the opposite is true as the days become shorter and the nights longer. For northern observers, the Plough is high up in the sky and Virgo is due south. Observers south of the equator are treated to the sight of the brilliant stars of Centaurus (the Centaur) and Crux (the Southern Cross) at their highest.

## NORTHERN LATITUDES

### THE STARS

The tip of the handle of the Plough lies on the north–south meridian this month. The second star in the handle, Mizar, has a fainter companion, Alcor, which is visible to the naked eye (see p.272). The curved handle of the Plough points towards orange Arcturus in Boötes, also high up. Almost due south is Spica, the brightest star in Virgo.

Gemini, the last of the winter constellations, begins to set in the northwest. As it departs, the stars of summer rise in the east, led by the brilliant blue-white star Vega (see p.249) in Lyra. For those observers at lower northerly latitudes, Antares and the stars of Scorpius begin to appear over the southeastern horizon.

### DEEP-SKY OBJECTS

Two large and relatively bright galaxies are well positioned for observation in May. South of the Plough's handle is the Whirlpool Galaxy (see p.305), or M51, while to the north of the handle is M101, which is larger but less prominent. On clear nights, each appears as a faint patch of light through binoculars; a telescope is needed to see their spiral structures. The fan-shaped Coma Star Cluster is well positioned, as is the Virgo Cluster of galaxies (see p.319).

### METEOR SHOWER

The Eta Aquarid meteor shower is visible this month, but because the radiant lies virtually on the celestial equator the shower is not seen well in far northerly latitudes.

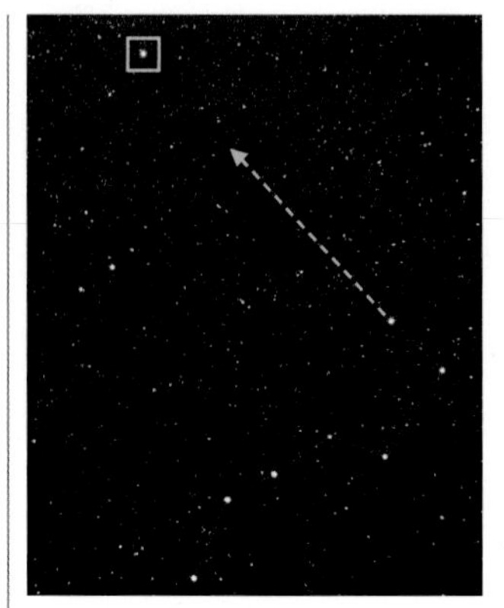

**FINDING THE POLE STAR**
Alpha (α) and Beta (β) Ursae Majoris, in the bowl of the Plough, point towards the north pole star, Polaris (in green box).

**NEPTUNE**

**URANUS**

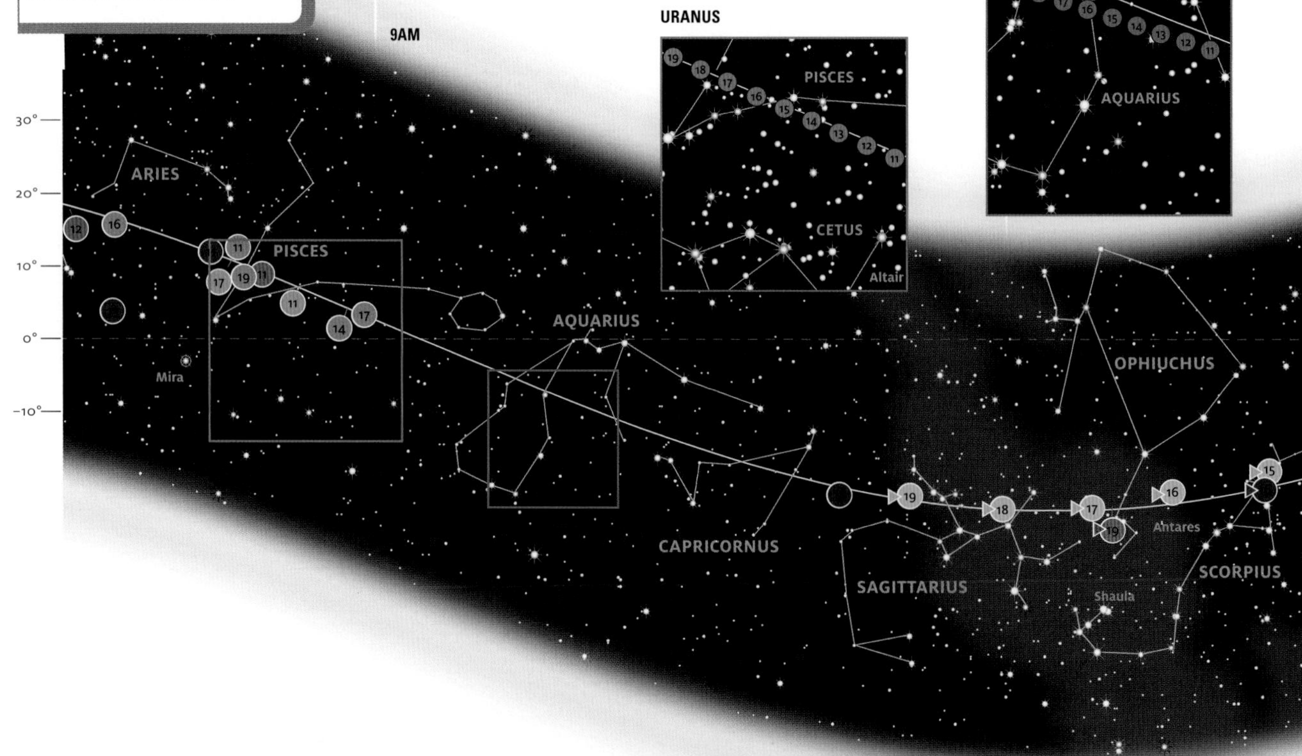


MORNING SKY

# SOUTHERN LATITUDES

## THE STARS

The constellation Crux and the two bright stars in Centaurus that act as a pointer to it, Alpha (α) Centauri – or Rigil Kentaurus (see p.248) – and Beta (β) Centauri – Hadar, are high in the southern sky in May. Crux is to the west of the north–south meridian, and Rigil Kentaurus and Hadar are on the eastern side. Although Rigil Kentaurus is usually described as the closest naked-eye star to the Sun, it actually consists of two yellowish stars, which form a double star that is easily divided through a small telescope. The brightest member of the Southern Cross, Acrux – Alpha (α) Crucis – is also a double star that is divisible with a small telescope, but its component stars are blue-white.

Spica, in Virgo, lies high overhead with orange Arcturus, in Boötes, in the north. Leo sinks towards the northwestern horizon, while in the southeast Scorpius and Sagittarius are coming into view – a sign that the southern winter is approaching.

## DEEP-SKY OBJECTS

The largest and brightest globular cluster in the sky, NGC 5139, or Omega (ω) Centauri, appears to the naked eye as a hazy 4th-magnitude star lying virtually on the north–south meridian this month. To the north of it lies NGC 5128, a peculiar radio-emitting galaxy also known as Centaurus A (see p.312), which is one of the easiest galaxies to find with binoculars. Another bright galaxy located near the meridian is M83, a spiral galaxy that is positioned face-on to the Earth.

In Crux, the dark Coalsack Nebula and the sparkling Jewel Box (see p.288) remain prominent.

## METEOR SHOWER

The Eta Aquarid meteor shower reaches its peak around 5–6 May, when 30 or so fast-moving meteors can be seen radiating each hour from near the star Eta (η) Aquarii, located almost exactly on the celestial equator. However, this part of the sky does not rise very high until around 3am, and the meteor shower is best seen from equatorial and southerly locations, where May nights are longer. The Eta Aquarids are caused by dust from Halley's Comet (see p.218).

**RICH STAR FIELDS**
Alpha (α) and Beta (β) Centauri (left) point towards the constellation Crux (right). The Coalsack nebula (bottom, right), most of which lies within Crux, obscures a large area of stars in the Milky Way.

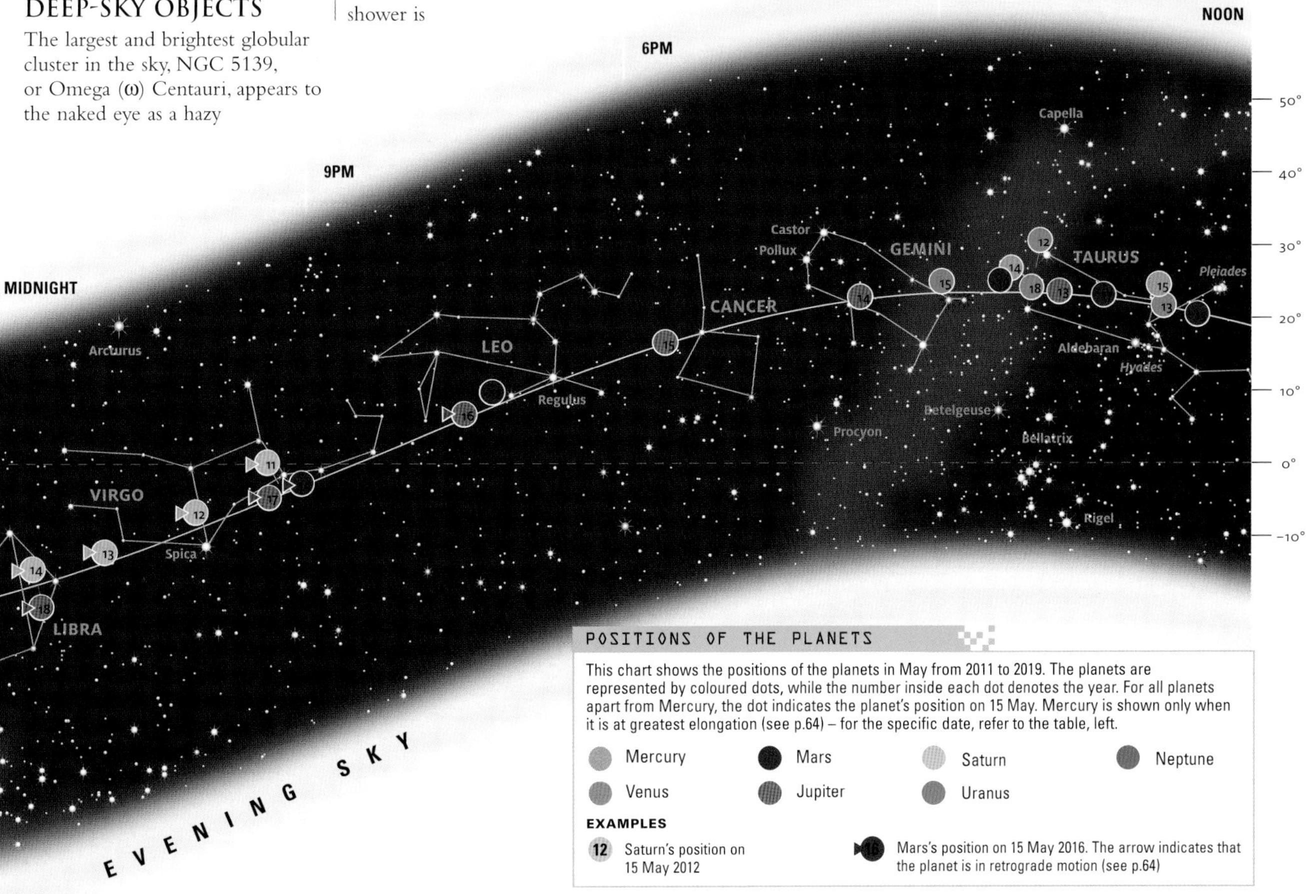


### POSITIONS OF THE PLANETS

This chart shows the positions of the planets in May from 2011 to 2019. The planets are represented by coloured dots, while the number inside each dot denotes the year. For all planets apart from Mercury, the dot indicates the planet's position on 15 May. Mercury is shown only when it is at greatest elongation (see p.64) – for the specific date, refer to the table, left.

Mercury, Venus, Mars, Jupiter, Saturn, Uranus, Neptune

**EXAMPLES**

12 Saturn's position on 15 May 2012

16 Mars's position on 15 May 2016. The arrow indicates that the planet is in retrograde motion (see p.64)

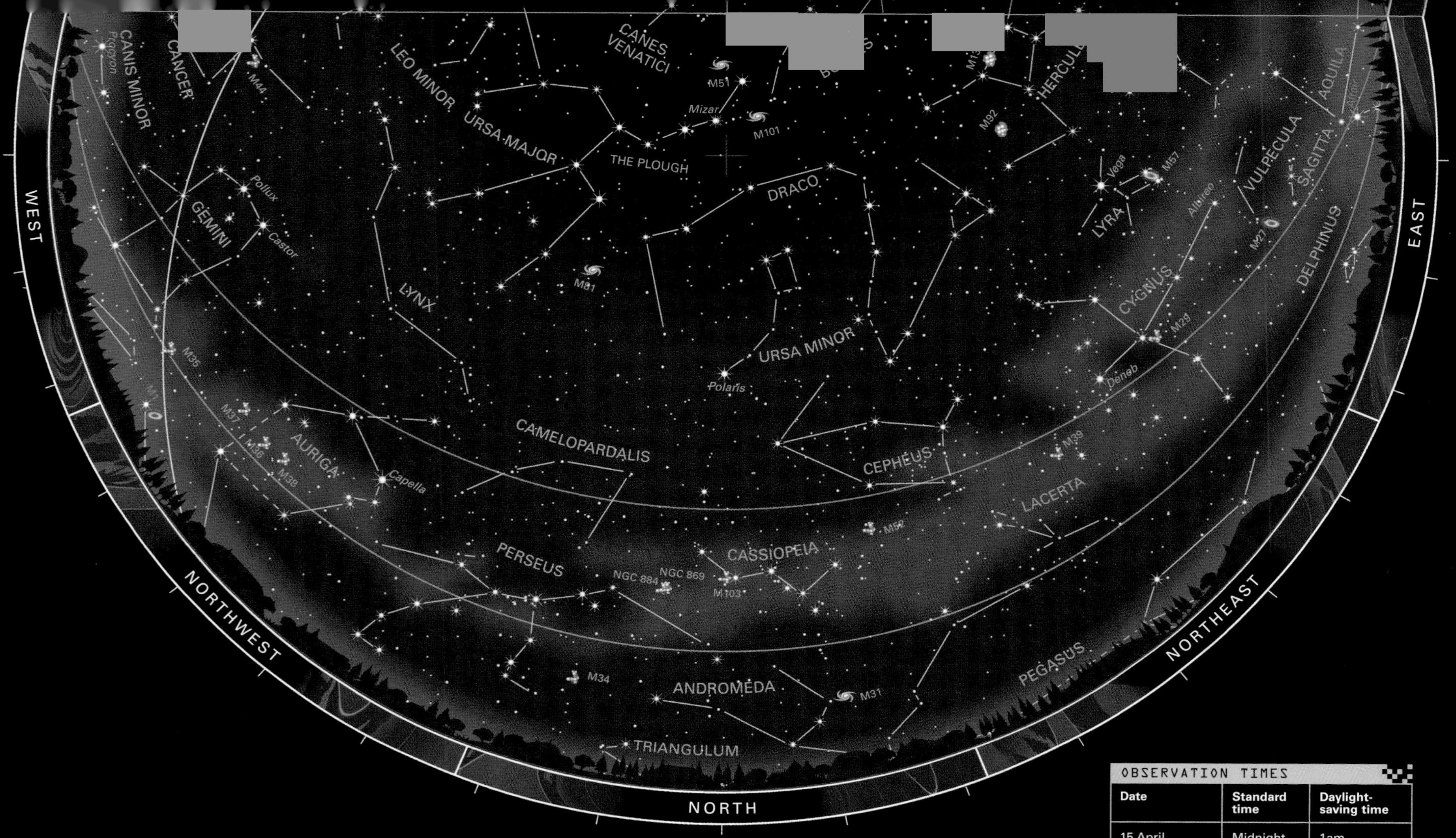

# MAY | NORTHERN LATITUDES

LOOKING NORTH

| OBSERVATION TIMES | | |
|---|---|---|
| Date | Standard time | Daylight-saving time |
| 15 April | Midnight | 1am |
| 1 May | 11pm | Midnight |
| 15 May | 10pm | 11pm |
| 1 June | 9pm | 10pm |
| 15 June | 8pm | 9pm |

STAR MAGNITUDES

-1 · 0 · 1 · 2 · 3 · 4 · 5 · Variable star

DEEP-SKY OBJECTS

Galaxy · Globular cluster · Open cluster · Diffuse nebula · Planetary nebula

POINTS OF REFERENCE

**Horizons** 60°N | 40°N | 20°N **Zeniths** 60°N | 40°N | 20°N | Ecliptic

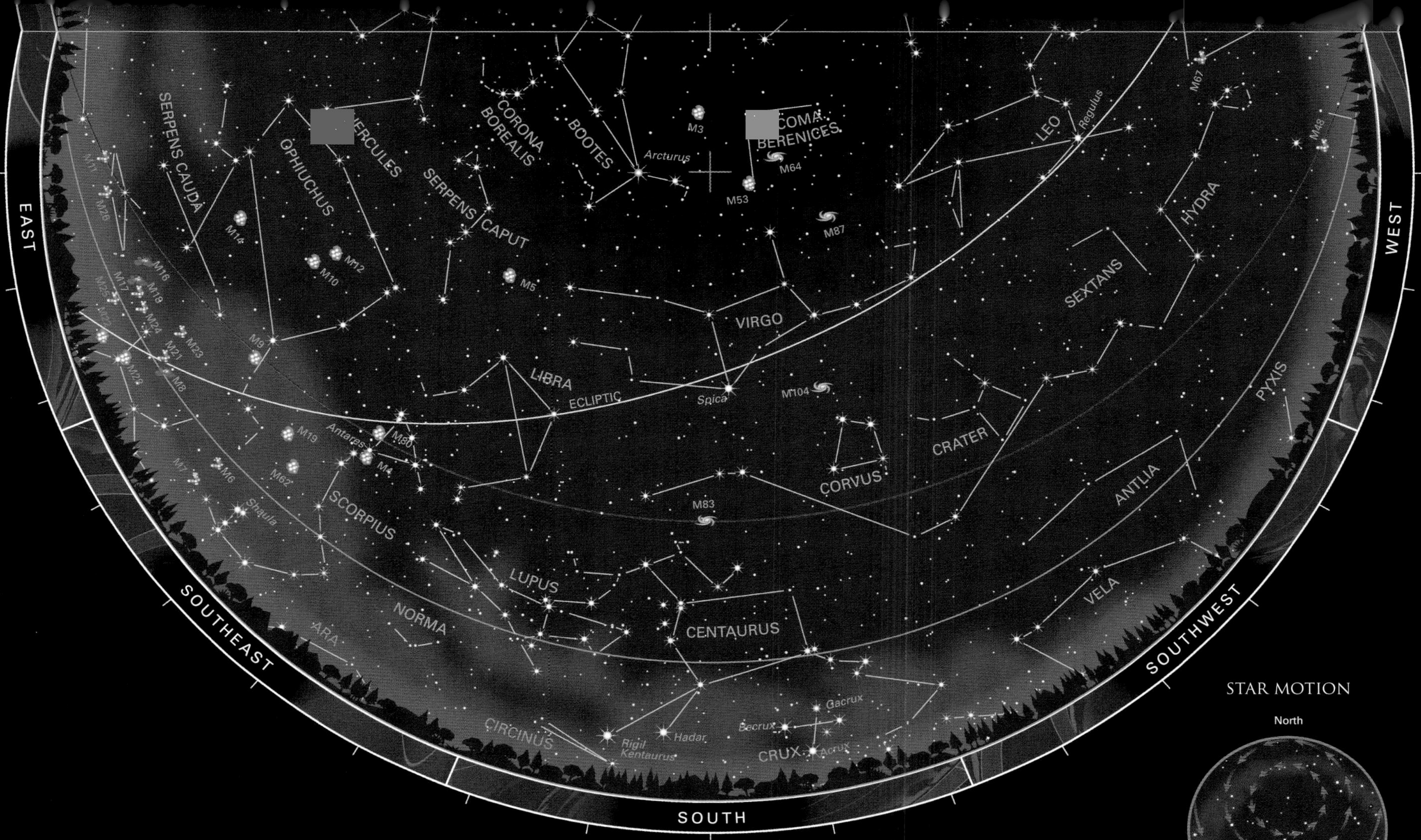

# MAY | NORTHERN LATITUDES

LOOKING SOUTH

EAST
SOUTHEAST
SOUTH
SOUTHWEST
WEST

STAR MOTION

North

South

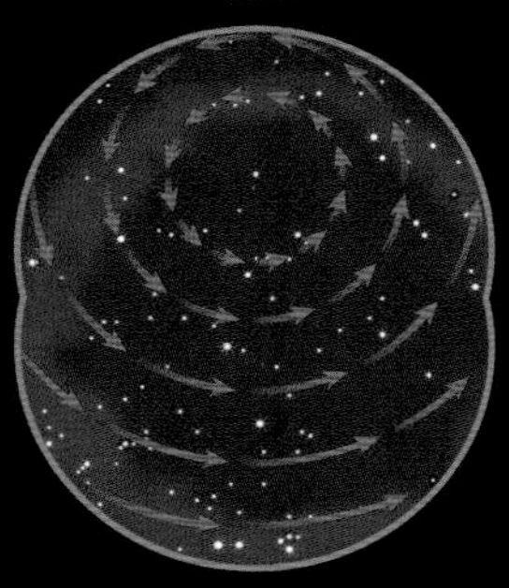

STAR MAGNITUDES

-1 · 0 · 1 · 2 · 3 · 4 · 5 · Variable star

DEEP-SKY OBJECTS

Galaxy · Globular cluster · Open cluster · Diffuse nebula · Planetary nebula

POINTS OF REFERENCE

Horizons: 60°N · 40°N · 20°N

Zeniths: 60°N · 40°N · 20°N

Ecliptic

LOOKING NORTH

# MAY | SOUTHERN LATITUDES

### OBSERVATION TIMES

| Date | Standard time | Daylight-saving time |
|---|---|---|
| 15 April | Midnight | 1am |
| 1 May | 11pm | Midnight |
| 15 May | 10pm | 11pm |
| 1 June | 9pm | 10pm |
| 15 June | 8pm | 9pm |

### STAR MAGNITUDES

-1 · 0 · 1 · 2 · 3 · 4 · 5 · Variable star

### DEEP-SKY OBJECTS

Galaxy · Globular cluster · Open cluster · Diffuse nebula · Planetary nebula

### POINTS OF REFERENCE

**Horizons** 0° | 20°S | 40°S **Zeniths** 0° | 20°S | 40°S | Ecliptic

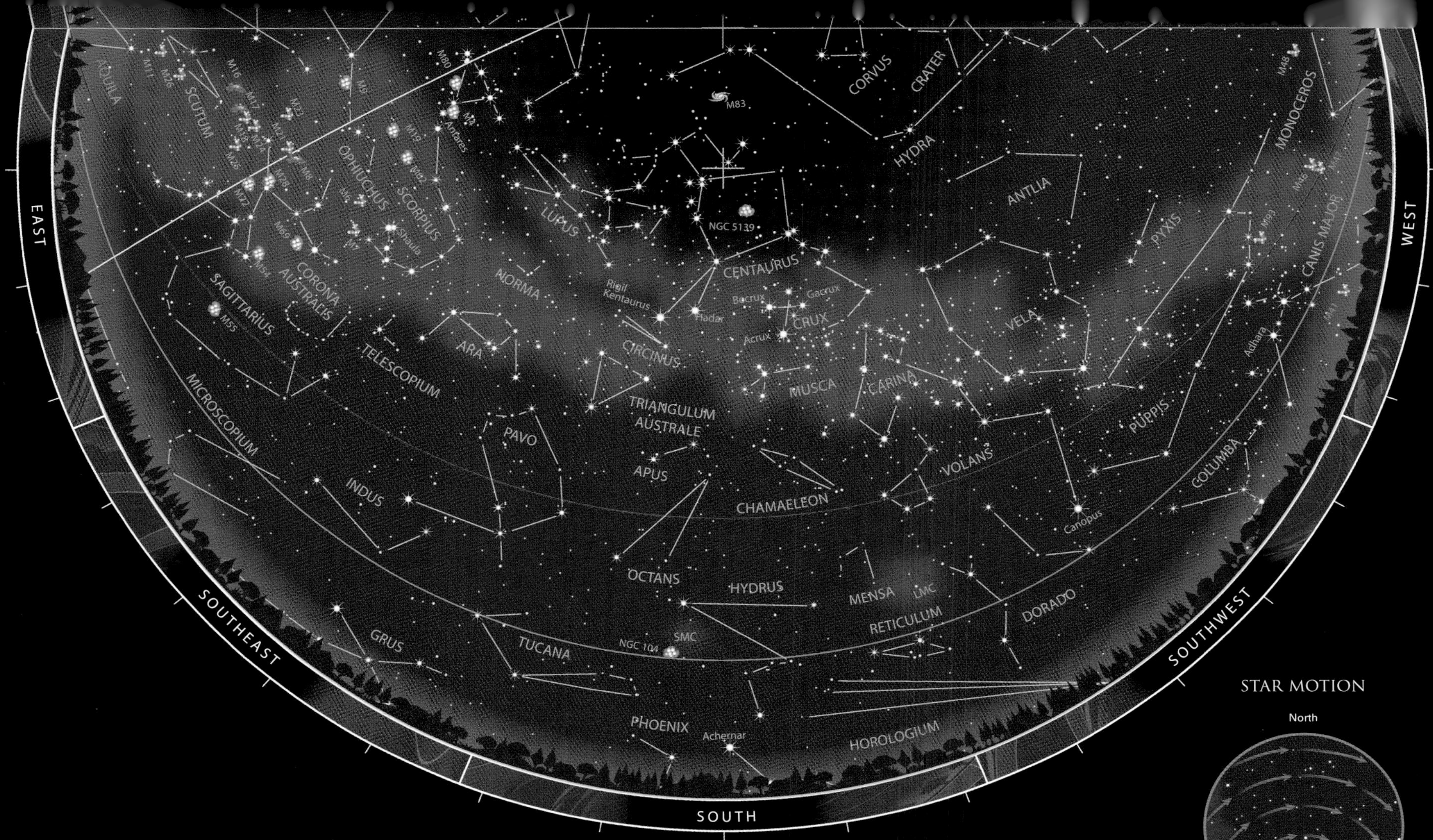

LOOKING SOUTH

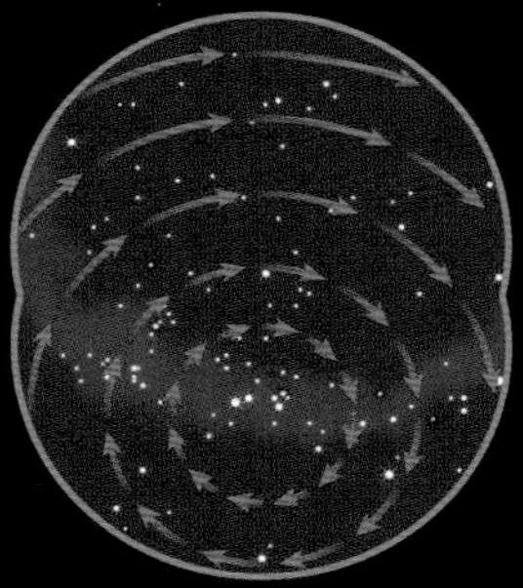

# MAY | SOUTHERN LATITUDES

STAR MOTION

North

South

**STAR MAGNITUDES**

-1 · 0 · 1 · 2 · 3 · 4 · 5 · Variable star

**DEEP-SKY OBJECTS**

Galaxy · Globular cluster · Open cluster · Diffuse nebula · Planetary nebula

**POINTS OF REFERENCE**

**Horizons** 0° | 20°S | 40°S **Zeniths** 0° 20°S 40°S | Ecliptic

# JUNE

Northern nights are at their shortest, and southern ones at their longest, around the solstice on 21 June, the date on which the Sun reaches its farthest point north of the celestial equator. In the northern sky, Arcturus and the other stars of Boötes stand high, and the giant Summer Triangle of Vega (in Lyra), Deneb (in Cygnus), and Altair (in Aquila) lies in the eastern half of the sky. Southern observers enjoy a rich band of constellations in the Milky Way during their long winter nights.

## SPECIAL EVENTS

**PHASES OF THE MOON**

| | FULL MOON | NEW MOON |
|---|---|---|
| **2011** | 15 June | 1 June |
| **2012** | 4 June | 19 June |
| **2013** | 23 June | 8 June |
| **2014** | 13 June | 27 June |
| **2015** | 2 June | 16 June |
| **2016** | 20 June | 5 June |
| **2017** | 9 June | 24 June |
| **2018** | 28 June | 13 June |
| **2019** | 17 June | 3 June |

**PLANETS**

**2013: 12 June** Mercury is at greatest evening elongation, magnitude 0.6.

**2015: 24 June** Mercury is at greatest morning elongation, magnitude 0.6.

**2016: 3 June** Saturn at opposition, magnitude 0.0.

**2016: 5 June** Mercury is at greatest morning elongation, magnitude 0.6.

**2017: 15 June** Saturn at opposition, magnitude -0.3.

**2018: 27 June** Saturn at opposition, magnitude -0.3.

**2019: 24 June** Mercury is at greatest evening elongation, magnitude 0.6.

**2019: 30 June** Jupiter at opposition, magnitude -2.6.

**ECLIPSES AND TRANSITS**

**2011: 1 June** A partial eclipse of the Sun is visible from Greenland, northern North America, and parts of northern and northeastern Asia.

**2011: 15 June** A total eclipse of the Moon is visible from Australia, southern Asia, the Indian Ocean, Africa, and Europe.

**2012: 4 June** A partial eclipse of the Moon is visible from western North and South America, the Pacific Ocean, Australasia, and eastern Asia.

**2012: 5–6 June** A transit of Venus across the Sun is visible from North America, the Pacific Ocean, Australasia, and Asia.

## NORTHERN LATITUDES

### THE STARS

The bowl of the Little Dipper, in Ursa Minor, stands high above the northern horizon with the sinuous body of Draco, the Dragon, winding around it. The horseshoe shape of Corona Borealis, the Northern Crown, lies on the north–south meridian with the head of Serpens, the Serpent, below it, while Arcturus, in Boötes, is high in the western half of the sky. In this area of sky, Arcturus is the base of a large Y-shaped pattern of bright stars completed by Epsilon (ε) and Gamma (γ) Boötis plus Alpha (α) Corona Borealis (also known as Alphecca). Leo is setting in the west, and Spica, in Virgo, is low in the southwest. In the eastern sky, the bright stars Vega (see p.249), Deneb, and Altair (see p.248) mark the corners of the Summer Triangle, best seen in late summer and autumn. Ruddy Antares and the stars of Scorpius twinkle low on the southern horizon – June and July are the best months of the year for far-northern observers to see Scorpius in the evening sky.

### DEEP-SKY OBJECTS

The brightest globular cluster in northern skies, M13, is high up on summer evenings. It can be found along one side of the Keystone of Hercules, a quadrangle of stars that form the torso of the constellation Hercules. M13 appears as a fuzzy 6th-magnitude star through binoculars, and it can be glimpsed by the naked eye under good conditions. It can be compared with M5, another 6th-magnitude globular cluster visible through binoculars. M5 lies in the head of Serpens and is usually regarded as the second-best northern globular cluster. Near the handle of the Plough, the spiral galaxies M51 and M101 remain well positioned for observation.

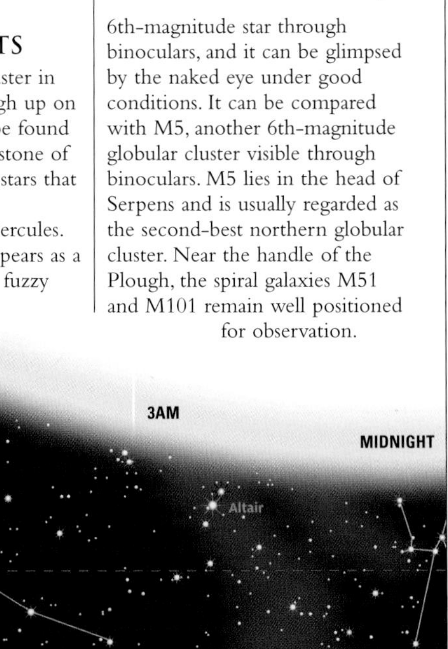

**NOCTILUCENT CLOUDS**
These high-altitude clouds can be seen on summer nights, illuminated by the Sun's rays that come over the horizon around midnight.

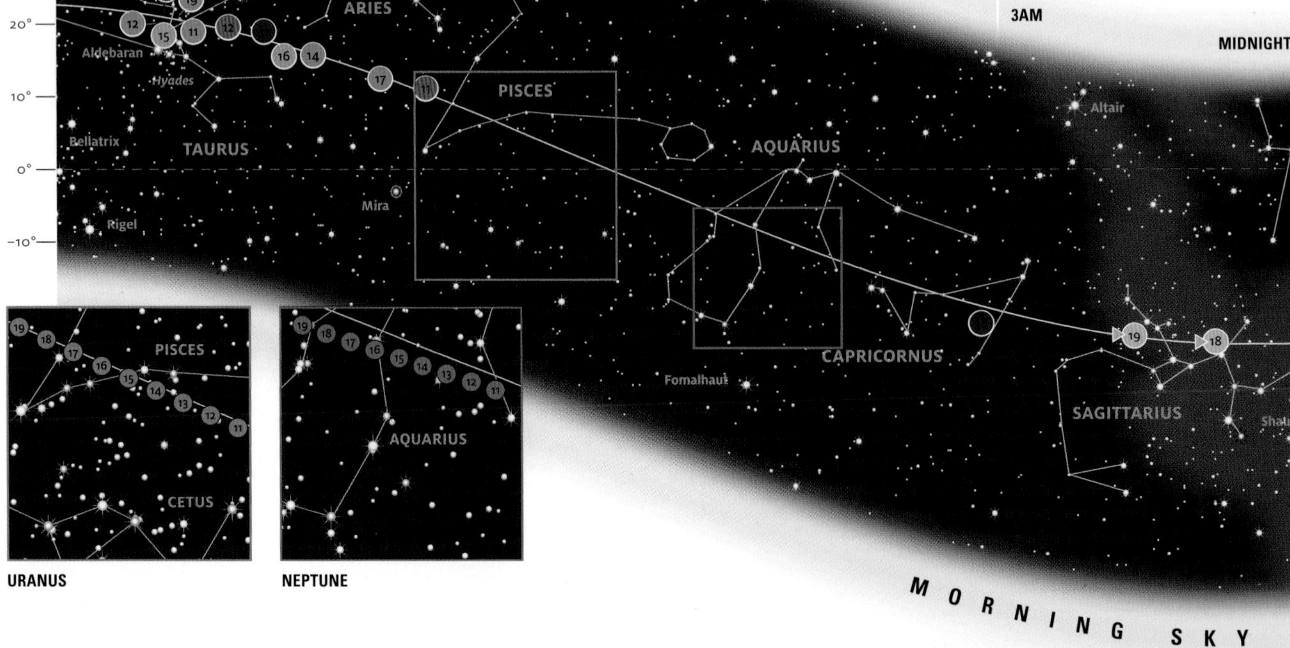

# SOUTHERN LATITUDES

## THE STARS

A rich band of constellations can be seen across the sky, from southwest to northeast, along the path of the Milky Way. Crux (the Southern Cross) and Centaurus (the Centaur), are in the southwest, to the right of the celestial meridian. The lesser-known constellations Lupus, Norma, and Triangulum Australe are on the meridian. Ruddy Antares (see p.252) is overhead, with the curving tail of Scorpius, the Scorpion, extending to the southeast. Next to its tail are the dense star fields of Sagittarius in the Milky Way. Along the Milky Way to the east is Altair (see p.248) in the constellation Aquila, while Vega (see p.249) is low in the northeast. Arcturus and Spica are high in the northwest.

## DEEP-SKY OBJECTS

Heading away from Scorpius and towards the Milky Way and the centre of the Galaxy, two magnificent open star clusters, M6 and M7, are positioned near the end of the Scorpion's tail. Both clusters are visible to the naked eye, and they appear magnificent through binoculars. M7 is the larger and brighter of the two; it appears twice the width of the full Moon. Another prominent open cluster in Scorpius is NGC 6231, positioned next to Zeta (ζ) Scorpii.

The globular cluster Omega (ω) Centauri, or NGC 5139, and the peculiar galaxy NGC 5128, or Centaurus A, remain well placed for observation this month, as do the Coalsack Nebula and the Jewel Box Cluster (see p.288), in Crux, and the spiral galaxy M83 (in the constellation Hydra).

**THE SCORPION'S LAIR**
Orange-red Antares, the star at the heart of Scorpius, and the curved line of stars marking the Scorpion's tail are distinctive sights in June skies. Hovering over the "sting" in the tail are two prominent star clusters, M6 and M7 (bottom, left).

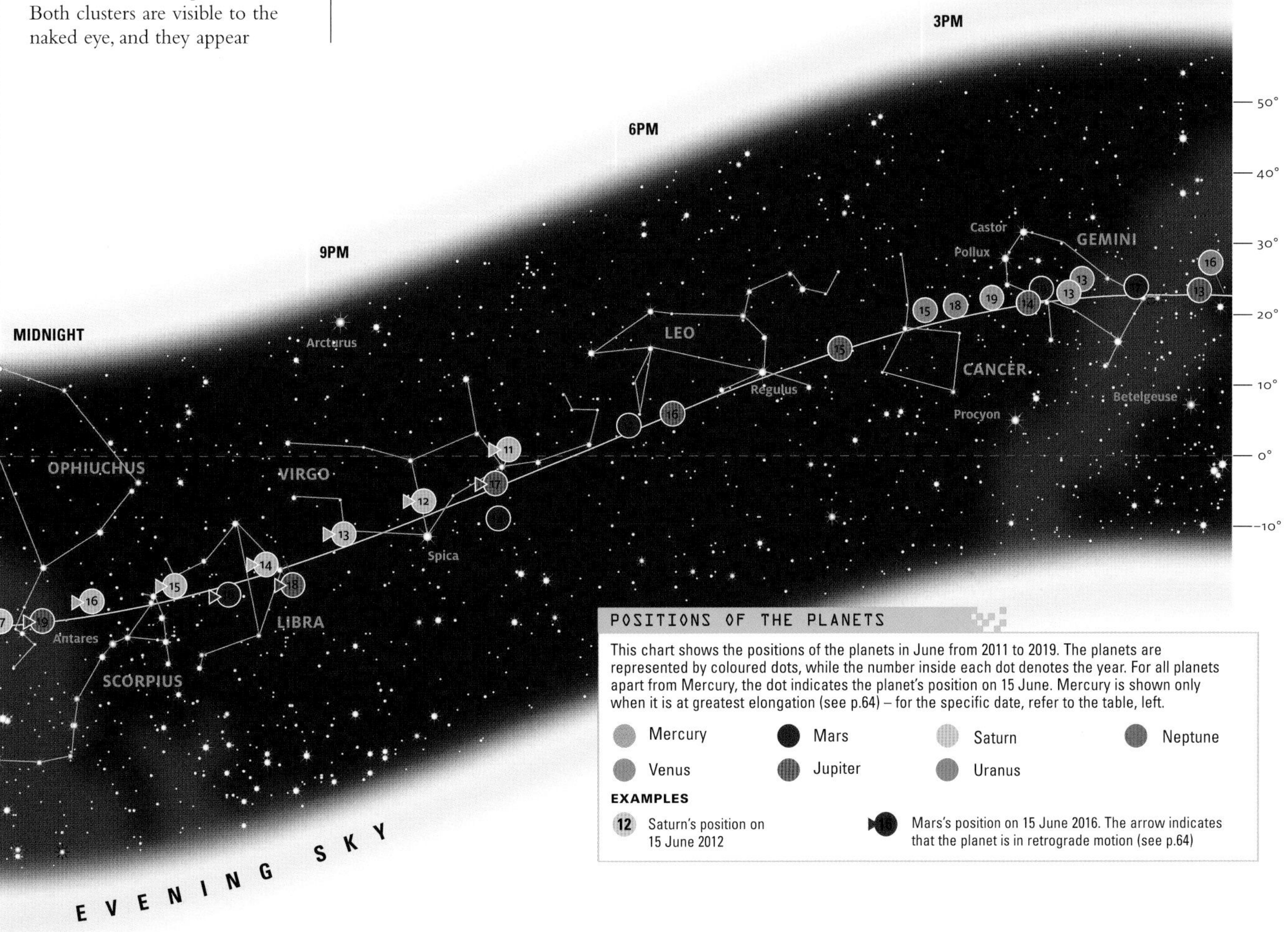


**POSITIONS OF THE PLANETS**

This chart shows the positions of the planets in June from 2011 to 2019. The planets are represented by coloured dots, while the number inside each dot denotes the year. For all planets apart from Mercury, the dot indicates the planet's position on 15 June. Mercury is shown only when it is at greatest elongation (see p.64) – for the specific date, refer to the table, left.

Mercury · Mars · Saturn · Neptune · Venus · Jupiter · Uranus

**EXAMPLES**

12 Saturn's position on 15 June 2012

Mars's position on 15 June 2016. The arrow indicates that the planet is in retrograde motion (see p.64)

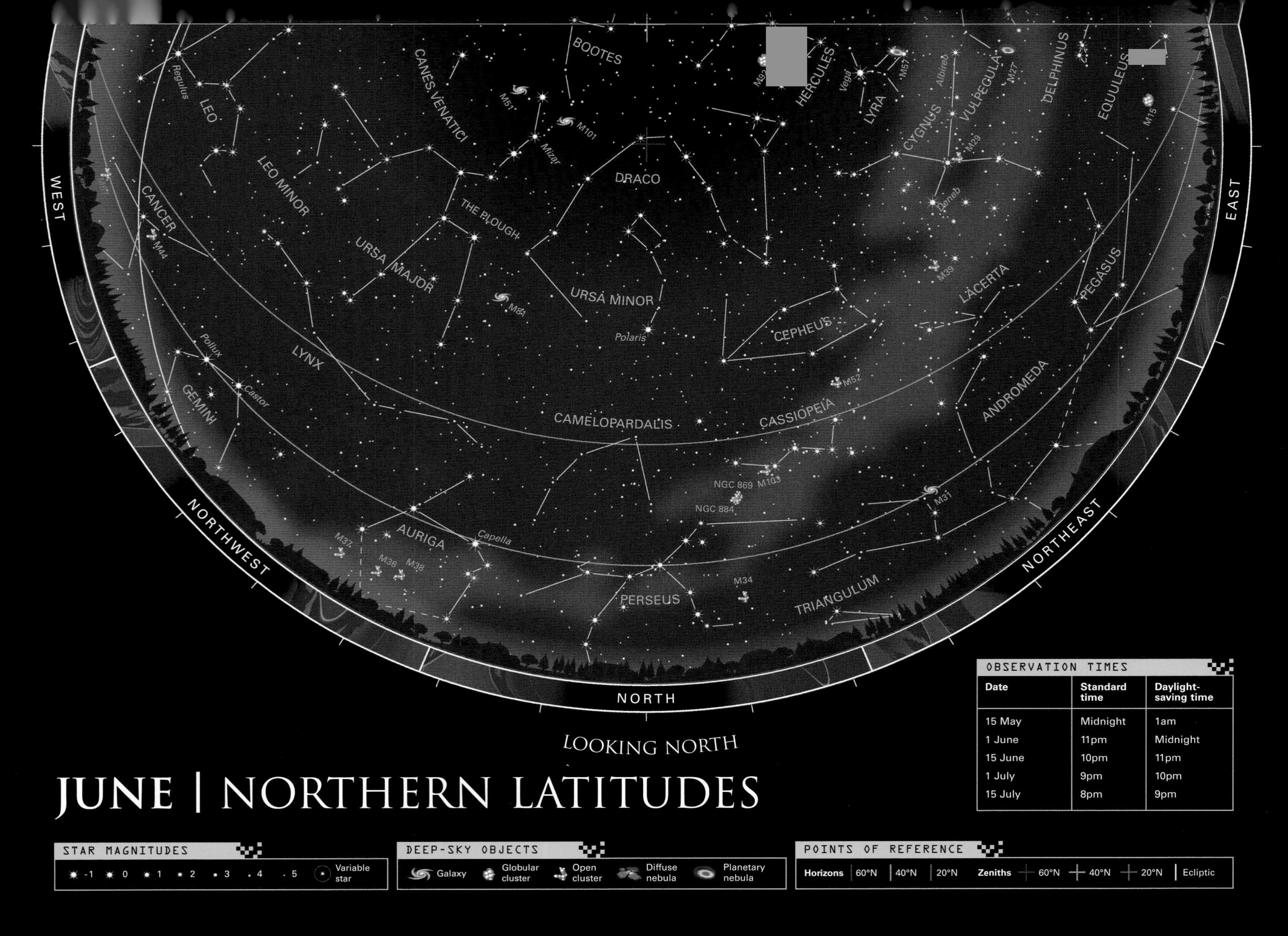


LOOKING NORTH

# JUNE | NORTHERN LATITUDES

**OBSERVATION TIMES**

| Date | Standard time | Daylight-saving time |
|---|---|---|
| 15 May | Midnight | 1am |
| 1 June | 11pm | Midnight |
| 15 June | 10pm | 11pm |
| 1 July | 9pm | 10pm |
| 15 July | 8pm | 9pm |

**STAR MAGNITUDES**

-1 0 1 2 3 4 5 Variable star

**DEEP-SKY OBJECTS**

Galaxy Globular cluster Open cluster Diffuse nebula Planetary nebula

**POINTS OF REFERENCE**

Horizons 60°N 40°N 20°N Zeniths 60°N 40°N 20°N Ecliptic

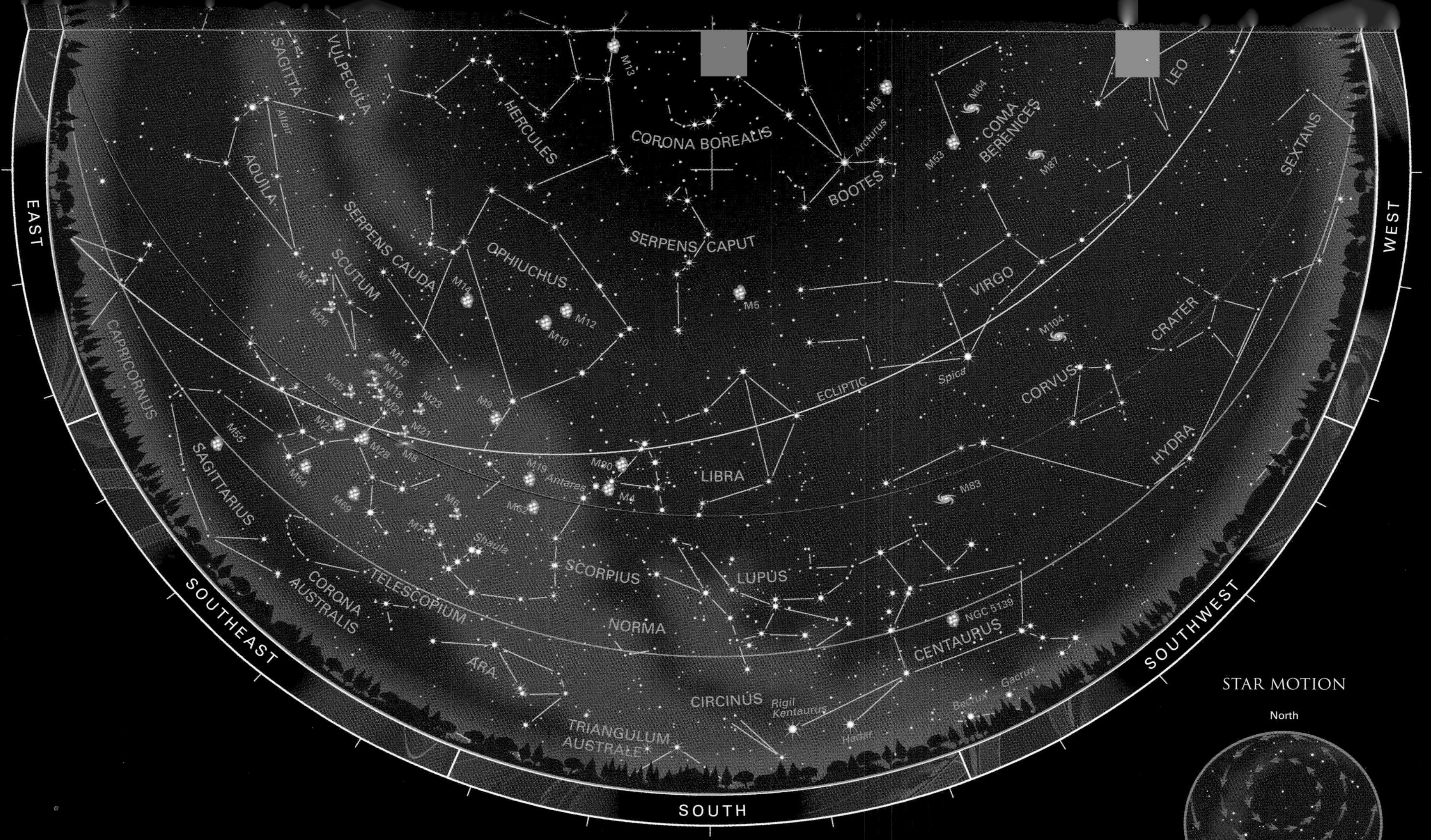

# JUNE | NORTHERN LATITUDES

LOOKING SOUTH

EAST
SOUTHEAST
SOUTH
SOUTHWEST
WEST

SAGITTA
VULPECULA
HERCULES
M13
CORONA BOREALIS
M3
Arcturus
M53
M64
COMA BERENICES
M87
LEO
SEXTANS
Altair
AQUILA
BOOTES
SERPENS CAPUT
SERPENS CAUDA
SCUTUM
M11
M26
OPHIUCHUS
M14
M12
M10
M5
VIRGO
CRATER
M104
Spica
CORVUS
ECLIPTIC
CAPRICORNUS
M16
M17
M18
M24
M25
M23
M21
M22
M28
M8
M9
M55
SAGITTARIUS
M54
M69
M19
Antares
M80
M4
LIBRA
HYDRA
M83
M6
M62
M7
Shaula
SCORPIUS
LUPUS
CORONA AUSTRALIS
TELESCOPIUM
NORMA
NGC 5139
CENTAURUS
ARA
CIRCINUS
Rigil Kentaurus
Hadar
Becrux
Gacrux
TRIANGULUM AUSTRALE

STAR MOTION

North

South

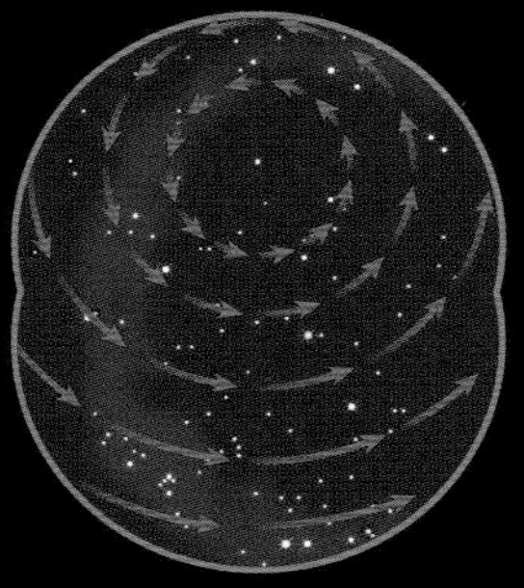

STAR MAGNITUDES

-1 0 1 2 3 4 5 Variable star

DEEP-SKY OBJECTS

Galaxy Globular cluster Open cluster Diffuse nebula Planetary nebula

POINTS OF REFERENCE

Horizons 60°N 40°N 20°N Zeniths 60°N 40°N 20°N Ecliptic

# JUNE | SOUTHERN LATITUDES

LOOKING NORTH

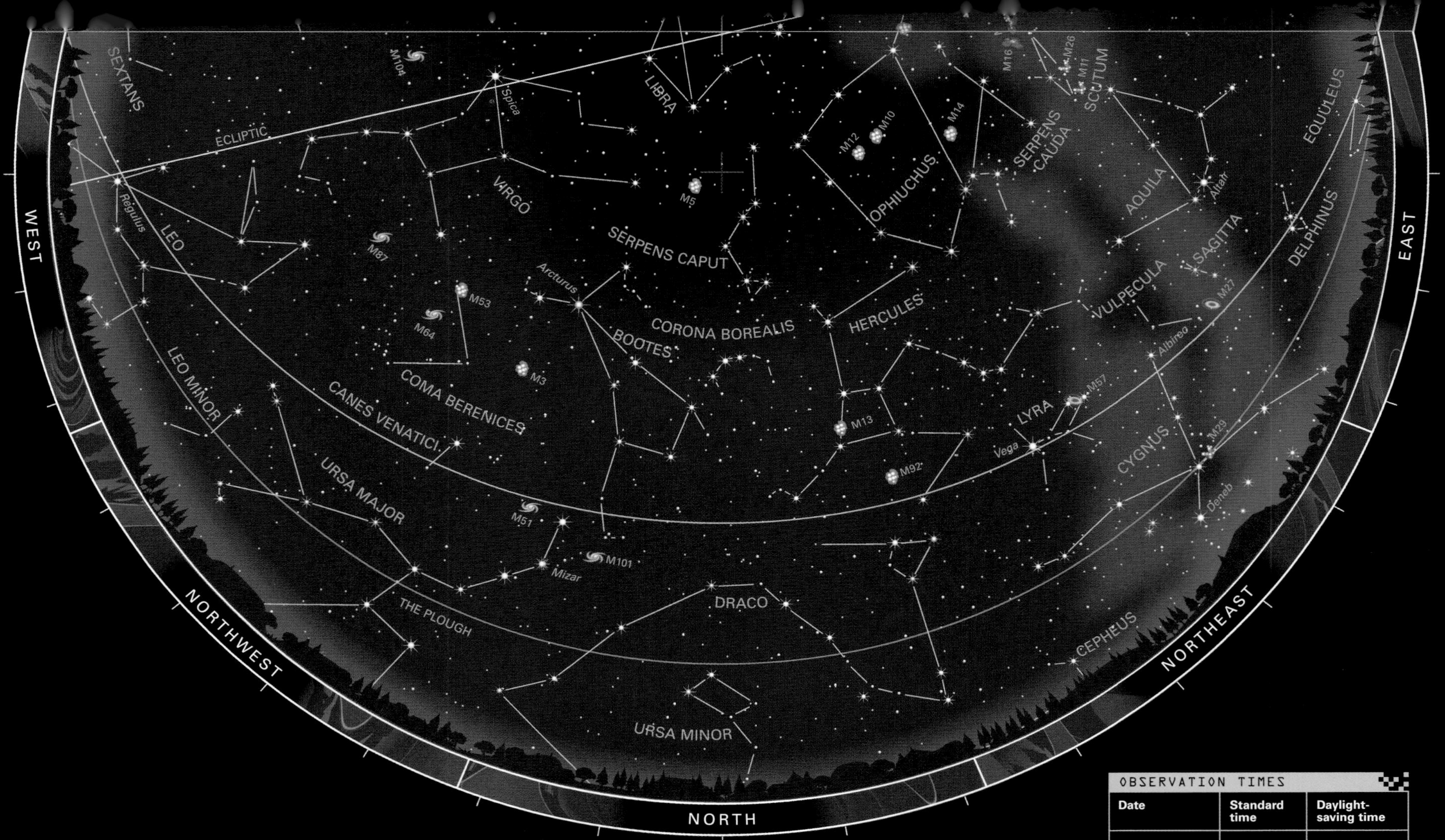

**OBSERVATION TIMES**

| Date | Standard time | Daylight-saving time |
|---|---|---|
| 15 May | Midnight | 1am |
| 1 June | 11pm | Midnight |
| 15 June | 10pm | 11pm |
| 1 July | 9pm | 10pm |
| 15 July | 8pm | 9pm |

**STAR MAGNITUDES**

-1 0 1 2 3 4 5 Variable star

**DEEP-SKY OBJECTS**

Galaxy Globular cluster Open cluster Diffuse nebula Planetary nebula

**POINTS OF REFERENCE**

Horizons 0° 20°S 40°S Zeniths 0° 20°S 40°S Ecliptic

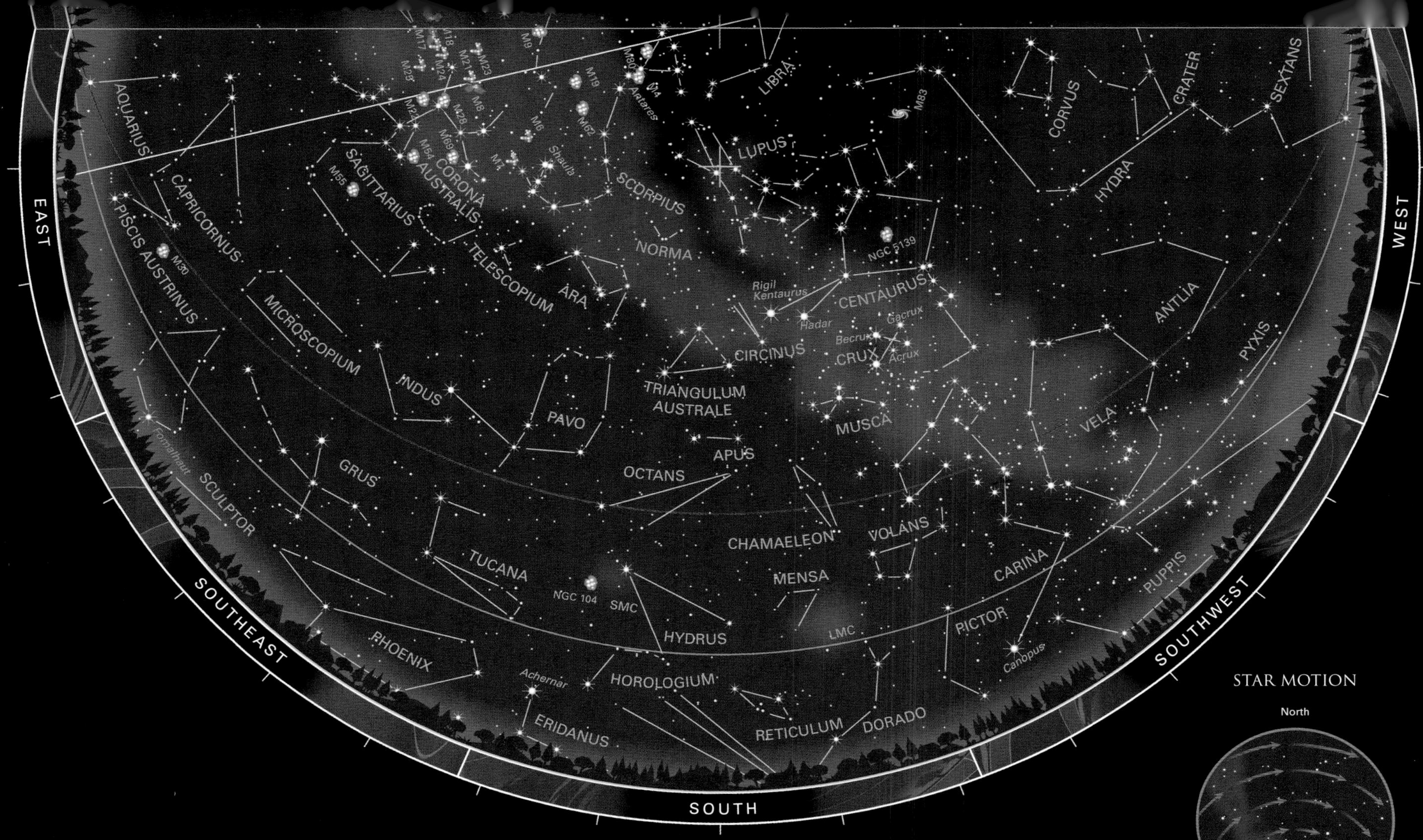

# JUNE | SOUTHERN LATITUDES

LOOKING SOUTH

EAST
SOUTHEAST
SOUTH
SOUTHWEST
WEST

AQUARIUS
CAPRICORNUS
PISCIS AUSTRINUS
M30
Fomalhaut
SCULPTOR
MICROSCOPIUM
SAGITTARIUS
M55
CORONA AUSTRALIS
M54
M69
M28
M22
M23
M24
M17
M18
M21
M23
M8
M9
M6
M7
Shaula
M19
M62
M80
M4
Antares
SCORPIUS
LIBRA
LUPUS
M83
CORVUS
CRATER
SEXTANS
HYDRA
TELESCOPIUM
ARA
NORMA
NGC 5139
Rigil Kentaurus
Hadar
CENTAURUS
Gacrux
Becrux
CRUX
Acrux
CIRCINUS
TRIANGULUM AUSTRALE
ANTLIA
PYXIS
INDUS
PAVO
MUSCA
VELA
GRUS
OCTANS
APUS
TUCANA
NGC 104
SMC
CHAMAELEON
VOLANS
MENSA
CARINA
PUPPIS
HYDRUS
LMC
PICTOR
Canopus
PHOENIX
Achernar
HOROLOGIUM
ERIDANUS
RETICULUM
DORADO

STAR MOTION
North
South

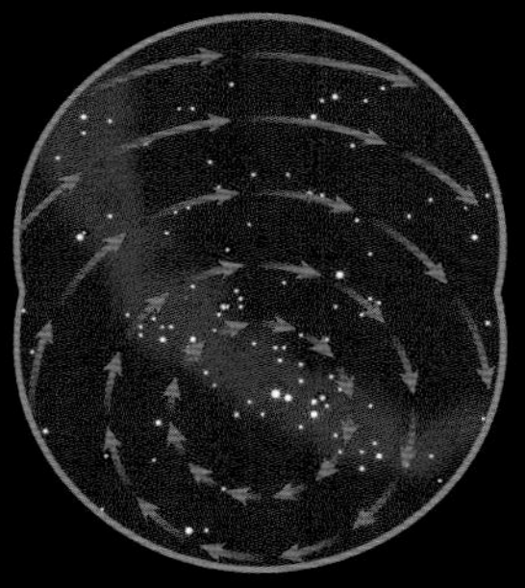

STAR MAGNITUDES
-1 0 1 2 3 4 5 Variable star

DEEP-SKY OBJECTS
Galaxy | Globular cluster | Open cluster | Diffuse nebula | Planetary nebula

POINTS OF REFERENCE
Horizons 0° | 20°S | 40°S Zeniths 0° | 20°S | 40°S | Ecliptic

# JULY

The strong man of Greek mythology, Hercules, lies overhead as seen from mid-northern latitudes, between the bright stars Vega (in Lyra) and Arcturus (in Boötes). South of Hercules is another large constellation, Ophiuchus, which represents a man encoiled by a serpent, Serpens. In southern skies, the Milky Way branches overhead from the southwest to the northeast. The zodiacal constellations Scorpius and Sagittarius stand high in the Milky Way's richest part.

## SPECIAL EVENTS

### PHASES OF THE MOON

| | FULL MOON | NEW MOON |
|---|---|---|
| **2011** | 15 July | 1, 30 July |
| **2012** | 3 July | 19 July |
| **2013** | 22 July | 8 July |
| **2014** | 12 July | 26 July |
| **2015** | 2, 31 July | 16 July |
| **2016** | 19 July | 4 July |
| **2017** | 9 July | 23 July |
| **2018** | 27 July | 13 July |
| **2019** | 16 July | 2 July |

### PLANETS

**2011: 20 July** Mercury is at greatest evening elongation, magnitude 0.5.

**2012: 1 July** Mercury is at greatest evening elongation, magnitude 0.5.

**2013: 30 July** Mercury is at greatest morning elongation, magnitude 0.3.

**2014: 12 July** Mercury is at greatest morning elongation, magnitude 0.3.

**2017: 30 July** Mercury is at greatest evening elongation, magnitude 0.4.

**2018: 12 July** Mercury is at greatest evening elongation, magnitude 0.5.

**2018: 27 July** Mars is at opposition, magnitude -2.8.

**2019: 9 July** Saturn is at opposition, magnitude -0.3.

### ECLIPSES

**2018: 13 July** A partial eclipse of the Sun is visible from southern Australia.

**2019: 2 July** A total eclipse of the Sun is visible from south Pacific, Chile, and Argentina. A partial solar eclipse is visible from south Pacific and South America.

**2019: 16 July** A partial eclipse of the Moon is visible from North America, South America, central Pacific, Europe, and Africa.

## NORTHERN LATITUDES

### THE STARS

Overhead lies Hercules, which is a large but not particularly striking constellation. Its most distinctive feature is a quadrangle formed by four stars, called the Keystone. North of Hercules lies the lozenge-shaped head of Draco, the Dragon. Between Draco and the north celestial pole is the bowl of the Little Dipper, in Ursa Minor.

Arcturus, in Boötes, remains prominent in the western sky. Spica, in Virgo, is lower in the southwest, and the Plough dips low in the northwest. In the eastern half of the sky, the stars of the Summer Triangle climb ever higher, while the Square of Pegasus appears closer to the eastern horizon.

Low in the south are the rich constellations Scorpius and Sagittarius. This is the best month for northern observers to see the two most southerly zodiacal figures in the evening sky.

### DEEP-SKY OBJECTS

Ophiuchus, the large constellation between Hercules and Scorpius, contains numerous globular clusters, although only two of them, M10 and M12, are of any note. The most impressive deep-sky objects in Ophiuchus are the open clusters IC 4665 and NGC 6633, both good binocular sights. The globular clusters M13, in Hercules, and M5, in the head of Serpens, remain well positioned this month.

**THE SUMMER TRIANGLE**
Deneb (left), Vega (top), and Altair (right) form a prominent triangle that remains visible well into autumn in northern skies.

9AM
6AM
3AM
MIDNIGHT
40°
30°
20°
10°
0°
-10°
Capella
Castor
GEMINI
TAURUS
Pleiades
ARIES
Aldebaran
Hyades
PISCES
Altair
Betelgeuse
Bellatrix
Mira
Rigel
AQUARIUS
CAPRICORNUS
Fomalhaut
SAGITTARIUS
PISCES
CETUS
AQUARIUS
URANUS
NEPTUNE
MORNING SKY

TOWARDS THE CENTRE OF THE GALAXY

The centre of the Galaxy cannot be seen directly, because it is obscured behind the dense Milky Way star fields of Sagittarius and Scorpius. The exact centre is thought to be marked by an intense radio source called Sagittarius A* (boxed).

# SOUTHERN LATITUDES

## THE STARS

The curved tail of Scorpius and the asterism known as the Teapot, formed from the main stars of Sagittarius, are virtually overhead for southern observers. The Milky Way is particularly dense and bright towards Sagittarius and Scorpius because this is the view towards the centre of the Galaxy.

Alpha (α) and Beta (β) Centauri – Rigil Kentaurus (see p.248) and Hadar – are in the southwest, pointing down to Crux, the Southern Cross. Spica (in Virgo) is in the eastern sky, Arcturus (in Boötes) in the northwest, and Vega (see p.249), in Lyra, is in the north. Altair (see p.248), in Aquila, is high in the northeast, and observers about 30°S or closer to the equator can see Deneb, in Cygnus, low in the northeast. In the southeast, 1st-magnitude Fomalhaut, in Piscis Austrinus, enters the scene.

## DEEP-SKY OBJECTS

Sagittarius is well stocked with outstanding deep-sky objects, among them the 5th-magnitude globular cluster M22, which is visible to the naked eye under good conditions. The Lagoon Nebula (see p.241), or M8, an elongated gas cloud containing the star cluster NGC 6530, can be seen well through binoculars. To the north, in Serpens Cauda, the tail of the Serpent, lies the cluster M16 – visible through binoculars – embedded in the much fainter Eagle Nebula (see pp.242–43).

Other famous deep-sky objects in Sagittarius, such as the Trifid Nebula, M20 (see p.244), need to be seen through a telescope. However, one particularly bright patch of the Milky Way, M24, is prominent to the naked eye. In adjoining Scorpius, the bright open clusters M6 and M7 remain high in the sky.

## METEOR SHOWER

The Delta Aquarids, the best southern meteor shower, is active in July and August, reaching a peak around 29 July. At best, perhaps 20 meteors an hour can be seen radiating from the southern half of Aquarius, but they are not particularly bright.

### POSITIONS OF THE PLANETS

This chart shows the positions of the planets in July from 2011 to 2019. The planets are represented by coloured dots, while the number inside each dot denotes the year. For all planets apart from Mercury, the dot indicates the planet's position on 15 July. Mercury is shown only when it is at greatest elongation (see p.64) – for the specific date, refer to the table, left.

Mercury · Mars · Saturn · Neptune · Venus · Jupiter · Uranus

EXAMPLES

12 Saturn's position on 15 July 2012

18 Mars's position on 15 July 2018. The arrow indicates that the planet is in retrograde motion (see p.64)

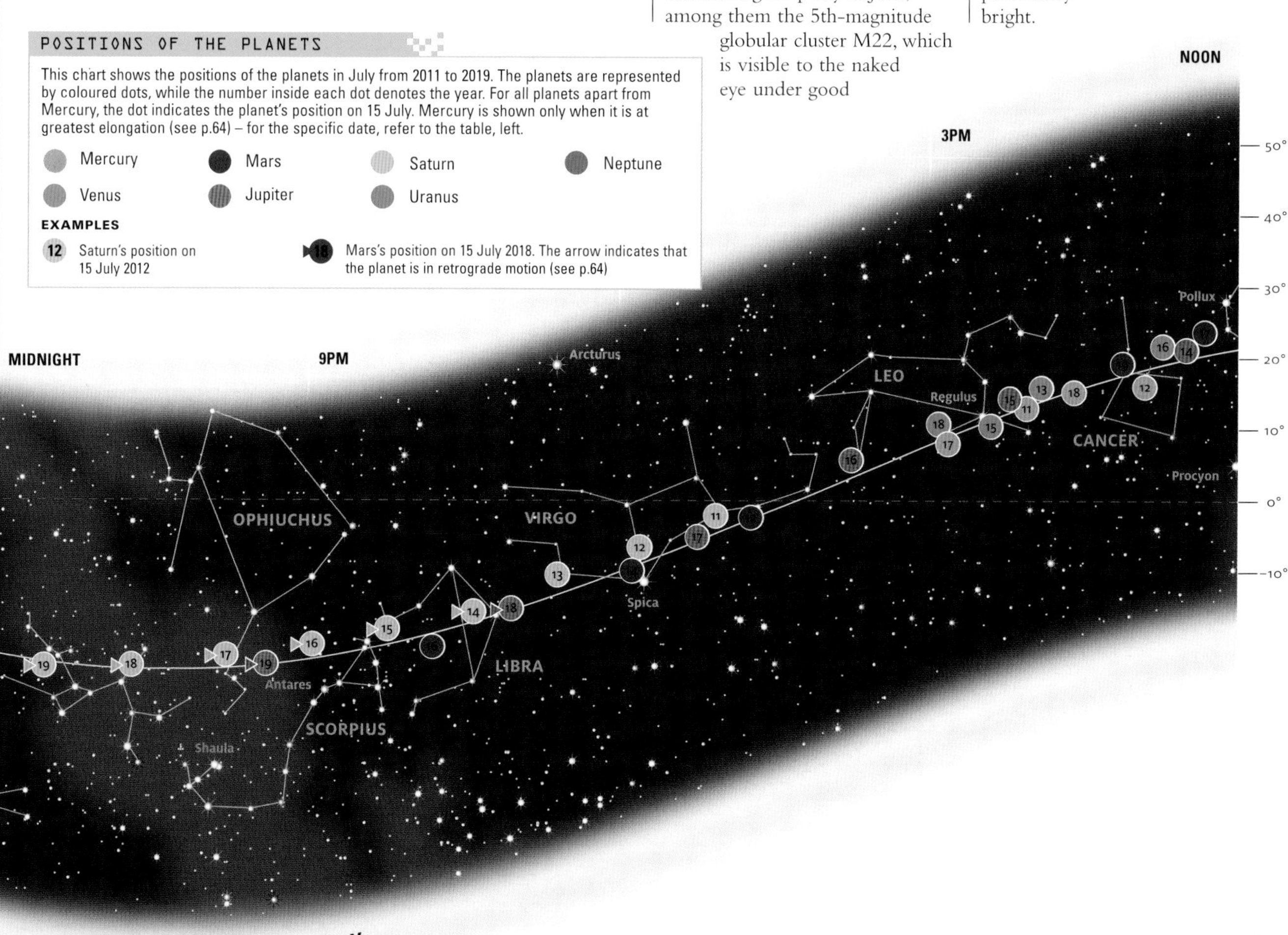

# JULY | NORTHERN LATITUDES

## LOOKING NORTH

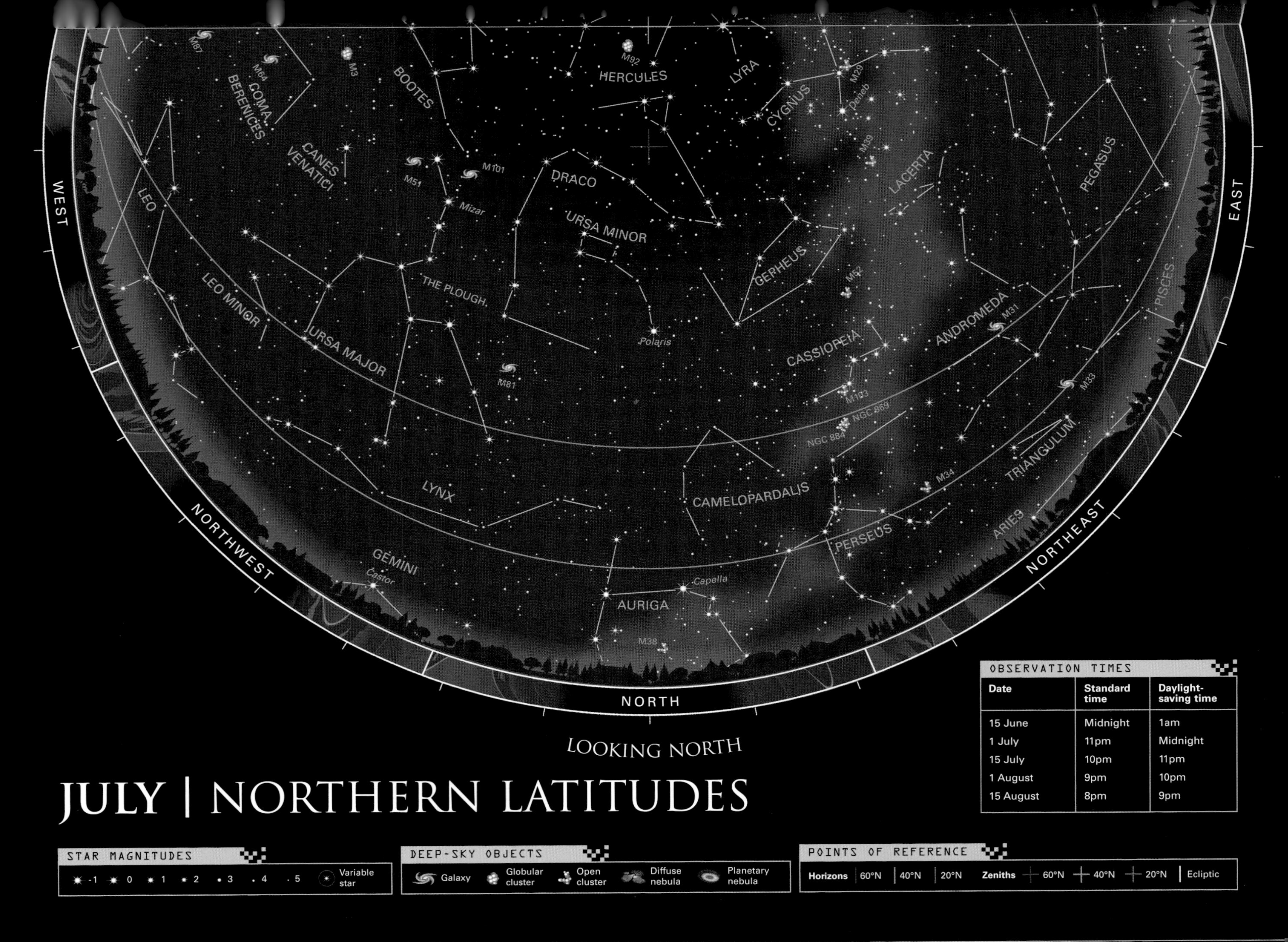


| OBSERVATION TIMES | | |
|---|---|---|
| **Date** | **Standard time** | **Daylight-saving time** |
| 15 June | Midnight | 1am |
| 1 July | 11pm | Midnight |
| 15 July | 10pm | 11pm |
| 1 August | 9pm | 10pm |
| 15 August | 8pm | 9pm |

STAR MAGNITUDES

-1 · 0 · 1 · 2 · 3 · 4 · 5 · Variable star

DEEP-SKY OBJECTS

Galaxy · Globular cluster · Open cluster · Diffuse nebula · Planetary nebula

POINTS OF REFERENCE

**Horizons** 60°N | 40°N | 20°N **Zeniths** 60°N 40°N 20°N | Ecliptic

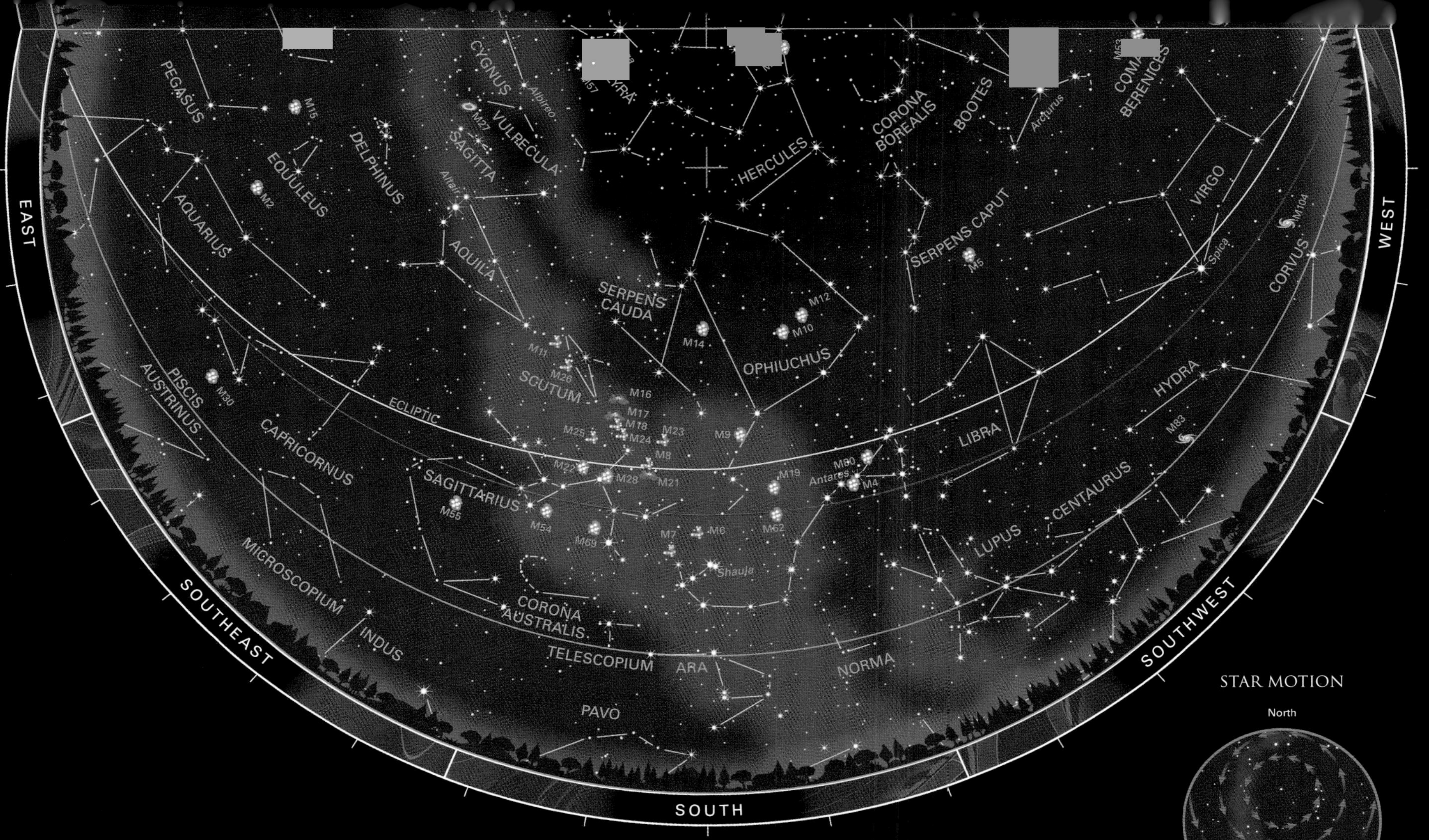

# JULY | NORTHERN LATITUDES

LOOKING SOUTH

EAST
SOUTHEAST
SOUTH
SOUTHWEST
WEST

PEGASUS
M15
EQUULEUS
M2
AQUARIUS
DELPHINUS
CYGNUS
Albireo
M27
VULPECULA
SAGITTA
Altair
AQUILA
LYRA
HERCULES
CORONA BOREALIS
BOOTES
Arcturus
COMA BERENICES
VIRGO
Spica
CORVUS
M104
SERPENS CAPUT
M5
SERPENS CAUDA
M14
M12
M10
OPHIUCHUS
M11
M26
SCUTUM
M16
M17
M18
M24
M25
M23
M8
M9
M22
M28
M21
M19
M80
Antares
M4
M62
LIBRA
HYDRA
M83
CENTAURUS
LUPUS
SAGITTARIUS
M55
M54
M69
M7
M6
Shaula
ECLIPTIC
PISCIS AUSTRINUS
M30
CAPRICORNUS
MICROSCOPIUM
INDUS
CORONA AUSTRALIS
TELESCOPIUM
ARA
NORMA
PAVO

STAR MOTION

North

South

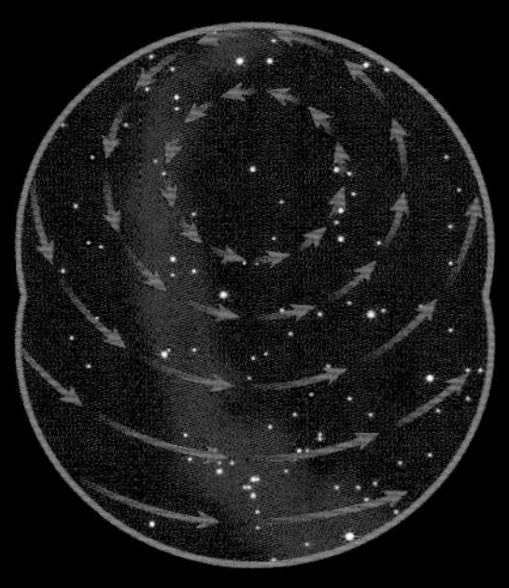

| STAR MAGNITUDES | | | | | | | |
|---|---|---|---|---|---|---|---|
| -1 | 0 | 1 | 2 | 3 | 4 | 5 | Variable star |

| DEEP-SKY OBJECTS | | | | |
|---|---|---|---|---|
| Galaxy | Globular cluster | Open cluster | Diffuse nebula | Planetary nebula |

| POINTS OF REFERENCE | | | | | | | |
|---|---|---|---|---|---|---|---|
| Horizons | 60°N | 40°N | 20°N | Zeniths | 60°N | 40°N | 20°N |
| Ecliptic | | | | | | | |

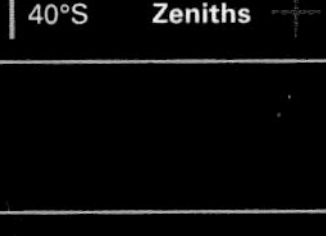

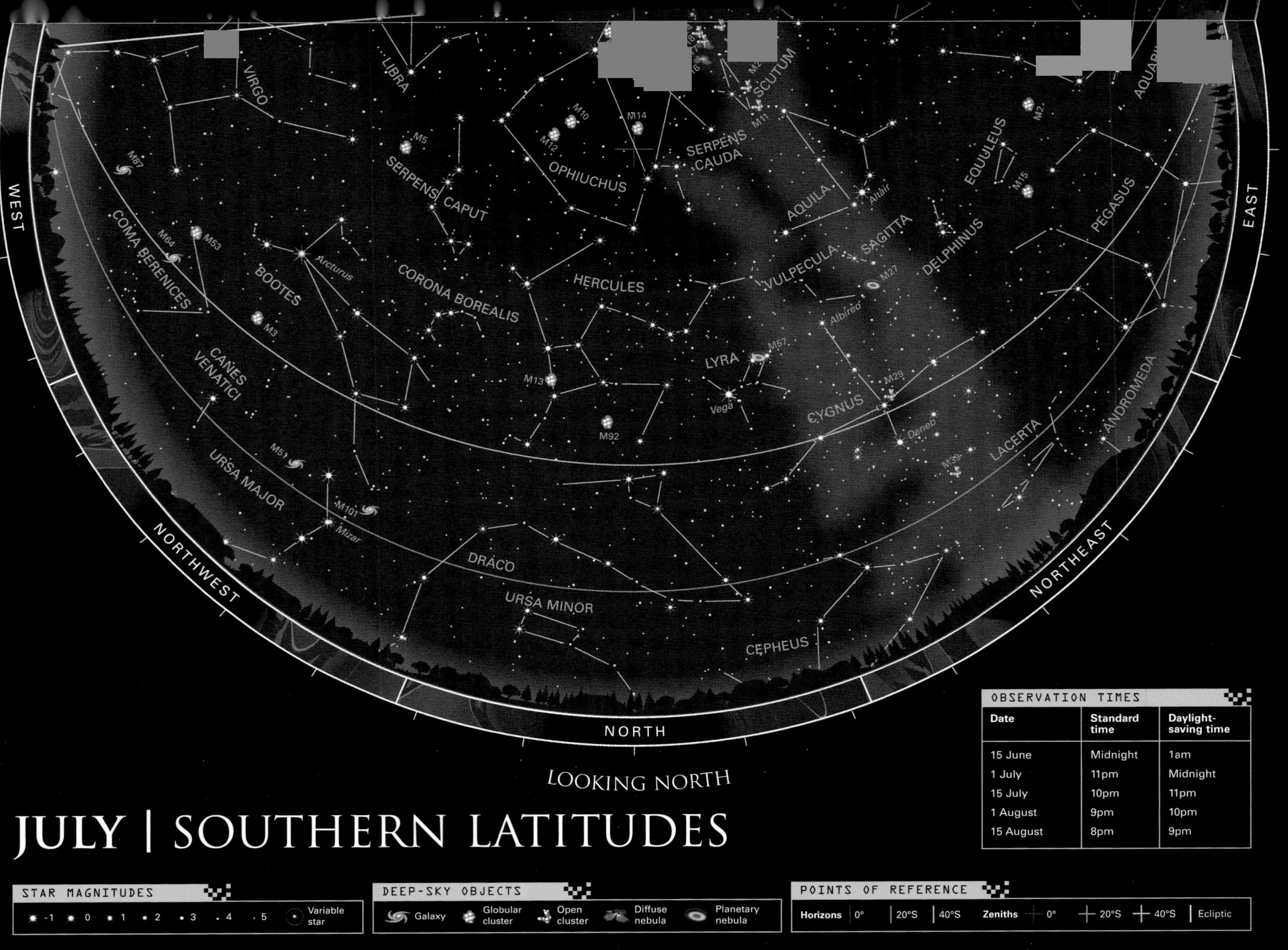


LOOKING NORTH

# JULY | SOUTHERN LATITUDES

OBSERVATION TIMES

| Date | Standard time | Daylight-saving time |
|---|---|---|
| 15 June | Midnight | 1am |
| 1 July | 11pm | Midnight |
| 15 July | 10pm | 11pm |
| 1 August | 9pm | 10pm |
| 15 August | 8pm | 9pm |

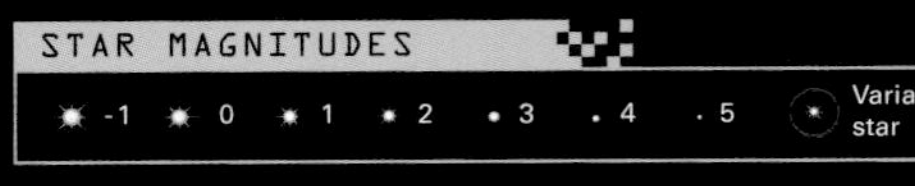

STAR MAGNITUDES

-1 · 0 · 1 · 2 · 3 · 4 · 5 · Variable star

DEEP-SKY OBJECTS

Galaxy · Globular cluster · Open cluster · Diffuse nebula · Planetary nebula

POINTS OF REFERENCE

**Horizons** 0° · 20°S · 40°S · **Zeniths** 0° · 20°S · 40°S · Ecliptic

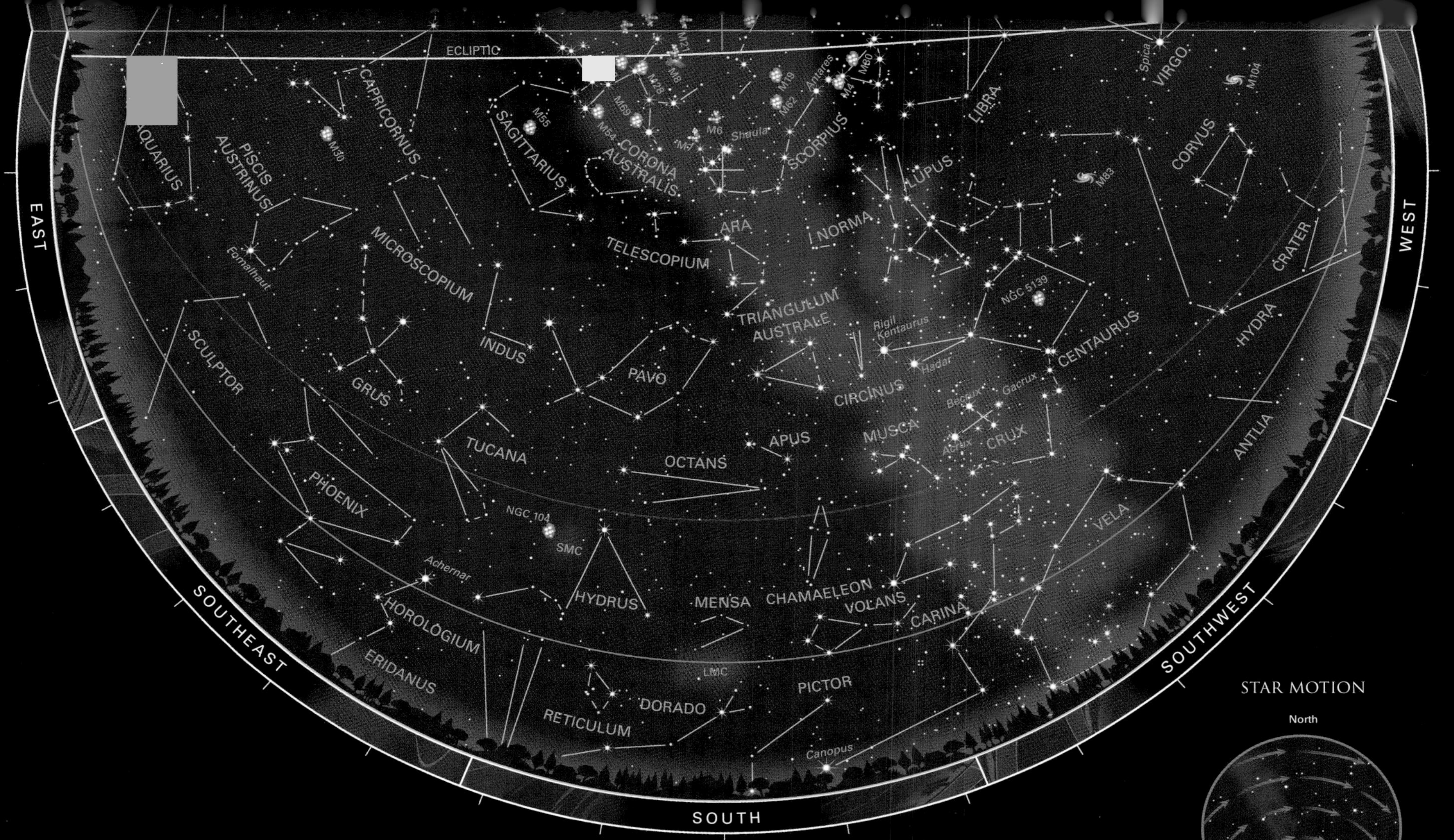

LOOKING SOUTH

# JULY | SOUTHERN LATITUDES

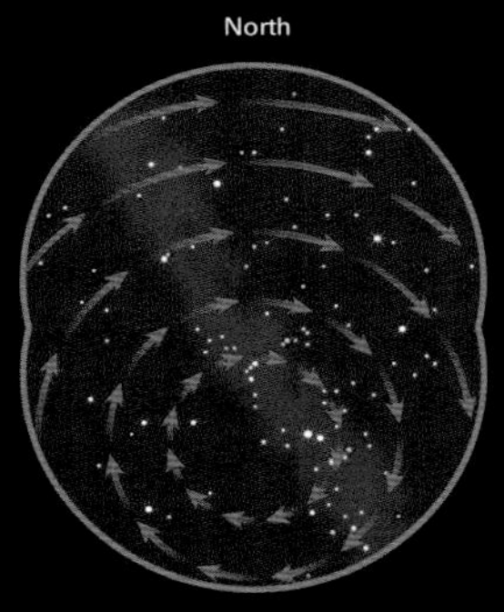

**STAR MAGNITUDES**

-1 · 0 · 1 · 2 · 3 · 4 · 5 · Variable star

**DEEP-SKY OBJECTS**

Galaxy · Globular cluster · Open cluster · Diffuse nebula · Planetary nebula

**POINTS OF REFERENCE**

Horizons: 0° · 20°S · 40°S
Zeniths: 0° · 20°S · 40°S
Ecliptic

# AUGUST

The Summer Triangle formed by the bright stars Vega (in Lyra), Deneb (in Cygnus), and Altair (in Aquila) lies on the north–south celestial meridian in the northern sky this month. The cross-shaped figure of Cygnus, the swan, stands out against the background of the Milky Way, which passes overhead in mid-northern latitudes. In the southern sky, the rich Milky Way star fields in Sagittarius and Scorpius, towards the centre of the Galaxy, remain well placed for observation.

## SPECIAL EVENTS

### PHASES OF THE MOON

| | FULL MOON | NEW MOON |
|---|---|---|
| **2011** | 13 August | 29 August |
| **2012** | 2, 31 August | 17 August |
| **2013** | 21 August | 6 August |
| **2014** | 10 August | 25 August |
| **2015** | 29 August | 14 August |
| **2016** | 18 August | 2 August |
| **2017** | 7 August | 21 August |
| **2018** | 26 August | 11 August |
| **2019** | 15 August | 1, 28 August |

### PLANETS

**2012: 15 August** Venus is at greatest morning elongation, magnitude -4.3.

**2012: 16 August** Mercury is at greatest morning elongation, magnitude 0.3.

**2016: 16 August** Mercury is at greatest morning elongation, magnitude 0.3.

**2018: 17 August** Venus is at greatest evening elongation, magnitude -4.4.

**2018: 27 August** Mercury is at greatest morning elongation, magnitude -0.1.

**2019: 9 August** Mercury is at greatest morning elongation, magnitude 0.3.

### ECLIPSES

**2017: 7 August** A partial eclipse of the Moon is visible from North and South America, Europe,Africa, and Asia.

**2017: 21 August** A total eclipse of the Sun is visible from North Pacific, USA, and South Atlantic. A partial solar eclipse is visible from North America and northern South America.

**2018: 11 August** A partial eclipse of the Sun is visible from northern Europe and northeast Asia.

## NORTHERN LATITUDES

### THE STARS

Blue-white Vega (see p.249), in the constellation Lyra, is the first bright star to appear overhead as the sky darkens on August evenings. Next to Lyra is Cygnus, popularly known as the Northern Cross. The star at the head of Cygnus, Albireo, is a beautifully coloured double star, easily divided by the smallest of telescopes. South of Cygnus is Aquila, the Eagle, from where the Milky Way continues, via Scutum, towards Sagittarius and Scorpius in the southwest. Hercules and Ophiuchus remain well placed in the southwest, and Arcturus, in Boötes, is lower in the west. In the east, the Square of Pegasus leads the stars of autumn into view.

### DEEP-SKY OBJECTS

The August skies are stocked with deep-sky objects for northern observers. The Milky Way is divided by a dark dust cloud known as the Cygnus Rift, which extends southwestwards from Cygnus into Ophiuchus. South of Cygnus, in the obscure constellation Vulpecula, is the planetary nebula M27, popularly known as the Dumbbell, the easiest such object to see through binoculars. Another celebrated planetary nebula, the Ring Nebula (see p.253) or M27, in Lyra, can be found with a telescope. The Wild Duck Cluster, or M11, in Scutum is a 6th-magnitude open cluster visible through binoculars.

### METEOR SHOWER

The year's top meteor shower, the Perseids, reaches a peak around 12 August, although some activity can be seen for a week or so either side of this date. Perseid meteors are bright: at best, an average of one a minute can be seen streaking away from northern Perseus. Most Perseids are seen after midnight, because Perseus does not rise high before then.

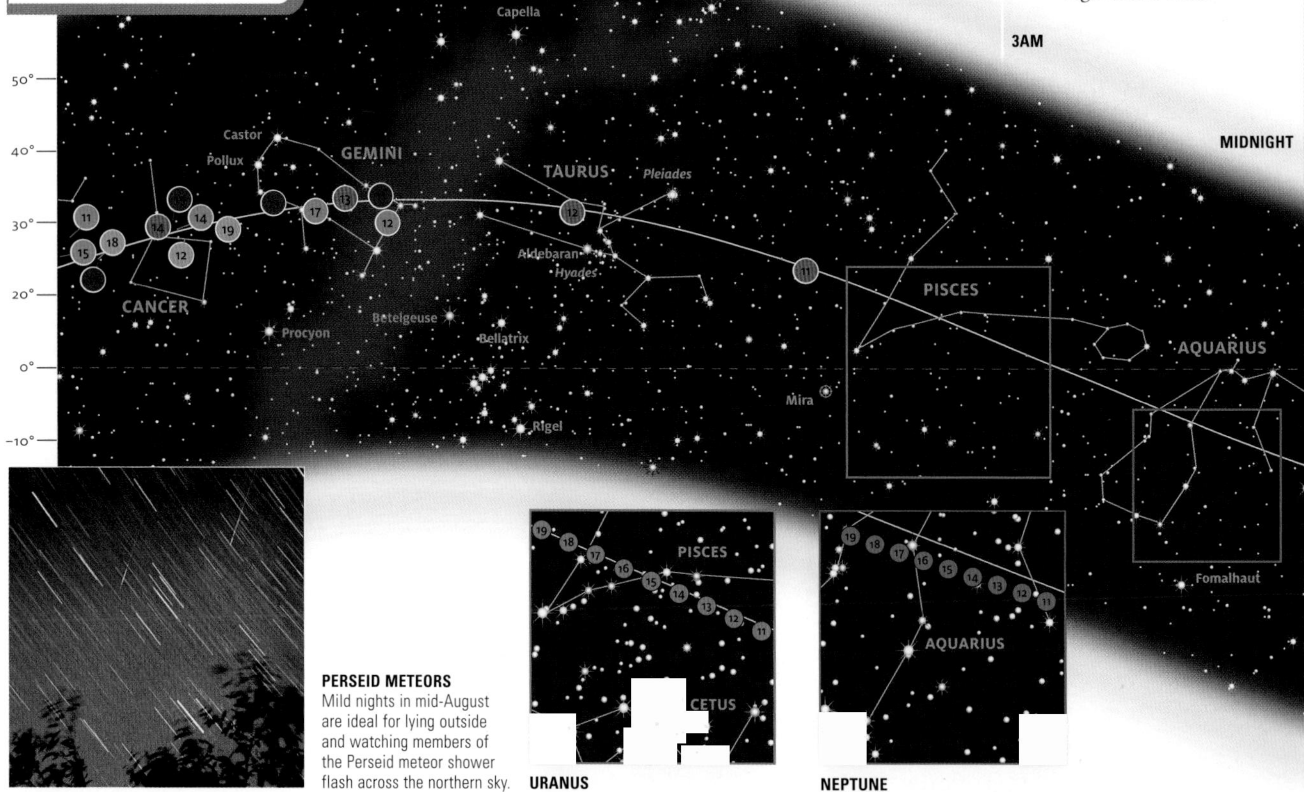


**PERSEID METEORS**
Mild nights in mid-August are ideal for lying outside and watching members of the Perseid meteor shower flash across the northern sky.

# SOUTHERN LATITUDES

## THE STARS

Sagittarius and its Milky Way star fields remain high overhead, with Scorpius to the southwest of it. Alpha (α) and Beta (β) Centauri – Rigil Kentaurus (see p.248) and Hadar – are low on the southwestern horizon. To the north are Altair (in Aquila), Vega (in Lyra), and Deneb (in Cygnus), the stars that form the northern Summer Triangle – this is the best time of year to see them in the evening sky from southern latitudes. The Square of Pegasus is rising in the northwest. Fomalhaut, in the constellation Piscis Austrinus, is high in the east, with Achernar, in Eridanus, lower in the southeast. The Small Magellanic Cloud (see p.301) is visible midway between Achernar and the south celestial pole.

## DEEP-SKY OBJECTS

The best deep-sky objects to view in the southern sky on August evenings are those that passed the celestial meridian earlier in the year, such as the Lagoon Nebula (see p.241), M22 in Sagittarius, M16 in Serpens Cauda, and M6 and M7 in Scorpius. In addition, this month southern observers can see the Wild Duck Cluster (M11) in Scutum and, looking north of the equator, the Dumbbell Nebula (M27) in Vulpecula, and the Ring Nebula (M57) in Lyra (see p.253).

**THE LAGOON NEBULA IN SAGITTARIUS**
Among the dense star fields of the Milky Way lies the Lagoon Nebula (bottom, right), also known as M8, in Sagittarius (right).

**SAGITTARIUS**
The Teapot asterism (bottom), formed by eight stars in Sagittarius, is a familiar pattern in summer skies.

**POSITIONS OF THE PLANETS**

This chart shows the positions of the planets in August from 2011 to 2019. The planets are represented by coloured dots, while the number inside each dot denotes the year. For all planets apart from Mercury, the dot indicates the planet's position on 15 August. Mercury is shown only when it is at greatest elongation (see p.64) – for the specific date, refer to the table, left.

Mercury · Mars · Saturn · Neptune · Venus · Jupiter · Uranus

**EXAMPLES**

12 Saturn's position on 15 August 2012

18 Mars's position on 15 August 2018. The arrow indicates that the planet is in retrograde motion (see p.64)

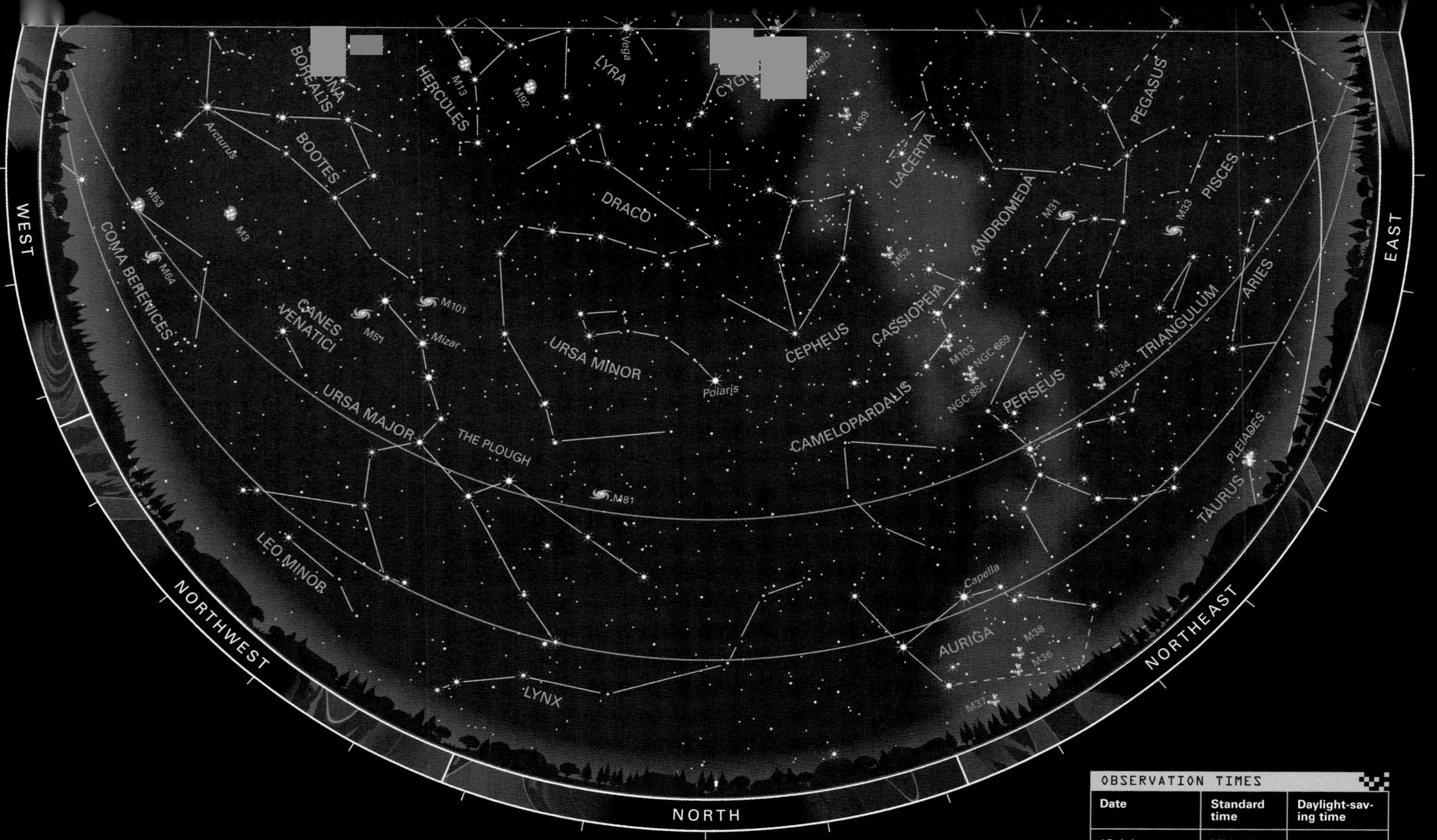

# AUGUST | NORTHERN LATITUDES

LOOKING NORTH

WEST
NORTHWEST
NORTH
NORTHEAST
EAST

CORONA BOREALIS
HERCULES
M13
M92
LYRA
Vega
CYGNUS
Deneb
M39
PEGASUS
LACERTA
PISCES
ANDROMEDA
M31
M33
Arcturus
BOOTES
DRACO
M53
M3
COMA BERENICES
M64
CANES VENATICI
M51
M101
Mizar
URSA MINOR
Polaris
CEPHEUS
M52
CASSIOPEIA
M103
NGC 869
NGC 884
PERSEUS
M34
TRIANGULUM
ARIES
URSA MAJOR
THE PLOUGH
CAMELOPARDALIS
PLEIADES
TAURUS
M81
LEO MINOR
Capella
AURIGA
M38
M36
M37
LYNX

| OBSERVATION TIMES | | |
|---|---|---|
| Date | Standard time | Daylight-saving time |
| 15 July | Midnight | 1am |
| 1 August | 11pm | Midnight |
| 15 August | 10pm | 11pm |
| 1 September | 9pm | 10pm |
| 15 September | 8pm | 9pm |

STAR MAGNITUDES

-1 · 0 · 1 · 2 · 3 · 4 · 5 · Variable star

DEEP-SKY OBJECTS

Galaxy · Globular cluster · Open cluster · Diffuse nebula · Planetary nebula

POINTS OF REFERENCE

Horizons: 60°N | 40°N | 20°N — Zeniths: 60°N, 40°N, 20°N — Ecliptic

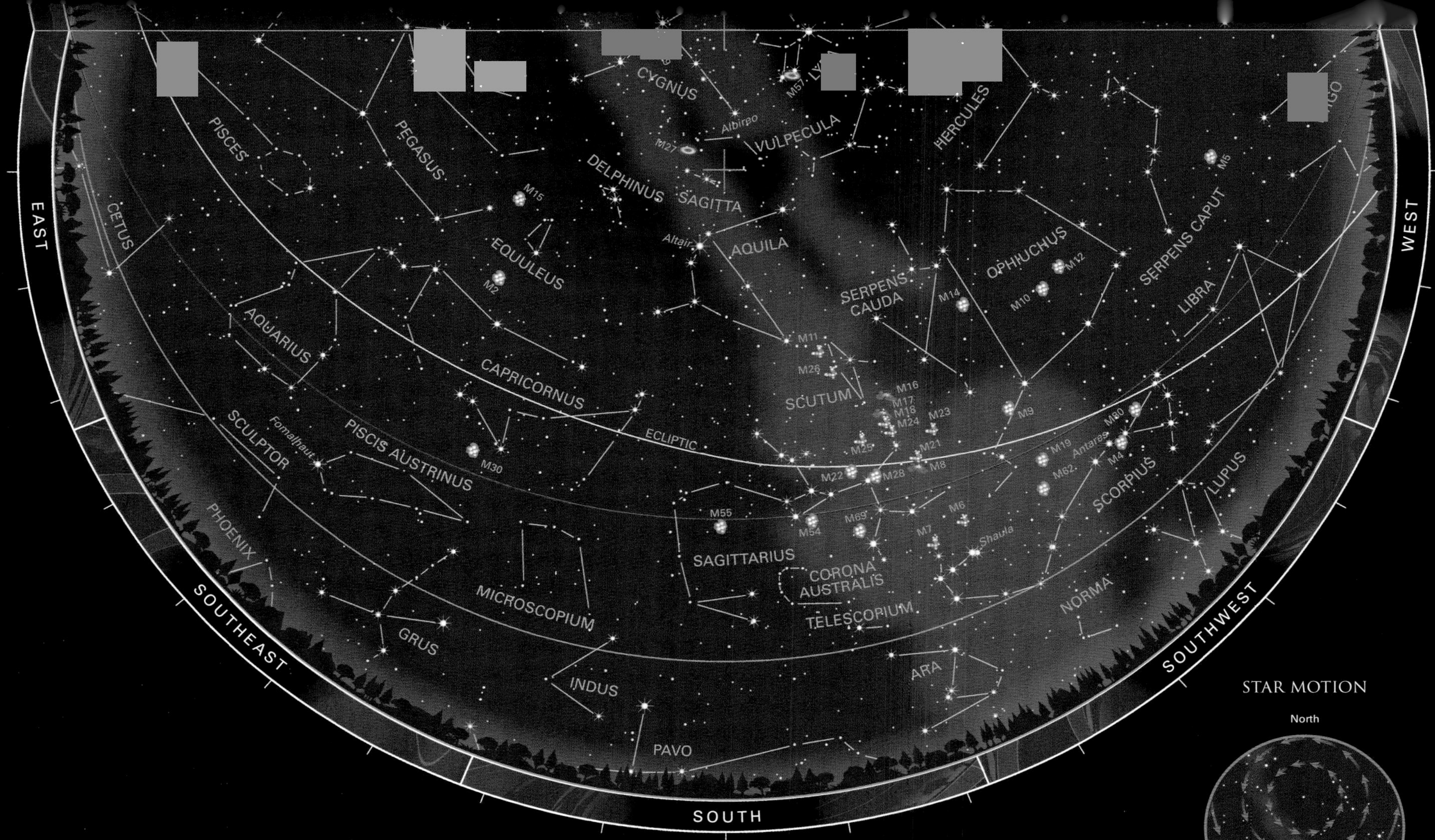

EAST
SOUTHEAST
SOUTH
SOUTHWEST
WEST

PISCES
CETUS
PEGASUS
M15
EQUULEUS
M2
DELPHINUS
SAGITTA
M27
CYGNUS
Albireo
VULPECULA
M57
HERCULES
M5
SERPENS CAPUT
OPHIUCHUS
M12
M10
M14
SERPENS CAUDA
AQUILA
Altair
LIBRA
AQUARIUS
CAPRICORNUS
M30
SCULPTOR
Fomalhaut
PISCIS AUSTRINUS
ECLIPTIC
M11
M26
SCUTUM
M16
M17
M18
M24
M23
M25
M21
M8
M22
M28
M9
M80
Antares
M4
M19
M62
SCORPIUS
LUPUS
M55
M54
M69
M7
M6
Shaula
SAGITTARIUS
CORONA AUSTRALIS
TELESCOPIUM
NORMA
PHOENIX
MICROSCOPIUM
GRUS
INDUS
ARA
PAVO

STAR MOTION
North
South

LOOKING SOUTH

# AUGUST | NORTHERN LATITUDES

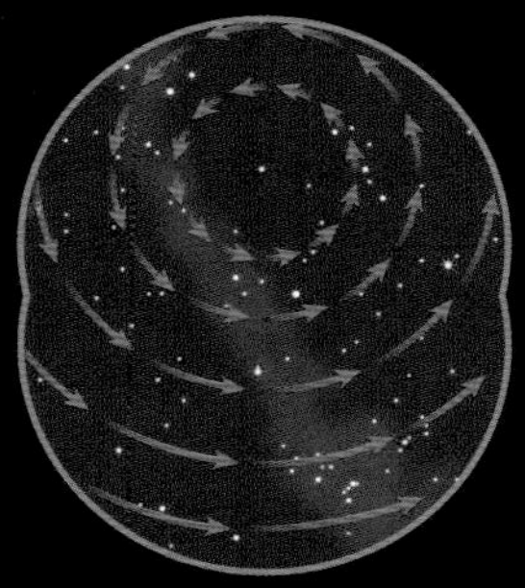

STAR MAGNITUDES

-1 · 0 · 1 · 2 · 3 · 4 · 5 · Variable star

DEEP-SKY OBJECTS

Galaxy · Globular cluster · Open cluster · Diffuse nebula · Planetary nebula

POINTS OF REFERENCE

Horizons: 60°N | 40°N | 20°N · Zeniths: 60°N | 40°N | 20°N · Ecliptic

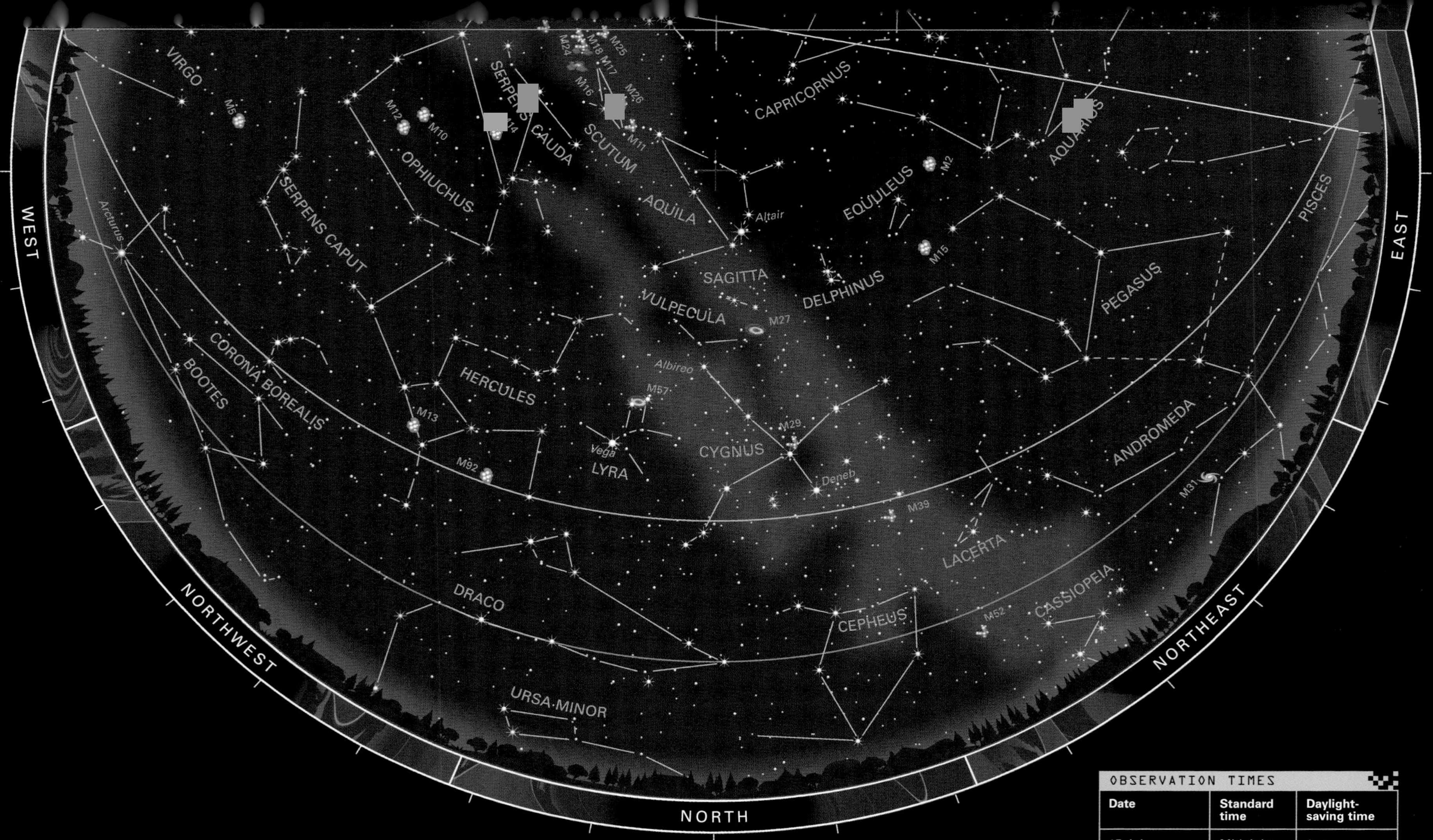

LOOKING NORTH

# AUGUST | SOUTHERN LATITUDES

OBSERVATION TIMES

| Date | Standard time | Daylight-saving time |
|---|---|---|
| 15 July | Midnight | 1am |
| 1 August | 11pm | Midnight |
| 15 August | 10pm | 11pm |
| 1 September | 9pm | 10pm |
| 15 September | 8pm | 9pm |

STAR MAGNITUDES

-1 · 0 · 1 · 2 · 3 · 4 · 5 · Variable star

DEEP-SKY OBJECTS

Galaxy · Globular cluster · Open cluster · Diffuse nebula · Planetary nebula

POINTS OF REFERENCE

Horizons 0° | 20°S | 40°S · Zeniths 0° | 20°S | 40°S · Ecliptic

EAST
SOUTHEAST
SOUTH
SOUTHWEST
WEST

CETUS
AQUARIUS
PISCIS AUSTRINUS
Fomalhaut
M30
CAPRICORNUS
M55
SAGITTARIUS
M22
M54
M28
M8
M21
M23
M69
M7
M6
M62
M19
M9
M4
M80
Antares
CORONA AUSTRALIS
Shaula
LIBRA
ECLIPTIC
VIRGO
Spica
SCULPTOR
MICROSCOPIUM
INDUS
TELESCOPIUM
ARA
SCORPIUS
NORMA
M83
HYDRA
GRUS
FORNAX
PHOENIX
TUCANA
PAVO
CIRCINUS
LUPUS
ERIDANUS
Achernar
NGC 104
SMC
Rigil Kentaurus
NGC 5139
Hadar
TRIANGULUM AUSTRALE
APUS
OCTANS
CENTAURUS
Becrux
Gacrux
Acrux
CRUX
HYDRUS
RETICULUM
MENSA
MUSCA
LMC
CHAMAELEON
HOROLOGIUM
DORADO
VELA
PICTOR
VOLANS
Canopus
CARINA

LOOKING SOUTH

# AUGUST | SOUTHERN LATITUDES

STAR MOTION
North
South

STAR MAGNITUDES
-1 0 1 2 3 4 5 Variable star

DEEP-SKY OBJECTS
Galaxy | Globular cluster | Open cluster | Diffuse nebula | Planetary nebula

POINTS OF REFERENCE
Horizons 0° | 20°S | 40°S Zeniths 0° 20°S 40°S | Ecliptic

# SEPTEMBER

Northern nights grow longer as the Sun approaches the celestial equator, but in the southern hemisphere the nights shorten. On 22–23 September, the Sun lies on the celestial equator, and day and night are of equal length worldwide. The rich band of constellations along the Milky Way, from Cygnus in the north to Sagittarius and Scorpius in the south, begin to give way this month to fainter constellations, many of them with watery associations, such as Capricornus, Aquarius, and Pisces.

## SPECIAL EVENTS

### PHASES OF THE MOON

| | FULL MOON | NEW MOON |
|---|---|---|
| **2011** | 12 September | 27 September |
| **2012** | 30 September | 16 September |
| **2013** | 19 September | 5 September |
| **2014** | 9 September | 24 September |
| **2015** | 28 September | 13 September |
| **2016** | 16 September | 1 September |
| **2017** | 6 September | 21 September |
| **2018** | 25 September | 11 September |
| **2019** | 14 September | 28 September |

### PLANETS

**2011: 3 September** Mercury is at greatest morning elongation, magnitude 0.0.

**2013: 20 September** Venus and Saturn are 3.5°apart in the western evening sky.

**2014: 21 September** Mercury is at greatest evening elongation, magnitude 0.1.

**2015: 4 September** Mercury is at greatest evening elongation, magnitude 0.3.

**2016: 28 September** Mercury is at greatest morning elongation, magnitude -0.4.

**2017: 12 September** Mercury is at greatest morning elongation, magnitude -0.2.

### ECLIPSES

**2015: 13 September** A partial eclipse of the Sun is visible from southern Africa and parts of Antarctica.

**2015: 28 September** A total eclipse of the Moon is visible from the Americas, Europe, and Africa.

**2016: 1 September** An annular eclipse of the Sun is visible from the Atlantic Ocean, central Africa, Madagascar, and the Indian Ocean.

## NORTHERN LATITUDES

### THE STARS

Cepheus, high up in the north, is best placed for evening observation this month and next. Its most celebrated star is Delta (δ) Cephei, the prototype of a class of pulsating variables. Deneb in Cygnus, Vega (see p.249) in Lyra, and Altair (see p.248) in Aquila, the stars of the Summer Triangle, remain high in the western half of the sky, while the Square of Pegasus is high in the east with Cassiopeia between it and the north celestial pole. The bright star Fomalhaut (see p.249) in Piscis Austrinus is low in the south with Aquarius above it. A cascade of faint stars suggests the flow of water from the water carrier's urn towards the southern fish, Piscis Austrinus. For observers at high northern latitudes, this is the best time of year to see the zodiacal constellation Capricornus in the evening sky, lying low in the south to the right of Fomalhaut.

### DEEP-SKY OBJECTS

Near Deneb in Cygnus lies one of the most distinctive nebulae in the sky, NGC 7000, popularly called the North America Nebula, on account of its shape. Under clear, dark skies, it can be detected with binoculars, but it is best seen on long-exposure photographs. Another object of note in Cygnus is the open star cluster M39, which is visible through binoculars. The 6th-magnitude globular cluster M15, also visible through binoculars, is not far from the star Enif – Epsilon (ε) Pegasi – which marks the horse's nose in Pegasus.

URANUS

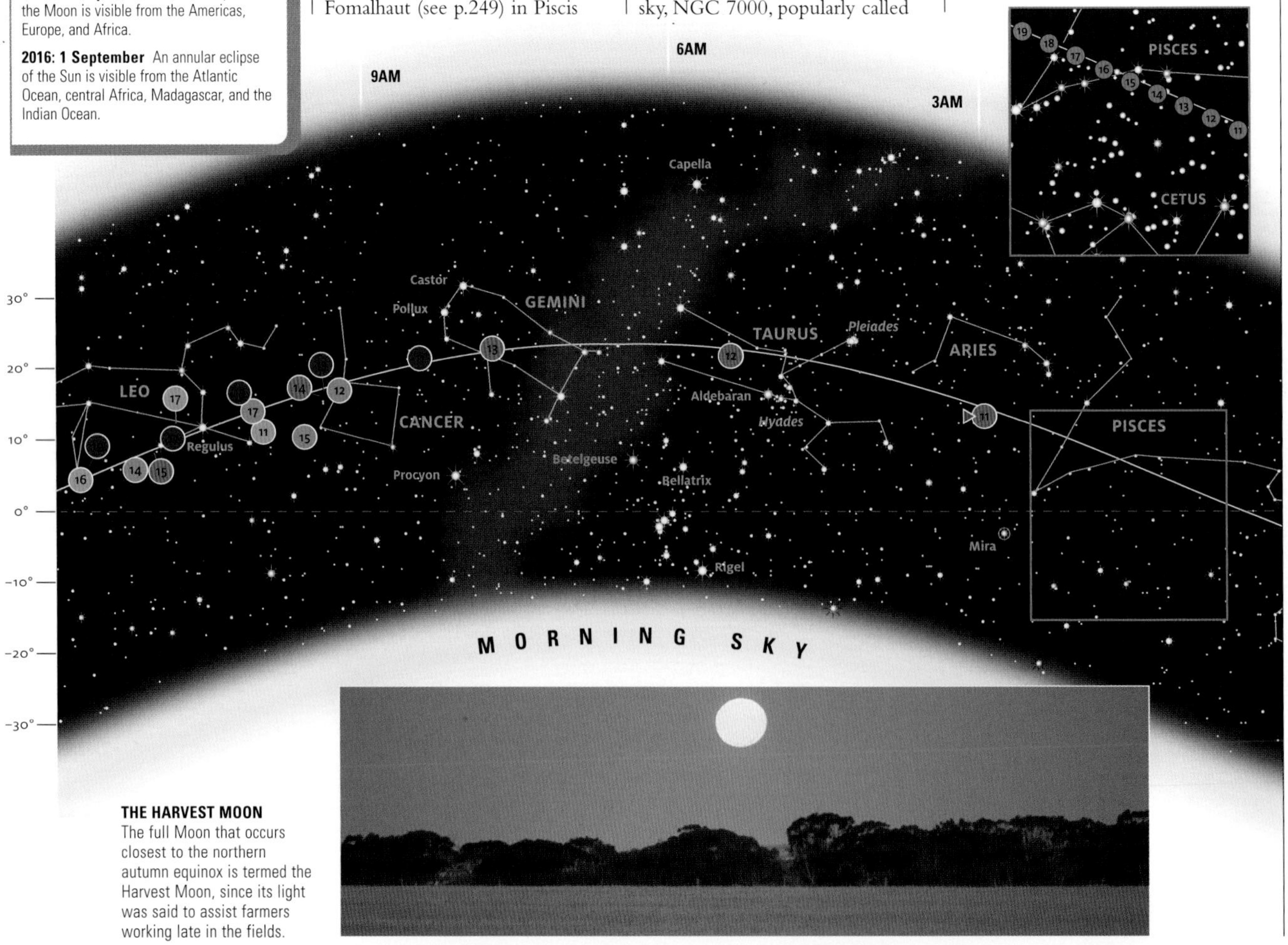

MORNING SKY

**THE HARVEST MOON**
The full Moon that occurs closest to the northern autumn equinox is termed the Harvest Moon, since its light was said to assist farmers working late in the fields.

# SOUTHERN LATITUDES

**THE SMALL MAGELLANIC CLOUD**
This small satellite galaxy (left) appears beside the globular cluster 47 Tucanae (right), which is in the foreground in our own galaxy.

## THE STARS

Scorpius is low in the west, with Sagittarius and the densest regions of the Milky Way above it. The large northern Summer Triangle of Altair, Vega, and Deneb is visible in the northwest, while in the southwest, Alpha (α) and Beta (β) Centauri – Rigil Kentaurus (see p.248) and Hadar – are visible from latitude 20°S and farther south. The Square of Pegasus dominates the northeastern sky.

First-magnitude Fomalhaut (see p.249) in Piscis Austrinus is almost overhead, along with Capricornus and Aquarius. Achernar, the bright star at the end of the celestial river Eridanus, is high in the southeast, as is the Small Magellanic Cloud (see p.301). A group of constellations with exotic names, such as Phoenix, Tucana, Grus, and Pavo, is spread across the southern half of the sky.

## DEEP-SKY OBJECTS

Aquarius contains two famous planetary nebulae, although neither is particularly easy to find through small instruments. The Helix Nebula (see p.253), or NGC 7293, is the nearest planetary nebula to us. Its size means that its light is spread out over such a large area that clear skies are essential to glimpse it through binoculars or a low-power telescope. The Saturn Nebula, NGC 7009, is so named because, when seen through a large telescope, it appears to have rings like the planet Saturn. A small telescope shows the Saturn Nebula simply as a greenish disc.

Also in Aquarius is the globular cluster M2, which resembles a fuzzy star when seen through binoculars. To the north of this is another globular cluster that can be viewed through binoculars, M15 in Pegasus.

**POSITIONS OF THE PLANETS**

This chart shows the positions of the planets in September from 2011 to 2019. The planets are represented by coloured dots, while the number inside each dot denotes the year. For all planets apart from Mercury, the dot indicates the planet's position on 15 September. Mercury is shown only when it is at greatest elongation (see p.64) – for the specific date, refer to the table, left.

Mercury, Venus, Mars, Jupiter, Saturn, Uranus, Neptune

**EXAMPLES**

12 Saturn's position on 15 September 2012

11 Jupiter's position on 15 September 2011. The arrow indicates the planet is in retrograde motion (see p.64).

**NEPTUNE**

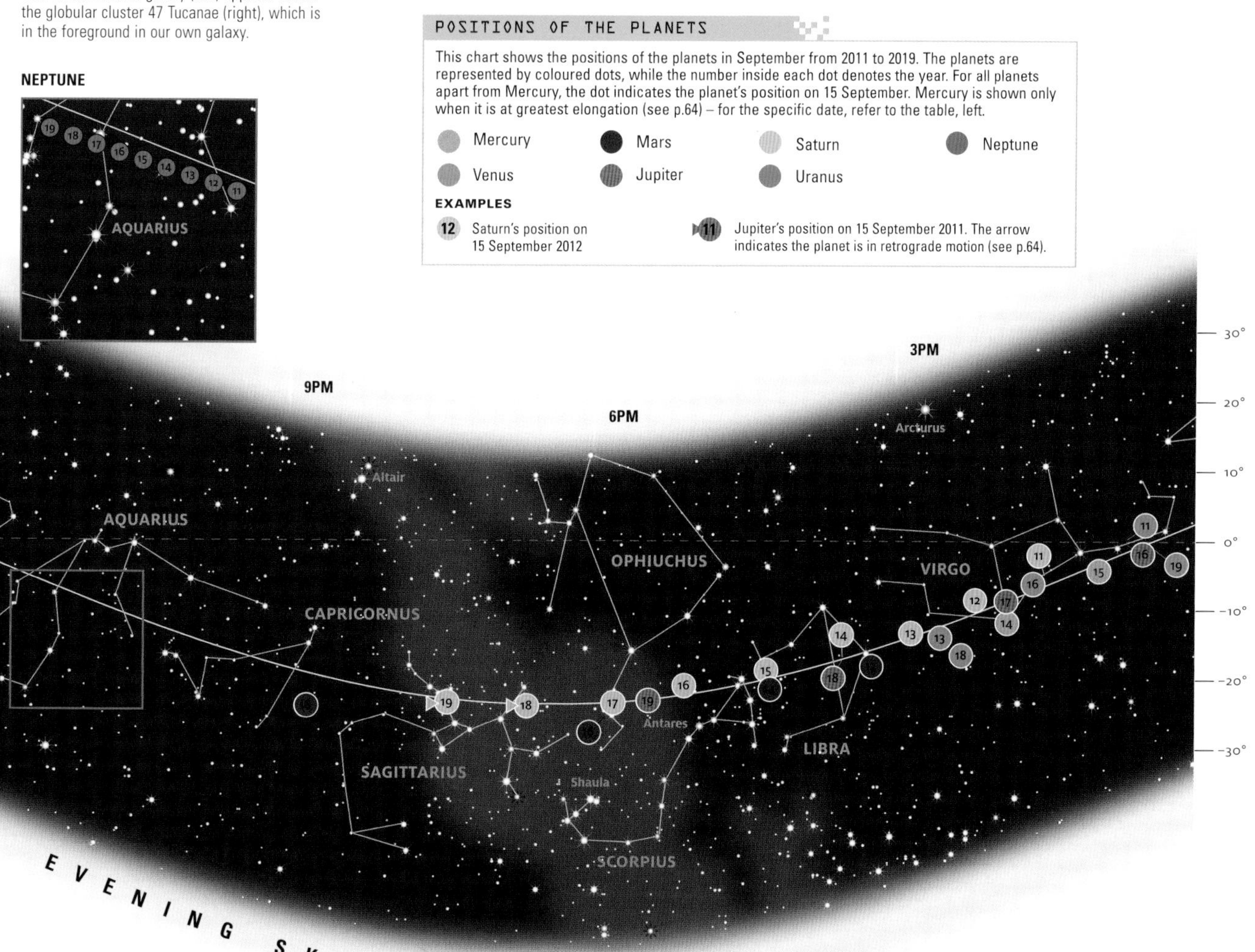

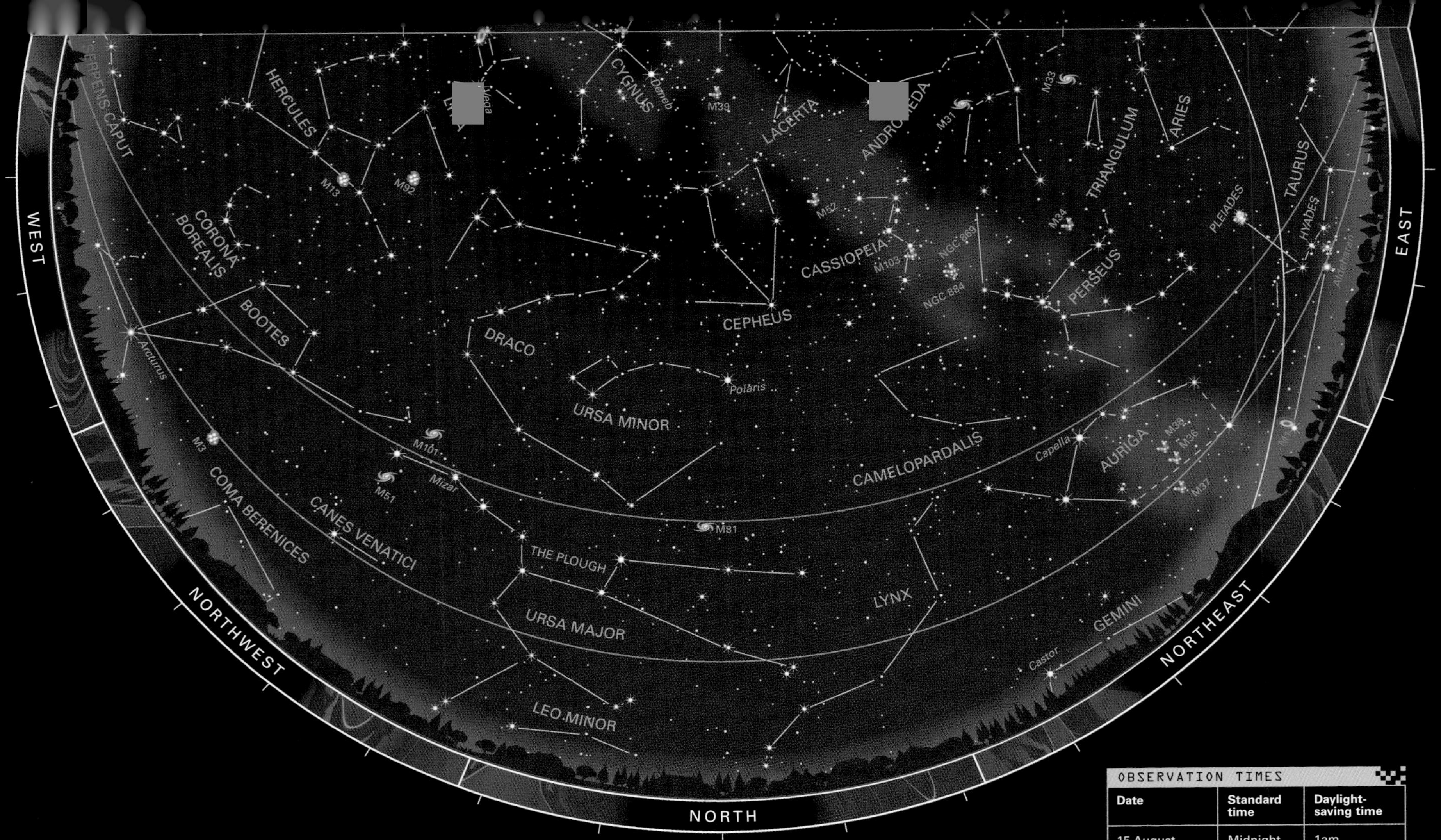

WEST
SERPENS CAPUT
HERCULES
LYRA
Vega
CYGNUS
Deneb
M39
LACERTA
ANDROMEDA
M31
M33
TRIANGULUM
ARIES
TAURUS
PLEIADES
HYADES
Aldebaran
EAST
M13
M92
CORONA BOREALIS
BOOTES
Arcturus
DRACO
CEPHEUS
M52
CASSIOPEIA
M103
NGC 869
NGC 884
M34
PERSEUS
URSA MINOR
Polaris
M3
M101
M51
Mizar
CAMELOPARDALIS
Capella
AURIGA
M38
M36
M37
M1
COMA BERENICES
CANES VENATICI
M81
THE PLOUGH
URSA MAJOR
LYNX
GEMINI
Castor
NORTHWEST
NORTHEAST
LEO MINOR
NORTH

LOOKING NORTH

# SEPTEMBER | NORTHERN LATITUDES

**OBSERVATION TIMES**

| Date | Standard time | Daylight-saving time |
|---|---|---|
| 15 August | Midnight | 1am |
| 1 September | 11pm | Midnight |
| 15 September | 10pm | 11pm |
| 1 October | 9pm | 10pm |
| 15 October | 8pm | 9pm |

**STAR MAGNITUDES**

-1 0 1 2 3 4 5 Variable star

**DEEP-SKY OBJECTS**

Galaxy Globular cluster Open cluster Diffuse nebula Planetary nebula

**POINTS OF REFERENCE**

**Horizons** 60°N 40°N 20°N **Zeniths** 60°N 40°N 20°N Ecliptic

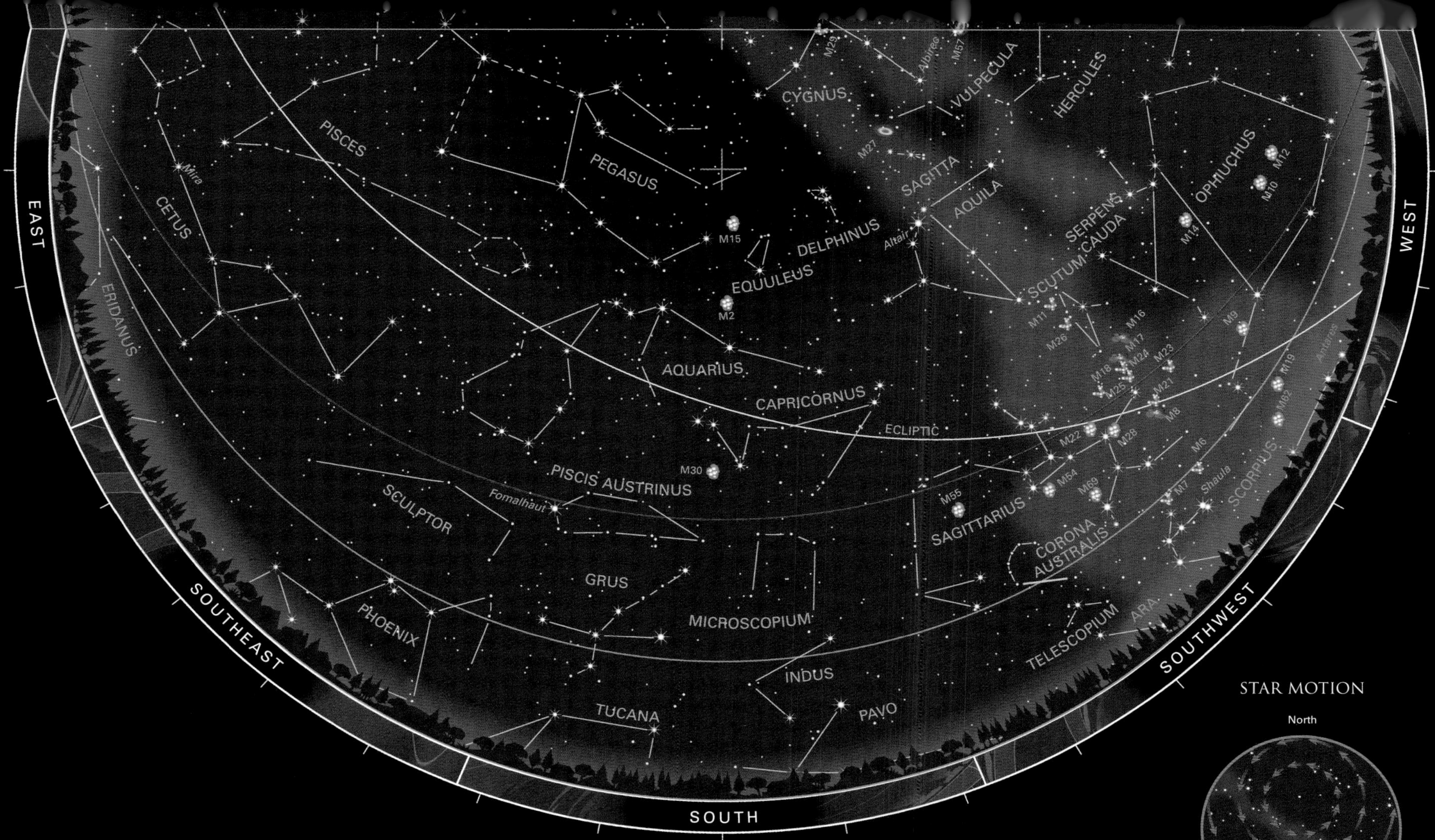

EAST
SOUTHEAST
SOUTH
SOUTHWEST
WEST

PISCES
PEGASUS
CETUS
Mira
ERIDANUS
CYGNUS
M29
Albireo
M57
VULPECULA
HERCULES
M27
SAGITTA
AQUILA
Altair
DELPHINUS
M15
EQUULEUS
M2
OPHIUCHUS
M12
M10
M14
SERPENS CAUDA
SCUTUM
M11
M26
M16
M17
M18
M24
M23
M25
M21
M8
M9
M19
M62
Antares
AQUARIUS
CAPRICORNUS
ECLIPTIC
M22
M28
M6
M7
Shaula
SCORPIUS
M54
M69
M30
PISCIS AUSTRINUS
Fomalhaut
SCULPTOR
M55
SAGITTARIUS
CORONA AUSTRALIS
GRUS
PHOENIX
MICROSCOPIUM
TELESCOPIUM
ARA
INDUS
PAVO
TUCANA

LOOKING SOUTH

# SEPTEMBER | NORTHERN LATITUDES

STAR MOTION

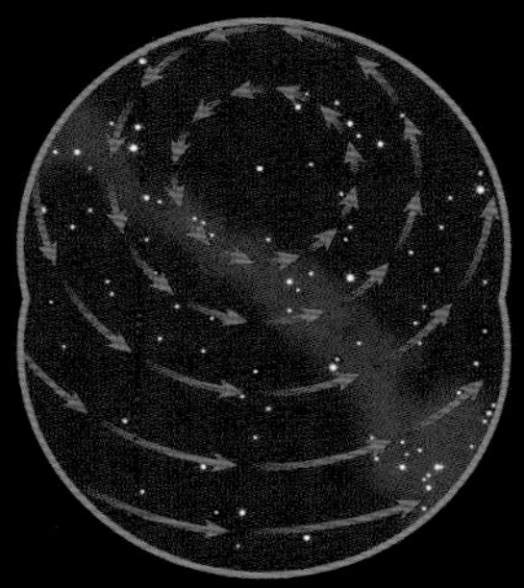

North

South

STAR MAGNITUDES

-1 · 0 · 1 · 2 · 3 · 4 · 5 · Variable star

DEEP-SKY OBJECTS

Galaxy · Globular cluster · Open cluster · Diffuse nebula · Planetary nebula

POINTS OF REFERENCE

**Horizons** 60°N | 40°N | 20°N **Zeniths** 60°N | 40°N | 20°N | Ecliptic

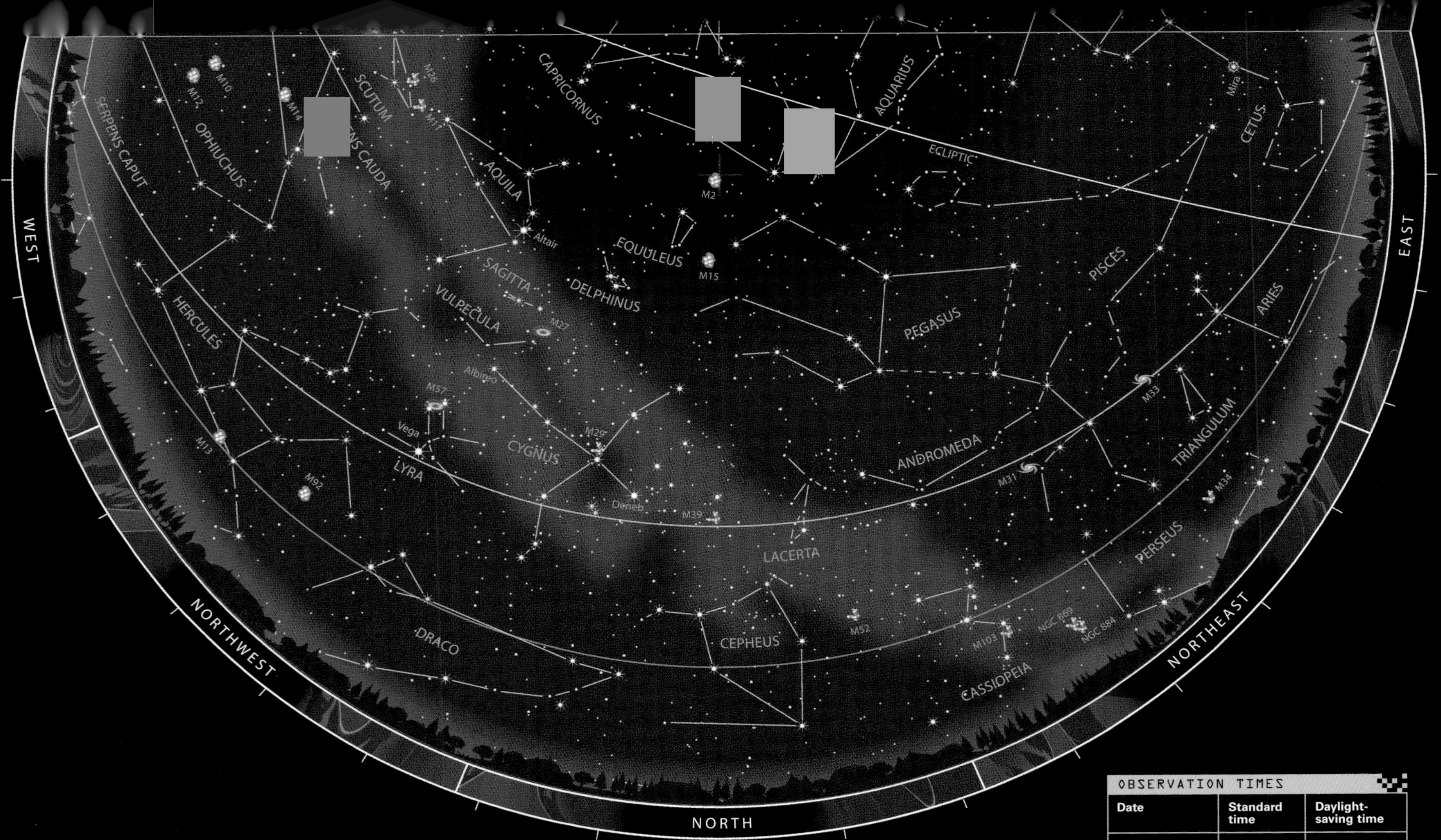

LOOKING NORTH

# SEPTEMBER | SOUTHERN LATITUDES

OBSERVATION TIMES

| Date | Standard time | Daylight-saving time |
|---|---|---|
| 15 August | Midnight | 1am |
| 1 September | 11pm | Midnight |
| 15 September | 10pm | 11pm |
| 1 October | 9pm | 10pm |
| 15 October | 8pm | 9pm |

STAR MAGNITUDES

-1 · 0 · 1 · 2 · 3 · 4 · 5 · Variable star

DEEP-SKY OBJECTS

Galaxy · Globular cluster · Open cluster · Diffuse nebula · Planetary nebula

POINTS OF REFERENCE

Horizons 0° | 20°S | 40°S · Zeniths 0° | 20°S | 40°S · Ecliptic

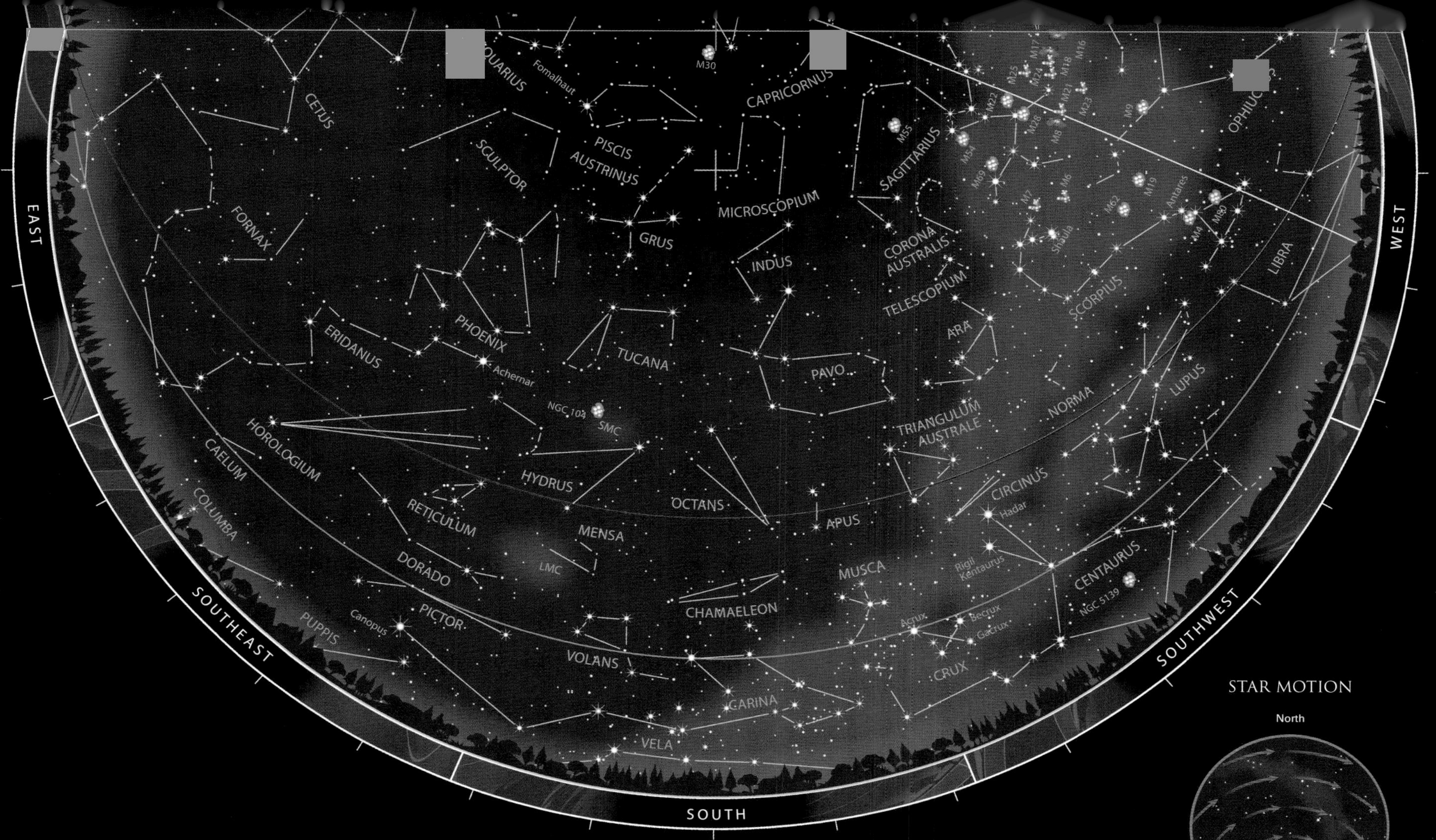

LOOKING SOUTH

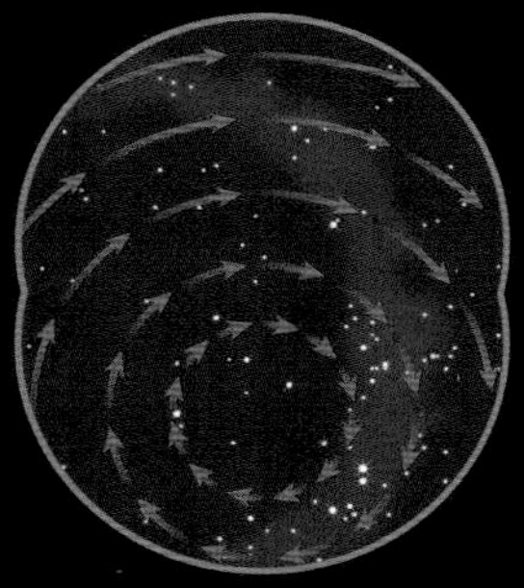

# SEPTEMBER | SOUTHERN LATITUDES

STAR MOTION

North

South

STAR MAGNITUDES

-1 · 0 · 1 · 2 · 3 · 4 · 5 · Variable star

DEEP-SKY OBJECTS

Galaxy · Globular cluster · Open cluster · Diffuse nebula · Planetary nebula

POINTS OF REFERENCE

Horizons 0° | 20°S | 40°S · Zeniths 0° | 20°S | 40°S · Ecliptic

# OCTOBER

The Square of Pegasus takes centre stage in the northern skies in both hemispheres, a sign of the arrival of the northern autumn and the southern spring. Northeast of it lies the Andromeda Galaxy, the nearest large galaxy to the Earth. South of the Square, a band of faint zodiacal constellations crosses the sky, from Aries in the east to Capricornus in the southwest.

## SPECIAL EVENTS

### PHASES OF THE MOON

| | FULL MOON | NEW MOON |
|---|---|---|
| **2011** | 12 October | 26 October |
| **2012** | 29 October | 15 October |
| **2013** | 18 October | 5 October |
| **2014** | 8 October | 23 October |
| **2015** | 27 October | 13 October |
| **2016** | 16 October | 1, 30 October |
| **2017** | 5 October | 19 October |
| **2018** | 24 October | 9 October |
| **2019** | 13 October | 28 October |

### PLANETS

**2011: 29 October** Jupiter is at opposition, magnitude -2.9.

**2011: 30–31 October** Mercury and Venus 2° apart in the western evening sky.

**2012: 4–5 October** Mercury and Saturn are 3.2° apart in the western evening sky.

**2012: 26 October** Mercury is at greatest evening elongation, magnitude -0.1.

**2013: 9 October** Mercury is at greatest evening elongation, magnitude 0.0.

**2013: 10 October** Mercury and Saturn are 5° apart in the southwestern evening sky.

**2015: 16 October** Mercury is at greatest morning elongation, magnitude -0.5.

**2015: 17–30 October** Venus, Jupiter, and Mars appear within 6° of each other in the eastern morning sky.

**2016: 30 October** Venus and Saturn are 3° apart in the southwestern evening sky.

**2019: 20 October** Mercury is at greatest evening elongation, magnitude -0.1.

### ECLIPSES

**2014: 8 October** A total eclipse of the Moon is visible from east Asia and North America.

**2014: 23 October** A partial eclipse of the Sun is visible from central and western USA, Canada, and Mexico.

## NORTHERN LATITUDES

### THE STARS

The Square of Pegasus lies high in the sky from mid-northern latitudes. From one corner of the Square, the stars of Andromeda extend northeastwards towards Perseus and Cassiopeia. Capella twinkles above the horizon in Auriga, lower in the northeast. In the north, the Plough is at its lowest, and it is below the horizon for observers south of about latitude 30°N.

Directly beneath the Square of Pegasus is a loop of stars known as the Circlet, representing the body of one of the fishes in the zodiacal constellation of Pisces. Fomalhaut (see p.249) in Piscis Austrinus is low on the southern horizon beneath the stars of Aquarius. In the western sky, the Summer Triangle lingers, while in the east Taurus leads the stars of winter into view.

### DEEP-SKY OBJECTS

October evenings are a good time to view the Andromeda Galaxy, M31 (see pp.302–303). It can be seen as an elongated misty patch with the naked eye, if skies are not too polluted, and it is easily visible through binoculars, spanning a greater width than the full Moon. High in the north, M52, an open cluster near Cassiopeia, is visible through binoculars. Between this and the Square of Pegasus lies an often-overlooked planetary nebula, NGC 7662, nicknamed the Blue Snowball. A small telescope is needed to see it.

### METEOR SHOWER

One of the year's lesser showers, the Orionids, reaches a peak of some 25 meteors an hour around 21 October. They radiate from northern Orion, near the border with Gemini. This area rises late, thus the meteors are best seen after midnight.

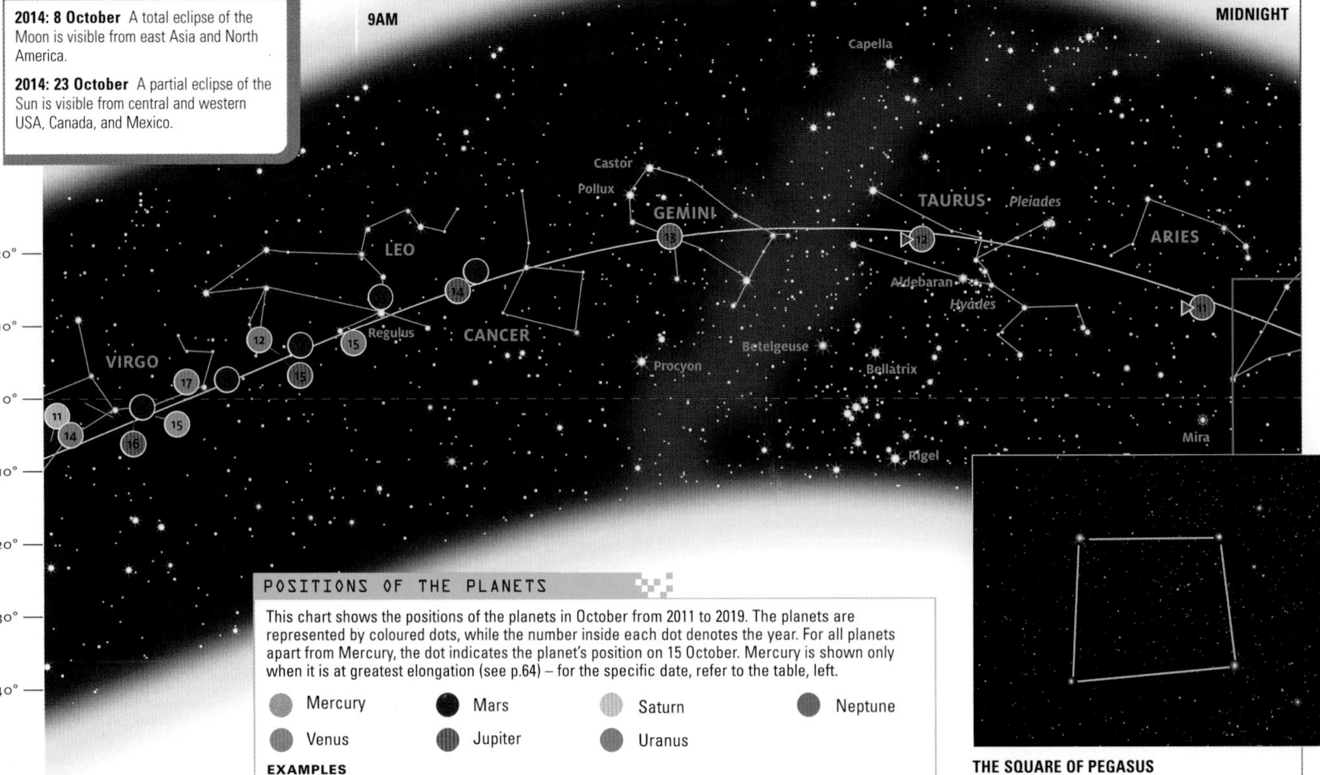

6AM
3AM
9AM
MIDNIGHT
Capella
Castor
Pollux
GEMINI
TAURUS
Pleiades
ARIES
LEO
Aldebaran
Hyades
Regulus
CANCER
VIRGO
Betelgeuse
Procyon
Bellatrix
Rigel
Mira
20°
10°
0°
-10°
-20°
-30°
-40°
POSITIONS OF THE PLANETS
This chart shows the positions of the planets in October from 2011 to 2019. The planets are represented by coloured dots, while the number inside each dot denotes the year. For all planets apart from Mercury, the dot indicates the planet's position on 15 October. Mercury is shown only when it is at greatest elongation (see p.64) – for the specific date, refer to the table, left.
Mercury
Mars
Saturn
Neptune
Venus
Jupiter
Uranus
EXAMPLES
Saturn's position on 15 October 2012
Jupiter's position on 15 October 2011. The arrow indicates the planet is in retrograde motion (see p.64)

**THE SQUARE OF PEGASUS**
This huge square in the northern autumn sky is composed of three stars in Pegasus and one in Andromeda (top, left).

# SOUTHERN LATITUDES

## THE STARS

In contrast to the sparkling skies of southern winter, the constellations of October evenings are mostly faint and unremarkable. One star that stands out is 1st-magnitude Fomalhaut (see p.249), almost overhead in the constellation Piscis Austrinus. In the northeast sky is Altair (see p.248) in Aquila and, high in the north, the Square of Pegasus. Between Pegasus and Fomalhaut lies Aquarius, the Water Carrier. More constellations with watery associations fill the western part of the sky – Pisces, the Fishes; Cetus, the Sea Monster or the Whale; and Eridanus, the River. The constellation Eridanus ends at the bright star Achernar, high in the south. The Small Magellanic Cloud (see p.301) is lower in the south, with the Large Magellanic Cloud (see p.301) now on view in the southeast. Canopus in Carina is also visible in the southeast, for those farther south of the equator than 20°S.

**FAMILIAR ASTERISMS**
The Circlet of Pisces (left) and the Y-shaped Water Jar of Aquarius (right) are two easily recognizable star patterns in the October evening sky.

## DEEP-SKY OBJECTS

Tucana contains the second-best globular cluster in the sky, 47 Tucanae, or NGC 104, which is visible to the naked eye as a fuzzy star and appears impressive through binoculars. It covers the same area of sky as the full Moon, near the Small Magellanic Cloud, but it lies much closer to us – about 15,000 light-years away – in our own galaxy. On the edge of the SMC, NGC 362 is another, fainter globular cluster, also in our galaxy.

October and November evenings are the best time for southern observers to view the Andromeda Galaxy, M31 (see pp.302–303), which lies low in the northern sky. Near it is another member of our Local Group of galaxies, M33, a smaller spiral galaxy that is less easy to see. In clear, dark skies, it can be glimpsed through binoculars or a low-power telescope as a large, rounded patch.

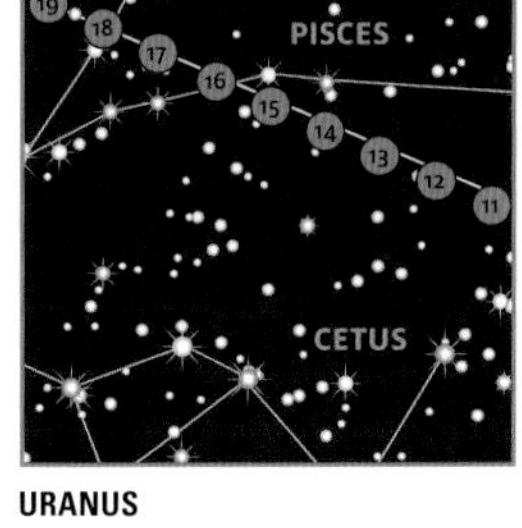


URANUS

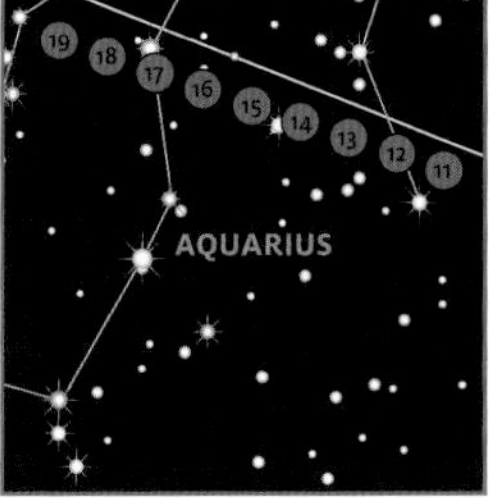


NEPTUNE

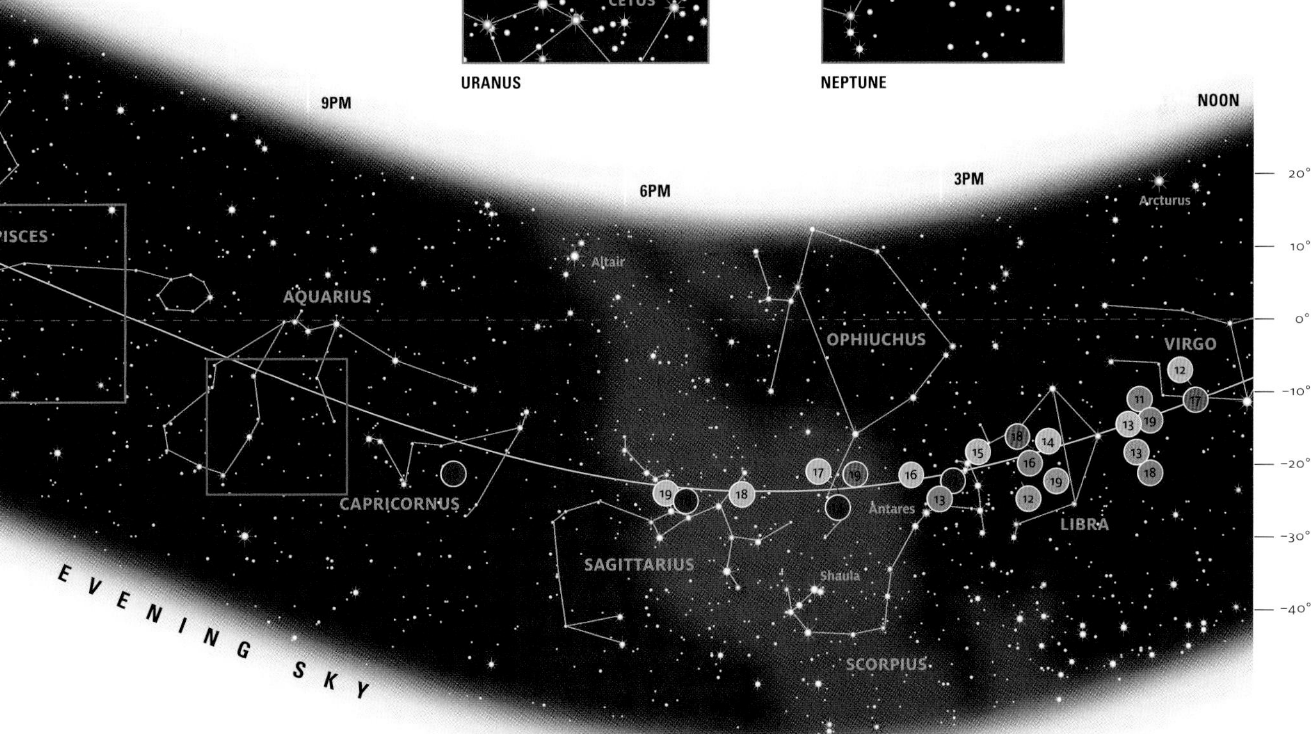

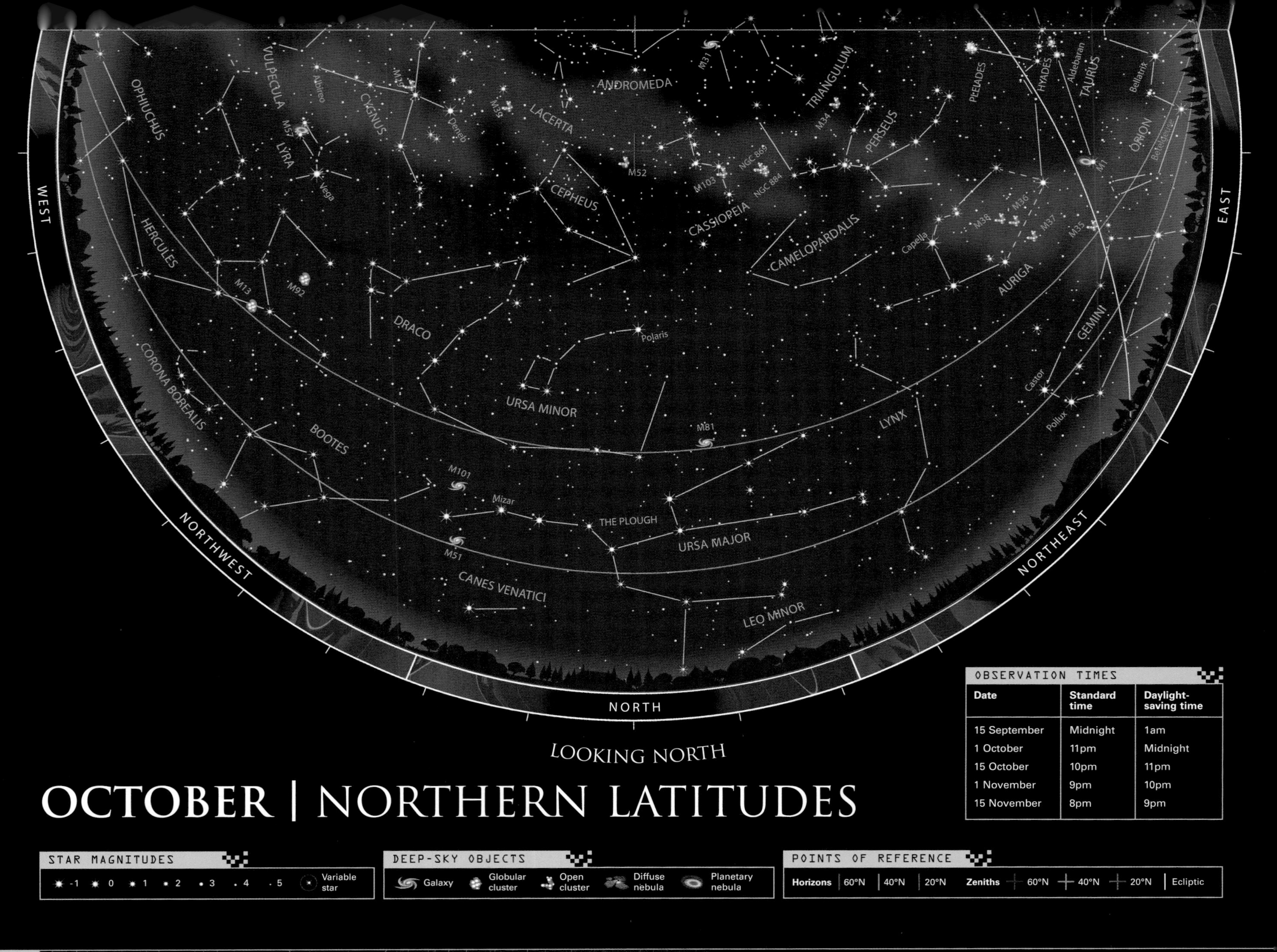


LOOKING NORTH

# OCTOBER | NORTHERN LATITUDES

**OBSERVATION TIMES**

| Date | Standard time | Daylight-saving time |
|---|---|---|
| 15 September | Midnight | 1am |
| 1 October | 11pm | Midnight |
| 15 October | 10pm | 11pm |
| 1 November | 9pm | 10pm |
| 15 November | 8pm | 9pm |

**STAR MAGNITUDES**

-1 · 0 · 1 · 2 · 3 · 4 · 5 · Variable star

**DEEP-SKY OBJECTS**

Galaxy · Globular cluster · Open cluster · Diffuse nebula · Planetary nebula

**POINTS OF REFERENCE**

**Horizons** | 60°N | 40°N | 20°N **Zeniths** 60°N 40°N 20°N | Ecliptic

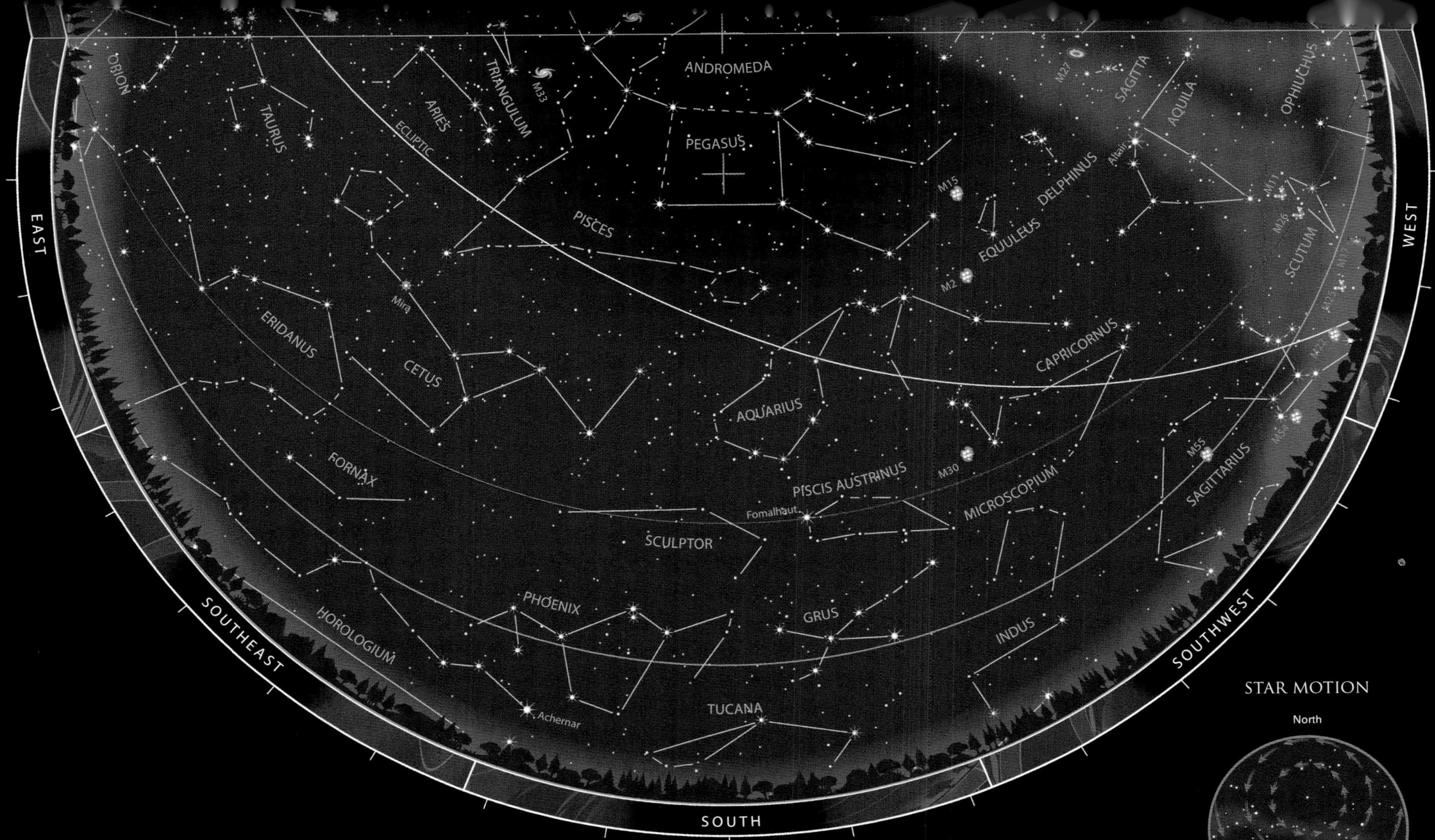

LOOKING SOUTH

# OCTOBER | NORTHERN LATITUDES

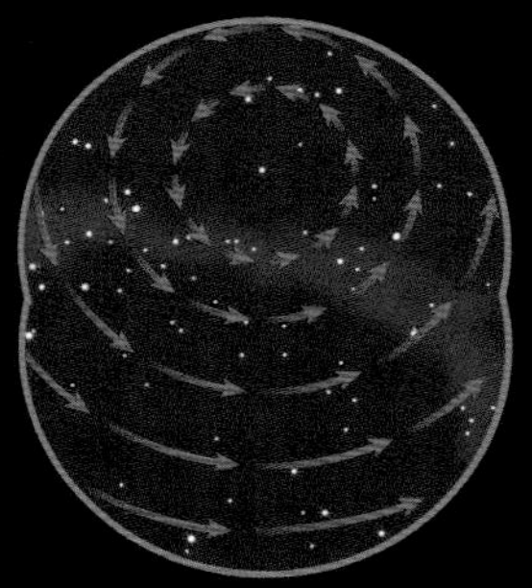

EAST
SOUTHEAST
SOUTH
SOUTHWEST
WEST

ORION
TAURUS
ARIES
TRIANGULUM
M33
ANDROMEDA
PEGASUS
PISCES
ECLIPTIC
ERIDANUS
Mira
CETUS
FORNAX
SCULPTOR
HOROLOGIUM
PHOENIX
Achernar
TUCANA
GRUS
AQUARIUS
PISCIS AUSTRINUS
Fomalhaut
M30
MICROSCOPIUM
INDUS
CAPRICORNUS
M2
M15
EQUULEUS
DELPHINUS
Altair
SAGITTA
M27
AQUILA
OPHIUCHUS
M11
M26
SCUTUM
M55
SAGITTARIUS
M54

STAR MOTION
North
South

STAR MAGNITUDES
-1 0 1 2 3 4 5 Variable star

DEEP-SKY OBJECTS
Galaxy | Globular cluster | Open cluster | Diffuse nebula | Planetary nebula

POINTS OF REFERENCE
Horizons 60°N | 40°N | 20°N
Zeniths 60°N | 40°N | 20°N
Ecliptic

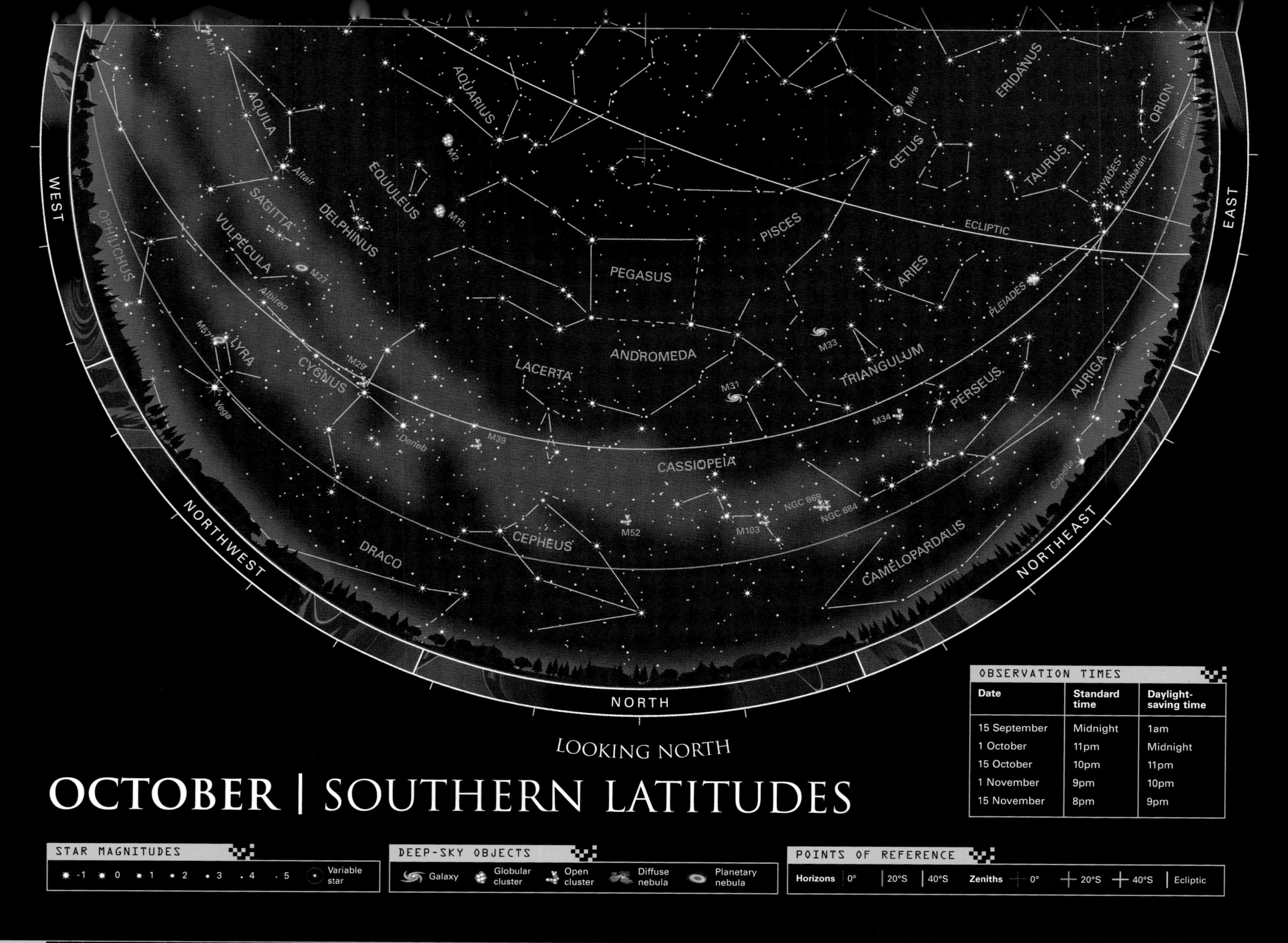


LOOKING NORTH

# OCTOBER | SOUTHERN LATITUDES

OBSERVATION TIMES

| Date | Standard time | Daylight-saving time |
|---|---|---|
| 15 September | Midnight | 1am |
| 1 October | 11pm | Midnight |
| 15 October | 10pm | 11pm |
| 1 November | 9pm | 10pm |
| 15 November | 8pm | 9pm |

STAR MAGNITUDES

-1 0 1 2 3 4 5 Variable star

DEEP-SKY OBJECTS

Galaxy Globular cluster Open cluster Diffuse nebula Planetary nebula

POINTS OF REFERENCE

**Horizons** 0° 20°S 40°S **Zeniths** 0° 20°S 40°S Ecliptic

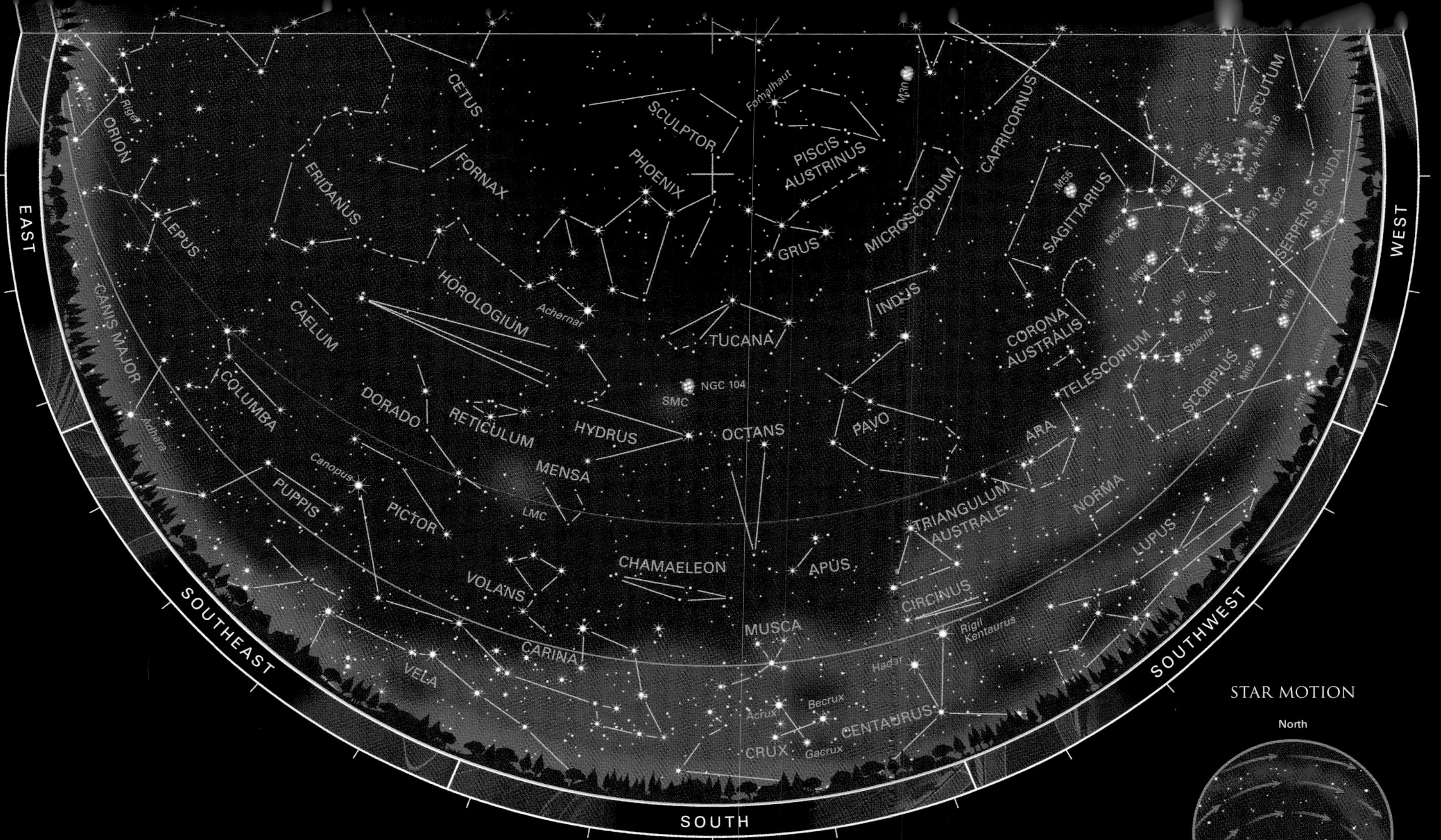

# OCTOBER | SOUTHERN LATITUDES

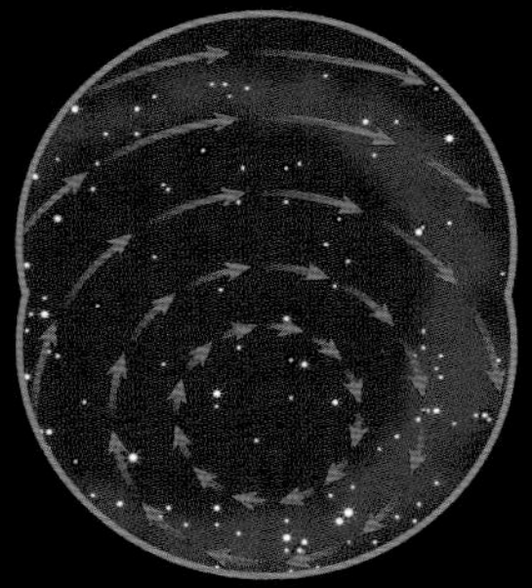

LOOKING SOUTH

EAST

SOUTHEAST

SOUTH

SOUTHWEST

WEST

STAR MOTION

North

South

**STAR MAGNITUDES**

-1 · 0 · 1 · 2 · 3 · 4 · 5 · Variable star

**DEEP-SKY OBJECTS**

Galaxy · Globular cluster · Open cluster · Diffuse nebula · Planetary nebula

**POINTS OF REFERENCE**

**Horizons** 0° | 20°S | 40°S

**Zeniths** 0° | 20°S | 40°S

Ecliptic

# NOVEMBER

Cassiopeia lies overhead for northern observers, as the Milky Way runs from Cygnus in the west to Gemini in the east. The large figures of Pisces, the Fishes, and Cetus, the Sea Monster or Whale, are spread across the equatorial region of the sky, while in the southern sky the Large and Small Magellanic Clouds are high up.

## SPECIAL EVENTS

### PHASES OF THE MOON

| | FULL MOON | NEW MOON |
|---|---|---|
| **2011** | 10 November | 25 November |
| **2012** | 28 November | 13 November |
| **2013** | 17 November | 3 November |
| **2014** | 6 November | 22 November |
| **2015** | 25 November | 11 November |
| **2016** | 14 November | 29 November |
| **2017** | 4 November | 18 November |
| **2018** | 23 November | 7 November |
| **2019** | 12 November | 26 November |

### PLANETS

**2011: 14 November** Mercury is at greatest evening elongation, magnitude -0.2.

**2012: 27 November** Venus and Saturn are 0.5° apart in the eastern dawn sky.

**2013: 18 November** Mercury is at greatest morning elongation, magnitude -0.5.

**2013: 26 November** Mercury and Saturn are 2° apart in the eastern morning sky.

**2014: 1 November** Mercury is at greatest morning elongation, magnitude -0.5.

**2017: 23 November** Mercury is at greatest evening elongation, magnitude -0.3.

**2018: 6 November** Mercury is at greatest evening elongation, magnitude -0.2.

**2019: 28 November** Mercury is at greatest morning elongation, magnitude -0.5.

### ECLIPSES AND TRANSITS

**2011: 25 November** A partial eclipse of the Sun is visible from the southern Indian Ocean and Antarctica.

**2012: 13–14 November** A total eclipse of the Sun is visible from northeastern Australia and the south Pacific. A partial eclipse is visible from New Zealand, the rest of Australia, and the Pacific Ocean.

**2013: 3 November** A total eclipse of the Sun is visible from the mid Atlantic Ocean and Central Africa.

**2019: 3 November** The transit of Mercury across the Sun is visible from North America, South America, Europe, Africa, and central Asia.

## NORTHERN LATITUDES

### THE STARS

All the main characters in the Perseus and Andromeda myth (see p.352) are on show in the November evening sky. Cetus contains a remarkable variable star, Mira (see p.281). It is easily visible to the naked eye when at maximum brightness, every 11 months or so, but the rest of the time it fades out of sight. High in the west is the Square of Pegasus, with the stars of the Summer Triangle lower in the northwest.

### DEEP-SKY OBJECTS

Two open star clusters, NGC 457 and NGC 663, are easy to see with binoculars in Cassiopeia. Even better are NGC 869 and 884, a pair known as the Double Cluster, embedded in the Milky Way between Perseus and Cassiopeia. The Andromeda Galaxy, M31 (see pp.302–303), remains high up this month.

### METEOR SHOWERS

The Taurids have a broad peak in the first week of the month, when around 10 meteors an hour may be seen coming from the region south of the Pleiades cluster. Although not numerous, the meteors are long-lasting and often bright. A second meteor shower in November, the Leonids, radiates from the head of Leo, reaching a peak around 17 November. Usually no more than 10 meteors per hour are seen, but surges of activity occur every 33 years or so. High activity is not expected again until around 2032.

**THE PRINCESS, HERO, KING, AND QUEEN** Joined in Greek myth, Andromeda (right), Perseus (bottom), Cepheus (top), and Cassiopeia (centre) appear together in northern skies in November.

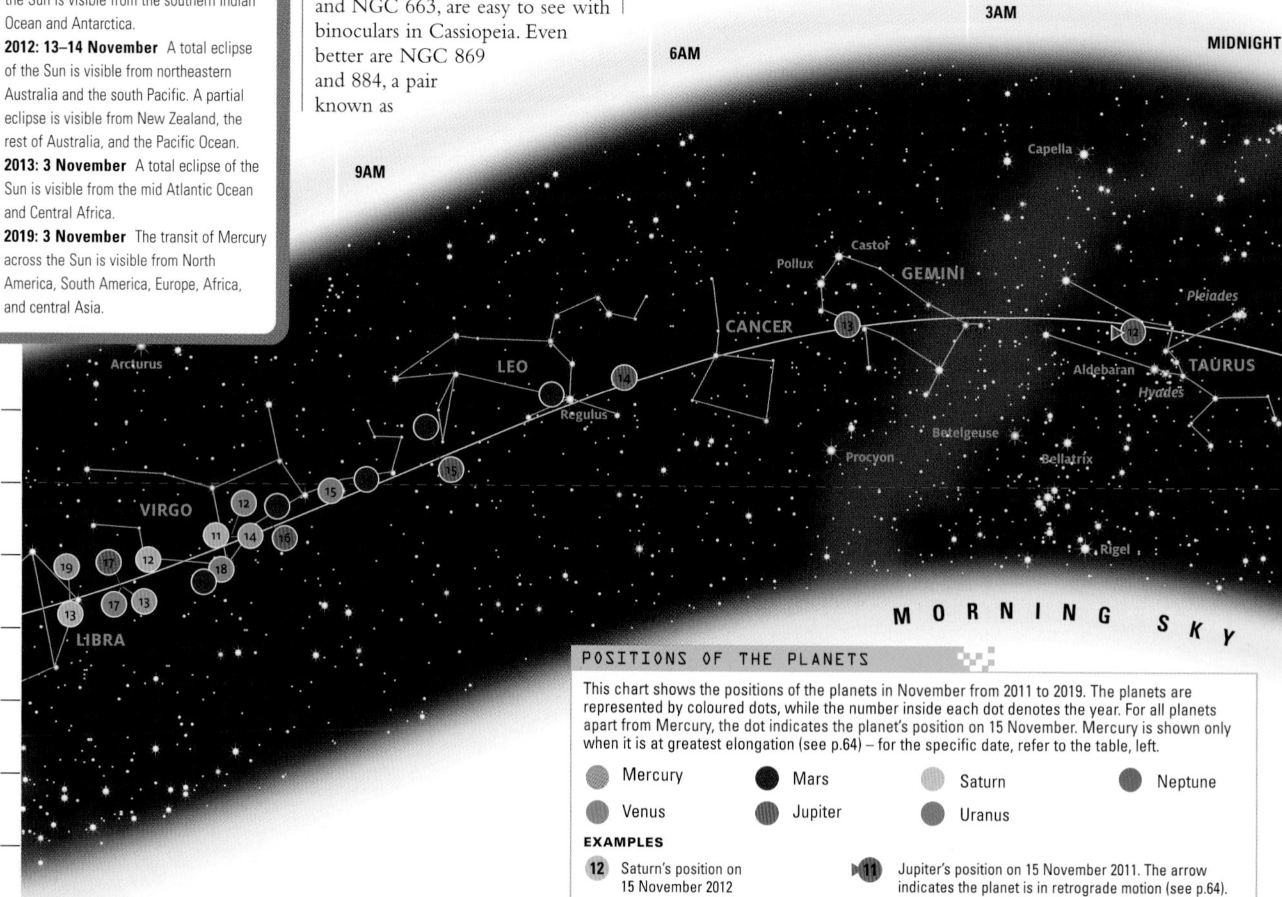


**POSITIONS OF THE PLANETS**

This chart shows the positions of the planets in November from 2011 to 2019. The planets are represented by coloured dots, while the number inside each dot denotes the year. For all planets apart from Mercury, the dot indicates the planet's position on 15 November. Mercury is shown only when it is at greatest elongation (see p.64) – for the specific date, refer to the table, left.

Mercury, Mars, Saturn, Neptune, Venus, Jupiter, Uranus

**EXAMPLES**

12 Saturn's position on 15 November 2012

11 Jupiter's position on 15 November 2011. The arrow indicates the planet is in retrograde motion (see p.64).

# SOUTHERN LATITUDES

## THE STARS

Achernar, the bright star at the end of Eridanus, lies high in the south on November evenings. The other stars of Eridanus extend to Orion, which is rising in the east. Aldebaran and the stars of Taurus are in the northeast, and the Square of Pegasus is high in the northwest. Aquarius is in the west, with Fomalhaut (see p.249) in Piscis Austrinus in the southwest. The Large and Small Magellanic Clouds (see p.300 and p.301) are high in the south. Brilliant Canopus in Carina is in the southeast, with Sirius (see p.264) in Canis Major rising in the east. Overhead is Cetus, containing the long-period variable star Mira.

## DEEP-SKY OBJECTS

South of the head of Cetus is M77, the brightest of the Seyfert type of galaxies (see p.310). Seyferts are spiral galaxies with unusually bright centres, caused by hot gas spiralling around a massive black hole. A telescope is required to see M77.

In the south, the globular cluster 47 Tucanae is still on view near the meridian. The Large Magellanic Cloud, with the Tarantula Nebula, NGC 2070, is in the southeast, but it is best seen in January. In the north, the galaxies M31 and M33 are visible, while the Pleiades (see p.287) and Hyades clusters (see p.286) are moving higher in the east.

**URANUS**

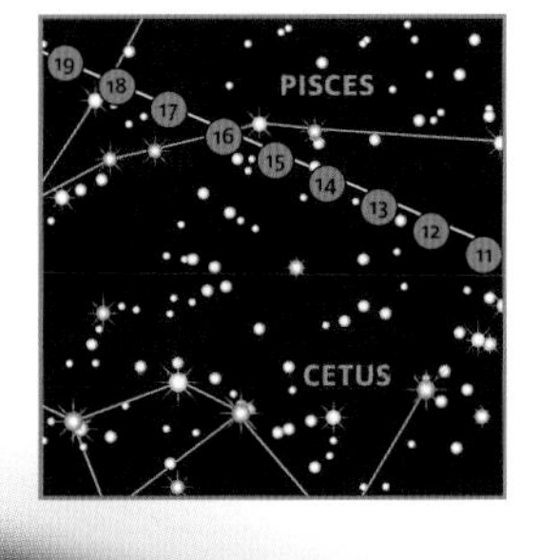


**NEPTUNE**

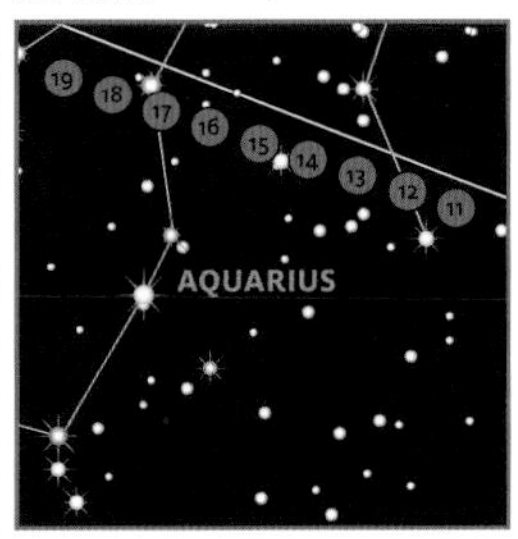


**CLASSIC VARIABLE**
The long-period variable star Mira (centre) appears strongly red when near maximum brightness. The 9th-magnitude star to its left is unrelated.

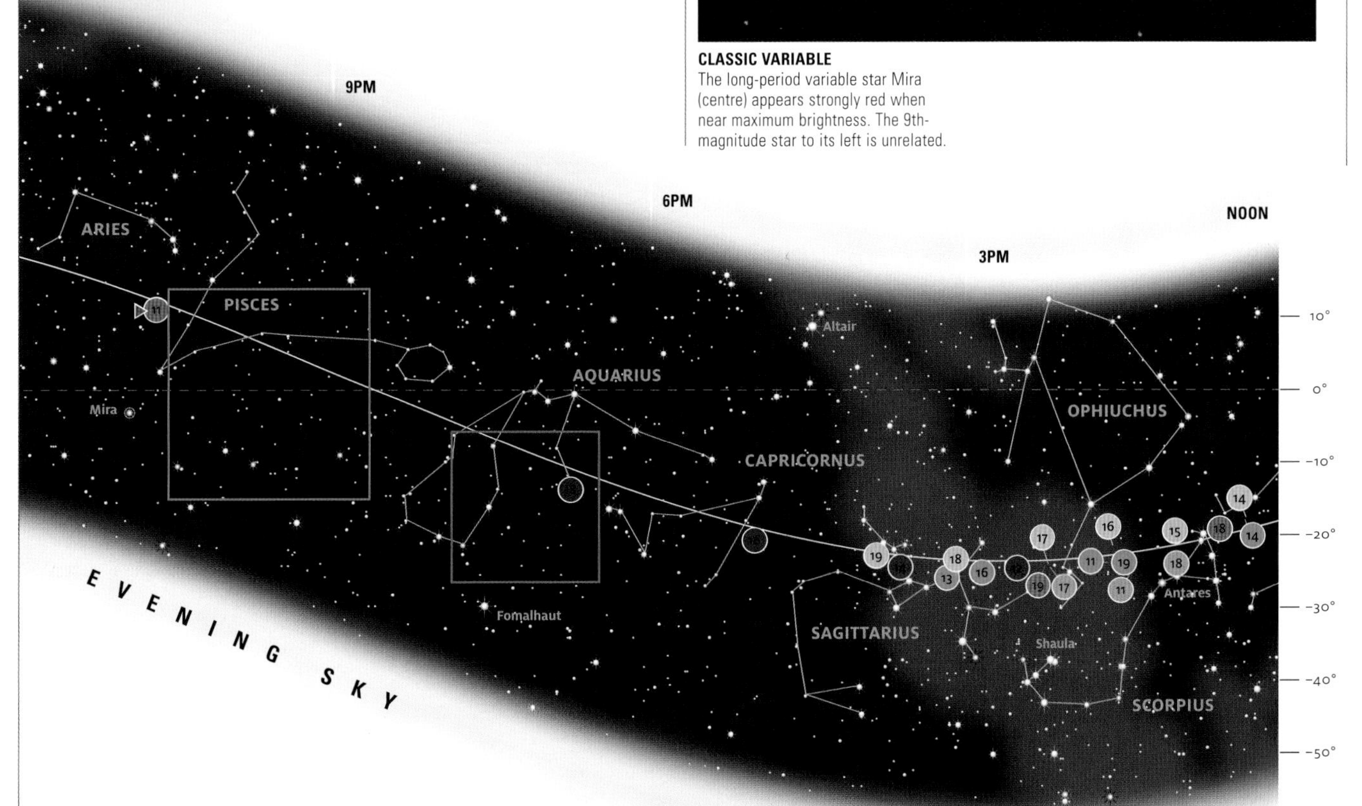

LOOKING NORTH

# NOVEMBER | NORTHERN LATITUDES

| OBSERVATION TIMES | | |
|---|---|---|
| **Date** | **Standard time** | **Daylight-saving time** |
| 15 October | Midnight | 1am |
| 1 November | 11pm | Midnight |
| 15 November | 10pm | 11pm |
| 1 December | 9pm | 10pm |
| 15 December | 8pm | 9pm |

**STAR MAGNITUDES**
-1 · 0 · 1 · 2 · 3 · 4 · 5 · Variable star

**DEEP-SKY OBJECTS**
Galaxy · Globular cluster · Open cluster · Diffuse nebula · Planetary nebula

**POINTS OF REFERENCE**
**Horizons** 60°N | 40°N | 20°N **Zeniths** 60°N | 40°N | 20°N | Ecliptic

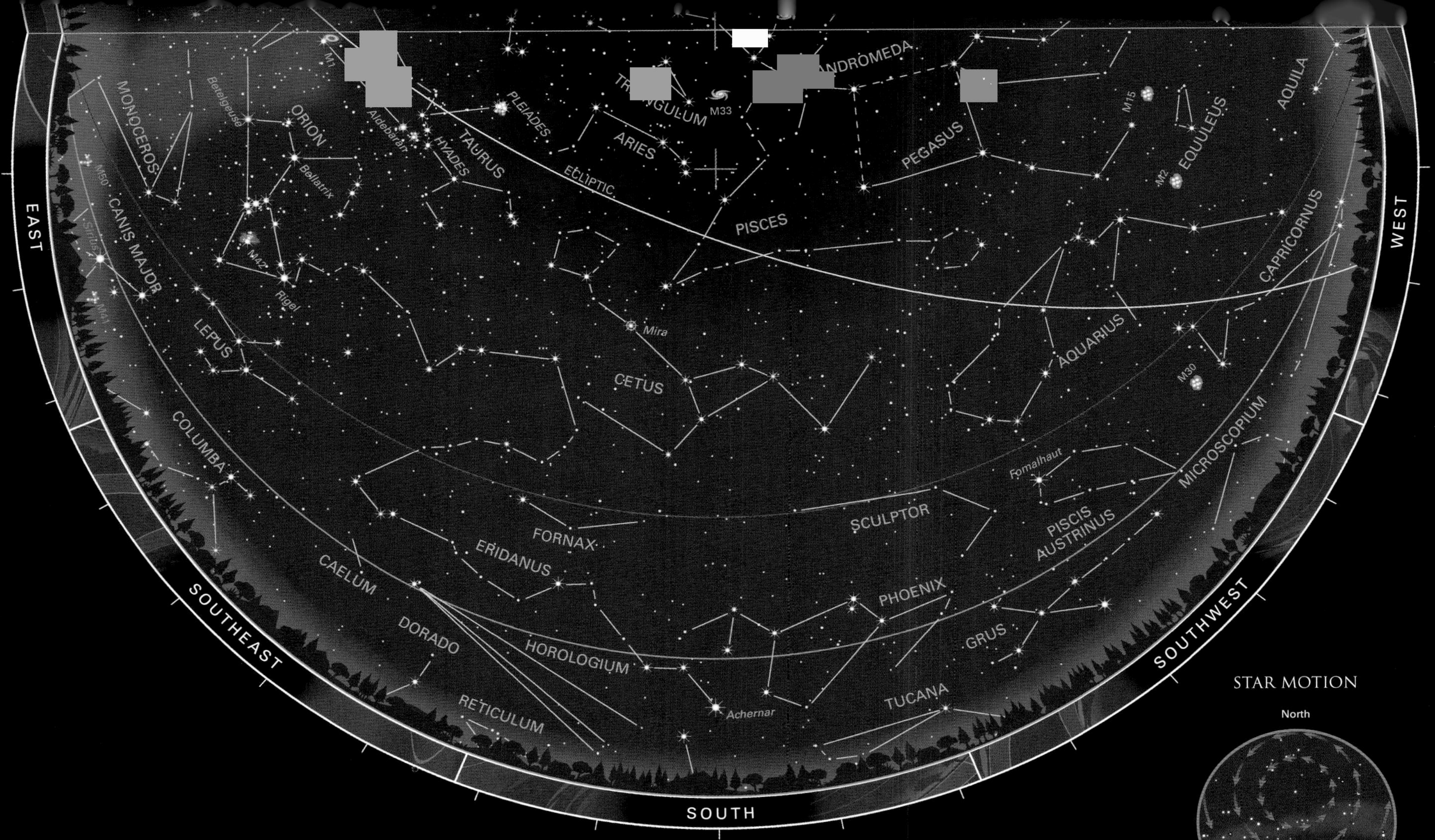

# NOVEMBER | NORTHERN LATITUDES

LOOKING SOUTH

EAST
SOUTHEAST
SOUTH
SOUTHWEST
WEST

MONOCEROS
CANIS MAJOR
M50
Sirius
M41
ORION
Betelgeuse
Bellatrix
M42
Rigel
M1
Aldebaran
HYADES
TAURUS
PLEIADES
ARIES
TRIANGULUM
M33
ANDROMEDA
PEGASUS
PISCES
ECLIPTIC
M15
M2
EQUULEUS
AQUILA
CAPRICORNUS
AQUARIUS
M30
MICROSCOPIUM
Fomalhaut
PISCIS AUSTRINUS
SCULPTOR
CETUS
Mira
LEPUS
COLUMBA
CAELUM
ERIDANUS
FORNAX
DORADO
HOROLOGIUM
RETICULUM
Achernar
PHOENIX
TUCANA
GRUS

STAR MOTION

North

South

STAR MAGNITUDES

-1 | 0 | 1 | 2 | 3 | 4 | 5 | Variable star

DEEP-SKY OBJECTS

Galaxy | Globular cluster | Open cluster | Diffuse nebula | Planetary nebula

POINTS OF REFERENCE

**Horizons** 60°N | 40°N | 20°N **Zeniths** 60°N | 40°N | 20°N | Ecliptic

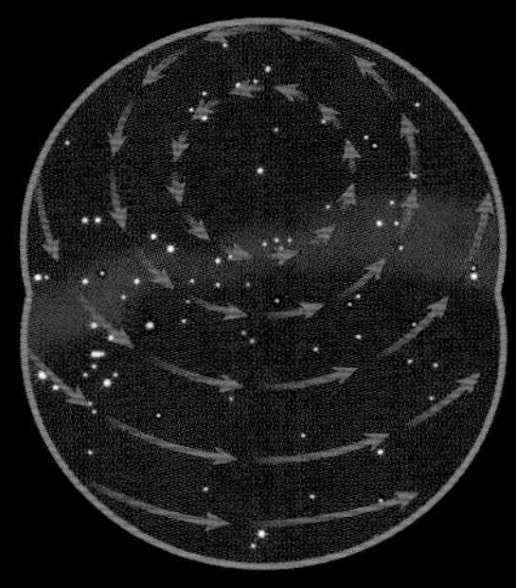

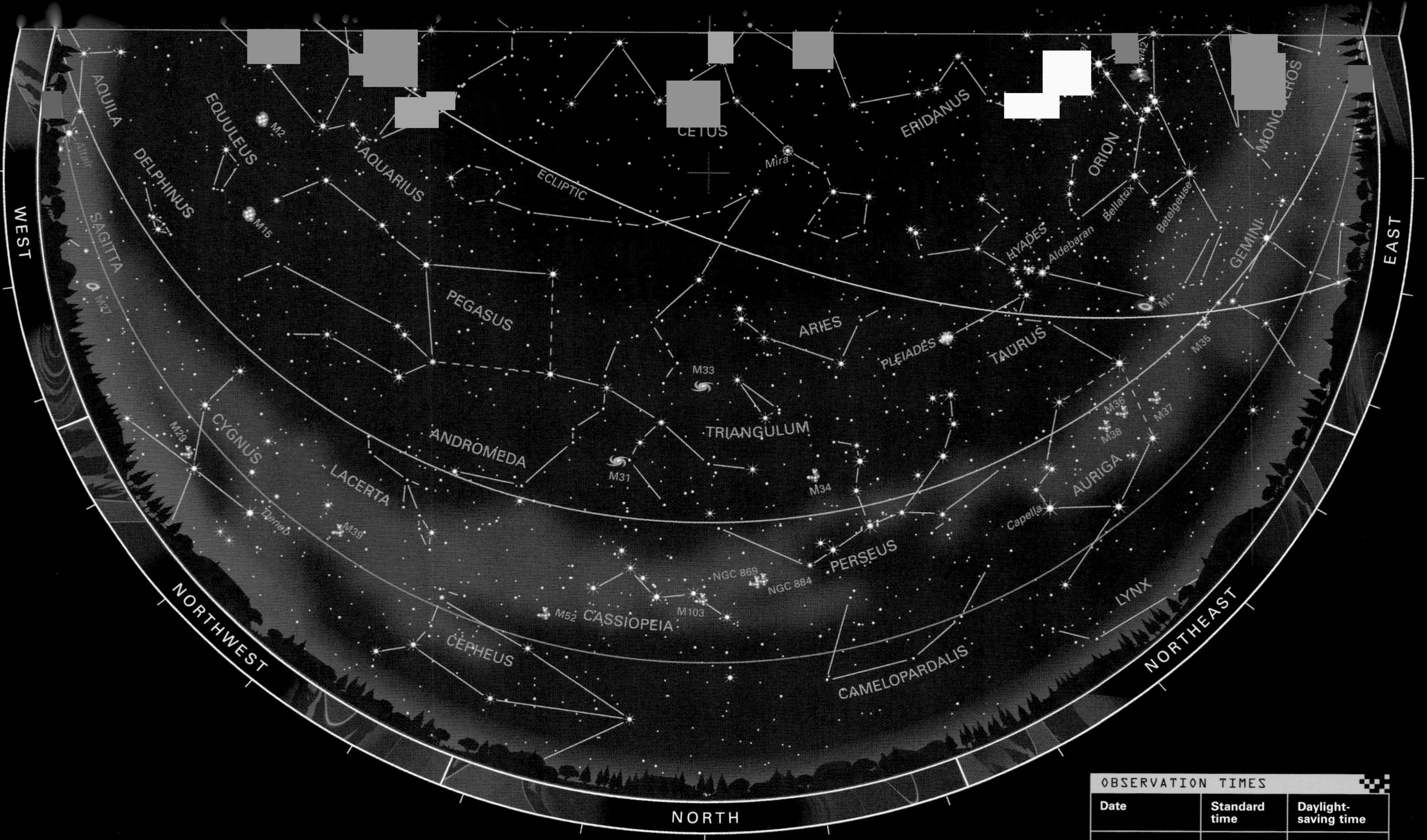

LOOKING NORTH

# NOVEMBER | SOUTHERN LATITUDES

OBSERVATION TIMES

| Date | Standard time | Daylight-saving time |
|---|---|---|
| 15 October | Midnight | 1am |
| 1 November | 11pm | Midnight |
| 15 November | 10pm | 11pm |
| 1 December | 9pm | 10pm |
| 15 December | 8pm | 9pm |

STAR MAGNITUDES

-1 · 0 · 1 · 2 · 3 · 4 · 5 · Variable star

DEEP-SKY OBJECTS

Galaxy · Globular cluster · Open cluster · Diffuse nebula · Planetary nebula

POINTS OF REFERENCE

**Horizons** 0° · 20°S · 40°S **Zeniths** 0° · 20°S · 40°S · Ecliptic

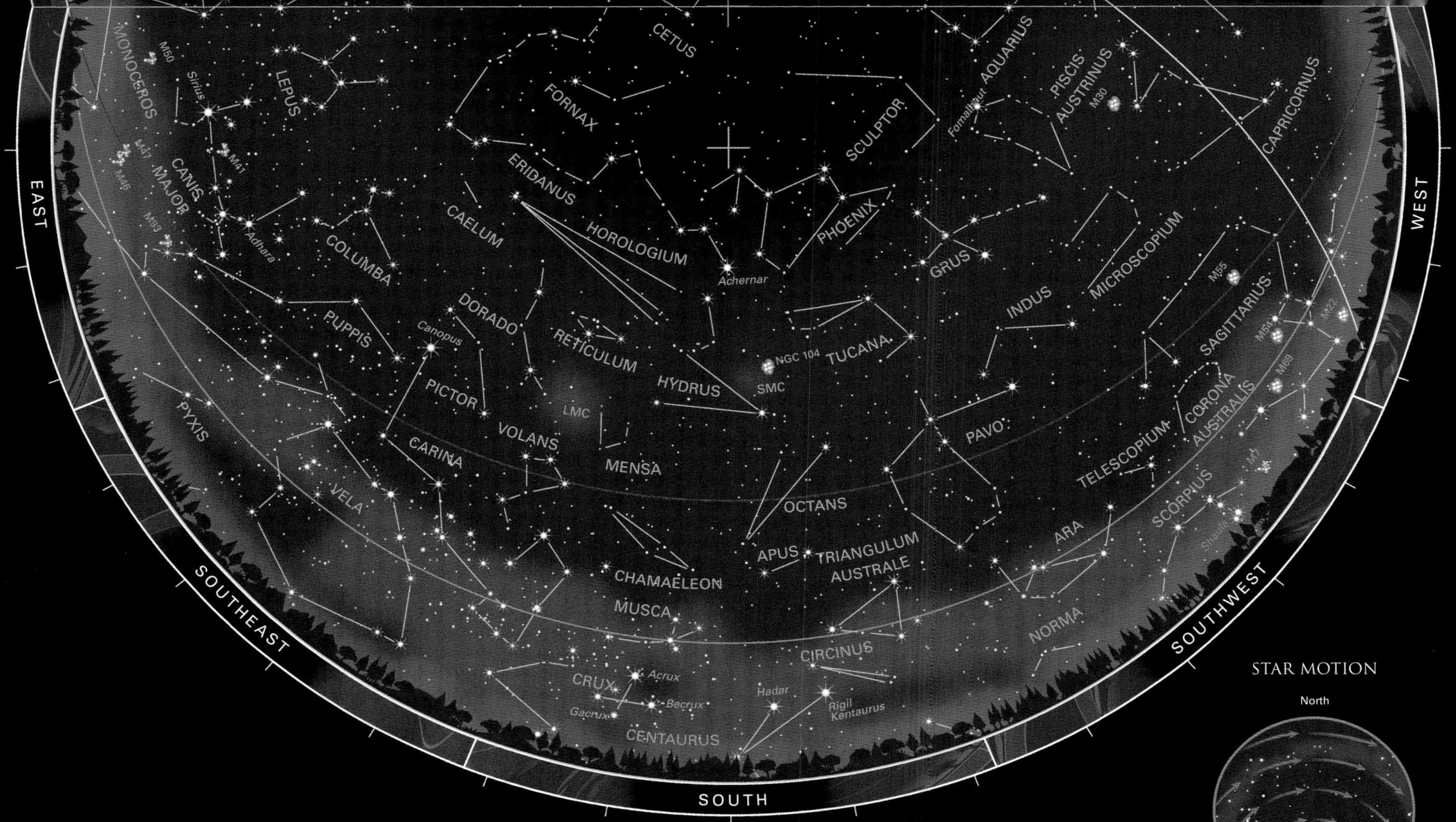

LOOKING SOUTH

# NOVEMBER | SOUTHERN LATITUDES

STAR MOTION

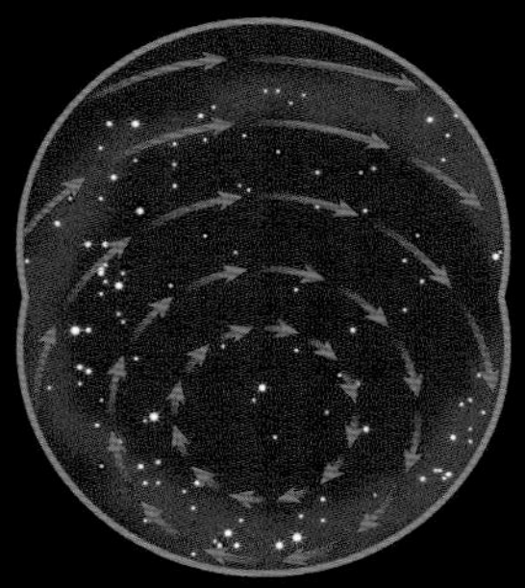

North

South

STAR MAGNITUDES

-1 · 0 · 1 · 2 · 3 · 4 · 5 · Variable star

DEEP-SKY OBJECTS

Galaxy · Globular cluster · Open cluster · Diffuse nebula · Planetary nebula

POINTS OF REFERENCE

**Horizons** 0° | 20°S | 40°S **Zeniths** 0° 20°S 40°S | Ecliptic

# DECEMBER

The Sun reaches its farthest point south of the celestial equator this month, on 21–22 December. As a result, northern hemisphere nights are the longest of the year, while in the southern hemisphere they are the shortest. The Earth has now completed another annual circuit of the Sun, and the evening stars end the year as they began, with the tableau of Orion and Taurus returning to centre stage.

## SPECIAL EVENTS

### PHASES OF THE MOON

| | FULL MOON | NEW MOON |
|---|---|---|
| **2011** | 10 December | 24 December |
| **2012** | 28 December | 13 December |
| **2013** | 17 December | 3 December |
| **2014** | 6 December | 22 December |
| **2015** | 25 December | 11 December |
| **2016** | 14 December | 29 December |
| **2017** | 3 December | 18 December |
| **2018** | 22 December | 7 December |
| **2019** | 12 December | 26 December |

### PLANETS

**2011: 23 December** Mercury is at greatest morning elongation, magnitude -0.3.

**2012: 3 December** Jupiter is at opposition, magnitude -2.8.

**2012: 4 December** Mercury is at greatest morning elongation, magnitude -0.3.

**2014: 30 December** Mercury and Venus are 3.8° apart in the southwestern evening sky.

**2015: 29 December** Mercury is at greatest evening elongation, magnitude -0.5.

**2016: 11 December** Mercury is at greatest evening elongation, magnitude -0.4.

**2018: 16 December** Mercury is at greatest morning elongation, magnitude -1.4.

### ECLIPSES

**2011: 10 December** A total eclipse of the Moon is visible from the Pacific Ocean, Australasia, and Asia.

**2019: 21 December** An annular eclipse of the Sun is visible from Saudi Arabia, India, Sumatra, and Borneo. A partial solar eclipse is visible from Asia and Australia.

## NORTHERN LATITUDES

### THE STARS

Overhead lies Perseus, containing the famous variable star Algol (see p.272). From Perseus, the Milky Way leads northwestwards to Cassiopeia and Cygnus, which is out of sight for those at around 20°N or closer to the equator. In the other direction, the Milky Way extends southeastwards via Auriga and past Taurus to Gemini and the northern arm of Orion. The Square of Pegasus is in the west, while the Winter Triangle of Betelgeuse (see p.252) in Orion, Procyon (see p.280) in Canis Minor, and Sirius (see p.264) in Canis Major dominates the southeast. By comparison with the richness of this southeastern part of the sky, the southwest seems dull and empty, as it is occupied by the faint constellations Aries, Pisces, and Cetus. As the year ends, Sirius lies due south around midnight.

### DEEP-SKY OBJECTS

Large, bright clusters of stars abound in the December evening sky. In central Perseus, a few dozen stars cluster around the constellation's brightest member, Alpha (α) Persei or Mirphak. They form a group known as the Alpha Persei cluster, which covers several diameters of the full Moon and is a fine sight through binoculars.

In Taurus lies probably the finest open cluster in the entire sky, the Pleiades or M45 (see p.287). At least six members are visible to normal eyesight, but binoculars bring dozens more into view. Taurus contains an even larger cluster, the Hyades (see p.286), a V-shaped grouping which outlines the Bull's face. In addition to these groupings, the Double Cluster in Perseus, NGC 869 and NGC 884, already encountered in November, remains well placed.

### METEOR SHOWER

The year's second-best meteor shower, the Geminids, reaches a peak around 13–14 December, when up to one meteor per minute can be seen radiating from a point near Castor in Gemini. Lesser activity is seen for a few days before the peak, but numbers fall off rapidly afterwards.

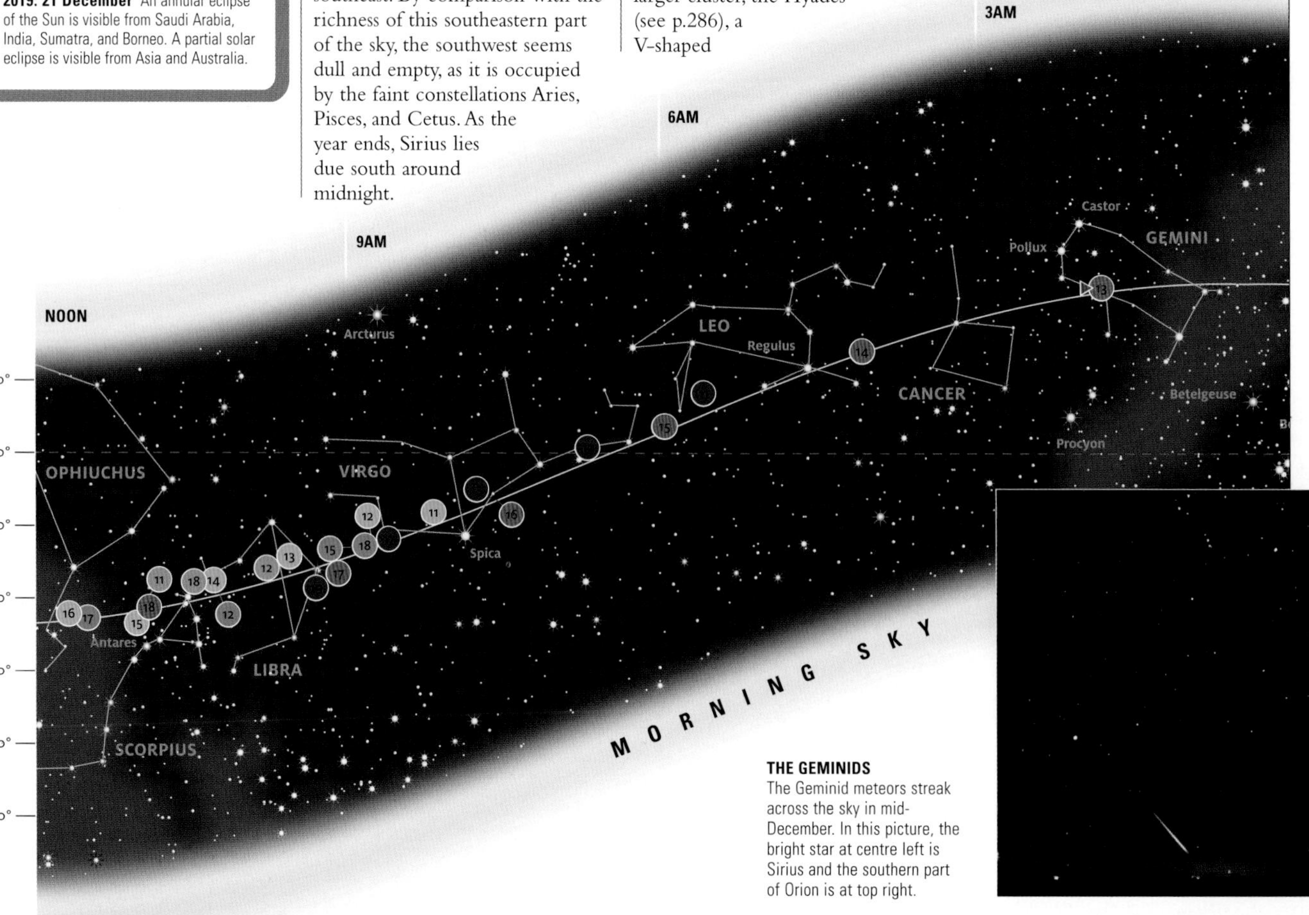


**THE GEMINIDS**
The Geminid meteors streak across the sky in mid-December. In this picture, the bright star at centre left is Sirius and the southern part of Orion is at top right.

# SOUTHERN LATITUDES

## THE STARS

The distinctive figures of Orion and Taurus are high in the northeast, with Gemini and Auriga closer to the horizon. Perseus lies low in the north, while the Square of Pegasus sets in the northwest, followed by Pisces. Fomalhaut (see p.249) in Piscis Austrinus is in the southwest.

Eridanus, the River, meanders southwestwards from the foot of Orion, ending at the bright star Achernar. Brighter Canopus is high in the southeast in Carina. The Large and Small Magellanic Clouds (see p.300 and p.301) lie high in the south, either side of the celestial meridian. In the east, Betelgeuse in Orion, Procyon in Canis Minor, and Sirius in Canis Major form a large triangle, which is a sign of the approaching southern summer.

## DEEP-SKY OBJECTS

December and January evenings are the best time for southern observers to see the Pleiades (see p.287) and Hyades (see p.286), two large and prominent open star clusters north of the equator in Taurus. The Large Magellanic Cloud, containing the Tarantula Nebula, NGC 2070, is high in the southeast but it is better seen in January. Overall, the southern evening sky is bereft of prominent deep-sky objects near the celestial meridian this month.

**THE LARGE MAGELLANIC CLOUD**
The LMC (bottom) lies deep in the southern sky between the bright stars Canopus (left) and Achernar (top right). The small pink patch on the LMC is the Tarantula Nebula.

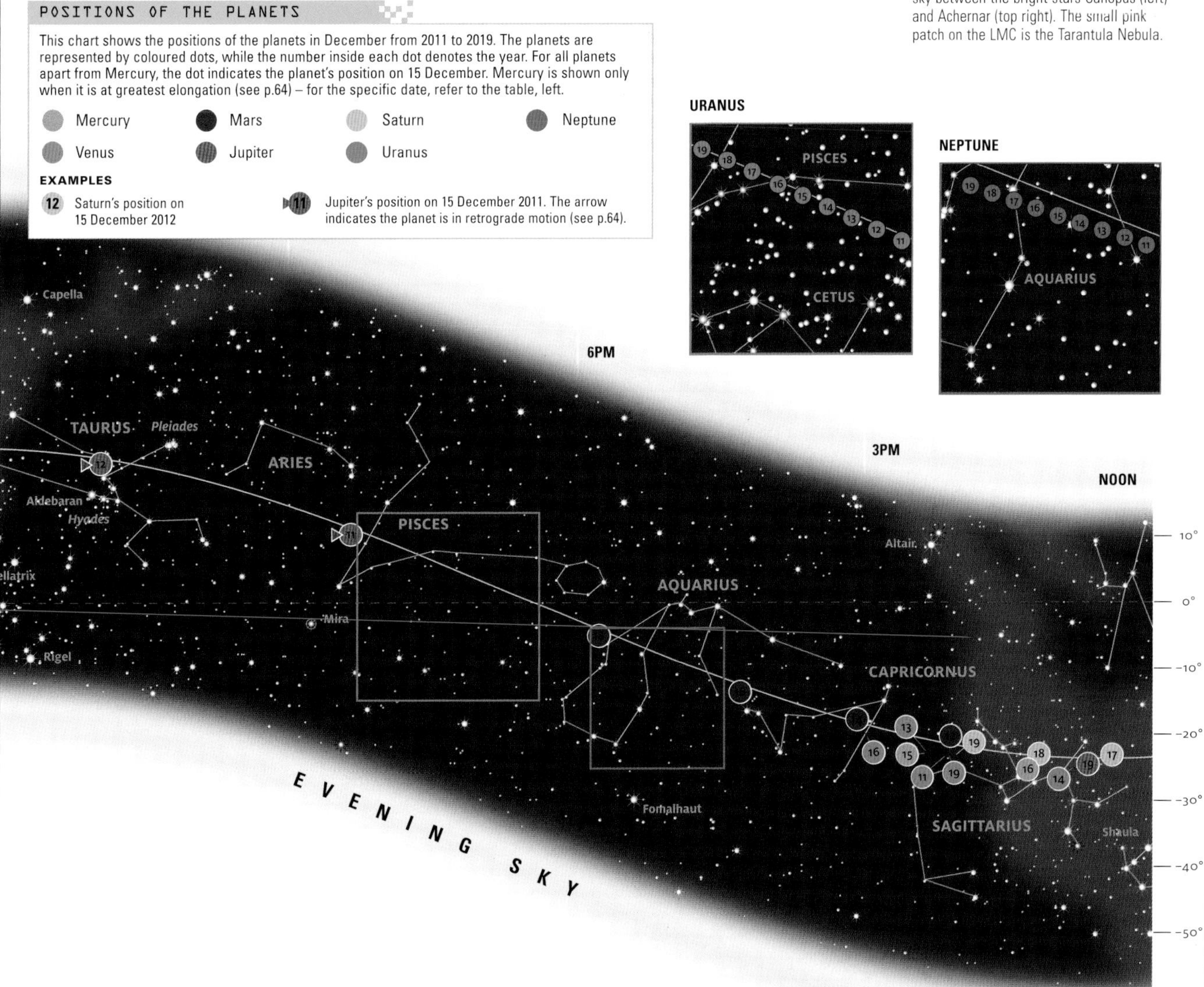
POSITIONS OF THE PLANETS
This chart shows the positions of the planets in December from 2011 to 2019. The planets are represented by coloured dots, while the number inside each dot denotes the year. For all planets apart from Mercury, the dot indicates the planet's position on 15 December. Mercury is shown only when it is at greatest elongation (see p.64) – for the specific date, refer to the table, left.
Mercury
Mars
Saturn
Neptune
Venus
Jupiter
Uranus
EXAMPLES
Saturn's position on 15 December 2012
Jupiter's position on 15 December 2011. The arrow indicates the planet is in retrograde motion (see p.64).
URANUS
PISCES
CETUS
NEPTUNE
AQUARIUS
Capella
TAURUS
Pleiades
ARIES
Aldebaran
Hyades
PISCES
Mira
Rigel
AQUARIUS
Altair
CAPRICORNUS
Fomalhaut
SAGITTARIUS
Shaula
6PM
3PM
NOON
10°
0°
-10°
-20°
-30°
-40°
-50°
EVENING SKY

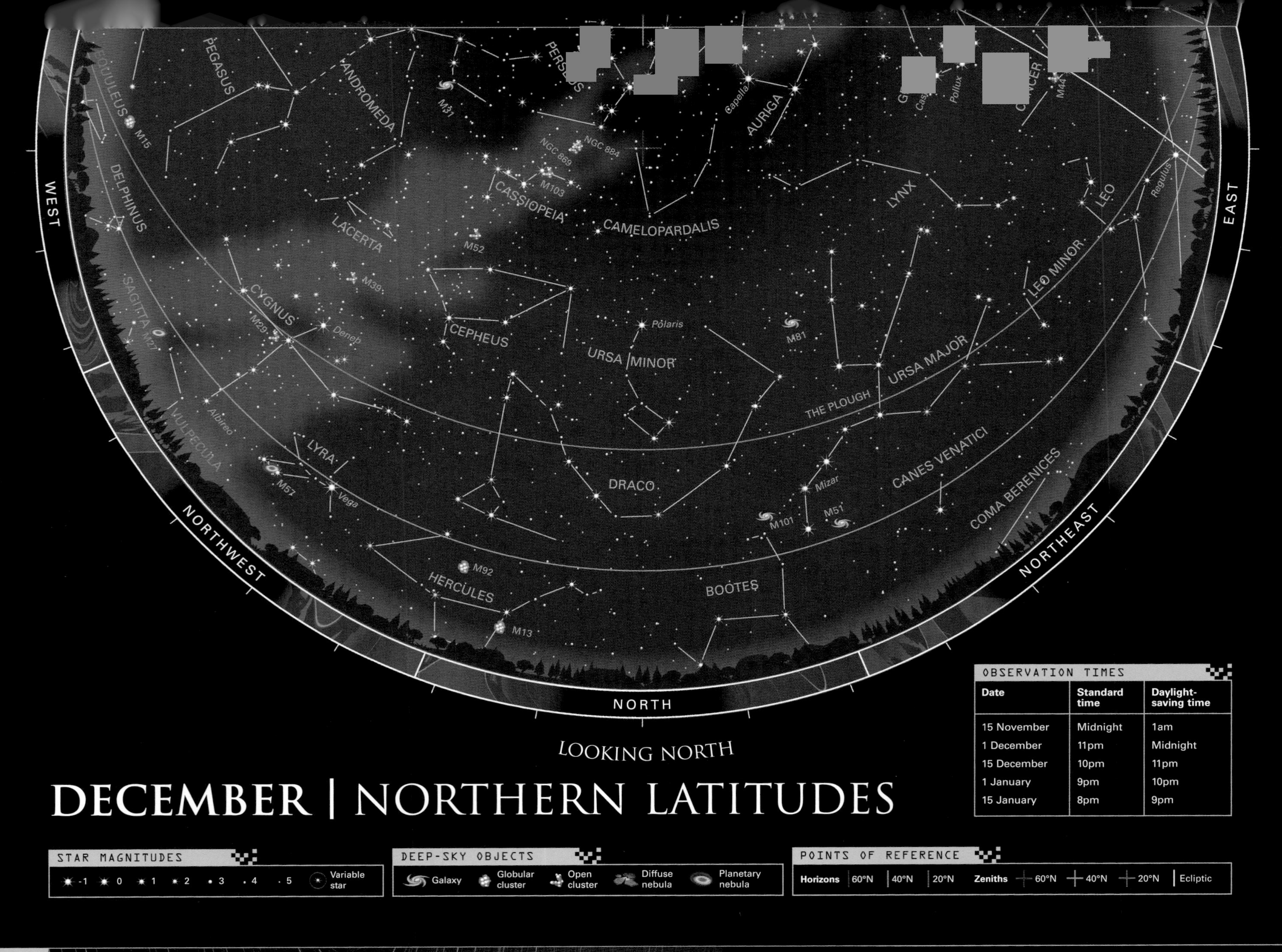


LOOKING NORTH

# DECEMBER | NORTHERN LATITUDES

OBSERVATION TIMES

| Date | Standard time | Daylight-saving time |
|---|---|---|
| 15 November | Midnight | 1am |
| 1 December | 11pm | Midnight |
| 15 December | 10pm | 11pm |
| 1 January | 9pm | 10pm |
| 15 January | 8pm | 9pm |

STAR MAGNITUDES

-1 · 0 · 1 · 2 · 3 · 4 · 5 · Variable star

DEEP-SKY OBJECTS

Galaxy · Globular cluster · Open cluster · Diffuse nebula · Planetary nebula

POINTS OF REFERENCE

Horizons 60°N 40°N 20°N · Zeniths 60°N 40°N 20°N · Ecliptic

# DECEMBER | NORTHERN LATITUDES

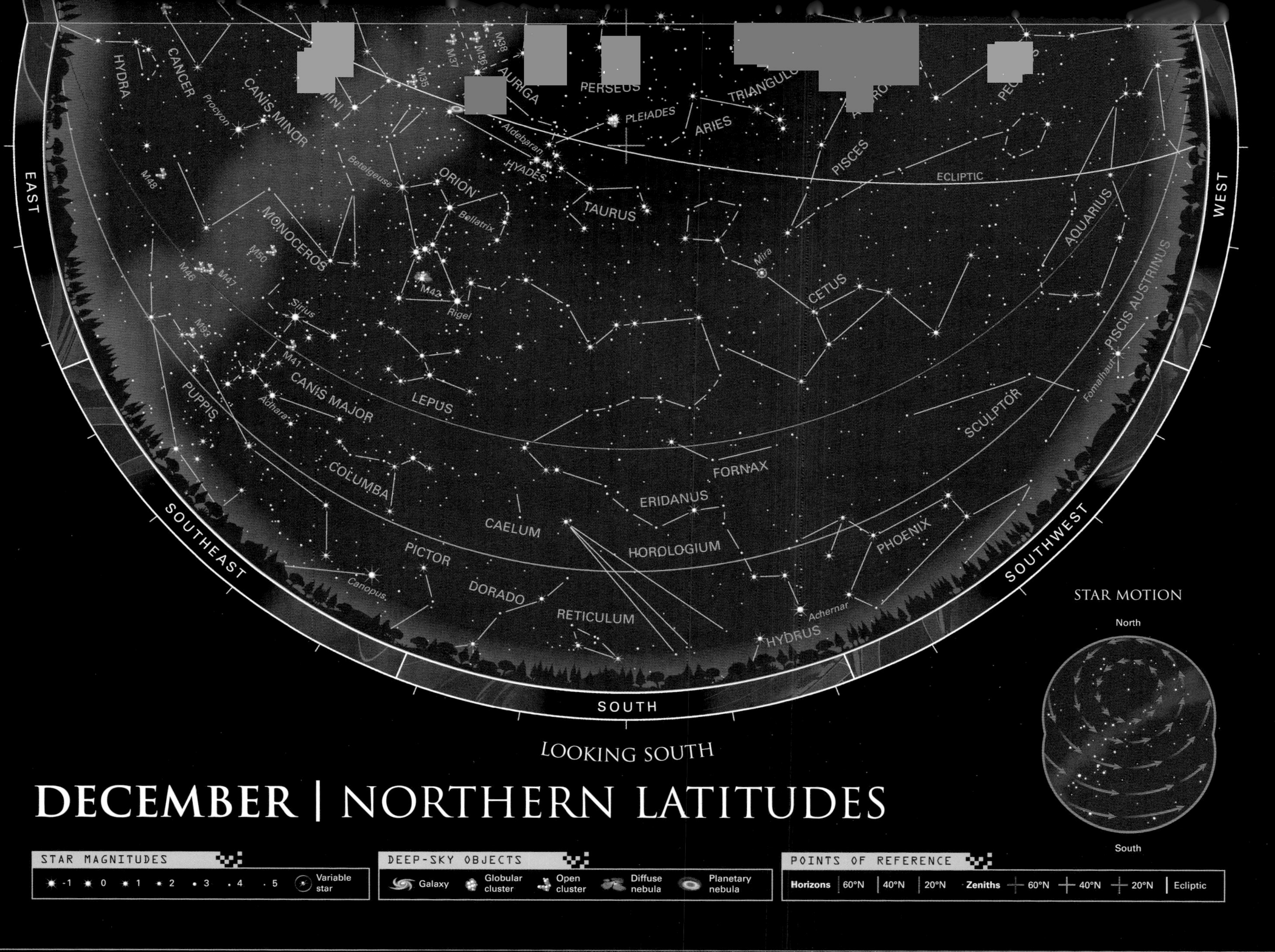

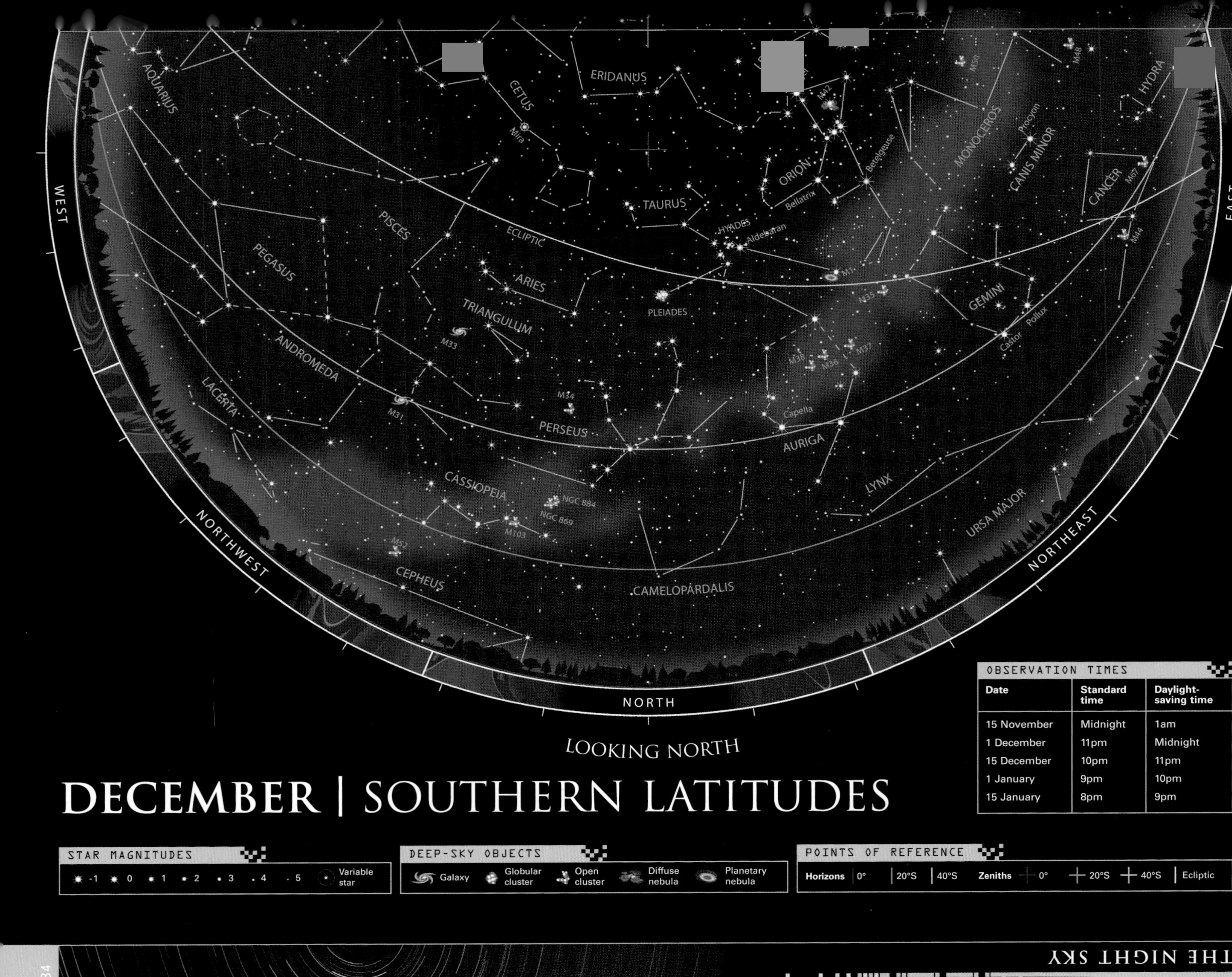


LOOKING NORTH

# DECEMBER | SOUTHERN LATITUDES

OBSERVATION TIMES

| Date | Standard time | Daylight-saving time |
|---|---|---|
| 15 November | Midnight | 1am |
| 1 December | 11pm | Midnight |
| 15 December | 10pm | 11pm |
| 1 January | 9pm | 10pm |
| 15 January | 8pm | 9pm |

STAR MAGNITUDES

-1 · 0 · 1 · 2 · 3 · 4 · 5 · Variable star

DEEP-SKY OBJECTS

Galaxy · Globular cluster · Open cluster · Diffuse nebula · Planetary nebula

POINTS OF REFERENCE

**Horizons** 0° · 20°S · 40°S · **Zeniths** 0° · 20°S · 40°S · Ecliptic

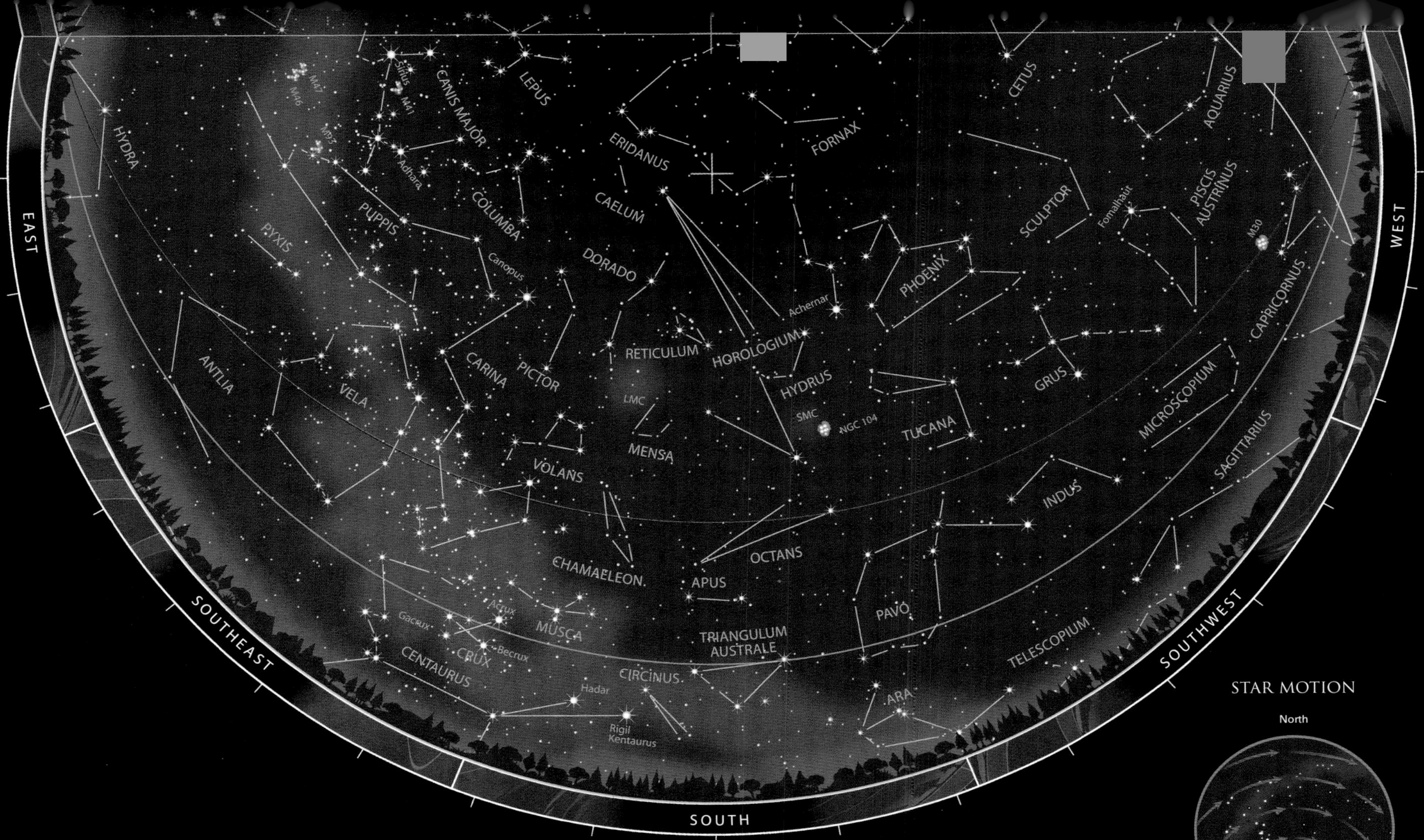

# DECEMBER | SOUTHERN LATITUDES

LOOKING SOUTH

EAST
SOUTHEAST
SOUTH
SOUTHWEST
WEST

HYDRA
M47
M46
M93
Sirius
M41
CANIS MAJOR
Adhara
LEPUS
ERIDANUS
FORNAX
CETUS
AQUARIUS
PISCIS AUSTRINUS
Fomalhaut
SCULPTOR
M30
CAPRICORNUS
PYXIS
PUPPIS
COLUMBA
Canopus
CAELUM
DORADO
Achernar
PHOENIX
ANTLIA
VELA
CARINA
PICTOR
RETICULUM
HOROLOGIUM
HYDRUS
LMC
SMC
NGC 104
TUCANA
GRUS
MICROSCOPIUM
SAGITTARIUS
MENSA
VOLANS
INDUS
CHAMAELEON
APUS
OCTANS
PAVO
Gacrux
Acrux
MUSCA
CRUX
Becrux
CENTAURUS
TRIANGULUM AUSTRALE
CIRCINUS
Hadar
Rigil Kentaurus
ARA
TELESCOPIUM

STAR MOTION

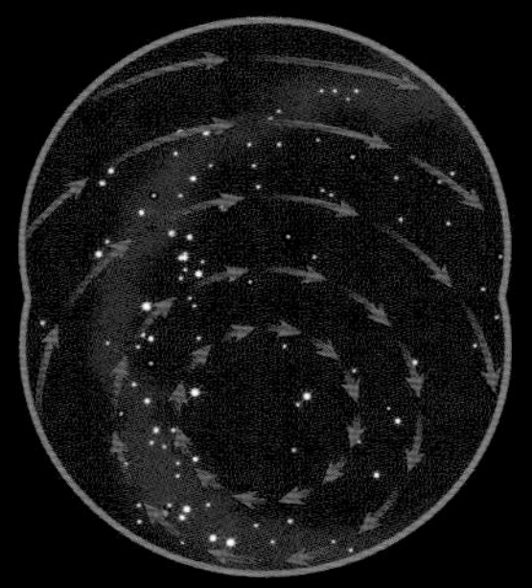

North

South

STAR MAGNITUDES

-1 0 1 2 3 4 5 Variable star

DEEP-SKY OBJECTS

Galaxy
Globular cluster
Open cluster
Diffuse nebula
Planetary nebula

POINTS OF REFERENCE

Horizons 0° | 20°S | 40°S

Zeniths 0° 20°S 40°S

Ecliptic

# GLOSSARY

## A

**absolute magnitude** see *magnitude.*

**absorption line** see *spectral line.*

**absorption nebula** see *nebula.*

**accelerating universe** A universe that expands at an accelerating rate. Current evidence indicates that the expansion of our Universe had been slowing down under the action of gravity until about 6 billion years ago, but that since then it has been accelerating. The acceleration is believed to be driven by the repulsive influence of dark energy. See also *dark energy.*

**accretion** **(1)** The colliding and sticking together of small, solid particles and bodies to make progressively larger ones. **(2)** The process whereby a body grows in mass by accumulating matter from its surroundings. An **accretion disc** is a disc of gas that revolves around a star or a compact object such as a white dwarf, neutron star, or black hole and which has been drawn in from a companion star or from neighbouring gas clouds.

**active galaxy** A galaxy that emits an exceptional amount of energy over a wide range of wavelengths, from radio waves to X-rays. An **active galactic nucleus (AGN)** is the compact, highly luminous core of an active galaxy that, in many cases, varies markedly in brightness over time, and is thought to be powered by the accretion of gas onto a supermassive black hole. See also *black hole, galaxy.*

**active prominence** see *prominence.*

**albedo** The ratio of the amount of light reflected by a body, such as a planet or a part of a planet's surface, to the amount of light that it receives from the Sun. Albedo values range from 0, for a perfectly dark object that reflects nothing, to 1, for a perfect reflector.

**altazimuth mounting** A mounting that enables a telescope to be rotated in altitude (around a horizontal axis) and in azimuth (around a vertical axis). Many large modern telescopes are mounted in this way, using computer-controlled motors to drive the telescope in altitude and azimuth so as to track the motion of an object across the sky. See also *altitude, azimuth, equatorial mounting.*

**altitude** The angular distance between the horizon and a celestial body. Altitude takes values from 0° (for an object on the horizon) to 90° (for an object that is directly overhead). See also *azimuth.*

**antimatter** Material composed of antiparticles. See *antiparticle.*

**antiparticle** An elementary particle that has the same mass as a particle of ordinary matter but exactly opposite values of other quantities such as spin and electrical charge. For example, the antiparticle of the negatively charged electron is the positively charged **positron**. If a particle and its antiparticle collide, both are annihilated and converted into energy.

**aperture** The clear diameter of the objective lens or primary mirror of a telescope or other optical instrument.

**aphelion** The point on its elliptical orbit at which a body such as a planet, asteroid, or comet is at its greatest distance from the Sun.

**apogee** The point on its elliptical orbit around the Earth at which a body such as the Moon or a spacecraft is at its greatest distance from the Earth. See also *perigee.*

**apparent magnitude** see *magnitude.*

**arachnoid** A type of structure, found on the surface of Venus, that consists of concentric circular or oval fractures or ridges, together with a complex network of fractures or ridges that radiate outwards. Its name derives from its superficial resemblance to a spider's web. Typical diameters range from 50 to 175km (30–110 miles).

**asterism** A conspicuous pattern of stars that is not itself a constellation. A well-known example is the Plough, or Big Dipper, which forms part of the constellation Ursa Major (the Great Bear). See also *constellation.*

**asteroid** One of the vast number of small bodies that revolve independently around the Sun. Their diameters range from a few metres (yards) to around 1,000km (600 miles). While the greatest concentration of asteroids is in the **Main Belt**, which lies between the orbits of Mars and Jupiter, asteroids are found throughout the Solar System. A **near-Earth asteroid (NEA)** is a body whose orbit comes close to, or intersects, the orbit of the Earth. Formally, a near-Earth asteroid is defined as one that has a perihelion distance of less than 1.3 times the Earth's mean distance from the Sun. See also *Kuiper Belt.*

**astronomical unit (AU)** A unit of distance measurement equal to the semimajor axis of the Earth's elliptical orbit, equivalent to the average of the maximum and minimum distances between the Earth and Sun. 1 AU = 149,598,000km (92,956,000 miles).

**atom** A basic building block of matter that is the smallest unit of a chemical element possessing the characteristics of that element. It consists of a nucleus of protons and neutrons, surrounded by a cloud of electrons. An atom has the same number of orbiting electrons as it has protons, so it is neutral (has no electrical charge). The chemical identity of an atom is determined by the number of protons in its nucleus (its **atomic number**). An atom of hydrogen (the simplest and lightest element) consists of a single proton and a single electron. See also *electron, neutron, proton.*

**aurora** A glowing, fluctuating display of light that is produced when charged particles entering a planet's upper atmosphere, usually in the vicinity of its north and south magnetic poles, collide with atoms and stimulate them to emit light.

**autumnal equinox** see *equinox.*

**azimuth** The angle between the north point on an observer's horizon and a celestial object, measured in a clockwise direction around the horizon. The azimuth of due north is 0°, due east 90°, due south 180°, and due west 270°. See also *altitude.*

## B

**background radiation** see *cosmic microwave background radiation.*

**barred spiral galaxy** A galaxy that has spiral arms emanating from the ends of an elongated, bar-shaped, nucleus. See also *galaxy, spiral galaxy.*

**baryon** A particle, composed of three quarks, that is acted on by the strong nuclear force. Examples include protons and neutrons, the building blocks of atomic nuclei.

**Big Bang** The event in which the Universe was born. According to Big Bang theory, the Universe originated a finite time ago in an extremely hot, dense initial state and ever since then has been expanding. The Big Bang was the origin of space, time, and matter.

**Big Crunch** The final state that will be reached by the Universe if it eventually ceases to expand and then collapses in on itself.

**Big Rip** The tearing apart of all forms of structure in the Universe – galaxy clusters, galaxies, stars, planets, atoms, and elementary particles – that is expected to occur should the repulsive effect of dark energy become infinitely strong in a finite time. See also *dark energy.*

**binary star** Two stars that revolve around each other under the influence of their mutual gravitational attraction. Each member star orbits the centre of mass of the system, a point that lies closer to the more massive of the two stars. A **spectroscopic binary** is a system in which the two stars are too close to be resolved into separate points of light, but whose binary nature is revealed by its spectrum. The combined spectrum of the two stars contains two sets of spectral lines that shift in wavelength as the stars revolve round each other. An **eclipsing binary** is a system in which each star alternately passes in front of the other, cutting off all or part of its light and causing a periodic variation in the combined light of the two stars. See also *Doppler effect, spectral line.*

**black body** An idealized body that absorbs and re-emits all the radiation that falls on its surface and which is a perfect radiator. A black body emits a continuous spectrum of radiation (**black-body radiation**) that peaks in brightness at a wavelength that depends on its surface temperature – the higher the temperature, the shorter the wavelength of peak brightness. See also *spectrum.*

**black-dwarf star** A white-dwarf star that has cooled to such a low temperature that it emits no detectable light. There has not been enough time since the origin of the Universe for any star to cool down enough to become a black dwarf. See also *brown-dwarf star, white-dwarf star.*

**black hole** A compact region of space, surrounding a collapsed mass, within which gravity is so powerful that no material object, light, or any other kind of radiation can escape to the outside Universe. The radius of a black hole is called the **Schwarzschild radius**, and its boundary is known as the **event horizon**. The greater its mass, the larger its radius. When a body collapses to form a black hole, all of its mass becomes compressed into a central point, a point of infinite density called a **singularity**. A **stellar-mass black hole** forms when the core of a high-mass star collapses; its mass is likely to be in the region of 3–100 times the mass of the Sun. A **super-massive black hole**, with a mass in the region of a few million to a few billion solar masses, is an object that forms when a very large mass collapses, or a number of black holes merge into one, in the core of a galaxy. See also *active galaxy, singularity.*

**blazar** The most variable type of active galaxy, which includes BL Lacertae objects and the most violently variable quasars. See also *active galaxy, BL Lacertae object, quasar.*

**BL lacertae object** A type of active galaxy that has no detectable absorption or emission lines in its spectrum but which is believed to be similar to a quasar. The name derives from an object in the constellation Lacerta that was at first thought to be a variable star. See also *quasar.*

**blue shift** The displacement of spectral lines to shorter wavelengths that occurs when a light source is approaching an observer. See also *Doppler effect, red shift, spectral line.*

**Bok globule** A compact dark nebula, roughly spherical in shape, which contains 1 to 1000 solar masses of gas and dust and has a diameter of between 0.1 and a few light-years. Globules of this kind are believed to be cool concentrations of gas and dust that eventually will collapse to form protostars. They are named after Dutch-born astronomer Bart Bok, who made a detailed study of these objects. See also *protostar.*

**brown-dwarf star** A body that forms out of a contracting cloud of gas in the same way as a star, but which, because it contains too little mass, never becomes hot enough to ignite the nuclear-fusion reactions that power a normal star. With less than 8 per cent of the Sun's mass, a brown dwarf glows dimly at infrared wavelengths, fading gradually as it cools down.

## C

**caldera** A bowl-shaped depression caused by the collapse of a volcanic structure into an emptied magma chamber. A caldera is usually found at the summit of shield volcanoes such as those on Venus and Mars.

**captured rotation** See *synchronous rotation.*

**carbonaceous chondrite** see *chondrite.*

**Cassegrain telescope** A type of reflecting telescope in which the converging cone of light reflected from a concave primary mirror is then reflected back from a convex secondary mirror, through a hole in the primary mirror, to a focus at the rear of the instrument. The design was devised by Guillaume Cassegrain in 1672. See also *reflecting telescope, Schmidt–Cassegrain telescope.*

**cataclysmic variable** see *variable star.*

**celestial equator** A great circle on the celestial sphere that is a projection of Earth's own equator onto the celestial sphere. See also *celestial sphere, great circle.*

**celestial poles** The two points at which the line of the Earth's axis, extended outwards, meets the celestial sphere and around which the stars appear to revolve. The north celestial pole lies directly above the Earth's North Pole and the south celestial pole directly above the Earth's South Pole. See also *celestial sphere.*

**celestial sphere** An imaginary sphere, that surrounds the Earth. As the Earth rotates from west to east, the sphere appears to rotate from east to west. In order to define the positions of stars and other celestial bodies, it is convenient to think of them as being attached to the inside surface of this sphere. See also *celestial equator, celestial poles.*

**centre of mass** The point within an isolated system of bodies around which those bodies revolve. Where the system consists of two bodies (for example, a binary star), it is located at a point on a line joining their centres. If both bodies have the same mass, the centre of mass lies midway between them, whereas if one body is more massive than the other, it lies closer to the more massive of the two.

**Cepheid variable** A type of variable star that increases and decreases in brightness in a regular, periodic way. Cepheids are pulsating variables, which vary in brightness as they expand and contract. The more luminous the Cepheid, the longer its period of variation. See also *variable star.*

**Chandrasekhar limit** The maximum possible mass for a white-dwarf star. If the mass of a white dwarf exceeds this limit, which is about 1.4 solar masses, gravity will overwhelm its internal pressure and it will collapse. The limit was first calculated by Indian astrophysicist Subrahmanyan Chandrasekhar in 1931. See also *white-dwarf star.*

**charge-coupled device (CCD)** An electronic imaging device that consists of a large array of tiny light-sensitive elements. The image of an object is constructed by reading off the electrical charges that accumulate in each element during an exposure.

**chondrite** A stony meteorite that contains a large number of small, spherical objects called **chondrules**. A **carbonaceous chondrite** is one that is rich in carbon, carbon compounds, and volatile materials. Carbonaceous chondrites are thought to be some of the least-altered primitive remnants of the protoplanetary disc from which the Solar System formed. See also *meteorite, protoplanetary disc.*

**chromosphere** The thin layer in the Sun's atmosphere that lies between the photosphere (the visible surface) and the corona. Its faint, reddish-pink light can be seen directly during a total eclipse of the Sun when the Moon hides the dazzling photosphere. See also *photosphere.*

**circumpolar** A term used to describe a star, or other celestial body, that remains above the horizon at all times when viewed from a particular place on the Earth's surface.

**circumstellar disc** A flattened, disc-shaped cloud of gas and dust that surrounds a star. A disc of this kind is usually associated with a young or newly forming star, in which case it is composed of material from the original dusty gas cloud that collapsed to form the central star. See also *protoplanetary disc.*

**closed universe** A universe that is curved in such a way that space is finite but has no discernable boundary (analogous to the surface of a sphere). A universe will be closed if its average density exceeds a particular value called the **critical density**. In the absence of a repulsive force, a closed universe will eventually cease to expand and will then collapse. See also *flat universe, open universe, oscillating universe.*

**coma** The cloud of gas and dust that surrounds the nucleus of a comet and which comprises its glowing "head". See also *comet.*

**comet** A small body composed mainly of dust-laden ice that revolves around the Sun, usually in a highly elongated orbit. Each time it approaches the Sun, gas and dust evaporate from its nucleus (the solid core of the comet) to form an extensive cloud, called the coma, and one or more tails. See also *coma, tail.*

**conjunction** A close alignment in the sky of two celestial bodies, which occurs when both bodies lie in the same direction as viewed from the Earth. When a planet lies directly on the opposite side of the Sun from the Earth, it is said to be at **superior conjunction**. If a planet passes between the Earth and the Sun (Mercury and Venus are the only planets that can do this), it is said to be at **inferior conjunction**. See also *opposition.*

**constellation** One of 88 regions of the celestial sphere. Each constellation contains a grouping of stars joined by imaginary lines to represent a figure. The constellations are officially referred to by the Latin names of these figures. Many have been named after mythological characters or creatures (such as Orion, the Hunter) but some after more mundane objects (for example, Sextans, the Sextant). See also *asterism.*

**continuous spectrum** see *spectrum.*

**convection** The transport of heat by rising bubbles or plumes of hot liquid or gas. In a **convection cell**, rising streams of hot material cool, spread out, and then sink down to be reheated, so maintaining a continuous circulation.

**core** **(1)** The dense central region of a planet. **(2)** The central region of a star within which energy is generated by means of nuclear-fusion reactions. **(3)** A dense concentration of material within a gas cloud.

**Coriolis effect** The tendency of a wind or current to be deflected from its initial direction as a consequence of a planet's rotation. In the case of the Earth, the deflection is to the right in the northern hemisphere and to the left in the southern hemisphere.

**corona** The outermost region of the atmosphere of the Sun or a star. The solar corona has an extremely low density and a very high temperature (1–5 million degrees Celsius/about 2–9 million degrees Fahrenheit). It cannot be observed except during a total eclipse of the Sun. See also *eclipse, solar wind.*

**coronal mass ejection** A huge, rapidly expanding bubble of plasma that is ejected from the Sun's corona. Containing billions of tons of material in the form of ions and electrons, together with associated magnetic fields, a typical coronal mass ejection propagates outwards through interplanetary space at a speed of several hundred kilometres (miles) per second. See also *corona, ion, plasma.*

**cosmic microwave background radiation (CMBR)** Remnant radiation from the Big Bang, which is detectable as a faint distribution of microwave radiation across the whole sky. See also *Big Bang.*

**cosmic rays** Highly energetic subatomic particles, such as electrons, protons, and atomic nuclei, that hurtle through space at speeds close to the speed of light.

**cosmological constant** An extra term in Einstein's relativity equations which, if it has a positive value, corresponds to a repulsive force that could cause the Universe to expand at an accelerating rate. Modern cosmologists associate the constant with a quantity called **vacuum energy** (residual energy that, according to quantum theory, exists even in a vacuum), one of the possible forms of the dark energy believed to permeate the Universe. See also *dark energy.*

**cosmological red shift** see *red shift.*

**cosmology** The study of the nature, structure, origin, and evolution of the Universe.

**crater** A bowl- or saucer-shaped depression in the surface of a planet or satellite, or at the summit of a volcano. Many have raised walls and some have a central peak. An **impact crater** is one excavated by an meteorite, asteroid, or comet impact, whereas a **volcanic crater** is the cavity from which a volcano discharges material. Raised walls are created by accumulation of ejected material.

**critical density** see *flat universe.*

**crust** The thin, rocky, outermost layer of a planet or major planetary satellite, which, like the Earth, has separated into several layers, with the densest material towards its centre and the least dense at its surface.

**dark energy** A little-understood form of energy that appears to comprise about 70 per cent of the total amount of mass and energy in the Universe. It exerts a repulsive effect and is believed to be causing the expansion of the Universe to accelerate. See also *accelerating universe.*

**dark matter** Matter that exerts a gravitational influence on its surroundings but does not emit detectable amounts of radiation. Dark matter appears to make up a large fraction of the total amount of mass contained in galaxies, galaxy clusters, and the Universe as a whole.

**dark-matter halo** see *halo.*

**dark nebula** see *nebula.*

**declination** The angular distance of a celestial body north or south of the celestial equator. Declination is positive (+) if the object is north of the celestial equator and negative (-) if it is south of the celestial equator. A star on the celestial equator has a declination of 0°, whereas a star at one of the celestial poles has a declination of 90°. See also *celestial equator, celestial sphere, right ascension.*

**declination axis** see *equatorial mounting.*

**diffuse nebula** A luminous cloud of gas and dust. The term "diffuse" refers to the cloud's fuzzy appearance and to the fact that it cannot be resolved into individual stars. See also *nebula.*

**direct motion** see *retrograde motion.*

**direct rotation** see *retrograde rotation.*

**Doppler effect** The observed change in the wavelength or frequency of radiation that is caused by the motion of its source towards or away from an observer. See also *blue shift, red shift.*

**double star** Two stars that appear close together in the sky. If the two stars revolve around each other, the system is called a binary. An **optical double star** consists of two stars that appear to be close together only because they happen to lie in almost exactly the same direction when viewed from the Earth; they lie at different distances and are not physically connected. See also *binary star.*

**dwarf planet** A celestial body that orbits the Sun and has sufficient mass and gravity to be spherical, but has not cleared the region around its orbit of other bodies, and is not a satellite.

**dwarf star** An alternative name for a main-sequence star that was originally devised to distinguish main-sequence stars, such as the Sun, from the much more luminous giant stars on the Hertzsprung–Russell diagram. See also *Hertzsprung–Russell diagram, main sequence.*

# E

**eccentricity (e)** A measure of how much an ellipse deviates from a perfect circle. Eccentricity takes a value between 0 and 1; a circle has eccentricity of 0, and the most elongated ellipses approach an eccentricity of 1. See also *ellipse.*

**eclipse** The passage of one celestial body into the shadow cast by another. A **lunar eclipse** occurs when the Moon passes into the Earth's shadow and a **solar eclipse** when part of the Earth's surface enters the shadow cast by the Moon. A **total lunar eclipse** takes place when the whole of the Moon lies within the dark cone of the Earth's shadow, and a **partial lunar eclipse** when only part of the Moon is in the shadow. During a **total solar eclipse**, the Sun is completely obscured by the dark disc of the Moon. A **partial solar eclipse** occurs when only part of the Sun's surface is hidden. If the Moon passes directly between the Sun and the Earth when it is close to apogee, it will appear smaller than the Sun, and its dark disc will be surrounded by a ring, or annulus, of sunlight; an event of this kind is called an **annular eclipse**. See also *apogee.*

**eclipsing binary** see *binary star.*

**ecliptic** The track along which the Sun appears to travel around the celestial sphere, relative to the background stars, in the course of a year. It is equivalent to the plane of the Earth's orbit.

**ejecta** Material thrown outwards by the blast of an impact. Ejecta, which is produced when a meteorite strikes the surface of a planet or moon and excavates a crater, consists of freshly exposed material that may be markedly brighter than the adjacent surface. Sometimes the ejected material forms extensive streaks, or rays, which radiate from the point of impact. An **ejecta blanket** is a continuous sheet of deposited ejecta that surrounds a crater. See also *crater.*

**electromagnetic (EM) radiation** Oscillating electric and magnetic disturbances that propagate energy through space in the form of waves (electromagnetic waves). Examples include light and radio waves.

**electromagnetic spectrum** The complete range of electromagnetic radiation from the shortest wavelengths (gamma rays) to the longest wavelengths (radio waves).

**electron** A lightweight fundamental particle with negative electrical charge. A cloud of electrons surrounds the nucleus of an atom. The number of orbiting electrons in an atom is the same as the number of protons in its nucleus.

**ellipse** An oval curve drawn around two points called **foci** (singular: **focus**) such that the total distance from one focus to any point on the curve and then back to the other focus is constant. The maximum diameter of an ellipse is the **major axis**, and half of this diameter is the **semimajor axis**. The two foci lie on the major axis; the greater their separation, the more elongated the ellipse. See also *eccentricity, orbit.*

**elliptical galaxy** A galaxy that appears round or elliptical in shape and normally contains very little gas or dust. See also *galaxy.*

**elongation** The angle between the Sun and a planet, or other Solar System body, when viewed from the Earth. The elongation of a planet is 0° when it is in conjunction with the Sun and 180° when it is at opposition. **Greatest elongation** is the maximum possible elongation of a body, such as Mercury or Venus, that lies inside the orbit of the Earth. See also *conjunction, opposition.*

**emission line** see *spectral line.*

**emission nebula** see *nebula.*

**equatorial mounting** A mounting that enables a telescope to be turned around two axes, one of which (the **polar axis**) is parallel to, and the other (the **declination axis**) perpendicular to, the Earth's axis of rotation. The telescope can follow the motion of a celestial object across the sky by being driven round the polar axis in the opposite direction to the Earth's rotation at a rate of one revolution per sidereal day. See also *declination, right ascension, sidereal time.*

**equinox** An occasion when the Sun is vertically overhead at a planet's equator, and day and night have equal duration for the whole planet. In the case of the Earth, the northern **vernal equinox** is the point at which the Sun crosses the celestial equator from south to north, on or around 21 March each year, and the northern **autumnal equinox** is the point at which the Sun crosses the celestial equator from north to south, on or around 22 September. See also *right ascension.*

**eruptive prominence** see *prominence.*

**eruptive variable** see *variable star.*

**escape velocity** The minimum speed at which a projectile must be launched in order to recede forever from a massive body and not fall back. The escape velocity at the Earth's surface is 40,320kph (25,200mph).

**event horizon** see *black hole.*

**extrasolar planet (exoplanet)** A planet that revolves around a star other than the Sun.

# F

**facula (plural: faculae)** A patch of enhanced brightness on the solar photosphere that may be seen in a white-light image of the Sun, usually near the edge of the Sun's visible disc where the background brightness is lower. Faculae correspond to regions that are hotter than their immediate surroundings. They are associated with active solar regions but may appear before, and persist after, any sunspots that develop in those regions. See also *photosphere, sunspot.*

**Fraunhofer line** One of the 574 dark absorption lines in the spectrum of the Sun that were identified by the 19th-century German optician and instrument maker Joseph von Fraunhofer. See also *spectral line.*

**flare star** A faint, cool, red-dwarf star that displays sudden, short-lived increases in luminosity caused by extremely powerful flares that occur above its surface. See also *red-dwarf star, solar flare.*

**flat universe** A universe in which the overall net curvature of space is zero. In such a universe, space is flat in the sense that, apart from localized distortions caused by massive bodies, its large-scale geometry is Euclidean and light rays travel in straight lines. A universe will be flat if its overall average density is equal to a particular value, called the **critical density**. See also *closed universe, open universe, oscillating universe.*

**focal length** The distance between the centre of a lens, or the front surface of a concave mirror, and the point at which it forms a sharp image of a very distant object.

**frequency** The number of wave crests of a wave motion that pass a given point in one second. In the case of an electromagnetic wave (for example, light) the frequency is equal to the speed of light divided by the wavelength. See also *electromagnetic radiation.*

**fusion (nuclear fusion)** The process whereby atomic nuclei are joined together during energetic collisions to form heavier atomic nuclei, with an associated release of large amounts of energy. Stars are powered by fusion reactions that take place in their central cores. In a main-sequence star such as the Sun, fusion reactions convert hydrogen into helium. See also *main sequence.*

**galactic cluster** see *open cluster.*

**galaxy** A large aggregation of stars and clouds of gas and dust. Galaxies, which may be elliptical, spiral, or irregular in shape, contain from a few million to several trillion stars and have diameters ranging from a few thousand to over a hundred thousand light-years. The Sun is a member of the **Milky Way galaxy**, which is also sometimes known as **the Galaxy**. See also *Milky Way.*

**galaxy cluster** An aggregation of galaxies held together by gravity. Clusters that contain up to a few tens of member galaxies are called groups. Larger clusters are divided into **regular** and **irregular clusters**, depending on their degree of structure. The most richly populated regular clusters (**rich clusters**) contain up to several thousand galaxies.

**galaxy supercluster** A cluster of galaxy clusters, which is a loose aggregation of up to about ten thousand galaxies, spread through a volume of space with a diameter of up to about 200 million light-years. See also *galaxy cluster.*

**Galilean moon** One of the four largest natural satellites of the planet Jupiter, which were discovered in 1610 by the Italian astronomer Galileo Galilei. In order of distance from the planet, they are Io, Europa, Ganymede, and Callisto.

**gamma radiation** Electromagnetic radiation with extremely short wavelengths (shorter than X-rays) and very high frequencies. Gamma radiation occupies the shortest-wavelength region of the spectrum. See also *electromagnetic radiation, electromagnetic spectrum.*

**gamma-ray burst (GRB)** A sudden burst of gamma radiation from a source in a distant galaxy. Gamma-ray bursts are the most powerful explosive events in the present-day Universe. They may be triggered by collisions between neutron stars or black holes or by an extreme version of a supernova called a **hypernova**.

**gas planet (gas giant)** A large planet that, like Jupiter or Saturn, consists predominantly of hydrogen and helium. Beneath its thick gaseous atmosphere, the pressure is so great that hydrogen and helium exist in liquid form. See also *rocky planet.*

**gegenschein** A very faint patch of light that sometimes may be seen on a clear, moonless night in the region of sky directly opposite the position of the Sun. It is caused by sunlight that has been reflected back towards the Earth by interplanetary dust particles lying beyond the orbit of the Earth. See also *zodiacal light.*

**general theory of relativity** see *relativity.*

**geocentric (1)** Treated as being viewed from the centre of the Earth. **(2)** Having the Earth at the centre (of a system). **Geocentric coordinates** are a system of positional measurements (such as right ascension and declination) that are treated as being measured from the centre of the Earth. A satellite that is travelling around the Earth is in a **geocentric orbit**. **Geocentric cosmology** was the ancient theory that the Sun, Moon, planets, and stars revolved around a central Earth. See also *heliocentric.*

**giant star** A star that is larger and much more luminous than a main-sequence star of the same surface temperature. See also *Hertzsprung–Russell diagram, main sequence, red giant.*

**globular cluster** A near-spherical cluster of between 10,000 and more than 1 million stars. Globular clusters, which consist of very old stars, are located predominantly in the halos of galaxies. See also *open cluster.*

**gravitation** see *gravity.*

**gravitational lens** A massive body, or a distribution of mass (such as a galaxy cluster), whose gravitational field deflects light rays from a more distant background object, thereby acting as a lens to produce a magnified or distorted image, or images, of that background object.

**gravitational wave** A wave-like distortion of space that propagates at the speed of light. Although waves of this kind have not yet been detected directly, there is strong, indirect evidence that they exist.

**gravity** The attractive force that acts between material bodies, particles, and photons. According to the theory of gravity developed in the 17th century by Isaac Newton (**Newtonian gravitation**), the force of gravity acting between two bodies is proportional to the product of their masses divided by the square of the distance between their centres. For example, if the distance between the bodies is doubled, the force of attraction is reduced to one quarter of its previous value. See also *relativity.*

**great circle** A circle on the surface of a sphere, the plane of which passes through the centre of the sphere and which exactly divides the sphere into two equal hemispheres. Its name derives from the fact that it is the largest circle that can be drawn on the surface of a sphere. See also *celestial equator, meridian.*

**greenhouse effect** The process by which atmospheric gases make the surface of a planet hotter than would be the case if the planet had no atmosphere. Incoming sunlight is absorbed at the surface of a planet and re-radiated as infrared radiation, which is then absorbed by **greenhouse gases** such as carbon dioxide, water vapour, and methane. Part of this trapped radiation is re-radiated back down towards the ground, so raising its temperature.

# H

**HII region** A glowing region of ionized hydrogen surrounding one or more hot, highly luminous stars. An HII region is often just a part of a more extensive cloud of gas and dust, the remainder of which has not been ionized and is not shining. See also *ion, nebula.*

**halo** A spherical region surrounding a galaxy that contains a distribution of globular clusters, thinly scattered stars, and some gas. A **dark-matter halo** is a distribution of dark matter within which a galaxy is embedded.

**heliocentric (1)** Treated as being viewed from the centre of the Sun. **(2)** Having the Sun at the centre (of a system). **Heliocentric coordinates** specify the position of an object as seen from the centre of the Sun. A body that is revolving round the Sun follows a **heliocentric orbit**. **Heliocentric cosmology** is a model of the Universe, such as the one proposed in 1543 by Nicolaus Copernicus, in which the planets revolve around a central Sun.

**heliosphere** The region of space around the Sun within which the solar wind and interplanetary magnetic field are confined by the pressure of the interstellar medium. Its boundary is called the **heliopause**. See also *interstellar medium, solar wind.*

**helium burning** The generation of energy by means of fusion reactions that convert helium into carbon and oxygen. Helium burning takes place in the core of a star that has left the main sequence and become a red giant, and it may occur again, later in a star's evolution, in a shell surrounding the core. See also *fusion, main sequence, red-giant star.*

**Hertzsprung–Russell (HR) diagram** A diagram on which stars are plotted as points according to their luminosity and surface temperature. Luminosity (or absolute magnitude) is plotted on the vertical axis, and surface temperature (or spectral class or colour) is plotted on the horizontal axis. Astrophysicists use the Hertzsprung–Russell diagram to classify stars. Depending on a star's position on the diagram, it may be classified as, for example, a main-sequence star, a giant, or a dwarf.

**Hubble constant** see *Hubble's law.*

**Hubble's law** The observed relationship between the red shifts in the spectra of remote galaxies and their distances, which implies that the speeds at which galaxies are receding are directly proportional to their distances. The **Hubble constant** (or **Hubble parameter**) – denoted by the symbol $H_0$ – is the constant of proportionality that relates speed of recession to distance.

**hydrogen burning** The generation of energy by means of fusion reactions that convert hydrogen into helium. Hydrogen burning takes place in the core of a main-sequence star. When a star has consumed all the available hydrogen in its core, the core contracts and hydrogen burning then continues in a thin shell surrounding the core. See also *fusion, main sequence, proton–proton reaction.*

**hypernova** see *gamma-ray burst.*

# I

**impact crater** see *crater.*

**inclination** The angle at which one plane is tilted relative to another. The inclination of a planetary orbit is the angle between its plane and the plane of the ecliptic (the plane of the Earth's orbit). The inclination of a planet's equator is the angle between the plane of its orbit and the plane of its equator. See also *ecliptic, orbit.*

**inferior conjunction** see *conjunction.*

**inferior planet** A planet that travels round the Sun on an orbit that is inside the orbit of the Earth. The two inferior planets are Mercury and Venus. See also *superior planet.*

**inflation** A sudden, short-lived episode of accelerating expansion thought to have occurred at a very early stage in the history of the Universe (about $10^{-35}$ seconds after the beginning of time). See also *Big Bang.*

**infrared radiation** Electromagnetic radiation with wavelengths longer than visible light but shorter than microwaves or radio waves. Infrared is the dominant form of radiation emitted from many cool astronomical objects, such as interstellar dust clouds. See also *electromagnetic radiation.*

**interstellar medium** The gas and dust that permeates the space between the stars within a galaxy.

**ion** A particle or system of particles with a net electrical charge. Positive ions are commonly formed when an atom loses one or more of its electrons, whereas negative ions result from an excess of electrons. Ions may form from complexes of former atoms. The process by which an atom or complex gains or loses an electron to become charged is called **ionization**. See also *electron, photon.*

**irregular cluster** see *galaxy cluster.*

**irregular galaxy** A galaxy that has no well-defined structure or symmetry.

**isotope** Any one of two or more forms of a particular chemical element, the atoms of which contain the same number of protons but different numbers of neutrons. For example, helium-3 and helium-4 are isotopes of helium; a nucleus of helium-4 (the heavier, and more common, isotope) contains two protons and two neutrons, whereas a nucleus of helium-3 contains two protons and one neutron. See also *atom, nucleus.*

# K

**Kepler's laws of planetary motion** Three laws, devised in the early 17th century by Johannes Kepler, that describe the orbital motion of planets around the Sun. In essence, the first law states that each planet's orbit is an ellipse, the second shows that a planet's speed varies as it travels around its orbit, and the third links its orbital period (the time taken to travel round the Sun) to its average distance from the Sun.

**Kuiper Belt (Edgeworth–Kuiper Belt)** A flattened distribution of icy planetesimals that orbit the Sun at distances in the region of 30–100 times the Earth's distance from the Sun and which is the source of many of the shorter-period comets. See also *Oort Cloud, planetesimal.*

# L

**lenticular galaxy** A galaxy that is shaped like a convex lens. It has a central bulge that merges into a disc, but no spiral arms. See also *galaxy, spiral galaxy.*

**lepton** A fundamental particle, such as an electron or a neutrino, that is not acted on by the strong nuclear force.

**light-year (ly)** A unit of distance equal to the distance light travels in one year – 9,460 billion km (5,878 billion miles).

**limb** The edge of the observed disc of the Sun, the Moon, or a planet.

**Local Group** The small cluster of more than 40 member galaxies to which the Milky Way galaxy belongs. The other major members are the spiral galaxies M31 (the Andromeda Galaxy) and M33. Most of the members are small (or dwarf) elliptical or irregular galaxies. See also *galaxy cluster*.

**local sidereal time** see *sidereal time*.

**luminosity** The total amount of energy emitted in one second by a source of radiation, such as the Sun or a star. The luminosity of a star can be expressed in watts or in units of solar luminosity (the luminosity of the Sun is 3.8 x $10^{26}$ watts). Stars are divided into luminosity classes denoted by Roman numerals. See also *magnitude*.

**lunar eclipse** see *eclipse*.

# M

**MACHO** An acronym for MAssive Compact Halo Object, a very low-luminosity object – such as a planet, brown dwarf, exceedingly dim white dwarf, or a black hole – that exists in the halo of a galaxy but is usually too faint to be seen directly. MACHOs are believed to account for a relatively small proportion of the unseen dark matter in a galaxy's halo. See also *dark matter, halo*.

**magnetic field** The region of space surrounding a magnetized body within which its magnetic influence affects the motion of an electrically charged particle.

**magnetosphere** The region of space around a planet within which the motion of charged particles is controlled by the planetary magnetic field rather than the solar wind and the associated interplanetary magnetic field. The shape of a planet's magnetosphere is influenced by the solar wind, which squeezes it inwards on the Sun-facing side and drags it out to form an elongated "tail" (a **magnetotail**) on the opposite, or downstream, side. See also *solar wind*.

**magnification** The increase in the apparent angular size of an object when viewed through an optical instrument, such as a telescope. The magnification of a telescope is equal to the focal length of its objective lens or primary mirror divided by the focal length of its eyepiece.

**magnitude (absolute and apparent)** **Apparent magnitude** is a measure of the apparent brightness of an object as seen in the sky. The fainter the object, the higher the numerical value of its magnitude. The faintest stars visible to the naked eye are of magnitude 6, whereas the brightest objects in the sky have negative apparent magnitudes. A star said to be of 1st magnitude has a magnitude of 1.49 or less, a star of 2nd magnitude has a value of 1.50 to 2.49, and so on. **Absolute magnitude** is the apparent magnitude a star would have if it were located at a standard distance of 10 parsecs (32.6 light-years) from Earth. See also *luminosity, parsec*.

**Main Belt** see *asteroid*.

**main sequence** A band that slopes diagonally from the upper left (hot, high-luminosity region) to the lower right (cool, low-luminosity region) of the Hertzsprung–Russell diagram and which contains about 90 per cent of stars. Main-sequence stars, such as the Sun, shine by converting hydrogen in their cores to helium. See also *dwarf star, Hertzsprung–Russell diagram*.

**major axis** see *ellipse*.

**mantle** The rocky layer that lies between the core and the crust of a rocky (Earth-like) planet or a major planetary satellite. See also *core, crust*.

**mare (plural: maria)** A relatively smooth, dark, lava-filled basin on the surface of the Moon. The name derives from the Latin for "sea".

**massive compact halo object** see *MACHO*.

**meridian** **(1)** A great circle on the surface of the Earth or another astronomical body that passes through the north and south poles and crosses the equator at right angles. **(2)** A great circle on the celestial sphere that passes through the north and south celestial poles and crosses the celestial equator at right angles. An observer's local meridian passes through the celestial pole, the zenith, and the north and south points of the horizon. See also *celestial sphere, great circle*.

**Messier catalogue** A widely used catalogue of nebulous objects (most of them nebulae, star clusters, and galaxies) that was published in 1781 by the French astronomer Charles Messier. Objects contained in this catalogue are designated by the letter "M" followed by a number. For example, M31 is the Andromeda Galaxy and M42 is the Orion Nebula. See also *New General Catalogue*.

**meteor** The short-lived streak of light seen when a meteoroid plunges into the Earth's atmosphere and is heated to incandescence by friction. A **sporadic meteor** is one that appears at a random time from a random direction. A **meteor shower** is a substantial number of meteors that appear to radiate from a common point in the sky (the **radiant**) when the Earth is passing through a stream of meteoroids. See also *meteorite, meteoroid*.

**meteorite** A rocky or metallic meteoroid that survives passage through the atmosphere and reaches the Earth's surface in one piece or in fragments. See also *meteor, meteoroid*.

**meteoroid** A lump or small particle of rock, metal, or ice orbiting the Sun in interplanetary space. Meteroid sizes range from a fraction of a millimetre (small fraction of an inch) to a few metres (yards). Some are debris from collisions between asteroids. Others are particles released by comets; these spread out along cometary orbits to form meteoroid streams. See also *asteroid, comet, meteor, meteorite*.

**Milky Way** **(1)** The spiral galaxy that contains the Sun, sometimes also referred to as the **Milky Way galaxy** or **the Galaxy**. **(2)** A faint, misty band of light that stretches across the night sky and consists of the combined light of vast numbers of stars and nebulae that lie in the disc and spiral arms of our galaxy. See also *galaxy*.

**Mira variable** A class of long-period variable star named after the star Mira – Omicron (o) Ceti – in the constellation Cetus. Mira variables are cool, giant pulsating stars that vary in brightness over periods ranging from 100 days to more than 500 days. See also *variable star*.

**molecular cloud** A cool, dense cloud of gas and dust in which the temperature is sufficiently low to enable atoms to join together to form molecules such as molecular hydrogen ($H_2$) or carbon monoxide (CO), and within which conditions are favourable for star formation.

**moon** Also known as a **natural satellite**, a body that orbits a planet. **The Moon** is the Earth's natural satellite. Orbiting the Earth at a mean distance of 384,000km (239,000 miles) in a period of 27.3 days, it has a diameter of 3,476km (2,159 miles). See also *satellite*.

**moon dog** See *sun dog*.

**multiple star** A system consisting of two or more stars bound together by gravity and revolving around each other (a system of just two stars is also called a binary). See also *binary star*.

# N

**near-Earth asteroid** see *asteroid*.

**nebula (plural: nebulae)** A cloud of gas and dust in interstellar space. The name derives from the Latin for "cloud". There are several types of luminous nebula (nebulae that shine). An **emission nebula** is a cloud of gas and dust that contains one or more extremely hot, young, high-luminosity stars; ultraviolet light emitted by these stars causes the surrounding gas to glow. Nebulae of this kind are also called HII regions because they contain a large proportion of ionized hydrogen. A **reflection nebula** is observed when the dust particles within a cloud are lit up by light from a neighbouring bright star. Other types of luminous nebulae include planetary nebulae (shells of gas puffed out by dying stars) and supernova remnants (the debris of exploded stars). A **dark nebula** (or **absorption nebula**) is a dust-laden cloud that blocks out light from background stars and appears as a dark patch in the sky. See also *diffuse nebula, HII region, planetary nebula, supernova*.

**neutrino** A fundamental particle of exceedingly low mass, which has zero electrical charge and which travels at very close to the speed of light.

**neutron** A particle, composed of three quarks, that has zero electrical charge and a mass fractionally greater than that of a proton. Neutrons are found in the nuclei of atoms. See also *atom*.

**neutron star** An exceedingly dense, compact star that is composed almost entirely of tightly packed neutrons. A typical neutron star has a diameter of around 10km (6 miles) yet has about the same mass as the Sun. A neutron star forms when the core of a high-mass star collapses, triggering a supernova explosion. See also *pulsar, supernova*.

**New General Catalogue (NGC)** A catalogue of nebulae, clusters, and galaxies that was published in 1888 by the Danish astronomer John L.E. Dreyer. Objects in this catalogue are denoted by "NGC" followed by a number. For example, the Andromeda Galaxy is NGC 224. See also *Messier catalogue*.

**Newton's laws of motion** Three laws describing the behaviour of moving bodies that were set out by Isaac Newton in 1687. Newton's first law states that a body continues to move in a straight line at a constant speed unless acted on by a force. The second law shows how a force causes a body to accelerate in the direction along which an applied force is acting. The third law states that for any force there is an equal and opposite reaction force.

**Newtonian gravity** see *gravity*.

**nova (plural: novae)** A star that suddenly brightens by a factor of thousands or more, then fades back to its original brightness over a period of weeks or months. The flare-up occurs when a fusion reaction is triggered on the surface of a white dwarf by gas flowing from a companion star. The name derives from the Latin for "new", because the rapid brightening produces what appears to be a new star. See also *white dwarf, fusion*.

**nuclear bulge** see *spiral galaxy*.

**nuclear fusion** see *fusion*.

**nucleus (plural: nuclei)** **(1)** The compact central core of an atom, which consists of a number of positively charged protons and neutral neutrons. The nucleus of a hydrogen atom consists of a single proton. **(2)** The solid, ice-rich body of a comet. **(3)** The central core of a galaxy, within which stars are relatively densely packed together.

**occultation** The passage of one body in front of another, which causes the more distant one to be wholly or partially hidden. The term is usually used to describe the passage of a body of larger apparent size in front of a body of smaller apparent size –

for example, when the Moon passes in front of a star or when a planet (such as Jupiter) passes in front of one of its moons.

**Oort Cloud (Oort–Opik Cloud)** A spherical distribution of trillions of icy planetesimals and cometary nuclei that surrounds the Solar System and extends out to a radius of about 1.6 light-years from the Sun. It provides the reservoir from which long-period and "new" comets originate. Its existence was proposed in 1950 by Dutch astronomer Jan H. Oort (a similar idea had also been suggested by Estonian astronomer Ernst J. Opik). See also *comet, planetesimal.*

**open cluster** A loose cluster of up to a few thousand stars that lies in or close to the plane of the Milky Way galaxy. Member stars of each cluster formed from the same cloud of gas and dust, and have closely similar ages and chemical compositions. Clusters of this kind are also known as **galactic clusters**. See also *globular cluster.*

**open universe** A universe in which the average density is less than the critical density that is needed to halt its expansion and which, therefore, will expand forever. See also *closed universe, flat universe, oscillating universe.*

**opposition** The position of a planet when it is exactly on the opposite side of the Earth from the Sun. Its elongation is then 180°, and it is highest in the sky at midnight. See also *conjunction, elongation.*

**optical double star** see *double star.*

**orbit** The path of a body that is moving within the gravitational field of another. The orbit of a planet around a star or a satellite around a planet will normally be an ellipse or, exceptionally, a circle (a circle is a special case of an ellipse).

**orbital period** The period of time during which a body travels once around its orbit. The **sidereal orbital period** is the time taken by one body to revolve around another (for example, the Moon around the Earth) measured relative to the background stars.

**oscillating universe** A universe that expands and contracts in a cyclic fashion. The collapse of such a universe at the end of one cycle triggers a new Big Bang that initiates the next cycle. See *closed universe, flat universe, open universe.*

# P

**parallax** The apparent shift in position of an object when it is observed from different locations. **Stellar parallax** is the apparent shift in position of a relatively nearby star when viewed from different points on the Earth's orbit. **Annual parallax** is the maximum angular displacement of a star from its mean position due to parallax. The greater the distance of a star, the smaller its parallax.

**parhelic circle** See *sun dog.*

**parsec (pc)** The distance at which a star would have an annual parallax of one second of arc (one second of angular measurement). One parsec is equivalent to 3.26 light-years, or 30,900 billion km (19,200 billion miles). See also *parallax.*

**parselene** See *sun dog.*

**penumbra** **(1)** The lighter, outer part of the shadow cast by an opaque body. An observer within the penumbra can see part of the illuminating source. See also *eclipse.* **(2)** The less dark and less cool outer region of a sunspot. See also *sunspot, umbra.*

**perigee** The point on its orbit at which a body that is revolving around the Earth is at its closest to the Earth. See also *apogee.*

**perihelion** The point on its orbit at which a planet, or other Solar System body, is at its closest to the Sun.

**phase** The proportion of the visible hemisphere of the Moon or a planet that is illuminated by the Sun at any particular instant.

**photon** An individual package, or quantum, of electromagnetic energy, which may be envisaged as a "particle" of light. The shorter the wavelength of the radiation and higher the frequency, the greater the energy of the photon. See also *electromagnetic radiation.*

**photosphere** The thin, gaseous layer at the base of the solar atmosphere, from which the Sun's visible light is emitted and which corresponds to the visible surface of the Sun.

**planet** A body that is much less massive than a star, revolves around a star, and shines by reflecting that star's light. As a general guide, an orbiting body is considered to be a planet (rather than a brown dwarf) if its mass is less than about 13 times the mass of Jupiter. See also *brown-dwarf star.*

**planetary nebula** A glowing shell of gas ejected by a star at a late stage in its evolution.

**planetesimal** One of the large number of small bodies, composed of rock or ice, that formed within the solar nebula and from which the planets were eventually assembled through the process of accretion.

**plasma** A completely ionized gas state of matter that consists of equal numbers of positively charged ions and negatively charged electrons. Plasmas usually have very high temperatures. Examples include the solar corona and solar wind, both of which consist predominantly of protons and electrons. See also *corona, solar wind.*

**polar axis** see *equatorial mounting.*

**positron** see *antiparticle.*

**precession** A slow change in the orientation of a rotating body's axis caused by the gravitational influence of neighbouring bodies. The Earth's axis precesses around in a conical pattern over a period of 25,800 years.

**prominence** A flame-like plume of gas that follows magnetic field lines in the solar atmosphere. An **active** or **eruptive prominence** undergoes rapid changes, whereas a **quiescent prominence** remains suspended in the solar atmosphere for a prolonged period.

**proper motion** The angular rate at which a star changes its observed position on the celestial sphere. **Annual proper motion** is the angle (seldom more than a small fraction of 1 second of angular measurement) through which a star appears to shift in the course of one year.

**protogalaxy** A progenitor of a normal galaxy. The building blocks from which galaxies were assembled through a process of collisions and mergers, protogalaxies are believed to have formed a few hundred million years after the Big Bang when clouds of gas collapsed under the action of gravity.

**proton** An elementary particle, composed of three quarks, that has a positive electrical charge and is a constituent of every atomic nucleus. See also *atom.*

**proton–proton chain (pp chain)** A sequence of reactions that fuse together hydrogen nuclei (protons) to create helium nuclei. The net result of the process is to convert four protons into one helium nucleus, which consists of two protons and two neutrons. The proton–proton reaction is the dominant hydrogen-burning process in stars similar to, or less massive than, the Sun. See also *fusion, hydrogen burning, neutron, proton.*

**protoplanetary disc** A flattened disc of dust and gas surrounding a newly formed star and within which matter may be aggregating together to form the precursors of planets. See also *planetesimal.*

**protostar** A star in the early stages of formation. It consists of the central part of a collapsing cloud that is heating up and is accreting matter from its surroundings, but within which hydrogen fusion reactions have not yet commenced.

**pulsar** A rapidly rotating neutron star from which we receive brief pulses of radiation, at short and precisely timed intervals, as it spins around its axis.

**pulsating variable** see *variable star.*

**quantum** see *photon.*

**quark** A fundamental particle, the main matter constituent of all atomic nuclei. Quarks join in bunches of three to make baryons (for example, protons and neutrons) or in quark–antiquark pairs to form particles called mesons. See also *antiparticle, baryon.*

**quasar** A very compact but extremely powerful source of radiation that is almost star-like in appearance but which is believed to be the most luminous kind of active galactic nucleus. The name is an abbreviation for quasi-stellar radio source, but is also applied to **quasi-stellar objects (QSOs)**, which are not strong radio emitters.

**quiescent prominence** see *prominence.*

# R

**radial velocity** The component of a body's velocity that is along the line of sight directly towards, or away from, an observer. The radial velocity of a celestial body can be obtained by measuring the Doppler effect in its spectrum. See also *Doppler effect, red shift, spectrum.*

**radiant** The point in the sky from which the tracks of meteors that are members of a particular meteor shower appear to radiate. See also *meteor.*

**radio galaxy** A galaxy that is exceptionally luminous at radio wavelengths. A typical radio galaxy contains an active galactic nucleus from which jets of energetic charged particles are being propelled towards huge clouds of radio-emitting material that in many cases are much larger than the visible galaxy. See also *active galaxy.*

**radio telescope** An instrument that is designed to detect radio waves from astronomical sources. The most familiar type is a concave dish that collects radio waves and focuses them onto a detector.

**red-dwarf star** A cool, red, low-luminosity star that, when plotted on a Hertzsprung–Russell diagram, is located towards the bottom end of the main sequence. See also *Hertzsprung–Russell diagram, main sequence.*

**red-giant star** A large, highly luminous star with a low surface temperature and a reddish colour. A red giant has evolved away from the main sequence, is "burning" helium in its core rather than hydrogen, and is approaching the final stages of its life. See also *helium burning, Hertzsprung–Russell diagram, main sequence.*

**red shift** The displacement of spectral lines to longer wavelengths that is observed when a light source is receding from an observer. The shift in wavelength is proportional to the speed at which the source is receding. **Cosmological red shift** is a wavelength shift that is caused by the expansion of the Universe. See also *blue shift, Doppler effect, spectral line.*

**red supergiant star** An extremely large star of very high luminosity and low surface temperature. Stars of this kind are located towards the top-right corner of the Hertzsprung–Russell diagram. See also *Hertzsprung–Russell diagram.*

**reflecting telescope (reflector)** A telescope that uses a concave mirror to collect light, reflect light rays to a focus, and form an image of a distant object.

**reflection nebula** see *nebula.*

**refracting telescope (refractor)** A telescope that uses a lens to refract (bend) light rays in order to bring them to a focus and form an image of a distant object.

**regolith** A layer of loose rock, rocky fragments, and dust that covers the surface of a planet or planetary satellite.

**regular cluster** see *galaxy cluster.*

**relativity** Theories developed in the early part of the 20th century by Albert Einstein to describe the nature of space and time and the motion of matter and light. The **special theory of relativity** describes how the relative motion of observers affects their measurements of mass, length, and time. One of its consequences is that mass and energy are equivalent. The **general theory of relativity** treats gravity as a distortion of space-time associated with the presence of matter or energy. One of its consequences is that massive bodies deflect rays of light. See also *gravitational lensing, space-time.*

**resonance** A gravitational interaction between two orbiting bodies that occurs when the orbital period of one is an exact, or nearly exact, simple fraction of the orbital period of the other. For example, Jupiter's moon Io is in a 1:2 resonance with another of Jupiter's moons, Europa (Io's period is half of Europa's period). When a small object is in resonance with a more massive one, it experiences a periodic gravitational tug each time one of the bodies overtakes the other, the cumulative effect of which gradually changes its orbit.

**retrograde motion** **(1)** The apparent backward motion of a planet, from east to west relative to the background stars. For most of the time, a planet such as Mars or Jupiter will move from west to east relative to the stars (**direct motion**), but it will appear to reverse direction each time it is being overtaken by the Earth (around the time of opposition). See also *opposition.* **(2)** Orbital motion in the opposite direction to that of the Earth and the other planets of the Solar System. **(3)** The motion of a satellite along its orbit in the opposite direction to that in which its parent planet is rotating.

**retrograde rotation** The rotation of a body around its axis in the opposite direction to the rotational motion of the Earth, the Sun, and the majority of the planets. Viewed from above its North Pole, the Earth rotates around its axis and revolves around the Sun, in an anticlockwise direction (**direct rotation**), whereas a planet with retrograde rotation spins in the opposite (clockwise) direction. The planets Venus, Uranus, and Pluto exhibit retrograde rotation.

**rich cluster** see *galaxy cluster.*

**right ascension (RA)** The angular distance, measured eastwards, between the **first point of Aries** (where the Sun's path around the sky crosses the celestial equator from south to north) and a celestial body. It is expressed in hours, minutes, and seconds of time, where 1 hour is equivalent to an angle of 15°. Together with declination, it specifies the position of a body on the celestial sphere. See also *celestial sphere, declination, ecliptic, equinox.*

**ring** A flat distribution of small particles and lumps of material that revolves around a planet, usually in the plane of its equator. A **ring system** consists of a number of concentric rings surrounding a planet. The planets Jupiter, Saturn, Uranus, and Neptune each have a ring system.

**rocky planet** A planet (also called a terrestrial planet) that is composed mainly of rocks and has similar basic characteristics to the Earth. Within the Solar System, there are four rocky planets: Mercury, Venus, Earth, and Mars. See also *gas planet.*

**rupes** Scarps or cliffs on the surface of a planet or a satellite. See also *moon.*

# S

**satellite** A body that revolves around a planet, otherwise known as a "moon". An **artificial satellite** is an object deliberately placed in orbit around the Earth or around another Solar System body.

**Schmidt–Cassegrain telescope** A telescope combining features of the Schmidt camera and the Cassegrain telescope. Light enters the telescope tube through a thin corrector lens and is reflected from a concave mirror at the bottom of the tube towards a small convex mirror fixed to the inner face of the correcting lens. It is then reflected back down the tube, through a hole in the concave mirror, to a focus. This is a popular, compact design for small and moderate-sized telescopes. See also *Cassegrain telescope.*

**Schwarzschild radius** see *black hole.*

**semimajor axis** see *ellipse.*

**Seyfert galaxy** A spiral galaxy with an unusually bright, compact nucleus that in many cases exhibits brightness fluctuations. First identified by American astronomer Carl Seyfert in 1943, Seyfert galaxies comprise one of the several categories of active galaxy. See also *active galaxy.*

**shepherd moon** A small natural satellite that, through its gravitational influence, confines orbiting particles into a well-defined ring around a planet. A pair of shepherd moons, where one is slightly closer to the planet than the other, can squeeze particles into particularly narrow rings.

**sidereal orbital period** see *orbital period.*

**sidereal time** A time system based on the apparent rotation of the celestial sphere. **Local sidereal time** is defined to be 0 hours at the instant the first point of Aries crosses an observer's meridian. The **sidereal day** corresponds to the Earth's axial rotation period measured relative to the background stars, and is equal to 23 hours 56 minutes 4 seconds of mean (civil) time. See also *equinox, right ascension.*

**singularity** A point of infinite density into which matter has been compressed by gravity, and a point at which the known laws of physics break down. Theory implies that a singularity exists at the centre of a black hole. See also *black hole.*

**solar cycle** A cyclic variation in solar activity (for example, the production of sunspots and flares), which reaches a maximum at intervals of about 11 years. Because the polarity pattern of magnetic regions on the Sun reverses every 11 years or so, the overall duration of the cycle is 22 years. The sunspot cycle is the 11-year variation in the number (and overall area) of sunspots. See also *solar flare, sunspot.*

**solar eclipse** see *eclipse.*

**solar flare** A violent release of huge amounts of energy – in the form of electromagnetic radiation, subatomic particles, and shock waves – from a site located just above the surface of the Sun.

**solar mass** A unit of mass equal to the mass of the Sun, which provides a convenient standard for comparing the masses of stars. One solar mass is equivalent to 1.989 x $10^{30}$ kg (1.96 x $10^{27}$ tons). Stellar masses range from about 0.08 solar masses up to about 100 solar masses.

**solar nebula** The cloud of gas and dust from which the Sun and planets formed. As the cloud collapsed, most of its mass accumulated at the centre to form the Sun, whereas the rest flattened out into a disc within which planets were assembled by the process of accretion. See also *accretion, protoplanetary disc.*

**Solar System** The Sun together with everything that revolves around it (the planets and their satellites, asteroids, comets, meteoroids, gas, and dust).

**solar wind** A stream of fast-moving, charged particles (predominantly electrons and protons) that escapes from the Sun and flows outwards through the Solar System like a wind.

**solstice** One of the two points on the ecliptic at which the Sun is at its maximum declination north or south of the celestial equator. On or around 21 June each year, the Sun reaches its greatest northerly declination. This is the northern-hemisphere **summer solstice** (the **winter solstice** in the southern hemisphere). On or around 22 December each year, the Sun reaches its greatest southerly declination. This is the northern-hemisphere winter solstice (the summer solstice in the southern hemisphere). See also *celestial equator, declination, ecliptic.*

**space-time** The four-dimensional combination of the three dimensions of space (length, breadth, and height) and the dimension of time. The concept that time and space are intimately linked, rather than (as Newton had believed) being separate entities, was proposed in 1908 by Hermann Minkowski and was incorporated into Albert Einstein's theories of relativity. See also *relativity.*

**special theory of relativity** see *relativity.*

**spectral class** A class into which a star is placed according to the lines that appear in its spectrum. The principal spectral classes, arranged in decreasing order of temperature, are labelled O, B, A, F, G, K, M and are subdivided into numbers from 0 to 9. For example, the spectral class of the Sun is G2. See also *luminosity, spectral line, spectrum.*

**spectral line** A feature that appears at a particular wavelength in a spectrum. An **emission line** is a bright feature corresponding to the emission of light at that wavelength, whereas an **absorption line** is a dark feature corresponding to the absorption of light at that wavelength. See also *spectrum.*

**spectroscopic binary** see *binary star.*

**spectroscopy** The science of obtaining and studying the spectra of objects. Because the detailed appearance of a spectrum is influenced by factors such as chemical composition, density, temperature, rotation, velocity, turbulence, and magnetic fields, spectroscopy can reveal a wealth of information about the physical and chemical properties of, and processes occurring in, planets, stars, gas clouds, galaxies, and other kinds of celestial bodies. See also *spectrum.*

**spectrum** A beam of electromagnetic radiation spread out into its constituent wavelengths. A **continuous spectrum** is the unbroken spread of wavelengths emitted by a hot solid or liquid or a dense gas (the continuous spectrum of sunlight appears to human eyes as a rainbow band of colours). A hot, low-density gas emits light at particular wavelengths only; the resulting spectrum consists of bright **emission lines**, each of which corresponds to one of the wavelengths at which emission takes place. If a low-density gas is silhouetted against a source of a continuous spectrum, it absorbs light at certain wavelengths to produce a series of dark **absorption lines**. A typical star has an **absorption-line spectrum** (a continuous spectrum with dark lines superimposed by its atmosphere), whereas an emission nebula has an **emission-line spectrum**. See also *spectral line.*

**spiral arm** A spiral-shaped structure extending outwards from the central bulge of a spiral or barred spiral galaxy. It consists of gas, dust, emission nebulae, and hot young stars,

**spiral galaxy** A galaxy that consists of a spheroidal central concentration of stars (the **nuclear bulge**) surrounded by a flattened disc composed of stars, gas, and dust, within which the major visible features are clumped together into a pattern of spiral arms. See also *galaxy, spiral arm.*

**star** A self-luminous body of hot plasma that generates energy by means of nuclear fusion reactions.

**starburst galaxy** A galaxy within which star formation is taking place at an exceptionally rapid rate.

**star cluster** A group of between a few tens and around 1 million stars held together by gravity. All the member stars of a particular cluster are thought to have formed from the same original massive cloud of gas and dust. There are two principal types of cluster: open clusters and globular clusters. See also *globular cluster, open cluster.*

**stellar-mass black hole** see *black hole.*

**stellar parallax** see *parallax.*

**stellar wind** An outflow of charged particles from the atmosphere of a star. See also *solar wind.*

**sun dog** One of a pair of coloured patches of light that sometimes may be seen on either side of the Sun, separated from the Sun by an angle of about 22°. Otherwise known as a **parhelion** or **mock sun**, a sun dog is formed when ice crystals in the Earth's atmosphere refract sunlight. A **moon dog**, or **parselene**, is a patch of light that sometimes forms by the same process on either side of the Moon. A **parhelic circle** is a large, faint ring of white light, produced by the reflection of sunlight from atmospheric ice crystals, which crosses the Sun, passes through a pair of sundogs, and extends around the sky. Although a complete circle may be seen occasionally, more usually it is only possible to see arcs of light extending outwards from the sundogs.

**sunspot** A patch on the surface of the Sun that appears dark because it is cooler than its surroundings. Sunspots occur in regions where localized concentrated magnetic fields impede the outward flow of energy from the solar interior. See also *solar cycle.*

**supergiant** An exceptionally luminous star with a very large diameter. Supergiant stars appear at the top of the Hertzsprung–Russell diagram. See also *Hertzsprung–Russell diagram.*

**superior conjunction** see *conjunction.*

**superior planet** A planet that travels around the Sun on an orbit that is outside the orbit of the Earth. The superior planets are Mars, Jupiter, Saturn, Uranus, and Neptune. See also *inferior planet.*

**supermassive black hole** see *black hole.*

**supernova (plural: supernovae)** A catastrophic event that destroys a star and causes its brightness to increase, temporarily, by a factor of around 1 million. A **type II supernova** occurs when the core of a massive star collapses and the rest of the star's material is blasted away; the collapsed core usually becomes a neutron star. A **type Ia supernova** involves the complete destruction of a white dwarf. The expanding cloud of debris from a supernova is called a **supernova remnant**. See also *neutron star, white dwarf.*

**synchrotron radiation** Electromagnetic radiation that is emitted when electrically charged particles (usually electrons) gyrate at very high speed around lines of force in a magnetic field. Synchrotron radiation has a characteristic continuous spectrum that is different from that which is emitted by a star or a black body. Astronomical sources of synchrotron radiation include supernova remnants and radio galaxies. See also *black body, electromagnetic radiation, spectrum.*

**synchronous rotation** The rotation of a body around its axis in the same period of time that it takes to orbit another body. Synchronous rotation, which is also known as **captured rotation**, is caused by tidal forces acting between the two bodies. Because its rotational and orbital periods are the same, the orbiting body always keeps the same face turned towards the object around which it is revolving. Like most of the planetary satellites, the Earth's moon displays synchronous rotation. See also *orbital period, satellite.*

# T

**tail (of a comet)** A stream, or streams, of ionized gas and dust that is swept out of the head of a comet (the coma) when it approaches, and begins to recede from, the Sun. A **type I tail** (or **gas tail**) consists of ionized gas driven out of the coma by the solar wind. A **type II tail** (or **dust tail**) is composed of dust particles that have been swept out of the coma by the pressure of sunlight. See also *comet.*

**tectonic plate** One of the large, rigid sections into which the Earth's lithosphere (which comprises the crust and the rigid uppermost layer of the Earth's mantle) is divided. Carried along by slow convection currents in the mantle, tectonic plates drift very slowly across the surface of the planet. Their relative motions give rise to phenomena such as earthquakes, volcanic activity, and mountain building. The term "tectonic" is sometimes also used to refer to large-scale geological structures, and features resulting from their movement, on planets other than the Earth. See also *convection, crust, mantle.*

**tektite** A small, rounded, glassy object formed when a large meteorite or asteroid strikes a rocky planet, melting the surface rocks and throwing molten drops of rock into the atmosphere. Typically a few centimetres (inches) across, tektites have been shaped by their flight through the atmosphere. On the Earth's surface, they are found in a number of specific locations, called **strewn fields**. See also *asteroid, meteorite.*

**terrestrial planet** see *rocky planet.*

**transit** **(1)** The passage of a particular celestial body across an observer's meridian. **(2)** The passage of a body in front of a larger one (for example, the passage of the planet Venus across the face of the Sun, or a satellite across the face of a planet).

**T Tauri star** A young star, surrounded by gas and dust, that varies in brightness and usually shows evidence of a strong stellar wind (a stream of gas flowing away from the star). T Tauri stars are believed still to be contracting towards the main sequence. They are named after the first star of this kind to be identified. See also *main sequence, protostar.*

# UV

**ultraviolet radiation** Electromagnetic radiation with wavelengths shorter than visible light but longer than X-rays. The hottest stars radiate strongly at ultraviolet wavelengths.

**umbra** **(1)** The dark, central cone of the shadow cast by an opaque body. The illuminating source will be completely hidden from view at any point within the umbra. **(2)** The darker, cooler central region of a sunspot, where the temperature is about 1,500–2,000°C (about 2,700–3,600°F) cooler than the average for the solar surface. See also *eclipse, penumbra, sunspot.*

**vacuum energy** see *cosmological constant.*

**Van Allen belts** Two concentric doughnut-shaped zones that contain charged particles (electrons and protons) trapped in the Earth's magnetic field. They were discovered in 1958 by American space scientist James Van Allen.

**variable star** A star that varies in brightness. A **pulsating variable** is a star that expands and contracts in a periodic way, varying in brightness as it does so. An **eruptive variable** is a star that brightens and fades abruptly. A **cataclysmic variable** is a star that suffers one or more major explosions (for example, a nova). See also *Cepheid variable, nova.*

**vernal equinox** see *equinox.*

**volcanic crater** see *crater.*

# W

**wavelength** The distance between two successive crests or between two successive troughs in a wave motion.

**WIMP** The acronym for **Weakly Interacting Massive Particle**, one of a range of postulated elementary particles that have high masses (tens or hundreds of times as great as that of a proton) but interact so exceedingly weakly with ordinary matter that they have not yet been directly detected. WIMPs are widely considered to comprise the major part of the dark-matter content of the Universe. See also *dark matter.*,

**white-dwarf star** A star of low luminosity but relatively high surface temperature that has ceased to generate energy by nuclear-fusion reactions, that has been compressed by gravity to a diameter comparable to that of the Earth, and that is slowly cooling and fading. See also *black dwarf, Hertzsprung–Russell diagram.*

**Wolf–Rayet star** A very hot star from which gas is escaping at an exceptionally rapid rate, which is surrounded by an expanding gaseous envelope, and which has emission lines in its spectrum. See also *emission line, spectrum.*

# XYZ

**X-ray burster** An object that emits strong bursts of X-rays, lasting from a few seconds to a few minutes. The bursts are believed to occur when gas drawn from an orbiting companion star accumulates on the surface of a neutron star and triggers a nuclear-fusion chain reaction. See also *fusion, neutron star.*

**X-ray radiation** Electromagnetic radiation with wavelengths shorter than ultraviolet radiation but longer than gamma rays. X-rays are emitted by extremely hot clouds of gas, such as the solar corona.

**zenith** The point on the sky directly above an observer (that is, 90° above the observer's horizon).

**zodiac** A band around the celestial sphere that extends for 9° on either side of the ecliptic, and through which the Sun, Moon, and naked-eye planets appear to travel. The zodiac contains part or all of 24 constellations. In the course of the year, the Sun passes through 13 of these constellations, 12 of which correspond to the astrological "signs of the zodiac". See also *ecliptic.*

**zodiacal light** A faint, cone-shaped glow that extends along the direction of the ecliptic from the western horizon after sunset or from the eastern horizon before sunrise. Most easily seen from tropical skies, it is caused by the scattering of sunlight by particles of interplanetary dust that lie close to the plane of the ecliptic.

# INDEX

Page numbers in **bold** indicate feature profiles or extended treatments of a topic. Page numbers in *italic* indicate pages on which the topic is illustrated. Celestial objects whose names begin with a number can be found at the end of the index.

## A


- A stars 231
- AASTO project 297
- AB Aurigae *233*
- Abell, George **321**
- Abell 400 *317*
- Abell 1060 (Hydra Cluster) **320**
- Abell 1656 (Coma Cluster) 316, *317*, **320**, *324*
- Abell 1689 *27*, *316*, **321**
- Abell 2029 *317*
- Abell 2065 (Corona Borealis Cluster) **321**
- Abell 2125 **321**
- Abell 2151 (Hercules Cluster) **321**, *348*
- Abell 2218 *23*, **322–23**
- Abell S 373 (Fornax Cluster) **319**
- absolute magnitude 231
  - absolute magnitude scale 67
  - Hertzsprung–Russell (H–R) diagram *230*
  - main-sequence stars 247
- absorption lines 33, *33*
  - Lyman Alpha lines 325, *325*
  - stellar classification 231
- Académie des Sciences 88
- accelerating motion 40, *40*
- accretion discs 245
  - black holes 263, 310
  - young stars *237*
- acetylene, on Jupiter 178
- Achernar (Alpha (α) Eridani) *246*, 390, *404*
  - Hertzsprung–Russell (H–R) diagram *230*
  - in monthly sky guides 421, 457, 463, 469, 475, 481
- Acheron Fossae (Mars) **168**
- achromatic telescopes 76
- Acidalia Planum (Mars) 170
- Acrux (Alpha (α) Crucis) 396, 439
- active galaxies **310–15**
  - BL Lacertae **315**
  - Centaurus A **312**
  - Circinus Galaxy **312**
  - Cygnus A **314**
  - Fried Egg Galaxy **313**
  - M87 **313**
  - NGC 1275 **314**
  - NGC 4261 **313**
  - NGC 5548 **313**
  - PKS 2349 **315**
  - "supermassive" black holes 297, *297*
  - types of 310
  - 3C 48 **315**
  - 3C 273 **315**
- Adams, John 90
- Adams, W.S. 94
- Adams ring (Neptune) 201, *201*
- Addams, Jane 137
- Addams Crater (Venus) **137**
- Adonis *209*
- Adrastea *178*, **180**
- AE Aurigae 392
- Aeneas Crater (Dione) *193*
- aerogel 219, *219*
- age
  - of Earth 92
  - of star clusters 285
  - of Universe 42
- Agena spacecraft 102, *102*
- Aglaonice Crater (Venus) 137
- Air Pump *see* Antlia
- Airy Crater (Mars) 171
- Aitken Basin Crater (Moon) **159**
- Aitne *179*
- Akna Montes (Venus) **132**
- Albiorix *189*
- Albireo (Beta (β) Cygni) *270*, **273**, 350, *350*, 456
- Alcmene 227
- Alcor (80 Ursae Majoris) **272**, 344, *345*, 438
- Alcott Crater (Venus) **137**
- Alcyone (Eta (η) Tauri) **273**, 287, 356
- Aldebaran (Alpha (α) Tauri) **252**, 356
  - classification 231
  - Hertzsprung–Russell (H–R) diagram *230*
  - and Hyades 286
  - in monthly sky guides 415, 475
  - naked-eye astronomy *73*
- Aldrin, Edwin "Buzz" 104, *104*, 154
- Alexandria 84, *84*
- algae *53*
- Algieba (Gamma (γ) Leonis) 361, *361*
- Algol (Beta (β) Persei) **272**, 354, *354*, 480
- ALH 81105 meteorite **223**
- aliens, search for 53, *53*
- alignments, planetary 65
- Alioth (Epsilon (ε) Ursae Majoris) *68*, 344
- Alkaid (Eta (η) Ursae Majoris) *68*, 344
- Allende meteorite **222**
- Almaak (Gamma (γ) Andromedae) **273**, 352
- Almaaz (Epsilon (ε) Aurigae) **277**, 279, *279*, 343
- Almach (Gamma (γ) Andromedae) **273**, 352
- Alnath (Beta (β) Tauri) *230*, 343, 356
- Alnilam (Epsilon (ε) Orionis) *230*
- Alnitak (Zeta (ζ) Orionis) *230*, 374, 375, *375*
- Alpha (α) Andromedae (Alpheratz) 352, 370
- Alpha (α) Aquilae (Altair) **248**, 350, 367, *367*
  - in monthly sky guides 445, 456, 457, 462, 463, 469
  - naked-eye astronomy 73
- Alpha (α) Arietis 355
- Alpha (α) Aurigae (Capella) 343
  - sky guides 414, 415, 432, 468
- Alpha (α) Boötis (Arcturus) 344, 347
  - Hertzsprung–Russell (H–R) diagram *230*
  - in monthly sky guides 426, 432, 433, 438, 439, 444, 445, 450, 451, 456
  - naked-eye astronomy *73*
- Alpha (α) Canis Majoris (Sirius A) **248**, 376
  - ancient astronomy 82
  - apparent magnitude 67
  - binary system 270
  - classification 231
  - Hertzsprung–Russell (H–R) diagram *230*
  - in monthly sky guides 414, 415, 421, 426, 427, 475, 481
  - naked-eye astronomy *73*
  - name, origin of 68
  - Winter Triangle 420, *420*, 480
- Alpha (α) Canis Minoris (Procyon) **280**, 376
  - classification 231
  - Hertzsprung–Russell (H–R) diagram *230*
  - in monthly sky guides 420, 481
  - naked-eye astronomy *73*
  - Winter Triangle 420, *420*, 480
- Alpha (α) Canum Venaticorum (Cor Caroli) 346, *346*
- Alpha (α) Capricorni 387
- Alpha (α) Centauri (Rigil Kentaurus) **248**, 382, *382*
  - apparent magnitude *67*
  - Hertzsprung–Russell (H–R) diagram *230*
  - in monthly sky guides 427, 432, 433, 439, *439*, 451, 457, 463
- Alpha (α) Ceti (Menkar) 373
- Alpha (α) Circini 397
- Alpha (α) Corona Borealis (Alpheca) 444
- Alpha (α) Corvi 381
- Alpha (α) Crucis (Acrux) 396, 439
- Alpha (α) Cygni (Deneb) 350
  - Hertzsprung–Russell (H–R) diagram *230*
  - luminosity *231*
  - in monthly sky guides 444, 451, 456, 457, 462, 463
  - naked-eye astronomy 73
- Alpha (α) Delphini (Sualocin) 369
- Alpha (α) Eridani (Achernar) *246*, 390, *404*
  - Hertzsprung–Russell (H–R) diagram *230*
  - in monthly sky guides 421, 457, 463, 469, 475, 481
- Alpha (α) Fornacis 389
- Alpha (α) Geminorum (Castor) **272**, 358
  - Hertzsprung–Russell (H–R) diagram *230*
  - in monthly sky guides 420, 421, 427
- Alpha (α) Herculis (Ras Algethi) **281**, 348
- Alpha (α) Horologii 403
- Alpha (α) Hydrae (Alphard) 378
  - Hertzsprung–Russell (H–R) diagram *230*
  - in monthly sky guides 426, 427
- Alpha (α) Leonis (Regulus) **249**, 361
  - Hertzsprung–Russell (H–R) diagram *230*
  - in monthly sky guides 427
  - naked-eye astronomy *73*
  - name, origin of 68
- Alpha (α) Librae (Zubenelgenubi) 363
- Alpha (α) Lyrae (Vega) **249**, 349, 350
  - circumstellar disc 290
  - Hertzsprung–Russell (H–R) diagram *230*
  - luminosity *231*
  - in monthly sky guides 432, 438, 444, 445, 450, 451, 456, 457, 462, 463
  - naked-eye astronomy 73
- Alpha (α) Mensae 406
- Alpha (α) Microscopii 387
- Alpha (α) Monocerotis 377
- Alpha (α) Orionis (Betelgeuse) *25*, **252**, 374, 376
  - apparent magnitude *67*
  - classification 231, *231*
  - Hertzsprung–Russell (H–R) diagram *230*
  - in monthly sky guides 415, 420, 481
  - naked-eye astronomy *73*
  - Winter Triangle 420, *420*, 480
- Alpha (α) Pavonis 408
- Alpha (α) Pegasi 370
- Alpha (α) Persei (Mirphak) *230*, 354, 480
- Alpha (α) Persei Cluster 354, 480
- Alpha (α) Piscis Austrini (Fomalhaut) **249**, 388, *388*
  - Hertzsprung–Russell (H–R) diagram *230*
  - in monthly sky guides 451, 457, 462, 463, 468, 475, 481
- Alpha (α) Piscium (Alrescha) 372, *372*
- Alpha (α) Scorpii (Antares) **252**, 365, 386
  - Hertzsprung–Russell (H–R) diagram *230*
  - in monthly sky guides 433, 438, 444, 445, *445*
- Alpha (α) Serpentis (Unukalhai) *364*
- Alpha (α) Tauri (Aldebaran) **252**, 356
  - classification 231
  - Hertzsprung–Russell (H–R) diagram *230*
  - and Hyades 286
  - in monthly sky guides 415, 475
  - naked-eye astronomy *73*
- Alpha (α) Triangulum Australis 398
- Alpha (α) Ursae Majoris (Dubhe) *68*, 344
  - Hertzsprung–Russell (H–R) diagram *230*
- Alpha (α) Ursae Minoris (Polaris) **274–75**, 338, *338*, 344
  - circumpolar stars 332
  - Hertzsprung–Russell (H–R) diagram *230*
  - in monthly sky guides 432, *438*
  - naked-eye astronomy 73, *73*
- Alpha (α) Virginis (Spica) *362*
  - Hertzsprung–Russell (H–R) diagram *230*
  - in monthly sky guides 426, 427, 432, 433, 438, 439, 444, 445, 450, 451
  - naked-eye astronomy *73*
- Alpha (α) Vulpeculae 368
- alphabet, Greek 7, 333
- Alphard (Alpha (α) Hydrae) 378
  - Hertzsprung–Russell (H–R) diagram *230*
  - in monthly sky guides 426, 427
- Alpheca (Alpha (α) Corona Borealis) 444
- Alpheratz (Alpha (α) Andromedae) 352, 370
- Alphonsus Crater (Moon) **155**
- Alrescha (Alpha (α) Piscium) 372, *372*
- Alshain (Beta (β) Aquilae) 367
- Altair (Alpha (α) Aquilae) **248**, 350, 367, *367*
  - in monthly sky guides 445, 456, 457, 462, 463, 469
  - naked-eye astronomy 73
- Altar *see* Ara
- altazimuth mountings, telescopes 76, *76*
- aluminium, properties *29*
- aluminium-26 222
- AM 0644-741 *292–93*
- Amalthea *178*, 179, **180**
- Amazon River (Earth) **146**
- American Association of Variable Star Observers 281, 283
- amino acids 52
- Ammavaru Volcano (Venus) *135*
- ammonia
  - interstellar medium 228
  - Jupiter 178, *178*
  - Neptune 200
  - Saturn 187, *187*, 188
  - Uranus 196, 197
- ammonium hydrosulphide, on Saturn 187
- Amor asteroids 208, *208*
- analemma, Sun's *60*
- Ananke *179*
- ancient astronomy **82–83**
- Andes (Earth) *143*
- Andromeda **352**
  - Almach (Gamma (γ) Andromedae) **273**, 352
  - Alpheratz (Alpha (α) Andromedae) 352, 370
  - in monthly sky guides 474, *474*

Andromeda *cont.*
Upsilon (υ) Andromedae 291, *291*
Andromeda Galaxy (M31, NGC 224) 301, **302–303**, 352, *352*
binocular astronomy *75*
Hooker Telescope 93
Local Group 318, *318*
in monthly sky guides 468, 469, 474, 475
radio waves *34*
angular diameter 73
angular momentum 37
animals 141
Annefrank **210**
annular eclipses 63
anorthite 223
anorthosite 125
Ant Nebula (Menzel 3) **255**
Antarctica
AASTO project 297
Ice-sheet **147**
meteorites 147, *221*
Antares (Alpha (α) Scorpii) **252**, 365, 386
Hertzsprung–Russell (H–R) diagram *230*
in monthly sky guides 433, 438, 444, 445, *445*
Antennae Galaxies (NGC 4038 and 4039) *35*, **307**, 308, 381, *381*
antielectrons *see* positrons
antimatter *31*, 311, *311*
antiparticles, Big Bang 46, 47, 48
antiquarks *31*
Big Bang *46*, *47*, 48
Antlia (the Air Pump) **380**
Zeta (ζ) Antliae 380
aperture
binoculars 74
telescopes 77, *77*
Aphrodite 372
Aphrodite Terra (Venus) *131*, 135
Apollinaris Patera (Mars) **168**
Apollo asteroids 208, *208*
Apollo missions **102–105**, *150*, 153, 154, 156, 249
Apollo–Soyuz mission 105, *105*
apparent magnitude 67, 231
April sky guide **432–37**
Apus (the Bird of Paradise) **407**
Delta (δ) Apodis 407
Theta (θ) Apodis 407
Aquarius (the Water Carrier) **371**
Eta (η) Aquarii 371, 439
Gamma (γ) Aquarii 371
Helix Nebula **253**, 371, *371*, 463
in monthly sky guides 469, *469*, 475
Pi (π) Aquarii 371
Saturn Nebula *251*, 371, *371*, 463
Zeta (ζ) Aquarii 371
Aquila (the Eagle) **367**
Alshain (Beta (β) Aquilae) 367
*see also* Altair (Alpha (α) Aquilae)
Eta (η) Aquilae **282**, 367
Lambda (λ) Aquilae *367*
sky guide 456
Tarazed (Gamma (γ) Aquilae) 367, *367*
15 Aquilae 367
57 Aquilae 367
Aquila Rift, Milky Way 229
Ara (the Altar) **399**
Ara (the Altar) *cont.*
Mu (μ) Arae 291
Stingray Nebula **260**
Arabs
constellations 330
early scientific astronomy 84, 85
mythology 275
star names 330
Arago ring (Neptune) 201
Aratus of Soli 330
Archer *see* Sagittarius
Arcturus (Alpha (α) Boötis) 344, 347
Hertzsprung–Russell (H–R) diagram *230*
in monthly sky guides 426, 432, 433, 438, 439, 444, 445, 450, 451, 456
naked-eye astronomy *73*
Arecibo radio telescope *53*, *97*
Arenal volcano (Earth) *143*
Argo Navis 393, 394, 395, 420
argon
Earth's atmosphere *140*
Moon's atmosphere *149*
Argonauts 394
Argyre Planitia (Mars) **173**
Ariadne 363
Ariel *197*, **199**
Ariel 1 observatory 96
Aries (the Ram) **355**
Alpha (α) Arietis 355
Beta (β) Arietis 355
Gamma (γ) Arietis 355, *355*
Lambda (λ) Arietis 355
Pi (π) Arietis 355
sky guide 480
Arion 369
Aristarchus 84, 85, 86
Aristarchus Crater (Moon) **154**
Aristotle 59, *59*, **84**, 85
arms, spiral galaxies 295
Armstrong, Neil 104, *104*, 154
Arp 220 *299*
Arp-Madore 1 (AM1) 403
Arrow *see* Sagitta
Arsia Mons (Mars) **164**
Artemis Chasma (Venus) *135*
Artemis Corona (Venus) **135**
Asclepius 365, *365*
Ascraeus Mons (Mars) **164**
Asellus Australis 359
Asellus Borealis 359
asterisms 68
asteroids 25, **208–13**
Annefrank **210**
asteroid belt 210
Ceres **211**
collisions 209, *209*
computerized telescopes 78
Eros **212–13**
formation of 233
formation of Moon 149, *149*
Gaspra **210**
Ida **211**
impact craters on Moon 151
Mathilde **210**
orbits 118, *119*, 208, *208–209*
space probes 111
structure 208
Toutatis **210**
Vesta **210**
Asterope 287, 357
astrobiology 97
astrolabes 85, *85*
astrology 60, 63, 83
astrometric binaries 270
astronauts
future missions 111
Moon landings **102–105**
Space Race 100, *100*
space stations 106
weightlessness *36*
astronomy
ancient astronomy **82–83**
binocular astronomy **74–75**
computerized telescopes **78–79**
early scientific astronomy **84–85**
18th- and 19th-century astronomy **90–91**
from space 96, 97
naked-eye astronomy **72–73**
radio astronomy 93, *93*
telescope astronomy **76–77**
20th-century astronomy **92–93**
astrophotography 79, *79*
astrophysics 90
Astroplanner *78*
Aten asteroids 208
Athos, Mount (Earth) *84*
Atlantic Ocean (Earth) 142
Atlas *188*, 287, 356
Atlas rockets 100
atlases, star 331
atmosphere (Earth) 140, *140*
aurorae 70, *70–71*, *123*
ice haloes 70, *70*
moving lights and flashes 71, *71*
noctilucent clouds 71, *71*, *444*
zodiacal light 71, *71*
atmospheres
formation of 233
Jupiter 178, *178*
Mars 161, *161*
Mercury 125, *125*
Moon 149, *149*
Neptune 200, *200*, 201, *201*
old stars 234
Pluto 204
Saturn 187, *187*
Sun 123, *123*
Titan 194
Uranus 197, *197*
Venus 129, *129*
atomic bomb *39*
atomic number, chemical elements 29
atoms 24, 28, *28–29*
after Big Bang 50
Big Bang 46
in chemical compounds 29, *29*
of chemical elements 29
emergence of matter 48, *49*
forces 30, *30*
ionization *28*
in molecules 29
nuclear fission and fusion 31, *31*
AU Microscopii *290*
August sky guide **456–58**
Augusta family, asteroids 210
Auriga (the Charioteer) **343**, 414, 420, 421, 426, 480, 481
AB Aurigae *233*
AE Aurigae 343, *343*, 392
Almaaz (Epsilon (ε) Aurigae) **277**, 279, *279*, 343
*see also* Capella (Alpha (α) Aurigae)
Zeta (ζ) Aurigae 343
Aurora Australis 70, *70*
Aurora Borealis 70, *70–71*
aurorae
Earth 70, *70–71*, *123*
Jupiter 177, *177*
Saturn *187*
Autonoe *179*
autumn equinox 61, *61*, *138*
Avebury 83
azimuth mountings, telescopes 76
Azophi *see* al-Sufi
Aztecs 83

# B

b Puppis *393*
B stars
classification 231
Regor (Gamma (γ) Velorum) 249
Wolf–Rayet stars 251
Babylonians
ancient astronomy 82, 83
constellations 330
Bach Crater (Mercury) **127**
bacteria 52, 53, 141
Baghdad 85
Baily, Francis 360
Baily's Beads *63*
Balch, Emily 136
Balch Crater (Venus) **136**
barium, formation of 51
Barnard, Edward 180, 256
Barnard 33 (Horsehead Nebula) **238**, 239, 375, *375*
Barnard 68 *24*
Barnard's Galaxy (NGC 6822) *318*
Barnard's Merope Nebula *287*
Barnard's Star 66, *230*, 365
barred spiral galaxies 26, *294*
NGC 1530 *26*
NGC 6782 **308**
baryons *31*
al-Battâni 85
Bayer, Johann 68, **331**, 333
Bayeux Tapestry *218*
Be stars 281
Beardmore Glacier (Earth) *147*
Becrux (Beta (β) Crucis) 396
Beehive Cluster (M44) **286**, 359, *359*
in monthly sky guides 420, 421, 426
Beethoven region (Mercury) 127
Belinda *197*
Bell, Jocelyn 94
Bellatrix (Gamma (γ) Orionis) *67*
Bellerophon 370
Belyaeyev, Pavel *101*
Belz Crater (Mars) **172**
Berenice's Hair *see* Coma Berenices
Bergerac, Cyrano de 98
Bessel, Friedrich 248
Beta (β) Aquilae (Alshain) 367
Beta (β) Arietis 355
Beta (β) Camelopardalis 342
Beta (β) Canum Venaticorum *346*
Beta (β) Capricorni 387
Beta (β) Centauri (Hadar) *248*, *382*
apparent magnitude *67*
in monthly sky guides 427, 432, 433, 439, *439*, 451, 457, 463
Beta (β) Corvi 381
Beta (β) Crucis (Becrux) 396
Beta (β) Cygni (Albireo) *270*, **273**, 350, *350*, 456
Beta (β) Delphini (Rotanev) 369
Beta (β) Doradus 405
Beta (β) Geminorum (Pollux) 358
Hertzsprung–Russell (H–R) diagram *230*
in monthly sky guides 420, 421, 427
Beta (β) Gruis 401
Beta (β) Leonis (Denebola) 68
Beta (β) Leonis Minoris 360
Beta (β) Librae (Zubeneschamali) 363
Beta (β) Lyrae (Sheliak) **277**, 349
Beta (β) Lyrae stars 277
Beta (β) Monocerotis **277**, 377
Beta (β) Orionis (Rigel) **277**, 374
classification 231, *231*
Hertzsprung–Russell (H–R) diagram *230*
in monthly sky guides 415
Beta (β) Pegasi 370
Beta (β) Persei (Algol) **272**, 354, *354*, 480
Beta (β) Pictoris *290*, 404, *404*
Beta (β) Piscis Austrini 388
Beta Regio (Venus) **133**
Beta (β) Sagittarii 384
Beta (β) Scorpii 386
Beta (β) Tauri (Alnath) *230*, 343, 356
Beta (β) Tucanae 402
Beta (β) Ursae Majoris (Merak) *68*, 73, 344
Betelgeuse (Alpha (α) Orionis) *25*, **252**, 374, 376
apparent magnitude *67*
classification 231, *231*
Hertzsprung–Russell (H–R) diagram *230*
in monthly sky guides 415, 420, 481
naked-eye astronomy *73*
Winter Triangle 420, *420*, 480
Bethe, Hans 95
BHR 71 **238**
Bianca *197*
Big Bang 22, **46–49**, 93, 96
aftermath of 50
cosmic microwave background radiation (CMBR) 34, 49, 50, *96*, 325
distribution of galaxies 298
expanding space 42
fate of Universe 54
galaxy superclusters 325
inflation theory 46, *46*
particle physics 31
recreating conditions 47
Big Chill 54, *54*, 55
Big Crunch 54, 55, *55*
Big Dipper *see* Plough
Big Rip *54*, 55
binary pulsars 270
binary stars **270**
Alpha (α) Herculis (Ras Algethi) 281
Beta (β) Lyrae (Sheliak) **277**
black holes 263
eclipsing binary stars 270, *270*, 354

binary stars *cont.*
Epsilon (ε) Aurigae (Almaaz) **277**
Eta (η) Geminorum (Propus) 280
Izar (Epsilon (ε) Bootis) **273**
Lambda (λ) Tauri 280
M40 **273**
novae 278, *278*
Polaris **274–75**
Porrima 249
Type I supernovae 279, *279*
Wolf–Rayet stars 251
Zeta (ζ) Boötis **273**
15 Monocerotis **276**
binocular astronomy **74–75**
biosphere, Earth 141
Biosphere 2 111, *111*
Bird of Paradise *see* Apus
BL Lacertae (BL Lac) **315**, 353
BL Lacertae objects *see* blazars
black dwarfs 233, *235*, 262
Black Eye Galaxy (M64, NGC 4826) **304**, 360, *360*
black holes 24, 25, **26**, 233, **263**
accretion discs 263, 310
active galaxies 310, *310–11*
Andromeda Galaxy 302, *302*
Big Chill 55
Cygnus X-1 **268**
discovery 94
event horizon *41*, *263*
formation 234, *234*, *235*
galaxies 297, *297*
GRO J1655-40 **268**
hypernovae *51*
lensing 263, 269
MACHO 96 **269**
matter 28
Milky Way *14*, 226
radiation 34
singularity 26, *41*
space-time *41*
SS 433 *26*
stellar black holes 26, *26*
supermassive black holes 26, *26*, 55, 297, *297*
black smokers *142*
Blake, William *89*
blazars 310, *310*
BL Lacertae (BL Lac) **315**, 353
distribution 311
superluminal jets 311, *311*
Blaze Star (T Coronae Borealis) **282**
blink comparators 205
Blinking Planetary 351
blue jets 71, *71*
blue light, photoelectric effect 32, *32*
Blue Planetary 382
blue shift 33, *33*
Blue Snowball (NGC 7662) 352, *352*, 468
blue supergiants
Eta (η) Carinae **258**
evolution *233*
HDE 226868 268, *268*
Sher 25 **261**
blue variable stars
Pistol Star **261**
blue-white stars
Regor (Gamma (γ) Velorum) **249**
Regulus (Alpha (α) Leonis) **249**
blueberries, Martian 175, *175*
BM Scorpii 286, 386
Bode, Johann Elert 304, 331, *344*
Bode's Galaxy (M81, NGC 3031) **304**, 344, *344*
Bohr, Niels **29**
Bok globules 236, *236*
BHR 71 238
Cone Nebula 240
Eagle Nebula 242
IC 2944 244, *244*
Lagoon Nebula 241, *241*
bolometric luminosity 231
bomb, atomic *39*
Bondi, Hermann 96
bonds, states of matter 30
Boötes (the Herdsman) **347**
*see also* Arcturus (Alpha (α) Boötis)
Gamma (γ) Boötis 444
Izar (Epsilon (ε) Boötis) *25*, **273**, 347, *347*, 444
Kappa (κ) Boötis 347
Mu (μ) Boötis 347
NGC 5548 **313**
Nu (ν) Boötis 347
Xi (ξ) Boötis 347
Zeta (ζ) Boötis **273**
Bopp, Thomas 218
Borrelly, Comet 215, *215*, **219**
bosons 30, *30*, *31*
Big Bang 46
bow shock
Orion Nebula *20–21*
solar wind *139*
Brahe, Tycho 87, *87*, 88, 268, *268*
Tycho's Supernova 268
Brahms Crater (Mercury) **127**
Braun, Wernher von 99, **102**
brightness, stars **67**
*see also* luminosity
Britain, observatories 88
Brocchi's Cluster 368, *368*
Broglie, Louis de **33**
bromine, properties *29*
Brontë Crater (Mercury) 127
Bronze Age 287
Brown, Mike 207
brown dwarfs 25
discovery 94, *94*
extra-solar planets *291*
formation *232*
Gliese 229b *25*
Bruno, Giordano *87*
Bubble Nebula *286*
bubble nebulae, Wolf–Rayet stars *260*
Bug Nebula (NGC 6302) **256–57**
Bull *see* Taurus
Burns Cliff (Mars) 174, *174*, 175
Butterfly Cluster (M6, NGC 6405) **286**, 386, *386*
in monthly sky guides 445, *445*, 451, 457
Butterfly Nebula (Hubble 5) *251*


# C


C153 321
Cacciatore, Niccolò **369**
Caelum (the Chisel) **389**
Gamma (γ) Caeli 389
Calabash Nebula (OH231.8+4.2) **258**
Calabi-Yau spaces *41*
calcium, on Mercury 125
calderas, Martian volcanoes *164*, *165*, 168, *168*
calendar, ancient astronomy 82, *83*
Caliban *197*, **199**
California Extremely Large Telescope (CELT) 35
Callirhoe *179*
Callisto *25*, *178*, **185**, 193
space probes 110
Callisto, in mythology 345
Caloris Basin (Mercury) 126, *126*, **127**
Calypso *188*, 192
Camelopardalis (the Giraffe) **342**
Beta (β) Camelopardalis 342
11 Camelopardalis 342
12 Camelopardalis 342
cameras 79, *79*
Cancer (the Crab) **359**
*see also* Beehive Cluster
Delta (δ) Canceri 359, *359*
Gamma (γ) Canceri 359, *359*
Iota (ι) Canceri 359
in monthly sky guides 420, 426, 433
Zeta (ζ) Canceri 359
Cancer, Tropic of 61
Candor Chasma (Mars) *166*, 167
Canes Venatici (the Hunting Dogs) **346**
Beta (β) Canum Venaticorum *346*
Cor Caroli (Alpha (α) Canum Venaticorum) 346, *346*
La Superba (Gamma (γ) Canum Venaticorum) 346
*see also* Whirlpool Galaxy
Canis Major (the Greater Dog) **376**
HD 56925 **260**
*see also* Sirius A (Alpha (α) Canis Majoris); Sirius B
Tau (τ) Canis Majoris 376, *376*
UW Canis Majoris 376
Canis Major Dwarf 300
Canis Minor (the Little Dog) *330*, **376**
*see also* Procyon (Alpha (α) Canis Minoris)
Canopus (Alpha (α) Carinae) 376, 395
Hertzsprung–Russell (H–R) diagram *230*
in monthly sky guides 415, 427, 433, 469, 475, 481
Canyon Diablo meteorite **222**
canyons, on Mars 166–67, *166–67*
Capella (Alpha (α) Aurigae) 343
in monthly sky guides 414, 415, 432, 468
Capricorn, Tropic of 61
Capricornus (the Sea Goat) **387**, 462
Alpha (α) Capricorni 387
Beta (β) Capricorni 387
carbon
atomic number 29
carbon cycle (CNO cycle) 246
dust 24
formation of 51, 95
interstellar medium 228
and life 52
carbon *cont.*
main-sequence stars 246
in meteorites 223
in old stars 234, 251
supergiant stars 250
Type I supernovae 279
Wolf–Rayet stars 251
carbon dioxide
atomic structure 29
in comets 215
interstellar medium 228
on Mars 161, 169, 171, *171*
on Venus 129
carbon monoxide
in comets 215
on Pluto 204
carbon stars 231, 252, *252*
carbonaceous (C-type) asteroids 208
carbonaceous chondrite (stony) meteorites 208
cardinal points 83
Carina (the Keel) **395**
*see also* Canopus (Alpha (α) Carinae)
Epsilon (ε) Carinae 395
Eta (η) Carinae 245, *252*, **258**, 395, *395*, 427, *433*
Iota (ι) Carinae 395
in monthly sky guides 415, 420, 426
Sher 25 **261**
Theta (θ) Carinae 395, 427
Carina Nebula (NGC 3372) *24*, **245**, 395, *395*
in monthly sky guides 427, 433, *433*
Carlyle, Thomas 329
Carme *179*
Cartwheel Galaxy (ESO 350-G40) **309**
Caspian Sea (Earth) **147**
Cassini, Giovanni 88
Saturn's moons 192, 193, 195
Cassini Regio (Iapetus) 195, *195*
Cassini spacecraft 110, *110*, 194, *194*
Cassiopeia **341**
Eta (η) Cassiopeiae 341
Gamma (γ) Cassiopeiae **281**, 341
M52 **286**, 341, *341*, 468
in monthly sky guides 414, 420, 474, 480
Phi (φ) Cassiopeiae 341
Rho (ρ) Cassiopeiae 341
Tycho's Supernova **268**
Cassiopeia, Queen 341, 352
Cassiopeia A (SN 1680) *51*, *264*, **269**
Castor (Alpha (α) Geminorum) **272**, 358
Hertzsprung–Russell (H–R) diagram *230*
in monthly sky guides 420, 421, 427
catalogues
nebulous objects 69
stars 68, 330
Cat's Eye Nebula (NGC 6543) **254**, 339, *339*
cD galaxies 296, 299, 316, *316*, 317
Celaeno 287
celestial coordinates 59
celestial cycles **60–63**, 82
celestial globes *330–31*
celestial meridian 59
celestial poles *421*
Celestial Police 209
celestial sphere **58–59**, *330*
constellations 68
mapping **332–37**
motion of planets 64–65
motion of stars 66
Centaur *see* Centaurus
Centaurs 206
Centaurus (The Centaur) **382**
IC 2944 **244**
*see also* Alpha (α) Centauri (Rigil Kentaurus); Hadar (Beta (β) Centauri)
in monthly sky guides 415, 421, 426, 433, 438, 445
Omega Centauri 75, *284*, *285*, *286*, **288**, 382, 402, 433, 439, 445
Proxima Centauri *22*, *230*, **248**, 382
RCW 49 **245**
Centaurus A (NGC 5128) *14*, **312**, 382
collision with spiral galaxy 308, *311*, 314
in monthly sky guides 439, 445
Cepheid variable stars 282
measuring distances with 42, 303, *303*
pulsation 278, *278*
in Small Magellanic Cloud 301
Cepheus **340**
Delta (δ) Cephei **282**, 340, *340*, 462
Epsilon (ε) Cepheus 340
IC 1396 **241**
Lambda (λ) Cepheus 340
Mu (μ) Cephei (Garnet Star) *230*, 241, *241*, *250*, **283**, 340, *340*
Zeta (ζ) Cepheus 340
Ceres 90, *208*, 209, **211**
CERN (European Centre for Nuclear Research) 47
Cernan, Eugene *104*, *105*
Cerro Tololo Inter-American Observatory 258
CETI (communication with extraterrestrial intelligence) 53
Cetus (the Sea Monster) **373**
Gamma (γ) Ceti 373
Menkar (Alpha (α) Ceti) 373
Mira (Omicron (ο) Ceti) **281**, 373, 474, 475, *475*
in monthly sky guides 469, 474, 475, 480
Tau (τ) Ceti *230*, 373
ZZ Ceti *230*
Chaffee, Roger 249
Chaldene *179*
Chamaeleon (the Chameleon) **407**
Delta (δ) Chamaeleontis 407
Chameleon *see* Chamaeleon
Chandra X-ray Observatory *35*, *268*
Chandrasekhar, S. 94
Chandrasekhar limit 262
charge-coupled device (CCD) detectors 79, *79*
charged particles
aurorae 70, *123*
ions 28
Jupiter 177
magnetic fields 247

charged particles *cont.*
pulsars *263*
solar wind 123, 139
Sun 122
Charioteer *see* Auriga
Charitum Montes (Mars) 173
Charles I, King of England 346
Charles II, King of England 88
Charon 204, 205, *205*
charts, star 331
Chasma Boreale (Mars) *169*
chasmata, on Mars 166–67
chemical compounds 29, *29*
chemical elements *see* elements
Chéseaux, Philippe Loys de 242
China
ancient astronomy *82*, 83
and Polaris 275
space programme 111
Chiron 206, *206*
in mythology 382
Chisel *see* Caelum
chondrites *222*, 223
chondrules *222*, 223, *223*
Christmas Tree Cluster 240
chromatic aberration 89
chromosphere (Sun) 123, *123*
Chryse Planitia (Mars) 166
Churyumov, Klim 219
Churyumov–Gerasimenko, Comet **219**
Cigar Galaxy (M82, NGC 3034) *69*, 297, **304**, 344
Circinus (the Compasses) **397**
Alpha (α) Circini 397
Circinus Galaxy (ESO 97-G13) **312**
Circlet 372, *372*, 468, *469*
circulation cells, Jupiter 178, *178*
Circumnuclear Disc, Milky Way 229
circumstellar discs, formation of planets 233, *233*, 290–91, *290*
CL0024+1654 *323*
CL-2244-02 *317*
Claritas Fossae (Mars) **168**
Clarke, Arthur C. 106, *106*
classification
galaxies 294, *294*
stars 231
Clementine space probe 151, *152*
Cleopatra Crater (Venus) **136**
climate, Earth 138
clocks, medieval astronomy *86*
closed universe 55, *55*
clouds
*see also* gas clouds
Jupiter 178, *178*, 179
lenticular clouds *71*
Mars 161
*see also* molecular clouds
Neptune 201
noctilucent clouds 71, *71*, *444*
Uranus 197, *197*
Venus 129, *129*
Clownface Nebula 358
clusters *see* galaxy clusters; galaxy superclusters; star clusters
CMBR *see* cosmic microwave background radiation
Coalsack Nebula 396, *396*, 397
in monthly sky guides 433, 439, *439*, 445
Coathanger 368, *368*
cobalt, formation of 95
CoKu Tau (τ) 4 290
Cold Bokkeveld meteorite **223**
Cold War 98
collapsing stars *235*, *262*
Collins, Michael 104, *104*
Collins, Peter 74, *74*, 283
collisions, galaxies 237, *237*, 299, 308, *308*
colour force *30*
colours, stars *66–67*, 231
Columba (the Dove) **392**
Mu (μ) Columbae 392
coma, comets 215
Coma Berenices (Berenice's Hair) **360**
Black Eye Galaxy **304**, 360, *360*
Gamma (γ) Comae Berenices 360
Malin 1 **309**
the Mice **308**
Coma Cluster (Abell 1656) 316, *317*, **320**, *324*
Coma Star Cluster (Melotte 111) 360, 432, 438
Comas Solá, Comet 219
"comet clouds" *309*
comets 25, **214–19**
binocular astronomy 74
Borrelly 215, *215*, **219**
Churyumov–Gerasimenko **219**
Comas Solá 219
computerized telescopes 78
Encke *214*, **217**
formation 233
Giacobini-Zinner **219**
Great Comet of 1680 **216**
Hale–Bopp *214*, *216*, **218**
Halley's Comet 89, 90, *90*, 111, *111*, *214*, *215*, 216, **218**, 439
Hyakutake 74, *207*, *214*, **217**, 218
Ikeya–Seki **216**
Ikeya–Zhang *25*
Kuiper Belt **206**
life cycles 215
meteoroids 220
Oort Cloud **207**
orbits 214, *214*
Shoemaker–Levy 9 179, *179*, **219**
Soho-6 *215*
space probes 111, *111*
structures 215, *215*
Swift–Tuttle *214*, **216**, 220
Tempel–Tuttle *214*, 220
West **217**
Wild 2 111, **219**
Wirtanen 111, *111*, 219
compact groups
Seyfert's Sextet **319**
Stephan's Quintet **320**
Compass *see* Pyxis
Compasses *see* Circinus
composite particles *31*
compounds, chemical 29, *29*
Compton Gamma Ray Observatory *35*
computerized telescopes **78–79**
Cone Nebula (NGC 2264) **240**, 276, 377, *377*
conjunction, planets *64*, 65, *65*
Constantine, Emperor 65
constellations 68, *68–69*, **328–409**
*see also individual constellations*
ancient astronomy 83
history **330–31**
mapping the sky **332–37**
zodiac 60, *61*
contact binary systems 270
convection 246
convection cells
red giants *250*
Sun *95*, 122
convection currents
Jupiter *178*
plate tectonics *140*
coordinates, celestial 59
Copernicus, Nicolaus **86–87**
Copernicus Crater (Moon) 149, *151*, **155**
Coprates Chasma (Mars) 167
Cor Caroli (Alpha (α) Canum Venaticorum) 346, *346*
Cordelia 197, *197*, **198**
core
Earth 138, *138*, 139
Jupiter 176, *176*
Mars 160, *160*
Mercury 125, *125*
Moon 148, *148*
Neptune 200, *200*
Pluto *204*
Saturn *186*
Uranus 196, *196*
Venus 128, *128*
Coriolis effect 140, *140*, 178
corona, Sun *10*, *63*, *122*, 123, *123*
Corona Australis (the Southern Crown) **399**
Gamma (γ) Coronae Australis 399
Kappa (κ) Coronae Australis 399
RX J1856.5-3754 **264**
Corona Borealis (the Northern Crown) **363**, 444
Abell 2065 (Corona Borealis Cluster) **321**
Alpheca (Alpha (α) Corona Borealis) 444
Nu (ν) Coronae Borealis 363
R Coronae Borealis 279, **283**, 363
Sigma (σ) Coronae Borealis 363
T Coronae Borealis (Blaze Star) **282**
Zeta (ζ) Coronae Borealis 363
coronal mass ejections (CMEs), Sun *122*, 123, *123*
Corvus (the Crow) **381**
Alpha (α) Corvi 381
Antennae Galaxies *35*, **307**, 308, 381, *381*
Beta (β) Corvi 381
Delta (δ) Corvi 381
Epsilon (ε) Corvi 381
Gamma (γ) Corvi 381
cosmic light horizon 23
cosmic microwave background radiation (CMBR) 34, 49, 50, *96*, 325
Sunyaev–Zel'dovich effect *322*, 323
cosmic rays 24, 32, 228
cosmological constant 54
cosmological red shift 33
cosmologists 22
cosmology, Earth-centred 85, *85*
covalent compounds 29
Crab *see* Cancer
Crab Nebula (M1, NGC 1952) 94, **266–67**, 356, *356*
Crane *see* Grus
Crater (the Cup) 68, **381**, 426
craters *see* impact craters; volcanoes
Crescent Nebula (NGC 6888) **255**
Cressida *197*
CRL 2688 (Egg Nebula) **254**
Crow *see* Corvus
crust
Earth 138, *138*, 140, *140*
Mars 160
Moon 148
Crux (the Southern Cross) **396**
Acrux (Alpha (α) Crucis) 396, 439
Becrux (Beta (β) Crucis) 396
Gacrux (Gamma (γ) Crucis) *230*, 396
*see also* Jewel Box (Kappa (κ) Crucis)
in monthly sky guides 421, 427, 432, 433, 438, 439, 445
Mu (μ) Crucis 396
naked-eye astronomy 73
Pointers *248*
Culann Patera (Io) *182*
Cunitz Crater (Venus) *136*
Cup *see* Crater
Curtis, Heber **92**, 93
cycles, celestial **60–63**, 82
Cygnus (the Swan) **350–51**
Albireo (Beta (β) Cygni) *270*, **273**, 350, *350*, 456
Crescent Nebula **255**
Cygnus A (3C 405) **314**, 351
Cygnus X-1 94, **268**, 351
*see also* Deneb (Alpha (α) Cygni)
DR 6 **241**
DR 21 **244**
in monthly sky guides 456, 474, 480
Nova Cygni 1992 **283**
Omicron (o) Cygni 350
TT Cygni **252**
56 Cygni 68
61 Cygni *230*, **248**, 351
Cygnus Loop (NGC 6960/95) *228*, **265**
Cygnus Rift 351, 456
Cygnus Star Cloud *268*

# D

Dactyl 211, *211*
Daedalia Planum (Mars) 168
Dali Chasma (Venus) **135**
Danilova Crater (Venus) 137
Dante Alighieri 182
dark ages 50
dark energy 27, 50, 54–55, *54*
dark galaxies 316
dark matter **27**, 28, 50
dwarf elliptical galaxies 296
galaxies 298
galaxy superclusters 325
gravitational lensing 323
Milky Way 229
dark nebulae 24, 228, 238
Barnard 68 *24*
BHR 71 **238**
Cone Nebula **240**
Horsehead Nebula **238**
Darwin, Charles 92
databases, computerized telescopes 78
days
and length of year 82
measuring 62, *62*
December sky guide **480–85**
declination 59, *59*, 73
Deep Space 1 mission 219
Degas Crater (Mercury) **127**
degrees of angle 73
Deimos 163, *163*
Delphinus (the Dolphin) **369**
Delta (δ) Delphini 369
Gamma (γ) Delphini 369
Rotaney (Beta (β) Delphini) 369
Sualocin (Alpha (α) Delphini) 369
Delta (δ) Apodis 407
Delta Aquarid meteor shower 451
Delta (δ) Canceri 359, *359*
Delta (δ) Cephei **282**, 340, *340*, 462
Delta (δ) Chamaeleontis 407
Delta (δ) Corvi 381
Delta (δ) Delphini 369
Delta (δ) Gruis 401, *401*
Delta (δ) Librae 363
Delta (δ) Lyrae 349
Delta (δ) Octantis 409
Delta (δ) Orionis 374
Delta (δ) Scorpii 386
Delta (δ) Scuti 366
Delta (δ) Serpentis 364
Delta (δ) Telescopii 400
Delta (δ) Ursae Majoris 344
Delta (δ) Velorum 394
Deneb (Alpha (α) Cygni) 350
Hertzsprung–Russell (H–R) diagram *230*
luminosity *231*
in monthly sky guides 444, 451, 456, 457, 462, 463
naked-eye astronomy 73
Denebola (Beta (β) Leonis) 68
density waves
formation of stars 232
spiral galaxies 227, 237, 295
Desdemona *197*
deserts, on Earth 141, *141*
Desktop Universe *78*
Despina *201*
deuterium 49
Devana Chasma (Venus) **133**
Dido Crater (Dione) *193*
differential rotation, spiral galaxies 294
digital astrophotography 79, *79*
dimensions
Calabi-Yau spaces *41*
space-time 39
Dione *188*, *189*, *190*, **193**
Discovery Rupes (Mercury) **127**
disrupted spiral galaxies
Antennae Galaxies **307**
Cartwheel Galaxy **309**
ESO 510-G13 **308**
the Mice **308**
distance
apparent magnitude 231
early scientific astronomy 84, *84*
expanding space 42–43
mapping the Universe 324
micrometers *91*
naked-eye astronomy 73, *73*
parallax shift 66, *66*
pulsating variable stars 278
size of Universe **22–23**
DNA 141
Dollond, John 91
Dolphin *see* Delphinus

Domovoy Crater (Ariel) 199
Doppler effect *33*
Dorado (the Goldfish) **405**
  Beta (β) Doradus 405
  *see also* Large Magellanic Cloud
  R Doradus 405
  *see also* Tarantula Nebula (30 Doradus)
double binary stars *270*
Double Cluster 354, *354*, 480
double-slit test *32*
double stars *see* binary stars
Dove *see* Columba
DR 6 **241**
DR 21 **244**
Draco (the Dragon) **339**, 444, 450
  Abell 2218 **322–23**
  Cat's Eye Nebula **254**, 339, *339*
  Etamin (Gamma (γ) Draconis) 339
  Mu (μ) Draconis 339
  Nu (ν) Draconis 339
  Omicron (ο) Draconis 339
  Psi (ψ) Draconis 339
  Spindle Galaxy **307**
  16 Draconis 339
  17 Draconis 339
  39 Draconis 339
  40 Draconis 339
  41 Draconis 339
Dragon *see* Draco
Dragon Storm (Saturn) *188*
Drake, Frank 53
Draper, Henry *239*
Dresden Codex *83*
Dreyer, J.L.E. **237**
Dubhe (Alpha (α) Ursae Majoris) *68*, 344
  Hertzsprung–Russell (H–R) diagram *230*
Dumbbell Nebula (M27) 253, 368, *368*, 456, 457
Dunlop, James 256
Dürer, Albrecht 331
dust
  interstellar medium 24, 228
  storms on Mars *167*
  zodiacal light 71, *71*
dusty elliptical galaxies 299, *299*
dwarf elliptical galaxies 296, *296*, 300
  Canis Major Dwarf 300
  galaxy clusters 316, *316*
  SagDEG **300**
dwarf stars
  black dwarfs 233, *235*, 262
  brown dwarfs 25, *25*, 94, *94*, *232*, *291*
  red dwarfs 25, *25*, *233*
  *see also* white dwarfs
dwarf planets **204–205**

# E

e Puppis 393
Eagle *see* Aquila
Eagle Crater (Mars) 171
Eagle Nebula (IC 4703) *236*, **242–43**, 364, *364*, 451
Earth *8*, *25*, **138–47**
  age 92
  asteroids 208
  atmosphere and weather 140, *140*
  aurorae 70, *123*
  axis of rotation 60, *60*
  celestial sphere **58–59**
Earth *cont.*
  as centre of cosmos 85, *85*
  circumference 84, *84*
  climate 138
  Copernican revolution **86–89**
  eclipses **63**
  features formed by water **146–47**
  life 52–53, 141, *141*
  lights in the sky **70–71**
  magnetic field 139, *139*
  meteorite craters 221, *221*, **222–23**
  meteorites 220
  the Moon 148, *149*, 150, *150*
  orbit and spin *118*, 138, *138*
  plate tectonics 140, *140*
  seasons 61, *61*, *138*
  size *22*
  structure 138, *138*
  surface features 141
  tectonic features 140, **142–45**
earthquakes
  Mercury 126
  meteorite impacts 221
eclipses **63**
eclipsing binary stars 270, *270*, 354
  Alpha (α) Herculis (Ras Algethi) **281**
  Eta (η) Geminorum (Propus) **280**
  Lambda (λ) Tauri **280**
ecliptic 60, *61*, 138
Eddington, Sir Arthur 92, *92*, **247**
Egg Nebula (CRL 2688) **254**
Egypt
  ancient astronomy 82, 83, *83*
  constellations 330
  early scientific astronomy 84
Eight-Burst Nebula (NGC 3132) *250–51*, 394, 427
Einstein, Albert 31, **38**, 92, *92*
  black holes 94
  cosmological constant 54
  energy and mass 39
  general theory of relativity **40–41**, 49
  mass and energy 54
  Mercury's orbit 124
  principle of equivalence 40, *40*
  special theory of relativity **38–39**
Eistla Regio (Venus) **133**
El Tajin, Mexico *82*
Elara *178*
Electra, in mythology 357
electromagnetic (EM) force 30, *30*, 47
electromagnetic (EM) radiation **32–35**
  "false colour" images 35, *35*
  observing **34–35**
  Sun 120
electron degeneracy pressure, white dwarfs 262
electrons 28, *28–29*
  Big Bang 47, 48, *48–49*
  Big Chill 55
  chemical elements 29
  forces 30, *30*
  molecules 29
  photoelectric effect 32, *32*
  plasma 30
  synchrotron mechanism *310*
electroweak era *46–47*
electroweak force 47
elements 29
  formation of 51, 262, *262*
  high-mass stars *234*
  planet formation 233
  spectroscopy 33, *33*
  star formation 232
  supergiant stars 250
Elephant's Trunk Nebula 241, *241*
ellipses, orbits 37, *37*, 87
elliptical galaxies 26, **296**
  classification 294, *294*
  distribution 298
  galaxy clusters 316, 317
  M60 **307**
  merger model 299, *299*
  SagDEG **300**
Eltanin 339
Elysium Planitia (Mars) **170**
emission nebulae 24, *33*, 228, 238
  Carina Nebula **245**
  DR 6 **241**
  DR 21 **244**
  Eagle Nebula **242–43**
  IC 1396 **241**
  IC 2944 **244**
  Lagoon Nebula **241**
  M43 *239*, 375
  NGC 604 301, *301*
  NGC 2359 260
  Omega Nebula *90*, *238*, 384, *385*
  Orion Nebula **239**
  RCW 49 **245**
  Trifid Nebula **244**
emission spectrum 33, *33*, 231
  planetary nebulae 251
  Wolf–Rayet stars 251
Enceladus *188*, *189*, **192**
Encke, Comet *214*, **217**
Encke, Johann 217
Encke gap, Saturn's rings *189*
Encounter 2001 message 53
end points, stellar **262–69**
Endurance Crater (Mars) **174–75**
Energetic Gamma (γ) Ray Experiment Telescope (EGRET) *35*
energy
  atomic bomb *39*
  atoms *28*
  Big Bang 46
  convection 246
  dark energy 27, 50, 54–55, *54*
  Einstein's theories 92
  electromagnetic (EM) radiation 32, *32*
  fate of Universe 54–55
  ionization *28*
  luminosity 231
  main-sequence stars 246
  mass 39, *39*
  nuclear fission and fusion 31, *31*
  photons 32
  protostars 237
  radiation 246
  rotation 37
  Saturn 187
  stars 230
  states of matter 30
  strong nuclear force *30*
  Sun 120
  supernovae 262
Enif (Epsilon (ε) Pegasi) 370, 462
Enki Catena (Ganymede) *215*
Enlightenment 90
Ensisheim meteorite **222**
Eos Chasma (Mars) 167
Epimetheus *188*, **190**
Epsilon (ε) Aurigae (Almaaz) **277**, 279, *279*, 343
Epsilon (ε) Bootis (Izar) *25*, **273**, 347, *347*, 444
Epsilon (ε) Carinae 395
Epsilon (ε) Cepheus 340
Epsilon (ε) Corvi 381
Epsilon (ε) Herculis 348
Epsilon (ε) Hydrae 378
Epsilon (ε) Indi 400
Epsilon (ε) Lupi 383
Epsilon (ε) Lyrae **272**, 349
Epsilon (ε) Normae 398
Epsilon (ε) Orionis (Alnilam) *230*
Epsilon (ε) Pegasi (Enif) 370, 462
Epsilon (ε) Sagittarii 384
Epsilon (ε) Sculptoris 388
Epsilon (ε) Ursae Majoris (Alioth) *68*, 344
Epsilon (ε) Volantis 406
equator, celestial sphere *58*, *59*
equatorial mountings, telescopes 76, *76*
equatorial sky charts *334–37*
equinoxes 61, *61*, *138*
  ancient astronomy 83
  Pisces 372
  precession 355
  sky guide 426
Equuleus (the Foal) **369**
  Gamma (γ) Equulei 369
  1 Equulei 369
Eratosthenes 84, *84*
Erichthonius 343
Eridanus (the River) **390**, 469, 481
  *see also* Achernar (Alpha (α) Eridani)
  Omicron (ο) Eridani **272**, 390, *390*
  Theta (θ) Eridani 390
  32 Eridani 390
  40 Eridani B *230*
Erie, Lake (Earth) 146
Erinome *179*
Eris 204, *205*, 206
Eros *13*, *111*, *208*, *210*, **212–13**
  mythology 372
erosion
  Mars 172, *172*
  Venus *131*
eruptive variable stars 258
  U Geminorum **280**
ESA *see* European Space Agency
Eskimo Nebula (NGC 2392) 358, *358*, **255**
ESO 97-G13 (Circinus Galaxy) **312**
ESO 350-G40 (Cartwheel Galaxy) 309
ESO 510-G13 **308**
Eta Aquarid meteor shower 371, 438, 439
Eta (η) Aquarii 371, 439
Eta (η) Aquilae **282**, 367
Eta (η) Carina Nebula *see* Carina Nebula
Eta (η) Carinae 245, *252*, **258**, 395, *395*, 427, *433*
Eta (η) Cassiopeiae 341
Eta (η) Geminorum (Propus) **280**, 358
Eta (η) Herculis 348
Eta (η) Lupi 383
Eta (η) Piscium 372
Eta (η) Tauri (Alcyone) **273**, 287, 356
Eta (η) Ursae Majoris (Alkaid) *68*, 344
Eta (η) Ursae Minoris 338
Etamin (Gamma (γ) Draconis) 339
ethane
  Jupiter 178
  Saturn *187*
"ether" 38
Euanthe *179*
Eudoxus 84, 330
Euporie *179*
Europa *25*, *178*, **180–81**
  possibility of life *53*
  space probes *110*
European Centre for Nuclear Research (CERN) 47
European Space Agency (ESA)
  Giotto mission 218
  Hipparcos satellite 66, *66*
  Rosetta mission *111*, 219
Eurydome *179*
evaporating gaseous globules (EGGs) *236*, 242
event horizon, black holes *41*, *263*
Everest, Mount (Earth) 144, *144*
evolution
  galaxies **298–99**
  galaxy clusters 317
  life 141
  multiple stars 270, *270*
  star clusters 285
  stars **233–35**
Ewen, Harold 95
exotic particles *31*, *46*
expanding space **42–43**, 54, 323
Explorer 1 satellite 99, 139
extra-solar planets 96, **290–91**
extraterrestrial life 53
extreme stars 94
Extreme Ultraviolet Explorer *35*
extremophile organisms 53
eyepieces, telescopes 77, *77*
eyes, adjusting to dark 72

# F

F stars 231
Fabricius, David 281
faculae, Sun *122*
false colour images, electromagnetic radiation 35, *35*
False Cross 394, 395
  in monthly sky guides 421, *427*
  naked-eye astronomy 73
Family Mountain (Moon) *156*
February sky guide **420–25**
fermions *31*
field equations 41
field galaxies 316
field of view, binoculars 75
filaments
  Big Bang *50*
  galaxy superclusters 325, *325*
films, astrophotography 79
filters 78, *78*
finder scopes 77, *77*
fire altars 83
fireballs 71, *220*

Fishes *see* Pisces
Fish's Mouth *239*
Flaming Star Nebula *343*
Flammarion, Camille *87*
Flamsteed, John 88
  star atlas 68, 331, *331*
flare stars 248
flares, solar *10*, *114–15*, 122, *246*
flat universe 54, 55, *55*
flocculent spiral galaxies 295, *295*, *301*
Florida Keys *9*
Fly *see* Musca
Flying Fish *see* Volans
flying saucers 71, *71*
Foal *see* Equuleus
focusing binoculars *75*
Fomalhaut (Alpha (α) Piscis Austrini) **249**, 388, *388*
  Hertzsprung–Russell (H–R) diagram *230*
  in monthly sky guides 451, 457, 462, 463, 468, 475, 481
force-carrier particles 30, *30*, *31*, 46
forces
  Big Bang 46
  electromagnetic (EM) force 30, *30*
  gravity 30, **36–37**, 89
  string theory 31
  strong nuclear force 30, *30*
  weak nuclear force 30, *30*
forests, on Earth 141, *141*
Fornax (the Furnace) **389**
  Alpha (α) Fornacis 389
  Fornax A 389
Fornax Cluster (Abell S 373) **319**, *324*, 389, *389*
Fornax dwarf galaxy *318*
Fortuna Tessera (Venus) **132**
Fox *see* Vulpecula
France, observatories 88, *88*
Frank, Anne 210
Fraunhofer, Joseph von 90, **123**
Fraunhofer lines 123
free fall 36, 37
Freedom 7 mission 100
Fried Egg Galaxy (NGC 7742) **313**
Fuji, Mount (Earth) *143*
fundamental particles *31*, 46, 48
fundamental strong nuclear force 30, *30*
fungi 141
Furnace *see* Fornax

## G

G stars 231
Gacrux (Gamma (γ) Crucis) *230*, 396
Gaea (Amalthea) *180*
Gagarin, Colonel Yuri 100, *100*
Galactic Centre, Milky Way 238
Galatea 201, *201*
galaxies *14–15*, 24, **26**, **292–325**
  *see also individual galaxies*
  active galaxies **310–15**
  barred spiral 26, *26*, *294*, 308
  Big Chill 55
  black holes 26, 297, *297*
  catalogues 69
  classification 294, *294*
  clusters *16–17*, *23*, 24
  collisions 237, *237*, 299, 308, *308*
  density waves 237
  distribution 298
  earliest 323
galaxies *cont.*
  *see also* elliptical galaxies
  evolution **298–99**
  expanding space 42, *42*
  formation 51, *51*, 298, *298*
  giant elliptical galaxies 300
  gravitational lensing 322–23
  Hubble Ultra-Deep Field *16*
  interstellar medium 228
  irregular galaxies 26, **297**
  lenticular galaxies 26, **296**
  merging 299, *299*, *317*
  radiation 34
  red shift 33, *33*
  rotation 37
  Seyfert galaxies 305, 310, *310*
  *see also* spiral galaxies
  star formation 237, *237*
  superclusters *16*, *23*
  20th-century astronomy 92, 93
  types of **294–97**
  wavelengths 297
galaxy clusters **27**, **316–21**
  Abell 1689 **321**
  Abell 2065 (Corona Borealis Cluster) **321**
  Abell 2125 **321**
  Abell 2218 **322–23**
  Coma Cluster **320**
  evolution 317
  Fornax Cluster **319**
  gravity bending light *41*
  Hercules Cluster **321**
  Hickson Compact Group *27*
  Hydra Cluster **320**
  Local Group 27, 316, *316*, **318**
  radiation 34
  Sculptor Group **319**
  Seyfert's Sextet **319**
  Stephan's Quintet **320**
  Virgo Cluster **319**
  X-rays 319, *319*
galaxy superclusters *16*, *23*, 24, 27, **324–25**
  filaments 325, *325*
  formation 50, *50*
  voids between 325
Galilean moons (Jupiter) *25*, 178, **180–85**
Galileo Galilei **88**
  Galilean moons *25*, 180–81
  mapping the Moon 151, *151*
  Saturn's rings 189
  study of gravity 36
  telescopes 76, *76*
Galileo space probe 110, *110*, 180, 181, *181*
Galle ring (Neptune) 201, *201*
Gamma (γ) Andromedae (Almach) **273**, 352
Gamma (γ) Aquarii 371
Gamma (γ) Aquilae (Tarazed) 367, *367*
Gamma (γ) Arietis 355, *355*
Gamma (γ) Boötis 444
Gamma (γ) Caeli 389
Gamma (γ) Canceri 359, *359*
Gamma (γ) Canum Venaticorum (La Superba) 346
Gamma (γ) Cassiopeiae **281**, 341
Gamma (γ) Ceti 373
Gamma (γ) Comae Berenices 360
Gamma (γ) Coronae Australis 399
Gamma (γ) Corvi 381
Gamma (γ) Crucis (Gacrux) *230*, 396
Gamma (γ) Draconis (Etamin) 339
Gamma (γ) Equulei 369
Gamma (γ) Leonis (Algieba) 361, *361*
Gamma (γ) Leporis 391
Gamma (γ) Lyrae 349
Gamma (γ) Orionis (Bellatrix) *67*
Gamma (γ) Pegasi 370
Gamma (γ) Piscis Austrini 388
gamma rays *31*, 32
  Geminga Pulsar 264
  main-sequence stars 246
  Milky Way *227*, 311
  observatories 35, *35*
Gamma (γ) Ursae Majoris (Phad) *68*, 344
Gamma (γ) Ursae Minoris 338
Gamma (γ) Veloram (Regor) 231, **249**, 394
Gamma (γ) Virginis (Porrima) **249**, 362
Gamma (γ) Volantis 406
Gamma-1 ($\gamma^1$) Normae 398
Gamma-2 ($\gamma^2$) Normae 398
Gamow, George **48**, 49
Ganymede *25*, *178*, **184**, 194
  impact craters *215*
  space probes 110
Ganymede, in mythology 371
Gaposchkin, Sergei 231
Garnet Star (Mu (μ) Cephei) *230*, 241, *241*, *250*, **283**, 340, *340*
gas clouds 24, *28*
  Eta (η) Carinae Nebula *24*
  formation of Solar System *116–17*
  galaxy collisions 299
gas-giant planets 25, 119
  extra-solar planets 291, *291*
  formation of Solar System *117*
  Jupiter **176–85**
  Neptune **200–203**
  Saturn **186–95**
  Uranus **196–99**
gases
  interstellar medium 228
  molecules 24
  novae 278
  spectroscopy *33*
  states of matter 30
  Sunyaev–Zel'dovich effect *322*
Gaspra **210**
gauge bosons *31*
gegenschein 71, *71*
Geminga Pulsar (SN 437) **264**
Gemini (the Twins) **358**
  *see also* Castor (Alpha (α) Geminorum) Eskimo Nebula **255**
  Eta (η) Geminorum (Propus) **280**, 358
  Geminga Pulsar **264**
  in monthly sky guides 420, 426, 432, 438, 474, 480, 481
  *see also* Pollux (Beta (β) Geminorum)
  U Geminorum **280**
  Zeta (ζ) Geminorum (Mekbuda) **282**, 358
Gemini missions *100*, 101, 102, *102*
Gemini North telescope *94*
Gemini South telescope *95*
Geminid meteor shower 358, 480, *480*
general theory of relativity **40–41**, 49
genes 52
geometry
  early scientific astronomy 84
  space-time curvature 55
Gerasimenko, Svetlana 219
Gertrude Crater (Titania) *199*
Ghost of Jupiter 378, *378*
Ghost Nebula (NGC 1977) *79*, 375
Giacobini–Zinner, Comet **219**
giant elliptical galaxies 296, *296*, 300
  galaxy clusters 316, 317
giant stars 25
  Aldebaran **252**
  classification 231
  evolution *233*, 234
  Hertzsprung–Russell (H–R) diagram 230, *230*, *251*
  multiple stars 270
  novae 278
  planetary nebulae 251
  red giants 25, **250**
  star life cycles *233–35*, 234
  *see also* supergiants
  TT Cygni **252**
  Type I supernovae 279, *279*
Giotto mission *111*, 218
Giraffe *see* Camelopardalis
Giza, pyramids *83*
glacial lakes, on Earth *144*
glaciers, on Earth *147*
glass
  impactites *221*
  volcanic glass 157, *157*
Glatton meteorite **222**
Glenn, John 100, *100*
Gliese 229 *25*
Gliese 229b *25*
Global Microlensing Alert Network 269
Global Positioning System (GPS) 39, 106
global warming, Earth 147
globes, celestial *330–31*
globular clusters 285, *285*, 286
  M4 **288**
  M12 **289**
  M14 **289**
  M15 **289**
  M68 **289**
  M107 **289**
  Milky Way 229, *229*
  NGC 3201 **288**
  NGC 4833 **289**
  Omega Centauri **288**
  47 Tucanae **288**
gluons 28, *29*, *31*
  Big Bang *46*, 48, *48*
  forces *30*
  recreating Big Bang 47
Goddard, Robert 98, *98*
gods and goddesses 83
Gold, Thomas 96
Golden Fleece 355
Goldfish *see* Dorado
Gomez, Arturo 258
Gomez's Hamburger Nebula (IRAS 18059-3211) **258**
Gossamer Ring (Jupiter) 180
GPS (Global Positioning System) 39, 78, 106
Grand Canyon (Earth) *146*, 166
"Grand Tour", space probes 109
Grand Unified Theory era *46*
granulation, Sun *122*
gravitational lensing
  black holes and 263, 269
  galaxy clusters 317, *317*, 322–23
gravitational waves, binary pulsars 270
gravitons 31, 46
gravity 24, 30, **36–37**, 89
  Big Bang 46
  Big Crunch 54, 55, *55*
  black holes 26, 263
  development of structures 50, *50*
  Einstein's theories 92
  expanding space 42, 54
  extra-solar planets *291*
  galaxy clusters 27
  galaxy superclusters 324, 325
  globular clusters 285
  light 40, *40*, *41*
  matter 28
  Moon 149, 150, *150*
  multiple stars 270, *270*
  Newton's laws 98
  orbits 37, *37*
  particle physics 31
  planet formation 233
  precession 60
  Principle of Equivalence 40, *40*
  protostars 237
  quantum gravity 41
  red giants 250
  Solar System 116
  space-time 40–41, *40–41*
  star formation 232, *232*, 236, 237
  stars 230, *230*
Great Attractor 324, *324*
Great Bear *see* Ursa Major
Great Comet of 1680 **216**
Great Dark Spot (Neptune) 201, *201*
Great Lakes (Earth) **146**
Great Red Spot (Jupiter) *12–13*, 90, 179, *179*
Great Rift Valley (Earth) **142**, 168
Great White Spots (Saturn) 188
Greater Dog *see* Canis Major
Greece
  ancient astronomy 83
  constellations 330
  early scientific astronomy 84, *84*
Greek alphabet, Bayer's system 68, 331, 333
greenhouse effect, on Venus *129*
Gregory, James 89
GRO J1655-40 **268**
Grus (the Crane) **401**, 463
  Beta (β) Gruis 401
  Delta (δ) Gruis 401, *401*
  Mu (μ) Gruis 401, *401*
Guardians of the Pole 338
Guardians of the Sky 252
Gula Mons (Venus) **133**
GUM 29 245
Gum Nebula *see* Vela Supernova
Guth, Alan 96

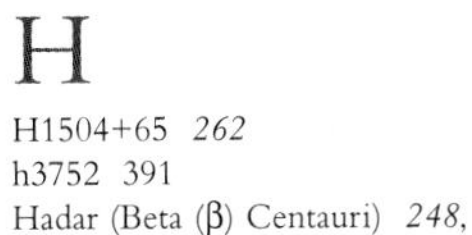

## H

H1504+65 *262*
h3752 391
Hadar (Beta (β) Centauri) *248*, *382*

apparent magnitude *67*
Hadar (Beta (β) Centauri) *cont.*
in monthly sky guides 427, 432, 433, 439, *439*, 451, 457, 463
Hadley, John 409
Hadriaca Patera (Mars) 171
Hadron Era *48*
hadrons *31*, *48*
Hahn, Friedrich von 253
Hale, Alan 218
Hale–Bopp, Comet *214*, *216*, **218**
Hale Crater (Mars) *172*
Hale Telescope 93, *93*
Hall, Asaph 163
Halley, Edmond 69, *90*, 91, 258
Halley's Comet **218**
discovery 90, *90*, 216
Eta Aquarid meteor shower 439
gravity 89
orbit *214*
space probes 111, *111*
tail *215*
halo stars 229
haloes
ice 70, *70*
Milky Way 229, 269
Hamlet Crater (Oberon) 199, *199*
Harch, Ann *211*
Harding, Karl Ludwig 253
Hare *see* Lepus
Harold II, King of England 218
Harpalyke *179*
Harvest Moon *462*
Hawking, Stephen 21, **94**
Hazard, Cyril 315
HD 23608 273
HD 44179 (Red Rectangle Nebula) **254**
HD 48915 B *see* Sirius B
HD 56925 **260**
HD 62166 (NGC 2440 nucleus) 264
HD 107146 *290*
HD 206267 241
HD 226868 268, *268*
heat 32
protostars 237
states of matter 30
*see also* temperature
Helen of Troy 358
Helene *188*, 193
Helios space probe 121, *121*
helioseismology 95, *95*
helium 24
Big Bang 48, *48–49*, 50
atomic number 29
burning in old stars 234, *234*, 251
carbon stars 252
first stars 51
helium flash 251
Jupiter 176, *176*, 178, *178*
main-sequence stars 246, 247
Mercury 125, *125*
Moon's atmosphere *149*
nebulae 236
Neptune 200, *201*
nuclear fusion 31, *31*, 95
red giants *250*
Saturn 186
star formation 232
stars 230, 232
Sun 120
supergiants 250
helium *cont.*
on Uranus *197*
Wolf–Rayet stars 251
Helix Nebula (NGC 7293) **253**, 371, *371*, 463
Hellas Basin (Mars) 162
Hellas Planitia (Mars) **173**
hematite, on Mars 171, 175
Hen-1357 (Stingray Nebula) 260
Heraclitus 84
Herbig Haro objects 240
Hercules 68, **348**
Epsilon (ε) Herculis 348
Eta (η) Herculis 348
Kappa (κ) Herculis 348
Keystone 348, 444, 450
in monthly sky guides 450, 456
mythology 227, 339, *339*, 359, 361, 378
Pi (π) Herculis 348
Ras Algethi (Alpha (α) Herculis) **281**, 348
Rho (ρ) Herculis 348
Zeta (ζ) Herculis 348
95 Herculis 348
100 Herculis 348
Hercules Cluster (Abell 2151) **321**, *348*
Herdsman *see* Boötes
Hermippe *179*
Herschel, Caroline 90, *91*, 217
Herschel, John 244
Herschel, William **90**
Cone Nebula 240
Eskimo Nebula 255
planetary nebulae 251
Polaris B 275
Saturn's moons 191
Sombrero Galaxy 306
telescopes 91, *91*
Uranus's moons 199
Herschel crater (Mars) *162*
Herschel Crateris (Mimas) 191, *191*
Herschel 36 241
Herschel's Garnet Star (Mu (μ) Cephei) *230*, 241, *241*, *250*, **283**, 340, *340*
Hertzsprung, Ejnar 230
Hertzsprung–Russell (H–R) diagram 94, 230, *230*
instability strip *251*
main-sequence stars 230, *230*, *247*
star classification 231
star evolution 233, *233*
Hevelius, Johannes 88, 330, **368**
Canes Venatici 346
Lacerta 353
Leo Minor 360
Lynx 343
Mira 281
Scutum 366
Sextans 380
Vulpecula 368
HH 320 238
HH 321 238
Hickson 92 (Stephan's Quintet) **320**
Hickson Compact Group *27*
Hidalgo *209*
Higgs bosons 46
high-mass stars
life cycle *233*, 234, *234*
nuclear reactions 246
structure *246*
supergiants **250**
supernovae 232
high-velocity stars 229
Himalayas (Earth) **144–45**
Himalia *178*
Himeros (Eros) *212*
Hinduism *83*
Hipparchus of Nicaea 66, 84, 330
Hipparcos satellite 66, *66*
Hiten space probe 153
Hoag's Object (PGC 54559) **309**
Hoba West meteorite **223**
Hodge 301 *301*
Hoffmeister, Cuno 315
Homer *286*, 374
Homunculus Nebula *258*
Hooke, Robert 216, 355
Hooker Telescope 93, *93*
Horologium (The Pendulum Clock) **403**
Alpha (α) Horologii 403
Horsehead Nebula (Barnard 33) **238**, 239, 375, *375*
Hourglass Nebula (MyCn18) 241, **259**
Houtman, Frederick de **400**
Apus 407
Chamaeleon 407
Dorado 405
Grus 401
Hydrus 403
Indus 400
Musca 397
Pavo 408
Phoenix 401
star catalogue 330
Triangulum Australe 398
Tucana 402
Volans 406
Hoyle, Fred 95, 96
H-R diagram *see* Hertzsprung–Russell diagram
Hubble, Edwin 43, *43*, 293
Andromeda Galaxy 302, 303
galaxy classification 294, *294*
Hooker Telescope 93, *93*
Hubble 5 (Butterfly Nebula) *251*
Hubble constant 42, *42*
Hubble Deep Field *96*, *298*
Hubble Space Telescope 43, 97, *107*
Pillars of Creation 242, *242*
Hubble Ultra-Deep Field *16*
Huggins, William 90, *90*
Hulst, Hendrick van der 95
Humboldt Crater (Moon) **155**
Hun Kal (Mercury) 126
Hunter *see* Orion
Hunting Dogs *see* Canes Venatici
Huron, Lake (Earth) 146, *146*
Huygens, Christiaan 115, 189, **194**
Huygens Crater (Mars) **172**
Huygens lander 110, *110*, 194, *194*
Hyades (MEL 25) **286**, *287*, 356, *356*
Aldebaran 252
in monthly sky guides 475, 480, 481
Hyakutake, Comet 74, *207*, *214*, **217**, 218
Hyakutake Yuji 74, 217
Hydra (the Water Snake) **378–79**
*see also* Alphard (Alpha (α) Hydrae)
Hydra (the Water Snake) *cont.*
Epsilon (ε) Hydrae 378
ESO 510-G13 **308**
M68 **289**, 378
in monthly sky guides 426, 432, 433
Mu (μ) Hydrae 378
Hydra Cluster (Abell 1060) **320**
hydrogen 24
absorption into galaxies 299
Big Bang 48, *48–49*, 50
atomic number 29
galaxy superclusters 325, *325*
Bug Nebula *256*
dark galaxies 316
first stars 51
intergalactic medium 317
interstellar medium 95, 228
Jupiter 176, *176*, 177, 178, *178*
Lyman Alpha lines 325, *325*
main-sequence stars 246, 247
Mercury 125
in meteorites 223
Moon *149*, 159
nebulae 236
Neptune 200, *201*
nuclear fusion 31, *31*, 95
planetary nebulae 251
properties *29*
re-ionization 51
red giants 250, *250*
Saturn 186
star formation 230, 232, *232*, 236
stellar evolution 233, 234, *234*
in Sun 120
supergiant stars 250
Uranus *197*
Wolf–Rayet stars 251
hydrothermal vents *142*
hydroxyl (OH) 217
Hydrus (the Little Water Snake) **403**
Pi (π) Hydri 403
Hyginus 330, *330*
hyperbolas, orbits 37
Hyperion *188*, **195**
hypernovae *35*, *51*

# I

Iapetus *188*, **195**
IC (Index Catalogue) 69
IC 349 *287*
IC 405 343
IC 434 375
IC 1396 **241**
IC 1590 *236*
IC 2163 *299*
IC 2391 394, *394*, 421
IC 2395 394, 421
IC 2602 (Southern Pleiades) 395, 427, 433
IC 2944 **244**
IC 4665 365, 450
IC 4703 (Eagle Nebula) *236*, **242–43**, 364, *364*, 451
IC 4756 364
Icarus *208*
ice
Callisto 185
Earth 141, *141*, 147, *147*
Europa 181
Ganymede 184, *184*
ice haloes 70, *70*
Mars 160, *160*, 163, *163*, **169**, 170, *170*, **171**
ice *cont.*
Neptune 200
Pluto 204, *204*
Uranus 196
ice ages 138
ice dwarfs 25
Ida *116*, *208*, **211**
Ijiraq *188*
Ikeya Kaoru 216
Ikeya–Seki, Comet **216**
Ikeya–Zhang, Comet *25*
Imbrium Basin (Moon) 154
impact craters
asteroids 209, *209*
Callisto 185
formation *119*
Mars 162, **172–75**
Mercury 126, *126*, 127, *127*
meteorites 221, *221*
Miranda *198*
Moon 149, 151, 154–55, 158–59
moons *215*
ray craters *151*
Venus 131, *131*, **136–37**
Vesta 210
impactites *221*
Incas 63
India, ancient astronomy 82, 83, *83*
Indian *see* Indus
Indus (the Indian) **400**
Epsilon (ε) Indi 400
Theta (θ) Indi 400
inferior planets, motion 64, *64*, 65
inflation theory, Big Bang 46, *46*
infrared 32
astronomy from space 96
galaxies *297*
telescopes 34, *34*
Infrared Space Observatory *107*
Inquisition 87
instability strip, Hertzsprung–Russell (H–R) diagram *251*
inter-continental ballistic missiles (ICBMs) 98, *98*, 99
interference, light waves 32, *32*
intergalactic medium 317
intermediate-period comets 214
International Astronomical Union 331
International Comet Explorer 219
International Space Station (ISS) *71*, 107, *107*
interstellar medium 24, 28
early Universe 51
Milky Way 228
radio astronomy 95, *95*
star formation 232, 237
inverse square law *67*
Io *13*, *25*, *178*, *180*, **182–83**
space probes *109*, 110, *110*
Iocaste *179*
ions 28, *28*
ionic compounds 29, *29*
plasma 30
Sun 122
Iota (ι) Canceri 359
Iota (ι) Carinae 395
Iota (ι) Librae 363
Iota (ι) Normae 398
Iota (ι) Orionis 375
Iota (ι) Pictoris 404
Iran 85
Iraq 85

IRAS telescope 249
IRAS 18059-3211 (Gomez's Hamburger Nebula) 258
iron
Earth 138
formation of 29, 51, 95
high-mass stars *234*
interstellar medium 228
Mercury 125
meteorites 223
old stars 234
supergiant stars 250
supernovae 262, *262*
iron meteorites 208, *220*
irregular clusters
Abell 2125 **321**
Hercules Cluster **321**
Local Group **318**
Sculptor Group **319**
Virgo Cluster **319**
irregular galaxies 26, **297**
Cigar Galaxy **304**
classification 294, *294*
distribution 298
Large Magellanic Cloud (LMC) **300–301**
Small Magellanic Cloud (SMC) **301**
Whirlpool Galaxy **305**
irregular variable stars
Gamma (γ) Cassiopeiae **281**
R Coronae Borealis **283**
Ishtar Terra (Venus) *131*, **132**
Isidis Planitia (Mars) 173
Islam
early scientific astronomy 84, 85, 86
zodiac *60*
islands, volcanic 142, *142*, 143
Isonoe *179*
Ithaca Chasma (Tethys) 192, *192*
Izar (Epsilon (ε) Bootis) *25*, **273**, 347, *347*, 444

# J

James Webb Space Telescope 97
January sky guide **414–19**
Janus **190**
Jason and the Argonauts 355, 394
Jason Crater (Phoebe) *195*
Jauslin, Karl *220*
Jeanne Crater (Venus) **136**
jets, superluminal 311, *311*
Jewel Box (Kappa (κ) Crucis, NGC 4755) **288**, 396, *396*
in monthly sky guides 433, 439, 445
Job's Coffin 369
Jodrell Bank *93*
John Sobieski, King of Poland 366
Juliet *197*
July sky guide **450–55**
June sky guide **444–49**
Jupiter **176–85**
atmosphere 178, *178*
Comet Shoemaker–Levy 9 219
formation of Solar System *117*
Great Red Spot *12–13*, 90
magnetic field 177, *177*
moons *13*, *25*, 178, *178–79*, **180–85**
occultations *65*
orbit and spin *118*, 176, *176*
Jupiter *cont.*
rings 179, *179*
short-period comets 214
space probes 109, 110
structure 176, *176*
Trojan asteroids 208, *208*
weather 179

# K

k Puppis 393
K stars 231
Kachina Chasmata (Ariel) 199
Kailas Range (Earth) *144*
Kale *179*
Kalyke *179*
Kant, Immanuel 116
Kappa (κ) Boötis 347
Kappa (κ) Coronae Australis 399
Kappa (κ) Crucis (Jewel Box, NGC 4755) **288**, 396, *396*
in monthly sky guides 433, 439, 445
Kappa (κ) Herculis 348
Kappa (κ) Leporis 391
Kappa (κ) Lupi 383
Kappa (κ) Pavonis 408
Kappa (κ) Tucanae 402
Kappa (κ) Velorum 394
Kapteyn, Jacobus Cornelius 229, *229*
Karatepe (Mars) 174
Kasei Vallis (Mars) **170**
Keck Telescope 93
Keel *see* Carina
Kemble, Lucian 342
Kemble's Cascade 342, *342*
Kennedy, John F. 102, *102*
Kennedy Space Center *9*
Kepler, Johannes **64**, *87*, 211, 268
Kepler's Star 269, 365
laws of planetary motion 87, *87*, 89, 118
Kepler's Star (SN 1604) *35*, **269**, 365
Keyhole Nebula 245, 395
Keyser, Pieter Dirkszoon **400**
Apus 407
Chamaeleon 407
Dorado 405
Grus 401
Hydrus 403
Indus 400
Musca 397
Pavo 408
Phoenix 401
star catalogue 330
Triangulum Australe 398
Tucana 402
Volans 406
Keystone 348, 444, 450
Kirch, Gottfried 216
Kiviuq *188*
Kiyotsugu, Hirayama 211
Kleinmann-Low Nebula 239
Köhler, Johann 307
Korolev, Sergei 100, *100*, 101, 158
Korolev Crater (Moon) **158**
Koronis family, asteroids 211
Kowal, Charles 180
Kuiper, Gerard **206**
Miranda 198
Neptune's moons 202
Kuiper Airborne Observatory 197, *197*
Kuiper Belt **206**
Pluto 204
space probes 111

# L

L Puppis 393
L² Puppis 393
Lacaille, Nicolas Louis de 330, **406**
Antlia 380
Caelum 389
Circinus 397
Horologium 403
Mensa 406
Microscopium 387
NGC 4833 289
Norma 398
Octans 409
Pictor 404
Pyxis 392
Reticulum 404
Sculptor 388
Telescopium 400
Lacerta (the Lizard) **353**
BL Lacertae (BL Lac) **315**, 353
Lada Terra (Venus) **135**
Laelaps 376
Lagoon Nebula (M8) **241**, 384, *384*
binocular astronomy 75
in monthly sky guides 451, 457, *457*
Lagrange points, orbits 192
Laika *99*
lakes, on Earth *144*, 146
Lakshmi Planum (Venus) **132**
Lalande, J.J. 381
Lambda (λ) Aquilae *367*
Lambda (λ) Arietis 355
Lambda (λ) Cepheus 340
Lambda (λ) Tauri **280**, 356
Lambda (λ) Velorum 394
Landsat satellites *106*
Langren, Arnold van *330–31*
Laplace, Pierre-Simon de 90, **116**
Large Magellanic Cloud (LMC) 297, **300–301**, 405, *405*
MACHO 96 269
Milky Way halo 229
in monthly sky guides 415, 421, 427, 469, 474, 475, 481, *481*
supernova 261
Larissa *201*, **202**
Larsen Ice-shelf (Earth) *147*
Lassell, William 199, 202, 203, *203*
Lassell ring (Neptune) 201, *201*
Latin names, constellations 68
latitude, and celestial sphere 58
lava flows
Io *182*
Mars 162, *165*, 168, *168*
Mercury 126
Moon 149, *150*, 154
Venus *130–35*
Le Verrier, Urbain 90, **118**
Le Verrier ring (Neptune) 201, *201*
lead, formation of 51
Leavitt, Henrietta **282**, 301, **340**
Leda *178*
Leda, Queen of Sparta 351, 358
Lemaître, Georges 48, 96
lenses
binoculars 74
telescopes 76
lensing *see* gravitational lensing
lenticular clouds *71*
lenticular galaxies 26, **296**
classification 294, *294*
Spindle Galaxy **307**
Leo (the Lion) 68, *331*, **361**
Algieba (Gamma (γ) Leonis) 361, *361*
Denebola (Beta (β) Leonis) 68
in monthly sky guides 420, 421, 426, *426*, 427, 432, 433, 439, 444
Zeta (ζ) Leonis 361
40 Leonis 361
*see also* Regulus (Alpha (α) Leonis)
Leo I galaxy *296*
Leo II groups, galaxy superclusters *324*
Leo Minor (the Little Lion) **360**
Beta (β) Leonis Minoris 360
46 Leonis Minoris 360
Leonid meteor shower 220, *220*, 361, *410–11*, 474
Leonov, Alexei 101, *101*, *105*
Lepton Era *48*
leptons *31*
after Big Bang 48, *48*
Lepus (the Hare) **391**
Gamma (γ) Leporis 391
Kappa (κ) Leporis 391
R Leporis 391
Levy, David 219
Libra (the Scales) **363**
Delta (δ) Librae 363
Iota (ι) Librae 363
Mu (μ) Librae 363
sky guide 433
Zubenelgenubi (Alpha (α) Librae) 363
Zubeneschamali (Beta (β) Librae) 363
life **52–53**
extra-solar planets 291, *291*
Mars *97*
search for 97
water and 141
life cycles, stars **232–35**
light 32
after Big Bang 50
analysing 33, *33*
black holes 263
emission nebulae 24, *24*
expanding space 43
galaxies *297*
gravitational lensing 263, 269, 317, *317*, 322–23
gravity 40, *40*, *41*
ice haloes 70, *70*
inverse square law *67*
light pollution *72*, *78*
observable Universe 23
Olbers' paradox *49*
optical telescopes 35, *35*
photoelectric effect 32, *32*
prisms 89, *89*
red shift 42
space and time 38–39
spectroscopy 90, *90*
stars 25
telescopes 76
velocity of 32, 38, 39
wave-like behaviour 32, *32*
light-year 22
lightning 71, *71*
lights in the sky **70–71**
line-of-sight binaries 270
Lion *see* Leo
Lippershey, Hans 76
liquids, states of matter 30
lithium 49
lithosphere, Earth 140, *140*
Little Bear *see* Ursa Minor
Little Dipper 338, 444, 450
Little Dog *see* Canis Minor
Little Lion *see* Leo Minor
Little Water Snake *see* Hydrus
Lizard *see* Lacerta
Lob Crater (Puck) *198*
Local Bubble, Milky Way 229, *229*
Local Group *23*, 27, 316, *316*, **318**
Andromeda Galaxy 302
galaxy superclusters 324, *324*
Local Interstellar Cloud 229
Local Supercluster *23*, 324
long-period comets 207, *207*, 214
lookback distance 43, *43*
Loop I, Milky Way 229
Loop II, Milky Way 229
Loop III, Milky Way 229
Lorentz contraction 39, *39*
Louis XIV, King of France 88
Lousma, Jack *106*
Lovell, Jim **103**
low-mass stars
life cycles *233*, 234, *234*
structure *246*
low-surface-brightness galaxies
Malin 1 **309**
Lowell, Percival 205
Lowell Crater (Mars) **173**
Lowell Observatory 306
luminosity 67, 231
Hertzsprung–Russell (H–R) diagram 230, *230*
main-sequence stars 247
pulsating variable stars 278
stellar classification 231
Type I supernovae 279
lunar eclipses 63, *63*
lunar month 62
Lunar Orbiter 103, *103*
Lunar Prospector 151, 153, 159, *159*
Lunar Rover *150*, *156*
Lunar Surveyor 103
Lunik (Luna) space probes 101, *101*, 108, 151, *151*, 153
Lunokhod rovers 105
Lupus (the Wolf) **383**
Epsilon (ε) Lupi 383
Eta (η) Lupi 383
Kappa (κ) Lupi 383
in monthly sky guides 445
Mu (μ) Lupi 383
Pi (π) Lupi 383
Xi (ξ) Lupi 383
Lyman Alpha lines 325, *325*
Lynx (the Lynx) **343**
12 Lyncis 343
19 Lyncis 343
38 Lyncis 343
Lyons, Harold 39
Lyra (the Lyre) **349**
Beta (β) Lyrae (Sheliak) **277**, 349
Delta (δ) Lyrae 349
Epsilon (ε) Lyrae **272**, 349
Gamma (γ) Lyrae 349
M40 **273**
Ring Nebula **253**, 349, *349*, 456, 457
RR Lyrae **282**
*see also* Vega (Alpha (α) Lyrae)
Zeta (ζ) Lyrae 349
Lyrid meteor shower 349, 432
Lysithea *178*

# M

M stars 231
M1 (Crab Nebula) 94, **266–67**, 356, *356*
M2 371, 463
M2-9 (Twin Jet Nebula) **253**
M3 *25*, 69, *346*
M4 268, **288**, 386
M5 364, *364*, 444, 450
M6 *see* Butterfly Cluster
M7 386, *386*, 445, *445*, 451, 457
M8 *see* Lagoon Nebula
M10 365, *365*, 450
M11 (Wild Duck Cluster) 366, *366*, 456, 457
M12 **289**, 365, 450
M13 53, 348, *348*, 444, 450
M14 **289**
M15 **289**, 370, *371*, 462, 463
M16 242, 364, 451, 457
M17 (Omega Nebula) *90*, *238*, 384, *385*
M20 (Trifid Nebula) **244**, 384, *384*, 451
M22 384, *384*, 451, 457
M23 384
M24 384, 451
M25 384
M27 (Dumbbell Nebula) 253, 368, *368*, 456, 457
M30 387, *387*
M31 *see* Andromeda Galaxy
M32 *294*, *303*, 318, 352, *352*
M33 (Triangulum Galaxy) *294*, **301**, *318*, 353, *353*, 469, 475
M34 354
M35 77, 358, *358*, 420
M36 343, 414, *414*
M37 343, 414, *414*
M38 343, 414, *414*
M39 *284*, 351, *351*, 462
M40 **273**
M41 376, 415, *415*, 421
M42 *see* Orion Nebula
M43 *239*, 375
M44 *see* Beehive Cluster
M46 393, *393*, 421
M47 393, 421
M48 378
M49 *296*, 319, 362
M50 377
M51 (Whirlpool Galaxy) *14*, *294*, **305**, 346, *346*, 438, 444
M52 **286**, 341, *341*, 468
M54 300
M57 (Ring Nebula) 253, 349, *349*, 456, 457
M59 *294*
M60 **307**
M63 (Sunflower Galaxy) 346, *346*
M64 (Black Eye Galaxy) **304**, 360, *360*
M65 361, *361*
M66 361, *361*
M67 359, *359*
M68 **289**, 378
M69 362
M71 366
M74 *35*, 372, *372*
M77 373, *373*, 475
M79 391, *391*
M81 (Bode's Galaxy) *26*, *69*, **304**, 344, *344*, 426, 432
M82 (Cigar Galaxy) *69*, 297, 304, 344
M83 (Southern Pinwheel) *294*, 378, *378*, 439, 445
M84 319, *319*, 362
M85 360
M86 319, *319*, 362
M87 *296*, *299*, **313**, 319, 362, *362*
M88 360
M89 *294*
M90 *35*
M92 348
M93 286, **286**, 393, *393*
M94 346
M95 361
M96 361
M97 *69*, 344, *344*
M99 360
M100 360
M101 (Pinwheel Galaxy) 306, 344, 438, 444
M102 (Spindle Galaxy) **307**, 380, *380*
M103 341, *341*
M104 (Sombrero Galaxy) **306**, 362, *362*
M105 361
M106 *310*
M107 **289**
M108 *69*
M109 *69*
M110 *294*, *303*, 318, 352, *352*
Maat Mons (Venus) 130, *130*, 134, *134*
MACHOs (massive compact halo objects) 27, 96 **269**
Maffei group, galaxy superclusters *324*
Maffei 1 319
Magellan, Ferdinand 300, 301, *301*
Magellan space probe *96*, 108, 130, 131
Magellanic Clouds *see* Large Magellanic Cloud (LMC); Small Magellanic Cloud (SMC)
Magellanic Stream 301
magma, plate tectonics 142
magnesium, on Earth 138
magnetic fields
  aurorae 70
  black holes 310
  Earth 139, *139*
  electromagnetic (EM) radiation 32
  Jupiter 177, *177*
  Mercury 125
  Milky Way 228
  Neptune 200
  neutron stars 263
  pulsars *263*
  Saturn 186
  stars 247
  Sun *10*, 122, *123*
  synchrotron mechanism *310*
  Uranus 196
magnetosphere
  Earth 139, *139*
  Jupiter 177, *177*
magnification
  binoculars 74, 75
  telescopes 77, *77*
magnitude *see* absolute magnitude; apparent magnitude
Maia 287
Main Belt, asteroids *119*, 208, *208*
main-sequence stars **246–49**
  Alpha (α) Centauri **248**
  Altair (Alpha (α) Aquilae) **248**
main-sequence stars *cont.*
  classification 231
  energy 246
  evolution 233, *233*, *234*
  Fomalhaut (Alpha (α) Piscis Austrini) **249**
  Hertzsprung–Russell (H–R) diagram 230, *230*, *247*
  magnetism 247
  Porrima (Gamma (γ) Virginis) **249**
  Proxima Centauri **248**
  Regor (Gamma (γ) Velorum) **249**
  Regulus (Alpha (α) Leonis) **249**
  rotation 247, *247*
  Sirius A (Alpha (α) Canis Majoris) **248**
  structure 246, *246*
  Vega (Alpha (α) Lyrae) **249**
  61 Cygni **248**
Malin 1 **309**
Manger Cluster *see* Beehive Cluster
mantle
  Earth 138, *138*
  Mars 160, *160*
  Mercury 125, *125*
  Moon 148, *148*
  Neptune 200, *200*
  Uranus 196, *196*
maps
  ancient astronomy 83
  mapping the sky **332–37**
  mapping the Universe 324
Maraghah 85
March sky guide **426–31**
Mare Crisium (Moon) **154**
Mare Imbrium (Moon) *151*
Mare Orientale Crater (Moon) **159**
Mare Serenitatis (Moon) 154, 156
Mare Tranquillitatis (Moon) **154**
maria, Moon 149
Mariner space probes
  Mars 101, 108, *108*, 109, 167
  Mercury 126, *126*
  Venus 101, 130
Marius, Simon 180–81, 182
Mars **160–75**
  asteroids 208, *208*
  atmosphere 161, *161*
  canyons *13*
  features formed by water **169–71**
  impact craters 162, **172–75**
  manned missions 111
  maps *162–63*, 163
  meteorites from *97*, 165, *165*, 222
  moons 163, *163*
  orbit and spin *118*, 160, *160*
  retrograde motion 64, *64*
  robot vehicles *97*
  search for life 53
  space probes 101, **108–10**, 162, 167
  structure 160, *160*
  surface features 162
  tectonic features 162, **164–68**
  water 163, *163*
Mars Exploration Rovers 110, *110*, 162, *162*
Mars Express 162, *162*, 163, 167, *167*
Mars Global Surveyor 110, 160, 162, *162*, 163, *163*, 167
Mars Pathfinder 110, *110*, 162
mass
  and energy 39, *39*
  fate of Universe 54–55
  galaxy clusters 317, 323, *323*
  gravitational lensing 317
  laws of gravity *36*
  and luminosity 231
  main-sequence stars 246, 247
  neutron stars 263
  nuclear reactions 230
  and space-time 40–41, *40–41*
  star evolution 233
  star formation 236
  stellar endpoints 262
  stellar structure *246*
  white dwarfs 262
massive stars, death of 262, *262*
Mathilde *111*, **210**
matter 24, **28–31**
  antimatter *31*, 311, *311*
  atoms 28, *28–29*
  Big Bang **46–49**
  Big Chill 55
  black holes 263
  chemical compounds 29, *29*
  chemical elements 29
  *see also* dark matter
  development of structures 50, *50*
  Einstein's theories 92
  forces 30, *30*
  particle physics 31
  states of matter 30
Mauna Kea Observatory, Hawaii *202*
Maximilian, Emperor 222
Maxwell, James Clerk 132
Maxwell Montes (Venus) 130, 131, **132**, 136
May sky guide **438–43**
Maya 83, *83*
Mayor, Michel 96
Mead Crater (Venus) **137**
measurements *see* distance
Mecca 85
Méchain, Pierre 69, 217
Medea, in mythology 355
medieval astronomy **86–87**
Medusa, in mythology 354, 370
Megaclite *179*
Mekbuda (Zeta (ζ) Geminorum) **282**, 358
MEL 25 *see* Hyades
Melas Chasma (Mars) 167
Melotte 20 354
Melotte 111 (Coma Star Cluster) 360, 432, 438
Menkar (Alpha (α) Ceti) 373
Mensa (the Table Mountain) **406**
  Alpha (α) Mensae 406
Menzel 3 (Ant Nebula) **255**
Merak (Beta (β) Ursae Majoris) *68*, 73, 344
Mercator, Gerardus 360
Mercury **124–27**
  atmosphere 125, *125*
  geography 126
  motion 64
  orbit and spin 41, 118, *118*, 124, *124*
  space probes 108, *108*
  structure 125, *125*
  surface features 126, *126*
  transits 65
Mercury programme 100, *100*, 102
meridian, celestial 59
Meridiani Planum (Mars) **171**, 174, 175
Merope 287, *287*
Merope, in mythology 357
mesas, on Mars *166*, *172*
mesons *31*, *48*
Mesopotamia, ancient astronomy 83, *83*
mesosphere, Earth's atmosphere *140*
Messier, Charles **69**
  catalogue 90
  Crab Nebula 267
  Eagle Nebula 242
  Pinwheel Galaxy 306
  Sombrero Galaxy 306
  Spindle Galaxy 307
  Whirlpool Galaxy 305
metallic (M-type) asteroids 208
Meteor Crater (Arizona) *221*, 222
meteor showers 220
  Delta Aquarid 451
  Eta Aquarid 371, 438, 439
  Geminid 358, 480, *480*
  Leonid 220, *220*, 361, *410–11*, 474
  Lyrid 349, 432
  Orionid 374, 468
  Perseid 216, 220, 456, *456*
  Quadrantid 347, 414
  Taurid 356, 474
meteorites 208, **220–23**
  in Antarctica 147, *221*
  impact craters *119*
  from Mars *97*, 165, *165*
  on Mars *171*
  Moon craters 149
  from Vesta 210, *210*
meteoroids 25, **220**
  *see also* meteor showers
meteors 71, **220**, *221*
methane
  atomic structure 29
  extra-solar planets 291
  Jupiter 178
  Neptune 200
  Pluto 204
  Saturn *187*
  Titan 194
  Uranus 196, 197, *197*
Methone *188*, 190
Methuselah 268
Metis *178*, **180**
Mexico, ancient astronomy *82*
the Mice (NGC 4676) **308**
Michell, John 94
Michigan, Lake (Earth) 146
micrometers *91*
Microscope *see* Microscopium
Microscopium (the Microscope) **387**
  Alpha (α) Microscopii 387
  AU Microscopii *290*
microwaves 32
  cosmic microwave background radiation (CMBR) 34, 49, 50, *96*, *322*, 325
  microwave observatories 34, *34*
Mid-Atlantic Ridge (Earth) **142**
Middle East
  ancient astronomy 82, 83
  early scientific astronomy 85, 86
midnight Sun *60–61*
Milk Dipper 384

Milky Way 26, **224–91**, *318*
  activity 311
  binocular astronomy *75*
  black hole *14*, 226
  Cygnus Rift 351, 456
  dark matter 229, 264
  galactic centre *34*, 229
  globular clusters 285, *285*
  halo 229, 269
  interstellar medium 228
  Local Group 318
  mapping *90*
  in monthly sky guides 414, 421, 456
  old stars 252
  Omega Centauri 288
  open star clusters 284, *285*
  size *22*
  sky guides 414, 421, 456, 480
  star clusters 286
  star formation 238
  stellar end points *262–63*
Miller, Stanley 52, *52*
Milton, John 225
Milvian Bridge, Battle of (AD 312) 65
Mimas *188*, 190, **191**
minerals, on Earth 138, 139
Mir space station 107, 111
Mira (Omicron (ο) Ceti) **281**, 373, 474, 475, *475*
Miralaidjii Corona (Venus) **135**
Miranda *197*, **198**
Mirphak (Alpha (α) Persei) *230*, 354, 480
mirrors, telescopes 76, 97
missiles 98, *98*, 99
Mizar (Zeta (ζ) Ursae Majoris) *68*, **272**, 344, *345*, 438
moldavite *221*
molecular clouds 238
  *see also* dark nebulae
  radio astronomy 95, *95*
  star formation 228, 232, *232*, 236
Molecular Ring, Milky Way *229*
molecules 29
monerans 141
Mongols 275
Monoceros (the Unicorn) **377**, 420
  Alpha (α) Monocerotis 377
  Beta (β) Monocerotis **277**, 377
  Cone Nebula **240**, 276, 377, *377*
  Red Rectangle Nebula **254**
  S Monocerotis 240, 377
  V838 Monocerotis **261**, *278–79*
  8 Monocerotis 377
  15 Monocerotis **276**
Montes Apenninus (Moon) **154**
Montes Cordillera (Moon) *159*
Montes Rook (Moon) *159*
monthly sky guides **410–85**
months, measuring 62
Moon *9*, **148–59**
  angular diameter 73
  Apollo missions **102–105**, 154
  astrology 60, 83
  atmosphere 149, *149*
  early scientific astronomy 84
  eclipses **63**
  far side *153*
  features **154–59**
  formation 149, *149*
  future missions 111
  Galileo's observations 88, *88*
Moon *cont.*
  gravity 36, 89
  Harvest Moon *462*
  history 149
  ice haloes 70, *70*
  impact craters 149, 151, 158–59
  influence on Earth 150, *150*
  maps 151
  meteorites from 222, 223, *223*
  movements across sky 59
  near side *152*
  occultations 65, *65*, *249*
  orbit and spin 37, 148, *148*
  phases 62, *62*
  size *22*
  space probes 101, *101*, 151, 153
  structure 148, *148*
  surface features 150, *150*
moon dogs 70, *70*
moons 25
  asteroids 211, *211*
  Jupiter *13*, *25*, 178, *178–79*, **180–85**
  Mars 163, *163*
  Neptune 201, *201*, **202–203**
  Pluto 204, *204*
  Saturn 188, *188–89*, **190–95**
  Uranus 197, **198–99**
Morecambe Bay (Earth) *150*
motion
  accelerating 40, *40*
  celestial sphere 58–59, *58–59*
  Newton's laws 36, *36*, 98
  planets **64–65**, 87
  retrograde 64, *64*
  stars 66
Mount Palomar, California 93
Mount Wilson, California 93
mountains
  Earth *143*, 144
  Moon 156
mountings, telescopes 76
moving clusters 344
moving lights, in sky 71
Mu (μ) Arae 291
Mu (μ) Boötis 347
Mu (μ) Cephei (Garnet Star) *230*, 241, *241*, *250*, **283**, 340, *340*
Mu (μ) Columbae 392
Mu (μ) Crucis 396
Mu (μ) Draconis 339
Mu (μ) Gruis 401, *401*
Mu (μ) Hydrae 378
Mu (μ) Librae 363
Mu (μ) Lupi 383
Mu (μ) Scorpii 386
multiple stars **270–77**
Mundilfari *189*
Mundrabilla meteorite **223**
Musca (the Fly) **397**
  BHR 71 **238**
  Hourglass Nebula **259**
  NGC 4833 **289**, 397
  Theta (θ) Muscae 397
MyCn18 (Hourglass Nebula) 241, 259
myths, Moon 150

# N

N44C *251*
Naiad *201*
naked-eye astronomy **72–73**
Nakhla meteorite **222**
names
  constellations 68
  stars 68, 330
Nanedi Vallis (Mars) **170**
Nansen, Fridtjof 173
Nansen Crater (Mars) **173**
Naos (Zeta (ζ) Puppis) 393
NASA 100
  Deep Space 1 mission 219
  Discovery programme 212
  Lunar Orbiters 151, *151*
  Moon landings **102–105**
  Skylab space station 106, *106*
  space probes 101, **108–10**
  Space Shuttle 107, *107*
  Stardust mission 111, 219
navigation, Pole Star 83, 275
Nazis 98
Near Earth Asteroid Rendezvous (NEAR) space probe 111, 212
Nebra Disc 287, *287*
Nebuchadnezzar, King of Babylon 150
nebulae 24
  BHR 71 **238**
  bubble nebulae *260*
  Carina Nebula **245**
  catalogues 69
  Cone Nebula **240**
  dark nebulae 24, 228, 238
  DR 6 **241**
  DR 21 **244**
  Eagle Nebula **242–43**
  emission nebulae 24, *33*, 228, 238
  Horsehead Nebula **238**
  IC 1396 **241**
  IC 2944 **244**
  Lagoon Nebula **241**
  Orion Nebula **239**
  *see also* planetary nebulae
  RCW 49 **245**
  reflection nebulae 237
  spectroscopy 33
  star-forming nebulae 25, 236, **238–45**
  Trifid Nebula **244**
neon, in Moon's atmosphere *149*
Neptune **200–203**
  atmosphere and weather 200, *200*, 201, *201*
  discovery 90
  and Kuiper Belt 206, *206*
  moons 201, *201*, **202–203**
  orbit and spin *119*, 200, *200*
  and Pluto 204
  rings 201, *201*
  space probes 109
  structure 200, *200*
Nereid 201, *201*, **202**
Nereidum Montes (Mars) 173
Net *see* Reticulum
neutrinos 28, *30*, *31*, 120
  after Big Bang 48, *48*, 50
  Big Chill 55
  detectors 27, *27*, 95
neutron stars 25, 233, **263**
  formation 94, 234, *234*, *235*
  gamma-ray astronomy *35*
  Geminga Pulsar **264**
  PSR B1620-26 **268**
  RX J1856.5-3754 **264**
  space-time *41*
neutrons 28, *28–29*
  after Big Bang 48, *48*
  forces 30, *30*
New General Catalogue *see* NGC
New Horizons space probe 111
Newgrange 83
Newton, Isaac **36**, *89*
  Great Comet of 1680 216
  law of universal gravitation 36, 64, 89, 98
  laws of motion 36, *36*, 98, 124
  telescope 89, *89*
Newtonian telescopes *76*
Newtonian universe, space and time 38
NGC (New General Catalogue) 26, 69, 237
NGC 55 319, 388
NGC 104 (47 Tucanae) **288**, 301, 402, *402*, *463*, 469, 475
NGC 224 *see* Andromeda Galaxy
NGC 253 319, *319*, 388
NGC 281 *236*
NGC 288 388
NGC 292 *see* Small Magellanic Cloud (SMC)
NGC 362 402, 469
NGC 383 *310*
NGC 457 341, 474
NGC 598 (Triangulum Galaxy) *294*, **301**, *318*, 353, *353*, 469, 475
NGC 604 301, *301*
NGC 660 *294*
NGC 663 341, 474
NGC 752 352
NGC 869 354, *354*, 474, 480
NGC 884 77, 354, *354*, 474, 480
NGC 1261 403, *403*
NGC 1275 **314**
NGC 1300 *294*, 390, *390*
NGC 1316 *299*, 319, 389
NGC 1365 *319*, 389, *389*
NGC 1399 *319*, *389*
NGC 1427A *237*
NGC 1435 *see* Pleiades
NGC 1502 342, *342*
NGC 1530 26, *26*
NGC 1851 392
NGC 1952 (Crab Nebula) 94, **266–67**, 356, *356*
NGC 1976 *see* Orion Nebula
NGC 1977 (Ghost Nebula) *79*, 375
NGC 1981 375
NGC 2017 391, *391*
NGC 2070 *see* Tarantula Nebula
NGC 2158 *358*
NGC 2207 *299*
NGC 2232 377
NGC 2244 377, *377*, 420
NGC 2264 (Cone Nebula) **240**, 276, 377, *377*
NGC 2264 IRS 240
NGC 2266 *250*, *285*
NGC 2359 260
NGC 2362 376, *376*
NGC 2392 (Eskimo Nebula) **255**, 358, *358*
NGC 2403 342
NGC 2440 nucleus (HD 62166) **264**
NGC 2447 286
NGC 2451 393, 421
NGC 2467 *236*
NGC 2477 393, *393*, 421
NGC 2516 395, 421
NGC 2547 394
NGC 2736 (Vela Supernova) **265**, 394
NGC 2755 *294*
NGC 2787 *296*
NGC 2841 *69*
NGC 2997 380, *380*
NGC 3031 (Bode's Galaxy) **304**, 344, *344*
NGC 3034 (Cigar Galaxy) *69*, 297, **304**, 344
NGC 3079 *69*
NGC 3114 395
NGC 3115 (Spindle Galaxy) **307**, 380, *380*
NGC 3132 (Eight-Burst Nebula) *250–51*, 394, 427
NGC 3195 407
NGC 3201 **288**
NGC 3242 (Ghost of Jupiter) 378, *378*
NGC 3309 *320*
NGC 3311 320, *320*
NGC 3312 *320*
NGC 3314 *320*
NGC 3370 *299*
NGC 3372 *see* Carina Nebula
NGC 3532 395, *395*
NGC 3603 *232*
NGC 3628 *361*
NGC 3766 382
NGC 3918 (Blue Planetary) 382
NGC 4038 and 4039 (Antennae Galaxies) *35*, **307**, 308, 381, *381*
NGC 4261 **313**
NGC 4414 *295*
NGC 4438 *26*
NGC 4449 *297*
NGC 4486 313
NGC 4526 *279*
NGC 4565 360, *360*
NGC 4590 289
NGC 4594 (Sombrero Galaxy) **306**, 362, *362*
NGC 4622 *294*
NGC 4649 307
NGC 4650A *297*
NGC 4676 (the Mice) 308
NGC 4755 (Jewel Box) **288**, 396, *396*, 433, 439, 445
NGC 4826 (Black Eye Galaxy) **304**, *360*
NGC 4833 **289**, 397
NGC 4881 *320*
NGC 4889 *316*, 320
NGC 5128 *see* Centaurus A
NGC 5139 *see* Omega Centauri
NGC 5194 and NGC 5195 (Whirlpool Galaxy) *14*, *294*, **305**, 346, *346*, 438, 444
NGC 5457 (Pinwheel Galaxy) **306**, 344, 438, 444
NGC 5460 382
NGC 5548 **313**
NGC 5822 383, *383*
NGC 5866 307
NGC 6025 398
NGC 6027 and NGC 6027A-C (Seyfert's Sextet) **319**
NGC 6041A 321
NGC 6050 321
NGC 6087 398, *398*
NGC 6121 288
NGC 6128 289
NGC 6171 289
NGC 6193 399
NGC 6231 386, 445
NGC 6302 (Bug Nebula) **256–57**
NGC 6397 *285*, 399, *399*
NGC 6402 289
NGC 6405 *see* Butterfly Cluster

NGC 6514 244, *244*
NGC 6523 241
NGC 6530 241, 384, 451
NGC 6541 399
NGC 6543 (Cat's Eye Nebula) **254**, 339, *339*
NGC 6633 365, 450
NGC 6744 408, *408*
NGC 6751 *25*, *251*
NGC 6752 408, *408*
NGC 6782 **308**
NGC 6822 (Barnard's Galaxy) *318*
NGC 6826 351
NGC 6888 (Crescent Nebula) **255**
NGC 6960/95 (Cygnus Loop) *228*, **265**
NGC 6992 351
NGC 7000 (North America Nebula) 351, *351*, 462
NGC 7009 (Saturn Nebula) *251*, 371, *371*, 463
NGC 7078 289
NGC 7293 (Helix Nebula) **253**, 371, *371*, 463
NGC 7320 320, *320*
NGC 7479 *294*
NGC 7654 286
NGC 7662 (Blue Snowball) 352, *352*, 468
NGC 7742 (Fried Egg Galaxy) **313**
Niagara Falls (Earth) *146*
Nicholson Regio (Ganymede) *184*
nickel
  Earth 138
  meteorites 222, 223
Nile, River 82
nitrogen
  Bug Nebula *256*
  Earth's atmosphere *140*
  main-sequence stars 246
  meteorites 223
  planetary nebulae 251
  Pluto 204, *204*
  Titan 194
  Wolf–Rayet stars 251
Nixon, Richard 104
Noah 392
noctilucent clouds 71, *71*, *444*
Noctis Labyrinthus (Mars) 166
Norma (the Set Square) **398**, 445
  Ant Nebula **255**
  Epsilon (ε) Normae 398
  Gamma-1 ($\gamma^1$) Normae 398
  Gamma-2 ($\gamma^2$) Normae 398
  Iota (ι) Normae 398
Norse mythology, Polaris 275
North America Nebula (NGC 7000) 351, *351*, 462
north celestial pole *58*
North Polar Region (Mars) **169**
North Polar sky *332*
Northern Coalsack 351
Northern Cross *see* Cygnus
Northern Crown *see* Corona Borealis
Northern Lights *123*
novae **278**
  binocular astronomy 74
  Nova Cygni 1992 *278*, **283**
  RS Ophiuchi **283**
  T Coronae Borealis (Blaze Star) **282**
November sky guide **474–79**
Nu (ν) Boötis 347
Nu (ν) Coronae Borealis 363
Nu (ν) Draconis 339
Nu (ν) Scorpii 386
nuclear fission 31
nuclear fusion 31, *31*
  discovery 92
  inside stars 95, 230, 232, *232*
  main-sequence stars 246
  protostars 237
  star formation 236
  Sun 120
nucleons *30*
Nucleosynthesis Era *48*
nucleus, atom 28, *29*
  after Big Bang 48
  forces 30, *30*

# O

O stars 231, 251
O3 stars 245
OB stars 239
Oberon *197*, **199**
Oberth, Hermann *98*
observable Universe 23
observatories 88, *88*
  early scientific astronomy 85, *85*
  *see also* telescopes
occultation 65, *249*
oceans (Earth) 141, *141*
  tides 150, *150*
Octans (the Octant) **409**
  Delta (δ) Octantis 409
  Sigma (σ) Octantis 409
Octant *see* Octans
October sky guide **468–73**
Odysseus Crater (Tethys) 192
OH231.8+4.2 (Calabash Nebula) 258
oil reserves, on Earth *147*
Okmok volcano (Earth) *143*
Olbers, Heinrich 211
Olber's paradox *49*
old stars **250–61**
  globular clusters 285
Olympus Mons (Mars) 162, *162*, 164, *164*, **165**
OMC-1 239
Omega Centauri (NGC 5139) *284*, *286*, **288**, 382
  binocular astronomy 75
  density *285*
  in monthly sky guides 402, 433, 439, 445
Omega Nebula (M17) *90*, *238*, 384, *385*
Omega (ω) Scorpii 386
Omicron (ο) Ceti (Mira) **281**, 373, 474, 475, *475*
Omicron (ο) Cygni 350
Omicron (ο) Draconis 339
Omicron (ο) Eridani **272**, 390, *390*
Omicron (ο) Velorum *394*
Ontario, Lake (Earth) 146
Oort, Jan Hendrik **207**
Oort Cloud **207**
Opaque Era *49*
open clusters **284**, 286
  Beehive Cluster **286**
  Butterfly Cluster **286**
  evolution 285, *285*
  Hyades **286**
  Jewel Box (Kappa (κ) Crucis) **288**
  M52 **286**
  M93 **286**
  Pleiades **287**
open universe 55, *55*
Ophelia *197*, **198**
Ophir Chasma (Mars) 167, *167*
Ophiuchus (the Serpent Holder) 364, **365**
  Barnard's Star 66, 365
  Cygnus Rift 351, 456
  Kepler's Star *35*, **269**, 365
  M12 **289**, 365, 450
  M14 **289**
  M107 **289**
  in monthly sky guides 450, 456
  Rho (ρ) Ophiuchi *290*, 365, *365*
  RS Ophiuchi **283**
  Twin Jet Nebula **253**
  Zeta (ζ) Ophiuchi 264
  36 Ophiuchi 365
  70 Ophiuchi 365
Opik, Ernst 207
Opportunity rover, on Mars 174–75
opposition, planets *64*
optical telescopes 35, *35*
orange stars
  Alpha (α) Centauri (Rigil Kentaurus) **248**
  61 Cygni **248**
orange-red stars
  Proxima Centauri **248**
orbits
  asteroids 208, *208–209*
  comets 214, *214*
  Earth-centred cosmos 85, *85*
  elliptical galaxies *296*
  globular clusters 285
  Jupiter 176, *176*
  Kuiper Belt objects 206, *206*
  Lagrange points 192
  laws of planetary motion 87
  Mars 160, *160*
  Mercury 41, 124, *124*
  Moon 148
  multiple stars 270, *270*
  Neptune 200, *200*
  Pluto 204, *204*
  Saturn 186, *186*
  Sedna *207*
  shapes of 37, *37*, 87
  Solar System **118–19**
  space-time 40–41, *40–41*
  spiral galaxies *294*
  synchronous rotation *148*
  Uranus 196, *196*
  Venus 128, *128*
Orion (the Hunter) 68, **374–75**
  Alnilam (Epsilon (ε) Orionis) *230*
  Alnitak (Zeta (ζ) Orionis) *230*, 374, 375, *375*
  Bellatrix (Gamma (γ) Orionis) *67*
  Delta (δ) Orionis 374
  Ghost Nebula *79*
  Horsehead Nebula **238**, 239, 375, *375*
  Iota (ι) Orionis 375
  mythology 374, 386
  in monthly sky guides 415, 420, 421, 426, 427, 480, 481
  naked-eye astronomy 73, *73*
  Orion's belt 68, 374, *415*
  Sigma (σ) Orionis 238, **277**, *374*, 375
  star colours *66–67*
  *see also* Betelgeuse (Alpha (α) Orionis); Rigel (Beta (β) Orionis)
Orion (the Hunter) *cont.*
  Trapezium (Theta (θ) Orionis) 239, *239*, *271*, *272*, **277**, 375, *375*
  42 Orionis 375
  45 Orionis 375
Orion Arm, Milky Way *227*, 229, *229*
Orion Nebula (M42, NGC 1976) *14–15*, **239**, 300, 374, 375, *375*
  binocular astronomy 75, *75*
  bow shock *20–21*
  in monthly sky guides 414, 415, *415*
  Theta (θ) Orionis *277*
  young stars *51*
Orionid meteor shower 374, 468
Orpheus, mythology 349
Orthosie *179*
Ovda Regio (Venus) **135**
Overwhelmingly Large Telescope (OWL) 35
Owl Nebula 344, *344*
oxygen
  Earth's atmosphere 140, *140*
  extra-solar planets 291
  formation 29, 51, 95
  main-sequence stars 246
  Mercury 125, *125*
  meteorites 223
  planetary nebulae 251
  silicates 24
  supergiants 250
  Type I supernovae 279
  Wolf–Rayet stars 251
Ozza Mons (Venus) 134

# P

Paaliaq *189*
Pacific Ocean (Earth) *8*, 83
Pacific Ring of Fire (Earth) **143**
Painter's Easel *see* Pictor
Palermo Circle 211, *211*
Palisa, Johann 211
Pallas 211
Pallene *188*, **190**
Palus Putredinis (Moon) 151
Pan (Amalthea) *180*
Pan (Saturn's moon) *188*, 387
Pandora *188*, *189*
Papin, Denis 380
parabolas, orbits 37
parallax shift 66, *66*
parhelia 70
Paris Observatory 260
Parsons, William 266, 305, *305*, 346, 356
partial eclipses 63
particle-like behaviour, electromagnetic (EM) radiation 32
particles
  in atoms 28
  aurorae 70
  Big Bang 46–49
  Big Chill 55
  cosmic rays 24, 228
  dark matter 27, 28
  electromagnetic (EM) force 30, *30*
  force-carrier particles 30, *30*
  and magnetic fields 247
  matter 28
  neutrinos 27, *27*
  particle accelerators 31, *31*, 47
particles *cont.*
  particle physics 31
  quantum mechanics 41
  radiation 32
  radioactive decay 32
  solar wind 123, 139
  states of matter 30
  string theory 31
  Sun 122
Pascal, Blaise 158
Pascal Crater (Moon) **158**
Pasiphae *179*
Pasithee *179*
Pavo (the Peacock) **408**, 463
  Alpha (α) Pavonis 408
  Kappa (κ) Pavonis 408
  NGC 6782 **308**
  Xi (ξ) Pavonis 408
Pavonis Mons (Mars) **164**
Payne-Gaposchkin, Cecilia **231**
Peacock *see* Pavo
peculiar (Pec) galaxies 297
Pegasus (the Winged Horse) **370**
  Alpha (α) Pegasi 370
  Beta (β) Pegasi 370
  Enif (Epsilon (ε) Pegasi) 370, 462
  Fried Egg Galaxy **313**
  Gamma (γ) Pegasi 370
  M15 **289**, 370, *371*, 462, 463
  *see also* Square of Pegasus
  Stephan's Quintet **320**
  Upsilon (υ) Pegasi 370
  51 Pegasi 96, 370
Pegasus, in mythology 370
Pele (Io) *183*
Pellepoix, Antoine Darquier de 253
Pendulum Clock *see* Horologium
penumbral eclipses 63
Penzias, Arno *49*
Perseid meteor shower 216, *216*, 220, 456, *456*
Perseus (the Victorious Hero) **354**
  Algol (Beta (β) Persei) **272**, 354, *354*, 480
  Mirphak (Alpha (α) Persei) *230*, 354, 480
  in monthly sky guides 414, 420, 421, 474, *474*, 480, 481
  NGC 1275 **314**
  Rho (ρ) Persei 354
Perseus, in mythology 352, 354
Pettifor, Arthur 222
PG 0052+251 *310*
PGC 54559 (Hoag's Object) 309
PGC 54876 321
Phad (Gamma (γ) Ursae Majoris) *68*, 344
Phaethon, in mythology 390
phases
  Moon 62, *62*
  planets 64
Phi (φ) Cassiopeiae 341
Philae lander 219
Phobos 163, *163*
Phoebe *189*, **195**
Phoenix **401**, 463
  Zeta (ζ) Phoenicis 401
photoelectric effect 32, *32*
photo-evaporation 242
photography 79, *79*, 91
photons *28*
  absorption lines 33
  after Big Bang 48, *48–49*, 50

photons *cont.*
Big Chill 55
charge-coupled devices (CCDs) 79
electromagnetic force *30*
energy 32
radiation 246
Sunyaev–Zel'dovich effect *322*
photosphere
stars 246, *246*
Sun 120, 122, *122*, 123
physics
astrophysics 90
gravity **36–37**
laws of motion 36, *36*
laws of planetary motion 64, 87
matter **28–31**
radiation **32–35**
space and time **38–41**
Pi (π) Aquarii 371
Pi (π) Arietis 355
Pi (π) Herculis 348
Pi (π) Hydri 403
Pi (π) Lupi 383
Piazzi, Giuseppe 369
Ceres 90, 209, 211
Pictor (the Painter's Easel) **404**
Beta (β) Pictoris *290*, 404, *404*
Iota (ι) Pictoris 404
Pillars of Creation 242, *242–43*
Pinwheel Galaxy (M101, NGC 5457) **306**, 344, 438, 444
Pioneer space probes 101, 108, *108*, *109*, 121, 130
pions *30*, *48*
Pisces (the Fishes) 68, **372**
Alrescha (Alpha (α) Piscium) 372, *372*
Eta (η) Piscium 372
in monthly sky guides 469, *469*, 474, 480, 481
PKS 2349 **315**
Psi-1 (ψ¹) Piscium 372
TX Piscium 372, *372*
Zeta (ζ) Piscium 372
Piscis Austrinus (the Southern Fish) **388**
Beta (β) Piscis Austrini 388
*see also* Fomalhaut (Alpha (α) Piscis Austrini)
Gamma (γ) Piscis Austrini 388
Pistol Nebula 261
Pistol Star **261**
Pius Institute, Pope 281
PKS 2349 **315**
Plancius, Petrus 330, **342**, 400
Camelopardalis 342
Columba 392
Monoceros 377
Planck era 46
Planet X 205
planetarium software 78, *78*
planetary nebulae 25, **251**, 252
Ant Nebula **255**
Bug Nebula **256–57**
Calabash Nebula **258**
Cat's Eye Nebula **254**
Crescent Nebula **255**
Egg Nebula **254**
Eskimo Nebula **255**
formation *233–35*, 234
Gomez's Hamburger Nebula **258**
Helix Nebula **253**
Hourglass Nebula **259**
planetary nebulae *cont.*
NGC 6751 *25*
NGC 7662 352, *352*, 468
Red Rectangle Nebula **254**
Ring Nebula **253**
Stingray Nebula **260**
Twin Jet Nebula **253**
planetesimals 117, *117*, 233
planets *12–13*, 25, **124–205**
ancient astronomy *83*
astrology 60
conjunction *64*, 65, *65*
Copernican revolution **86–89**
Earth **138–47**
Earth-centred cosmos 85, *85*
extra-solar planets 96, **290–91**
formation 116, *117*, 233, *233*
gas giants 119
gravity 89
Jupiter **176–85**
Mars **160–75**
Mercury **124–27**
moons 25
motion 59, **64–65**, 64, 87
Neptune **200–203**
orbits **118–19**
protoplanetary discs 25
rocky planets 119
rotation 37
Saturn **186–95**
search for life *53*
searching for 205
Solar System 25
space probes **108–11**
space-time 40–41, *40–41*
transits 65, *65*
Uranus **196–99**
Venus **128–37**
zodiac 65
planispheres 72, *72*
plants 141
Planum Australe (Mars) **171**
Planum Boreum (Mars) **169**
plasma *30*
magnetic fields 247
recreating Big Bang 47
states of matter 30
Sun 122, 123, *123*
plasma balls *30*
plate tectonics, on Earth 140, *140*
Plato 84
Pleiades (NGC 1435) **287**, 356, *356*
Alcyone (Eta (η) Tauri) **273**
Aldebaran 252
binocular astronomy *75*
"missing" Pleiad 357
in monthly sky guides 475, 480, 481
Pleione 287, 356
Plough **344–45**
changing shape *66*
in monthly sky guides 414, 420, 426, 432, *432*, 433, 438, 450, 468
naked-eye astronomy 73, *73*
naming stars *68*
pattern 68, *68*
Pluto **204**
atmosphere 204
formation of Solar System 117
and Kuiper Belt *206*
moons 204, *205*
and Neptune 204
orbit and spin 37, 118, *119*, 204, *204*
Pluto *cont.*
space probes *111*
structure 204, *204*
Pointers *248*
polar ring galaxies *297*
polar sky charts *332–33*
Polaris (Alpha (α) Ursae Minoris) **274–75**, 338, *338*, 344
circumpolar stars 332
Hertzsprung–Russell (H–R) diagram *230*
in monthly sky guides 432, *438*
naked-eye astronomy 73, *73*
Pole Star
navigation 83, *83*
*see also* Polaris (Alpha (α) Ursae Minoris)
Vega as 249
poles
magnetic poles 139
celestial poles *58*, *421*
Poliakov, Valeri 111, *111*
pollution, light *72*, *78*
Pollux (Beta (β) Geminorum) 350
Hertzsprung–Russell (H–R) diagram *230*
in monthly sky guides 420, 421, 427
Polydeuces *188*, 193
Polynesians, navigation 83
Pons, Jean Louis 217
Pope, Alexander 198
Population I stars 227
Population II stars 227, 285
populations, stars 227
Porrima (Gamma (γ) Virginis) **249**, 362
Portia *197*
positrons *31*
Big Bang 47
Big Chill 55
emergence of *48*
Milky Way 311, *311*
potassium, on Mercury 125
Praesepe 286, 359, 420
Praxidike *179*
precession 60, *60*, 138
pressure
Earth's atmosphere 140
star formation *232*
stars 230, *230*
pressure stripping, galaxy collisions 299
Principle of Equivalence 40, *40*
Principle of Relativity 38
prisms, analysing light 33, *33*
Procyon (Alpha (α) Canis Minoris) **280**, 376
classification 231
Hertzsprung–Russell (H–R) diagram *230*
in monthly sky guides 420, 481
naked-eye astronomy *73*
Winter Triangle 420, *420*, 480
Procyon B *230*
Project Gemini 102, *102*
Promethei Terra (Mars) 171
Prometheus (Io) *182*
Prometheus (Saturn's moon) *188*, *189*, **190**
Prometheus space probe 110
prominences, Sun *10*, *122*
propane, on Jupiter 178
proper motion, stars 66
Propus (Eta (η) Geminorum) **280**, 358
Prospero *197*
Proteus *201*, **202**
protists 141
protons 28
after Big Bang 48
Big Chill 55
in chemical elements 29
forces 30, *30*
proton–proton chain reaction (pp chain) *31*, 246
protoplanetary discs 25
protoplanets 117
protostars
brown dwarfs *232*
evolution to main-sequence stars 230
formation 232, *232–33*, 236, 237
protosun *117*
Proxima Centauri *22*, *230*, **248**, 382
Psi (ψ) Draconis 339
Psi-1 (ψ¹) Piscium 372
PSR B1620-26 **268**
PSR 0531 +21 267, *267*
PSR 1257 291
Ptolemy 57, **85**
astrology 83
Cetus 373
Corona Australis 399
Delphinus 369
Earth-centred cosmos 59, 85, 86, 87
Equuleus 369
Piscis Austrinus 388
star catalogue 85, 330, 405
Puck *197*, **198**
Pulcherrima (Epsilon (ε) Boötis) *see* Izar
pulsars 94, 263, *263*
binary systems 270
extra-solar planets 291
Geminga Pulsar **264**
PSR B1620-26 **268**
PSR 0531 +21 267, *267*
rotation 37
Vela Pulsar 265, *265*
pulsating variable stars 278
Delta (δ) Cephei **282**
Eta (η) Aquilae **282**
Mira (Omicron (ο) Ceti) **281**
Mu (μ) Cephei (Garnet Star) **283**
RR Lyrae **282**
W Virginis **282**
Zeta (ζ) Geminorum (Mekbuda) **282**
Puppis (the Stern) **393**, 420, 421
b Puppis *393*
Calabash Nebula **258**
e Puppis 393
k Puppis 393
L Puppis 393
L² Puppis 393
M93 **286**
Naos (Zeta (ζ) Puppis) 393
NGC 2440 nucleus **264**
Xi (ξ) Puppis 393
Purcell, Edward 95
Pwyll Crater (Europa) *181*
pyramids, astronomical orientation 82, *82*, 83, *83*
Pythagoras 84
Pyxis (the Compass) **392**
T Pyxidis 392

# Q

*Quadrans Muralis* 347, 414
Quadrantid meteor shower 347, 414
quadruple stars 270
Alcor (80 Ursae Majoris) **272**
Alcyone (Eta (η) Tauri) **273**
Algol (Beta (β) Persei) **272**
Almach (Gamma (γ) Andromedae) **273**
Epsilon (ε) Lyrae **272**
Mizar (Zeta (ζ) Ursae Majoris) **272**
Trapezium (Theta (θ) Orionis) **277**
quanta 32
quantum mechanics 41
Quaoar *206*
Quark era *46–47*
quarks 28, *29*, *31*
Big Bang *46–48*, 47, 48
forces *30*
recreating Big Bang 47
quasars 310, *310*
BL Lac objects 353
discovery 94, *94*
distribution 311
Lyman Alpha (α) lines 325, *325*
PKS 2349 **315**
superluminal jets 311
3C 48 **315**
3C 273 **315**, 362
Queloz, Didier 96
quintuple stars
Sigma (σ) Orionis **277**

# R

R Coronae Borealis 279, **283**, 363
R Leporis 391
R Scuti 366
Rabinowitz, David 207
radar, space probes *96*
radial velocity, stars 66
radiation
Big Bang 22
black holes and 263
cosmic background microwave radiation (CMBR) 34, 49, 50, *96*, *322*, 325
*see also* electromagnetic (EM) radiation
main-sequence stars 246
red shift and blue shift 33, *33*
radiation belts
Jupiter 177
Van Allen radiation belts (Earth) 139
radio astronomy 93, *93*
telescopes 34, *34*, *97*
radio galaxies 310, *310*
Centaurus A **312**
Cygnus A **314**
distribution 311
M87 **313**
NGC 1275 **314**
NGC 4261 **313**
Radio Lobe, Milky Way *229*
radio pulsars 94
radio waves 32
interstellar medium 95
Milky Way 229, *229*
radio window *34*
radioactive decay 30, *30*, 32
radioactivity 31, 92

Ram *see* Aries
Ramsden, Jesse 211
random walk, radiation 246
Ranger space probes *101*, 103, 153, 155
Ras Algethi (Alpha (α) Herculis) **281**, 348
ray craters, Moon *151*
Rayet, Georges 251, 260, *260*
RCW 49 **245**
red dot finders 77
red dwarfs 25
evolution *233*
Gliese 229 *25*
red giants 25, **250**
Aldebaran **252**
Hertzsprung–Russell (H–R) diagram 230, *230*, *251*
multiple stars 270
planetary nebulae 251
star life cycles *233–35*, 234
TT Cygni **252**
red light, photoelectric effect 32, *32*
Red Rectangle Nebula (HD 44179) **254**
red shift 33, *33*
cosmological red shift 33
expanding space 42, 323
Lyman Alpha forest 325, *325*
mapping the Universe 324
red sprites 71
red supergiant stars 250
Antares (Alpha (α) Scorpii) **252**
Betelgeuse (Alpha (α) Orionis) **252**
evolution *233*
V838 Monocerotis **261**
Redstone rockets 100
reflecting telescopes 76, *76*, 89, *89*, *91*, 93
reflection nebulae 228, 237
reflex finders 77
refracting telescopes 76, *76*, 91
refraction phenomena 70, *70*
Regor (Gamma (γ) Velorum) 231, **249**, 394
regular clusters
Abell 1689 **321**
Abell 2065 (Corona Borealis Cluster) **321**
Abell 2218 **322–23**
Coma Cluster **320**
Fornax Cluster **319**
Hydra Cluster **320**
Regulus (Alpha (α) Leonis) **249**, 361
Hertzsprung–Russell (H–R) diagram *230*
in monthly sky guides 427
naked-eye astronomy *73*
name 68
relativity
general theory of relativity **40–41**, 49
special theory of relativity **38–39**
Renoir region (Mercury) 127
replication, and life 52
residual strong nuclear force 30, *30*
Reticulum (The Net) **404**
Zeta (ζ) Reticuli 404
retrograde motion 64, *64*
Reull Vallis (Mars) *169*, **171**
Rhea *188*, **193**
Rhea Mons (Venus) 133
Rho (ρ) Cassiopeiae 341
Rho (ρ) Herculis 348
Rho (ρ) Ophiuchi *290*, 365, *365*
Rho (ρ) Persei 354
rifting, plate tectonics 142
Rigel (Beta (β) Orionis) **277**, 374
classification 231, *231*
Hertzsprung–Russell (H–R) diagram *230*
in monthly sky guides 415
Right Ascension 59, *59*, 73
Rigil Kentaurus *see* Alpha (α) Centauri
Riley, Margaretta 136
Riley Crater (Venus) **136**
ring galaxies
Hoag's Object **309**
Ring Nebula (M57) **253**, 349, *349*, 456, 457
Ring of Fire (Earth) **143**
rings
gravity *36–37*
Jupiter 179, *179*
Neptune 201, *201*
Saturn *13*, *36–37*, 186, 189, *189*
Uranus 197, *197*
River *see* Eridanus
rivers, on Earth 146, *146*
robotic space exploration *97*, 101
rockets **98–100**
Apollo missions *102*, 103, *103*
rocks
Earth 138
Mars 175, *175*
Moon 150, *150*, 156, *157*
rocky planets 119
Roman Catholic church 87
Romans, constellations 330
Romulus and Remus Crater (Dione) *193*
Rosalind *197*
Rosetta space probe 111, *111*, 219
Rosette Nebula 377, *377*, 420
Rotanev (Beta (β) Delphini) 369
rotating variable stars
Procyon (Alpha (α) Canis Minoris) **280**
rotation
angular momentum 37, *37*
neutron stars 263, *263*
spiral galaxies 294
stars 247
synchronous rotation *148*
Rotten Egg Nebula 258
Royal Greenwich Observatory 88, *88*
Royal Society 88
Royal Stars 252
RR Lyrae **282**
RS Ophiuchi **283**
Ruapehu, Mount (Earth) *143*
Rupes Altai (Moon) **155**
Russell, Henry 230
RX J1856.5-3754 **264**

# S

S Monocerotis 240, 377
S Sagittae 366
S1986/U10 *197*
S/2000 J11 *178*
S2001/U2 *197*
S2001/U3 *197*
S/2002 N1 *201*, **202**
S2002 N2 *201*
S2002 N3 *201*
S2002 N4 *201*
S/2003 J1 *178*, *179*
S/2003 J2 *179*
S/2003 J3 *179*
S/2003 J4 *179*
S/2003 J5 *179*
S/2003 J6 *179*
S/2003 J7 *179*
S/2003 J8 *179*
S/2003 J9 *178*
S/2003 J10 *179*
S/2003 J11 *178*
S/2003 J12 *179*
S/2003 J13 *179*
S/2003 J14 *179*
S/2003 J15 *179*
S/2003 J16 *179*
S/2003 J17 *179*
S/2003 J18 *179*
S/2003 J19 *178*
S/2003 J20 *179*
S/2003 J21 *179*
S/2003 S1 *189*
S2003/U1 *197*
S2003/U2 *197*
S2003/U3 *197*
Sacajawea Patera (Venus) **133**
Sachs Patera (Venus) **133**
Sagan, Carl 109, *109*
SagDEG (Sagittarius Dwarf Elliptical Galaxy) **300**, 318
Sagitta (the Arrow) **366**
S Sagittae 366
WZ Sagittae 366
Zeta (ζ) Sagittae 366
Sagittarius (the Archer) **384–85**
Beta (β) Sagittarii 384
Epsilon (ε) Sagittarii 384
Gomez's Hamburger Nebula **258**
Lagoon Nebula **241**
MACHO 96 **269**
in monthly sky guides 439, 445, 450, 451, 456, 457
Pistol Star **261**
Teapot 384, 451, *457*
Trifid Nebula **244**
WR 104 **255**
WR 124 **260**
9 Sagittarii 384
15 Sagittae *94*
Sagittarius A 229, 384
Sagittarius A* 229, *229*, 384, *451*
Sagittarius A West 229, *229*
Sagittarius Arm, Milky Way *227*
Sails *see* Vela
salts 29, *29*
Salyut space stations 106, *106*
Samarkand 85, *85*
Sandage, Allan **315**
Sapas Mons (Venus) **134**
Saskia Crater (Venus) **137**
satellites 106, *106*
astronomy from space 96
observing 71
Space Race 99–100, *99*
Saturn **186–95**
atmosphere 187, *187*
formation of Solar System *117*
moons 188, *188–89*, **190–95**
orbit and spin *119*, 186, *186*
rings *13*, *36–37*, 186, 189, *189*
space probes 109, *109*, 110, 194, *194*
Saturn *cont.*
structure 186, *186*
weather 188, *188*
Saturn Nebula (NGC 7009) *251*, 371, *371*, 463
Saturn V rockets *102*, 103, *103*, 105
Scales *see* Libra
Schiaparelli, Giovanni 172, 216, **220**
Schiaparelli Crater (Mars) **172**
Schmidt–Cassegrain telescopes *76*, *78*
Schmitt, Harrison "Jack" 156, *156*
Schwarzschild, Martin 95
Scooter (Neptune) *201*
Scorpion *see* Scorpius
Scorpius (the Scorpion) **386**
*see also* Antares (Alpha (α) Scorpii)
Beta (β) Scorpii 386
BM Scorpii 286, 386
Bug Nebula **256–57**
Butterfly Cluster **286**
Delta (δ) Scorpii 386
GRO J1655-40 **268**
M4 **288**
Mu (μ) Scorpii 386
in monthly sky guides 439, 445, 450, 451, 456, 457
Nu (ν) Scorpii 386
Omega (ω) Scorpii 386
PSR B1620-26 **268**
Scorpius X-1 386
Xi (ξ) Scorpii 386
Zeta (ζ) Scorpii 386, 445
Scorpius–Centaurus Association 229, *229*
Sculptor **388**
Cartwheel Galaxy **309**
Epsilon (ε) Sculptoris 388
Sculptor Group **319**, *324*
Scutum (the Shield) **366**, 456
Delta (δ) Scuti 366
R Scuti 366
Wild Duck Cluster 366, *366*, 456, 457
Scutum Star Cloud 366
Sea Goat *see* Capricornus
Sea Monster *see* Cetus
Sea of Tranquillity (Moon) **154**
seas, on Earth 147
seasons
Earth 61, *61*, *138*
Mars 160, *160*
Neptune 200
Uranus 196
Secchi, Father Angelo 281
Second World War 98
Sedan Crater (Nevada Desert) *158*
Sedna **207**
Seki Tsutomu 216
September sky guide **462–67**
Serpens (the Serpent) **364**, 444, 450
Delta (δ) Serpentis 364
Eagle Nebula **242–43**
Hoag's Object **309**
Seyfert's Sextet **319**
Theta (θ) Serpentis 364
Unukalhai (Alpha (α) Serpentis) *364*
Serpens Cauda 451
Serpent *see* Serpens
Set Square *see* Norma
Setebos *197*
SETI (search for extraterrestrial intelligence) 53
Seven Sisters *see* Pleiades
Sextans (the Sextant) **380**
sky guide 426
17 Sextantis 380
18 Sextantis 380
Sextant *see* Sextans
sextuple systems
Castor **272**
Seyfert, Carl **313**, 314
Seyfert galaxies 305, 310, *310*
Circinus Galaxy **312**
distribution 311
Fried Egg Galaxy **313**
M77 373, *373*, 475
NGC 1275 **314**
NGC 5548 **313**
Seyfert's Sextet (NGC 6027 and NGC 6027A-C) **319**
Shackleton, Ernest 174
Shakespeare, William 198, 199
Shakespeare region (Mercury) 127
Shapley, Harlow **92**
sheets, galaxy superclusters 325
Sheliak (Beta (β) Lyrae) **277**, 349
Shepard, Alan 100, *104*
shepherd moons, Saturn *189*
Sher 25 **261**
Shergotty meteorite *165*
Shield *see* Scutum
Shoemaker, Carolyn 219, *219*
Shoemaker, Eugène (Gene) 151, *151*, 219
Shoemaker–Levy 9, Comet 179, *179*, **219**
shooting stars 71, 220
short-period comets 214
Shorty Crater (Moon) *156*, 157
Siarnaq *189*
Sickle 361
sidereal day 62, *62*
sidereal month 62, 65
Sif Mons (Venus) 133
Sigma (σ) Coronae Borealis 363
Sigma (σ) Octantis 409
Sigma (σ) Orionis 238, **277**, *374*, 375
Sigma (σ) Tauri 356
silicaceous (S-type) asteroids 208
silicates
dust 24
interstellar medium 228
silicon, formation of 51
Silverstein, Dr. Abe 102
singularity, black holes 26, *41*
Sinope *179*
Sinus Iridum (Moon) 151
Sippar Sulcus (Ganymede) *184*
Sirius A (Alpha (α) Canis Majoris) **248**, 376
ancient astronomy 82
apparent magnitude 67
binary system 270
classification 231
Hertzsprung–Russell (H–R) diagram *230*
in monthly sky guides 414, 415, 421, 426, 427, 475, 481
naked-eye astronomy *73*
name 68
Winter Triangle 420, *420*, 480

Sirius B (HD 48915 B) **264**, 376
binary system 270
Hertzsprung–Russell (H–R) diagram *230*
as white dwarf 94, 262
Sirrah 352
Sk-69 202 261
sky guides **410–85**
Skylab space station 105, 106, *106*
Slayton, Donald "Deke" *105*
Slipher, Vesto 205, 306, *306*
Sloan Digital Sky Survey 324, *325*
Small Magellanic Cloud (SMC, NGC 292) *288*, *294*, 297, **301**, 402, *402*
Milky Way halo 229
in monthly sky guides 415, 457, 463, *463*, 469, 474, 475, 481
SMART-1 spacecraft *111*, *151*
SN 437 (Geminga Pulsar) **264**
SN 1572 (Tycho's Supernova) 268
SN 1604 (Kepler's Star) *35*, **269**, 365
SN 1680 (Cassiopeia A) *51*, 269
SN 1987A 261
Society for Space Travel (VfR) 98, *98*
sodium, on Mercury 125, *125*
sodium chloride *29*
software, astronomical 78
SOHO solar observatory *120–21*, 121, *122*, *123*
Soho-6, Comet *215*
Sojourner 110, 162, *162*
solar day 62, *62*
solar eclipses 63, *63*
solar flares *10*, *114–15*, 122, *246*
Solar Maximum Mission 121, *121*
solar nebulae, formation of Solar System 116
solar quakes *122*
Solar System 25, **114–223**
asteroids **208–13**
comets **214–19**
Earth **138–47**
history **116–17**
Jupiter **176–85**
Kuiper Belt **206**
life, search for 53
Mars **160–75**
Mercury **124–7**
meteors and meteorites **220–23**
in Milky Way *229*
Moon **148–59**
Neptune **200–3**
Oort Cloud **207**
orbits 37, **118–19**
planets *12–13*
Pluto **204**
Saturn **186–95**
size *22*
space probes **108–11**
Sun **120–23**
Uranus **196–9**
Venus **128–37**
solar systems, formation 233, *233*, 290–91
solar wind *10*, 123
aurorae 70
bow shock *139*
charged particles 139
Jupiter 177
solids, states of matter 30
Solis Planum (Mars) 168
solstices 61, *61*, 83, *138*, 444
Sombrero Galaxy (M104, NGC 4594) **306**, 362, *362*
sound waves, helioseismology 95, *95*
south celestial pole *58*, *421*
South Polar Group 319
South Polar Region (Mars) **171**
South Polar sky *333*
South Pole, AASTO project 297
South Pole-Aitken Basin Crater (Moon) **159**
Southern Cross *see* Crux
Southern Crown *see* Corona Australis
Southern Fish *see* Piscis Austrinus
Southern Pinwheel (M83) *294*, 378, *378*, 439, 445
Southern Pleiades (IC 2602) 395, 427, 433
Soviet Union
Moon landings *102*, 103, 105
space probes 101, 108
Space Race **98–101**
space stations 106, *106*, 107, 111
Soyuz spacecraft 100, 101
space
astronomy from 96, 97
Big Bang 46
Einstein's theories 92
expanding **42–43**, 54
space and time **38–41**, 55
*see also* Universe
space probes 101, **108–11**
asteroids 111
astronomy from space 96
Jupiter 109, 110
Jupiter's moons 110
Mars 101, **108–10**, 162, 167
Mercury 108, *108*
Moon 101, *101*, 151, 153
Neptune 109
Pluto *111*
Saturn 109, *109*, 110, 194, *194*
Sun 121, *121*
Uranus 109
Venus *96*, 101, 108, *108*, 130
Space Race **98–101**
Space Shuttle *8*, *80–81*, 107, *107*
space stations 106, *106*, 107, 111
spacecraft **98–101**
Moon landings **102–105**
*see also* space probes
special theory of relativity **38–39**
spectra 33, *33*
astrophysics 90, *90*
circumstellar discs 290
identifying binary stars 270
spectroscopy 33, *33*
star classification 231
Wolf–Rayet stars 251
spectroscopic binaries 270
speed of light 32, 38, 39
Spencer, Dr L.J. *223*
Spica (Alpha (α) Virginis) *362*
Hertzsprung–Russell (H–R) diagram *230*
in monthly sky guides 426, 427, 432, 433, 438, 439, 444, 445, 450, 451
naked-eye astronomy *73*
spicules, Sun *95*, *122*, 123
spin
Earth 138, *138*
Jupiter 176, *176*
Mars 160, *160*
Mercury 124, *124*
Moon 148, *148*
Neptune 200, *200*
Pluto 204
Saturn 186, *186*
Uranus 196, *196*
Venus 128, *128*
Spindle Galaxy (M102, NGC 3115) **307**, 380, *380*
spiral galaxies 26, **294–95**
Andromeda Galaxy (M31, NGC 224) **302–303**
Antennae Galaxies **307**
barred spiral galaxies 26, *294*
Black Eye Galaxy **304**
Bode's Galaxy **304**
Cartwheel Galaxy **309**
classification 294, *294*
density waves 227, 237, 295
ESO 510-G13 **308**
galaxy clusters 317
merger model 299, *299*
the Mice **308**
Milky Way **226–29**
NGC 6782 **308**
Pinwheel Galaxy **306**
Sombrero Galaxy **306**
Triangulum Galaxy **301**
Whirlpool Galaxy **305**
Spitzer Space Telescope 26, 34, 51, 245, *245*
Sponde *179*
spring equinox 61, *61*, *138*
Sputnik 1 satellite 99, *99*
Sputnik 2 satellite *99*, 100
Square of Pegasus 68, 352, 370
in monthly sky guides 414, 450, 456, 457, 462, 463, 468, *468*, 469, 474, 475, 480, 481
SS 433 *26*
star clusters
Beehive Cluster **286**
Butterfly Cluster **286**
catalogues 69
Christmas Tree Cluster 240
evolution 285, *285*
Hyades **286**
Jewel Box (Kappa (κ) Crucis) **288**
M4 **288**
M12 **289**
M14 **289**
M15 **289**
M52 **286**
M68 **289**
M93 **286**
M107 **289**
moving clusters 344
NGC 3201 **288**
NGC 4833 **289**
Omega Centauri **288**
open clusters **284**, 286
Pleiades **287**
Trapezium 239, *239*, **277**
47 Tucanae **288**
starburst galaxies 297, 304
Stardust mission 111, 219
starquakes 263
stars *14–15*, **230–91**
accretion discs 242, 245
ancient astronomy 83
apparent magnitude 231
Arabic names 330
stars *cont.*
asterisms 68
Big Chill 55
binary stars *25*, 270, 272
brightness **67**
brown dwarfs 25, *25*, 94, *94*
carbon stars 252, *252*
catalogues 68, 330
celestial coordinates 59, *59*
celestial sphere **58–59**
Cepheid variable stars 42, 301, *303*
charts and atlases 331
classification 231
collapsing *235*, *262*
colours *66–67*
Copernican revolution 87
death 25
18th-century astronomy 90
evolution **233–35**
extreme stars 94
first stars 51
formation 25, 230, 232, **236–45**
giant stars 25
Hertzsprung–Russell (H–R) diagram 94, 230, *230*
hypernovae *35*, *51*
interstellar medium 228
life cycles 230, **232–35**
light 25
luminosity 230, *230*, 231
magnetic fields 247
main-sequence stars **246–49**
mapping the sky **332–37**
mass 230
Milky Way **226–29**
molecular clouds 228
motion and patterns **66–69**
multiple stars **270–77**
names 68
neutron stars 25, 94
nuclear fusion 31, *31*, 92
old stars **250–61**
planet formation 233, *233*
plasma 30
populations 227
radio pulsars 94
red dwarfs 25, *25*
rotation 37, 247
sidereal day 62, *62*
space-time *41*
spectroscopy 33, *33*, 90, *90*
stellar end points **264–69**
structure 246
Sun **120–23**
supergiants 25, *25*
temperature 230, *230*
variable stars 258, **278–83**
white dwarfs 25, *25*, 94
Wolf–Rayet stars 245, **251**, 252, 260
*see also* constellations; galaxies; star clusters *and individual stars*
states of matter 30
Stein Crater Field (Venus) **137**
stellar black holes 26, *26*
stellar end points **262–69**
stellar nurseries 236
stellar winds 236, 237
Stephano *197*
Stephan's Quintet (Hickson 92) **320**
Stern *see* Puppis
Stingray Nebula (Hen-1357) **260**
Stone Age 83
stone circles *82*, 83
Stonehenge *82*, 83
stony-iron meteorites 208, *220*
stony meteorites *220*
storms
Jupiter 179, *179*
Mars 161, *161*, *167*
Neptune 201
Saturn 188, *188*
stratosphere, Earth's atmosphere *140*
string theory 31, *31*, 41
stromatolites *52*
strong nuclear force 30, *30*, 46
Struve, F. 273
Struve 747 375
Struve 2725 369
Sualocin (Alpha (α) Delphini) 369
subatomic particles *see* particles
al-Sufi 302, 330, 405, *405*
sulphur, properties *29*
sulphuric acid, on Venus 129
Sumerians
ancient astronomy 83
constellations 330
summer solstice 61, *61*, *138*
Summer Triangle 444, 450, *450*, 456, 457, 462, 463, 468, 474, 481
*see also* Altair (Alpha (α) Aquilae); Deneb (Alpha (α) Cygni)
naked-eye astronomy 73
*see also* Vega (Alpha (α) Lyrae)
Sun *10–11*, **120–23**
analemma *60*
angular diameter 73
astrology 60, 83
atmosphere 123, *123*
celestial cycle 60
classification 231, *231*
comets 215
Copernican revolution **86–89**
corona *10*, *63*
early scientific astronomy 84, *84*
Earth-centred cosmos 85, *85*
eclipses **63**
energy 92
formation of Solar System 116, *117*
helioseismology 95, *95*
helium production 246
Hertzsprung–Russell (H–R) diagram *230*
ice haloes 70
internal structure **120**
laws of planetary motion 87
luminosity 231
as main-sequence star 247
midnight Sun *60–61*
in Milky Way 229, *229*
movements across sky 59
nuclear fusion *31*
photosphere 120, 122, *122*, 123
plasma loops 122
prominences *10*
solar day 62, *62*
solar flares *10*, *114–15*, 122, *246*
Solar System 25
space probes 108, 121, *121*
space-time *40–41*
sunspots *10–11*, 122, *122*, *247*
surface 122, *122*
temperature 95, *122*, 123
transits by planets 65, *65*
zodiac 60, *61*

sun dogs 70
sundials *82*
Sunflower Galaxy (M63) 346, *346*
sunlight
  noctilucent clouds 71, *71*
  zodiacal light 71, *71*
suns *see* stars
Sunyaev–Zel'dovich effect *322*, 323
La Superba (Gamma (γ) Canum Venaticorum) 346
superclusters *see* galaxy superclusters
supergiants 25, **250**
  Antares (Alpha (α) Scorpii) **252**
  Betelgeuse (Alpha (α) Orionis) **252**
  Eta (η) Carinae **258**
  evolution *233*
  Hertzsprung–Russell (H–R) diagram 230, *230*, *251*
  Sher 25 **261**
  star life cycles 233, *233–35*, 234
  stellar black holes 26
  V838 Monocerotis **261**
Superior, Lake (Earth) 146, *146*
superior planets, motion 64, *64*
superluminal jets 311, *311*
supermassive black holes 26, *26*, 55, 297, *297*
Supernova 1987A *262*, 300, 405
Supernova 1994D *279*
supernova remnants 25, *25*
  Crab Nebula **266–67**
  Cygnus Loop **265**
  Vela Supernova **265**
supernovae 25, 250
  and black holes 263
  Cassiopeia A **269**
  dark energy 54, *54*
  formation of 234, *235*
  formation of elements in *234*
  Kepler's Star **269**
  life cycles of stars 232
  and meteorites 222
  and neutron stars 263
  radiation 34
  star evolution 233, *233*
  star formation 236, 237, *237*
  Tycho's Supernova **268**
  Type I supernovae 279
  Type II supernovae 262, *263*
Surtsey (Earth) *142*
Surveyor space probes 153
Suttung *189*
Swan *see* Cygnus
Swift, Lewis 216
Swift–Tuttle, Comet *214*, **216**, 220
Sycorax *197*, 199
Syene 84, *84*
synchronous rotation *148*
synchrotron mechanism 310, *310*


## T


T Coronae Borealis (Blaze Star) **282**
T Pyxidis 392
T Tauri *237*
Table Mountain *see* Mensa
the Tadpole 27
Tagish Lake meteorite **222**
tails, comets 214, 215, *215*
Tarantula Nebula (30 Doradus) *301*, 405
  brightness 300, *300*
  in monthly sky guides 415, 421, 427, 475, 481
Tarazed (Gamma (γ) Aquilae) 367, *367*
Tarvos *189*
Tau (τ) Canis Majoris 376, *376*
Tau (τ) Ceti *230*, 373
Taurid meteor shower 356, 474
Taurus (the Bull) **356–57**
  Alcyone (Eta (η) Tauri) **273**, 287, 356
  *see also* Aldebaran (Alpha (α) Tauri)
  Alnath (Beta (β) Tauri) *230*, 343, 356
  Crab Nebula **266–67**
  Hyades **286**
  Lambda (λ) Tauri **280**, 356
  in monthly sky guides 420, 421, 426, 480, 481
  Pleiades **287**
  Sigma (σ) Tauri 356
  T Tauri *237*
  Theta (θ) Tauri 356
  Zeta (ζ) Tauri 356
Taurus-Littrow Valley (Moon) **156–57**
Taurus Molecular Cloud 290
Taygeta 287
Taygette *179*
Teapot 384, 451, *457*
tectonic features
  Earth 140, **142–45**
  Mars 162, **164–68**
  Venus 130, *130*, **132–35**
tektite *221*
Telescope *see* Telescopium
telescopes **76–77**, 331
  astrophotography 79, *79*
  computerized telescopes **78–79**
  early astronomy 88, *88*
  18th- and 19th-century astronomy 90, **91**
  filters 78, *78*
  Galileo's 76, *76*
  Hubble Space Telescope 43, 97, *107*
  infrared astronomy 34, *34*
  Newton's 89, *89*
  optical telescopes 35, *35*
  planet-hunting 291
  radio astronomy 34, *34*, 93, *93*
  reflecting telescopes 76, *76*, 89, *89*, *91*, 93
  refracting telescopes 76, *76*, 91
  size 97
  space telescopes 107, *107*
  Spitzer telescope 245, *245*
  20th-century astronomy 93
Telescopium (the Telescope) **400**
  Delta (δ) Telescopii 400
Telesto *188*, **192**
Tempel–Tuttle, Comet *214*, 220
temperature
  Big Bang 46, 49, 50
  gas giants 291
  Hertzsprung–Russell (H–R) diagram 230, *230*
  interstellar medium 228
  on Io 182
  Jupiter 176, 179
  main-sequence stars 246, *247*
temperature *cont.*
  Mars 160, 161
  Mercury 124, 125
  Moon 149
  old stars 251
  Pluto 204
  red giants 250
  Saturn 187
  star classification 231
  star formation *232*, 236
  Sun 95, *122*, 123
  Uranus 197
  Venus 129
Tethys *188*, *190*, **192**, 193
Teviot Vallis (Mars) *171*
Thackeray, A.D. 244
Thalassa *201*
Tharsis Bulge (Mars) 162, 164, 166, 168
Thebe *178*, 179, **180**
Theia Mons (Venus) 133
Themisto *178*, **180**
thermosphere, Earth's atmosphere *140*
Theta (θ) Apodis 407
Theta (θ) Carinae 395, 427
Theta (θ) Eridani 390
Theta (θ) Indi 400
Theta (θ) Muscae 397
Theta (θ) Orionis (Trapezium) 239, *239*, *271*, *272*, **277**, 375, *375*
Theta (θ) Serpentis 364
Theta (θ) Tauri 356
Thor's Helmet *260*
Thrym *189*
Thyone *179*
Tibetan Plateau (Earth) 144, *145*
tides, and gravity 89, 150, *150*
Tigre, River (Earth) *146*
Tikhonravov Crater (Mars) **172**
time and space **38–41**
  ancient astronomy 82, *82*
  Big Bang 46
  celestial cycles 60, 62
  Einstein's theories 92
  expanding space 43
  lunar month 62
  medieval astronomy *86*
  sidereal day 62, *62*
  sidereal month 62
  solar day 62, *62*
  space-time 39, *39*, 40–41, *40–41*, 55
  time dilation 39, *39*
Titan 188, *188*, **194**
  search for life 53
  space probes 109, 110, *110*
Titania *197*, **199**
titanium, on Moon *154*
Titanomachia 399
Titans, in mythology 399
Tito, Dennis 107, *107*
TMR-1C *291*
Tohil Mons (Io) *182*
Tombaugh, Clyde **205**
total eclipses 63, *63*
Toucan *see* Tucana
Toutatis **210**
Tr37 star cluster 241
TRACE satellite 121, *121*, *123*
transition region, Sun 123
transits, planets 65, *65*
transverse velocity, stars 66
Trapezium (Theta (θ) Orionis) 239, *239*, *271*, *272*, **277**, 375, *375*
Triangulum (the Triangle) **353**
Triangulum (the Triangle) *cont.*
  Local Group **318**
  3C 48 **315**
  6 Trianguli 353
Triangulum Australe (the Southern Triangle) **398**, 445
  Alpha (α) Triangulum Australis 398
Triangulum Galaxy (M33, NGC 598) *294*, **301**, *318*, 353, *353*, 469, 475
Trifid Nebula (M20) **244**, 384, *384*, 451
trigonometry, early scientific astronomy 84, 85
Trinculo *197*
triple stars
  Albireo (Beta (β) Cygni) **273**
  Beta (β) Monocerotis **277**
  Omicron (o) Eridani **272**
  Rigel (Beta (β) Orionis) **277**
Triton 201, *201*, **202–203**, 204, 205
Trojan asteroids 208, *208–209*
Tropic of Cancer 61
Tropic of Capricorn 61
troposphere, Earth's atmosphere 140, *140*
Trujillo, Chad 207
Trumpler 14 *245*
Trumpler 16 *245*
Tsiolkovsky, Konstantin 81, 98, *98*, 158
Tsiolkovsky Crater (Moon) **158**
TT Cygni **252**
Tucana (the Toucan) **402**, 463
  Beta (β) Tucanae 402
  Kappa (κ) Tucanae 402
  *see also* Small Magellanic Cloud (SMC)
  47 Tucanae **288**, 301, 402, *402*, *463*, 469, 475
Tuttle, Horace 216
TWA-5B *291*
Twin Jet Nebula (M2-9) **253**
Twins *see* Gemini
Tycho catalogue 66
Tycho Crater (Moon) *151*, **155**, 157
Tycho's Supernova (SN 1572) **268**
Type I supernovae 279
Type II supernovae 262, *263*


## U


U Geminorum **280**
UFOs (unidentified flying objects) 71, *71*
Uhuru satellites 96
ultraviolet radiation 32, *35*
  astronomy from space 96
  first stars 51
  galaxies *297*
  observatories 35, *35*
  photoelectric effect *32*
Ulugh Beg 85, *85*, 86
Ulysses space probe 121, 217
Umbriel *197*, **199**
Unicorn *see* Monoceros
United Kingdom Infrared Telescope (UKIRT) 34
United States of America
  future missions 111
  Moon landings **102–105**
  Skylab space station 106, *106*
  space probes 101, **108–10**
  Space Race **98–101**
United States of America *cont.*
  Space Shuttle 107, *107*
Universe
  age 42
  Big Bang 22, **46–49**
  constituent parts **24–25**
  Copernican revolution **86–89**
  dark ages 50
  early models of 59, *59*
  Earth-centred cosmos 85, *85*
  expanding space **42–43**, 54, 323
  fate of **54–55**
  general theory of relativity 41, 49
  geometry of 55
  life in **52–53**
  mapping 324
  matter **28–31**
  observable Universe 23
  radiation **32–35**
  scale of **22–23**
  space and time **38–41**
Unukalhai (Alpha (α) Serpentis) *364*
Upsilon (υ) Andromedae 291, *291*
Upsilon (υ) Pegasi 370
Uranus **196–99**
  atmosphere and weather 197, *197*
  discovery 90
  and Kuiper Belt *206*
  moons 197, **198–99**
  orbit and spin *118*, 196, *196*
  rings 197, *197*
  space probes 109
  structure 196, *196*
Ursa Major (the Great Bear) **344–45**, 432
  Alcor (80 Ursae Majoris) **272**, 344, *345*, 438
  Alioth (Epsilon (ε) Ursae Majoris) *68*, 344
  Alkaid (Eta (η) Ursae Majoris) *68*, 344
  Bode's Galaxy **304**, 344, *344*
  Cigar Galaxy 297, **304**, 344
  Delta (δ) Ursae Majoris 344
  *see also* Dubhe (Alpha (α) Ursae Majoris)
  Merak (Beta (β) Ursae Majoris) *68*, 73, 344
  Mizar (Zeta (ζ) Ursae Majoris) *68*, **272**, 344, *345*, 438
  naked-eye astronomy 73, *73*
  Phad (Gamma (γ) Ursae Majoris) *68*, 344
  Pinwheel Galaxy **306**, 344, 438, 444
  star chart *68–69*
  Xi (ξ) Ursae Majoris 344
  47 Ursae Majoris 291
Ursa Minor (the Little Bear) **338**
  Abell 2125 **321**
  Eta (η) Ursae Minoris 338
  Gamma (γ) Ursae Minoris 338
  *see also* Polaris (Alpha (α) Ursae Minoris)
  11 Ursae Minoris 338
  19 Ursae Minoris 338
US Explorer satellites 96
US Navy 99
Utopia Planitia (Mars) **170**
UW Canis Majoris 376
Uzbekistan 85

# V


V*V1033 Sco 268
V2 rockets 98, *99*
V647 Tau (τ) 273
V838 Monocerotis **261**, *278–79*
Valhalla Basin (Callisto) 185, *185*
Valles Marineris (Mars) 109, *161*, 162, *162*, **166–67**
Van Allen, James **139**
Van Allen radiation belts (Earth) 139
Van De Graaff Crater (Moon) **158**
Vanguard rocket 99
variable stars 258, **278–83**
  bizarre variables 279
  Cepheid variable stars 42, 278, *278*, 282, 301, *303*
  Delta (δ) Cephei **282**
  Eta (η) Aquilae **282**
  Gamma (γ) Cassiopeiae **281**
  Mira (Omicron (ο) Ceti) **281**
  Mu (μ) Cephei (Garnet Star) **283**
  Pistol Star **261**
  Procyon (Alpha (α) Canis Minoris) **280**
  pulsating variable stars 278
  R Coronae Borealis **283**
  RR Lyrae **282**
  U Geminorum **280**
  W Virginis **282**
  Zeta (ζ) Geminorum (Mekbuda) **282**
Vatican Observatory 281
Vega (Alpha (α) Lyrae) **249**, 349, 350
  circumstellar disc 290
  Hertzsprung–Russell (H–R) diagram *230*
  luminosity *231*
  in monthly sky guides 432, 438, 444, 445, 450, 451, 456, 457, 462, 463
  naked-eye astronomy 73
Veil Nebula *25*, 265, 351
Vela (the Sails) **394**, 420, 421
  Delta (δ) Velorum 394
  Kappa (κ) Velorum 394
  Lambda (λ) Velorum 394
  NGC 3201 **288**
  Omicron (ο) Velorum *394*
  Regor (Gamma (γ) Veloram) 231, **249**, 394
Vela Pulsar 265, *265*
Vela Supernova (NGC 2736) **265**, 394
velocity
  light 32, 38, 39
  motion of stars 66
Venera space probes 108, *108*, 130, 131
Venus **128–37**
  atmosphere 129, *129*
  formation of Solar System *117*
  Galileo's observations 88
  impact craters 131, *131*, **136–37**
  maps *130–31*, 131
  motion 64, *64*
  occultations 65
  orbit and spin *118*, 128, *128*
  phases 64
Venus *cont.*
  space probes *96*, 101, 108, *108*, 130
  structure 128, *128*
  tectonic features 130, *130*, **132–35**
  transits 65, *65*
vernal equinox 61, *61*, *138*, 355, 372
Verne, Jules *98*
Very Large Array, New Mexico 34, *34*
Very Large Telescope 51
Vespucci, Amerigo 396, *396*
Vesta **210**
vibrations, string theory 31
Viking space probes 109, 162
Virgin *see* Virgo
Virgo (the Virgin) **362**, 438
  Abell 1689 **321**
  M60 **307**
  M87 *296*, *299*, **313**, 319, 362, *362*
  NGC 4261 **313**
  Porrima (Gamma (γ) Virginis) **249**, 362
  Sombrero Galaxy **306**, 362, *362*
  *see also* Spica (Alpha (α) Virginis)
  W Virginis **282**
  3C 273 **315**
Virgo Cluster *23*, 27, **319**, 360, 362
  central region *317*
  dark galaxies 316
  galaxy superclusters 324, *324*
  in monthly sky guides 432, 433, 438
Virgo III groups, galaxy superclusters *324*
Virgo Supercluster 324
viruses 52, *52*
visual binaries 270
voids, galaxy superclusters 325
Volans (the Flying Fish) **406**
  Epsilon (ε) Volantis 406
  Gamma (γ) Volantis 406
volcanoes
  Earth 140, 142, *142*, 143, *143*
  Io 182, *182–83*
  Mars 162, 164–65, *164–65*, 168, *168*
  Moon 149
  Venus 130, *130*, 133, *133*, 134, *134*
Volga Delta (δ) (Earth) *147*
Voskhod spacecraft 101, *101*
Vostok missions 100, *100*
Voyager space probes
  Grand Tour 109, *109*
  Neptune 109, 200
  Uranus 109, 196
Vulcan 124
Vulpecula (the Fox) **368**, 456
  Alpha (α) Vulpeculae 368
  Dumbbell Nebula 368, *368*, 456, 457


# W


W Virginis **282**
Wanda Crater (Venus) **136**
water
  atomic structure 29
  Earth *139*, *140*, 141, *141*
  extra-solar planets 291
  features formed on Earth **146–47**
water *cont.*
  features formed on Mars **169–71**
  interstellar medium 228
  Jupiter 178
  and life 52
  Mars 163, *163*
  the Moon 159
  Neptune 200
  Pluto 204
  Saturn 187
  Saturn's rings 189
  states of matter *30*
  Uranus 196, 197
Water Carrier *see* Aquarius
Water Jar 371
Water Snake *see* Hydra
wave-like behaviour, electromagnetic (EM) radiation 32, *32*
wavelengths
  analysing light 33, *33*
  celestial objects 34
  electromagnetic (EM) radiation 32
  galaxies 297
  luminosity 231
  photons 32
  red shift and blue shift 33, *33*
WC stars 251
weak interaction, Big Bang 47
weak nuclear force 30, *30*
weather
  Earth 140, *140*
  Jupiter 179, *179*
  Mars 161
  Neptune 201
  Saturn 188, *188*
  Uranus 197
weight
  and gravity 36
  weightlessness 36, *36*
Weinberg, Steven **30**
werewolves 150, *150*
West, Comet **217**
West, Richard 217
Whipple, Fred 215, *215*
Whirlpool Galaxy (M51, NGC 5194, NGC 5195) *14*, *294*, **305**, 346, *346*, 438, 444
White, Ed *100*
white dwarfs 25, *25*, **262**
  Big Chill 55
  classification 231
  Hertzsprung–Russell (H–R) diagram 230, *230*
  mass 94
  multiple stars 270
  NGC 2440 nucleus **264**
  novae 278
  planetary nebulae 251
  Sirius B 94, **264**, 270, 376, 421, 427, 475
  space-time *41*
  star life cycles 233, *233*, 234, *235*
  Type I supernovae 279, *279*
white stars
  Altair (Alpha (α) Aquilae) **248**
  Fomalhaut (Alpha (α) Piscis Austrini) **249**
  Sirius A (Alpha (α) Canis Majoris) **248**
  Vega (Alpha (α) Lyrae) **249**
Wild 2, Comet 111, **219**
Wild Duck Cluster (M11) 366, *366*, 456, 457
Wilkinson Microwave Anisotropy Probe (WMAP) 34, *34*
William the Conqueror, King of England 218
Wilson, Robert *49*
WIMPs (weakly interacting massive particles) 27, 28
wind erosion
  Mars 172, *172*
  Venus *131*
winds
  Jupiter 179
  Mars 161, *161*
  Saturn 188
  stellar winds 236, 237
Winged Horse *see* Pegasus
winter solstice 61, *61*, *138*
Winter Triangle
  in monthly sky guides 414, 420, *420*, 480
  naked-eye astronomy 73, *73*
Wirtanen, Comet 111, *111*, 219
WN stars 245, 251
WO stars 251
Wolf *see* Lupus
Wolf, Charles 251, 260
Wolf, Max 211
Wolf–Rayet stars 245, **251**, 252, 260
  HD 56925 **260**
  Regor (Gamma (γ) Velorum) **249**
  WR 104 **255**
  WR 124 **260**
Wunda Crater (Umbriel) 199, *199*
WZ Sagittae 366


# X


X-bosons 47
X-rays 32
  black holes 94, 263, 310
  galaxy clusters 319, *319*
  gravitational lensing 323
  intergalactic medium 317
  observatories 35, *35*, 96
  Sunyaev–Zel'dovich effect *322*
  supermassive black holes *297*
Xanadu (Titan) *194*
Xi (ξ) Boötis 347
Xi (ξ) Lupi 383
Xi (ξ) Pavonis 408
Xi (ξ) Puppis 393
Xi (ξ) Scorpii 386
Xi (ξ) Ursae Majoris 344


# Y


year, length of 82
yellow stars
  Alpha (α) Centauri (Rigil Kentaurus) **248**
yellow-white stars
  Porrima **249**, 362
Ymir *189*
Yohkoh space probe 121, *121*


# Z


Zach, Franz Xaver von **209**
Zeta (ζ) Antliae 380
Zeta (ζ) Aquarii 371
Zeta (ζ) Aurigae 343
Zeta (ζ) Boötis **273**
Zeta (ζ) Canceri 359
Zeta (ζ) Cepheus 340
Zeta (ζ) Coronae Borealis 363
Zeta (ζ) Geminorum (Mekbuda) **282**, 358
Zeta (ζ) Herculis 348
Zeta (ζ) Leonis 361
Zeta (ζ) Lyrae 349
Zeta (ζ) Ophiuchi 264
Zeta (ζ) Orionis (Alnitak) *230*, 374, 375, *375*
Zeta (ζ) Phoenicis 401
Zeta (ζ) Piscium 372
Zeta (ζ) Puppis (Naos) 393
Zeta (ζ) Reticuli 404
Zeta (ζ) Sagittae 366
Zeta (ζ) Scorpii 386, 445
Zeta (ζ) Ursae Majoris (Mizar) *68*, **272**, 344, *345*, 438
Zeus 338, *338*, 345, 351, 399
zodiac *61*, 65
  ancient astronomy 83
  astrology 60
  Islamic *60*
zodiacal light 71, *71*
Zond 3 spacecraft *102*
Zubenelgenubi (Alpha (α) Librae) 363
Zubeneschamali (Beta (β) Librae) 363
ZZ Ceti *230*

1 Ceres 90, *208*, 209
1 Zwicky 18 *298*
1ES 1853-37.9 264
2 Pallas 211
2M1207 *291*
3C 31 *310*
3C 48 **315**
3C 273 **315**, 362
3C 279 *310*
3C 405 (Cygnus A) **314**
4 Vesta **210**
9 Sagittarii 241
15 Monocerotis **276**
24 Tau (τ) 273
30 Doradus *see* Tarantula Nebula
47 Tucanae (NGC 104) **288**, 301, 402, *402*, *463*, 469, 475
61 Cygni **248**
243 Ida *116*, *208*, **211**
253 Mathilde *111*, **210**
433 Eros *13*, *111*, *208*, *210*, **212–13**
951 Gaspra **210**
1992 QB1 206
4179 Toutatis **210**
5535 Annefrank **210**

# ACKNOWLEDGMENTS

**Dorling Kindersley** would like to thank the following people for their help in the preparation of this book: Anne Brumfitt and her colleagues at the European Space Agency for editorial advice; Stephen Hawking for permission to reproduce the quotation on p.21; Giles Sparrow for advice on the contents list; Gillian Tester and Andrew Pache for DTP support; Dave Ball, Sunita Gahir, and Marilou Prokopiou for additonal artwork; Malcolm Godwin of Moonrunner Design; Rajeev Doshi of Combustion Design and Advertising; Philip Eales and Kevin Tildsley of Planetary Visions; Tim Brown and Giles Sparrow of Pikaia Imaging; Tim Loughhead of Precision Illustration; John Plumer of JP Map Graphics; Richard Tibbitts of Antbits; and Greg Whyte of Fanatic Design.

**PICTURE CREDITS**

Dorling Kindersley would like to thank the following for their help in supplying images: Till Credner; Robin Scagell at Galaxy Picure Library; Romaine Werblow in the DK Picture Library; Anna Bond at Science Photo Library.

**Key:**
t = top; b = bottom; c = centre; l = left; r = right; A = above; B = below.

**Abbreviations:**
**AAO** = Anglo Australian Observatory; **ASU** = Arizona State University; **BAL** = Bridgeman Art Library (www.bridgeman.co.uk); **Caltech** = California Institute of Technology; **Chandra** = Chandra X-Ray Observatory; **Credner** = Till Credner www.allthesky.com; **DSS** = Digitized Sky Survey; **ESA** = European Space Agency; **ESO** = © European Southern Observatory; **GPL** = Galaxy Picture Library; **GSFC** = Goddard Space Flight Center; **HHT** = The Hubble Heritage Team; **HST** = Hubble Space Telescope; **JHU** = John Hopkins University; **JPL** = Jet Propulsion Laboratory; **JSC** = Johnson Space Center; **KSC** = Kennedy Space Center; **DMI** = David Malin Images; **MSFC** = Marshall Space Flight Center; **NASA** = National Aeronautics and Space Administration; **NOAO** = National Optical Astronomy Observatory/Association of Universities for Research in Astronomy/National Science Foundation; **NRAO** = Image courtesy of National Radio Astronomy Observatory/AUI; **NSSDC** = National Space Science Data Center; **SPL** = Science Photo Library; **SOHO** = Courtesy of SOHO/EIT Consortium. SOHO is a project of international cooperation between ESA and NASA; **STScI** = Space Telescope Science Institute; **TRACE** = Image courtesy of the Lockheed Martin team of NASA's TRACE Mission; **USGS** = U.S. Geological Survey.

SIDEBAR IMAGES
**© CERN Geneva** (Introduction); **SOHO** (The Solar System); **NASA:** HST/ESA, HEIC and HHT (STScI/AURA) (Milky Way); HST/HHT (STScI/AURA) (Beyond our Galaxy); **SPL:** Kaj R. Svensson (The Night Sky).

**1 NASA:** HST/ESA, HEIC and HHT (STScI/AURA).
**2–3 GPL:** NASA/STScI/AURA.
**4–5 Corbis:** Roger Ressmeyer tcl; **NASA:** JPL tr; JPL/STScIcrA; **NOAO:** T.A. Rector (NRAO/AVI/NSF and NOAO) and B.A. Wolpa (NOAO) b.
**6–7 NOAO:** Adam Block (background).
**8–9 Corbis:** cAr; Digital Image © 1996 Corbis; Original image courtesy of NASA l; **Landsat 7 satellite image courtesy of NASA Landsat Project Science Office and USGS National Center for Earth Resources Observation Science:** tcr; **NASA:** JSC br.
**10–11 Corbis:** Roger Ressmeyer clA; **SPL:** ESA tl; Jisas/Lockheed cl; Scharmer et al/Royal Swedish Academy of Sciences r; **SOHO:** clB.
**12–13 ESA:** DLR/FU Berlin (G. Neukum) tr; **GPL:** JPL l; **NASA:** JPL crA, crB; JPL/STScI trB.
**14–15 Chandra:** NASA/CXC/MIT/F.K. Baganoff et al tl; **© 2005 Russell Cromon** (www.rc-astro.com)**:** r; **NOAO:** Eric Peng (JHU), Holland Ford (JHU/STScI), Ken Freeman (ANU), Rick White (STScI) clA; T.A. Rector and Monica Ramirez clB.
**16–17 2MASS:** T.H. Jarrett, J. Carpenter, & R. Hurt clA; **Chandra:** X-Ray: NASA/CXC/ESO/P. Rosati et al; **Optical:** ESO/VLT/P. Rosati et al r; **NASA:** HST/ESA, S. Beckwith (STScI) and the HUDF Team clB; **SPL:** Carlos Frenk, Univ. of Durham tl.
**18–19 Corbis:** Roger Ressmeyer.
**20–21 NASA:** HST/HHT (STScI/AURA).
**22–23 Credner:** cBl; **NASA:** HST/Dr Michael S. Vogeley – Princetown Univ. Obs. bcl; HST/ESA, Richard Ellis (Caltech) and Jean-Paul Kneib (Observatoire Midi-Pyrenees, France) tcr.
**24–25 Corbis:** tcr (Cluster); **GPL:** Andrea Dupree, Ronald Gilliland (STScI)/NASA/ESA tcr (Supergiant); Damian Peach tcr (Double); Nigel Sharp, NSF REU/AURA/NOAO cr; STScI tr; **NASA:** GSFC bcr; HST/ESA and J. Hester (ASU) bl; HST, HHT (STScI/AURA) c; JPL cBr (Europa), cBr (Ganymede), cBr (Io); JPL/DLR (German Aerospace Center) cBr (Callisto); **NOAO:** Nathan Smith, Univ. of Minnesota tcl; **SPL:** J-C Cuillandre/Canada–France–Hawaii Telescope clA; Pekka Parviainen crB.
**26–27 Chandra:** NASA/CXC/U. Amsterdam/S. Migliari et al bl; **Gemini Observatory/Association of Universities for Research in Astronomy:** GMOS–South Commissioning Team tcr; **NASA:** HST/H. Ford (JHU), G. Illingworth (UCSC/LO), M. Clampin (STScI), G. Hartig (STScI), the ACS Science Team and ESA c; HST/Jeffrey Kenney and Elizabeth Yale (Yale Univ.) bcl; HST/N. Benitez (JHU), T. Broadhurst (The Hebrew Univ.), H. Ford (JHU), M. Clampin (STScI), G. Hartig (STScI), G. Illingworth (UCO/Lick Obs.), the ACS Science Team and ESA cAr; JPL – Caltech/ASU/Harvard–Smithsonian Center for Astrophysics/NOAO cl; **SPL:** Los Alamos National Laboratory br; Max-Planck-Institut für Astrophysik cr; NOAO cAl; STScI/NASA tl.
**28–29 DK Images:** Andy Crawford cr (aluminium); Clive Streeter/Courtesy of the Science Museum, London cr (bromine); Colin Keates/Courtesy of the Natural History Museum, London cr (sulphur); Harry Taylor cr (hydrogen); **NOAO:** Todd Boroson cAl; **SPL:** Lawrence Berkeley Laboratory trB; Philippe Plailly cl.
**30–31 Corbis:** Raymond Gehman tlB; **SPL:** Alfred Pasieka tc; CERN bcl, cAr; **SOHO:** bcr.
**32–33 DK Images:** clB; **NASA:** HST/HHT (STScI/AURA) tc; **SPL:** bcr.
**34–35 2MASS:** cl (Infrared); **Chandra:** NASA/SAO/CXC/G. Fabbiano et al crB (X-Ray); NGST crA (Chandra); **GPL:** EGRET Team crB (Gamma-Ray); Rainer Beck/Philipp Hoernes/MPIFR clB; Robin Scagell tcr; **NASA:** Compton Gamma Ray Obs. crA (CO); HST/ESA, R. Sankrit and W. Blair (JHU) tr; Ultraviolet Imaging Telescope crB (Ultraviolet); **NOAO:** crB (Visible); **SPL:** David Nunak cl; Dr Fred Espenak tcl; NASA crA (EUVE); courtesy of **NASA/WMAP Science Team:** cl (microwaves), tclB.
**36–37 Alamy Images:** Kolvenbach br; **Corbis:** bl; **NASA:** JSC bcl; JPL tr.
**38–39 Corbis:** bl, cAr; Bettmann bcr; Lester Lefkowitz cl.
**40–41 SPL:** W. Couch and R. Ellis/NASA bc.
**42–43 NASA:** HST/ESA, J. Blakeslee and H. Ford (JHU) tcr; **NOAO:** Todd Boroson bcl; **SPL:** Sanford Roth crA.
**44–46 NASA:** HST/H. Ford (JHU), G. llingworth (UCSC/LO), M. Clampin (STScI), G. Hartig (STScI), the ACS Science Team, and ESA; **© CERN Geneva:** tcr.
**48–49 Corbis:** Bettmann tcl, tr; **NASA:** HST/HHT (STScI/AURA) tcr.
**50–51 Chandra:** NASA/CXC/GSFC/U. Hwang et al br; Image courtesy of **Andrey Kravstov:** Simulations were performed at the National Center for Supercomputing Applications (Urbana-Champaign, Illinois) by Andrey Kravtsov (The Univ. of Chicago) and Anatoly Klypin (New Mexico State Univ.). Visualizations by Andrey Kravtsov bl; **NASA:** HST/ESA, A.M. Koekemoer (STScI), M. Dickinson (NOAO) and the GOODS Team tr; HST/K.L. Luhman (Harvard–Smithsonian Center for Astrophysics, Cambridge, Mass.); and G. Schneider, E. Young, G. Rieke, A. Cotera, H. Chen, M. Rieke, and R. Thompson (Steward Obs., ASU, Tuscon, Ariz.) tc; **SPL:** NASA crA, crB; courtesy of **NASA/WMAP Science Team:** cAl.
**52–53** Courtesy of the **NAIC–Arecibo Observatory, a facility of the NSF:** bcr; **Corbis:** Roger Ressmeyer bcl; **NASA:** JPL/ASU cr; Provided by the SeaWiFS Project, NASA/GSFC, and ORBIMAGE tc; **SPL:** Dr Linda Stannard, UCT cl; John Reader bl; MSFC/NASA clB; **SETI League photo, used by persmission:** br.
**54–55 SPL:** Royal Obs., Edinburgh/AATB br; courtesy of **Saul Perlmutter and The Supernova Cosmology Project:** bl.
**56–57 Corbis:** Roger Ressmeyer.
**58–59 BAL:** Bibliothèque des Arts Decoratifs, Paris, France/Archives Charmet cr; **SPL:** David Nunuk tcr.
**60–61 British Library, London:** shelfmark: Or.5259, folio: f.29 c; **Corbis:** Paul A Souders b; **The Picture Desk:** The Art Archive/British Library, London cl; **SPL:** Frank Zullo tcl.
**62–63 Alamy Images:** Robert Harding Picture Library cl; **Corbis:** JeffVanuga tr; Royalty–Free cBr; **DMI:** Akiri Fujii cr; **The Picture Desk:** The Art Archive/Biblioteca d'Ajuda, Lisbon/Dagli Orti cAr; **SPL:** John Sanford bl.
**64–65 akg-images:** Bibliothèque Nationale, Paris bcrr; **GPL:** Jon Harper cr; **SPL:** Eckhard Slawik tcr, tr; John Sanford bcr; Pekka Parviainen tcl; Sheila Terry clB; **Tunc Tezel:** bcl.
**66–67 AAO:** Photograph by David Malin c; **SPL:** ESA cl; John Chumack crr; Rev. Ronald Royer cr.
**68–69 BAL:** Private Collection/Archives Charmet bcr; courtesy of the **Archives, California Institute of Technology:** bl; **Corbis:** Stapleton Collection tl; **NOAO:** bcrr (M97), br; Jeff Hageman/Adam Block cr; Joe Jordan/Adam Block bcrr (M82); N.A. Sharp bcrr (M81); Peter Kukol/Adam Block tr; Yon Ough/Adam Block bcrr (M108).
**70–71 Corbis:** Digital image © 1996 Corbis; original image courtesy of NASA clA; **Credner:** bcrr, tcr; **NAOJ:** H. Fukushima, D. Kinoshita, and J. Watanabe tr; **Nature Publishing Group** (www.nature.com): Victor Pasko bcrA; **Polar Image/Pekka Parviainen:** cr; **SPL:** Chris Madeley c; Magrath/Folsom br; Stephan J Krasemann bl.
**72–73 DK Images:** bl; Andy Crawford tr; **GPL:** Dave Tyler cAl; Robin Scagell c; **NASA:** C. Mayhew and R. Simmon (NASA/GSFC), NOAA/NGDC, DMSP Digital Archive cl; **SPL:** Frank Zullo clB.
**74–75 Credner:** cr (Telescope); **DK Images:** cl, tcr, tr; **GPL:** Robin Scagell bcr, bcrr, br, cr (Binoculars), crr (Focusing 1), crr (Focusing 2), crr (Focusing 3), crr (Focusing 4), crr (Focusing 5); courtesy of **John W. Griese:** bl; **SPL:** Frank Zullo c.
**76–77 akg-images:** Museo delle Scienze, Florence bcl; **DK Images:** clA, clB, tclB; Andy Crawford bcr, bl; **GPL:** Optical Vision bcll; Robin Scagell bcrr, br, cr (Nagler), cr (Plossl), crrA, crrB, tr.
**78–79 DK Images:** bcr, br (inset), bl, cBl; Andy Crawford cr, crr; **GPL:** Celestron International tcl; Philip Perkins br; Robin Scagell bcl, bcll, bcrr, tcr; courtesy of **iLanga, Inc.** (www.ilangainc.com): clA; courtesy of **Main-Sequence Software Inc.:** clB.
**80–81 NASA:** JSC.
**82–83 akg-images:** Bibliothèque Municipale, Boulogne-sur-Mer tr; Erich Lessing crr; **Alamy Images:** Tor Eigeland cAr; ViewStock bl; **BAL:** Bibliothèque National, Paris, France cr; British Library, London, UK bcr; **Corbis:** Danny Lehman bcl; Jason Hawkes tcl; **DK Images:** The British Museum, London br; The British Museum, London/Alan Hills tcr; **GPL:** Robin Scagell cl; Tatsuo Nakagawa cArr.
**84–85 akg-images:** Musée du Louvre, Paris crr; **BAL:** British Library, London, UK c; Stapleton Collection, UK cr; **Corbis:** Bettmann cl; Roger Ressmeyer crB; **DK Images:** Courtesy of the National Maritime Museum, London/Tina Chambers br; **The Picture Desk:** The Art Archive/Museo Nazionale Roman, Rome/Dagli Orti cll.
**86–87 akg-images:** bcrA, crr; **Alamy Images:** images-of-france tlB; **BAL:** Orlicka Galerie, Rychnov, and Kneznou, Czech Republic brA; **Corbis:** Archivo Iconografica, S.A. bcrB; Carl and Ann Purcell tclB; Paul Almasy bl; The Stapleton Collection cr; **GPL:** Richard Hook tcrB; **The Picture Desk:** The Art Archive/British Library, London, UK cl; The Art Archive/Maritiem Museum Prins Hendrik, Rotterdam/Dagli Orti tr.
**88–89 Alamy Images:** Adrian Chinery bl; **BAL:** Courtesy of the Warden and Scholars of New College, Oxford tr; **Corbis:** Arthur Thévenart bcl; Bettmann bcr, cArr; **DK Images:** Courtesy of the Science Museum, London/Clive Streeter cAr; courtesy of the Science Museum, London/Dave King crB, tcl; **© National Maritime Museum, London:** (D7061) blr; **The Picture Desk:** The Art Archive/National Palace Mexico City/Dagli Orti tl; The Art Archive/Private Collection/Eileen Tweedy cll; **Science and Society Picture Library:** Science Museum, London cAl, cl.
**90–91 BAL:** Philip Mould, Historical Portraits Ltd call; **Corbis:** Christel Gerstenberg bcr; Hulton–Deutsch Collection cBl; **DK Images:** cAl; courtesy of the National Maritime Museum, London/Tina Chambers cr, tcr; courtesy of the Science Museum, London/Clive Streeter bcll; **GPL:** bl, r; **SPL:** Dr Jeremy Burgess clB; Freeman D Miller bclA; Harvard College Obs. tclB.
**92–93 Alamy Images:** ImageState cl; Popperfoto tcl; **Corbis:** Bettmann bl; Dennis di Cicco c; Hulton–Deutch Collection cAl; London Aerial Company crB; Roger Ressmeyer br; **SPL:** bclA; David Parker cr; Hale Observatories tr.
**94–95 AAO:** Photography from UK Schmidt plates by David Malin crr; **Corbis:** Michael S Yamashita bl; **ESO:** (VLT KUEYEN + FORSZ) tcl; **Gemini Observatory/Association of Universities for Research in Astronomy:** b, cAl; Univ. of Hawaii Institute of Astronomy/Michael Liu/NSF clA; **A.S.Brun (CEA/Saclay), M.S. Miesch (NCAR), and J. Toomre (JILA/Univ. of Colorado):** tcr; **SST, Royal Swedish Academy of Sciences, LMSAL:** cr; **SPL:** Fred Espenak trB.
**96–97** Courtesy of the **NAIC–Arecibo Observatory, a facility of the NSF:** tr; **DMR, COBE, NASA, Two-Year Sky Map:** bl; **Corbis:** Roger Ressmeyer tcr; **ESO:** (VLT) bcl; **NASA:** HST/Susan Tereby (Extrasolar Research Corp.) cl; Jet Propulsion Laboratory/California Institute of Technology/Mars Exploration Rover br; JPL cll, tl; **SPL:** NASA cBr.
**98–99 akg-images:** tl; **Corbis:** Bettmann bcl; Michael Nicholson cBr; Schenectady Museum; Hall of Electrical History Foundation tr; **DK Images:** crB; **Getty Images:** Hulton Archive bl; **SPL:** Detlev van Ravenswaay cBl, cl; NASA cAl; Novosti brl.
**100–101 Alamy Images:** Popperfoto tclB; **Corbis:** Leonard de Selva cAl; Roger Ressmeyer tl; **NASA:** Asif Siddiqi cl, crA; JSC/James McDivit c; JPL cr; KSC bl; Langley Research Center bcl; **NSSDC/GSFC/NASA:** Ranger 8 br; **SPL:** Novosti crB, tlB, tr.
**102–103 Corbis:** Bettmann tclB; **Getty Images:** MPI crB; **NASA:** JSC bcl, bcr, crA; KSC c; Langley Research Center tcr; MSFC br, clB; **NSSDC/GSFC/NASA:** Lunar Orbiter 3 tcrB; **SPL:** NASA cBr; Novosti cl.
**104–105 Alamy Images:** Popperfoto tcl; **DK Images:** Courtesy of Bob Gathany/Andy Crawford bcr; **Getty Images:** NASA tclB; NASA; Photo by Time Life Pictures/Life Magazine, Copyright Time Inc./Time Life Pictures cBl; **NASA:** JSC bcll, bl, brl, cBr, clA, clB, crB, tlB, tr; KSC bcl, cr; **NSSDC/GSFC/NASA:** Apollo 11 brA.
**106–107 Corbis:** Reuters crB; **ESA:** tr; ISO/ISOPHOT and M. Haas, D. Lemke, M. Stickel, H. Hippelein, et al CAr; **Getty Images:** Luis Enrique Ascui bl; **Landsat 7 satellite image courtesy of NASA Landsat Project Science Office and USGS National Center for Earth Resources Observation Science:** bcl, cBBl; **NASA:** HST tcr; JSC b, clA; MSFC cr; Skylab cAl; **SPL:** A Sokolov/ASAP tcl; Space Imaging cBl.
**108–109 Corbis:** Jeff Albertson cr; **NASA:** Ames Research Center cBr; Ames Research Center/Pioneer Project clA; JPL br, tcl; JPL – Caltech crr; **NSSDC/GSFC/NASA:** Mariner 10 cBl, cBll; Mariner 4 cAl, tclB; Mariner 9 tcr, tr; Venera 13 bcl, clB.
**110–111 Corbis:** Joseph Sohm; ChromoSohm Inc. cr; Roger Ressmeyer bcr; Sygma/Georges de Keerle br; **ESA:** crr; AOES Medialab tcrrB; MPAE, Courtesy of Dr H.U. Keller crA; **NASA:** ESA/ASU clB; JHU Applied Physics Laboratory/Southwest Research Institute tr; JPL cAl, tcl; JPL/Cassini is a cooperative project of NASA, the ESA, and the Italian Space Agency. The JPL, a division of Caltech, manages the Cassini mission for NASA's Office of Space Science, Washington, D.C. bcl; JPL/Cornell tclB, tl; JPL/Galileo Mission cll; JPL/JHU/Applied Physics Laboratory tcrB; JPL/STScI bl; JPL/ASU cBl.
**112–113 NASA:** JPL/STScI.

**114–115 TRACE**.
**116–117 akg-images:** cl; **NASA:** JPL bcl, cr.
**118–119 Corbis:** Yann Arthus-Bertrand crA; **NASA:** Erich Karkoschka (ASU Lunar and Planetary Lab) and NASA tcr; **SPL:** tcl.
**120–121 NASA:** KSC trB; MSFC crA; **SPL:** Julian Baum crB; **SOHO:** brA, c; **TRACE:** br.
**122–123 Alamy Images:** Steve Bloom Images bcr; **Science and Society Picture Library:** Science Museum, London cAr; **SPL:** Chris Butler trB; Jerry Rodriguess tr; John Chumack cBl; NOAO tcl; **SOHO:** bcl, cBr, l; **TRACE:** crr; A. Title (Stanford Lockheed Institute) cr.
**124–125** Courtesy of **Andrew E. Potter:** cBr; **SPL:** A.E. Potter and T.H. Morgan cr; Fred Espenak cAl; NASA c.
**126–127 GPL:** NASA/JPL/Northwestern Univ. br, tcl, tr; **NASA:** JPL/Northwestern Univ. cr, crr; **NSSDC/GSFC/NASA:** Mariner 10 bcr, bcrr, bl, cl; **SPL:** NASA brl; **USGS:** Mariner 10 Image Project bcl.
**128–129 NASA:** JPL bcr; **SPL:** NASA c.
**130–131 NASA:** Ames Research Center clll; JPL cl, tcl, tcr, tcrB, cll; **NSSDC/GSFC/NASA:** Magellan tclB, trl; Venera 13 clB; Venera 4 tl; **SPL:** Julian Baum bl; NASA cr, trB.
**132–133 NASA:** JPL bcr, bl, br, cAr, cArr, cBr, cl, clll, tcl; **NSSDC/GSFC/NASA:** Magellan cArrr, cll, tclB, trB; **SPL:** David P. Anderson, SMU/NASA bcl.
**134–135 NASA:** JPL bcr, bcrr, br, cAr, tcl, tcr, tl crB; **SPL:** David P. Anderson, SMU/NASA bl, crA.
**136–137 NASA:** JPL bcl, bcr, br, cAr, cArr, cl, clll, tl, tr; **NSSDC/GSFC/NASA:** Magellan bl, cbr, cll.
**138–139 NASA:** GSFC. Image by Reto Stöckli, enhancements by Robert Simmon c; **SPL:** Emilio Segre Visual Archive/American Institute of Physics crB.
**140–141 Corbis:** Jamie Harron/Papillio tcr; **DK Images:** bcr (fungi); Andrew Butler bcr (plant); Geoff Brightling bcr (animal); M.I. Walker bcr (protist); **FLPA – Images of Nature:** Frans Lanting tc; **SPL:** Scimat bcr (moneran).
**142–143 Alamy Images:** FLPA bclA; **Corbis:** br; image by Digital image © 1996 Corbis; original image courtesy of NASA cAl; Jon Sparks bcr; Kevin Schafer tr; Lloyd Cluff tl; Michael S Yamashita crB; Robert Gill/Papilio cAll; Sygma/Pierre Vauthey bcl; **NASA:** ASF/JPL cr; **National Geographic Image Collection:** Image from *Volcanoes of the Deep*, a giant screen motion picture, produced for IMAX Theaters by the Stephen Low Company in association with Rutgers Univ. Major funding for the project is provided by the National Science Foundation. bl.
**144–145 Corbis:** Craig Lovell cBl, tcl; Macduff Everton bl; **Landsat 7 satellite image courtesy of NASA Landsat Project Science Office and USGS National Center for Earth Resources Observation Science:** clA; **NASA:** JSC – Earth Sciences and Image Analysis r.
**146–147 Corbis:** tcr; Elio Ciol cAll; Galen Rowell br, crr; image by Digital image © 1996 Corbis; original image courtesy of NASA bl; Layne Kennedy bcl; Marc Garanger tr; Tom Bean tl; **NASA:** GSFC/JPL, MISR Team cl; JSC – Earth Sciences and Image Analysis cAl; JPL cr.
**148–149 Michael Light** (www.projectfullmoon.com): c.
**150–151 akg-images:** clA; **Corbis:** Roger Ressmeyer cAl, crr; **ESA:** Space-X, Space Exploration Institute br; **Galaxy Contact:** NASA cr; **GPL:** Thierry Legault tr; **NASA:** JSC cl; MSFC bl; **NSSDC/GSFC/NASA:** Lunar 3 cBrr; **Scala Art Resource:** Biblioteca Nazionale, Florence, Italy cBr; **SPL:** ESA, Eurimage tclB; **USGS:** crB.
**152–153 Berkeley Cosmology Group:** tcl.
**154–155 NASA:** JSC bl; JPL cAr, tl; **NSSDC/GSFC/NASA:** Apollo 11 bcl; Apollo 12 bcrr; Apollo 17 tcr; Galileo cBl; Lunar Orbiter 5 br, cl; Ranger 9 crA; USGS/Clementine crB; **SPL:** John Sanford bcr, cAl.
**156–157 Michael Light** (www.projectfullmoon.com): b; **NASA:** JSC cAl, cAr, tcl; **NSSDC/GSFC/NASA:** Apollo 17 tr.
**158–159 U.S. Department of Energy:** bcl; **ESA:** Space-X, Space Exploration Institute clll; **GPL:** NASA bcll, bcr, cl; **NASA:** JPL/USGS tr; Lunar Prospector br; The Clementine Project crB; **NSSDC/GSFC/NASA:** Apollo 15 tcl; Lunar Orbiter 3 cll; Lunar Orbiter 4 tcr.
**160–161 NASA:** JPL br; JPL/ASU cr; **USGS:** c.
**162–163 ESA:** DLR/FU Berlin (G. Neukum) cAr (Phobos), cl; illustration by Medialab, ESA 2001 clllB; **NASA:** JPL cll, clllA, tcl, tclB; JPL/Cornell Univ./Mars Digital bl; JPL/MSSS tcr, tr; **NSSDC/GSFC/NASA:** JPL/Mars Global Surveyor clll; Viking Orbiter 1 tl; Viking Orbiter 2 cAr (Deimos).
**164–165 ESA:** DLR/FR Berlin (G. Neukum) bcr; DLR/FU Berlin (G. Neukum) tcl; **NASA:** JPL brA, trB; JPL/ASU cl, tcr; JPL/MSSS bcl, bl, cAl, cAll, cAr, cArr, cBl.
**166–167 ESA:** DLR/FU Berlin (G. Neukum) cAr, tcr, trB; **NASA:** JPL/MSSS cAl, tcrB, tl; JPL/USGS b.
**168–169 ESA:** DLR/FU Berlin (G. Neukum) bcl, br, cAl, cl, tcl, tcll; **NASA:** JPL/ASU tlB; JPL/MSSS bl, cll, cr, crA; JPL/USGS tr.
**170–171 ESA:** DLR/FU Berlin (G. Neukum) bcr, bl, cl, crB; OMEGA br; **NASA:** JPL cll; JPL/Cornell cAr, tcr; JPL/MSSS bcl, tcl.
**172–173 ESA:** DLR/FU Berlin (G. Neukum) bcl, bcrr, crA, tcl; **NASA:** JPL/ASU blA; JPL/MSSS br, cAl, cll, crB, tr; JPL/USGS tcr; Mars Global Surveyor/USGS bcr; Mars Orbiter Laser Altimeter (MOLA) Science Team cl; **NSSDC/GSFC/NASA:** Viking Orbiter 1 clll.
**174–175 NASA:** JPL tcr; JPL/Cornell b, clA, tcl, tcrB, tr.
**176–177 GPL:** NASA/JPL/ASU c; **NASA:** HHT (STScI/AURA); NASA/ESA, John Clarke (Univ. of Michigan) cr.
**178–179 NASA:** HST/Dr. Hal Weaver and T. Ed Smith (STScI) cBr; HST/ESA, and E.Karkoschka (ASU) cBl; JPL tr; JPL – Caltech cAr; JPL/Cornell crB; JPL/STScI tl. **180–181 DK Images:** Andy Crawford crA; **GPL:** NASA/JPL/DLR (German Aerospace Center)/ASU br; **Laurie Hatch Photography/Lick Observatory:** cBl; **NASA:** JPL/Cornell Univ. bl, cAll, cl, cll; JPL/DLR (German Aerospace Center) cAr, tcr, tcrr; JPL/Lowell Obs. cAl; JPL/ASU tl; courtesy of **Scott S. Sheppard, University of Hawaii:** bcl.
**182–183 GPL:** NASA/JPL tcl; NASA/JPL/USGS r; **NASA:** JPL/PIRL/ASU cl; JPL/ASU bl; JPL/ASU/LPL bcl.
**184–185 BAL:** Private Collection crB; **GPL:** NASA/JPL br; NASA/JPL/DLR (German Aerospace Center) tl; **NASA:** JPL bl; JPL/ASU tr; JPL/Brown Univ. bcl, cBl; JPL/DLR (German Aerospace Center) cAr, tcr.
**186–187 GPL:** NASA/JPL/USGS c; **NASA:** HST/ESA, J. Clarke (Boston Univ.), and Z. Levay (STScI) bcr.
**188–189 NASA:** JPL tl, trB; JPL/STScI bcrA, c, cBl, cBr, crB, tcl; JPL/Univ. of Colorado tr.
**190–191 NASA:** JPL tcl; JPL/STScI bcl, bl, cAl, cAr, cBr, cl, cll, r.
**192–193 NASA:** JPL bcl, bcrrA, br, tcl, tcr; JPL/STScI bcll, bcr, bl, cl, clA, cr, tclB, tr.
**194–195 akg-images:** Huygens Museum Hofwijck/Nimatallah bl; **NASA:** ESA/JPL/ASU cAl; JPL/Cassini is a cooperative project of NASA, the ESA, and the Italian Space Agency. The JPL, a division of Caltech, manages the Cassini mission for NASA's Office of Space Science, Washington, D.C. tcl; JPL/STScI bcl, br, cBr, cl, clA, crA, tcr, trB; JPL/ASU bcr, tl.
**196–197 Corbis:** Roger Ressmeyer crB; **GPL:** JPL/STScI c; **W. M. Keck Observatory:** Courtesy Lawrence Sromovsky, UW-Madison Space Science and Engineering Center tr; **NASA:** JPL cBr.
**198–199 Corbis:** Sygma cAr; **Brett Gladman, Paul Nicholson, Joseph Burns, and JJ Kavelaars**, using the 200 inch Hale Telescope: br; **NASA:** HST/Erich Karkoschka (ASU) tl; JPL bcl, bcr, bcrr, bl, cBr, cl, cll, tcr, tr; JPL/USGS cBl; **NSSDC/GSFC/NASA:** Voyager 2 cAl.
**200–201 GPL:** NASA/JPL c; **NASA:** JPL crB; JPL/HST cr.
**202–203 Liverpool Astronomical Society:** With thanks to Mike Oates br; **Corbis:** Roger Ressmeyer bcll; **NASA:** JPL bcl, bl, tl; JPL/USGS cBr, tr; **NSSDC/GSFC/NASA:** Voyager 2 cl, cll; courtesy of **A. Tayfun Oner:** bcr; **SPL:** NASA cAl.
**204–205 Getty Images:** Michal Cizek/ Stringer/ AFP bl; **W.M. Keck Observatory:** Marcos van Dam bc; **NASA:** ESA, H. Weaver (JHU/APL), A. Stern (SwRI), and the HST Pluto Companion Search Team; ESA, J. Parker (Southwest Research Institute), P. Thomas (Cornell University), L. McFadden (University of Maryland, College Park), and M. Mutchler and Z. Levay (STScI) cr
**206–207 Corbis:** Bettmann bcl; Jonathan Blair crA; **GPL:** Michael Stecker cAr; **NASA:** HST/M. Brown (Caltech) bl; JPL – Caltech br, cBr; **NSSDC/GSFC/NASA:** Denis Bergeron, Canada cl.
**208–209 DK Images:** tr; **NASA:** HST/R. Evans and K. Stapelfeldt (JPL) cl.
**210–211 Corbis:** R Kempton cBl; **GPL:** NASA/JPL br; **NASA:** HST/Ben Zellner (Georgia Southern Univ.), Peter Thomas (Cornell Univ.) bl, blAr; JPL cr; JPL – Caltech cll, cill; JPL/JHU/APL bcl, tcl; JPL/USGS clll; **NSSDC/GSFC/NASA:** Goldstone DSC antenna-radar cl; courtesy of **Osservatorio Astronomico di Palermo Giuseppe S. Vaiana:** trB; **SPL:** Dennis Milon cAr; Mark Garlick cArr.
**212–213 GPL:** NASA/JPL/JHU/APL r; **NASA:** JPL/JHU/APL bl, clA, clB; **SPL:** NASA tcl.
**214–215 Corbis:** Jonathan Blair crA; **DK Images:** br; **NASA:** JPL/Brown Univ. cr; JPL/USGS cAr; **SPL:** Pekka Parviainen cAl; **SOHO:** cBr.
**216–217 akg-images:** cll; **DK Images:** cl; **ESO:** Peter Stättmayer of the Munich Public Obs. bcr; **DMI:** Akira Fujii cr; **NOAO:** Roger Lynds bl; **SPL:** Detlev van Ravenswaay tcl; John Thomas tr; Pekka Parviainen bcl; Rev. Ronald Royer cBl; **James V. Scotti, Spacewatch Project of the Lunar and Planetary Laboratory, ASU**. © 1994 by the Arizona Board of Regents. Reproduced by permission: br.
**218–219 Corbis:** Gianni Dagli Orti cl; **ESA:** MPAE, 1986, 1996 cll; **Rolando Ligustri/CAST Circolo AStrofili Talmassons, Italy:** cArr, tcr; courtesy of **Lowell Observatory:** br; **NASA:** JPL crA; JPL – Caltech bcr, cBr; **SPL:** cAr; Frank Zullo bl; Richard J. Wainscoat, Peter Arnold Inc. tcl; STScI/NASA bcrr.
**220–221 Corbis:** Jonathan Blair bcr; **DK Images:** Harry Taylor bcrr, cBr; **GPL:** Arne Danielsen cll; **Getty Images:** NASA/AFP tr; **NASA:** Carnegie Mellon Univ./Robotic Antarctic Explorer (LORAX) br; **© The Natural History Museum, London:** bcl, bcll, cBl; **SPL:** bl, cl; David McLean cAl.
**222–223 Alamy Images:** H.R. Bramaz cAr; **Corbis:** Matthew McKee/Eye Ubiquitous bl; **DK Images:** courtesy of the Natural History Museum, London/Colin Keates cll; **GPL:** UWO/Univ. of Calgary clll; **Muséum National d'Histoire Naturelle, Paris:** Département Histoire de la Terre bcll; **NASA:** JSC br; KSC crB; **© The Natural History Museum, London:** bcl, bcr, tr; **SPL:** D. van Ravenswaay cBl; Michael Abbey cl; Pascal Goetgheluck/Francois Robert tcl.
**224–225 SPL:** Tony and Daphne Hallas.
**226–227 Corbis:** Image by © National Gallery Collection; by kind permission of the Trustees of the National Gallery, London cr; **NASA:** D. Dixon (UCR), D. Hartmann (Clemson), E. Kolaczyk (U. Chicago) cBr; **SPL:** Chris Butler cAl; Tony and Daphne Hallas tcl; Milky Way map and 3-D view by **Planetary Visions**.
**228–229** Reprinted by permission of **American Scientist**, magazine of Sigma Xi, the Scientific Research Society: cAr; **NASA:** HST/Jeff Hester (ASU) tcl; **NOAO:** Adam Block bl; **NRAO:** cr, crr; **SPL:** bcr; **B.J. Mochejska (CfA), J. Kaluzny (CAMK)**, 1m Swope Telescope: bcrr.
**230–231** Courtesy of **Andy Steere:** bcr; **Corbis:** Bettmann br; **GPL:** Andrea Dupree, Ronald Gilliland (STScI)/NASA/ESA crA; Robin Scagell crB; **SOHO:** tc.
**232–233 NASA:** HST/C.A. Grady (NOAO, NASA, GSFC), B. Woodgate (NASA, GSFC), F. Bruhweiler and A. Boggess (Catholic Univ. of America), P. Plait and D. Lindler (ACC, Inc., GSFC), and M. Claupin (STScI) br; HST/Wolfgang Brandner (JPL/IPAC), Eva K. Grebel (Univ. Washington), You-Hua Chu (Univ. Illinois Urbana-Champaign) tcl.
**234–235** Courtesy of **Andy Steere:** bl; **Chandra:** NASA/STScI/R. Gilliand et al tcr.
**236–237 AAO:** Photograph by David Malin cAl; courtesy of **Armaugh Observatory:** bcr; **NASA:** HST/ESA and HHT (STScI/AURA) cr; HST/J. Hester and P. Scowen (ASU) cl; HST/J. Hester (ASU) tr; HST/Kirk Borne (STScI) tcr; **NOAO:** Gemini Obs./Travis Rector, Univ. of Alaska, Anchorage bl; **C. and F. Roddier (IfA, Hawaii), CFHT:** brA; **Chris Sauer:** tcl.
**238–239 ESO:** J. Alves (ESO), E. Tolstoy (Groningen), R. Fosbury (ST–ECF), and R. Hook (ST–ECF) (VLT) cl; Leonardo Testi (Arcetri Astrophysical Obs., Florence, Italy (NTT + SOFI) tl; Mark McCaughrean (Astrophysical Institute, Potsdam, Germany (VLT, ANTU, and ISAAC) tr; **NOAO:** T.A. Rector (NOAO) and HHT (STScI/AURA/NASA) bl; **SPL:** Chris Madeley c; **© Smithsonian Institution:** br.
**240–241 Richard Crisp** (www.narrowbandimaging.com): tcr; **NASA:** HST/H. Ford (JHU), G. Illingworth (UCSC/LO), M. Clampin (STScI), G. Hartig (STScI), the ACS Science Team and ESA tcl; JPL – Caltech/S. Carey (Caltech) bcr; **NOAO:** Michael Gariepy/Adam Block bcl; N.A. Sharp, REU Program cBr; T.A. Rector (NRAO/AUI/NSF and NOAO) and B.A. Wolpa (NOAO) cl; **SPL:** J-C Cuillandre/Canada–France–Hawaii Telescope tr; Mount Stromlo and Siding Spring Observatories bcrr.
**242–243 ESO:** (VLT, ANTU + ISAAC) r; **NASA:** HST/ESA, STScI, J. Hester, and P. Scowen (ASU) bl; **NOAO:** T.A. Rector (NRAO/AUI/NSF and NOAO) and B.A. Wolpa (NOAO) tcl.
**244–245 2MASS:** E.Kopan (IPAC)/Univ. of Massachusetts tcr; **NASA:** HST/HHT (AURA/STScI) bcr; HST/HHT (STScI/AURA) cAl; JPL – Caltech tcl; JPL – Caltech/Spitzer Space Telescope br; JPL – Caltech/Univ. of Wisconsin tr; **NOAO:** Todd Boroson blA; **SPL:** National Optical Astronomy Observatories bcl.
**246–247 Corbis:** Bettmann cr; **GPL:** Gordon Garradd cl; **SOHO:** bc, tr; **TRACE:** tcl.
**248–249 GPL:** Deep Sky Survey cBr; Duncan Radbournel bl; **W. Holland (JAC) et al:** tcr; **DMI:** Akira Fujii tcl, tr, cAr; **NASA:** HST/HHT (AURA/STScI) clll; courtesy of **Joe Orman:** bcrA; **SPL:** Dr. Fred Espenak cl; Eckhard Slawik bcl, bcr; NOAO cll.
**250–251 Matt BenDaniel** (http://starmatt.com): bl; **Credner:** clA; **NASA:** HST/Bruce Balick (Univ. of Washington), Jason Alexander (Univ. of Washington), Arsen Hajian (U.S. Naval Obs.), Yervant Terzian (Cornell Univ.), Mario Perinotto (Univ. of Florence, Italy), Patrizio Patriarchi (Arcetri Obs. Italy) bcr; HST/Bruce Balick (Univ. of Washington), Vincent Icke (Leiden Univ., The Netherlands), Garrett Mellema (Stockholm Univ.) crB; HST/HHT (STScI/AURA) c; HST/HHT (STScI/AURA) cr; HHT (STScI/AURA); D. Garnett (University of Arizona) crA.
**252–253 NASA:** HST/Andrea Dupree (Harvard–Smithsonian CfA), Ronald Gilliland (STScI) and ESA cAl; HST/Bruce Balick (Univ. of Washington), Vincent Icke (Leiden Univ., The Netherlands), Garrett Mellema (Stockholm Univ.) br; HST/Jon Morse (Univ. of Colorado) tl; HST/NOAO, ESA, the Hubble Helix Nebula Team, M. Meixner (STScI), and T.A. Rector (NRAO) cr, tr; **NOAO:** Adam Block bcr; **H. Olofsson (Stockholm Observatory) et al:** bcl; **SPL:** Eckhard Slawik clA; John Chumack cl; Royal Obs., Edinburgh/AAO bl.
**254–255 R. Corradi (Isaac Newton Group), D. Goncalves (Inst. Astrofisica de Canarias):** clB; **W. M. Keck Observatory:** U.C. Berkeley Space Sciences Laboratory cBr; **NASA:** HST/Andrew Fruchter and ERO Team (Sylvia Baggett (STScI), Richard Hook (ST–ECF), and Zoltan Levay (STScI)) br; HST/ESA/Hans van Winckel (Catholic Univ. of Leuven, Belgium) and Martin Cohen (Univ. of California Berkely) tcl; HST/ESA, HEIC, and HHT (STScI/AURA) bcll; HST/HHT (STScI/AURA); W. Sparks (STScI) and R. Sahai (JPL) bcl; STScI cAr; **SPL:** NOAO crA.
**256–257 ESO:** (ANTU UT1 + TC) bl; **NASA:** HST/ESA and A. Zijlstra (UMIST, Manchester, UK) r.
**258–259 Chandra:** NASA/HST/J. Morse/K. Davidson bcl; **NASA:** ESA and Valentin Bujarrabal (Observatorio Astronomico Nacional, Spain) clA; HST/Raghvendra Sahai and John Trauger (JPL), the WFPC2 Science Team r; HST/HHT (STScI/AURA) tcl; **SPL:** Dr Kris Davidson cBl.
**260–261 ESO:** W. Brandner (UIUC) et al, ESO, 1.54-m Telescope, Chile bcr; **NASA:** HST br; HST/Matt Bobrowsky (Orbital Sciences Corporation) bcl; HST/HHT (AURA/STScI) tr; HST/Yves Grosdidier (Univ. of Montreal and Observatoire de Strasbourg), Anthony Moffat (Univ. of Montreal), Gilles Joncas (Univ. Laval), Agnes Acker (Observatoire de Strasbourg) cl; **NOAO:** Peter and Suzie Erickson/Adam Block tcl; **© Observatoire de Paris:** bl; **SPL:** Celestial Image Co. tl.
**262–263 Chandra:** NASA/U. Mass/D. Wang et al c; Univ. of Leicester tcl; **NASA:** HST/Peter Challis and Robert Kirshner (Harvard–Smithonian Center for Astrophysics), Peter Garnavich (Univ. of Notre Dame), and the SINS Collaboration bl.
**264–265 AAO:** Royal Obs., Edinburgh. Photograph from UK Schmidt plates by David Malin crB; **Chandra:** G. Pavlov, M. Teter, O. Kargaltsev, D. Sanwal (PSU), CXC, NASA br; NASA/CXC/SAO tcl; NASA/SAO/CXC cll; **ESO:** M. van Kerkwijk (Institute of Astronomy, Utrecht), S. Kulkarni (Caltech), VLT Kueyen cl; **NASA:** Compton Gamma Ray Obs. blAr; HST/Fred Walter (State Univ. of New York at Stony Brook) cAl; HST/Jeff Hester (ASU) crA; HST/HHT (AURA/STScI) bclA; William P. Blair and Ravi Sankrit (JHU) tcr.
**266–267 ESO:** (VLT KUEYEN + FORS2) l; **NASA:** HST/Jeff Hester and Paul Scowen (ASU) bcr, cBr; **SPL:** Dr S. Gull and Dr J. Fielden cr; GSFC/NASA tr.
**268–269 Chandra:** NASA/CXC/GSFC/U. Hwang et al bcr; **Corbis:** bl; **GPL:** Michael Stecker cAl; **NASA:** ESA, R.Sankrit, and W. Blair (JHU) tr; HST/Dave Bennett (Univ. of Notre Dame, Indiana) br; HST/ESA and HHT (STScI/AURA) tcr; HST/ESA, CXO, and P. Ruiz-Lapuente (Univ. of Barcelona) bcl; HST/H. Richer (Univ. of British Columbia) clA; HST/NOAO, Cerro Tololo Inter-American Obs. bcrr; **NOAO:** Doug Matthews and Charles Betts/Adam Block cr; **SPL:** Dr S. Gull and Dr J. Fielden cBl; Royal Greenwich Obs. cAl (inset).
**270–271 ESO:** Mark McCaughrean (Astrophysical Institute Potsdam, Germany) (VLT ANTU + ISAAC) r; **Astronomische Vereinigung Kärnten, Austria:** cAl.

**272–273 GPL:** Damian Peach cAl, cBl, crr; Robin Scagell bcl; courtesy of **Padric McGee, University of Adelaide:** cll; **NASA:** HST/K.L. Luhman (Harvard–Smithsonian Center for Astrophysics, Cambridge, Mass.), G. Schneider, E. Young, G. Rieke, A.Cotera, H. Chen, M. Rieke, and R. Thompson (Steward Obs., ASU) tl; **NOAO:** crA; **Astronomische Vereinigung Kärnten, Austria:** br; **Johannes Schedler, Panther Observatory, Austria:** cr; **SPL:** Dr Fred Espenak bcr; Eckhard Slawik bl, cl; John Sanford tr.
**274–275 AAO:** Photograph by David Malin l; **GPL:** Damian Peach cBr; **The Picture Desk:** The Art Archive/National Library, Cairo/Dagli Orti bcr; **SPL:** Tony and Daphne Hallas cr.
**276–277 AAO:** Photograph by David Malin bcrr; **GPL:** Duncan Radbourne bcr; **SPL:** Celestial Image Co. bl, tcl; George Fowler brl; John Sanford br, cr; **Matthew Spinelli:** crr; courtesy of **Thomas Williamson, New Mexico Museum of Natural History and Science:** crrr.
**278–279 Credner:** tcl; **NASA:** HST tr; HST/ESA and HHT (STScI/AURA) b; HST/F. Paresce, R. Jedrzejewski (STScI) and ESA cl; **SPL:** Mark Garlick cr.
**280–281 Credner:** tr; courtesy of **Mark Crossley:** cAr (1); **GPL:** Damian Peach cAr (2); DSS (N) tl; **infoastro.com/Victor R. Ruiz:** cl; **NASA:** HST/Margarita Karovska (Harvard–Smithsonian Center for Astrophysics) crA; **NOAO:** Tom Bash and John Fox/Adam Block br; **Johannes Schedler, Panther Observatory, Austria:** cBl; **SPL:** Eckhard Slawik crB; John Sanford cll, cll (insert); courtesy of **Jerry Xiaojin Zhu, Carnegie Melon University:** bcl.
**282–283 Matt BenDaniel** (http://starmatt.com): tr; **GPL:** DSS (N) cll; DSS (S) clll; Martin Mobberley crB; Robin Scagell cBl; **Sean Lockwood and David Yeaton-Massey, Leuschner Observatory, Lafayette, CA:** cl; **NASA:** HST/F. Paresce, R. Jedrjejewski (STScI), ESA brA; **SPL:** tclB; NOAO bcr.
**284–285 Credner:** bcr; ESO: (ANTU UT1 + TC) tcr; **© 2005 Loke Tan** (www.starryscapes.com): bl; **NASA:** HST/HHT (AURA/STScI) cAr; **NOAO:** Heidi Schweiker tcl.
**286–287 ESO:** Das Universumisteine Scheibe crr; **NASA:** HST/HHT (STScI/AURA) tcr; **NOAO:** bcl, bl; N.A. Sharp, REU Program cBl; **SPL:** Celestial Image Co. cAl; Eckhard Slawik cll, cr; Jerry Lodriguss cl; Tony and Daphne Hallas br; **P. Seitzer (Univ. Michigan):** tl.
**288–289 AAO:** Photograph by David Malin bcl; **ESO:** crr; courtesy of **William Keel:** tclB; **© 2005 Loke Tan** (www.starryscapes.com): cBl; **NASA:** HST/HHT (STScI/AURA) bcll, brA; **NOAO:** bcrr, br, clA, tr; Bruce Hugo and Leslie Gaul/Adam Block bcr; Michael Gariepy/Adam Block cr; **SPL:** Dr. Fred Espenak tcl; Mount Stromlo and Siding Spring Observatories tlBr.
**290–291 Chandra:** NASA/CXC/Chuo U./Y. Tsuboi et al cBr; **ESO:** (NACO + VLT) tr; (SOFI + NTT) tcl; Jean-Luc Beuzit, Anne-Marie Lagrange (Observatoire de Grenoble, France), and David Mouillet (Observatoire de Paris-Meudon, France) bl; **NASA:** HST/ESA, C. Beichman (JPL), D. Ardila (JHU), and J. Krist (STScI/JPL) cBl, bclA; HST/S. Terebey (Extrasolar Research Corp) cr.
**292–293 NASA:** HST/ESA and HHT (AURA/STScI).
**294–295 NASA:** HST/HHT (STScI/AURA) bcll, tr; HST/HHT (STScI/AURA) cl (Sb), tc; **NOAO:** cAl, cl (E0), cl (E6), cll; Adam Block cl (E2), cl (Sa), cl (SBa), cl (SBb); Allan Cook/Adam Block bcl; Jeff Newton/Adam Block cl (S0); Jon and Bryan Rolfe/Adam Block cl (Sc); Nicole Bies and Esidro Hernandez/Adam Block cl (SBc); P. Massey (Lowell), N. King (STScI), S. Holmes (Charleston), G. Jacoby (WIYN) cllB; **SPL:** Royal Obs., Edinburgh cllA.
**296–297 AAO:** Photograph by David Malin clll, cl; **Chandra:** NASA/SAO/G. Fabbiano et al br; courtesy of **D. A. Harper, University of Chicago:** bcr; **NASA:** HST/ESA and D. Maoz (Tel-Aviv Univ. and Columbia Univ.) tr; HST/R. de Grijs (Institute of Astronomy, Cambridge, UK) crr; HST/HHT (STScI/AURA) bcl, cr; **NOAO:** cll, tcl; John and Christie Connors/Adam Block tc.
**298–299** Courtesy of **Daisuke Kawata and Brad K. Gibson, Swinburne University of Technology:** bcl; **NASA:** HST tc; HST/E. Shaya, D. Dowling/U. of Maryland, the WFPC Team crA; HST/ESA and HHT (STScI/AURA) brA; HST/ESA, Y. Izotov (Main Astronomical Obs., Kyiv, UA) and T. Thuan (Univ. ofVirginia) cl; HST/Rogier Windhorst and Sam Pascarelle (ASU) cAl; HST/HHT and A. Riess (STScI) crB; **NOAO:** Adam Block br.
**300–301 AAO:** Photograph by David Malin cr; **Mary Evans Picture Library:** crA; **NASA:** HST/HHT (STScI/AURA) cBr; HST/HHT (STScI/AURA) br; **NOAO:** bc, tl; P. Massey (Lowell), N. King (STScI), S. Holmes (Charleston), and G. Jacoby (WIYN) crr; **SPL:** Max-Planck-Institut für Radioastronomie bl; Royal Obs., Edinburgh tcr; courtesy of **www.seds.org:** cAl.
**302–303 Chandra:** NASA/CXC/SAO bcl; **NASA:** HST/T.R. Lauer (NOAO) bl; **SPL:** STScI/NASA bcr; Tony and Daphne Hallas t.
**304–305 Chandra:** NASA/CXC/SAO/PSU/CMU blAr; **NASA:** HST/HHT (STScI/AURA) bcl; **NOAO:** Jon and Bryan Rolfe/Adam Block br; Mark Westmoquette (Univ. College London), Jay Gallagher (Univ. of Wisconsin–Madison), Linda Smith (Univ. College London), WIYN, ESA, NASA clB; N.A. Sharp bl; **NRAO:** tcr; **SPL:** crA; George Bernard trB; GSFC clA; Los Alamos National Laboratory tcrB; Max-Planck-Institut für extraterrestrische Physik tl; Tony and Daphne Hallas tcl.
**306–307 Corbis:** Bettmann bcl; **NASA:** HST/Brad Whitmore (STScI) br; HST/HHT (STScI/AURA) cl; **NOAO:** cAr, crA; George Jacoby, Bruce Bohamanm, and Mark Hanna tcl; **SPL:** Celestial Image Co. bcr; Dr Rudolph Schild, Smithsonian Astrophys–ICAL Obs. cBl; Kapteyn Laboratorium tcll; NOAO cr.
**308–309 DMI:** Photograph by David Malin br; **NASA:** HST/H. Ford (JHU), G. Illingworth (UCSC/LO), M. Clampin (STScI), G. Hartig (STScI), the ACS Science Team and ESA bcl; HST/Kirk Borne (STScI) crA; HST/HHT (STScI/AURA) bcr, bl, cAl; **SPL:** Max-Planck-Institut für Astrophysik cBl; NOAO bcll; STScI/NASA cAr.
**310–311 Chandra:** NASA/MIT/F. Bayanoff et al br; **Credner:** c(background); **NASA:** W. Purcell (NWU) et al, OSSE, Compton Obs. crB; **NOAO:** Adrian Zsilavee and Michelle Qualls/Adam Block bcl; Eric Peng, Herzberg Institute of Astrophysics/NRAO/AUI tr; **NRAO:** clB, cr; **SPL:** Jodrell Bank bl; Space Science Telescope Institute cBl.
**312–313 Chandra:** X-ray (NASA/CXC/M. Karovska et al); radio 21-cm image (NRAO/VLA/J. Van Gorkom/Schminovich et al); radio continuum image (NRAO/VLA/J.Condon et al); optical (DSS U.K. Schmidt Image/STScI) cl; courtesy of **Vanderbilt Dyer Observatory:** trB; **GPL:** STScI br; **NASA:** HST/Andrew S. Wilson (Univ. of Maryland); Patrick L. Shopbell (Caltech); Chris Simpson (Subaru Telescope); Thaisa Storchi-Bergmann and F. K. B. Barbosa (UFRGS, Brazil) and Martin J. Ward (Univ. of Leicester, U.K.) cAl; HST/E.J. Schreier (STScI) bcl; HST/L. Ferrarese (JHU) bcrr; HST/HHT (STScI/AURA) crr, tcrB; HST/Walter Jeffe/Leiden Obs., Holland Ford/JHU/STScI; **NOAO:** Adam Block cr; **NRAO:** tl.
**314–315 GPL:** DSS crB; STScI crA; **NASA:** HST/A, Martel (JHU), H. Ford (JHU), M. Clampin (STScI), G. Hartig (STScI), G. Illingworth (UCO/Lick Obs.), the ACS Science Team and ESA bcrA; HST/J. Holtzman cAl; HST/HHT (STScI/AURA) tl; **NRAO:** cBl; courtesy of **Cormac Reynolds, Joint Institute for VLBI in Europe, The Netherlands:** cAr; **SPL:** © Estate of Francis Bello br; Jodrell Bank bcr.
**316–317 AAO:** AURA/Royal Obs., Edinburgh/UK Schmidt Telescope, Skyview bcr; Royal Obs. Edinburgh. Photograph from UK Schmidt plates by David Malin cBr; **Chandra:** NASA/CXC/UCI/A. Lewis et al tcrB; Pal. Obs. DSS tcr; **ESO:** (VLT UT1 + ISAAC) cr; **GPL:** DSS/California Institute of Technology/Palomar Obs. crB; **NASA:** HST/N. Benitez (JHU), T. Broadhurst (The Hebrew Univ.), H. Ford (JHU), M. Clampin (STScI), G. Hartig (STScI), G. Illingworth (UCO/Lick Obs.), the ACS Science Team, ESA c; **NRAO:** F.N. Owen, C.P. O'Dea, M. Inoue, and J. Eilek br; **SPL:** Dr. Rudolph Schild bcl; Royal Obs., Edinburgh bcll.
**318–319 Matt BenDaniel** (http://starmatt.com): bl; **Chandra:** NASA/CXC/Columbia U./C. Scharf et al bcr; **ESO:** FORS Team, 8.2 meter VLT Antu cBr; **GPL:** Robin Scagell clA; **NASA:** HST/J. English (U. Manitoba), S. Hunsberger, S. Zonak, J. Charlton, S. Gallagher (PSU), and L. Frattare (STScI) br; **NOAO:** Doug Matthews/Adam Block cAr; Local Group Galaxies Survey Team cAl; N.A. Sharp tl; **SPL:** Celestial Image Co. bcrr, cl; Jerry Lodriguss tcrB; Tony and Daphne Hallas crA.
**320–321 AAO:** Photograph by David Malin cl; **Dr. Victor Andersen (University of Alabama, KPNO),** courtesy of **W. Keel:** cAr; courtesy of **the Archives, California Institute of Technology:** crA; **Chandra:** NASA/CXC/U. Mass/Q.D. Wang et al bcrr; NASA/STScI and NOAO/Kitt Peak brAl; **NASA:** HST/J. English (Univ. of Manitoba), S. Hunsberger (Pennsylvania State Univ.), Z. Levay (STScI), S. Gallagher (Pennsylvania State Univ.), and J. Charlton (Pennsylvania State Univ.) tcl; HST/N. Benitez (JHU), T. Broadhurst (The Hebrew Univ.), H. Ford (JHU), M. Clampin (STScI), G. Hartig (STScI), G. Illingworth (UCO/Lick Obs.), the ACS Science Team, ESA cArr; HST/HHT (STScI/AURA) clB; HST/W. Keel (Univ. Alabama), F. Owen (NRAO), M. Ledlow (Gemini Obs.), and D. Wang (Univ. Mass.) br; HST/WFPC Team/STScI bcl; **NOAO:** Jack Burgess/Adam Block bcr; N.A. Sharp cBl.
**322–323 Bell Labs, Lucent Technologies:** Greg Kockanski, Ian Dell'Antonio, and Tony Tyson br; **GPL:** NASA – MSFC/Chandra/M. Bonamente et al bcl; **NASA:** HST/Andrew Fruchter and the ERO Team – Sylvia Baggett (STScI), Richard Hook (ST–ECF), Zoltan Levay (STScI) t.
**324–325 ESO:** (MPG/ESO 2.2-m + WFI) bl; Tom Theuns at the Max-Planck-Institut für Astrophysik, Garching, Germany tc; courtesy of **Sloan Digital Sky Survey:** Fermilab Visual Media Services bcl; NASA/NSF/DOE cr; **J. Shalf, Y. Zhang (UIUC) et al, GCCC:** cBr; **© Smithsonian Institution:** clB.
**326–327 NASA:** JPL.
**328–329 Credner.**
**330–331 BAL:** Private Collection, The Stapleton Collection bcr, tc; **British Library, London:** shelfmark: Harley 647, folio: f.13 cl; shelfmark: Maps.C.10.c.10, folio: 5 tr; **Corbis:** The Stapleton Collection br; **DK Images:** courtesy of the National Maritime Musem, London/James Stevenson bcl; courtesy of the National Maritime Museum, London/Tina Chambers bl.
**338–339 BAL:** National Gallery of Art, Washington D.C., USA/Lauros/Giraudon bcl; Palazzo Vecchio (Palazzo della Signoria) Florence, Italy br; **Credner:** tr; **GPL:** Damian Peach cl; Robin Scagell tl; **NOAO:** Adam Block cr.
**340–341 Credner:** br, tcl; **GPL:** Michael Stecker tl; **NOAO:** tcrB; Hillary Matthis, N.A. Sharp tcr; courtesy of **Ian Ridpath:** crA; **SPL:** Harvard College Obs. bcl.
**342–343 Credner:** bl, br, cr; **Digital Library of Dutch Literature** (www.dbnl.org): bcl; **GPL:** Robin Scagell tcl; **NOAO:** Adam Block crr; Fred Calvert/Adam Block cBl.
**344–345 British Library, London:** shelfmark: Or. 8210/S. 3326 br; **Credner:** r; **GPL:** Damian Peach tcr; **NOAO:** Gary White and Verlenne Monroe/Adam Block tcl; Jeff Cremer/Adam Block bcl; Joe Jordan/Adam Block cl.
**346–347 Corbis:** The Stapleton Collection tr; **Credner:** bcl, br; **GPL:** Damian Peach cr; **NOAO:** Elliot Gellam and Duke Creighton/Adam Block bl; Jon and Bryan Rolfe/Adam Block tl; N.A. Sharp cBl.
**348–349 akg-images:** Hessisches Landesmuseum bcr; **Credner:** br, tcl; **GPL:** Eddie Guscott cBl; **NOAO:** Adam Block cr; Burt May/Adam Block bcl.
**350–351 Corbis:** Allinari Archives/Mauro Magliani tr; **Credner:** b; **GPL:** Damian Peach tcl; Philip Perkins cr; **NOAO:** Adam Block, Jeff and Mick Stuffings, Brad Ehrhorn, Burt May, and Jennifer and Louis Goldring br; Heidi Schweiker cAr.
**352–353 Corbis:** Archivo Iconografico, S.A. bcl; **Credner:** bcr, bl, tcr; **NOAO:** Adam Block cl; T.A. Rector (NRAO/AUI/NSF and NOAO) and M. Hanna br; **SPL:** Tony Hallas cll.
**354–355 akg-images:** bcr; **Corbis:** Massimo Listri tcl; **Credner:** cl, r; **GPL:** Robin Scagell tcr; **SPL:** Jerry Lodriguss bcl.
**356–357 akg-images:** © Sotheby's br; **Credner:** tr; **NOAO:** Adam Block bl; **SPL:** John Sanford tcl.
**358–359 BAL:** Private Collection, The Stapleton Collection trB; **Credner:** bl, br; **GPL:** Robin Scagell tcrB; **NOAO:** N.A. Sharp tclB; Nigel Sharp, Mark Hanna cAr; Sharon Kempton and Karen Brister/Adam Block tcl.
**360–361 Corbis:** Arte and Immagini sr bcrr; **Credner:** bcl, bcr, tcl; **GPL:** Damian Peach cBr; Nik Szymanek/Ian King cAl; **NOAO:** cBl; REU Program cr.
**362–363 akg-images:** Museum of Fine Arts Boston/Erich Lessing br; **Corbis:** Archivo Iconografico, S.A. tcl; **Credner:** bcl, bcr, cAr; **NOAO:** Adam Block bl; Morris Wade/Adam Block clB.
**364–365 Corbis:** Gianni Dagli Orti br; **Credner:** bcl, bcr; **GPL:** Michael Stecker cBr; **NOAO:** Bill Schoening bl; Hillary Matthis, REU Program cl; N.A. Sharp, Vanessa Harley/REU Program bcrr.
**366–367 akg-images:** Erich Lessing tr; **Credner:** bcl, br, tcl; **GPL:** Robin Scagell cAr, cr; **SPL:** John Sanford cl.
**368–369 Corbis:** Bettmann bcl; **Credner:** bl, br, tr; **GPL:** Damian Peach tcr; Nik Szymanek clA; Robin Scagell cl; courtesy of **Osservatorio Astronomico di Palermo Giuseppe S. Vaiana:** cr.
**370–371 Corbis:** Richard T. Nowitz bcl; **Credner:** br, tcl; courtesy of **William McLaughlin:** cBr, crB; **NOAO:** bl.
**372–373 BAL:** Palais du Luxembourg, Paris, France/Giraudon cBl; **Corbis:** The Stapleton Collection br; **Credner:** bcr, cl; **GPL:** Robin Scagell tcl; **NOAO:** Francois and Shelley Pelletier tcr; Todd Boroson bcl.
**374–375 Credner:** l; **GPL:** Duncan Radbourne cl; Michael Stecker br, tr; **NOAO:** Jim Rada/Adam Block tcr; **The Picture Desk:** The Art Archive/Bodleian Library, Oxford bl.
**376–377 Credner:** bcl, bcr, cl; **GPL:** Michael Stecker cr; Pedro Rè tcl; **NOAO:** Michael Gariepy/Adam Block tr; **The Picture Desk:** The Art Archive/Private Collection/Marc Charmet cll.
**378–379 Corbis:** Todd Gipstein tcl; **Credner:** r; **NOAO:** Adam Block cBl; Allan Cook/Adam Block bl.
**380–381 Daniel Verschatse** (www.astrosurf.com): clA; **Credner:** br, cAr, cBl; **Mary Evans Picture Library:** crA; **GPL:** Gordon Garradd bcl; Yoji Hirose cAl; **NOAO:** Bob and Bill Twardy/Adam Block cBr.
**382–383 BAL:** Bibliothèque Nationale, Paris, France/Archives Charmet bcr; **Corbis:** Araldo de Luca cl; **Credner:** br; **GPL:** Chris Picking bl; Gordon Garradd cAr, cBl.
**384–385 Credner:** r; **GPL:** Michael Stecker clB, tcl; **NOAO:** bcr; Todd Boroson bl.
**386–387 Corbis:** Andrew Cowin cr; Archivo Iconografico, S.A. bcl; **Credner:** bcr, bl, cAr; **GPL:** Pedro Rè tcr; **SPL:** Rev. Ronald Royer tcl.
**388–389 AAO:** Royal Obs. Edinburgh. Photograph from UK Schmidt plates by David Malin tcr; **Credner:** bl, br, cAl, cr; **ESO:** (VLT UT1 + FORS1) trB; **NOAO:** T.A Rector bcl.
**390–391 BAL:** Musée Conde, Chantilly, France/Giraudon br; **Credner:** bcr, bl; **GPL:** DSS cr; Gordon Garadd tcl; **NOAO:** Adam Block cAr; Nicole Bies and Esidro Hernandez/Adam Block bcl.
**392–393 akg-images:** Museo Capitular de la Catedral, Gerona/Erich Lessing cl; **Credner:** bcl, bcr, tcl; **GPL:** Michael Stecker br; Pedro Rè cAr; **NOAO:** crB.
**394–395 Credner:** r, tcl; **GPL:** Chris Picking bcl; Gordon Garradd tl; Robin Scagell cBr; **NOAO:** bcr; **The Picture Desk:** The Art Archive/Museo Civico Padua/Dagli Orti bl.
**396–397 Corbis:** Bettmann bcl; **Credner:** bcr, cl, tr; **GPL:** Gordon Garradd tcr; Yoji Hirose cBl; **NOAO:** bl.
**398–399 akg-images:** Pergamon Museum, Berlin/Erich Lessing cr; **Credner:** bcl, br, cAl, tr; **GPL:** Gordon Garadd crr; Gordon Garradd tl.
**400–401 akg-images:** Coll. Archiv f. Kunst and Geschichte crB; **BAL:** Cheltenham Art Gallery and Museums, Gloucestershire, UK cl; **Credner:** bcr, bl, tcl, tr.
**402–403 Credner:** bcl, br, tr; **GPL:** Chris Livingstone tcl; Gordon Garradd bcr; Michael Stecker cAl.
**404–405 BAL:** © The Trustees of the Chester Beatty Library, Dublin bcr; **Credner:** bcl, br, tl; **ESO:** Jean-Luc Beuzit, Anne-Marie Lagrange (Observatoire de Grenoble, France), and David Mouillet (Observatoire de Paris-Meudon, France) cl; **GPL:** Chris Livingstone tr; **NOAO:** Marcelo Bass/CTIO tcr.
**406–407 Alamy Images:** Chris Cameron bcl; **Credner:** bl, br, tcl, tr; **SPL:** bclA.
**408–409 Alamy Images:** Adam van Bunnens tr; **Credner:** bcr, bl; **DK Images:** Courtesy of the Science Museum, London/Dave King crB; **GPL:** Gordon Garradd cl; **Volker Wendel and Bernd Flach-Wilken** (www.spiegelteam.de): tclB.
**410–411 GPL:** Juan Carlos Casado.
**412–413 DK Images:** tr.
**414–415 GPL:** Robin Scagell tr, cl; **NOAO:** Ryan Steinberg and family tcr.
**420–421 Credner:** tr; **DMI:** Akira Fujii cl.
**426–427 Corbis:** Roger Ressmeyer cl; **GPL:** Gordon Garradd tcr.
**432–433 Credner:** cAl; **GPL:** Yoji Hirose cr; **NOAO:** tcr.
**438–439 Credner:** cl; **DMI:** Akira Fujii crA.
**444–445 Alamy Images:** Pixonnet.com cAl; **DMI:** Akira Fujii tr.
**450–451 DMI:** Akira Fujii cl, tcl.
**456–457 Corbis:** Reuters/Ali Jarekji bl; **Credner:** tr; **NOAO:** Svend and Carl Freytag/Adam Block cAr.
**462–463 Alamy Images:** Gondwana Photo Art bcl; **GPL:** Chris Livingstone tcr.
**468–469 GPL:** Robin Scagell cAr; Yoji Hirose bcl.
**474–475 Credner:** tcl; **GPL:** Robin Scagell tr.
**480–481 Credner:** bcl; **GPL:** Yoji Hirose tr.

ENDPAPERS **NOAO:** Nathan Smith, Univ. of Minnesota.

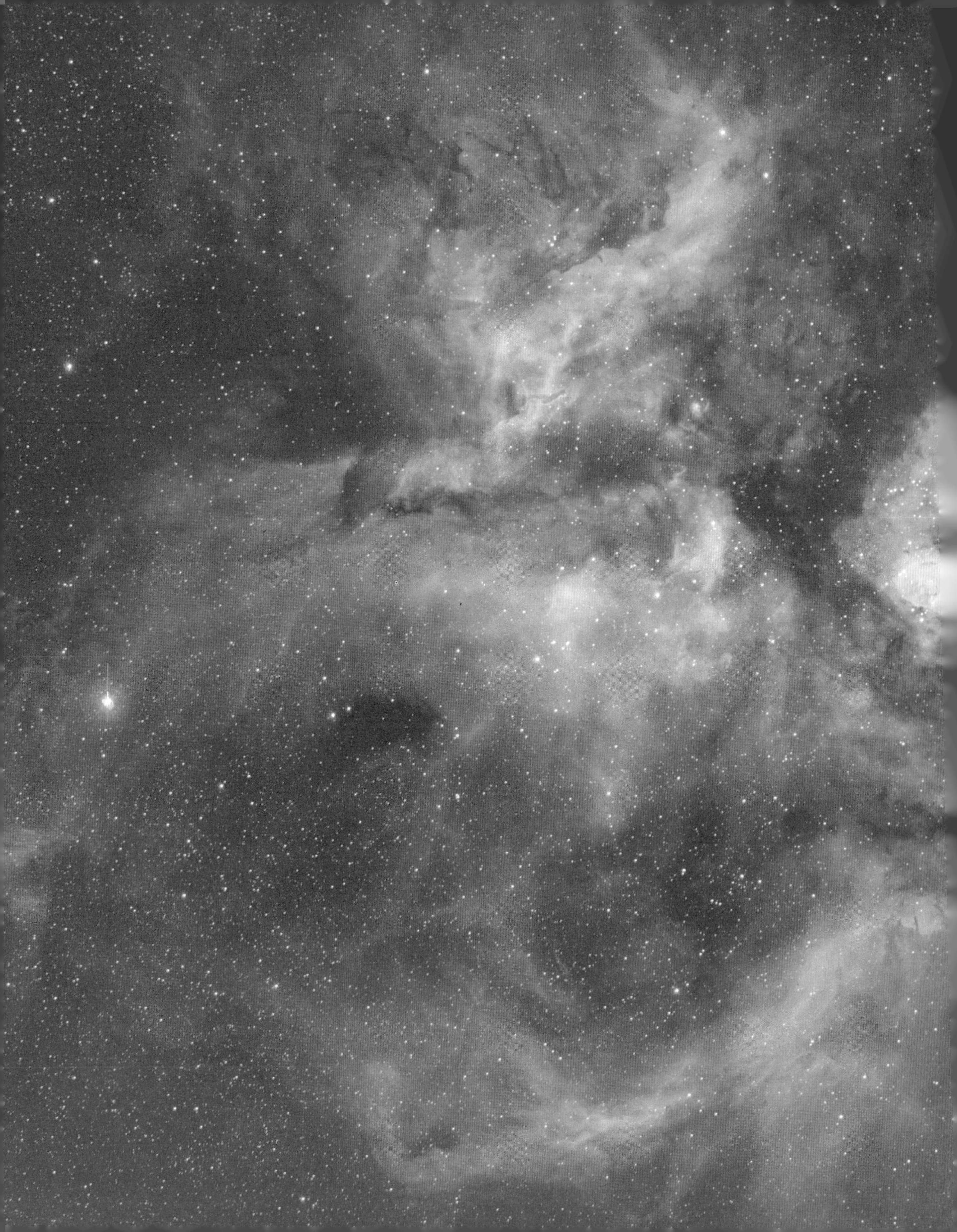